35
AF855462

# CONTENTS

Copyright© 1996

by

Canadian Prestressed Concrete Institute

All rights reserved. This book or any part thereof
must not be reproduced in any form without the written permission of the
Canadian Prestressed Concrete Institute.

Third Edition
First Printing 1996

ISBN 0-9691816-7-1

Printed in Canada

# DESIGN MANUAL

## PRECAST AND PRESTRESSED CONCRETE


CANADIAN PRESTRESSED
CONCRETE INSTITUTE

196 BRONSON AVENUE
SUITE 100, OTTAWA
ONTARIO K1R 6H4

Ph (613) 232-2619
Fax (613) 567-3064
email cpci@fox.nstn.ca


STRUCTURAL CPCI ARCHITECTURAL

# FOREWORD

The Canadian Prestressed Concrete Institute is a non-profit corporation founded in 1961 for the purpose of advancing the design, manufacture and use of precast, and prestressed concrete throughout Canada.

Although relatively young in years, CPCI represents a growing part of the Canadian construction industry. The first prestressed concrete structure was erected in Canada in 1952, a bridge in North Vancouver. In succeeding years, the use of prestressed concrete spread into the fields of buildings and virtually all types of engineering structures. This "new material" - structural and architectural, pretensioned and post-tensioned, precast concrete - was successfully and economically utilized in and ever expanding variety of applications.

CPCI developed early and naturally into a unique combination of trade and professional association, comprising a representative mix of companies and individuals. Accordingly, its membership includes producers (Active Members), suppliers (Associate and Supporting Members), engineers and architects (Professional Members), plus Affiliate and Student Members.

From the beginning, CPCI established a working arrangement with its counterpart in the United States, the Prestressed Concrete Institute. As an early organization member of PCI, the Canadian Prestressed Concrete Institute continues to enjoy a mutually beneficial relationship with that Institute, resulting in a meaningful sharing of experiences, knowledge and developments. In this respect, a dual-membership agreement exists, whereby Professional and Producer members of CPCI hold similar membership in PCI.

The focus of CPCI's current activities are primarily external and include the delivery of state of the art information about the industry, its products and services to the design and construction community in Canada. To this end, the Institute involves itself in specifications, codes and standards, liaison with technical schools and universities, awards programs, seminars, trade shows and conventions of user groups.

A further important focus for the Institute has been, and continues to be, speaking to the federal government on behalf of the industry in two principal areas. The Institute assists the producer members in marketing their products and services to government. Secondly, an important dialogue has been established in providing the government with information about the industry as it may effect the formulation of government policy.

CPCI is increasing the quality and quantity of its communications with other cement based trade and technical organizations, particularly the Canadian Portland Cement Association. As an example, CPCA participated with CPCI in the publication of "Prestressed Concrete Basics", a text book for students, and in the publication of this manual.

The Active Membership in the Institute represents over 75% of the industry's capacity in Canada. As the recognized spokesman for the prestressed and precast concrete industry in Canada, CPCI continually disseminates information on design data, production practice and field techniques, via regional/national programs and technical publications - all directed towards advancing the state of the art for the entire Industry.

In recent years CPCI has sponsored the development of software to assist designers with the design of precast and prestressed concrete. CONCISE - BEAM has become the industry standard for the analysis and design of precast prestressed concrete members. CONNECT-IT is particularly useful for the design of connections between precast members. CONCISE - BRIDGE is focused on the computer analysis and design of pretensioned and post-tensioned bridges. During the useful life of this manual, there will no doubt be other programs developed to assist the design community.

Interested engineers, architects, owners, developers and contractors are invited to visit or contact CPCI headquarters in Ottawa for additional information on the design and construction of prestressed and precast concrete and on membership in CPCI.

# PREFACE

The Canadian Prestressed Concrete Institute embarked on its first venture into the publishing of comprehensive design manuals in 1964, with the publication of the first design manual in North America applicable to precast, prestressed concrete. This formidable task was a major undertaking for a fledgling industry, not even 15 years old. The 1964 CPCI Handbook consisted of 4 parts and was based on CSA-A135-1962. Part 1 contained the CSA Standard and a commentary; Part 2 contained design theory and practice: Part 3 dealt extensively with connection design: and Part 4 contained sections on specifications and plant standards. The handbook was well received with orders placed from around the world.

During the 1970's, the standard design references for precast, prestressed concrete in North America were published by the Chicago based Prestressed Concrete Institute; the first edition of the PCI Design Handbook in 1971 was followed by a second edition in 1978.

The Canadian Government's decision in the mid 70's to take Canada into SI units reduced the relevance of the American publications in this country. Consequently, in 1978, the CPCI undertook what was to become the single largest undertaking in its history; the publishing of the First Edition of the CPCI Metric Design Manual. This publication, which was four years in the making, was based on a combination of the PCI Design Handbook and the PCI Structural Design of Architectural Precast Concrete Handbook, was written entirely in SI units with extensive reference to Canadian design codes, and contained design information on post-tensioned concrete.

In 1984, it was decided to embark upon the publication of a Second Edition of the CPCI Metric Design Manual. This decision was based upon the extensive revisions to A23.3.

In 1985, PCI published the Third Edition of its Design Handbook.

In the period since the Second Edition of the CPCI Metric Design Manual was first published, there have been significant changes in the state of the art for precast, prestressed concrete. In addition there have been important changes in A23.3 that recognize the benefits of quality control in certified precast concrete plants. These developments have brought about the need for this, the Third Edition of the CPCI Design Manual.

The chapter on methods and materials has been updated to reflect current applications of precast, prestressed concrete. New applications are presented. Previous applications have become more sophisticated and have kept pace with modern architectural trends.

The chapter on Design of Structures has been modified to reflect the recent changes in the philosophy in earthquake design. The changes in the 1995 National Building Code of Canada have had a major impact on the design examples in this chapter.

The chapter on Element Design has required the greatest degree of revision due to changes in A23.3-94. The concrete resistance factor for precast concrete produced in a CSA certified precast plant has increased from 0.60 to 0.65. The expression for the modulus of elasticity has been modified to give somewhat lower values. This results in increased deflection and camber. The General Method for design of elements for shear and torsion was modified to suit the new requirements of A23.3-94. The design procedures for compression members have also been updated.

The chapter on Design of Connections has been revised to suit the new A23.3-94 formulations for strut and tie models, and shear friction. Several of the connection details have been modified to reflect current practice.

In addition to updating the technical content of the chapter on Related Considerations, sections have been added to deal with vibrations, and with the integration of structural, mechanical, and electrical systems. The chapter on Architectural Precast Concrete has been updated to reflect current practice.

The chapter containing Product Information and Capability, has undergone extensive revision. The load capacity tables have been generated using A23.3-94 design equations. In general, the capacity of elements either remains the same, or improves slightly. Load tables for solid slabs have been added.

There are minor modifications to the chapter on General Design Information. Development and lap lengths required by A23.3-94 are given. The post-tensioning anchorage section has been updated to reflect currently available systems.

Finally, this Manual now includes a copy of CSA Standard A23.4-94, Precast Concrete - Materials and Construction.

# ACKNOWLEDGMENTS

The first individuals to be acknowledged are those that produced the Second Edition of the Design Manual. Readers should refer to the Second Edition for the names of the more than 40 individuals responsible. They laid the foundation for this the Third Edition.

The work of coordinating the preparation of new material and the revision of sections of the Second Edition was executed by an Editorial Committee consisting of the following persons:

J. P. Audet, Con-Force Structures Limited, Calgary

R. Dozzi, Harris P/T, Stoney Creek

J. Fowler[3], Canadian Prestressed Concrete Institute, Ottawa

R. Golec, Pre-Con Company, Brampton

L. Grant, Strescon Limited, Saint John

P. Kluchert, Kluchert & Associates, Toronto

D. Morse[2], Lafarge Construction Materials, Edmonton

D. Rogowsky[1], University of Alberta, Edmonton

A wide range of consultants, university professors and industry personnel gave generously of their time to prepare material and review draft copies and final gally proofs for this manual. The CPCI would like to express its sincere appreciation for the effort expended by the following persons:

P. Adebar, University of British Columbia, Vancouver
D. Allen, Consultant, Ottawa
J. P. Audet, Con-Force Structures Limited, Calgary
Y. Beaudoin, Betcon Inc., Longueil
P. Breeze, American Concrete Institute, Detroit
W. Brown, National Research Council, Ottawa
P. Cicuto, Architectural Precast Systems, Newmarket
A. Cornelissen, Canadian Portland Cement Association, Ottawa
R. Craik, Coreslab Structures, Burlington
D. Deare, Pre-Con Company, Brampton
R. Dozzi, Harris P/T, Stoney Creek
R. Erfani, University of Alberta, Edmonton
J. Fowler, Canadian Prestressed Concrete Institute, Ottawa
R. Golec, Pre-Con Company, Brampton
L. Grant, Strescon Limited, Saint John
M. Green, Queen's University, Kingston
S. Heuvel, ARC & CADD, Winnipeg
W. Houston, W.D. Houston & Associates Ltd., Willowdale
N. Huszar, Nandor H. Huszar Ltée., Montréal
D. Isnor, Strescon Limited, Saint John
W. Kassian, Con-Force Structures Limited, Calgary
P. Kluchert, Kluchert & Associates, Toronto
M. Lau, Manitoba Highways and Transportation, Winnipeg
W. LeBlanc, Con-Force Structures Ltd., Winnipeg
J. Lockwood, Western Concrete Products, Armstrong
R. Loov, University of Calgary, Calgary
A. Mehta, Prestressed Systems Inc., Windsor
R. McGrath, Canadian Portland Cement Association, Ottawa
D. Morse, Lafarge Construction Materials, Edmonton
D. Olynick, Lafarge Construction Materials, Winnipeg
A. Perry, CBCL Limited, Halifax
R. Petanca, Strescon Limited, Saint John
R. Riffel, Con-Force Structures Ltd, Vancouver
D. Rogowsky, University of Alberta, Edmonton
C. Seaborn, Seaborn Engineering, Calgary
L. Seig, Harris P/T, Delta
G. Shelestynsky, Procon Engineering Inspections Inc., Inglewood
C. Shirtliffe, National Research Council, Ottawa
D. Simms, Pre-Con Company, Belleville
W. Slater, Slater Associates, Toronto
W. Stewart, Coreslab Structures, Burlington
A. Warnock, National Research Council, Ottawa
C. Weber, Lafarge Construction Materials, Calgary
D. White, Canadian Prestressed Concrete Institute, Ottawa
R. Wong, Pre-Con Company, Brampton

CPCI acknowledges the generosity of the Precast/Prestressed Concrete Institute for permission to use sections of their excellent publications as a basis for this manual.

Acknowledgment is also extended to the Canadian Portland Cement Association for their financial contribution and permission to use various figures and material from their Concrete Design Handbook.

CPCI is indebted to the National Research Council of Canada for permission to reproduce certain tables and figures from the National Building Code of Canada, 1995 and the supplement to the National Building Code of Canada, 1995.

Acknowledgment is also extended to the Canadian Standards Association for licensing CPCI to reproduce CSA Standard A23.4-94

Manual design and production by K. Chalmers and J. Morris, Boldface Technologies Inc., Edmonton. Cover Photo: J. Fowler

---

[1] Editor in Chief

[2] Chair, CPCI Technical Committee

[3] CPCI President

# CHAPTER 1
# METHODS AND MATERIALS

Page No.

# PRECAST PRESTRESSED CONCRETE METHODS 1.1

# PRODUCT MANUFACTURE 1.1.1

Long span 305 mm and 203 mm deep pretensioned hollow-core. Slabs are usually continuously extruded on 100 to 150 m long casting beds and saw cut after curing to the exact length required. Slabs are also rip and angle cut to suit building geometry.

The use of innovative forming materials and techniques (like these urethane molds) provide high quality finishes and an endless variety of shapes and sizes for both architectural and structural applications.

Double tee load test at the McGill University structural laboratory. These tests helped to develop design recommendations for precast pretensioned elements cast with high performance concrete.

Scientists test precast rain screen insulated wall panels in the dynamic wall test facility at the National Research Council in Ottawa. The experiments demonstrated that architectural precast wall panels can be manufactured to control rain penetration using the pressure equalized rainscreen principle. An effective air barrier is an inherent feature of these panels.

Long-line double tee casting bed. A wide variety of widths, depths and span ranges can be provided by most manufacturers. Strictly controlled work in a precast plant and an industry supported certification program ensure high quality and a strict adherence to construction industry standards.

Solid and voided precast pretensioned piles can be manufactured and driven in lengths up to 40 m to eliminate or minimize mechanical splices. Precast piles are durable and perform well in soil and marine environments.

# WAREHOUSES & INDUSTRIAL BUILDINGS 1.1.2

The ability of precast prestressed concrete to span long distances (hollow-core 9 to 15 m, double tees 20 to 30 m, single tees 25 to 40 m, bridge sections 50 m and up) and carry heavy loads with a maximum span/depth ratio is particularly useful in the construction of warehouses and industrial buildings.

Precast floor and roof framing can be designed to accommodate a variety of mechanical systems and support hanging loads and bridge cranes (heavy industrial uses). High strength precast concrete resists fire, damp conditions and a variety of chemical substances. The clean, smooth surfaces obtainable in a precast concrete factory make this material ideal for food processing, building computer components, manufacturing and wet processing operations.

Solid precast concrete panels or insulated sandwich panels can be readily used for load bearing exterior walls. Roof and floor elements can be supported on wall panels bearing directly on haunches provided on the inside faces of the wall panels (allowing the latter to extend above the roof to form a parapet) or by bearing directly on the top of the wall panels (projecting, if desired, beyond the walls to form a decorative or protective overhang).

Attractive, durable exterior walls can be formed using standard shapes economically prestressed in long line production facilities. Custom shapes are produced in architectural molds with smooth, sandblasted or exposed aggregate surfaces. Sandwich insulation can be incorporated to provide required RSI-values.

PRECAST RAILWAY REPAIR AND MAINTENANCE FACILITY. Long span precast prestressed concrete roof and heavy crane beams provide a clean, fire resistant, wide open building for efficient operation.

Precast warehouse uses long-span double tees for the roof and hollow-core slabs for the mezzanine. The I-girder roof beams have web openings for mechanical piping in the main plant. Inverted tee beams are used to increase headroom in the mezzanine area.

# SCHOOLS UNIVERSITIES COLLEGES 1.1.3

Wallace Building, University of Manitoba, Winnipeg

Precast prestressed concrete has been a favoured material for school, college and university building structures, providing design flexibility and fast construction time.

The superior finishes achievable in a precast plant have enabled many designers to expose the structure in the finished building. Durable, good looking exterior finishes using architectural precast panels provide years of maintenance free use.

In addition to classroom and office facilities; student residences, auditoriums, gymnasiums and school swimming pools have been constructed using long span precast concrete floor and roof members.

University of Lethbridge, Lethbridge

Mount Royal College, Calgary

Lethbridge Community College, Lethbridge

La Cité Collegiale, Ottawa

# SHOPPING CENTRES 1.1.4

Precast prestressed concrete components can be quickly fabricated and erected to provide early occupancy of retail stores and shopping centres. The long spans achievable using precast prestressed floor and roof slabs provide column free retail areas. Solid concrete construction minimizes floor vibrations and provides built-in fire resistance.

The use of architectural precast concrete projects a quality appearance and offers years of maintenance free operation. Warehouse retailers use prestressed sandwich panels for fast all-weather construction, economy, low maintenance and the desired corporate image.

WESTMINSTER MALL, VANCOUVER

EATON'S RETAIL STORE, CALGARY

EDMONTON CENTRE

# OFFICE BUILDINGS 1.1.5

MANITOBA SPORTS FEDERATION, WINNIPEG. Eight storey total precast retail and office complex. A four level parking garage underneath uses long span double tee floor slabs.

CABLE REGINA. Total precast one and two storey building housing offices, retail, studio and production facilities was built in record time during winter conditions.

REVENUE CANADA, WINNIPEG. Total precast six storey office building has an attractive exterior finish which consists of rustication joints and sandblasting to expose the local aggregates in the precast concrete panels.

The combination of high quality architectural load bearing exterior walls with mass produced structural precast floor and roof members and precast concrete framing members can produce open, attractive, fire resistant, economical office buildings.

Long span single tee, double tee or hollow-core floors reduce interior framing, providing large column free areas. Architectural finishes can be used in the interior of an office building for columns, atrium framing, entrance and elevator shaft walls. Interior or exterior shear wall systems and rigid frame column beam jointing have all been successfully used to resist lateral forces.

High quality components can be readily manufactured under controlled conditions in a precast prestressed concrete plant.

Significant time savings usually result from the choice of a total precast concrete structure. The superstructure is prefabricated while the onsite foundations are being built. Scheduling and coordination of the work is simplified and potential delays are reduced when the complete building system is being supplied under one contract.

The erection of large precast components can proceed during all weather conditions to quickly enclose the structure. The prestressed floors provide an immediate working platform to allow the interior tradesmen an early start on the mechanical, electrical and interior finishing work.

Quality finishes and fast schedules result in early occupancy, tenant satisfaction and reduced financing costs. These factors make a precast prestressed office building system a choice investment.

REVENUE CANADA, WINNIPEG. Long span hollow-core floors and load bearing exterior spandrel beams provide plenty of wide open areas for flexible space planning.

# RESIDENTIAL BUILDINGS 1.1.6

FORT GARRY PLACE, WINNIPEG. Three 20 storey condominium towers use precast load bearing interior and exterior walls, hollow-core floors and precast balcony slabs. The exterior panels have exposed Tyndall stone aggregates and a stone jointing pattern to match the natural stone used on the 4 storey podium around the whole complex.

CHIMO INN, OTTAWA. Six storey hotel uses untopped hollow core slabs supported on load-bearing masonry walls.

Precast prestressed concrete enjoys wide use in all types of residential structures, particularly multiple units where the repetitive use of standard components manufactured in a factory can be fully utilized.

The significant benefits of superior fire resistance, sound control, durability and low maintenance, rapid construction and attractive exterior treatments are recognized by owners and designers alike.

The most common floor and roof elements employed are 205 mm deep untopped hollow-core units. These slabs can span up to 9 m or more without intermediate supports. Longer spans can be achieved by using 254 mm or 305 mm deep hollow-core units.

For hotels, motels and apartments the hollow-core slabs are oriented to either span between load bearing shear walls or to span from the central corridor to an exterior wall. Slabs can be cantilevered up to 1.8 m to form exterior balconies.

Bearing walls may be of masonry, precast concrete or cast-in-place construction. The use of precast walls speeds the erection process. Construction rates of 1 floor per week and better are often achieved. Architectural insulated panels for the exterior of a building can provide a durable, attractive, energy efficient envelope. Ancillary recreation, parking, and convention facilities are commonly constructed using precast concrete framing with long span roof and floor members.



SINGLE FAMILY HOMES, WHITBY. Single span hollow-core slabs provide a finished column free basement and a solid, sound and fire resistant ground floor.

# PARKING GARAGES 1.1.7

Most parking structures are either open structures or are ventilated with outside air. The speed of construction, versatility of design, attractive exterior finishes, durability and economy of precast prestressed concrete make this material a popular choice for commercial, municipal and institutional clients.

## Loads and Forces

Allowances must be made in the design to accommodate volume changes resulting from creep, shrinkage and extreme temperature differences. Lateral design loads for wind, earthquake or earth pressures (for in-ground or partially buried structures) can be resisted in a precast concrete structure by transferring loads through the floor diaphragm to either shear walls or to column beam frames. Care in locating shear walls, the adequate isolation of shear walls and the introduction of adequate isolation (expansion) joints will prevent cracking of the floor diaphragm.

Connections between shear walls to resist uplift forces can be minimized by loading the walls with framing beams or floor members. The torsional resistance of eccentrically loaded beams and spandrel panels must be considered. Connections should be designed to prevent beam rotation and absorb bumper loads (if applicable) without undue restraint against volume changes.

CANADA POST HEADQUARTERS GARAGE, OTTAWA. Five level precast garage has white architectural exterior framing and interior bearing walls with punched openings to support long span double tee floors.

MISSISSAUGA HOSPITAL GARAGE, MISSISSAUGA. Long clear spans make parking easy and very efficient. Precast construction can provide durable, low maintenance, functional structures which make good long term investments.

PEARSON INTERNATIONAL AIRPORT, TERMINAL 3 PARKADE, TORONTO. Four storey 70,000 m² total precast parking garage is crescent shaped to match the terminal building. The long clear spans offer wider aisles and faster, safer, maneuvering and parking within the structure.

# PARKING GARAGES 1.1.7

EMILY STREET PARKADE, WINNIPEG. Open design, long spans, high visibility, excellent lighting and glass enclosed stairwells act together to give increased visibility and security for parkers working at and visitors to the Health Science Centre.

## Bay Sizes

Bay sizes should be as large as possible and a module of the standard precast concrete floor elements selected. For clear span parking, the bay size selected need not be a multiple of the width of the parking stall. Cranked (bent up or down) double tees can be used to accommodate complex geometric layouts.

## Drainage

Sloping the structure to achieve good drainage is essential to quickly remove salt laden water from the structure. The drainage pattern selected should repeat for all floors to allow for repetition in the manufacturing of the precast elements. Locate isolation(expansion) joints at high points to minimize possible leakage. Slope the floors away from columns, walls and spandrels where standing water and leakage can corrode connections.

## Durability

Except for column base plates, all connections and exposed hardware should be hot dipped galvanized or stainless steel. Where connections are subsequently welded, the welds should be minimal and located where they can be easily maintained.

High strength factory produced precast reinforced and pretensioned concrete components have been found to be highly resistant to attack by chloride ions. Where cast-in-place composite topping is used over precast floor members, wire mesh reinforcement should be incorporated in the topping. Good results have been achieved by providing a high strength concrete topping having a water cement ratio of 0.40 or less. Concrete containing 6% entrained air and five days of curing under wet burlap will produce the best results.

A series of control joints should be tooled into the topping above all joints in the precast members below. A V-shaped trowel will produce a 25 mm deep by 12 mm wide joint. Later, this joint is prepared by grinding with a V-shaped abrasive wheel and filled with a polyurethane sealant.

The applications of a penetrating sealant to the concrete surfaces is usually a good investment to help inhibit water and chloride ion penetration. A regular maintenance program is essential to keep the parking structure long lasting and trouble free.

CSA Standard S413 specifies requirements for low-permeablity concrete, acceptable protection systems and concrete cover for reinforcement and prestressing tendons.



CANADA POST HEADQUARTERS GARAGE, OTTAWA. Cranked (bent up and down) double tees are used to form the interior driving ramps to provide the required interior geometry. All floors are sloped to allow for proper drainage and to eliminate any ponding.

# PUBLIC BUILDINGS 1.1.8

STATION EST TRAITMENT DES EAUX USÉES, QUÉBEC. Precast columns, beams and double tees proved to be the ideal way to enclose the treatment tanks for this waste water treatment facility. The building envelope is clad using 800 architectural precast concrete panels, half of which have sandwich insulation.

Designers strive to create public buildings (airports, theaters, museums, galleries, libraries, convention centres, bus and train stations) which are open, functional and inviting.

The use of precast prestressed concrete will contribute in a number of ways:

- The massive effect of exposing the concrete in the interior of public buildings can produce a dramatic facade.
- Rugged exterior and interior surface treatments look good and are long lasting.
- Integral insulation in exterior walls will conserve energy and lower operating costs.

Precast concrete construction allows the plastic nature of concrete to be fully realized in a unique way with the highest quality achieved in a controlled factory environment.

OTTAWA CARLETON REGIONAL HEADQUARTERS, OTTAWA Prestressed I-girders with a cast-in-place deck were used to provide long span support for a landscaped public square over the parking garage in front of this building. This is a good example of how standard precast bridge sections can often be used to support heavy loads in buildings.

CALGARY INTERNATIONAL AIRPORT. Architectural and structural precast concrete was used boldly and extensively for the terminal building, elevated roadway and parking garage. All the architectural panels and most of the exposed structural elements were manufactured using a light brown structural mix and sandblasted to achieve a rough, even texture.

MOUNT SINAI RESEARCH CENTRE, TORONTO. Precast concrete framing was used to construct an addition on top of the existing structure. The addition consisted of a full bay extending from the 3rd to 10th floors plus a 9th and 10th floor and a 2 storey penthouse. Work was completed on a tight schedule without interrupting the day to day activities at the hospital. Exterior panels were cast using Georgian Bay quartz to match the existing structure.

# STADIUMS/ ARENAS 1.1.9

New Montreal Forum. Folded plate bleacher units speed the construction of stadiums and arenas throughout North America. Bleachers are usually prestressed longitudinally. Seating units are bolted directly to the concrete.

Large stadiums and arenas are impressive structures. Often these projects are built on tight schedules to accommodate some important sporting event. Precast prestressed concrete has been the overwhelming choice for many of these projects.

The technique of post-tensioning precast segments together has allowed this versatile material to form complex cantilever arm and ring beam construction to support the roofs of these structures. Post-tensioning is also commonly employed to reinforce precast concrete cantilevered raker beams which carry the seating and provide unobstructed viewing of the playing surface.

Mass produced seating units have been manufactured in a variety of configurations and spans to provide for quick installation and long lasting service. Pedestrian ramps, concession, toilet, and dressing room areas can all be framed and constructed using precast prestressed concrete elements.

Saddledome, Calgary. Erection of some of the 391 - 6 m x 6m low density roof panels on a stressed net of cables suspended from the ring beam. When the joints between panels are concreted in, the entire roof behaves as a hyperbolic paraboloid shell.

Kingdome, Seattle Three step precast concrete folded plate bleacher units were cast with a bend at mid span. This allowed the bleachers in plan to closer approximate a smooth curve and increased seating capacity. These horizontally kinked bleacher sections were precast with deflected prestressing strands – the first project ever to employ this innovation.

Olympic Oval, Calgary. Built for the XV Winter Olympics, this innovative structure is a precast concrete, 88 m x 199 m clear span, lattice arch grid. Hollow trapezoidal precast beams were erected on temporary scaffolding. Reinforcing steel was added, post-tensioning ducts were coupled, joints were concreted and beams were post-tensioned together. A well planned sequential jacking down of the temporary supports transferred the roof loads to thrust bearings at the 28 buttresses.

Saddledome, Calgary. From a structural standpoint, the Saddledome is divided into five independent parts: two grandstands having three tiers each, two grandstands having two tiers, and a roof constrained at two ends by four stability A-frames. The outer rib-shaped precast columns, arranged along the perimeter, follow the meridians on the original sphere which generates the stadium geometry. The sphere is divided into 32 equal parts by these columns. The precast segmental ring beam at the border, to which the roof is anchored, rests on the columns. The grandstand structure is formed from precast concrete interior framing (H-frames, beams and columns), bleacher support raker beams, double tee floor slabs, bleacher slabs, exterior rib-shaped columns and A-frames. Extensive post-tensioning was used throughout to provide continuous reinforcement and moment connections in the structure.

# SPECIAL STRUCTURES 1.1.10

A high degree of design flexibility makes prestressed concrete ideal for a wide variety of innovative structures.

Properties such as corrosion resistance (piling), durability (railway ties), fire resistance (pipe racks), tight tolerances (tunnel liners), architectural finishes (chimney stacks), strength (silos) and fast installation and economy (water tanks), are all used to good advantage.

Where repetition and standardization exist, precast components can economically provide quality, plant manufactured products and eliminate expensive and risky field procedures.

New applications await the imagination of creative designers.

**Effluent Treatment Tanks.** Precast prestressed clarifier tanks have built-in launder troughs which can be cast on either the inside or outside of the main tank walls.

**Hexagonal Prestressed Concrete Piles** Piles are driven to support a massive sewage treatment plant in Winnipeg.

**Effluent Treatment Tanks .** Precast wall segments are pretensioned vertically. Segments are erected at the jobsite on steel shims and temporarily braced in position until the horizontal post-tensioning ducts are spliced and the vertical joints are concreted. The post-tensioning resists the hydraulic pressure of the liquid in the tank.

**Kil-Cona Park Pedestrian Bridge, Winnipeg.** Multi-span concrete arch pedestrian bridges were designed to set a rustic tone for the park, yet meet the stringent requirements of a modern bridge.

Prestressed concrete switch ties for CN Rail.

# MODULAR SANDWICH WALL PANELS 1.1.11

Insulated precast sandwich wall panels consist of two reinforced or prestressed concrete wythes with a continuous layer of rigid insulation sandwiched between. They are mass produced in standard widths on long line casting beds. With attractive sculptured exterior surfaces and steel form-finished interior faces, they inexpensively provide a strong, durable, energy efficient, fire resistant cladding system.

The type and thickness of insulation contained in sandwich panels varies with the RSI-value required for each building. The insulation is installed under controlled factory conditions and is well protected by the concrete.

Erected at the rate of up to 120 linear metres per day on concrete or steel frame buildings, precast sandwich wall panels are available in a range of widths, lengths, thicknesses and exterior finishes. Designed to stand without intermediate supports, they are used for both load bearing and non-load bearing applications.

**Firewall - Canadian Tire Corp., Brampton.** Two layers of panels, stacked two high, form a four hour fire wall to protect the contents of this high level stacking warehouse.

Site storage is not required. Panels are quickly hoisted directly from trucks and placed on the building.

**Leon's Centre, Winnipeg.** Picturesque single-storey retail shopping centre has two attractive buildings clad with precast concrete sandwich panels. The architect wanted a durable, energy efficient wall panel and specified precast panels with 75 mm exterior concrete, 50 mm insulation (RSI 1.44) and 75 mm interior concrete. The rolled finish white concrete precast panels provide a durable finish which can withstand the elements, physical impact and retain their appearance over the lifespan of the building.

# ARCHITECTURAL WALL PANELS 1.1.12

AARP BUILDING, WASHINGTON. The return sections of this precast concrete column cover are first poured flat, then placed in the form, and cast integrally with the face pour. This assures a uniform, highly detailed finish on all three faces of the panel.

Architectural precast cladding was developed to combine the maximum freedom of architectural expression with the economies of mass production of repetitive precast elements. This concept functions most effectively when architects consult with local CPCI architectural precast concrete producers in the early design stages, and through the complete development of the contract documents.

With this precast-specialist help, proper functional, structural and mechanical features can then be detailed. The results will accurately reflect material characteristics, manufacturing and erection efficiencies, cost factors, quality control standards, and local trade practices. A continuing dialogue between designer and precaster will ensure the optimum quality of the product, and its economical, efficient installation at the construction site.

Architectural precast concrete can be cast in almost any colour, form, or texture (within budget guidelines) to meet aesthetic and practical requirements. Special sculptured effects can provide such visual expressions as strength and massiveness or grace and openness. The appearance of properly designed shapes is often enhanced by the long term effects of weathering.

1501 M STREET, WASHINGTON. Spectacular facade is stacked vertically, bearing on the foundation, and laterally supported behind by the structural frame. Column sections were precast vertically as single pieces. The precast concrete finish is acid etched to simulate limestone.

AARP BUILDING, WASHINGTON. Office project has a simulated stone pattern and classic architectural shapes. Pieces were produced with a rough textured concrete mix coloured to the exact shade required by the project designers.

# ARCHITECTURAL WALL PANELS 1.1.12

ROBERT O. PICKARD ENVIRONMENTAL CENTRE, OTTAWA. Precast concrete was used extensively for several new and expanded buildings during the expansion and rehabilitation of the regional waste water treatment centre.

GRAND HARBOUR, ETOBICOKE. Waterfront condominium project on the shore of Lake Ontario contains over 7500 pieces of architectural precast concrete ranging from 3 to 23 m$^2$ in size and 230 kg to 9 tonnes in mass. Shapes included Ionic columns, pilasters, straight and curved entable units, arched cornices, single, double and triple open ended window panels, sills, copings and many other decorative pieces.

CHURCH OF JESUS CHRIST OF LATTER DAY SAINTS, TORONTO. Church and assembly hall are clad using textured white architectural precast panels. The project is beautifully detailed to provide a clean, simple, elegant, place of worship.

Design flexibility is achieved by varying aggregates and matrix colour, the size of aggregates, finishing processes, and depth of exposure. Combining colour with texture accents the natural beauty of aggregates.

Panel geometry (shape details) have a major influence on fabrication economy and engineering requirements, with overall size and configuration being the most important elements. Generally, the smaller the elements, the higher the cost.

Surface textures range from smooth, to ribbed, sandblasted or exposed aggregate finishes. Texture allows the natural concrete ingredients to be expressed, provides the mass with scale, expresses the plasticity of the material and improves its weathering characteristics.

Supplementing the freedom of aesthetic expression and economic features achievable with the use of architectural concrete, are a number of other important functional and construction advantages.

- Architectural cladding panels are sometimes used as permanent concrete exterior formwork, thus becoming an integral part of the structure.
- Insulated architectural wall panels contribute substantially to the overall thermal efficiency of a building. Since the thermal resistance of glass is less than that of concrete, reducing window areas minimizes heat gains and losses. In suitable environments, properly designed precast sun screens help to reduce air conditioning loads.
- Pre-assembly in a precast plant of all the components comprising a total precast wall system, including insulation, air gap, flashing, window washing tracks, window sash and glazing can be very cost effective.

For more information see Reference R1.19.

LES VERRIÈRS SUR LE FLEUVE, MONTRÉAL. Luxury condominium project, consists of a 27 storey central tower surrounded by four, 18 storey residences. The rapid all weather construction and attractive finish of the precast concrete panels increased sales and allowed the construction of the five phases during a four year period without interruption.

# VENEER FACED WALL PANELS 1.1.13

125 Summer Street, Boston. This magnificent entranceway is constructed using a combination of exquisitely detailed, acid etched, architectural precast concrete and thermal finished granite panels anchored to the precast.

To supplement the incredible variety of colours and textures available with exposed aggregate architectural finishes, various other materials may be cast on the face of precast concrete panels. Natural stone (natural and polished granite, marble and various other types of stone) are attached to precast panels. Panels are constructed to exacting tolerances as either solid or insulated units.

Some very complex brick faced panels have been precast in plants to produce results virtually impossible using field-set masonry. Attaching large or small ceramic tiles to the face of architectural panels provides an additional choice of exterior or interior finish.

Veneer faced architectural panels are produced in precast plants under tightly controlled conditions. The panels are quickly erected at the site, rapidly enclosing a building and speeding the construction process.

World Exchange Plaza, Ottawa. The precast window panels were stored along side the site and erected during a second shift using the project's tower crane. Work on the project was speeded up with finishing trades able to comfortably work inside the closed in building all through the winter.

World Exchange Plaza, Ottawa. Pressure equalized rainscreen insulated granite faced window panels were delivered daily to the jobsite. The complete, sealed, multi-pane, window unit was cast into the panel at the precast plant.

Mackenzie Health Sciences Building, Edmonton. The precast spandrel panels incorporate a brick face, an air gap, rigid insulation and the interior concrete wythe. The precast envelope is very energy efficient with no evidence of condensation or air leakage associated with the precast panel system. Good looks and superior performance have made this system an excellent choice.

# POST-TENSIONED CONCRETE METHODS 1.2

## 1.2.1 INTRODUCTION

The advantages of prestressing materials for greater strength have been recognized for centuries. The classic example is that of a barrel (wooden staves prestressed by the application of iron bands). The principles, however, were not applied to concrete construction until 1886, when a patent for tightening steel tie rods in artificial stones and concrete arches was obtained.

Developments since then have been numerous. Of note were the contributions of Eugene Freyssinet who started using high strength steel wires for post-tensioning concrete beams in 1928. Among his innovative developments were conical wedges for anchoring wires at the end of prestressed members, and special jacks to stress and anchor the wires.

For all practical purposes, post-tensioning in North America began in the early 1950's. Although experiencing a slow start, the use of post-tensioning in concrete continues to increase each decade. This growth can be attributed to the growing awareness of its advantages by architects and engineers. Advantages include design flexibility, reduced structural depth, control of deflection, longer spans with reduced construction costs, and crack free construction.

The increased use of post-tensioning is due in a large part to the fabricators of post-tensioning products. Expertise, gained through years of experience, allows them to provide services and equipment to meet any job requirement.

## 1.2.2 BUILDING CONSTRUCTION

In the construction of offices, condominiums, warehouses and parking garages, the use of post-tensioned concrete is widespread. Post-tensioned concrete offers increased usable floor space as the increased span/depth ratio requires fewer intermediate columns. The application of prestressing will result in shallower more slender members which reduces concrete volumes and lowers building heights.

The photographs depict a few of the many buildings constructed with post-tensioned concrete with emphasis on the use of the unbonded, monostrand system.

## 1.2.3 MISCELLANEOUS STRUCTURES

The versatility of post-tensioned concrete is demonstrated in its wide range of uses, from the slender grace of the CN Tower to more common uses such as bridges. Experienced designers can create structures as diverse as circular slip formed water towers and coal silos; and the Skydome in Toronto.

Post-tensioning tendons are also applied to soil and structure stabilization in the form of rock and soil anchors.

**CN Tower and Skydome, Toronto.** The CN tower, at 533 m, is the world's tallest free-standing structure. Both its foundation and superstructure were prestressed using in-place post-tensioning to make the concrete portion of the tower a fully prestressed concrete structure. The 442 m concrete shaft has a three-legged cross section which tapers upward. The central shaft was slipformed and post-tensioned in about 8 months.

**Skydome, Toronto.** Skydome has a seating capacity of 54,000 (football) and a fully retractable space frame roof whose apex is 95 m above the playing field. The main concrete structure consists of 48 post-tensioned frames in an approximately circular arrangement (inner radius 83 m, outer radius 106 m, height 50m). The concrete frames are typically 900 mm thick and 15 m wide with the inner and outer column cross-sections being 900 mm x 1600 mm and 900 mm x 2200 mm respectively. Five basic levels of frame beams were used in conjunction with three diagonals. The main members were post-tensioned and anchored to shale bedrock with post-tensioned rock anchors where required by the loading conditions. In all, 800 tonnes of low-relaxation strand was used for post-tensioning the frames and rock anchors.

**Cooperators Building, Scarborough.** Project uses unbonded post-tensioned reinforcement in both the beam to column strips and transversely in the floor and roof slabs.

# 1.3 PRECAST PRESTRESSED CONCRETE MATERIALS

This section of the manual provides a brief review of the materials used in precast and prestressed concrete. For more complete information, see references R1.1 to 1.4.

## 1.3.1 CONCRETE

The 28-day design strength of concrete used in precast and prestressed products is usually in the 35 MPa to 45 MPa range. The strength at which prestress force is transferred to concrete is usually about 25 MPa and may be more or less as required by design. However, a practical limit is that which can be attained in about 16 hours to allow the removal of the product from the forms on a daily basis.

For post-tensioned construction, the concrete strength at stressing should be specified by the post-tensioning contractor or by the designer.

Aggregates for structural products are usually the same as those used for other high-quality concrete in the local area. Semi-low density structural concrete is also used for prestressed products and appropriate mix designs should be obtained from local plants.

Aggregates commonly selected for exposed concrete facings are quartz, granite or marble. These offer a wide variety of colour and texture. Lower cost sand and gravel aggregates may also be used to produce architectural concrete. Special attention should be paid to sand and gravel aggregates to determine that they do not rust or stain when the aggregate is exposed to the environment.

During production, architectural precast concrete panels generally do not receive accelerated heat curing, as do precast, prestressed concrete structural members. Architectural precast panels are removed from forms at an age of about 16 hours after the concrete has reached a strength adequate to withstand stripping and handling.

### Compressive Strength

The compressive strength of concrete made with aggregate of adequate strength is governed by either the strength of the cement paste or the bond between the paste and the aggregate particles. At early ages, the bond strength is lower than the paste strength. At later ages, the reverse may be the case.

For a given cement and acceptable aggregates, the strength that may be developed by a workable, properly placed mixture of cement, aggregate, and water (under the same mixing, curing, and testing conditions) is influenced by:

1. The ratio of water to cement.
2. The ratio of cement to aggregate.
3. Grading, surface texture, shape, strength, and stiffness of aggregate particles.
4. Maximum size of the aggregate.

Mix factors, partially or totally independent of water/cement ratio, which affect the strength are:

1. Type and brand of cement.
2. Amount and type of admixture or pozzolan.
3. Mineral composition of the aggregate.

Compressive strength is measured by testing 150 $\phi$ x 300 mm or 100 $\phi \times$ 200 mm cylinders in accordance with standard CSA procedures. The precast concrete industry generally uses 100 $\phi$ x 200 mm cylinders and 50 or 100 mm cube specimens. These specimens must be correlated with the standard 150 $\phi$ x 300 mm cylinders.

The use of "zero" slump concrete in hollow-core slabs has resulted in the use of special 150 mm x 150 mm cylinders. These are either made in the same manner as asphaltic concrete cylinders or by vibrating the concrete in a manner similar to the product being manufactured. Either method can be used and correlated with cubes cut from the actual product.

Because of the need for early strength gain, Type 30 cement is often used by precasters so that molds may be reused daily. Structural precast concrete and much architectural concrete is made with gray cement that meets CAN/CSA-A5/A8/A362. Type 30 and Type 10 white portland cements are frequently used in architectural products and are usually assumed to have the same characteristics (other than colour) as gray cement. Pigments are also available to colour concrete, and, at the recommended dosages, have little or no effect on strength. Cement type and colour should be chosen with the help of local producers who may have experience with the proposed mix.

Concrete mixes with strengths of 45 MPa or more are available in most areas. Local suppliers should be contacted to furnish mix and design information.

Initial curing of precast concrete takes place in the form, usually by covering to prevent loss of moisture and, in many instances (particularly for structural products), by the application of radiant heat or live steam. Additional curing is rarely necessary to attain the specified strength. Control techniques for the most effective and economical accelerated curing have been the subject of recent research.

Most concrete subjected to freezing and thawing should be air-entrained, although the dense mixes used in precast products have a high resistance to freezing and thawing even without the addition of entrained air. For some precast concrete mixes it is sometimes difficult to obtain air contents as large as those normally specified for the leaner mixes most often used in cast-in-place flatwork. For precast concrete elements located above grade in a vertical position, air contents as low as 2 to 3% will usually provide the required durability. The precast concrete industry does not use air-entraining portland cements. Instead, admixtures are added to the concrete during the mixing cycle to entrain the air. A slight reduction of strength should be anticipated when concrete is air entrained.

### Tensile Strength

A critical measure of the performance of architectural precast concrete is its resistance to cracking, which is a function of tensile strength. Reinforcement does not prevent cracking, but controls crack widths after cracking has occurred. Tensile stresses which would theoretically result in cracking are permitted by CSA-A23.3 in prestressed concrete.

The flexural tensile strength is called the modulus of rupture. It can be determined by test, but for structural design the modulus of rupture is generally assumed to be a function of compressive strength as given by:

$$f_r = 0.6\lambda\sqrt{f'_c}$$

where:

$f_r$ = modulus of rupture (MPa)
$f'_c$ = compressive strength (MPa)
$\lambda$ = 1.0 for normal density concrete
= 0.85 for structural semi-low density concrete
= 0.75 for structural low density concrete

### Shear Strength

The shear (or diagonal tension) strength of concrete is also a function of compressive strength. The equations for shear strength specified in CSA-A23.3 are given in Chap. 3. The shear strength of semi-low density and low-density concrete is determined with the use of the factor, $\lambda$, as described above.

### Modulus of Elasticity

The modulus of elasticity, E, is the ratio of normal stress to corresponding strain for tensile or compressive stress. It is the material property which determines the deformability under load. Thus it is used to calculate deflections, axial shortening and elongation, buckling and relative distribution of applied forces in composite and non-homogeneous structural elements.

The modulus of elasticity of concrete and other masonry materials is not as well defined as, for example, steel. It is therefore defined by an approximate slope, such as the "secant modulus". Calculations which involve its use have an inherent imprecision, but this seldom affects practical performance. While it may be desirable in rare instances to determine the modulus of elasticity by test, particularly with some low density concretes, the equation given in CSA-A23.3 is usually adequate for design:

$$E_c = (3300\sqrt{f'_c} + 6900)\left(\frac{\gamma_c}{2300}\right)^{1.5}$$

where:

$E_c$ = modulus of elasticity (MPa)
$\gamma_c$ = density of concrete (kg/m$^3$)

### Poisson's Ratio

Poisson's ratio is the ratio of transverse strain to axial strain resulting from uniformly distributed axial load. Values generally range between 0.11 and 0.27, and are usually assumed to be 0.20 for both normal and low density concrete.

### Volume Changes

Volume changes of precast concrete members are caused by variations in temperature, shrinkage due to air-drying, and by creep caused by sustained stress. If precast concrete is free to deform, volume changes are of little consequence. If elements are restrained by foundations, connections, steel reinforcement, or connecting elements, significant stresses may develop over time.

The volume changes due to temperature variations and creep can be positive (expansion) or negative (contraction), while volume changes from shrinkage are typically only negative.

Because precast concrete elements are usually 30 days or older when erected, much of the creep and shrinkage will have taken place during yard storage. Connection details and joints must be designed to accommodate the volume changes which will occur after the precast element has been erected and connected to the structure. Hollow-core slabs are often shipped at an early age, but are not normally rigidly connected to the supporting structure.

Typical creep, shrinkage, and temperature strains and design examples are given in Chap. 2.

*Temperature effects:*

The coefficient of thermal expansion of concrete varies with the aggregate used as shown in Fig. 1.3.1. Ranges for normal density concrete are 9 to 13 × $10^{-6}\cdot{}^\circ C^{-1}$ when made with siliceous aggregates and 6 to 9 x $10^{-6}\cdot{}^\circ C^{-1}$ when made with calcareous aggregates. The approximate values for structural low density concretes are 6.5 to 11 × $10^{-6}\cdot{}^\circ C^{-1}$, depending on the

**Fig. 1.3.1 Average coefficients of linear thermal expansion of rock (aggregate) and concrete**

| Type of Rock (Aggregate) | Average Coefficient Of Thermal Expansion [$10^{-6}\cdot{}^\circ C^{-1}$)] | |
|---|---|---|
| | Aggregate | Concrete* |
| Quartzite, Cherts | 11.0 - 12.6 | 11.9 - 12.8 |
| Sandstones | 10.0 - 12.0 | 10.0 - 11.7 |
| Quartz Sands & Gravels | 9.9 - 12.8 | 11.0 - 15.7 |
| Granites & Gneisses | 5.8 - 9.5 | 6.8 - 9.5 |
| Syenites, Diorites, Andesite, Gabbros, Diabas, Basalt | 5.4 - 8.1 | 8.0 - 9.5 |
| Limestones | 3.6 - 6.5 | 6.1 - 9.2 |
| Marbles | 4.0 - 7.0 | 4.1 |
| Dolomites | 7.0 - 9.9 | — |
| Expanded Shale, Clay & Slate | — | 6.5 - 7.7 |
| Expanded Slag | — | 7.0 - 11.2 |
| Blast-Furnace Slag | — | 9.2 - 10.6 |
| Pumice | — | 9.4 - 11.0 |
| Perlite | — | 7.6 - 11.7 |
| Vermiculite | — | 8.3 - 14.2 |
| Barite | — | 18.0 |
| Limonite, Magnetite | — | 8.3 - 11.0 |
| None (Neat Cement) | — | 18.5 |
| Cellular Concrete | — | 9.0 - 12.6 |
| 1 : 1 (Cement:Sand) † | | 13.5 |
| 1 : 3 † | | 11.2 |
| 1 : 6 † | | 10.0 |

* Coefficients for concretes made with aggregates from different sources vary from these values, especially those for gravels, granites, and limestones. Fine aggregates are generally the same material as coarse aggregates.

† Tests made on 2-year old samples.

type of aggregate and amount of natural sand. Coefficients of $11 \times 10^{-6} \cdot °C^{-1}$ for normal density and $9 \times 10^{-6} \cdot °C^{-1}$ for semi-low density concrete are frequently used. If greater accuracy is needed, tests should be conducted on the specific concrete.

Since the thermal coefficient for steel is also about $11 \times 10^{-6} \cdot °C^{-1}$, the steel reinforcement does not produce significant stresses in the concrete.

*Shrinkage and creep:*

Precast concrete elements are subject to air-drying as soon as they are removed from the molds. As a result of this drying, the concrete slowly loses some of its original water causing shrinkage to occur.

When concrete is subjected to a sustained load, the deformation may be divided into two parts: (1) an elastic deformation which occurs immediately, and (2) a time-dependent deformation called creep which begins immediately and continues over time.

Creep and shrinkage strains vary with relative humidity, volume/surface ratio (ratio of area to perimeter - see Fig.1.3.2), level of sustained load including prestress, concrete strength at time of load application, amount and location of steel reinforcement, and other characteristics of the material and design.

**Fig.1.3.2 Volume/surface ratios for precast structural concrete elements**

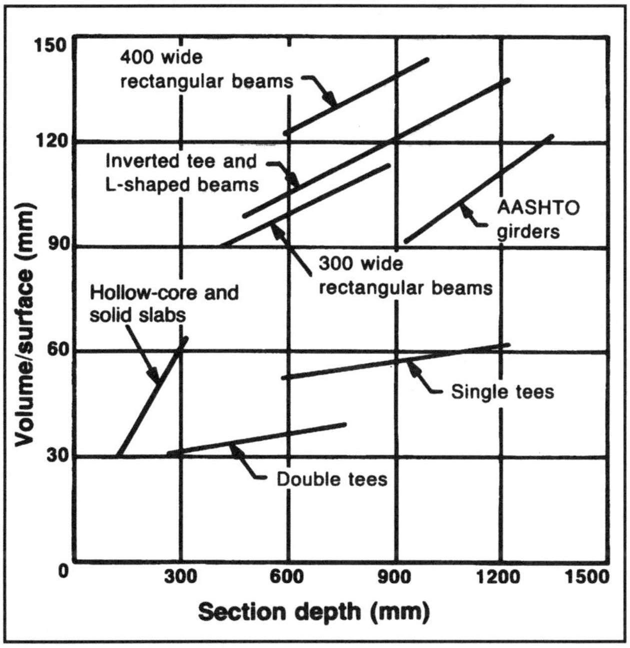


### Freeze-Thaw and Chemical Resistance

Cycles of freezing and thawing can cause damage to concrete ranging from minor surface scaling to severe disintegration. Corrosion of reinforcement, prestressing strand or connection hardware can also result, which may affect the integrity of the structure.

The effects of freezing and thawing can be resisted by high quality concrete and air entrainment. Adequate concrete cover over reinforcement and surface drainage is essential in structures exposed to weather.

Freeze-thaw damage is accelerated by deicing chemicals. Deicers can be applied indirectly in various ways; such as by dripping from the undersides of vehicles or by splash-water which can be carried a considerable distance by the wind. Some proprietary surface treatments have been found to be effective deterrents to freeze-thaw, deicer and other chemical damage (see CSA Standard S413-94 Parking Structures).

Other foreign materials, such as sulphates in soils or ground water and industrial acids, can damage concrete. The former can be resisted by specifying cements with low $C_3A$ contents. The presence of acids generally requires a membrane or a topping of concrete or other material. When aggregates or cement with a high alkali content are used in a moist environment the danger of alkali-aggregate reactivity (AAR) should be considered.

## 1.3.2 REINFORCEMENT

Reinforcement used in structural and architectural precast concrete includes prestressing tendons, deformed steel bars, and welded wire fabric.

### Prestressing Tendons

Tendons for prestressing concrete may be wires, strands, or bars. In precast, prestressed structural concrete, nearly all tendons are 7-wire strands conforming to CSA-G279. The strands are pretensioned (tensioned prior to concrete placement). After the concrete has reached a predetermined strength, the strands are cut and the prestress force is transferred to the concrete through bond.

Years ago, most strand was stress-relieved, but low-relaxation strand is now almost universally used. Low-relaxation strand as specified in CSA-G279 differs from stress-relieved strand in two respects: first, it meets more restrictive relaxation loss requirements, and second, the minimum yield strength at an extension of 1% is 90% of the specified minimum tensile strength, as opposed to 85% for stress-relieved strand. The load tables in Chap. 7 are based on low-relaxation strand.

Architectural precast concrete is sometimes prestressed. Depending on the facilities available at the plant, the prestressing tendons may be either pretensioned or post-tensioned.

Prestressing wire or bars are rarely used as primary reinforcement in precast products, but weldable-grade bars meeting CSA-G30.18 are frequently used in connections. The properties of prestressing strand, wire and bars are given in Chap. 8.

### Deformed Reinforcing Bars

Hot-rolled reinforcing bars are required to meet CSA-G30.18. This specification covers both weldable steel and regular steel which has less stringent control of the steel chemistry. Bars are usually specified to have a minimum yield strength of 400

MPa (Grade 400R and 400W). Grade 300R bars are available in sizes No. 10 and No. 15 only. Grade 500R and 500W steel is typically available only on special orders. The yield strength of 400W and 500W bars is limited to 525 MPa and 625 MPa respectively. The W in the grade designation indicates a weldable bar with controlled chemistry and a maximum carbon equivalent of 0.55.

Some precasting operations have found it economically advantageous to use weldable steel (400W) for all reinforcement. One of the advantages is a reduction of inventory and possibilities for errors. Another advantage is that bar ends can be used for welded connections instead of being scrapped. For welding of reinforcing bars, see Sect. 4.16 and CSA-W186.

For a reinforcing bar to develop its full strength in concrete, a minimum length of embedment or a hook is required. Information on bar sizes, bend and hook dimensions, standard lap and development length are given in Figs. 8.2.9 to 8.2.12, and R1.6.

### Welded Wire Fabric

Welded wire fabric is a prefabricated reinforcement consisting of parallel cold-drawn wires welded together in square or rectangular grids. Each wire intersection is electrically resistance-welded by a continuous automatic welder. Pressure and heat fuse the intersecting wires together and fix all wires in their proper position.

Smooth wires, deformed wires or a combination of both may be used in welded wire fabric. Wire sizes are denoted by their area in $mm^2$ prefixed with the letters MW or MD for smooth or deformed wire, respectively. Welded wire fabric styles are designated by the spacing and wire sizes as shown in Fig. 8.2.13 and 8.2.14.

Smooth wire fabric bonds to concrete by the mechanical anchorage at each welded wire intersection. Deformed wire fabric utilizes wire deformations plus welded intersections for bond and anchorage.

Welded wire fabric for architectural precast concrete is normally supplied in flat sheets. Use of fabric from rolls, particularly in thin precast sections, is not recommended because the rolled fabric cannot be flattened to the required tolerance.

Many plants have equipment for bending welded wire fabric into various shapes such as U-shaped stirrups, four-sided cages, etc. Designers are cautioned to use stress-relieved mesh as shear reinforcement because of special requirements for ductility when used in this application (A23.1 Clause 7.1.5).

### Protection of Reinforcement

Reinforcement is protected from corrosion by embedment in concrete. A protective iron oxide film forms on the surface of bars, wires and tendons as a result of the high alkalinity of the cement paste. As long as the alkalinity is maintained, this film is effective in preventing corrosion.

The protective high alkalinity of the cement paste is usually lost only by leaching or carbonation. Sufficient cover over the reinforcement using concrete of low permeability will usually provide adequate protection. Low permeability is obtained in well-consolidated concrete having a low water-cement ratio and high cement content, both characteristics of precast concrete. Permeability can be further decreased with the addition of small amounts of silica-fume to the concrete mix.

Cracking may allow oxygen and moisture to reach the embedded steel, providing conditions where rusting of the steel and staining of the surface may occur. A sufficient amount of closely spaced reinforcement limits the width of cracks, hence minimizing the intrusion of water, and maintaining the protection of the steel. Prestressing may also be used to limit, or eliminate cracking.

Concrete cover is the minimum clear distance from the reinforcement to the surface of the concrete. For exposed aggregate surfaces, the concrete cover is measured from the deformations of the reinforcement to the deepest point of the exposed aggregate surface. Allowance must also be given to scoring, false joints, and drips, as these reduce the cover.

In determining cover, consideration should be given to the following:

1. Structural or nonstructural use of precast element.
2. Maximum aggregate size (cover should always be greater than the nominal maximum aggregate size, particularly if a face mix is used).
3. The means of securing the reinforcement in a controlled position and maintaining this control during placement of concrete.
4. Accessibility for placement of concrete, and the proportioning of the concrete mix relative to the structural environment.
5. The type of finish treatment of the concrete surface.
6. The environment at the concrete surface: interior or exposed to weather, ocean atmosphere, or corrosive industrial fumes.
7. Fire resistance requirements.

The minimum cover requirements for precast concrete manufactured under plant control conditions as specified in CSA-A23.4 are shown in Fig. 1.3.3.

For exposure to earth or weather, the cover is required to be at least 1.5 times the nominal maximum aggregate size.

Cover requirements over reinforcement should be increased when the precast elements are acid treated, exposed to a corrosive environment, or subjected to other severe exposure conditions. For these conditions the cover should be greater than twice the nominal maximum aggregate size (CSA-A23.1 Clause 15.1.7.1).

Galvanized reinforcing bars or welded wire fabric are sometimes used when minimum cover requirements cannot be achieved, or when the concrete is exposed to a particularly severe environment. However, a detrimental chemical reaction can take place when the concrete is damp and chlorides are present. Therefore, galvanizing is not recommended for elements subjected to deicing salts or similar treatment.

Galvanized welded wire fabric is usually available as a stock item in some sizes. Individual wires are galvanized before they are welded together to form the fabric. Zinc at each wire inter-

**Fig. 1.3.3 Minimum cover requirements for precast concrete**

| Condition | Minimum Cover (mm) |
|---|---|
| **Exposed to earth or weather:** | |
| Piles: | |
| Principal reinforcement, No. 35 and smaller | 50 |
| Ties and spirals | 40 |
| Beams, girders, and columns: | |
| Principal reinforcement, No. 35 and smaller | 40 |
| Ties stirrups, and spirals | 30 |
| Slabs, joists, shells, and folded plates: | |
| No. 20 and smaller | 25 |
| Wall panels: | |
| No. 15 and smaller | 25 |
| For bars with a diameter $d_b$ larger than listed above, the cover shall be at least 1.5 $d_b$ but not necessarily more than 50 mm. The ratio of the cover to the nominal maximum aggregate size shall be at least 1.5. | |
| **Not exposed to earth or weather:** | |
| Beams, girders, and columns: | |
| Principal reinforcement, No. 35 and smaller | 30 |
| Ties stirrups, and spirals | 20 |
| Slabs, joists, shells, and folded plates: | |
| No. 20 and smaller | 20 |
| Wall panels: | |
| No. 15 and smaller | 20 |
| For bars with a diameter $d_b$ larger than listed above, the cover shall be at least 1.0 $d_b$ but not necessarily more than 50 mm. The ratio of the cover to the nominal maximum aggregate size shall be at least 1.5. | |

1. $d_b$ = diameter of reinforcing bar
2. Bar diameters apply to both prestressed and reinforced concrete.
3. The cover for a bundle of bars shall be the same as that for a single bar with an equivalent area.

section is burned off during welding, but the resulting black spots have not caused appreciable problems. After welding, the fabric is normally shipped to the plant without further treatment. There is no CSA specification for galvanized welded wire fabric.

When galvanized reinforcement is used in concrete, it should not be coupled directly to ungalvanized steel reinforcement or placed close to non-galvanized metal forms. Polyethylene and similar tapes can be used to provide local insulation between dissimilar materials. Galvanized reinforcement should be fastened with ties of soft stainless steel, zinc coated wire or non-metallic coated tie wire.

### Epoxy Coated Reinforcement

Epoxy coated reinforcing bars and welded wire fabric are available for use in products which require special corrosion protection. Epoxy coated reinforcing bars should conform to ASTM A775M and epoxy coated welded wire fabric should conform to ASTM A884. These standards describe steel surface protection, coating application, thickness, continuity and adhesion of coating and quality control requirements.

Epoxy coating reduces bond strength (see CSA A23.3 Sect. 12.2). Similarly, the requirements for bond strength may need to be modified.

Bars are generally coated when straight. Subsequent shearing, bending and handling should conform to ASTM D3963M. Damaged or removed epoxy coating should be patched using the manufacturer's approved patch compound. All tie wire, chairs and bar supports used for the installation of coated rebar should be non-metallic materials or materials with an acceptable protective coating. Epoxy coated reinforcing bars should be handled with nylon slings.

### Epoxy Coated Strand

Epoxy coated strand is a relatively new product proposed for use in unusually aggressive environments. This material is described in the standard ASTM A882 "Standard Specification for Epoxy-Coated Seven-Wire Prestressing Steel Strand". For bonded construction, the epoxy coating is impregnated with a grit to develop a bond with the concrete.

The behavior of epoxy coated strand at elevated temperatures is of concern due to the softening of the epoxy. Pull-out tests show there is a progressive reduction in bond strength initiated at about 50°C with a virtual complete loss of bond occurring at about 90°C. This behavior necessitates a careful monitoring of concrete temperature at transfer of prestress.[R1•20]

Because of the uncertainties in the properties noted above, particularly the behavior under elevated temperatures, it is recommended that epoxy coated strand not be used for pretensioned precast concrete products.

## 1.3.3 GROUT, MORTAR AND DRYPACK

When water, sand and a cementitious material are mixed together without coarse aggregate, the result is called mortar, grout, or drypack, depending on consistency. These materials have numerous applications in precast concrete construction: sometimes only for fire or corrosion protection, or for cosmetic treatment, other times to transfer loads in horizontal and vertical joints.

Different cementitious materials are used:

1. Portland cement.
2. Shrinkage-compensating portland cement.
3. Expansive portland cement made with special additives.
4. Gypsum or gypsum/portland cements.
5. Epoxy resins.

### Sand-Cement Mixtures

Most grout is a simple mixture of portland cement, sand, and water. Proportions are usually one part cement to 2.25 to 3 parts sand. The amount of water depends on the method of placement.

Flowable grouts are high-slump mixes used to fill voids that are either formed in the field or cast into the precast element. They are used at joints that are heavily congested but not confined, thus requiring some formwork. These grouts usually have a high water-cement ratio, resulting in low strength and high shrinkage. There is a tendency for the solids to settle, leaving a layer of water on the top. Special ingredients or treatments can improve the characteristics of grout.

For very small spaces in confined areas, grouts may be pumped or pressure injected. The confinement must be sufficiently strong to resist the pressure. Less water can be used than for flowable grouts, hence less shrinkage and higher strength can be expected.

A stiffer grout, or mortar, is used when the joint is not totally confined, for example in vertical joints between wall panels. This material will usually develop strengths of 20 MPa to 45 MPa, and have much less shrinkage than flowable grout.

Drypack is the name used for very stiff sand-cement mixes. It is used if a relatively high strength is desired, for example, under column base plates. Compaction is by hand tamping, using a rod or stick.

When freeze-thaw durability is a factor, grout should be air-entrained. An air content of 9 or 10% may be required for adequate protection.

Typical portland cement mortars have very slow early strength gain when placed in cold weather. Heating is usually not effective, because the heat is rapidly dissipated to the surrounding concrete. Thus, unless a heated enclosure can be provided, special proprietary mixes, usually containing gypsum, may be required. However, these mixes may deteriorate under prolonged exposure to water.

### Non-Shrink Grout

Shrinkage of sand-cement grout can be reduced by using proprietary non-shrink mixes, or by adding aluminum powder to the mix. Non-shrink grouts can be classified by the method of expansion:

1. Gas-liberating.
2. Metal-oxidizing.
3. Gypsum-forming.
4. Expansive cement.

Some expansive ingredients may cause undesirable effects in some applications, so manufacturers' recommendations should be followed.

Aluminum powder added to ordinary sand-cement grout forms a gas-liberating mixture. Extremely small amounts of powder are required (about a teaspoonful per bag of cement). Thus these mixes are very sensitive and trial mixes should be tested.

### Epoxy Grouts

Epoxy grouts are used when very high strength is desired, or positive bonding to the concrete is necessary. They are mixtures of epoxy resins and a filler material, usually sand.

The physical properties of epoxy compounds vary widely. The user should be familiar with the compound to be used, either through experience or test[R1.8]. Of particular importance in some applications is the thermal expansion, which can be up to 7 times that of concrete.

Epoxy resins without fillers are sometimes pressure-injected into cracked concrete as a repair measure.

## 1.4 POST-TENSIONED CONCRETE MATERIALS

Post-tensioned construction differs from pre-tensioned construction in the timing of the application of stress to the prestressing tendons. With pre-tensioned elements, stress is applied to the steel before concrete is placed and has cured, while in the post-tensioned elements, the stress is applied to the steel after the concrete has been placed and has cured to an acceptable compressive strength.

Post-tensioned construction offers many of the benefits of pre-tensioned construction:

1. Effective use of high strength materials.
2. Shallow, more attractive structures.
3. Control of deflections.
4. Virtual elimination of cracks.
5. Long, economical spans.
6. Low cost fire resistant construction.

and offers other benefits as well:

7. Prestressing of field construction.
8. Connections for precast elements
9. Stage prestressing.
10. Watertight construction.
11. Reduced prestress losses.
12. Structural continuity.

Post-tensioning methods and materials may be applied to both precast and cast-in-place concrete construction as well as rehabilitation and construction of wood and steel structures and elements.

### 1.4.1 POST-TENSIONING MATERIALS

Post-tensioning hardware generally consists of sheathing, tendons and anchors. Although there are several different post-tensioning "systems" in use, these three basic components are common to all.

#### Sheathing

Sheathing may be fabricated from either steel or plastic. In bonded post-tensioned construction where a bond between the concrete outside the sheath and the grout inside the sheath is required, steel sheathing is employed. Steel sheath may be made from strip steel formed into a corrugated, helical tube, or from thin walled steel tube.

In unbonded post-tensioned construction with monostrand tendons, plastic sheath is employed as it does not completely bond to the concrete. Inside the sheath, the post-tensioned strand is lubricated with a non-corrosive grease for additional corrosion protection.

### Tendons

Post-tensioning tendons can be of two different categories: strand or wire. These designations also determine the general post-tensioning system.

Wire and strand properties are given in Fig. 8.2.3.

Of the several grades and sizes of seven wire strand available, the most commonly used is 16 mm diameter Grade 1760 strand. Less common, but still in use is 13 mm diameter Grade 1860 strand. Other sizes and grades are not extensively used for post-tensioning in Canada.

Prestressing wire is made of high tensile steel and is available in several diameters. Although popular in the 1960's and 1970's, wire post-tensioning is no longer common, as strand systems are more economical.

### Anchorages

The anchorages for post-tensioning tendons are specially designed for the type of tendon that they are anchoring. Several types of anchorages are shown in Figs. 8.2.5 to 8.2.7. The information is presented in these diagrams in "generic" form. While dimensions may vary from actual sizes by 10%, the data provided will allow for adequate detailing of structural elements. Post-tensioning subcontractors should be consulted for actual sizes.

Anchorage systems can be used for live ends (i.e. the end of the tendon to which force is applied) or dead ends (i.e. the end of the tendon which is fixed within the concrete, or to which no direct force is applied-also called "fixed" end). In many cases, live end anchorages may be used as dead end anchorages, or may be used for stressing the opposite end of a tendon where friction losses are high.

Seven wire strand systems are perhaps the most versatile, with various anchorages for multistrand (2 or more strands per tendon), live and dead end conditions, as well as monostrand systems (1 strand per tendon).

Multistrand live end anchors consist of an anchor head which grips the individual strands by means of tapered wedges encircling the strand, inserted in conical holes in the anchor head (Fig. 8.2.5). Dead end or fixed anchors can be made of individual strands with the ends crimped or "flowered" to form a bulb. The concrete flows into this bulb and cures, effectively anchoring it (Fig. 8.2.6).

The monostrand system consists of a steel casting which can be used as a live, intermediate or dead end anchor. Used mostly for unbonded suspended slabs, monostrand tendons may be bonded by providing a steel tube sheath and grout vents. The strand in this case is not greased (Fig. 8.2.7).

### Protection of Tendons

The minimum clear cover requirements of Clause 12.6.3.1 of CSA-A23.1 for post-tensioning tendons are summarized in Fig. 1.4.1.

## 1.4.2 PRESTRESSED ROCK AND SOIL ANCHORS

### Description and Uses

Prestressed ground anchors, used in all types of civil engineering, provide a constant active force as illustrated in Fig. 1.4.2 for:

a) Anchoring of support structures.
b) Tying down cantilever structures.
c) Stabilization of uplift.

**Fig. 1.4.1 Minimum cover requirements for post-tensioned concrete**

| Exposure to earth or weather | Exposed | Not exposed |
|---|---|---|
| (a) When cast against and permanently exposed to earth | 75 mm | |
| (b) For<br>(i) Beams, girders, columns and piles;<br>Principal reinforcement, No. 35 and smaller | 50 mm | 40 mm |
| Ties, stirrups, and spirals | 40 mm | 30 mm |
| (ii) Slabs, walls, and joists, shells and folded plates: No. 20 and smaller | 30 mm | 20 mm |
| (c) For bars with a diameter $d_b$ larger than listed above, the cover shall be at least but need not be more than 60 mm | 1.5 $d_b$ | 1.0 $d_b$ |
| (d) The ratio of the cover to the nominal maximum aggregate size shall be at least | 1.5 $d_b$ | 1.0 $d_b$ |
| (e) The cover for a bundle of bars shall be the same as that for a single bar with an equivalent area. | | |

**Fig. 1.4.2 Applications of prestressed rock and soil anchors**

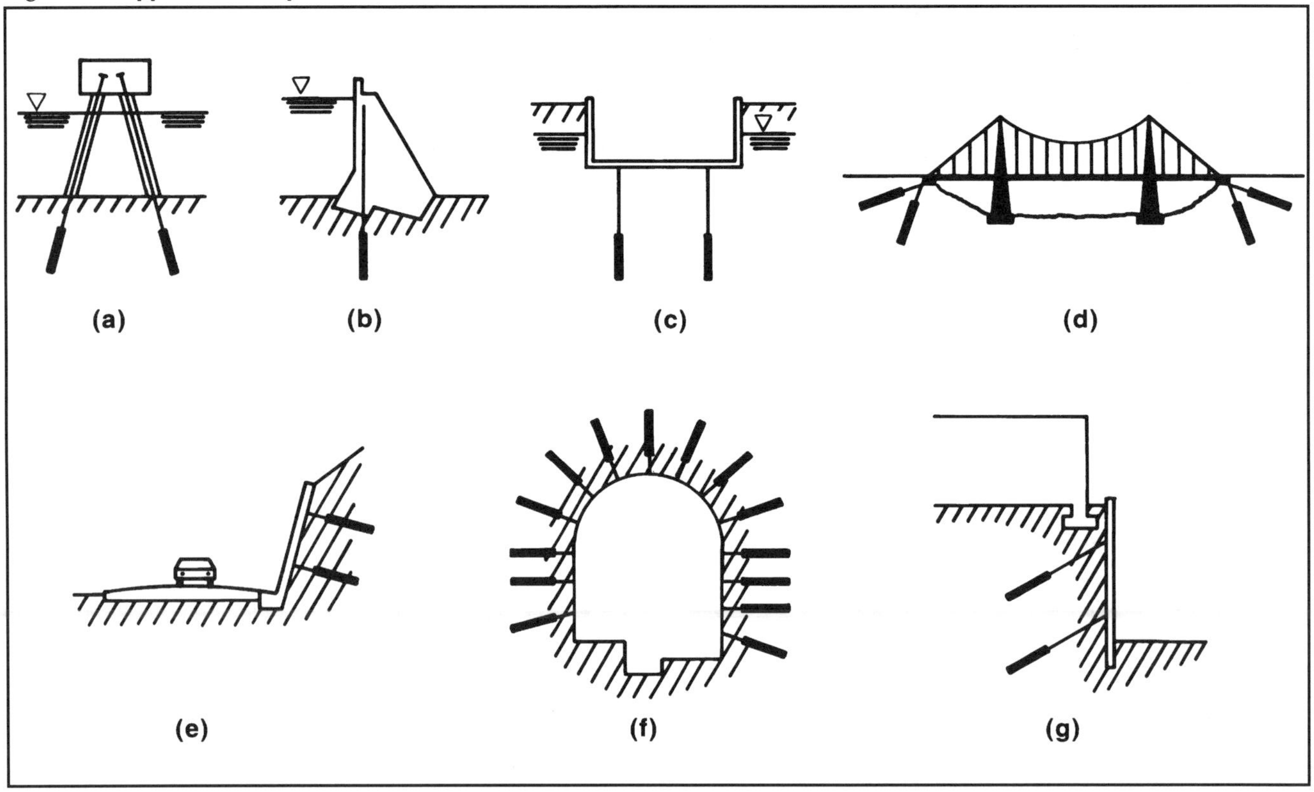


d) Transfer of external tensile forces to the ground.
e) Stabilization of retaining walls.
f) Rock stabilization and roof support in mining.
g) Foundation stabilization and tunnel work.

Ground anchors consist of one or more tendons grouted in a hole of suitable diameter and length. The bond length is that portion of the anchor that permits a transfer of force from the strands to the ground. The free stressing length refers to that portion of the anchor free to elongate during stressing (see Fig. 1.4.3). Ground anchors are categorized as either rock or soil anchors.

Rock anchors are installed in holes drilled to the ground and into solid rock. The anchor force is transmitted to the rock mainly through friction between the rock and cement grout or resin.

Soil anchors transmit the anchor force to the soil or loose rock in two ways: through friction between the soil and cement grout, or through an enlarged pressure grout bulb creating an end bearing force component.

### Installation Methods

Four types of installation methods are described as follows (see Fig. 1.4.4):

*a) Gravity grouted, straight shaft bore holes:*

This type of anchor is employed in rock and very stiff cohesive deposits. The borehole may be lined or unlined, depending on the hole stability, and resistance to withdrawal is dependent on side shear (friction) at the grout/ground interface.

*b) Low pressure (1-2 MPa) grouted bore holes:*

These are employed most commonly in soft fissured rock and coarse alluvial deposits, as well as in fine grained cohesionless soil. Because the grout permeates through fissures and pores in rock or soil, the pullout capacity may include an end bearing component with the side shear (friction) component.

*c) High pressure (2-4 MPa) grouted bore holes:*

The strength of this anchor is derived using the assumption of a uniform shear component along the fixed anchor length. Primarily used in cohesionless soils, the grouted fixed anchor is increased in size due to compaction of the soil or hydrofracturing due to pressure.

*d) Gravity grouted bore hole with mechanical enlargements:*

The mechanical enlargement or bell gives this anchor an end bearing component as well as a side shear (friction) component. It is most commonly employed in stiff or hard cohesive soils.

### Further Information

An extensive treatment of ground anchor design, testing, and corrosion protection is beyond the scope of this work. Further information on these and other subjects may be found in the reference material and by consulting suppliers.

Fig. 1.4.3 Typical ground anchor

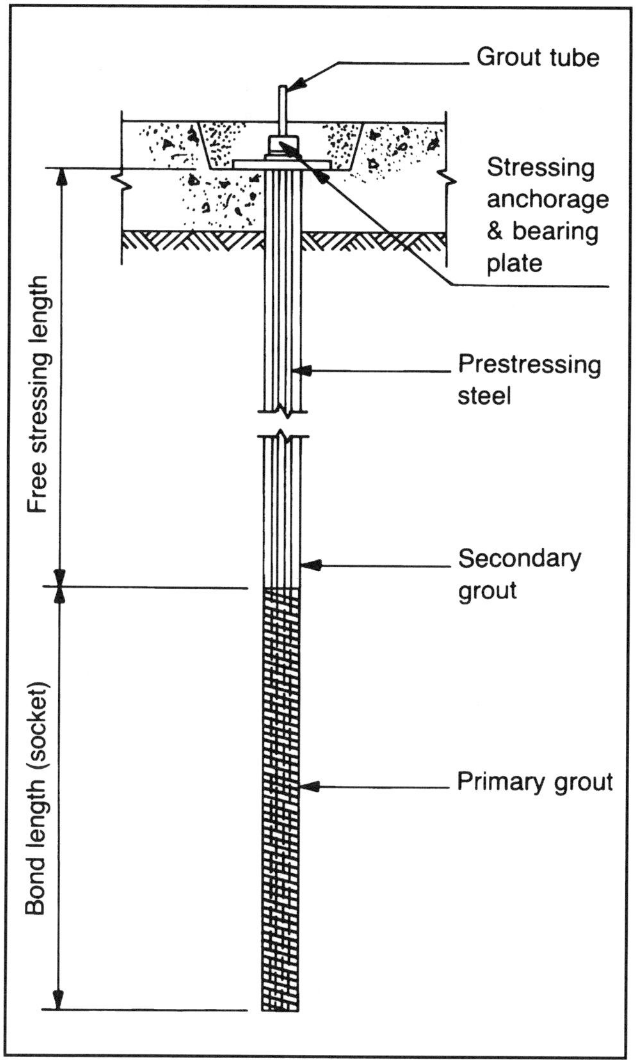


Fig. 1.4.4 Ground anchor installation methods

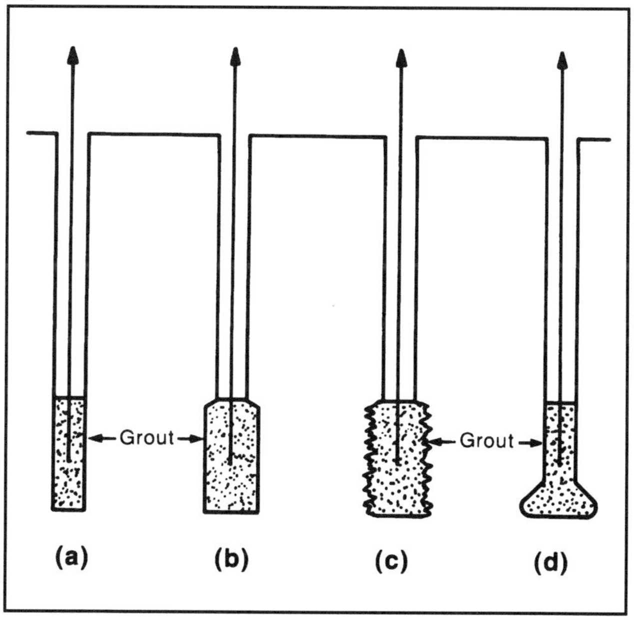


## 1.4.3 SEGMENTAL CONSTRUCTION

Post-tensioning products and techniques are used extensively in segmental construction. Here, the primary load carrying elements consist of precast (sometimes pretensioned) or cast-in-place concrete segments post-tensioned together for structural integrity. A familiar example is a precast segmental bridge.

Various post-tensioning components are used as follows:

### Anchorages

Post-tensioning tendon anchors are located in end block or anchorage areas. Their purpose is to safely transfer anchorage forces into the structure during all loading stages, including time of initial post-tensioning.

When all tendons extend the full length of the structure, the end blocks are located in end segments of segmental structures, which, in a simply supported beam system, will also be the bearing or the support segments. In some cases, however, anchorage areas may be located in intermediate segments (continuous beam or cantilever structures with segmental tendons distributed according to the moment or shear diagram) or in nonbearing end segments (cantilever structures).

Special attention should be given to proper reinforcement of bearings and anchorages. Anchorage zones should contain sufficient horizontal and vertical stirrups or grillage reinforcement placed in the plane parallel to the end surface.

The anchor plate is the component of a post-tensioning system which transmits the prestress force from the tendon anchoring device directly to the concrete. Its function is to distribute the concentrated force from the anchoring device over a larger bearing area to the concrete.

The anchor plate may be placed in any of the following ways:

1. Cast into precast segment.
2. Placed against precast surface.
3. Placed against cushioning materials.

### Sheathing

Sheathing is used to form the holes or enclose the space in which the prestressing steel is located. The prestressing steel is installed after the concrete is placed. The cross section of the sheathing must be adequate to allow proper installation of the prestressing steel and to provide enough passage area for filling the tube with grout after stressing.

### Grouting

The purpose of grouting is to provide corrosion protection to the prestressing steel, and to develop bond between the prestressing steel and the surrounding concrete. To accomplish this, the grout has to fill all the voids in and around the post-tensioning tendon for its entire length.

### Unbonded Tendons

Unbonded tendons may be used in segmental construction provided that the strength of the anchorage is not less than the strength of the tendon. In unbonded post-tensioning, a corrosion protection system must be provided which ensures at least the same degree of corrosion protection as grout. This can be achieved by using a protective coating on the prestressing steel and enclosing the coated tendon in an encasement to protect the coating during handling, installation and stressing of the tendon.

### Placement and Stressing of Tendons

When tendons are installed in the segments before casting, they are subsequently coupled together at each joint. This construction method permits stressing of part of a tendon, after installing one or more segments, before the full length is completely installed.

Tendons may also be installed after casting and erection and may be coupled together at the joint. Usually, however, these tendons are installed full length. It may be noted that this procedure permits intermediate stressing of portions of the structure by using tendons of variable lengths, stressing the short ones first and the long ones later.

Special attention must be given to the corrosion protection of the post-tensioning steel which, at any stage of construction, must remain unbonded. If a tendon which contains couplers is to be stressed over its full length, the couplers must be able to accommodate the movement anticipated at that location.

### References:


R1.1 Fintel, M., Editor, *Handbook of Concrete Engineering,* Van Nostrand Reinhold Company, New York, NY, Second Edition, 1984.

R1.2 *Design and Control of Concrete Mixtures,* Sixth Canadian Edition, Canadian Portland Cement Association, Ottawa, 1995.

R1.3 *ACI Manual of Concrete Practice* (5 Volumes), American Concrete Institute, Detroit, MI, 1996.

R1.4 *Concrete Manual, A Water Resources Technical Publication,* Eighth Edition, U.S. Department of Interior, Bureau of Reclamation, Engineering and Research Center, Denver, CO, 1975.

R1.5 Pfeifer, D.W., Marusin, Stella, and Langren, J.R., "Energy Efficient Accelerated Curing of Concrete", *Technical Report No. 1,* Prestressed Concrete Institute, Chicago, IL, 1981.

R1.6 *Manual of Standard Practice,* Reinforcing Steel Institute of Canada, Second Edition, 1995.

R1.7 *Corps of Engineers Specification for Non-shrink Grout,* CRD-C588-78A, U.S. Army Corps of Engineers, 1978.

R1.8 ACI Committee 503, "Use of Epoxy Compounds with Concrete", *ACI Journal,* V. 70, No. 9, Sept., 1973.

R1.9 Martin, L.D. and Pellow, D.L. "Low-Relaxation Strand-Practical Applications in Precast Prestressed Concrete", *PCI Journal,* V. 28, No. 4, July-August, 1983.

R1.10 *Post-Tensioning Manual,* Fifth Edition, Post-Tensioning Institute, Phoenix, AZ, 1990.

R1.11 Guyon, Y., *Limit-State Design of Prestressed Concrete,* Vol. 2, The Design of the Member, Translated by F. Turner, John Whlie & Sons, Inc., New York, 1974.

R1.12 Leonhardt, F., *Prestressed Concrete,* Wilhelm Ernst & Son, (Berlin), Second Edition, 1964.

R1.13 Construction Industry Research & Information Asoc., *A Guide to the Design of Anchor Blocks for Post-tensioned Prestressed Concrete,* Ciria (London), 1976.

R1.14 Canadian Standards Association, *Design of Concrete Structures,* A23.3-94, Toronto, 1994.

R1.15 Ministry of Transportation of Ontario, *Ontario Highway Bridge Design Code,* Third Edition, 1991.

R1.16 Lin., T.Y. Burns, N.H., *Design of Prestressed Concrete Structures,* Third Edition, John Wiley & Son, 1981.

R1.17 *Recommendations for Prestressed Concrete Ground Anchors,* Third Edition, Post-Tensioning Institute, 1996.

R1.18 The Concrete Society, *Post-tensioned Flat Slab Design Handbook, Technical Report No. 25,* The Concrete Society, 1984.

R1.19 *Architectural Precast Concrete* - Second Edition, Precast/Prestressed Concrete Institute, 1989.

R1.20 "Guidelines for the Use of Epoxy Coated Strand", JR383, 1993, Precast/Prestressed Concrete Institute.

R1.21 Recomendations for Prestressed Rock and Soil Anchors, 3rd Edition, Post-Tensioning Institute, Phoenix, 1996.

R1.22 Field Procedures Manual for Unbonded Single Strand Tendons, 2nd Edition, Post-Tensioning Institute.

# CHAPTER 2
# ANALYSIS AND DESIGN OF PRECAST AND PRESTRESSED CONCRETE STRUCTURES

Page No.

## 2.0 NOTATION

Note: Notation for earthquake design is in Sect. 2.8.0.

$A$ = area (with subscripts)

$A_{vf}$ = area of shear friction reinforcement

$b$ = width of a section or structure

$C$ = compressive force

$C$ = coefficient of thermal expansion

$C_f$ = factored compressive force

$C_m$ = a factor relating actual moment to equivalent uniform moment

$D$ = dead load

$E$ = modulus of elasticity of a beam (with subscripts)

$E_c$ = modulus of elasticity of column concrete

$e$ = eccentricity of axial load

$F_b$ = degree of base fixity (decimal)

$F_i$ = lateral force at bay i or shear wall i

$f'_c$ = specified concrete compressive strength

$f'_m$ = specified compressive strength of masonry

$f_t$ = unfactored tensile stress

$f_{ft}$ = factored tensile stress

$f_y$ = specified yield strength of steel

$G$ = shear modulus of elasticity (modulus of rigidity)

$g$ = assumed length over which elongation of the anchor bolt takes place

$H_f$ = total factored lateral force within a storey

$h$ = column width in direction of bending

$h_s$ = storey height

$I$ = moment of inertia

$I_b$ = moment of inertia of a beam

$I_{bp}$ = moment of inertia of base plate (vertical cross-section dimensions)

$I_c$ = moment of inertia of a column

$I_f$ = moment of inertia of the footing (plan dimensions)

$I_g$ = uncracked moment of inertia

$K$ = stiffness (with subscripts)

$K$ = constant used for the calculation of equivalent creep and shrinkage shortening

$K_r$ = relative stiffness

$K_t$ = constant used for the calculation of equivalent temperature shortening

$k$ = effective length factor

$k_b$, $k_f$, $k_m$ = coefficients used to determine forces and moments in beams and columns

$k_s$ = coefficient of subgrade reaction

$L$ = live load

$\ell$ = length of span or structure

$\ell_n$ = clear span

$\ell_s$ = distance from column to centre of stiffness

$\ell_u$ = unsupported length of a compression element

$\ell_w$ = length of weld

$M$ = unfactored moment

$M_t$ = volume change moment

$N$ = unfactored horizontal force

$N_f$ = factored horizontal force

$P$ = shrinkage modification factor (with subscripts)

$P$ = unfactored axial load

$P_c$ = critical load

$P_o$ = prestressing force after assumed initial loss

$P_f$ = factored axial load

$Q$ = creep modification factor (with subscripts)

$Q$ = stability index

$Q$ = statical moment

$r$ = radius of gyration

$T$ = unfactored tensile force

$T_f$ = factored tensile force

$t$ = thickness

$t$ = time in days

$V$ = shear at support (with subscripts)

$V_f$ = factored shear force

$V_r$ = factored shear resistance

$V_w$ = total wind shear

$v_r$ = unit shear on panel edge

$v_f$ = factored unit shear

$W$ = total lateral load

$x_1$ = distance from face of column to centre of anchor bolts

$x_2$ = distance from face of column to base plate anchorage

$\alpha$ = load factor (with subscripts)

$\alpha_1$ = ratio of average stress in rectangular compression block to the specified concrete strength

$\beta_1$ = stress block depth factor

$\gamma$ = flexibility coefficient (with subscripts)

$\Delta$ = total equivalent shortening or column deflection

$\Delta_f$ = deflection due to factored loads

$\delta$ = moment magnifier

$\delta$ = volume-change shortening (with subscripts)

$\epsilon$ = unit strain (with subscripts)

$\theta$ = rotation (with subscripts)

$\mu_e$ = effective shear friction coefficient

$\phi$ = resistance factor

$\psi$ = ratio of column to beam stiffnesses

## 2.1 INTRODUCTION

This chapter provides guidelines for the analysis and design of structures that are comprised wholly or partially of precast and/or precast, prestressed components. The primary advantages of precast concrete include:

1. Construction speed.
2. Plant-controlled quality control.
3. Fire resistance and durability.
4. With prestressing: greater span-depth ratios, more controllable performance, less material usage.
5. Architectural precast concrete provides a wide variety of highly attractive surfaces and shapes.
6. Thermal and acoustical control.

To fully realize these benefits and thereby gain the most economical and effective use of the material, the following general principles are offered:

1. Precast concrete is basically a "simple-span" material, but continuity is often effectively achieved with properly conceived connection details.
2. Sizes and shapes of elements are often a function of production, hauling and erection considerations.
3. Concrete is a massive material. This is an advantage for such things as stability under wind loads, acoustical and vibration control, reduction in heating and cooling loads, and fire resistance. Also, the high dead to live load ratio will provide a greater safety factor against gravity overloads. However, framing details, or loading conditions such as earthquake, which result in eccentrically loaded supports, need careful attention.
4. Maximum economy is achieved with maximum repetition, and standard sections should be used whenever possible.
5. Successful use is largely dependent on carefully conceived connection design and details.
6. The effects of restraint of volume changes caused by creep, shrinkage and temperature change must be considered in every structure.
7. While architectural panels are often used only as cladding, the inherent load-carrying capacity of these products should not be overlooked.
8. Prestressing improves the economy and performance of precast members, but is usually only feasible when they are standard shapes and capable of being cast in "long-line" beds.

## 2.2 PRELIMINARY ANALYSIS

Maximum economy occurs when the building is laid out to take advantage of the principles discussed above. The primary considerations in preliminary analysis of the total structure are:

1. Framing dimensions
2. Span to depth ratios
3. Lateral load resisting systems
4. Control of volume change deformations and restraint forces
5. Connection concepts.

## 2.2.1 FRAMING DIMENSIONS

When possible, bays should be laid out to fit the module of the components selected. Standard section dimensions are shown in Chap. 7, but others may be available in a given geographical area. Height and width of wall and deck units may be limited by over the road hauling regulations.

It is often feasible to cast wall panels and columns in multi-storey units, and economy is achieved with fewer pieces to handle. Length and weight limitations for hauling and erection, and stability during erection, are items to be considered in this decision.

### 2.2.2 Span to Depth Ratios

Selection of floor to floor dimensions should consider the practical span to depth ratio of the horizontal framing elements, allowing adequate space for mechanical ductwork.

Typical span to depth ratios of flexural precast, prestressed concrete elements are:

| | |
|---|---|
| Hollow core floor slabs | 30 to 40 |
| Hollow core roof slabs | 40 to 50 |
| Stemmed floor slabs | 25 to 35 |
| Stemmed roof slabs | 35 to 40 |
| Beams | 10 to 20 |

These values are intended as guidelines, not limits. The required depth of a beam or slab is influenced by the ratio of live load to total load. Where this ratio is high, deeper sections may be needed.

For reinforced flexural elements, span-depth ratios are given in Table 9-1, CSA-A23.9-94.

## 2.2.3 LATERAL LOAD RESISTING SYSTEMS

The most important task in the preliminary analysis is the selection of the lateral load resisting system. Methods used to resist lateral loads, in the approximate order of economy, include:

1. Shear walls. These can be of precast concrete, cast-in-place concrete, or masonry. These are discussed in more detail in Sect. 2.7. When architectural or structural precast elements are used for the exterior cladding, they can often be used as shear walls.
2. Cantilevered columns or wall panels. This is usually only feasible in low-rise buildings. Base fixity can be attained by fixing the column to the footing. A detailed analysis of the footing rotation can be made as described in Sect. 2.6.2.
3. Steel or concrete X-bracing. This has been used effectively in some mid-rise buildings, and is largely dependent on connection details. A related resistance system, in terms of analysis procedures, occurs naturally in parking structures with sloped decks in the direction of traffic flow.
4. Moment resisting frames. Only the minimum number of frames should be used to avoid building up substantial volume change forces. However, building function sometimes makes it necessary. It is sometimes feasible to provide a moment connection at only one end of an element, or a connection that will resist moments with lateral forces in one direction but not in the other, to reduce the buildup of restraint forces. To reduce the number of moment resisting frames required, a combined shear wall-frame system may be used. Moment resisting frames are discussed in more detail in Sect. 2.6.

All of the above systems depend on distribution of lateral loads through diaphragm action of the roof and floor systems (see Sect. 2.5).

## 2.2.4 CONTROL OF VOLUME CHANGE DEFORMATIONS AND RESTRAINT FORCES

Volume changes of concrete are those resulting from creep, shrinkage and temperature change. Creep and shrinkage cause a shortening of the element, so the critical condition is creep shrinking and temperature drop.

It is usually desirable to design connections so that volume change shortening is accommodated. Sect. 2.4 provides data and guidelines for estimating the amount of shortening that may take place.

Problems that have been caused by restraint of volume change movements have appeared when relatively long elements, usually stemmed deck units, were welded to their supports at the bottom on both ends. When such elements are connected only at the top, experience has shown that volume changes are adequately accommodated. Top connections may attract negative moments if compression resistance is encountered at the bottom. This may be difficult to accommodate.

Connections using cast-in-place concrete have exhibited few volume change problems. This is probably because micro-cracking and creep in the cast-in-place portion effectively relieve the restraint.

Long buildings may require full height expansion joints. This is discussed in Sect. 2.3.

## 2.2.5 CONNECTION CONCEPTS

The types of connections to be used should be determined during the preliminary analysis, as this may have an effect on the component dimensions, the overall structural behavior, as discussed above, and on the erection procedure. Chap. 4 and some PCI publications are devoted entirely to connections.

# 2.3 EXPANSION JOINTS

## 2.3.1 GENERAL

Joints are placed in structures to limit the magnitude of forces which result from volume change deformations (temperature changes, shrinkage and creep), and to permit movements (volume change deformations) of structural elements. If the forces generated by temperature rise are significantly greater than shrinkage and creep forces, a true "expansion joint" is needed. However, in concrete structures, true expansion joints are seldom required. Instead, joints that permit contraction of the structure are needed to relieve the strains caused by temperature drop and restrained creep and shrinkage, which are additive. Such joints are properly called contraction or control joints but are commonly referred to as expansion joints.

It is desirable to have as few expansion joints as possible. Expansion joints are often located by "rules of thumb" without considering the structural framing method. The purpose of this section is to present guidelines for determining the spacing and width of expansion joints.

## 2.3.2 SPACING OF EXPANSION JOINTS

There is a wide divergence of opinion concerning the spacing of expansion joints. Typical practice in concrete structures, prestressed or reinforced, is to locate expansion joints at distances between 45 m and 90 m. However, reinforced

**Fig. 2.3.1 General guidelines for determination of building length without use of expansion joints**

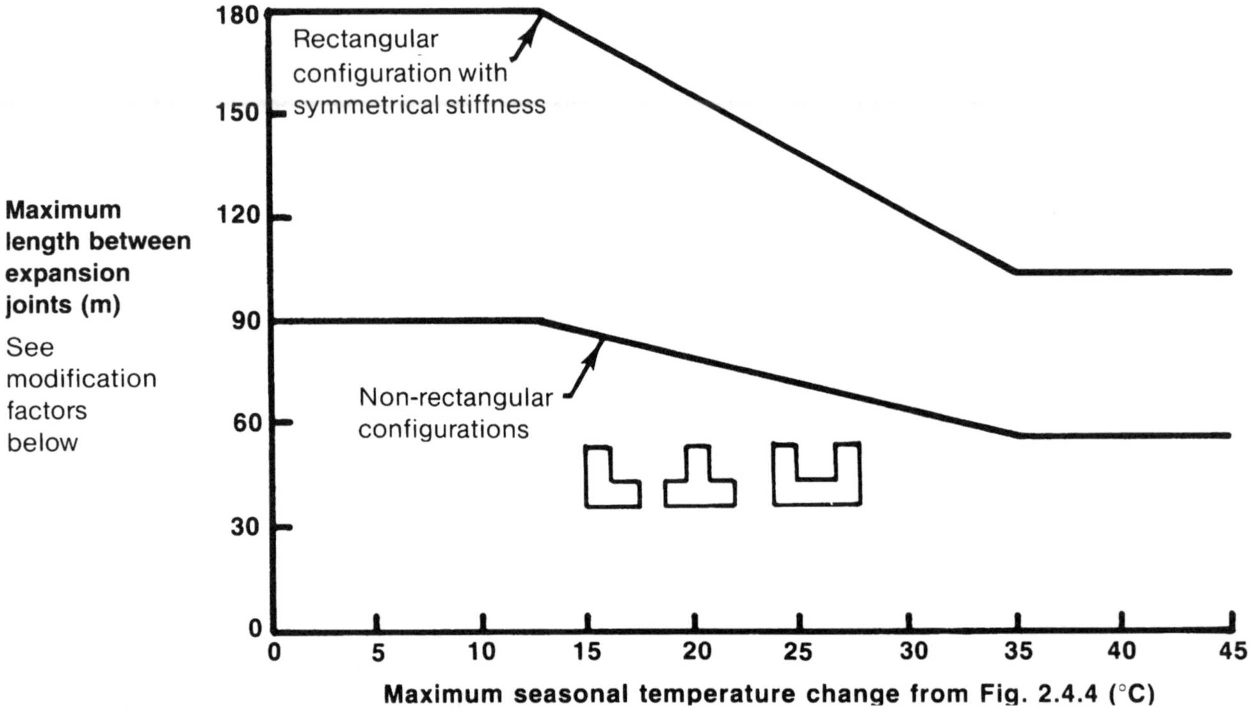


These curves are directly applicable to buildings of beam-and-column construction, hinged at the base, and with heated interiors. When other conditions prevail, the following rules are applicable:

(a) If the building will be heated only and will have hinged column bases, use the allowable length as specified;

(b) If the building will be air conditioned as well as heated, increase the allowable length by 15 percent (provided the environmental control system will run continuously);

(c) If the building will be unheated, decrease the allowable length by 33 percent;

(d) If the building will have fixed-column bases, decrease the allowable length by 15 percent;

(e) If the building will have substantially greater stiffness against lateral displacement at one end of the plan dimension, decrease the allowable length by 25 percent.

When more than one of these design conditions prevail in a building, the percentile factor to be applied should be the algebraic sum of the adjustment factors of all the various applicable conditions.

*Note:* A detailed volume change frame analysis will provide more accurate expansion joint requirements.

*Source:* "Expansion Joints in Buildings", *Technical Report No. 65,* National Research Council, National Academy of Sciences, 1974.

concrete buildings exceeding these limits have performed well without expansion joints. Recommended joint spacings for precast concrete buildings are generally based on experience and may not consider several important items. Among these items are the types of connections used, the column stiffness in simple span structures, the relative stiffness between beams and columns in framed structures, and the weather exposure conditions. Non-heated structures such as parking garages, are subjected to greater temperature changes than occupied structures, so shorter distances between expansion joints are warranted.

Sects. 2.4 and 2.6 present methods for analyzing the potential movement of framed structures, and the effect of restraint of movement on the connections and structural frame. This information along with the connection design methods in Chap. 4 can aid in determining the spacing of expansion joints.

**Fig. 2.4.1 Creep and shrinkage modification factors for non-standard conditions***

$$Q_{cr} = Q_a Q_h Q_f Q_r Q_{sl} Q_v$$
$$P_{sh} = P_c P_h P_f P_r P_{sl} P_v$$

| Age at loading (days) | $Q_a$ Moist cured | $Q_a$ Accelerated cured | Relative humidity[4] (%) | $Q_h$ | $P_h$ |
|---|---|---|---|---|---|
| 1 | 1.25 | 1.00 | 40 | 1.00 | 1.00 |
| 7 | 1.00 | 0.94 | 60 | 0.87 | 0.80 |
| 20 | 0.87 | 0.85 | 80 | 0.73 | 0.60 |
| 60 | 0.77 | 0.76 | 100 | 0.60 | 0.00 |
| **Ratio of fine to total aggr.[3]** | **$Q_f$** | **$P_f$** | **Volume/surface ratio (mm)[2]** | **$Q_r$** | **$P_r$** |
| 0.30 | 0.95 | 0.72 | 38 | 1.00 | 1.00 |
| 0.40 | 0.98 | 0.86 | 75 | 0.82 | 0.84 |
| 0.50 | 1.00 | 1.00 | 150 | 0.70 | 0.59 |
| 0.70 | 1.05 | 1.04 | 250 | 0.67 | 0.37 |
| **Slump (mm)** | **$Q_{sl}$** | **$P_{sl}$** | **Air (%)** | **$Q_v$** | **$P_v$** |
| 0 | 0.82 | 0.85 | ≤6 | 1.00 | 1.00 |
| 50 | 0.95 | 0.97 | 8 | 1.18 | 1.01 |
| 70 | 1.00 | 1.00 | 10 | 1.36 | 1.03 |
| 125 | 1.15 | 1.09 | | | |

| | Cement content (kg/m³) | | |
|---|---|---|---|
| | 225 | 300 | 410 |
| $P_c$ | 0.89 | 0.93 | 1.00 |

(1) Standard conditions produce modification factors of 1.0.

(2) Volume/surface ratio of a rectangular member having a × b cross-section is ab/(2a + 2b).

(3) Ratio of fine aggregate to total aggregate is expressed as the ratio of the masses.

(4) For average ambient relative humidity, see Fig. 2.4.2.

* Above table based on CPCA Concrete Design Handbook, Second Edition.

Fig. 2.3.1 shows recommended expansion joint spacing. Note that the spacings obtained from the graph in Fig. 2.3.1 should be modified for various conditions as shown in the notes beneath the graph. Values for the design temperature change can be obtained from Sect. 2.4.

When expansion joints are required in non-rectangular structures, they should always be located at locations where the plan or elevation dimensions change radically.

### 2.3.3 WIDTH OF EXPANSION JOINTS

The width of the joint can be calculated theoretically using a coefficient of expansion of $10 \times 10^{-6}$ mm/(mm · °C) for normal density concrete and $9 \times 10^{-6}$ mm/(mm · °C) for low density concrete. The report referenced in Fig. 2.3.1 recommends a minimum width of 25 mm. However, since the primary problem in concrete buildings is contraction rather than expansion, joints that are too wide may result in problems with reduced bearing or loss of filler material.

## 2.4 VOLUME CHANGES

The strains resulting from creep, shrinkage, and temperature change, and the forces caused by restraining these strains have important effects on connections, service load behaviour, and ultimate capacity of precast prestressed structures. Consequently, these strains and forces must be considered in the design.

Vertical elements, such as load bearing wall panels, are also subject to volume change strains. The approximate magnitude can be calculated using Figs. 2.4.1 to 2.4.5. The effects will only be significant in high rise buildings, and then only differential movements between elements will significantly affect the performance of a structure. This can occur, for example, at the corner of a building where load bearing and non-load bearing panels meet.

### 2.4.1 VOLUME CHANGE DATA

#### Creep

Creep of concrete is defined as the time dependent increase in strain under sustained loading. Other time dependent deformations are generally attributed to shrinkage and temperature changes in concrete.

Creep is usually expressed in terms of the creep coefficient, $C_u$, defined as the ratio of creep strain $\epsilon_{cr}$ to initial immediate strain $\epsilon_i$. Creep strain increases with time at a decreasing rate. Creep strain $\epsilon_{cr}$ is determined using the expression:

$$\epsilon_{CR} = \epsilon_i \left( \frac{t^{0.6}}{10 + t^{0.6}} \right) C_u Q_{cr}$$

where:

$$\epsilon_i = \frac{P_i}{AE_{ci}}$$

t = time in days

**Fig. 2.4.2 Annual average ambient relative humidity (%)**

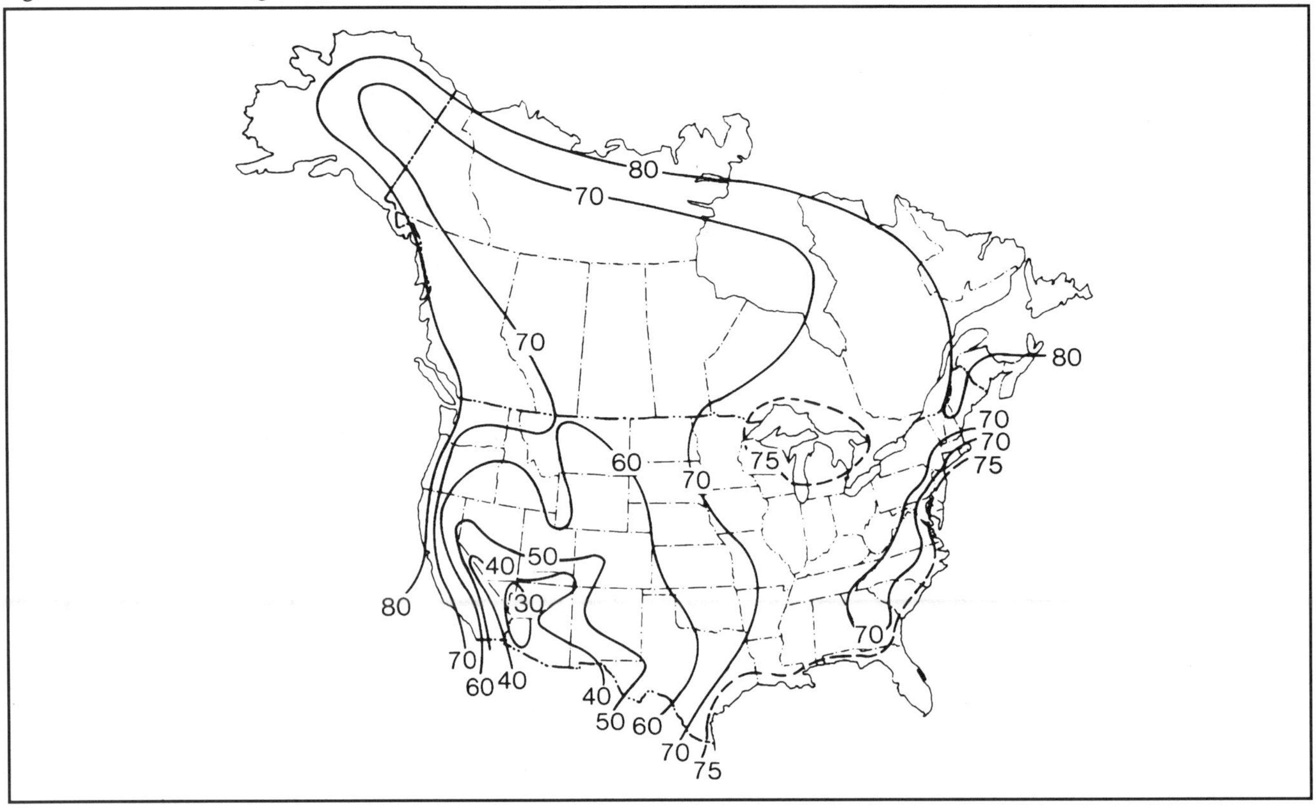


**Fig. 2.4.3 Approximate creep and shrinkage strains**

| Time t (days) | Creep[1] $\epsilon_{cr}=\epsilon_i\left(\frac{t^{0.6}}{10+t^{0.6}}\right)C_uQ_{cr}$ | | Shrinkage[2] $\epsilon_{sh}=\left(\frac{t}{C_s+t}\right)\epsilon_{shu}P_{sh}$ | |
|---|---|---|---|---|
| | Normal density | Low density | Moist cure | Accelerated cure |
| 1 | 59 | 88 | 23 | 16 |
| 3 | 106 | 158 | 62 | 39 |
| 5 | 137 | 204 | 101 | 62 |
| 7 | 160 | 238 | 133 | 86 |
| 10 | 188 | 280 | 172 | 117 |
| 20 | 246 | 367 | 281 | 210 |
| 30 | 286 | 426 | 359 | 273 |
| 40 | 314 | 468 | 413 | 328 |
| 50 | 336 | 501 | 460 | 374 |
| 60 | 356 | 516 | 491 | 406 |
| 90 | 395 | 589 | 562 | 483 |
| 120 | 420 | 626 | 600 | 538 |
| 180 | 456 | 679 | 655 | 601 |
| 1 yr. | 510 | 760 | 710 | 679 |
| 2 yr. | 552 | 822 | 741 | 725 |
| 5 yr. | 594 | 885 | 764 | 757 |
| 10 yr | 613 | 913 | 772 | 764 |
| Final | 630 | 940 | 780 | 780 |

1. Based on $P_i/A = 7.0$ MPa, $f'_{ci} = 25$ MPa, $E_{ci} = 25\,300$ MPa, $C_u = 2.35$, $Q_{cr} = 1.0$. For other values of $P_i/A$ and $E_{ci}$, multiply creep strains by $\frac{3615P_i}{AE_{ci}}$
2. Based on $C_s = 35$ for moist cure, $C_s = 55$ for accelerated cure, Î$_{shu} = 780 \times 10^{-6}$ mm/mm, $P_{sh} = 1.0$.

$C_u$ = ultimate creep coefficient which varies between 1.30 and 4.15

= 2.35 in the absence of specific creep data for local aggregates and conditions

$Q_{cr}$ = modifications factor for non-standard conditions (Fig. 2.4.1)

**Shrinkage**

Shrinkage is the decrease in the volume of hardened concrete with time. Unlike creep, shrinkage is independent of externally applied loads. The decrease in volume is mainly attributed to the moisture loss caused by drying and hydration as well as the chemical changes that result in the carbonation of cement hydration products.

Shrinkage strains start taking place immediately after concrete is exposed to a drying environment.

The shrinkage strain $\in_{sh}$ is determined using the expression:

$$\in_{sh} = \in_{shu}\left(\frac{t}{C_s + t}\right)P_{sh}$$

where:

$\in_{shu}$ = ultimate shrinkage strain

= 780 × $10^{-6}$ mm/mm in the absence of specific shrinkage data for local conditions

$C_s$ = Shrinkage coefficient

= 35 if concrete moist cured for 7 days

= 55 if steam cured for 1-3 days

**Fig. 2.4.4 Maximum seasonal climatic temperature change (°C)**

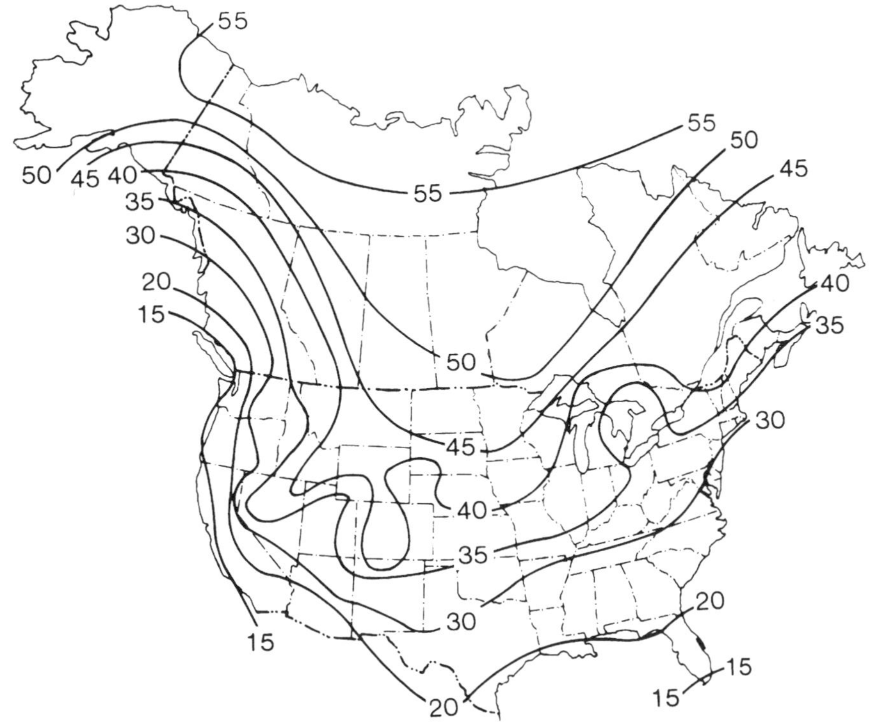


**Fig. 2.4.5 Design temperature strains* ($10^{-6}$)**

| Temperature zone (from Fig. 2.4.4) | Normal density | | Low density | |
|---|---|---|---|---|
| | Heated | Unheated | Heated | Unheated |
| 5 | 27 | 41 | 23 | 34 |
| 10 | 54 | 81 | 45 | 68 |
| 15 | 81 | 122 | 68 | 101 |
| 20 | 108 | 162 | 90 | 135 |
| 25 | 135 | 203 | 113 | 169 |
| 30 | 162 | 243 | 135 | 203 |
| 35 | 189 | 284 | 158 | 236 |
| 40 | 216 | 324 | 180 | 270 |
| 45 | 243 | 365 | 203 | 304 |
| 50 | 270 | 405 | 225 | 338 |

*Based on accepted coefficients of thermal expansion, reduced to account for thermal lag (See referenced committee report, *PCI Journal*, September-October, 1977).

$P_{sh}$ = shrinkage modification factor for non-standard conditions (Fig. 2.4.1)

### Effect of Time on Creep and Shrinkage

The influence of time in the above equation for creep and shrinkage is demonstrated in Fig. 2.4.3, where values of creep and shrinkage strain for a prestressed element under standard conditions are presented.

### Temperature strains

Design values for maximum seasonal climatic temperature for various geographic locations are shown in Fig. 2.4.4. Design values for temperature strains associated with these temperature zones are given in Fig. 2.4.5.

### Approximate Values of Creep and Shrinkage

The behaviour of actual structures indicates that reasonable estimates of volume change characteristics are satisfactory for the design of most structures even though test data relating volume changes to the variables shown in Fig. 2.4.1 exhibit a considerable scatter. Therefore, it is possible to reduce the variables and use approximate values as shown in Figs. 2.4.6 and 2.4.7.

**Example 1 - Calculation of volume change shortening by detailed method**

*Given:*

300 × 700 rectangular concrete beam.

Heated structure in Calgary, Alberta.

Initial accelerated cure.

Prestress transfer at one day.

12-13 mm diameter, 1860 MPa strands.

Assume initial prestress loss = 10%

Length = 7.5 m

**Fig. 2.4.6 Volume change strains for typical non-prestressed building elements ($10^{-6}$)**

| Temp. zone (Fig. 2.4.4) | Normal density concrete | | | | | Low density concrete | | | | |
|---|---|---|---|---|---|---|---|---|---|---|
| | Avg. relative humidity (Fig. 2.4.2) | | | | | Avg. relative humidity (Fig. 2.4.2) | | | | |
| | 40 | 50 | 60 | 70 | 80 | 40 | 50 | 60 | 70 | 80 |
| **Heated structures** | | | | | | | | | | |
| 0 | 320 | 290 | 260 | 220 | 190 | 320 | 290 | 260 | 220 | 190 |
| 5 | 350 | 320 | 290 | 250 | 220 | 340 | 310 | 280 | 240 | 210 |
| 10 | 370 | 340 | 310 | 270 | 240 | 360 | 330 | 300 | 260 | 230 |
| 15 | 400 | 370 | 340 | 300 | 270 | 390 | 360 | 330 | 290 | 260 |
| 20 | 430 | 400 | 370 | 330 | 300 | 410 | 380 | 350 | 310 | 280 |
| 25 | 460 | 430 | 400 | 360 | 330 | 430 | 400 | 370 | 330 | 300 |
| 30 | 480 | 450 | 420 | 380 | 350 | 460 | 430 | 400 | 360 | 330 |
| 35 | 510 | 480 | 450 | 410 | 380 | 480 | 450 | 420 | 380 | 350 |
| 40 | 540 | 510 | 480 | 440 | 410 | 500 | 470 | 440 | 400 | 370 |
| 45 | 560 | 530 | 500 | 460 | 430 | 520 | 490 | 460 | 420 | 390 |
| 50 | 590 | 560 | 530 | 490 | 460 | 550 | 520 | 490 | 450 | 420 |
| **Unheated structures** | | | | | | | | | | |
| 0 | 320 | 290 | 260 | 220 | 190 | 320 | 290 | 260 | 220 | 190 |
| 5 | 360 | 330 | 300 | 260 | 230 | 350 | 320 | 290 | 250 | 220 |
| 10 | 400 | 370 | 340 | 300 | 270 | 390 | 360 | 330 | 290 | 260 |
| 15 | 440 | 410 | 380 | 340 | 310 | 420 | 390 | 360 | 320 | 290 |
| 20 | 480 | 450 | 420 | 380 | 350 | 460 | 430 | 400 | 360 | 330 |
| 25 | 520 | 490 | 460 | 420 | 390 | 490 | 460 | 430 | 390 | 360 |
| 30 | 560 | 530 | 500 | 460 | 430 | 520 | 490 | 460 | 420 | 390 |
| 35 | 600 | 570 | 540 | 500 | 470 | 560 | 530 | 500 | 460 | 430 |
| 40 | 640 | 610 | 580 | 540 | 510 | 590 | 560 | 530 | 490 | 460 |
| 45 | 690 | 660 | 630 | 590 | 560 | 620 | 590 | 560 | 520 | 490 |
| 50 | 720 | 690 | 660 | 620 | 590 | 660 | 630 | 600 | 560 | 530 |

Above table is based on:

1. Moist cured
2. Ratio of fine to total aggregates = 0.50
3. Volume to surface ratio = 38 mm
4. Slump = 70 mm
5. Air % ≤ 6
6. Cement content = 410 kg/m$^3$
7. Age at erection = 50 days

*Concrete properties:*

| | |
|---|---|
| $f'_{ci}$: | 30 MPa |
| Air content: | 5% |
| Slump: | 100 mm |
| Fine aggregate: | 670 kg/m$^3$ |
| Coarse aggregate: | 1000 kg/m$^3$ |
| Type 10 cement: | 410 kg/m$^3$ |

*Problem:*

Determine the actual shortening that can be anticipated from:

a) Casting to erection at 50 days
b) Erection to the end of service life
c) Temperature variations

*Solution:*

*a) At 50 days:*

Calculate initial creep:

$$A_p = (12)(99) = 1188 \text{ mm}^2$$

$$P_i = (1188)(1860)(0.70)(0.90)/10^3 = 1392 \text{ kN}$$

$$\in_i = \frac{P_i}{AE_{cr}} = \frac{1392 \times 10^3}{(300)(700)(27700)}$$

$$= 239 \times 10^{-6} \text{ mm/mm}$$

Calculate creep and shrinkage modification factors (Fig. 2.4.1):

Accelerated cure for 1 day:
$Q_a = 1.00$
Cement content = 410 kg/m$^3$
$P_c = 1.00$
Relative humidity = 70% (Fig. 2.4.2):
$Q_n = 0.80$, $P_h = 0.70$

Ratio of fine to total aggregates

$$= \frac{670}{670 + 1000} = 0.40$$

$Q_f = 0.98$, $P_f = 0.86$

Volume to surface ratio

$$= \frac{(300)(700)}{(2)(300 + 700)} = 105 \text{ mm}$$

$Q_r = 0.77$, $P_r = 0.74$
Slump = 100 mm
$Q_{sl} = 1.08$, $P_{sl} = 1.05$
Air = 5%
$Q_v = 1.00$, $P_v = 1.00$

Creep modification factor:

$$Q_{cr} = Q_a Q_h Q_f Q_r Q_{s\ell} Q_v$$
$$= (1.00)(0.80)(0.98)(0.77)(1.08)(1.00)$$
$$= 0.65$$

Shrinkage modification factor:

$$P_{sh} = P_c P_h P_f P_r P_{s\ell} P_v$$
$$= (1.00)(0.70)(0.86)(0.74)(1.05)(1.00)$$
$$= 0.47$$

Calculate total strains and displacement:

$$\in_{cr} = \in_i \left(\frac{t^{0.6}}{10 + t^{0.6}}\right) C_u Q_{cr}$$

$$= (239 \times 10^{-6}) \left(\frac{50^{0.6}}{10 + 50^{0.6}}\right)(2.35)(0.65)$$

$$= 187 \times 10^{-6} \text{ mm/mm}$$

$$\in_{sh} = \in_{shu} \left(\frac{t}{C_s + t}\right) P_{sh}$$

$$= (780 \times 10^{-6}) \left(\frac{50}{55 + 50}\right)(0.47)$$

$$= 174 \times 10^{-6} \text{ mm/mm}$$

$$\Delta_{tot} = (187 \times 10^{-6} + 174 \times 10^{-6})(7500)$$
$$= 2.7 \text{ mm}$$

*b) From 50 days to final:*

Calculate total final strains and displacements:

$$\in_{cr} = (239 \times 10^{-6})(1.0)(2.35)(0.65)$$
$$= 365 \times 10^{-6} \text{ mm/mm}$$

$$\in_{sh} = (780 \times 10^{-6})(1.0)(0.47)$$
$$= 367 \times 10^{-6} \text{ mm/mm}$$

$$\Delta_{tot} = (365 \times 10^{-6} + 367 \times 10^{-6})(7500)$$
$$= 5.5 \text{ mm}$$

Difference from 50 days to final:
$$\Delta_{diff} = 5.5 - 2.7 = 2.8 \text{ mm}$$

*c) Temperature effects:*

From Fig. 2.4.4:

$\Delta_t$ = 39°C for Calgary:

From Fig. 2.4.5:

$$\in_t = 211 \times 10^{-6} \text{ mm/mm}$$

$$\therefore \Delta_t = (211 \times 10^{-6})(7500)$$
$$= 1.6 \text{ mm}$$

## 2.4.2 VOLUME CHANGE CONSIDERATIONS

Load bearing exterior wall panels are subjected to concrete volume change deformations, which are defined as movements due to concrete creep, shrinkage and temperature change. Creep and shrinkage result in shortening while thermal movements can cause shortening or expansion. The amount of movement anticipated must be determined to properly design joints and connections.

### Typical Volume Change Movements

a) Creep strains of load bearing exterior wall panels can be estimated at $120 \times 10^{-6}$ mm/mm for design, considering the usual magnitudes of loadings for panels supporting floors. Stacked exterior panels, supporting only the weight of exterior wall panels and not floors, can be estimated to have a design creep strain in the range of $30 \times 10^{-6}$ mm/mm.

b) Shrinkage strains for structural wall panels, considering usual panel volume to surface ratios and erection at an age of 90 days, can be estimated to be $200 \times 10^{-6}$ mm/mm for normal density concrete and $250 \times 10^{-6}$ mm/mm for low density concrete.

c) Thermal movements can be determined based on the map shown in Fig. 2.4.4. In concrete elements, particularly beams, there is a certain amount of thermal lag which reduces the temperature shortening effect. Also, it is apparent that elements in heated buildings will not undergo as large a temperature differential as those in unheated ones. Therefore, the calculated temperature strains are reduced by a factor of 0.5 and 0.75 for heated and unheated buildings, respectively. If the anticipated construction schedule is such that the elements are likely to be exposed through a full annual temperature cycle, the value for the unheated building should be used.

The preceding volume change movement values are approximate and suitable for most designs. However, where volume change movements may be a significant design factor, it is recommended an in-depth determination be made to establish more exact values compatible with the specific design parameters.

**Example 2 - Shortening of load bearing panels**

Estimate the total shortening of the load bearing panels in a 10 storey structure. Panels are normal density concrete 2750 mm high. Structure is located in Kingston, Ontario:

| | |
|---|---|
| Creep strain | $= 120 \times 10^{-6}$ mm/mm |
| Shrinkage strain | $= 200 \times 10^{-6}$ mm/mm |
| Design temperature | $= 35°$C (from Fig. 2.4.4) |
| Coefficient of | |
| expansion | $= 10 \times 10^{-6}$ mm/(mm · °C) |
| Temperature strain | $= (0.75)(10 \times 10^{-6})(35)$ |
| | $= 262 \times 10^{-6}$ |
| Total strain | $= (120 + 200 + 262)(10^{-6})$ |
| | $= 582 \times 10^{-6}$ |
| Total shortening | $= (582 \times 10^{-6})(2750)(10)$ |
| | $= 16$ mm |

Estimates of building movement must be tempered with engineering judgement. Floors and interior walls attached to the exterior load bearing panel will tend to restrain vertical movement; also a heavily loaded element will tend to distribute load to less heavily loaded ones. Thus, calculated values will tend to be on the high side.

### Movements of Building Corners

Where load bearing and non-load bearing panels meet, e.g., at the corners of a building, it is possible that differential volume change movements may occur. If connections are employed which restrain these movements, it is likely that the connections will attempt to transfer significant vertical forces. Structural behaviour of the building at corners where panels meet requires specific attention and design for volume change movement forces as well as for other design structural forces.

Design consideration of panels meeting at corners should consider the influence of temperature differentials because of sun exposure. Depending upon the exterior panel finishes and plan orientation of the building, 5°C to 8°C temperature differentials may develop. As previously discussed, the interaction of corner panel connections needs consideration when temperature differentials at building corners can develop.

### Influence Upon Non-structural Components

Previous concerns with volume change deformations have related to structural factors. Of equal concern is the influence of volume change movement upon non-structural items.

Exterior sealants used to prevent water penetration into the building must be able to accommodate movements caused by volume changes. Sealants subjected to differential volume change movements, either horizontally or vertically at building corners, at non-precast adjacent construction or at windows not having similar movements must be studied.

As the height and length of a building increase, the cumulative movements at the top or ends of the structure increase. The movements of exterior walls can affect the interior partitions on upper floors resulting in distress to or cracking of the partitions. Non-structural components at the building interior must be detailed to allow for volume change movements of exterior precast structural walls.

## 2.4.3 THERMAL BOWING

Temperature difference between the inside and outside of a wall panel, especially composite insulated sandwich panels, or between the top and underside of an uninsulated roof deck, can cause the elements to bow. The theoretical magnitude of bowing (see Fig. 2.4.8) can be determined by:

$$\Delta = \frac{C(T_1 - T_2)\ell^2}{8h}$$

**Fig. 2.4.8 Thermal bow of uninsulated wall panel**

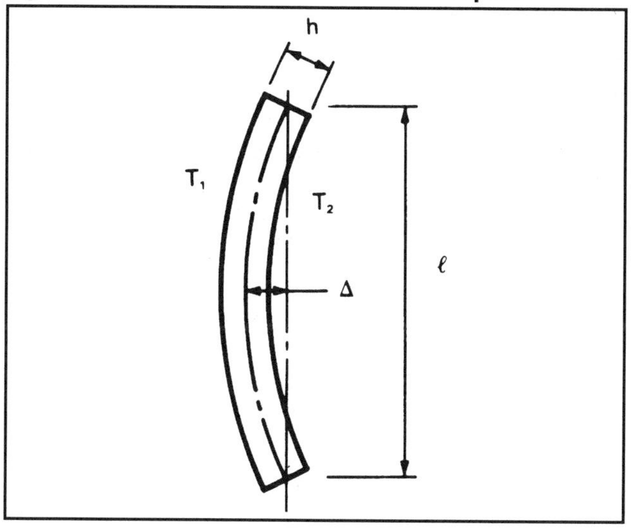


where:

C = coef. of thermal expansion

$T_2$, $T_1$ = inside and outside temperature

ℓ = distance between supports

h = element thickness

Limited records of temperature measurements indicate that in open structures, such as the roofs of parking decks, the temperature differential ($T_1 - T_2$) seldom exceeds 16 to 22°C. In an insulated sandwich wall panel, the theoretical difference can be higher, but this is tempered by "thermal lag" due to the mass of the concrete.

Moisture differences between the inside and outside of an enclosed building can also cause bowing, however, calculation is much less precise and involves more variables. The exterior layer of the concrete panel absorbs moisture from the atmosphere and periodic precipitation, while the interior layer is relatively dry, especially when the building is heated. This causes the inside layer to shrink more than the outside, causing an outward bow. This would tend to balance the theoretical inward thermal bowing in cold weather, which is believed to explain the observation that "wall panels always bow out".

**Example 3 - Thermal bow in an uninsulated wall panel**

*Given:*

A 6 metre high, 150 mm thick wall panel as shown below. Coefficient of thermal expansion:

$C = 10.8 \times 10^{-6}$ mm/(mm · °C)

Temperature differential $T_1 - T_2 = 20$°C

$E_c$ = 29 900 MPa

*Problem:*

Determine the potential thermal bow $\Delta_1$, the force, P, required at midheight to restrain the bowing, the stress in the panel caused by the restraint, and the residual bow, $\Delta_2$.

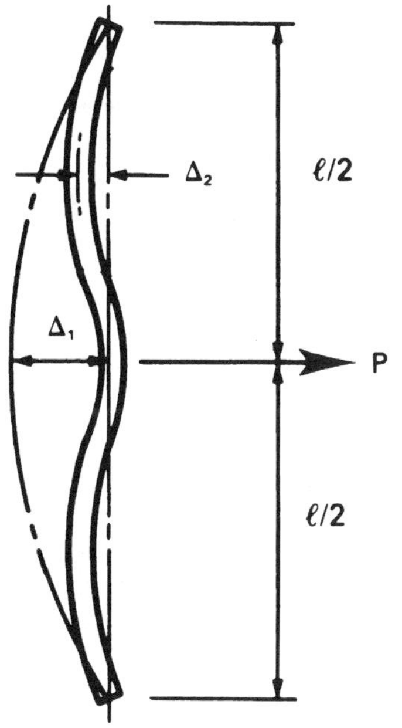


*Solution:*

$$\Delta_1 = \frac{(10.8 \times 10^{-6})(20)(6000)^2}{(8)(150)} = 6.5 \text{ mm}$$

From Fig. 2.4.10:

$$E_t = (0.75)(29\,900) = 22\,400 \text{ MPa}$$

$$I = \frac{bh^3}{12} = \frac{(1000)(150^3)}{12} = 281 \times 10^6 \text{ mm}^4/\text{m}$$

From Fig. 2.4.10, Case (a):

$$P = \frac{48E_tI\Delta}{\ell^3} = \frac{(48)(22\,400)(281 \times 10^6)(6.5)}{(6000)^3}$$

$$= 9.09 \text{ kN/m width}$$

$$M = \frac{P\ell}{4} = \frac{(9.09)(6)}{4} = 13.6 \text{ kN} \cdot \text{m/m width}$$

$$\text{Panel Stress} = \frac{My}{I} = \frac{(13.6)(10^6)(75)}{(281 \times 10^6)} = 3.6 \text{ MPa}$$

The residual bow can be calculated by adjusting the equation in Fig. 2.4.10. Case (e) to read:

$$\Delta = \frac{M\ell^2}{16E_tI}$$

Substituting ℓ/2 for ℓ:

$$\Delta_2 = \frac{(13.6 \times 10^6)(3000)^2}{(16)(22400)(281 \times 10^6)} = 1.2 \text{ mm}$$

While the magnitude of bowing is usually not very significant, in the case of wall panels it may cause unacceptable separation at the corners (see Fig. 2.4.9), and possible damage to joint sealants. It may therefore be desirable to restrain bowing with one or more connectors between panels. Fig. 2.4.10 gives equations for calculating the required restraint and the moments this would cause in the panel.

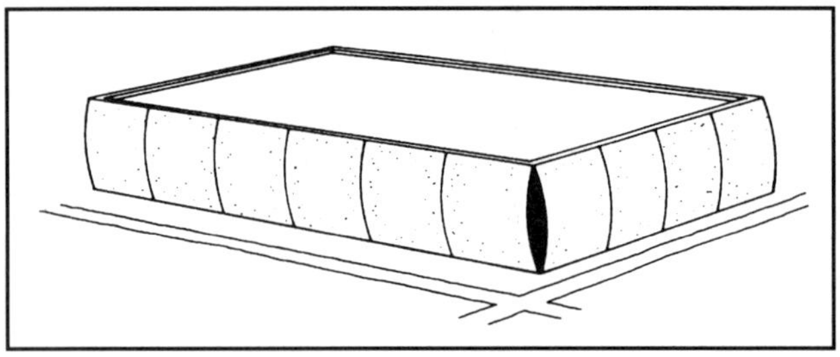

**Fig. 2.4.9 Corner separation due to thermal bow**

Similarly, differential temperature can cause upward bowing in roof elements, especially in open structures such as parking decks. If these elements are restrained from rotations at the ends, positive moments (bottom tension) can develop at the support, as shown in Cases (d) and (e), Fig. 2.4.10. The bottom tension can cause severe cracking, but once the cracks occur, the tension is relieved. Examination of the equations shows that the thermal induced positive moments are independent of the span length.

Note from Fig. 2.4.10 that if only one end is restrained, as is sometimes done to relieve axial volume change force, the restraint moment is doubled. Also note that, since thermal bow occurs with daily temperature changes, the cyclical effects could magnify the potential damage.

**Example 4 - Thermal bow in a roof element**

*Given:*

An inverted tee beam supporting the double tees of the upper level of a parking deck, as shown in Fig. 2.4.11, is welded at both ends at the bottom as part of a moment-resisting frame:

$C = 10.8 \times 10^{-6}$ mm/(mm · °C)

$T_1 - T_2 = 20°C$

$E_c = 29\ 900$ MPa

**Fig. 2.4.10 Force required to restrain bowing**

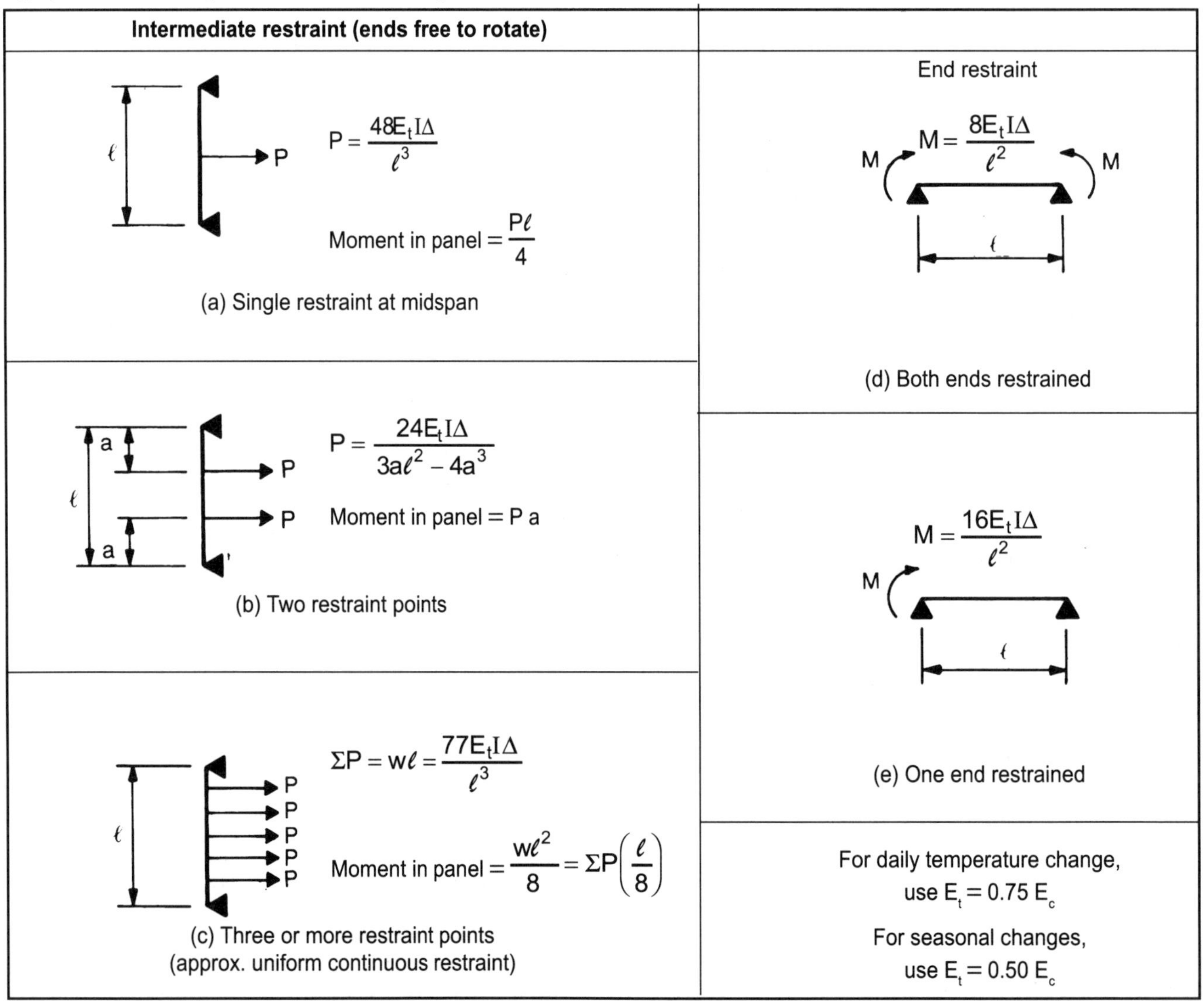

**Fig. 2.4.11 Inverted tee beam**

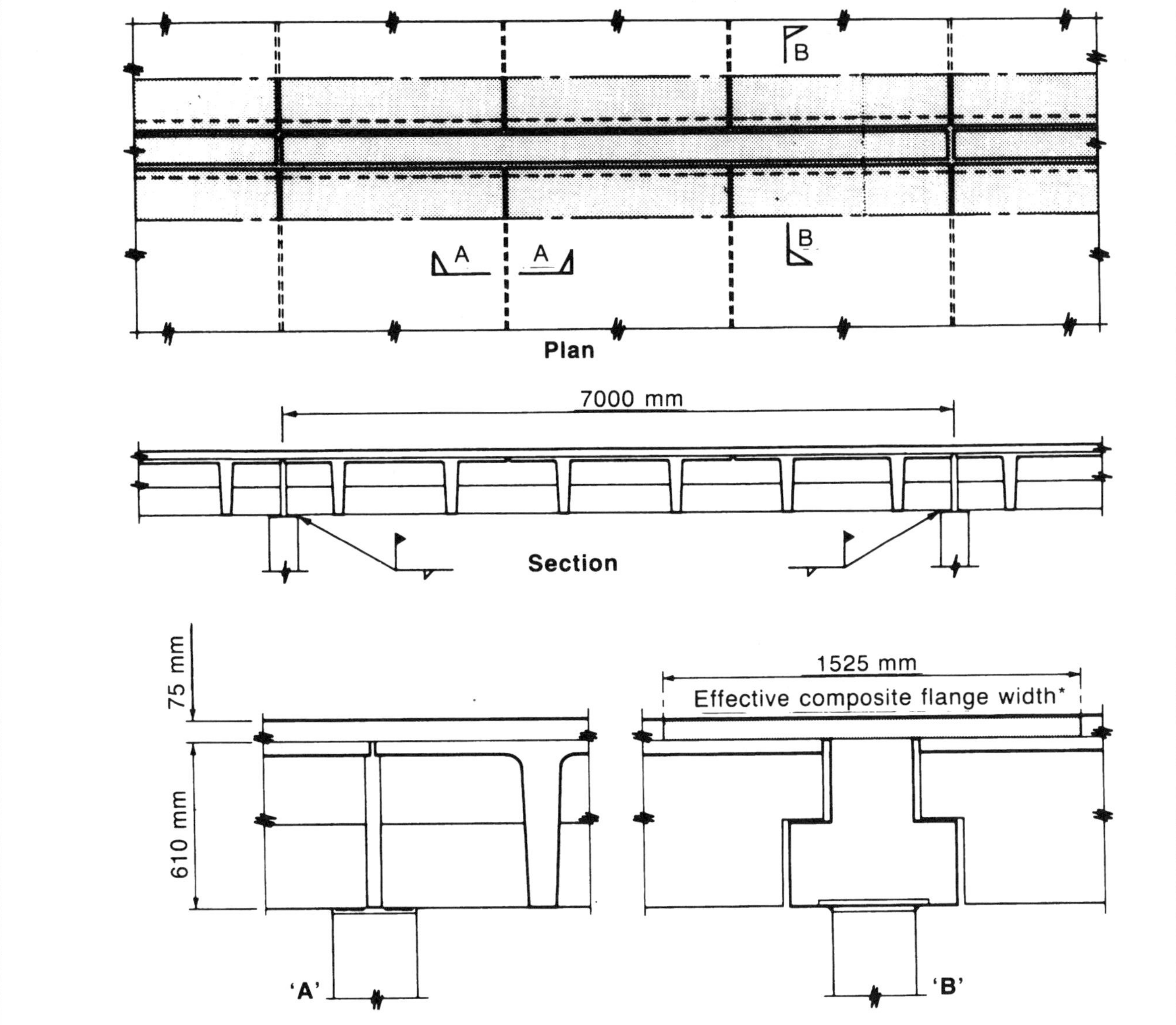


*Note: Cracking of the composite topping at the junction of the beam and double tee will occur due to volume change effects. Adequate transverse reinforcement must be provided to ensure integrity of the assumed effective composite flange width.

Composite I = 20 700 × $10^6$ $mm^4$

*Problem:*

Find the tensile force developed at the support.

*Solution:*

$$\Delta = \frac{(10.8 \times 10^{-6})(20)(7000)^2}{(8)(686)} = 1.93 \text{ mm}$$

From Fig. 2.4.10, Case (d):

$$E_t = (0.75)(29\,900) = 22\,400 \text{ MPa}$$

$$M = \frac{(8)(22\,400)(20\,700)(1.93)}{(7000)^2} = 146 \text{ kN·m}$$

$$T \approx \frac{(146)(10^3)}{(686)} = 213 \text{ kN}$$

## 2.5 DIAPHRAGM DESIGN

Horizontal loads from wind or earthquake are usually transmitted to shear walls or moment-resisting frames through the roof and floors acting as horizontal diaphragms.

### 2.5.1 METHOD OF ANALYSIS

The diaphragm is analyzed by considering the roof or floor as a deep horizontal beam, analogous to a plate girder or I-beam. The shear walls or structural frames are the supports for this analogous beam. Thus, the lateral loads are transmitted to these supports as reactions. As in a beam, tension and compression are induced in the chords or "flanges" of the analogous I-beam as shown in Fig. 2.5.1.

When precast concrete elements which span parallel to the supporting shear walls or frames are used for the diaphragm, it is apparent that the shear in the analogous beam must be transferred between adjacent elements and also to the supporting elements. The "web" shear must also be transferred to the chord elements. Thus the design of a diaphragm is essentially a connection design problem.

## 2.5.2 SHEAR TRANSFER BETWEEN ELEMENTS

In floors or roofs without composite topping, the shear transfer between elements is usually accomplished by grout keys, depending on the element.

Weld plates may be analyzed as illustrated in Fig. 2.5.2. In addition to the hardware details shown, many others are used by precast concrete manufacturers.

For elements connected by grout keys not intentionally roughened, a conservative value of 0.3 MPa can be used for the average shear resistance of the grouted key. A value of 0.6 MPa may be used for grouted joints with intentionally roughened surfaces. If necessary, reinforcement placed as shown in Fig. 2.5.3 can be used to transfer the shear. This steel is designed by the shear-friction principles discussed in Chap. 4.

In floors or roofs with composite topping, the topping itself can act as the diaphragm, if it is adequately reinforced. Reinforcement requirements can be determined by shear-friction analysis.

It should be noted that the connections between elements often serve functions in addition to the transfer of shear for lateral loads. For example, weld plates in flanged elements are often used to adjust differential camber. Grout keys may be called upon to distribute concentrated loads.

Connections which transfer shear from the diaphragm to the shear walls or moment-resisting frame are analyzed in the same manner as the connections between elements.

## 2.5.3 CHORD FORCES

Chord forces are calculated as shown in Fig. 2.5.1. For roofs with intermediate supports as shown, the shear force is carried across the beam with weld plates or bars in grout keys as shown in Section A. Bars are designed by shear-friction. Forces are usually quite low, and only as many bars or weld plates as required should be used.

**Fig. 2.5.1 Analogus beam design of a diaphragm**

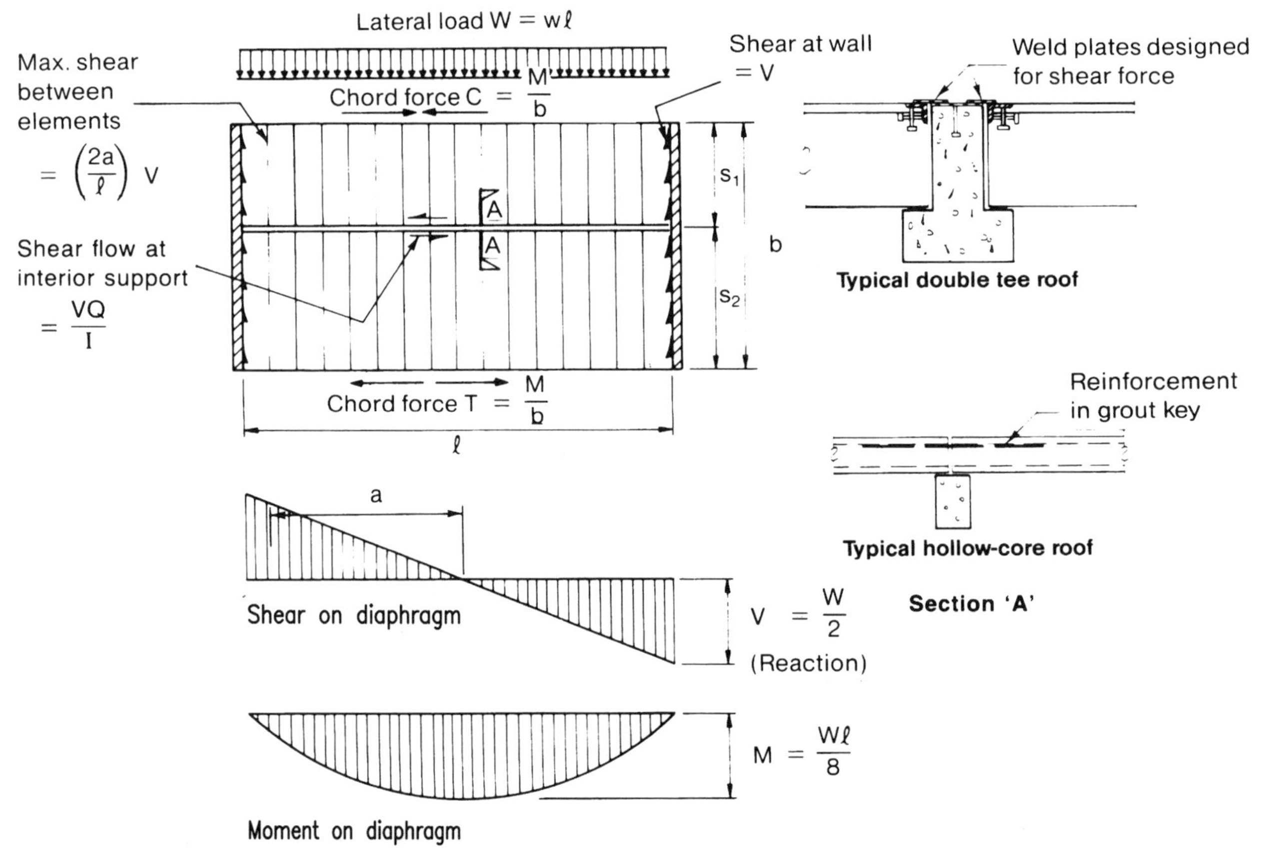


Notes:
1. Grouted joints cannot be considered effective unless prevented from separating. Chord forces for diaphragms involving grouted joints must be based on shear forces in addition to bending.
2. $T_f = \frac{M_f}{b} + \frac{V_f}{\mu_e}$

In flanged deck elements, the chord tension at the perimeter of the building is usually transferred between elements by the same type of connection used for shear transfer (Fig. 2.5.2). When forces are high, such as in design for earthquake, transverse reinforcing bars may be placed in the flange and attached

**Fig. 2.5.2 Typical flange weld plate details**

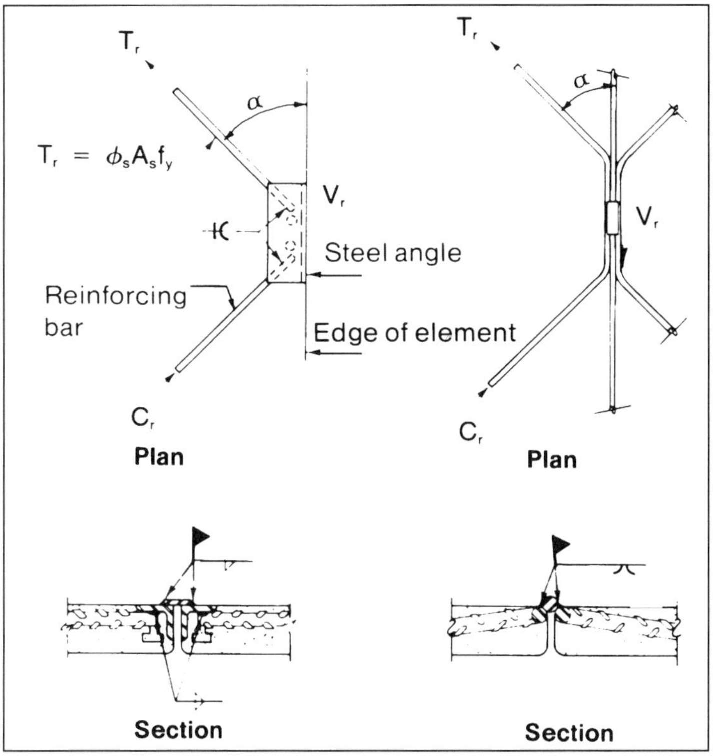


Note: The above assumption would only be used under factored loading conditions

**Fig. 2.5.3 Use of perimeter reinforcement as shear-friction reinforcement**

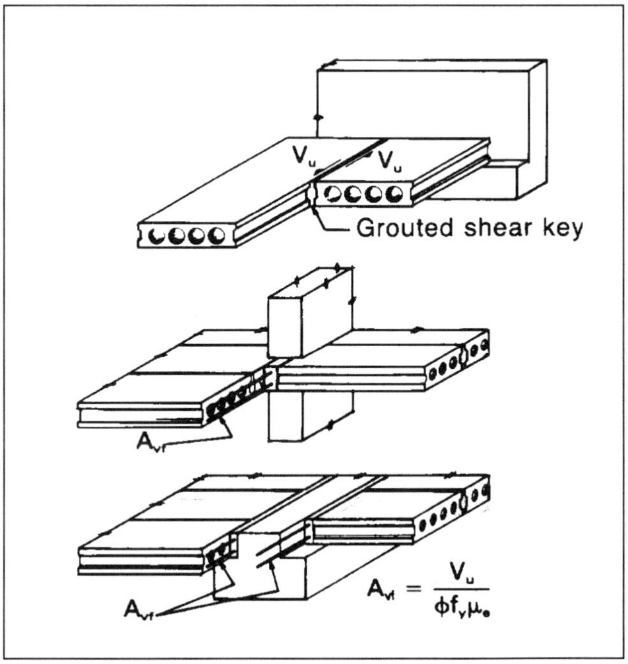


to the connection device by welding or lapping with the connection bars. Examples of each of these are given in Sects. 2.7 and 2.8.

In bearing wall buildings, perimeter reinforcement is required for structural integrity (CSA A23.3-94 Clause 16.5),[R2.5.1] and must be checked very carefully, particularly for earthquake loads.

## 2.6 BUILDINGS WITH MOMENT-RESISTING FRAMES

### 2.6.1 GENERAL

Precast, prestressed concrete beam and slab elements are usually most economical when they can be designed and connected into a structure as simple-span elements. This is because:

1. Positive moment capacity is much easier and less expensive to attain with pretensioned elements than negative moment capacity at supports.
2. Connections which achieve continuity at the supports are sometimes complicated and costly.
3. The restraint to volume changes that occurs in rigid connections may cause serious cracking and unsatisfactory performance.

Therefore, it is simpler when designing precast, prestressed concrete structures to have connections which allow lateral movement and rotation, i.e., pinned ends, and achieve lateral stability through the use of floor and roof diaphragms and shear walls.

However, in some structures, adequate shear walls interfere with the function of the building, or are more expensive than alternate solutions. In these cases, the lateral stability of the structure depends on the moment-resisting capacity of either the column bases, a beam-column frame, or both.

When moment connections between beams and columns are required to resist lateral loads, it is very important that the amount of beam-column moment framing be kept to a minimum and that it be located centrally so as to reduce volume change effects.

Also, when possible, in order to reduce the magnitude of the connections, make the moment connection after most of the dead loads have been applied. This requires careful detailing, specification of the construction process, and inspection. If such details are possible, the moment connections need only resist the negative moments from live load, lateral loads and volume changes, and will then be less costly.

### 2.6.2 MOMENT RESISTANCE OF COLUMN BASES

Single-storey and some low-rise buildings without shear walls may depend on the fixity of the column base to resist lateral loads. The ability of a spread footing to resist moments caused by lateral loads is dependent on the rotational characteristics of the base. The total rotation of the column base is a function

**Fig. 2.6.1 Assumptions used in derivation of rotational coefficients for column bases**

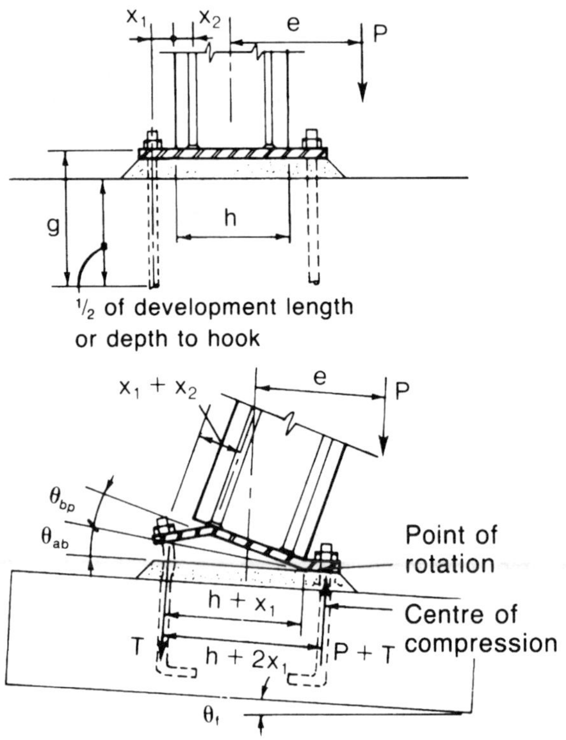


*Note: The above assumptions would only be used under factored loading conditions.*

of rotation between the footing and soil, bending in the base plate, and elongation of the anchor bolts, as shown in Fig. 2.6.1.

The total rotation of the base is:

$$\theta_b = \theta_f + \theta_{bp} + \theta_{ab}$$

If the axial load is large enough so that there is no tension in the anchor bolts, $\theta_{bp}$ and $\theta_{ab}$ are zero, and:

$$\theta_b = \theta_f$$

Rotational characteristics can be expressed in terms of flexibility or stiffness coefficients:

$$\theta = \gamma M = M/K$$

where:

$M$ = applied moment = Pe

$e$ = eccentricity of the applied load, P

$\gamma$ = flexibility coefficient

$K$ = stiffness coefficient = $1/\gamma$

If bending of the base plate and strain in the anchor bolts are assumed as shown in Fig. 2.6.1, the flexibility coefficients for the base can be derived,[R2.6.1] and the total rotation of the base becomes:

$$\theta_b = M(\gamma_f + \gamma_{bp} + \gamma_{ab}) = Pe(\gamma_f + \gamma_{bp} + \gamma_{ab})$$

$$\gamma_f = 1/(k_s I_f)$$

$$\gamma_{bp} = \frac{(x_1 - x_2)^3[2e/(h+2x_1) - 1]}{6eE_s I_{bp}(h + x_1)} \geq 0$$

$$\gamma_{ab} = \frac{g[2e/(h+2x_1) - 1]}{2eA_b E_s(h + x_1)} \geq 0$$

where:

$\gamma_f, \gamma_{bp}, \gamma_{ab}$ = flexibility coefficients of the footing/soil interaction, the base plate and the anchor bolts, respectively

$k_s$ = coefficient of subgrade reaction from Fig. 2.6.2.

$I_f$ = moment of inertia of the footing (plan dimensions)

$E_s$ = modulus of elasticity of steel

$I_{bp}$ = moment of inertia of the base plate (vertical cross-section dimensions)

$A_b$ = total area of anchor bolts which are in tension

$h$ = width of the column in the direction of bending

$x_1$ = distance from face of column to the centre of the anchor bolts, positive when anchor bolts are outside the column, and negative when anchor bolts are inside the column

$x_2$ = distance from the face of the column to base plate anchorage

$g$ = assumed length over which elongation of the anchor bolt takes place = 1/2 of development length + projection for deformed anchor bolts or the length to the hook + projection for smooth anchor bolts (see Fig. 2.6.1.)

**Fig. 2.6.2 Approximate relationship between allowable soil bearing value and coefficient of subgrade reaction, $k_s$**

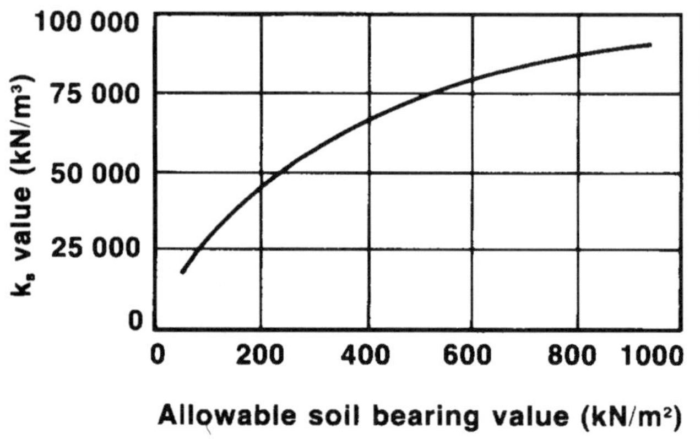


Rotation of the base may cause an additional eccentricity of the loads on the columns, causing moments which must be added to the moments induced by the lateral loads.

Note that if the eccentricity, e, is less than $h/2 + x_1$ (inside the centre of compression), $\gamma_{bp}$ and $\gamma_{ab}$ are less than zero, meaning that there is no rotation between the column and the footing, and only the rotation from soil deformation need be considered.

Values are tabulated for typical cases in Figs. 2.6.3 and 2.6.4.

**Fig. 2.6.3 Flexibility coefficients for footing soil interaction**

Flexibility of base $= \gamma_b = \gamma_f + \gamma_{ab} + \gamma_{bp}$

Rotation of base $= \gamma_b Pe$

Stiffness of base $= K_b = 1/\gamma_b$

Fixity of base $= K_b/(K_c + K_b)$

$K_c$ = column stiffness = $4E_cI_c/h_s$

$E_c$ = modulus of elasticity of column concrete

$I_c$ = moment of inertia of column

$h_s$ = storey height

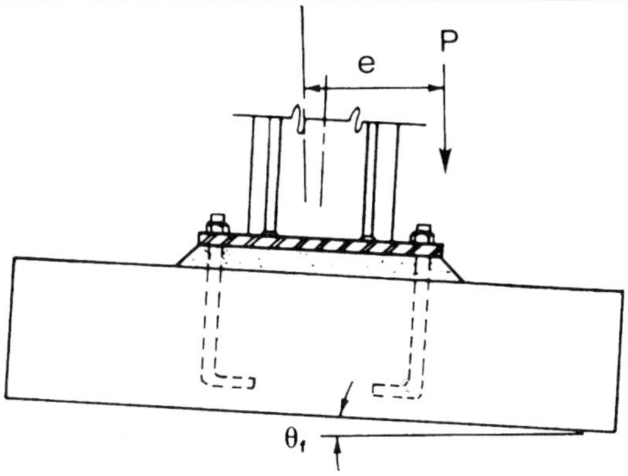


$\gamma_f$ [$10^{-6}$/(kN · m)] for square footings

| Footing size (mm) | $k_s$ (kN/m$^3$) | | | | |
|---|---|---|---|---|---|
| | 20 000 | 40 000 | 60 000 | 80 000 | 100 000 |
| 600 × 600 | 4630 | 2315 | 1543 | 1157 | 926 |
| 750 × 750 | 1896 | 948 | 632 | 474 | 380 |
| 900 × 900 | 915 | 457 | 305 | 229 | 183 |
| 1050 × 1050 | 494 | 247 | 165 | 123 | 99 |
| 1200 × 1200 | 289 | 145 | 97 | 72 | 58 |
| 1350 × 1350 | 181 | 90 | 60 | 45 | 36 |
| 1500 × 1500 | 119 | 59 | 40 | 30 | 24 |
| 1650 × 1650 | 81 | 41 | 27 | 20 | 16 |
| 1800 × 1800 | 57 | 29 | 19 | 14 | 11 |
| 1950 × 1950 | 42 | 21 | 14 | 10 | 8 |
| 2100 × 2100 | 31 | 15 | 10 | 8 | 6 |
| 2250 × 2250 | 23 | 12 | 8 | 6 | 5 |
| 2400 × 2400 | 18 | 9 | 6 | 5 | 4 |
| 2700 × 2700 | 11 | 6 | 4 | 3 | 2 |
| 3000 × 3000 | 7 | 4 | 3 | 2 | 2 |
| 3300 × 3300 | 5 | 3 | 2 | 1 | 1 |
| 3600 × 3600 | 4 | 2 | 1 | 1 | 1 |

**Example 1 - Stability analysis of an unbraced frame**

*Given:*

The column shown in Fig. 2.6.5

Soil bearing capacity = 240 kN/m$^2$

P = 355 kN dead load, 135 kN live load

W = 9 kN wind load

*Problem:*

Determine the column design loads and moments for stability as an unbraced frame.

*Solution:*

The column is designed for the following conditions:

1. 1.25D + 1.5L
2. 1.25D + 0.7(1.5L + 1.5W)
3. 0.85D + 1.5W

The maximum eccentricity would occur when (3) is applied.

Moment at base of column:

$M = (9)(5) = 45\text{kN} \cdot \text{m}$

$P_f = (0.85)(355) = 302 \text{ kN}$

$M_f = (1.5)(45) = 67.5 \text{ kN} \cdot \text{m}$

Eccentricity due to wind load:

$$= \frac{M_f}{P_f} = \frac{67.5}{302} = 0.224\text{m} = 224 \text{ mm}$$

To determine the moments caused by base rotation, an iterative procedure is required.

Estimate eccentricity due to rotation = 9 mm:

$e = 224 + 9 = 233 \text{ mm}$

Check rotation between column and footing:

$h/2 + x_1 = 500/2 + (-50) = 200 \text{ mm} < 233$

**Fig. 2.6.4 Flexibility coefficients for anchor bolts and base plates**

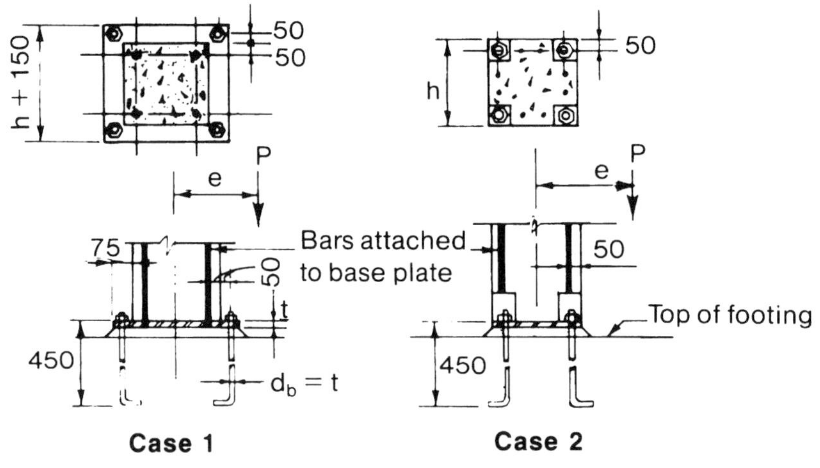


| Column size, h (mm) | e (mm) | $\gamma_{ab} + \gamma_{bp}$ $[10^{-6}/(kN \cdot m)]$ for typical details — Case 1 : Exterior anchor bolts — Base plate thickness & anc. bolt diam. (mm) | | | | Case 2 : Interior anchor bolts — Base plate thickness & anc. bolt diam. (mm) | | | |
|---|---|---|---|---|---|---|---|---|---|
| | | 20 | 25 | 30 | 35 | 20 | 25 | 30 | 35 |
| 300 × 300 | 100 | 0 | 0 | 0 | 0 | 0 | 0 | 0 | 0 |
| | 150 | 0 | 0 | 0 | 0 | 23.9 | 15.3 | 10.6 | 7.8 |
| | 200 | 0 | 0 | 0 | 0 | 35.8 | 22.9 | 15.9 | 11.7 |
| | 250 | 13.1 | 7.3 | 4.6 | 3.2 | 43.0 | 27.5 | 19.1 | 14.0 |
| | 300 | 21.8 | 12.2 | 7.7 | 5.3 | 47.8 | 30.6 | 21.2 | 15.6 |
| | 350 | 28.0 | 15.7 | 9.9 | 6.8 | 51.2 | 32.7 | 22.7 | 16.7 |
| | 400 | 32.6 | 18.3 | 11.6 | 7.9 | 53.7 | 34.4 | 23.8 | 17.5 |
| | 450 | 36.3 | 20.4 | 12.9 | 8.8 | 55.7 | 35.7 | 24.8 | 18.2 |
| 400 × 400 | 150 | 0 | 0 | 0 | 0 | 0 | 0 | 0 | 0 |
| | 200 | 0 | 0 | 0 | 0 | 8.5 | 5.5 | 3.8 | 2.8 |
| | 250 | 0 | 0 | 0 | 0 | 13.6 | 8.7 | 6.1 | 4.5 |
| | 300 | 6.0 | 3.4 | 2.2 | 1.5 | 17.1 | 10.9 | 7.6 | 5.6 |
| | 350 | 10.3 | 5.9 | 3.7 | 2.6 | 19.5 | 12.5 | 8.7 | 6.4 |
| | 400 | 13.5 | 7.7 | 4.9 | 3.4 | 21.3 | 13.6 | 9.5 | 7.0 |
| | 450 | 16.1 | 9.1 | 5.8 | 5.8 | 22.7 | 14.6 | 10.1 | 7.4 |
| | 500 | 18.1 | 10.3 | 6.5 | 6.5 | 23.9 | 15.3 | 10.6 | 7.8 |
| 500 × 500 | 200 | 0 | 0 | 0 | 0 | 0 | 0 | 0 | 0 |
| | 250 | 0 | 0 | 0 | 0 | 4.0 | 2.6 | 1.8 | 1.3 |
| | 300 | 0 | 0 | 0 | 0 | 6.6 | 4.2 | 3.0 | 2.2 |
| | 350 | 3.2 | 1.8 | 1.2 | 1.2 | 8.5 | 5.5 | 3.8 | 2.8 |
| | 400 | 5.6 | 3.2 | 2.1 | 2.1 | 10.0 | 6.4 | 4.4 | 3.3 |
| | 450 | 7.5 | 4.3 | 2.8 | 2.8 | 11.1 | 7.1 | 4.9 | 3.6 |
| | 500 | 9.0 | 5.2 | 3.3 | 3.3 | 11.9 | 7.6 | 5.3 | 3.9 |
| | 550 | 10.2 | 5.9 | 3.8 | 3.8 | 12.7 | 8.1 | 5.6 | 4.1 |
| 600 × 600 | 250 | 0 | 0 | 0 | 0 | 0 | 0 | 0 | 0 |
| | 300 | 0 | 0 | 0 | 0 | 2.2 | 1.4 | 1.0 | 0.7 |
| | 350 | 0 | 0 | 0 | 0 | 3.7 | 2.4 | 1.7 | 1.2 |
| | 400 | 1.9 | 1.1 | 0.7 | 0.5 | 4.9 | 3.1 | 2.2 | 1.6 |
| | 450 | 3.4 | 2.0 | 1.3 | 0.9 | 5.8 | 3.7 | 2.6 | 1.9 |
| | 500 | 4.6 | 2.6 | 1.7 | 1.2 | 6.5 | 4.2 | 2.9 | 2.1 |
| | 550 | 5.5 | 3.2 | 2.1 | 1.4 | 7.1 | 4.6 | 3.2 | 2.3 |
| | 600 | 6.3 | 3.7 | 2.4 | 1.6 | 7.6 | 4.9 | 3.4 | 2.5 |

**Fig. 2.6.5 Column stability and base fixity for Examples 1 and 2**

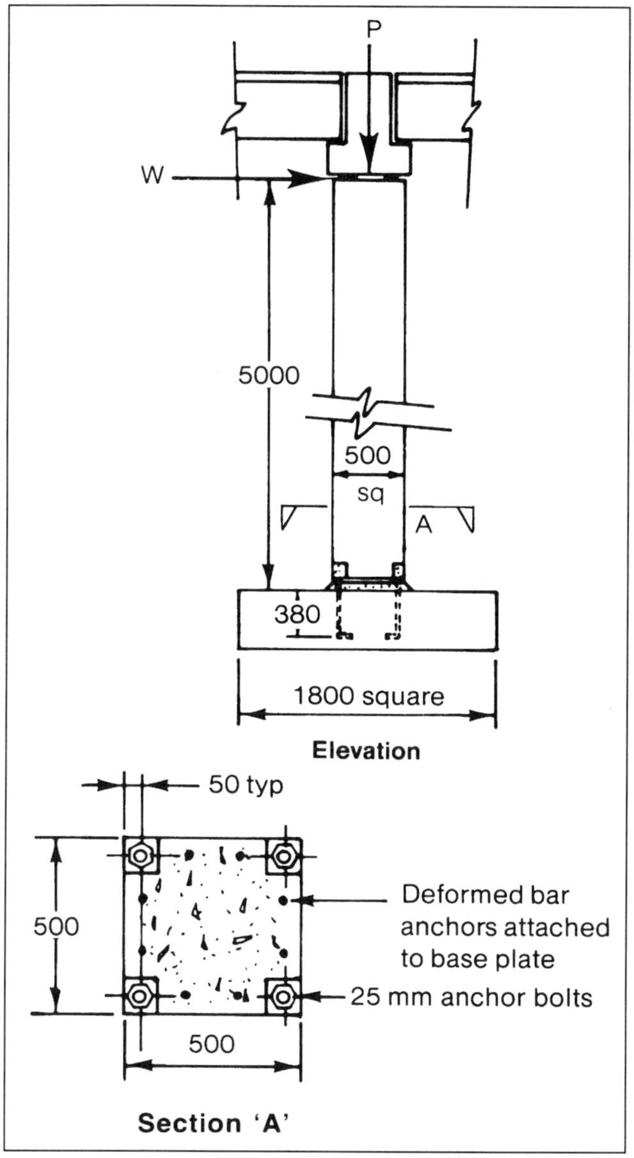


Thus there is tension in the anchor bolts and rotation between the column and footing:

$$I_f = (1800)^4/12 = 875\,000 \times 10^{-6}\ \text{mm}^4$$

From Fig. 2.6.2:

$$k_s \approx 55\,000\ \text{kN/m}^3$$

$$\gamma_f = 1/(k_s\,I_f) = 1/[(55\,000)(875)]$$

$$= 20.8 \times 10^{-6}/(\text{kN} \cdot \text{m})$$

$$\gamma_{ab} = \frac{g\left[\dfrac{2e}{h+2x_1} - 1\right]}{2eE_sA_b(h+x_1)}$$

$$= \frac{(465)\left[\dfrac{(2)(233)}{(500)+(2)(-50)} - 1\right]}{(2)(233)(193)(349)(500-50)(10^3)}$$

$$= 5.4 \times 10^{-6}/(\text{kN}\cdot\text{m})$$

$$\gamma_{bp} = \frac{(x_1 - x_2)^3[2e/(h+2x_1) - 1]}{6eE_sI_{bp}(h+x_1)}$$

$$= 0\ (\text{since } x_1 + x_2 = -50 + 50 = 0)$$

$$M_f = (302)(0.233) = 70.3\ \text{kN} \cdot \text{m}$$

$$\theta_b = (\gamma_f + \gamma_{ab})\,M_f$$

$$= (20.8 + 5.4)(10^{-6})(70.3)$$

$$= 0.00184$$

Eccentricity caused by rotation:

$$= \theta_b\,h_s = (0.00184)(5000) = 9.2\ \text{mm}$$

Therefore no further trial is required.

Design requirements for 0.85D – 1.5W:

$$P_f = 302\ \text{kN}$$

$$M_f = 70.3\ \text{kN} \cdot \text{m}$$

Check for 1.25D + 0.7(1.5L + 1.5W):

$$P_f = (1.25)(355) + (1.05)(135) = 585\ \text{kN}$$

$$M_f = (0.7)(1.5W) = (0.7)(1.5)(9)(5) = 47.2\ \text{kN} \cdot \text{m}$$

$$e = \frac{47.2}{585} = 0.0807\ \text{m} = 81\ \text{mm}$$

Estimate eccentricity due to rotation = 6 mm:

$$M_f = (585)(81 + 6)/10^3 = 50.9\ \text{kN} \cdot \text{m}$$

$$\theta_b = \gamma_f M_f = (20.8 \times 10^{-6})(50.9) = 0.00106\ \text{rad}$$

$$\theta_b h_s = (0.00106)(5000) = 5.3\ \text{mm} \quad \text{OK}$$

Design requirements for 1.25D + 0.7(1.5L + 1.5W):

$$P_f = 585\ \text{kN}$$

$$M_f = 50.9\ \text{kN} \cdot \text{m}$$

It is recommended that the moment caused by a minimum eccentricity of 15 + 0.03h be considered when designing for 1.25D + 1.5L:

$$P_f = (1.25)(355) + (1.5)(135) = 646\ \text{kN}$$

$$e = 15 + 0.03h = 15 + (0.03)(500) = 30\ \text{mm}$$

Estimate eccentricity due to rotation = 3 mm:

$$M_f = P_f e = (646)(30 + 3)/10^3 = 21.3\ \text{kN} \cdot \text{m}$$

$$\theta_b = (20.8 \times 10^{-6})(21.3) = 0.00044\ \text{rad}$$

$$\theta_b h_s = (0.00044)(5000) = 2.2\ \text{mm} \approx 3\ \text{mm} \quad \text{OK}$$

Design requirements for 1.25D + 1.5L:

$P_f$ = 646 kN

$M_f$ = 21.3 kN · m

## 2.6.3 FIXITY OF COLUMN BASES

The degree of fixity of a column base is the ratio of the rotational stiffness of the base to the sum of the rotational stiffnesses of the column plus the base:

$$F_b = \frac{K_b}{K_b + K_c}$$

where:

$F_b$ = degree of base fixity, expressed as a decimal

$K_b$ = $1/\gamma_b$

$K_c$ = $\frac{4E_cI_c}{h_s}$

$E_c$ = modulus of elasticity of the column concrete

$I$ = moment of inertia of the column

$h_s$ = column height

**Example 2 - Calculation of degree of fixity**

*Given:*

The column base shown in Example 1 on Page 2-19.

$E_c$ = 29 900 MPa

*Problem:*

Determine the degree of fixity.

*Solution:*

$K_b = 1/\gamma_b = 1/(20.8 \times 10^{-6}) = 48.1 \times 10^3$ kN · m

$I_c = bd^3/12 = 500^4/12 = 5208 \times 10^6$ mm$^4$

$$K_c = \frac{4E_cI_c}{h_s} = \frac{(4)(29\,900)(5208 \times 10^6)}{(5000)(10^6)}$$

$$= 124.6 \times 10^3 \text{ kN}\cdot\text{m}$$

$$F_b = \frac{K_b}{K_b + K_c} = \frac{48.1}{48.1 + 124.6} = 0.28$$

## 2.6.4 MODELING PARTIALLY FIXED BASES

In lieu of using a computer program that accepts column base stiffness as input, a simple way to model base fixity for computer analyses is to incorporate an "imaginary column" below the actual column base. If the bottom of the imaginary column is modeled as pinned, then the expression for its rotational stiffness is:

$K_{ci} = 3E_{ci}\,I_{ci} / h_{ci}$

where the subscript "ci" denotes the properties of the imaginary column (Fig. 2.6.6) and $K_{ci} = K_b$ as calculated in Sect. 2.6.3. Or, the degree of base fixity, $F_b$, can be determined or estimated, and $K_b$ calculated from Sect. 2.6.3.

For the computer model, either $I_{ci}$ or $h_{ci}$ may be varied for different values of $K_{ci}$, with the other terms left constant for a given problem. It is usually preferable to use $E_{ci} = E_c$. For the assumptions of Fig. 2.6.6:

$h_{ci} = 3E_{ci}I_{ci} / K_{ci}$

or:

$I_{ci} = K_{ci}\,h_{ci} / (3E_{ci})$

**Fig. 2.6.6 Model for partially fixed column base**

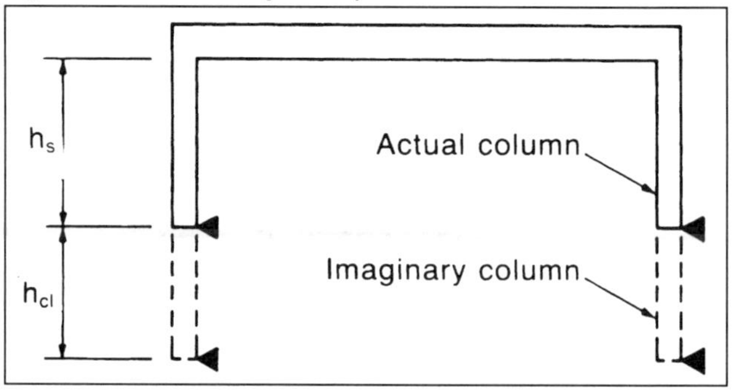


**Example 3 – Imaginary column for computer model**

*Given:*

The column base shown in Examples 1 and 2 on Pages 2-19 and 2-22 respectively.

*Problem:*

Determine the length of an imaginary column to model fixity of the column base.

*Solution:*

$$h_{ci} = \frac{3E_{ci}I_{ci}}{K_{ci}}$$

$$= \frac{(3)(29\,900)(5208)}{(48.1 \times 10^3)} = 9712 \text{ mm}$$

## 2.6.5 VOLUME CHANGE EFFECTS IN MOMENT-RESISTING FRAMES

The restraint of volume changes in moment-resisting frames causes tension in the girders and deflections and moments in the columns. The magnitude of these tensions, moments and deflections is dependent on the distance from the *centre of stiffness* of the frame.

The centre of stiffness is that point of a frame, which is subject to a uniform unit shortening, at which no lateral movement will occur. For frames which are symmetrical with respect to bay

**Fig. 2.6.7 Effect of volume change restraints in building frames**

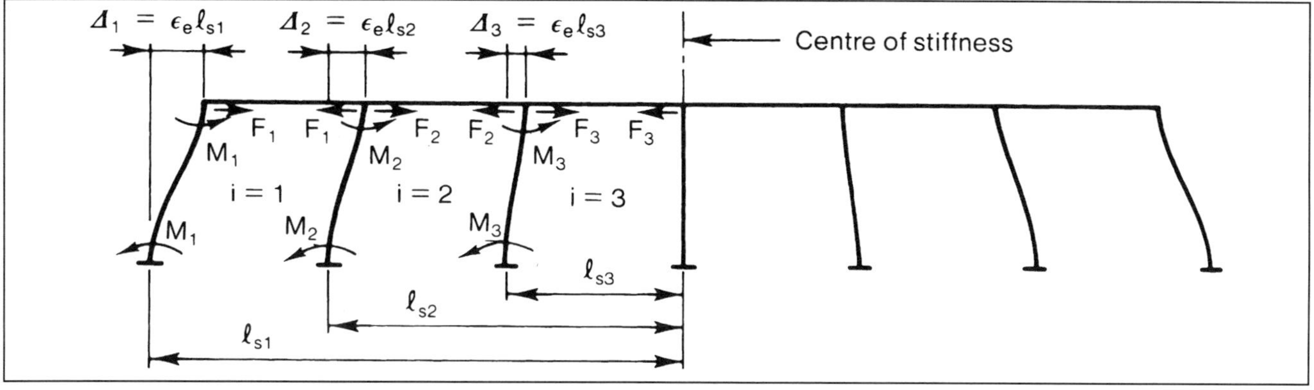


sizes, storey heights and element stiffnesses, the centre of stiffness is located at the midpoint of the frame, as shown in Fig. 2.6.7.

Tension in girders is maximum in the bay nearest the centre of stiffness. Deflections and moments in columns are maximum furthest from the centre of stiffness. Thus in Fig. 2.6.7:

$F_1 < F_2 < F_3$

$\Delta_1 > \Delta_2 > \Delta_3$

$M_1 > M_2 > M_3$

The degree of fixity of the column base, as described in Sect. 2.6.3, has a great effect on the magnitude of the forces and moments caused by volume change restraint. An assumption of a fully fixed base in the analysis of the structure may result in significant overestimation of the restraint forces, whereas, assuming a pinned base may have the opposite effect. The degree of fixity used in the volume change analysis should be consistent with that used in the analysis of the column for other loadings, and determination of slenderness effects.

### Calculation of Volume Change Restraint Forces In Moment-resisting Frames

Several computer analysis programs are available that will allow the input of not only gravity and lateral loads, but also the shortening strains of elements from volume changes. The equivalent strains as described below can be input directly into such programs. While these programs have methods of inputting partially fixed bases, they are somewhat complex. The designer may prefer to make two computer runs, one with the base pinned and one with the base fixed and then interpolate between the results, or use the method outlined in Sect. 2.6.4.

For frames that are approximately symmetrical, the coefficients from Figs. 2.6.8 and 2.6.10 can be used with small error. The use of these tables is described in Fig. 2.6.9.

**Fig. 2.6.8 Build-up of restraint forces in beams ($k_b$)**

| Total number of bays (n) | Number of bays from end (i) | | | | | | | |
|---|---|---|---|---|---|---|---|---|
| | 1 | 2 | 3 | 4 | 5 | 6 | 7 | 8 |
| 2 | 1.00 | | | | | | | |
| 3 | 1.00 | 4.00 | | | | | | |
| 4 | 1.00 | 3.00 | | | | | | |
| 5 | 1.00 | 2.67 | 9.00 | | | | | |
| 6 | 1.00 | 2.50 | 6.00 | | | | | |
| 7 | 1.00 | 2.40 | 5.00 | 16.00 | | | | |
| 8 | 1.00 | 2.33 | 4.50 | 10.00 | | | | |
| 9 | 1.00 | 2.29 | 4.20 | 8.00 | 25.00 | | | |
| 10 | 1.00 | 2.25 | 4.00 | 7.00 | 15.00 | | | |
| 11 | 1.00 | 2.22 | 3.86 | 6.40 | 11.67 | 36.00 | | |
| 12 | 1.00 | 2.20 | 3.75 | 6.00 | 10.00 | 21.00 | | |
| 13 | 1.00 | 2.18 | 3.67 | 5.71 | 9.00 | 16.00 | 49.00 | |
| 14 | 1.00 | 2.17 | 3.60 | 5.50 | 8.33 | 13.50 | 28.00 | |
| 15 | 1.00 | 2.15 | 3.55 | 5.33 | 7.86 | 12.00 | 21.00 | 64.00 |
| 16 | 1.00 | 2.14 | 3.50 | 5.20 | 7.50 | 11.00 | 17.50 | 36.00 |

**Fig. 2.6.9 Use of Fig. 2.6.10**

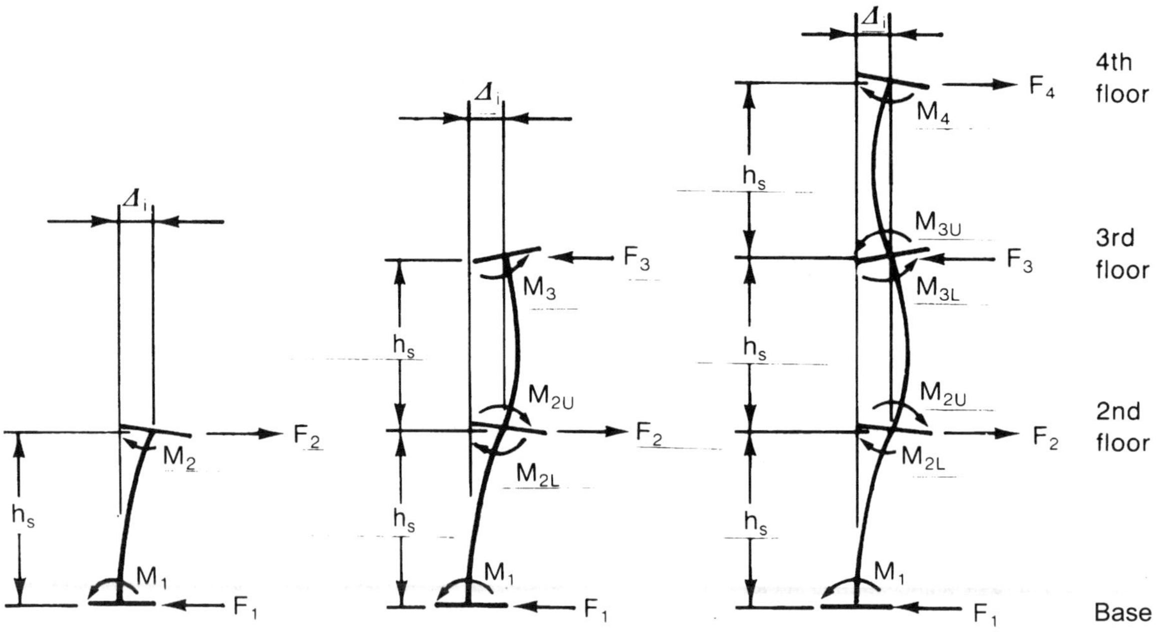


$\Delta_i \ = \epsilon_e \, \ell_s$

$F_i \ = k_f k_b \Delta_i E_c I_c / h_s^3$

$M_i \ = k_m \Delta_i E_c I_c / h_s^2$

where:

$\epsilon_e$ = equivalent unit strain

$\ell_s$ = distance from column to centre of stiffness

$F_i$ = $F_1$, $F_2$, etc., as shown above

$k_f$, $k_m$ = coefficients from Fig. 2.6.10

$k_b \ = i\left(\dfrac{n+1-i}{n+2-2i}\right)$ (or from Fig. 2.6.8)

$n$ = number of bays

$i$ = as shown in Fig. 2.6.7

$E_c$ = modulus of elasticity of the column concrete

$I_c$ = moment of inertia of the column

### Equivalent Volume Change

If a horizontal framing element is connected at the ends, such that the volume change shortening is restrained, a tensile force is built up in the element and transmitted to the supporting elements. However, since the shortening takes place gradually over a period of time, the effect of the shortening on the shears and moments of the support is lessened because of creep and micro-cracking of the element and its support.

For ease of design, the volume change shortenings can be treated in the same manner as short term elastic deformations by using a concept of "equivalent" shortening.

Thus, the following relations can be assumed:

$\delta_{ec} \ = \delta_c / K_\ell$

$\delta_{es} \ = \delta_s / K_\ell$

where:

$\delta_{ec}$, $\delta_{es}$ = equivalent creep and shrinkage shortenings, respectively

$\delta_c$, $\delta_s$ = calculated creep and shrinkage shortenings, respectively

$K_\ell$ = a constant for design purposes which varies from 3 to 5

The value of $K_\ell$ will be near the lower end of the range when the elements are heavily reinforced, and near the upper end when they are lightly reinforced. For most common structures, a value of $K_\ell = 4$ is sufficiently conservative.

Shortening due to temperature change[R2.6.2] will be similarly modified. However, the maximum temperature change will usually occur over a much shorter time, probably within 60 to 90 days. Thus:

$\delta_{et} \ = \delta_t / K_t$

where:

$\delta_{et}$ and $\delta_t$ = the equivalent and calculated temperature shortening, respectively

$K_t$ = a constant; recommended value = 1.5

The total equivalent shortening to be used for design is:

$$\Delta \ = \delta_{ec} + \delta_{es} + \delta_{et}$$

$$= \frac{\delta_c + \delta_s}{K_\ell} + \frac{\delta_t}{K_t}$$

**Fig. 2.6.10 Coefficients $k_f$ and $k_m$ for forces and moments caused by volume change restraint (see Fig. 2.6.9 for notation)**

| No. of stories | $K_r = \frac{\Sigma E_b I_b / \ell}{\Sigma E_c I_c / h_s}$ | Base fixity | Values of $k_f$ | | | | Values of $k_m$ | | | | | |
|---|---|---|---|---|---|---|---|---|---|---|---|---|
| | | | | | | | Base | 2nd floor | | 3rd floor | | 4th |
| | | | $F_1$ | $F_2$ | $F_3$ | $F_4$ | $M_1$ | $M_{2L}$ | $M_{2U}$ | $M_{3L}$ | $M_{3U}$ | $M_4$ |
| 1 | 0 | Fixed | 3.0 | 3.0 | | | 3.0 | 0 | | | | |
| | | Pinned | 0 | 0 | | | 0 | 0 | | | | |
| | 0.5 | Fixed | 6.0 | 6.0 | | | 4.0 | 2.0 | | | | |
| | | Pinned | 1.2 | 1.2 | | | 0 | 1.2 | | | | |
| | 1.0 | Fixed | 7.5 | 7.5 | | | 4.5 | 3.0 | | | | |
| | | Pinned | 1.7 | 1.7 | | | 0 | 1.7 | | | | |
| | 2.0 | Fixed | 9.0 | 9.0 | | | 5.0 | 4.0 | | | | |
| | | Pinned | 2.2 | 2.2 | | | 0 | 2.2 | | | | |
| | 4.0 or more | Fixed | 10.1 | 10.1 | | | 5.4 | 4.7 | | | | |
| | | Pinned | 2.5 | 2.5 | | | 0 | 2.5 | | | | |
| 2 | 0 | Fixed | 6.8 | 9.4 | 2.6 | | 4.3 | 2.6 | 2.6 | 0 | | |
| | | Pinned | 1.5 | 3.0 | 1.5 | | 0 | 1.5 | 1.5 | 0 | | |
| | 0.5 | Fixed | 8.1 | 10.7 | 2.6 | | 4.7 | 3.4 | 2.1 | 0.4 | | |
| | | Pinned | 1.9 | 3.4 | 1.4 | | 0 | 1.9 | 1.2 | 0.2 | | |
| | 1.0 | Fixed | 8.9 | 11.2 | 2.3 | | 4.9 | 3.9 | 1.8 | 0.5 | | |
| | | Pinned | 2.1 | 3.4 | 1.3 | | 0 | 2.1 | 1.0 | 0.3 | | |
| | 2.0 | Fixed | 9.7 | 11.6 | 1.9 | | 5.2 | 4.5 | 1.4 | 0.5 | | |
| | | Pinned | 2.4 | 3.4 | 1.0 | | 0 | 2.4 | 0.8 | 0.3 | | |
| | 4.0 or more | Fixed | 10.4 | 11.9 | 1.4 | | 5.5 | 5.0 | 1.0 | 0.4 | | |
| | | Pinned | 2.6 | 3.4 | 0.8 | | 0 | 2.6 | 0.5 | 0.2 | | |
| 3 or more | 0 | Fixed | 7.1 | 10.6 | 4.1 | 0.7 | 4.4 | 2.8 | 2.8 | 0.7 | 0.7 | 0 |
| | | Pinned | 1.6 | 3.6 | 2.4 | 0.4 | 0 | 1.6 | 1.6 | 0.4 | 0.4 | 0 |
| | 0.5 | Fixed | 8.2 | 11.1 | 3.5 | 0.5 | 4.7 | 3.5 | 2.2 | 0.7 | 0.4 | 0.09 |
| | | Pinned | 1.9 | 3.6 | 1.9 | 0.3 | 0 | 1.9 | 1.2 | 0.4 | 0.2 | 0.05 |
| | 1.0 | Fixed | 8.9 | 11.4 | 2.9 | 0.4 | 5.0 | 3.9 | 1.9 | 0.7 | 0.3 | 0.09 |
| | | Pinned | 2.2 | 3.5 | 1.6 | 0.2 | 0 | 2.2 | 1.0 | 0.4 | 0.2 | 0.05 |
| | 2.0 | Fixed | 9.7 | 11.7 | 2.2 | 0.2 | 5.2 | 4.7 | 1.4 | 0.6 | 0.2 | 0.06 |
| | | Pinned | 2.4 | 3.5 | 1.2 | 0.1 | 0 | 2.4 | 0.8 | 0.3 | 0.1 | 0.03 |
| | 4.0 or more | Fixed | 10.4 | 11.9 | 1.5 | 0.04 | 5.5 | 5.0 | 1.0 | 0.5 | 0.04 | 0.01 |
| | | Pinned | 2.6 | 3.4 | 0.8 | 0.02 | 0 | 2.6 | 0.5 | 0.2 | 0.02 | 0.00 |

When the equivalent shortening is used in the frame analysis for determining shears and moments in the supporting elements, the actual modulus of elasticity of the elements is used, rather than a reduced modulus as used in other methods.

Figures 2.6.11 and 2.6.12 provide equivalent volume change strains for typical building frames.

**Example 4 - Calculation of column moment caused by volume change shortening of a beam**

*Given:*

The beam in Example 1 on Page 2-19 is supported and attached to two 400 × 400 columns as shown in the sketch.

$f'_c(\text{col}) = 35$ MPa

$E_c = 29\,900$ MPa

**Fig. 2.6.11 Equivalent volume change strains for typical continuous building frames with prestressed elements ($10^{-6}$)**

| Temp. zone (Fig. 2.4.4) | Normal density concrete | | | | | Low density concrete | | | | |
|---|---|---|---|---|---|---|---|---|---|---|
| | Avg. relative humidity (Fig. 2.4.2) | | | | | Avg. relative humidity (Fig. 2.4.2) | | | | |
| | 40 | 50 | 60 | 70 | 80 | 40 | 50 | 60 | 70 | 80 |
| **Heated structures** | | | | | | | | | | |
| 0 | 140 | 130 | 120 | 100 | 90 | 170 | 160 | 140 | 130 | 110 |
| 5 | 160 | 150 | 130 | 120 | 110 | 180 | 170 | 160 | 140 | 130 |
| 10 | 180 | 160 | 150 | 140 | 130 | 200 | 190 | 170 | 160 | 140 |
| 15 | 190 | 180 | 170 | 160 | 150 | 210 | 200 | 190 | 170 | 160 |
| 20 | 210 | 200 | 190 | 180 | 160 | 230 | 220 | 200 | 190 | 170 |
| 25 | 230 | 220 | 210 | 190 | 180 | 240 | 230 | 220 | 200 | 190 |
| 30 | 250 | 240 | 220 | 210 | 200 | 260 | 250 | 230 | 220 | 200 |
| 35 | 270 | 250 | 240 | 230 | 220 | 270 | 260 | 250 | 230 | 220 |
| 40 | 280 | 270 | 260 | 250 | 240 | 290 | 280 | 260 | 250 | 230 |
| 45 | 300 | 290 | 280 | 270 | 250 | 300 | 290 | 280 | 260 | 250 |
| 50 | 320 | 310 | 300 | 280 | 270 | 320 | 310 | 290 | 280 | 260 |
| **Unheated structures** | | | | | | | | | | |
| 0 | 140 | 130 | 120 | 100 | 90 | 170 | 160 | 140 | 130 | 110 |
| 5 | 170 | 160 | 140 | 130 | 120 | 190 | 180 | 170 | 150 | 140 |
| 10 | 190 | 180 | 170 | 160 | 150 | 210 | 200 | 190 | 170 | 160 |
| 15 | 220 | 210 | 200 | 190 | 170 | 240 | 220 | 210 | 190 | 180 |
| 20 | 250 | 240 | 220 | 210 | 200 | 260 | 250 | 230 | 220 | 200 |
| 25 | 280 | 260 | 250 | 240 | 230 | 280 | 270 | 260 | 240 | 230 |
| 30 | 300 | 290 | 280 | 270 | 250 | 300 | 290 | 280 | 260 | 250 |
| 35 | 330 | 320 | 310 | 290 | 280 | 330 | 310 | 300 | 280 | 270 |
| 40 | 360 | 340 | 330 | 320 | 310 | 350 | 340 | 320 | 310 | 290 |
| 45 | 380 | 370 | 360 | 350 | 340 | 370 | 360 | 350 | 330 | 320 |
| 50 | 410 | 400 | 390 | 370 | 360 | 390 | 380 | 370 | 350 | 340 |

Above table based on the same parameters as Fig. 2.4.6 with the addition that $K_\ell = 5$ and $K_t = 1.5$.

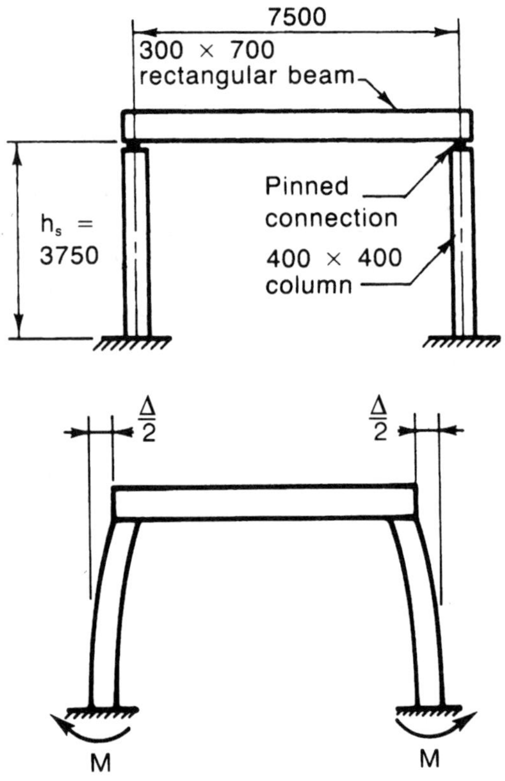


*Problem:*

Determine the horizontal force at the top of the column caused by volume change shortening of the beam.

*Solution:*

$I_c = bh^3/12 = 400^4/12 = 2130 \times 10^6 \text{ mm}^4$

From Example 1 on Page 2-19

Total volume change shortening from erection to final

$= 5.6$ mm or 2.8 mm each end

Calculate the equivalent shortening:

$$\Delta = \frac{\delta_c + \delta_s}{K_\ell} + \frac{\delta_t}{K_t}$$

$$= \left[\frac{(365 - 187 + 367 - 174)}{5} + \frac{211}{1.5}\right]$$

$$\times (10^{-6})(7500)$$

$$= 1.6 \text{ mm}$$

$\Delta/2 = 1.6 / 2 = 0.8$ mm each end

$N = 3E_c I_c (\Delta / 2) / h_s^3$

**Fig. 2.6.12 Equivalent volume change strains for typical continuous building frames with Non-prestressed elements ($10^{-6}$)**

| Temp. zone (Fig. 2.4.4) | Normal density concrete, Avg. relative humidity (Fig. 2.4.2) | | | | | Low density concrete, Avg. relative humidity (Fig. 2.4.2) | | | | |
|---|---|---|---|---|---|---|---|---|---|---|
| | 40 | 50 | 60 | 70 | 80 | 40 | 50 | 60 | 70 | 80 |
| **Heated structures** | | | | | | | | | | |
| 0 | 70 | 60 | 50 | 40 | 40 | 70 | 60 | 50 | 40 | 40 |
| 5 | 80 | 80 | 70 | 60 | 60 | 80 | 70 | 70 | 60 | 50 |
| 10 | 100 | 90 | 90 | 80 | 70 | 90 | 90 | 80 | 70 | 70 |
| 15 | 120 | 110 | 110 | 100 | 90 | 110 | 100 | 100 | 90 | 80 |
| 20 | 140 | 130 | 120 | 120 | 110 | 120 | 120 | 110 | 100 | 100 |
| 25 | 160 | 150 | 140 | 130 | 130 | 140 | 130 | 130 | 120 | 110 |
| 30 | 180 | 170 | 160 | 150 | 150 | 150 | 150 | 140 | 130 | 130 |
| 35 | 190 | 180 | 180 | 170 | 160 | 170 | 160 | 160 | 150 | 140 |
| 40 | 210 | 200 | 200 | 190 | 180 | 180 | 180 | 170 | 160 | 160 |
| 45 | 230 | 220 | 210 | 210 | 200 | 200 | 190 | 190 | 180 | 170 |
| 50 | 250 | 240 | 230 | 220 | 220 | 210 | 210 | 200 | 190 | 190 |
| **Unheated structures** | | | | | | | | | | |
| 0 | 70 | 60 | 50 | 40 | 40 | 70 | 60 | 50 | 40 | 40 |
| 5 | 90 | 80 | 80 | 70 | 60 | 90 | 80 | 80 | 70 | 60 |
| 10 | 120 | 110 | 110 | 100 | 90 | 110 | 100 | 100 | 90 | 80 |
| 15 | 150 | 140 | 130 | 130 | 120 | 130 | 120 | 120 | 110 | 110 |
| 20 | 170 | 160 | 160 | 150 | 150 | 150 | 150 | 140 | 130 | 130 |
| 25 | 200 | 190 | 190 | 180 | 170 | 180 | 170 | 170 | 160 | 150 |
| 30 | 230 | 220 | 210 | 210 | 200 | 200 | 190 | 190 | 180 | 170 |
| 35 | 250 | 250 | 240 | 230 | 230 | 220 | 210 | 210 | 200 | 200 |
| 40 | 280 | 270 | 270 | 260 | 250 | 240 | 240 | 230 | 220 | 220 |
| 45 | 310 | 300 | 300 | 290 | 280 | 270 | 260 | 250 | 250 | 240 |
| 50 | 330 | 330 | 320 | 310 | 310 | 290 | 280 | 280 | 270 | 260 |

Above table based on the same parameters as Fig. 2.4.7 with the addition that $K_\ell = 5$ and $K_t = 1.5$.

$= (3)(29\,900)(2130 \times 10^6)(0.8) / 3750^3$

$= 2800 \text{ N} = 2.8 \text{ kN}$

$M = Nh_s$

$= (2.8)(3750) / 10^3$

$= 10.5 \text{ kN} \cdot \text{m}$

**Example 5 - Volume change restraint forces**

*Given:*

The 4-bay, 2-storey heated frame as shown

Beam modulus of elasticity = $E_b$ = 29 900 MPa

Column modulus of elasticity = $E_c$ = 32 000 MPa

Column bases 20% fixed (see Subsect. 2.6.3)

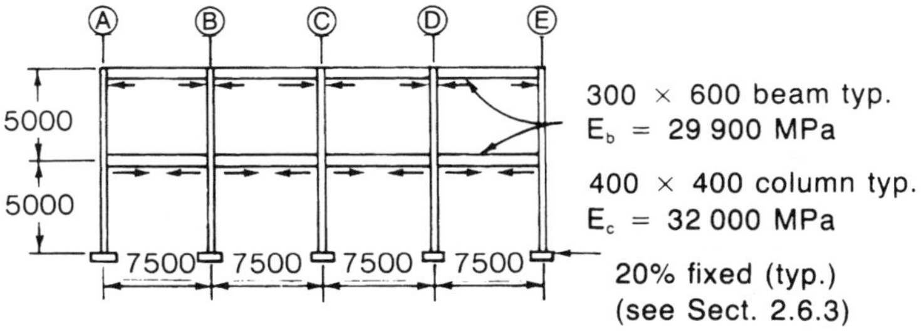


Design R.H. = 70%

Design temperature change = 39°C

*Problem:*

Determine the maximum tension in the beams and the maximum moment in the columns caused by volume change restraint.

*Solution:*

1. Determine relative stiffness between columns and beams:

$I_b = (300)(600)^3 / 12 = 5400 \times 10^6 \text{ mm}^4$

$E_b I_b / \ell = (29.9)(5400) / 7500 = 21.5 \times 10^3 \text{ kN} \cdot \text{m}$

$I_c = 400^4/12 = 2130 \times 10^6 \text{ mm}^4$

$E_c I_c / h_s = (32.0)(2130) / 5000 = 13.6 \times 10^3 \text{ kN} \cdot \text{m}$

$$K_r = \frac{E_b I_b / \ell}{E_c I_c / h_s} = 1.6$$

2. Determine deflections:

From Fig. 2.6.11:

$\epsilon_e = 246 \times 10^{-6}$ mm/mm

**Fig. 2.6.13 Use of Fig. 2.6.14**

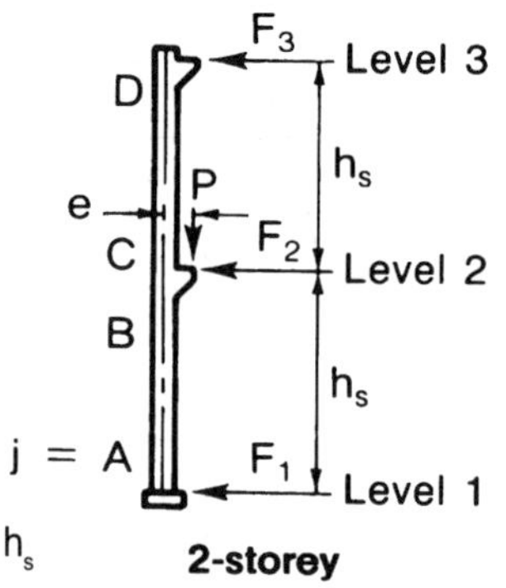


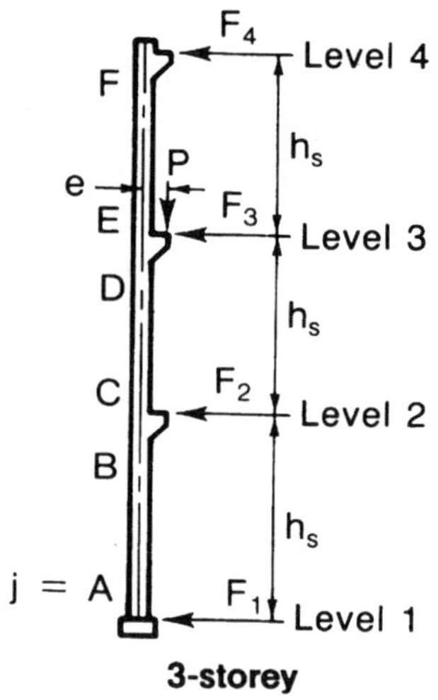


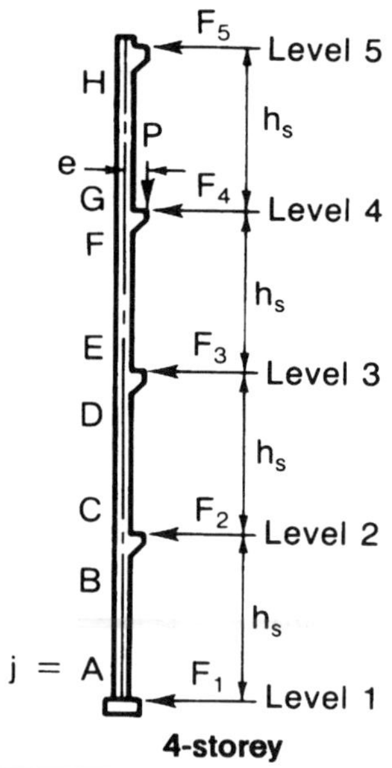


$F_i \quad = k_f\, P\, e / h_s$

$M_j \quad = k_m\, P\, e$

where:

$F_i$ = restraining force at level I

$m_j$ = moment at point j

$k_f$, $k_m$ = coefficients from Fig. 2.6.14

P = vertical load acting at eccentricity e

Columns braced against sidesway

See text for limitations and design example

$\Delta_B \quad = \epsilon_e \ell = (246 \times 10^{-6})(7500) = 1.8$ mm

$\Delta_A \quad = \epsilon_e(2\ell) = 3.6$ mm

3. Determine maximum beam tension which occurs nearest the centre of stiffness, i.e., beams BC and CD, 2nd floor:

From Fig. 2.6.8:

For n = 4 and i = 2, $k_b = 3.00$

From Fig. 2.6.10:

For $K_r = 1.0$, fixed base, $k_f = 11.2$

For $K_r = 2.0$, fixed base, $k_f = 11.6$

Therefore for $K_r = 1.5$, $k_f = 11.4$

For pinned base, $k_f = 3.4$ (for $K_r = 1.0$ and 2.0)

For 20% fixed:

$k_r \quad = 3.4 + 0.20\,(11.4 - 3.4) = 5.0$

$F_2 \quad = k_f k_b \Delta_i E_c I_c / h_s^3$

$= (5.0)(3.0)(1.8)(32.0)(2130 \times 10^6) / 5000^3$

$= 14.7$ kN

4. Determine maximum column moments:

For base moment, $M_1$:

From Fig. 2.6.10 by interpolation similar to above:

$k_m$ (fixed) $= (4.9 + 5.2)/2 = 5.05$

$k_m$ (pinned) $= 0$

$k_m$ (20% fixed) $= 0 + 0.20\,(5.05) = 1.0$

$M_1 \quad = k_m \Delta_i E_c I_c / h_s^2$

$= (1.0)(3.6 \times 10^{-3})(32.0)(2130 \times 10^6) / 5000^2$

$= 9.8$ kN · m

For second floor moment, $M_{2L}$:

$k_m$ (fixed) $= (3.9 + 4.5) / 2 = 4.2$

$k_m$ (pinned) $= (2.1 + 2.4)/2 = 2.25$

$k_m$ (20% fixed) $= 2.25 + 0.20\,(4.20 - 2.25) = 2.64$

$M_{2L} \quad = (2.64)(3.6 \times 10^{-3})(32.0)(2130 \times 10^6) / 5000^2$

$= 25.9$ kN · m

## 2.6.6 ECCENTRICALLY LOADED COLUMNS

Many precast concrete structures utilize multi-storey columns with simple-span beams resting on haunches. Figs. 2.6.13 and 2.6.14 are provided as aids for determining the various combinations of load and moment that can occur with such columns.

The following conditions and limitations apply to Figs. 2.6.13 and 2.6.14:

1. The coefficients are only valid for braced columns. Lateral stability must be achieved by other shear walls or moment resisting frames.
2. For partially fixed column bases (see Sect. 2.6.3), a straight line interpolation between the coefficients for pinned and fixed bases can be used with small error.
3. For higher columns, the coefficients for the 4-storey columns can be used with small error.

**Fig. 2.6.14 Coefficients $k_f$ and $k_m$ for determining momemts and restaining forces on eccentrically loaded columns braced against sidesway**

+ indicates clockwise moments on the columns and compression in the restraining beam

| No. of stories | Base fixity | P acting at level | $k_f$ at level 1 | 2 | 3 | 4 | 5 | $k_m$ at point A | B | C | D | E | F | G | H |
|---|---|---|---|---|---|---|---|---|---|---|---|---|---|---|---|
| 2 | Pinned | 3 | +0.25 | −1.50 | +1.25 | | | 0 | −0.25 | +0.25 | +1.00 | | | | |
| | | 2 | −0.50 | 0 | +0.50 | | | 0 | +0.50 | +0.50 | 0 | | | | |
| | | Σ Max | ±0.75 | ±1.50 | ±1.75 | | | 0 | ±0.75 | ±0.75 | ±1.00 | | | | |
| | | Σ One side | −0.25 | −1.50 | +1.75 | | | 0 | +0.25 | +0.75 | +1.00 | | | | |
| | Fixed | 3 | +0.43 | −1.72 | +1.29 | | | −0.14 | −0.29 | +0.29 | +1.00 | | | | |
| | | 2 | −0.86 | +0.43 | +0.43 | | | +0.29 | +0.57 | +0.43 | 0 | | | | |
| | | Σ Max | ±1.29 | ±2.15 | ±1.72 | | | ±0.43 | ±0.86 | ±0.72 | ±1.00 | | | | |
| | | Σ One side | −0.43 | −1.29 | +1.72 | | | +0.15 | +0.28 | +0.72 | +1.00 | | | | |
| 3 | Pinned | 4 | −0.07 | +0.40 | −1.60 | +1.27 | | 0 | +0.07 | −0.07 | −0.27 | +0.27 | +1.00 | | |
| | | 3 | +0.13 | −0.80 | +0.20 | +0.47 | | 0 | −0.13 | +0.13 | +0.53 | +0.47 | 0 | | |
| | | 2 | −0.47 | −0.20 | +0.80 | −0.13 | | 0 | +0.47 | +0.53 | +0.13 | −0.13 | 0 | | |
| | | Σ Max | ±0.67 | ±1.40 | ±2.60 | ±1.87 | | 0 | ±0.67 | ±0.73 | ±0.93 | ±0.87 | ±1.00 | | |
| | | Σ One side | −0.41 | −0.60 | −0.60 | +1.61 | | 0 | +0.40 | +0.60 | +0.40 | +0.60 | +1.00 | | |
| | Fixed | 4 | −0.12 | +0.47 | −1.62 | +1.27 | | +0.04 | +0.08 | −0.08 | −0.27 | +0.27 | +1.00 | | |
| | | 3 | +0.23 | −0.92 | +0.23 | +0.46 | | −0.08 | −0.15 | +0.15 | +0.54 | +0.46 | 0 | | |
| | | 2 | −0.81 | +0.23 | +0.70 | −0.12 | | +0.27 | +0.54 | +0.46 | +0.12 | −0.12 | 0 | | |
| | | Σ Max | ±1.16 | ±1.62 | ±2.55 | ±1.85 | | ±0.38 | ±0.77 | ±0.69 | ±0.92 | ±0.85 | ±1.00 | | |
| | | Σ One side | −0.70 | −0.22 | −0.69 | +1.61 | | +0.23 | +0.46 | +0.54 | +0.38 | +0.62 | +1.00 | | |
| 4 | Pinned | 5 | +0.02 | −0.11 | +0.43 | −1.61 | +1.27 | 0 | −0.02 | +0.02 | +0.07 | −0.07 | −0.27 | +0.27 | +1.00 |
| | | 4 | −0.04 | +0.22 | −0.86 | +0.22 | +0.46 | 0 | +0.04 | −0.04 | −0.14 | +0.14 | +0.54 | +0.46 | 0 |
| | | 3 | +0.13 | −0.75 | 0 | +0.75 | −0.12 | 0 | −0.13 | +0.13 | +0.50 | +0.50 | +0.12 | −0.12 | 0 |
| | | 2 | −0.46 | −0.22 | +0.86 | −0.22 | +0.04 | 0 | +0.46 | +0.54 | +0.14 | −0.14 | −0.04 | +0.04 | 0 |
| | | Σ Max | ±0.65 | ±1.30 | ±2.15 | ±2.80 | ±1.89 | 0 | ±0.64 | ±0.72 | ±0.86 | ±0.86 | ±0.97 | ±0.89 | ±1.00 |
| | | Σ One side | −0.35 | −0.86 | +0.43 | −0.86 | +1.65 | 0 | +0.35 | +0.65 | +0.57 | +0.43 | +0.35 | +0.65 | +1.00 |
| | Fixed | 5 | +0.03 | −0.12 | +0.43 | −1.61 | +1.27 | −0.01 | −0.02 | +0.02 | +0.07 | −0.07 | −0.27 | +0.27 | +1.00 |
| | | 4 | −0.06 | +0.25 | −0.87 | +0.22 | +0.46 | +0.02 | +0.04 | −0.04 | −0.14 | +0.14 | +0.54 | +0.46 | 0 |
| | | 3 | +0.22 | −0.87 | +0.03 | +0.74 | −0.12 | −0.07 | −0.14 | +0.14 | +0.51 | +0.50 | +0.12 | −0.12 | 0 |
| | | 2 | −0.80 | +0.21 | +0.74 | −0.18 | +0.03 | +0.27 | +0.54 | +0.46 | +0.12 | −0.12 | −0.03 | +0.03 | 0 |
| | | Σ Max | ±1.11 | ±1.45 | ±2.07 | ±2.75 | ±1.88 | ±0.37 | ±0.74 | ±0.67 | ±0.84 | ±0.83 | ±0.96 | ±0.88 | ±1.00 |
| | | Σ One side | −0.61 | −0.53 | +0.33 | −0.83 | +1.64 | +0.21 | +0.41 | +0.59 | +0.56 | +0.44 | +0.36 | +0.64 | +1.00 |

4. The coefficients in the "Σ Max" line will yield the maximum required restraining force, $F_i$, and column moments caused by loads (equal at each level) which can occur on either side of the column, for example, live loads on interior columns. The maximum force will not necessarily occur with the same loading pattern that causes the maximum moment.

5. The coefficients in "Σ One Side" line will yield the maximum moments which can occur if the column is loaded on only one side, such as the end column in a bay.

**Example 6 - Use of Figs. 2.6.13 and 2.6.14**

*Given:*

3-storey frame with beam reactions to column haunch at each level:

Dead load = 220 kN

Live load = 90 kN

Eccentricity, e = 350 mm

Storey height, $h_s$ = 5000 mm

Column base is 65% fixed

*Problem:*

Using Fig. 2.6.14, determine the maximum restraining force and moment in the lowest storey for:

a) An interior column in a multi-bay frame

b) An exterior column

*Solution:*

Factored load = (1.25)(220) + (1.5)(90)

= 275 + 135 = 410 kN

a) *For the interior column:*

The dead load reaction would be the same on either side, thus no moment results. The live load could occur on any one side at any floor, hence use the coefficients in the "Σ Max" line:

$P_f e = (135)(350 \times 10^{-3}) = 47.3 \text{ kN} \cdot \text{m}$

To determine the maximum moment at point B:

For a pinned base, $k_m$ = 0.67

For a fixed base, $k_m$ = 0.77

For 65% fixed, $k_m$ = 0.67+ (0.65)(0.77 – 0.67)

= 0.74

$M_f = k_m P_f e = (0.74)(47.3) = 35.0$ kN · m

Maximum restraining force at level 2:

$k_f = 1.40 + (0.65)(1.62 - 1.40) = 1.54$

$F_f = k_f P_f e / h_s = (1.54)(47.3) / 5$

= 14.6 kN (tension or compression)

*b) For the exterior column:*

The total load is eccentric on the same side of the column, hence use the coefficients in the "Σ One Side" line:

$P_f e = (410)(350 \times 10^{-3}) = 143.5$ kN · m

To determine the maximum moment at point B:

For a pinned base, $k_m$ = 0.40

For a fixed base, $k_m$ = 0.46

For a 65% fixed, $k_m$ = 0.40 + 0.65)(0.46-0.40)

= 0.44

$M_f = k_m P_f e = (0.44)(143.5) = 63.1$ kN·m

Maximum restraining force at level 2:

$k_f = -0.60 + (0.65)(-0.60 + 0.22) = -0.35$

$F_f = k_f P_f e/h_s = (-0.35)(143.5) / 5$

= –10.0 kN (tension)

## 2.6.7 SLENDERNESS EFFECTS IN COMPRESSION ELEMENTS

### Approximate Evaluation of Slenderness Effects

The Code permits an approximate evaluation of slenderness. Application of this section of the Code for elements braced against sidesway is shown in Sect. 3.8.3 and for unbraced frames in the Example in Sect. 2.6.8. A more rigorous approach is discussed briefly later in this section.

**Fig. 2.6.15 Alignment charts for determining effective length factors**

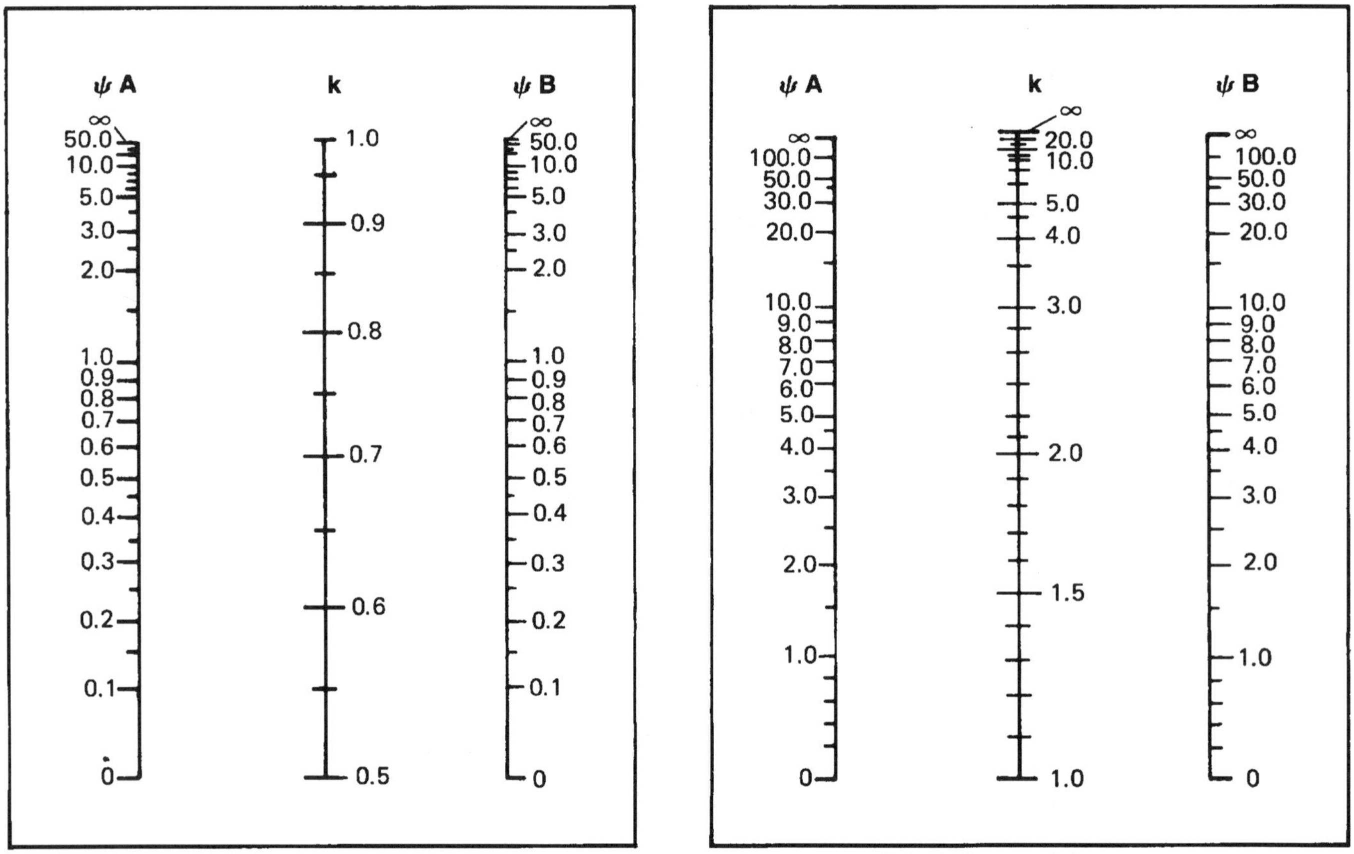


**(a) Braced frames**

ψ = ratio of Σ K of compression elements to Σ K of flexural elements in a plane at one end of a compression element

**(b) Unbraced frames**

ψ base = $K_c/K_b$ (see Sect. 2.6.2) 10 ≥ ψ base ≥ 1.0

k = effective length factor

The effective length factor, k, can be determined from the alignment charts, Fig. 2.6.15. For column bases, the value of $\psi$ for use in these charts can be calculated from the rotational stiffness coefficients described in Sect. 2.6.2 with $\psi$ base = $K_c/K_b$. For most structures, $\psi$ base should not be taken less than 1.0. For column bases which are assumed pinned in the frame analysis, $\psi$ base can be assumed equal to 10 when using Fig. 2.6.15.

### Slenderness Effects by Structural Analysis

In lieu of the approximate evaluation of slenderness effects, the requirements of the Code can be met by using an iterative P Δ analysis that is suitable for computer solution[R2.6.3], or by direct P Δ analysis[R2.6.4].

An iterative P Δ analysis accounts for the effects of frame drift due to slenderness by designing for frame moments which correspond to the calculated shears, which are required for equilibrium for an initial sway.

A short explanation of this procedure follows:

1. Select an initial frame deflection to account for out-of-plumbness. (1/4 of 1% has been suggested in the past.)
2. Calculate approximate storey shears resulting from this initial deflection. See Fig. 2.6.16.
3. Determine the frame deflection corresponding to the above shears. Note that this deflection must be based on a reduced EI to account for creep due to sustained loads and cracking.
4. Calculate the shears resulting from the increased deflection determined in Step 3. Once again determine the corresponding deflection. (This may be quickly done using a ratio of shears.)
5. The addition to the initial shears reduces after each successive iteration (if it does not, the structure has a stability problem and element dimensions should be reconsidered) and generally an upper bound value for the design storey shear can be selected following several iterations.
6. The computed forces corresponding to these storey shears account for the effects of column slenderness on the frame. Now each column may be designed without moment magnifiers for the moments resulting from gravity loads acting at an eccentricity of Δ.
7. When designing for the case of wind, an additional storey shear results from the wind deflection and must be accounted for. For this case, use the actual EI, as wind effects are short term. Due to the nature of volume change effects, generally no additional storey shear will result from these forces as they tend to cancel.

Storeys in structures shall be designated as non-sway if $Q \le 0.05$ where:

$$Q = \frac{\Sigma P_f \Delta_o}{V_f \ell_c}$$

$\Sigma P_f$ = summation of factored loads in a given storey

**Fig. 2.6.16 P Δ effects in a frame**

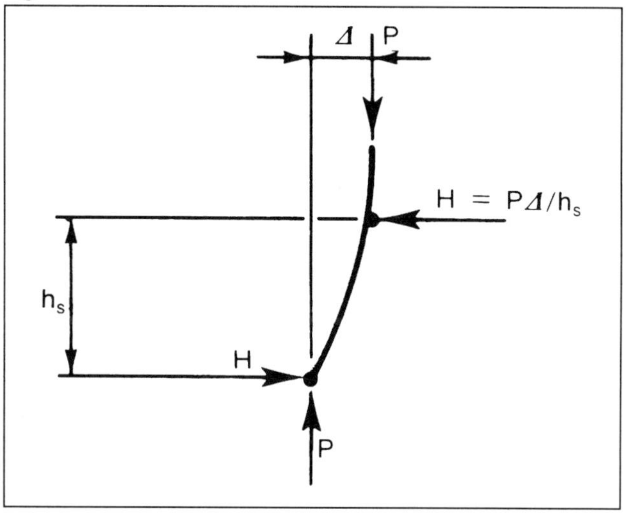


$\Delta_o$ = elasticity-computed first order lateral deflection due to $H_f$ (neglecting PΔ effects) at the top of the storey relative to the bottom of the storey

$V_f$ = total factored lateral force within the storey

$\ell_c$ = height of storey, centre-to-centre of floors or roof

## 2.6.8 EXAMPLE 7 – SEVEN STOREY PARKING STRUCTURE

*Given*:

A multi-level parking structure located in Medicine Hat, Alberta, illustrated by the schematic drawings shown in Fig. 2.6.17. Deck elements are low density concrete single tees with 75 mm (min.) normal density concrete topping. Columns and beams are also precast, prestressed elements. Unfactored loads are given as follows:

Dead loads:

| | |
|---|---|
| Tees and topping | = 4.5 kN/m² |
| Beams, columns, misc. | = 3.6 kN/m² |
| | 8.1 kN/m² |

Live loads:

| | |
|---|---|
| Roof | = 3.6 kN/m² |
| Floor | = 2.4 kN/m² |
| Wind | = 1.0 kN/m² |
| | (Gross projected area) |

*Problem*:

Analyze the structure for volume change effects and for wind loads.

*Solution*:

*Wind load analysis, North and South direction:*

Lateral stability in the North-South direction is achieved by providing moment connections between the spandrel beams and columns on lines 1 and 16.

*Storey wind shears per frame:*

Roof: $P = (1.0)(2.6)(59.2/2)$

$= 77$ kN

Floor: $P = (1.0)(3.0)(59.2/2)$

$= 89$ kN/floor

The wind analysis can be done by computer. Output moments are shown in Fig. 2.6.20. (Note: because of the size of the beams in this structure, uncracked section properties are used in the analysis.)

*Wind drift:*

From the computer analysis, deflection at the first floor due to wind is 2.5 mm. The "drift index" is then:

$$\frac{2.5}{3\text{x}10^3} = 0.00083\,\text{mm} / \text{mm} < 0.002^1$$

*Wind load analysis, East-West direction*:

The full analysis for wind in the East-West direction is not shown here, but the assumptions used are shown in Fig. 2.6.17(c).

*Volume change analysis, North-South direction*:

From Figs. 2.4.4 and 2.4.2, the structure should be designed for a temperature change of 44°C and an average ambient R.H. of 70%.

Beams and columns are prestressed, normal density concrete, so use Fig. 2.4.6 and Fig. 2.6.11 to find volume change strains. For an unheated structure:

Actual strain $= 872 \times 10^{-6}$

Equivalent strain $= 344 \times 10^{-6}$

Because of the stiffening effect of the scissor ramps on the interior columns at line D, and the deck diaphragm, the centre of stiffness of the structure can be assumed to be approximately at line D.

From Fig. 2.6.10 for a structure with 3 or more floors, pinned column base and $K_r > 4$:

$k_m$ at second floor $= 2.6$

Column moment $= k_m \Delta_i E_c I_c / h_s^2$

For column A, 19 250 mm from stiffness centre:

$\Delta_A = (344\times10^{-6})(19\,250) = 6.58$ mm

$M_A = 2.6(6.58)(29\,900)(7813) / 3000^2$

$= 445$ kN · m

$\Delta_B = 4.70$ mm $\quad M_B = 716$ kN · m

$\Delta_C = 1.92$ mm $\quad M_C = 293$ kN · m

$\Delta_E = 2.77$ mm $\quad M_E = 422$ kN · m

$\Delta_F = 4.69$ mm $\quad M_F = 317$ kN · m

*Volume change analysis, East-West direction*:

$\Delta_{max} = (344\times10^{-6})(58\,000 / 2) = 9.91$ mm

(at lines 1 and 16)

From Fig. 2.6.10 with $K_r = 0$:

$k_m$ at second floor $= 1.6$

For columns at lines A and F:

$I = 17\,580 \times 10^6$ mm$^4$

(bending in 750 mm direction)

$M = (1.6)(9.91)(29\,900)(17\,580) / 3000^2$

$= 926$ kN · m

For columns at lines B, C, D, and E:

$I = 7813 \times 10^6$ mm4

(bending in 500 mm direction)

$M = 412$ kN · m

Moments at upper levels can be calculated in a similar manner.

*Column slenderness effects – North and South bending*:

Using the approximate evaluation, it is necessary to calculate two moment magnifiers, one for moments caused by gravity loads (in this case the required minimum eccentricity), another for moments caused by wind.

Sum of all columns loads:

| | $P_{storey}$ (kN) | | |
|---|---|---|---|
| **Storey** | **D** | **L** | **D+L** |
| 7 | 11 710 | 5 200 | 16 910 |
| 6 | 15 600 | 4 620 | 20 220 |
| 5 | 15 600 | 4 620 | 20 220 |
| 4 | 15 600 | 4 620 | 20 220 |
| 3 | 15 600 | 4 620 | 20 220 |
| 2 | 15 600 | 4 620 | 20 220 |
| 1 | 15 600 | 4 620 | 20 220 |
| Total | 105 310 | 32 920 | 138 230 |

$P_f = (1.25)(105\,310) + (1.5)(32\,920)$

$= 131\,640 + 49\,380$

$= 181\,020$ kN

[1] *The NBC of Canada requirement for wind drift is that the value not exceed 0.002.*

**Fig. 2.6.17 7-storey structure of Example 7**

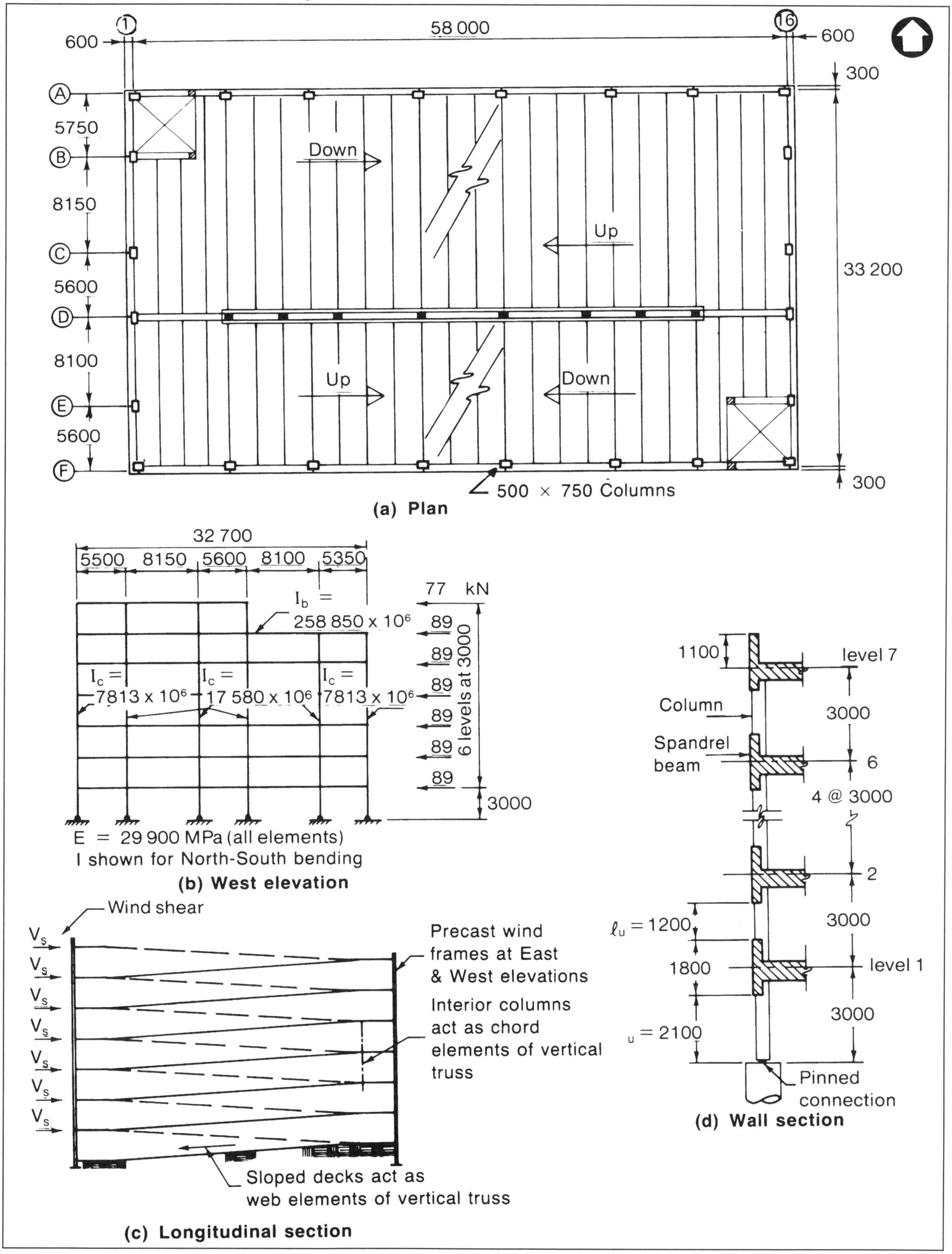

*First level moment magnifier*:

$\ell_u$ = 2100 mm (see Fig. 2.6.17d)

For moments caused by gravity:

$\beta_d$ = 131 640 / 181 020 = 0.73

For moments caused by wind:

$\beta_d$ = 0

For columns at lines A and F:

$E_c = E_b$ = 29 900 MPa

$I_c = 7813 \times 10^6$ mm$^4$

$I_b = 258\,850 \times 10^6$ mm$^4$

From Fig. 2.6.15.

$\psi_A$ (base) = 10.0 (see Sect. 2.6.7)

In using Fig. 2.6.15, $I_b$ can be considered equal to 0.5 $I_g$:

$$\psi_B = \frac{\Sigma E_c I_c / h_s}{\Sigma E_b I_b / \ell} = \frac{(2)(7813)/3000}{(0.5)(258\,850)/5500} = 0.23$$

k = 1.75 (unbraced frame)

$EI = (E_c I_c / 2.5) / (1 + \beta_d)$

Gravity loads:

$EI = (29.9)(7813)/(2.5 \times 1.73)$

= 54 000 kN · mm$^2$

$P_c = \pi^2 EI / (k\ell_u)^2 = \pi^2 (54\,000) / (1.75 \times 2.1)^2$

= 39 500 kN

Wind loads:

$EI = (29.9)(7813) / (2.5 \times 1.0)$

= 93 400 kN · mm$^2$

$P_c = \pi^2(93\,400)/(1.75 \times 2.1)^2$

= 68 250 kN

For columns at lines B, C, D, and E, assuming column B as typical:

$I_c = 17\,580 \times 10^6$ mm$^4$

$\psi_A$ = 10.0

$$\psi_B = \frac{(2)(17\,580)/3000}{0.5(258\,850/5500 + 258\,850/8150)} = 30$$

k = 1.80

Gravity loads:

$EI = (29.9)(17\,580)/(2.5 \times 1.73)$

= 122 000 kN · mm$^2$

$P_c = \pi(122\,000) / (1.80 \times 2.1)^2$

= 84 200 kN

Wind loads:

$EI = (29.9)(17\,580) / (2.5 \times 1.0)$

= 211 000 kN · mm$^2$

$P_c = \pi^2 (211\,000) / (1.80 \times 2.1)^2$

= 145 800 kN

Calculate the sum of $P_c$ of all columns which resist lateral loads:

Gravity loads:

$\Sigma P_c = (4)(39\,500) + (8)(84\,200) = 832\,000$ kN

Note: This neglects any contribution toward stability of the interior columns, which is theoretically correct, since they are assumed pinned at each end. Actual connections seldom behave as full pins, so the above assumption is conservative.

$$\delta = \frac{C_m}{1 - \dfrac{\Sigma P_f}{\phi_m(\Sigma P_c)}} = \frac{1.0}{1 - \dfrac{181020}{(0.65)(832\,000)}} = 1.50$$

Wind loads:

$\Sigma P_c = (4)(68\,250) + (8)(145\,800) = 1\,440\,000$ kN

$$\delta = \frac{1.0}{1 - \dfrac{(0.7)(181020)}{(0.65)(1\,440\,000)}} = 1.16$$

The magnifier for moments caused by wind loads can also be computed as follows:

$\Delta$ = first level wind drift from the computer analysis

= 2.5 mm

$\Delta_f$ = (0.70)(1.5)(2.5) = 2.63 mm

$\Sigma P_f$ = (0.70)(181 020) = 126 710 kN

V = (6)(89) + 77 = 611 kN

$V_f$ = (0.70)(1.5)(611) = 642 kN

$\ell_c$ = storey height = 3000 mm

$$Q = \frac{\Sigma P_f \Delta_o}{V_f \ell_c} = \frac{(126\,710)(2.63)}{(642)(3000)} = 0.17$$

$$\delta = \frac{1}{1-Q} = \frac{1}{1-0.17} = 1.2$$

To complete the analysis for slenderness effects, the moment magnifier for the columns assumed braced should also be calculated. In this case, it can be shown that for braced conditions slenderness can be neglected.

Lines B, C, D, and E upper level effective length factors:

$\psi_A = \psi_B = 0.30$

From Fig. 2.6.15, k = 1.10:

$\ell_u$ = 1200 mm (see Fig. 2.6.17d)

$k\ell_u / r = (1.10)(1200)/[(0.3)(750)] = 5.9 < 22$

Lines A and F upper level effective length factors:

$\psi_A = \psi_B = 0.23$

**Fig. 2.6.18 Summary of column A-16 design forces – North-South bending**

| Load effect | | Floor | | | | | | |
|---|---|---|---|---|---|---|---|---|
| | | 1 | 2 | 3 | 4 | 5 | 6 | 7 |
| Unfactored forces | Axial dead load, P, (kN) | 630 | 540 | 450 | 360 | 270 | 180 | 90 |
| | Axial live load, P, (kN) | 190 | 165 | 140 | 115 | 90 | 65 | 40 |
| | Wind moment, W, (kN • m) | 178 | 74 | 59 | 48 | 34 | 22 | 15 |
| | Volume change moment, $M_t$, (kN • m) | 445 | 84 | 4 | – | – | – | – |
| 1.25 D + 1.5 L | Calculated $M_f$, (kN • m) | 40 | 27 | 27 | 27 | 27 | 27 | 21 |
| | Minimum $M_f$, (kN • m) | 32 | 28 | 23 | 19 | 14 | 10 | 5 |
| | Moment magnifier, δ | 1.50 | 1.0 | 1.0 | 1.0 | 1.0 | 1.0 | 1.0 |
| | Design $M_f$, (kN • m) | 60 | 28 | 27 | 27 | 27 | 27 | 21 |
| | Design $P_f$, (kN) | 1073 | 923 | 773 | 623 | 473 | 323 | 173 |
| 1.25 D + 0.70 (1.5 L + 1.5 Q) (Unbraced) | Moment magnifier, δ | 1.16 | 1.0 | 1.0 | 1.0 | 1.0 | 1.0 | 1.0 |
| | Design $M_f$, (kN • m) | 257 | 104 | 86 | 75 | 60 | 47 | 35 |
| | Design $P_f$, (kN) | 987 | 848 | 710 | 571 | 432 | 293 | 155 |
| 0.85 D + 1.5 W (Unbraced) | Moment magnifier, δ | 1.11 | 1.0 | 1.0 | 1.0 | 1.0 | 1.0 | 1.0 |
| | Design $M_f$, (kN • m) | 313 | 127 | 101 | 84 | 63 | 46 | 35 |
| | Design $P_f$, (kN) | 536 | 459 | 383 | 306 | 230 | 153 | 77 |
| 1.25 (D + T) (Braced) | Design $M_f$, (kN • m) | 578 | 129 | 23 | 18 | 18 | 18 | 14 |
| | Design $P_f$, (kN) | 788 | 675 | 563 | 450 | 338 | 225 | 113 |
| 1.25 D + 0.70 (1.25T + 1.5L) (Braced) | Design $M_f$, (kN • m) | 424 | 99 | 28 | 24 | 24 | 24 | 19 |
| | Design $P_f$, (kN) | 987 | 848 | 710 | 571 | 432 | 293 | 155 |

**Fig. 2.6.19 Summary of column A-16 design forces – East-West bending**

| Load effect | | Floor | | | | | | |
|---|---|---|---|---|---|---|---|---|
| | | 1 | 2 | 3 | 4 | 5 | 6 | 7 |
| Unfactored forces | Axial dead load, P, (kN) | 4960 | 4220 | 3480 | 2780 | 2000 | 1260 | 520 |
| | Axial live load, P, (kN) | 1474 | 1254 | 1034 | 814 | 594 | 374 | 154 |
| | Volume change moment, $M_t$, (kN • m) | 412 | 400 | 100 | – | – | – | – |
| 1.25 D + 1.5 L | Calculated $M_f$, (kN • m) | 154 | 285 | 189 | 220 | 250 | 189 | 358 |
| | Minimum $M_f$, (kN • m) | 315 | 268 | 221 | 176 | 127 | 80 | 33 |
| | Design $M_f$, (kN • m) | 315 | 285 | 221 | 220 | 250 | 189 | 358 |
| | Design $P_f$, (kN) | 8411 | 7156 | 5901 | 4696 | 3391 | 2136 | 881 |
| 1.25 (D + T) (Braced) | Design $M_f$, (kN • m) | 627 | 724 | 269 | 163 | 185 | 139 | 264 |
| | Design $P_f$, (kN) | 6200 | 5275 | 4350 | 3475 | 2500 | 1575 | 650 |
| 1.25 D + 0.70 (1.25T + 1.5L) (Braced) | Design $M_f$, (kN • m) | 502 | 617 | 263 | 202 | 230 | 174 | 330 |
| | Design $P_f$, (kN) | 7748 | 6372 | 5436 | 4330 | 3124 | 1968 | 812 |

$k = 1.08$

$k\ell_u / r = (1.08)(1200)/[(0.3)(500)] = 8.6 < 22$

Therefore slenderness may be neglected at upper levels.

*Column slenderness effects – East-West bending:*

Braced frame, $k \leq 1.0$ (see Fig. 2.6.17d)

Max. $k\ell_u / r = (1.0)(2100)/[(0.3)(750)] = 9.3$

Slenderness may be neglected in East and West direction.

*Gravity loads and moments on columns:*

A detailed analysis of the gravity loads and moments on the columns is not shown here, but is summarized in Figs. 2.6.18 and 2.6.19 for typical columns. The moments in the North-South direction are calculated using moment distribution or other suitable indeterminate structural analysis, and are caused by the dead load of the spandrel beam. The gravity moments in the East-West direction are caused by the eccentric reaction of the interior beams on column corbels, and may be calculated using the coefficients from Fig. 2.6.14.

Thus for the columns on lines A and F, minimum eccentricity to be considered:

$e_{min}$ (North-South) $= 15 + (0.03)(500) = 30$ mm

$e_{min}$ (East-West) $= 15 + (0.03)(750) = 37.5$ mm

*Moment connections at lines 1 and 16:*

Negative beam moments can be determined by a computer analysis or by approximate methods shown here.

Volume change moments ($M_t$):

From Fig. 2.6.10, $k_m$ for the column above the second floor is 0.5. Thus the net moment to be distributed to the beams is (2.6 + 0.5) / 2.6 times the calculated column moment below the second floor. This is distributed to the beam on each side of the column in proportion to the stiffness. For example at column E:

$$M_E = 422 \text{ kN} \cdot \text{m (see Page 2-32)}$$

$$\Sigma M_{Beams} = \left(\frac{2.6+0.5}{2.6}\right)(422) = 503 \text{ kN} \cdot \text{m}$$

$$M_{EF} = \left(\frac{8.10}{5.60+8.10}\right)(503) = 297 \text{ kN} \cdot \text{m}$$

$$M_{ED} = 503 - 297 = 206 \text{ kN} \cdot \text{m}$$

Gravity load moments:

In lieu of a more precise analysis, the approximate moment coefficients given in the Code can be used to calculate the gravity load moment. For example, at column E, with the spandrel dead load equal to 21 kN/m:

Avg. clear span $= \ell_n \approx 6200$ mm

$$M_{EF} = \frac{w\ell_n^2}{10} = \frac{(21)(6200 \times 10^{-3})^2}{10} = 81 \text{kN} \cdot \text{m}$$

$$M_{ED} = \frac{w\ell_n^2}{11} = \frac{(21)(6200 \times 10^{-3})^2}{11} = 73 \text{ kN} \cdot \text{m}$$

Other moments:

In this example, the wind load moments in the beams can be taken from the computer output shown in Fig. 2.6.20. In some cases, other moments, such as those caused by differential foundation settlement should also be considered. In this example, an analysis of the effects of differential settlement, which included the moderating effect of creep, resulted in an additional negative moment of 407 kN · m. The wind moment in beam EF (see Fig. 2.6.20) is 309 kN · m. Thus the total factored negative moment at beam EF is the greater of:

$$M_f = 1.25\, M_d + 1.5\, M_w$$
$$= (1.25)(81) + (1.5)(309) = 565 \text{ kN} \cdot \text{m}$$

or:

$$M_f = 1.25\,(M_d + M_t)$$
$$= (1.25)(81 + 297 + 407)$$
$$= 981 \text{ kN} \cdot \text{m (controls)}$$

The moment connection used in this example is shown conceptually in Fig. 2.6.21.

## 2.6.9 COMPUTER MODELS FOR FRAME ANALYSES

When precast concrete framing elements are modeled for computer analyses as "sticks", as is usually done with steel frames, the results are often very misleading. For example, the structure in Fig. 2.6.22, when assumed connected at the centreline as in (a), will indicate more flexibility than is actually true.

Thus, wind drift will be overestimated, and the moments caused by axial shortening may be underestimated. Fig. 2.6.22(b) shows a suggested model that will more nearly estimate the true condition.

## 2.6.10 SHEAR WALL-FRAME INTERACTION

Rigid frames and shear walls exhibit different responses to lateral loads, which may be important, especially in high-rise structures. This difference is illustrated in Fig. 2.6.23.

A frame bends predominantly in a shear mode as shown in Fig. 2.6.23(a), while a shear wall deflects predominantly in a

**Fig. 2.6.20 Wind moments in elements**

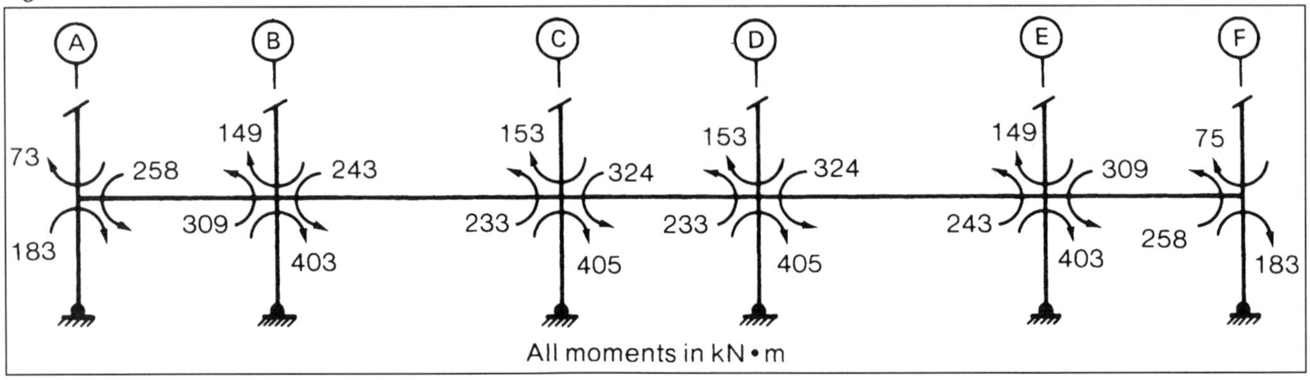

**Fig. 2.6.21 Moment connection details for Example 7**

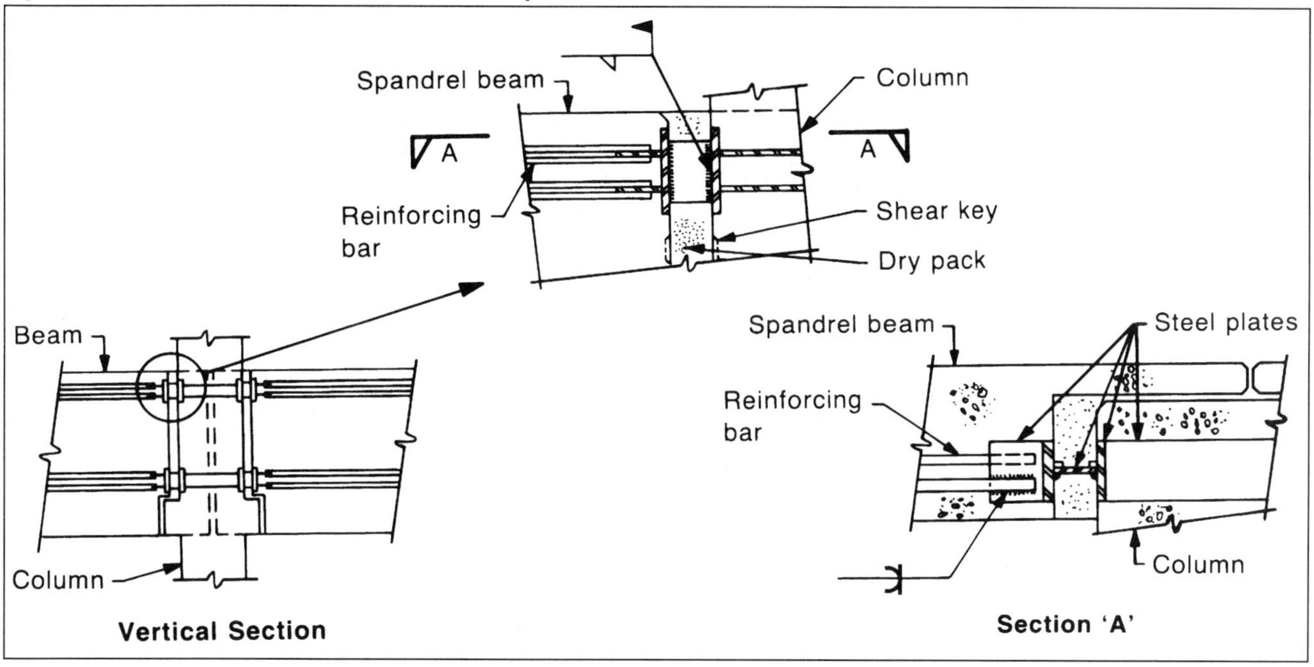


**Fig. 2.6.22 Computer models**

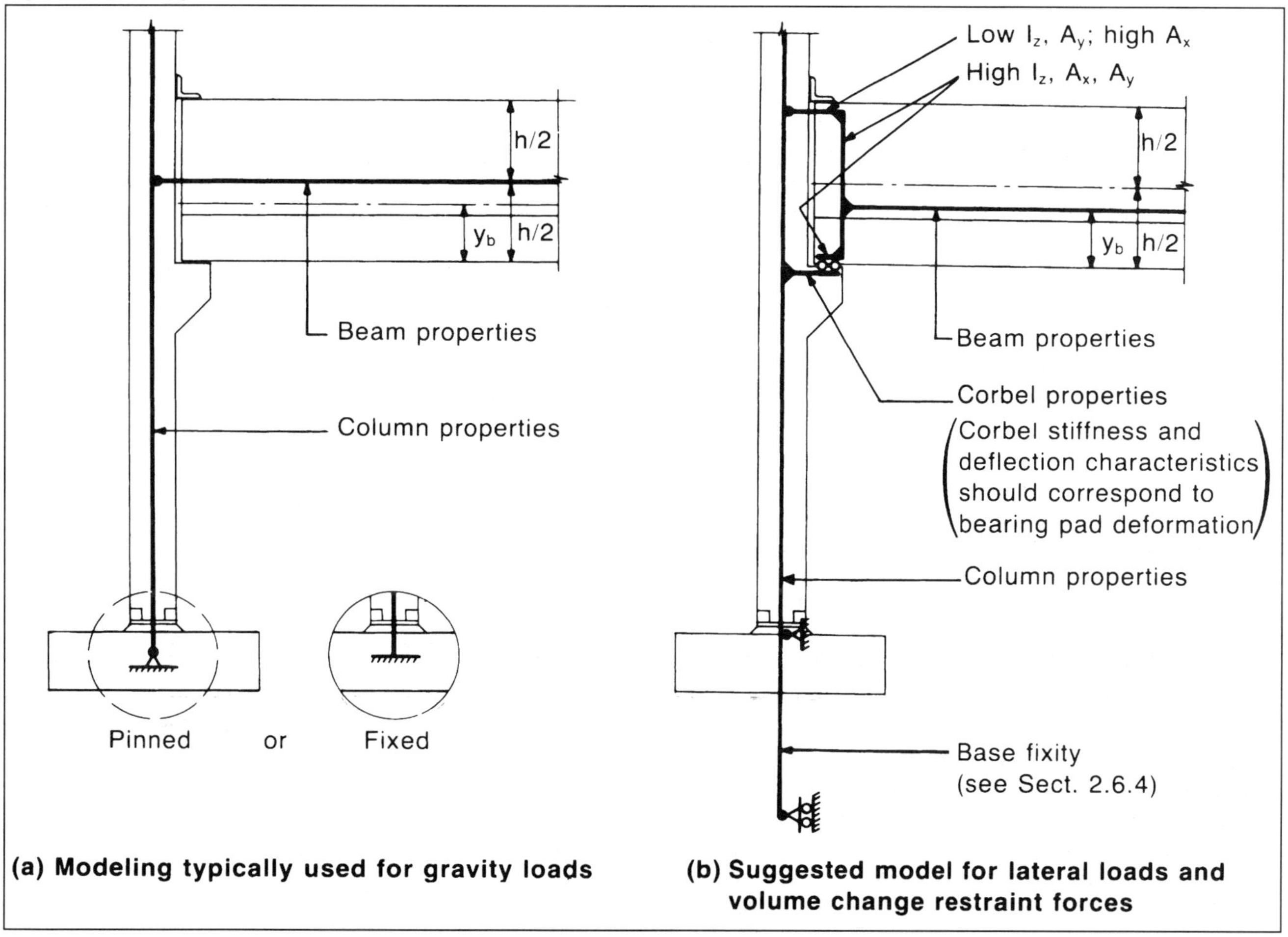

cantilever bending mode, Fig. 2.6.23(b). Elevator shafts, stairwells, and concrete walls normally exhibit this behavior.

**Fig. 2.6.23 Deformation modes**

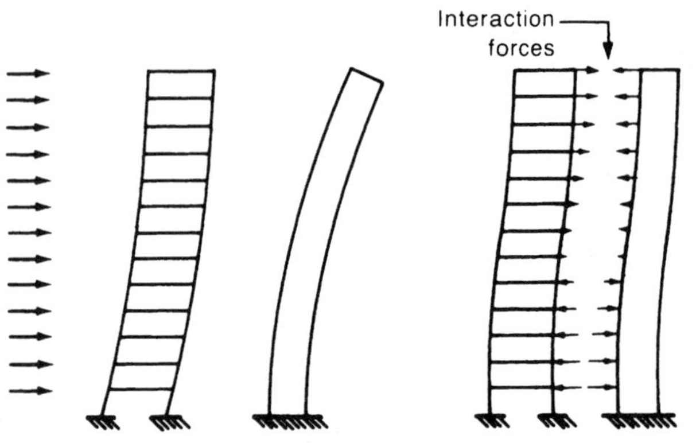


It is not always easy to differentiate between modes of deformation. For example, a shear wall weakened by a row, or rows of openings may tend to act like a frame, and an infilled frame will tend to deflect in a bending mode. Also, shear deformation of a shear wall can be more important than bending deformation if the height to length ratio is low, as discussed in Sect. 2.7.2.

If all vertical elements of a structure exhibit the same behavior under load, that is, if they are all frames or all shear walls, the load can be distributed to the units in proportion to their stiffnesses (Sect. 2.7.3). However, because of the difference in bending modes, the load distribution in structures with both frames and shear walls is considerably more complex and beyond the scope of this manual.

# 2.7 SHEAR WALLS

## 2.7.1 GENERAL

In most precast, prestressed concrete buildings, it is desirable to resist lateral loads with shear walls of precast or cast-in-place concrete, or unit masonry. Shear walls are usually the exterior wall system, interior walls, or walls of elevator, stairway, mechanical shafts or cores. The transfer of load from horizontal diaphragm to shear walls, or walls of elevator and stairway cores or mechanical shafts, can be achieved either through connections or by direct bearing.

Shear walls act as vertical cantilever beams which transfer lateral forces from the superstructure to the foundation. Most structures contain a number of walls which resist lateral load in two orthogonal directions. The portion of the total lateral force which each wall resists depends on the bending and shear resistance of the wall, the participation of the floor, and the characteristics of the foundation. It is common practice to assume that floors act as rigid elements for loads in the plane of the floor, and that the deformations of the footings and soil can be neglected. Thus, for most structures, lateral load distribution is based only on the properties of the walls.

If the floor is considered to be a rigid body, it will translate in a direction parallel to the applied load an amount related to the flexural and shear rigidity of the participating shear walls (Fig. 2.7.1 (a)). If the centre of rigidity is not coincident with the line of action of the applied loads, the floor will tend to rotate about the centre of rigidity, introducing additional forces (Fig. 2.7.1(b)). Therefore, the load on each shear wall will be determined by combining the effects produced by rigid body translation and rotation.

**Fig. 2.7.1 Translation and rotation of rigid floors**

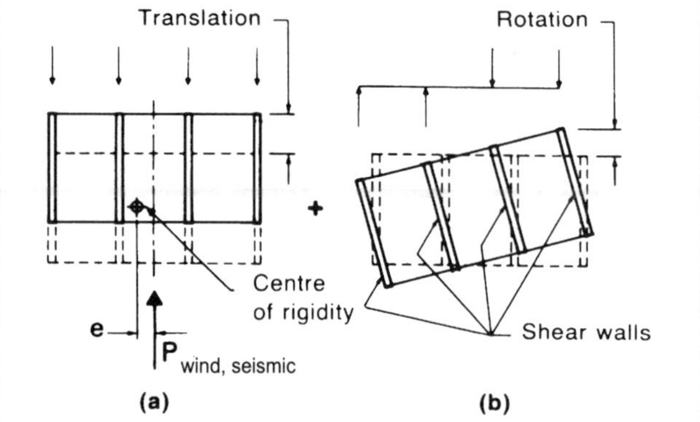


A shear wall need not consist of a single element. It can be composed of independent units such as double tee, hollow-core, or architectural precast wall panels. If such units have adequate shear ties between them, they can be designed to act as a single unit, greatly increasing their shear resistance. Connecting such units can, however, result in a buildup of volume change forces, so it is usually desirable to connect only as many units as necessary to resist the overturning moment. Connecting as few units as necessary near mid-length of the wall will minimize the volume change restraint forces.

Connection of rectangular wall units to form "T" or "L" shaped walls increases their flexural rigidity, but has little effect on shear rigidity. The effective flange width that can be assumed for such walls is illustrated in Fig. 2.7.2.

In some structures, it may be desirable to provide shear connections between non-load bearing and load bearing shear walls in order to increase the dead load resistance to moments caused by lateral loads.

## 2.7.2 RIGIDITY OF SOLID SHEAR WALLS

In order to determine the distribution of lateral loads, the relative rigidity of all shear walls must be established. Rigidity is defined as:

$$r = 1 / \Delta$$

where:

$\Delta$ = sum of flexure and shear deflections

**Fig. 2.7.2 Effective width of walls perpendicular to shear walls**

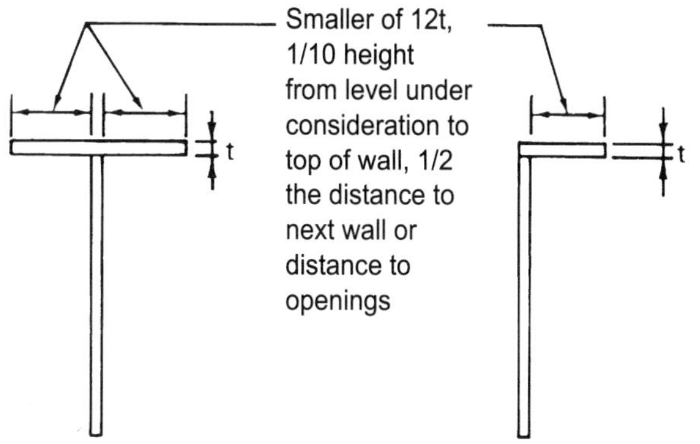


Note: The above flange width recommendations will normally be satisfactory for single storey buildings or isolated walls but may underestimate the effective flange width and resulting shear wall rigidity in multi-storey buildings. Shear lag must be considered in determining the effective flange width to be used in these cases.

For a structure with rectangular shear walls of the same material, flexural deflections can be neglected when the wall height to length ratio is less than about 0.3. The rigidity of the element is then directly proportional to its web cross-sectional area. When the wall height to length ratio is greater than about 3.0, shear deflections can be neglected, and the rigidity is proportional to the moment of inertia (plan dimensions). When the height to length ratio is between 0.3 and 3.0, an equivalent moment of inertia, $I_{eq}$, can be derived for simplifying the calculation of wall rigidity. $I_{eq}$ is an approximation of the moment of inertia that would result in a flexural deflection equal to the combined flexural and shear deflections of the wall. Fig. 2.7.3 compares the deflections and $I_{eq}$ for several load and restraint conditions.

Connecting or coupling of shear walls and large openings in walls also affect stiffness, as discussed in Sects. 2.7.5 and 2.7.6.

## 2.7.3 LATERAL LOAD DISTRIBUTION TO SHEAR WALLS

Lateral loads are distributed to each shear wall in proportion to its rigidity. It is usually considered sufficient to design for lateral loads in only two orthogonal directions.

When the shear walls are symmetrical with respect to the centre of load application, the force resisted by any shear wall is:

$$F_i = r_i / \Sigma r$$

where:

$F_i$ = the force resisted by an individual shearwall i

$r_i$ = the rigidity of wall i

$\Sigma r$ = sum of rigidities of all shear walls

## 2.7.4 UNSYMMETRICAL SHEAR WALLS

Structures which have shear walls placed unsymmetrically with respect to the centre of the lateral load, should take the torsional effect into account. Typical examples are shown in Fig. 2.7.4. For wind loading on most structures, a simplified method of determining the torsional resistance may be used in lieu of more exact design. The method is similar to the design of bolt groups in steel connections, and is illustrated in the following example.

**Example – Design of unsymmetrical shear walls**

*Given:*

The structure of Fig. 2.7.4. All walls are 2400 mm high and 200 mm thick.

**Fig. 2.7.3 Shear wall deflections**

| Case | Deflection due to | | Equivalent moment of inertia, $I_{eq}$ | |
|---|---|---|---|---|
| | Flexure | Shear | Single storey | Multi-storey |
| ①  | $\frac{Ph^3}{3EI}$ | $\frac{2.78Ph}{A_wE}$ | $\frac{I}{1+\frac{8.34I}{A_wh^2}}$ | $\frac{I}{1+\frac{13.4I}{A_wh^2}}$ |
| ②  | $\frac{Wh^3}{8EI}$ | $\frac{1.39Wh}{A_wE}$ | — | $\frac{I}{1+\frac{23.6I}{A_wh^2}}$ |
| ③ 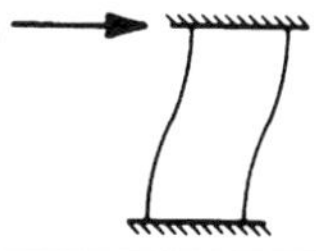 | $\frac{Ph^3}{12EI}$ | $\frac{2.78Ph}{A_wE}$ | $\frac{I}{1+\frac{33.4I}{A_wh^2}}$ | — |

**Fig. 2.7.4 Unsymmetrical shear walls**

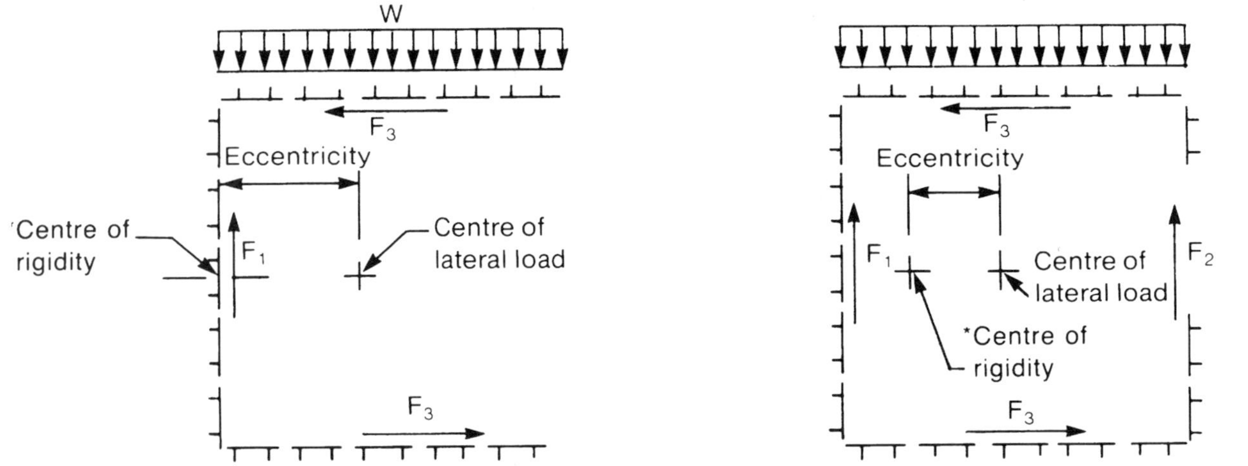


(a) Frequently occurs in buildings with expansion joints

(b) Frequently occurs in buildings with large door openings

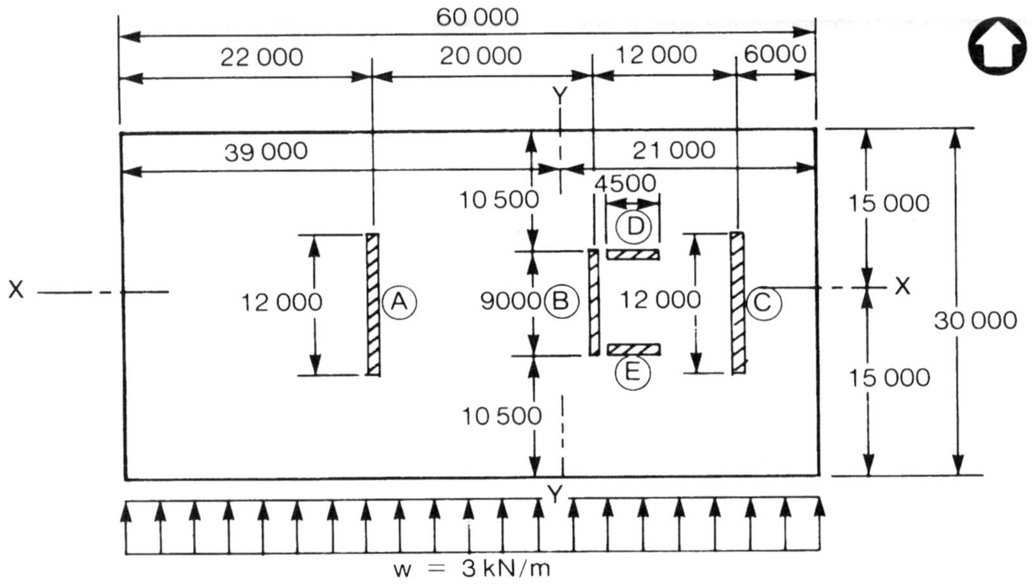


(c) Building for Example in Sect. 2.7.4

*Note: For tall structures, the shear centre should replace the centre of rigidity in determining the torsional effects.

*Problem* :

Determine the shear in each wall, assuming the floors and roof are rigid diaphragms. Walls D and E are not connected to wall B.

*Solution*:

Maximum height-to-length ratio of North-South walls = 2.4 / 9 < 0.3. Thus for distribution of the direct wind shear, neglect flexural stiffness. Since walls are the same thickness and material, distribute in proportion to length.

Total lateral load, W = (60)(3) = 180 kN

Determine centre of rigidity:

$$\bar{X} = \frac{(12\,000)(22\,000)+(9000)(42\,000)+(12\,000)(54\,000)}{12\,000+9000+12\,000}$$

= 39 000 mm from left

$\bar{Y}$ = centre of building, since walls D and E are placed symmetrically about the centre of the building in the North-South direction

Torsional moment:

$T = (180)(39\,000 - 60\,000/2)/10^3$

$= 1620 \text{ kN} \cdot \text{m}$

Determine the polar moment of inertia of the shear wall group about the centre of rigidity:

$I_p = I_{xx} + I_{yy}$

$I_{xx} = \sum \ell y^2$ of East-West walls

$= (2)(4500)(4500)^2 = 182\,250 \times 10^6 \text{ mm}^3$

$I_{yy} = \sum \ell x^2$ of North-South walls

$= (12\,000)(39\,000 - 22\,000)^2$

$+ (9000)(42\,000 - 39\,000)^2$

$+ (12\,000)(54\,000 - 39\,000)^2$

$= 6\,249\,000 \times 10^6 \text{ mm}^3$

$I_p = 182\,250 + 6\,249\,000$

$= 6\,431\,000 \times 10^6 \text{ mm}^3$

Shear in North-South walls $= \dfrac{W\ell}{\Sigma\ell} + \dfrac{Tx\ell}{I_p}$

Wall A =

$$\frac{(180)(12\,000)}{(33\,000)} + \frac{(1620)(39\,000 - 22\,000)(12\,000)(10^3)}{(6\,431\,000 \times 10^6)}$$

$= 65.5 + 51.4 = 116.9 \text{ kN}$

Wall B =

$$\frac{(180)(9\,000)}{(33\,000)} + \frac{(1620)(-3\,000)(9\,000)(10^3)}{(6\,431\,000 \times 10^6)}$$

$= 49 - 6.8 = 42.2 \text{ kN}$

Wall C =

$$\frac{(180)(12\,000)}{(33\,000)} + \frac{(1620)(-15\,000)(12\,000)(10^3)}{(6\,431\,000 \times 10^6)}$$

$= 65.5 - 45.3 = 20.2 \text{ kN}$

Shear in East-West walls $= \dfrac{Ty\ell}{I_p}$

$$= \frac{(1620)(4500)(4500)(10^3)}{(6\,431\,000 \times 10^6)}$$

$= 5.1 \text{ kN}$

## 2.7.5 COUPLED SHEAR WALL

Fig. 2.7.5 shows two examples of coupled shear walls. The effect of coupling two walls is to increase the stiffness by transfer of shear through the coupling beam. The wall curvatures are altered from that of a cantilever because of the frame action developed. Fig. 2.7.6 shows how the deflected shapes differ in response to lateral loads.

Several approaches may be used to analyze the response of coupled shear walls. A simple approach is to ignore the coupling effect by considering the walls as independent cantilevers. This method results in a conservative wall design. However, if the coupling beam is rigidly connected, significant shears

**Fig. 2.7.5 Coupled shear walls**

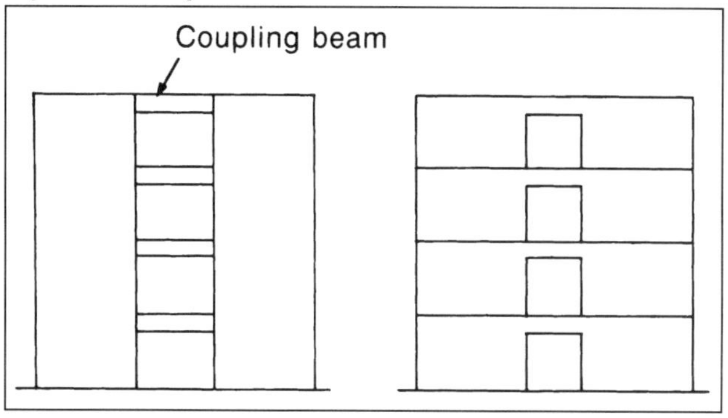


**Fig. 2.7.6 Response to lateral loads**

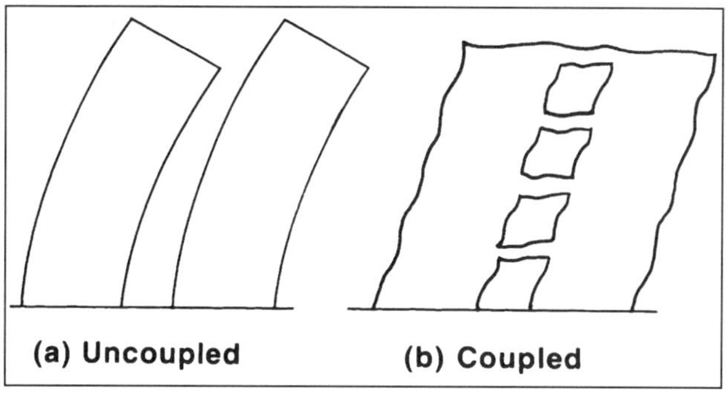


and moments will occur in the beam that may cause unsightly and possibly dangerous cracking. To avoid the problem, the beam to panel connection can be detailed for little or no rigidity, or the beam can be designed to resist the actual shears and moments.

Finite element analysis may be used to determine the stiffness and the distribution of shears and moments within a coupled shear wall. The accuracy (and cost) of such an analysis is a function of the element size used. This method is usually reserved for very complex structures.

A "plane frame" computer analysis will be sufficiently precise for the great majority of structures. In modelling the coupled shear wall as a frame, the element dimensions must be considered, as a centerline analysis may yield very inaccurate results. A suggested model is shown in Fig. 2.7.7(a).

Either a finite element or frame analysis may be used to determine the deflection of a coupled shear wall, and hence its equivalent moment of inertia. This may then be used to determine distribution of shears in a building which contains both solid and coupled shear walls. Most frame analysis programs do not calculate shear deformations so, if significant, shear deformations may have to be manually calculated.

## 2.7.6 SHEAR WALLS WITH LARGE OPENINGS

Window panels and other wall panels with large openings may also be analyzed with a plane frame computer program. Fig. 2.7.7(b) shows suggested models. Where length to depth ratios for vertical and horizontal segments are similar, a frame model based on segment centerlines will be reasonably accurate. Otherwise, an analysis similar to that described for coupled shear walls may be used.

**Fig. 2.7.7 Computer models**

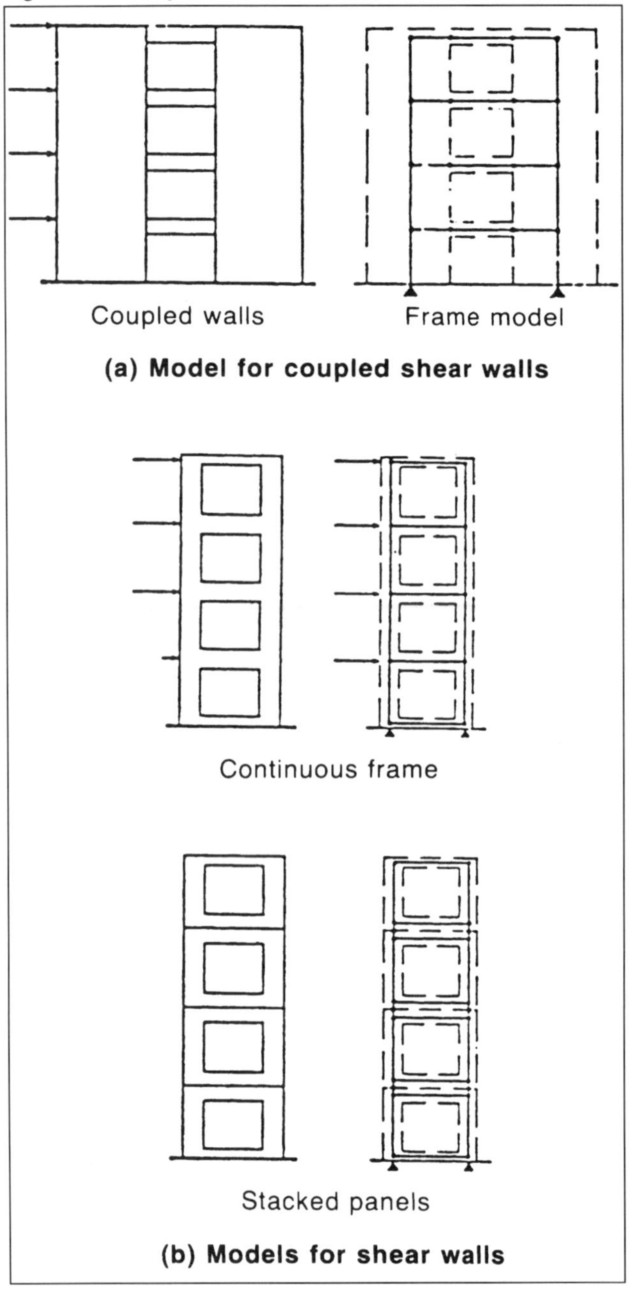


As with coupled shear walls, the deflections yielded by the computer analysis may be used to determine equivalent stiffness for determining lateral load distribution. Again, shear deflections, if significant, may have to be hand calculated and added to the flexural stiffnesses from the frame analysis.

In very tall structures, vertical shear and axial deformations influence the rigidity of panels with large openings, so a more rigorous analysis may be required.

## 2.7.7 EXAMPLE – ONE STOREY BUILDING

*Given*:

A typical one-storey industrial type building illustrated by the structure shown in Fig. 2.7.8. 2400 mm wide double tees are used for both the roof and walls. The local building code specifies that a wind load of 1.2 kN/m$^2$ be used for buildings of this height.

**Fig. 2.7.8 Example of one-storey building**

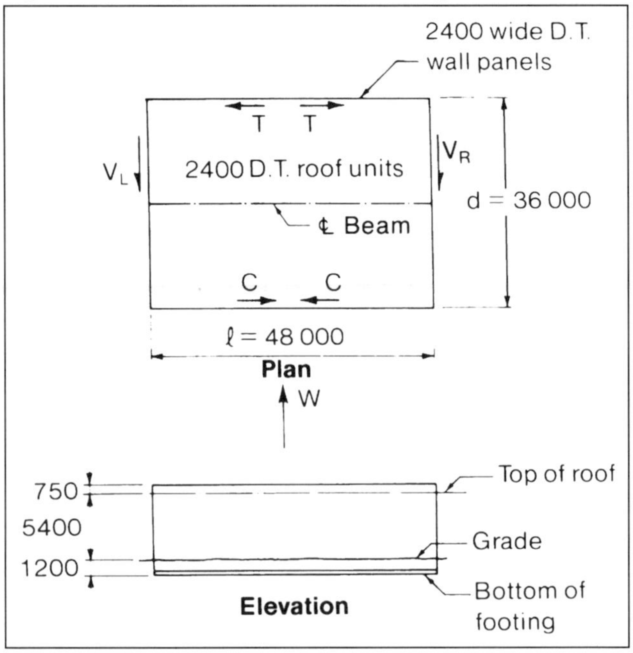


*Problem*:

Analyze the structure for the effects of wind loads.

*Solution:*

1. Calculate forces, reactions, shears and moments:

Total wind force to roof:

$$W = (1.2)(48)(5.4/2 + 0.75)$$

$$= 200 \text{ kN}$$

$$V_L = V_R = 100 \text{ kN}$$

$$\text{Diaphragm moment} = \frac{w\ell}{8} = \frac{(200)(48)}{8}$$

$$= 1200 \text{ kN} \cdot \text{m}$$

2. Check sliding resistance of the shear wall:

Determine dead load on the footing:

| | |
|---|---|
| DT wall: (7.05)(36)(1.9) | = 482 kN |
| Footing: (0.3)(0.45)(36)(24) | = 117 kN |
| Backfill: (0.5)(0.45)(36)(16) | = 130 kN |
| Total | = 729 kN |

Assume coefficient of friction against granular soil:

$\mu_s$ = 0.5

Sliding resistance = $\mu_s$ N = (0.5)(729)

= 365 kN

Factor of safety = 365 / 100 = 3.65 OK

(Note: A factor of 2.0 is specified by many building codes.)

3. Check overturning resistance:

Applied overturning moment = (100)(1.2 + 5.4)

= 660 kN · m

Resistance to overturning:

Assume axis of rotation at leeward edge of the building. (Note: Some engineers prefer to use the more conservative assumption of an axis at d/5, d/4 or d/3 from the leeward edge, depending on the foundation conditions.)

Resisting moment = (729)(36/2)

= 13 120 kN · m

Factor of safety = 13 120 / 660

= 19.9 > 2.0 OK

4. Analyze connections:

a) Shear ties in double tee roof joint:

(Maximum load at next to last joint)

Applied shear = [(24 − 2.4) / 24](100)

= 90 kN

Load factor $\alpha_L$ = 1.5

$V_f$ = (90)(1.5) = 135 kN

$v_f$ = 135 / 36 = 3.8 kN/m

Use ties as shown in Fig. 2.5.2. (Note: Most engineers and precasters prefer a maximum connection spacing of about 2500 mm to 3000 mm.)

b) Shear ties at the shear walls:

$V_f$ = (100)(1.5) = 150 kN

$v_f$ = 150 / 36 = 4.2 kN/m

A connection as shown in Fig. 2.7.9 is designed similar to the shear tie between tees.

In order to distribute the load to the wall panels, at least one connection per panel is required. From Fig. 2.7.9(a), it is apparent that these connections should occur at the tee stems. Thus a spacing of 1200 mm or 2400 mm would be used in this case.

Other types of connections using short welded headed studs are commonly used for this application.

In some cases, the designer may find it necessary to provide a connection that permits vertical movement of the roof element. Such a connection is illustrated in Fig. 2.7.9(b).

**Fig. 2.7.9 Connection of roof tee to wall**

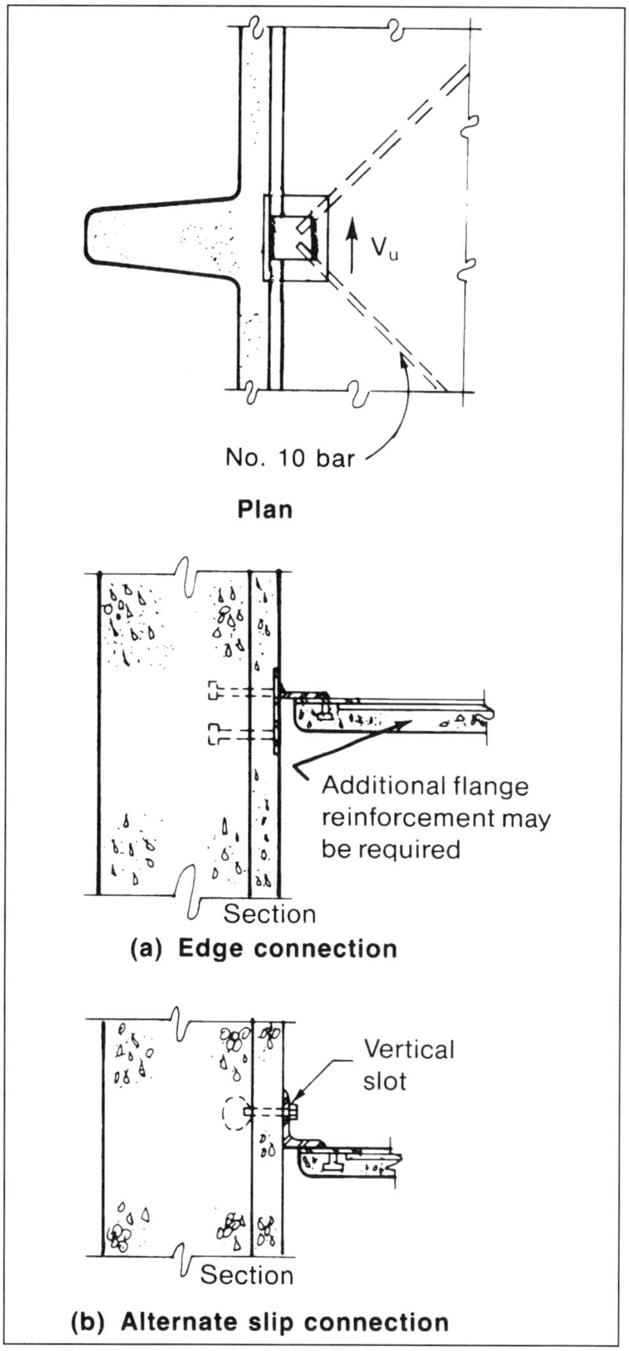


c) Chord force (see Fig. 2.7.10):

T = C = M / d = 1200 / 36 000 × $10^{-3}$

= 33.3 kN

$T_f$ = (1.5)(33.3) = 50.0 kN

This force can be transmitted between elements by ties at the roof tees, wall panels or a combination, as illustrated in Fig. 2.7.10.

For higher forces, or where more ductility is required, reinforcing bars can be placed in the flanges. Design procedures are shown in Sect. 2.8.

**Fig. 2.7.10 Chord forces**

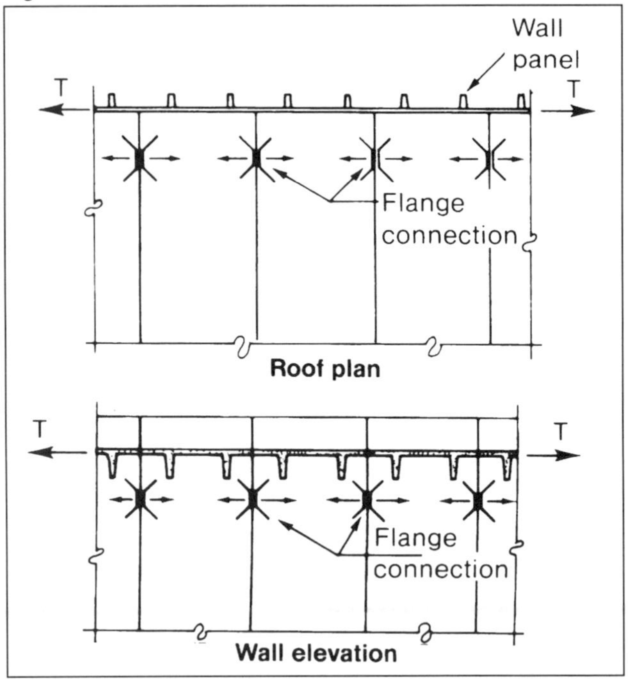


d) Wall panel connections:

This shear wall may be designed to act as a series of independent units, without ties between the panels. The shear force is assumed to be distributed equally among the wall panels (see Fig. 2.7.11).

**Fig. 2.7.11 Panels acting as idividual units in a shear wall**

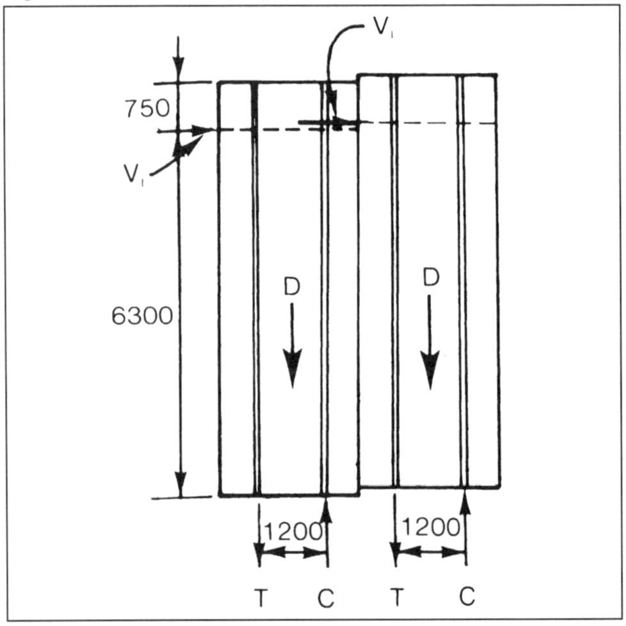


n = 36 000/2400 = 15 panels

V = $V_R$ /n = 100/15 = 6.67 kN

D = $(1.9\ kN/m^2)(2400)(7050) \times 10^{-6}$

= 32.1 kN

Design base connection for 1.5Q – 0.85D

$T_f$ = [(1.5)(6.67)(6300) – (0.85)(32.1)(600)] / 1200

= 38.7 kN tension

As an alternative, the shear wall may be designed with 2 or more panels connected together. The following illustrates the connection design if all panels are connected (see Fig. 2.7.12).

**Fig. 2.7.12 Panels connected together as a monolithic shear wall**

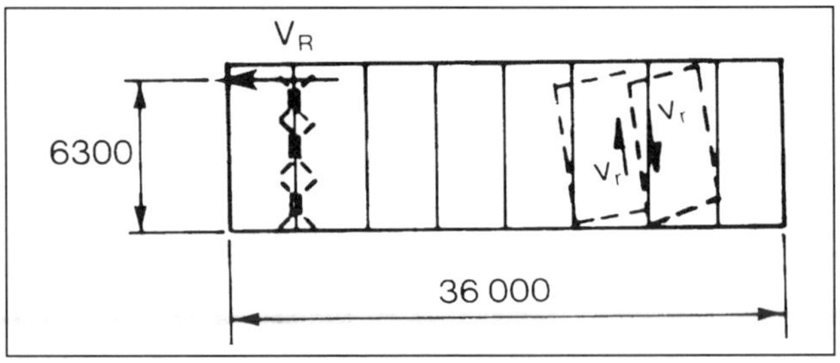


Shear ties between panels:

v = $V_R$ / d = 100 / 36 = 2.78 kN/m

The unit shear stress, $v_f$, is equal on all sides of the panel (see Subsection 2.8.13).

$v_f$ = (2.78)(1.5) = 4.2 kN/m

(Note: Most designers prefer a minimum of 2 or 3 ties per panel, at a maximum spacing of 2400 to 3000 mm).

Check for tension using factored loads:

By the Code, the required load factor equation to use under this condition is 0.85D – 1.5Q. The tensile stress would be:

$$f_t = \frac{P}{A} - \frac{M}{S}$$

S = $\ell^2 / 6 = 36^2 / 6 = 216\ mm^2$

A = $\ell$ = 36 m

M = $V_r h_s$ = (100)(6.3) = 630 kN · m

P = D.L. of wall = 482 kN

$$f_{ft} = \frac{0.85P}{A} - \frac{1.5M}{S}$$

$$= \left[\frac{(0.85)(482)}{36} - \frac{(1.5)(630)}{216}\right]$$

= 7.0 kN / m (compression)

Thus no tension connections are required for earthquake loads. However minimum tension connections are required for structural integrity.

**Fig. 2.7.13 Four-storey building Example**

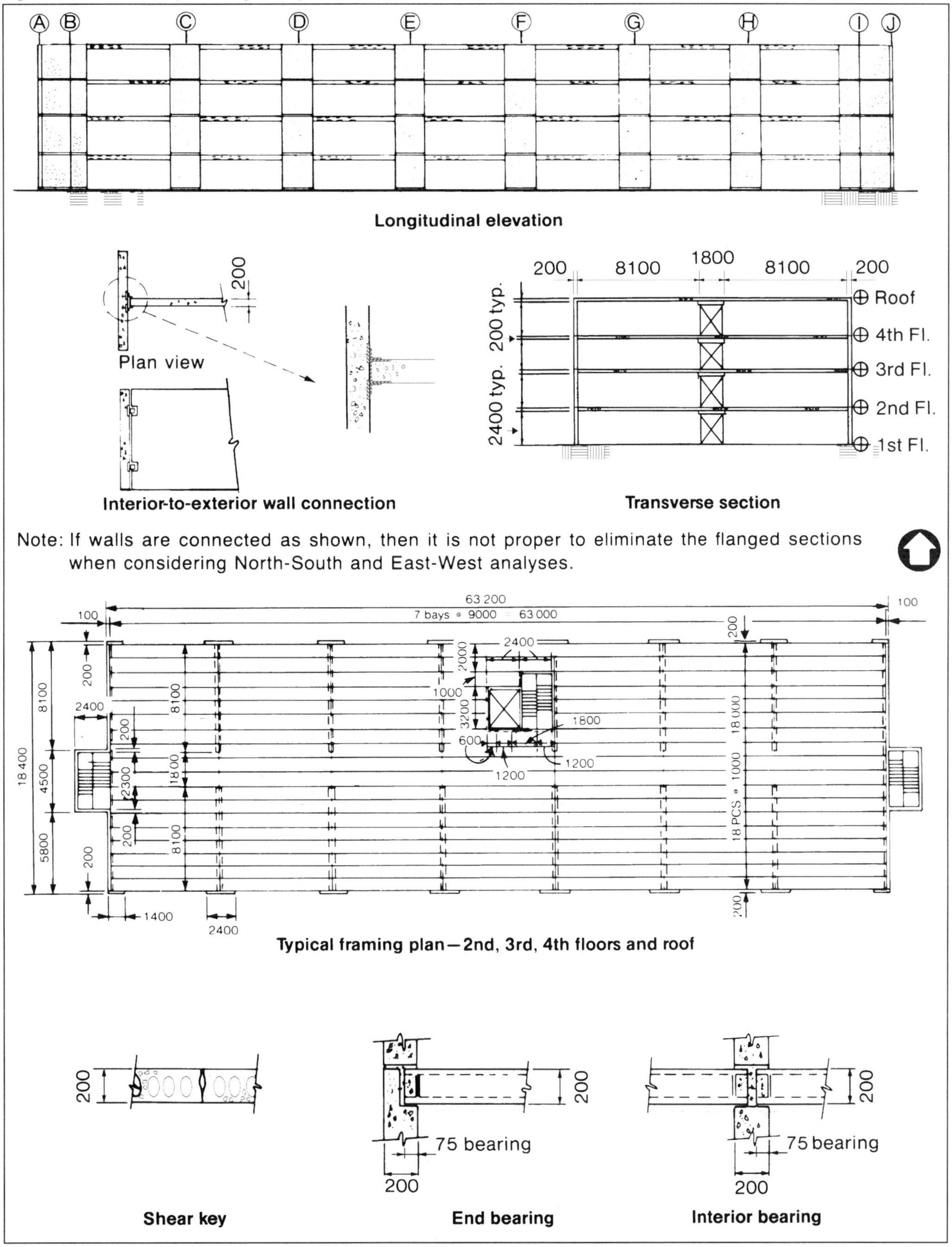

## 2.7.8 EXAMPLE - FOUR STOREY BUILDING

*Given*:

A typical four-storey residential building illustrated by the structure shown in Fig. 2.7.13. 200 mm deep hollow-core units are used for the floors and roof, and 200 mm thick precast concrete walls are used for all walls shown. Unfactored loads are given as follows:

| | L.L. | D.L. |
|---|---|---|
| Gravity loads: | | |
| Roof: | 1.44 | |
| Roofing, mechanical, etc. | | 0.48 |
| Hollow-core slabs | | 3.06 |
| | | 3.54 kN/m² |
| Typical floor: | | |
| Living areas | 1.92 | |
| Corridors & stairs | 4.80 | |
| Partitions | | 0.48 |
| Hollow-core slabs | | 3.06 |
| | | 3.54 kN/m² |
| Walls: | | 4.8kN/m² |
| Stairs: | 4.8 | 6.2 kN/m² |
| Wind loads: | | |
| 0 to 9 m above grade | | 1.2kN/m² |
| 9 m to 10.4 m above grade | | 1.4 kN/m² |

*Problem*:

Analyze and design the structure for wind in the North-South and East-West directions.

*Solution:*

*Wind in transverse North-South direction*:

For wind in the transverse direction, normal practice for this structure would be to conservatively neglect the resistance provided by the stairs, elevator and longitudinal walls. However, as the cross walls are so rigidly connected (see interior to exterior wall connection, Fig. 2.7.13), flanged walls are considered in this solution. Fig. 2.7.14 shows the wind resisting elements for North-South wind and a summary of their properties is shown in Fig. 2.7.15.

A sample calculation of these properties is given below for Element ④:

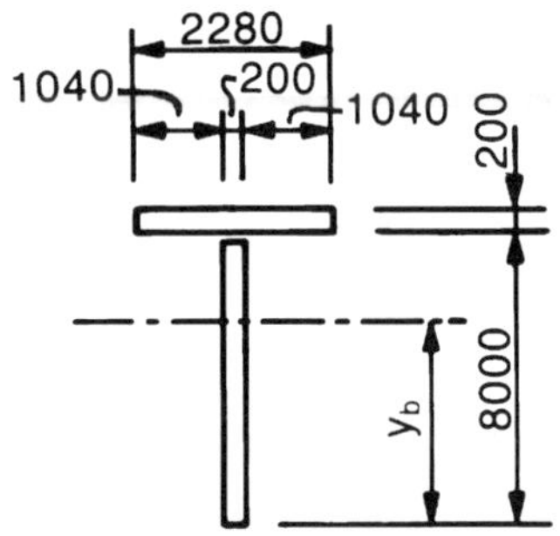


**Fig. 2.7.14 Wind resisting elements for North-South wind**

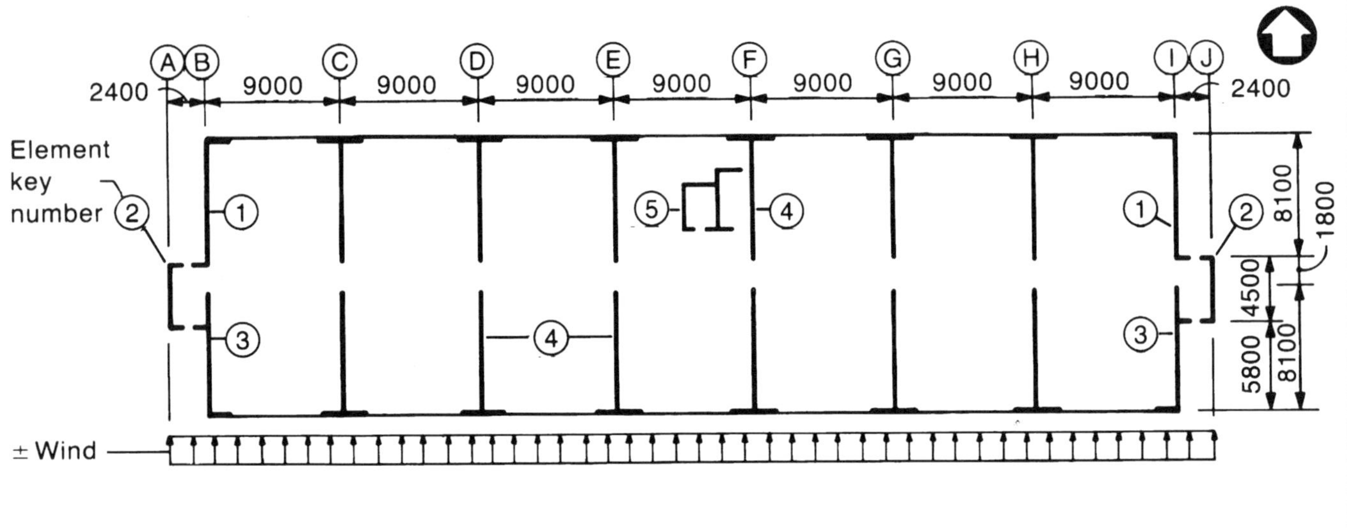


**Shear wall— resisting element diagram**

Note: For tall structures it would be inaccurate to consider wall elements 1, 2 & 3 and 4 & 5 individually as shown above. Consideration of shear lag would be necessary to determine to what extent these elements would interact as a composite wall element.

**Fig. 2.7.15 Properties of resisting elements for wind in transverse North-South direction**

| Element | $A_w$ ($mm^2 \times 10^3$) | I ($mm^4 \times 10^9$) | $I_{eq}$ ($mm^4 \times 10^9$) | No. of elements | n $I_{eq}$ ($mm^4 \times 10^9$) | $\frac{I_{eq}}{\Sigma n I_{eq}}$(100%) |
|---|---|---|---|---|---|---|
| ① 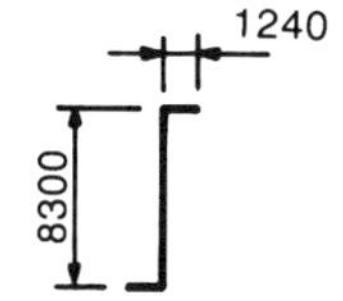  | 1 660 | 16 355 | 800 | 2 | 1 600 | 5.7 |
| ② 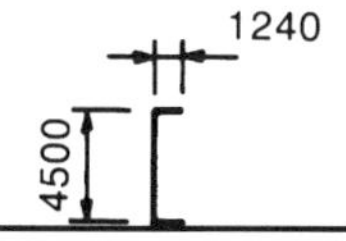  | 900 | 3 442 | 400 | 2 | 800 | 2.8 |
| ③ 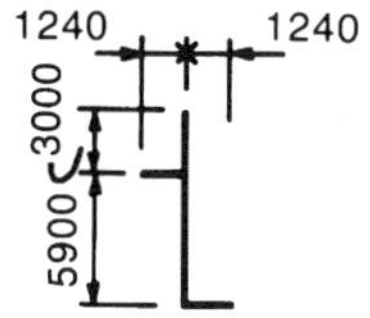  | 1 780 | 15 956 | 850 | 2 | 1 700 | 6.0 |
| ④ 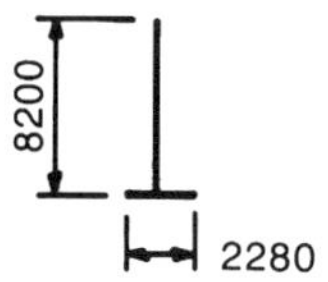  | 1 640 | 14 500 | 780 | 12 | 9 360 | 5.5 |
| ⑤ 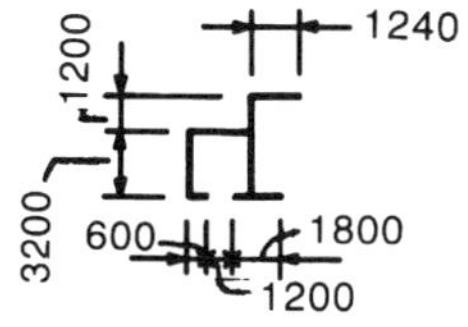  | 1 520 | 5 321 | 670 | 1 | 670 | 4.7 |
| | | | | | | $\Sigma$ n $I_{eq}$ = 14 130 |

The effective flange projection of the longitudinal wall (see Fig. 2.7.2) is the smaller of:

$$12t = (12)(200) = 2400$$

$$h/10 = 10\,400/10 = 1040 \text{ (Controls)}$$

The sectional properties of this T section are:

$$A_w = (8200)(200) = 1640 \times 10^3 \text{ mm}^2$$

$$A_f = (2080)(200) = 416 \times 10^3 \text{ mm}^2$$

$$y_b = \frac{(1640\,000)(4000) + (416\,000)(8100)}{1\,640\,000 + 416\,000} = 4910 \text{ mm}$$

$$y_t = 8200 - 4910 = 3290 \text{ mm}$$

$$I = \frac{(200)(8200)^3}{12} + (1640\,000)(4100 - 4910)^2 + (416\,000)(8100 - 4910)^2 = 14\,500 \times 10^9 \text{ mm}^4$$

The equivalent stiffness is calculated using the Case 1 multi-storey formula from Fig. 2.7.3:

$$I_{eq} = \frac{I}{1 + \frac{13.4I}{A_w h^2}} = \frac{14\,500 \times 10^9}{1 + \frac{(13.4)(14\,500 \times 10^9)}{(1640 \times 10^3)(2600)^2}} = 780 \times 10^9 \text{ mm}^4$$

$I_{eq}$ is essentially a relative stiffness:

$$K_r = 1/\Delta$$

$$\Delta = \frac{Ph^3}{3EI_{eq}}$$

$$\therefore K_r = \frac{3EI_{eq}}{Ph^3}$$

Since 3, E, P, and h are all constants when comparing stiffnesses, $K_r$ varies directly with $I_{eq}$.

Distribution of wind load to Element ④, based on its relative stiffness, is (Fig. 2.7.15):

**Fig. 2.7.16 Loading on all Element ④**

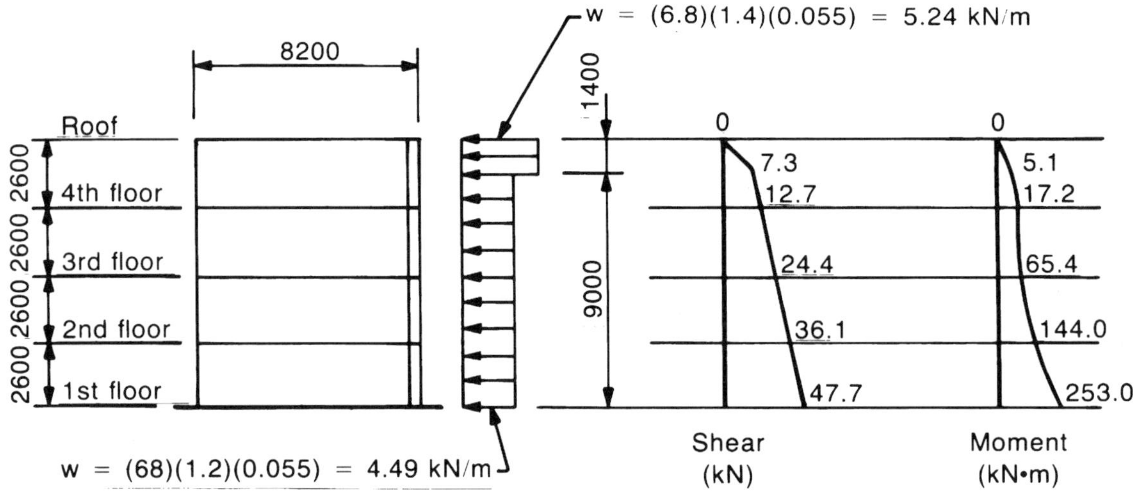


**(a) North-South lateral wind loading**

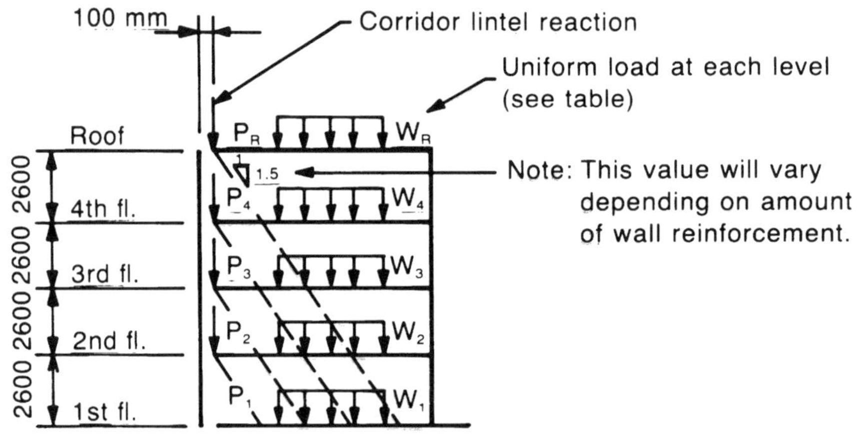


**(b) Gravity loads on bearing wall**

**Summary of gravity loads**

| Load mark | Tributary area | Unit loads ($kN/m^2$) | | Wall weight (kN/m) | Total unfactored load | | |
|---|---|---|---|---|---|---|---|
| | | L.L. | D.L. | | L.L. | D.L. | T.L. |
| $P_R$ | 8.1 $m^2$ | 1.44 | 3.54 | – | 11.7 kN | 28.7 kN | 40.4 kN |
| $P_4$ | 8.1 $m^2$ | 4.80 | 3.06 | – | 38.9 kN | 24.8 kN | 63.7 kN |
| $P_3$ | 8.1 $m^2$ | 4.80 | 3.06 | – | 38.9 kN | 24.8 kN | 63.7 kN |
| $P_2$ | 8.1 $m^2$ | 4.80 | 3.06 | – | 38.9 kN | 24.8 kN | 63.7 kN |
| $W_R$ | 9.0 $m^2$ | 1.44 | 3.54 | – | 13.0 kN/m | 31.9 kN/m | 44.9 kN/m |
| $W_4$ | 9.0 $m^2$ | 0.77* | 3.54 | 11.7 | 7.0 kN/m | 43.6 kN/m | 50.6 kN/m |
| $W_3$ | 9.0 $m^2$ | 0.77* | 3.54 | 11.7 | 7.0 kN/m | 43.6 kN/m | 50.6 kN/m |
| $W_2$ | 9.0 $m^2$ | 0.77* | 3.54 | 11.7 | 7.0 kN/m | 43.6 kN/m | 50.6 kN/m |
| $W_1$ | N/A | – | – | 11.7 | 0 | 11.7 kN/m | 11.7 kN/m |

*Includes live load reduction allowed by codes

**Fig. 2.7.17 Wind resisting elements for East-West wind**

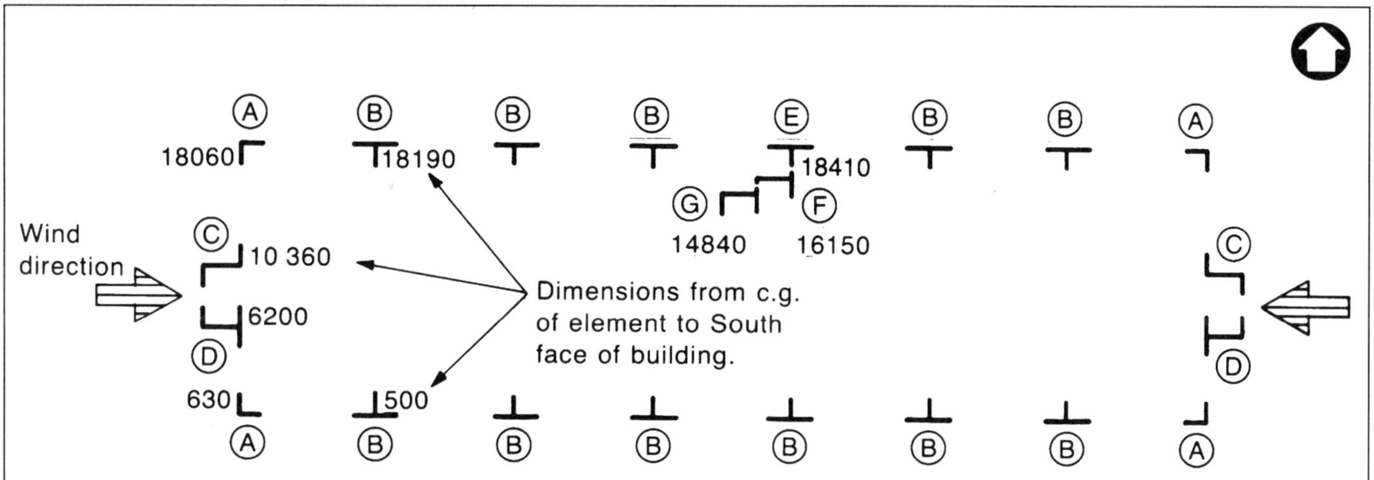


Note: For tall structures it would be inaccurate to consider wall elements C & D and E, F & G individually as shown above. Consideration of shear lag would be necessary to determine to what extent these elements would interact as a composite wall element.

**Fig. 2.7.18 Properties of resisting elements for wind in longitudinal East-West direction**

| Element | $A_w$ ($mm^2 \times 10^3$) | I ($mm^4 \times 10^9$) | yb (mm) | $I_{eq}$ ($mm^4 \times 10^9$) | No. of elem. | $n\,I_{eq}$ ($mm^4 \times 10^9$) | $\frac{I_{eq}}{\Sigma n I_{eq}}$ (100%) | $\Sigma \bar{y}$ ($mm \times 10^3$) | $I_{eq}(\Sigma \bar{y})$ ($mm^5 \times 10^9$) |
|---|---|---|---|---|---|---|---|---|---|
| A 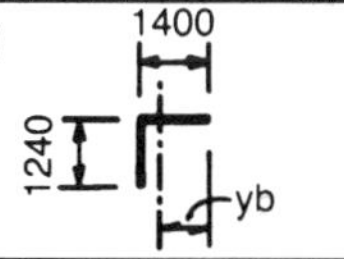 | 280 | 101 | 956 | 54.5 | 4 | 218 | 1.90 | 37.4 | 2 038 |
| B 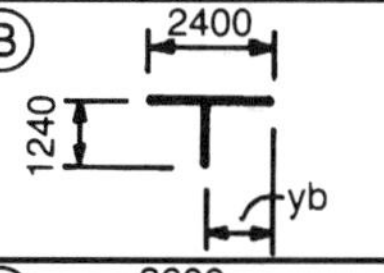 | 480 | 230 | 1 200 | 118 | 11 | 1 298 | 4.10 | 93.5 | 11 033 |
| C 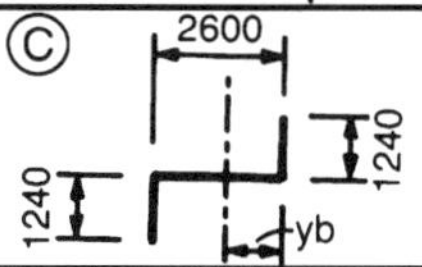 | 520 | 892 | 1 300 | 203 | 2 | 406 | 7.06 | 20.7 | 4 202 |
| D 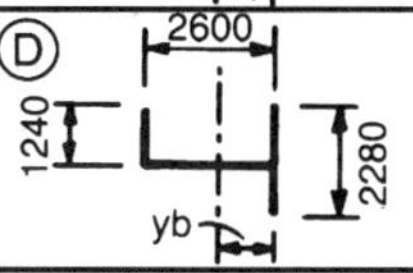 | 520 | 1 137 | 1 082 | 213 | 2 | 426 | 7.41 | 12.4 | 2 641 |
| E 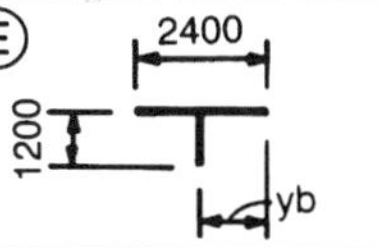 | 480 | 231 | 1 200 | 117 | 1 | 117 | 4.07 | 18.4 | 2 153 |
| F 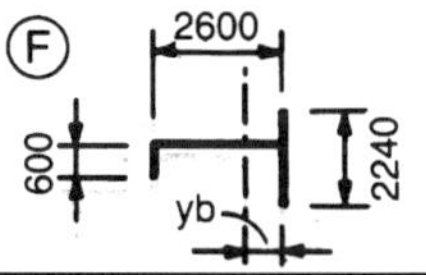 | 520 | 851 | 810 | 200 | 1 | 200 | 6.96 | 16.2 | 3 240 |
| G 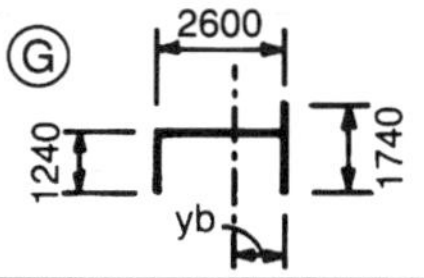 | 520 | 1 022 | 1 184 | 209 | 1 | 209 | 7.27 | 14.8 | 3 093 |
| | | | | | | $\Sigma n I_{eq}$=2 874 | | | $\Sigma$ = 28 400 |

Centre of rigidity = 28 400 / 2 874 = 9.88 m

Note: The East-West wind load is slightly eccentric by 9.8 – 18.4 / 2 = 0.68 m Torsion due to this eccentricity is neglected in calculating shears and moments in Fig. 2.7.19

$$\frac{I}{\Sigma n I_{eq}} = \frac{(780)(100)}{14\,130} = 5.5\%$$

Fig. 2.7.16 shows the load summary for wall Element ④ for gravity and transverse wind loading. Note that 5.5% of the total wind load has been assumed to be carried by this element. Concentrated loads from the corridor lintels can be assumed to be distributed as shown in this figure. In this example, these loads are conservatively neglected to simplify the calculations.

Check overturning of shear wall dead load resisting moment about toe of wall:

$$= (8.1)\frac{(8.1)}{2}[31.9 + (3)(43.6) + 11.7]$$

$$= 5720 \text{ kN·m}$$

Factor of safety:

$$= \frac{\text{dead load resisting moment}}{\text{wind moment}}$$

$$= \frac{5720}{253} = 22.6 > 2.0 \quad \text{OK}$$

Check for tension using factored loads:

Dead load on wall:

$$P = (8.1)\,[31.9 + (3)(43.6) + (11.7)]$$

$$= 1410 \text{ kN}$$

Maximum wind moment at foundation:

$$M = 253 \text{ kN} \cdot \text{m}$$

$$f = \frac{0.85P}{\ell} - \frac{1.5M}{(\ell^2/6)}$$

$$= \frac{(0.85)(1410)}{8.1} - \frac{(1.5)(253)}{(8.1^2/6)}$$

$$= +113 \text{ kN/m} \quad \text{(compression)}$$

No tension connections are required between panels and the foundation. Thus, the building is stable under wind loads in North-South direction. (Note: Other design considerations may dictate the use of minimum vertical ties.[R2.7.1])

**Fig. 2.7.19 Wind load in East-West direction**

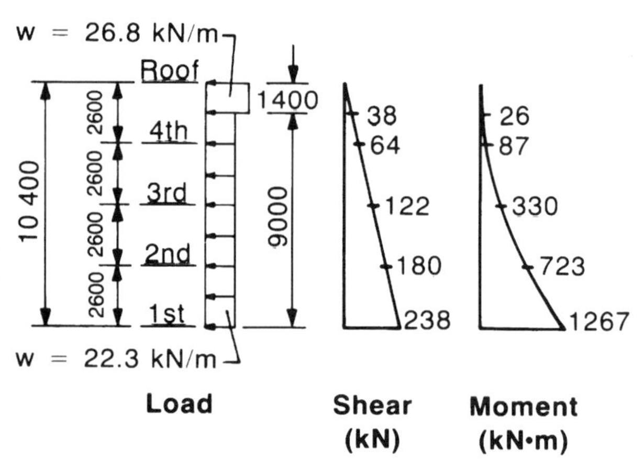


**Fig. 2.7.20 Distribution of wind shears and moments in longitudinal East-West direction**

| Element | Dist. (%) | 4th floor | | 3rd floor | | 2nd floor | | 1st floor | |
|---|---|---|---|---|---|---|---|---|---|
| | | Shear (kN) | Moment (kN · m) | Shear (kN) | Moment (kN · m) | Shear (kN) | Moment (kN · m) | Shear (kN) | Moment (kN · m) |
| | | **64** | **87** | **122** | **330** | **180** | **723** | **283** | **1267** |
| A | 1.90 | 1.2 | 1.7 | 2.3 | 6.3 | 3.4 | 13.7 | 5.4 | 24.1 |
| B | 4.10 | 2.6 | 3.6 | 5.0 | 13.5 | 7.4 | 29.6 | 11.6 | 51.9 |
| C | 7.06 | 4.5 | 6.1 | 8.6 | 23.3 | 12.7 | 51.0 | 20.0 | 89.5 |
| D | 7.41 | 4.7 | 6.5 | 9.0 | 24.5 | 13.3 | 53.5 | 21.0 | 94.0 |
| E | 4.07 | 2.6 | 3.5 | 5.0 | 13.4 | 7.3 | 29.4 | 11.5 | 51.6 |
| F | 6.96 | 4.5 | 6.0 | 8.5 | 23.0 | 12.5 | 50.3 | 19.7 | 88.2 |
| G | 7.27 | 4.7 | 6.3 | 8.9 | 24.0 | 13.1 | 52.6 | 20.6 | 92.1 |

Note: The relative stiffness and percent distribution for the elements in this table are assumed the same for all stories. The exact values may be slightly different for each storey because the values change due to the reduced flange width (see Fig. 2.7.2).

*Wind in the longitudinal East-West direction:*

For wind in the longitudinal direction, the shear walls will be connected to the load bearing walls. The assumed resisting elements are shown in Fig. 2.7.17 and a summary of the properties is shown in Fig. 2.7.18.

The shears and moments in the East-West direction are shown in Fig. 2.7.19, and the distributions are shown in Fig. 2.7.20.

To check overturning due to wind in the East-West direction, consider Element B at the first floor. From Fig. 2.7.16, the dead loads on the 1040 mm long portion of Element B:

$$= 31.9 + (3)(43.6) + 11.7 = 174 \text{ kN/m}$$

The dead load on the 2400 mm long portion of Element B is the weight of the wall:

$$= (10.4)(4.8) = 50 \text{ kN/m}$$

The dead load resisting moment for 11 elements is therefore:

$$M = (11)\,[(174)(1.04)(1.2) + (50)(2.4)(1.2)]$$
$$= 3970 \text{ kN} \cdot \text{m}$$

The factor of safety against overturning can conservatively be calculated neglecting the contribution of the other elements:

$$\text{Factor of safety} = \frac{3970}{1267} = 3.1 > 2.0 \quad \text{OK}$$

To check for tension, also consider Element B:

Total dead load on wall:

$$P = (174)(1.04) + (50)(2.4) = 301 \text{ kN}$$

Total wall area:

$$A = (2400 + 1040)(200) = 688\,000 \text{ mm}$$

The wind moment carried by Element Ⓑ (Fig. 2.7.20):

$$M = 51.9 \text{ kN} \cdot \text{m}$$

The factored stress is therefore:

$$f = \frac{0.85P}{A} - \frac{1.5M}{S}$$
$$= \frac{(0.85)(301 \times 10^3)}{688\,000} - \frac{(1.5)(51.9 \times 10^6)}{(230 \times 10^9 / 1200)}$$
$$= -0.035 \text{ MPa} \quad \text{(tension)}$$

Minor tension connections are required between panels and the foundation. Thus the building is stable under wind loads in the East-West direction.

The connections required to assure that the elements will act in a composite manner as assumed can be designed by considering Element Ⓐ. The unit stress at the interface is determined using the classic equation for horizontal shear:

$$v_h = \frac{VQ}{I}$$
$$Q = (1040)(200)(1400 - 956 - 100)$$
$$= 71.6 \times 10^6 \text{ mm}^3$$
$$v_h = \frac{(5.4)(71.6)}{101} = 3.8 \text{ kN/m}$$

Total shear = (3.8)(2.4) = 9.1 kN/level

Connections similar to those shown in Fig. 2.7.13 can be designed using the principles outlined in Chap. 4.

*Design of floor diaphragm:*

Analysis procedures for the floor diaphragm are described in Sect. 2.5. For this example, refer to Fig. 2.7.21.

The factored wind load for a typical floor is:

$$w_f = (1.5)(1.2)(2.6) = 4.68 \text{ kN/m}$$

For wind from the North or South:

$$V_{fR} = \frac{(4.68)(9)}{2} = 21 \text{ kN}$$

$$C_f = T_f = \frac{M_f}{\ell} = \frac{(4.68)(9)^2}{(8)(17.0)} = 2.8 \text{ kN}$$

The reaction V is transferred to the shear wall by static friction:

| | | |
|---|---|---|
| DL of floor: (9/2)(18)(3.54) | = | 287 kN |
| DL of wall: (2.4)(16.2)(4.8) / 2 | = | 93 kN |
| Total | = | 300 kN |

Static coefficient of friction from Fig. 4.4.1 (hardboard to concrete) = 0.5. Reduce by factor of 5 as recommended in Sect. 4.4:

$$\mu = 0.5 / 5 = 0.10$$

Resisting force = (0.10)(380)

= 38 > 21 OK

The chord tension, $T_f$, is resisted by the tensile resistance of the floor slab. The grout key between slabs must also resist approximately the same force:

**Fig. 2.7.21 Diaphragm analysis**

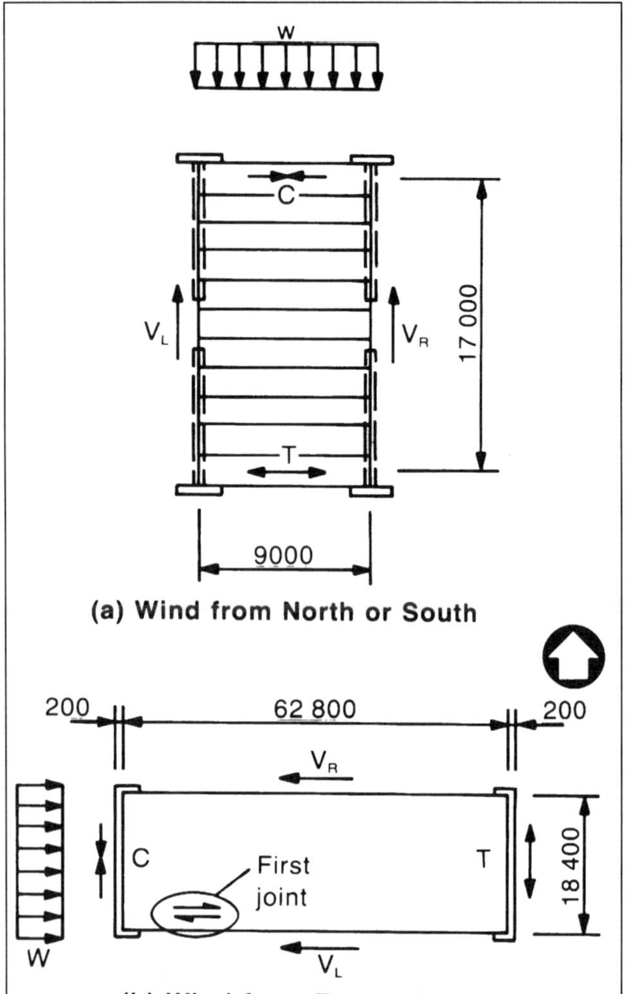


**(a) Wind from North or South**

**(b) Wind from East or West**

Assume area of exterior slab = 140 × 10³ mm²

Grout key = 75 mm deep

Concrete $f'_c$ = 35 MPa

The tensile resistance of the exterior floor slab units is:

$$T_r = 0.6\phi_c\sqrt{f'_c}A_c$$
$$=(0.6)(0.6)(1.0)\sqrt{35}(140)$$
$$= 298 \text{ kN} > 2.8 \quad \text{OK}$$

The shear resistance of the grout key is (see Sect. 2.5.2):

$$V_r = 0.3\,A = (0.3)(9000)(75) / 10^3$$
$$= 202 \text{ kN} > 2.8 \quad \text{OK}$$

For wind from the East or West (Fig. 2.7.21(b)):

$$V_{fR} = \frac{(4.68)(18.4)}{2} = 43 \text{ kN}$$

$$C_f = T_f = \frac{(4.68)(18.4)^2}{(8)(62.8)} = 3.2 \text{ kN}$$

Shear resistance of first joint:

$$V_r = (0.3)(62\,800)(75) / 10^3$$
$$= 1410 \text{ kN} > 43 \quad \text{OK}$$

In this example, only the resistance to wind loading has been analyzed. Any other required loading (including "abnormal" loads) must be reviewed for a complete analysis.

## 2.7.9 ARCHITECTURAL PANELS AS SHEAR WALLS

In many structures it is economical to take advantage of the inherent strength and rigidity of exterior panels, and design them to serve as the lateral load resisting system. The effectiveness of such a system is largely dependent on the panel-to-panel connections.

**Fig. 2.7.22 Foundation reaction distributions resulting from lateral loads**

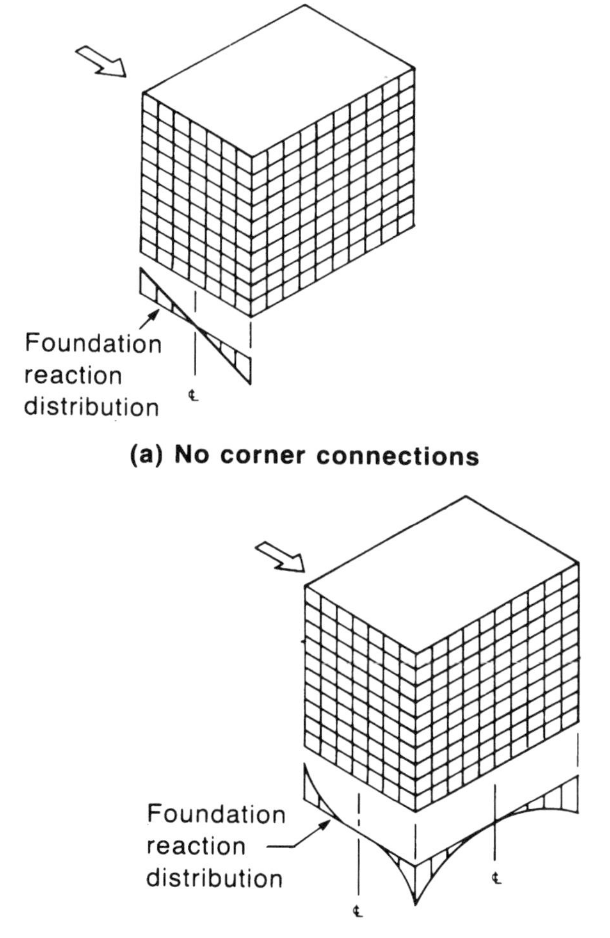


**(a) No corner connections**

**(b) Connections at corner**

**Fig. 2.7.23 Influence of shear lag on tube behaviour**

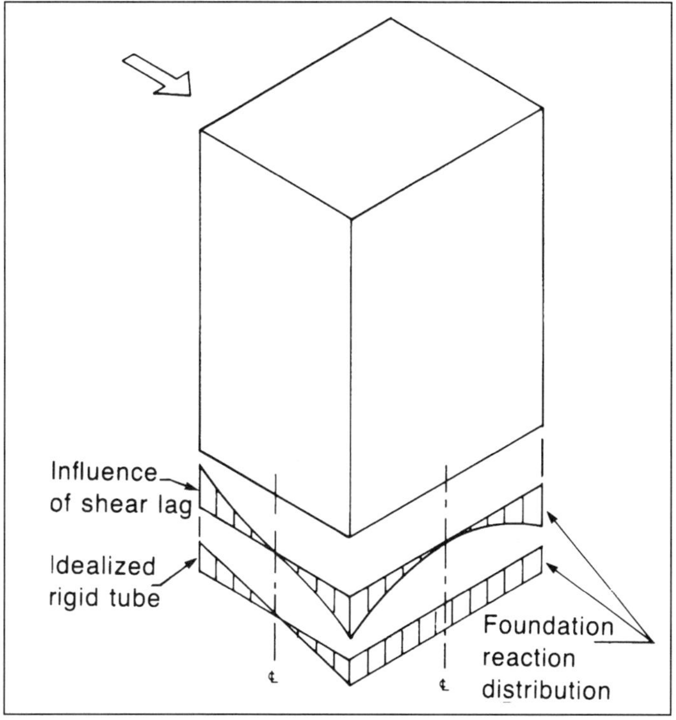


Fig. 2.7.22 illustrates the foundation reaction distributions of exterior architectural precast shear wall systems under the action of lateral load, with and without connections between the shear walls and the windward or leeward walls. The structure with corner connections is structurally more efficient for resisting lateral loads.

The lateral load resisting system shown in Fig. 2.7.22(b) is frequently labeled a "tube". However, because the components and the connections are not perfectly rigid, full tube behavior does not develop. Fig. 2.7.23 illustrates the difference. The peaking of the foundation reaction at the corner results from shear lag, which limits the effective width of the "flange". Accurate evaluation of shear lag is difficult, but analytical and research studies have indicated the following limitations on the effective flange are sufficiently accurate for most structures:

1. One-half the length of the shear wall.
2. One-third the length of the windward or leeward wall.
3. One-tenth the height of the building.
4. Twelve times the thickness of the "flange" wall.
5. Distance to nearest major opening.
6. One-half the distance to the nearest shear wall.

# 2.8 EARTHQUAKE DESIGN AND ANALYSIS

## 2.8.0 NOTATION

$A_{cv}$ = area of concrete section resisting shear transfer
$A_{vf}$ = area of shear friction reinforcement
$A_\ell$ = cross-sectional area in linear measure
$A_s$ = area of reinforcement
$b$ = width of panel
$b_v$ = width of cross section at contact surface being investigated for longitudinal shear
$C_a$ = snow accumulation factor
$C_f$ = factored compressive force
$C_{pR}$ = probable compressive chord force in roof
$D$ = dimension of building in direction parallel to applied lateral force
$D_n$ = plan dimension of the *building* in the direction of the computed eccentricity
$D_p$ = probable resistance of a dowel
$D_r$ = factored resistance of a dowel
$D_s$ = dimension of wall of braced frame which constitutes the main lateral load-resisting system in a direction parallel to the applied forces
$d$ = dimension of building; distance from extreme compression fibre to centroid of tension reinforcement
$e$ = distance between the location of the resultant of all forces at and above the level being considered and the centre of rigidity at the level being considered
$e_x$ = design eccentricity at level x
$F$ = foundation factor
$F_{fh}$ = factored horizontal force
$F_i$, $F_x$ = force at level x
$F_t$ = that portion of V considered concentrated at the top of the structure, level n
$F_{fv}$ = factored vertical force
$F_x$ = lateral force applied to level x
$f$ = fibre stress
$f'_c$ = specified compressive strength of concrete
$f_y$ = specified yield strength of reinforcement
$H$ = horizontal force needed to overcome friction
$H_{fP}$ = factored sliding force acting on a wall panel
$H_{pP}$ = probable sliding resistance of a wall panel
$H_{rP}$ = factored sliding resistance of a wall panel
$h$ = height of element
$h_i, h_n, h_x$ = height above base level to level "i", "n", and "x", respectively
$h_s$ = interstorey height ($h_i - h_{i-1}$)

$I$ = seismic importance factor of the structure
$J$ = numerical reduction coefficient for base overturning moment
$J_x$ = numerical reduction coefficient for moment at level "x"
Level i = any level of the *building*, i = 1 for first level above the base
Level n = that level which is uppermost in the main portion of the structure
Level x = that level which is under design consideration
$\ell$ = length of building or element
$M_{pP}$ = probable flexural resistance of wall panel fixed at the foundation
$M_{rP}$ = factored flexural resistance of a wall panel fixed at the foundation
$M_{tx}$ = torsional moment at level x
$M_x$ = moment at level x
$N$ = total number of *storeys* above exterior *grade* to level "n" (N is usually numerically equal to n)
$n$ = uppermost level in the structure
$R$ = force modification factor that reflects the capability of a structure to dissipate energy through inelastic behaviour
$R_{B,T}$ = reaction at bottom, top
$S$ = seismic response factor, for unit value of zonal velocity ratio or specified snow loading
$S_\ell$ = section modulus in linear measure
$S_p$ = horizontal force factor for part or portion of structure
$s$ = spacing of weld clips
$T$ = fundamental period of vibration of the building in the direction under consideration; total tensile capacity or force
$T_f$ = factored tensile force
$T_{pR}$ = probable tensile chord force in roof
$T_r$ = factored tensile resistance
$U$ = factor representing level of protection based on experience
$V$ = minimum lateral seismic force at the base of the structure, to be used with a load factor $\alpha_Q = 1.0$
$V_e$ = equivalent lateral force at the base of the structure representing elastic response
$V_{fP}$ = factored shear force on a wall panel
$V_{fW}$ = factored shear force on a wall
$V_p$ = specified lateral force on the part of the structure and in the direction under consideration
$V'_p$ = lateral force due to non-structural wythe and insulation
$V_{pP}$ = probable shear resistance of a wall panel
$V'_{pP}$ = probable shear resistance of a wall panel – seismic force due to panel weight
$V_{pR}$ = probable shear force transferred from wall to roof
$V_{rC}$ = factored shear resistance of a connection
$V_{rP}$ = factored shear resistance of a wall panel
$V_{rW}$ = factored shear resistance of a wall
$V_W$ = specified shear force on a wall
$v$ = zonal velocity ratio
$v_c$ = factored shear resistance per unit length due to concrete alone
$v_f$ = factored shear force per unit length
$v_r$ = factored shear resistance per unit length
$W$ = total dead load of building plus snow and other loads (see NBCC)
$W_p$ = total weight of a part or portion of a structure
$W_x$ = portion of weight of building assigned to level "x"
$W_i$, $W_x$ = that portion of "W" which is located at or is assigned to level "i" or "x" respectively
$Z_a$ = acceleration related seismic zone
$Z_v$ = velocity related seismic zone
$\lambda$ = factor to account for low density concrete
$\mu$ = shear-friction coefficient
$\mu_s$ = static coefficient of friction
$\Sigma$ = sum of values
$\phi$ = resistance factor (with subscripts)

## 2.8.1 GENERAL

There is a critical difference between earthquake design and gravity and wind load design. The seismic forces that may be expected during the life of a structure are dependent not only on the earthquake intensity but also on the strength, stiffness and ductility of the structure. They may be much larger than the factored design forces given in building codes and it is expected that the structure will suffer damage during a severe earthquake. To protect life and property, it is important that this damage not cause collapse. This is accomplished by designing elements that can deform in the nonlinear range without failing. Because the strength of these elements limits the forces that act on other parts of the structure, it is important that their real or probable strength, as opposed to their design or factored strength, should not be excessive. This will ensure that those parts of the structure that are assumed by the designer to remain elastic, and which may not be detailed for ductility, are not loaded beyond their elastic limit.

Load tests of prestressed concrete elements have consistently shown that large deflections occur as the design strength is approached. Because of prestress, the transition from linear to nonlinear response is gradual and smooth. Cyclic load tests have shown that prestressed concrete beams can undergo several cycles of intense load reversals and still maintain their design strengths.

Prestressed concrete structures can be designed to withstand the effects of earthquakes in accordance with the requirements of building codes. For some types of buildings, box-type structures offer an economical solution. In other cases, ductile moment-resisting frames can be used.

This section deals with the design of precast concrete buildings located in Seismic Zones 1 to 6. Because the actual seismic forces during an earthquake may exceed the factored forces found using the Code, proper detailing for seismic forces should not be ignored even when the factored forces found for wind exceed the factored seismic forces. The discussion that follows deals mainly with the design of critical connections and components to provide adequate resistance to seismic forces, and the design procedures necessary to ensure that the complete structure will behave as intended during an earthquake.

Earthquakes generate horizontal and vertical ground movement. When an earthquake passes beneath a structure, the foundation will tend to move with the ground, while the superstructure will tend to remain in its original position. The lag between foundation and superstructure movement will cause distortions and develop forces in the structure. These distortions and forces can best be investigated using non-linear dynamic analysis taking into account the properties of the soil, the intensity and duration of the ground shaking, and the physical properties of the structure. A linear or elastic dynamic analysis can be used to predict response to moderate earthquakes and can also provide an insight into which elements are likely to yield and what the maximum deflection is likely to be in a more severe earthquake. Important structural properties include mass, strength, stiffness and ductility, and their distribution throughout the structure. In lieu of dynamic analysis, building codes present formulae for equivalent static forces and force distributions which can, if the underlying assumptions are understood and respected, be used to design acceptable structures in the majority of cases.

The current philosophy for the design of earthquake-resistant structures permits minor damage for moderate earthquakes, and accepts major damage for severe earthquakes, provided that collapse is prevented and occupants can leave the building without serious injury. Large non-linear or inelastic deformations may be expected in some connections and other elements as a result of the earthquake induced forces. In precast structures, these are accommodated by providing ductility in these connections and elements. While this ductility limits the earthquake forces in the building and prevents total collapse, the resultant distortions will usually lead to significant damage to connections and other structural elements that yield and may also result in damage to non-structural elements and mechanical and electrical systems.

Buildings may be designed as either flexible or rigid. Flexible structures will develop large deflections and small inertial forces; conversely, rigid structures will develop large inertial forces but small deflections. Either type may be designed to be safe against total collapse. However, experience demonstrates that a rigid structure, properly designed to account for the large inertia forces, will incur significantly less damage to architectural, mechanical, and electrical elements.

To limit the damage to non-structural or architectural elements in a flexible building, these elements may be uncoupled from the structural system so that they are not forced to undergo as much deformation as the supporting structure. This requires connections between the individual elements and the supporting structure that can withstand large distortions without fracture. As explained above, damage can also be reduced by building a more rigid building to minimize the deformations imposed on the architectural elements. This will require a stronger structure to reduce the amount of yielding and consequent inelastic deformation that occurs. In most buildings, non-structural architectural wall panels should be uncoupled from the lateral force resisting structure.

Since ground motion is random in direction, a structure which is shaped so as to be equally resistant in any direction is the optimum solution. Experience has shown that structures that are symmetrical in plan, with minimum torsional eccentricity, generally behave better in earthquakes than structures that are unsymmetrical and have their centres of mass and rigidity well separated. Furthermore, closed sections (i.e., boxes or tubes) have demonstrated markedly improved behaviour when compared with open sections, because: (1) closed sections provide a high degree of torsional resistance, and (2) the higher axial stresses and resultant deformations in the exterior columns provide significant energy absorption.

An architectural precast concrete facade can be used to provide the earthquake resistant structural elements. This can be in the form of either shear walls or spandrel beams and closely spaced columns (mullions) which simultaneously provide the exterior skin and the structural tube. It is also often possible to use these elements to create a structure which is symmetrical in plan.

The energy dissipation characteristics of precast walls will depend to a great extent upon the behaviour of the connections. In addition, precast walls with large openings can behave in a flexurally ductile manner, similar to a beam-column system. However, unless comprehensive test data is available, precast wall type structures should be treated as specified by codes for shear wall structures.

## 2.8.2 BUILDING CODE REQUIREMENTS

In the example problems that follow, the seismic requirements of the National Building Code of Canada are used, except as noted.

The response of a building to earthquake motion depends on its structural properties including mass, strength, stiffness, and damping and their distribution throughout the structure, as well as the characteristics of the ground motion and the properties of the foundation material. Although a non-linear dynamic analysis is required to predict the structural response to strong earthquakes, the Code provides an equivalent static force method for design which takes into account many of these factors. While this method is satisfactory for the design of the majority of buildings, calculations must be supplemented with engineering judgment. This should be based on an understanding of how buildings respond to earthquakes.

**Fig. 2.8.1 Maps of contours[*] of peak horizontal ground accelerations and velocities[†]**

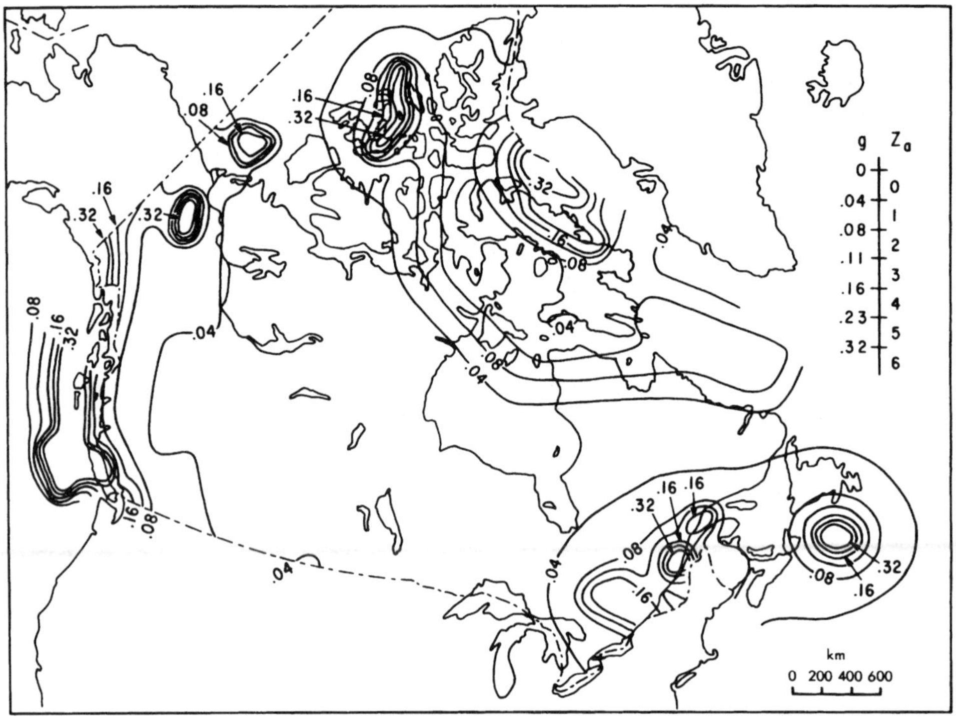


(a) Contours of peak horizontal ground accelerations, in units of g, having a probability of exceedance of 10 per cent in 50 years.

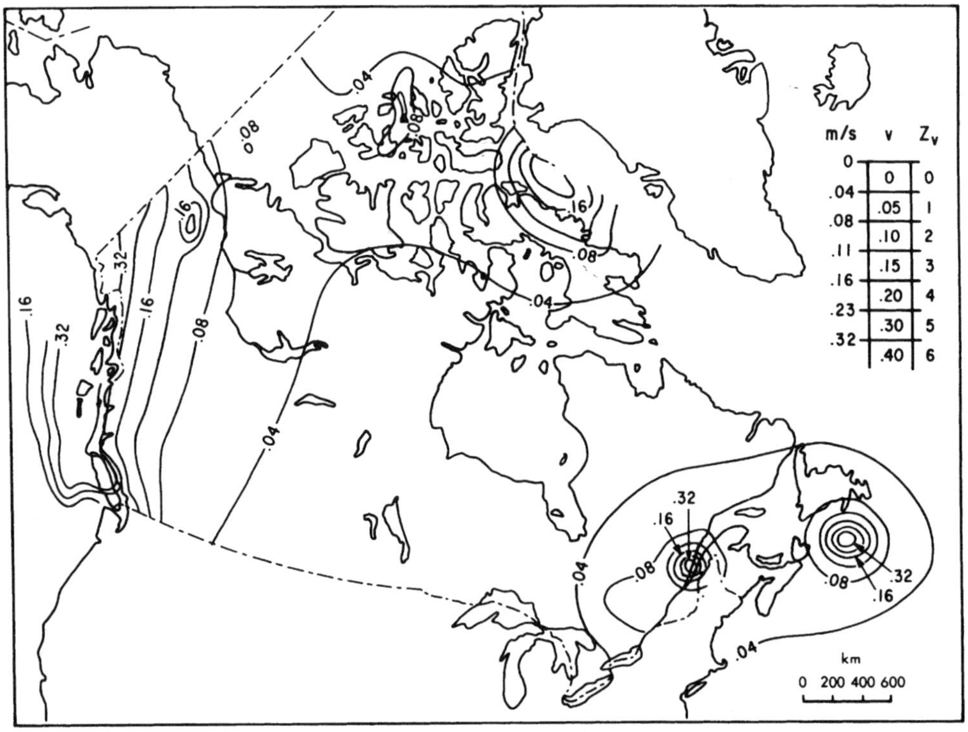


(b) Contours of peak horizontal ground velocities, in m/s having a probability of exceedance of 10 per cent in 50 years.

*For exact values for specific locations, refer to "Chapter 1, The Supplement to the National Building Code of Canada, 1995."

†Source: "The Supplement to the National Building Code of Canada, 1995"

For non-structural elements, or structural elements which do not participate in the lateral load resistance of the building, special load factors are applied to the connections, to ensure that these elements will perform satisfactorily. The forces on these connections are treated locally; that is, the reactions are carried to the support but not through the entire structure. The connections must also allow for distortion of the building, including relative movement between storeys (drift).

## 2.8.3 LATERAL FORCE RESISTING ELEMENTS

The lateral force resisting system of a building must be designed to resist a total force equal to the base shear V. This force is distributed over the height of the structure and is then assigned to the load bearing wall panels and shear walls according to their relative lateral stiffnesses. In addition, torsional moments in the horizontal plane, overturning, drift and building separations must be investigated.

The minimum lateral seismic force, V, shall be calculated in accordance with the following formula:

$$V = (V_e / R)\,U$$

where: U = 0.6

The equivalent lateral seismic force representing elastic response, $V_e$, shall be calculated in accordance with the following formula:

$$V_e = v \cdot S \cdot I \cdot F \cdot W$$

where v is the zonal velocity ratio determined in conformance with Fig. 2.8.1(b), except when $Z_v = 0$ and $Z_a > 0$ the value of $Z_v$ shall be taken as 1 and v as 0.05.

The seismic response factor, S, shall conform to Fig. 2.8.2.

The fundamental period T, shall be determined by:

(a) the formula 0.1N where the lateral force-resisting system consists of a moment-resisting space frame which resists 100 per cent of the required lateral forces and the frame is not enclosed by or adjoined by more rigid elements that would tend to prevent the frame from resisting lateral forces,

(b) the formula $0.09h_n / \sqrt{D_s}$ for other structures, where $H_n$ and $D_s$ are in metres and
$D_s$ = length of wall or braced frame which constitutes the main lateral load resisting system in the direction parallel to the applied forces; if the main lateral load resisting system does not have a well-defined length then D shall be used in lieu of $D_s$, or

(c) other established methods of mechanics provided they do not result in a value that exceeds 1.2 times the period calculated in Clause (a) or (b).

Values of the force modification factor, R, shall conform to Fig. 2.8.3.

The seismic importance factor, I, shall conform to Fig. 2.8.4.

The foundation factor, F, shall conform to Fig. 2.8.5 except that the product F•S need not exceed 3.0 where $Z_a$ does not exceed $Z_v$ and need not exceed 4.2 where $Z_a$ is greater than $Z_v$.

The weight, W, of the structure shall be calculated in accordance with the following formula:

$$W = \sum_{i=1}^{n} W_i$$

The total lateral seismic force, V, shall be distributed as follows:

(a) a portion, $F_t$, shall be assumed to be concentrated at the top of the structure and equal to 0.07 T V, except that $F_t$ need not exceed 0.7 s; the remainder, V-$F_t$, shall be distributed along the height of the *building*, including the top level, in accordance with the formula:

$$F_x = (V - F_t) W_x h_x \Big/ \left( \sum_{i=1}^{n} W_i h_i \right); \text{ or}$$

(b) by dynamic analysis.

The total shear in any horizontal plane shall be distributed to the various elements of the lateral force-resistant system in proportion to their rigidities according to rational analysis, with due regard to the capacities and stiffnesses of the non-structural elements and to the effects of torsion as required by Sentence 4.1.9.1(8) NBCC.

## 2.8.4 NON-LOAD RESISTING ELEMENTS

Precast non-load bearing, non-shear wall elements and their supporting connections should be designed for a lateral force, $V_p$, to be applied at the centre of gravity of the panel in any direction, given by:

$$V_p = v\, I\, S_p\, W_p$$

where:

v = zonal velocity ratio

$S_p$ = horizontal force factor for part of a building as given in Fig.2.8.6

$W_p$ = the weight of a part of the building

**Fig 2.8.2 Seismic Response Factors**

| T | $Z_a / Z_v$ | S |
|---|---|---|
| ≤ 0.25 | > 1.0 | 4.2 |
| | 1.0 | 3.0 |
| | < 1.0 | 2.1 |
| > 0.25 but < 0.50 | > 1.0 | 4.2–8.4(T–0.25) |
| | 1.0 | 3.0–3.6(T–0.25) |
| | < 1.0 | 2.1 |
| ≥ 0.50 | All values | $1.5 / \sqrt{T}$ |

**Fig 2.8.4 Importance Factor**[R2.8.1]

| Type of Occupancy | I |
|---|---|
| Post-disaster buildings | 1.5 |
| schools | 1.3 |
| Other buildings | 1.0 |

**Fig. 2.8.3 Force Modification Factors**[R2.8.1]

| Case | Type of Lateral-Force-Resisting System | R |
|---|---|---|
| | Steel Structures Designed and Detailed According to CAN/CSA-S16.1-M | |
| 1 | ductile moment-resisting frame | 4.0 |
| 2 | ductile eccentrically braced frame | 4.0 |
| 3 | ductile steel plate shear wall | 4.0 |
| 4 | ductile braced frame | 3.0 |
| 5 | moment-resisting frame with nominal ductility | 3.0 |
| 6 | nominally ductile steel plate shear wall | 3.0 |
| 7 | braced frame with nominal ductility | 2.0 |
| 8 | ordinary steel plate shear wall | 2.0 |
| 9 | other lateral-force-resisting systems not defined in Cases 1 to 8 | 1.5 |
| | Reinforced Concrete Structures Designed and Detailed According to CSA A23.3 | |
| 10 | ductile moment-resisting frame | 4.0 |
| 11 | ductile coupled wall | 4.0 |
| 12 | other ductile wall systems | 3.5 |
| 13 | moment-resisting frame with nominal ductility | 2.0 |
| 14 | wall with nominal ductility | 2.0 |
| 15 | other lateral-force-resisting systems not defined in Cases 10 to 14 | 1.5 |
| | Timber Structures Designed and Detailed According to CSA O86.1 | |
| 16 | nailed shear panel with plywood, waferboard or OSB | 3.0 |
| 17 | concentrically braced heavy timber frame with ductile connections | 2.0 |
| 18 | moment-resisting wood frame with ductile connections | 2.0 |
| 19 | other systems not included in Cases 16 to 18 | 1.5 |
| | Masonry Structures Designed and Detailed According to CSA S304.1 | |
| 20 | reinforced masonry wall with nominal ductility | 2.0 |
| 21 | reinforced masonry | 1.5 |
| 22 | unreinforced masonry | 1.0 |
| 23 | Other Lateral-force-resisting Systems not Defined in Cases 1 to 22 | 1.0 |

**Fig. 2.8.5 Foundation Factors**[R2.8.1]

| Categories | Type and Depth of Soil Measured from the Foundation or Pile Cap Level | F |
|---|---|---|
| 1 | Rock, dense and very dense coarse-grained soils, very stiff and hard fine-grained soils; compact coarse-grained soils and firm and stiff fine-grained soils from 0 to 15 m deep | 1.0 |
| 2 | Compact coarse grained soils, firm and stiff fine-grained soils with a depth greater than 15 m; very loose and loose coarse grained soils and very soft and soft fine-grained soils from 0 to 15 m deep | 1.3 |
| 3 | Very loose and loose coarse-grained soils with depth greater than 15 m | 1.5 |
| 4 | Very soft and soft fine-grained soils with depth greater than 15 m | 2.0 |

## 2.8.5 CONCEPT OF BOX-TYPE BUILDINGS

A box-type building consists of a roof diaphragm, floor diaphragms, and shear walls. These are connected where they meet, and if properly designed, the resulting structure can resist lateral forces.

Since earthquake forces result from accelerations applied to building elements, the equivalent lateral forces are proportional to the masses of the building parts and are applied at their centroids. These forces must be transferred to the foundation through a complete load path. The lateral force resisting elements share in transferring these forces in direct proportion to their contribution to the lateral stiffness of the building. With box-type buildings of moderate height-to-width ratio, the stiffnesses of the diaphragms and walls in their own planes are usually high compared to other stiffnesses. Forces originating in the diaphragms are transferred through connections into the walls, and through the walls to the footings, rather than through moment-resisting frames.

In a multi-storey box-type building, the forces originating in each floor diaphragm are transferred to the supporting walls. After a force has entered a wall, it will not leave that wall unless the wall is discontinued or its stiffness is substantially reduced. If this occurs, there must be an adequate load carrying system to transfer that force to another lateral load resisting wall, and then on down to the foundation. The designer should try and make the load path from each diaphragm down to the foundation as direct as possible.

A diaphragm made up of precast elements requires special consideration in order to develop resistance to shear and moment. Shear walls may have door and window openings. The designer must judge if the wall should be considered as a shear wall or a moment-resisting frame, depending on the number, size and location of the openings. It is also important to remember that although the building is designed for earthquake by applying forces to the roof and floor diaphragms and other elements, and transferring these forces to the foundation, in an earthquake, the motion is actually transferred

**Fig. 2.8.6 Horizontal force factors for part of a building**[R2.8.1]

| Category | Architectural Part or Portion of *Building* | Direction of Force | Value of $S_p$ |
|---|---|---|---|
| 1 | All exterior and interior walls except those of Categories 2 and 3 | Normal to flat surface | 1.5 |
| 2 | Cantilever parapet and other cantilever walls except retaining walls | Normal to flat surface | 6.5 |
| 3 | Exterior and interior ornamentations and appendages | Any direction | 6.5 |
| 4 | Connections/attachments for Categories 1, 2 and 3 | | |
| | The body of ductile connections/attachments | Any direction | 2.5 |
| | All fasteners and anchors in the ductile connection, such as bolts, inserts, welds, or dowels | Any direction | (1) |
| | Non-ductile connections/attachments | Any direction | 15.0 |
| 5 | Floors and roofs acting as diaphragms(2) | Any direction | 0.7 |
| 6 | Towers, chimneys, smokestacks and penthouses when connected to or forming part of a building(3) | Any direction | 4.5 |
| 7 | Horizontally cantilevered floors, balconies, beams etc. | Vertical | 4.5 |
| 8 | Suspended ceilings, light fixtures and other attachments to ceilings with independent vertical support | Any direction | 2.0 |
| 9 | Masonry veneer connections | Normal to flat surface | 5.0 |

(1) All fasteners and anchors in a ductile connection, such as bolts, inserts, welds, or dowels, shall be capable of developing 3 times the yield load of the body of the connection.

(2) Floors and roofs acting as diaphragms shall be designed for a minimum force corresponding to a value of $S_p$ equal to 0.7 applied to loads tributary from that *storey*, unless a greater force $F_x$ is assigned to the level under consideration.

(3) Lower values of $S_p$ may be used for towers, chimneys, smokestacks and penthouses when connected to or forming part of a building if the lower values can be proven by analysis. Information on the seismic design of towers, chimneys, smokestacks and penthouses can be found in the Commentary on Effects of Earthquakes in Structural Commentaries on the National Building Code of Canada 1995.

from the ground up into the building. The forces that act on the upper parts of the building in a strong earthquake will be determined by the actual lateral resistance of the system that supports these elements and will therefore exceed the factored code forces.

## 2.8.6 STRUCTURAL LAYOUT AND CONNECTIONS

A box-type structure may have a large number of precast concrete elements that are assembled into walls, floors, roof and possibly frames. Properly designed connections between the many pieces create the diaphragms and shear walls, and the connections between these, in turn, create the box-type structure. Some connections must be designed to transmit forces between the elements within a horizontal diaphragm or shear wall, others to transmit forces between these diaphragms and shear walls. In addition, some parts of the building, such as the shear walls, must be designed to absorb energy through non-linear deformation, while other parts, such as the horizontal diaphragms, will normally be intended to remain elastic during an earthquake.

In seismic design, forces must be positively transmitted. Load paths must be as direct as possible. Anchors should be attached to or hooked around reinforcing bars or otherwise terminated so as to effectively transfer forces to the bars. Reinforcement in the vicinity of the anchors should be designed to distribute the forces so as to preclude local failure. Concrete dimensions and reinforcement must be adequate so that the hardware of the connection is confined, and the connection can thus transmit accidental forces that are normal to the usual plane of the load path. Connections should be designed so that they can yield in a ductile manner without loss of load carrying capacity as the connection and surrounding concrete is damaged. The ductility of each connection must be adequate for its location and intended role in resisting seismic forces.

## 2.8.7 DESIGN GUIDELINES FOR PANELS

The Code provides sufficient information for the determination of forces on, and the structural design of, load bearing walls subjected to seismic forces. With respect to behaviour of structures subjected to the large lateral forces associated with seismic activity, the following comments are pertinent:

1. Exterior walls, normally perforated for windows, will act somewhere between an unperforated wall and a flexible frame. For tall buildings, this will result in a non-linear distribution of forces, due to the effect of shear lag, as indicated in Fig. 2.8.7.[R2.8.2] This figure indicates the relative differences, and the base stress patterns, due to lateral loading on a 23-storey building (plan dimension 22 m by 30 m).

2. Portions of walls with openings can be subjected to significant axial loads. These portions may require reinforcement with closely spaced ties, in accordance with the Code.

3. Connected walls may act as coupled walls, the connection achieved either by a lintel or by the floor construction. Provisions must be made for reversible shears and moments in the connecting element.

**Fig. 2.8.7 Forces and deflections due to lateral load**

| Forces and deflection | 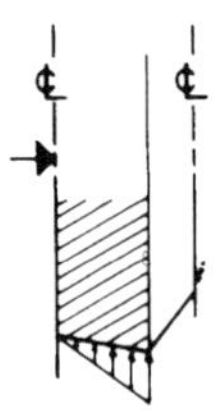 Side wall only | 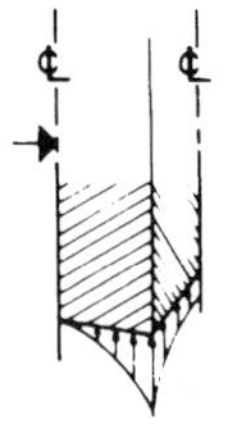 Computer solution | 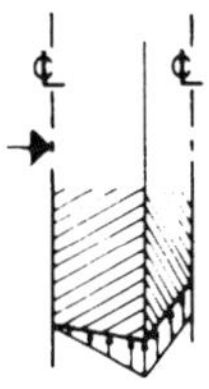 Flexural tube |
|---|---|---|---|
| Half sidewall shear | 1.00 | 1.00 | 1.00 |
| Half sidewall moment | 2.19 | 1.00 | 0.43 |
| Half endwall axial force | 0 | 1.00 | 1.53 |
| Building deflection | 2.87 | 1.00 | 0.89 |

4. Walls will be subjected to lateral loads perpendicular to the plane of the wall (wind, seismic) in combination with loads in the plane.

5. Design should consider the eccentricities produced by storey drift. These are combined with the eccentricities due to manufacturing and erection tolerances. Drift is defined as the relative movement of one storey with respect to the stories immediately above or below the level under consideration. Between points of connection, non-load bearing panels should be separated from the building frame to avoid contact under seismic action. In the immediate area of connections, the panel will tend to distort the same amount as the supporting frame. Internal forces induced due to a statically indeterminate support system should be checked. Even in a statically determinate panel there may be some built-in restraint at the connections, so that some allowance for internal forces should be considered. Fig. 2.8.8 illustrates the effects of storey drift.

**Fig. 2.8.8 Drift response**

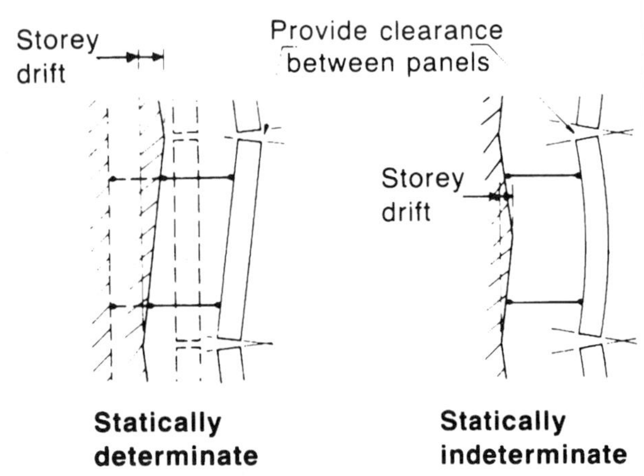


6. Under severe earthquake, large deflection may be anticipated. The investigation of individual walls and of the entire structure should include the consideration of deflection (P-$\Delta$ effect).

7. Accidental torsion may exist for any element subjected to seismic forces. Arrangement of reinforcement should consider this possibility.

8. Seismic-induced forces are reversible. This is particularly important at joints.

9. The best energy absorbing elements are those with high moment-rotation capabilities. The energy absorbing capacity of a flexural element is measured by the area under the moment-rotation curve. Correctly reinforced concrete can exhibit high ductility. Reference should be made to the Code, for proper methods or reinforcing to achieve ductility.

10. Joints represent discontinuities, and may be the location of stress concentrations. Reinforcing or mechanical anchorage must be provided through the joint to fully transmit the horizontal shear and flexure developed during seismic activity. See Chap. 4 and Sect. 2.8.8 for a discussion on connections. In zones of high seismicity, cast-in-place reinforced concrete in combination with precast concrete has proved successful in economically transferring the seismic forces[R2.8.3].

## 2.8.8 DESIGN GUIDELINES FOR PANEL CONNECTIONS

When choosing the number, location, and degree of restraint of connections, the following guidelines are recommended:

1. Wherever possible, make panel connections to the supporting structure statically determinate, in order to permit a more accurate determination of force distribution.

2. Choose the number and location of connections to minimize internal stresses and permit movements in the plane of the panel to accommodate storey drift and volume changes.
3. Locate connections to minimize torsional moments on supporting spandrel beams, particularly if the beams are structural steel.
4. Provide separation between non-load bearing panels and the building frame to prevent contact during an earthquake.

A generalized arrangement for connecting a non- load bearing panel to the supporting structure is shown in Fig. 2.8.9 and a specific connection is shown in Fig. 2.8.10.

**Fig. 2.8.9 Typical connection arrangement**

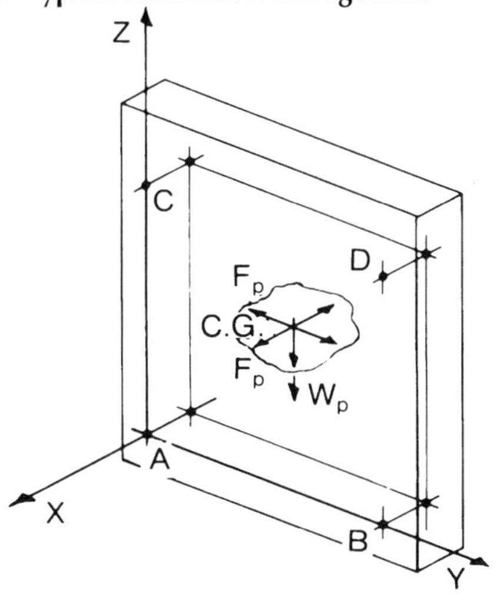


| Connection | Restraint* | | |
|---|---|---|---|
| | X | Y | Z |
| A | 1 | 1 | 1 |
| B | 1 | 0 | 1 |
| C | 1 | 1 | 0 |
| D | 1 | 0 | 0 |

* Restrained = 1 Free = 0

## 2.8.9 ANALYSIS OF PANELS

The analysis of panels should be done in accordance with established methods of structural mechanics. The following systematic approach is recommended:

1. Start with a free-body diagram for every load condition.
2. Solve for reactions using equations of equilibrium for statically determinate systems, or equations of equilibrium and compatibility for statically indeterminate systems.
3. Tabulate loads and reactions and identify critical design values.
4. Compute the internal forces (moments, shears, axial) due to the critical values. For complicated architectural shapes, consider use of computer analysis using methods such as finite element analysis.

A free-body diagram of a typical panel is shown in Fig. 2.8.11.

**Fig. 2.8.11 Free body diagrams**

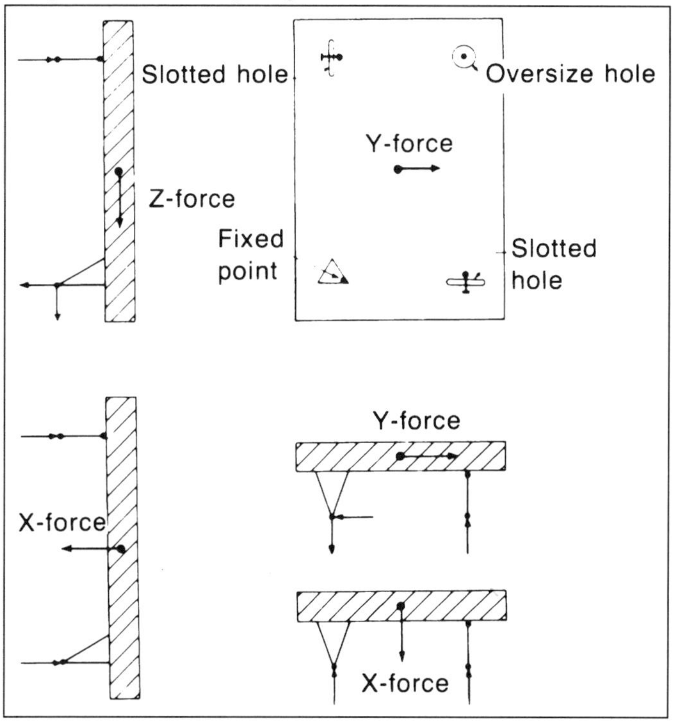


### Sandwich Panels

There are several additional design considerations relating to sandwich panels (see Chaps. 3 and 6 for a more complete discussion):

1. The structural wythe resists all loads, contains all connections, and transfers loads through these connections to the building frame (see Fig. 2.8.12).

**Fig. 2.8.12 Details of a typical sandwich panel**

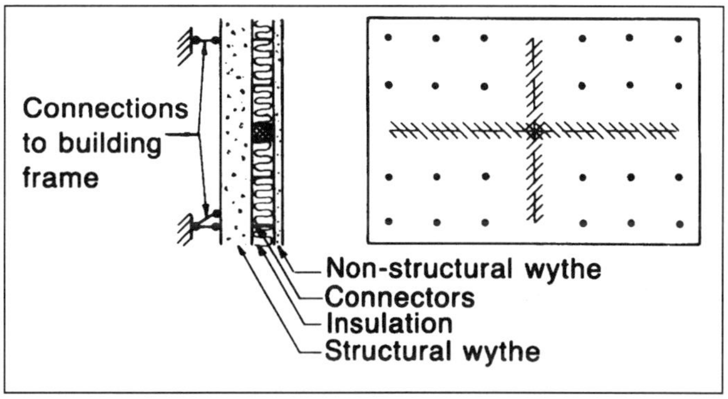


2. Seismic loads, analysis, and design of connections between structural wythe and building frame is identical to that of non-sandwich panels. The value $W_p$ used to determine the seismic force $V_p$ will include the weight of all elements of the panel.

**Fig. 2.8.10 A spandrel connection**

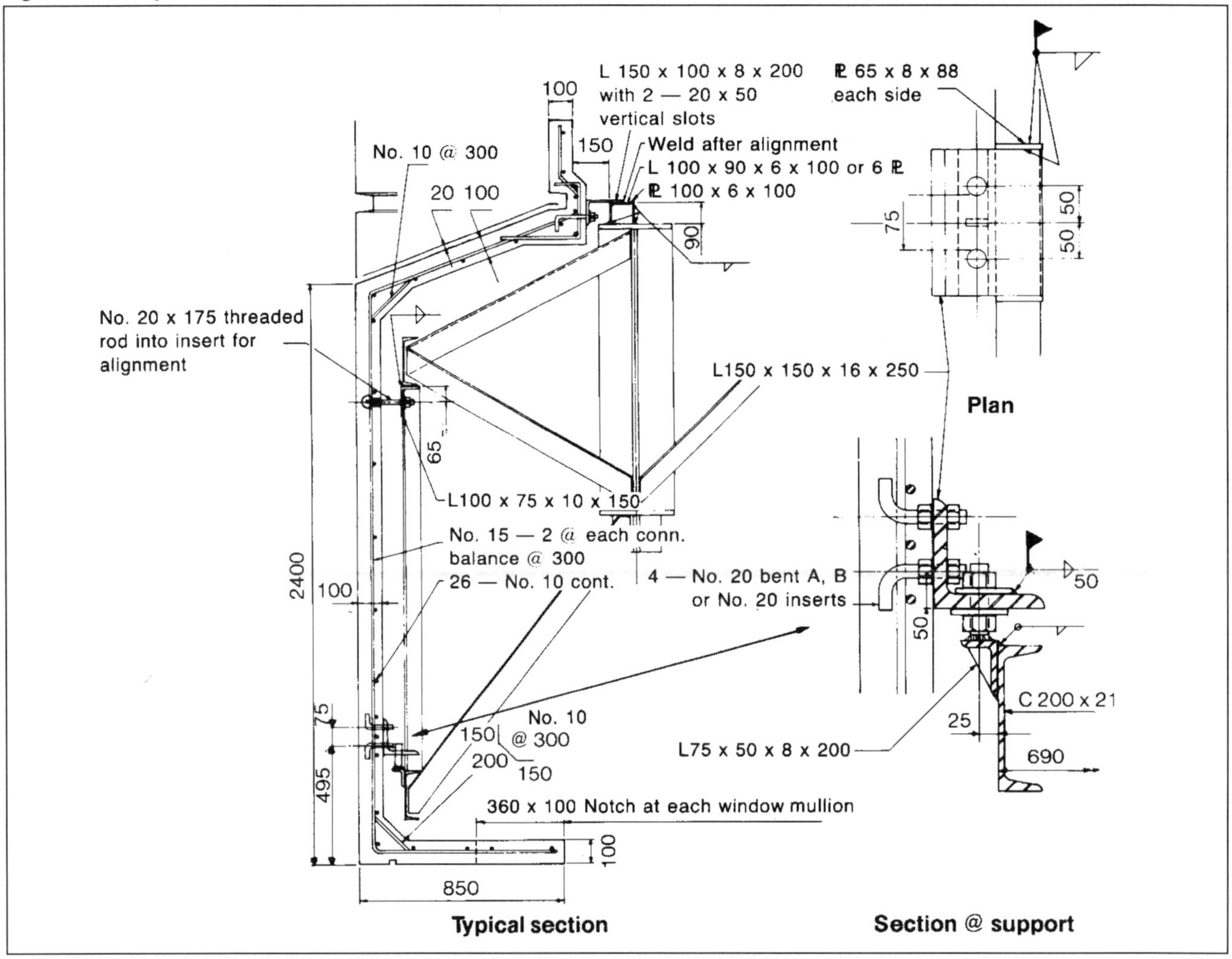


3. Connection between the two wythes should be capable of transferring a seismic force $V'_p$, where $V'_p$ is computed using the weight of the non-structural wythe and insulation.
4. The seismic forces $V_p$ and $V'_p$ should be assumed to act in any horizontal direction in conjunction with the weight of the panel or non-structural wythe.
5. The non-structural wythe should be proportioned and reinforced to resist the stresses induced by the seismic force $V'_p$ plus its self weight.

## 2.8.10 SEISMIC STRENGTHENING OF EXISTING STRUCTURES

Buildings in seismic zones may require strengthening due to seismic damage or due to upgrading of a seismic zone. One method of upgrading is to install stiffening walls as indicated in Fig. 2.8.13.

The following are some general guidelines which have been developed:[R2.8.3]

1. The factored shear resistance of panels may be taken as:

$$V_r = 0.2\lambda\phi_c\sqrt{f'_c}bd$$

where:

$\lambda$ = factor to account for low density concrete
$\phi_c$ = resistance factor for concrete
b = panel width
d = distance from extreme compression fibre to centroid of tension reinforcement

2. The system stiffness should be taken as either the stiffness of the single panel or the sum of the stiffnesses of the multiple panels, neglecting the contribution of the existing frame.

**Fig. 2.8.13 Infill stiffening walls**

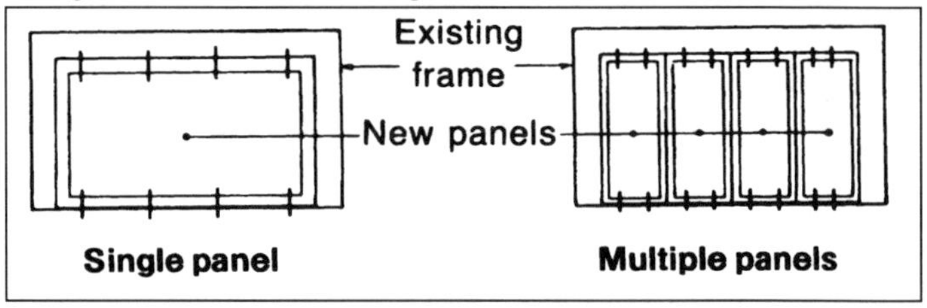

3. Connection between panel and existing frame should be at the top and bottom only. Sufficient room should be provided between the vertical legs of the frame and the panel to prevent contact during seismic loading.
4. Single panels of the multiple panel system should be designed as beams fixed top and bottom. Horizontal reinforcement should be in the form of closely spaced stirrups, spaced in accordance with the requirements of the Code.
5. Panel anchors should be mild steel bolts or threaded rods, located so as to minimize the eccentricity of shear transfer between existing frame and panel.

## 2.8.11 EXAMPLE-EXTERIOR NON-LOAD BEARING PANEL

*Given*:

A typical exterior non-load bearing panel used on a multi-storey structure in Seismic Zone 4. The supporting structure is composed of rigid shear walls, and storey drift is neglected. The concrete density of the panel is 1900 kg/m³.

*Problem*:

Determine the forces for which the connections must be designed.

*Solution*:

Reactions at connections:

1. Due to gravity:

$W_P = 63\text{ kN}$ including tile facing

$$W_p = \frac{63}{(1.8)(5.675)} = 6\text{ kN/m}^2$$

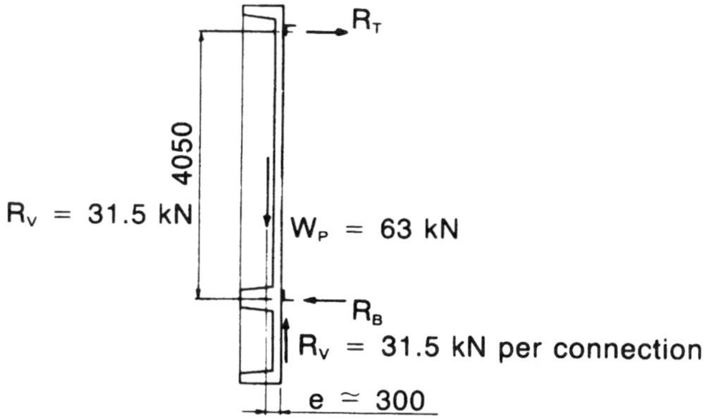


$$R_T = R_B = \frac{(63)(300)}{(4050)(2)} = 2.3\text{ kN per connection}$$

2. Horizontal seismic force normal to panel:

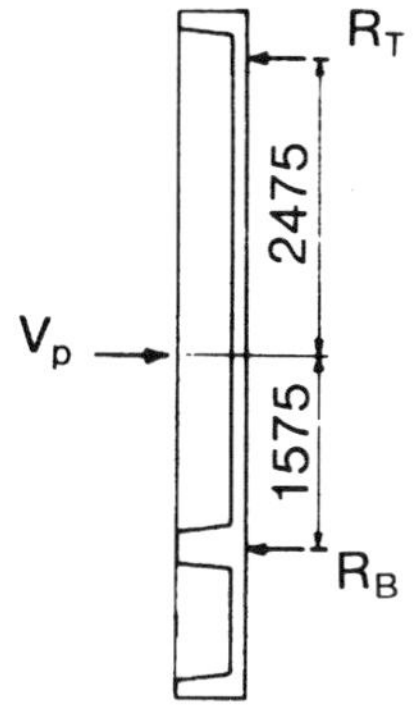


$S_p = 15.0$

$v = 0.2$

$V_p = (15)(0.2)(63) = 189\text{ kN}$

$$R_T = \pm\frac{(189)(1575)}{(4050)(2)} = 36.8\text{ kN per connection}$$

$$R_B = \pm\frac{(189)(2475)}{(4050)(2)} = 57.8\text{ kN per connection}$$

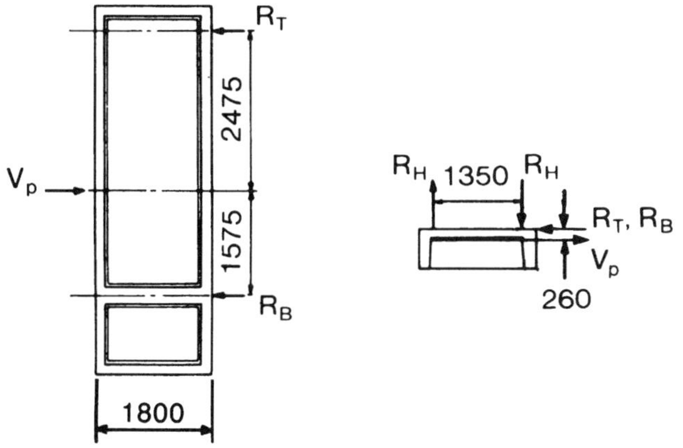


3. Horizontal seismic force in plane of panel:

$$@R_T: \quad R_H = \frac{(260)(36.8)}{1350} = 7.1\text{ kN}$$

$$@R_B: \quad R_H = \frac{(260)(57.8)}{1350} = 11.1\text{ kN}$$

Forces on panel normal to panel:

Following are forces for design of panel. For connection forces, see above:

1. Wind (1.2 kN/m²):

$$R_T = \frac{(2.16)(5.675)(1.575)}{(4.050)(2)} = 2.4\text{ kN per connection}$$

$$R_B = \frac{(2.16)(5.675)(2.475)}{(4.050)(2)} = 3.8\text{ kN per connection}$$

2. Seismic:

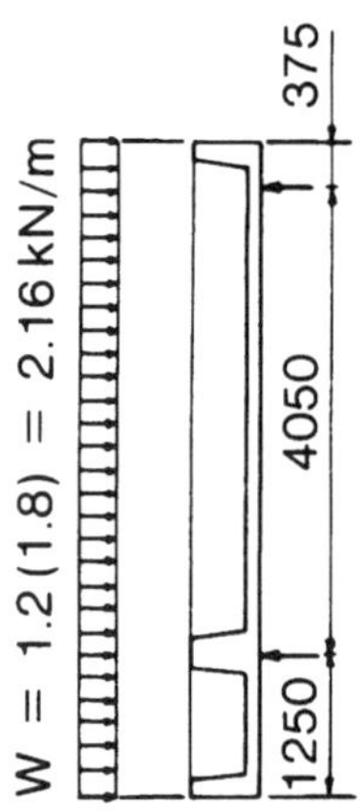


v = 0.2

$S_p$ = 1.5

w = (1.5)(0.2)(6) = 1.8 kN/ $m^2$, 50% more than wind

### 2.8.12 EXAMPLE – ONE STOREY BUILDING

*Given:*

The one storey building illustrated in Fig. 2.8.14. It is 38 400 mm × 48 000 mm in plan, and has 4800 mm clear height inside. It is entirely precast above the floor, using 400 mm double tees for the walls and 600 mm double tees for the roof. The double tees are 2400 mm wide, with stems 1200 mm on centres. The flanges of the roof tees are 50 mm thick, and of the wall tees, 100 mm thick. The concrete strength is 40 MPa and all reinforcement is Grade 400 weldable steel conforming to CSA-G30.16. The building is assumed to be in Zone 5 for both Seismic Acceleration and Seismic Velocity.

*Problem:*

Determine the loads due to an earthquake in the N-S direction and design the East and West wall connections, roof connections and roof to wall connections.

*Solution*:

As pointed out earlier, earthquake design differs from design for wind or gravity loads. Actual dynamic earthquake loads are dependent not only on the earthquake but also on the design of the building. For this reason, the lateral loads to be used in design are always related to assumptions about the stiffness, strength and ductility of the building. If these assumptions are incorrect for the building being designed, then the lateral forces that are used in design will also be incorrect. In addition, the dynamic loads resulting from strong earthquake motion can far exceed the factored lateral forces recommended for design. It is expected that buildings will deflect beyond their elastic limit during a severe earthquake and suffer damage as a result. The designer must identify the elements that are in-

**Fig. 2.8.14 One-storey example building**

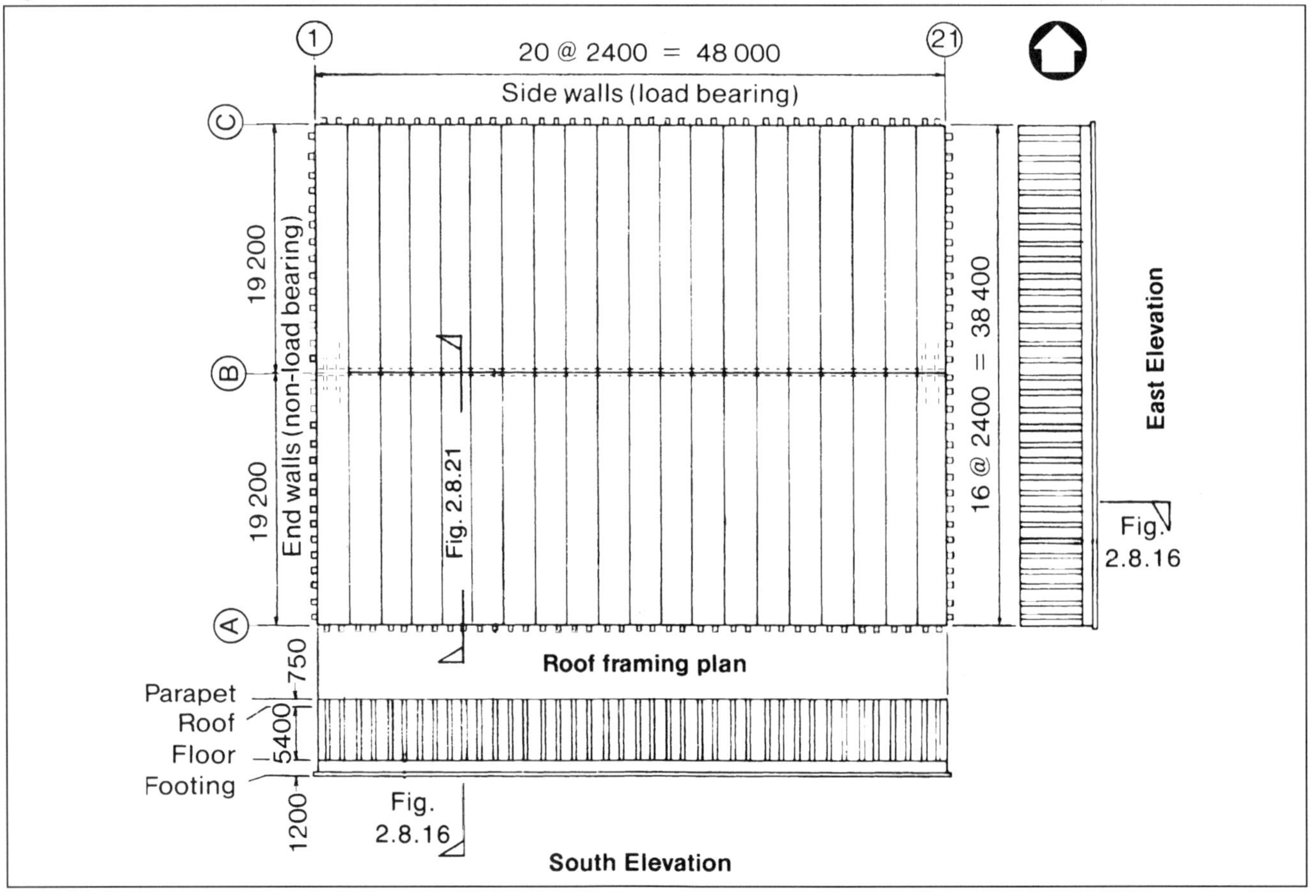

tended to yield during a severe earthquake and be satisfied that they can sustain the anticipated deformations without failure. The building must be able to undergo the expected deflections and suffer the consequent damage without collapsing.

The real earthquake accelerations applied to the foundation are essentially independent of the strength and stiffness of the building. It follows that if parts of the building are stiffer or stronger than intended, they may be subject to much higher forces than those assumed in design. This in turn may lead to higher than expected forces in other parts of the building. Therefore, elements that are intended to yield must be checked not only for adequate ductility but also to ensure that their strength is not excessive. The yielding of these elements should limit the forces that act in the building. If they are too strong, then other elements may fail in ways that are not intended, leading to a design that is unsafe for severe earthquake loading, in spite of the fact that the building would be more than capable of carrying the factored lateral forces used in design.

The nonlinear response of the example building to earthquake excitation has been studied and is discussed elsewhere.[R2.8.4]

*Load analysis*:

The example building resists lateral forces by diaphragm and shear wall action. Earthquake induced forces (mass × acceleration) are assumed to act on the roof diaphragm which acts like a plate girder laid flat, spanning between the shear walls. To find the total base shear V, the mass of the roof diaphragm, the mass tributary to it and the snow load must be considered. This is done in terms of the dead load, as follows:

Wall panels:

Unit weight of wall panels = 3.3 kN/m²

Height = 6150 mm

Half height + parapet = 3450 mm

N or S wall:

Total length of one wall = 48.0 m

Weight above mid-height = (3.45)(48.0)(3.3) = 545 kN

Weight of complete wall = (6.15)(48.0)(3.3) = 971 kN

E or W wall:

Total length of one wall = 38.4 m

Weight above mid-height = (3.45)(38.4)(3.3) = 436 kN

Weight of complete wall = (6.15)(38.4)(3.3) = 777 kN

Roof:

Weight of a roof panel (with waterproof membrane) = 2.8 kN/m²

Weight of roof = (48.0)(38.4)(2.8) = 5179 kN

Weight of building (roof + all walls) = 8675 kN

Snow load:

(see NBCC 4.1.7 for details of this calculation)

$S_o = 2.5$ kPa.

$S = S_o C_b C_w C_s C_a = (2.5)(0.8)(1.0)(1.0)(1.0) = 2.0$ kPa

Effective snow load = (0.25)(48.0)(38.4)(2) = 922 kN

The minimum lateral seismic force is given by:

$$V = (V_e / R)U$$

where the equivalent lateral seismic force representing elastic response is:

$$V_e = v \cdot S \cdot I \cdot F \cdot W$$

In this example, v is taken as 0.3, the value for Seismic Velocity-related Zone 5. The period of the building is estimated as $T = 0.09h_n / \sqrt{D_s}$, where $h_n$ is the height to the roof level in metres and $D_s$ is taken as the length of the building in the N-S direction, giving:

$$T = (0.09)(5.4)\sqrt{38.4} = 0.08 \text{ s}$$

The table in Fig. 2.8.2 gives S = 3.0 when $Z_a = Z_b$. The importance factor I and the foundation factor F are both taken as unity.

The choice of R requires some discussion. Fig. 2.8.3 includes a brief definition of buildings conforming to Case 15, for which R = 1.5. These include those of "continuously reinforced concrete". The discussion of Case 15 buildings in the Supplement to the NBCC explains that Case 15 includes "structures having nominal ductility, such as those of continuously reinforced concrete, reinforced masonry shear wall buildings....". Continuously reinforced concrete is defined to mean normal reinforced concrete, except that it must be detailed "so that, in regions of anticipated plastic hinging, the flexural steel can develop its full yield strength". Although precast concrete is not specifically included in Case 11, it can be considered to qualify provided that the precast building is detailed so that it can develop ductility comparable to that of a continuously reinforced structure of the type described. Therefore, R will be taken to be 1.5. Note that yielding and consequent damage, probably to the connections, must be anticipated if R = 1.5.

The weight W is the weight assigned to all levels above the base. For this single storey building, only the roof level needs to be considered. W includes the roof, the upper half of each wall and 25% of the snow load:

$$W = 5179 + (2)(545) + (2)(436) + 922 = 8063 \text{ kN}$$

The equivalent lateral seismic force representing elastic response is:

$$V_e = v \cdot S \cdot I \cdot F \cdot W$$
$$= (0.3)(3.0)(1.0)(1.0)(8063)$$
$$= 7258 \text{ kN}$$

The minimum lateral seismic force is:

$$V = (V_e / R)U$$
$$= \frac{(7258)}{1.5}(0.6)$$
$$= 2903 \text{ kN}$$

In comparison, the specified shear due to wind load is 175 kN. This building in this location is definitely governed by earthquake and not by wind.

*Overall stability of building*:

The overall stability of the complete building requires adequate resistance to sliding and overturning. This is considered using specified or unfactored forces. The force tending to cause sliding is the base shear V = 2903 kN. The weight of the building is 8675 kN. Assuming a granular soil with a coefficient of sliding friction $\mu_s$ = 0.5, the force resisting sliding is:

$$\mu_s W = (0.5)(8675) = 4338 \text{ kN}$$

This gives a factor of safety of:

4338/2903 = 1.49

The overturning moment:

= (base shear V) × (height to roof)

= (2903)(5.4 + 1.2) = 19 160 kN · m

The resistance to overturning can be taken as:

= (weight) × (half length of East wall)

= (8675)(38.4) / 2 = 166 000 kN · m

The factor of safety against overturning:

= 166 000 / 19 160 = 8.66

These factors of safety are adequate.

*East and West wall connections:*

The East and West walls act as shear walls for earthquake loading in the N-S direction, and must resist the entire lateral force V, assumed to act at roof level. To guard against torsion resulting from accidental eccentricity, a minimum 10% eccentricity must be assumed between the centre of resistance and the centre of mass as shown in Fig. 2.8.15. The specified lateral force due to earthquake, $V_W$, assumed to act on each of the East and West walls at roof level is:

$$V_W = 0.6\,V = (0.6)(2903) = 1740 \text{ kN}$$

The factored horizontal force acting at roof level to be resisted by each wall is:

**Fig. 2.8.15 Specified forces acting on roof & walls**

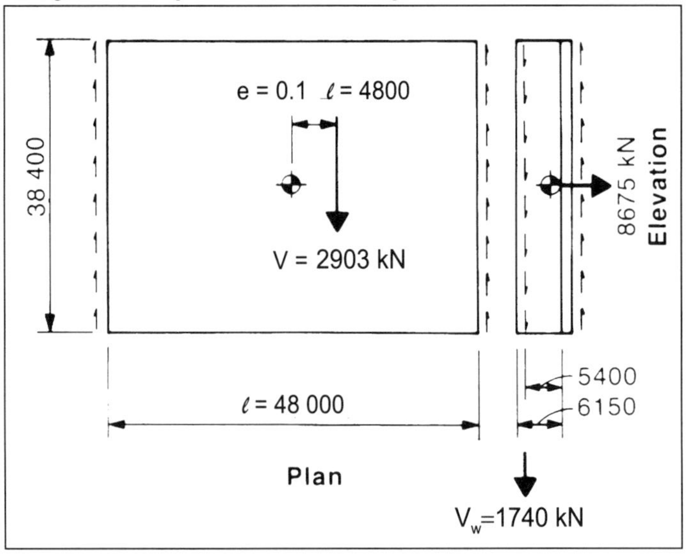


$$V_{fW} = 1.0\,V_W = (1.0)(1740) = 1740 \text{ kN}$$

Each wall panel contributes to the strength of the wall. If the panels are similar, the factored horizontal force to be resisted by each panel is:

$$V_{fP} = V_{fW} / (\text{number of panels}) = 1740/16 = 109 \text{ kN}$$

The building should be designed so that the yielding which is anticipated during a severe earthquake occurs in the shear walls and not in the roof diaphragm. This means that for the shear walls, both the factored resistance, found using material resistance factors, and the probable resistance, found using probable values of the material properties, must be considered. The design of the roof diaphragm will depend on the probable strength of the walls.

**Fig. 2.8.16 Wall-footing connection**

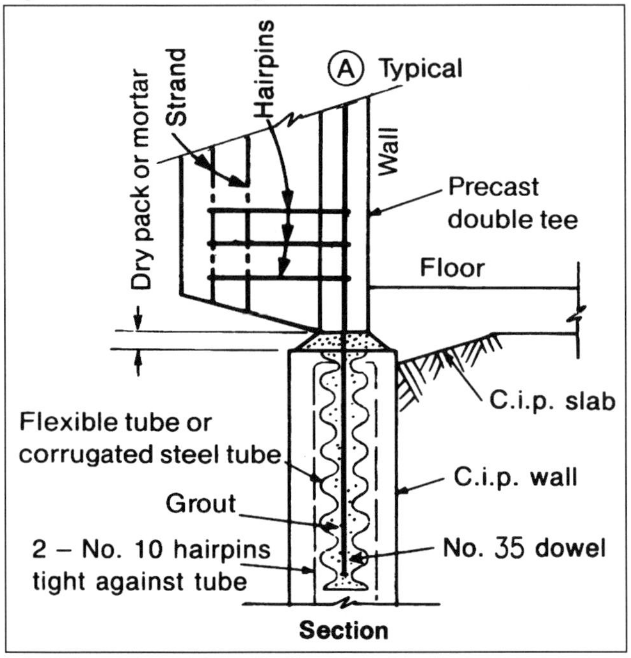

The factored resistance of the East and West shear walls will be considered first. It should be at least equal to the factored earthquake force $V_{fW}$, and the wall should be able to maintain this resistance through several cycles of loading into the inelastic range.

The required factored resistance will be provided by dowels projecting from the stems of the double tees. These dowels are anchored into grouted sleeves in the footing to develop their full tension resistance, as shown in Fig. 2.8.16. They must have an embedment length in the panel sufficient to transfer their full yield strength to the prestressing strand in the stem. Shear reinforcement should be provided to transfer this force, as shown in Fig. 2.8.16. The dowels provide resistance to rotation of the panels about their bottom corners and they resist sliding of the panels through shear-friction.

**Fig. 2.8.17 Assembly of double tee panels into wall**

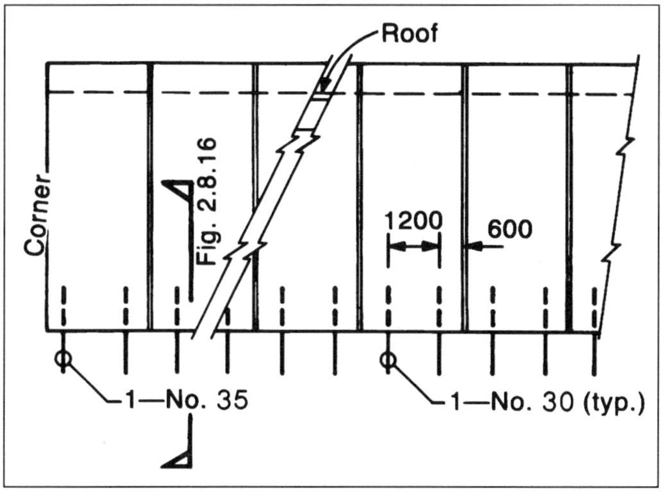


**Fig. 2.8.18 Forces acting on a wall panel**

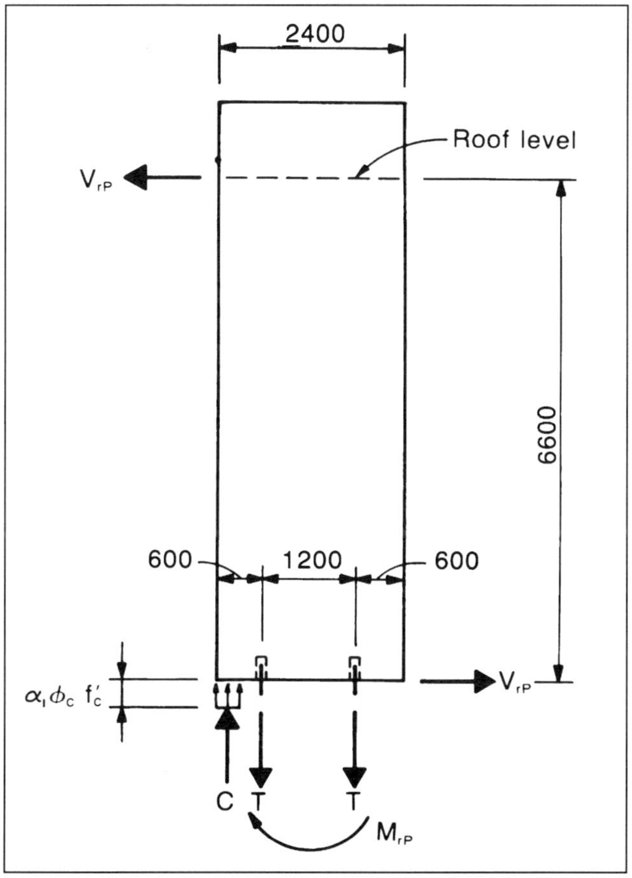


For simplicity, it is assumed that the walls have no openings. A wall is assembled as shown in Fig. 2.8.17. Each panel must be in equilibrium under the action of the forces acting on it, as shown in Fig. 2.8.18.

Each wall panel behaves like a cantilever beam fixed at the foundation and loaded by a horizontal force acting at roof level. If each panel has a factored flexural resistance at the base of $M_{rP}$, the factored shear resistance can be taken as:

$$V_{rP} = M_{rP} / (\text{height to roof})$$

The panel must be designed to ensure that a shear failure does not occur before this value is reached. The shear resistance of a wall, $V_{rW}$, is then the sum of the shear resistances of the individual panels.

The dowels are analogous to the flexural reinforcement in a beam. Try 1–No. 35 dowel per stem. The factored resistance of a dowel in tension is:

$$D_r = \phi_s f_y A_b = (0.85)(400)(1000) / 10^3 = 340 \text{ kN}$$

The depth of the rectangular stress block, a, and the distance of the neutral axis, c, from the corner of the panel when both dowels are yielding and the compression force C is acting, can be estimated with the normal beam analysis procedure:

$$\alpha_1 = 0.79$$

$$\beta_1 = 0.87$$

$$a = 2(D_r)/(\alpha_1 \phi_c f'_c b) = \frac{(2)(340)(10^3)}{(0.79)(0.65)(40)(100)} = 331 \text{mm}$$

where b is the thickness of the flange. Checking the dowel nearest to the compression force for flexural yielding, the maximum permissible value of c, with d = 600 mm for the dowel nearest to corner of the panel, is:

$$c_b = 700\, d / (700 + f_y)$$

$$= (700)(600) / (700 + 400) = 382 \text{ mm}$$

Therefore both dowels will yield when $V_{rP}$ causes the panel to rotate.

Ignoring the vertical force due the weight of the panel and considering moment equilibrium, it can be seen from Fig. 2.8.18 that for each panel:

$$M_{rP} = D_r(600 - a/2) + D_r(1800 - a/2)$$

$$= \left[340\left(600 - \frac{331}{2}\right) + 340\left(1800 - \frac{331}{2}\right)\right] / 10^3$$

$$= 703 \text{kN·m}$$

$$V_{rP} = 703/5.4 = 130 \text{ kN}$$

This is more than the required value of 109 kN. Try 1–No. 30 per stem. This gives $V_{rP}$ = 95 kN which is less than required. Because the wall should not be stronger than necessary, use 8 panels with No. 35 dowels and 8 with No 30 dowels. The factored shear resistance of the wall is then:

$$V_{rW} = (8)(95.5) + (8)(130.3) = 1806 \text{ kN} > V_{fW}$$

Use 4 panels with No. 35 dowels at each end of the wall.

For the panels with No. 35 dowels, the flexural compressive force acting at the corner of a panel is 680 kN. Because the flange thickness is only 100 mm, transverse reinforcement must be provided in the flange close to the end of the double tee to prevent this force from pushing off the corner of the flange. By shear-friction, the required area is:

$$A_{vf} = \frac{1}{A_{cv} f_y f'_c}\left(\frac{V_f}{0.5\lambda\phi_c}\right)^2$$

$$= \frac{1}{(100)(2400)(400)(40)}\left[\frac{(680)(10^3)}{(0.5)(1.0)(0.65)}\right]^2$$

$$= 1140 \text{ mm}^2$$

Use 4–No. 20 bars, giving $A_{vf}$ = 1200 mm² and locate these bars close to the bottom of the panel.

Each panel must also resist sliding on the foundation. The factored sliding force that must be resisted with No. 35 dowels:

$$H_{fP} = V_{rP} = 130 \text{ kN}$$

Taking c as 0.25 MPa and μ as 0.60, as recommended in CSA A23.3, Sec. 11.6.2, (a), and with 2–No. 35 dowels in each panel, the factored resistance of each panel to sliding on the foundation is:

$$H_{rP} = \lambda\phi_c\left(c + \mu\frac{A_{vf} f_y}{A_{cv}}\right)A_{cv}$$

$$= (1.0)(0.60)\left[0.25 + (0.60)\frac{(2000)(400)}{(100)(2400)}\right]$$

$$\times (100)(2400)/10^3$$

$$= 324 \text{ kN}$$

The No. 35 dowels provided to resist rotation will provide adequate shear-friction resistance. Similar checks must be made for the panels with No. 30 dowels.

To ensure that the footing will not fail by sliding under the action of the shear-friction force, similar dowels should be provided between the footing and the floor slab. To ensure the overall structural integrity of the building, the floor slab should have welded wire fabric reinforcement providing at least this much steel area in each direction.

*Probable resistance of East and West walls*:

During an earthquake, the base of the building is subject to varying acceleration. Forces are transferred from the foundation to the upper parts of the building to make them respond to the earthquake motion. For this building, the force that can be transferred to the roof is not limited to the factored force found from the Code, but is determined by the actual strength of the shear walls.

The probable lateral resistance of a shear wall is an estimate of the actual force acting at roof level when it begins to yield. It is therefore also an estimate of the maximum horizontal force that can be transferred to the roof by the shear wall during an earthquake. It is important that both the connections between the roof and the shear walls, and the entire roof diaphragm, should be able to remain elastic under the action of this force so that any yielding will occur in the walls and not in the roof or its connections.

The probable resistance of a No. 35 dowel in tension is:

$$D_p = 1.25\, f_y A_b = (1.25)(400)(1000)/10^3 = 500 \text{ kN}$$

The probable resistance of a No. 30 dowel is:

$$D_P = (1.25)(400)(700)/10^3 = 350 \text{ kN}$$

Using the probable resistance of the dowels and taking the compressive strength of the concrete as $f'_c$ gives a probable flexural resistance for a panel with No. 35 dowels rotating about its base:

$$M_{pP} = 1042 \text{ kN}\cdot\text{m}$$

and a corresponding probable shear strength:

$$V_{pP} = M_{pP}/(\text{height to roof}) = 1042/5.4 = 193 \text{ kN}$$

The probable sliding resistance between the wall panel and the foundation is:

$$H_{pP} = \lambda\phi_c\left(c + \mu\frac{A_{vf} 1.25 f_y}{A_{cv}}\right)A_{cv}$$

$$= (1.0)(0.60)\left[0.25 + (0.6)\frac{(2000)(1.25)(400)}{(100)(2400)}\right]$$

$$\times (100)(2400)/10^3$$

$$= 396 \text{ kN}$$

This is greater than $V_{pP}$, as required to ensure that a flexural failure will occur at the base of the panel before a sliding failure.

Each panel must also resist some earthquake force arising from its own mass. This reduces the earthquake force required to act at roof level to yield the dowels at the base of the wall panel, and so reduces the force that could be transfered from the wall through the connections to the roof. The reduced force can be conservatively estimated as:

$$V'_{pP} = V_{pP} - [(v\,S\,I\,F/R)\,U]\,(\text{weight of panel})$$

$$= 193 - (0.36)(27.3) = 183 \text{ kN}$$

The corresponding force for a panel with No. 30 dowels is:

$$V'_{pP} = 141 - (0.36)(27.3) = 131 \text{ kN}$$

The total force which the connections between the shear wall and the roof must be able to transfer to the roof is therefore:

$$V_{pR} = \Sigma V'_{pP} = (8)(183) + (8)(131) = 2512 \text{ kN}$$

Note that the specified force that would be obtained by multiplying the weight of the roof, North and South walls and effective snow load by [(v S I F / R) U] is 2546 kN. Allowing for 10% eccentricity would give a specified force on the connections between the roof and an East or West wall of 1566 kN, and a factored force of 2291 kN. This value is lower than the force of 2512 kN found above using the probable resistance of the East and West walls.

The expected inelastic behaviour of the wall is rotation of the wall panels, with yielding of the dowels acting as flexural reinforcement. This is an acceptable mode of inelastic behaviour, and supports the use of R = 1.5 in the equation for base shear.

**Fig. 2.8.19 Probable forces on roof**

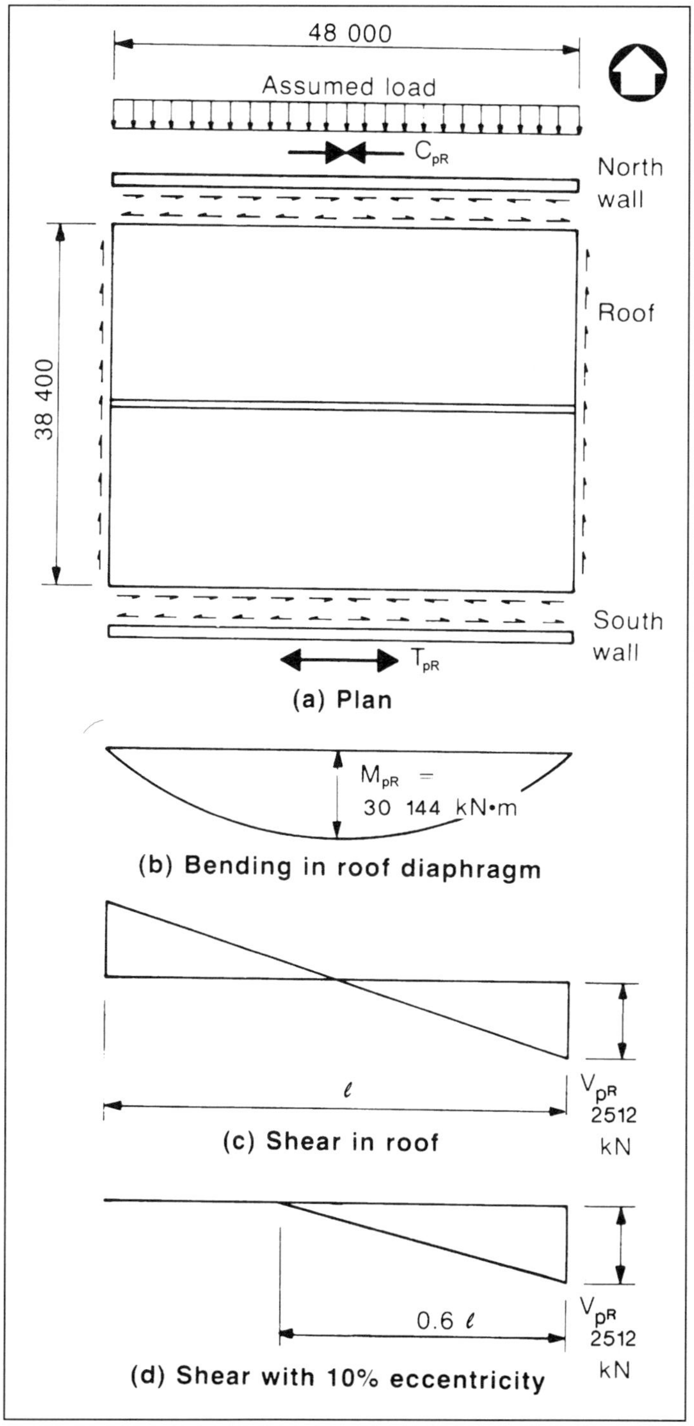


*East and West wall to roof connections:*

The roof and the wall to roof connections must now be designed to remain elastic when the force $V_{pR}$ = 2512 kN is applied along the East and West edges. This force will acceler-

**Fig. 2.8.20 Connections between flanges of roof double tees and between roof and E & W walls**

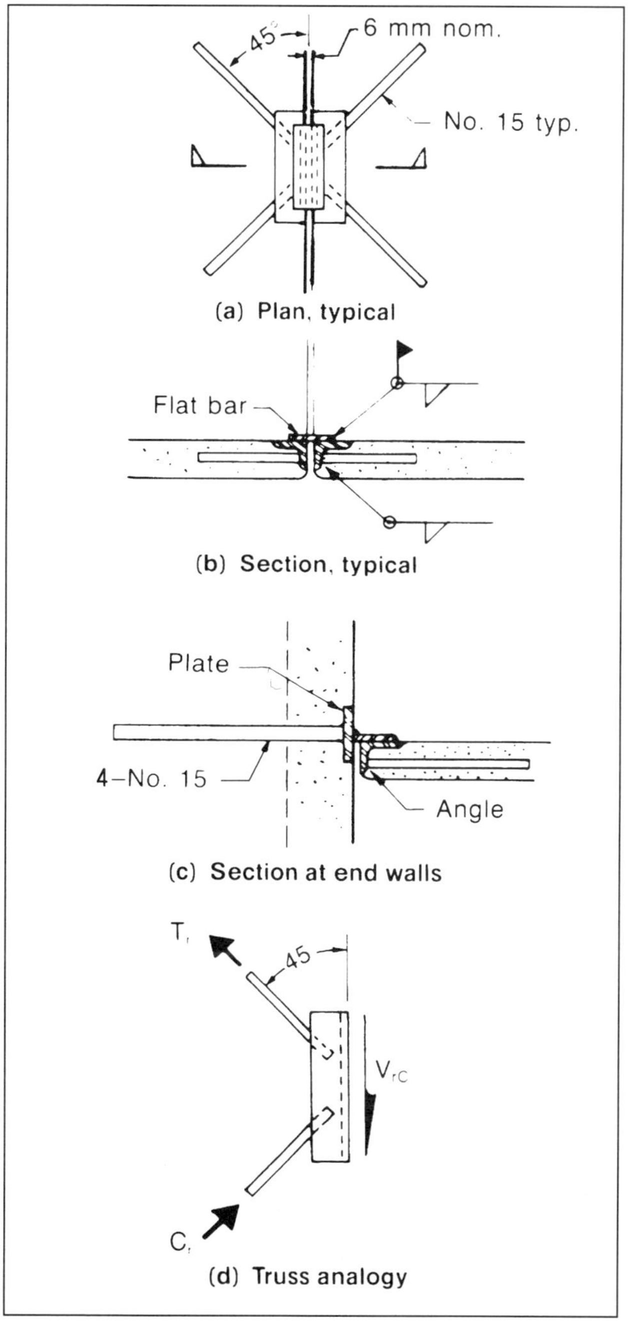

ate the roof which is assumed to act like a plate girder laid flat and subject to a uniformly distributed horizontal force.

The probable forces internal to the roof diaphragm are shown in Fig 2.8.19. The shear forces are analogous to those in the web of a plate girder. The chord forces are analogous to those in the flanges of a plate girder.

Weld clip connections with No. 15 bars are used in the flanges of the roof double tees. The factored shear resistance of a weld clip is found using a truss analogy, illustrated in Fig. 2.8.20. The material resistance factor for reinforcing steel, $\phi_s = 0.85$, is used to find the force in the reinforcing bars. If these connections were intended to yield during an earthquake, an allowance would be made for loss of strength during cycles of inelastic loading by using an additional member resistance factor, $\phi_m = 0.7$, when calculating the factored resistance of the connection.

Since these connections will not be designed to yield, the member resistance factor can be taken as $\phi_m = 1$. The factored resistance of a connection is:

$$C_r = T_r = \phi_s f_y A_s = (0.85)(400)(200) / 10^3 = 68 \text{ kN}$$

$$V_{rC} = (C_r + T_r)\cos 45° = 96 \text{ kN}$$

The required spacing for the panels with No. 35 dowels is:

$$s = (V_r / V'_{pP})(\text{panel width})$$
$$= (96.2 / 183)(2400) = 1260 \text{ mm}$$

A suitable spacing would be 1200 mm on centres, allowing the connections in the wall panels to be anchored in the webs. The factored horizontal force to be transferred by each connection is thus:

$$F_{fh} = (V_{pC})(s) / (\text{panel width})$$
$$= (183)(1.2) / 2.4 = 91.5 \text{ kN}$$

which is less than the factored resistance of 96 kN.

The exterior flange of the outside roof tees must also be checked for a shear failure along a plane parallel to the web. The factored shear resistance of the concrete alone is:

$$v_c = 0.17\phi_c\sqrt{f'_c}bd$$
$$= (0.17)(0.65)\sqrt{40}(1000)(50) / 10^3$$
$$= 35 \text{ kN/m}$$

The factored shear force per unit length is found from $V'_{pP}$ for a panel with No. 35 dowels to be:

$$v_f = V_{pC} / (\text{panel width}) = \frac{183}{2.4} = 76 \text{ kN/m}$$

The welded wire fabric used to reinforce the flange should be used as shear friction reinforcement to prevent a shear failure. The factored shear friction resistance will be:

$$v_r = 0.5\lambda\phi_c\sqrt{\frac{A_{vf}f_y f'_c}{A_{cv}}}$$

therefore:

$$A_{vf} = \frac{A_{cv}}{f_y f'_c}\left(\frac{v_f}{0.5\lambda\phi_c}\right)^2$$
$$= \frac{(1000)(50)}{(400)(40)}\left[\frac{76/50}{(0.5)(1.0)(0.65)}\right]^2$$
$$= 68 \text{ mm}^2/\text{m}$$

This can be provided with WWF 152 × 152 MW13.3 × MW13.3, with $A_s = 87.5$ mm²/m in each direction.

Vertical acceleration during an earthquake may cause bouncing of the double tees. Since vertical motion of the edge double tee relative to the wall cannot be accommodated, the connections should be able to yield the double tee flange as a cantilever where it meets the first interior stem. The welded wire fabric in the flanges has 13 mm clear top cover, giving a maximum flexural resistance of approximately 1.4 kN·m/m. To yield the flange, weld clips spaced at 1200 mm on centres and acting at approximately 500 mm from the web will have to apply a vertical force:

$$F_{fv} = (1.4)(1200) / (500) = 3.4 \text{ kN}$$

Although the weld clip connection is not designed for loading normal to the plane of the flange, it should be able to carry this relatively small force. However, the presence of this force might be expected to reduce the shear resistance somewhat.

Note that these connections will have to be checked for a different load case when the earthquake acts in the East-West direction. It should also be pointed out that the flange is only 50 mm thick in the roof double tees. Weld clip connections embedded in these thin flanges will almost certainly perform badly if they are allowed to yield, but the design approach used will ensure that they have adequate strength and remain elastic.

The connection in the wall double tee is shown in Fig. 2.8.20. Shear-friction reinforcement is required to carry the forces $F_{fh}$ and $F_{fv}$. It is conservative to design for the sum of these forces. The minimum reinforcement area required is:

$$A_{vf} = \frac{F_{fh} + F_{fv}}{\lambda\mu\phi_c f_y}$$
$$= \frac{(91.5 + 3.4)(10^3)}{(1.0)(0.6)(0.65)(400)}$$
$$= 608 \text{ mm}^2$$

Use 4–No. 15 bars ($A_{vf} = 800$ mm²) for each connection. Provide full development of the bars within the 400 mm depth of the stem.

*North and South wall to roof connections:*

The ends of the roof double tees must be connected to the North and South walls in order to transfer the "VQ/Ib" web shear to the chords. This is analogous to the connection between the web and flange of a plate girder. This same connection must also function as a tension tie, holding the wall panels onto the roof against wind and earthquake, and holding the building together against shrinkage and related forces.

The specified earthquake tension force, $V_p$, acting between a North or South wall panel and the roof is:

$$V_p = v\,S_p\,W_p$$

where v is the zonal velocity factor, $S_p$ is a horizontal force factor from Fig. 2.8.6 and $W_p$ is the tributary weight of this part of the building. In this case, v is 0.3 and $S_p$ is 15.0.

The tributary weight of a 2400 mm wide wall panel is the weight of half the height plus the parapet

$$W_p = (3.45)(2.4)(3.29) = 27.2 \text{ kN}$$

$$V_p = v\,S_p\,W_p = (0.3)(15.0)(27.2) = 122 \text{ kN}$$

With one connection per stem, a specified force of 61 kN acts on each connection.

As this is equivalent to a specified wind load of 10.8 kPa, wind does not govern. The connection must transmit the diaphragm shear and accommodate live load end rotation of the simple span roof tees. This is done by welding the roof tees to the top of the wall tees , but not at the bearing (Fig. 2.8.21). The bearing will then slide under live load end rotation.

**Fig. 2.8.21 Connections between roof double tees at (a) end walls and (b) ledger beam**

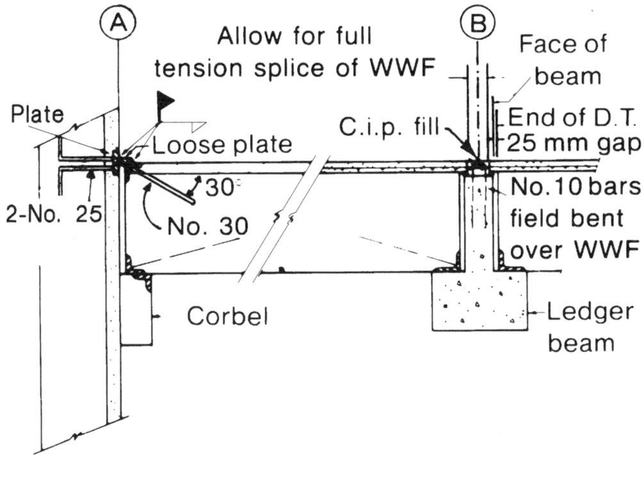


**(a) Section at wall** **(b) Section at ledger beam**

The horizontal friction force at each bearing and the corresponding tension tie resistance required, assuming a coefficient of friction of 0.6 (see CSA A23.3-M94, Sec. 11.6.2 (d)) and the dead load reaction of the roof = 32.4 kN, is:

$$H = \mu_s N = (0.6)(32.4) = 19.4 \text{ kN}$$

This value of H is smaller than the $V_p$ = 61 kN required for earthquakes. The wall panel to roof connection is thus designed for a specified horizontal pullout force of 61 kN. The reinforcement anchorage is therefore sized for a factored tension force:

$$T_f = 1.0\,V_p = (1.0)(61) = 61 \text{ kN/stem}$$

$$A_s = \frac{T_f}{\phi_s f_y} = \frac{(61)(10^3)}{(0.85)(400)} = 180 \text{ mm}^2 \text{ / stem}$$

These connections must also be able to transfer the shear between the diaphragm and a North or South wall. At the ends of the walls, adjacent to an East or West wall, the shear will be equal to the shear acting between an East or West wall and the roof. The shear will diminish toward the centre of the North or South wall. The maximum factored shear force per unit length to be transferred is $v_f$ = 76 kN/m. If the connections are located at each stem, the factored shear force to be transferred by a connection closest to the end of the wall where the shear is a maximum is:

$$V_f = (76)(1200) / 10^3 = 91 \text{ kN / stem}$$

The shear friction reinforcement required to carry the force $V_f$, with $\mu$ = 0.6 from CSA-A23.3-M94, Sec. 11.6.2 (d), is:

$$A_{vf} = \frac{V_f}{\mu\phi_s f_y} = \frac{(91)(10^3)}{(0.6)(0.85)(400)} = 446 \text{ mm}^2 \text{ / stem}$$

The total reinforcement area required is $A_s$ = 180 + 446 = 626 mm². The connections used are shown in Fig. 2.8.21. The embedment in the wall uses 2–No. 25 bars ($A_s$ = 1000 mm²) welded to a steel plate, and anchored in the stem to develop their full yield strength.

The embedment in the stem of each roof double tee has 1–No. 30 bar ($A_s$ = 700 mm²) welded to an angle and placed at 30° to the horizontal as shown. This is equivalent to a reinforcement area $A_s$ = (700) cos 30° = 606 mm² placed horizontally.

This is sufficient to carry the direct tension force $T_f$ and develop the required shear-friction force. The embedment length should be sufficient to develop the full tension capacity of the bar. Vertical reinforcement in the end of the stem, over the length of the No. 30 bar, should be capable of transferring its probable tension resistance into the welded wire fabric in the flange.

*Diaphragm shear connections*:

The double tee flanges must also be connected to each other along their edges using weld clips. This is analogous to providing shear strength along vertical joints in a plate girder. The number of connections required will be proportional to the shear which diminishes towards the centre of the diaphragm. In an earthquake, as a result of torsion or other effects, one wall may yield while the other remains elastic. Therefore, it would be a mistake to assume zero shear force at the centre of the diaphragm as is implied by the plate girder analogy with uniform load. The resultant force on the diaphragm should be assumed to act with at least 10% eccentricity relative to the centre. The shear force can therefore be assumed to reach zero at (0.60)(48 000) = 28 800 mm from the wall that is yielding (see Fig. 2.8.19).

Using simple proportion, the required spacing of the connections along the interior edge of the first double tee would be:

$$s = (2512)(28\,800) / (28\,800 - 2400) = 2740 \text{ mm}$$

The spacing should not be increased beyond a reasonable maximum, say 3000 mm, along any connection line.

*Diaphragm chord reinforcement*:

Maximum bending moment in the diaphragm:

$$M_{pR} = \frac{2V_{pR}\ell}{8} = \frac{(2)(2512)(48)}{8} = 30\,144 \text{ kN·m}$$

Maximum chord forces at the North and South edges of the diaphragm:

$$C_{pR} = T_{pR} = \frac{M_{pR}}{d} = \frac{30\,144}{38.4} = 785 \text{ kN}$$

The flange or "chord" bar size is sized for tension as follows:

$$A_s = \frac{T_{pR}}{\phi_s f_y} = \frac{(785)(10^3)}{(0.85)(400)} = 2310 \text{ mm}^2$$

Use 2–No. 45 bars ($A_s = 3000 \text{ mm}^2$)

These bars are located at the top of the wall panels and must be spliced at the vertical joints as shown in Fig. 2.8.22. Their development length is 1200 mm. This will ensure that the bars in the wall panels nearest to the East and West walls can develop their full yield strength. The cross sectional area of the angle should be sufficient to assure prior tensile yield in the bar, and the angle should be of a size to provide reasonable concentricity between bar and angle. Field welding is designed in accordance with current welding standards.

**Fig. 2.8.22 Flange or "chord" reinforcement at top of wall panels**

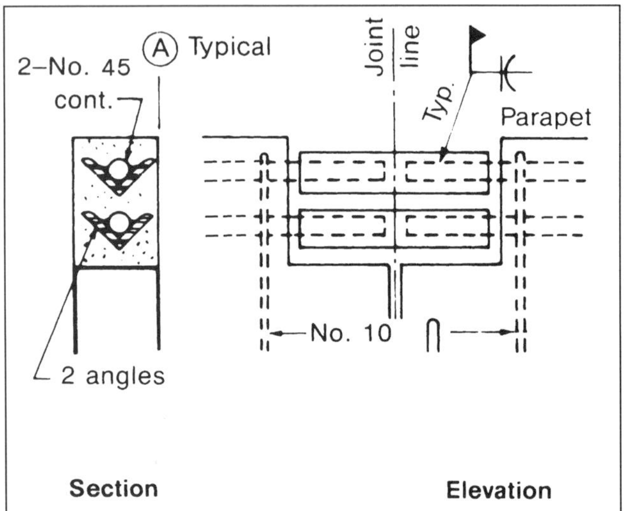


*Note: This example assumes a mild climate. In colder climates, it would be preferable to locate the tensile chord in a location less exposed to the elements - say, under the roof insulation at the roof-to-wall junction.*

The hairpin bars shown must provide enough shear-friction capacity to transmit the shear acting between the end of the double tees and the North and South walls into the 2–No. 45 chord bars. For the maximum shear force per unit length, $v_f = 76$ kN/m, assumed to act over the width of the panel nearest to an East or West wall, the required shear-friction reinforcement is:

$$A_{vf} = \frac{(2400)(100)}{(400)(40)}\left[\frac{76/100}{(0.5)(1.0)(0.65)}\right]^2$$

$$= 82 \text{ mm}^2$$

One No. 10 hairpin gives $A_{vf} = 200 \text{ mm}^2$. Hairpins should be provided in all panels and should extend an anchorage length below the double tee to wall tension ties. This provides a load path for the diaphragm shear between the chord bars and the roof weld clips.

*Roof tee to ledger beam connection:*

The diaphragm must be able to transfer the same shear forces over the centreline beam as those transferred between the roof and the North and South walls, using the connection shown in Fig. 2.8.21. Therefore, the required area of shear-friction reinforcement is governed by the shear force per unit length close to the East and West walls, 76 kN/m, and is:

$$A_{vf} = \frac{(76)(10^3)}{(1.0)(0.65)(400)} = 290 \text{ mm}^2/\text{m}$$

ignoring the resistance given by the cohesion of concrete, which is conservative.

This reinforcement should be provided over the full length of the ledger beam to ensure adequate performance of the diaphragm.

The shear-friction requirement governs close to the East and West walls, while the tension reinforcement requirement will govern toward the centre of the roof. Use a 152 × 152 MW47.6 × MW47.6 welded wire fabric with an area of 313 mm²/m.

This completes the design of the building for earthquake in the N-S direction. Note that all connections must also be checked for earthquake loads acting in the E-W direction, and this load case may govern.

## 2.8.13 EXAMPLE - FOUR STOREY BUILDING

*Given*:

The four storey rectangular building shown in Figs. 2.8.23 and 2.8.24, in which the structural system resisting lateral forces is of the box type, that is, floors and roof are horizontal diaphragms, and exterior walls are shear walls. Floors and roof are made up of 2400 mm wide precast, prestressed concrete double tees, which are supported by exterior bearing walls and a central interior frame of precast, reinforced concrete beams and columns. There are smaller frames at the two elevator-stairway shafts, one at each end of the building. The interior frames are made up of full height columns and short beams. The exterior walls are 2400 mm wide precast, prestressed concrete double tees placed on end, and are continuous from top of footing to top of parapet.

The unit dead loads are:

| | |
|---|---|
| Roof: | 4.8 kN/m² |
| Floors: | 5.3 kN/m² |
| Walls: | 4.3 kN/m² max., 3.2 kN/m² average |

**Fig. 2.8.23 Four-storey example building - elevations**

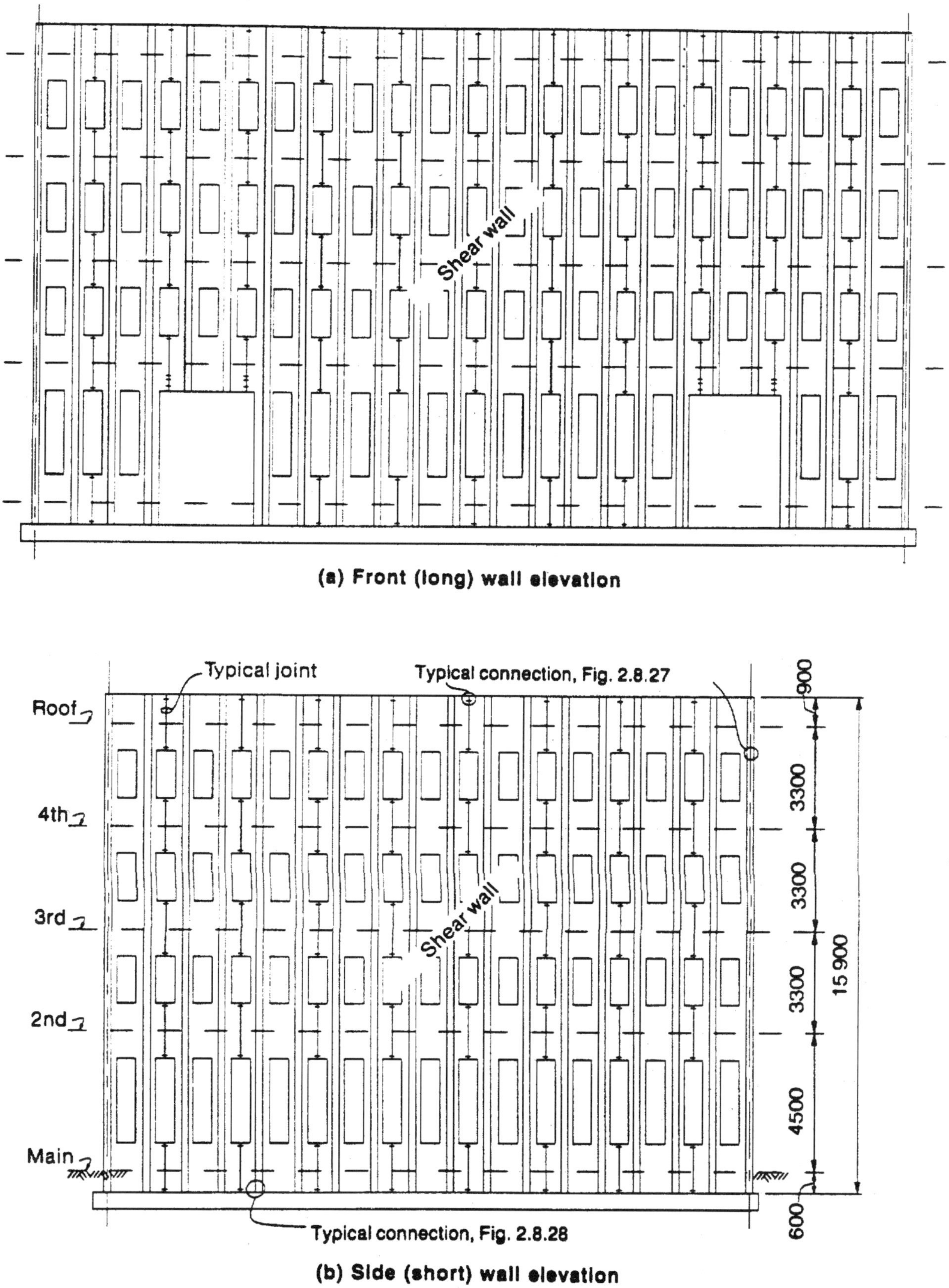


(b) Side (short) wall elevation

**Fig. 2.8.24 Four storey example building - typical floor plan**

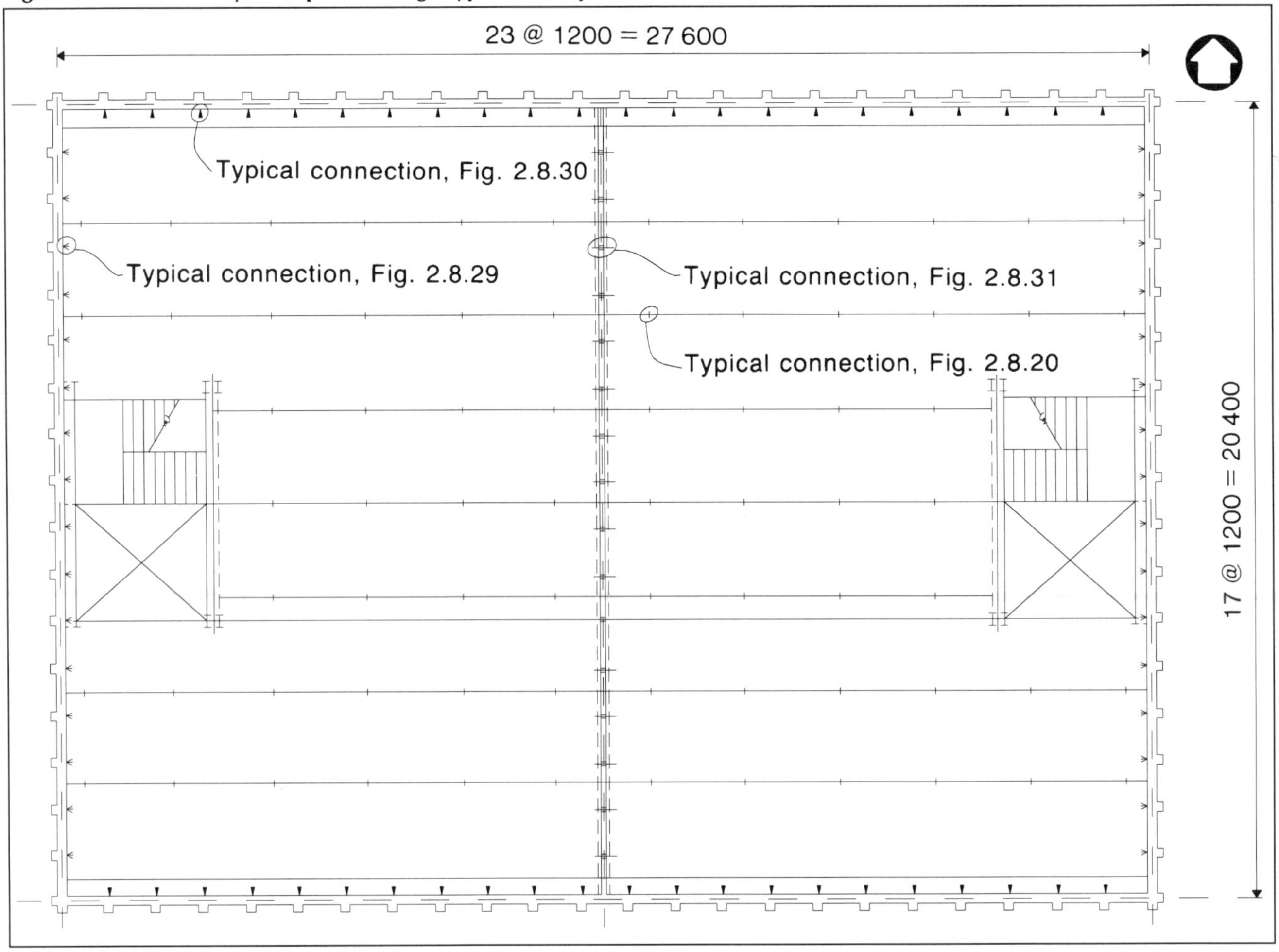


*Problem*:

Determine the earthquake forces and design the shear walls, diaphragms and connections.

*Solution*:

In the example, several approximations are made, but they are all on the conservative side. The designs are not complete in all respects but in the case of several similar parts, the more critical part is analyzed to illustrate the procedure. Some items, such as number or spacing of connections, are in some cases determined by judgement.

The dead load assigned to each level of the building is the weight of the roof or floor plus the walls from mid-storey to mid-storey (only walls for first level).

This gives values of $W_x$ as listed in Fig. 2.8.25, Column 3.

*Equivalent static loads*:

The building is assumed to be located in Ottawa, with $Z_a = 4$, $Z_v = 2$ and the zonal velocity factor v = 0.1.

The structural system is a box-type and, for example purposes, assume R = 1.5.

The fundamental period of vibration in the longitudinal direction of the building is:

$$T = \frac{0.09h_n}{\sqrt{D_s}} = \frac{(0.09)(15.0)}{\sqrt{27.6}} = 0.26\text{ s}$$

The corresponding Seismic Response Factor is:

$$S = 4.2 - (8.4)(T - 0.25) = 4.12$$

**Fig. 2.8.25**

| (1) Level | (2) $h_x$ (mm) | (3) $W_x$ (kN) | (4) $F_x$ (kN) | (5) $V_p$ (kN) |
|---|---|---|---|---|
| Roof | 15 000 | 3 690 | 983* | 258 |
| 4th | 11 700 | 3 998 | 831* | 280 |
| 3rd | 8 400 | 3 998 | 596* | 280 |
| 2nd | 5 100 | 4 182 | 379* | 293 |
| 1st | 600 | 783 | 8 | 55* |
| Totals | | 16 651 | 2797 | |

* These are used for design of diaphragms.

In the transverse direction, T = 0.30 s and S = 3.78.

For compact coarse grained soils F = 1.3, but FS ≤ 4.2 (when $Z_a$ is greater than $Z_v$) governs. The total dead load of the building is W = 16 651 kN, as shown in Fig. 2.8.25. The base shear is thus:

$$V = [(v \cdot S \cdot I \cdot F \cdot W)/R]U$$
$$= \frac{(0.1)(4.2)(1.0)(16\,651)}{1.5}(0.6)$$
$$= 2797 \text{ kN}$$

The base shear is distributed over the height of the building. In this example, the extra lateral force at the top level is $F_t = 0$, since T < 0.7 s. The distribution to the several levels is then:

$$F_x = \frac{(V - F_t)W_x h_x}{\left(\sum_{i=1}^{n} W_i h_i\right)} = \frac{2797 W_x h_x}{157\,508}$$

which gives the values of $F_x$ listed in Fig. 2.8.25, Column 4. These lateral loads are used for design of the building as a whole, like shear in and overturning of walls. They are also used for design of building parts unless they are exceeded by the loads developed in the following:

$$V_p = v\, S_p\, W_p$$

When applied to the design of the roof and floor diaphragms, use $S_p = 0.7$ which gives:

$$V_p = (0.1)(0.7)\, W_p = 0.07\, W_p$$

The resulting values of $V_p$ are listed in Fig. 2.8.25, Column 5. However, at each level the larger of the loads $F_x$ or $V_p$ must be used; this is indicated by asterisks in Fig. 2.8.25. Note that these loads include the effects of the weights of walls that are parallel to the direction of motion, and although this is not really required, the design is somewhat conservative.

*Shear wall-general*:

The exterior walls of the building provide the horizontal reactions for the diaphragms. These forces are in the plane of the wall and become horizontal shears, therefore the term shear wall. The end walls in the example building do double duty in that they are also bearing walls.

The design of the individual wall elements can be important, but will not be covered here. For the example, next examine the shear in the joints, and overall stability. The equivalent static loads, which are assumed to act horizontally at each floor level, cause a tendency for the wall to overturn. Gravity loads and footing will stabilize the wall and prevent overturning, but internal shear and direct stresses will develop. In this case, with T < 0.5 secs, the Code does not permit any reduction in overturning moment, and the full value of the cantilever moment will be used for each wall.

Results of the calculations are shown in Fig. 2.8.26 for one wall.

*Shear in shear wall*:

The maximum specified shear in the shorter wall due to the equivalent static loads and the minimum 10% eccentricity is:

$$V = (0.6)(2797) = 1678 \text{ kN}$$

The maximum factored load is (1.0)(1678) = 1678 kN, and the corresponding unit shear at the base of the wall is:

**Fig. 2.8.26**

| Level | 0.6 Σ $F_x$ (kN) | $M_x$ (kN · m) |
|---|---|---|
| Roof | | |
| | 590 | |
| 4th | | 1 947 |
| | 1088 | |
| 3rd | | 5 537 |
| | 1446 | |
| 2nd | | 10 309 |
| | 1673 | |
| 1st | | 17 838 |
| | 1678 | |
| Top of footing | | 18 845 |

$$v_f = 1678 / 20.4 = 82 \text{ kN/m}$$

This shear acts in both horizontal and vertical directions.

A typical connection for the vertical joints between wall elements is shown in Fig. 2.8.27. In each edge there is a small angle anchored with 2 - No. 15 bars. With 150 mm thick walls, it is possible to drypack the joints effectively. Apply the shear-friction concept, with the anchor bars placed at 90° to the joint at a distance of 50 mm from the interior face of the wall. The required spacing of the anchor bars is:

$$s \geq \frac{A_{vf} f_y f'_c}{b_v}\left(\frac{0.5\lambda\phi_c}{v_f}\right)^2$$
$$= \frac{(200)(400)(40)}{50}\left[\frac{(0.5)(1.0)(0.65)}{82/50}\right]^2$$
$$= 2500 \text{ mm}$$

Near the bottom, the average vertical spacing should be not more than 2500 mm. Use three connections in each storey. Since windows cross the joints, be sure to allow for this when locating the connections.

*Moment in shear wall:*

The factored moment at the top of the footing due to the specified horizontal forces with 10% arbitrary eccentricity is 18 845 kN · m. The factored gravity load is:

| | | |
|---|---|---|
| End wall | = (1.25)(20.4)(15.9)(3.2) | = 1297kN |
| 1/4 of roof | = (1.25)(20.4)(27.6)(4.8)/4 | = 845kN |
| 1/4 of roof | = (1.25)(20.4)(27.6)(5.3)(3)/4 | = 2798 kN |
| | Total W | = 4940 kN |

With wall ribs neglected, section properties in linear measure are, area $A_\ell = 20.4$ m and section modulus $S_\ell = (20.4)^2 / 6 = 69.4$ m². Fibre stresses at bottom of wall are:

$$f_f = \frac{W_f}{A_\ell} \pm \frac{M_f}{S_\ell} = \frac{4940}{20.4} \pm \frac{18\,845}{69.4} = \begin{matrix} 514 \text{ kN/m} \\ -29 \text{ kN/m} \end{matrix}$$

These values are a good measure of the forces in the gross wall section, and can also be used to calculate the necessary width of wall footing. Additional refinement of stress calculations is necessary at window openings.

**Fig. 2.8.27 Typical connection, wall to wall**

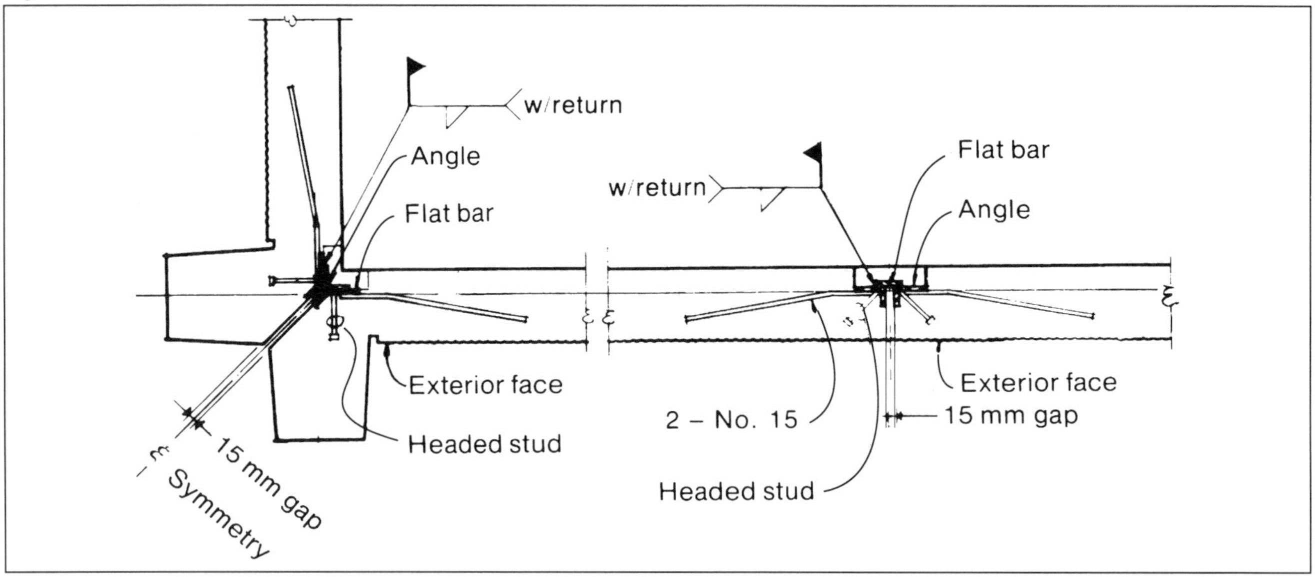


*Other shear walls*:

The long shear walls, which provide the resistance to earthquake in the longitudinal direction of the building, are less critical than the short walls, and they will not be discussed. Details of design will be the same as for the short walls.

There is a special situation at the building's entrance doors. The wall panel above the doorway is supported by connections to neighbouring units. The necessary resistance can be provided by two additional connections in each joint above the doorway.

*Connections of walls to footings*:

The footings can be proportioned on the basis of reactions from the walls, as indicated above. Each wall element can be connected to the footing by projecting the main reinforcing bars of the ribs downward into blockouts in the footings, and filling the blockouts with concrete. The space between wall element and footing must be dry-packed as shown in Fig. 2.8.28.

*Design of diaphragms*:

As pointed out in the one storey building example, the forces that can act on the diaphragm and its connections are determined by the probable resistance of the shear walls, and so depend only indirectly on the specified or factored horizontal forces. Since a detailed design is not presented here, it will be assumed that the walls have a probable resistance, based on the probable strength of the steel of 1.25 $f_y$, of 1.25 / $\phi_s$ = 1.25 / 0.85 = 1.5 times their required resistance to the factored forces. In a complete design, the probable resistance of each connection or element would be calculated and these values would be used to find the probable resistance of the walls.

**Fig. 2.8.28 Typical connection of wall to footing**

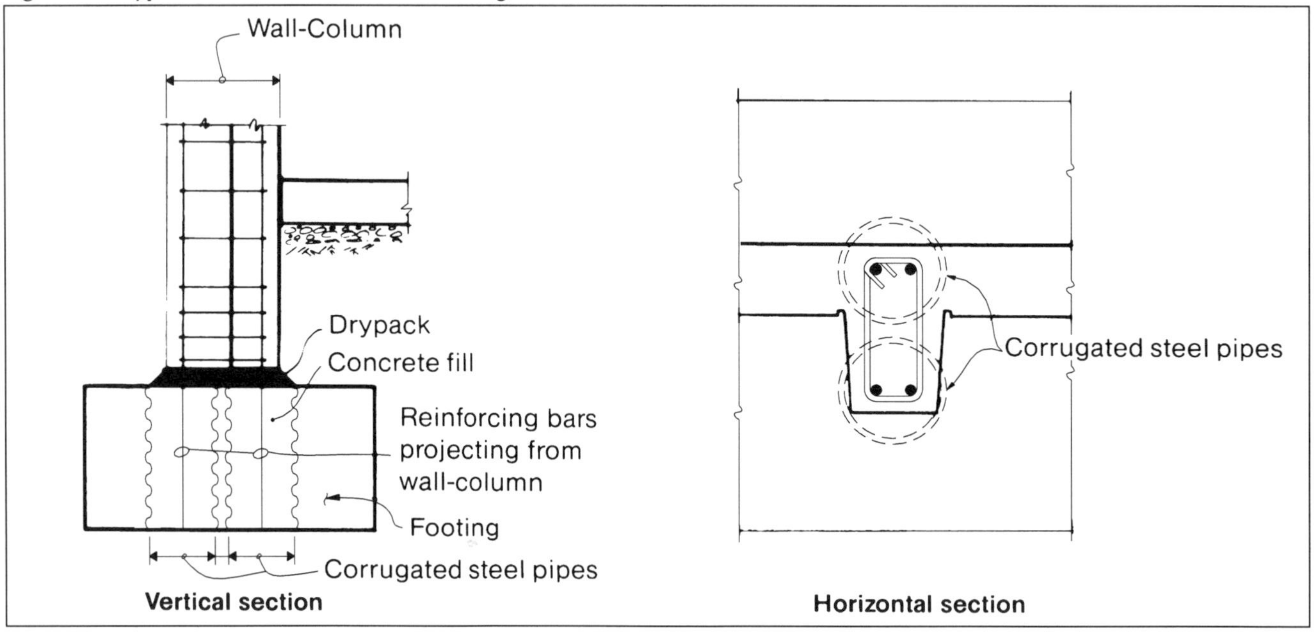

*Chord forces in diaphragm*:

The shear force acting between a wall and a diaphragm is assumed to be 1.25 / 0.85 times the factored force acting on the wall at that level. At roof level:

$$V_{pR} = (1.25 / 0.85)(0.6)(983) = 833 \text{ kN.}$$

The maximum bending moment in the diaphragm:

$$M_{pR} = \frac{2V_{pR}\ell}{8} = \frac{(2)(833)(27.6)}{8} = 5745 \text{ kN} \cdot \text{m}$$

Maximum chord forces at the North and South edges of the diaphragm:

$$C_{pR} = T_{pR} = \frac{M_{pR}}{d} = \frac{5745}{(20.4 - 0.6)} = 290 \text{ kN}$$

Design of the chord reinforcement:

$$A_s = \frac{T_{pR}}{\phi_s f_y} = \frac{(290)(10^3)}{(0.85)(400)} = 853 \text{ mm}^2$$

Use 2- No. 25 bars placed as shown in Figs. 2.8.29 and 2.8.30.

**Fig. 2.8.29 Typical connection between diaphragm and wall at end of double tee**

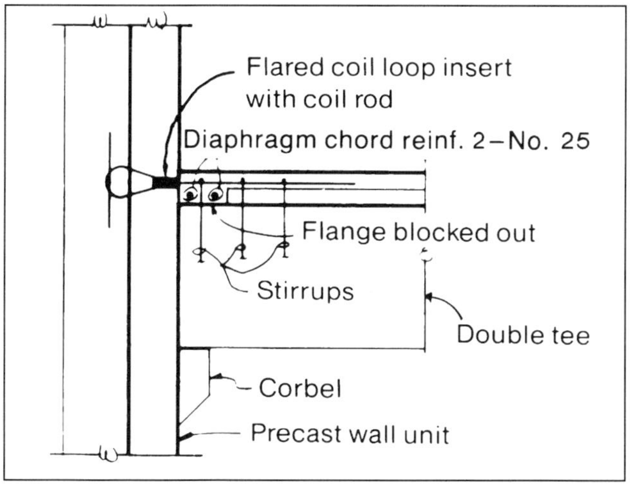


**Fig. 2.8.30 Typical connection between diaphragm and wall at edge of double tee**

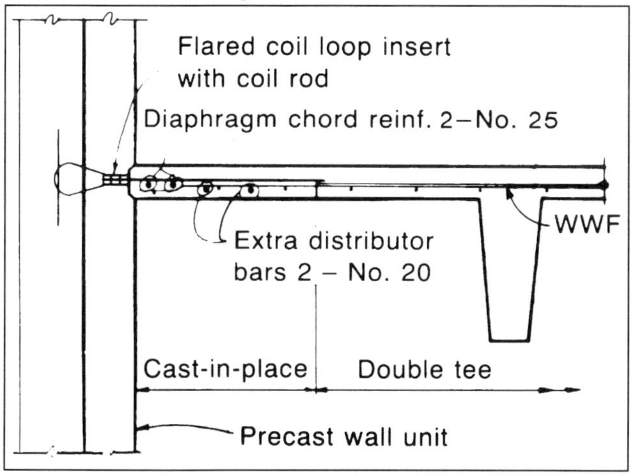


Requirements at lower levels and for the diaphragm spanning in the short direction are less, but for simplicity the same diaphragm chord reinforcement is used throughout.

*Shear stresses in diaphragm*:

This example building is assumed to be symmetrical, and the centres of mass and rigidity coincide. It is then necessary to include the effect of an arbitrary torsion, that is, the lateral load is offset from the centre of rigidity by 10 percent of the maximum dimension of the diaphragm.

The torsional moment is resisted by all four exterior walls, depending on rigidities and distances from the centre of rigidity. However, it is conservative to assume that the torsional moment is resisted by the two walls that are parallel to the direction of motion. For earthquake in the transverse direction, the maximum shear to one wall is $V_{fW} = 625$ kN (from the assumed probable strength of the wall). Deducting the length of the service shafts, the maximum unit shear is $v_{fW} = V_{fW} / \ell = 625 / (20.4 - 6.0) = 43.4$ kN/m.

For earthquake in the longitudinal direction, the maximum unit shear to the long walls is less.

For simplicity in production of elements and erection at the site, the 43.4 kN/m is used throughout all diaphragms and for their connections to the walls.

*Typical connections for interior of diaphragm*:

A typical connection transmitting shear between double tees is illustrated in Fig. 2.8.20. In each edge there is a small angle anchored with two reinforcing bars at 45°. The shear resistance is calculated from force components in the anchor bars, as in the previous example:

$$V_r = 96 \text{ kN}$$

To equalize cambers and deflections in neighbouring elements, a maximum spacing should not exceed about 2000 mm. This provides a resisting unit shear of $v_r = 96/2.0 = 48$ kN/m which compares to the requirement of 43.4 kN/m.

The cast-in-place strip along the longitudinal wall is reinforced with welded wire fabric with $A_s$ =184 mm²/m projecting out of the double tee flange. By shear-friction, this has a resistance of:

$$\begin{aligned} V_r &= 0.5\lambda\phi_c\sqrt{A_{vf} f_y f_c' A_{cv}} \\ &= (0.5)(1.0)(0.65) \\ &\quad \times\sqrt{(184)(400)(40)(1000)(50)} / 10^3 \\ &= 125 \text{ kN / m which is larger than that required.} \end{aligned}$$

Across the centre of the building, the double tee elements are supported by a beam, and they meet end to end. It is desirable to make the shear path stay at the level of the flange, that is, to make the path as direct as feasible. The connections to take care of this may be angle splices on reinforcing bars projecting out of the tops of the legs. The gap is filled with concrete, and shear-friction if applicable. Such a connection is shown in Fig. 2.8.31.

Fig. 2.8.31 Typical end connection between double tees

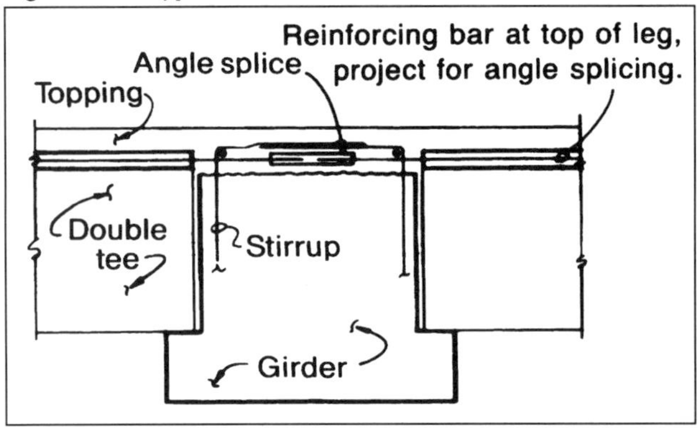


It should be noted that structural topping containing a rather light welded wire fabric could be used in place of the direct connections between neighbouring elements. Even so, it may be desirable to use some of the hardware-type connections to make the building resistant to lateral loads during construction.

*Connection of diaphragm wall:*

First consider the connections between the ends of the double tees and the short wall. The legs of the double tees and of the wall elements are aligned, and the vertical loads are carried by corbels projecting from the interior wall surface. Two requirements must be satisfied:

1. Horizontal shear between diaphragm and wall.
2. Direct horizontal tension caused by the tendency of the wall to fall away from the building.

The two requirements will be superimposed because earthquake motion may occur in any horizontal direction, and is not limited to the principal axes of the building.

It is convenient to space the connections 1200 mm apart so that they will occur at wall ribs. At each connection the forces are:

1. Horizontal shear:

$V_f = (43.4)(1200) / 10^3 = 52.1$ kN

This can be handled by shear-friction. Then the factored tensile force is:

$T_f = V_f / \mu = 52.1 / 1.0 = 52.1$ kN

2. Direct horizontal tension:

$V_p = v\, S_p\, W_p$

where:

$v = 0.10$

$S_p = 15.0$

$W_p = (4.3)(1200)(400) / 10^6 = 21.7$ kN

thus:

$V_p = (0.10)(15.0)(21.7) = 33$ kN

$T_f = (1.0)(33) = 33$ kN

A choice must be made of the type and size of connection. The coil-loop insert with continuous threaded coil rod is a simple and useful type, and there are several proprietary makes. The installation is illustrated in Fig. 2.8.29. A note of caution is in order: the coil-loop insert and the coil rod should each be amply anchored in their respective parts so that failure by yielding in the coil rod is assured.

The connections between the edge of the double tee and the long wall are of the same kind, and are calculated in the same way. The primary difference is caused by the cast-in-place strip of diaphragm adjacent to the wall. The threaded coil rods extend into this strip and overlap the welded wire fabric that projects out of the edge of the double tee flange. See Fig. 2.8.30.

*Special situation at service shafts*:

From the typical floor plan, Fig. 2.8.24, it is seen that several wall ribs are not tied to the floor, and the wall, because of vertical joints, is not suited to span horizontally past the shaft openings. This situation is readily alleviated with a strong-back type beam placed horizontally at each floor level spanning the shaft opening. The adjacent connections at each side of the shaft will have to be stronger than the standard connection previously designed.

## 2.8.14 EXAMPLE - MULTI-STOREY BUILDING

For box-type buildings over 60 m in height, the Code requires that the value of V shall be increased by 50 % in velocity-re-

Fig. 2.8.32 Multi-storey building - elevation

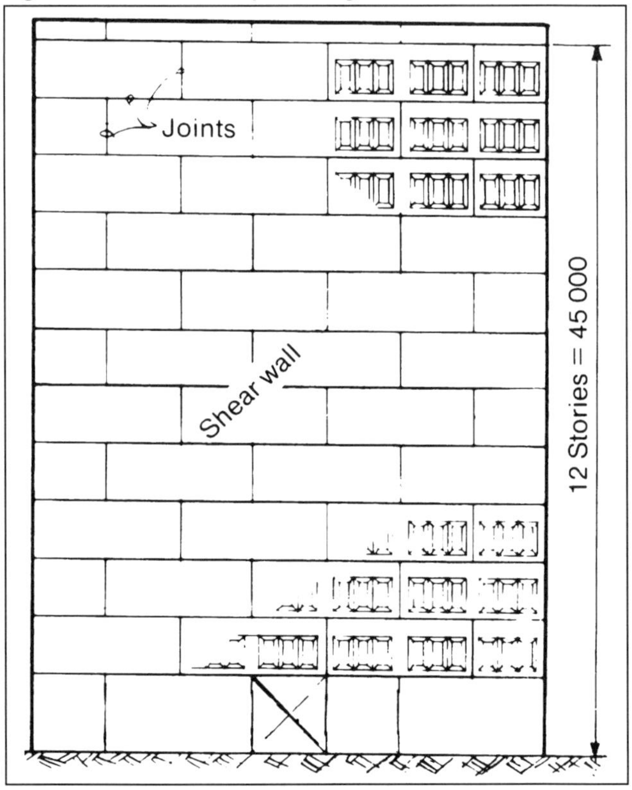

**Fig. 2.8.33 Typical connections for multi-storey example building - seismic loads**

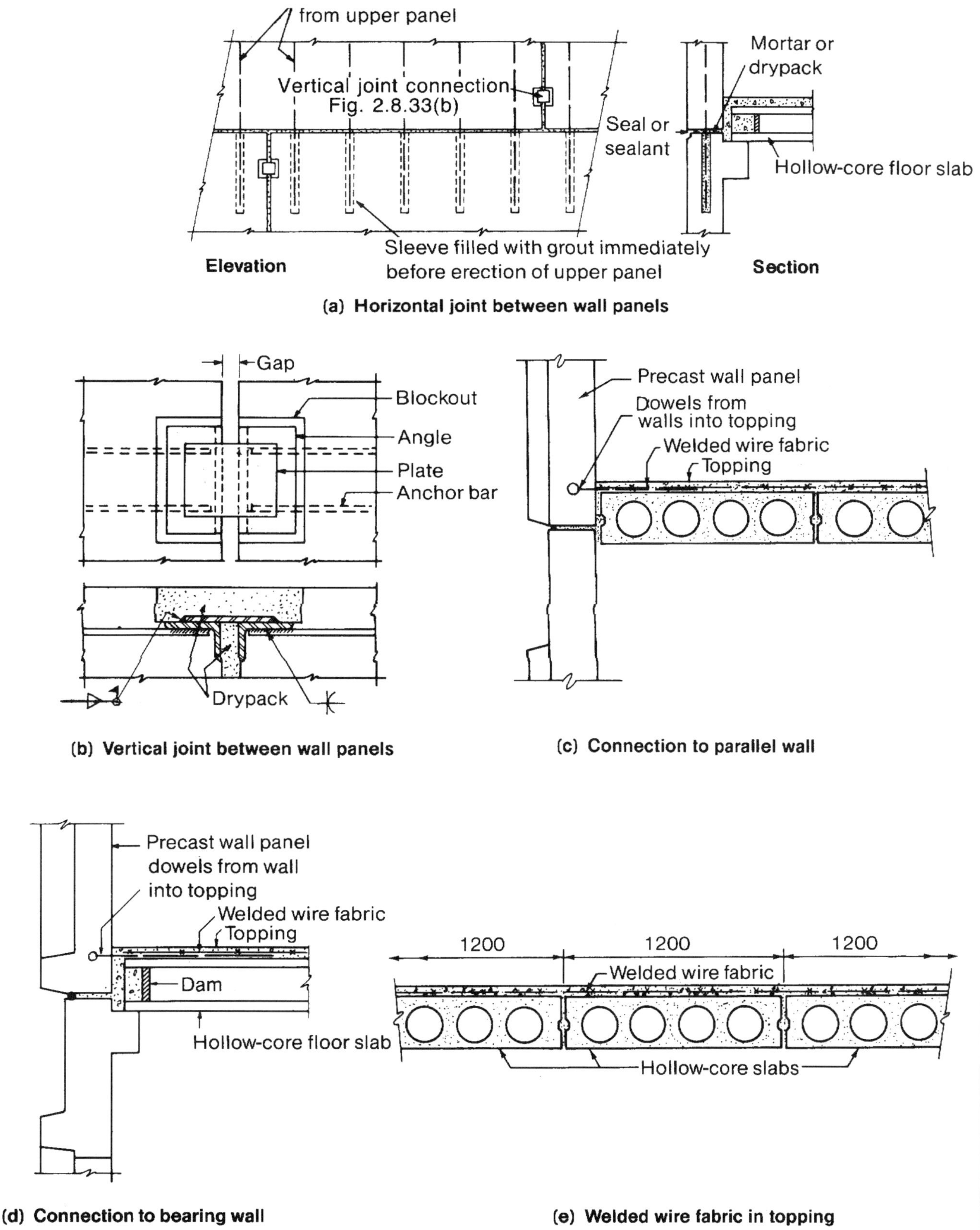

lated seismic zones of 4 and higher. For buildings approaching that height, the structural layout and connection details may differ from those of the preceding example, but the design procedure is the same. The base shear, its distribution as equivalent static loads over the height of the building, and the load paths down to the foundations, are calculated as illustrated in Sect. 2.8.13.

Floors consist of precast, prestressed hollow-core slabs, 300 mm deep and 1200 mm wide. They bear on continuous corbels on the walls and on the interior beams.

The connections shown in Fig. 2.8.33 are illustrations of methods of providing the diaphragms and continuous load paths discussed in the previous examples. Many other details have been used successfully on similar structures in high-intensity earthquake areas.

# 2.9 SEGMENTAL CONSTRUCTION

## 2.9.1 GENERAL

Segmental construction[R2.9.1] is defined as a method of construction for buildings and other structures in which primary load carrying elements are composed of individual segments post-tensioned together.

Segmental construction requires the Engineer to consider the following:

1. Choice of element size (dimension and weight).
2. Configuration and structural response of the joint between elements.
3. Construction sequence, loads and deflections imposed at various stages.
4. Recognition of tolerances and the effect of these upon the joint.

Segmental construction provides all the advantages associated with precast concrete and, in addition, permits the economical use of precast for the solution of specific problems:

1. Reduction of erection weights, permitting the use of smaller capacity erection equipment.
2. Ability to manufacture large elements in plants of limited size.
3. Multiple re-use of forms, permitting the economical construction of complicated shapes.
4. The ability of constructing deep elements (by keeping the width of individual elements within the allowable for transportation), and thus permitting the economical spanning of long distances.
5. Various structural configurations, such as trusses with inclined or parallel chords, may be readily constructed.
6. Temporary falsework can be minimized by use of cantilever construction for columns, similar to familiar bridge construction methods.

## 2.9.2 JOINTS AND JOINT CONSTRUCTION

Joints are of two types: either open, to permit completion by a field pour, or narrow, where the joint may be completed either dry or by use of a thin layer of adhesive (Fig. 2.9.1). Both methods, which have been successfully used in the past, are discussed below.

**Fig. 2.9.1 Types of joints**

Wide or Open Joints

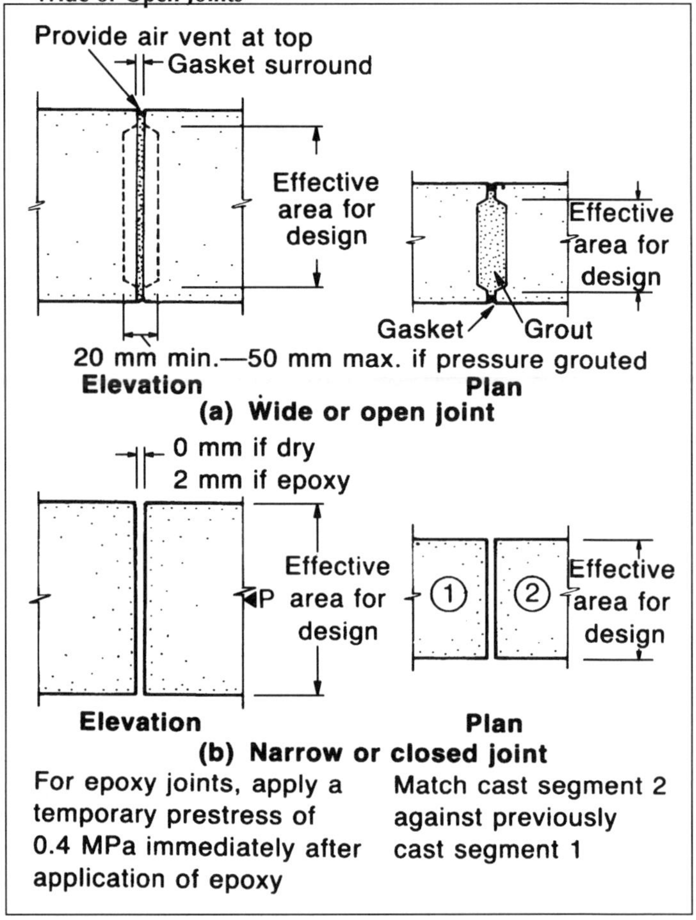


The width of these joints vary from 50 to 300 mm. They can be filled with concrete, grout or dry pack.

Concrete filled joints require forming around the joint and should be about 100 mm wide to allow proper concrete placement and compaction. High range water reducing agents mixed in the joint concrete will substantially reduce the problem of compaction. The joining of the tendon ducts can be well supervised and inspected with wide joints.

The compressive strength of the concrete at a specified age should be equal to the strength of the concrete in the adjacent precast segments. High early strength portland cement may be used. Aggregate size should be selected to ensure maximum compaction.

The height of each concrete placement or lift must be limited so that the concrete can be properly consolidated. Ports are normally provided for inspection.

Formwork must prevent leakage of concrete during and after its placement. Adequate curing is necessary to reach the design strength of the concrete.

Grouted joints are usually filled using pressure grouting. The perimeter of the joint is sealed with a compressible gasket, the compression achieved either by use of external devices or by introducing a small amount of prestress. Gaskets are also provided around the individual post-tensioning elements to prevent leakage into the ducts, blocking passage of the tendons. The joint detail should provide for vents at the top to permit escape of entrapped air during grouting. At the conclusion of the grouting operation, vents should be closed and pressure increased to a minimum of 0.1 MPa at the vent, so as to ensure full grout intrusion. Within a few days after grouting, vents should be re-opened and, if consolidation is evident, the vent filled.

The width of the pressure grouted joint should be not greater than 50 mm. The compressive strength of the grout should equal that of the concrete in adjacent segments, but not less than 30 MPa. Admixtures, such as water reducing agents or expansive agents, if used, should be of the non-staining type.

Dry-packed joints are used in conjunction with smaller elements having good access to all parts of the joint. Dry packing allows earlier application of the final post-tensioning.

Dry-pack mortar must have a compressive strength equal to the concrete in adjacent segments, or at least 30 MPa. Good mortar should be thoroughly mixed and have zero slump. Maximum aggregate size normally does not exceed 5 mm. Mortar should be rammed into place using a heavy hammer and a wood ram. The width of dry-packed joints should not exceed approximately 50 mm. Mortar should be introduced into the joint in small quantities or batches not exceeding a mass of 5 kg. Each batch must be thoroughly tamped and packed before the next batch is placed. Containment may be necessary, particularly at the bottom of the joint.

All wide joints require preparatory treatment of the joint surfaces prior to filling. The joint surfaces must be clean, free from grease and oil, etc. and preferably wire brushed or sandblasted. Prior to construction of the joint, the adjacent concrete surface should be kept thoroughly wet for approximately 6 hours, or bonding agent should be applied.

### Narrow or Closed Joints

Narrow joints can be dry or epoxy bonded. The joint width varies from practically zero for dry joints to about 2 mm for epoxy bonded joints.

In both cases, a perfect match of the joint surfaces is essential. This can be achieved through match-casting or the use of precision steel bulkheads.

Match-casting requires that the new segment is cast against its previously cast neighbour. The use of bond breaking agents allows the segments to be separated. Match-casting involves careful organization and additional handling of the segments.

The use of precision bulkheads provides more flexibility in production and ties up less floor space, but its success hinges on machined steel bulkheads fabricated to very tight tolerances.

Match-cast segments are normally joined by coating abutting surfaces with a thin (1mm) layer of epoxy adhesive, and then drawing together and holding in position the precast elements. An epoxy is not always necessary, however, it does provide the following advantages:

1. During placement, the epoxy will act as a lubricant, which will assist in the alignment process.
2. The epoxy will even out any small contact surface differences and thus provide a uniform bearing surface.
3. The epoxy will provide a joint with tensile capacity greater than the concrete, and thus assure monolithic behaviour.
4. The epoxy will provide water tightness and durability at the joints.

Epoxy bonding agents for match cast joints should be thermosetting, 100% solid compositions that do not contain solvents or any non-reactive organic ingredient, except for pigments required for colouring. Epoxy bonding agents should be of two components, a resin and a hardener. The two components should be distinctly pigmented, so that mixing produces a third colour similar to the concrete in the segments to be joined, and should be packaged in pre-portioned, labeled, ready-to-use containers.

Epoxy bonding agents should be formulated to provide application temperature ranges which will permit erection of match-cast segments at ambient temperatures from 5°C to 45°C. If two surfaces to be bonded have different temperatures, the adhesive applicable at the lower temperature should be used.

If the project requires or would benefit from erection at a concrete temperature lower than 5°C, the temperature of the concrete within the vicinity of the joint should be elevated to at least 5°C to ensure effective wetting of the surface by the epoxy compound in a reasonable length of time. An artificial environment will have to be provided to accomplish this elevation in temperature, and should be created by an enclosure heated by circulating warm air or by radiant heaters. In any event, localized heating should be avoided, and the heat should be provided in a manner that prevents surface temperatures greater than 45°C during the epoxy hardening process. Direct flame heating of concrete surfaces should be prohibited.

Surfaces of the match-cast joints to be bonded should be sound, and clean. All traces of mould release agents, curing compounds, laitance, oil, dirt and loose concrete should be removed from surfaces to be bonded by proper washing or sandblasting.

The concrete surfaces which are to be bonded should not be wet; a damp but not saturated surface is permissible. To get rid of a wet surface, the concrete may be dried with hot air just before applying the adhesive.

The adhesive should be applied in a uniform thickness to both surfaces. Care must be taken that no epoxy mix enters the ducts for the tendons. After joining the segments, the ducts must be checked to be sure they are not blocked by the adhesive. Some post-tensioning should be applied within 45 minutes after application of the epoxy material and a minimum average temporary prestress of 0.4 MPa over the cross-section should be applied within 70% of the open time of the epoxy material, and should be maintained until the permanent tendons are stressed. At no point of the cross-section should the temporary prestress be less than 0.2 MPa. If the correct amount of adhesive has been used, a small amount will extrude from the joint when pressure is applied.

In case of unforeseen interruptions, and 70% of the open time expires before the segments are fully joined (an approximate determination of the open time can be noted from behaviour of lap joint samples spread on small cement-asbestos boards), the epoxy should be removed and any remainder washed off with solvent, according to the instructions of the manufacturer, followed by sandblasting. Particular care is required in cold weather.

Epoxy bonding agents should be relatively insensitive to damp conditions during application and, after curing, should exhibit high bonding strength to cured concrete, good water resistivity, low creep characteristics, and tensile strength greater than the concrete. The compressive strength of the epoxy should equal that of the concrete in adjacent segments under any environmental condition that may be encountered during the life of the structure.

Epoxy bonding agents should be tested to determine their workability, set time, open time, bond, compression and shear strength, and working temperature range. The frequency of tests should be stated in the Contract.

If desired, test specimens can be made at the site to later verify the properties of the epoxy. These may include prisms of adhesive to test the compressive strength, and diagonally-cut concrete cylinders or prisms bonded together with the epoxy to indirectly test shear strength by compressive loading. If necessary, concrete cores may be drilled through the bond line and tested in compression.

Dry joints are fast to handle during construction and, providing there is a perfect match of segments, very little can go wrong. Post-tensioning should be designed in such a way that after all losses a compressive stress of about 1.0 MPa remains in all dry joints.

#### Joint Surfaces

Requirements concerning surface quality must be stricter for narrow joints than for wide joints filled with mortar or concrete.

*Orientation*:

Surfaces should be oriented perpendicular to the main post-tensioning tendons, to minimize shearing forces and dislocation in the plane of the joint during post-tensioning. Inclination with respect to a plane perpendicular to the longitudinal axis is permitted for joints with assured friction resistance. The inclination should generally not exceed 20°. Larger inclination, but not more than approximately 30°, may be permitted if the inclined surface area is located close to the neutral axis and does not exceed 25 percent of the total joint's surface area.

*Quality*:

For wide joints, rough surfaces are preferable, as they produce better bond between segment and filling material. Since it is difficult in normal practice to produce perfect sharp edges, it is advisable to make joint surface edges slightly rounded or chamfered. Although this will tend to make joints visible, it will also reduce the contrast if neighbouring segments have slight colour variance. Rounding or chamfering of edges should not decrease the joint surface area by more than approximately 2.5 percent.

For narrow joints, the surface, including formed keys, should be even and smooth, to avoid point contact and surface crushing or chipping off of edges during post-tensioning.

#### Holes for Tendons and Couplers

Holes or sheathing for tendons must be located very precisely, such as by using steel templates or precision steel bulkheads, when producing segments joined by post-tensioning. Care is required to prevent leakage or penetration of joint-filling materials into the duct, thus blocking passage of the tendons.

Joints are always crossed by post-tensioning ducts and tendons. Two basic methods are known:

1. Tendons in ducts through the segments.
2. Tendons placed outside the segment.

The second case does not constitute any problems for the joints. The first case is discussed in detail in the PTI Post-Tensioning Manual.[R2.9.2]

### 2.9.3 DESIGN CONSIDERATIONS

Design procedures for precast segmental structures are essentially the same as for monolithic prestressed concrete structures.[R2.9.3]

Settlement and shortening of scaffolding due to dead load of segments as well as construction loads must be considered. Segments need to be carefully aligned and leveled before forming joints and post-tensioning.

Shortening of the segments and jointing materials due to temperature, settlement, or change in loading conditions should be checked before post-tensioning of the structure. If joints separate due to the above causes, the post-tensioning may cause uneven distribution of stresses or grout leakage.

Elements which are post-tensioned require the following additional considerations:

#### Flexural Stresses

1. For elements which are assembled using a dry joint, tension should not be permitted between segments under any stage of erection or service loading.
2. For elements which are assembled using an epoxy joint, the allowable tensile stress at the joint may be considered the same as that for monolithic construction.

3. For elements which are assembled using a grout joint, the allowable tensile stress at the joint should be taken as zero unless tests indicate otherwise.

### Joint Shear

With joints composed of an epoxy or grout with a tensile strength at least equal to the concrete, the element may be designed as a monolith. When a dry joint is used, shear should be considered transferred by friction, using a friction factor of 0.7. (This value is given as a guide. The actual value to be used requires engineering judgement.) At working loads, care should be taken that there is no relative displacement at the joint. At ultimate, the available shear resistance $V_r$ should be at least equal to $V_f$, where $V_f$ is the factored shear.

### Reinforcement

Reinforcement to at least meet minimum requirements for flexural elements in accordance with the Code, is normally required in segments for:

1. Transverse bending moments.
2. Shear.
3. Torsion due to eccentric loading.
4. Thermal and volume change forces.
5. Temporary forces imposed during fabrication, transportation, or erection.

### Bearing and Anchorage

Bearing areas are subjected to large concentrated forces; they may also need to accommodate substantial movement due to volume changes and thermal effects.

Generally, four different areas of diagonal splitting and cracking can be identified in anchorage or bearing areas which will require reinforcement:

1. Under end surfaces, not more than 20 mm deep, to control possible surface cracking around anchorages.
2. Internally, to prevent splitting of separate anchorages. Size and location of this area and of the magnitude of splitting (bursting) force depends on the type of anchorage and the force in the post-tensioning tendon, and should be investigated individually for each type of anchorage.
3. Internally, to prevent splitting between groups of anchorages (not distributed on bearing surface uniformly). For each case, the splitting force and necessary reinforcement should be defined separately for the vertical and horizontal directions.
4. To decrease the possibility of damage to segments during post-tensioning due to unintended stress concentrations, or during handling of segments, supplemental reinforcement may be desirable immediately adjacent to the joint surface.

Concrete which is placed around anchorages, after post-tensioning, should be provided with reinforcement to ensure that it will not spall.

### Couplers

Couplers should be designed to develop the full ultimate resistance of the tendons they connect. Adjacent to the coupler, the tendons should be straight for a minimum length of 12 times the diameter of the coupler. Adequate provisions should be made to ensure that couplers can move during prestressing. It is particulary important that the void areas around a coupler be deducted from gross section areas when computing stresses at the time of prestressing.

### Deflections

Deflection calculations should consider dead load, prestress, erection loads, concrete creep and shrinkage, and steel relaxation. For joints composed of an epoxy or grout with a tensile strength at least equal to the concrete, the elements may be considered monolithic. When a dry joint is used, deflection should be computed based on a bilinear moment-deflection relationship, using a transformed cracked section when tensile stresses exist in the precompressed tensile zone.

## 2.9.4 POST-TENSIONING CONSIDERATIONS

Background information on various post-tensioning systems and their applications is given in Chap. 1 and 8 and in the PTI Post-Tensioning Manual.[R2.9.2]

### References:


R2.5.1 Speyer, I.J., Considerations for the Design of Precast Concrete Bearing Wall Buildings to Withstand Abnormal Loads, PCI Journal, Vol. 21, No. 2, Mar-Apr 1976, pp. 18-51.

R2.6.1 Martin, L.D., Background and Discussion on PCI Design Handbook Second Edition, PCI Journal, Vol. 25, No. 1, Jan-Feb 1980, pp. 24-41.

R2.6.2 Temperature change is, of course, a reversible effect; increases cause expansion and are important in design and location of expansion joints (see Sect. 2.3 ). Temperature differential in roof and wall elements should also be considered.

R2.6.3 Gouwens, A.J., Lateral Load Analysis of Multistorey Frames with Shear Walls, Bulletin AEC2, Portland Cement Association Computer Program.

R2.6.4 MacGregor, J.G., and Hage, S.E., Stability Analysis and Design of Concrete,Proceedings, Journal of the Structural Division, ASCE, October, 1977.

R2.7.1 Speyer, I.J., Consideration for the Design of Precast Concrete Bearing Wall Buildings to Withstand Abnormal Loads, PCI Journal, vol. 21, No. 2, Mar-Apr 1976, pp. 18-51

R2.8.1 Source: The National Building Code of Canada.

R2.8.2 Design Considerations for Precast Prestressed Apartment Building, Prestressed Concrete Institute,, 1975, 224 pp.

R2.8.3 Kahn, L.F., Reinforced Concrete Infilled Shear Walls for Aseismic Strengthening, Report UMEE 76R1, University of Michigan, Ann Arbor, Jan. 1976.

R2.8.4 Spencer, R.A. and Tong, W.K.T., Design of a One-story Precast Concrete Building for Earthquake Loading, Proc., Eighth World Conf. on Earthquake Engineering, San Francisco, 1984, Vol. V, pp. 653-660.

R2.9.1 PCI Committee on Segmental Construction, Recommended Practice for Segmental Construction in Prestressed Concrete, PCI Journal, Vol. 20, No. 2, Mar-Apr 1975, pp. 22-41.

R2.9.2 Post-Tensioning Manual, Fourth Edition, Post-Tensioning Institute Phoenix, Arizona, 1985.

R2.9.3 Martynowicz, A. and McMillan, C.B. , Large Precast Prestressed Vierendeel Trusses Highlight Multistorey Building, PCI Journal, Vol. 20, No. 6, Nov-Dec 1975, pp. 50-65.

# CHAPTER 3
# DESIGN OF ELEMENTS

## 3.0 NOTATION

$A$ = effective tension area of concrete surrounding the flexural tension reinforcement

$A_c$ = concrete cross sectional area, not including area of reinforcement

$A_{comp}$ = cross sectional area of the equivalent rec-tangular stress block

$A_{cr}$ = area of cracked section

$A_{cv}$ = area of concrete section resisting shear transfer

$A_g$ = gross area of section

$A_o$ = area enclosed by shear flow path

$A_{oh}$ = area enclosed by stirrup centreline

$A_p$ = area of prestressed reinforcement in tension zone

$A'_p$ = area of prestressed reinforcement in compression zone

$A_r$ = concrete area in tension below the neutral axis

$A_s$ = area of non-prestressed tension reinforcement

$A'_s$ = area of non-prestressed compression reinforcement

$A_{st}$ = $A_p + A_s E_s/E_p$

$A_t$ = area of one leg of closed stirrup

$A_{top}$ = effective area of cast-in-place composite topping

$A_{tr}$ = area of uncracked transformed section

$A_v$ = area of shear reinforcement

$A_{vf}$ = area of shear friction reinforcement

$A_{v\ell}$ = area of shear ties

$a$ = depth of equivalent rectangular stress block

$a$ = width of a panel

$b$ = length of a panel

$b$ = width of compression face of element

$b_t$ = width of tension zone of section

$b_v$ = width of cross section at contact surface being investigated for longitudinal shear

$b_w$ = minimum effective web width within depth d

$C$ = compressive force

$C$ = cross sectional constant used in definition of equivalent column torsional properties

$C_c$ = compressive force capacity of composite topping

$C_m$ = factor relating the actual moment diagram to an equivalent uniform moment diagram

$C_s$ = shrinkage coefficient

$C_t$ = creep coefficient

$C_u$ = ultimate creep coefficient for standard conditions

$c$ = distance from extreme compression fibre to neutral axis

$c_1$ = width of column in direction moments taken

$c_2$ = width of column in direction perpendicular to moments taken

$c_u$ = neutral axis depth assuming $f_{pr} = f_{pu}$

$c_y$ = neutral axis depth assuming $f_{pr} = f_{py}$

$D$ = dead load

$d$ = distance from extreme compression fibre to centroid of prestressed and non-prestressed tension reinforcement

$d$ = distance from extreme compression fibre to centroid of non-prestressed tension reinforcement

$d'$ = distance from extreme compression fibre to centroid of non-prestressed compression reinforcement

$d_b$ = nominal diameter of reinforcing bar or prestressing strand

$d_c$ = concrete cover to centre of reinforcement

$d_p$, $d'_p$ = distance from extreme compression fibre to centroid of stressed reinforcement

$d_v$ = effective shear depth; taken as flexural lever arm but need not be taken less than 0.9 d

$E$ = modulus of elasticity

$E_c$ = modulus of elasticity of concrete

$E_{cc}$ = modulus of elasticity of concrete for column

$E_{co}$ = modulus of elasticity of concrete at time of prestress transfer

$E_{cs}$ = modulus of elasticity of concrete for slab

$E_p$ = modulus of elasticity of prestressed reinforcement

$E_s$ = modulus of elasticity of non-prestressed reinforcement

$e$ = eccentricity of design load or prestress force parallel to axis measured from the centroid of the section

$e'$ = distance between centroid of prestress force at end and centroid of prestress force at lowest point $= e_c - e_e$

$e_c$ = eccentricity of prestress force from the centroid of the section at the centre of the span

$e_e$ = eccentricity of prestress force from the centroid of the section at the end of the span

$e_{tr}$ = eccentricity of prestress force in uncracked transformed section

$e_x$ = eccentricity of prestress force from x axis

$e_y$ = eccentricity of prestress force from y axis

$F$ = force as defined in section used (with subscripts)

$F_{eq}$ = equilibrium forces in anchorage zone

$F_{sp}$ = spalling force

$f_{all}$ = allowable compressive stress

$f_b$ = stress in the bottom fibre of the cross section

$f_c$ = unit stress in concrete

$f'_c$ = specified compressive strength of concrete

$f'_{cc}$ = specified compressive strength of composite topping

$f_{ce}$ = stress in concrete at the level of prestressing tendons after all losses

$f_{ce}$ = compressive stress in concrete due to prestress only after all losses, at the extreme fibre of a section at which tensile stresses are caused by applied loads

$f_{ci}$ = compressive stress in concrete at time of prestress transfer

$f'_{ci}$ = compressive strength of concrete at time of prestress transfer

$f_{co}$ = concrete compressive stress at centroid of tendon at critical section immediately after transfer

$f_{cp}$ = compressive stress in concrete (after allowance for all prestress losses) at the centroid of the cross section

$f_{c1}$ = concrete stress at centroid of tendon at the critical section caused by sustained loads not included in the calculation of $f_{co}$ (tension negative)

$f_\ell$ = calculated stress due to live load

$f_{pe}$ = effective stress in prestressed reinforcement after allowance for all prestress losses

$f_{pi}$ = stress in tendon after transfer

$f_{po}$ = stress in pretensioned tendon immediately before transfer

$f_{pp}$ = stress in tendon at a distance y from the jacking end

$f_{pr}$ = stress in prestressed reinforcement

$f_{pr}$ = stress in prestressed reinforcement at factored resistance

$f_{pu}$ = tensile strength of tendon

$f_{px}$ = stress in tendon at a distance x from the jacking end

$f_{py}$ = yield strength of tendon

$f_r$ = modulus of rupture of concrete

$f'_r$ = allowable flexural tension stress computed on the basis of gross concrete section

$f_{re}$ = intrinsic relaxation loss in prestressing steel

$f_s$ = stress in the non-prestressed tension reinforcement due to specified loads

$f_s$ = increase in stress in the prestressed and non-prestressed reinforcement beyond state of decompression

$f'_s$ = stress in the non-prestressed compression reinforcement due to specified loads

$f_{sd}$ = stress due to service dead load

$f_t$ = stress in the top fibre of the cross section

$f_{t\ell}$ = final total stress in the element

$f_y$, $f'_y$ = specified yield strength of non-prestressed reinforcement

$h$ = overall depth of element

$h_c$ = centre to centre column height

$h_f$ = thickness of compression flange of an element

$h_1$ = distance from centroid of tensile reinforcement to neutral axis

$h_2$ = distance from extreme tensile fibre to neutral axis

$I$ = moment of inertia

$I_c$ = moment of inertia of concrete section

$I_{cr}$ = moment of inertia of the cracked trans-formed section

$I_e$ = effective moment of inertia

$I_g$ = moment of inertia of gross concrete section

$I_{minor}$ = moment of inertia about the minor axis

$I_{major}$ = moment of inertia about the major axis

$I_s$ = moment of inertia of reinforcement

$I_{tr}$ = moment of inertia of uncracked transformed section

$I_x$ = moment of inertia about x axis

$I_y$ = moment of inertia about y axis

$I_{xy}$ = $A_g (x - x_L) (y - y_b)$

$J$ = polar moment of inertia

$j$ = ratio of distance between centroid of compression and centroid of tension to the depth, d, for a cracked section at service moment

$K$ = wobble coefficient for a post-tensioning tendon

$K_c$ = flexural stiffness of column

$K_{ec}$ = flexural stiffness of equivalent column

$K_r$ = moment coefficient

$K_s$ = flexural stiffness of slab

$K_t$ = torsional stiffness of torsion element

$k$ = effective length factor for compression elements

$k$ = ratio of depth to the neutral axis to the depth, d, for a cracked section at service moment

$k_c$ = stiffness coefficient

$k_1$, $k_p$,

$k_r$ = factor for type of prestressing steel

$k_p$ = coefficient used in calculating critical load, $P_c$

L = live load due to intended use

$\ell$ = span length

$\ell_1$ = span in direction moments considered

$\ell_2$ = span in direction perpendicular to moments considered

$\ell_b$ = length of bar

$\ell_c$ = vertical distance between supports

$\ell_d$ = development length

$\ell_e$ = length of tendon between anchors divided by the number of plastic hinges required to develop a failure mechanism in the span under consideration

$\ell_n$ = length of clear span of element

$\ell_u$ = unsupported length of a compression element

$\ell_v$ = horizontal shear length as defined in Fig. 3.6.6

M = moment due to specified loads

M′ = secondary moment

$M_a$ = total moment at the section

$M_{bal}$ = balance load moment in post-tensioned element

$M_c$ = moment used for design of compression elements

$M_{cr}$ = cracking moment

$M_{CS}$ = moment in column strip

$M_d$ = moment due to specified dead load

$M_{dc}$ = decompression moment

$M_{DS}$ = moment in design strip

$M_f$ = moment due to factored loads

$M_\ell$ = moment due to specified live loads

$M_n$ = net moment at a section

$M_{n\ell}$ = net load moment in post-tensioned element

$M_{ns}$ = moments from a first order analysis due to gravity loads, which do not contribute to sway

$M_{ocr}$ = pure flexural cracking moment

$M_r$ = factored flexural resistance

$M_{ro}$ = factored flexural resistance of a compression element with zero axial load

$M_{rb}$ = factored flexural resistance under balanced conditions

$M_s$ = factored sway moment

$M_s$ = moment due to all specified loads

$M_{sd}$ = moment due to all sustained load except element mass

$M_{sw}$ = moment due to self weight

$M_v$ = moment transferred by eccentricity of shear

$M_{vcr}$ = total moment due to factored loads plus eccentric shear

$M_x$, $M_y$,

$M_z$ = service load moments about x, y and z axes

$M_1$ = smaller factored end moment on a compression element, positive if bent in single curvature, negative if double curvature

$M_2$ = larger factored end moment, always positive

$M_{2ns}$ = factored end moment on a compression member at the end at which $M_2$ acts, due to loads that cause no appreciable sway, calculated using a fist-order elastic frame analysis

$M_{2s}$ = factored end moment on a compression member at the end at which $M_2$ acts, due to loads that cause appreciable sway, calculated using a fist-order elastic frame analysis

m = unfactored mass

N = unfactored axial load

N′ = allowable axial load

$N_f$ = factored axial load occuring with $V_f$

$N_v$ = equivalent factored axial load caused by shear and torsion

n = modular ratio

n = number of bars

$n_o$ = $E_p/E_{co}$

$n\alpha$ = $\dfrac{M_s}{bd^2}\dfrac{E_p}{E_c}$

$n\rho$ = $\dfrac{A_pE_p + A_sE_s}{E_cbd}$

$n_1$ = $E_p / E_{c1}$

$P_{bu}$ = final force in upper layer of bottom prestressed reinforcement

$P_{b\ell}$ = final force in lower layer of bottom prestressed reinforcement

$P_c$ = critical axial load

$P_c$ = shrinkage coefficient to account for cement content

$P_{dc}$ = decompression force

$P_e$ = prestress force after all losses

$P_f$ = shrinkage coefficient to account for fine to total aggregate ratio

$P_f$ = factored axial load

$P_h$ = shrinkage coefficient to account for relative humidity

$P_i$ = prestress force immediately after transfer

$P_o$ = initial prestress force immediately before transfer

$P_R$ = anchorage zone equilibrium force reaction

$P_r$ = factored axial load resistance

$P_r$ = shrinkage coefficient to account for volume to surface ratio

$P_{rb}$ = factored axial resistance under balanced conditions

$P_{ro}$ = factored axial resistance of a compression element with zero eccentricity

$P_s$ = force in non-prestressed reinforcement due to time dependent strain

$P_{sh}$ = long term shrinkage modification factor

$P_{s\ell}$ = shrinkage coefficient to account for slump

$P_t$ = final force in top prestressed reinforcement

$P_v$ = shrinkage coefficient to account for air content

$p$ = $P_{dc}\,d_p/M_s$

$p_c$ = outside perimeter of cross section

$p_h$ = perimeter of stirrup centreline

$p_o$ = perimeter of shear flow path

$Q$ = stability index of a storey

$Q_a$ = creep coefficient to account for concrete age at time of loading and curing conditions

$Q_{cp}$ = creep modification factor for non-standard conditions

$Q_f$ = creep coefficient to account for the fine to total aggregate ratio

$Q_h$ = creep coefficient to account for relative humidity

$Q_r$ = creep coefficient to account for the volume to surface ratio

$Q_{s\ell}$ = creep coefficient to account for concrete slump

$Q_v$ = creep coefficient to account for concrete air content

$r$ = radius of gyration

$S$ = section modulus

$S_b$ = section modulus with respect to the bottom fibre of a cross section

$S_t$ = section modulus with respect to the top fibre of a cross section

$s$ = shear or torsion reinforcement spacing in a direction parallel to the longitudinal reinforcement

$T$ = tensile force

$T$ = cumulative effects of temperature, creep, shrinkage and differential settlement

$T_{cr}$ = pure torsional cracking resistance

$T_f$ = factored torsional moment on a section

$t$ = thickness

$t$ = time

$V_c$ = factored shear resistance of the concrete

$V_{cw}$ = factored shear resistance provided by concrete when diagonal cracking results from excessive principal tensile stress in web

$V_f$ = factored shear force

$v_f$ = factored shear stress

$V_{f\ell}$ = factored longitudinal shear force

$V_p$ = component in the direction of the applied shear of the effective prestressing force or, for variable depth members, the sum of the component of the prestressing force and the components of flexural compression and tension in the direction of the applied shear, positive if resisting applied shear

$V_r$ = factored shear resistance

$V_{r\ell}$ = factored longitudinal shear resistance

$V_s$ = factored shear resistance provided by shear reinforcement

$v_s$ = factored shear stress resisted by shear reinforcement

$V_{se}$ = shear due to specified loads

$W$ = live load due to wind or internal moment or forces

$w$ = specified load per unit length of beam or per unit area of slab

$w$ = maximum crack width at extreme tension fibre

$w_d$ = specified dead load per unit length

$w_\ell$ = specified live load per unit of length

$w_{n\ell}$ = net service load per unit of length (unfactored)

$w_p$ = vertical distributed load exerted on an ele-ment by a post-tensioning tendon

$w_{sd}$ = specified superimposed dead load

$x$ = anchorage length of a tension tie of torsion resisting cross section

$x_L$ = distance from left edge to y axis

$y$ = larger overall dimension of rectangular part of torsion resisting cross section

$y'$ = distance from top to centroid of $A_{comp}$

$y_b$ = distance from bottom fibre to centroid of the section

$y_{cr}$ = distance to tension fibre being considered from centroid of cracked section

$y_t$ = distance from top fibre to centroid of cracked section

$y_{tr}$ = distance to extreme tension fibre from centroid of uncracked transformed section

Z = total bursting force

z = a quantity limiting distribution of flexural reinforcement

$\alpha$ = angle change in post-tensioning tendon

$\alpha$ = load factor (with subscripts) as defined in Fig. 3.2.1

$\alpha_1$ = ratio of average stress in rectangular compression block to the specified concrete strength

$\alpha_b$ = $b_w/b$

$\alpha_f$ = $h_f/h$

$\alpha_r$ = relaxation reduction coefficient (Fig. 3.4.4)

$\beta$ = angle used in post-tensioning friction calculations

$\beta$ = factor accounting for shear resistance of cracked concrete

$\beta_1$ = ratio of depth of rectangular compression block to depth to the neutral axis

$\beta_d$ = ratio of factored dead load moment to total load moment

$\beta_d$ = ratio of the maximum factored sustained shear within a storey to the maximum factored shear in that storey

$\gamma$ = importance factor

$\Delta$ = deflection (with subscripts)

$\Delta$ = post-tensioning anchorage set

$\Delta f_p$ = change in prestress due to creep, shrinkage and relaxation

$\Delta f_{po}$ = change in prestress due to elastic loss

$\Delta f_s$ = change in stress in reinforcement due to creep and shrinkage

$\Delta_i^{su}$ = initial deflection due to sustained load

$\Delta_t^{su}$ = time dependent deflection due to sustained load

$\Delta_i^{P}$ = initial deflection due to prestressing

$\Delta_t^{P}$ = time dependent deflection due to prestres-sing

$\Delta P$ = change in prestressing force (loss)

$\Delta P_o$ = force in prestressed reinforcement corresponding to $f_c$ due to dead load and effective prestress

$\Delta P_s$ = force in non-prestressed reinforcement corresponding to $f_c$ due to dead load and effective prestress

$\delta$ = strand elongation due to post-tensioning

$\delta$ = moment magnifier (with subscripts)

$\epsilon_{ce}$ = the strain in concrete corresponding to a stress of $f_{pe}$ in the steel

$\epsilon_{cp}$ = strain due to creep

$\epsilon_i$ = initial immediate elastic strain

$\epsilon_p$ = strain at level of prestressed reinforcement

$\epsilon_{pr}$, $\epsilon'_{pr}$ = strain in prestressed reinforcement

$\epsilon_s$, $\epsilon'_s$ = strain in non-prestressed reinforcement

$\epsilon_{pe}$ = strain in prestressed reinforcement after losses

$\epsilon_{sh}$ = shrinkage strain

$\epsilon_{shu}$ = ultimate shrinkage strain

$\epsilon_x$ = longitudinal strain of flexural tension chord of the member

$\epsilon_y$ = yield strain of non-prestressed reinforcement

$\theta$ = angle of minor axis of a section

$\theta$ = angle of inclination of diagonal compressive stresses to the longitudinal axis of member

$\lambda$ = factor to account for low density concrete

$\mu$ = shear friction coefficient

$\mu$ = coefficient of curvature friction (post-tensioned tendon)

$\rho$ = reinforcement ratio

$\rho_p$ = $A_p/bd$ = reinforcement ratio for prestressed reinforcement

$\rho_s$ = $\dfrac{A_s E_s}{A_c E_p}$

$\sigma_o$ = bursting stress in anchorage zone at $x = 0$

$\sigma_x$ = bursting stress in anchorage zone at "x"

$\sigma_y$ = bursting stress in anchor zone at "y"

$\phi$ = resistance factor (with subscripts) as defined in Fig. 3.2.3

$\phi_m$ = member resistance factor used in moment magnification calculation

$\psi$ = load combination factor (see Fig. 3.2.2)

$\psi$ = curvature

$\omega$ = $\rho f_y / f'_c$

$\omega_p$ = $\rho_p f_{pr} / f'_c$

$\omega_{pu}$ = $\rho_p f_{pu} / f'_c$

$\Omega$ = $\Delta f_p / f_{pi}$

$*$ = superscript used to identify terms used in the creep transformed section method (see Sec. 3.4.7)

# 3.1 INTRODUCTION

This Chapter of the Manual provides a summary of procedures for the design of precast concrete elements, reinforced with prestressed and non-prestressed reinforcement, and post-tensioned elements. Post-tensioning is a method of prestressing in which the tendons are tensioned after the concrete has reached the predetermined strength. Pretensioning is a method of prestressing in which the tendons are tensioned before the concrete is placed. Unless otherwise noted, in this Chapter it is assumed that precast concrete elements are pretensioned and that cast-in-place concrete elements are post-tensioned.

No attempt has been made in this Chapter to differentiate between the design of architectural and structural elements or between the design of load bearing and non-load bearing elements, as the design approach is common to all cases.

# 3.2 LOAD AND RESISTANCE FACTORS

## 3.2.1 LOAD COMBINATIONS

In accordance with CSA Standard A23.3 - 94:

(a) for load combinations not including earthquake, the factored load combination shall be taken as:

$\alpha_D D + \gamma\psi(\alpha_L L + \alpha_W W + \alpha_T T)$

(b) for load combinations including earthquake, the factored load combinations shall be taken as:

$1.0D + \gamma(1.0E)$

and either:

(i) for storage and assembly occupancies

$1.0D + \gamma(1.0L + 1.0E)$ ; or

(ii) for all other occupancies

$1.0D + \gamma(0.5L + 1.0E)$

## 3.2.2 LOAD FACTORS

The load factors are shown in Fig. 3.2.1. Moments due to the reactions induced by prestressing (secondary moments) have a load factor of 1.0.

**Fig. 3.2.1 Load factors**

| Load type | Factor |
|---|---|
| Dead loads resisting overturning uplift or stress reversal | $\alpha_D = 0.85$ |
| Dead loads not included above | $\alpha_D = 1.25$ |
| Live loads due to use and occupancy | $\alpha_L = 1.50$ |
| Live loads due to wind | $\alpha_W = 1.50$ |
| Cumulative effects of temperature, creep, shrinkage and differential settlement | $\alpha_T = 1.25$ |

## 3.2.3 LOAD COMBINATION FACTORS

The load combination factors are shown in Fig. 3.2.2.

**Fig. 3.2.2 Load combination factors**

| Load combination | Factor |
|---|---|
| When only one of the loads, L, W or T is included | $\psi = 1.0$ |
| When two of the loads, L, W or T are included. Eg. L + W or L + T or W + T | $\psi = 0.7$ |
| When all three and the loads, L, W, and T are included | $\psi = 0.6$ |

## 3.2.4 IMPORTANCE FACTOR

The importance factor, $\gamma$, shall not be less than 1.0 except for buildings where collapse is unlikely to cause injury or other serious consequences, where it shall be taken not less than 0.8.

## 3.2.5 LOAD COMBINATION FOR SERVICEABILITY CHECKS

The load combination for checking serviceability requirements shall be taken as:

$$D + \psi(L + W + T)$$

where $\psi$ is per Fig. 3.2.2.

## 3.2.6 MATERIAL RESISTANCE FACTORS

The material resistance factors are as shown in Fig. 3.2.3. Clause 16.1.3 of CSA Standard A23.3 - 94 states that for elements produced in manufacturing plants certified in the appropriate categories by a Certification Organization accredited by the Standards Council of Canada, the concrete material resistance factor, $\phi_c$, may be taken as 0.65.

**Fig. 3.2.3 Material resistance factors**

| Materials | Factor |
|---|---|
| Cast in place concrete | $\phi_c = 0.60$ |
| Precast concrete non certified | $\phi_c = 0.60$ |
| Precast concrete certified | $\phi_c = 0.65$ |
| Reinforcing bars | $\phi_s = 0.85$ |
| Prestressing tendons | $\phi_p = 0.90$ |
| Structural steel | $\phi_a = 0.90$ |

# 3.3 FACTORED FLEXURAL RESISTANCE

## 3.3.1 DESIGN PROCEDURE

The factored flexural resistance of an element must be equal to or greater than the effect of factored load:

$M_r \geq M_f$

The flexural resistance can be determined for any section based on procedures which take into account equilibrium and strain compatibility. For many prestressed elements, the stress in the prestressed reinforcement at factored resistance, $f_{pr}$, can be obtained using the approximate equations given in the Code. These equations are particularly suitable for rectangular elements or for T-shaped sections in which the depth of the compressive stress block is less than the flange thickness. These equations are different from those used in ACI-318.

The equations of equilibrium using the rectangular stress block are shown in Fig. 3.3.1.

**Fig. 3.3.1 Factored flexural resistance**

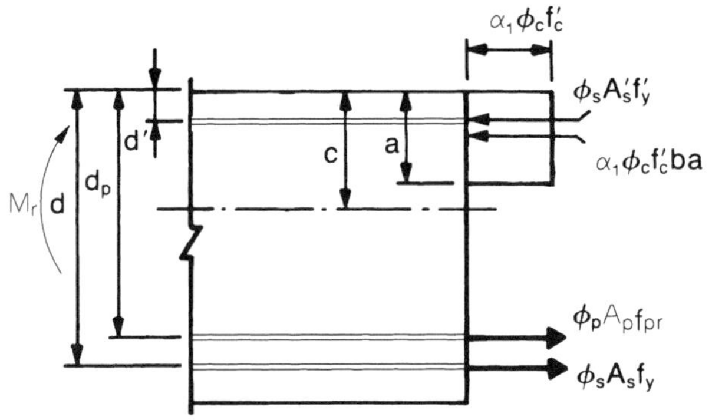


For equilibrium at factored resistance:

$$\alpha_1\phi_c f'_c ba = \phi_p A_p f_{pr} + \phi_s A_s f_y - \phi_s A'_s f'_y$$

$$M_r = \phi_p A_p f_{pr}(d_p - a/2) + \phi_s A_s f_y (d - a/2) - \phi_s A'_s f'_y (d' - a/2)$$

### Depth of Stress Block

The depth "a" of a rectangular stress block is related to the depth to the neutral axis "c" by the equation:

$a = \beta_1 c$

where:

$\beta_1 = 0.97 - 0.0025 f'_c \geq 0.67$

| $f'_c$ (MPa) | $\beta_1$ | $f'_c$ (MPa) | $\beta_1$ |
|---|---|---|---|
| 30 | 0.895 | 60 | 0.820 |
| 35 | 0.883 | 65 | 0.808 |
| 40 | 0.870 | 70 | 0.795 |
| 45 | 0.858 | 75 | 0.783 |
| 50 | 0.845 | 80 | 0.770 |
| 55 | 0.833 | | |

A flow chart illustrating the application of design calculations for the computation of flexural resistance is shown in Fig. 3.3.2.

### Flanged Elements

Clause 10.3.3, CSA-A23.3-94 limits the overhanging flange width on either side of the web to be used in the design of simply supported, symmetrical T-beams to 12 times the flange thickness, 1/2 the clear distance to the next web or 1/5 the span length. These empirical provisions were developed for conventionally reinforced concrete and do not necessarily apply to prestressed concrete.

The equations for flexural resistance given in Fig. 3.3.1 apply only to rectangular cross sections and flanged sections in which the stress block lies entirely within the depth of the flange, $h_f$. The depth of the stress block, a, can be found from the equations of equilibrium given in Fig. 3.3.1:

$$a = \frac{\phi_p A_p f_{pr} + \phi_s A_s f_y - \phi_s A'_s f'_y}{\alpha_1 \phi_c f'_c b}$$

where:

$\alpha_1 = 0.85 - 0.0015 f'_c \geq 0.67$

| $f'_c$ (MPa) | $\alpha_1$ | $f'_c$ (MPa) | $\alpha_1$ |
|---|---|---|---|
| 30 | 0.805 | 60 | 0.760 |
| 35 | 0.798 | 65 | 0.753 |
| 40 | 0.790 | 70 | 0.745 |
| 45 | 0.783 | 75 | 0.737 |
| 50 | 0.775 | 80 | 0.730 |
| 55 | 0.768 | | |

If $a > h_f$ the force required to develop the compressive strength of the overhanging flanges is deducted from the total force in the tension reinforcement as shown in Fig. 3.3.2.

### Limitations on Reinforcement

For reinforced flexural elements, the minimum reinforcement shall be proportioned so that:

**Fig. 3.3.2 Flow chart for flexural resistance calculations**

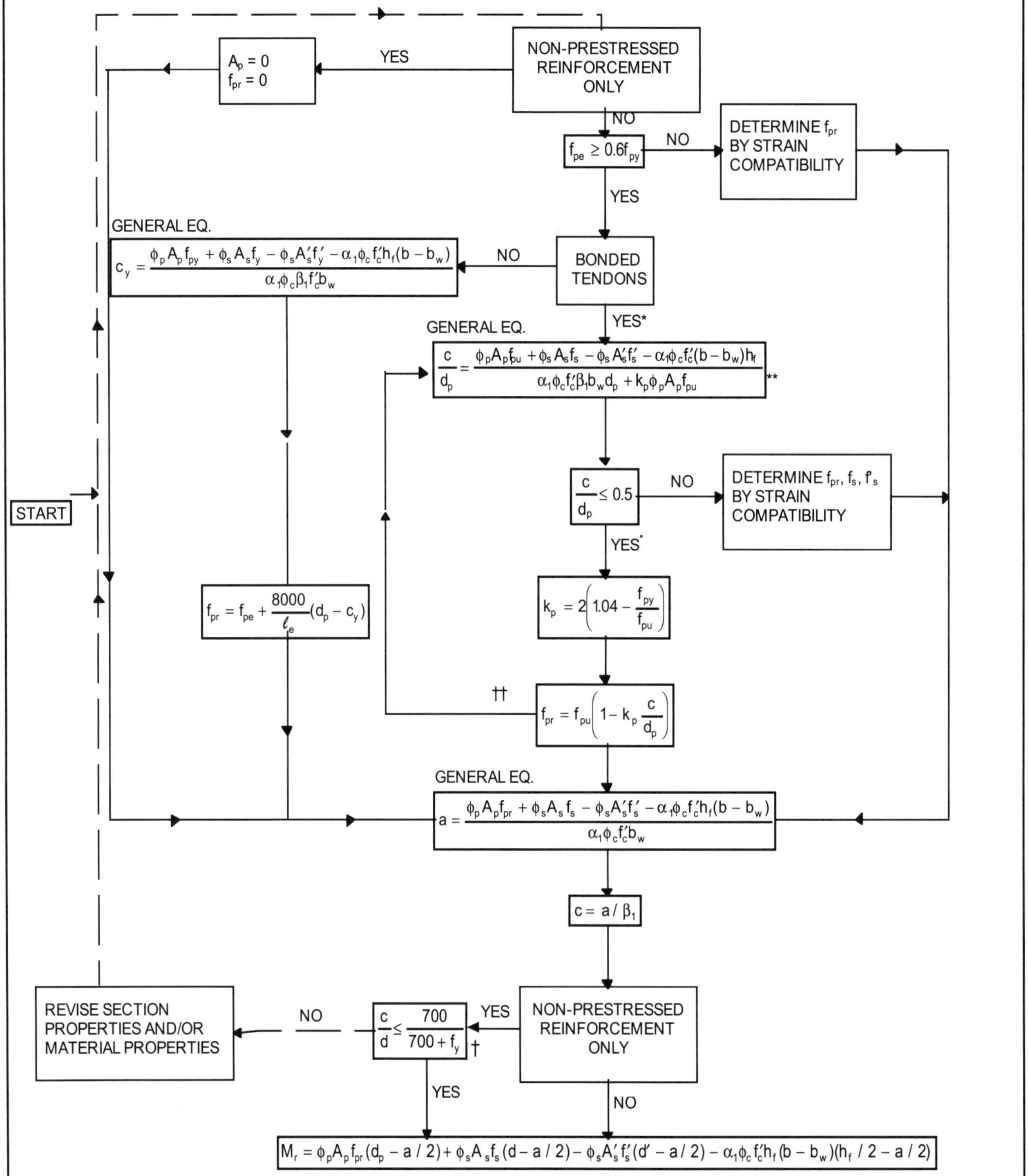


* Provided that the above equation is satisfied, the following approximate values of $f_{pr}$ may be used.

** As a first iteration, the upper bound value of c/d can be determined by assuming $f_{pr} = f_{pu}$. Also $f_s$ and $f'_s$ can be taken equal to $f_y$ and $f'_y$ provided they are located 0.75c from the neutral axis, otherwise strain compatibility must be used.

† Preferred values of c/d lie in the range 0.23 to 0.37.

†† Iterate until values of $f_{pr}$ converge.

**Fig. 3.3.3 Minimum Area of Bonded Reinforcement**

| Type of member | Concrete stress | | | |
|---|---|---|---|---|
| | Tensile stress $\leq 0.5\lambda\sqrt{f'_c}$ | | Tensile stress $> 0.5\lambda\sqrt{f'_c}$ | |
| | Type of tendon | | Type of tendon | |
| | Bonded | Unbonded | Bonded | Unbonded |
| Beams | 0 | 0.004 A | 0.003 A | 0.005 A |
| One-way slabs | 0 | 0.003 A | 0.002 A | 0.004 A |
| Two-way slabs: Negative moment regions | 0 | 0.0006 h $\ell_n$ | 0.00045 h $\ell_n$ | 0.00075 h $\ell_n$ |
| Positive moment regions, concrete stress $> 0.2\lambda\sqrt{f'_c}$ | 0 | 0.004 A | 0.003 A | 0.005 A |
| Positive moment regions, concrete tensile stress $\leq 0.2\lambda\sqrt{f'_c}$ | 0 | 0 | – | – |

$$M_r \geq 1.2\ M_{cr}$$

where $M_{cr}$ is calculated using:

$$f_r = 0.6\lambda\sqrt{f'_c}$$

or, except for two-way slabs and footings, minimum reinforcement may be determined from:

$$A_{smin} = \frac{0.2\sqrt{f'_c}}{f_y} b_t h$$

where $b_t$ is the width of the tension zone of the section considered.

One of the two methods shown above shall be used in calculating the minimum reinforcement unless $M_r \geq 1.33M_f$.

The maximum reinforcement ratio for reinforced elements shall be such that the following expression is satisfied:

$$\frac{c}{d} \leq \frac{700}{700 + f_y}$$

Compression reinforcement, in conjunction with additional tension reinforcement may be used to increase the strength of flexural elements.

For prestressed elements, the minimum reinforcement requirements will be satisfied provided:

$M_r \geq 1.20\ M_{cr}$ unless

$M_r \geq 1.33\ M_f$

where:

$M_{cr}$ = cracking moment

$$= \left(\frac{P_e}{A_g} + \frac{P_e e}{S_b} + 0.6\lambda\sqrt{f'_c}\right) S_b$$

where:

$A_g$ = gross area of section

The minimum quantity of bonded non-prestressed reinforcement is dependent on the degree of prestress (see Sect. 3.4.3) and the bonding conditions of the tendons, which is summarized in Fig. 3.3.3.

**Fig. 3.3.4 Critical sections for flexural design**

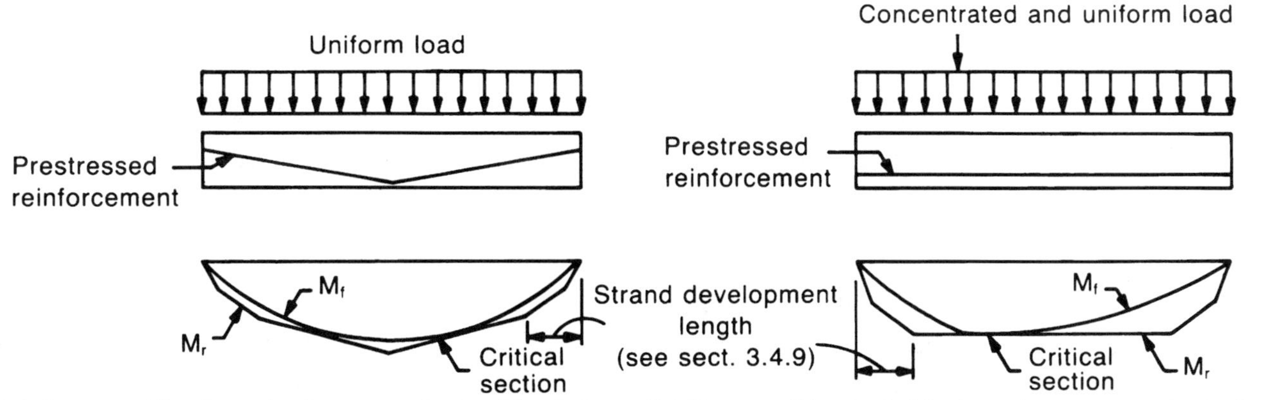


**(a) Beam with strands depressed at midpoint uniform load** **(b) Beam with straight strands – concentrated a uniform load**

### Critical Section

For simply supported, uniformly loaded, prismatic reinforced elements, the critical section for flexural design will occur at midspan. Provided that reinforcement is properly developed and adequate shear reinforcement is provided, the amount of flexural reinforcement may be reduced in areas of lower moment towards the support.

For prestressed elements, because of the limitation on end stresses at the time of prestress transfer (see Sect. 3.4.3), tendons are often draped or depressed, producing a varying effective depth, $d_p$ along the length of the element. For draped tendons, and also for non-uniform loading, it may be necessary to compare the factored moment ($M_f$) with the factored flexural resistance ($M_r$) at points other than at midspan. This is illustrated in Fig. 3.3.4. For uniform loads with straight tendons, it is sufficient to check the midspan condition. For uniform loads with single point depressed tendons, the critical section is usually near $0.4\ell$ and it is common practice to check only at midspan and at $0.4\ell$.

**Fig. 3.3.5 Flexural resistance design aid for rectangular section with non-prestressed reinforcement only produced in CSA certified plants ($\varnothing_c$ = 0.65)**

Procedure:

Design

1. Determine $K_r = M_f / (f'_c bd^2)$
2. Calculate $\omega$ from the table
3. Determine $\rho = \omega f'_c / f_y$
4. Determine $A_s = \rho bd$

Analysis

1. Determine $\rho = A_s / bd$
2. Determine $\omega = \rho f_y / f'_c$
3. Calculate $K_r$ from the table
4. Determine $M_r = K_r f'_c bd^2$

| | $\omega_{max.}$ | | | | | | | | | | | |
|---|---|---|---|---|---|---|---|---|---|---|---|---|
| | $f'_c$(MPa) | 30 | 35 | 40 | 45 | 50 | 55 | 60 | 65 | 70 | 75 | 80 |
| | $\beta_1$ | 0.895 | 0.883 | 0.870 | 0.858 | 0.845 | 0.833 | 0.820 | 0.808 | 0.795 | 0.783 | 0.770 |
| $f_y$ (MPa) | $\alpha_1$ | 0.805 | 0.798 | 0.790 | 0.783 | 0.775 | 0.768 | 0.760 | 0.753 | 0.745 | 0.738 | 0.730 |
| 300 | | 0.386 | 0.377 | 0.368 | 0.359 | 0.351 | 0.342 | 0.334 | 0.325 | 0.317 | 0.309 | 0.301 |
| 400 | | 0.351 | 0.342 | 0.334 | 0.327 | 0.319 | 0.311 | 0.303 | 0.296 | 0.288 | 0.281 | 0.274 |

| $f'_c$ | $\omega$ | $K_r$ |
|---|---|---|
| 30 MPa | $0.615 - \sqrt{0.3786 - 1.4482K_r}$ | $0.85\omega - 0.6907\omega^2$ |
| 35 | $0.610 - \sqrt{0.3716 - 1.4347K_r}$ | $0.85\omega - 0.6967\omega^2$ |
| 40 | $0.604 - \sqrt{0.3647 - 1.4212K_r}$ | $0.85\omega - 0.7038\omega^2$ |
| 45 | $0.598 - \sqrt{0.3578 - 1.4077K_r}$ | $0.85\omega - 0.7101\omega^2$ |
| 50 | $0.592 - \sqrt{0.3509 - 1.3942K_r}$ | $0.85\omega - 0.7174\omega^2$ |
| 55 | $0.587 - \sqrt{0.3442 - 1.3807K_r}$ | $0.85\omega - 0.7240\omega^2$ |
| 60 | $0.581 - \sqrt{0.3375 - 1.3672K_r}$ | $0.85\omega - 0.7316\omega^2$ |
| 65 | $0.575 - \sqrt{0.3309 - 1.3537K_r}$ | $0.85\omega - 0.7384\omega^2$ |
| 70 | $0.570 - \sqrt{0.3246 - 1.3405K_r}$ | $0.85\omega - 0.7460\omega^2$ |
| 75 | $0.564 - \sqrt{0.3185 - 1.3278K_r}$ | $0.85\omega - 0.7531\omega^2$ |
| 80 | $0.558 - \sqrt{0.3116 - 1.3135K_r}$ | $0.85\omega - 0.7613\omega^2$ |

## 3.3.2 ANALYSIS USING CODE EQUATIONS

### Reinforced Elements

Fig. 3.3.5 and 3.3.6 are a convenient aid for the design and analysis of rectangular sections with non-prestressed reinforcement. Values of $\omega_{max}$ are also listed which correspond to the maximum c/d ratio.

The use of this design aid is illustrated in Example 1.

### Prestressed elements

The following conservative values of $f_{pr}$ may be used if $f_{pe}$ is not less than 0.6 $f_{py}$:

1. For elements with bonded tendons, provided $c/d_p$ is not greater than 0.5:

$$f_{pr} = f_{pu}\left(1 - k_p \frac{c}{d_p}\right)$$

or considering that $c = c_u \frac{f_{pr}}{f_{pu}}$:

**Fig. 3.3.5 Continued**

**Table is based on:**
$f'_c$ = 40 MPa
$\alpha_1$ = 0.79
The error for other concrete strengths does not exceed 3.2%.

| | Values of $K_r$ | | | | | | | | | |
|---|---|---|---|---|---|---|---|---|---|---|
| ω | 0.000 | 0.001 | 0.002 | 0.003 | 0.004 | 0.005 | 0.006 | 0.007 | 0.008 | 0.009 |
| 0.00 | 0.0000 | 0.0008 | 0.0017 | 0.0025 | 0.0034 | 0.0042 | 0.0051 | 0.0059 | 0.0068 | 0.0076 |
| 0.01 | 0.0084 | 0.0093 | 0.0101 | 0.0109 | 0.0118 | 0.0126 | 0.0134 | 0.0142 | 0.0151 | 0.0159 |
| 0.02 | 0.0167 | 0.0175 | 0.0184 | 0.0192 | 0.0200 | 0.0208 | 0.0216 | 0.0224 | 0.0232 | 0.0241 |
| 0.03 | 0.0249 | 0.0257 | 0.0265 | 0.0273 | 0.0281 | 0.0289 | 0.0297 | 0.0305 | 0.0313 | 0.0321 |
| 0.04 | 0.0329 | 0.0337 | 0.0345 | 0.0352 | 0.0360 | 0.0368 | 0.0376 | 0.0384 | 0.0392 | 0.0400 |
| 0.05 | 0.0407 | 0.0415 | 0.0423 | 0.0431 | 0.0438 | 0.0446 | 0.0454 | 0.0462 | 0.0469 | 0.0477 |
| 0.06 | 0.0485 | 0.0492 | 0.0500 | 0.0508 | 0.0515 | 0.0523 | 0.0530 | 0.0538 | 0.0545 | 0.0553 |
| 0.07 | 0.0561 | 0.0568 | 0.0576 | 0.0583 | 0.0590 | 0.0598 | 0.0605 | 0.0613 | 0.0620 | 0.0628 |
| 0.08 | 0.0635 | 0.0642 | 0.0650 | 0.0657 | 0.0664 | 0.0672 | 0.0679 | 0.0686 | 0.0694 | 0.0701 |
| 0.09 | 0.0708 | 0.0715 | 0.0722 | 0.0730 | 0.0737 | 0.0744 | 0.0751 | 0.0758 | 0.0765 | 0.0773 |
| 0.10 | 0.0780 | 0.0787 | 0.0794 | 0.0801 | 0.0808 | 0.0815 | 0.0822 | 0.0829 | 0.0836 | 0.0843 |
| 0.11 | 0.0850 | 0.0857 | 0.0864 | 0.0871 | 0.0878 | 0.0884 | 0.0891 | 0.0898 | 0.0905 | 0.0912 |
| 0.12 | 0.0919 | 0.0925 | 0.0932 | 0.0939 | 0.0946 | 0.0953 | 0.0959 | 0.0966 | 0.0973 | 0.0979 |
| 0.13 | 0.0986 | 0.0993 | 0.0999 | 0.1006 | 0.1013 | 0.1019 | 0.1026 | 0.1032 | 0.1039 | 0.1046 |
| 0.14 | 0.1052 | 0.1059 | 0.1065 | 0.1072 | 0.1078 | 0.1085 | 0.1091 | 0.1097 | 0.1104 | 0.1110 |
| 0.15 | 0.1117 | 0.1123 | 0.1129 | 0.1136 | 0.1142 | 0.1148 | 0.1155 | 0.1161 | 0.1167 | 0.1174 |
| 0.16 | 0.1180 | 0.1186 | 0.1192 | 0.1199 | 0.1205 | 0.1211 | 0.1217 | 0.1223 | 0.1229 | 0.1236 |
| 0.17 | 0.1242 | 0.1248 | 0.1254 | 0.1260 | 0.1266 | 0.1272 | 0.1278 | 0.1284 | 0.1290 | 0.1296 |
| 0.18 | 0.1302 | 0.1308 | 0.1314 | 0.1320 | 0.1326 | 0.1332 | 0.1338 | 0.1343 | 0.1349 | 0.1355 |
| 0.19 | 0.1361 | 0.1367 | 0.1373 | 0.1378 | 0.1384 | 0.1390 | 0.1396 | 0.1401 | 0.1407 | 0.1413 |
| 0.20 | 0.1419 | 0.1424 | 0.1430 | 0.1436 | 0.1441 | 0.1447 | 0.1452 | 0.1458 | 0.1464 | 0.1469 |
| 0.21 | 0.1475 | 0.1480 | 0.1486 | 0.1491 | 0.1497 | 0.1502 | 0.1508 | 0.1513 | 0.1519 | 0.1524 |
| 0.22 | 0.1530 | 0.1535 | 0.1540 | 0.1546 | 0.1551 | 0.1556 | 0.1562 | 0.1567 | 0.1572 | 0.1578 |
| 0.23 | 0.1583 | 0.1588 | 0.1593 | 0.1599 | 0.1604 | 0.1609 | 0.1614 | 0.1619 | 0.1625 | 0.1630 |
| 0.24 | 0.1635 | 0.1640 | 0.1645 | 0.1650 | 0.1655 | 0.1660 | 0.1665 | 0.1670 | 0.1675 | 0.1680 |
| 0.25 | 0.1685 | 0.1690 | 0.1695 | 0.1700 | 0.1705 | 0.1710 | 0.1715 | 0.1720 | 0.1725 | 0.1730 |
| 0.26 | 0.1734 | 0.1739 | 0.1744 | 0.1749 | 0.1754 | 0.1758 | 0.1763 | 0.1768 | 0.1773 | 0.1777 |
| 0.27 | 0.1782 | 0.1787 | 0.1792 | 0.1796 | 0.1801 | 0.1805 | 0.1810 | 0.1815 | 0.1819 | 0.1824 |
| 0.28 | 0.1828 | 0.1833 | 0.1838 | 0.1842 | 0.1847 | 0.1851 | 0.1856 | 0.1860 | 0.1864 | 0.1869 |
| 0.29 | 0.1873 | 0.1878 | 0.1882 | 0.1887 | 0.1891 | 0.1895 | 0.1900 | 0.1904 | 0.1908 | 0.1913 |
| 0.30 | 0.1917 | 0.1921 | 0.1925 | 0.1930 | 0.1934 | 0.1938 | 0.1942 | 0.1946 | 0.1951 | 0.1955 |
| 0.31 | 0.1959 | 0.1963 | 0.1967 | 0.1971 | 0.1975 | 0.1979 | 0.1984 | 0.1988 | 0.1992 | 0.1996 |
| 0.32 | 0.2000 | 0.2004 | 0.2008 | 0.2012 | 0.2015 | 0.2019 | 0.2023 | 0.2027 | 0.2031 | 0.2035 |
| 0.33 | 0.2039 | 0.2043 | 0.2047 | 0.2050 | 0.2054 | 0.2058 | 0.2062 | 0.2066 | 0.2069 | 0.2073 |
| 0.34 | 0.2077 | 0.2080 | 0.2084 | 0.2088 | 0.2091 | 0.2095 | 0.2099 | 0.2102 | 0.2106 | 0.2110 |
| 0.35 | 0.2113 | 0.2117 | 0.2120 | 0.2124 | 0.2127 | 0.2131 | 0.2134 | 0.2138 | 0.2141 | 0.2145 |
| 0.36 | 0.2148 | 0.2152 | 0.2155 | 0.2158 | 0.2162 | 0.2165 | 0.2169 | 0.2172 | 0.2175 | 0.2179 |
| 0.37 | 0.2182 | 0.2185 | 0.2188 | 0.2192 | 0.2195 | 0.2198 | 0.2201 | 0.2205 | 0.2208 | 0.2211 |
| 0.38 | 0.2214 | 0.2217 | 0.2220 | 0.2224 | 0.2227 | 0.2230 | 0.2233 | 0.2236 | 0.2239 | 0.2242 |
| 0.39 | 0.2245 | 0.2248 | 0.2251 | 0.2254 | 0.2257 | 0.2260 | 0.2263 | 0.2266 | 0.2269 | 0.2272 |

$$f_{pr} = f_{pu}\left(\frac{1}{1+k_p c_u / d_p}\right)$$

where $k_p = 2(1.04 - f_{py} / f_{pu})$ and c is determined using a stress of $f_{pr}$ in the tendons.

2. For elements with unbonded tendons:

$$f_{pr} = f_{pe} + 8000(d_p - c_y) / \ell_e$$

but shall not exceed $f_{py}$ and $c_y$ is determined assuming a stress of $f_{py}$ in the tendons.

The above term $k_p$ accounts for the shape of the tendon stress-strain curve. Values of $k_p$ are given below:

**Fig. 3.3.7 Tendon stress-strain curve factor**

| Tendon Type | $k_p$ |
|---|---|
| Low relaxation strand or wire | 0.28 |
| Plain prestressing bars | 0.38 |
| Deformed prestressing bars | 0.48 |

Example 2 shows the method of calculating the flexural resistance of a prestressed element with bonded tendons using Fig. 3.3.1 and the appropriate equation for $f_{pr}$.

**Fig. 3.3.6 Flexural resistance design aid for rectangular section with non-prestressed reinforcement only produced in non-CSA certified plants ($\O_c = 0.60$)**

Procedure:

Design

1. Determine $K_r = M_f / (f'_c bd^2)$
2. Calculate ω from the table
3. Determine $\rho = \omega f'_c$
4. Determine $A_s = \rho bd$

Analysis

1. Determine $\rho = A_s / bd$
2. Determine $\omega = \rho f_y / f'_c$
3. Calculate $K_r$ from the table
4. Determine $M_r = K_r f'_c bd^2$

| | $\omega_{max}$ | | | | | | | | | | | |
|---|---|---|---|---|---|---|---|---|---|---|---|---|
| | $f'_c$ MPa) | 30 | 35 | 40 | 45 | 50 | 55 | 60 | 65 | 70 | 75 | 80 |
| | $\beta_1$ | 0.895 | 0.883 | 0.870 | 0.858 | 0.845 | 0.833 | 0.820 | 0.808 | 0.795 | 0.783 | 0.770 |
| $f_y$ (MPa) | $\alpha_1$ | 0.805 | 0.798 | 0.790 | 0.783 | 0.775 | 0.768 | 0.760 | 0.753 | 0.745 | 0.738 | 0.730 |
| 300 | | 0.356 | 0.348 | 0.340 | 0.332 | 0.324 | 0.316 | 0.308 | 0.301 | 0.293 | 0.285 | 0.278 |
| 400 | | 0.324 | 0.317 | 0.308 | 0.302 | 0.294 | 0.287 | 0.280 | 0.273 | 0.266 | 0.259 | 0.252 |

| $f'_c$ | ω | $K_r$ |
|---|---|---|
| 30 MPa | $0.568 - \sqrt{0.3229 - 13370K_r}$ | $0.85\omega - 0.7479\omega^2$ |
| 35 | $0.563 - \sqrt{0.3173 - 13254K_r}$ | $0.85\omega - 0.7545\omega^2$ |
| 40 | $0.558 - \sqrt{0.3110 - 13121K_r}$ | $0.85\omega - 0.7621\omega^2$ |
| 45 | $0.555 - \sqrt{0.3078 - 13055K_r}$ | $0.85\omega - 0.7660\omega^2$ |
| 50 | $0.547 - \sqrt{0.2993 - 12872K_r}$ | $0.85\omega - 0.7769\omega^2$ |
| 55 | $0.542 - \sqrt{0.2939 - 1.2756K_r}$ | $0.85\omega - 0.7840\omega^2$ |
| 60 | $0.536 - \sqrt{0.2878 - 12623K_r}$ | $0.85\omega - 0.7922\omega^2$ |
| 65 | $0.532 - \sqrt{0.2825 - 1.2507K_r}$ | $0.85\omega - 0.7996\omega^2$ |
| 70 | $0.526 - \sqrt{0.2765 - 12373K_r}$ | $0.85\omega - 0.8082\omega^2$ |
| 75 | $0.521 - \sqrt{0.2714 - 1.2258K_r}$ | $0.85\omega - 0.8158\omega^2$ |
| 80 | $0.515 - \sqrt{0.2655 - 12124K_r}$ | $0.85\omega - 0.8248\omega^2$ |

For elements with bonded prestressed reinforcement only, the calculations shown in Fig. 3.3.2 can be simplified by the use of the coefficients given in Fig. 3.3.8 and 3.3.9, as illustrated in Example 3.

It should be noted that if $c/d_p > 0.5$ or if $f_{pe} < 0.6\, f_{py}$, then the approximate equation for $f_{pr}$ should only be used as the first iteration of a strain compatibility analysis. Similarly, if $c/d_p$ is greater than 0.5, the coefficients in Fig. 3.3.8 and Fig. 3.3.9 should only be used to estimate $A_p$ or $M_r$.

For sections with bonded prestressed tendons only, the coefficients are derived on the following basis:

From Fig. 3.3.1:

$$\alpha_1\, \phi_c\, f'_c\, b\, a = \phi_p\, A_p\, f_{pr}$$

Dividing by $b\, d_p\, f'_c$ and defining the term:

$$\omega_p = \rho \frac{f_{pr}}{f'_c} = \frac{A_p f_{pr}}{b d_p f'_c}$$

then:

**Fig. 3.3.6 Continued**

**Table based on:**
$f'_c$ = 40 MPa
$\alpha_1$ = 0.79

The error for other concrete strengths does not exceed 3.2%.

| | Values of $K_r$ | | | | | | | | | |
|---|---|---|---|---|---|---|---|---|---|---|
| ω | 0.000 | 0.001 | 0.002 | 0.003 | 0.004 | 0.005 | 0.006 | 0.007 | 0.008 | 0.009 |
| 0.00 | 0.0000 | 0.0008 | 0.0017 | 0.0025 | 0.0034 | 0.0042 | 0.0051 | 0.0059 | 0.0068 | 0.0076 |
| 0.01 | 0.0084 | 0.0093 | 0.0101 | 0.0109 | 0.0118 | 0.0126 | 0.0134 | 0.0142 | 0.0151 | 0.0159 |
| 0.02 | 0.0167 | 0.0175 | 0.0183 | 0.0191 | 0.0200 | 0.0208 | 0.0216 | 0.0224 | 0.0232 | 0.0240 |
| 0.03 | 0.0248 | 0.0256 | 0.0264 | 0.0272 | 0.0280 | 0.0288 | 0.0296 | 0.0304 | 0.0312 | 0.0320 |
| 0.04 | 0.0328 | 0.0336 | 0.0344 | 0.0351 | 0.0359 | 0.0367 | 0.0375 | 0.0383 | 0.0390 | 0.0398 |
| 0.05 | 0.0406 | 0.0414 | 0.0421 | 0.0429 | 0.0437 | 0.0444 | 0.0452 | 0.0460 | 0.0467 | 0.0475 |
| 0.06 | 0.0483 | 0.0490 | 0.0498 | 0.0505 | 0.0513 | 0.0520 | 0.0528 | 0.0535 | 0.0543 | 0.0550 |
| 0.07 | 0.0558 | 0.0565 | 0.0572 | 0.0580 | 0.0587 | 0.0595 | 0.0602 | 0.0609 | 0.0617 | 0.0624 |
| 0.08 | 0.0631 | 0.0638 | 0.0646 | 0.0653 | 0.0660 | 0.0667 | 0.0675 | 0.0682 | 0.0689 | 0.0696 |
| 0.09 | 0.0703 | 0.0710 | 0.0717 | 0.0725 | 0.0732 | 0.0739 | 0.0746 | 0.0753 | 0.0760 | 0.0767 |
| 0.10 | 0.0774 | 0.0781 | 0.0788 | 0.0795 | 0.0802 | 0.0808 | 0.0815 | 0.0822 | 0.0829 | 0.0836 |
| 0.11 | 0.0843 | 0.0850 | 0.0856 | 0.0863 | 0.0870 | 0.0877 | 0.0883 | 0.0890 | 0.0897 | 0.0904 |
| 0.12 | 0.0910 | 0.0917 | 0.0924 | 0.0930 | 0.0937 | 0.0943 | 0.0950 | 0.0957 | 0.0963 | 0.0970 |
| 0.13 | 0.0976 | 0.0983 | 0.0989 | 0.0996 | 0.1002 | 0.1009 | 0.1015 | 0.1021 | 0.1028 | 0.1034 |
| 0.14 | 0.1041 | 0.1047 | 0.1053 | 0.1060 | 0.1066 | 0.1072 | 0.1079 | 0.1085 | 0.1091 | 0.1097 |
| 0.15 | 0.1104 | 0.1110 | 0.1116 | 0.1122 | 0.1128 | 0.1134 | 0.1141 | 0.1147 | 0.1153 | 0.1159 |
| 0.16 | 0.1165 | 0.1171 | 0.1177 | 0.1183 | 0.1189 | 0.1195 | 0.1201 | 0.1207 | 0.1213 | 0.1219 |
| 0.17 | 0.1225 | 0.1231 | 0.1237 | 0.1242 | 0.1248 | 0.1254 | 0.1260 | 0.1266 | 0.1272 | 0.1277 |
| 0.18 | 0.1283 | 0.1289 | 0.1295 | 0.1300 | 0.1306 | 0.1312 | 0.1317 | 0.1323 | 0.1329 | 0.1334 |
| 0.19 | 0.1340 | 0.1345 | 0.1351 | 0.1357 | 0.1362 | 0.1368 | 0.1373 | 0.1379 | 0.1384 | 0.1390 |
| 0.20 | 0.1395 | 0.1401 | 0.1406 | 0.1411 | 0.1417 | 0.1422 | 0.1428 | 0.1433 | 0.1438 | 0.1444 |
| 0.21 | 0.1449 | 0.1454 | 0.1459 | 0.1465 | 0.1470 | 0.1475 | 0.1480 | 0.1486 | 0.1491 | 0.1496 |
| 0.22 | 0.1501 | 0.1506 | 0.1511 | 0.1516 | 0.1522 | 0.1527 | 0.1532 | 0.1537 | 0.1542 | 0.1547 |
| 0.23 | 0.1552 | 0.1557 | 0.1562 | 0.1567 | 0.1572 | 0.1577 | 0.1582 | 0.1586 | 0.1591 | 0.1596 |
| 0.24 | 0.1601 | 0.1606 | 0.1611 | 0.1615 | 0.1620 | 0.1625 | 0.1630 | 0.1635 | 0.1639 | 0.1644 |
| 0.25 | 0.1649 | 0.1653 | 0.1658 | 0.1663 | 0.1667 | 0.1672 | 0.1677 | 0.1681 | 0.1686 | 0.1690 |
| 0.26 | 0.1695 | 0.1699 | 0.1704 | 0.1708 | 0.1713 | 0.1717 | 0.1722 | 0.1726 | 0.1731 | 0.1735 |
| 0.27 | 0.1739 | 0.1744 | 0.1748 | 0.1752 | 0.1757 | 0.1761 | 0.1765 | 0.1770 | 0.1774 | 0.1778 |
| 0.28 | 0.1782 | 0.1787 | 0.1791 | 0.1795 | 0.1799 | 0.1803 | 0.1808 | 0.1812 | 0.1816 | 0.1820 |
| 0.29 | 0.1824 | 0.1828 | 0.1832 | 0.1836 | 0.1840 | 0.1844 | 0.1848 | 0.1852 | 0.1856 | 0.1860 |
| 0.30 | 0.1864 | 0.1868 | 0.1872 | 0.1876 | 0.1880 | 0.1884 | 0.1887 | 0.1891 | 0.1895 | 0.1899 |
| 0.31 | 0.1903 | 0.1906 | 0.1910 | 0.1914 | 0.1918 | 0.1921 | 0.1925 | 0.1929 | 0.1932 | 0.1936 |
| 0.32 | 0.1940 | 0.1943 | 0.1947 | 0.1950 | 0.1954 | 0.1957 | 0.1961 | 0.1965 | 0.1968 | 0.1972 |
| 0.33 | 0.1975 | 0.1979 | 0.1982 | 0.1985 | 0.1989 | 0.1992 | 0.1996 | 0.1999 | 0.2002 | 0.2006 |
| 0.34 | 0.2009 | 0.2012 | 0.2016 | 0.2019 | 0.2022 | 0.2025 | 0.2029 | 0.2032 | 0.2035 | 0.2038 |
| 0.35 | 0.2041 | 0.2045 | 0.2048 | 0.2051 | 0.2054 | 0.2057 | 0.2060 | 0.2063 | 0.2066 | 0.2069 |
| 0.36 | 0.2072 | 0.2075 | 0.2078 | 0.2081 | 0.2084 | 0.2087 | 0.2090 | 0.2093 | 0.2096 | 0.2099 |
| 0.37 | 0.2102 | 0.2104 | 0.2107 | 0.2110 | 0.2113 | 0.2116 | 0.2119 | 0.2121 | 0.2124 | 0.2127 |
| 0.38 | 0.2129 | 0.2132 | 0.2135 | 0.2138 | 0.2140 | 0.2143 | 0.2145 | 0.2148 | 0.2151 | 0.2153 |
| 0.39 | 0.2156 | 0.2158 | 0.2161 | 0.2163 | 0.2166 | 0.2168 | 0.2171 | 0.2173 | 0.2176 | 0.2178 |

$$\frac{\alpha_1\phi_c f_c' ba}{bd_p f_c'} = \frac{\phi_p A_p f_{pr}}{bd_p f_c'} = \phi_p \omega_p$$

$$\therefore a = \frac{\phi_p \omega_p d_p}{\alpha_1 \phi_c}$$

$$M_r = \phi_p A_p f_{pr}(d_p - a/2)$$

$$= \frac{\phi_p A_p f_{pr} bd_p^2 f_c'(1 - a/2d_p)}{bd_p f_c'}$$

$$= \phi_p f_c' bd_p^2 \omega_p \left(1 - \frac{\phi_p \omega_p}{2\alpha_1 \phi_c}\right)$$

Multiplying the equation for $f_{pr}$ by $\frac{A_p}{bd_p f_c'}$:

$$\frac{A_p f_{pr}}{bd_p f_c'} = \frac{A_p f_{pu}(1 - k_p c/d_p)}{bd_p f_c'}$$

and defining the term $\omega_{pu}$:

$$\omega_{pu} = \rho_p \frac{f_{pu}}{f_c'} = \frac{A_p f_{pu}}{bd_p f_c'}$$

then:

$$\frac{c}{d_p} = \frac{1}{\left(\frac{\alpha_1 \phi_c \beta_1}{\phi_p \omega_{pu}} + k_p\right)}$$

and:

$$\omega_p = \omega_{pu} \frac{1}{\left(1 + k_p \frac{\phi_p \omega_{pu}}{\alpha_1 \phi_c \beta_1}\right)}$$

By substituting the equation for $\omega_p$ in the equation for $M_r$ and introducing the coefficient:

$$K_r = \phi_p \omega_{pu} \frac{1}{\left(1 + k_p \frac{\phi_p \omega_{pu}}{\alpha_1 \phi_c \beta_1}\right)} \times \left[1 - \frac{\phi_p \omega_{pu}}{2\alpha_1 \phi_c \left(1 + k_p \frac{\phi_p \omega_{pu}}{\alpha_1 \phi_c \beta_1}\right)}\right]$$

leads to:

$$M_r = K_r f_c' bd_p^2$$

For flanged sections, the design aid only applies if the compression stress block is confined to the flange, that is, where:

$$\frac{\phi_p d_p \omega_{pu}}{\alpha_1 \phi_c \left(1 + k_p \frac{\phi_p \omega_{pu}}{\alpha_1 \phi_c \beta_1}\right)} \le h_f$$

If the stress block is deeper than the flange, the equations shown in the flow chart in Fig. 3.3.2 should be used.

**Example 1 – Use of Fig. 3.3.5 coefficients for determination of non-prestressed reinforcement**

*Given:*

300/450 × 1000 L-shaped beam produced in a CSA certified plant.

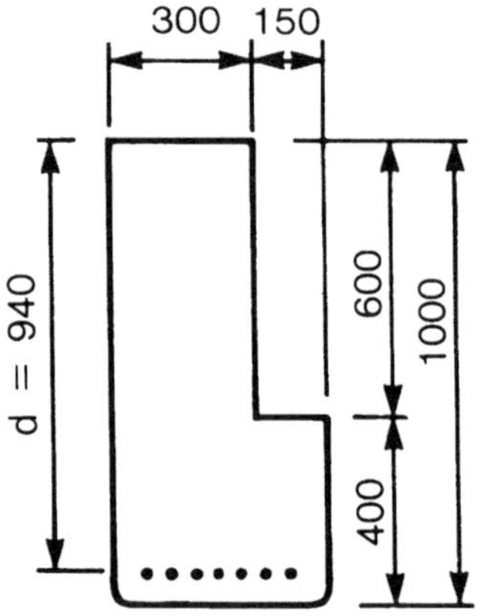


**Fig. 3.3.8 Flexural resistance coefficients for elements with bonded prestressed reinforcement only, produced in CSA certified plant ($Ø_c$ = 0.65)**

| Procedure: | |
|---|---|
| Design | Analysis |
| 1. Determine $K_r = \frac{M_f}{f_c' bd_p^2}$ | 1. Determine $\rho_p = A_p / bd_p$ |
| 2. Calculate $\omega$ from the table | 2. Determine $\omega_{pu} = \rho_p f_{pu} / f_c'$ |
| 3. Determine $\rho_p = \omega_{pu} f_c' / f_{pu}$ | 3. Calculate $K_r$ from the table |
| 4. Determine $A_p = \rho_p bd_p$ | 4. Determine $M_r = K_r f_c' bd_p^2$ |

**Fig. 3.3.8 Continued**

$\phi_c$ = 0.65

$\phi_p$ = 0.90

| | $k_p$ = 0.28 | | | | | | $k_p$ = 0.38 | | | | | | $k_p$ = 0.48 | | | | | |
|---|---|---|---|---|---|---|---|---|---|---|---|---|---|---|---|---|---|---|
| f'c | 30 | 35 | 40 | 45 | 50 | 55 | 30 | 35 | 40 | 45 | 50 | 55 | 30 | 35 | 40 | 45 | 50 | 55 |
| $\beta_1$ | 0.8950 | 0.8825 | 0.8700 | 0.8575 | 0.8450 | 0.8325 | 0.8950 | 0.8825 | 0.8700 | 0.8575 | 0.8450 | 0.8325 | 0.8950 | 0.8825 | 0.8700 | 0.8575 | 0.8450 | 0.8325 |
| $\alpha_1$ | 0.8050 | 0.7975 | 0.7900 | 0.7825 | 0.7750 | 0.7675 | 0.8050 | 0.7975 | 0.7900 | 0.7825 | 0.7750 | 0.7675 | 0.8050 | 0.7975 | 0.7900 | 0.7825 | 0.7750 | 0.7675 |
| $\omega_{pu}$ | $K_r$ | | | | | | $K_r$ | | | | | | $K_r$ | | | | | |
| 0.005 | 0.0045 | 0.0045 | 0.0045 | 0.0045 | 0.0045 | 0.0045 | 0.0045 | 0.0045 | 0.0045 | 0.0045 | 0.0045 | 0.0045 | 0.0045 | 0.0045 | 0.0045 | 0.0045 | 0.0045 | 0.0045 |
| 0.010 | 0.0089 | 0.0089 | 0.0089 | 0.0089 | 0.0089 | 0.0089 | 0.0089 | 0.0089 | 0.0089 | 0.0089 | 0.0088 | 0.0088 | 0.0088 | 0.0088 | 0.0088 | 0.0088 | 0.0088 | 0.0088 |
| 0.015 | 0.0132 | 0.0132 | 0.0132 | 0.0132 | 0.0132 | 0.0132 | 0.0132 | 0.0132 | 0.0132 | 0.0132 | 0.0132 | 0.0132 | 0.0131 | 0.0131 | 0.0131 | 0.0131 | 0.0131 | 0.0131 |
| 0.020 | 0.0175 | 0.0175 | 0.0175 | 0.0175 | 0.0175 | 0.0175 | 0.0174 | 0.0174 | 0.0174 | 0.0174 | 0.0174 | 0.0174 | 0.0174 | 0.0174 | 0.0174 | 0.0173 | 0.0173 | 0.0173 |
| 0.025 | 0.0217 | 0.0217 | 0.0217 | 0.0217 | 0.0217 | 0.0217 | 0.0216 | 0.0216 | 0.0216 | 0.0216 | 0.0216 | 0.0216 | 0.0215 | 0.0215 | 0.0215 | 0.0215 | 0.0215 | 0.0214 |
| 0.030 | 0.0259 | 0.0259 | 0.0259 | 0.0258 | 0.0258 | 0.0258 | 0.0258 | 0.0257 | 0.0257 | 0.0257 | 0.0257 | 0.0257 | 0.0256 | 0.0256 | 0.0256 | 0.0255 | 0.0255 | 0.0255 |
| 0.035 | 0.0300 | 0.0300 | 0.0300 | 0.0299 | 0.0299 | 0.0299 | 0.0298 | 0.0298 | 0.0298 | 0.0297 | 0.0297 | 0.0297 | 0.0296 | 0.0296 | 0.0296 | 0.0295 | 0.0295 | 0.0295 |
| 0.040 | 0.0341 | 0.0340 | 0.0340 | 0.0340 | 0.0339 | 0.0339 | 0.0338 | 0.0338 | 0.0337 | 0.0337 | 0.0337 | 0.0336 | 0.0336 | 0.0335 | 0.0335 | 0.0334 | 0.0334 | 0.0334 |
| 0.045 | 0.0380 | 0.0380 | 0.0380 | 0.0379 | 0.0379 | 0.0379 | 0.0377 | 0.0377 | 0.0377 | 0.0376 | 0.0376 | 0.0375 | 0.0374 | 0.0374 | 0.0373 | 0.0373 | 0.0372 | 0.0372 |
| 0.050 | 0.0420 | 0.0419 | 0.0419 | 0.0419 | 0.0418 | 0.0418 | 0.0416 | 0.0416 | 0.0415 | 0.0415 | 0.0414 | 0.0413 | 0.0412 | 0.0412 | 0.0411 | 0.0411 | 0.0410 | 0.0409 |
| 0.055 | 0.0459 | 0.0458 | 0.0458 | 0.0457 | 0.0457 | 0.0456 | 0.0454 | 0.0454 | 0.0453 | 0.0452 | 0.0452 | 0.0451 | 0.0450 | 0.0449 | 0.0448 | 0.0448 | 0.0447 | 0.0446 |
| 0.060 | 0.0497 | 0.0496 | 0.0496 | 0.0495 | 0.0494 | 0.0494 | 0.0492 | 0.0491 | 0.0490 | 0.0490 | 0.0489 | 0.0488 | 0.0487 | 0.0486 | 0.0485 | 0.0484 | 0.0483 | 0.0482 |
| 0.065 | 0.0535 | 0.0534 | 0.0533 | 0.0533 | 0.0532 | 0.0531 | 0.0529 | 0.0528 | 0.0527 | 0.0526 | 0.0525 | 0.0524 | 0.0523 | 0.0522 | 0.0521 | 0.0520 | 0.0519 | 0.0518 |
| 0.070 | 0.0572 | 0.0571 | 0.0570 | 0.0569 | 0.0569 | 0.0568 | 0.0565 | 0.0564 | 0.0563 | 0.0562 | 0.0561 | 0.0560 | 0.0558 | 0.0557 | 0.0556 | 0.0555 | 0.0554 | 0.0553 |
| 0.075 | 0.0609 | 0.0608 | 0.0607 | 0.0606 | 0.0605 | 0.0604 | 0.0601 | 0.0600 | 0.0599 | 0.0598 | 0.0596 | 0.0595 | 0.0593 | 0.0592 | 0.0591 | 0.0590 | 0.0588 | 0.0587 |
| 0.080 | 0.0645 | 0.0644 | 0.0643 | 0.0642 | 0.0641 | 0.0639 | 0.0636 | 0.0635 | 0.0634 | 0.0632 | 0.0631 | 0.0630 | 0.0628 | 0.0626 | 0.0625 | 0.0623 | 0.0622 | 0.0620 |
| 0.085 | 0.0680 | 0.0679 | 0.0678 | 0.0677 | 0.0676 | 0.0674 | 0.0671 | 0.0669 | 0.0668 | 0.0667 | 0.0665 | 0.0664 | 0.0661 | 0.0660 | 0.0658 | 0.0657 | 0.0655 | 0.0653 |
| 0.090 | 0.0716 | 0.0714 | 0.0713 | 0.0712 | 0.0710 | 0.0709 | 0.0705 | 0.0703 | 0.0702 | 0.0700 | 0.0699 | 0.0697 | 0.0694 | 0.0693 | 0.0691 | 0.0689 | 0.0688 | 0.0686 |
| 0.095 | 0.0750 | 0.0749 | 0.0747 | 0.0746 | 0.0744 | 0.0743 | 0.0738 | 0.0737 | 0.0735 | 0.0733 | 0.0732 | 0.0730 | 0.0727 | 0.0725 | 0.0723 | 0.0721 | 0.0719 | 0.0717 |
| 0.100 | 0.0784 | 0.0783 | 0.0781 | 0.0780 | 0.0778 | 0.0776 | 0.0772 | 0.0770 | 0.0768 | 0.0766 | 0.0764 | 0.0762 | 0.0759 | 0.0757 | 0.0755 | 0.0753 | 0.0751 | 0.0749 |
| 0.105 | 0.0818 | 0.0816 | 0.0815 | 0.0813 | 0.0811 | 0.0809 | 0.0804 | 0.0802 | 0.0800 | 0.0798 | 0.0796 | 0.0794 | 0.0791 | 0.0788 | 0.0786 | 0.0784 | 0.0782 | 0.0779 |
| 0.110 | 0.0851 | 0.0849 | 0.0848 | 0.0846 | 0.0844 | 0.0842 | 0.0836 | 0.0834 | 0.0832 | 0.0830 | 0.0827 | 0.0825 | 0.0822 | 0.0819 | 0.0817 | 0.0814 | 0.0812 | 0.0809 |
| 0.115 | 0.0884 | 0.0882 | 0.0880 | 0.0878 | 0.0876 | 0.0874 | 0.0868 | 0.0865 | 0.0863 | 0.0861 | 0.0858 | 0.0856 | 0.0852 | 0.0850 | 0.0847 | 0.0844 | 0.0842 | 0.0839 |
| 0.120 | 0.0916 | 0.0914 | 0.0912 | 0.0910 | 0.0907 | 0.0905 | 0.0899 | 0.0896 | 0.0894 | 0.0891 | 0.0889 | 0.0886 | 0.0882 | 0.0879 | 0.0877 | 0.0874 | 0.0871 | 0.0868 |
| 0.125 | 0.0948 | 0.0946 | 0.0943 | 0.0941 | 0.0939 | 0.0936 | 0.0929 | 0.0927 | 0.0924 | 0.0921 | 0.0919 | 0.0916 | 0.0911 | 0.0909 | 0.0906 | 0.0903 | 0.0899 | 0.0896 |
| 0.130 | 0.0979 | 0.0977 | 0.0974 | 0.0972 | 0.0969 | 0.0967 | 0.0959 | 0.0957 | 0.0954 | 0.0951 | 0.0948 | 0.0945 | 0.0940 | 0.0937 | 0.0934 | 0.0931 | 0.0928 | 0.0924 |
| 0.135 | 0.1010 | 0.1008 | 0.1005 | 0.1002 | 0.0999 | 0.0997 | 0.0989 | 0.0986 | 0.0983 | 0.0980 | 0.0977 | 0.0974 | 0.0969 | 0.0966 | 0.0962 | 0.0959 | 0.0955 | 0.0952 |
| 0.140 | 0.1041 | 0.1038 | 0.1035 | 0.1032 | 0.1029 | 0.1026 | 0.1018 | 0.1015 | 0.1012 | 0.1009 | 0.1005 | 0.1002 | 0.0997 | 0.0993 | 0.0990 | 0.0986 | 0.0982 | 0.0979 |
| 0.145 | 0.1071 | 0.1068 | 0.1065 | 0.1062 | 0.1058 | 0.1055 | 0.1047 | 0.1044 | 0.1040 | 0.1037 | 0.1033 | 0.1030 | 0.1024 | 0.1021 | 0.1017 | 0.1013 | 0.1009 | 0.1005 |
| 0.150 | 0.1100 | 0.1097 | 0.1094 | 0.1091 | 0.1087 | 0.1084 | 0.1075 | 0.1072 | 0.1068 | 0.1064 | 0.1061 | 0.1057 | 0.1052 | 0.1048 | 0.1044 | 0.1040 | 0.1035 | 0.1031 |
| 0.155 | 0.1129 | 0.1126 | 0.1123 | 0.1119 | 0.1116 | 0.1112 | 0.1103 | 0.1099 | 0.1096 | 0.1092 | 0.1088 | 0.1084 | 0.1078 | 0.1074 | 0.1070 | 0.1066 | 0.1061 | 0.1057 |
| 0.160 | 0.1158 | 0.1154 | 0.1151 | 0.1147 | 0.1144 | 0.1140 | 0.1130 | 0.1127 | 0.1123 | 0.1119 | 0.1114 | 0.1110 | 0.1104 | 0.1100 | 0.1096 | 0.1091 | 0.1087 | 0.1082 |
| 0.165 | 0.1186 | 0.1183 | 0.1179 | 0.1175 | 0.1171 | 0.1167 | 0.1157 | 0.1153 | 0.1149 | 0.1145 | 0.1141 | 0.1136 | 0.1130 | 0.1126 | 0.1121 | 0.1116 | 0.1111 | 0.1106 |
| 0.170 | 0.1214 | 0.1210 | 0.1206 | 0.1202 | 0.1198 | 0.1194 | 0.1184 | 0.1180 | 0.1175 | 0.1171 | 0.1166 | 0.1162 | 0.1155 | 0.1151 | 0.1146 | 0.1141 | 0.1136 | 0.1131 |
| 0.175 | 0.1241 | 0.1237 | 0.1233 | 0.1229 | 0.1225 | 0.1221 | 0.1210 | 0.1206 | 0.1201 | 0.1196 | 0.1192 | 0.1187 | 0.1180 | 0.1175 | 0.1170 | 0.1165 | 0.1160 | 0.1155 |
| 0.180 | 0.1269 | 0.1264 | 0.1260 | 0.1256 | 0.1251 | 0.1247 | 0.1236 | 0.1231 | 0.1226 | 0.1222 | 0.1216 | 0.1211 | 0.1205 | 0.1200 | 0.1194 | 0.1189 | 0.1184 | 0.1178 |
| 0.185 | 0.1295 | 0.1291 | 0.1286 | 0.1282 | 0.1277 | 0.1272 | 0.1261 | 0.1256 | 0.1251 | 0.1246 | 0.1241 | 0.1236 | 0.1229 | 0.1224 | 0.1218 | 0.1213 | 0.1207 | 0.1201 |
| 0.190 | 0.1321 | 0.1317 | 0.1312 | 0.1307 | 0.1303 | 0.1298 | 0.1286 | 0.1281 | 0.1276 | 0.1271 | 0.1265 | 0.1260 | 0.1253 | 0.1247 | 0.1241 | 0.1236 | 0.1230 | 0.1223 |
| 0.195 | 0.1347 | 0.1343 | 0.1338 | 0.1333 | 0.1328 | 0.1322 | 0.1311 | 0.1305 | 0.1300 | 0.1294 | 0.1289 | 0.1283 | 0.1276 | 0.1270 | 0.1264 | 0.1258 | 0.1252 | 0.1246 |
| 0.200 | 0.1373 | 0.1368 | 0.1363 | 0.1358 | 0.1352 | 0.1347 | 0.1335 | 0.1329 | 0.1324 | 0.1318 | 0.1312 | 0.1306 | 0.1299 | 0.1293 | 0.1287 | 0.1280 | 0.1274 | 0.1268 |
| 0.205 | 0.1398 | 0.1393 | 0.1387 | 0.1382 | 0.1377 | 0.1371 | 0.1359 | 0.1353 | 0.1347 | 0.1341 | 0.1335 | 0.1329 | 0.1322 | 0.1315 | 0.1309 | 0.1302 | 0.1296 | 0.1289 |
| 0.210 | 0.1423 | 0.1417 | 0.1412 | 0.1406 | 0.1401 | 0.1395 | 0.1382 | 0.1376 | 0.1370 | 0.1364 | 0.1358 | 0.1351 | 0.1344 | 0.1337 | 0.1331 | 0.1324 | 0.1317 | 0.1310 |
| 0.215 | 0.1447 | 0.1441 | 0.1436 | 0.1430 | 0.1424 | 0.1418 | 0.1405 | 0.1399 | 0.1393 | 0.1386 | 0.1380 | 0.1373 | 0.1366 | 0.1359 | 0.1352 | 0.1345 | 0.1338 | 0.1331 |
| 0.220 | 0.1471 | 0.1465 | 0.1459 | 0.1453 | 0.1447 | 0.1441 | 0.1428 | 0.1421 | 0.1415 | 0.1408 | 0.1402 | 0.1395 | 0.1387 | 0.1380 | 0.1373 | 0.1366 | 0.1359 | 0.1351 |
| 0.225 | 0.1495 | 0.1489 | 0.1483 | 0.1476 | 0.1470 | 0.1464 | 0.1450 | 0.1444 | 0.1437 | 0.1430 | 0.1423 | 0.1416 | 0.1408 | 0.1401 | 0.1394 | 0.1386 | 0.1379 | 0.1371 |
| 0.230 | 0.1518 | 0.1512 | 0.1505 | 0.1499 | 0.1492 | 0.1486 | 0.1472 | 0.1465 | 0.1458 | 0.1451 | 0.1444 | 0.1437 | 0.1429 | 0.1422 | 0.1414 | 0.1407 | 0.1399 | 0.1391 |
| 0.235 | 0.1541 | 0.1534 | 0.1528 | 0.1521 | 0.1515 | 0.1508 | 0.1494 | 0.1487 | 0.1480 | 0.1472 | 0.1465 | 0.1457 | 0.1449 | 0.1442 | 0.1434 | 0.1426 | 0.1418 | 0.1410 |
| 0.240 | 0.1563 | 0.1557 | 0.1550 | 0.1543 | 0.1536 | 0.1529 | 0.1515 | 0.1508 | 0.1501 | 0.1493 | 0.1485 | 0.1478 | 0.1470 | 0.1462 | 0.1454 | 0.1446 | 0.1437 | 0.1429 |
| 0.245 | 0.1585 | 0.1579 | 0.1572 | 0.1565 | 0.1558 | 0.1550 | 0.1536 | 0.1529 | 0.1521 | 0.1513 | 0.1505 | 0.1497 | 0.1489 | 0.1481 | 0.1473 | 0.1465 | 0.1456 | 0.1448 |
| 0.250 | 0.1607 | 0.1600 | 0.1593 | 0.1586 | 0.1579 | 0.1571 | 0.1557 | 0.1549 | 0.1541 | 0.1533 | 0.1525 | 0.1517 | 0.1509 | 0.1501 | 0.1492 | 0.1484 | 0.1475 | 0.1466 |
| 0.255 | 0.1629 | 0.1622 | 0.1614 | 0.1607 | 0.1599 | 0.1592 | 0.1577 | 0.1569 | 0.1561 | 0.1553 | 0.1545 | 0.1536 | 0.1528 | 0.1520 | 0.1511 | 0.1502 | 0.1493 | 0.1484 |
| 0.260 | 0.1650 | 0.1643 | 0.1635 | 0.1628 | 0.1620 | 0.1612 | 0.1597 | 0.1589 | 0.1581 | 0.1572 | 0.1564 | 0.1555 | 0.1547 | 0.1538 | 0.1529 | 0.1520 | 0.1511 | 0.1502 |
| 0.265 | 0.1671 | 0.1663 | 0.1656 | 0.1648 | 0.1640 | 0.1632 | 0.1617 | 0.1608 | 0.1600 | 0.1591 | 0.1583 | 0.1574 | 0.1566 | 0.1557 | 0.1548 | 0.1538 | 0.1529 | 0.1519 |
| 0.270 | 0.1692 | 0.1684 | 0.1676 | 0.1668 | 0.1659 | 0.1651 | 0.1636 | 0.1628 | 0.1619 | 0.1610 | 0.1601 | 0.1592 | 0.1584 | 0.1575 | 0.1565 | 0.1556 | 0.1546 | 0.1536 |
| 0.275 | 0.1712 | 0.1704 | 0.1696 | 0.1687 | 0.1679 | 0.1670 | 0.1655 | 0.1646 | 0.1638 | 0.1628 | 0.1619 | 0.1610 | 0.1602 | 0.1592 | 0.1583 | 0.1573 | 0.1563 | 0.1553 |
| 0.280 | 0.1732 | 0.1723 | 0.1715 | 0.1707 | 0.1698 | 0.1689 | 0.1674 | 0.1665 | 0.1656 | 0.1647 | 0.1637 | 0.1628 | 0.1619 | 0.1610 | 0.1600 | 0.1590 | 0.1580 | 0.1570 |
| 0.285 | 0.1751 | 0.1743 | 0.1734 | 0.1726 | 0.1717 | 0.1708 | 0.1692 | 0.1683 | 0.1674 | 0.1664 | 0.1655 | 0.1645 | 0.1637 | 0.1627 | 0.1617 | 0.1607 | 0.1597 | 0.1586 |
| 0.290 | 0.1771 | 0.1762 | 0.1753 | 0.1744 | 0.1735 | 0.1726 | 0.1711 | 0.1701 | 0.1692 | 0.1682 | 0.1672 | 0.1662 | 0.1654 | 0.1644 | 0.1634 | 0.1623 | 0.1613 | 0.1602 |
| 0.295 | 0.1790 | 0.1781 | 0.1772 | 0.1763 | 0.1753 | 0.1744 | 0.1728 | 0.1719 | 0.1709 | 0.1699 | 0.1689 | 0.1679 | 0.1671 | 0.1661 | 0.1650 | 0.1640 | 0.1629 | 0.1618 |
| 0.300 | 0.1808 | 0.1799 | 0.1790 | 0.1781 | 0.1771 | 0.1761 | 0.1746 | 0.1736 | 0.1726 | 0.1716 | 0.1706 | 0.1695 | 0.1687 | 0.1677 | 0.1666 | 0.1656 | 0.1645 | 0.1633 |
| 0.305 | 0.1827 | 0.1817 | 0.1808 | 0.1798 | 0.1789 | 0.1779 | 0.1763 | 0.1753 | 0.1743 | 0.1733 | 0.1722 | 0.1712 | 0.1704 | 0.1693 | 0.1682 | 0.1671 | 0.1660 | 0.1649 |
| 0.310 | 0.1845 | 0.1835 | 0.1826 | 0.1816 | 0.1806 | 0.1796 | 0.1780 | 0.1770 | 0.1760 | 0.1749 | 0.1738 | 0.1728 | 0.1720 | 0.1709 | 0.1698 | 0.1687 | 0.1675 | 0.1664 |
| 0.315 | 0.1863 | 0.1853 | 0.1843 | 0.1833 | 0.1823 | 0.1812 | 0.1797 | 0.1787 | 0.1776 | 0.1765 | 0.1754 | 0.1743 | 0.1736 | 0.1725 | 0.1713 | 0.1702 | 0.1690 | 0.1678 |
| 0.320 | 0.1880 | 0.1870 | 0.1860 | 0.1850 | 0.1839 | 0.1829 | 0.1814 | 0.1803 | 0.1792 | 0.1781 | 0.1770 | 0.1759 | 0.1751 | 0.1740 | 0.1728 | 0.1717 | 0.1705 | 0.1693 |
| 0.325 | 0.1897 | 0.1887 | 0.1877 | 0.1866 | 0.1856 | 0.1845 | 0.1830 | 0.1819 | 0.1808 | 0.1797 | 0.1785 | 0.1774 | 0.1766 | 0.1755 | 0.1743 | 0.1731 | 0.1719 | 0.1707 |
| 0.330 | 0.1914 | 0.1904 | 0.1893 | 0.1883 | 0.1872 | 0.1861 | 0.1846 | 0.1835 | 0.1824 | 0.1812 | 0.1801 | 0.1789 | 0.1781 | 0.1770 | 0.1758 | 0.1746 | 0.1734 | 0.1721 |
| 0.335 | 0.1931 | 0.1920 | 0.1910 | 0.1899 | 0.1888 | 0.1876 | 0.1862 | 0.1850 | 0.1839 | 0.1827 | 0.1815 | 0.1803 | 0.1796 | 0.1784 | 0.1772 | 0.1760 | 0.1748 | 0.1735 |
| 0.340 | 0.1947 | 0.1937 | 0.1926 | 0.1915 | 0.1903 | 0.1892 | 0.1877 | 0.1866 | 0.1854 | 0.1842 | 0.1830 | 0.1818 | 0.1811 | 0.1799 | 0.1786 | 0.1774 | 0.1761 | 0.1748 |
| $\omega_{pu (max)}$ | 0.303 | 0.296 | 0.289 | 0.282 | 0.275 | 0.268 | 0.321 | 0.314 | 0.306 | 0.299 | 0.292 | 0.285 | 0.342 | 0.334 | 0.327 | 0.319 | 0.311 | 0.304 |

Concrete:

$f'_c$ = 40 MPa

Normal Density

Factored Moment $M_f$ = 950 kN · m

*Problem:*

Find the required amount of non-prestressed reinforcement with $f_y$ = 400 MPa.

*Solution:*

To use Fig. 3.3.5, determine:

$$K_r = \frac{M_f(10^6)}{f'_c bd^2} = \frac{(950)(10^6)}{(40)(300)(940^2)} = 0.0896$$

From Fig. 3.3.5, for $K_r$ = 0.0896:

$\omega$ = 0.117 which is less than the maximum value of 0.334

$$A_s = \frac{\omega f'_c bd}{f_y} = \frac{(0.117)(40)(300)(940)}{(400)} = 3299 \text{ mm}^2$$

Check minimum reinforcement required:

$$A_{smin} = \frac{0.2\sqrt{f'_c}}{f_y} b_t h = \frac{0.2\sqrt{40}}{(400)}(450)(1000)$$

$$= 1423 \text{ mm}^2 < A_s$$

Therefore, use 7 No. 25 Grade 400 bars

($A_s$ = 3500 mm²)

**Example 2 – Factored flexural resistance by use of approximate equation for $f_{pr}$**

*Given:*

1220 × 203 hollow-core slab produced in a CSA certified plant.

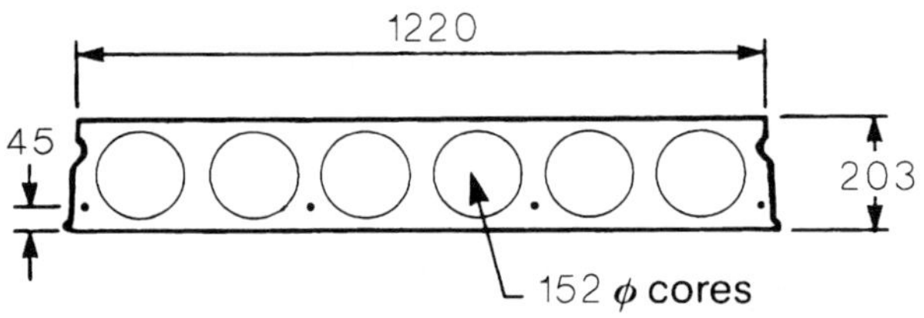


Concrete:

$f'_c$ = 45 MPa  $\alpha_1$ = 0.78  $\beta_1$ = 0.86  $\phi_c$ = 0.65

Normal Density

Prestressed reinforcement:

6 - 9 mm 1860 MPa low relaxation strands

$k_p$ = 0.28

$A_p$ = (6) (55) = 330 mm²

Section Properties:

$A_g$ = 138 700 mm²

$S_b$ = 6825 × 10³ mm³

*Problem:*

Find the factored flexural resistance $M_r$

*Solution:*

Check that the depth of compression block is less than top flange thickness of (203 – 152)/2 = 25 mm:

$$\omega_{pu} = \frac{A_p f_{pu}}{bd_p f'_c} = \frac{(330)(1860)}{(1220)(158)(45)} = 0.0708$$

**Fig. 3.3.9 Flexural resistance coefficients for elements with bonded prestressed reinforcement only, produced in non-CSA certified plant ($Ø_c$ = 0.60)**

| Procedure: | |
|---|---|
| **Design** | **Analysis** |
| 1. Determine $K_r = \frac{M_f}{f'_c bd_p^2}$ | 1. Determine $\rho_p = A_p / bd_p$ |
| 2. Calculate $\omega$ from the table | 2. Determine $\omega_{pu} = \rho_p f_{pu} / f'_c$ |
| 3. Determine $\rho_p = \omega_{pu} f'_c / f_{pu}$ | 3. Calculate $K_r$ from the table |
| 4. Determine $A_p = \rho_p bd_p$ | 4. Determine $M_r = K_r f'_c bd_p^2$ |

**Fig. 3.3.9 Continued**

$\phi_c$ = 0.60

$\phi_p$ = 0.90

| | $k_p$ = 0.28 | | | | | | $k_p$ = 0.38 | | | | | | $k_p$ = 0.48 | | | | | |
|---|---|---|---|---|---|---|---|---|---|---|---|---|---|---|---|---|---|---|
| $f'_c$ | 30 | 35 | 40 | 45 | 50 | 55 | 30 | 35 | 40 | 45 | 50 | 55 | 30 | 35 | 40 | 45 | 50 | 55 |
| $\beta_1$ | 0.8950 | 0.8825 | 0.8700 | 0.8575 | 0.8450 | 0.8325 | 0.8950 | 0.8825 | 0.8700 | 0.8575 | 0.8450 | 0.8325 | 0.8950 | 0.8825 | 0.8700 | 0.8575 | 0.8450 | 0.8325 |
| $\alpha_1$ | 0.8050 | 0.7975 | 0.7900 | 0.7825 | 0.7750 | 0.7675 | 0.8050 | 0.7975 | 0.7900 | 0.7825 | 0.7750 | 0.7675 | 0.8050 | 0.7975 | 0.7900 | 0.7825 | 0.7750 | 0.7675 |
| $\omega_{pu}$ | $K_r$ | | | | | | $K_r$ | | | | | | $K_r$ | | | | | |
| 0.005 | 0.0045 | 0.0045 | 0.0045 | 0.0045 | 0.0045 | 0.0045 | 0.0045 | 0.0045 | 0.0045 | 0.0045 | 0.0045 | 0.0045 | 0.0045 | 0.0045 | 0.0045 | 0.0045 | 0.0045 | 0.0045 |
| 0.010 | 0.0089 | 0.0089 | 0.0089 | 0.0089 | 0.0089 | 0.0089 | 0.0088 | 0.0088 | 0.0088 | 0.0088 | 0.0088 | 0.0088 | 0.0088 | 0.0088 | 0.0088 | 0.0088 | 0.0088 | 0.0088 |
| 0.015 | 0.0132 | 0.0132 | 0.0132 | 0.0132 | 0.0132 | 0.0132 | 0.0132 | 0.0132 | 0.0131 | 0.0131 | 0.0131 | 0.0131 | 0.0131 | 0.0131 | 0.0131 | 0.0131 | 0.0131 | 0.0131 |
| 0.020 | 0.0175 | 0.0175 | 0.0174 | 0.0174 | 0.0174 | 0.0174 | 0.0174 | 0.0174 | 0.0174 | 0.0174 | 0.0174 | 0.0173 | 0.0173 | 0.0173 | 0.0173 | 0.0173 | 0.0173 | 0.0173 |
| 0.025 | 0.0217 | 0.0217 | 0.0216 | 0.0216 | 0.0216 | 0.0216 | 0.0216 | 0.0215 | 0.0215 | 0.0215 | 0.0215 | 0.0215 | 0.0215 | 0.0214 | 0.0214 | 0.0214 | 0.0214 | 0.0214 |
| 0.030 | 0.0258 | 0.0258 | 0.0258 | 0.0258 | 0.0257 | 0.0257 | 0.0257 | 0.0256 | 0.0256 | 0.0256 | 0.0256 | 0.0255 | 0.0255 | 0.0255 | 0.0255 | 0.0254 | 0.0254 | 0.0254 |
| 0.035 | 0.0299 | 0.0299 | 0.0298 | 0.0298 | 0.0298 | 0.0298 | 0.0297 | 0.0297 | 0.0296 | 0.0296 | 0.0296 | 0.0295 | 0.0295 | 0.0294 | 0.0294 | 0.0294 | 0.0293 | 0.0293 |
| 0.040 | 0.0339 | 0.0339 | 0.0338 | 0.0338 | 0.0338 | 0.0337 | 0.0336 | 0.0336 | 0.0336 | 0.0335 | 0.0335 | 0.0334 | 0.0334 | 0.0333 | 0.0333 | 0.0332 | 0.0332 | 0.0332 |
| 0.045 | 0.0379 | 0.0378 | 0.0378 | 0.0377 | 0.0377 | 0.0377 | 0.0375 | 0.0375 | 0.0374 | 0.0374 | 0.0373 | 0.0373 | 0.0372 | 0.0372 | 0.0371 | 0.0370 | 0.0370 | 0.0369 |
| 0.050 | 0.0417 | 0.0417 | 0.0417 | 0.0416 | 0.0416 | 0.0415 | 0.0413 | 0.0413 | 0.0412 | 0.0412 | 0.0411 | 0.0411 | 0.0410 | 0.0409 | 0.0408 | 0.0408 | 0.0407 | 0.0406 |
| 0.055 | 0.0456 | 0.0455 | 0.0455 | 0.0454 | 0.0454 | 0.0453 | 0.0451 | 0.0450 | 0.0450 | 0.0449 | 0.0448 | 0.0448 | 0.0446 | 0.0446 | 0.0445 | 0.0444 | 0.0443 | 0.0443 |
| 0.060 | 0.0494 | 0.0493 | 0.0492 | 0.0492 | 0.0491 | 0.0490 | 0.0488 | 0.0487 | 0.0486 | 0.0486 | 0.0485 | 0.0484 | 0.0483 | 0.0482 | 0.0481 | 0.0480 | 0.0479 | 0.0478 |
| 0.065 | 0.0531 | 0.0530 | 0.0529 | 0.0528 | 0.0528 | 0.0527 | 0.0524 | 0.0523 | 0.0523 | 0.0522 | 0.0521 | 0.0520 | 0.0518 | 0.0517 | 0.0516 | 0.0515 | 0.0514 | 0.0513 |
| 0.070 | 0.0567 | 0.0567 | 0.0566 | 0.0565 | 0.0564 | 0.0563 | 0.0560 | 0.0559 | 0.0558 | 0.0557 | 0.0556 | 0.0555 | 0.0553 | 0.0552 | 0.0551 | 0.0549 | 0.0548 | 0.0547 |
| 0.075 | 0.0603 | 0.0602 | 0.0601 | 0.0600 | 0.0599 | 0.0598 | 0.0595 | 0.0594 | 0.0593 | 0.0592 | 0.0590 | 0.0589 | 0.0587 | 0.0586 | 0.0585 | 0.0583 | 0.0582 | 0.0580 |
| 0.080 | 0.0639 | 0.0638 | 0.0637 | 0.0636 | 0.0634 | 0.0633 | 0.0630 | 0.0628 | 0.0627 | 0.0626 | 0.0624 | 0.0623 | 0.0621 | 0.0619 | 0.0618 | 0.0616 | 0.0615 | 0.0613 |
| 0.085 | 0.0674 | 0.0673 | 0.0671 | 0.0670 | 0.0669 | 0.0668 | 0.0664 | 0.0662 | 0.0661 | 0.0659 | 0.0658 | 0.0656 | 0.0654 | 0.0652 | 0.0650 | 0.0649 | 0.0647 | 0.0645 |
| 0.090 | 0.0708 | 0.0707 | 0.0706 | 0.0704 | 0.0703 | 0.0701 | 0.0697 | 0.0695 | 0.0694 | 0.0692 | 0.0690 | 0.0689 | 0.0686 | 0.0684 | 0.0682 | 0.0681 | 0.0679 | 0.0677 |
| 0.095 | 0.0742 | 0.0741 | 0.0739 | 0.0738 | 0.0736 | 0.0734 | 0.0730 | 0.0728 | 0.0726 | 0.0724 | 0.0723 | 0.0721 | 0.0718 | 0.0716 | 0.0714 | 0.0712 | 0.0710 | 0.0708 |
| 0.100 | 0.0776 | 0.0774 | 0.0772 | 0.0771 | 0.0769 | 0.0767 | 0.0762 | 0.0760 | 0.0758 | 0.0756 | 0.0754 | 0.0752 | 0.0749 | 0.0747 | 0.0745 | 0.0742 | 0.0740 | 0.0738 |
| 0.105 | 0.0808 | 0.0807 | 0.0805 | 0.0803 | 0.0801 | 0.0799 | 0.0794 | 0.0792 | 0.0790 | 0.0787 | 0.0785 | 0.0783 | 0.0780 | 0.0777 | 0.0775 | 0.0772 | 0.0770 | 0.0767 |
| 0.110 | 0.0841 | 0.0839 | 0.0837 | 0.0835 | 0.0833 | 0.0831 | 0.0825 | 0.0823 | 0.0820 | 0.0818 | 0.0816 | 0.0813 | 0.0810 | 0.0807 | 0.0805 | 0.0802 | 0.0799 | 0.0797 |
| 0.115 | 0.0873 | 0.0871 | 0.0868 | 0.0866 | 0.0864 | 0.0862 | 0.0856 | 0.0853 | 0.0851 | 0.0848 | 0.0846 | 0.0843 | 0.0839 | 0.0836 | 0.0834 | 0.0831 | 0.0828 | 0.0825 |
| 0.120 | 0.0904 | 0.0902 | 0.0899 | 0.0897 | 0.0895 | 0.0892 | 0.0886 | 0.0883 | 0.0880 | 0.0878 | 0.0875 | 0.0872 | 0.0868 | 0.0865 | 0.0862 | 0.0859 | 0.0856 | 0.0853 |
| 0.125 | 0.0935 | 0.0932 | 0.0930 | 0.0927 | 0.0925 | 0.0922 | 0.0915 | 0.0913 | 0.0910 | 0.0907 | 0.0904 | 0.0901 | 0.0897 | 0.0893 | 0.0890 | 0.0887 | 0.0884 | 0.0881 |
| 0.130 | 0.0965 | 0.0963 | 0.0960 | 0.0957 | 0.0955 | 0.0952 | 0.0944 | 0.0941 | 0.0938 | 0.0935 | 0.0932 | 0.0929 | 0.0924 | 0.0921 | 0.0918 | 0.0914 | 0.0911 | 0.0907 |
| 0.135 | 0.0995 | 0.0992 | 0.0990 | 0.0987 | 0.0984 | 0.0981 | 0.0973 | 0.0970 | 0.0967 | 0.0963 | 0.0960 | 0.0957 | 0.0952 | 0.0948 | 0.0945 | 0.0941 | 0.0938 | 0.0934 |
| 0.140 | 0.1024 | 0.1022 | 0.1019 | 0.1016 | 0.1012 | 0.1009 | 0.1001 | 0.0998 | 0.0994 | 0.0991 | 0.0988 | 0.0984 | 0.0979 | 0.0975 | 0.0971 | 0.0968 | 0.0964 | 0.0960 |
| 0.145 | 0.1053 | 0.1050 | 0.1047 | 0.1044 | 0.1041 | 0.1037 | 0.1029 | 0.1025 | 0.1022 | 0.1018 | 0.1014 | 0.1011 | 0.1005 | 0.1001 | 0.0998 | 0.0993 | 0.0989 | 0.0985 |
| 0.150 | 0.1082 | 0.1079 | 0.1075 | 0.1072 | 0.1068 | 0.1065 | 0.1056 | 0.1052 | 0.1049 | 0.1045 | 0.1041 | 0.1037 | 0.1031 | 0.1027 | 0.1023 | 0.1019 | 0.1014 | 0.1010 |
| 0.155 | 0.1110 | 0.1107 | 0.1103 | 0.1099 | 0.1096 | 0.1092 | 0.1083 | 0.1079 | 0.1075 | 0.1071 | 0.1067 | 0.1062 | 0.1057 | 0.1053 | 0.1048 | 0.1044 | 0.1039 | 0.1034 |
| 0.160 | 0.1138 | 0.1134 | 0.1130 | 0.1126 | 0.1123 | 0.1119 | 0.1109 | 0.1105 | 0.1101 | 0.1097 | 0.1092 | 0.1088 | 0.1082 | 0.1077 | 0.1073 | 0.1068 | 0.1063 | 0.1058 |
| 0.165 | 0.1165 | 0.1161 | 0.1157 | 0.1153 | 0.1149 | 0.1145 | 0.1135 | 0.1131 | 0.1126 | 0.1122 | 0.1117 | 0.1113 | 0.1107 | 0.1102 | 0.1097 | 0.1092 | 0.1087 | 0.1082 |
| 0.170 | 0.1191 | 0.1187 | 0.1183 | 0.1179 | 0.1175 | 0.1171 | 0.1160 | 0.1156 | 0.1151 | 0.1147 | 0.1142 | 0.1137 | 0.1131 | 0.1126 | 0.1121 | 0.1116 | 0.1110 | 0.1105 |
| 0.175 | 0.1218 | 0.1214 | 0.1209 | 0.1205 | 0.1200 | 0.1196 | 0.1185 | 0.1181 | 0.1176 | 0.1171 | 0.1166 | 0.1161 | 0.1155 | 0.1149 | 0.1144 | 0.1139 | 0.1133 | 0.1128 |
| 0.180 | 0.1244 | 0.1239 | 0.1235 | 0.1230 | 0.1226 | 0.1221 | 0.1210 | 0.1205 | 0.1200 | 0.1195 | 0.1190 | 0.1184 | 0.1178 | 0.1173 | 0.1167 | 0.1161 | 0.1156 | 0.1150 |
| 0.185 | 0.1269 | 0.1265 | 0.1260 | 0.1255 | 0.1250 | 0.1245 | 0.1234 | 0.1229 | 0.1224 | 0.1218 | 0.1213 | 0.1207 | 0.1201 | 0.1195 | 0.1190 | 0.1184 | 0.1178 | 0.1172 |
| 0.190 | 0.1294 | 0.1289 | 0.1284 | 0.1280 | 0.1274 | 0.1269 | 0.1258 | 0.1252 | 0.1247 | 0.1241 | 0.1236 | 0.1230 | 0.1223 | 0.1218 | 0.1212 | 0.1206 | 0.1199 | 0.1193 |
| 0.195 | 0.1319 | 0.1314 | 0.1309 | 0.1304 | 0.1298 | 0.1293 | 0.1281 | 0.1276 | 0.1270 | 0.1264 | 0.1258 | 0.1252 | 0.1245 | 0.1239 | 0.1233 | 0.1227 | 0.1221 | 0.1214 |
| 0.200 | 0.1343 | 0.1338 | 0.1333 | 0.1327 | 0.1322 | 0.1316 | 0.1304 | 0.1298 | 0.1292 | 0.1286 | 0.1280 | 0.1274 | 0.1267 | 0.1261 | 0.1255 | 0.1248 | 0.1241 | 0.1235 |
| 0.205 | 0.1367 | 0.1362 | 0.1356 | 0.1350 | 0.1345 | 0.1339 | 0.1327 | 0.1321 | 0.1315 | 0.1308 | 0.1302 | 0.1296 | 0.1289 | 0.1282 | 0.1275 | 0.1269 | 0.1262 | 0.1255 |
| 0.210 | 0.1390 | 0.1385 | 0.1379 | 0.1373 | 0.1367 | 0.1361 | 0.1349 | 0.1343 | 0.1336 | 0.1330 | 0.1323 | 0.1317 | 0.1310 | 0.1303 | 0.1296 | 0.1289 | 0.1282 | 0.1275 |
| 0.215 | 0.1413 | 0.1408 | 0.1402 | 0.1396 | 0.1390 | 0.1383 | 0.1371 | 0.1364 | 0.1358 | 0.1351 | 0.1344 | 0.1337 | 0.1330 | 0.1323 | 0.1316 | 0.1309 | 0.1302 | 0.1294 |
| 0.220 | 0.1436 | 0.1430 | 0.1424 | 0.1418 | 0.1411 | 0.1405 | 0.1392 | 0.1385 | 0.1379 | 0.1372 | 0.1365 | 0.1358 | 0.1350 | 0.1343 | 0.1336 | 0.1329 | 0.1321 | 0.1313 |
| 0.225 | 0.1458 | 0.1452 | 0.1446 | 0.1439 | 0.1433 | 0.1426 | 0.1413 | 0.1406 | 0.1399 | 0.1392 | 0.1385 | 0.1378 | 0.1370 | 0.1363 | 0.1355 | 0.1348 | 0.1340 | 0.1332 |
| 0.230 | 0.1480 | 0.1474 | 0.1467 | 0.1461 | 0.1454 | 0.1447 | 0.1434 | 0.1427 | 0.1420 | 0.1412 | 0.1405 | 0.1397 | 0.1390 | 0.1382 | 0.1375 | 0.1367 | 0.1359 | 0.1350 |
| 0.235 | 0.1502 | 0.1495 | 0.1489 | 0.1482 | 0.1475 | 0.1468 | 0.1454 | 0.1447 | 0.1440 | 0.1432 | 0.1424 | 0.1417 | 0.1409 | 0.1401 | 0.1393 | 0.1385 | 0.1377 | 0.1369 |
| 0.240 | 0.1523 | 0.1516 | 0.1509 | 0.1502 | 0.1495 | 0.1488 | 0.1474 | 0.1467 | 0.1459 | 0.1451 | 0.1443 | 0.1435 | 0.1428 | 0.1420 | 0.1412 | 0.1403 | 0.1395 | 0.1386 |
| 0.245 | 0.1544 | 0.1537 | 0.1530 | 0.1523 | 0.1515 | 0.1508 | 0.1494 | 0.1486 | 0.1478 | 0.1470 | 0.1462 | 0.1454 | 0.1447 | 0.1438 | 0.1430 | 0.1421 | 0.1413 | 0.1404 |
| 0.250 | 0.1565 | 0.1557 | 0.1550 | 0.1542 | 0.1535 | 0.1527 | 0.1513 | 0.1505 | 0.1497 | 0.1489 | 0.1481 | 0.1472 | 0.1465 | 0.1456 | 0.1448 | 0.1439 | 0.1430 | 0.1421 |
| 0.255 | 0.1585 | 0.1577 | 0.1570 | 0.1562 | 0.1554 | 0.1546 | 0.1532 | 0.1524 | 0.1516 | 0.1507 | 0.1499 | 0.1490 | 0.1483 | 0.1474 | 0.1465 | 0.1456 | 0.1447 | 0.1438 |
| 0.260 | 0.1605 | 0.1597 | 0.1589 | 0.1581 | 0.1573 | 0.1565 | 0.1551 | 0.1543 | 0.1534 | 0.1525 | 0.1517 | 0.1508 | 0.1500 | 0.1491 | 0.1482 | 0.1473 | 0.1464 | 0.1454 |
| 0.265 | 0.1624 | 0.1616 | 0.1608 | 0.1600 | 0.1592 | 0.1583 | 0.1569 | 0.1561 | 0.1552 | 0.1543 | 0.1534 | 0.1525 | 0.1518 | 0.1508 | 0.1499 | 0.1490 | 0.1480 | 0.1470 |
| 0.270 | 0.1643 | 0.1635 | 0.1627 | 0.1619 | 0.1610 | 0.1601 | 0.1587 | 0.1579 | 0.1570 | 0.1561 | 0.1551 | 0.1542 | 0.1535 | 0.1525 | 0.1516 | 0.1506 | 0.1496 | 0.1486 |
| 0.275 | 0.1662 | 0.1654 | 0.1645 | 0.1637 | 0.1628 | 0.1619 | 0.1605 | 0.1596 | 0.1587 | 0.1578 | 0.1568 | 0.1559 | 0.1551 | 0.1542 | 0.1532 | 0.1522 | 0.1512 | 0.1502 |
| 0.280 | 0.1681 | 0.1672 | 0.1663 | 0.1655 | 0.1646 | 0.1636 | 0.1622 | 0.1613 | 0.1604 | 0.1594 | 0.1585 | 0.1575 | 0.1568 | 0.1558 | 0.1548 | 0.1538 | 0.1528 | 0.1517 |
| 0.285 | 0.1699 | 0.1690 | 0.1681 | 0.1672 | 0.1663 | 0.1654 | 0.1640 | 0.1630 | 0.1621 | 0.1611 | 0.1601 | 0.1591 | 0.1584 | 0.1574 | 0.1564 | 0.1553 | 0.1543 | 0.1532 |
| 0.290 | 0.1717 | 0.1708 | 0.1699 | 0.1689 | 0.1680 | 0.1670 | 0.1656 | 0.1647 | 0.1637 | 0.1627 | 0.1617 | 0.1607 | 0.1600 | 0.1589 | 0.1579 | 0.1569 | 0.1558 | 0.1547 |
| 0.295 | 0.1734 | 0.1725 | 0.1716 | 0.1706 | 0.1697 | 0.1687 | 0.1673 | 0.1663 | 0.1653 | 0.1643 | 0.1633 | 0.1622 | 0.1615 | 0.1605 | 0.1594 | 0.1584 | 0.1573 | 0.1562 |
| 0.300 | 0.1751 | 0.1742 | 0.1733 | 0.1723 | 0.1713 | 0.1703 | 0.1689 | 0.1679 | 0.1669 | 0.1659 | 0.1648 | 0.1637 | 0.1631 | 0.1620 | 0.1609 | 0.1598 | 0.1587 | 0.1576 |
| 0.305 | 0.1768 | 0.1759 | 0.1749 | 0.1739 | 0.1729 | 0.1719 | 0.1705 | 0.1695 | 0.1685 | 0.1674 | 0.1663 | 0.1652 | 0.1646 | 0.1635 | 0.1624 | 0.1613 | 0.1601 | 0.1590 |
| 0.310 | 0.1785 | 0.1775 | 0.1765 | 0.1755 | 0.1745 | 0.1734 | 0.1721 | 0.1710 | 0.1700 | 0.1689 | 0.1678 | 0.1667 | 0.1660 | 0.1649 | 0.1638 | 0.1627 | 0.1615 | 0.1603 |
| 0.315 | 0.1801 | 0.1791 | 0.1781 | 0.1771 | 0.1760 | 0.1750 | 0.1736 | 0.1726 | 0.1715 | 0.1704 | 0.1693 | 0.1681 | 0.1675 | 0.1664 | 0.1652 | 0.1641 | 0.1629 | 0.1617 |
| 0.320 | 0.1817 | 0.1807 | 0.1797 | 0.1786 | 0.1776 | 0.1765 | 0.1751 | 0.1741 | 0.1730 | 0.1718 | 0.1707 | 0.1695 | 0.1689 | 0.1678 | 0.1666 | 0.1654 | 0.1642 | 0.1630 |
| 0.325 | 0.1833 | 0.1823 | 0.1812 | 0.1801 | 0.1791 | 0.1779 | 0.1766 | 0.1755 | 0.1744 | 0.1733 | 0.1721 | 0.1709 | 0.1703 | 0.1692 | 0.1680 | 0.1668 | 0.1656 | 0.1643 |
| 0.330 | 0.1849 | 0.1838 | 0.1827 | 0.1816 | 0.1805 | 0.1794 | 0.1781 | 0.1770 | 0.1758 | 0.1747 | 0.1735 | 0.1723 | 0.1717 | 0.1705 | 0.1693 | 0.1681 | 0.1669 | 0.1656 |
| 0.335 | 0.1864 | 0.1853 | 0.1842 | 0.1831 | 0.1820 | 0.1808 | 0.1795 | 0.1784 | 0.1772 | 0.1760 | 0.1748 | 0.1736 | 0.1731 | 0.1719 | 0.1706 | 0.1694 | 0.1681 | 0.1669 |
| 0.340 | 0.1879 | 0.1868 | 0.1856 | 0.1845 | 0.1834 | 0.1822 | 0.1809 | 0.1798 | 0.1786 | 0.1774 | 0.1762 | 0.1749 | 0.1744 | 0.1732 | 0.1719 | 0.1707 | 0.1694 | 0.1681 |
| $\omega_{pu(max)}$ | 0.279 | 0.273 | 0.266 | 0.260 | 0.254 | 0.248 | 0.296 | 0.290 | 0.283 | 0.276 | 0.269 | 0.263 | 0.316 | 0.309 | 0.301 | 0.294 | 0.287 | 0.280 |

$$a = \frac{\phi_p d_p \omega_{pu}}{\alpha_1 \phi_c \left(1 + k_p \dfrac{\phi_p \omega_{pu}}{\alpha_1 \phi_c \beta_1}\right)}$$

$$= \frac{(0.9)(158)(0.0708)}{(0.78)(0.65)\left(1 + (0.28)\dfrac{(0.9)(0.0708)}{(0.78)(0.65)(0.86)}\right)}$$

$$= 19 \text{ mm} < 25$$

Determine:

$$c/d_p = \frac{1}{\left(\dfrac{\alpha_1 \phi_c \beta_1}{\phi_p \omega_{pu}} + k_p\right)}$$

$$= \frac{1}{\left[\dfrac{(0.78)(0.65)(0.86)}{(0.9)(0.0708)} + 0.28\right]} = 0.140 < 0.5$$

$$f_{pr} = f_{pu}(1 - k_p c/d_p) = (1860)[1 - (0.28)(0.140)]$$
$$= 1790 \text{ MPa}$$

$$M_r = \phi_p A_p f_{pr}(d - a/2)$$
$$= (0.9)(330)(1790)(158 - 19/2)/10^6$$
$$= 78.9 \text{ kN}\cdot\text{m}$$

Check the ductility requirement that $M_r \geq 1.2\, M_{cr}$. Assuming that the effective prestress is $0.56\, f_{pu}$:

$$f_{pe} = (0.56)(1860) = 1042 \text{ MPa}$$

$$P_e = f_{pe} A_p = (1042)(330)/10^3 = 344 \text{ kN}$$

$$1.2M_{cr} = 1.2\left(\frac{P_e}{A_g} + \frac{P_e e}{S_b} + 0.6\lambda\sqrt{f'_c}\right) S_b$$

$$= 1.2\left[\frac{(344)(10^3)}{138\,700} + \frac{(344)(56.5)(10^3)}{6825 \times 10^3} + (0.6)(1.0)\sqrt{45}\right] \times \frac{(6825 \times 10^3)}{10^6}$$

$$= 76.6 \text{ kN}\cdot\text{m} < 78.9 \quad \text{OK}$$

**Example 3 – Use of Fig. 3.3.7 coefficients for determination of bonded prestressed reinforcement**

*Given:*

400 × 600 rectangular beamproduced in a CSA certified plant

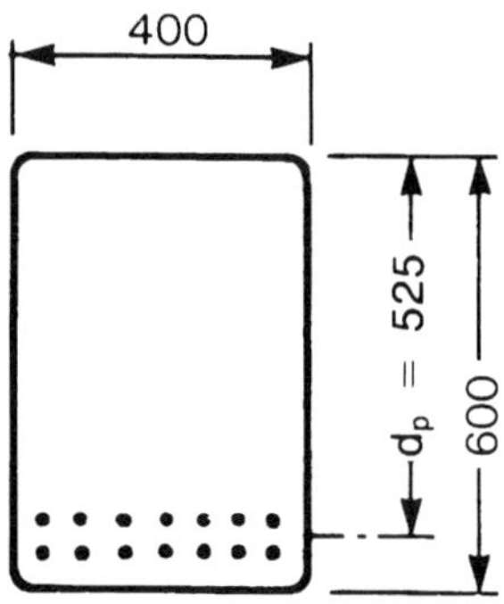


Concrete:

$f'_c$ = 35 MPa

Normal Density

Factored Moment: 350 kN · m

*Problem:*

Find the amount of prestressed reinforcement based on $f_{pu}$ = 1860 MPa and $k_p$ = 0.28 (low relaxation strand).

*Solution:*

To use Fig. 3.3.5, determine:

$$K_r = \frac{M_f}{f'_c b d_p^2}$$

$$= \frac{(350)(10^6)}{(35)(400)(525)^2} = 0.0907$$

For $K_r$ = 0.0907, $f'_c$ = 35 MPa and $k_p$ = 0.28:

$$\omega_{pu} = 0.1189$$

The required area of prestressed reinforcement is:

$$A_p = \omega_{pu} f'_c b d_p / f_{pu}$$
$$= (0.1189)(35)(400)(525)/1860$$
$$= 470 \text{ mm}^2$$

Provide 5 - 13 mm 1860 MPa strands ($A_p$ = 495 mm²)

**Example 4 – Factored flexural resistance of stadia unit**

*Given:*

1600 × 800 stadia unit produced in a CSA certified plant

Length = 10.0 m

Span = 9.7 m

Load:

Superimposed dead load: 0.2 kN/m² = 0.32 kN/m

Live Load:

| | | | | |
|---|---|---|---|---|
| Vertical | 4.80 | kN/m² | = | 7.68 kN/m |
| Horizontal | (0.15)(2) | kN/m | = | 0.30 kN/m |
| Axial | (0.3)(2)(10) | kN | = | 6.00 kN |

(Axial load acts independent of horizontal load)

Section Properties:

$A_g$ = 298 500 mm²

**Stadia Unit - dimensions**

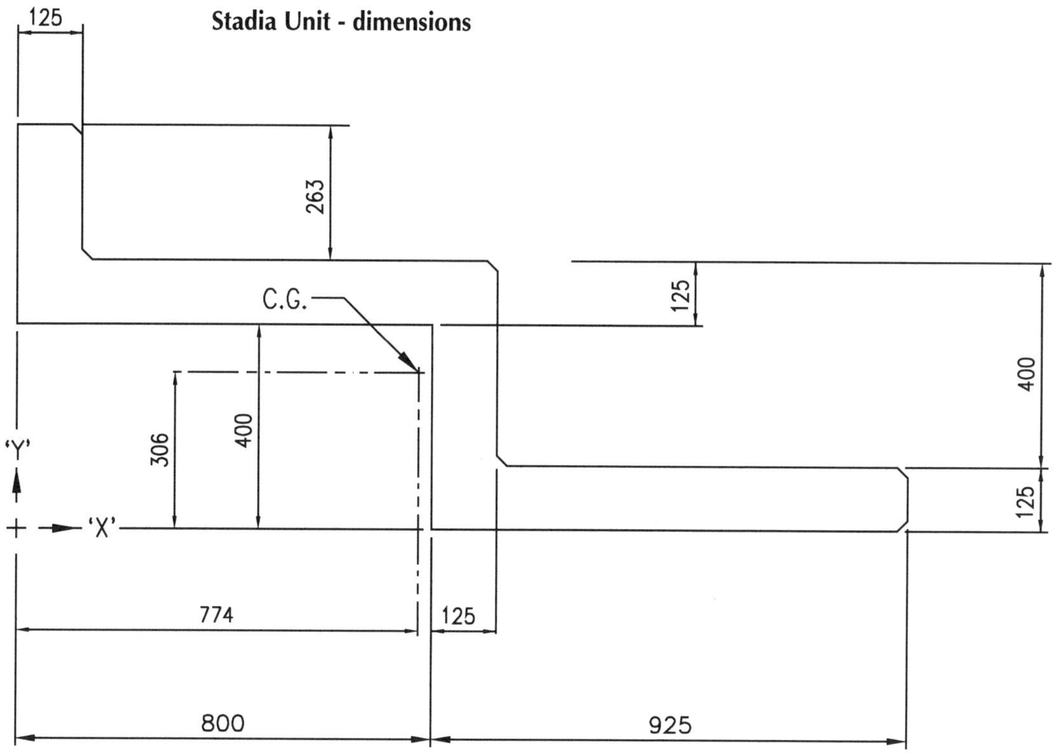


Weight $= 7.0$ kN/m

$y_b = 306$ mm

$x_L = 774$ mm

$I_x = 14\ 500 \times 10^6$ mm$^4$

$I_y = 72\ 300 \times 10^6$ mm$^4$

$I_{xy} = A_g\,(x - x_L)(y - y_b) = 27\ 720 \times 10^6$ mm$^4$

from $\tan 2\theta = \dfrac{2I_{xy}}{I_x - I_y}$:

$\theta = 21.9^\circ =$ angle of minor axis

$I_{minor} = 3350 \times 10^6$ mm$^4$

$I_{major} = 83\ 400 \times 10^6$ mm$^4$

Concrete:

$f'_c = 35$ MPa  $\alpha_1 = 0.80$  $\beta_1 = 0.88$  $\phi_c = 0.65$

$f'_{ci} = 25$ MPa

Normal density

Prestressed reinforcement:

9 – 13 mm 1860 MPa low relaxation strands

$k_p = 0.28$

$A_p = 99$ mm$^2$/strand

$P_o = 0.75\,A_p\,f_{pu} = 1243$ kN

$P_i = 0.95\,P_{jacking} = 1181$ kN

$P_e = 0.85\,P_{jacking} = 1057$ kN

*Problem:*

Determine the factored flexural resistance.

*Solution:*

Due to the shape of the compression block and to the fact that the failure will occur in bending around the minor axis, use a graphical solution to this problem.

By trial and error a value of "a" was found that simultaneously solved the following two equations:

area of compression block of depth "a"

$$a = \frac{\phi_p A_p f_{pr}}{\alpha_1 \phi_c f'_c b}$$

Stress in prestressing steel at factored resistance:

$$f_{pr} = f_{pu}(1 - k_p c / d_p)$$

where:

$\phi_p = 0.9$

$d_p = (363 + 345 + 326 + 297 + 436 + 418 + 380 + 399) / 8 = 370$ mm

Note: one strand in the compression zone was not included in the determination of $M_r$

By trial and error:

$a = 165$ mm

**Stadia Unit - prestressing strand locations**

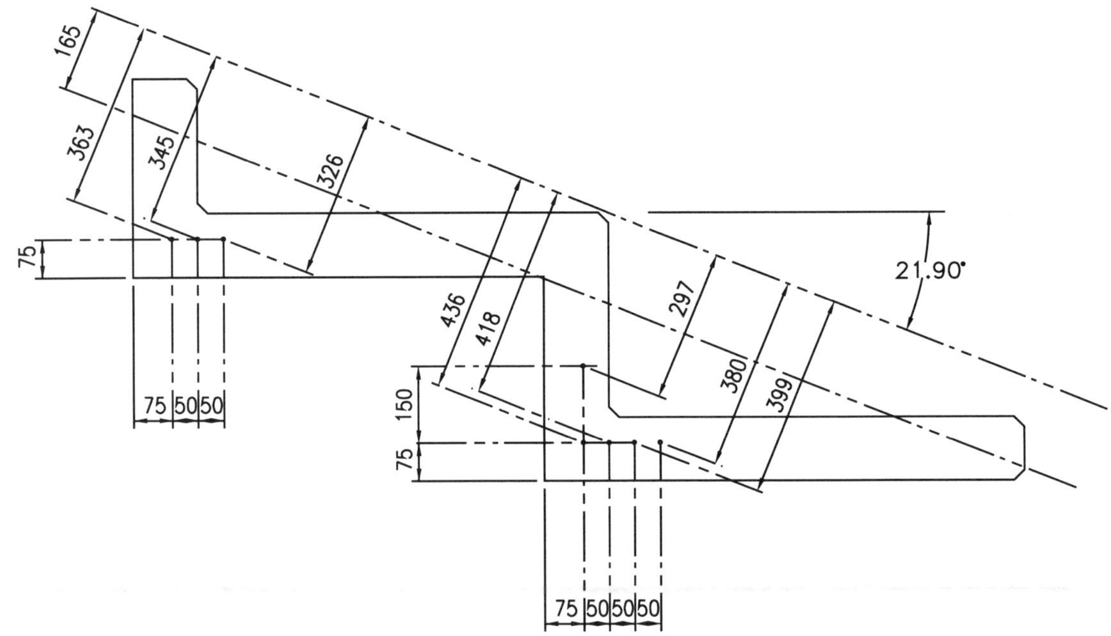


$$c = \frac{165}{0.88} = 187\,\text{mm}$$

$$c/d_p = 187/370 = 0.505 \approx 0.5$$

$$f_{pr} = 1860[1-(0.28)(0.505)] = 1597\,\text{MPa}$$

Required area of compression block

$$= \frac{(0.9)(8)(99)(1597)}{(0.80)(0.65)(35)} = 62\,703\,\text{mm}^2$$

From Fig. 3.3.2, in order for the area of the compression block to equal 62 703 mm$^2$:

$$a = 165\,\text{mm}$$

$$c = a/\beta_1 = 165/0.88 = 187\,\text{mm}$$

The centre of the equivalent compression block is conservatively estimated at 2/3 $\beta_1 c$ from the top fibre.

$$2/3\,\beta_1 c = (2/3)(0.88)(187) = 110\,\text{mm}$$

$$M_r = \phi_p A_p f_{pr}(d_p - a/2)$$
$$= (0.9)(8)(99)(1597)(370-110)/10^3$$
$$= 296\,\text{kN}\cdot\text{m}$$

$$M_f = \frac{(9.7)^2}{8}\{[(7)(1.25)+(0.32)(1.25)$$
$$+(7.68)(1.5)]\cos 21.9^\circ + (0.3)(1.5)\sin 21.9^\circ\}$$
$$= 228\,\text{kN}\cdot\text{m} < 296\,\text{kN}\cdot\text{m}$$

The bending moment around the major axis is small compared to the section depth. The axial load is small compared to the horizontal load. Both of these effects can safely be ignored.

## 3.3.3 ANALYSIS USING STRAIN COMPATIBILITY

The fundamental strain compatibility approach is a convenient method for analyzing fully and partially prestressed flexural elements (see Sect. 3.4.3 for definitions). Example 5 demonstrates the use of this method of analysis.

**Example 5 – Strain compatibility analysis of a partially prestressed flanged section**

*Given:*

600 × 550 inverted tee beam produced in a CSA certified plant with 50 mm composite topping as shown

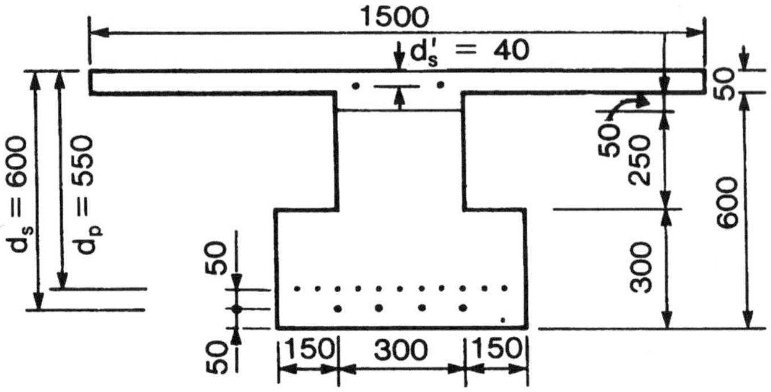


Concrete:

$f'_c$ precast = 35 MPa  $\alpha_1$ = 0.80  $\phi_c$ = 0.65

$f'_c$ topping = 25 MPa  $\alpha_1$ = 0.81  $\phi_c$ = 0.60

Normal density

Prestressed reinforcement:

11 - 13 mm 1860 MPa strands (low relaxation)

$A_p = (11)(99) = 1089\ mm^2$

$E_p = 190\ 000$ MPa

Non-prestressed reinforcement:

4 - No. 25 bottom, $f_y = 400$ MPa

$A_s = (4)(500) = 2000\ mm^2$

2 – No. 20 top, $f_y = 400$ MPa

$A'_s = (2)(300) = 600\ mm^2$

$E_s = 200\ 000$ MPa

Composite Section Properties:

$A_g = 333\ 400\ mm^2$

$I = 14\ 700 \times 10^6\ mm^4$

$y_b = 321$ mm

*Problem:*

Find factored flexural resistance $M_r$

*Solution:*

$$C1 + C2 + C3 + C4 = T1 + T2$$

$$C1 = \alpha_1 \phi_c f'_c h_f b = (0.81)(0.6)(25)(50)(1500 - 300)/10^3 = 732\ kN$$

$$C2 = (0.81)(0.6)(25)(100)(300)/10^3 = 366\ kN$$

$$C3 = (0.8)(0.65)(35)(300)(0.883c - 100)/10^3 = (4.81c - 545)\ kN$$

$$C4 = \phi_s A'_s f_y = (0.85)(600)(400)/10^3 = 204\ kN$$

$$f_{pr} = f_{pu}(1 - k_p c/d_p)$$

$$T1 = \phi_p A_p f_{pr} = (0.9)(1089)(1860)(1 - k_p c/d_p)/10^3 = (1823)(1 - k_p c/d_p)\ kN$$

$$T2 = \phi_s A_s f_y = (0.85)(2000)(400)/10^3 = 680\ kN$$

According to Fig. 3.3.2:

$$c/d_p = \frac{\phi_p A_p f_{pu} + \phi_s A_s f_s - \phi_s A'_s f'_s - \alpha_1 \phi_c f'_c (b - b_w) h_f}{\alpha_1 \phi_c f'_c \beta_1 b_w d_p + k_p \phi_p A_p f_{pu}}$$

hence:

$$c/d_p = \frac{1823 + 680 - 732 - 366 + 545 - 204}{(4.81)(550) + (1823)(0.28)} = 0.553$$

Because $c/d_p > 0.5$, use strain compatibility analysis. Use the approximate equation for $f_{pr}$ for first iteration:

$$f_{pr} = (1860)[1 - (0.28)(0.553)] = 1572\ MPa$$

$$T1 = (0.9)(1089)(1572)/10^3 = 1541\ kN$$

Rearranging terms from relationship

$$T1 + T2 = C1 + C2 + C3 + C4$$

$$c = \frac{T1 + 680 - 732 - 366 + 545 - 204}{4.81} = \frac{1541 - 77}{4.81} = 304\ mm$$

$\epsilon'_s = (0.0035)(304 - 40)/304 = 0.00304$

$\epsilon_y = 400/200\ 000 = 0.002$ ($A'_s$ has yielded)

$\epsilon_s = (0.0035)(296)/304 = 0.0034$ ($A_s$ has yielded)

$\epsilon_p = (0.0035)(246)/304 = 0.00283$

The prestressed reinforcement has an additional prestrain corresponding to the elastic strain in the steel when the concrete stress is zero at the level of the prestress.

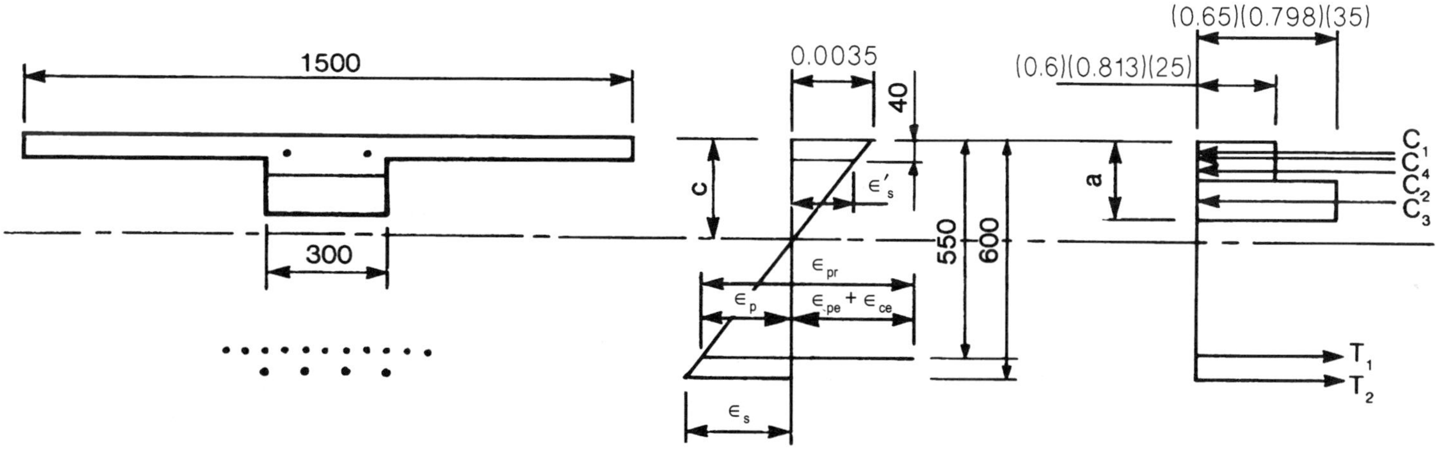


Strain diagram

Force diagram

Assume:

$f_{pe} = 0.6f_{pu} = (0.6)(1860) = 1116\,MPa$

hence:

$\in_{pe} = f_{pe} / E_p = 1116 / 190\,000 = 0.00587$

Concrete stress at level of tendons:

$e = 321 - 100 = 221\,mm$

$$f_{ce} = P_e\left(\frac{1}{A_g} + \frac{e^2}{I}\right)$$

$$= (1116)(1089)\left(\frac{1}{333\,400} + \frac{221^2}{14\,700 \times 10^6}\right)$$

$= 7.7\ MPa$

$\in_{ce} = f_{ce} / E_c = 7.7 / 26\,600 = 0.00029$

$\in_{pe} + \in_{ce} = 0.00587 + 0.00029 = 0.00616$

Total strain in tendons $= 0.00616 + 0.00283 = 0.00899$

From the stress-strain curve in Fig. 8.2.4:

$f_{pr} = 1645\ MPa > 1572\ MPa$

try:

$f_{pr} = (1572 + 1645) / 2 = 1609\,MPa$

$T1 = (0.9)(1089)(1609) / 10^3 = 1577\,kN$

2nd iteration:

$$c = \frac{1577 - 77}{4.81} = 312\,mm$$

$\in'_s = (0.0035)(312 - 40) / 312 = 0.00305$

$\in_s = (0.0035)(600 - 312) / 312 = 0.00323 > 0.002$

$\in_p = (0.0035)(550 - 312) / 312 = 0.00267$

$\in_{pr} = 0.00616 + 0.00267 = 0.00883$

From the stress-strain curve:

$f_{pr} = 1635\ MPa > 1609\ MPa$

try:

$f_{pr} = 1620\,MPa$

$T1 = (0.9)(1089)(1620) / 10^3 = 1588\,kN$

3rd iteration:

$$c = \frac{1588 - 77}{4.81} = 314\ mm$$

$\in_p = (0.0035)(550 - 314) / 314 = 0.00263$

$\in_{pr} = 0.00613 + 0.00263 = 0.00876$

From the stress-strain curve:

$f_{pr} = 1625\ MPa \approx 1620\ MPa$

This matches the trial value so that this is the correct solution.

$a = \beta_1 c = (0.8)(314) = 251\,mm$

$C3 = 4.81c - 545 = (4.81)(314) - 545 = 965\,kN$

$T1 = (0.9)(1089)(1625) / 10^3 = 1593\,kN$

therefore:

$$M_r = [(1593)(550 - 314) + (680)(600 - 314) + (732)(314 - 25) + (366)(314 - 75) + (965)(314 - 100 - 151/2) + (204)(314 - 40)]/10^3$$

$M_r = 1059\,kN \cdot m$

## 3.4 FLEXURAL DESIGN AT SERVICE LIMIT STATE

### 3.4.1 DESIGN PROCEDURE

For precast concrete elements, the two distinct phases which require design consideration are:

1. The manufacturing through erection process.
2. In-service conditions, defined as those conditions imposed upon an element after it is permanently connected to the supporting structure.

The specified in-service loads, which are determined by the designer, may not produce the critical stress condition. The forces imposed during the manufacturing or erection process can control design. These stresses occur primarily as a result of the element being manufactured in an orientation differing from that of its final position in the structure, and the limitations of these stresses are controlled by the fact that the concrete strengths at time of stripping are lower than when the element is in service.

The forces imposed on an element during stripping and handling are affected by:

1. The orientation of the element.
2. External restraints such as form suction and impact.
3. Location and number of handling inserts.
4. Dead load of element and any additional items which must be lifted (such as forms which remain with the element during stripping).
5. Location of temporary points of support during storage.

The forces imposed on an element during transportation are affected by:

1. Orientation of the element.
2. Location of the supports.
3. Condition of the transporting vehicle, road and site.

Since it is the manufacturer's option as to which production and transportation methods will be employed, it is also the manufacturer's responsibility to verify satisfactory behaviour of the precast element during these processes.

The Contract Documents should make reference to The CSA Certification Program for Architectural and Structural Precast Concrete and CSA-A23.4, "Precast Concrete – Materials and Construction".

Specific requirements for stripping, handling and transportation should be clearly set forth in the Contract Documents. Some examples of specific requirements include: type and location of handling inserts, methods of curing, protection during storage and transportation, and limits on when elements may be transported.

Structural design criteria related to the critical manufacturing processes are discussed in the following sections.

## 3.4.2 REINFORCED ELEMENT DESIGN

Non-prestressed flexural elements are normally proportioned, and reinforcement selected, on the basis of the procedures described in Section 3.3. However, when reinforcement with a yield strength exceeding 300 MPa is used, or when the precast element is exposed to view, it is necessary to ensure that excessive crack widths do not occur. In general, a larger number of closer spaced bars will lead to smaller crack widths than fewer large diameter bars with an equivalent area.

The detailing of reinforcement for elements not exposed to view is governed by a Code limitation on the maximum value of the quantity z as shown in Fig. 3.4.1, in which z is calculated from the equation:

$$z = f_s \sqrt[3]{d_c A}$$

where:

$f_s$ = reinforcement stress due to specified loads (MPa)

= $0.6 f_y$ unless otherwise determined

$d_c$ = concrete cover to the centre of the reinforcement (mm)

$A$ = average effective area around one reinforcing bar ($mm^2$)

= $2b\, d_c/n$

**Fig. 3.4.1 Recommended maximum values of z and corresponding crack widths**

| Type of element and exposure | Maximum value of z (N/mm) | Corresponding value of w (mm) |
|---|---|---|
| **Critical appearance** | | |
| Exterior exposure | 9 000 | 0.12 |
| Interior exposure | 18 000 | 0.24 |
| **Prestressed elements** | | |
| Exterior exposure | 15 000 | 0.20 |
| Interior exposure | 20 000 | 0.27 |
| **Reinforced elements** | | |
| Exterior exposure | 25 000 | 0.33 |
| Interior exposure | 30 000 | 0.40 |

$b$ = width of tension face (mm)

$n$ = number of reinforcing bars

The above equation is derived from the Gergely-Lutz[R3.4.1] expression:

$$w = 11 \times 10^{-6} \frac{h_2}{h_1} f_s \sqrt[3]{d_c A}$$

with $h_2/h_1$ taken equal to 1.2

where:

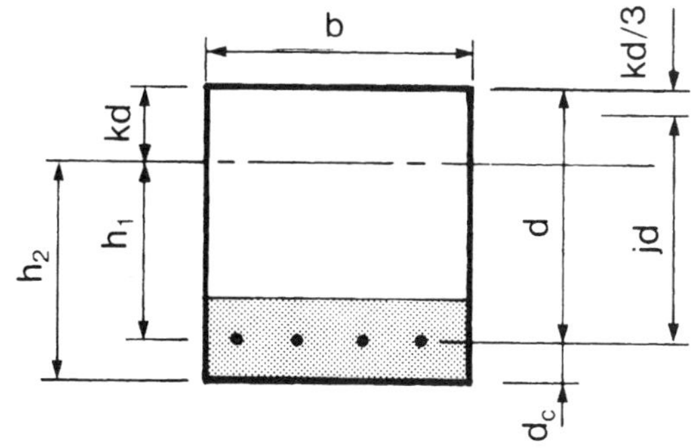


$w$ = maximum crack width at extreme tension fibre (mm)

$h_1$ = distance from centroid of tensile reinforcement to neutral axis (mm)

$h_2$ = distance from extreme tensile fibre to neutral axis (mm)

For precast elements where appearance is critical, it is recommended that the Code requirement for crack control be extended to prevent the formation of unsightly cracks[R3.4.2]. Recommended values for the maximum crack width are given in Fig. 3.4.1.

For additional recommendations on the design of water retaining sanitary structures, reference should be made to "Application of Strength Design Methods to Sanitary Structures" by F.Klein, E.S. Hoffman and P.F. Rice, Concrete International, April 1981.

In certain circumstances, it may be desirable for surfaces exposed to view to remain free from discernable cracks. This can be achieved by ensuring that flexural tensile stresses during handling and under service loads remain less than the modulus of rupture by a sufficient margin. Dividing the modulus of rupture of normal density concrete, taken as:

$0.6\lambda\sqrt{f'_{ci}}$, by 1.5, leads to:

$$f'_r \le 0.4\lambda\sqrt{f'_{ci}}$$

where:

$f'_r$ = allowable flexural tension, computed on the basis of the gross concrete section

$f'_{ci}$ = concrete strength at the time considered

$\lambda$ = factor for low density concrete (see Sect 1.3.1)

If values of $f_s$ under service load conditions are required to be less than 0.6 $f_y$ to satisfy crack control requirements, reinforcement should be provided equal to:

$$A_s = \frac{M}{0.9 f_s d}$$

This equation is based on working stress design principles with the assumption that j = 0.9 and k = 0.3.

**Example 1 – Reinforced panel design**

*Given:*

A 150 mm thick architectural panel exposed to the weather.

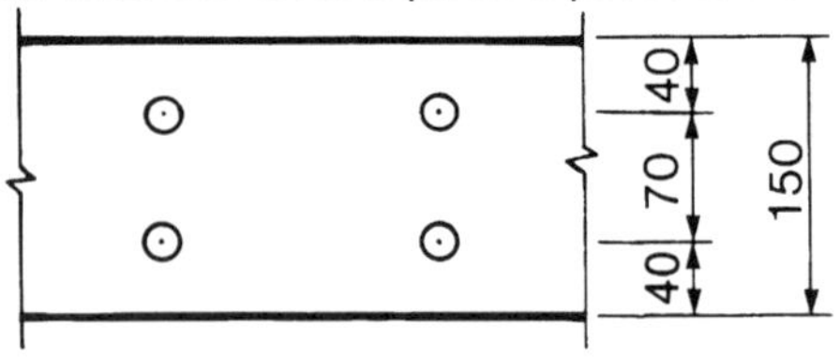


Concrete:

$f'_c$ = 35 MPa

Normal Density

Specified load moment M = 60 kN · m/m

*Problem:*

Find the required amount of non-prestressed 300 grade reinforcement with 40 mm cover to the centre of the reinforcement.

*Solution:*

For a panel with critical appearance exposed to the weather, the recommended maximum value of w from Fig. 3.4.1 is:

w = 0.12 mm

Assuming j = 0.9 and k = 0.3, calculate:

$d = 150 - 40 = 110$ mm

$kd = (0.3)(110) = 33$ mm

$h_1 = 110 - 33 = 77$ mm

$h_2 = 150 - 33 = 117$ mm

$$\frac{h_2}{h_1} = \frac{117}{77} = 1.52$$

$d_c = 40$ mm

Try a bar spacing of 150 mm:

$A = (80)(150) = 12\,000$ mm²

$$f_s = \frac{w}{(11 \times 10^{-6})\frac{h_2}{h_1}\sqrt[3]{d_c A}} = \frac{0.12}{(11 \times 10^{-6})(1.52)\sqrt[3]{(40)(12\,000)}} = 92 \text{ MPa}$$

$$A_s = \frac{M}{0.9 f_s d} = \frac{(6.0)(10^6)}{(0.9)(92)(110)} = 658 \text{ mm}^2/\text{m}$$

Use No. 10 bars at 150 mm top and bottom ($A_s$ = 666 mm²/m)

### 3.4.3 PRESTRESSED ELEMENT DESIGN - GENERAL

Prestressed concrete elements are classified as either fully or partially prestressed depending on the value of the calculated tensile stresses due to the effects of specified loads. The permissible stresses for fully prestressed elements are summarized as follows:

**Permissible Concrete Stresses:**

1. At release (transfer) of prestress, before time-dependent losses:
   a) Compression: $0.6 f'_{ci}$
   b) Tension (except at ends): $0.25\lambda\sqrt{f'_{ci}}$
   c) Tension at ends of simply supported elements: $0.5\lambda\sqrt{f'_{ci}}$
   d) Architectural elements exposed to view: $0.4\lambda\sqrt{f'_{ci}}$

2. Due to prestress and specified loads, after all losses:
   a) Compression due to sustained loads: $0.45 f'_c$
   b) Compression due to total load: $0.6 f'_c$

**Fig. 3.4.2 Calculation of concrete stresses due to specified loads**

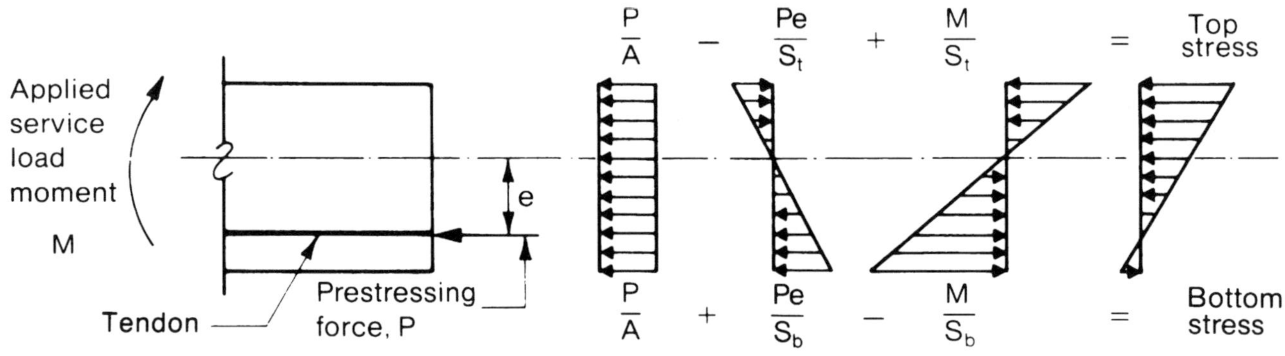

c) Tension in precompressed tensile zone (except corrosive environments): $0.5\lambda\sqrt{f'_c}$

d) Tension in precompressed tensile zone exposed to corrosive environments: $0.25\lambda\sqrt{f'_c}$

e) Tension in precompressed tensile zone for architectural elements exposed to view: $0.4\lambda\sqrt{f'_c}$

The stresses given in 1(d) and 2(e) are recommended values and are not specified in CSA-A23.3. Calculation of stresses at critical points follows classical straight line theory as illustrated in Fig. 3.4.2. The calculation of these stresses is illustrated in Example 2 at the end of this Subsection.

Prestressed elements which do not comply with 2(c) are defined as partially prestressed and are required to meet the crack control criteria given in Fig. 3.4.1.

### Permissible Stresses in Prestressing Tendons

1. During jacking:
   a) For post-tensioning tendons: $0.85\ f_{pu}$ but not greater than $0.94\ f_{py}$
   b) For pretensioning tendons: $0.80\ f_{pu}$
2. Tension immediately after prestress transfer:
   a) Except at post-tensioning anchorages and couplers: $0.82\ f_{py}$ but not greater than $0.74\ f_{pu}$
   b) At anchorages and couplers of post-tensioning tendons: $0.70\ f_{pu}$

In accordance with CSA-G279, "Steel for Prestressed Concrete Tendons", the specified yield strength, $f_{py}$, of prestressing tendons is based on the following:

1. Low relaxation strand or wire: $0.90\ f_{pu}$
2. Stress relieved strand or wire: $0.85\ f_{pu}$
3. Plain bars: $0.85\ f_{pu}$
4. Deformed bars: $0.80\ f_{pu}$

The maximum jacking stress given in 1(a) is based solely on the stress-strain characteristics of the tendons; this maximum value should only be used after consultation with the manufacturer of the anchorages or the supplier of the post-tensioning system.

### Critical Sections for Pretensioned Elements

For stresses immediately after transfer, the most critical section is usually near the end of the element, although in elements with single-point depressed tendons, the release stresses at midspan may also be critical and should be checked. The actual critical end stress is at the point where the prestressing force has been completely transferred to the concrete, usually assumed to be 50 strand diameters from the end. For convenience, using hand calculation, it is normal practice to calculate the stress at the end (assuming full transfer), and check the transfer point only if necessary to meet Code requirements.

If release stresses are higher than allowed by Code, it may be necessary to either increase the specified release strength, or provide supplemental tensile reinforcement. In short span, heavily loaded elements, such as beams, it is usually more practical to reinforce for the release tension. This is illustrated in Example 2.

Under uniform service loads, the critical section is at midspan for elements with straight tendons and near $0.4\ell$ from the end ($\ell$ = span) for elements with tendons depressed at midpoint. The exact critical point can be determined by a detailed analysis, but designing for critical stresses at the midspan and $0.4\ell$ points will usually determine the capacity within 1 or 2 percent.

*Tendon profiles other than straight or single-point depressed are uncommon in building products and should be avoided unless absolutely necessary.* Straight tendons are used in flat deck elements and beams and either straight or depressed tendons are used in stemmed deck elements.

For unusual loading conditions, such as concentrated loads, other sections may be critical. In these cases, it is common to estimate where the critical section may be, and check two or three points near that estimated position.

### Composite Elements

It is usually more economical to place cast-in-place composite topping without shoring the elements, especially for deck elements. This means that the mass of the topping must be carried by the precast element alone. Additional superimposed dead and live loads are carried by the composite section.

The following examples illustrate a tabular form of superimposing the stresses caused by the prestress force and the dead and live load moments.

### Sign Convention

The customary sign convention used in the design of precast, prestressed concrete elements for service load stresses is positive (+) for compression and negative (–) for tension. Bending moment diagrams are drawn on the tension side of flexural elements. This convention is used throughout this Manual.

**Example 2 – Calculation of critical stresses – single point depressed strand**

*Given:*

2400 × 600 double tee

Span = 21.0 m

| | | |
|---|---|---|
| Superimposed dead load | = 0.5 kN/m² | = 1.2 kN/m |
| Superimposed live load | = 1.75 kn/m² | = 4.2 kN/m |

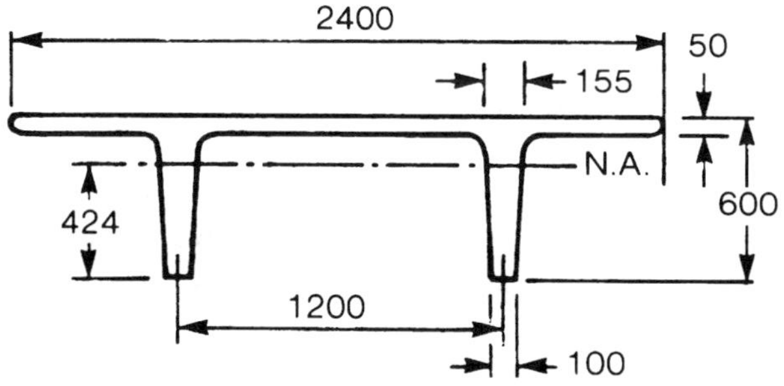

Concrete:

$f'_c$ = 35 MPa
$f'_{ci}$ = 25 MPa

Normal Density

Prestressed reinforcement:

14 - 13 mm 1860 MPa stress-relieved strands

$A_p$ = (14) (99) = 1386 mm$^2$

Section properties:

$A_g$ = 260 000 mm$^2$
$I$ = 8 580 × 10$^6$ mm$^4$

$y_b$ = 424 mm

$y_t$ = 176 mm
$S_b$ = 20 200 × 10$^3$ mm$^3$
$S_t$ = 48 800 × 10$^3$ mm$^3$

m = 260 kg/m$^2$ = 624 kg/m

w = 2.6 kN/m$^2$ = 6.24 kN/m

Eccentricities, single point depression:

*Problem:*

Find critical service load stresses.

*Solution:*

Prestress force:

$P_o$ = (1386)(0.70)(1860)/10$^3$ = 1800 kN

$P_i$ = (Assume 10% initial loss)
= (0.90)(1800) = 1620 kN

$P_e$ = (Assume 24% total loss)
= (0.76)(1800) = 1370 kN

Service load moments:

at midspan:

$M_d$ = (6.24)(21)$^2$/8 = 344 kN · m
$M_{sd}$ = (1.20)(21)$^2$/8 = 66 kN · m
$M_\ell$ = (4.20)(21)$^2$/8 = 232 kN · m

at 0.4 $\ell$

$M_d$ = (344)(0.96) = 330 kN · m
$M_{sd}$ = (66)(0.96) = 63 kN · m
$M_\ell$ = (232)(0.96) = 223 kN · m

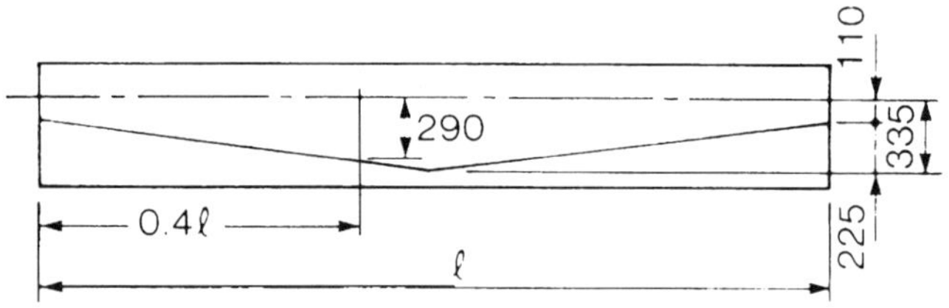


In this example, a release strength of:

$$f'_{ci} = \frac{16.1}{0.6} = 26.8\,\text{MPa}$$

should be provided and, because the extreme fibre stress in the precompressed tensile zone exceeds $0.5\lambda\sqrt{f'_c}$ (5.4 > 3.0), the element is partially prestressed (see Sect. 3.4.8).

| Load | Support at release P = $P_i$ | | Midspan at release P = $P_i$ | | 0.4 $\ell$ at service load P = $P_e$ | |
|---|---|---|---|---|---|---|
| | $f_b$ | $f_t$ | $f_b$ | $f_t$ | $f_b$ | $f_t$ |
| $P/A_g$ | + 6.2 | + 6.2 | + 6.2 | + 6.2 | + 5.3 | + 5.3 |
| Pe / S | + 8.8 | − 3.7 | + 26.9 | − 11.1 | + 19.7 | + 8.1 |
| $M_d$ / S | | | − 17.0 | + 7.1 | − 16.3 | + 6.8 |
| $M_{sd}$ / S | | | | | − 3.1 | + 1.3 |
| $M_\ell$ / S | | | | | − 11.0 | + 4.6 |
| Stresses | + 15.0 | + 2.5 | + 16.1 | + 2.2 | − 5.4 | + 9.9 |
| Allowable stresses | $0.6f'_{ci}$ | $0.6f'_{ci}$ | $0.6f'_{ci}$ | $0.6f'_{ci}$ | $0.5\lambda\sqrt{f'_{ci}}$ | $0.6f'_{ci}$ |
| | + 15.0 | + 15.0 | + 15.0 | + 15.0 | − 3.0 | + 21.0 |
| | OK | OK | HIGH | OK | HIGH | OK |

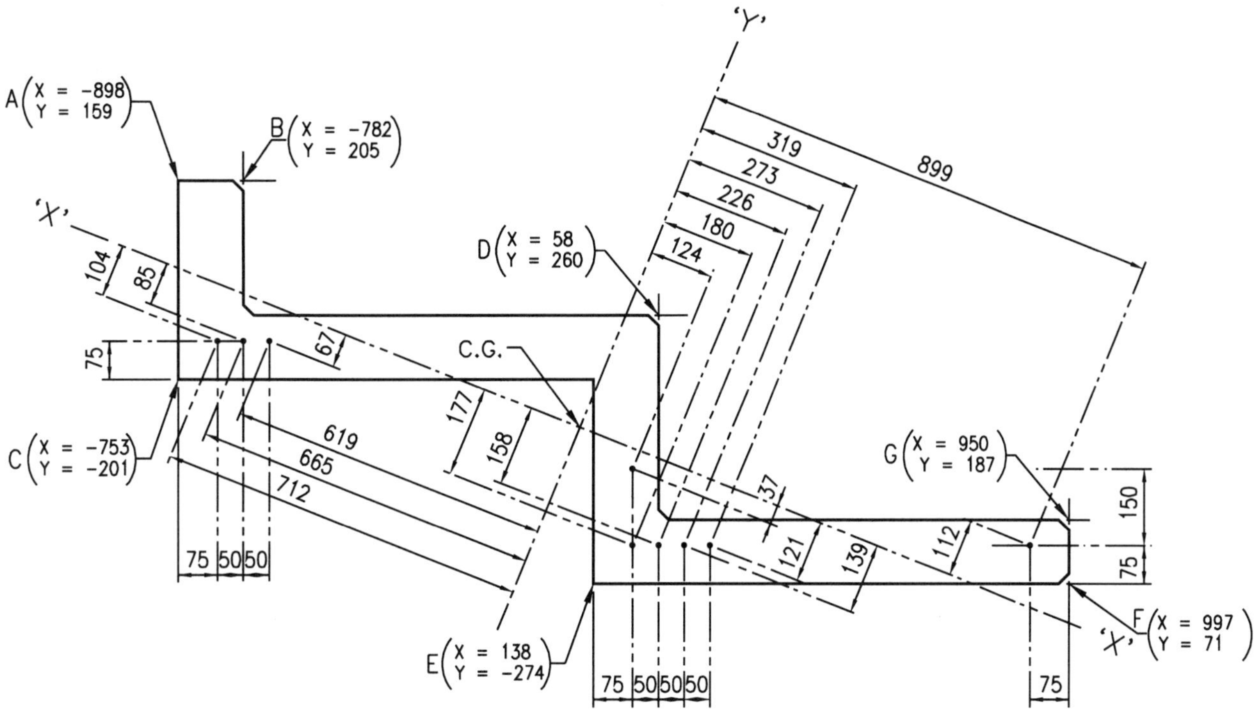


**Example 3 – Serviceability check of a stadia unit**

*Given:*

The stadia unit and loads described in Example 4, Section 3.3.2.

*Problem:*

Check service limits, final stresses, release stresses and cracking moment.

*Solution:*

*Final Stresses*

$$e_y = (-104 - 85 - 67 - 37 - 177 - 158 - 139 - 121 + 112)/9 = -86 \text{ mm}$$

$$e_x = (-712 - 665 - 619 + 124 + 180 + 226 + 273 + 319 + 899)/9 = 3 \text{ mm}$$

$$M_x = [(7 + 0.32 + 7.68)\cos 21.9^\circ + (0.3)\sin 21.9^\circ](9.7)^2/8 = 165.0 \text{ kN}\cdot\text{m}$$

$$M_y = [(7 + 0.32 + 7.68)\sin 21.9^\circ + (0.3)\cos 21.9^\circ](9.7)^2/8 = 69.1 \text{ kN}\cdot\text{m}$$

Final stress at point E

$$= \frac{P}{A_g} + \frac{Pe_y}{S_x} + \frac{Pe_x}{S_y} - \frac{M_x}{S_x} - \frac{M_y}{S_y}$$

$$= \frac{(1057)(10^3)}{298\,500} + \frac{(1057)(10^3)(86)(274)}{(3350)(10^6)} + \frac{(1057)(10^3)(3)(138)}{(83\,400)(10^6)} - \frac{(165)(10^6)(274)}{(3350)(10^6)} - \frac{(69.1)(10^6)(138)}{(83\,400)(10^6)}$$

$$= 3.54 + 7.43 + 0.01 - 13.50 - 0.11$$

$$= -2.63 \text{ MPa} < 2.96 \text{ MPa} = 0.5\sqrt{f'_c}$$

The final stresses at other points are summarized in the figure on next page.

*Release Stresses*

Check stresses at 50 strand diameters from end of unit.

$$M_{sw} = (7)(0.65/2)(10 - 0.65) = 21.3 \text{ kN}\cdot\text{m}$$

$$M_x = (21.3)\cos 21.9^\circ = 19.8 \text{ kN}\cdot\text{m}$$

$$M_y = (21.3)\sin 21.9^\circ = 7.9 \text{ kN}\cdot\text{m}$$

Release stress at point D

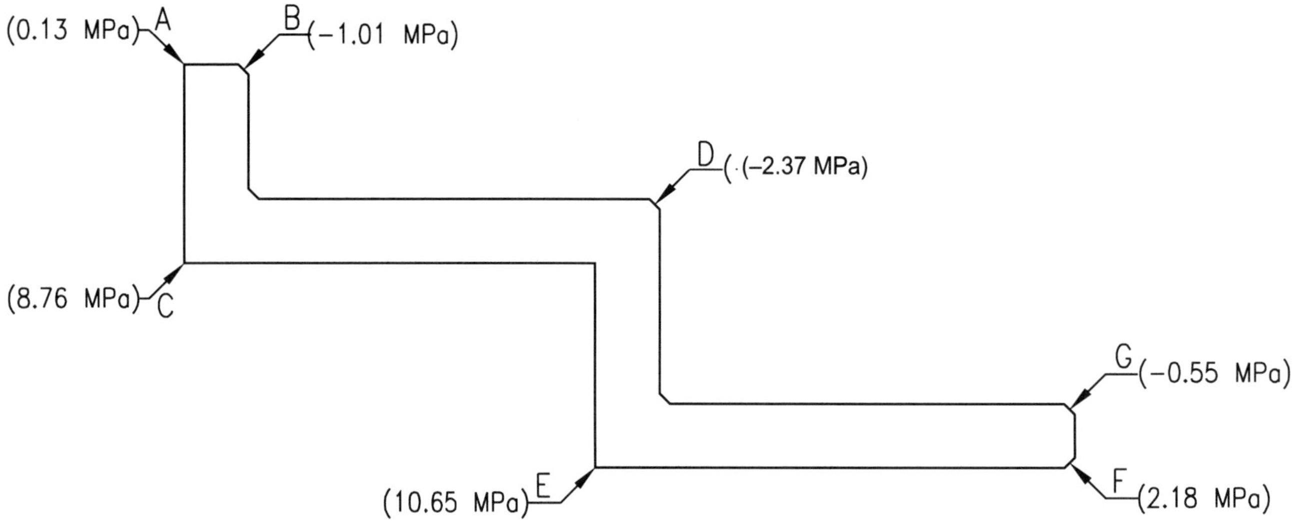


**RELEASE STRESSES AT 650 mm FROM ENDS**

$$= \frac{(1181)(10^3)}{298\,500} - \frac{(1181)(10^3)(86)(260)}{(3350)(10^6)}$$

$$+ \frac{(1181)(10^3)(3)(58)}{(83\,400)(10^6)} + \frac{(19.8)(10^6)(260)}{(3350)(10^6)}$$

$$+ \frac{(7.9)(10^6)(58)}{(83\,400)(10^6)}$$

$$= 3.96 - 7.88 + 0 + 1.54 + 0.01$$

$$= -2.37\text{ MPa} < 2.5\text{ MPa} = 0.5\sqrt{f'_c}$$

The release stresses at other points are summarized in the figure on this page.

Cracking Moment:

$$M_{cr} = \frac{I}{y_t}(f_{ce} + f_r)$$

$$f_r = 0.6\sqrt{f'_c} = (0.6)\sqrt{35} = 3.55\text{ MPa}$$

$$f_{ce} = 3.54\text{ MPa}$$

$$M_{cr} = \frac{(3350)(10^6)}{274}(3.55 + 3.54)/10^6 = 87\text{ kN·m}$$

$$M_r = 296\text{ kN·m} > (1.2)(87)$$

Although not part of this example, the designer should do a deflection and vibration analysis to complete the serviceability check.

## 3.4.4 PRESTRESS LOSS - GENERAL

### Sources of Prestress Loss

Loss of prestress is the reduction of tensile stress in prestressing tendons due to shortening of the concrete around the tendons, relaxation of stress within the tendons, and external factors which reduce the total initial force before it is applied to the concrete. The Code identifies the following sources of loss of prestress:

1. Anchorage seating loss.
2. Elastic shortening of concrete.
3. Friction loss due to intended or unintended curvature in post-tensioning tendons.
4. Creep of concrete.
5. Shrinkage of concrete.
6. Relaxation of tendon stress.

Accurate determination of losses is more important in some prestressed concrete elements than in others. Losses have little effect on the flexural resistance of an element unless the tendons are unbonded or unless the final stress after losses is less than 0.50 $f_{pu}$. Underestimation or overestimation of losses can affect service conditions such as camber, deflection and cracking.

*Anchorage seating loss and friction:*

These two sources of loss are mechanical. They represent the difference between the tension applied to the tendon by the jacking unit and the initial tension available for application

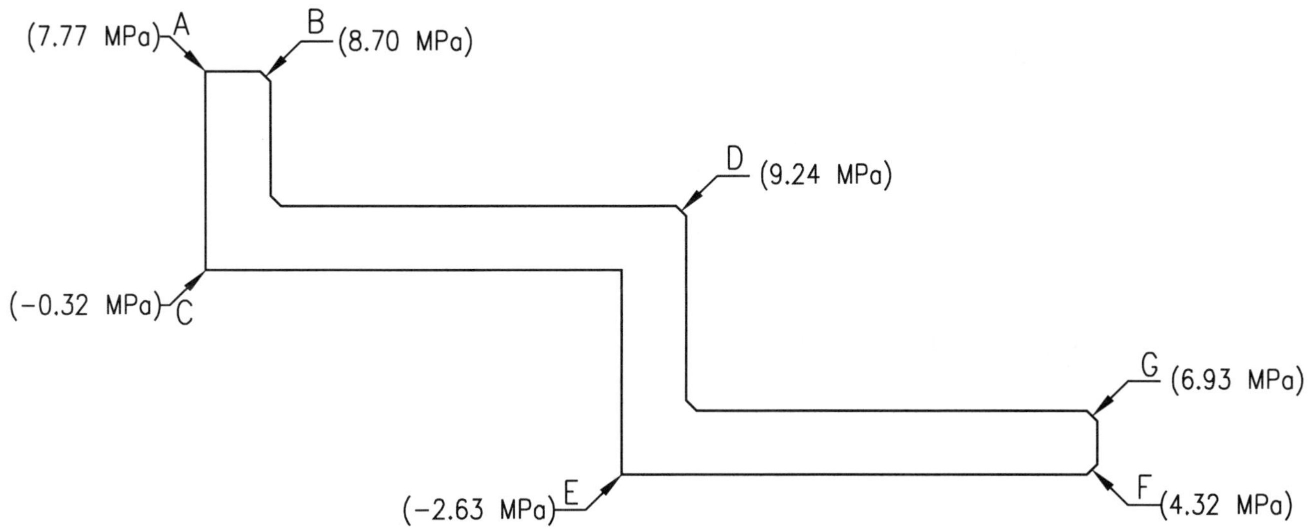


**FINAL STRESSES AT MID SPAN**

to the concrete by the tendon. Their magnitude can be determined with reasonable accuracy and, in many cases, they are fully or partially compensated for by overjacking.

*Elastic shortening of concrete:*

The concrete around the tendons shortens as the prestressing force is applied to it. Those tendons which are already bonded to the concrete shorten with it.

*Shrinkage of concrete:*

Loss of stress in the tendon due to shrinkage of the concrete surrounding it is proportional to that part of the shrinkage that takes place after the transfer of prestress force to the concrete.

*Creep of concrete and relaxation of tendons:*

Losses due to creep of concrete and relaxation of tendons complicate stress loss calculations. The rate of loss due to each of these factors changes when the stress level changes and the stress level is changing constantly throughout the life of the structure. Therefore, the rates of loss due to creep and relaxation are constantly changing.

### Range of Values for Total Prestress Loss

Due to the reduced availability of stress relieved strands, all loss calculations are based on low relaxation strands in this manual edition.

The total loss of prestress in typical elements will range from about 200 to 350 MPa for normal density concrete elements, and from about 250 to 400 MPa for semi-low density concrete. The value of 240 MPa (corresponding to a 21% loss) which had been recommended in the 1958 ACI-ASCE committee report should only be used for preliminary design. Two methods for loss calculation are discussed in the following:

1. The Simplified Method proposed by the PCI Committee.
2. The Detailed Method[R3.4.3]

The Simplified Method takes into account the level of concrete stress, type of prestressed reinforcement and volume to surface ratio. The Detailed Method takes into account all the major variables affecting creep and shrinkage and includes the effect of non-prestressed reinforcement located in the tension zone of the element. The presence of non-prestressed reinforcement in partially prestressed elements affects the losses in a significant way and should not be neglected in the loss calculations.

## 3.4.5 PRESTRESS LOSS BY SIMPLIFIED PCI COMMITTEE METHOD

"Recommendations for Estimating Prestress Losses", by the PCI Committee on Prestress Losses, appears in the PCI Journal for July/August 1975. Reader's Comments on the Recommendations appear in the PCI Journal for March/April 1976.

Using the Simplified Method, stress loss is determined by computing the value of $f_{co}$ and $f_{ci}$ and substituting them in the appropriate empirical equations. These equations are used to compute total loss, $\Delta f_p$, in MPa. Total loss is the sum of losses due to shrinkage, elastic shortening and creep of concrete plus loss due to relaxation of tendons.

*For normal density concrete:*

$$\Delta f_p = 137 + 16.3\, f_{co} + 5.4\, f_{c1}$$

*For semi-low density concrete:*

$$\Delta f_p = 121 + 20.4\, f_{co} + 4.8\, f_{c1}$$

The above empirical equations apply to pretensioned elements only. For typical elements it was found that the only variable that is not included in the equations but which could make an appreciable difference in the net result is volume/surface ratio. A correction factor is applied for that:

| V/S ration (mm) | 25 | 50 | 75 | 100 |
|---|---|---|---|---|
| Adjustment (percent) | +3.2 | 0 | -3.8 | -7.6 |

Example: For V/S = 75 reduce losses by 3.8%

The equations above are based on the initial tension, after reduction for anchor slip, normally used in pretensioned elements, i.e. 0.75 $f_{pu}$ for low-relaxation strand. Use of a higher or lower initial tension will result in an appreciable change in net losses.

Use of the equations requires calculation of the stresses $f_{co}$ and $f_{c1}$:

$$f_{co} = \frac{P_i}{A_g} + \frac{P_i e^2}{I} - \frac{M_d e}{I}$$

$$f_{c1} = \frac{M_{sd} e}{I}$$

where:

- $A_g$ = gross area of the precast section
- $e$ = eccentricity of the strand at the critical section
- $f_{c1}$ = concrete stress at centroid of tendon at the critical section caused by sustained loads not included in the calculation of $f_{co}$ (tension negative)
- $f_{co}$ = concrete compressive stress at centroid of tendon at critical section immediately after transfer
- $I$ = moment of inertia of the section
- $M_d$ = moment due to mass of the element
- $M_{sd}$ = moment due to all sustained loads except the element mass
- $P_i$ = prestress force immediately after transfer, after initial loss. It is within reasonable accuracy to assume 7.5% initial loss for low-relaxation strand

**Example 4 – Loss of prestress – by the Simplified Method**

*Given:*

2400 × 800 double tee with 50 mm composite topping

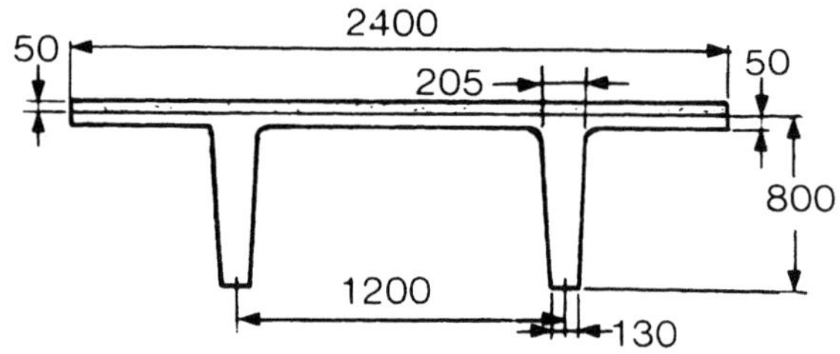


Span = 21.0 m

No superimposed dead load except topping

Section properties (untopped):

- $A_g$ = 371 000 mm²
- $I$ = 22 800 × 10⁶ mm⁴
- $S_b$ = 43 700 × 10³ mm³
- $y_b$ = 523 mm
- V/S = 48 mm
- w = 3.6 kN/m² = 8.6 kN/m
- w of topping = 2.9 kN/m

Precast concrete:

- $f'_c$ = 35 MPa
- $f'_{ci}$ = 25 MPa
- Normal Density

Topping concrete:

- Normal Density

Prestressed reinforcement:

- 12 - 13 mm 1860 MPa
- $A_p$ = (12)(99) = 1188 mm²
- Depressed at mid-span:
- $e_e$ = 307 mm
- $e_c$ = 441 mm

*Problem:*

Determine total prestress loss and effective prestress by PCI Committee simplified method.

*Solution:*

For depressed strand, critical section is at 0.4ℓ. Determine moments, eccentricity, and prestress force:

$$M\ @\ 0.4\ell = \frac{wx}{2}(\ell - x) = \frac{w(0.4\ell)}{2}(\ell - 0.4\ell)$$

$$= 0.12\ w\ell^2$$

$$M_d = (0.12)(8.6)(21)^2 = 455\ \text{kN} \cdot \text{m}$$

$$M_{sd} = (0.12)(2.9)(21)^2 = 154\ \text{kN} \cdot \text{m}$$

$$e\ \text{at}\ 0.4\ell = 307 + (0.8)(441 - 307)$$

$$= 414\ \text{mm}$$

$$P_o = 0.75\ A_p\ f_{pu} = (0.75)(1188)(1860)/10^3$$

$$= 1660\ \text{kN}$$

$$P_i = 0.925\ P_o = (0.925)(1660) = 1536\ \text{kN}$$

Determine $f_{co}$ and $f_{ci}$ :

$$f_{co} = \frac{P_i}{A_g} + \frac{P_i e^2}{I} - \frac{M_d e}{I}$$

$$= \frac{(1536)(10^3)}{371\,000} + \frac{(1536)(414)^2(10^3)}{(22\,800 \times 10^6)}$$

$$- \frac{(455)(414)(10^6)}{(22\,800 \times 10^6)}$$

$$= 4.1 + 11.5 - 8.3 = 7.3\ \text{MPa}$$

$$f_{c1} = \frac{-M_{sd}e}{I} = \frac{-(154)(414)(10^6)}{(22\,800 \times 10^6)} = -2.8\ \text{MPa}$$

For normal density concrete and low relaxation strand:

$$\Delta f_p = 137 + 16.3\ f_{co} + 5.4\ f_{c1}$$

$$= 137 + (16.3)(7.3) + (5.4)(-2.8) = 241\ \text{MPa}$$

Adjust for V/S ratio (interpolate between 25 and 50):

$$\text{Adjustment factor} = 3.2 - \frac{(23)}{25}(3.20 - 0)$$

$$= +\ 0.26\%$$

(+ means additional loss)

$$\frac{(0.26)}{100}(241) = 0.6\ \text{MPa}$$

Final loss is 241 + 0.6 = 242 MPa or 17.3% of 1395 MPa

Final prestress force after all losses:

$$P_e = 1660 - \frac{(242)(1188)}{10^3} = 1373\ \text{kN}$$

## 3.4.6 PRESTRESS LOSS BY DETAILED METHOD

In this method, creep and shrinkage of the concrete and the relaxation of the prestressed reinforcement have to be determined initially.

### Creep

If data about creep and shrinkage are not available from tests or observations on structural elements, the method of ACI Committee 209 is used to predict the time-dependent properties. For use in the Detailed Method, creep is expressed in terms of the creep coefficient $C_t$, defined as the ratio of creep strain $\epsilon_{cp}$ to initial immediate elastic strain $\epsilon_i$ at time of application of stress.

The following expression is applicable to normal, semi-low and low density concretes:

$$C_t = \frac{t^{0.6}}{10 + t^{0.6}} C_u Q_{cp}$$

where:

- $t$ = time in days after loading
- $C_u$ = ultimate creep coefficient
- $Q_{cp}$ = creep modification factor for non-standard conditions

In the absence of specific creep data for local aggregates and conditions, the average value suggested for $C_u$ is 2.35.

The above equation was developed for sustained compressive stress not exceeding 50% of concrete strength. It consists of an expression for creep under standard conditions multiplied by the correction factor $Q_{cp}$ to modify for non-standard conditions. The standard conditions and the correcting factor $Q_{cr}$ are given in Fig. 2.4.1.

For long term creep (t = ∞), the above equation reduces to:

$$C_\infty = C_u Q_{cp}$$

### Shrinkage

Shrinkage strains start taking place immediately after concrete is exposed to the drying environment. According to ACI Committee 209, the shrinkage strain $\epsilon_{sh}$ is determined using the following equation which is applicable to normal, semi-low and low density concretes:

$$\epsilon_{sh} = \frac{t}{C_s + t} \epsilon_{shu} P_{sh}$$

where:

- $t$ = time starting immediately after the initial wet curing (days)
- $C_s$ = 35 if concrete is moist cured for 7 days
  = 55 if steam cured for 1-3 days
- $\epsilon_{shu}$ = ultimate shrinkage strain

In the absence of specific shrinkage data for local conditions, the average value of $\epsilon_{shu}$ suggested for use is $780 \times 10^{-6}$. $P_{sh}$ is a correction factor for conditions that are other than the standard condition specified in Fig. 2.4.1.

For long term shrinkage, the above equation reduces to:

$$\epsilon_{sh} = \epsilon_{shu} P_{sh}$$

## Relaxation of prestressed reinforcement

In lieu of detailed information from the steel manufacturer, the intrinsic relaxation of prestressing tendons may be predicted as:

$$f_{re}(t) = \frac{\log t}{45}\left(\frac{f_{pi}}{f_{py}} - 0.55\right) f_{pi}$$

The term "intrinsic" relaxation signifies the loss of stress under constant strain. In this equation:

$f_{re}(t)$ = intrinsic relaxation of prestressing steel at time t

$f_{pi}$ = initial stress in prestressed reinforcement

$f_{py}$ = 0.90 $f_{pu}$ for low relaxation strands

$t$ = time since prestressing (hours)

Assuming 50 years as the lifetime of the structure, the final value for the intrinsic relaxation is:

$$f_{re} = 0.125\left(\frac{f_{pi}}{f_{py}} - 0.55\right) f_{pi}$$

### Fig. 3.4.3 Intrinsic relaxation of low relaxation strand

These equations are valid for temperatures up to 20°C. At higher temperatures, the relaxation increases (see Fig. 3.4.3). There is also a substantial loss of prestress due to relaxation before transfer of prestress. This loss may be high because of the elevated temperature during steam curing, even though the duration may be only about one day. Fig. 3.4.3 may be useful in estimating the relaxation loss at higher temperature.

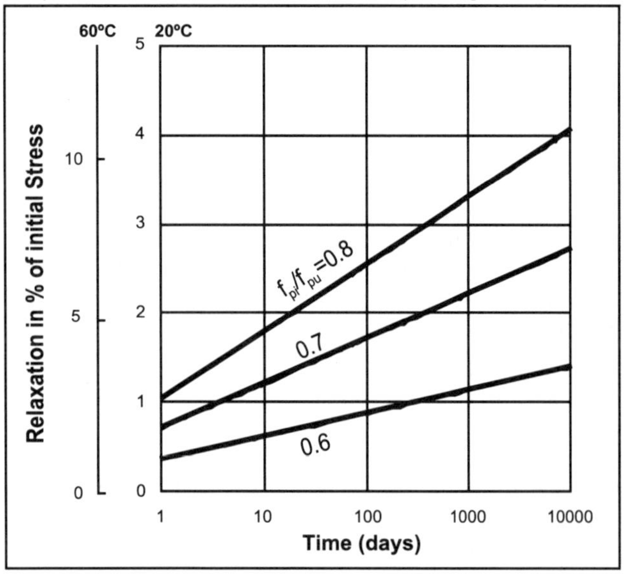


## Elastic loss

At the time of transfer, the concrete stress in any fibre at a distance $y_{tr}$ from the centroid of the transformed section, is determined by:

$$f_{co} = \frac{P_o}{A_{tr}} + \left(\frac{P_o e_{tr} - M_d}{I_{tr}}\right) y_{tr}$$

At the level of the tendons, $y_{tr}$ is equal to $e_{tr}$, and the change in stress in the prestressed reinforcement at transfer is accurately determined by the relation:

$$\Delta f_{po} = n_o \left[\frac{P_o}{A_{tr}} + \left(\frac{P_o e_{tr} - M_d}{I_{tr}}\right) e_{tr}\right]$$

In these equations:

$P_o$ = prestress force before transfer (after de-ducting relaxation losses occurring prior to transfer)

$A_{tr}$ = area of transformed section (including transformed areas of prestressed and non-prestressed reinforcement

$I_{tr}$ = moment of inertia of transformed section

$e_{tr}$ = eccentricity of tendons in transformed section

$M_d$ = moment due to dead load

$n_o$ = $E_p / E_{co}$

$E_p$ = modulus of elasticity of prestressed reinforcement

$E_{co}$ = modulus of elasticity of concrete at time of prestress transfer

$A_p$ = area of prestressed reinforcement

Note: For most pretensioned elements, the use of the gross section properties is sufficiently accurate in the above equations, except when additional non-prestressed reinforcement is present.

The prestressing force after prestress transfer is:

$$P_i = P_o - \Delta f_{po} A_p$$

Moments M, applied after transfer of prestress produce the following flexural stresses in concrete:

$$f_c = \frac{M}{I_{tr}} y_{tr}$$

and in the prestressed or non-prestressed reinforcement:

$$f_s = n\frac{M}{I_{tr}}e_{tr}$$

The modular ratio n is determined for concrete strength at the time of application of M. Moment $M_{sd}$ due to superimposed dead load increases the stress in the prestressed reinforcement by:

$$f_{pr} = n\frac{M_{sd}}{I_{tr}}e_{tr}$$

### Time-dependent Loss

The loss of prestress for an element with one layer of tendons and non-prestressed reinforcement at about the same level can be estimated from the relation:

$$\Delta f_p = \frac{n_o f_{co} C_t + \epsilon_{sh} E_p + f_{re}}{1 + n_o(\rho_p + \rho_s)(1 + e^2/r^2)(1 + 0.8C_t)}$$

where:

- $f_{co}$ = concrete compressive stress at level of tendon after transfer (use previous equation for $f_{co}$ with $y_{tr} = e_{tr}$)
- $C_t$ = creep coefficient for age of concrete at transfer of prestress
- $\epsilon_{sh}$ = shrinkage
- $f_{re}$ = relaxation of prestressed reinforcement
- $\rho_p$ = $A_p / A_c$
- $\rho_s$ = $\frac{A_s}{A_c}\frac{E_s}{E_p}$
- $r^2$ = $I_c / A_c$

The factor applied to $C_t$ in the denominator is the so-called aging coefficient which can be assumed to be 0.8 for practical creep calculations.

In the above equations, the theoretically correct concrete area is the net concrete area not including the reinforcement area. However, considering the uncertainties involved in the various concrete properties, the gross concrete area is normally used.

If a moment $M_{sd}$ due to a superimposed load is applied at a later stage, the additional time-dependent change in the prestressed or non-prestressed reinforcement stress is:

$$\Delta f_{p1} = \frac{n_1 f_{c1} C_{t1}}{1 + n_1(\rho_p + \rho_s)(1 + e^2/r^2)(1 + 0.8C_{t1})}$$

where:

- $f_{c1}$ = $\frac{M_{sd}}{I_{tr}}e_{tr}$
- $n_1$ = $E_p / E_{c1}$
- $E_{c1}$ = modulus of elasticity of concrete at age $t_1$ of application of moment $M_{sd}$
- $C_{t1}$ = creep coefficient for concrete loaded at age $t_1$

The time-dependent stress $\Delta f_{p1}$ is a tensile stress reducing the loss $\Delta f_p$. The modulus of elasticity of the concrete at age $t_1$ may be considerably higher than the value at transfer of prestress and the creep coefficient is smaller than the value $C_t$ determined for the age at transfer.

The relaxation loss of the prestressed reinforcement is considerably reduced by the effects of creep and shrinkage of the concrete. This may be included in the calculations by replacing the intrinsic loss $f_{re}$ by a reduced value of $\alpha_r f_{re}$. The coefficient $\alpha_r$ is determined from Fig. 3.4.4 using the parameters $\beta = f_{pi}/f_{pu}$ and $\Omega = \Delta f_p / f_{pi}$. The stress $f_{pi}$ is the initial stress applied to the tendon, $f_{pu}$ is the tensile strength and $\Delta f_s$ is the loss of prestress due to creep and shrinkage only:

$$\Delta f_p = \frac{n f_c C_t + \epsilon_{sh} E_p}{1 + n(\rho_p + \rho_s)(1 + e^2/r^2)(1 + 0.8C_t)}$$

If a superimposed dead load is present, the stress $\Delta f_{p1}$ is deducted from $\Delta f_p$.

The value $\Delta f_p$ also is the time-dependent stress in the non-prestressed reinforcement, the centroid of which is assumed to coincide with that of the prestressed reinforcement. Total time-dependent reduction of compression acting on the concrete is:

$$\Delta P = \Delta f_p A_p + \Delta f_s A_s + \Delta f_{s1}(A_p + A_s)$$

In this equation, $\Delta f_p$ and $\Delta f_s$ are compressive stresses while $\Delta f_{s1}$ is a tensile stress less than $\Delta f_s$ in magnitude. The result of this stress change will be a reduction of the tensile stress in the prestressed reinforcement and a compressive stress in the non-prestressed reinforcement.

**Fig. 3.4.4 Relaxation reduction coefficient $a_r$**

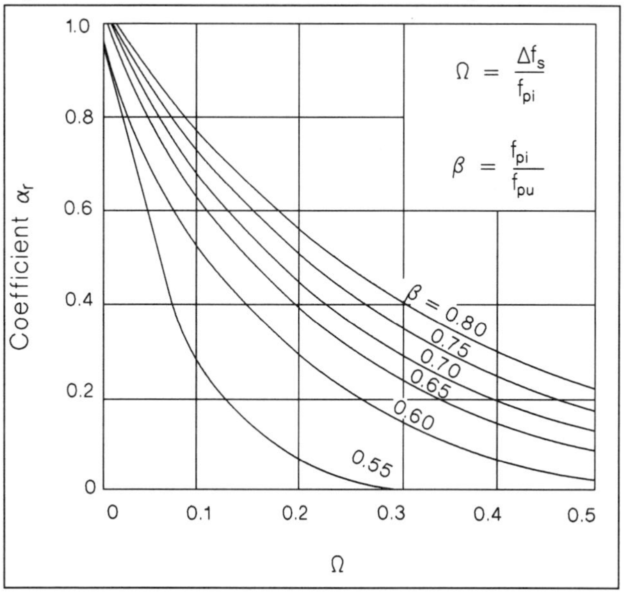

**Example 5 – Loss of prestress for element without non-prestressed reinforcement - Detailed Method**

*Given:*

The 2400 × 800 double tee of Example 4 on Page 3-32

Concrete:

$$E_{co} = \left(3300\sqrt{f'_{ci}} + 6900\right)\left(\frac{2400}{2300}\right)^{1.5}$$

= 24 900 MPa at transfer

$$E_{c1} = \left(3300\sqrt{f'_{c}} + 6900\right)\left(\frac{2400}{2300}\right)^{1.5}$$

= 28 200 MPa at 40 days

Section Properties:

$$r = \sqrt{\frac{I_c}{A_c}} = \sqrt{\frac{228\,000 \times 10^6}{371000}} = 248 \text{ mm}$$

*Problem:*

Determine total loss of prestress by Detailed Method.

*Solution:*

It is assumed that the tendons are pretensioned to $0.75 f_{pu} = 1395$ MPa and that the topping does not contribute to the section properties. A relaxation loss prior to transfer of prestress is calculated assuming an average temperature of 60°C during steam curing for a period of 24 hours. According to Fig. 3.4.3, $\Delta f_{re} = (0.025)(1395) = 35$ MPa.

Prestress immediately before transfer:

$$f_{po} = 1395 - 35 = 1360 \text{ MPa}$$

Elastic loss (using transformed section properties):

$$\Delta f_{po} = n_o f_{co} = n_o\left[\frac{P_o}{A_{tr}} + \left(\frac{P_o e_{tr} - M_d}{I_{tr}}\right)e_{tr}\right]$$

with:

$$n_o = \frac{E_p}{E_{co}} = \frac{190\,000}{24\,900} = 7.6$$

$$P_o = f_{po} A_p = (1360)(1188)/10^3 = 1616 \text{ kN}$$

At 0.4ℓ, the transformed section properties are:

$$A_{tr} = A_g + (n_o - 1) A_p$$

$$= 371\,000 + (7.6 - 1)(1188)$$

$$= 379\,000 \text{ mm}^2$$

From Example 3:

$$e = 414 \text{ mm}$$

$$d_p = h - y_b + e$$

$$= 800 - 523 + 414 = 691 \text{ mm}$$

$$y_{tr} = \frac{A_g y_b + (n_o - 1)A_p(h - d_p)}{A_{tr}}$$

$$= \frac{(371\,000)(523) + (7.6 - 1)(1188)(800 - 691)}{379\,000}$$

$$= 514 \text{ mm}$$

$$e_{tr} = y_{tr} - (h - d_p)$$

$$= 514 - (800 - 691)$$

$$= 405 \text{ mm}$$

$$I_{tr} = I_g + A_g (y - y_{tr})^2 + (n_o - 1) A_p e^2_{tr}$$

$$= (22\,800 \times 10^6) + (371\,000)(523 - 514)^2$$

$$+ (7.6 - 1)(1188)(405)^2$$

$$= 24\,100 \times 10^6 \text{ mm}^4$$

Concrete stress at level of prestressed reinforcement after transfer:

$$f_{co} = \frac{P_o}{A_{tr}} + \left(\frac{P_o e_{tr} - M_d}{I_{tr}}\right)e_{tr}$$

$$= \frac{(1616)(10^3)}{379\,000}$$

$$+ \left[\frac{(1616)(405)(10^3) - (455)(10^6)}{24\,100 \times 10^6}\right](405)$$

$$= 7.6 \text{ MPa}$$

Elastic prestress loss:

$$\Delta f_{po} = n_o f_{co}$$

$$= (7.6)(7.6) = 58 \text{ MPa}$$

Prestress after transfer:

$$f_{pi} = f_{po} - \Delta f_{po}$$

$$= 1360 - 58 = 1302 \text{ MPa}$$

$$P_i = A_p f_{pi}$$

$$= (1188)(1302)/10^3 = 1547 \text{ kN}$$

The value of $P_i$ is slightly higher than that obtained by assuming a 7.5% loss in Example 4.

Concrete stress at level of prestressed reinforcement due to $M_{sd}$:

$$f_{c1} = \frac{M_{sd} e_{tr}}{I_{tr}} = \frac{(154)(405)(10^6)}{(24\,100 \times 10^6)}$$

$= 2.6$ MPa (tension)

Corresponding change in tendon stress:

with:

$$n_1 = \frac{E_p}{E_{c1}} = \frac{190\,000}{28\,200} = 6.7$$

$n_1 f_{c1} = (6.4)(2.6) = 17$ MPa (tension)

Creep and shrinkage coefficients from Fig. 2.4.1:

| | | |
|---|---|---|
| 1 day at prestress transfer | $Q_a$ | = 1.00 |
| 40 days when topping is applied | $Q_{s1}$ | = 0.80 |
| Cement content = 410 kg/m | $P_c$ | = 1.00 |
| Relative humidity 70% | $Q_h$<br>$P_h$ | = 0.80<br>= 0.70 |
| Ratio of fine to total aggregates = 0.40 | $Q_f$<br>$P_f$ | = 0.98<br>= 0.86 |
| V/S = 48 mm | $Q_r$<br>$P_r$ | = 0.95<br>= 0.96 |
| Slump 50 mm | $Q_{s\ell}$<br>$P_{s\ell}$ | = 0.95<br>= 0.97 |
| Air 7% | $Q_v$<br>$P_v$ | = 1.09<br>= 1.01 |

Long term creep due to dead load and prestressing:

$$Q_{cp} = Q_a Q_h Q_f Q_r Q_{s\ell} Q_v$$

$$= (1.0)(0.80)(0.98)(0.95)(0.95)(1.09) = 0.77$$

With the ultimate creep coefficient, $C_u = 2.35$

$$C_\infty = C_u Q_{cp} = (2.35)(0.77) = 1.81$$

Long term creep due to superimposed dead load:

$$Q_{cp1} = (0.80)(0.80)(0.98)(0.95)(0.95)(1.09)$$

$$= 0.62$$

$$C_{\infty 1} = C_u Q_{cp1} = (2.35)(0.62) = 1.45$$

Long term shrinkage:

$$P_{sh} = P_c P_h P_f P_r P_s P_v$$

$$= (1.0)(0.70)(0.86)(0.96)(0.97)(1.01)$$

$$= 0.566$$

With the ultimate shrinkage strain, $\epsilon_{shu} = 780 \times 10^{-6}$

$$\epsilon_{shu} = \epsilon_{shu} P_{sh} = (780 \times 10^{-6})(0.566)$$

$$= 442 \times 10^{-6}$$

Time-dependent loss not including relaxation, precast section:

$$\Delta f_p = \frac{n_o f_{co} C_\infty + \epsilon_{sh} E_p}{1 + n(\rho_p + \rho_s)(1 + e^2/r^2)(1 + 0.8C_\infty)}$$

with:

$$\rho_p = \frac{A_p}{A_c} = \frac{1188}{371000} = 0.0032$$

$$\Delta f_p = \frac{n f_c C_t + \epsilon_{sh} E_p}{1 + n(\rho_p + \rho_s)(1 + e^2/r^2)(1 + 0.8C_t)}$$

$$= \frac{(7.6)(7.6)(1.81) + (442 \times 10^{-6})(190\,000)}{1 + (7.6)(0.0032 + 0)\left[1 + \left(\frac{414}{248}\right)^2\right][1 + (0.8)(1.81)]}$$

$$= \frac{188.5}{1.225} = 154 \text{ MPa}$$

Reduction in time-dependent loss due to application of $M_{sd}$:

$$\Delta f_{p1} = \frac{n_1 f_{c1} C_{t1}}{1 + n_1(\rho_p + \rho_s)(1 + e^2/r^2)(1 + 0.8C_{t1})}$$

$$= \frac{(6.7)(2.6)(145)}{1 + (6.7)(0.0032 + 0)\left[1 + \left(\frac{414}{248}\right)^2\right][1 + (0.8)(1.45)]}$$

$$= \frac{25.3}{1.175} = 22 \text{ MPa} \quad \text{(tension)}$$

Intrinsic relaxation:

$$f_{re} = 0.125\left(\frac{f_{pi}}{f_{py}} - 0.55\right) f_{pi}$$

with:

$f_{pi} = 1360 - 58 + 17 = 1319$ MPa[R3.4.4]

$$f_{re} = (0.125)\left[\frac{1319}{(0.9)(1860)} - 0.55\right](1319)$$

$$= 39 \text{ MPa}$$

Reduced relaxation taking into account creep and shrinkage:

with

$$\Omega = \frac{\Delta f_p - \Delta f_{p1}}{f_{pi}} = \frac{154 - 22}{1319} = 0.10$$

$$\beta = \frac{f_{pi}}{f_{pu}} = \frac{1319}{1860} = 0.71$$

from Fig. 3.4.4:

$\alpha_r = 0.73$

$\alpha_r f_{re} = (0.73)(39) = 29$ MPa

Therefore, loss due to relaxation only:

$$\Delta f_p = \frac{\alpha_r f_{re}}{1 + n_o(\rho_p + \rho_s)(1 + e^2/r^2)(1 + 0.8C_\infty)}$$

$$= \frac{29}{1.225} = 23 \text{ MPa}$$

Prestress loss summary (MPa):

| | |
|---|---|
| Relaxation prior to transfer: | 35 |
| Elastic losses: | |
| – precast section only | 58 |
| – due to $M_{sd}$ | –17 |
| Time-dependent losses: | |
| – creep and shrinkage | 154 |
| – due to $M_{sd}$ | –22 |
| – relaxation | 23 |
| Total prestress loss: | 231MPa |

This value is 11 MPa (5%) less than that obtained by the simplified method. However, for different conditions this loss may vary by as much as ± 35 percent.

Final effective prestress after all losses:

$f_{pe} = 1395 - 231 = 1164$ MPa

which corresponds to a final stress of 0.62 $f_{pu,}$ and:

$P_e = (1164)(1188)/10^3 = 1382$ kN

**Example 6 – Loss of prestress for element with non-prestressed reinforcement – Detailed Method**

*Given:*

Same as Example 4, except that two of the 13 mm strands are replaced by 2 – No. - 25 bars (partially prestressed element).

Concrete:

$E_{co} = 24\,900$ MPa at transfer

$E_{c1} = 28\,200$ MPa at 40 days

Prestressed reinforcement:

$A_p = (10)(99) = 990$ mm$^2$

$E_p = 190\,000$ MPa

$d_p = 691$ mm (at $0.4\ell$)

Non-prestressed reinforcement:

$A_s = (2)(500) = 1000$ mm$^2$

$E_s = 200\,000$ MPa

$d_s = 750$ mm

Section Properties (including all reinforcement at $0.4\ell$):

$A_{tr} = 385\,000$ mm$^2$

$I_{tr} = 25\,400 \times 10^6$ mm$^4$

$e_{trps} = 398$ mm (prestressed reinforcement)

$e_{trs} = 457$ mm (non-prestressed reinforcement)

$e_{tr} = 428$ mm (both reinforcements)

$y_{tr} = 507$ mm

$$r = \sqrt{\frac{228\,000 \times 10^6}{371\,000}} = 248 \text{ mm}$$

*Problem:*

Determine the total prestress loss and effective prestress by Detailed Method.

*Solution:*

Elastic loss:

From Example 4, the relaxation loss prior to transfer is 35 MPa.

Hence:

$f_{po} = 1360$ MPa

$P_o = (1360)(990)/10^3 = 1346$ kN

Concrete stress after transfer at level of centroid of both prestressed and non-prestressed reinforcement:

$$f_{co} = \frac{P_o}{A_{tr}} + \left(\frac{P_o e_{trp} - M_d}{I_{tr}}\right) e_{tr}$$

$$= \frac{(1346)(10^3)}{385\,000}$$

$$+ \left[\frac{(1346)(398)(10^3) - (455)(10^6)}{25\,400 \times 10^6}\right](428)$$

$$= 3.5 + 1.4 = 4.9\,\text{MPa}$$

Elastic prestress loss:

$$\Delta f_{po} = n_o f_{co} = (7.6)(4.9) = 37 \text{ MPa}$$

Prestress after transfer:

$$f_{pi} = f_{po} - \Delta f_{po}$$

$$= 1360 - 37 = 1323 \text{ MPa}$$

Concrete stress at level of centroid of steel due to $M_{sd}$:

$$f_{c1} = \frac{M_{sd}}{I_{tr}} e_{tr}$$

$$= \frac{(154)(10^6)}{25\,400 \times 10^6}(428) = 2.6 \text{ MPa} \quad \text{(tension)}$$

Corresponding change in reinforcement stress

$$n_1 f_{c1} = (6.7)(2.6) = 17 \text{ MPa (tension)}$$

Time-dependent change in stress at centroid of both prestressed and non-prestressed reinforcement (without relaxation):

with

$$\rho_p + \rho_s = \frac{990}{371\,000} + \left(\frac{1000}{371\,000}\right)\left(\frac{200 \times 10^3}{190 \times 10^3}\right) = 0.0055$$

$$\Delta f_p = \frac{n_o f_{co} C_\infty + \epsilon_{sh} E_p}{1 + n_o(\rho_p + \rho_s)(1 + e^2/r^2)(1 + 0.8C_\infty)}$$

$$= \frac{(7.6)(4.9)(1.81) + (442 \times 10^{-6})(190\,000)}{1 + (7.6)(0.0055)\left[1 + \left(\frac{414}{248}\right)^2\right][1 + (0.8)(1.81)]}$$

$$= \frac{151}{1.387} = 109 \text{ MPa}$$

$$\Delta f_{p1} = \frac{n_1 f_{c1} C_{t1}}{1 + n_1(\rho_p + \rho_s)(1 + e^2/r^2)(1 + 0.8C_{t1})}$$

$$= \frac{(6.7)(2.6)(1.45)}{1 + (6.7)(0.0055)\left[1 + \left(\frac{414}{248}\right)^2\right][1 + (0.8)(1.45)]}$$

$$= \frac{25.3}{1.301} = 19\,\text{MPa} \quad \text{(tension)}$$

For stress-relieved strands intrinsic relaxation:

$f_{pi} = 1360 - 37 + 17 = 1340$ MPa[R3.4.4]

$$f_{re} = 0.125\left(\frac{f_{pi}}{f_{py}} - 0.55\right) f_{pi}$$

$$= 0.125\left[\frac{1340}{(0.9)(1860)} - 0.55\right](1340)$$

$$= 42 \text{ MPa}$$

Reduced relaxation takes into account creep and shrinkage.

with:

$$\Omega = \frac{\Delta f_p - \Delta f_{p1}}{f_{pi}} = \frac{109 - 19}{1340} = 0.067$$

$$\beta = \frac{f_{pi}}{f_{pu}} = \frac{1340}{1860} = 0.72$$

from Fig. 3.4.4:

$$\alpha_r = 0.80$$

$$\alpha_r f_{re} = (0.80)(42) = 34 \text{ MPa}$$

Therefore, loss due to relaxation only:

$$\Delta f_p = \frac{\alpha_r f_{re}}{1 + n_o(\rho_p + \rho_s)(1 + e^2/r^2)(1 + 0.8C_\infty)}$$

$$= \frac{34}{1.387} = 24 \text{ MPa}$$

Prestress loss summary (MPa):

| | |
|---|---|
| Relaxation prior to transfer: | 35 |
| Elastic losses: | |
| – precast section only | 37 |
| – due to $M_{sd}$ | –17 |
| Time-dependent losses: | |
| – creep and shrinkage | 109 |
| – due to $M_{sd}$ | –19 |
| – relaxation | 24 |
| Total prestress loss | 169 MPa |

Final forces:

In prestressed reinforcement:

$$P_e = (1395 - 169)(990)/10^3$$

$$= 1214 \text{ kN (tension)}$$

In non-prestressed reinforcement:

$$P_s = (37 - 17 + 90)(1000)/10^3$$

$$= 110 \text{ kN (compression)}$$

Net force acting on concrete:

$$P_e - P_s = 1214 - 110 = 1104 \text{ kN}$$

### 3.4.7 PRESTRESS LOSSES AND DEFLECTION IN COMPOSITE ELEMENTS

Analysis of time-dependent effects in composite members involves the time-dependent properties of the two different concretes and the relaxation of the prestressing steel. In addition, non-prestressed reinforcement may be present. The problem can be solved by the "Creep-transformed Section Method" developed by Dilger[R3.4.5]. The method involves the following steps (see Fig. 3.4.5).

*Step 1:* Determine concrete stresses $f_{ci}$ (see Fig. 3.4.5(b)) in the precast girder (Section 1) after transfer of the prestressing force (see Sect. 3.4.6). At level i:

$$f_{ci} = \frac{P_i}{A_{tr}} + \frac{M_d + P_i e_{tr}}{I_{tr}} y_{tr}$$

(For elements without non-prestressed reinforcement, the gross properties of the section may be used.)

The following sign convention is strictly adhered to in this analysis: compression and shortening are positive, eccentricity e and ordinate y above the centroid are positive, a sagging moment is positive.

*Step 2:* Determine the creep, shrinkage and $E_c$ of the precast girder (Section 1) and the deck (Section 2).

The following data are required:

a) Creep coefficients of girder concrete $C(t_\infty, t_o)$, $C(t_{ca}, t_o)$ and $C(t_\infty, t_{ca})$ where:

$t_\infty$ = time infinity (days)

$t_o$ = time at transfer of prestress (days)

$t_{ca}$ = time at the beginning of the composite action (days)

b) Creep coefficient of deck concrete $C_{deck}(t_\infty, t_{ca} - t_{cc})$ where $t_{cc}$ is the time at casting of topping (days).

For unshored construction, $t_{ca} - t_{cc}$ is usually 1 to 3 days.

c) Shrinkage of girder concrete:

$\epsilon_{sh}(t_\infty, t_o)$ = ultimate shrinkage

$\epsilon_{sh}(t_{ca}, t_o)$ = shrinkage occurring between time, $t_o$ and $t_{ca}$

d) Shrinkage of deck concrete:

$\epsilon_{shdeck}(t_\infty, t_{ca} - t_{cc})$ = ultimate shrinkage of deck concrete

e) Relaxation of prestressing steel, $f_{re}$

f) $E_c(t_o)$, $E_c(t_{ca})$, $E_{cdeck}(t_{ca} - t_{cc})$

Calculate:

$$E^*_1 = E_c(t_1)/[1 + 0.8C(t_\infty, t_{ca})]$$

$$E^*_2 = E_{cdeck}(t_{ca})/[1 + 0.7C_{deck}(t_\infty, t_{ca})]$$

*Step 3:* Calculate prestress loss at time $t_{ca}$ when composite action begins (see Sect. 3.4.6), neglecting relaxation of the steel:

$$\Delta f_p(t_{ca}) = \frac{n_o f_{co} C(t_{ca}, t_o) + \epsilon_{sh}(t_{ca}, t_o) E_p}{1 + n_o(\rho_p + \rho_s)(1 + e^2/r^2)[1 + 0.8C(t_{ca}, t_o)]}$$

Loss of prestressing force:

$$\Delta P_{ca} = \Delta f_p(t_{ca}) A_p$$

*Step 4:* Calculate stress in precast section due to $\Delta P_{ca}$. Referring to Fig. 3.4.5(c), stress at level i (i = 2, ... 5):

$$f_{ci} = -\frac{\Delta P_{ca}}{A_g} - \frac{\Delta P_{ca} e_3}{I} y_i$$

Use properties of concrete section (not including $A_p$), or gross section properties.

*Step 5:* Calculate stresses due to weight of deck (or topping) in fibres i (i = 2, ... 5):

a) For unshored construction the stresses are:

$$f_{sdi} = \frac{M_{sd}}{I} y_i$$

b) For shored construction the stresses are:

$$f'_{sdi} = \frac{M_{sd}}{I'} y'_i$$

The prime indicates properties of the transformed composite section, determined by multiplying the width of the deck by the ratio $E_{cdeck}(t_{ca} - t_{cc})/E_c(t_{ca})$.

*Step 6:* Calculate strains:

In fibre 2:

$$\epsilon^*_2 = \frac{f_{c2}}{E_c(t_o)}[C(t_\infty, t_o) - C(t_{ca}, t_o)] + \frac{f_{sd2}}{E_c(t_{ca})} C(t_\infty, t_{ca}) + \epsilon_{sh}(t_\infty, t_o) - \epsilon_{sh}(t_{ca}, t_o) - \epsilon_{shdeck}(t_\infty, t_{ca} - t_{cc})$$

where $f_{c2}$ is the sum of the stress components of steps 1 and 4 at level 2.

The strain $\epsilon^*_2$ represents the difference in free strain between the precast and deck concrete at level of fibre 2. For shored construction the term:

$\frac{f_{sd2}}{E_c(t_o)} C(t_\infty, t_{ca})$ is replaced by:

$$\frac{f'_{sd2}}{E_c(t_{ca})}[C(t_\infty, t_{ca}) - C_{deck}(t_\infty, t_{ca} - t_{cc})]$$

In fibre 3:

$$\epsilon^*_3 = \frac{f_{c3}}{E_c(t_o)}[C(t_\infty, t_o) - C(t_{ca}, t_o)] + \frac{f_{sd3}}{E_c(t_{ca})} C(t_\infty, t_{ca}) + \epsilon_{sh}(t_\infty, t_{ca}) - \epsilon_{sh}(t_{ca}, t_o)$$

For fibres 4 and 5, $f_{c3}$ and $f_{sd3}$ are replaced, respectively, by $f_{c4}$ and $f_{sd4}$, and $f_{c5}$ and $f_{sd5}$.

**Fig. 3.4.5 Prestress loss in composite elements**

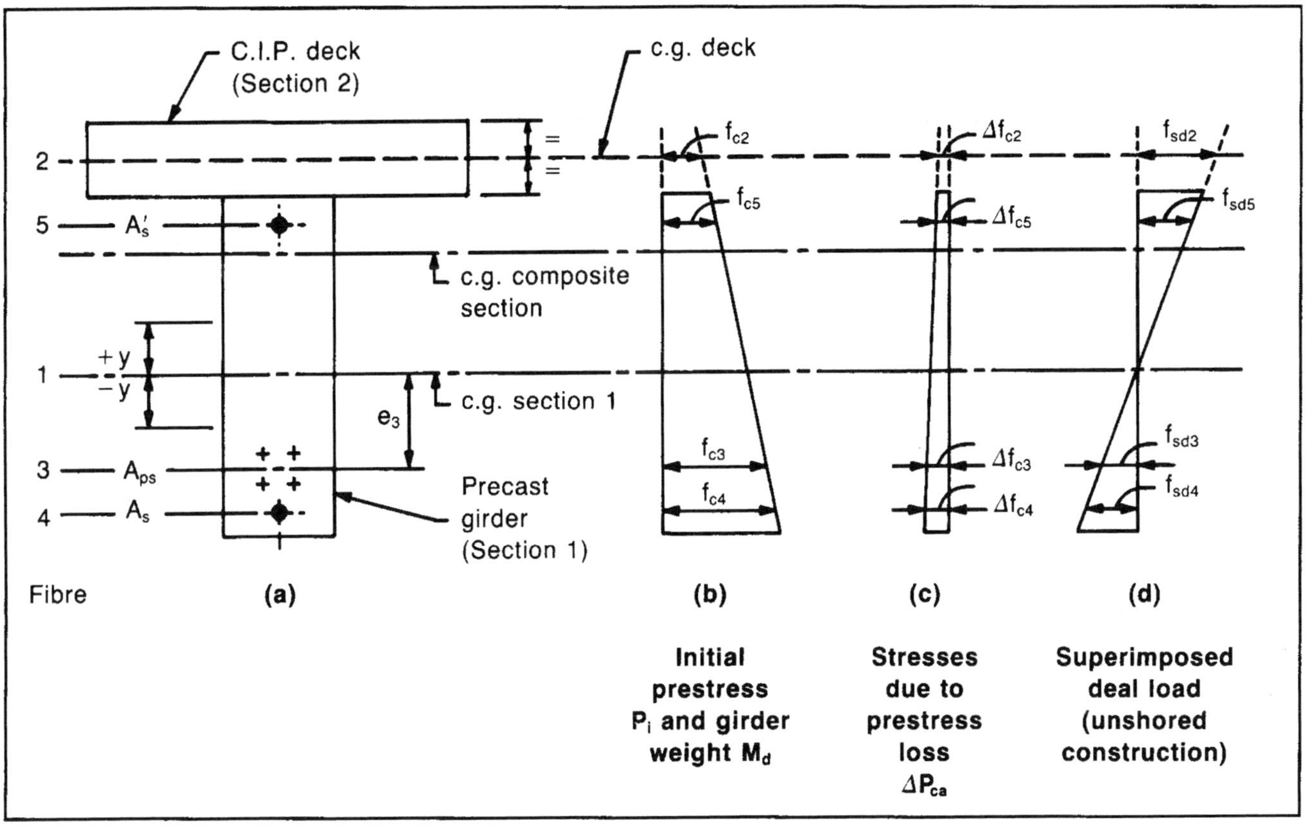


*Step 7:* Determine properties of creep-transformed section. Multipliers for the calculation of the properties of the creep-transformed section are:

For the steel:

$(n^* - 1)$ where $n^* = E_p / E_1^*$

For the deck concrete:

$E^*_2 / E^*_1$

*Step 8:* Calculate normal forces and bending moments:

| | | |
|---|---|---|
| $N^*_2$ | $= \epsilon^*_2 E^*_2 A_{deck}$ | $M^*_2 = N^*_2 y^*_2$ |
| $A_{deck}$ | = Area of concrete deck (Section 2) | |
| $N^*_3$ | $= (\epsilon^*_3 E^*_p + f_{re}) A_p$ | $M^*_3 = N^*_3 y^*_3$ |
| $N^*_4$ | $= \epsilon^*_4 E_s A_{s4}$ | $M^*_4 = N^*_4 y^*_4$ |
| $N^*_5$ | $= \epsilon^*_5 E_s A_{s5}$ | $M^*_5 = N^*_5 y^*_5$ |
| $N^*$ | $= \Sigma N^*_i$ | $M^* = \Sigma N_{Ni} y^*_i$ |

The distances y* are the distances from the centroid of the creep-transformed section.

Note: The reduced relaxation may be determined as shown in Example 5.

*Step 9:* Calculate stress due to N* and M*: Concrete stress in the top fibre of the deck for unshored construction:

$$f^*_t = -\left(\frac{N^*}{A^*} + \frac{M^*}{I^*} y^*_t\right)\frac{E^*_2}{E^*_1} + \frac{N^*_2}{A_{deck}}$$

A*, I* and y* are properties of the creep transformed section (see Example 5).

Note that y* above the centroid is positive, below negative.

For shored construction:

$$f = f_{sd} + f^*$$

where $f_{sd}$ is the flexural stress in the top fibre due to the weight of the deck (see Step 4).

Time dependent stress in the bottom fibre:

$$f^*_b = -\left(\frac{N^*}{A^*} + \frac{M^*}{I^*} y^*_b\right)$$

This stress is to be added to the stresses due to prestressing force P at the time $t_{ca}$, and moments $M_d$ and $M_{sd}$ at the bottom fibre.

The time-dependent change due to composite action in the prestressed reinforcement:

$$f^*_3 = -\left(\frac{N^*}{A^*} + \frac{M^*}{I^*} y^*_3\right) n^* + \epsilon^*_3 E_p + f_{re}$$

and in the non-prestressed reinforcement, if any:

$$f^*_4 = -\left(\frac{N^*}{A^*} + \frac{M^*}{I^*} y^*_4\right) n^* + \epsilon^*_4 E_s$$

The stresses f* have to be added to the stresses at the beginning of the composite action (time $t_{ca}$).

*Step 10:* Calculation of deformations:

Initial plus time-dependent curvature (starting from age $t_o$):

$$\psi_\infty = \psi(t_o)[1 + C(t_\infty, t_o)] + \psi_{\Delta p} + \psi(t_{ca})[1 + C(t_\infty, t_{ca})] - \frac{M^*}{I^* E^*_1}$$

where:

$$\psi(t_o) = \frac{M_d + P_o e_3}{EI_{tr}}$$

= elastic curvature of precast girder at transfer

$$\psi\Delta_p = \frac{\Delta P_{ca} e_3}{EI}$$

= elastic curvature due to prestress loss before composite action begins.

$$\psi(t_{ca}) = \frac{M_{sd}}{EI_{tr}}$$

= elastic curvature due to super-imposed load for unshored construction.

or:

$$\psi(t_{ca}) = \frac{M_{sd}}{EI'}$$

= elastic curvature due to super-imposed load for shored construction.

To calculate the deflection, curvatures should be established at $0.4\ell$ for draped tendons and the support and an appropriate distribution assumed in between for integration.

**Example 7– Prestress loss in a composite element using Detailed Method**

*Given:*

The 2400 × 800 double tee of Example 4 on Page 3-32 but assuming that the topping acts compositely.

$f'_c$ precast = 35 MPa

$f'_c$ topping = 25 MPa

The topping is applied at a precast double tee age of 40 days. Assume that composite action begins at age 43 days and that the concrete strength of the topping (after moist curing) has reached 17.5 MPa at the beginning of the composite action.

*Problem:*

Determine the total loss of prestress using the detailed method.

*Solution:*

*Step 1:* Concrete stress after transfer:

At c.g. of topping (level 2):

$f_{c2}$ = 1.7 MPa

At level of prestressing steel (level 3)

$f_{c3}$ = 7.6 MPa

At other levels, see Fig. 3.4.6(b).

*Step 2:* Creep and shrinkage properties:

Precast section:

$C(t_\infty, 1)$ = 1.81 (see Example 4)

$C(t_\infty, 43)$ = 1.45 (see Example 4)

$$C(43,1) = \frac{t^{0.6}}{10+t^{0.6}} C_\infty$$

$$= \left[\frac{(43-1)^{0.6}}{10+(43-1)^{0.6}}\right](1.81) = 0.88$$

$\in_{sh}(t_\infty, 1)$ = $442 \times 10^{-6}$ (see Example 4)

$$\in_{sh}(43, 1) = \left(\frac{t}{C_s + t}\right) \in_{sh}$$

$$= \left[\frac{(43-1)}{55+(43-1)}\right](442 \times 10^{-6})$$

$$= 191 \times 10^{-6}$$

Cast-in-place topping:

$C_{deck}(t_\infty, 43)$ = 2.40

$\in_{shdeck}(t_\infty, 43)$ = $600 \times 10^{-6}$

Moduli of elasticity:

$E_c(t_o)$ = 24 900 MPa for $f'_{ci}$ = 25 MPa

$E_c(t_{ca})$ = 28 200 MPa for $f'_c$ = 35 MPa

$E_{cdeck}(t_{ca} - t_{cc})$ = 22 100 MPa for $f'_c$ = 17.5 MPa and $\gamma_c = 2400$ kg / m$^3$

Age adjusted moduli:

$$E^*_1 = \frac{E_c(t_{ca})}{1+0.8C(t_\infty, t_{ca})}$$

$$= \frac{28\,200}{1+(0.8)(1.45)} = 13\,100 \text{ MPa}$$

$$E^*_2 = \frac{E_{cdeck}(t_{ca} - t_{cc})}{1+0.7C_{deck}(t_\infty, t_{ca})}$$

$$= \frac{22\,100}{1+(0.7)(1.45)} = 8200 \text{ MPa}$$

| | | |
|---|---|---|
| $E^*_2 / E^*_1$ | = 8200/13 100 | = 0.63 |
| $n^*$ | = 190 000/13 100 | = 14.5 |
| $n_o$ | = 190 000/24 900 | = 7.6 (at transfer) |
| $n$ | = 190 000/28 200 | = 6.7 (at time $t_{ca}$) |

*Step 3:* Loss of prestress at time $t_{ca}$ (neglecting relaxation):

$$\Delta f_p(t_{ca}) = \frac{n_o f_{c3} C(t_{ca}, t_o) + \in_{sh}(t_{ca}, t_o) E_p}{1+n_o(\rho_p + \rho_s)(1+e^2/r^2)[1+0.8C(t_{ca}, t_o)]}$$

$$= \frac{(7.6)(7.6)(0.88) + (191 \times 10^{-6})(190\,000)}{1+(7.6)(0.0032)\left[1+\left(\frac{414}{248}\right)^2\right][1+(0.8)(0.88)]}$$

$$= 75 \text{ MPa}$$

$$\Delta P_{ca} = \Delta f_p(t_{ca}) A_p$$

$$= (75)(1188)/10^3 = 89 \text{ kN}$$

*Step 4:* Stresses due to $\Delta P_{ca}$ = 89 kN:

$$f_{ci} = -\frac{\Delta P_{ca}}{A_g} - \frac{\Delta P_{ca} e_3}{I} y_i$$

At c.g. of topping (level 2):

$$f_{c2} = \frac{(-89)(10^3)}{371\,000} + \left[\frac{(-89)(10^3)(-414)}{(22.8)(10^9)}\right](287+25)$$

$$= 0.3 \text{ MPa}$$

At level of prestressing steel (level 3):

$f_{c3}$ = – 0.9 MPa (tension)

For other levels, see Fig. 3.4.6(c)

*Step 5:* Stress due to superimposed dead load of topping:

$$f_{sdi} = \frac{M_{sd}}{I} y_i$$

$$f_{sd2} = \left[\frac{(154)(10^6)}{24.1 \times 10^9}\right](287+25) = 2.0 \text{ MPa}$$

$$f_{sd3} = \left[\frac{(154)(10^6)}{24.1 \times 10^9}\right](-405) = -2.6 \text{ MPa} \quad \text{(tension)}$$

Increase in steel stress:

$= n_1 f_{sd3} = (6.7)(-2.6) = -17$ MPa (tension)

Steel stress at the beginning of composite action:

$f_{pr}$ = 1360 – (7.6)(7.6) – 75 + 17 = 1244 MPa

*Step 6:* Strain:

$$\in^*_2 = \frac{f_{c2}}{E_c(t_o)}[C(t_\infty, t_o) - C(t_{ca}, t_o)] + \frac{f_{sd2}}{E_c(t_{ca})} C(t_\infty, t_{ca}) + \in_{sh}(t_\infty, t_o) - \in_{sh}(t_{ca}, t_o) - \in_{shdeck}(t_\infty, t_{ca} - t_{cc})$$

$$= \left(\frac{1.7+0.3}{24\,900}\right)(1.81 - 0.88) + \left(\frac{2.0}{28\,200}\right)(1.45) + (442 - 191 - 600)(10^{-6})$$

$$= -171 \times 10^{-6}$$

$$\in^*_3 = \frac{f_{c3}}{E_c(t_o)}[C(t_\infty, t_o) - C(t_{ca}, t_o)] + \frac{f_{sd3}}{E_c(t_{ca})} C(t_\infty, t_{ca}) + \in_{sh}(t_\infty, t_o) - \in_{sh}(t_{ca}, t_o)$$

$$= \left(\frac{7.6 - 0.9}{24\,900}\right)(1.81 - 0.88)$$

$$+\left(\frac{-2.6}{28\,200}\right)(145) + (442 - 191)(10^{-6})$$

$$= 368 \times 10^{-6}$$

*Step 7:* Properties of creep-transformed section:

From Step 2, the multiplier for the deck concrete at level 2 is:

$$= \frac{E_2^*}{E_1^*} = 0.63$$

and the multiplier for the prestressing steel at level 3 is:

$$= n^*_1 - 1 = 14.5 - 1 = 13.5$$

$A^* = 463 \times 10^3\ mm^2$

$\bar{y} = \dfrac{16.29 \times 10^6}{463 \times 10^3} = 35\ mm$

$I^* = (9.11 + 22.8)(10^9) = 31.9 \times 10^9\ mm^4$

*Step 8:* Normal force and bending moments:

$N^*_2 = \in^*_2 E^*_2 A_2$

$= (-171 \times 10^{-6})(8200)(120 \times 10^3)/10^3$

$= -168.3\ kN$

$N^*_3 = (\in^*_3 E_p + f_{re}) A_p$

$= (368 \times 10^{-6})(190\,000)(1188)/10^3$

$= 83.1\ kN$

$N^* = \Sigma N^*_i = -168.3 + 83.1 = -85.2\ kN$

$M^*_2 = N^*_2 y^*_2$

$= (-168.3)(268)/10^3 = -45.1\ kN{\cdot}m$

$M^*_3 = N^*_3 y^*_3$

$= (83.1)(-449)/10^3 = -37.3\ kN{\cdot}m$

$M^* = M^*_2 + M^*_3 = -45.1 - 37.3$

$= -82.4\ kN{\cdot}m$

Reduced relaxation:

Time-dependent change in prestress due to composite action without relaxation:

$$f^*_3 = -\left[\frac{N^*}{A^*} + \frac{M^*}{I^*} y^*_3\right] n^*_1 + \in^*_3 E_p$$

$$= -\left[\frac{(-85.2)(10^3)}{463 \times 10^3} + \frac{(-82.4)(10^6)}{31.9 \times 10^9}(-449)\right](14.5)$$

$$+(368 \times 10^{-6})(190\,000)$$

$$= 56\ MPa$$

Intrinsic relaxation with:

$f_{pi} = 1244\ MPa$

$$f_{re} = 0.125\left(\frac{f_{pi}}{f_{py}} - 0.55\right) f_{pi}$$

$$= (0.125)\left(\frac{1244}{(0.9)(1860)} - 0.55\right)(1244)$$

$= 30.0\ MPa$

$\Omega = \dfrac{56}{1244} = 0.045$

$\beta = \dfrac{1244}{1860} = 0.67$

From Fig. 3.4.4:

$\alpha_r = 0.86$

$\alpha_r f_{re} = (0.86)(30) = 26\ MPa$

Add forces due to reduced relaxation to above values of N* and M*. The new values are:

$N^* = -85.2 + (26)(1188)/10^3 = -54.3\ kN$

$M^* = -82.4 + (26)(1188)(-449)/10^6$

$= -96\ kN \cdot m$

*Step 9:* Stresses:

Final concrete stress in top and bottom fibres of composite section:

$$f_t = f^*_t = -\left(\frac{N^*}{A^*} + \frac{M^*}{I^*} y_t^*\right)\frac{E_2^*}{E_1^*} + \frac{N_2^*}{A_{c2}}$$

$$= -\left[\frac{(-54.3)(10^3)}{(463 \times 10^3)} + \frac{(-96)(10^6)}{(31.9 \times 10^9)}(293)\right](0.63)$$

$$+\frac{(-168.3)(10^3)}{(120 \times 10^3)} = -0.8\ MPa \quad \text{(tension)}$$

| Level | Area ($mm^2 \times 10^3$) | Multiplier | A* ($mm^2 \times 10^3$) | y | A*y ($mm^3 \times 10^6$) | y* = y – y (mm) | A*(y*)² ($mm^4 \times 10^9$) | I ($mm^4 \times 10^9$) | I* ($mm^4 \times 10^9$) |
|---|---|---|---|---|---|---|---|---|---|
| 1 | 371 | 1.0 | 371.0 | 0 | 0 | – 35 | 0.45 | 22.8 | 23.25 |
| 2 | 120 | 0.63 | 75.6 | 303 | 22.91 | 268 | 5.43 | | 5.43 |
| 3 | 1.188 | 13.5 | 16.0 | – 414 | – 6.62 | – 449 | 3.23 | – | 3.23 |
| Total | | | 462.6 | | 16.29 | | 9.11 | 22.8 | 31.9 |

**Fig. 3.4.6 Stress distributions for Example 6**

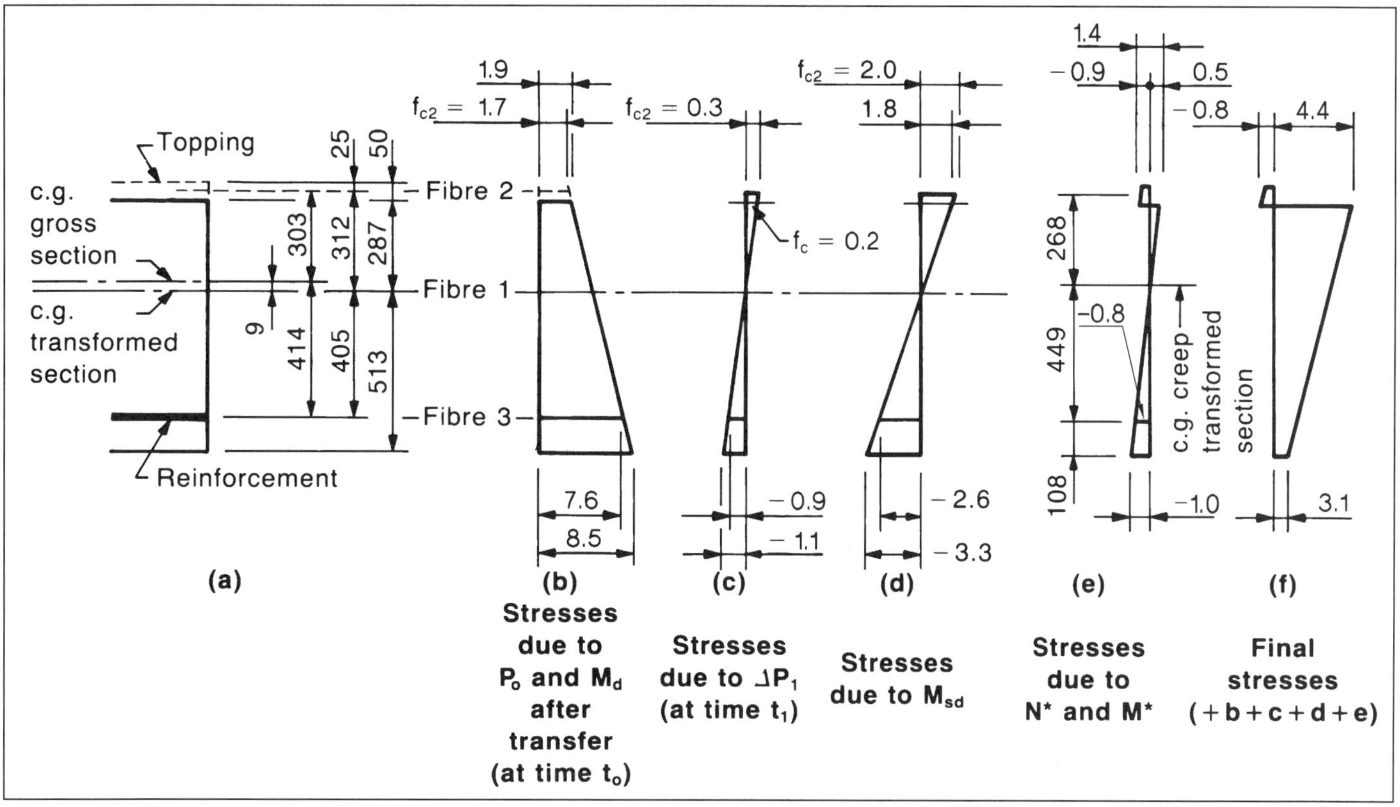


and from Fig 3.4.6:

$$f_b = f^*_b$$

$$= -\left[\frac{(-54.3)(10^3)}{(463\times10^3)} + \frac{(-96)(10^6)}{(31.9\times10^9)}(-557)\right](0.63)$$

$$+8.5 - 1.1 - 3.3 = 3.1\,\text{MPa} \quad \text{(compression)}$$

Effective prestress:

$$f^*_3 = -\left(\frac{N^*}{A^*} + \frac{M^*}{I^*}y^*_3\right)n_1 + \epsilon^*_3 E_p + \alpha_r f_{re}$$

$$= -\left[\frac{(-54.3)(10^3)}{(463\times10^3)} + \frac{(-96)(10^6)}{(31.9\times10^9)}(-449)\right](14.5)$$

$$+(368\times10^{-6})(190\,000) + 26$$

$$= 78\,\text{MPa}$$

$$P_e = (1244 - 78)(1188)/10^3 = 1385\,\text{kN}$$

The distribution of the final concrete stresses is shown in Fig. 3.4.6(f).

## 3.4.8 PARTIALLY PRESTRESSED ELEMENT DESIGN

In many prestressed concrete structures it is not likely that the full specified load will be applied during the lifetime of the structure. It is therefore possible to design the structural elements such that some cracking may occur under full specified load should it ever be applied. Within the range of specified dead and live loads found in most building structures, elements should be proportioned such that:

$$M_{cr} > M_d + M_{sd}$$

This is not a requirement of the Standard but is good design practice; it also assists the designer in the selection of prestressed and non-prestressed reinforcement.

The advantages of partial prestressing are:

1. A reduction in camber.
2. A reduction in prestress force which may allow and increase in tendon eccentricity.
3. A reduction in prestress losses due to creep and relaxation (where partial prestressing is achieved using lower tendon stresses).
4. A reduction of cracking in the end zones of post-tensioned elements.

Partial prestressing may be achieved in two ways:

1. By providing non-prestressed reinforcement in addition to the prestressing tendons. This would normally be utilized in post-tensioned construction. However, this may also provide greater economy in long-line pretensioned fabrication when it is desirable to produce elements of dissimilar strength.
2. By reducing the effective prestress below the maximum and relying on the increase in tendon stress after decompression and cracking to resist the increase in moment. This is a useful option in pretensioned

construction where there may be difficulties in placing the non-prestressed reinforcement.

The crack control criteria for partially prestressed elements, given in Fig. 3.4.1, are more conservative than for reinforced elements, due to the potential effects of corrosion on small diameter tendons.

For the calculation of z, the same equation for reinforced concrete elements is used:

$$z = f_s \sqrt[3]{d_c A}$$

where:

$f_s$ = the increase in steel stress in the non-prestressed reinforcement beyond the state of decompression

$d_c$ = concrete cover to the centre of the reinforcement (mm)

$A$ = effective tension area of concrete surrounding the flexural tension reinforcement and having the same centroid as that reinforcement, divided by the number of bars or wires ($mm^2$). When the flexural reinforcement consists of different bar or wire sizes, the number of bars or wires used to compute A is taken as the total area of reinforcement divided by the area of the largest bar or wire used.

Note: This definition was developed for elements with a rectangular tension zone. For elements with a large tension flange (eg. a box section) this definition may not be applicable and $w_{max}$ should be calculated as shown in Step 5 of the Detailed Method on Page 3-50.

The value of $f_s$ may be determined by either a simplified or a detailed method. The first method is rather conservative, particularly for elements with a small reinforcement ratio. The second is a more rigorous method and is given here, together with some recent research results concerning maximum crack widths in partially prestressed elements with different types of prestressed and non-prestressed reinforcement. The detailed method follows the paper by Suri and Dilger published in the PCI Journal in May/June 1986. In this paper, design aids to calculate $f_s$ are presented which are used here.

*Note:* The CPCI computer software program CONCISE BEAM uses the detailed method and calculates crack widths directly.

### Simplified Method

The stress, $f_s$, may be calculated as the stress corresponding to the difference between the moment due to specified loads, $M_s$, and the decompression moment, $M_{dc}$, according to:

$$f_s = \frac{M_s - M_{dc}}{(A_p + A_s)d}$$

The decompression moment, $M_{dc}$, is the moment which produces zero stress at the extreme concrete fibre of the precompressed tensile zone. The moment does not result in complete decompression of the whole section; this would require the removal of the bending moment due to applied loads and the application of a fictitious decompression force. The advantages of the equation are that it eliminates the complicated calculations required to compute the decompression force $P_{dc}$ and it facilitates a quick determination of the stresses due to specified loads.

### Detailed Method

The reference point for the calculation of the incremental steel stress corresponds to zero stress throughout the concrete section. The state of zero concrete stress is achieved by applying a so-called decompression force $P_{dc}$ to the concrete section.

The step-by-step procedure is summarized as follows:

*Step 1:*

Determine the effective prestress $P_e$ and the time-dependent change in stress in the non-prestressed reinforcement, $\Delta f_s$ according to the procedure given in Sect. 3.4.6. Note that $\Delta f_s = \Delta f_p$.

*Step 2:*

Determine the decompression force $P_{dc}$:

$$P_{dc} = P_e + \Delta P_p - P_s + \Delta P_s$$

where:

$P_e$ = effective prestress (after all losses)

$$\Delta P_p = \frac{f_c}{E_c} E_p A_p$$

= force in tendons corresponding to the concrete stress $f_c$ at the level of the tendon under dead load and effective prestress

$$P_s = A_s \Delta f_s$$

= force in the non-prestressed reinforcement due time-dependent strain

$$\Delta P_s = \frac{f_c}{E_c} E_s A_s$$

= force in non-prestressed reinforcement corresponding to the concrete stress $f_c$ at the level of the non-prestressed reinforcement under dead load and effective prestress.

The forces $\Delta P_p$ and $\Delta P_s$ are normally small and are often neglected.

*Step 3:*

Determine parameters:

$$d = \frac{E_p A_p d_p + E_s A_s d_s}{E_p A_p + E_s A_s}$$

$$p = \frac{P_{dc} d_p}{M_s}$$

$$n\rho = \frac{A_p E_p + A_s E_s}{b d E_c}$$

$$\alpha_b = \frac{b_w}{b}$$

**Fig. 3.4.7 (a) T-beam design coefficients**

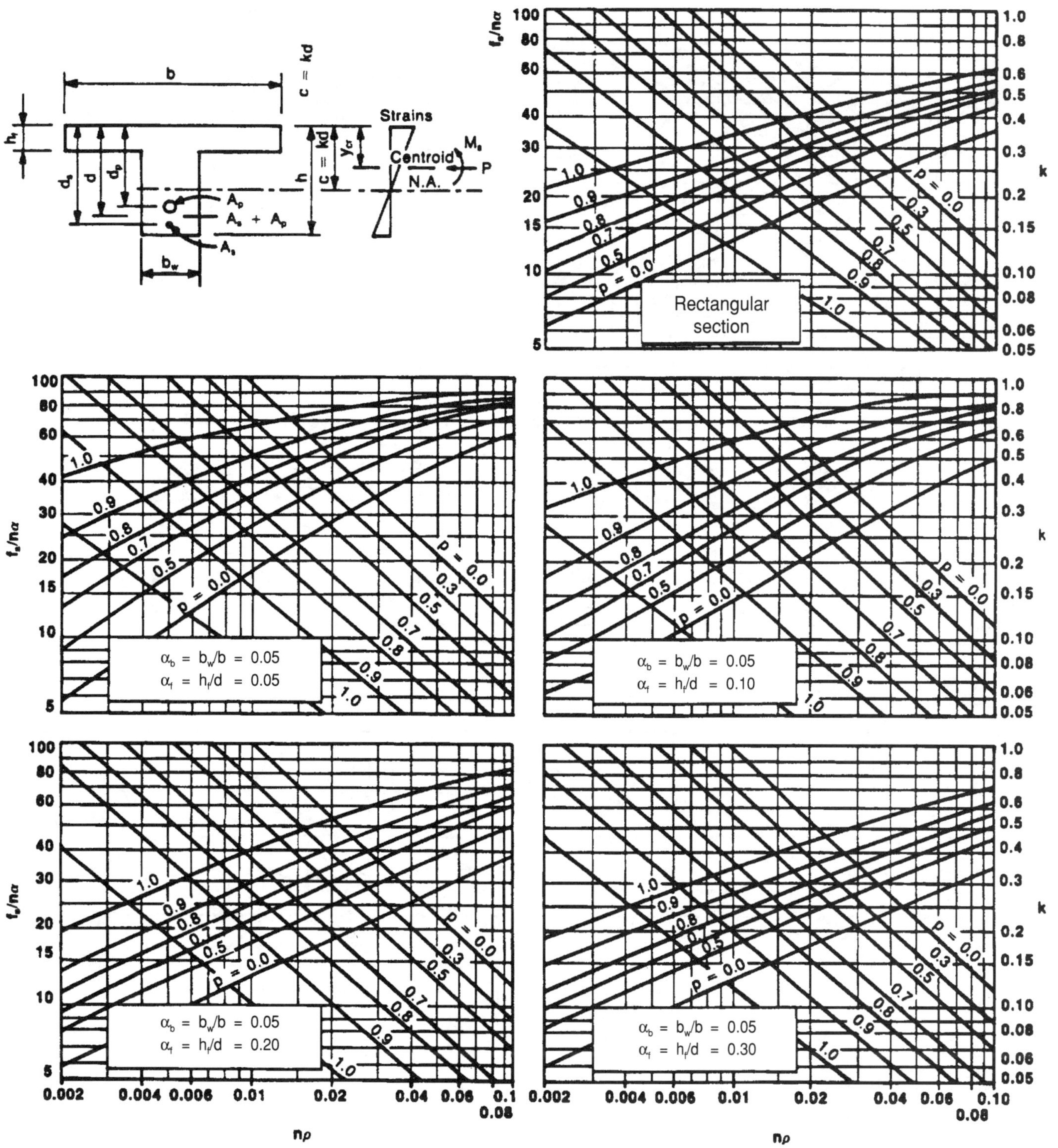

**Fig. 3.4.7 (b) T-beam design coefficients**

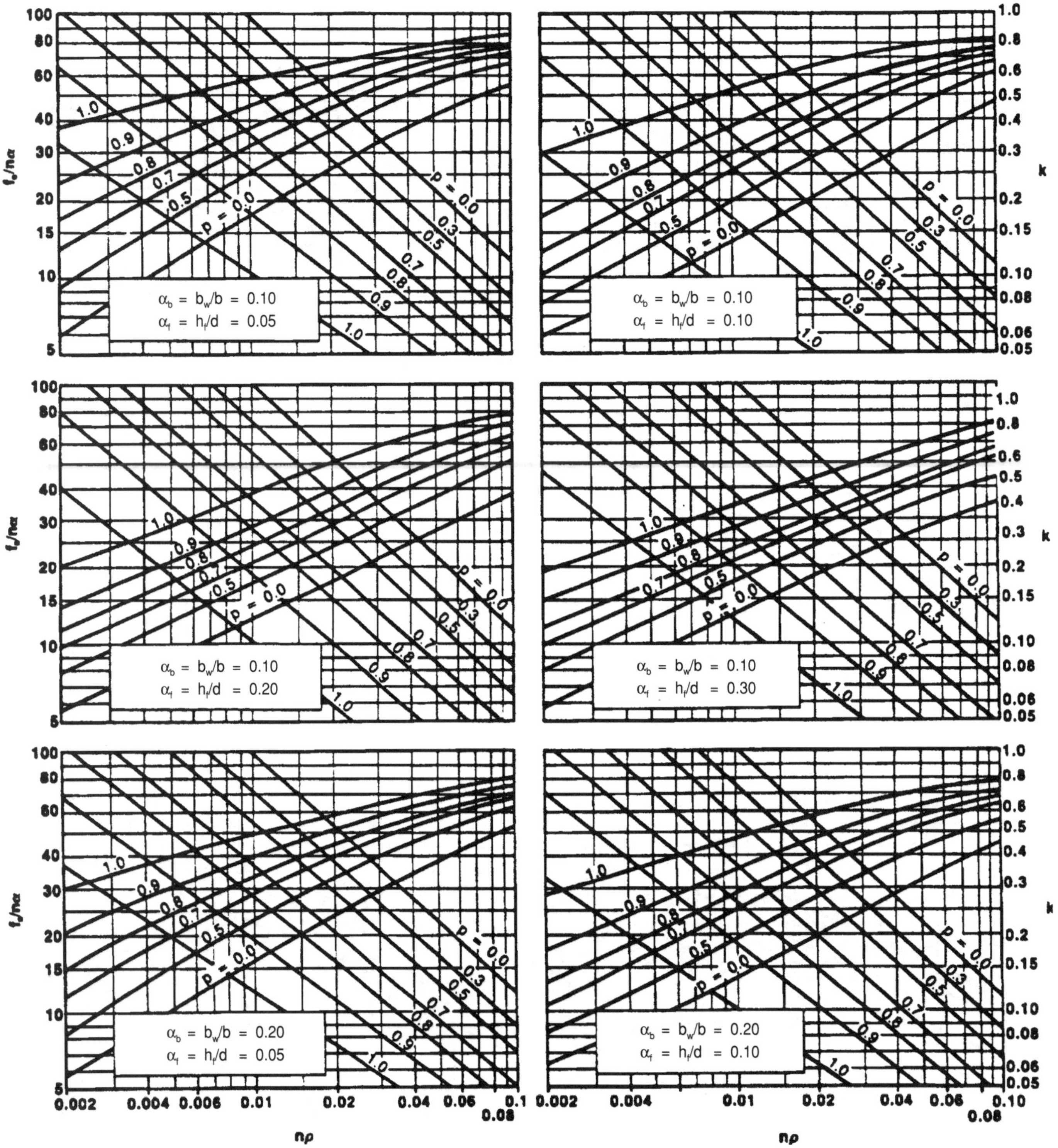

**Fig. 3.4.7 (c) T-beam design coefficients**

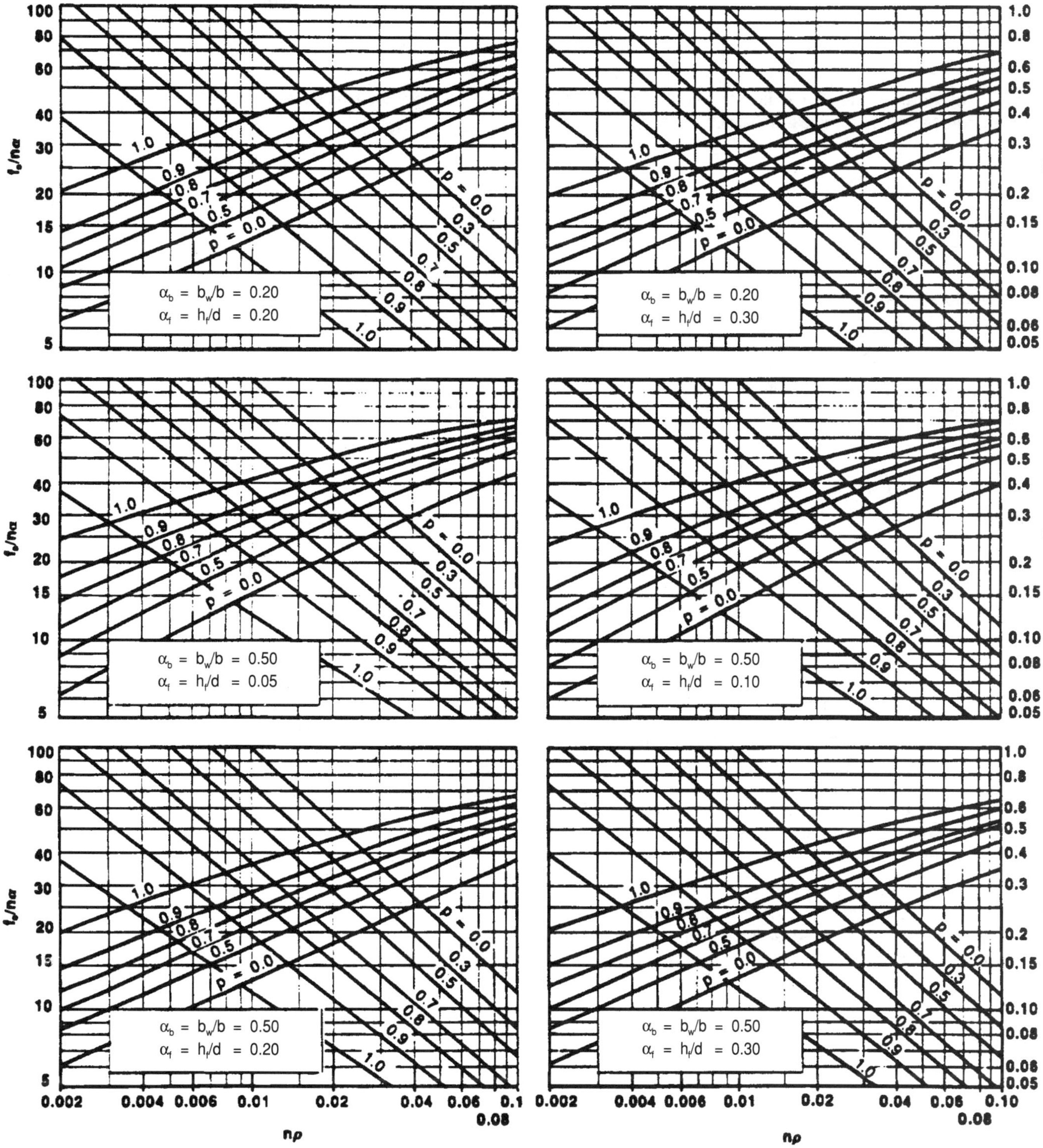

$$\alpha_f = \frac{h_f}{h}$$

For rectangular sections $\alpha_b = \alpha_f = 1.0$

*Step 4:*

From the charts in Fig. 3.4.7, find the coefficient for the appropriate parameters and use it to calculate the incremental steel stress:

$$f_s = n\alpha \times (\text{coefficient})$$

where:

$$n\alpha = \frac{E_p}{E_c}\frac{M_s}{bd^2}$$

The charts also provide the neutral axis depth coefficient k. (See Example 8)

*Step 5:*

Calculate the maximum crack width or crack control parameter z and compare with allowable values.

Maximum crack width[R3.4.6]:

$$w_{max} = k_1 f_s d_c \sqrt{A_t / A_{st}}$$

where:

$k_1$ = $3 \times 10^{-6}$ for a combination of deformed bars and strands and for strands only

$k_1$ = $4 \times 10^{-6}$ for a combination of deformed bars and wires

$k_1$ = $5 \times 10^{-6}$ for wires only

$d_c$ = concrete cover to centre of the reinforcement (mm)

$A_t$ = concrete area in tension below the neutral axis

$A_{st}$ = $A_p + A_s E_s/E_p$

For unbonded tendons the prestressed reinforcement is not considered when calculating $w_{max}$. Recommended values of $w_{max}$ are given in Fig. 3.4.1.

If only the non-prestressed reinforcement is considered to contribute to crack control, or if there is a combination of strands and deformed non-prestressed reinforcement, the crack control parameter z may be used and the limiting values for Fig. 3.4.1 apply:

$$z = f_s \sqrt[3]{d_c A}$$

*Step 6:*

Calculate compressive stress in top fibre:

$$f_c = \frac{1}{n} f_s \left(\frac{kd}{d - kd}\right)$$

where:

n = $E_p/E_c$

k = neutral axis depth coefficient

d = average depth as defined in Step 3

**Example 8 – Crack width analysis of partially prestressed element using detailed method**

*Given:*

3000 × 1200 single tee

Span = 27.0 m

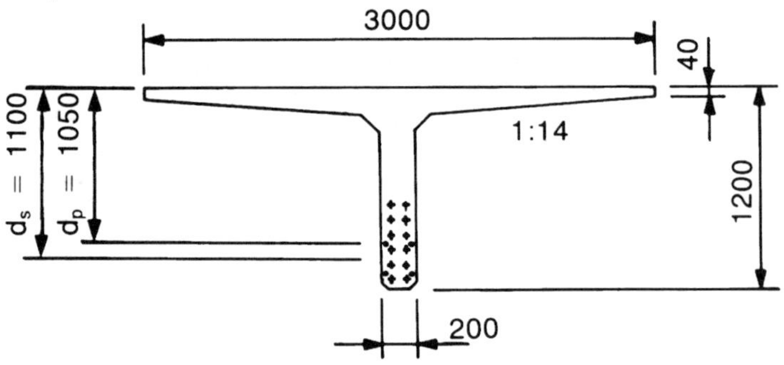


Concrete:

$f'_{ci}$ = 25 MPa

$f'_c$ = 35 MPa

$E_{co}$ = 24 900 MPa at transfer

$E_c$ = 28 200 MPa at 40 days

Normal Density

Prestressed reinforcement:

12 – 13 mm 1860 MPa low relaxation, bonded strands

$A_p$ = 1188 mm$^2$

$E_p$ = 190 000 MPa

Non-prestressed reinforcement:

4 – No. 25 Grade 400 bars

$A_s$ = (4)(500) = 2000 mm$^2$

$E_s$ = 200 000 MPa

Section Properties:

$A_c$ = 492 000 mm$^2$

$I_g$ = 66 300 × 10$^6$ mm$^4$

$y_b$ = 882 mm

$e_c$ = 731 mm

$A_{tr}$ = 513 900 mm$^2$

$I_{tr}$ = 79 150 × 10$^6$ mm$^4$

$e_{trp}$ = 699 mm (prestressed reinforcement)

$e_{trs}$ = 749 mm (non-prestressed reinforcement)

$e_{tr}$ = 731 mm (both reinforcements)

$y_{tr}$ = 849 mm

r = 367 mm

Creep, shrinkage and relaxation:

Creep Coefficient $C_t = 1.76$

Shrinkage $\epsilon_{sh} = 430 \times 10^{-6}$

Intrinsic Relaxation $f_{re} = 50$ MPa

Service Moments:

$M_d = 1066$ kN·m

$M_{sd} = 410$ kN·m

$M_\ell = 546$ kN·m

$$M_s = M_d + M_{sd} + M_\ell$$
$$= 1066 + 410 + 546 = 2022 \text{ kN·m}$$

*Problem:*

Calculate z and check the crack width at midspan for exterior exposure conditions.

*Solution:*

*Step 1(a):* Elastic loss:

Concrete stress after transfer at level of centroid of both pre-stressed and non-prestressed reinforcement:

$$f_{co} = \frac{P_o}{A_{tr}} + \left(\frac{P_o e_{trp} - M_d}{I_{tr}}\right) e_{tr}$$

where:

$$P_o = f_{po} A_p$$

with:

$$f_{po} = (0.75)(f_{pu})$$
$$= (0.75)(1860) = 1395 \text{ MPa}$$

$$P_o = (1395)(1188)/10^3 = 1657 \text{ kN}$$

$$f_{co} = \frac{(1657)(10^3)}{513\,900} + \left[\frac{(1657)(699)(10^3) - (1066)(10^6)}{79\,150 \times 10^6}\right](731)$$
$$= 3.2 + 0.9 = 4.1 \text{ MPa}$$

Elastic prestress loss:

$$n_o = \frac{E_p}{E_{co}} = \frac{190\,000}{24\,900} = 7.6$$

$$\Delta f_{po} = n_o f_{co} = (7.6)(4.1) = 31 \text{MPa}$$

Prestress after transfer:

$$f_{pi} = f_{po} - \Delta f_{po}$$
$$= 1395 - 31 = 1364 \text{ MPa}$$

Concrete stress at level of centroid of steel due to $M_{sd}$:

$$f_{c1} = \frac{M_{sd}}{I_{tr}} e_{tr}$$
$$= \frac{(410)(10^6)}{79\,150 \times 10^6}(731) = 3.8 \text{ MPa} \quad \text{(tension)}$$

Corresponding change in reinforcement stress:

$$n_1 = \frac{E_p}{E_{c1}} = \frac{190\,000}{28\,200} = 6.7$$

$$n_1 f_{c1} = (6.7)(3.8) = 25 \text{ MPa} \quad \text{(tension)}$$

*Step 1(b):* Time dependent losses:

with:

$$\rho_p + \rho_s = \frac{A_p}{A_c} + \frac{A_s}{A_c}\frac{E_s}{E_p}$$
$$= \frac{1188}{492 \times 10^3} + \frac{(2000)(200\,000)}{(492 \times 10^3)(190\,000)}$$
$$= 0.0024 + 0.0043 = 0.0067$$

The dependent change in stress at centroid of reinforcement (without relaxation):

$$\Delta f_p = \frac{n_o f_{co} C_\infty + \epsilon_{sh} E_p}{1 + n_o(\rho_p + \rho_s)(1 + e^2/r^2)(1 + 0.8C_\infty)}$$
$$= \frac{(7.6)(4.1)(1.76) + (430 \times 10^{-6})(190\,000)}{1 + (7.6)(0.0067)\left[1 + \left(\frac{731}{367}\right)^2\right][1 + (0.8)(1.76)]}$$
$$= \frac{136.5}{1.609} = 85 \text{ MPa}$$

Assuming the superimposed dead load is applied at 40 days, then from Fig. 2.4.1:

$$Q_{a1} = 0.80$$

$$C_{t1} = 0.80\, C_\infty$$
$$= (0.80)(1.76) = 1.41$$

$$\Delta f_{p1} = \frac{n_1 f_{c1} C_{t1}}{1 + n_1(\rho_p + \rho_s)(1 + e^2/r^2)(1 + 0.8C_{t1})}$$
$$= \frac{(6.7)(3.8)(1.41)}{1 + (6.7)(0.0067)\left[1 + \left(\frac{731}{367}\right)^2\right][1 + (0.8)(1.41)]}$$
$$= \frac{35.9}{1.475} = 24 \text{ MPa} \quad \text{(tension)}$$

Reduced relaxation taking into account creep and shrinkage:

$$f_{pi} = 1395 - 31 + 25 = 1389 \text{ MPa}^{R3.4.4}$$

$$\Omega = \frac{\Delta f_p - \Delta f_{p1}}{f_{pi}} = \frac{85 - 24}{1389} = 0.044$$

$$\beta = \frac{f_{pi}}{f_{pu}} = \frac{1389}{1860} = 0.75$$

From Fig. 3.4.4:

$\alpha_r \quad = 0.90$

Therefore, loss due to relaxation only:

$$\Delta f_p = \frac{\alpha_r f_{re}}{1+n_o(\rho_p+\rho_s)(1+e^2/r^2)(1+0.8C_\infty)} = \frac{(0.90)(50)}{1.609} = 28\ \text{MPa}$$

*Step 2:* Decompression force:

$$P_{dc} = P_e - P_s + \Delta P_p + \Delta P_s$$

with:

$$\Delta f_p = -25 + 85 - 24 + 28 = 64\ \text{MPa}$$

$$P_e = (f_{pi} - \Delta f_p)\,A_p = (1364 - 64)(1188)/10^3 = 1544\ \text{kN}$$

with:

$$\Delta f_s = 31 - 25 + 85 - 24 = 67\ \text{MPa}$$

$$P_e = \Delta f_s A_s = (67)(2000)/10^3 = 134\ \text{kN}$$

with:

$$f_c = f_{co} - f_{c1} = 4.1 - 3.8 = 0.3\ \text{MPa}$$

$$\Delta P_p = \frac{f_c}{E_{c1}} E_p A_p = \frac{(0.3)(190\,000)(1188)}{(28\,200)(10^3)} = 2\ \text{kN}$$

$$\Delta P_s = \frac{f_c}{E_{c1}} E_s A_s = \frac{(0.3)(200\,000)(2000)}{(28\,200)(10^3)} = 4\ \text{kN}$$

$$P_{dc} = 1544 - 134 + 2 + 4 = 1416\ \text{kN}$$

It is apparent that the terms $\Delta P_p$ and $\Delta P_s$ could be neglected.

*Step 3:*

Parameters to find $f_s$:

$$d = \frac{A_s E_s d_s + A_p E_p d_p}{A_s E_s + A_p E_p} = \frac{(2000)(200\,000)(1100) + (1188)(190\,000)(1050)}{(2000)(200\,000) + (1188)(190\,000)} = 1082\ \text{mm}$$

$$n\rho = \frac{A_p E_p + A_s E_s}{E_{c1} bd} = \frac{(1188)(190\,000) + (2000)(200\,000)}{(28\,200)(3000)(1082)} = 0.00684$$

$$p = \frac{P_{dc} d_p}{M_s} = \frac{(1416)(1050)(10^3)}{(2022)(10^6)} = 0.735$$

$$\alpha_b = \frac{b_w}{b} = \frac{200}{3000} = 0.0667$$

$$\alpha_f = \frac{h_f}{d} = \frac{90}{1082} = 0.0832$$

*Step 4:*

With these parameters, by interpolation from Fig.3.4.7:

$f_s/n\alpha \quad = 47$

Steel stress at the level of the centroid of reinforcement:

$$f_s = \frac{47 n_1 M_s}{bd^2} = \frac{(47)(6.7)(2022)(10^6)}{(3000)(1082)^2} = 181\ \text{MPa}$$

*Step 5*: Crack control parameter and crack width:

Disregarding the bundled prestressed reinforcement for the calculation of A, the effective concrete area surrounding each bar:

$$A = \frac{(200)(2)(100)}{4} = 10\,000\ \text{mm}^2$$

$$d_c = 50\ \text{mm}$$

$$z = f_s\sqrt[3]{d_c A} = (181)\sqrt[3]{(50)(10\,000)} = 14\,400\ \text{N/mm}$$

For exterior exposure, $z \le 15\,000$ N/mm (Fig. 3.4.1). Hence, 12 - 13 mm diameter strands plus 4 - No. 25 bars satisfy crack control under specified loads.

With $d_c = 32$ mm (concrete cover) and using the parameters calculated in step 3 and Fig. 3.4.7, by interpolation k = 0.25.

The area of concrete in tension is:

$$A_t = (h - kd)\,b_w = [1200 - (0.25)(1082)](200) = 185\,900\ \text{mm}^2$$

and:

$$A_{st} = A_p + A_s\frac{E_s}{E_p} = 1188 + \frac{(2000)(200\,000)}{190\,000} = 3300\ \text{mm}^2$$

The maximum crack width is:

$$w_{max} = k_1 f_s d_c\sqrt{A_t/A_{st}} = (3.0\times10^{-6})(181)(32)\sqrt{\frac{185\,900}{3300}} = 0.13\ \text{mm}$$

This is less than 0.20 mm (see Fig. 3.4.1) which is assumed to be satisfactory for exterior exposure of prestressed concrete elements.

*Step 6:* Check concrete stress in top fibre (compression):

$$f_c = \frac{f_s}{n_1}\left(\frac{kd}{d-kd}\right)$$

$$= \frac{181}{6.7}\left[\frac{(0.25)(1082)}{1082-(0.25)(1082)}\right]$$

$$= 9.0 \text{ MPa} < 0.45f_c' = 15.7 \text{ MPa} \quad \text{OK}$$

**Buckling of Non-Prestressed Reinforcement**

In the previous example, a compressive stress of 67 MPa is introduced in the non-prestressed reinforcement by creep and shrinkage. At this low stress, buckling of the bars will not occur. For compressive stresses exceeding 100 MPa it is recommended that some transverse reinforcement be provided.

The minimum shear reinforcement is considered adequate to prevent the buckling of the non-prestressed reinforcement up to a compressive stress of 200 MPa. Experiments by Hutton and Loov[R3.4.7] on beams without transverse reinforcement showed no sign of buckling after a compressive stress of about 200 MPa had been reached in the non-prestressed reinforcement.

## 3.4.9 PRESTRESS TRANSFER AND STRAND DEVELOPMENT

In a pretensioned element, the prestress force is transferred to the concrete by bond. The length required to accomplish this transfer is called the transfer length, and is assumed to be approximately 50 times the nominal diameter of the strand. The transfer length should be doubled for pretensioned strands that are partially debonded because the bond develops more slowly under such circumstances.

The development length required to develop the full design strength of the strand $f_{pr}$ is much longer, and is specified by the Code as:

$$\ell_d = 0.145(f_{pr} - 0.67f_{pe})d_b$$

**Fig. 3.4.8 Variation of strand stress with distance from free end**

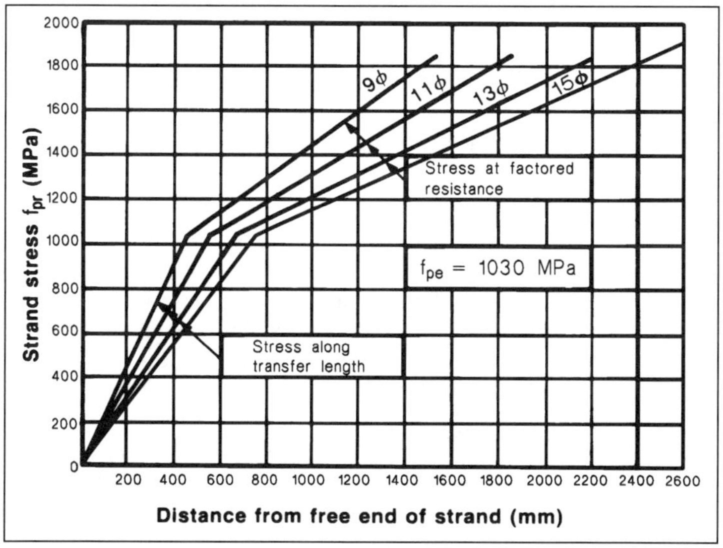


The approximate variation of strand stress along the transfer length and the maximum stress that can be developed near the end of a strand is shown in Fig. 3.4.8. Where bonding of a strand does not extend to the end of a member, and the design includes tension at specified loads, the development length of $\ell_d$ should be doubled.

**Example 9 – Debonding strands**

*Given:*

400 × 600 rectangular beam, normal density concrete with

$\gamma_c$ = 24 kN / m³

Span = 6 m plus a 2 m cantilever

**Fig. 3.4.9**

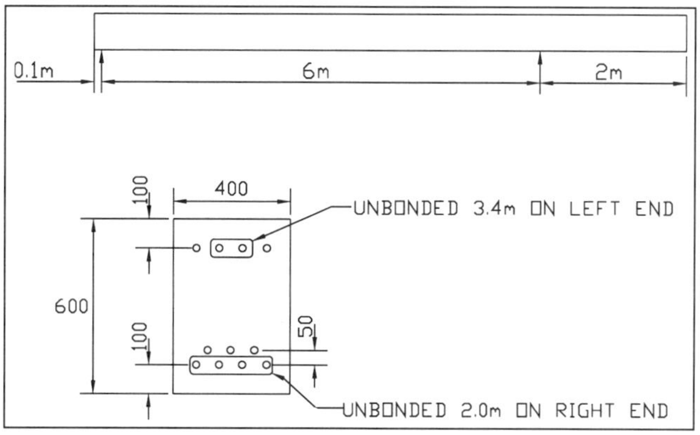


Prestress:

4 – 13 mm 1860 MPa strands in top

7 – 13 mm 1860 MPa strands in bottom

$P_i$ = (99)(0.7)(1860) /10³ = 128.9 kN/strand

$P_e$ = (128.9)(0.87) = 112.2 kN/strand

Section Properties:

$A_g$ = 240 000 mm²

$I$ = 7200 ×10⁶ mm⁴

$y_t$ = 300 mm

$y_b$ = 300 mm

$S_t$ = 24 000 ×10³ mm³

$S_b$ = 24 000 ×10³ mm³

$e_t$ = – (300 – 100) = – 200 mm

(for 4 top strands)

$e_{bu}$ = 300 – 150 = 150 mm

(for 3 upper strands in bottom)

$e_{b\ell}$ = 300 – 100 = 200 mm

(for 4 lower strands in bottom)

Loading:

$w \quad = (400)(600)(10^{-6}) = 5.76$ kN/m

$w_d \quad = 35$ kN/m (not including self weight of beam)

$w_\ell \quad = 26$ kN/m

*Problem:*

Choose appropriate debond lengths and determine stresses at critical sections to determine the required concrete strength, $f'_{ci}$, at transfer and required $f'_c$ at service load. Check factored flexural resistance at critical sections as limited by strand development length.

*Discussion:*

In a cantilever beam such as this, some of the bottom strands will usually be debonded because they increase the stresses produced by the cantilever moment and increase the deflection of the cantilever. A few strands should, however, continue through to the end for reinforcement and crack control during transfer, stripping, storage, transportation, erection and construction.

In this example the 4 strands in the bottom layer are debonded over the full 2 m length of the cantilever. The development length of these strands needs to be adequate so they can provide their full flexural resistance near the mid span.

From Fig. 3.4.8 we can see that, based on Eq. 12–5 of the Code, a development length of approximately 2 m is required for 13 mm strands. Debonded strands normally need a development length which is twice as long as bonded.

Top strands are necessary along the length of the cantilever, but near the centre of the span they increase the stresses produced by the positive moment. Normally, some of the top strands should be debonded over most of the span. Caution is needed because the full flexural resistance of these strands may be needed to resist the factored negative moment over the right support.

In this example, 2 of the 4 top strands have been debonded over a length of 3.4 m from the left end of the beam. A length of 2.7 m is therefore available for development of $f_{pr}$ of these strands at the right support.

The variation of prestress force in each layer of strands is shown below. These forces have been labeled as $P_t$, $P_{bu}$ and $P_{bl}$ to designate their location in the beam. A transfer length of 650 mm at the ends and 1300 mm for the debonded strands has been assumed.

**Fig. 3.4.10**

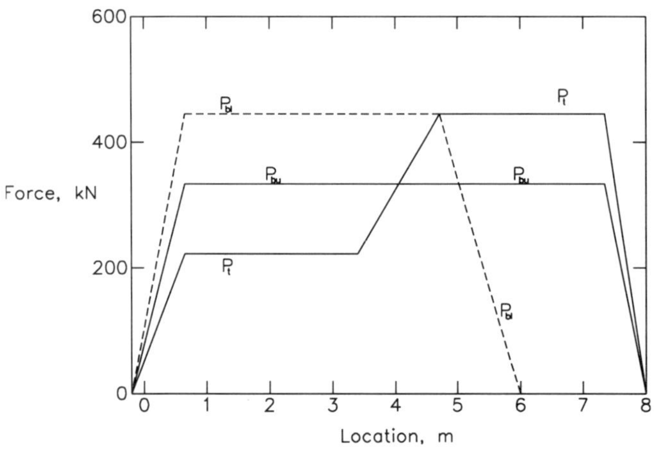


*Solution:*

To show the variation of stresses along the beam under different loads, the stresses at the top and bottom of the beam at transfer and the stresses in the beam under service loads have been plotted. To compute the stresses at transfer the beam is assumed to be supported at the ends. The stresses at service load are based on $P_e$ and two loading cases: (a) with no live load on the cantilever and (b) with no live load on the 6 m span. The following equations with appropriate variations in P and M have been used to determine the stresses:

$$\sigma_t = P_t\left(\frac{1}{A_g} - \frac{e_t}{S_t}\right) + P_{bu}\left(\frac{1}{A_g} - \frac{e_{bu}}{S_t}\right) + P_{b\ell}\left(\frac{1}{A_g} - \frac{e_{b\ell}}{S_t}\right) + \frac{M}{S_t}$$

$$\sigma_b = P_t\left(\frac{1}{A_g} + \frac{e_t}{S_b}\right) + P_{bu}\left(\frac{1}{A_g} + \frac{e_{bu}}{S_b}\right) + P_{b\ell}\left(\frac{1}{A_g} + \frac{e_{b\ell}}{S_b}\right) - \frac{M}{S_b}$$

From these figures it is apparent that the sections which generally need to be checked include:

(a) The transfer point, 650 mm from each end.

(b) The location of maximum positive moment near mid span.

(c) The right support.

Other locations, which may need to be checked, are the lifting and support points used during stripping, storage, transportation and erection.

*Stresses:*

At the transfer point, 650 mm from the left support, $P_{ti} = 257.8$ kN, $P_{bui} = 386.7$ kN, and $P_{b\ell i} = 515.6$ kN. The moment at this point, assuming the beam is simply supported 100 mm from each end, is M = 11. 6 kN·m.

The compressive stress at transfer is therefore:

$\sigma_b \quad = 8.91$ MPa

**Fig. 3.4.11**

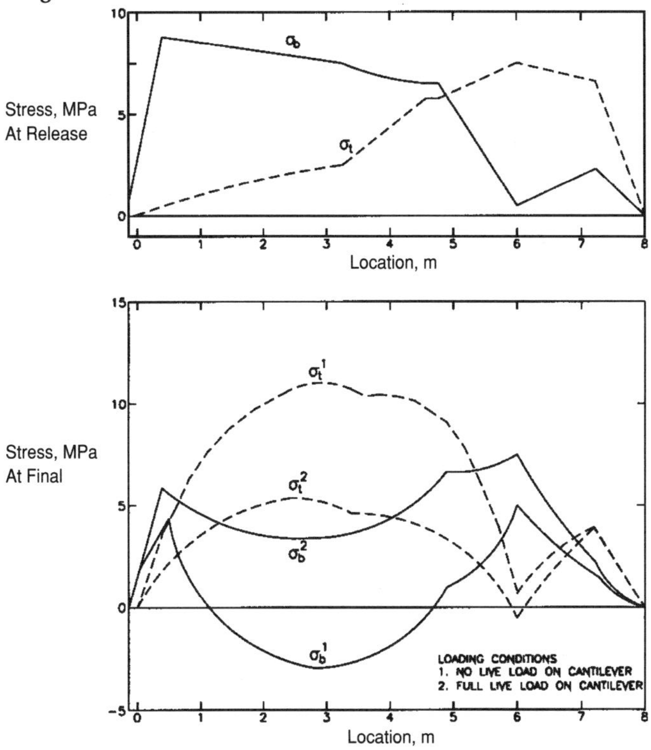


*Since design includes tension at these points, the development length of debonded strand must be doubled (see Clause 12.9.2 - CSA A23.3 - 94)*

The required concrete strength at transfer is therefore:

$$f'_{ci} \geq 8.91 / 0.60 = 14.9 \text{ MPa}$$

The maximum positive moment occurs with no live load on the cantilever. The left reaction under this condition is 186.7 kN so that zero shear and corresponding maximum moment occur at 2.80 m from the left support. The maximum service load moment at this section is 261.0 kN·m. (The use of the centrespan moment would have introduced only negligible error). The forces in the different layers of prestressed steel are: $P_t = 224.4$ kN, $P_{bu} = 336.6$ kN, and $P_{b\ell} = 448.8$ kN.

The maximum stresses at this section are:

$$\sigma_t = 11.11 \text{ MPa and } \sigma_b = -2.70 \text{ MPa.}$$

The maximum negative moment, M = – 133.5 kN·m, occurs at the right support with the cantilever fully loaded. The forces in the prestressed steel at this section are: $P_t = 448.8$ kN, $P_{bu} = 336.6$ kN, and $P_{b\ell} = 0$.

The maximum stresses at this section are:

$$\sigma_t = -0.65 \text{ MPa and } \sigma_b = 7.20 \text{ MPa.}$$

The required concrete strength for service conditions will therefore be based on the stresses from the positive moment:

$$f'_c \geq 11.11 / 0.45 = 24.7 \text{ MPa}$$

and

$$f'_c \geq [2.70 / (0.5\lambda)]^2 = 29.1 \text{ MPa}$$

A transfer strength of $f'_{ci} = 20$ MPa and an $f'_c = 30$ MPa will thus be adequate.

*Check factored flexural resistance at critical sections.*

In positive moment region:

$$M_{f\,max} = (2.81)^2(0.5)[(5.76 + 35.0)(1.25) + (26)(1.5)] = 355.4 \text{ kN} \cdot \text{m}$$

$$d_p = [(3)(450) + (4)(500)] / 7 = 478.6 \text{ mm}$$

$$\phi_p A_p f_{pu} = (0.9)(7)(99)(1860) / 10^3 = 1160 \text{ kN}$$

$$\frac{c}{d_p} = \frac{\phi_p A_p f_{pu}}{\phi_c \alpha_1 \beta_1 f'_c b d_p + k_p \phi_p A_p f_{pu}}$$

$$= \frac{(1160)(10^3)}{(0.65)(0.805)(0.895)(30)(400)(478.6) + (0.28)(1160)(10^3)}$$

$$= 0.385$$

$$f_{pr} = f_{pu}[1 - k_p(c / d_p)] = (1860)[1 - (0.28)(0.385)] = 1660 \text{ MPa}$$

$$a = \frac{\phi_p A_p f_{pr}}{\alpha_1 \phi_c f'_c b} = \frac{(0.9)(7)(99)(1660)}{(0.805)(0.65)(30)(400)} = 164.9 \text{ mm}$$

$$M_r = \phi_p A_p f_{pr}(d_p - a/2) = (0.9)(7)(99)(1660)(478.6 - 164.9/2) = 410.1 \text{ kN·m}$$

Adjust moment resistance based on actual development length provided.

For $f_{pr} = 1660$ MPa

$\ell_d = 1700$ mm for bonded strands

$= 3400$ mm for debonded strands

From left end:

$$\ell_{actual} = 2810 + 100 = 2910 \text{ mm} > 1700$$

From right end:

for bottom upper $\ell_{actual} = 5190$ mm > 1700 mm

for bottom lower $\ell_{actual} = 3190$ mm < 3400 mm

No. of strands effective

$$= (3) + (4)(3190)/(3400) = 6.75$$

$$d_p = \frac{(3)(450) + (4)(500)\frac{(3190)}{(3400)}}{(6.75)}$$

$$= 478.0$$

$$a = \frac{(0.9)(6.75)(99)(1660)}{(0.805)(0.65)(30)(400)}$$
$$= 159.0 \text{ mm}$$

$$M_r = (0.9)(6.75)(99)(1660)\left(476.7 - \frac{151.5}{2}\right)$$
$$= 397.8 \text{ kN}\cdot\text{m} > 355.4$$

In negative moment region:

$$M_{f\,max} = (2)^2(0.5)[(5.76 + 35.0)(1.25) + (26)(1.5)]$$
$$= 179.9 \text{ kN}\cdot\text{m}$$

$$d_p = 500 \text{ mm}$$

$$\phi_p A_p f_{pu} = (0.9)(4)(99)(1860)/10^3$$
$$= 663 \text{ kN}$$

$$\frac{c}{d_p} = \frac{(663)(10^3)}{(0.65)(0.805)(0.895)(30)(400)(500) + (0.28)(663)(10^3)}$$
$$= 0.2213$$

$$f_{pr} = (1860)[1 - (0.28)(0.2213)]$$
$$= 1745 \text{ MPa}$$

$$a = \frac{(0.9)(4)(99)(1745)}{(0.805)(0.65)(30)(400)}$$
$$= 99.0 \text{ mm}$$

$$M_r = (0.9)(4)(99)(1745)\left(500 - \frac{99.0}{2}\right)$$
$$= 179.9 \text{ kN}\cdot\text{m}$$

Adjust moment resistance based on actual development length provided on either side of critical section.

$$\ell_d = (0.145)[1745 - (0.67)(1133)](13)$$
$$= 1860 \text{ mm}$$

$$\ell_d = 1860 \text{ mm for fully bonded strand}$$
$$\ell_d = (2)(1860) = 3720 \text{ mm for debonded strand}$$

From left end:

$$\ell_{actual} = 6100 \text{ mm for bonded strands}$$
$$= 2700 \text{ mm for debonded strands}$$

Conservatively assuming that $f_{pr}$ is proportional to the length, then:

$$a = \frac{(0.9)[2 + (2)(2.7)/3.86](99)(1745)}{(0.805)(0.65)(30)(400)}$$
$$= 84.2 \text{ mm}$$

$$M_r = (0.9)(3.4)(99)(1745)(500 - 84.2/2)$$
$$= 242.1 \text{ kN}\cdot\text{m} > 179.9$$

From right end:

$$\ell_{actual} = 2000 \text{ mm} > 1860$$

No adjustment to moment resistance is required

## 3.5 CAMBER AND DEFLECTION

The deflection of precast, prestressed and reinforced elements under the application of dead and live loads is an important design consideration. For floor and roof elements, excessive

**Fig. 3.5.1 Maximum allowable computed deflections**

| Type of Element | Deflection to be Considered | Deflection limitation |
|---|---|---|
| Flat roofs not supporting or attached to non-structural elements likely to be damaged by large deflections | Immediate deflection due to live load | $\ell_n/180$ (1) |
| Floors not supporting or attached to non-structural elements likely to be damaged by large deflections | Immediate deflection due to live load | $\ell_n/360$ |
| Roof or floor construction supporting or attached to non-structural elements likely to be damaged by large deflections | That part of the total deflection which occurs after attachment of the non-structural elements, the sum of the long-time deflection due to all sustained loads and the immediate deflection due to any additional live load(2) | $\ell_n/480$ (3) |
| Roof or floor construction supporting or attached to non-structural elements likely to be damaged by large deflections | | $\ell_n/240$ (4) |

(1) This limit is not intended to safeguard against ponding. Ponding should be checked by suitable calculations of deflection including the added deflection due to ponded water, and considering-long time effects of all sustained loads, camber, construction tolerances, and reliability of provisions for drainage.

(2) The long-time deflection may be reduced by the amount of deflection which occurs before attachment of the non-structural elements. This amount shall be determined on the basis of accepted engineering data relating to the time-deflection characteristics of members similar to those being considered.

(3) This limit may be exceeded if adequate measures are taken to prevent damage to supported or attached elements.

(4) But not greater than the tolerance provided for the non-structural elements. This limit may be exceeded if camber is provided so that the total deflection minus the camber does not exceed the limitation.

deflection can lead to possible damage of interior finishes and equipment; excessive deflection of roof elements can lead to water ponding problems. Inadequate stiffness of wall panels may cause damage to the glazing and lead to sealant failure.

While the emphasis in this section is on prestressed floor and roof elements, the same principles are applicable to the design of reinforced elements and wall panels.

Most precast, prestressed concrete flexural elements will have a net positive (upward) camber at the time of transfer of prestress, caused by the eccentricity of the prestressing force. This camber may increase or decrease with time, depending on the stress distribution across the element under sustained loads.

Limitations on instantaneous deflections and time-dependent cambers and deflections are specified in the Code, and are reprinted in Fig. 3.5.1 for reference.

The following sections contain suggested methods for computing cambers and deflections. There are many inherent variables that affect camber and deflection, such as concrete mix, storage method, time of release of prestress, time of erection and placement of superimposed loads, relative humidity, etc. **Because of this, calculated long-time values should never be considered any better than estimates**. Non-structural components attached to elements which could be affected by camber variations, such as partitions or folding doors, should be placed with adequate allowance for these variations. Calculation of topping quantities should also recognize the imprecision of camber calculations.

## 3.5.1 INITIAL CAMBER

Initial camber can be calculated using conventional moment-area equations.

Figs. 8.1.3 and 8.1.4 provide deflection equations for typical loading conditions and camber equations for common tendon profiles.

**Example 1 – Calculation of initial camber**

*Given:*

2400 × 600 double tee of Example 2, Page 3-27.

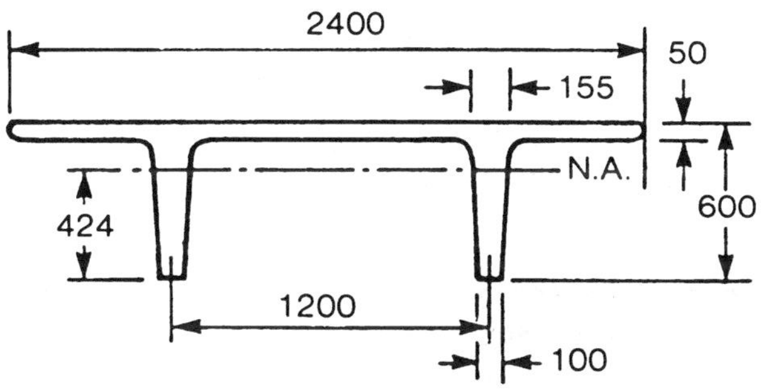


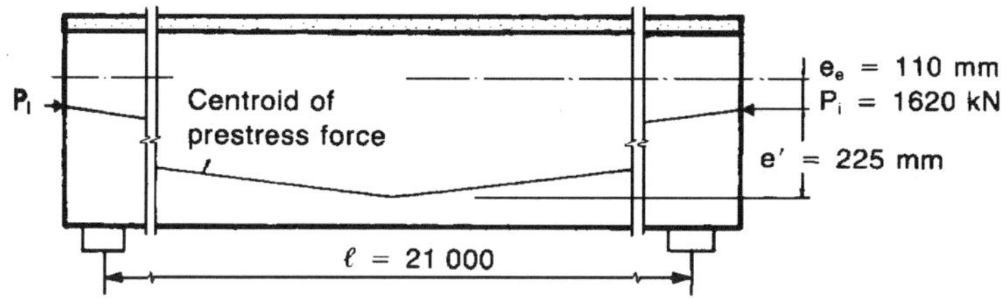


Section properties:

$A_g$ = 260 000 mm$^2$

I = 8 580 × 10$^6$ mm$^4$

$y_b$ = 424 mm

$y_t$ = 176 mm

$S_b$ = 20 200 × 10$^3$ mm$^3$

$S_t$ = 48 800 × 10$^3$ mm$^3$

m = 260 kg/m$^2$ = 624 kg/m

w = 2.6 kN/m$^2$ = 6.2 kN/m

Concrete:

$f'_c$ = 35 MPa

Normal Density

$$E_c = (3300\sqrt{f'_c} + 6900)\left(\frac{\gamma_c}{2300}\right)^{1.5}$$

$$= (3300\sqrt{35} + 6900)\left(\frac{2400}{2300}\right)^{1.5} = 28\,200 \text{ MPa}$$

$f'_{ci}$ = 25 MPa

$$E_{ci} = (3300\sqrt{25} + 6900)\left(\frac{2400}{2300}\right)^{1.5} = 24\,900 \text{ MPa}$$

(Note: The values of $E_c$ and $E_{ci}$ could also be read from Fig. 8.2.2.)

*Problem:*

Find the initial camber at time of transfer of prestress.

*Solution:*

The prestress force immediately after transfer and strand eccentricities are calculated in Example 2, Page 3-27, and are shown in the illustration on Page 3-28.

Calculate the upward component:

$$\Delta\uparrow = \frac{P_i e_e \ell^2}{8E_{ci}I} + \frac{P_i e' \ell^2}{12E_{ci}I}$$

$$= \frac{(1620)(110)(21)^2(10^9)}{(8)(24\,900)(8580 \times 10^6)}$$

$$+ \frac{(1620)(225)(21)^2(10^9)}{(12)(24\,900)(8580 \times 10^6)}$$

$$= 46 + 63 = 109 \text{ mm}\uparrow$$

Deduct deflection caused by self weight of element:

$$\Delta\downarrow = \frac{5w\ell^4}{384E_{ci}I}$$

$$= \frac{(5)(6.2)(21)^4(10^{12})}{(384)(24\,900)(8580 \times 10^6)} = 73 \text{ mm}\downarrow$$

Net camber at release = 109 – 73 = 36 mm ↑

### 3.5.2 ELASTIC DEFLECTIONS

Calculation of instantaneous deflections caused by superimposed service loads follows classical methods of mechanics. Design equations for various load conditions are given in Chapter 8 of this Manual. If the bottom tension in a simple span element does not exceed the modulus of rupture, the deflection is calculated using the uncracked moment of inertia of the section. The modulus of rupture of concrete is defined in the Code as:

$$f_r = 0.6\lambda\sqrt{f'_c}$$

see Sect. 1.3.1 for a definition of λ.

#### Bilinear Behaviour

It is normal practice that "bilinear moment-deflection relationships" be used to calculate instantaneous deflections when the bottom tension exceeds $0.6\lambda\sqrt{f'_c}$. This means that the deflection before the element has cracked is calculated using the gross (uncracked) moment of inertia, $I_g$, and the additional deflection after cracking is calculated using the moment of inertia of the cracked section, $I_{cr}$. This is illustrated graphically in Fig. 3.5.2.

**Fig. 3.5.2 Bilinear and effective moment of inertia moment-deflection relationship**

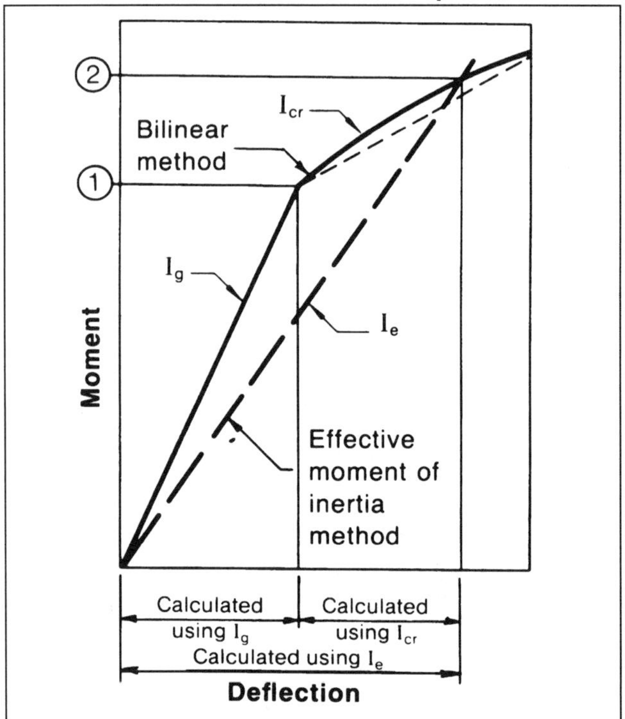


*① Moment which results in bottom tension of $f_r$*

*② Moment corresponding to final bottom tension*

*(Note: Prestress effects not included in illustration.)*

In lieu of a more exact analysis, the relationship:

$$I_{cr} = n\rho\, bd^3 (1 - k)(1 - k/3)$$

may be used to determine the cracked moment of inertia. In this equation:

$$d = \frac{A_s E_s d + A_p E_p d_p}{A_s E_s + A_p E_p}$$

$$k = \sqrt{(n\rho)^2 + 2n\rho} - n\rho$$

$$n\rho = \frac{A_s E_s + A_p E_p}{E_c bd}$$

Fig. 3.5.3 provides coefficients for use in solving this equation. For elements in which there is only prestressed reinforcement, the above expression may be simplified to:

$$I_{cr} = nA_p d_p^2 (1 - 1.67\sqrt{n\rho_p})$$

A more detailed procedure for calculating $I_{cr}$, which includes the effect of the prestressing force, is given in Sect. 3.5.6.

#### Effective Moment of Inertia

For reinforced elements, the Code allows an alternative to the method of calculation described in the previous section. An effective moment of inertia, $I_e$, can be determined and the deflection then calculated by substituting $I_e$ for $I_g$ in the deflection calculation.

The equation for effective moment of inertia is:

$$I_e = \left(\frac{M_{cr}}{M_a}\right)^3 I_g + \left[1 - \left(\frac{M_{cr}}{M_a}\right)^3\right] I_{cr}$$

$$= I_{cr} + (I_g - I_{cr})\left(\frac{M_{cr}}{M_a}\right)^3 \le I_g$$

The difference between the bilinear method and the $I_e$ method is illustrated in Fig. 3.5.2.

The use of $I_e$ with prestressed concrete elements is described in a paper by Branson[R3.5.1]. The value of $M_{cr}/M_a$ for use in determining live load deflections can be expressed as:

$$\frac{M_{cr}}{M_a} = 1 - \left(\frac{f_{t\ell} - f_r}{f_\ell}\right)$$

where:

$f_{t\ell}$ = final calculated total stress in the element

$f_\ell$ = calculated stress due to live load

A more accurate application of the I-Effective method is described in a paper by Branson and Trost[R3.5.2]

**Example 2 - Deflection calculation using bilinear moment-deflection relationships**

*Given:*

2400 × 600 double tee of Example 2, Page 3-27 and Example 1, Page 3-57.

**Fig. 3.5.3 Momemt of inertia of cracked transformed section**

$$d = \frac{A_s E_s d + A_p E_p d_p}{A_s E_s + A_p E_p}$$

$$I_{cr} = n\rho b d^3 (1-k)(1-k/3) = C \text{ (from table) } bd^3$$

$$k = \sqrt{(n\rho)^2 + 2n\rho} - n\rho$$

$$n\rho = \frac{A_s E_s + A_p E_p}{E_c b d}$$

| | Values of coefficient C | | | | | | | | |
|---|---|---|---|---|---|---|---|---|---|
| **nρ** | **0.0010** | **0.0020** | **0.0030** | **0.0040** | **0.0050** | **0.0060** | **0.0070** | **0.0080** | **0.0090** |
| 0.0000 | 0.0009 | 0.0018 | 0.0027 | 0.0036 | 0.0044 | 0.0052 | 0.0060 | 0.0068 | 0.0075 |
| 0.0100 | 0.0090 | 0.0098 | 0.0105 | 0.0112 | 0.0119 | 0.0127 | 0.0133 | 0.0140 | 0.0147 |
| 0.0200 | 0.0161 | 0.0167 | 0.0174 | 0.0180 | 0.0187 | 0.0193 | 0.0199 | 0.0206 | 0.0212 |
| 0.0300 | 0.0224 | 0.0230 | 0.0236 | 0.0242 | 0.0248 | 0.0254 | 0.0260 | 0.0266 | 0.0271 |
| 0.0400 | 0.0283 | 0.0288 | 0.0294 | 0.0299 | 0.0305 | 0.0310 | 0.0316 | 0.0321 | 0.0327 |
| 0.0500 | 0.0337 | 0.0343 | 0.0348 | 0.0353 | 0.0358 | 0.0363 | 0.0369 | 0.0374 | 0.0379 |
| 0.0600 | 0.0389 | 0.0394 | 0.0399 | 0.0404 | 0.0408 | 0.0413 | 0.0418 | 0.0423 | 0.0428 |
| 0.0700 | 0.0437 | 0.0442 | 0.0447 | 0.0451 | 0.0456 | 0.0461 | 0.0465 | 0.0470 | 0.0474 |
| 0.0800 | 0.0483 | 0.0488 | 0.0492 | 0.0497 | 0.0501 | 0.0506 | 0.0510 | 0.0514 | 0.0519 |
| 0.0900 | 0.0527 | 0.0532 | 0.0536 | 0.0540 | 0.0544 | 0.0548 | 0.0553 | 0.0557 | 0.0561 |
| 0.1000 | 0.0569 | 0.0573 | 0.0577 | 0.0581 | 0.0585 | 0.0589 | 0.0594 | 0.0597 | 0.0601 |

*Problem:*

Determine the total instantaneous deflection caused by the specified live load.

*Solution:*

$$f_r = 0.6\lambda\sqrt{f'_c}$$

From Example 2, Page 3-27, the final tensile stress is 5.4 MPa, which is more than 3.6 MPa, so bilinear behaviour must be considered.

Determine $I_{cr}$ from Fig. 3.5.3:

$A_p = 1386 \text{ mm}^2$

$d_p$ at midspan = $e_c + y_t = 335 + 176 = 511$ mm[R3.5.3]

$$n\rho = \frac{A_p E_p}{E_c b d} = \frac{(1386)(190\,000)}{(28\,200)(2400)(511)} = 0.0076$$

From Fig. 3.5.3, C = 0.0065

hence:

$$I_{cr} = Cbd_p^3 = (0.0065)(2400)(511)^3 = 2082 \times 10^6 \text{ mm}^4$$

Determine the portion of the live load that would result in a bottom tension of 3.6 MPa:

5.4 – 3.6 = 1.8 MPa

The tension caused by live load alone is 11.0 MPa, therefore, the portion of the live load that would result in a bottom tension of 3.6 MPa is:

$$\left(\frac{11.0 - 1.8}{11.0}\right)(4.2) = 3.5 \text{ kN/m}$$

and:

$$\Delta_g = \frac{5w\ell^4}{384 E_c I_g} = \frac{(5)(3.5)(21)^4(10^{12})}{(384)(28\,200)(8580 \times 10^6)} = 37 \text{ mm}$$

$$\Delta_{cr} = \frac{(5)(0.7)(21)^4(10^{12})}{(384)(28\,200)(2082 \times 10^6)} = 30 \text{ mm}$$

Total instantaneous deflection = 37 + 30 = 67 mm

**Example 3 - Deflection calculation using effective moment of inertia**

*Given:*

Same section and loading conditions of Example 2 of page 3-27.

*Problem:*

Determine the instantaneous deflection caused by live load using the $I_e$ method.

*Solution:*

From the table of stresses in Example 2, Page 3-27.

$f_{t\ell}$ = 5.4 MPa (tension)

$f_{\ell}$ = 11.0 MPa (tension)

$f_r = 0.6\lambda\sqrt{f'_c} = 3.6$ MPa (tension)

$$\frac{M_{cr}}{M_a} = 1 - \left(\frac{5.4 - 3.6}{11.0}\right) = 0.83$$

$$\left(\frac{M_{cr}}{M_a}\right)^3 = 2082 + (8580 - 2082)(0.57)(10^6)$$

From Example 2, $I_{cr} = 2082 \times 10^6$ mm$^4$

$$I_e = I_{cr} + (I_g - I_{cr})\left(\frac{M_{cr}}{M_a}\right)^3$$

$$= 2082 + (8580 - 2082)(0.57)(10^6)\ )^6)$$

$$= 5786 \times 10^6 \text{ mm}^4$$

$$\Delta_{\ell} = \frac{5w\ell^4}{384E_cI_e}$$

$$= \frac{(5)(4.2)(21)^4(10^{12})}{(384)(28\,200)(5786 \times 10^6)} = 65 \text{ mm} \downarrow$$

## 3.5.3 LONG-TIME CAMBER/DEFLECTION

The Code provides a convenient multiplier for estimating the additional long-time deflection of reinforced concrete elements:

$$\left[1 + \frac{S}{1 + 50\rho'}\right]$$

where $\rho'$ is $A'_s/bd$ at midspan for simple and continuous spans and at the support for cantilevers. The time dependent factor, S, for sustained loads may be taken equal to:

| | |
|---|---|
| 5 years or more: | 2.0 |
| 12 months: | 1.4 |
| 6 months: | 1.2 |
| 3 months: | 1.0 |

No such convenient guide is given for prestressed concrete.

The determination of long-time cambers and deflections in precast, prestressed elements is somewhat more complex because of (1) the effect of prestress and the loss of prestress over time, (2) the strength gain of concrete after release of prestress, and because (3) the camber or deflection is important not only at the "initial" and "final" stages, but also at erection, which occurs at some intermediate stage, usually from 30 to 60 days after casting.

## 3.5.4 USE OF MULTIPLIERS FOR DETERMINING LONG-TIME CAMBER AND DEFLECTION

It has been customary in the design of precast, prestressed concrete to estimate the camber of an element after a period of time by multiplying the initial calculated camber by some factor, usually based on the experience of the designer. To properly use these "multipliers," the upward and downward component of the initial calculated camber should be separated in order to take into account the effects of loss of prestress, which only affect the upward component.

Fig. 3.5.4 provides suggested multipliers which can be used as a guide in estimating long-time cambers and deflections for typical elements, i.e., those elements which are within the span-depth ratios recommended in this Manual. Derivation of these multipliers is contained in a paper by Martin.[R3.5.5]

Long-time effects can be substantially reduced by adding non-prestressed reinforcement in prestressed concrete elements. The reduction effects proposed by Shaikh and Branson[R3.5.6] can be applied to the approximate multipliers of Fig. 3.5.4 as follows:

$$C_2 = \frac{C_1 + A_s / A_p}{1 + A_s / A_p}$$

where:

$C_1$ = multiplier from Fig. 3.5.4

$C_2$ = revised multiplier

$A_s$ = area of non-prestressed reinforcement

$A_p$ = area of prestressed reinforcement

**Example 4 - Use of multipliers for determination of long-time cambers and deflections**

*Given:*

2400 × 600 double tee of Example 2, Page 3-27, and Examples 1, 2 and 3 in this Subsection. Non-structural elements are attached, but not likely to be damaged by deflections (light fixtures, etc.).

*Problem:*

Estimate the camber and deflection and determine if they meet the requirements of the Code.

*Solution:*

Calculate the instantaneous deflections caused by the superimposed dead and live loads:

**Fig. 3.5.4 Suggested multipliers to be used as a guide in estimating long-time cambers and deflections for typical lements**

| | Without composite topping | With composite topping |
|---|---|---|
| **At erection:** | | |
| (1) Deflection (downward) component - apply to the elastic deflection due to the element mass ar release of prestress | 1.85 | 1.85 |
| (2) Camber (upward) component - apply to the elastic camber due to prestress at the time of release of prestress | 1.80 | 1.80 |
| **Final:** | | |
| (3) Deflection (downward) component - apply to the elastic deflection due to the element mass ar release of prestress | 2.70 | 2.40 |
| (4) Camber (upward) component - apply to the elastic camber due to prestress at the time of release of prestress | 2.45 | 2.20 |
| (5) Deflection (downward) - apply to elastic deflection due to superimposed dead load only | 3.00 | 3.00 |
| (6 Deflection (downward) - apply to elastic deflection caused by the composite topping) | – | 2.30 |

**Example - Use of multiipliers for determining long-time cambers and deflections**

| | (1) Release | Multiplier | (2) Erection | Multiplier | (3) Final |
|---|---|---|---|---|---|
| Prestress | 109 ↑ | 1.80 × (1) | 196 ↑ | 2.45 × (1) | 267 ↑ |
| $w_d$ | 73 ↓ | 1.85 × (1) | 135 ↓ | 2.70 × (1) | 197 ↓ |
| Resultant | 36 ↑ | | 61 ↑ | | 68 ↑ |
| $w_{sd}$ | | | 13 ↓ | 3.00 × (1) | 39 ↓ |
| Resultant | | | 48 ↑ | | 32 ↑ |
| $w_\ell$ | | | | | 67 ↓ |
| Resultant | | | | | 36 ↓ |

$$\Delta_{sd} = \frac{5w\ell^4}{384E_cI}$$

$$= \frac{(5)(1.2)(21)^4(10^{12})}{(384)(28\,200)(8580\times10^6)} = 13\text{ mm}\downarrow$$

$\Delta_\ell$ = 67 mm ↓ (see Example 2, Page 3.58)

For convenience, a tabular format is shown above.

The estimated critical cambers and deflections would then be:

At erection of the element,

after $w_{sd}$ is applied = 48 mm

"Final" long-time camber = 32 mm

The deflection limitation of Fig. 3.5.1 for the above condition is $\ell/240$:

(21)(1000)/240 = 88 mm

Total deflection occurring after attachment of non-structural elements:

$\Delta_{t\ell}$ = (47 – 32) + 67 = 83 mm < 88 OK

## 3.5.5 DETAILED METHOD FOR DETERMINING LONG-TIME CAMBER AND DEFLECTION - FULLY PRESTRESSED ELEMENTS

As discussed in Sections 3.5.1 and 3.5.2, the elastic cambers and deflections for uncracked elements are calculated by conventional methods of analysis; the downward deflection due to gravity load is reduced by the upward deflection due to prestressing. The initial deflection due to prestressing is calculated with $P = P_o$, the initial prestressing force.

The time-dependent deflections of uncracked elements due to prestressing can be obtained from the following equation:

$$\Delta_t^p = \Delta_i^p\left[C_t - \frac{\Delta P}{P_o}(1+0.8C_t)\right]$$

where:

$C_t$ = creep coefficient, see Sect. 3.4.6

$\Delta_i^p C_t$ = time-dependent deflection (camber) assuming $P = P_0$

$\Delta P$ = average prestress loss along the tendon, see Sect. 3.4.6

The factor 0.8 is the aging coefficient[R3.5.6] and this value can be used for practical creep computations.

The time-dependent deflections of uncracked elements due to sustained loads can be obtained from the following expression:

$$\Delta_t^{su} = \Delta_i^{su} C_t$$

where:

$\Delta_i^{su}$ = initial deflection

The values obtained from the above expressions are additive to the elastic values.

## 3.5.6 DETAILED CALCULATION OF LIVE LOAD DEFLECTION - PARTIALLY PRESTRESSED ELEMENTS

As discussed in Section 3.5.2, the deflection of partially prestressed (cracked) elements may be calculated by using an effective moment of inertia, $I_e$, or by bilinear moment-deflection relationships. The calculation of $I_{cr}$ by Fig. 3.5.3 ignores the effect of the prestressing force which is of particular importance in partially prestressed elements.

The cracked section properties for a given section are dependent on the decompression load, $P_{dc}$, and the moment due to specified loads, $M_s$.

The calculation of the cracked section properties involves the determination of the neutral axis depth, c, the centroidal depth, $y_{cr}$, the area, $A_{cr}$, and the moment of inertia, $I_{cr}$.

The following parameters are required to determine the neutral axis depth, c:

$\alpha_b = b_w/b$

$\alpha_f = h_f/d$

$p = P_{dc}\, d_p/M_s$

$$n\rho = \frac{A_s E_s + A_p E_p}{E_c bd}$$

where:

$P_{dc}$ = decompression force (see Sect. 3.4.8)

$M_s$ = moment due to specified loads

$$d = \frac{A_s E_s d_s + A_p E_p d_p}{A_s E_s + A_p E_p}$$

Having determined k from Fig. 3.4.7:

$c = kd$

The properties of the cracked section can be calculated using the following equations:

Centroidal depth:

$$y_{cr} = \frac{0.5bh_f^2 + 0.5b_w(c^2 - h_f^2) + n_p A_p d_p + n_s A_s d_s}{bh_f + b_w(c - h_f) + n_p A_p + n_s A_s}$$

Transformed area:

$$A_{cr} = bh_f + b_w(c - h_f) + n_p A_p + n_s A_s$$

Moment of inertia:

$$I_{cr} = \frac{1}{12}bh_f^3 + bh_f\left(y_{cr} - \frac{h_f}{2}\right)^2 + \frac{1}{12}b_w(c - h_f)^3 + b_w(c - h_f)\left(y_{cr} - \frac{c + h_f}{2}\right)^2 + n_p A_p(d_p - y_{cr})^2 + n_s A_s(d - y_{cr})^2$$

In these equations:

b = effective flange width

$b_w$ = web width

c = kd = neutral axis depth (k from Fig. 3.4.7)

$n_p = E_p/E_c$

$n_s = E_s/E_c$

$A_p$ = area of prestressing steel

$A_s$ = area of reinforcing steel

$h_f$ = flange thickness

Alternatively, $I_{cr}$ and $y_{cr}$ may be determined according to Tadros et al[R3.5.7] from Figs. 3.5.5 and 3.5.6 respectively. (See Example 5)

According to a recent study[R3.5.8], the most reliable predictions for the live load deflections of partially prestressed members are obtained by the method proposed by Tadros et al[R3.5.7] which takes into account the eccentricity of the tendons relative to the centroid of the cracked section.

For the calculation of the deflection, the curvatures have to be determined at selected points (normally at midspan for beams with straight strands and an additional value is needed at $0.4\ell$ for draped strands). The effective moment of inertia for calculating curvatures is:

$$I'_e = I_{cr} + \left(\frac{M_{cr}}{M_s - M_{dc}}\right)^4 (I_g - I_{cr})$$

The terms $M_{cr}$, $M_s$ and $M_{dc}$ are defined in Sect. 3.4.8.

After cracking, the effective eccentricity of the tendons is $(d_p - y_e)$, where:

$$y_e = y_{cr} + \left(\frac{M_{cr}}{M_s - M_{dc}}\right)^4 (y_t - y_{cr})$$

In this equation $y_t$ and $y_{cr}$ are the distance between the centroid and the top fibre of the uncracked section and the cracked section respectively.

**Example 5 - Deflection calculation using Detailed Method**

*Given:*

Same section and loading condition as for Example 2 in this Subsection.

**Fig. 3.5.5 Moment of inertia of cracked transformed section**

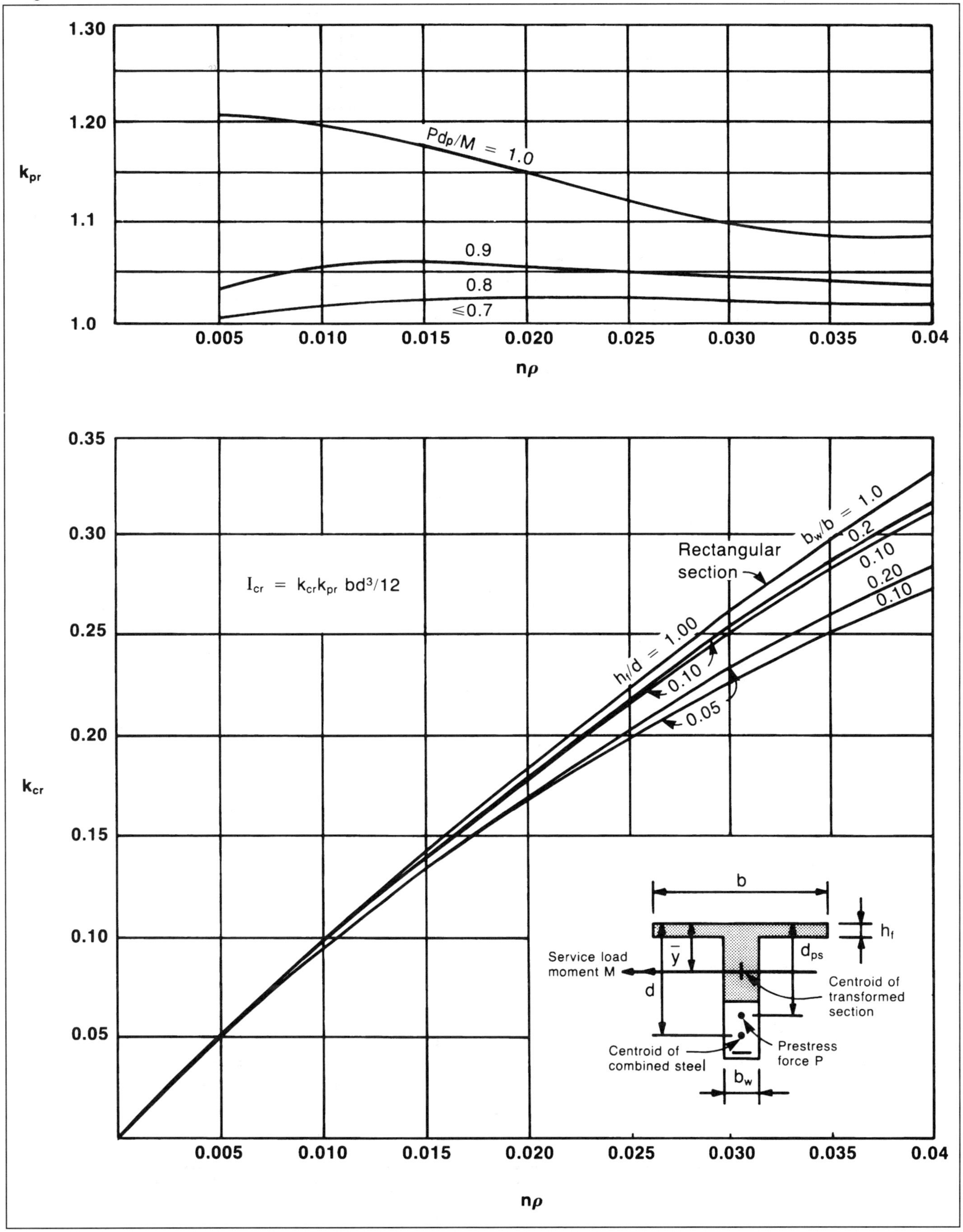

**Fig. 3.5.6 Centroidal axis depth of cracked transformed section**

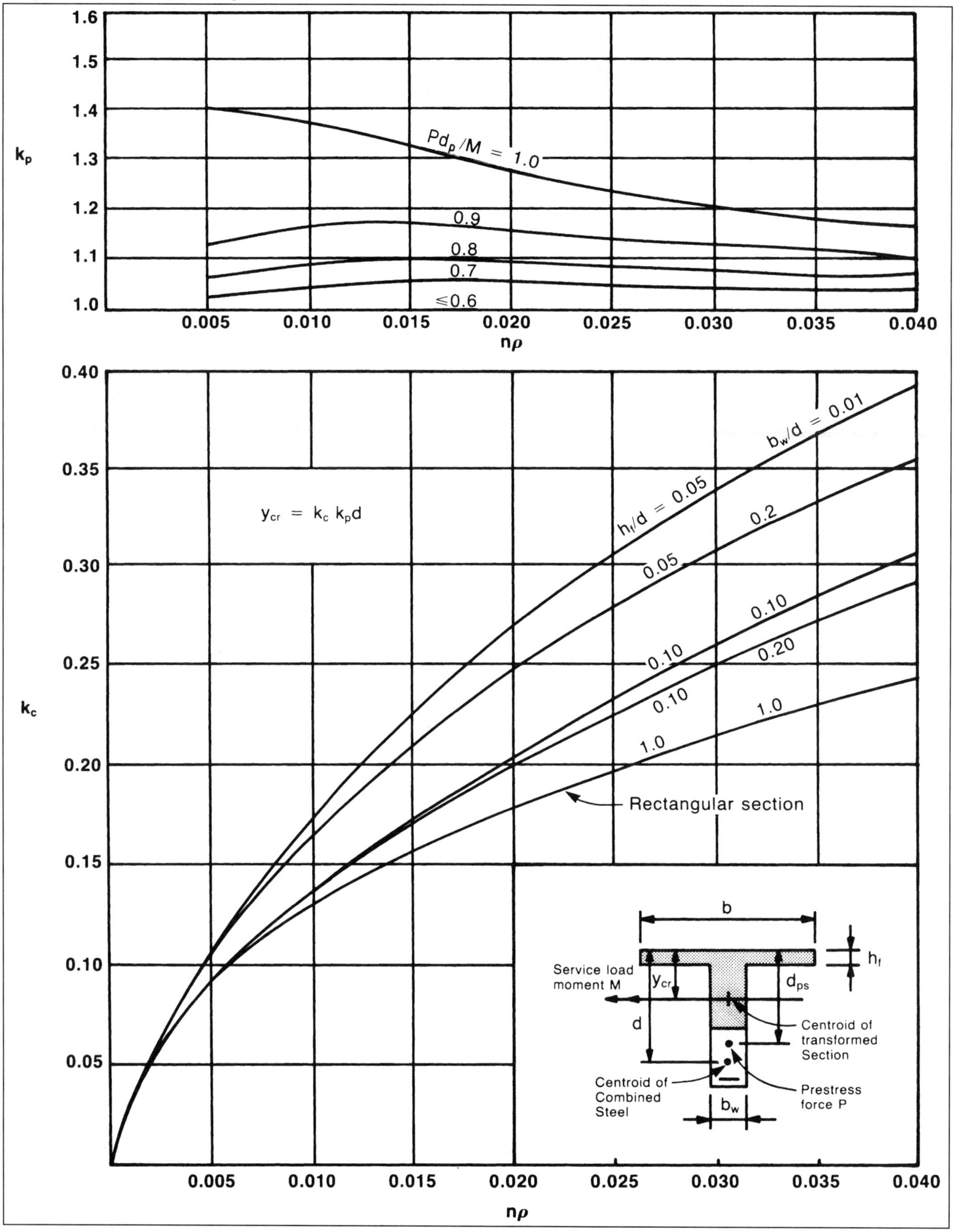

*Problem:*

Determine instantaneous deflection caused by live load using detailed method.

*Solution:*

Determine $I_{cr}$ and $y_{cr}$ using Fig. 3.5.5 and 3.5.6 (or using Equations):

At $x = 0.4\ell$, no non-prestressed reinforcement:

$\therefore A_s = 0$

$P_{dc} = P_e = 1370$ kN

$d = d_p = e_{0.4\ell} + y_t = 290 + 176 = 466$ mm

$$p = \frac{P_{dc}d_p}{M_s} = \frac{(1370)(466)(10^3)}{(616)(10^6)} = 1.04$$

$$n\rho = \frac{A_sE_s + A_pE_p}{E_cbd} = \frac{(1386)(190\,000)}{(28\,200)(2400)(466)}$$

$= 0.00835$

$$b_w/b = \frac{100+155}{2400} = 0.11$$

$h_f/d = 50/466 = 0.107$

With these parameters:

$k_{cr} = 0.085$, $k_{pr} = 1.21$ from Fig. 3.5.5

$k_c = 0.125$, $k_p = 1.39$ from Fig. 3.5.6

thus:

$$I_{cr} = \frac{1}{12}k_{cr}k_{pr}bd^3 = \frac{1}{12}(0.085)(1.21)(2400)(466)^3$$

$= 2082 \times 10^6$ mm$^4$

$y_{cr} = k_c k_p d = (0.125)(1.39)(466) = 81$ mm

$$M_{cr} = \frac{f_rI_g}{y} = \frac{(3.6)(8580\times10^6)}{(424)(10^6)} = 72 \text{ kN·m}$$

$$M_{dc} = \frac{f_{pe}I_g}{y_t} = \frac{(24.9)(8580\times10^6)}{424} = 504 \text{ kN·m}$$

where:

$$f_{ce} = \frac{P_e}{A} + \frac{P_ee}{I}y$$

$$= \frac{(1370)(10^3)}{260\,000} + \frac{(1370)(10^3)(290)}{8580\times10^6}(424)$$

$= 24.9$ MPa

Effective I for curvature:

$$I'_e = I_{cr} + (I_g - I_{cr})\left(\frac{M_{cr}}{M_s - M_{dc}}\right)^4$$

$$= (2082\times10^6) + (8580 - 2082)(10^6)\left(\frac{72}{616-504}\right)^4$$

$= 3192\times10^6$ mm$^4$

Effective distance of centroid from top fibre:

$$y_e = 81 + \left(\frac{72}{616-504}\right)^4(176-81) = 97 \text{ mm}$$

Curvature due to live load at $0.4\ell$:

$$\psi_e = \frac{M_s - P_e(d_p - y_e)}{E_cI'_e}$$

$$= \frac{(616)(10^6) - (1370)(466-97)(10^3)}{(28\,200)(3192\times10^6)}$$

$= 1.23 \times 10^{-6}$

Curvature due to live load at $0.4\ell$:

$$\psi_\ell = \psi_e + \left(\frac{M_d + M_{sd} - P_ee}{E_cI_g}\right)$$

$$= 1.23 \times 10^{-6} - \left[\frac{(330+63)(10^6) - (1370)(10^3)(290)}{(28\,200)(8580\times10^6)}\right]$$

$= 1.25 \times 10^{-6}$

Repeating the same procedure for midspan:

$\psi_\ell = 0.96 \times 10^{-6}$

Deflection due to live load:

$$\Delta_{\ell\ell} = \frac{5}{48}\left(\frac{\psi_\ell^{0.5\ell} + \psi_\ell^{0.4\ell}}{2}\right)\ell^2$$

$$= \frac{5}{48}\left(\frac{0.96+1.25}{2}\right)(10^{-6})(21000)^2 = 51 \text{ mm}$$

### 3.5.7 CAMBER AND DEFLECTION OF COMPOSITE ELEMENTS

For camber and deflection of composite elements, refer to Sect. 3.4.7 (Step 10).

## 3.6 SHEAR

### 3.6.1 DESIGN PROCEDURES

The Canadian Concrete Code CSA A23.3-94 contains three different methods that can be used to design structural concrete elements for shear and torsion.

The Simplified Method (Clause 11.3) and the General Method (Clause 11.4) are sectional procedures that can be used for "flexural" regions where it is reasonable to assume that plane sections remain plane and that the shear stresses are uniform over the depth of the member. For regions near statical or geometrical discontinuities, the Strut-and-Tie Model (Clause 11.5) must be used. The Strut-and-Tie Model, which considers the flow of forces in a structural element, can also be used for the "flexural" regions of a member.

The Simplified Method is the traditional 45° truss model with an empirical concrete contribution to shear resistance. It can only be used for members not subjected to significant axial tension. While the Simplified Method can be used to design members subjected to torsion, it does not account for the beneficial effects of prestressing and therefore the General Method should be used. The Simplified Method is described in this section.

The General Method (Clause 11.4) is a variable angle truss model approach. Unlike the General Method of the 1984 Canadian Concrete Code, the new approach includes a rational concrete contribution to shear resistance and uses design tables and charts to simplify the calculations. For prestressed concrete, the General Method would typically require less transverse shear reinforcement than the Simplified Method. The General Method for shear and torsion design is described in section 3.7.

The Strut-and-Tie Model is not discussed in this Manual.[R3.6.1]

## 3.6.2 DESIGN BY THE SIMPLIFIED METHOD

The shear resistance of structural concrete elements must meet the requirement:

$$V_f \leq V_r$$

where:

$$V_r = V_c + V_s$$

but $V_r$ shall not exceed:

$$V_r = V_c + 0.8\lambda\phi_c\sqrt{f'_c}\; b_w d$$

and:

$V_f$ = factored shear force

$V_r$ = factored shear resistance

$V_c$ = factored shear resistance of concrete

$V_s$ = factored shear resistance of shear reinforcement

For reinforced elements complying with Clause 11.3.2, sections located within a distance d from the face of the support may be designed for the same shear $V_f$ as computed at distance d, provided that concentrated loads do not occur within this distance. For prestressed elements, this provision applies for a distance of h/2.

## 3.6.3 SHEAR RESISTANCE OF REINFORCED CONCRETE

The nominal shear resistance, $V_c$, provided by reinforced concrete for sections having either:

(a) at least the minimum amount of transverse reinforcement given by

$$A_v = 0.06\sqrt{f'_c}\,\frac{b_w s}{f_y}$$

or

(b) an effective depth not exceeding 300 mm,

is:

$$V_c = 0.2\lambda\phi_c\sqrt{f'_c}\; b_w d$$

For sections with effective depths greater than 300 mm and with no transverse reinforcement or less transverse reinforcement than the minimum required, $V_c$ shall be computed from:

$$V_c = \left(\frac{260}{1000+d}\right)\lambda\phi_c\sqrt{f'_c}\; b_w d > 0.10\lambda\phi_c\sqrt{f'_c}\; b_w d$$

## 3.6.4 SHEAR RESISTANCE OF PRESTRESSED CONCRETE USING CODE EQUATIONS

For prestressed elements, the shear strength is:

$$V_c = 0.06\lambda\phi_c\sqrt{f'_c}\; b_w d + \frac{V_f}{M_f} M_{cr}$$

where:

$$M_{cr} = (I_g / y_t)(0.6\lambda\phi_c\sqrt{f'_c} + \phi_p f_{ce})$$

but:

$$0.17\lambda\phi_c\sqrt{f'_c}\; b_w d < V_c < V_{cw}$$

and:

$$V_{cw} = 0.4\lambda\phi_c\sqrt{f'_c}\left(\sqrt{1+\frac{\phi_p f_{cp}}{0.4\lambda\phi_c\sqrt{f'_c}}}\right) b_w d + \phi_p V_p$$

The distance, d, is the effective depth from the extreme compression fibre to the centroid of the longitudinal tension reinforcement, but need not be taken less than 0.80 h.

**Fig. 3.6.1 Shear diagram**

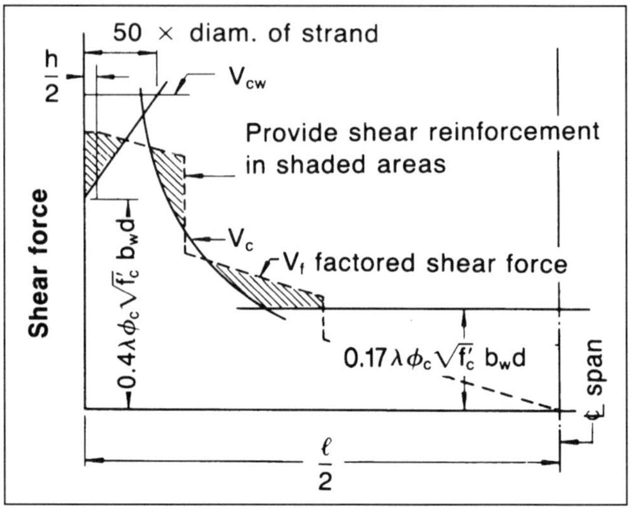

It is usually helpful to construct a shear resistance diagram and superimpose the factored shear force diagram in order to determine how much transverse reinforcement is required and where it is required. This is illustrated in Fig. 3.6.1 and is described below.

The steps for constructing the shear resistance diagram are as follows:

1. Draw a horizontal line at a value of $0.17\lambda\phi_c\sqrt{f'_c}\ b_w d$
2. Construct the curved portion of the diagram using the equation for $V_c$.
3. Draw upper limit line, $V_{cw}$.
4. The diagonal line at the upper left of Fig. 3.6.1 delineates the upper limit of the shear resistance diagram in the prestress transfer zone. The line starts at a value of $0.17\lambda\phi_c\sqrt{f'_c}\ b_w d$ at the end of the element which is obtained by substituting $f_{cp} = 0$ and $V_p = 0$ in the equation for $V_{cw}$. The diagonal line intersects the $V_{cw}$ line at 50 strand diameters from the end of the element for multi-wire strand.

**Example 1 - Construction of a shear diagram**

*Given:*

2400 × 1000 single tee with 50 mm composite topping

Span = 24.0 m

| | | |
|---|---|---|
| Self weight and topping | = | 11.7 kN/m |
| Superimposed dead load | = | 4.3 kN/m |
| Superimposed live load | = | 12.0 kN/m |

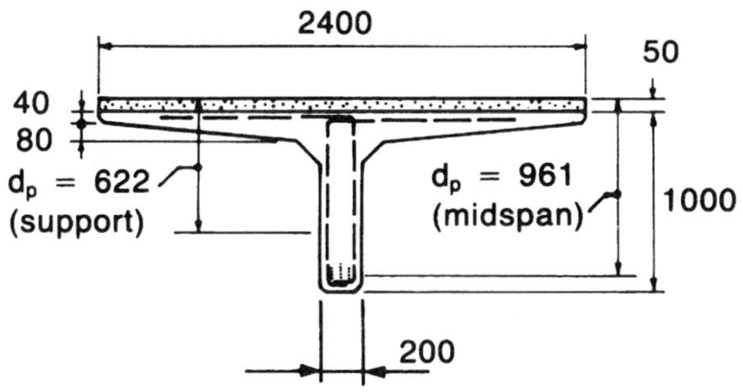


Section properties:

(a) without topping

$A_g$ = 376 000 mm²

$I_g = 36\ 300 \times 10^6$ mm⁴

$y_b$ = 714 mm

(b) with topping

$I_g = 44\ 200 \times 10^6$ mm⁴

$y_b$ = 782 mm

Concrete:

Precast: $f'_c$ = 35 MPa
Normal Density

Topping: $f'_c$ = 30 MPa
Normal Density

Prestressed reinforcement:

18-13 mm (special) 1860 MPa strands

$A_p = (18)(107.7) = 1939\ \text{mm}^2$

$$P_e = \frac{(18)(107.7)(0.79)(0.70)(1860)}{10^3} = 1994\ \text{kN}$$

Single point depression.

Shear reinforcement:

$f_y$ = 400 MPa

*Problem:*

Construct the shear diagram

*Solution:*

1. Determine factored loads:

| | | |
|---|---|---|
| Uniform dead = (1.25)(4.3 + 11.7) | | = 20.0 kN/m |
| Uniform live | = (1.5)(12.0) | = 18.0 kN/m |

2. Construct the applied shear force diagram as shown in Fig. 3.6.2.

3. Construct the factored shear resistance diagram as described in the previous section:

(a) Construct line at $0.17\lambda\phi_c\sqrt{f'_c}\ b_w d$

$$= (0.17)(1.0)(0.65)\sqrt{35}(200)(840)/10^3$$
$$= 109.8\ \text{kN}$$

(b) Construct $V_c$ line:

$$V_c = 0.06\lambda\phi_c\sqrt{f'_c}\ b_w d + \frac{V_f}{M_f}M_{cr}$$

$$= (0.06)(1.0)(0.65)\sqrt{35}(200)(840)/10^3 + \frac{V_f}{M_f}M_{cr}$$

$$= 38.8 + \frac{V_f}{M_f}M_{cr}$$

$e_{end}$ = 714 − 428 = 286 mm

$e_{centreline}$ = 714 − 89 = 625 mm

At ends (neglecting transfer effects):

$$f_{ce} = \frac{P_e}{A_g} + \frac{P_e e y_b}{I}$$

$$= \frac{(1994)(10^3)}{376\ 000} + \frac{(1994)(286)(714)(10^3)}{(36\ 300 \times 10^6)}$$

= 5.3 + 11.2 = 16.5 MPa

| Point in Fig. 3.6.2 | x (m) | $M_{cr}$ (kN·m) | $V_f$ (kN) | $M_f$ (kN·m) | $\frac{V_f}{M_f}M_{cr}$ (kN) | $V_c$ (kN) |
|---|---|---|---|---|---|---|
| 1 | 1 | 1026.2 | 418 | 437 | 981.6 | 1020.4 |
| 2 | 2 | 1082.6 | 380 | 836 | 492.1 | 530.9 |
| 3 | 4 | 1195.3 | 304 | 1520 | 239.1 | 277.9 |
| 4 | 6 | 1308.1 | 228 | 2052 | 145.3 | 184.1 |
| 5 | 8 | 1420.8 | 152 | 2432 | 88.8 | 127.6 |
| 6 | 10 | 1533.6 | 76 | 2660 | 43.8 | 82.6 |
| 7 | 12 | 1646.3 | 0 | 2736 | 0 | 38.8 |

At centreline:

$$f_{ce} = \frac{(1994)(10^3)}{376\,000} + \frac{(1994)(10^3)(625)(714)}{(36\,300 \times 10^6)}$$
$$= 5.3 + 24.5 = 29.8 \text{ MPa}$$

At ends:

$$M_{cr} = \frac{I_g}{y_B}(0.6\lambda\phi_c\sqrt{f'_c} + \phi_p f_{ce})$$
$$= \frac{(44\,200)(10^6)}{782}[(0.6)(1.0)(0.65)\sqrt{35}$$
$$+(0.9)(16.5)]$$
$$= 969.8 \text{ kN·m}$$

At centreline:

$$M_{cr} = \frac{(44\,200)(10^6)}{782}[(0.6)(1.0)(0.65)\sqrt{35}$$
$$+(0.9)(29.8)]$$
$$= 1646.3 \text{ kN·m}$$

(c) Construct upper limit line corresponding to $V_{cw}$:

$$V_{cw} = 0.4\lambda\phi_c\sqrt{f'_c}\left(\sqrt{1+\frac{\phi_p f_{cp}}{0.4\phi_c\sqrt{f'_c}}}\right)b_w d + \phi_p V_p$$

with:

$$0.4\lambda\phi_c\sqrt{f'_c} = (0.4)(1.0)(0.65)\sqrt{35} = 1.54 \text{ MPa}$$

$$f_{cp} = \frac{P_e}{A_g} = \frac{(1994)(10^3)}{376\,000} = 5.3 \text{ MPa}$$

$$V_{cw} = (1.54)\left[\sqrt{1+\frac{(0.9)(5.3)}{1.54}}\right](200)(840)/10^3$$
$$+(0.9)(1994)\left(\frac{961-622}{12\,000}\right)$$
$$= 523.7 + 50.7$$
$$= 574.4 \text{ kN}$$

(d) Construct upper limit line in transfer zone from:

$$0.4\lambda\phi_c\sqrt{f'_c}\; b_w d = (0.4)(1.0)(0.65)\sqrt{35}(200)(840)/10^3$$
$$= 258.4 \text{ kN}$$

at the end of the element to $V_{cw} = 574.4$ kN

at 50 $d_b$ = 650 mm from end

(e) Superimpose the factored shear force diagram on the shear resistance diagram.

(f) Design shear reinforcement for shaded area.

**Fig. 3.6.2 Solution of Example 1**

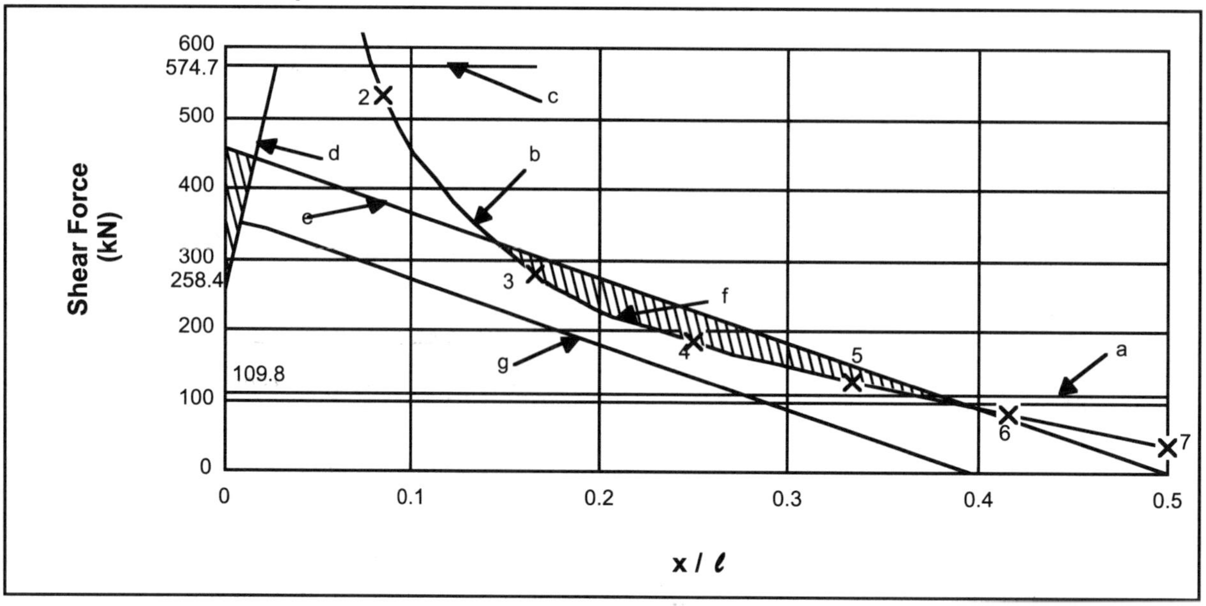

## 3.6.5 TRANSVERSE SHEAR REINFORCEMENT

### Minimum Transverse Shear Reinforcement

A minimum area of shear reinforcement is required in all precast elements except:

1. Slabs
2. Beams with a total depth not greater than 250 mm
3. Where $V_f$ is less than $V_c/2 + \phi_p V_p$
4. If shown not to be required by test

For stirrups normal to the axis of reinforced and prestressed elements, the minimum area $A_v$ required is:

$$A_v = 0.06\sqrt{f'_c}\,\frac{b_w s}{f_y}$$

### Spacing

The maximum spacing of stirrups allowed is:

(1) 600 mm or 0.7d if $(V_f - \phi_p V_p)/(b_w d) < 0.1\lambda\phi_c f'_c$

(2) 300 mm or 0.35d if $(V_f - \phi_p V_p)/(b_w d) \geq 0.1\lambda\phi_c f'_c$

### Resistance of Shear Reinforcement

For shear reinforcement perpendicular to the axis of an element:

$$V_s = \phi_s \frac{A_v f_y d}{s} \leq 0.8\lambda\phi_c\sqrt{f'_c}\ b_w d$$

**Example 2 - Design of shear reinforcement**

*Given:*

The single tee of Example 1, Page 3-67.

Stirrups:

No. 10, 2 legs

$A_v = (2)(100) = 200\ \text{mm}^2$

$f_y = 400\ \text{MPa}$

*Problem:*

Determine the required shear reinforcement spacing.

*Solution:*

Calculate the maximum spacings based on the minimum area requirements:

$$s = \frac{A_v f_y}{0.06\sqrt{f'_c}\ b_w} = \frac{(200)(400)}{(0.06)\sqrt{35}(200)} = 1127\ \text{mm}$$

At 525 mm (h/2) from support:

$$\frac{V_f - \phi_p V_p}{b_w d} = \frac{[436 - (0.9)(56)](10^3)}{(200)(840)} = 2.3\ \text{MPa}$$

$$0.1\lambda\phi_c f'_c = (0.1)(1.0)(0.65)(35) = 2.3\ \text{MPa}$$

$$\frac{V_f - \phi_p V_p}{b_w d} = 0.1\lambda\phi_c f'_c$$

$0.7\,d = (0.7)(840) \approx 600$ mm

$s \leq 600$ mm

Therefore, 600 mm requirement controls. Shear stress resisted by minimum shear reinforcement:

$$V_s = \frac{\phi_s f_y A_v d}{s} = \frac{(0.85)(400)(200)(840)}{600}/10^3 = 95.2\ \text{kN}$$

Draw line g parallel to line e in Fig. 3.6.2 at a distance of 95.2 kN. By inspection, the $V_c$ (line b) does not go below the g line and no additional transverse shear reinforcement is required.

The CSA A23.3–94 Code simplified shear design procedure requires an explicit check of the longitudinal reinforcement at the supports of simple spans (Clause 11.3.8.2). The longitudinal reinforcement at the inside edge of the bearing area must be capable of resisting a force of $V_f - 0.5V_s$.

$$V_f - 0.5V_s = (12.0 - 0.1)(38.0) - (0.5)(95.2) = 404.6\ \text{kN}$$

Resistance of strands at $50d_b = 650$ mm:

$f_{pe} = 1029$ MPa

$$T_r = \phi_p f_{pe} A_p = (0.9)(1029)(1939) = 1795\ \text{kN}$$

At face of support (edge of bearing area):

$$T_r = \frac{(300)(1795)}{650} = 828.5\ \text{kN} > 404.6\ \text{kN}$$

Longitudinal reinforcement is adequate.

## 3.6.6 LONGITUDINAL SHEAR TRANSFER IN COMPOSITE ELEMENTS

In order for a precast prestressed element with topping to behave compositely, full transfer of the longitudinal shear force must be assured at the interface of the precast element and the cast-in-place topping. The procedure recommended in this section is based on Clause 17.5.4 of CSA-A23.3. An alternative procedure is given in Clause 17.5.3.

The longitudinal shear force, $V_{f\ell}$, which must be resisted is the total force in the topping; compression in positive moment regions, and tension in negative moment regions, as shown in Fig. 3.6.3.

**Fig. 3.6.3 Longitudinal shear in composite section**

**(a) Positive moment section**

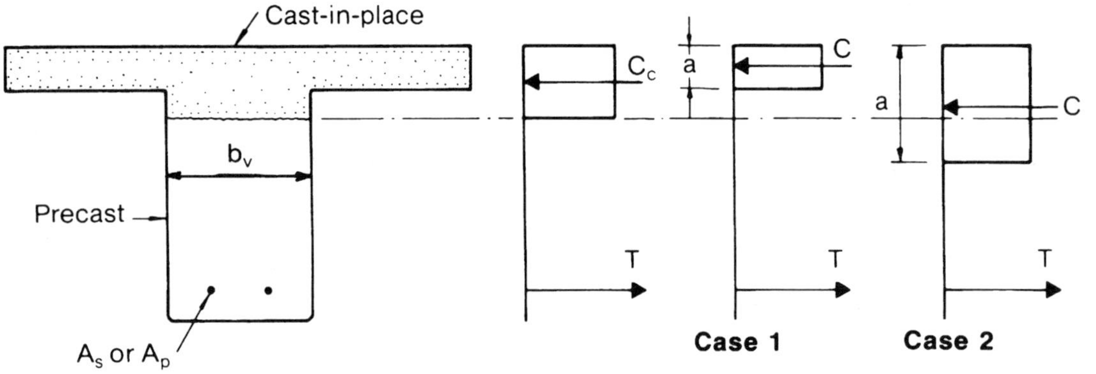


$A_{top}$ = effective area of the cast-in-place composite topping
$C_c$ = normal compressive force resistance of the composite topping
= $\alpha_1 \phi_c f'_{cc} A_{top}$
C = total compressive force
T = total tensile force = $\phi_s A_s f_s + \phi_p A_{ps} f_{ps}$
$f'_{cc}$ = compressive strength of the topping
$V_{f\ell}$ = factored longitudinal shear force

Case 1: $C < C_c$
$V_{f\ell} = C = T$
Case 2: $C > C_c$
$V_{f\ell} = C_c < T$

**(b) Negative moment section**

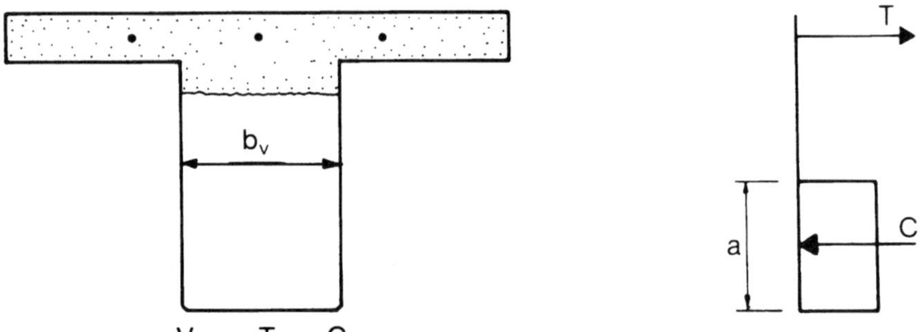


$V_{f\ell} = T = C$

**Fig. 3.6.4 Longitudinal shear length**

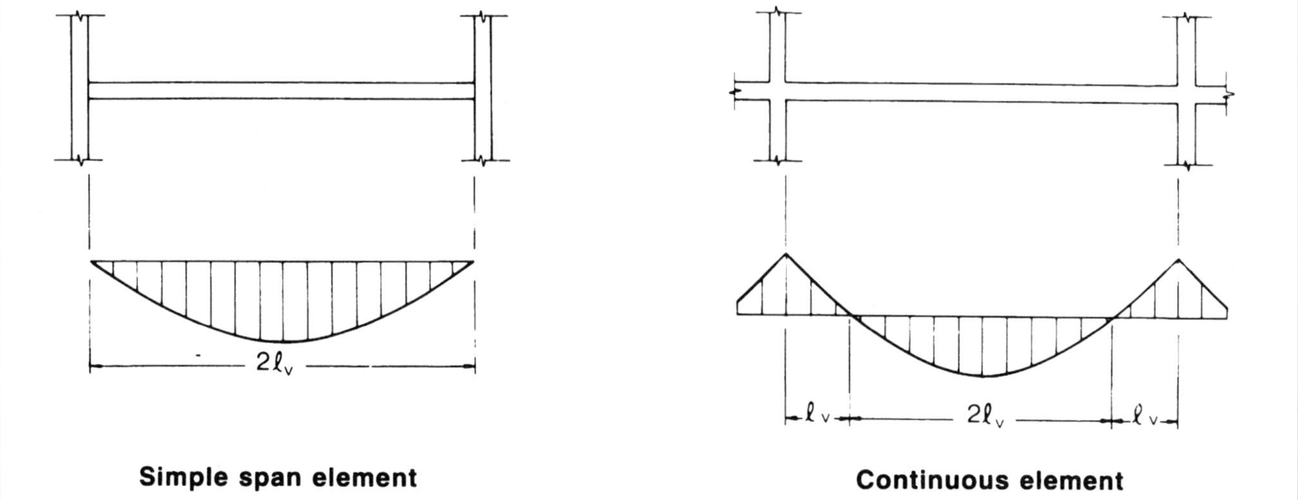

When the contact surfaces are clean, free of laitance, and intentionally roughened (but without ties), the calculated factored longitudinal shear resistance is $V_{r\ell} = 0.7\,\phi_c b_v \ell_v$, where $b_v$ is the width of the interface surface and $\ell_v$ is the longitudinal shear length as defined in Fig. 3.6.4.

When minimum ties are provided but the interface is not intentionally roughened, the same limit applies.

Where $V_{f\ell}$ exceeds $0.7\phi_c b_v \ell_v$, the design for longitudinal shear shall be based on the interface shear transfer provisions of Clause 11.6 of CSA-A23.3.

For concrete placed against hardened concrete with the surface clean and intentionally roughened to a full amplitude of at least 5 mm, the factored shear resistance may be computed using Code Equation 11-33 with k = 0.5[R3.6.2]. A k = 0.6 is specified for monolithic concrete. For other situations, Code Equation 11-32 must be used.

The stirrups are almost invariably placed with $\alpha_f = 90°$ therefore $\sin \alpha_f = 1$ and $\cos \alpha_f = 0$. The permanent load perpendicular to the shear plane is generally negligible. For these circumstances Code Equation 11-33 simplifies to:

$$v_r = 0.5\lambda\phi_c\sqrt{\sigma f'_c}$$

where:

$$\sigma = \frac{A_{vf} f_y}{A_{cv}}$$

For uniformly spaced ties:

$$\sigma = \frac{A_v f_y}{b_v s}$$

The required tie spacing is therefore:

$$s \le \frac{A_v f_y f'_c}{b_v}\left(\frac{0.5\lambda\phi_c}{v_r}\right)^2$$

The shear stress is also limited to:

$$v_r \le 0.25\phi_c f'_c \quad \text{when } f'_c \le 28\text{ MPa}$$

$$v_r \le 7.0\phi_c \quad \text{when } f'_c > 28\text{ MPa}$$

In these equations

- $A_{cv}$ = area of shear interface
- $A_{vf}$ = area of shear ties crossing shear interface
- $A_v$ = area of one shear tie
- $b_v$ = width of shear interface
- $f'_c$ = specified compressive strength of the precast concrete or the topping whichever has the lower strength.
- $f_y$ = specified yield strength of shear ties
- $s$ = tie spacing
- $v_r$ = horizontal shear resistance
- $\lambda$ = factor to account for low density concrete
- $\phi_c$ = resistance factor for concrete (use 0.60 not 0.65 because topping is cast-in-place)

When ties are required, their spacing must be not greater than four times the thickness of the topping nor more than 600 mm, and from Code Equation 11-1:

$$s \le \frac{A_v f_y}{0.06\sqrt{f'_c}\,b_v}$$

**Example 3 - Horizontal shear design for composite beam**

*Given:*

Inverted tee beam with 50 mm composite topping (100 mm over web, see Example 5, Page 3-22)

Span length = 6.0 m

$b_v$ = 300 mm

Concrete:

Precast: $f'_c$ = 35 MPa

Topping: $f'_c$ = 25 MPa

Prestressed reinforcement:

11 - 13 mm 1860 MPa strands

$A_p = (11)(99) = 1089\text{ mm}^2$

Tie reinforcement:

No. 10 ties ($A_v = 200\text{ mm}^2$)

$f_y$ = 400 MPa

*Problem:*

Determine the tie requirements to transfer horizontal shear force. Note that this force must include the force in the compression steel which is located in the topping.

$$V_{f\ell} = C_1 + C_2 + C_4 = 732 + 366 + 204 = 1302\text{ kN}$$

*Solution:*

Check maximum interface shear without ties:

$$\ell_v = \ell / 2 = 6000 / 2 = 3000\text{ mm}$$

$$V_{r\ell} = 0.7\phi_c b_v \ell_v = (0.7)(0.6)(300)(3000) / 10^3 = 378\text{ kN}$$

This is less than the factored shear, therefore ties are necessary.

Check maximum interface shear with ties:

$$V_{r\ell} \le 0.25\phi_c f'_c b_v \ell_v$$
$$= (0.25)(0.6)(25)(300)(3000)/10^3 = 3375 \text{ kN}$$

The horizontal shear force is less than this, so it can be resisted by providing sufficient ties across the interface.

Maximum tie spacing:

$$s \le 600 \text{ mm}$$
$$s \le 4h_t = (4)(100) = 400 \text{ mm}$$
$$s \le \frac{A_v f_y}{0.06\sqrt{f'_c}\,b_v} = \frac{(200)(400)}{(0.06)\sqrt{25}(300)} = 890 \text{ mm}$$
$$v_{f\ell} = \frac{V_{f\ell}}{b_v \ell_v} = \frac{(1302)(10^3)}{(300)(3000)} = 1.44 \text{ MPa}$$
$$s \le \frac{(200)(400)(25)}{300}\left[\frac{(0.5)(1.0)(0.6)}{1.44}\right]^2 = 289 \text{ mm}$$

The spacing based on shear friction governs. Use No. 10 ties @ 250 mm.

Check shear on a vertical plane through the flange at the face of the web.

$$b_v = 50 \text{ mm}$$
$$V_{f\ell} = C_1/2 = 732/2 = 366 \text{ kN}$$

Check maximum interface shear with ties:

$$V_{f\ell} \le 0.25\phi_c f'_c b_v \ell_v$$
$$= (0.25)(0.6)(25)(50)(3000)/10^3$$
$$= 562 \text{ kN} > 365 \text{ kN} \quad \text{OK}$$

Maximum tie spacing:

$$s \le 600 \text{ mm}$$
$$s \le 4h_f = (4)(50) = 200 \text{ mm}$$
$$s \le \frac{(100)(400)}{(0.06)\sqrt{25}(50)} = 2670 \text{ mm}$$
$$v_{f\ell} = \frac{V_{f\ell}}{b_v \ell_v} = \frac{(366)(10^3)}{(50)(3000)} = 2.43 \text{ MPa}$$

Because the topping concrete is placed against hardened concrete across this joint, use k = 0.5.

$$s \le \frac{(100)(400)(25)}{50}\left[\frac{(0.5)(1.0)(0.6)}{2.43}\right]^2 = 305 \text{ mm}$$

Use No. 10 bars @ 200 mm governed by 4 times the topping thickness.

**Example 4 - Horizontal shear design for composite inverted T-beam**

*Given:*

Inverted T-beam produced in a CSA certified plant with composite topping as shown

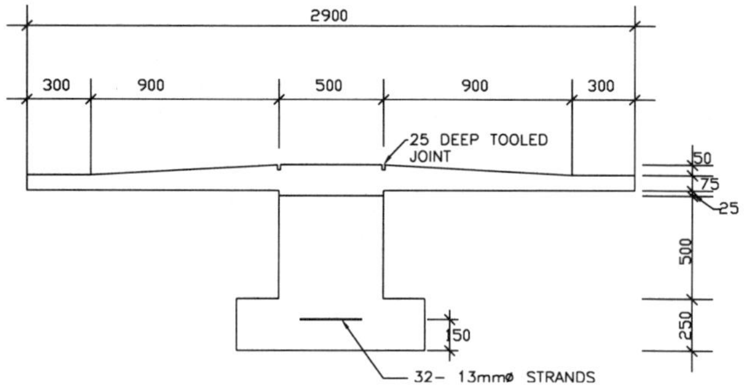


Span length = 10.0 m
$b_v$ = 500 mm

Concrete:

Precast: $f'_c$ = 35 MPa
Topping: $f'_c$ = 35 MPa

Prestressed reinforcement:

32 - 13 mm 1860 MPa strands
$A_p = (32)(99) = 3168 \text{ mm}^2$

Tie reinforcement:

No. 10 ties ($A_v = 200 \text{ mm}^2$)
$f_y$ = 400 MPa

*Problem:*

Determine the tie requirements to transfer the horizontal shear force.

*Solution:*

In order to determine the horizontal shear force to be transferred, the size of the compression block must be determined.

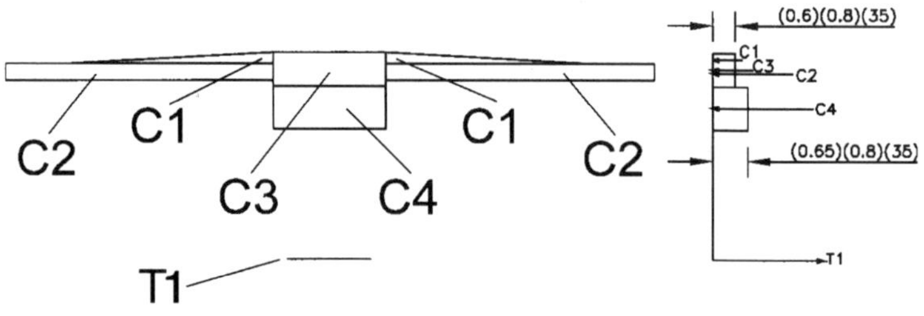


$$C_1 + C_2 + C_3 + C_4 = T_1$$
$$C_1 = (0.80)(0.6)(35)(0.5)(1200)(50)(2)/10^3$$
$$= 1008 \text{ kN}$$
$$C_2 = (0.8)(0.6)(35)(1200)(75)(2)/10^3 = 3024 \text{ kN}$$
$$C_3 = (0.80)(0.6)(35)(500)(150)/10^3 = 1260 \text{ kN}$$
$$C_4 = (0.80)(0.65)(35)(500)(0.88c - 150)/10^3$$
$$C_4 = (8.01c - 1365) \text{ kN}$$

$$T_1 = \phi_p A_p f_{pu}\left(1 - k_p \frac{c}{d_p}\right)$$

$$= (0.9)(3168)(1860)\left(1 - 0.28\frac{c}{d_p}\right)$$

$$T_1 = (5303)\left(1 - 0.28\frac{c}{d_p}\right)$$

$$\frac{c}{d_p} = \frac{5303 - 1008 - 3024 - 1260 + 1365}{(8.01)(750) + (0.28)(5303)}$$

$$= \frac{1376}{7492} = 0.184$$

$$c = (0.184)(750) = 138\,\text{mm}$$

$$a = \beta_1 c = (0.88)(138) = 121\,\text{mm}$$

This is slightly less than the flange thickness. Redesign as rectangular flange with triangular part removed:

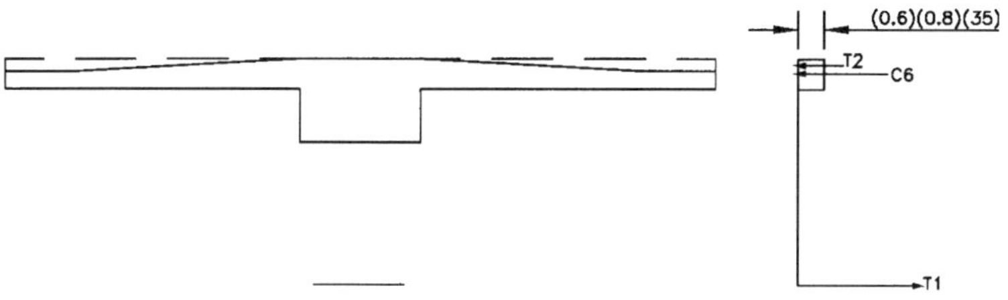


$$C_6 = T_1 + T_2$$

$$T_2 = C_1 = 1008\text{ kN (Force on negative area)}$$

$$C_6 = \alpha_1 \phi_c f'_c b \beta_1 c$$

$$= (0.80)(0.60)(35)(2900)(0.88)c/10^3 = 42.9c$$

$$\frac{c}{d_p} = \frac{5303 + 1008}{(42.9)(750) + (0.28)(5303)} = 0.187$$

$$c = (0.187)(750) = 141\,\text{mm}$$

$$a = (0.883)(141) = 124\text{ mm}$$

$$f_{pr} = (1860)[1 - (0.28)(0.187)] = 1763\text{ MPa}$$

$$T_1 = (0.9)(3168)(1763)/10^3 = 5027\text{ kN}$$

Total horizontal shear force to be transferred across beam topping interface

$$= 5027\text{ kN}$$

Check maximum interface shear without ties:

$$b_v = 500\text{ mm}$$

$$\ell_v = \ell/2 = 10\,000/2 = 5000\,\text{mm}$$

$$V_{r\ell} = 0.7\phi_c b_v \ell_v = (0.7)(0.6)(500)(5000)/10^3$$

$$= 1050\text{ kN} < 5027\text{ kN}$$

therefore ties are necessary.

Check maximum interface shear with ties:

$$V_{r\ell} \le 7.0\phi_c b_v \ell_v = (7.0)(0.6)(500)(5000)/10^3$$

$$= 10\,500\text{ kN}$$

This is greater than $V_{f\ell}$, so compute tie spacing.

Maximum tie spacing:

$$s \le 600\text{ mm}$$

$$s \le (4)(150) = 600\text{ mm}$$

$$s \le \frac{A_v f_y}{0.06\sqrt{f'_c}\,b_v} = \frac{(200)(400)}{(0.06)\sqrt{35}(500)} = 450\text{ mm}$$

$$v_{f\ell} = \frac{(5027)(10^3)}{(500)(5000)} = 2.01\,\text{MPa}$$

$$s \le \frac{(200)(400)(35)}{500}\left[\frac{(0.5)(1.0)(0.6)}{2.01}\right]^2 = 125\text{ mm}$$

This spacing governs. Consideration might be given to the use of No. 15 ties at 250 mm but caution is necessary because of the short anchorage length available[R3.6.3].

Use No. 10 ties @ 125 mm.

Check interface shear in a vertical section through the flange at the face of web. The 25 mm deep tooled joint at this location creates a crack control joint and also reduces the shear area.

$$b_v = 75 + 50 - 25 = 100\,\text{mm}$$

$$V_{f\ell} = C_6(1200/2900) - T_2 = 1495\,\text{kN}$$

Check maximum interface shear with ties:

$$V_{f\ell} \le (7)(0.6)(100)(5000)/10^3$$

$$= 2100\,\text{kN} > 1495\,\text{kN} \quad \text{OK}$$

Maximum tie spacing:

$$s \le 600\text{ mm}$$

$$s \le (4)(100) = 400\text{ mm}$$

$$s \le \frac{A_v f_y}{0.06\sqrt{f'_c}\;b_v} = \frac{(100)(400)}{(0.06)\sqrt{35}(100)} = 1127\,\text{mm}$$

$$v_{f\ell} = \frac{(1495)(10^3)}{(100)(5000)} = 2.99\text{ MPa}$$

Because the concrete is monolithic across this joint, use $k = 0.6$.

$$s \le \frac{(100)(400)(35)}{100}\left[\frac{(0.6)(1.0)(0.6)}{2.99}\right]^2 = 203\text{ mm}$$

Use No. 10 bars @ 200 mm. These bars may also be used to resist moment.

# 3.7 DESIGN FOR SHEAR AND TORSION ACCORDING TO GENERAL METHOD

## 3.7.1 INTRODUCTION

The 1994 Canadian Concrete Code contains general provisions for shear and torsion design which are based on the modified compression field theory. The method is a variable angle truss model with a rational concrete contribution. The same general procedure is used for prestressed and reinforced members and unlike the Simplified Method, the procedure is also valid for members subjected to significant axial tension.

In the Simplified Method, the shear resisted by the stirrups crossing a 45° diagonal crack is added to an empirical concrete contribution equal to the shear that causes initial diagonal cracking. The General Method accounts for the fact that the critical diagonal crack may be inclined at angles other than 45°. For example, a member with significant prestressing will usually have diagonal cracks that are considerably flater than 45°, while in a member subjected to axial tension, or in the flexure dominated regions of a beam, the diagonal cracks will tend to be steeper than 45°. The effect of variables which influence the crack inclination such as the amount of prestressing, the magnitude of the applied axial load and bending moment, and the stiffness of the longitudinal reinforcement are quantified in terms of the axial strain $\epsilon_x$ of the tension cord of the member.

The concrete contribution, $V_c$, in the general method, is the shear that can be transmitted across diagonal cracks by concrete stresses (e.g. shear stress transferred by aggregate interlock). The ability of a diagonal crack to resist shear stress is controlled mostly by the width of the crack, and the width of the diagonal cracks is strongly influenced by the axial strain $\epsilon_x$.

To simplify the general shear design procedure, design tables are provided in the 1994 Canadian Concrete Code, as well as equivalent figures which may be used as an alternative to the tables. Two different tables (and two different equivalent figures) are provided - one is for members with at least minimum amount of transverse reinforcement, while the other is for members with less than the minimum amount of transverse reinforcement. The reason for the two different tables is that the width of diagonal cracks in members without minimum transverse reinforcement depends greatly on the depth of the member (deeper members have wider diagonal cracks). Also, the shear stress may be considerably larger in members with significant transverse reinforcement.

In addition to accounting for the influence of the applied bending moment and amount of longitudinal reinforcement on the transverse reinforcement design, in the general shear design method the longitudinal reinforcement required for shear is considered in an explicit way.

The design for shear by the General Method is described in Section 3.7.2, while the design for combined shear and torsion by the General Method is described in Section 3.7.3.

**Fig. 3.7.1 Effective shear area**

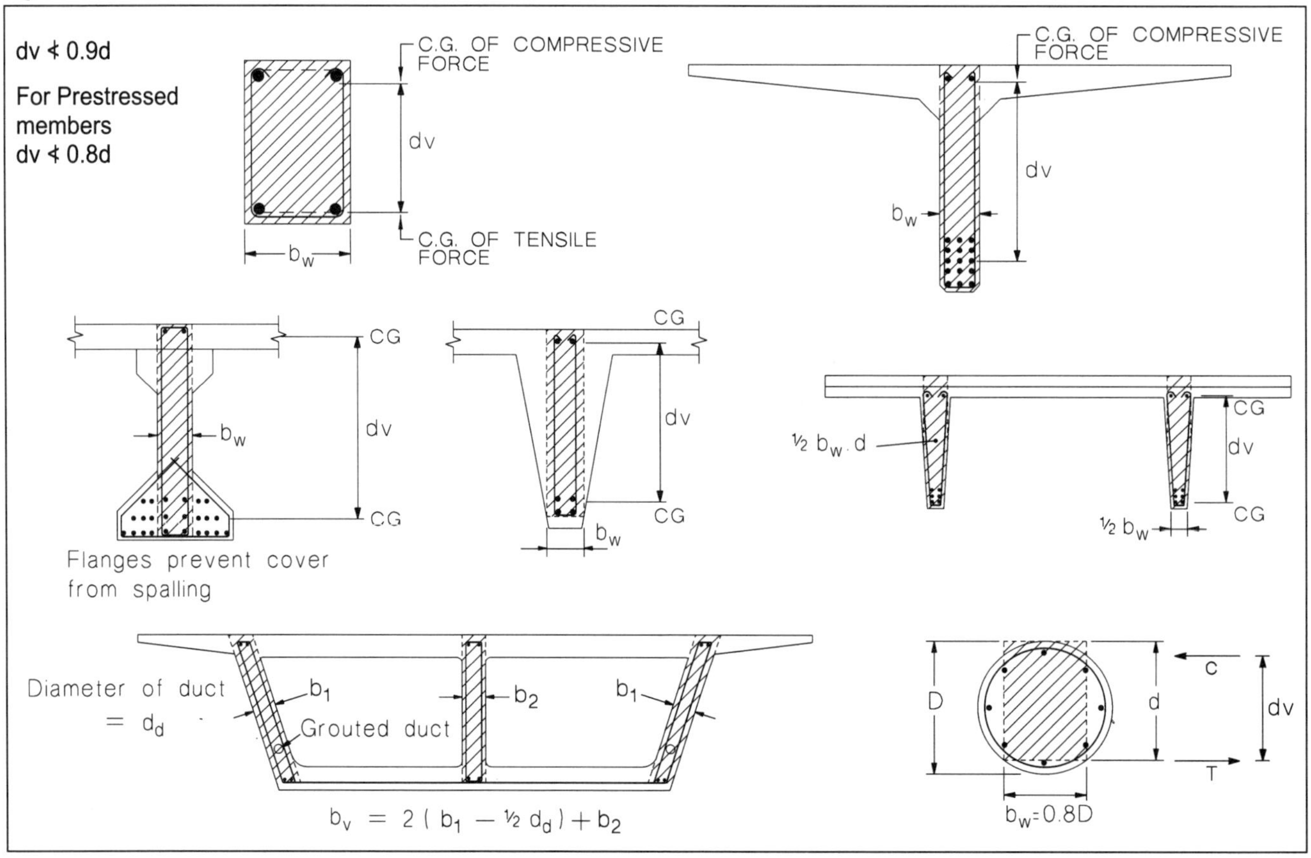

## 3.7.2 DESIGN FOR SHEAR BY THE GENERAL METHOD

In the General Method, the factored shear resistance is given by:

$$V_{rg} = V_{cg} + V_{sg} + \phi_p V_p$$

but $V_{rg}$ shall not exceed:

$$V_{rg} = 0.25\, \phi_c f'_c\, b_w d_v + \phi_p V_p$$

For both reinforced and prestressed members, a section located within $d_v$ from the face of the support may be designed for the same shear as computed at a distance of $d_v$ if the support reaction introduces compression into the end region of the member.

As shown in Fig. 3.7.1, the effective shear depth, $d_v$, over which the shear stresses are reasonably uniform, shall be taken as the distance between the flexural tension and flexural compression (d – a / 2), but need not be taken as less than 0.9 d. As d need not be taken less than 0.8 h for prestressed members, $d_v$ need not be less than 0.72 h.

The shear resistance of both reinforced and prestressed concrete is given by:

$$V_{cg} = 1.3\lambda\phi_c\beta\sqrt{f'_c}\; b_w d_v$$

where β is determined from Fig. 3.7.2 if the member has at least the minimum amount of transverse reinforcement, or from Fig. 3.7.3 if the member has no transverse reinforcement.

In both tables the main parameter which influences the β value is the maximum axial strain which can be calculated from:

$$\epsilon_x = \frac{\dfrac{M_f}{d_v} + \dfrac{N_f}{2} + \dfrac{V_f \cot\theta}{2} - A_p f_{po}}{E_s A_s + E_p A_p}$$

**Fig. 3.7.2 Values of β and θ for sections with at least minimum transverse reinforcement**

| $v_f / (\lambda\, \phi_c\, f'_c)$ | | Longitudinal strain, $\epsilon_x$ | | | | | | |
|---|---|---|---|---|---|---|---|---|
| | | ≤ 0.0000 | ≤ 0.00025 | ≤ 0.0005 | ≤ 0.00075 | ≤ 0.0010 | ≤ 0.0015 | ≤ 0.0020 |
| ≤ 0.050 | β<br>θ | 0.405<br>27.0° | 0.290<br>28.5° | 0.208<br>29.0° | 0.197<br>33.0° | 0.185<br>36.0° | 0.162<br>41.0° | 0.143<br>43.0° |
| ≤ 0.075 | β<br>θ | 0.405<br>27.0° | 0.250<br>27.5° | 0.205<br>30.0° | 0.194<br>33.5° | 0.179<br>36.0° | 0.158<br>40.0° | 0.137<br>42.0° |
| ≤ 0.100 | β<br>θ | 0.271<br>23.5° | 0.211<br>26.5° | 0.200<br>30.5° | 0.189<br>34.0° | 0.174<br>36.0° | 0.143<br>38.0° | 0.120<br>39.0° |
| ≤ 0.125 | β<br>θ | 0.216<br>23.5° | 0.208<br>28.0° | 0.197<br>31.5° | 0.181<br>34.0° | 0.167<br>36.0° | 0.133<br>37.0° | 0.112<br>38.0° |
| ≤ 0.150 | β<br>θ | 0.212<br>25.0° | 0.203<br>29.0° | 0.189<br>32.0° | 0.171<br>34.0° | 0.160<br>36.0° | 0.125<br>36.5° | 0.103<br>37.0° |
| ≤ 0.200 | β<br>θ | 0.203<br>27.5° | 0.194<br>31.0° | 0.174<br>33.0° | 0.151<br>34.0° | 0.131<br>34.5° | 0.100<br>35.0° | 0.083<br>36.0° |
| ≤ 0.250 | β<br>θ | 0.191<br>30.0° | 0.167<br>32.0° | 0.136<br>33.0° | 0.126<br>34.0° | 0.116<br>35.5° | 0.108<br>38.5° | 0.104<br>41.5° |

**Fig. 3.7.3 Values of β and θ for sections with less than minimum transverse reinforcement**

| $s_z$ | | Longitudinal strain, $\epsilon_x$ | | | | | |
|---|---|---|---|---|---|---|---|
| | | ≤ 0.0000 | ≤ 0.00025 | ≤ 0.0005 | ≤ 0.0010 | ≤ 0.0015 | ≤ 0.0020 |
| ≤ 125 | β<br>θ | 0.406<br>27° | 0.309<br>29° | 0.263<br>32° | 0.214<br>34° | 0.183<br>36° | 0.161<br>38° |
| ≤ 250 | β<br>θ | 0.384<br>30° | 0.283<br>34° | 0.235<br>37° | 0.183<br>41° | 0.156<br>43° | 0.138<br>45° |
| ≤ 500 | β<br>θ | 0.359<br>34° | 0.248<br>39° | 0.201<br>43° | 0.153<br>48° | 0.127<br>51° | 0.108<br>54° |
| ≤ 1000 | β<br>θ | 0.335<br>37° | 0.212<br>45° | 0.163<br>51° | 0.118<br>56° | 0.095<br>60° | 0.080<br>63° |
| ≤ 2000 | β<br>θ | 0.306<br>41° | 0.171<br>53° | 0.126<br>59° | 0.084<br>66° | 0.064<br>69° | 0.052<br>72° |

The numerator of this equation is the axial force applied to the longitudinal reinforcement in the tension zone of the cross section, while the denominator is the axial stiffness of that reinforcement. The decompression stress, $f_{po}$, is the result of prestressing, i.e., the stress in the prestressing when the stress in the surrounding concrete is zero, may be taken as 1.1 $f_{pe}$.

For members with minimum transverse reinforcement, the β value also depends on the shear stress ratio

$$\frac{V_f}{\lambda\phi_c f_c'(b_v d_v)}$$

while for members without minimum transverse reinforcement, the β value depends on the crack spacing parameter, $s_z$, which shall be taken as $d_v$ or the maximum distance between layers of longitudinal crack control reinforcement with an area of at least $0.004b_w s_z$.

The stirrup contribution is given by:

$$V_{sg} = \phi_s A_v f_y \frac{d_v \cot\theta}{s}$$

where θ is given in the same tables as the β values.

Longitudinal reinforcement shall be proportioned so that at all sections the factored resistance of the tension reinforcement, taking account of the stress that can be developed in that reinforcement, shall be greater than or equal to:

$$\frac{M_f}{d_v} + 0.5N_f + (V_f - 0.5V_{sg} - \phi_p V_p)\cot\theta$$

### Example 1 - Design of Single Tee

The shear design of the 2400 × 1000 single tee with 50 mm composite topping given in Example 1 and 2 of Section 3.6 will be redesigned using the general method.

*Solution:*

Determine location where no stirrups are required (see Fig. 3.7.3).

From CSA A23.3-94 (Clause 11.2.8), stirrups are required if $V_f$ exceeds $0.5V_c + \phi_p V_p$, where $V_c$ may be taken as $0.2\lambda\phi_c\sqrt{f_c'}\, b_w d$ .

$$V_c = (0.2)(1.0)(0.65)\sqrt{35}(200)(840) / 10^3 = 129.2 \text{ kN}$$

$$V_f = (0.5)(129.2) + 50.4 = 115.0 \text{ kN}$$

at $\frac{115.0}{38.0} = 3.03$ m from midspan.

Critical section at $d_v$ from face of support:

$$V_f = (456)\frac{(12\,000 - 100 - 756)}{12\,000} = 423.5 \text{ kN}$$

$$M_f = (38.0)(12)(0.856) - (38.0)(0.856)^2 / 2 = 376.4 \text{ kN·m}$$

$$v_f = \frac{V_f - \phi_p V_p}{b_w d_v} = \frac{(423.5 - 50.7)(10^3)}{(200)(756)} = 2.47 \text{ MPa}$$

$$\frac{v_f}{\lambda\phi_c f_c'} = \frac{2.47}{(1.0)(0.65)(35)} = 0.108$$

$$A_p f_{po} = (1939)(1.1)(0.79)(0.7)(1860) / (10^3) = 2194 \text{ kN}$$

$$\epsilon_x = \frac{\frac{M_f}{d_v} + \frac{N_f}{2} + \frac{V_f \cot\theta}{2} - A_p f_{po}}{E_s A_s + E_p A_p} = \frac{\frac{376.4}{0.756} + 0 + \frac{(423.5)\cot\theta}{2} - 2194}{0 + (200\,000)(1939)/(10^3)} = \frac{(212)\cot\theta - 1696}{387\,800}$$

A conservative estimate of $\epsilon_x$ results from a smaller value for θ.

From Fig. 3.7.2, assuming θ = 23.5°:

$$\epsilon_x = -0.0031 < 0$$

Use $\epsilon_x = 0$:

$$\theta = 23.5°$$

$$\beta = 0.253$$

$$V_{cg} = 1.3\lambda\phi_c\beta\sqrt{f_c'}\, b_w d_v = (1.3)(1.0)(0.65)(0.253)\sqrt{35}(200)(756)/10^3 = 191.2 \text{ kN}$$

$$s = \frac{\phi_s A_v f_y d_v \cot\theta}{V_f - \phi_p V_p - V_{cg}} = \frac{(0.85)(2 \times 100)(400)(756)\cot 23.5°}{(423.5 - 50.7 - 191.2)(10^3)} = 651 \text{ mm}$$

Repeating these calculations for a number of locations, gives the values in Fig. 3.7.4 and summarized in the shear diagram in Fig. 3.7.6.

The requirement for minimum stirrups (maximum spacing) are similar to what is given in Example 2 of Section 3.6. Thus No. 10 stirrups at 600 mm are to be provided where stirrups are required.

Check longitudinal reinforcement:

$$T_f = \frac{M_f}{d_v} + 0.5N_f + (V_f - 0.5V_{sg} - \phi_p V_p)\cot\theta$$

**Fig. 3.7.4 Spacing required for single tee stirrups Example 1**

| No. | x (m) | $V_f$ (kN) | $M_f$ (kN · m) | $\frac{V_f}{\lambda\phi_c f'_c}$ | θ (deg) | $\epsilon_x \times 10^3$ | β | $V_{cg}$ (kN) | s (mm) |
|---|---|---|---|---|---|---|---|---|---|
| 1 | 0.856 | 423.5 | 376 | 0.108 | 23.5 | < 0.000 | 0.253 | 191.2 | 651 |
| 2 | 2.200 | 372.4 | 911 | 0.094 | 24.3 | < 0.000 | 0.303 | 229.0 | 1229 |
| 3 | 3.550 | 321.1 | 1379 | 0.079 | 26.4 | < 0.000 | 0.384 | 290.3 | – |
| 4 | 4.900 | 269.8 | 1778 | 0.064 | 35.0 | 0.904 | 0.187 | 141.3 | 944 |
| 5 | 6.250 | 218.5 | 2108 | 0.049 | 42.4 | 1.841 | 0.149 | 112.6 | 1020 |
| 6 | 7.600 | 167.2 | 2368 | 0.034 | 43.0 | > 2.000 | 0.143 | 108.1 | 6563 |

**Fig. 3.7.5 Axial tension required for single tee longitudinal reinforcement Example 1**

| x (m) | d–a/2 (m) | $\frac{M_f}{d-a/2}$ (kN) | $(V_f-0.5V_{sg}$ $-\phi_p V_p)\cot\theta$ (kN) | $T_f$ (kN) |
|---|---|---|---|---|
| 0.856 | 0.600 | 627 | 404 | 1031 |
| 2.200 | 0.637 | 1430 | 292 | 1722 |
| 3.550 | 0.675 | 2043 | 197 | 2240 |
| 4.900 | 0.713 | 2494 | 138 | 2632 |
| 6.250 | 0.752 | 2803 | 81 | 2884 |
| 7.600 | 0.790 | 2997 | 26 | 3024 |
| 9.600 | 0.846 | 3105 | – | 3105 |
| 12.000 | 0.914 | 2995 | – | 2993 |

At face of support, $M_f \approx 0$.

$$T_f = (V_f - 0.5V_{sg} - \phi_p V_p)\cot\theta$$
$$= [452.2 - (0.5)(85.68)\cot 23.5^\circ$$
$$-50.7]\cot 23.5^\circ$$
$$= 696.8 \text{ kN}$$

The axial tension required at a number of locations along the span are given in Fig. 3.7.5.

Resistance of strands:

At $50d_b = 650$ mm

$$f_{pe} = 1029 \text{ MPa}$$
$$T_r = \phi_p f_{pe} A_p$$
$$= (0.9)(1029)(1939)/10^3$$
$$= 1795 \text{ kN}$$

At development length $\ell_d = 2040$ mm

$$c_u = \frac{\phi_p A_p f_{pu}}{\beta_1 \alpha_1 \phi_c f'_c b}$$
$$= \frac{(0.9)(1939)(1860)}{(0.895)(0.805)(0.60)(30)(2400)}$$
$$= 104.3 \text{ mm}$$

$$f_{pr} = \frac{f_{pu}}{1 + k_p c_u / d_p}$$
$$= \frac{1860}{1 + (0.28)(104.3/680)}$$
$$= 1783 \text{ MPa}$$

$$T_r = \phi_p f_{pr} A_p$$
$$= (0.9)(1783)(1939)/10^3$$
$$= 3112 \text{ kN}$$

At face of support:

$$T_r = \frac{(300)}{650}(1795)$$
$$= 828.5 \text{ kN} > 697 \text{ kN} \quad \text{OK}$$

Longitudinal reinforcement is adequate.

**Example 2 - Prestressed concrete spandrel beam**

*Given:*

Typical prestressed spandrel panel beam shown in Fig. 3.7.7 supporting an 18 m span double tee parking deck.

D.L. of deck = 4.4 kN/m$^2$

L.L. of deck = 2.4 kN/m$^2$

*Beam properties:*

$A_c$ = 435 000 mm$^2$

$p_c$ = 4550 mm

$w_d$ = 10.2 kN/m

$f'_c$ = 35 MPa, Normal Density

$f_y$ = 400 MPa

$y_b$ = 827 mm

$I$ = $141 \times 10^9$ mm$^4$

**Fig. 3.7.6 Force diagrams for single tee shear Example 1 by General Method**

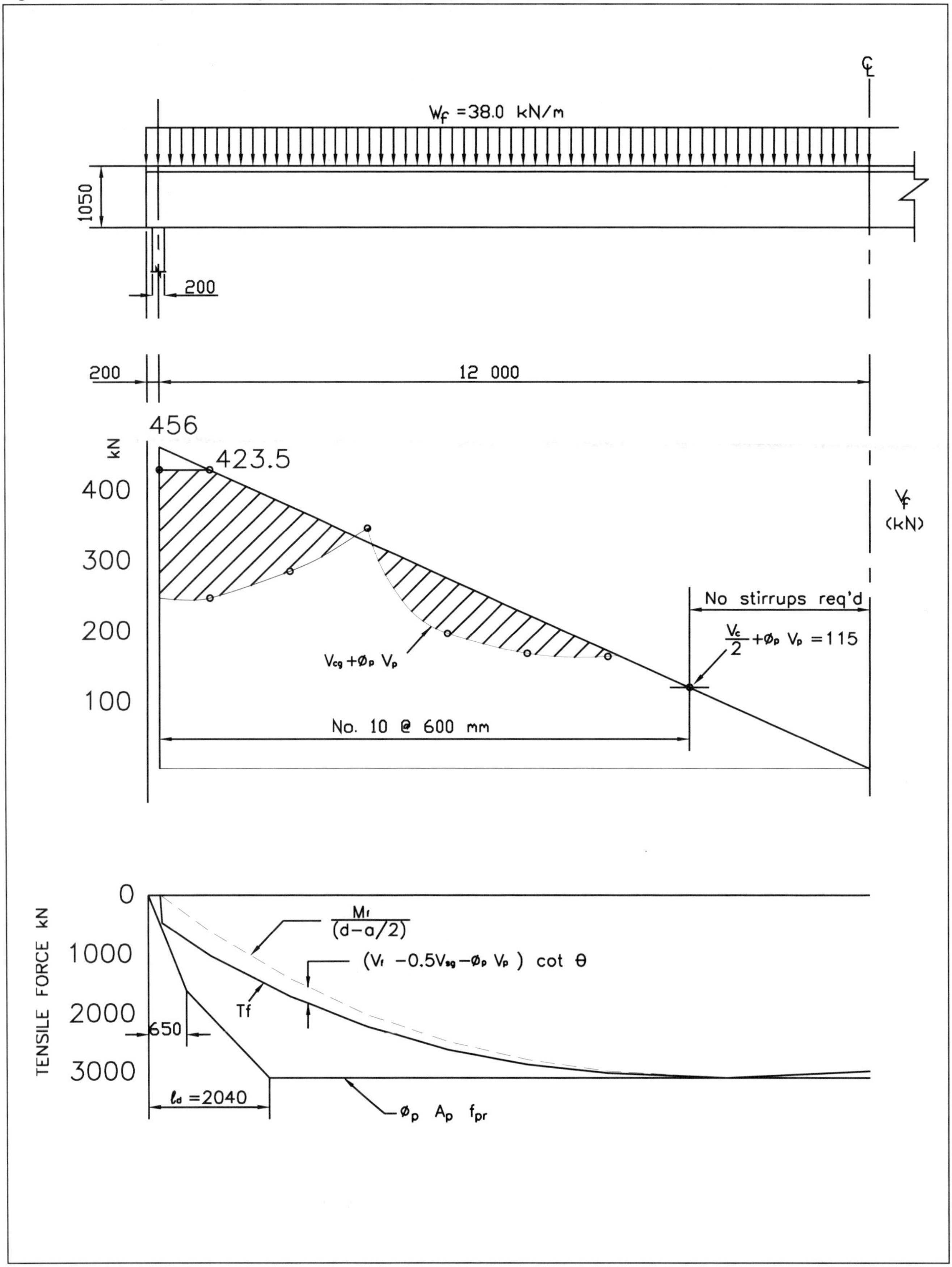

$S_b$ = 170 × $10^6$ $mm^3$

Cover to stirrups = 25 mm

Prestressed reinforcement:

4 – 13 mm 1860 MPa low relaxation straight strands

$A_p$ = (4)(99) = 396 $mm^2$

$d_p$ = 1725 mm

e = 677 mm

$f_{pe}$ = 1030 MPa

*Problem:*

Find required torsion reinforcement for the spandrel.

*Solution:*

1. Determine factored loads on spandrel:

D.L. of Beam = (1.25)(10.2)
= 12.75 kN/m

D.L. of deck = (1.25)(4.4)(1.2)(18)/2
= 59.4 kN/stem

L.L. = (1.5)(2.4)(1.2)(18)/2
= 38.9 kN/stem

Concentrated load:

$P_f$ = 59.4 + 38.9 = 98.3 kN/stem

The loads and the resulting values of $M_f$, $V_f$ and $T_f$ are shown in Fig. 3.7.9.

**Fig. 3.7.7 Structural framing for Example 2**

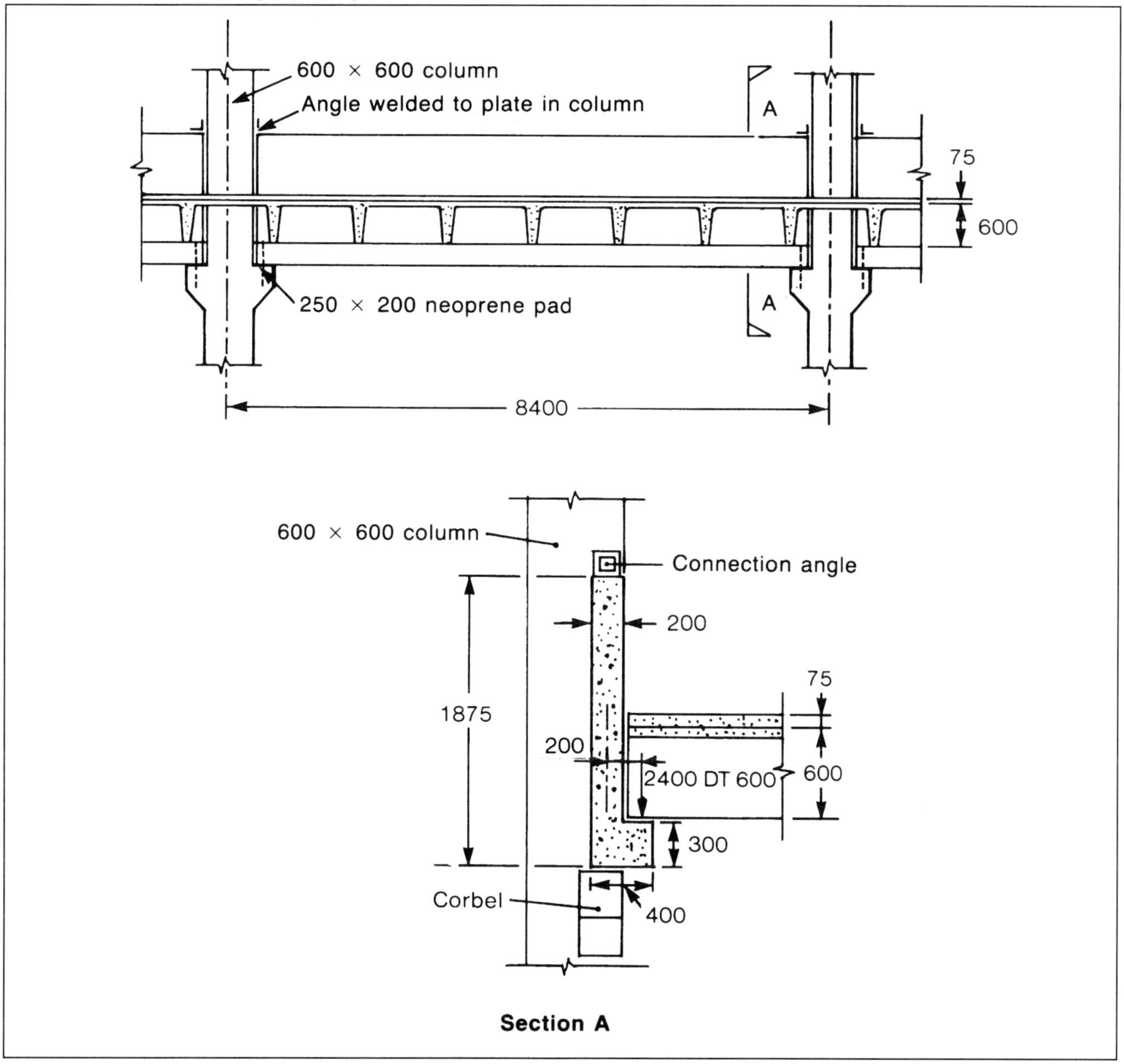

**Fig. 3.7.8 Force diagrams for Example 2**

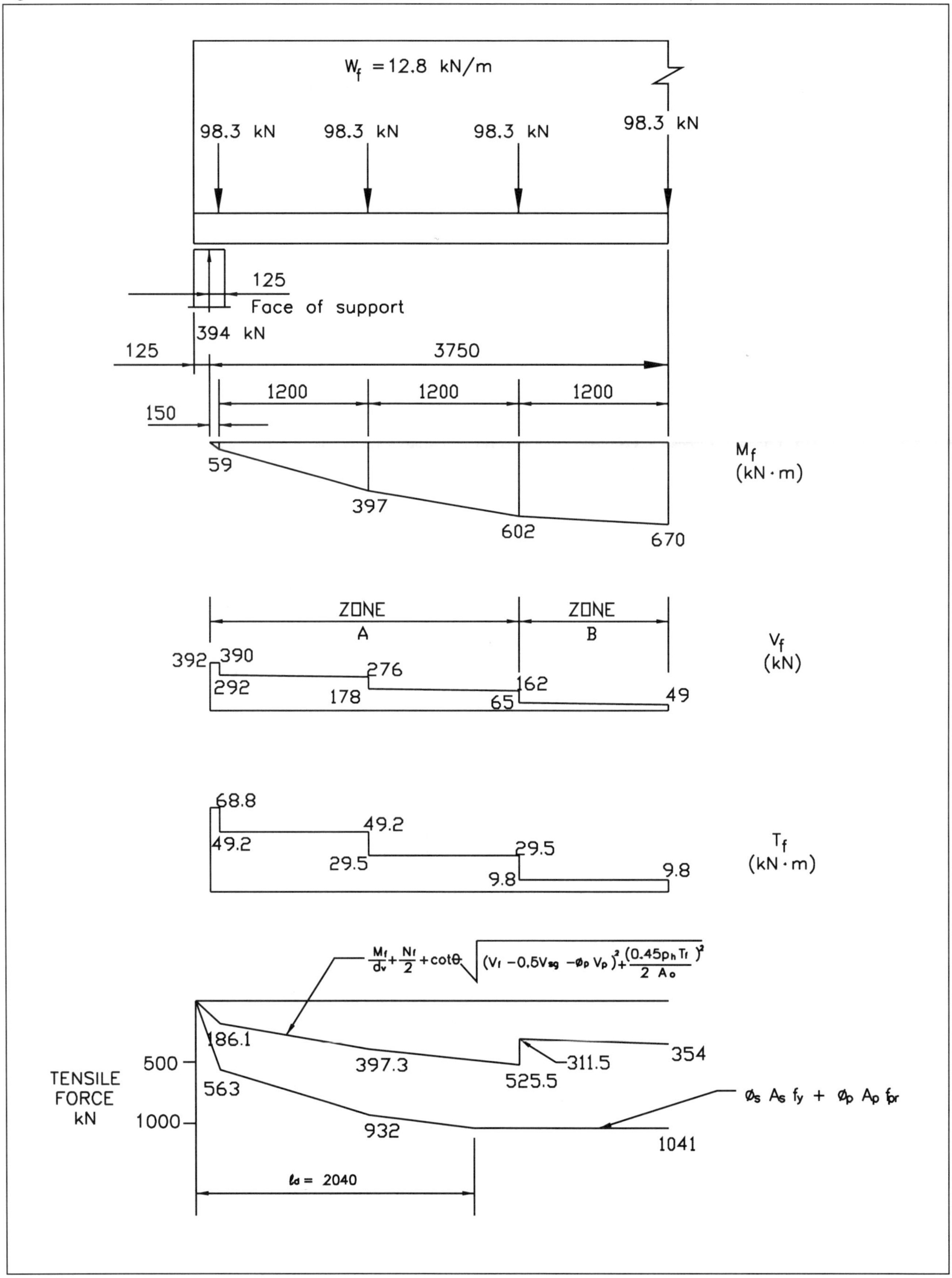

2. Minimum reinforcement requirements.

(a) Check minimum flexural reinforcement:

$$P_e = A_p f_{pe} = (396)(1030)/10^3 = 408\,\text{kN}$$

$$f_{ce} = \frac{P_e}{A_c} + \frac{P_e e}{S_b}$$

$$= \frac{(408)(10^3)}{435000} + \frac{(408)(677)(10^3)}{170 \times 10^6} = 2.56\,\text{MPa}$$

$$M_{ocr} = S_b(0.6\lambda\sqrt{f'_c} + f_{cp})$$

$$= \frac{170 \times 10^6}{10^6}[(0.6)(1.0)\sqrt{35} + 2.56]$$

$$= 1039\ \text{kN·m}$$

$$1.2M_{cr} = 1246.8\ \text{kN·m}$$

Check flexural resistance:

$$\alpha_1 = 0.85 - (0.0015)(35) = 0.80$$

$$\beta_1 = 0.97 - (0.0025)(35) = 0.88$$

$$\frac{c}{d_p} = \frac{\phi_p A_p f_{pu} + \phi_s A_s f_y - \phi_s A'_s f'_y}{\alpha_1 \phi_c f'_c \beta_1 b_w d_p + k_p \phi_p A_p f_{pu}}$$

$$= \frac{(0.9)(396)(1860) + (0.85)(1200)(400) - (0.85)(400)(400)}{(0.88)(0.8)(0.65)(35)(200)(1725) + (0.28)(0.9)(396)(1860)}$$

$$= 0.16$$

$$f_{pr} = f_{pu}(1 - k_p c/d_p)$$

$$= (1860)[1 - (0.28)(0.16)]$$

$$= 1775\,\text{MPa}$$

$$a = \frac{\phi_p A_p f_{pr} + \phi_s A_s f_y - \phi_s A'_s f'_y}{\alpha_1 \phi_c f'_c b_w}$$

$$= \frac{(0.9)(396)(1775) + (0.85)(1200)(400) - (0.85)(400)(400)}{(0.8)(0.65)(35)(200)}$$

$$= 250\,\text{mm}$$

At centreline:

$$M_r = \phi_p A_p f_{pr}(d_p - a/2) + \phi_s A_s f_y (d - a/2)$$

$$= (0.9)(396)(1775)(1775 - 250/2)$$

$$+ (0.85)(1200)(400)(1750 - 250/2)$$

$$= 10120 + 6630$$

$$= 16750\ \text{kN·m} > 1.2M_{cr}$$

At prestress transfer point (650 mm from end):

$$M_r \approx \frac{1030}{1775}(1012) + 663$$

$$= 1250.2\ \text{kN·m} > 1.2Mcr$$

(b) Minimum shear reinforcement:

$$\frac{A_v}{s} = 0.06\sqrt{f'_c}\frac{b_w}{f_y}$$

$$= 0.06\sqrt{35}\frac{200}{400} = 0.177\ \text{mm}^2/\text{mm}$$

Using No. 10 closed stirrups ($A = 100\ \text{mm}^2$), the spacing is:

$$s = \frac{(2)(100)}{0.177} = 1130\ \text{mm}$$

3. Determine need for torsional reinforcement:

$$T_f = 49.2\ \text{kN·m}$$

$$f_{cp} = \frac{(408)(10^3)}{435\,000} = 0.94\ \text{MPa}$$

$$p_c = 1875 + 200 + 1575 + 200 + 300 + 400$$

$$= 4550\ \text{mm}$$

$$T_{cr} = \frac{0.4 A_c^2 \lambda \phi_c \sqrt{f'_c}}{p_c}\left(\sqrt{1 + \frac{\phi_p f_{cp}}{0.4\lambda\phi_c\sqrt{f'_c}}}\right)$$

$$= \left[\frac{(0.4)(435\,000)^2(10)(0.65)\sqrt{35}}{(4550)(10^6)}\right]$$

$$\times\left[\sqrt{1 + \frac{(0.9)(0.94)}{(0.4)(10)(0.65)\sqrt{35}}}\right] = 79.6\ \text{kN·m}$$

$$0.25\,T_{cr} = 19.9\ \text{kN·m} < T_f$$

Torsion reinforcement is required in zone A.

4. Determine section parameters:

$$d_v = d_p - \frac{a}{2} = 1725 - \frac{250}{2} = 1600\ \text{mm}$$

$$A_{oh} = (138)[1875 - (2)(31)] + (200)(238)$$

$$= 297\,000\ \text{mm}^2$$

$$p_h = (138 + 1813)(2) + (200)(2) = 4302\ \text{mm}$$

$$A_o = 0.85 A_{oh}$$

$$= (0.85)(297\,000) = 252\,500\ \text{mm}^2$$

5. Design shear and torsion reinforcement:

Zone A, at a distance d = 1600 mm from face of support:

*Note:* For precast framing, the face of the support is taken to be the centreline of the support.

$$V_f = 174.7\ \text{kN}$$

$$T_f = 29.5\ \text{kN·m}$$

Determine angle $\theta$ and factor $\beta$:

$$v_f = \sqrt{\left(\frac{V_f - \phi_p V_p}{b_w d_v}\right)^2 + \left(\frac{T_f p_h}{A_{oh}^2}\right)^2}$$

$$= \sqrt{\left[\frac{(174.7)(10^3)}{(200)(1600)}\right]^2 + \left[\frac{(29.5)(10^6)(4302)}{(297\,000)^2}\right]^2}$$

$$= 1.54\ \text{MPa}$$

$$\frac{v_f}{\lambda\phi_c f'_c} = \frac{1.54}{(10)(0.65)(35)} = 0.068$$

From design aid, Fig. 3.7.2, assuming $\epsilon_x = 0.002$:

$\theta = 42.3°$

$\beta = 0.139$

Transverse reinforcement for torsion:

$$\frac{A_t}{s} = \frac{T_f \tan\theta}{\phi_s 2A_o f_y}$$

$$= \frac{(29.5)(10^6)\tan 42.3°}{(0.85)(2)(252\,500)(400)} = 0.156 \text{ mm}^2/\text{mm}$$

Transverse reinforcement for shear:

$$V_{cg} = 1.3\lambda\phi_c\beta\sqrt{f'_c}b_w d_v$$

$$= (1.3)(1.0)(0.65)(0.139)\sqrt{35}(200)(1600)/10^3$$

$$= 222.4 \text{ kN}$$

$$\frac{A_v}{s} = \frac{V_f - V_{cg} - \phi_p V_p}{\phi_s f_y d_v \cot\theta}$$

$$= \frac{(174.7 - 222.4)(10^3)}{(0.85)(400)(1600)\cot 42.3°}$$

$$= 0$$

Total area of transverse reinforcement (for one leg):

$$= \frac{A_t}{s} + \frac{A_v}{2s} = 0.156 + 0$$

$$= 0.156 \text{ mm}^2/\text{mm}$$

Using No. 10 closed stirrups ($A = 100 \text{ mm}^2$), the spacing is:

$$s = \frac{100}{0.156} = 641 \text{ mm}$$

Check minimum shear reinforcement:

$$\frac{A_v}{s} = 0.06\sqrt{f'_c}\frac{b_w}{f_y}$$

$$= (0.06)\sqrt{35}\frac{(200)}{400} = 0.177 \text{ mm}^2/\text{mm}$$

For No. 10 stirrups:

$$s = \frac{200}{0.177} = 1127 \text{ mm}$$

$$\frac{V_f - \phi_p V_p}{b_w d}$$

$$= \frac{(174.7)(10^3)}{(200)(1725)}$$

$$= 0.51 \text{ MPa} < 0.1\lambda\phi_c f'_c = 2.3 \text{ MPa}$$

$$s = 0.7d = (0.7)(1725) = 1207 \text{ mm}$$

Use No. 10 stirrups at 600 mm o.c.

$$\frac{A_v}{s} = 0.333 \text{ mm}^2/\text{mm}$$

Check longitudinal reinforcement:

$$V_{sg} = \phi_s \frac{A_v}{s} f_y d_v \cot\theta$$

$$= (0.85)(0.333)(400)(1600)(\cot 42.3°)/10^3$$

$$= 199.1 \text{ kN}$$

$$\frac{M_f}{d_v} + 0.5N_f + \cot\theta\sqrt{(V_f - 0.5V_{sg} - \phi_p V_p)^2 + \left(\frac{0.45 p_h T_f}{2A_o}\right)^2}$$

$$= \frac{439.7}{1.6} + 0$$

$$+ \cot 42.3° \sqrt{[174.7 - (0.5)(199)]^2 + \left[\frac{(0.45)(4302)(29.5)(10^3)}{(2)(252\,500)}\right]^2}$$

$$= 274.8 + 149.2$$

$$= 424.0 \text{ kN}$$

$$\phi_s A_s f_y + \phi_p A_p f_{pr}$$

$$= [(0.85)(1200)(400) + (0.9)(396)(1605)]/10^3$$

$$= 980 \text{ kN} > 424 \text{ kN}$$

No additional longitudinal reinforcement is required.

Zone B:

No torsion design is required since $T_f < 0.25\,T_{cr}$.

$$v_f = \frac{V_f - \phi_p V_p}{b_w d_v}$$

$$= \frac{(65)(10^3)}{(200)(1600)} = 0.203 \text{ MPa}$$

$$\frac{v_f}{\lambda\phi_c f'_c} = \frac{0.203}{(1.0)(0.65)(35)} = 0.0089$$

Try $\epsilon_x = 0.002$

and using Fig. 3.7.3 (sections without transverse reinforcement):

$\theta > 38°$

$\beta = 0.161$

$$V_{cg} = 1.3\lambda\phi_c\beta\sqrt{f'_c}b_w d_v$$

$$= (1.3)(1.0)(0.65)(0.161)\sqrt{35}(200)(1600)/10^3$$

$$= 313.9 \text{ kN} > 2V_f$$

Although no shear ties are required for shear design, the requirements for hanger steel and ledge reinforcement at point loads would govern making it virtually impossible to eliminate transverse reinforcing in this type of member.

With minimum shear reinforcement:

$s = 1127 \text{ mm}$ for No. 10 stirrups.

Use $s = 600 \text{ mm o.c.}$

**Fig. 3.7.9 Additional longitudinal reinforcing calculations for Example 2**

| Distance from end of beam | mm | <275 | >275 | <1475 | >1475 | <2675 | >2675 | 3875 |
|---|---|---|---|---|---|---|---|---|
| $M_f$ | kN · m | 59 | 59 | 397 | 397 | 602 | 602 | 670 |
| $V_f$ | kN | 390 | 292 | 276 | 178 | 162 | 65 | 49 |
| $T_f$ | kN · m | 68.8 | 49.2 | 49.2 | 29.5 | 29.5 | 9.8 | 9.8 |
| $V_{sg}$ | kN | 199.1 | 199.1 | 199.1 | 199.1 | 199.1 | 199.1 | 199.1 |
| $\frac{M_f}{d_v} + 0.5N_f + \cot\theta\sqrt{(V_f - 0.5V_{sg} - \phi_p V_p)^2 + \left(\frac{0.45 p_h T_f}{2A_o}\right)^2}$ | kN | 186.1 | 186.1 | 397.3 | 397.3 | 525.5 | 311.5 | 354 |
| $\phi_s A_s f_y + \phi_p A_p f_{pr}$ | kN | 563 | 563 | 932 | 932 | 1041 | 1041 | 1041 |

$$V_{sg} = \phi_s \frac{A_v}{s} f_y d_v \cot\theta$$
$$= (0.85)\frac{(200)}{600}(400)(1600)(\cot 38^\circ)$$
$$= 199.1\,\text{kN}$$

The reinforcement required for shear and torsion is summarized in Fig. 3.7.10. Additional reinforcing may be required for the following:

(a) end bearing reactions

(b) end torsional equilibrium reactions

(c) ledge point loads

(d) hanger steel at ledge point loads

(e) horizontal reinforcing

(f) anchorage of longitudinal bars at ends

The quantity of reinforcement should be calculated by the Strut-and-Tie modeling methods outlined in Chapter 4.

*Note:* In Fig. 3.7.10, the stirrup spacing of 215 mm and 600 mm would require longitudinal 15M and 45 M corner bars respectively. The maximum spacing for the 15M longitudinal bars shown is 240 mm. Some adjustment is required of either or both the longitudinal reinforcement and the transverse reinforcement after reinforcement (a) to (d) has been considered.

## 3.8 COMPRESSION

Precast and prestressed concrete columns and load-bearing wall panels are usually proportioned to resist factored loads. Stresses under service conditions, particularly during handling and erection (especially for wall panels) must be considered. The procedures in this section are based on the Code and on the recommendations of the PCI Committee on Prestressed Concrete Columns[R3.8.1] (Referred to in this section as "the Recommended Practice").

### 3.8.1 FACTORED RESISTANCE

The factored resistance of a reinforced concrete compression element with eccentric loads is most easily determined by constructing an interaction curve. Points on this curve are calculated using the compatibility of strains and solving equations of equilibrium as prescribed in the Code. The solution of these equations is illustrated in Fig. 3.8.1.

**Fig. 3.7.10 Reinforcement for Example 2**

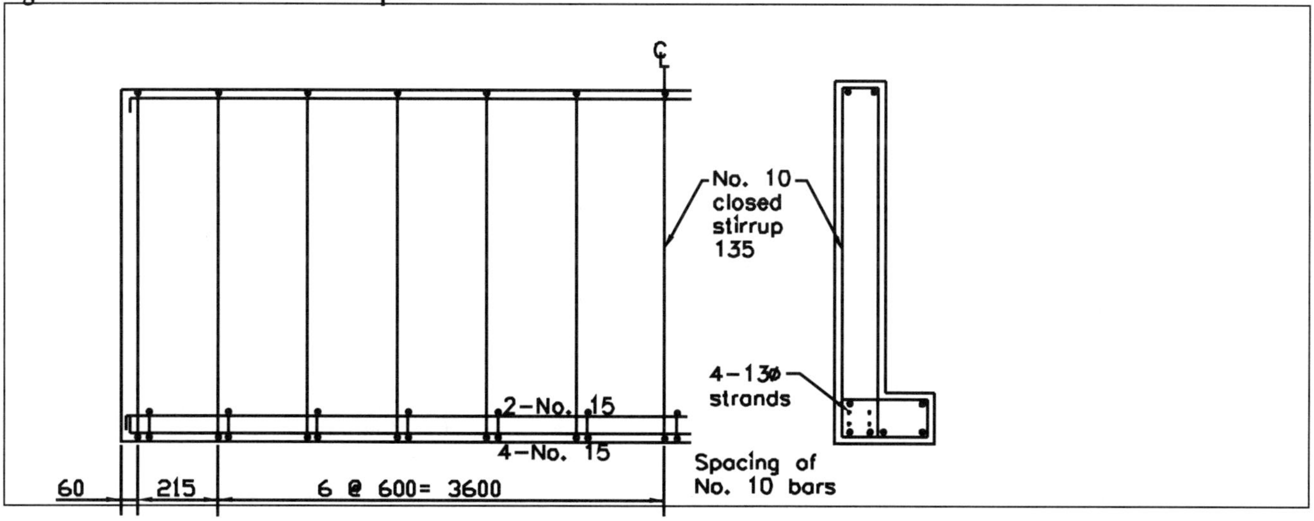

**Fig. 3.8.1 Design relationships for prestressed and reinforced compression elements**

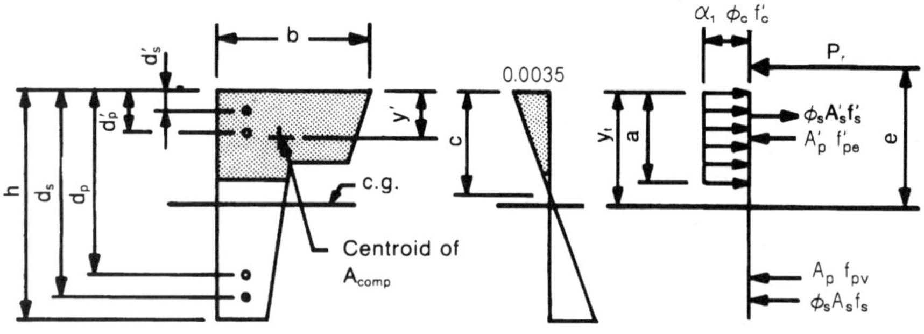


$$\epsilon'_s = \frac{0.0035}{c}(c - d'_s) \qquad \text{(Comp.)}$$
$$\epsilon_s = \frac{0.0035}{c}(d_s - c) \qquad \text{(Ten.)}$$
$$\epsilon'_{pv} = \frac{f_{pe}}{E_p} + \frac{f_{cp}}{E_c} - \frac{0.0035}{c}(c - d'_p) < 0.035 \qquad \text{(Ten.)}$$
$$\epsilon_{pv} = \frac{f_{pe}}{E_p} + \frac{f_{cp}}{E_c} + \frac{0.0035}{c}(d_p - c) < 0.035 \qquad \text{(Ten.)}$$
$$f'_s = \epsilon'_s E_s \leq f_y \qquad \text{(Comp.)}$$
$$f_s = \epsilon_s E_s \leq f_y \qquad \text{(Ten.)}$$

$f'_{pr}$ from stress-strain diagram $< f_{pu}$
$f_{pr}$ from stress-strain diagram $< f_{pu}$

$$P_r = \phi_c A_{comp}(\alpha_1 f'_c) + (\phi_s f'_s - \alpha_1 \phi_c f'_c)A'_s - (\phi_p f'_{pr} + \alpha_1 \phi_c f'_c)A'_p - \phi_s A_s f_s - A_p f_{pr}$$
$$M_r = \phi_c A_{comp}(\alpha_1 f'_c)(y_t + y') + (\phi_s f'_s - \alpha_1 \phi_c f'_c)A'_s(y_t - d'_s) - (f'_{pr} + \alpha_1 \phi_c f'_c)A'_p(y_t - d'_p) + \phi_s A_s f_s(d_s - y_t) + A_p f_{pr}(d_p - y_t)$$

**(a) Basic relationships**

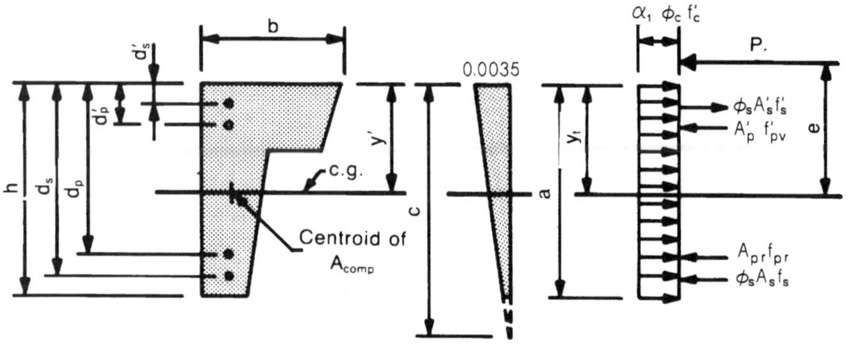


$A_{comp} \approx A$ if $a > h$

$$\epsilon'_s = \frac{0.0035}{c}(c - d'_s) \qquad \text{(Comp.)}$$
$$\epsilon_s = \frac{0.0035}{c}(c - d_s) \qquad \text{(Comp.)}$$
$$\epsilon'_{pr} = \frac{f_{pe}}{E_p} + \frac{f_{cp}}{E_c} - \frac{0.0035}{C}(c - d'_p) < 0.035 \qquad \text{(Ten.)}$$
$$\epsilon_{pr} = \frac{f_{pe}}{E_p} + \frac{f_{cp}}{E_c} - \frac{0.003}{C}(c - d_p) < 0.035 \qquad \text{(Ten.)}$$

Remaining equations same as above.

**(b) Special case with neutral axis outside of the section**

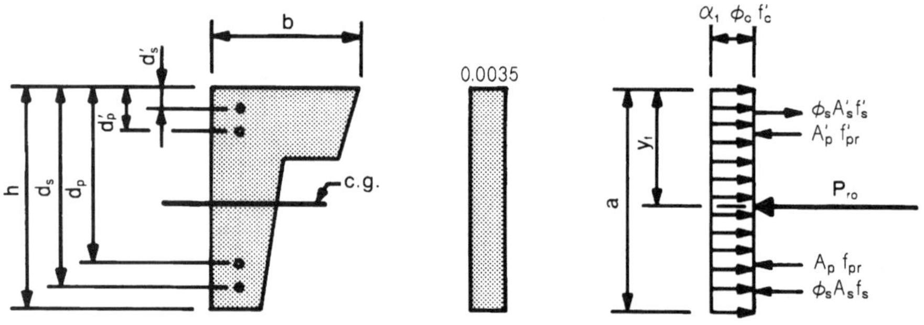


$$P_{ro} = \phi_c(A - A'_s - A'_p - A_p - A_s)(\alpha_1 f'_c) - \phi_p(A'_p + A_p)\left(f_{pe} + \frac{f_{cp}E_p}{E_c} - 0.0035 E_p\right) + \phi_s(A'_s + A_s) f_y$$

(Multiply by 0.85 if no ties are used)

**(c) Special case when $M_r = O$, $P_r = P_{ro}$**

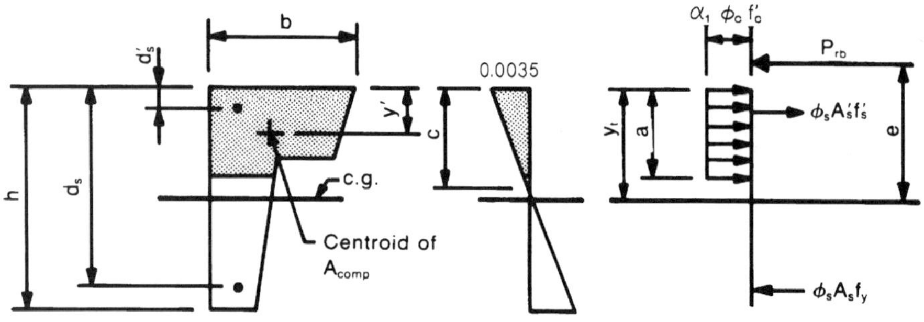


$$c = \frac{0.0035\ d_s}{0.0035 + f_y/E_s} = \frac{700\ d_s}{700 + f_y}$$
$$f'_s = E_s\left[\frac{0.0035}{c}(c - d'_s)\right] \leq f_y$$
$$= \frac{700\ (c - d_s)}{c} \leq f_y$$

$$P_{rb} = \phi_s A_{comp}(\alpha_1 f'_c) + (\phi_s f'_s - \alpha_1 \phi_c f'_c) A'_s - \phi_s A_s f_y$$
$$M_{rb} = P_{rb} e = \phi_c A_{comp}(\alpha_1 f'_c)(y_t - y') + (\phi_s f'_s - \alpha_1 \phi_c f'_c) A'_s (y_t - d'_s) + \phi_s A_s f_y (d_s - y_t)$$

**(d) Special case at balance point for non-prestressed elements**

The Code waives the lateral reinforcement requirement for walls if the concrete is prestressed to at least an average of 1.5 MPa after all losses (Clause 18.12.2.2). In addition, the Recommended Practice permits the elimination of lateral reinforcement in flat walls, if the factored resistance is multiplied by 0.85 (Clause 12.4.4).

Interaction curves for typical prestressed square columns and wall panels are provided in Chapter 7.

Construction of an interaction curve usually follows these steps:

*Step 1:*

Determine $P_r$ for $M_r = 0$ (see Fig. 3.8.1(c)).

*Step 2:*

Determine $M_r$ for $P_r = 0$. This is normally done by neglecting the reinforcement above the neutral axis and determining the moment capacity by one of the methods described in Section 3.3.

*Step 3:*

For reinforced columns, $P_b$ and $M_b$ at the balance point may be determined (see Fig. 3.8.1(d)). For prestressed columns, the yield point of the prestressed reinforcement is not well defined and the stress strain relationship is non-linear over a broad range (see Fig. 8.2.4).

*Step 4:*

For each additional point on the interaction curve, proceed as follows:

a) Select a value of "c" and calculate $a = \beta_1 c$.

b) Determine the value of and the centroid of $A_{comp}$ from the geometry of the section (shaded portion in Fig. 3.8.1(a)).

c) Determine the strain in the reinforcement assuming that $\in = 0.0035$ at the compression face of the column. For prestressed reinforcement, add the strain due to the prestress, corresponding to zero strain in the concrete:

$$\epsilon_{pe} - \epsilon_{ce} = \frac{f_{pe}}{E_p} - \frac{f_{cp}}{E_c}$$

(tensile strain is called positive here).

d) Determine the stress in the reinforcement. For non-prestressed reinforcement, $f_s = \epsilon_s E_s < f_y$. For prestressed reinforcement, the stress is determined by stress-strain relationship (see Fig. 8.2.4). When these stresses are used to find forces, the stress times the area is multiplied by the appropriate $\phi$-factor to account for uncertainties in the dimensions and properties of the materials. However the $\phi_p$ factor is not applied to prestressed reinforcement as this generally adds axial load to the section and should be unreduced. If the maximum factored moment occurs near the end of a prestressed element where the strand is not fully developed, an appropriate reduction in the value of $f_{pr}$ should be made as described in Sect. 3.8.2, Fig. 3.8.2.

e) Calculate $P_r$ and $M_r$ by statics.

*Step 5:*

Calculate the maximum factored axial resistance, specified by the Code as 0.80 $P_{ro}$ for tied columns and 0.85 $P_{ro}$ for spiral columns.

**Fig. 3.8.2 End reinforcement in a precast, prestressed concrete column**

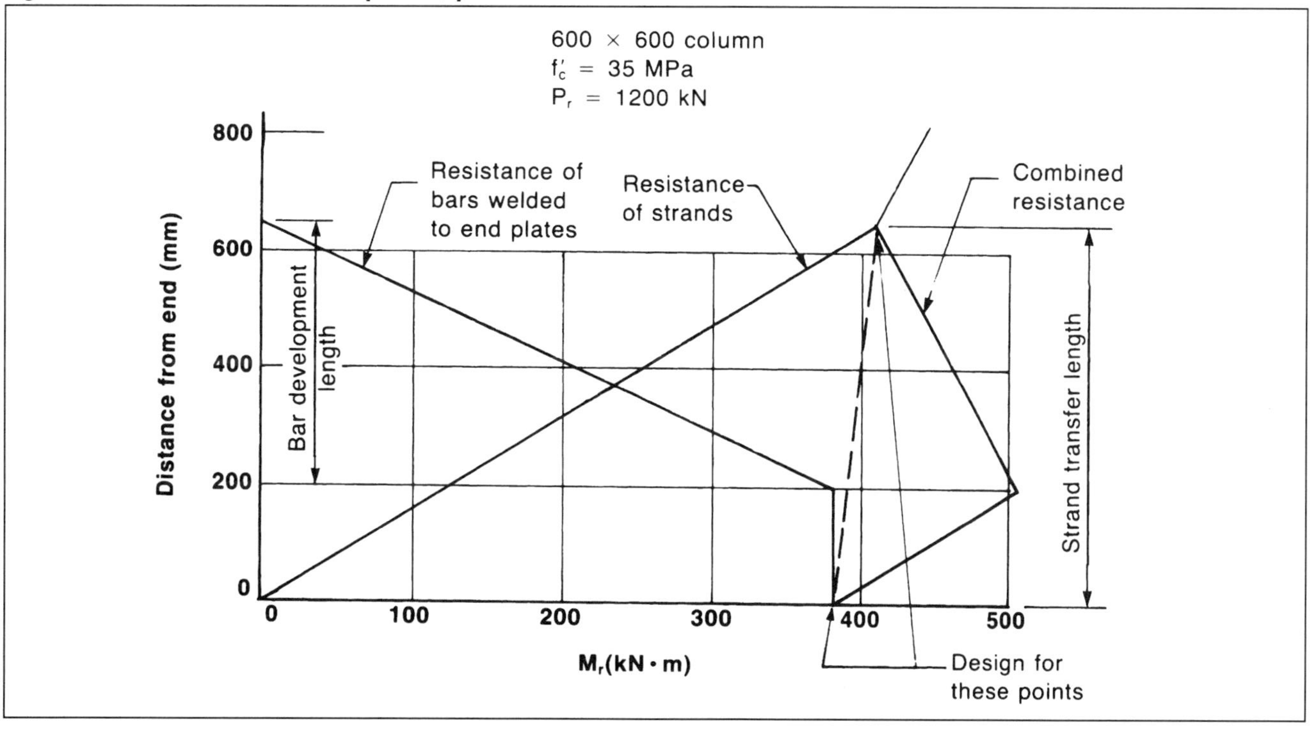

## 3.8.2 DEVELOPMENT LENGTH

As noted in Step 4(d) above, the factored flexural resistance is reduced for prestressed elements at locations within a distance equal to the strand development length from each end. The factored flexural resistance of the prestressed reinforcement in this zone can be supplemented by non-prestressed reinforcement that is anchored to end plates, or otherwise developed.

The interaction curves for prestressed columns in Chap. 7 are based on a maximum value of $f_{pr} = f_{pe}$, which is valid at a development length equal to the assumed transfer length from the end of the element. The required area of end reinforcement can be determined by matching interaction curves, or can be approximated by the following equation, if the bar locations approximately match the strand locations:

$$A_s = \frac{A_p f_{pe}}{f_y}$$

where:

$A_s$ = required area of bars

$f_{pe}$ = strand stress after losses

$f_y$ = yield strength of bars

The effects of adding end reinforcement to a 600 x 600 mm prestressed concrete column to improve flexural resistance in the end 600 mm are shown in Fig. 3.8.2.

**Example 1 - Construction of interaction curve for a reinforced column produced in a CSA certified plant**

*Given:*

300 × 500 column cross-section shown.

Concrete:

$f'_c = 35$ MPa $\quad \alpha_1 = 0.80 \quad \beta_1 = 0.88 \quad \phi_c = 0.65$

Normal Density

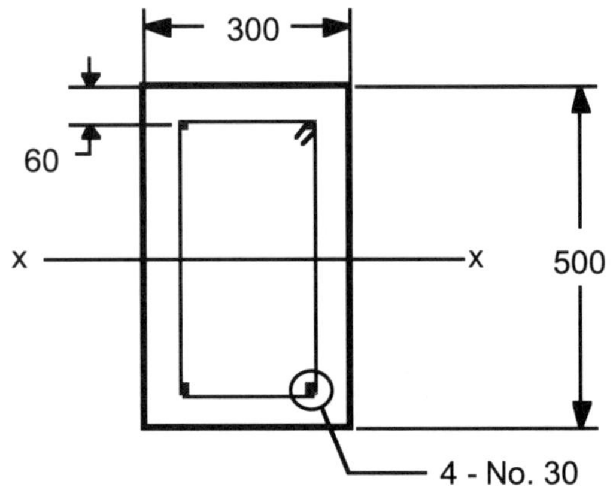


Reinforcement:

$f_y$ = 400 MPa

$E_s$ = 200 000 MPa

*Problem:*

Construct interaction curve for bending about x-x axis.

*Solution:*

Determine following parameters:

$d = 500 - 60 = 440$ mm

$d' = 60$ mm

$y_t = 250$ mm

$A_g = (300)(500) = 150\,000\ \text{mm}^2$

$A_s = A'_s = (2)(700) = 1400\ \text{mm}^2$

1. Determine $P_{ro}$ from Fig. 3.8.1(c):

$$P_{ro} = (\alpha_1 \phi_c f'_c)(A - A'_s - A_s) + \phi_s (A'_s + A_s) f_y$$
$$= [(0.8)(0.65)(35)(150\,000 - 2800) + (0.85)(2800)(400)] / 10^3$$
$$= 3631\ \text{kN}$$

2. Determine $M_{ro}$ - use conservative solution neglecting compressive reinforcement:

$$a = \frac{\phi_s A_s f_y}{\alpha_1 \phi_c f'_c b} = \frac{(0.85)(1400)(400)}{(0.8)(0.65)(35)(300)} = 87\ \text{mm}$$

$$M_{ro} = \phi_s A_s f_y \left(d - \frac{a}{2}\right)$$
$$= (0.85)(1400)(400)\left(440 - \frac{87}{2}\right) / 10^6$$
$$= 189\ \text{kN}\cdot\text{m}$$

3. Determine $P_{rb}$ and $M_{rb}$ from Fig. 3.8.1(d):

$$c = \frac{(0.0035)(440)}{0.0035 + 400/200\,000} = 280\ \text{mm}$$

$$f'_s = (200\,000)\frac{(0.0035)}{280}(280 - 60) = 550\ \text{MPa} > f_y$$

therefore:

$$f'_s = f_y = 400\ \text{MPa}$$

$$A_{comp} = ab = \beta_1 cb = (0.88)(280)(300)$$
$$= 73\,920\ \text{mm}^2$$

$$y' = \frac{a}{2} = \frac{(0.88)(280)}{2} = 123\ \text{mm}$$

$$P_{rb} = (0.8)(0.65)(35)(73\,920)/10^3 + [(0.85)(400) - (0.8)(0.65)(35)](1400)/10^3 - (0.85)(1400)(400)/10^3 = 1320 \text{ kN}$$

$$M_{rb} = (0.8)(0.65)(35)(73\,920)(250-123)/10^6 + [(0.85)(400) - (0.8)(0.65)(35)](1400)(250-60)/10^6 + (0.85)(1400)(400)(440-250)/10^6 = 347 \text{ kN·m}$$

4. To determine intermediate points on the curve:

a) Set a = 150 mm, $c = \dfrac{150}{0.88} = 170$ mm

b) $A_{comp} = (150)(300) = 45\,000 \text{ mm}^2$

c) Use Fig. 3.8.1(a):

$$\epsilon'_s = \frac{(0.0035)(170-60)}{170} = 0.0023$$

$$\epsilon_s = \frac{(0.0035)(440-170)}{170} = 0.0056$$

d) $f'_s = (0.0023)(200\,000) = 460 \text{ MPa} > f_y$

$f_s = (0.0056)(200\,000) = 1120 \text{ MPa} > f_y$

Use $f_s = f'_s = f_y = 400$ MPa

e) $$P_r = (0.8)(0.65)(35)(45\,000)/10^3 + [(0.85)(400) - (0.8)(0.65)(35)](1400)/10^3 - (0.85)(1400)(400)/10^3 = 794 \text{ kN}$$

$$M_r = (0.8)(0.65)(35)(45\,000)(250-75)/10^6 + [(0.85)(400) - (0.8)(0.65)(35)](1400)(250-60)/10^6 + (0.85)(1400)(400)(440-250)/10^6 = 319 \text{ kN·m}$$

Note: Step 4 can be repeated for as many points as desired. A plot of such points is shown in Fig. 3.8.3.

5. Maximum axial load:

$= 0.8P_{ro} = (0.8)(3631) = 2905$ kN

**Fig. 3.8.3 Interaction curve for Example 1**

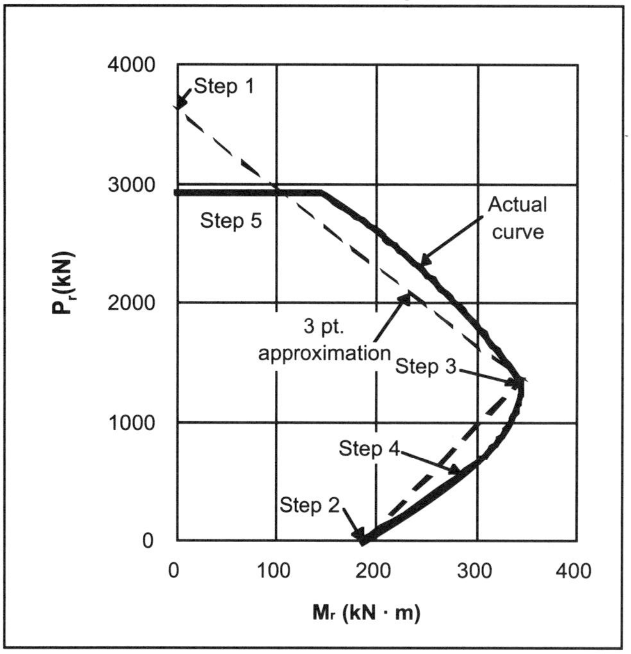


**Example 2 - Calculation of interaction points for pre-stressed concrete compression element**

*Given:*

1220 × 203 hollow-core wall panel shown.

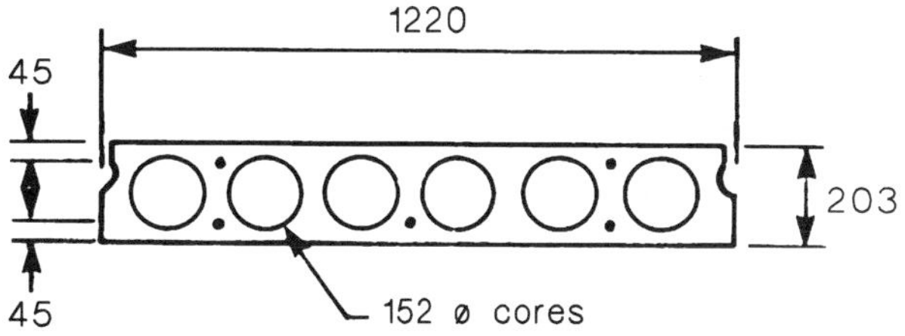


Concrete:

$f'_c = 40$ MPa $\alpha_1 = 0.79$ $\beta_1 = 0.87$ $\phi_c = 0.65$

Normal Density

$A_g = 138\,700 \text{ mm}^2$

$E_c = 29\,600$ MPa

Prestressed reinforcement:

$f_{pu} = 1860$ MPa

$E_p = 190\,000$ MPa

$f_{pe} = 1030$ MPa

5-13 mm 1860 MPa strands

$A_p = (3)(99) = 297 \text{ mm}^2$

$A'_p = (2)(99) = 198 \text{ mm}^2$

$$f_{cp} = -\frac{(1030)(297+198)}{138\,700}$$

$$\approx 3.7 \text{ MPa (not quite uniform)}$$

*Problem:*

Calculate a point on the design interaction curve for $a = 50$ mm.

*Solution:*

*Step a:*

$$a = 50 \text{ mm}$$

$$c = \frac{50}{0.87} = 58 \text{ mm}$$

*Step b:*

From Fig. 8.3, calculate area and centroid of core segments excluded from compression zone (Fig. 3.8.11 can be used):

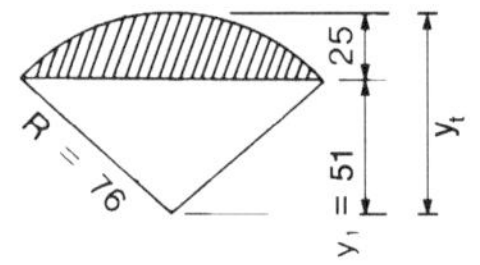


$$A = \frac{\pi R^2}{2} - y_1\sqrt{R^2 - y_1^2} - R^2 \sin^{-1}\left(\frac{y_1}{R}\right)$$

$$= \frac{(\pi)(76)^2}{2} - 51\sqrt{76^2 - 51^2} - 76^2 \sin^{-1}\left(\frac{51}{76}\right)$$

$$= 1950 \text{mm}^2$$

$$\bar{y} = \frac{2(R^2 - y_1^2)^{3/2}}{3A} = \frac{(2)(76^2 - 51^2)^{3/2}}{(3)(1908)} = 61 \text{mm}$$

$$A_{comp} = (1220)(50) - (6)(1950) = 49\,300 \text{ mm}^2$$

$$y' = \frac{(1220)(52^2/2) - (6)(1950)(41)}{49\,300}$$

$$= 22 \text{ mm from top}$$

*Step c:*

Prestrain at zero concrete stress:

$$= \frac{1030}{190\,000} + \frac{3.7}{29\,600} = 0.00555$$

From Fig. 3.8.1(a):

$$\epsilon'_p = 0.00555 - \frac{(0.0035)(58-45)}{58} = 0.00477$$

$$\epsilon_p = 0.00555 + \frac{(0.0035)(158-58)}{58} = 0.01158$$

*Step d:*

From Fig. 8.2.4, $\epsilon'_p$ is on the linear portion of the curve and $\epsilon_p$ is on the non-linear portion:

$$f'_{pr} = \epsilon'_p E_p = (0.00477)(190\,000) = 906 \text{ MPa}$$

$$f_{pr} = 1703 \text{ MPa}$$

*Step e:*

From Fig. 3.8.1(a):

$$P_r = (0.79)(0.65)(40)(49\,512)/10^3$$
$$-(198)(906)/10^3 - (297)(1703)/10^3$$
$$= 1017 - 179 - 506 = 332 \text{ kN}$$

(The concrete area displaced by the compression-side tendons is ignored here.)

$$M_r = (1017)(102-22)/10^3$$
$$-(179)(102-45)/10^3$$
$$+(506)(158-102)/10^3$$
$$= 81 - 10 + 28 = 99 \text{ kN·m}$$

Since no lateral ties are used in this element, the values should be multiplied by 0.85:

$$P_r = (0.85)(332) = 282 \text{ kN}$$

$$M_r = (0.85)(99) = 84 \text{ kN·m}$$

Note that this is for *fully developed* strand. If the capacity at a point near the end of the transfer zone is desired, then $f_{pr} \le f_{pe}$ = 1030 MPa. Take $f_{pr}$ = 1030 and adjust the prestress contribution accordingly:

$$P_r = (0.85)[1017 - 179 - (297)(1030)/10^3]$$
$$= 452 \text{ kN}$$

$$M_r = (0.85)[81 - 10$$
$$+(297)(1030)(158-102)/10^6]$$
$$= 75 \text{ kN·m}$$

Note: In prestressed wall panels, the effects of unsymmetrical prestress should also be investigated.

### 3.8.3 SLENDERNESS EFFECTS

For compression elements braced against sideway, effects of slenderness may be neglected when $k\ell_u/r$ is less than $\frac{25 - 10(M_1/M_2)}{\sqrt{P_f/(f'_c A_g)}}$. Observe that $M_1/M_2$ is taken to be positive if the element is bent in single curvature, negative if it is bent in double curvature and is not taken less than -0.5. $M_2$ is the larger and $M_1$ is the smaller end moment.

The radius of gyration, r, may be taken as 0.3 times the overall dimension in the direction that stability is being considered for rectangular elements, 0.25 times the diameter for circular elements, or computed from the gross concrete section for other shapes:

**Fig. 3.8.4 Slenderness effects**

**Notation**

- $\ell_u$ = unsupported length of compression element
- r = radius of gyration
- k = effective length factor
- $M_1$ = smaller factored end moment, positive if bent in single curvature, negative if bent in double curvature
- $M_2$ = larger factored end moment always positive (if less than minimum Code values, use Code values)
- $M_c$ = moment to be used in design = $\delta_b M_2$
- $I_g$ = gross moment of inertia
- $P_c$ = critical load

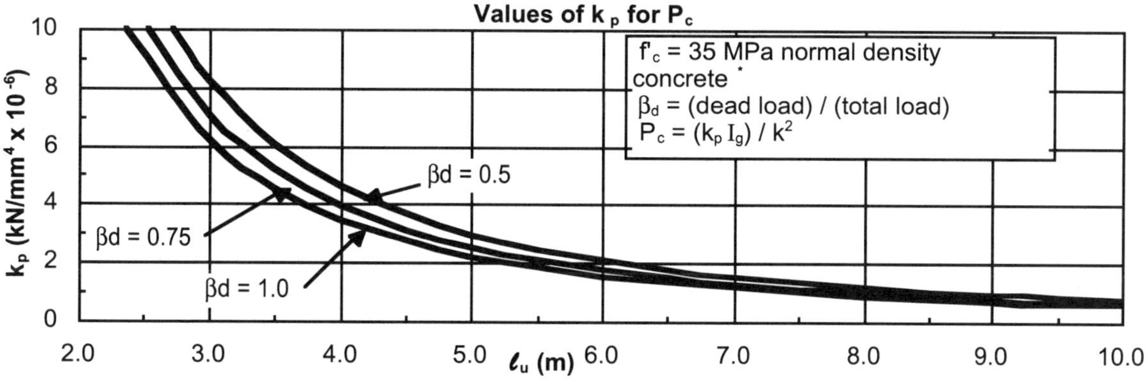


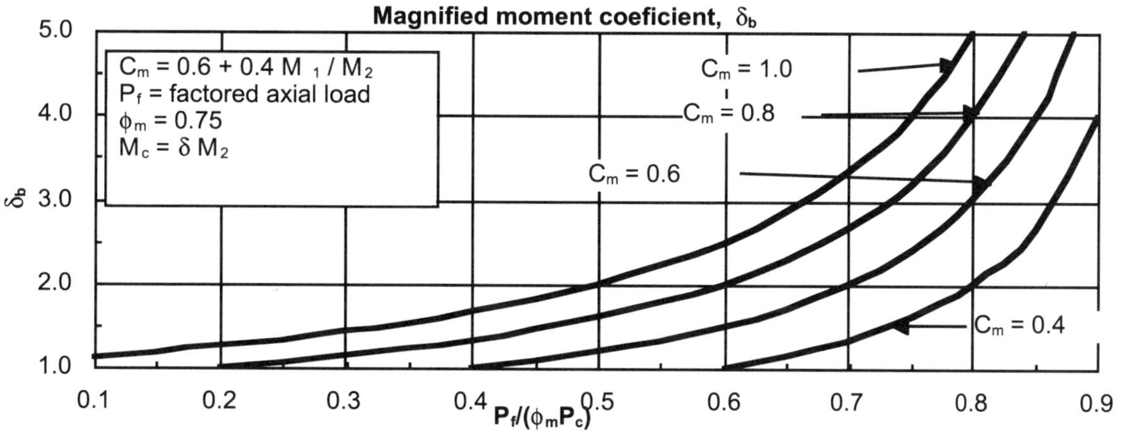


* For other concretes, $P_c = \frac{k_p I_g}{k^2}\left(\frac{E_c}{28\,200}\right)$, $E_c$ = modulus of elasticity, MPa

$$r = \sqrt{\frac{I}{A_g}}$$

The Code contains provisions for evaluating slenderness effects (buckling) of compression elements based on the following three approaches:

1. The approximate magnified moment approach.
2. Empirical design for certain load bearing walls.
3. A more rigorous PΔ approach as described in Section 2.6.7, where the first order lateral storey deflection is computed using an elastic first order frame analysis with the section properties calculated taking into account the influence of axial loads, the presence of cracked regions along the length of the member, the member curvature and effects of duration of the loads.

Alternatively, the following properties may be used for the members in the structure:

a) Modulus of elasticity

$$= E_c = (3300\sqrt{f'_c} + 6900)\left(\frac{\gamma_c}{2300}\right)^{1.5}$$

b) Moment of inertia

| Type of element | Moment of inertia |
|---|---|
| Beams | $0.35I_g$ |
| Columns | $0.70I_g$ |
| Walls | |
| -Uncracked | $0.70I_g$ |
| -Cracked | $0.35I_g$ |
| Flat plates and flat slabs | $0.25I_g$ |

c) Area $= A_g$

## Magnified Moment Approach

The principles of the moment magnifier method are illustrated in Fig. 3.8.5. The magnified moment approach is valid for compression members with $k\ell_u / r$ not greater than 100.

The approximate evaluation of slenderness effects in this Manual involves:

**Fig. 3.8.5 Slenderness effects**

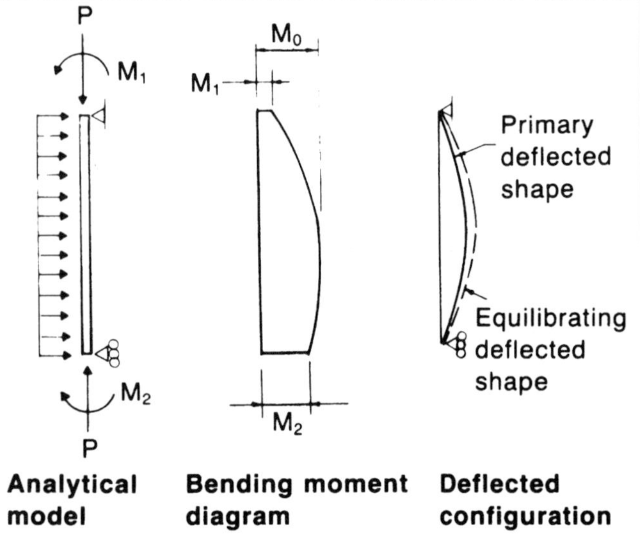


where:

$M_1$ = value of smaller design end moment on compression member calculated from a conventional elastic analysis - positive if member is bent in single curvature, otherwise negative

$M_2$ = value of larger design end moment, always positive

$M_o$ = first order bending moment at critical section of compression members as obtained from elastic analysis

1. The determination of the effective length factor, k, and end moments, $M_1$ and $M_2$, as described in Section 2.6.7.
2. The calculation of the moment magnifier, $\delta_b$, as shown in the Example 3 on Page 3-91.
3. The magnified moment in combination with the factored axial load is compared with the resistance of the element as determined by an interaction diagram in accordance with Section 3.8.1.

### Hinged Columns and Columns in Braced Frames

The magnified moment is taken as:

$$M_c = \delta_b M_2 = \frac{C_m M_2}{1 - \dfrac{P_f}{\phi_m P_c}}$$

where $\phi_m = 0.75$ and $C_m M_2$ shall not be taken as less than $P_f(15 + 0.03\ h)$ about each axis separately.

For members without transverse loads between supports, $C_m$ shall be taken as

$$C_m = 0.6 + 0.4\frac{M_1}{M_2} \geq 0.4$$

When there are lateral loads, $C_m = 1.0$.

$P_f$ is the factored axial load, and $P_c$ is the critical buckling load, calculated as:

$$P_c = \frac{\pi^2 EI}{(k\ell_u)^2}$$

where $\ell_u$ is the unsupported height of the element.

The effective length factor, k, for pin ended, laterally restrained elements with various edge conditions can be determined from Fig. 3.8.6 as follows:

1. For columns or panels braced against sidesway and free along both the vertical edges:

   For all values of $\ell_u / b$ $\quad k = 1.0$

2. For panels braced against sidesway and restrained along both vertical edges:

   $\ell_u / b < 1/2 \qquad k = 1.0$

   $1/2 \leq \ell_u / b \leq 1 \qquad k = 1.5 - \ell_u / b$

   $\ell_u / b > 1 \qquad k = 1.0/[1 + (\ell_u / b)^2]$

3. For panels braced against sidesway and restrained along one vertical edge:

   $\ell_u / b < 1 \qquad k = 1.0$

   $1/2 \leq \ell_u / b \leq 2 \qquad k = 1.0 - 0.423[(\ell_u / b) - 1]$

   $\ell_u / b > 2 \qquad k = 1.0 / \sqrt{1 + (\ell_u / b)^2 / 2}$

The value EI may be determined from the following:

1. For columns or panels with a double layer of reinforcement, the value of EI may be taken as:

**Fig. 3.8.6 Effective length factor k**

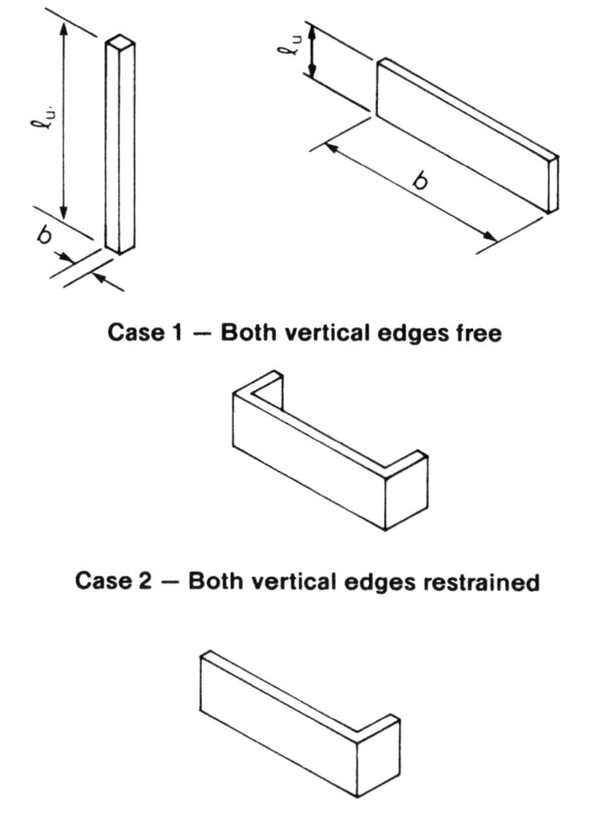

Case 1 — Both vertical edges free

Case 2 — Both vertical edges restrained

Case 3 — One vertical edge restrained

$$EI = \frac{0.2E_cI_g + E_sI_{st}}{1+\beta_d}$$

Approximate values for lightly reinforced elements are:

$$EI = 0.25E_cI_g$$

or:

$$EI = \frac{0.4E_cI_g}{1+\beta_d}$$

2. For wall panels with a single layer of reinforcement centered in the panel, the value of EI may be taken as[R3.8.2]:

$$EI = \frac{E_cI_g}{\beta_d}\left(0.5 - \frac{e}{h}\right) \geq \frac{0.10\ E_cI_g}{\beta_d}$$

3. The above equations were developed for reinforced concrete columns with at least 1% reinforcement; since most prestressed compression elements and precast load bearing wall panels have much less reinforcement than this, modified equations for EI are recommended. Fig. 3.8.7 presents one set of such equations. Others have also been used. Such approximations are necessarily conservative, and therefore second order analysis procedures are strongly recommended for slender prestressed elements.

4.

(a) For non-sway frames and for stability checks of sway frames in accordance with Code Clause 16.16.5, $\beta_d$ is the ratio of the maximum factored dead load to the total factored axial load.

(b) For sway frames [except as noted in (a)], $\beta_d$ is the ratio of the maximum factored sustained shear within a storey to the maximum total factored shear in that storey.

For wall panels subject to transverse loading, the maximum moment can occur at a section away from the end of the element. In this case, the value of the largest calculated moment occurring anywhere along the element is used for the value of $M_2$. $C_m$ is taken as 1.0 for this case.

If computations show that there is no moment at either end of a compression element or that computed end eccentricities are less than 15 + 0.03 h (mm), $M_2$ should be based on a minimum eccentricity of 15 + 0.03 h, where h is the thickness of the panel or column. The ratio $M_1/M_2$ should be determined by either of the following:

1. When computed end eccentricities are less than 15 + 0.03 h, computed end moments may be used to evaluate $M_1/M_2$.
2. If computations show that there is essentially no moment at either end of a compression element, the ratio $M_1/M_2$ may be taken equal to one.

For elements with $k\ell_u/r > 100$, the design of compression elements shall be based on the factored forces and moments from a second order analysis considering material nonlinearity and cracking, effects of member curvature and lateral drift, duration of the loads, shrinkage and creep, and interaction with the supporting foundation.

### Columns in Sway Frames

Stories in structures shall be designated as sway frames if:

$$Q = \frac{\Sigma P_f \Delta_o}{V_f \ell c} > 0.05$$

where $\Sigma P_f$ and $V_f$ are the total factored vertical load and the factored storey shear in the storey in question and $\Delta_o$ is the first order relative deflection of the top and bottom of that storey due to $V_f$.

The moments from a first order analysis are divided into two parts: $M_{ns}$ due to gravity loads, which do not contribute to sway, and $M_s$, the moments associated with sway. Gravity load moments causing deflections greater than $\ell_u$ / 1500, due to lack of symmetry, should be classed as $M_s$ moments.

The $M_s$ moments are magnified by $\delta_s$, given by:

$$\delta_s = \frac{1}{1 - \dfrac{\Sigma P_f}{\phi_m \Sigma P_c}}$$

where the summations are over all the columns of the storey, and the other quantities are as described in the previous section for braced columns. The critical load, $P_c$, is based on the unbraced effective length, which may be determined from Fig. 2.6.15 and E and I values in accordance with Standard Clause 10.14.1. For moments associated with sway, $\beta_d$ is calculated in

**Fig. 3.8.7 Coefficients, λ, for modified EI**

$\eta = 2.5 + \dfrac{1.6}{P_f / P_o}$

$6 \le \eta \le 70$

$\lambda = \eta\theta$

For $P_c$ equation:

$EI = \dfrac{E_c I_c / \lambda}{1 + \beta_d}$

**(a) Compression flange**

$\theta = \dfrac{35}{k\ell_u / r} - 0.09$

**(b) No compression flange**

$\theta = \dfrac{27}{k\ell_u / r} - 0.05$

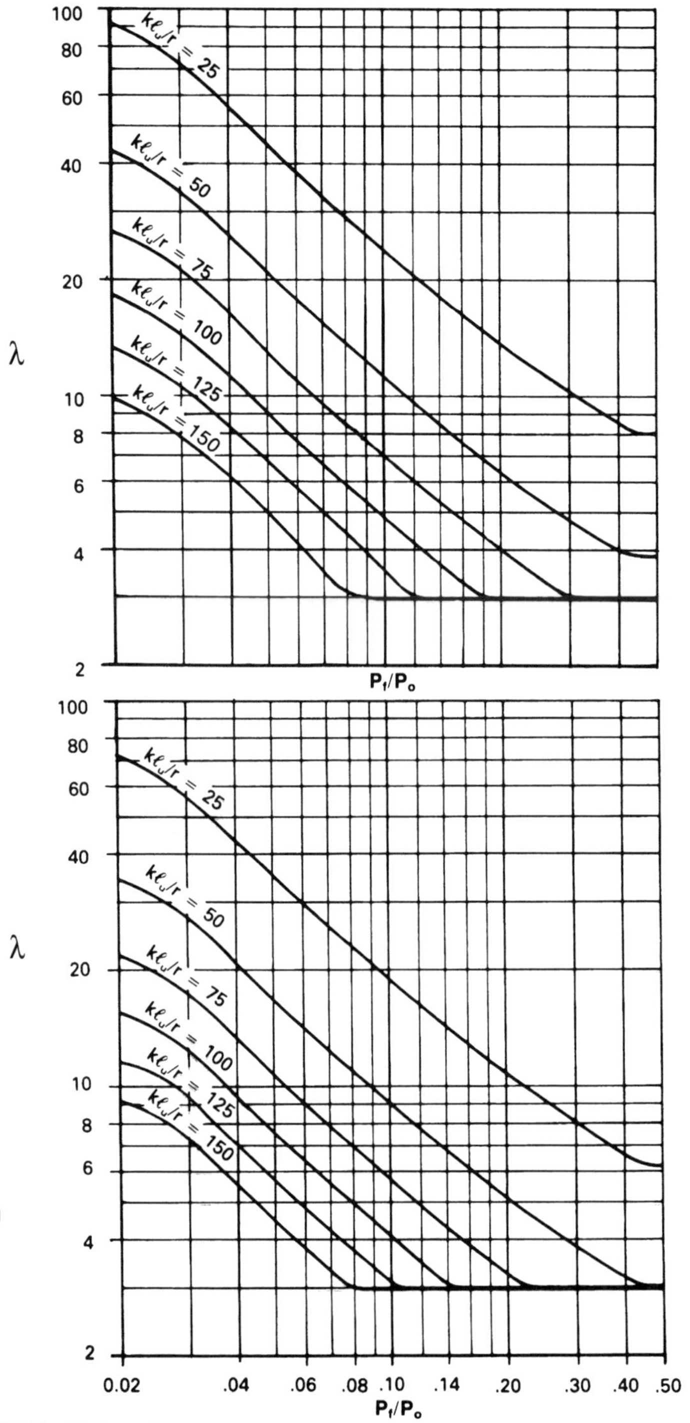


accordance with 4(b). If the stability index of a storey, Q, is less than 1/3, then $\delta_s$ can be computed as:

$$\delta_s = \frac{1}{1 - Q}$$

The final end moments are then computed from:

$$M_{ns} + \delta_s M_s$$

and the design moment is:

$$M_c = M_2$$

where $M_2$ is the value of $(M_{ns} + \delta_s M_s)$ at the top or bottom of the column, whichever is greater.

In addition to the above requirements for columns in sway frames, individual compression members which have $\dfrac{\ell_u}{r} > \dfrac{35}{\sqrt{P_f / (f'_c A_g)}}$ shall also be designed for:

$$M_c = \frac{C_m M_2}{1 - \dfrac{P_f}{\phi_m P_c}}$$

where:

$$M_2 = M_{2ns} + \delta_s M_{2s}$$

and $\delta_s$ is calculated as above except $\beta_d$ according to 4(a) is used and $k \le 1.0$.

In addition to load cases involving lateral loads, the strength and stability of the structure as a whole should be considered (see Code Clause 10.16.5).

Alternative methods of accounting for the magnification of sway moments are given in Section 2.6.7.

## Empirical Design of Wall Panels

If the resultant of all factored axial loads including the effects (moments and sway) of lateral loads is located within the kern dimension of the panel (i.e., middle third of the thickness of a flat panel), the strength of the wall may be approximated as:

$$P_r = \frac{2}{3}\alpha_1 \phi_c f'_c A_g \left[1 - \left(\frac{k\ell_c}{32h}\right)^2\right]$$

where:

$\alpha_1$ = 0.85 - (0.0015)$f'_c \ge 0.67$

$\phi_c$ = 0.65 (for CSA certified products)

$A_g$ = gross area

$\ell_c$ = vertical distance between supports

$h$ = thickness of panel

For panels which are of an irregular cross-section, an equivalent h may be determined from:

$$h = \sqrt[3]{\frac{12I_g}{b}}$$

where:

$I_g$ = moment of inertia based on gross section

$b$ = width of section

**Example 3 - Moment magnifier for a column in a non-sway frame**

*Given:*

Pin ended column in a braced frame bent in single curvature:

| | |
|---|---|
| Column size | = 350 × 350 mm |
| $P_f$ dead load | = 1200 kN |
| $P_f$ live load | = 600 kN |
| Eccentricity at top | = 350 mm |
| Eccentricity at base | = 35 mm |
| $\ell_u$ | = 4800 mm |
| $f'_c$ | = 35 MPa |
| $E_c$ | = 28 200 MPa |

*Problem:*

Determine non-sway moment magnifier $\delta_b$.

*Solution:*

$$M_1 = \frac{(1200+600)(35)}{10^3} = +63 \text{ kN·m}$$

$$M_2 = \frac{(1200+600)(350)}{10^3} = +630 \text{ kN·m}$$

$$\frac{M_1}{M_2} = \frac{+63}{+630} = 0.10$$

Note that this ratio is positive because the element is bent in single curvature.

$$r = (0.3)(350) = 105 \text{ mm}$$

$$\frac{k\ell_u}{r} = \frac{(1.0)(4800)}{105} = 45.7$$

Slenderness may be neglected when $k\ell_u/r$ is less than:

$$\frac{25 - 10(M_1/M_2)}{\sqrt{P_f/(f'_c \cdot A_g)}} = \frac{25-(10)(0.1)}{\sqrt{(1800)(10^3)/(35)(122\,500)}}$$

$$= 37.0$$

Therefore slenderness must be considered.

$$\beta_d = \frac{\text{factored dead load}}{\text{factored total load}} = \frac{1200}{1200+600} = 0.67$$

Note that $\beta_d$ is a factor that takes into account creep due to sustained loads. When the moment to be magnified is caused by short-term loads, such as wind or seismic, $\beta_d$ may be taken as zero. However, if the more conservative, simplified expression for EI is used, an allowance for $\beta_d$ is included in the formula:

$$EI = 0.25E_cI_g = \frac{(0.25)(28\,200)(350)^4}{12}$$

$$= 8.82 \times 10^{12} \text{ kN·mm}^2$$

$$P_c = \frac{\pi^2 EI}{(k\ell_u)^2} = \frac{(\pi^2)(8.82\times10^{12})}{[(1.0)(4.8)]^2(10^9)} = 3778 \text{ kN}$$

$$C_m = 0.6 + 0.4M_1/M_2 = 0.6 + (0.4)(0.1) = 0.64$$

$$\delta_b = \frac{C_m}{1 - \dfrac{P_f}{\phi_m P_c}} = \frac{0.64}{1 - \dfrac{(1200+600)}{(0.75)(3778)}} = 175$$

Fig. 3.8.4 could also be used for this example.

## 3.8.4 EFFECTIVE WIDTH OF WALL PANELS

In lieu of a detailed analysis, the Code suggests the following for effective width (the portion of the wall assumed to resist the loads):

Flat wall panels (solid or hollow-core): length of loaded area plus four times the thickness of the wall on either side (see Fig. 3.8.8).

Ribbed wall panels: Width of the rib at the flange plus four times the flange thickness on either side (See Fig. 3.8.8).

**Fig. 3.8.8 Effective width of wall panels**

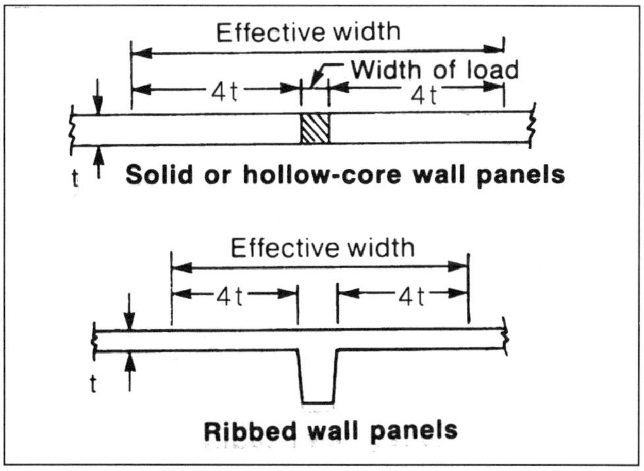


In no case should the effective width exceed the centre-to-centre distance between loads.

Note that these effective widths are for longitudinal compression loads. The effective widths for flexure are discussed in Section 3.3.1.

### 3.8.5 INSULATED LOAD-BEARING WALL PANELS

Insulated wall panels, or sandwich panels of the type shown in Fig. 3.8.9, are usually designed assuming only one wythe as structurally effective, unless it can be shown by experience, test, or calculation that both wythes are fully or partially effective. The method of transmitting the superimposed loads to the panel (corbels, for example) should be carefully detailed to assure that the non-structural elements do not carry loads they are not designed for.

**Fig. 3.8.9 Typical precast concrete load-bearing insulated wall panels**

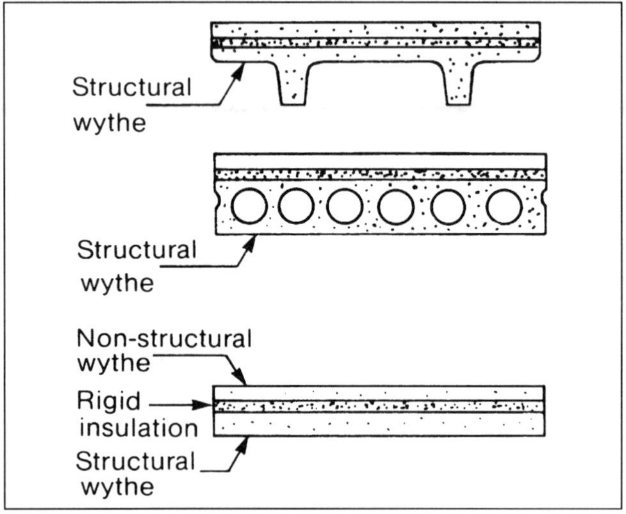


### 3.8.6 NON-UNIFORM MOMENT OF INERTIA

An approximate analysis of non-uniform cross section compression elements can be obtained by computing an equivalent moment of inertia, $I_{equiv}$, which would lead to the same mid-height deflection as the actual non-prismatic element when considered as a simply supported beam subjected to a uniformly distributed lateral load.

Once $I_{equiv}$ has been determined, $P_c$ can be found for the determination of the moment magnifier. The effective length factor, k, typically would be selected as 1 for most usual design conditions for slenderness about the x-x axis. In certain design cases, slenderness design about the y-y axis may control. The following design example illustrates a typical calculation for slenderness effects in the centre mullion of a load bearing panel.

**Example 4 - Varying I for slenderness design**

*Given:*

The load bearing panel shown in Fig. 3.8.10.

**Fig. 3.8.10 Varying I Example 4**

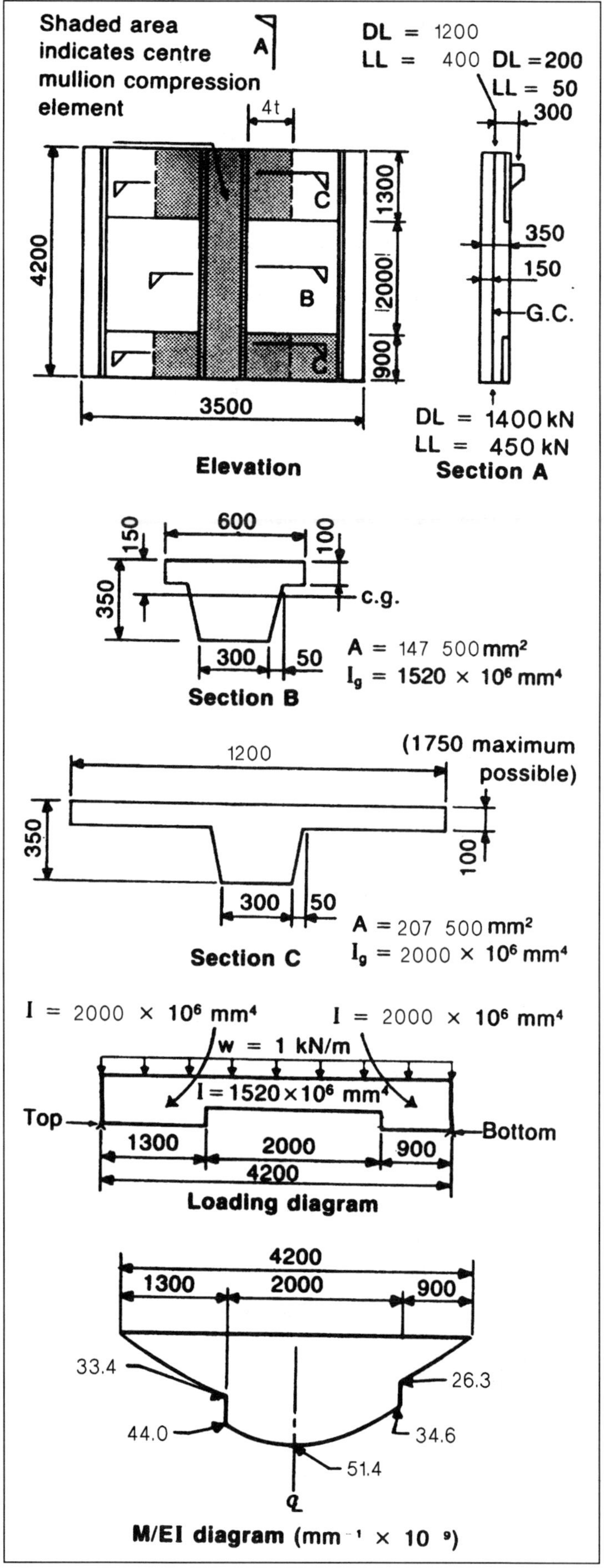

*Concrete:*

$f'_c$ = 35 MPa

Normal Density

$E_c$ = 28 200 MPa

*Problem:*

Calculate magnified moment for design.

*Solution:*

Calculate factored loads and moments:

$$P_f = (1.25)(1200 + 200) + (1.5)(400 + 50) = 2425 \text{ kN}$$

$$M_1 = 0$$

$$M_2 = [(1.25)(200) + (1.5)(50)](0.3) = 98 \text{ kN·m}$$

$$e_{min} = 15 + (0.03)(350) = 26 \text{ mm}$$

$$M_{min} = (2425)(0.026) = 63 \text{ kN·m}$$

Find an equivalent moment of inertia, $I_{equiv}$, by applying a 1.0 kN/m uniform lateral load to the centre mullion, constructing the M/EI diagram, and calculating the mid-height deflection using the moment-area method:

$$\Delta = 0.0900 \text{ mm}$$

Substituting into the deflection equation for a simply supported beam gives:

$$I_{equiv} = \frac{5w\ell^4}{384E_c\Delta} = \frac{(5)(1.0)(4200)^4}{(384)(28\ 200)(0.09)} = 1596 \times 10^6 \text{ mm}^4$$

Check whether slenderness must be considered. Assume k = 1 (braced against sidesway) and calculate the equivalent radius of gyration:

$$A_{avg} = \frac{147\ 500 + 207\ 500}{2} = 177\ 500 \text{ mm}^2$$

$$r = \sqrt{\frac{1596 \times 10^6}{177\ 500}} = 95 \text{ mm}$$

$$\frac{25 - 10(M_1 / M_2)}{\sqrt{P_f / (f'_c A_g)}} = \frac{25}{\sqrt{2425 \times 10^3 / (35)(177\ 500)}} = 40.0$$

$$\frac{k\ell_u}{r} = \frac{(10)(4200)}{95} = 44.2 > 40$$

Therefore slenderness effects must be considered.

$$\text{For } \beta_d = \frac{(1.25)(1000 + 200)}{2425} = 0.62$$

and:

$\ell_u = 4.2$ m, from Fig. 3.8.4, find:

$$k_p = 3.9 \times 10^{-6} \text{ kN / mm}^4$$

Calculate:

$$P_c = \frac{k_p I_g}{k^2} = \frac{(3.9 \times 10^{-6})(1596 \times 10^6)}{(10)^2} = 6224 \text{ kN}$$

$$C_m = 0.6 + (0.4)(0) = 0.6$$

$$C_m M_2 = (0.6)(98) = 58.8 \text{ kN·m}$$

$$P_f(15 + 0.03h) = (2425 \times 10^3)[15 + (0.03)(350)](10^{-6}) = 61.8 \text{ kN·m} > 58.8$$

$$M_f = \frac{61.8}{1 - \dfrac{2425}{(0.75)(6224)}} = 129 \text{ kN·m}$$

The short column interaction diagram for this mullion cross section should now be checked for the following axial load and magnified moment:

$P_f$ = 2425 kN

$M_f$ = 129 kN·m

## 3.8.7 SERVICE LOAD STRESSES

There are no limitations in the Code on service load stresses in compression elements subject to bending. The Recommended Practice suggests that the tension limitations of CSA - A23.3 Clause 18.4 be applied. If the surface is to remain free of discernable cracks, it is suggested that the allowable tensile stress be limited to $0.4\lambda\sqrt{f'_c}$ .

## 3.8.8 PILES

### General

This section refers only to the structural design of piles. It is assumed that the capacity to transfer load to the soil has been established by competent geotechnical engineers.

The design for factored load conditions follows exactly the procedures discussed above for columns. The necessary investigation of service load conditions during transporting, handling, and driving also follows standard procedures based on first principles. Stresses which should be allowed under these conditions are discussed in Ref. R3.8.3. Nevertheless, there are three reasons why piles merit special attention:

1. The tensile stresses induced by driving can cause severe cracking problems.
2. The seismic response of piles involves different problems from those of other structural elements.
3. Piles are often produced in standard shapes whose geometry can be dealt with in generalized form.

The first two problems will be briefly outlined and references will be given for fuller discussion and for design procedures. The third point will be illustrated by an example.

### Driving Stresses

When the hammer strikes a pile, a compression wave travels down the length of the pile, moving at the speed of sound (In concrete of good quality, this is between 3600 m/s and 4000 m/s). The length, and therefore the intensity, of this stress wave depends upon:

1. The ratio of the hammer mass to the pile mass.
2. The velocity of the hammer at impact.
3. The stiffness of the cushion block.
4. The stiffness of the pile.
5. The driving resistance of the soil.

When the compression wave reaches the tip of the pile, it may simply be transmitted to the soil. If the tip has encountered very hard material, however, the wave may be reflected back up the pile as a compression wave; or, in very easy driving conditions, the reflected wave may be tensile. In either case, the reflected wave passes up through the trailing end of the downward traveling wave, and the stresses are momentarily superimposed. If the reflected wave is compressive (hard-driving conditions) the stresses are additive, and compression failure may occur near the pile tip; if it is tensile (easy-driving conditions, pile longer than about 15 m), cracking may occur further up the pile when the reflected wave clears the end of the downward-traveling compression wave.

The stress waves tend to be dissipated by friction, damping and tip resistance, but, in hard-driving conditions, a significant compression wave may reach back up to the pile head. Here it will again be reflected, as a tension wave, and cracking may occur near the pile head.

The parameters that are most readily controlled are the ratio of hammer mass to pile mass, the stiffness of the cushion block, and the capacity of the pile to resist the imposed stresses. The parameters are selected to increase the length of the traveling wave, since the intensity of the stress is reduced thereby.

Detailed discussions of these problems, of driving stress calculations, and of required prestress levels may be found in Refs. R3.8.3 and R3.8.4.

Note that piles tend to rotate slightly as they are driven. If they are restrained by the driving leads or by a tightly fitting helmet or pile cap, torsional stresses will be induced which will add to the tensile stresses discussed above. Also, spalling of concrete at the pile head may be caused by high or irregular compressive stress concentrations during driving. These points, too, are considered in Ref. R3.8.3.

## Seismic Response of Piles

Piles reduce the rocking motions of foundations in earthquakes, but not the lateral motion. Installation of battered piles is more likely to lead to failure of the piles than to reduction of lateral motion. Piles essentially follow the horizontal motion of the surrounding soil, which undergoes horizontal shear deformation as the seismic motion travels upward from the bedrock. Therefore, piles may be subjected to significant curvature, particularly just below the pile cap, at the interface between hard and soft-soil layers, and at the point within the pile length where the shear force is zero. Thus, the ideal pile from the seismic point of view is flexible and ductile.

Discussion of the curvatures to be expected is found in Ref. R3.8.5. Design for acceptance of this curvature is generally a matter of achieving the correct quantity and distribution of spiral reinforcement. This is discussed in Refs. R3.8.6 and R3.8.7.

For further information on the response of hollow piles in seismic regions, see Ref. R3.8.8.

**Example 5 – Interaction curve for octagonal hollow bearing pile**

*Given:*

500 mm diameter octagonal pile with 275 mm diameter circular hollow-core.

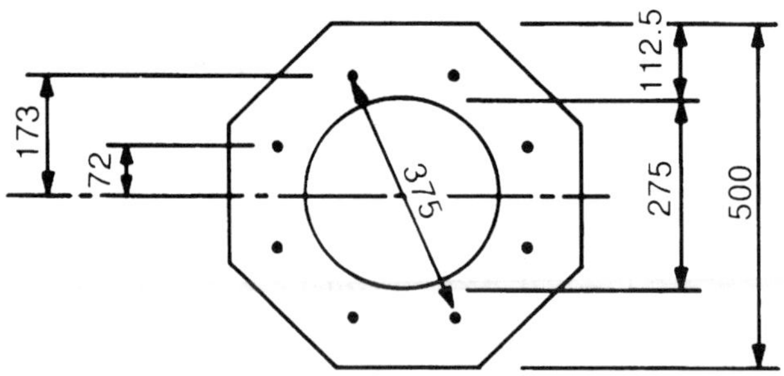


$\gamma_c = 2400$ kg/m³

$f'_c = 50$ MPa

$f_{pu} = 1860$ MPa

8 - 13 mm strands

*Problem:*

Construct the factored resistance interaction diagram. Determine the service load moment capacity at a load of 800 kN if the concrete stresses are limited to 22.5 MPa (compression) and 3.5 MPa (tension).

*Solution:*

$$\alpha_1 = 0.85 - (0.0015)(50) = 0.78$$

$$\beta_1 = 0.97 - (0.0025)(50) = 0.85$$

$$\phi_c = 0.65$$

$$E_c = 32\,200 \text{ MPa}$$

$$A_c = (0.8284)(500)^2 - (\pi)(275)^2/4 = 147\,000 \text{ mm}^2$$

$$A_p = (8)(99) = 792 \text{ mm}^2$$

Assuming $f_{pe} = 1100$ MPa:

$$f_{cp} = \frac{(1100)(792)}{147\,000} = 5.9 \text{ MPa}$$

Prestrain in strand at zero concrete strain:

$$= \frac{1100}{190\,000} + \frac{5.9}{32\,200} = 0.00597$$

Stress in strand at maximum concrete compression:

$= (190\,000)(0.00597 - 0.0035) = 469$ MPa

$P_{ro} = \alpha_1 \phi_c \omega f'_c (A_c - A_p) - f_{pr} A_p$

$= [(0.78)(0.65)(50)(147\,000 - 792)$

$-(469)(792)] / 10^3$

$= 3335$ kN

Factored axial load resistance:

$= 0.85P_{ro} = (0.85)(3335) = 2835$ kN

To determine points on interaction curve, select values of neutral axis depth, say 500, 400, 300, 250, 200 mm, and compute strains, stresses, and forces. As an example, the calculations for a depth of 400 mm are as follows:

c $= 400$ mm

a $= \beta_1 c = (0.85)(400) = 340$ mm

a/D for the octagon $= 340 / 500 = 0.680$

a/D for the hollow core $= (340 - 112.5) / 275$

$= 0.827$

From Fig 3.8.11:

Positive concrete area (octagon)

$= (0.594)(500)^2 = 148\,500\ \text{mm}^2$

Centroidal distance from mid-depth axis

$= (0.125)(500) = 63$ mm

Negative concrete area (hollow circular core)

$= (0.695)(275)^2 = 52\,560\ \text{mm}^2$

Centroidal distance from mid-depth axis

$= (0.045)(275) = 12$ mm

Concrete strain at steel levels:

1. $\frac{(400 - 77)}{400}(0.0035) = 0.00283$
2. $\frac{(400 - 178)}{400}(0.0035) = 0.00194$
3. $\frac{(400 - 322)}{400}(0.0035) = 0.00068$
4. $\frac{(400 - 423)}{400}(0.0035) = -0.00020$

Total strain of steel = prestrain - concrete strain:

**Fig. 3.8.11 Area and centroid coefficients for octagonal and circular piles**

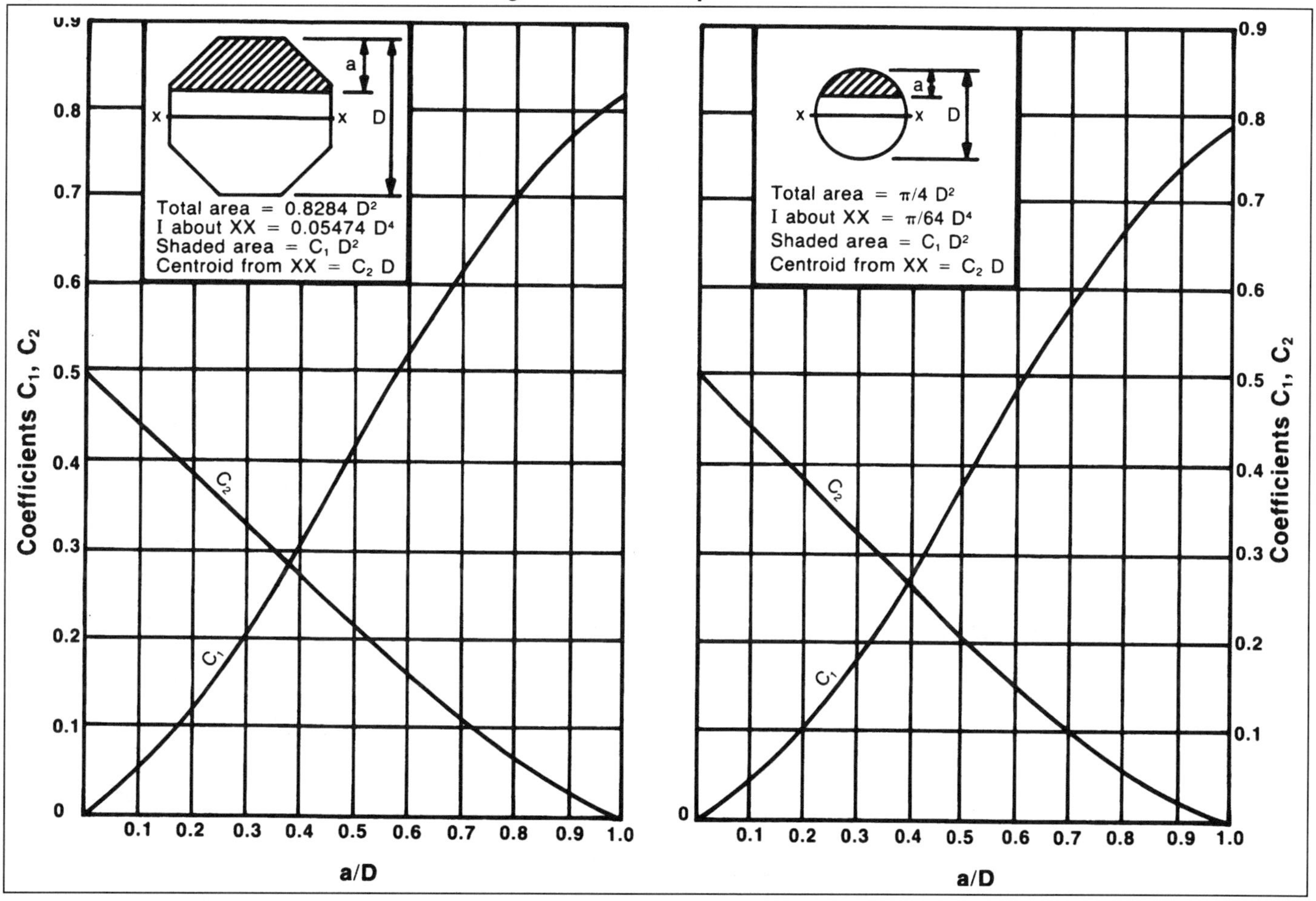

1. 0.00597 – 0.00283 = 0.00314
2. 0.00597 – 0.00194 = 0.00403
3. 0.00597 – 0.00068 = 0.00529
4. 0.00597 – (–0.0002) = 0.00616

Steel stresses from Fig. 8.2.4:

1. 597 MPa
2. 766 MPa
3. 1005 MPa
4. 1172 MPa

Concrete force = (0.78)(0.65)(50)(148 500) – (0.78)(0.65)(50)(52 560)/10³

= 3764 - 1332 = 2432 kN

Steel forces = $f_p A_p$:

1. (597)(198)/10³ = 118 kN
2. (766)(198)/10³ = 152 kN
3. (1005)(198)/10³ = 199 kN
4. (1172)(198)/10³ = 232 kN

Total factored axial resistance:

= 2432 – 118 – 152 – 199 – 232

= 1731 kN (compression)

Concrete moment:

= (3764)(0.063) – (1332)(0.012)

= 221 kN·m

Steel moments:

1. (118)(–0.173) = –20
2. (152)(–0.072) = –11
3. (199)(0.072) = +14
4. (232)(0.173) = +40

Total factored resisting moment:

= 221 – 20 – 11 + 14 + 40

= 244 kN·m

Values for other depths of neutral axis are:

| c | $P_r$(kN) | $M_r$(kN·m) |
|---|---|---|
| ≥ 509 | 2835 | 0 to 80 |
| 500 | 2572 | 132 |
| 400 | 1731 | 244 |
| 300 | 1089 | 284 |
| 250 | 756 | 279 |
| 200 | 390 | 260 |

These values are plotted in Fig. 3.8.12. The value of $M_{ro}$ for prestressed elements is found most easily by extending the plot or investigating further values of c until values of $P_r$ either side of zero are obtained.

To check the service conditions, determine the moment capacity, M, at a service axial load P of 800 kN:

F = prestress force at zero concrete strain

$= \epsilon_p E_p A_p$

= (0.00597)(190 000)(792)/10³

= 898 kN

$A_{tr}$ = transformed area (see Fig. 3.8.11)

$$= (0.8284)(500)^2 - \frac{(\pi)(275)^2}{4} + \frac{(190\,000)(792)}{32\,200} = 152\,400\text{ mm}^2$$

$S_{tr}$ = transformed section modulus (see Fig. 3.8.11)

$$= \left[(0.05474)(500)^4 - \frac{(\pi)(275)^4}{64} + \frac{(190\,000)(396)\cdot(72^2 + 173^2)}{32\,200}\right] / 250$$

$$= 12.9 \times 10^6 \text{ mm}^3$$

(In practice, the error introduced by using the area and section modulus of the gross concrete area may be acceptable.)

Equate the extreme fibre stresses to the allowable values and solve for the smaller value of M:

$$\frac{P}{A_{tr}} + \frac{F}{A_{tr}} + \frac{M}{S_{tr}} = 22.5 \text{ MPa}$$

$$\frac{(800)(10^3)}{(152\,400)} + \frac{(898)(10^3)}{(152\,400)} - \frac{M}{(12.9 \times 10^6)} = 22.5$$

**Fig. 3.8.12 Interaction curve for bearing pile example**

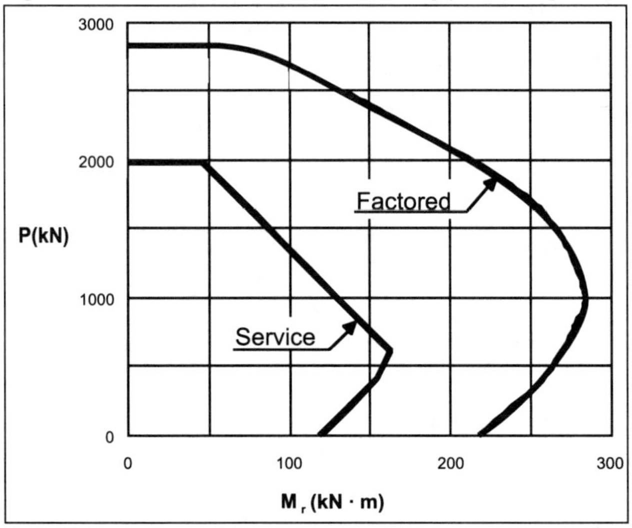

$\therefore\ M = 146\ kN{\cdot}m$

$$\frac{P}{A_{tr}} + \frac{F}{A_{tr}} - \frac{M}{S_{tr}} = -3.5\ MPa$$

$$\frac{(800)(10^3)}{(152\,400)} + \frac{(898)(10^3)}{(152\,400)} - \frac{M}{(12.9 \times 10^6)} = -3.5$$

$\therefore\ M = 189\ kN{\cdot}m$

The service load moment capacity at an axial load of 800 kN is thus 146 kN·m.

A plot of the complete curve of P vs. M at these service load conditions is shown on Fig. 3.8.12.

## 3.9 HANDLING CONSIDERATIONS

The art of prefabrication requires careful attention to loads and stresses during the handling stages between fabrication and erection.

### 3.9.1 FORM SUCTION AND IMPACT FACTORS

Elements will be subjected to forces during stripping, in addition to the mass of the element, which will depend upon methods of forming, surface treatment, and element geometry. During the handling process, elements may be subjected to dynamic forces. For purposes of determining concrete stresses and reinforcement, it is common practice to use multipliers on the panel mass, and treat the resulting force as an equivalent static dead load. There are no fixed values which can be quantitatively derived; values which have been used in the past are generally based on the experience of the manufacturer. Fig 3.9.1 provides one set of multipliers which may be used to determine equivalent static load.

#### Lifting Devices

For simplicity, it has become practice to design handling and erection devices for a factored resistance of four times the mass supported by the device.

### 3.9.2 FLEXURE

The arrangement of lifting devices will be determined by the plant engineer, since such arrangement is a function of the manufacturing process and the strength of concrete at time of stripping. The magnitude of stresses generated at time of stripping will determine whether a design based on no cracking, or controlled cracking, is most feasible. When the maximum moment at time of stripping has been determined, the stresses may be evaluated based on the gross section properties; if they are excessive, an increase in the number of lifting points may be required. The embedded devices used for stripping may be used again for erection; however, additional devices may be required to rotate the panel from the stripped and handled position to the erection position.

#### Flat Panels

Panels which are stripped by rotating the panel about one edge, with lifting devices attached near the opposite edge, will develop moments in accordance with Fig. 3.9.2(a) and (b). For determining stresses, it may be assumed that the calculated moment is resisted by a width as shown in the figure. When elements are stripped in this fashion, care should be taken to prevent spalling of the edge along which rotation is accomplished. Use of a compressible filler, or sand bed, are two of the more common ways to protect this edge.

Elements which are stripped flat from the mould will develop moments in accordance with Fig. 3.9.3 (a) and (b). As above, the calculated moment may be assumed resisted by the width as shown.

Lift lines which are inclined to the plane of the panel will induce in-plane compression. When the point of load application is not coincident with the middle plane of the panel, additional flexure will be introduced. Design considerations similar to those discussed under Ribbed Panels are applicable.

When available, the use of a tilt table will significantly reduce stripping stresses. The slight inclination of the panel as it is lifted from the tilted position to the vertical may be neglected when determining stripping forces and stresses (Fig. 3.9.4).

**Fig. 3.9.1 Equivalent static load multipliers[(1)] to account for stripping and dynamic forces**

| Stripping | | |
|---|---|---|
| | Finish | |
| Element type | Exposed aggregate with retarder | Smooth form (form oil only) |
| Flat, with removable side forms, no false joints or reveals | 1.2 | 1.3 |
| Flat, with false joints and/or reveals | 1.3 | 1.4 |
| Fixed forms with proper draft[(4)] | 1.4 | 1.6 |
| Sculptured | 1.5 | 1.7 |
| **Yard handling[(2)] and erection[(3)]** | | |
| All elements | 1.2 | |
| **Transportation[(2)]** | | |
| All elements | 1.5 | |

(1) These factors are used in flexural design of elements. At stripping, suction between product and form introduces forces, which are treated here by introducing a multiplier on product mass. It would be more accurate to establish these multipliers based on the actual contact area and a suction factor independent of product mass.

(2) Certain unfavourable conditions in road surface, equipment, etc., may require use of higher values. See Sect. 3.9.5.

(3) Under certain circumstances may be higher.

(4) For example, Tees, channels and fluted panels.

**Fig. 3.9.2 Moments developed in panels stripped by rotating about one edge**

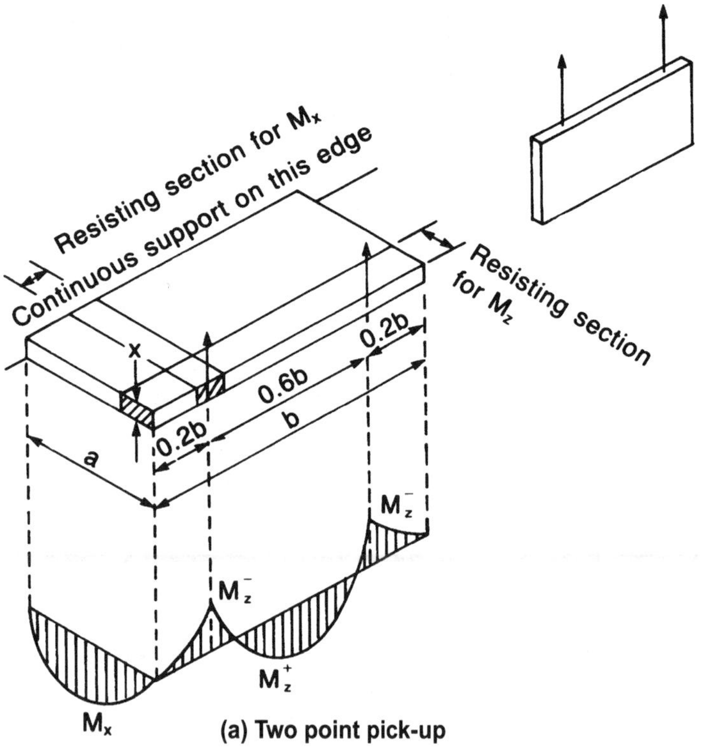


(a) Two point pick-up

Maximum moments (approximate)

$M_x = 0.0625\ wa^2b$

$M_z^- = M_z^+ = 0.0125\ wab^2$

$M_x$ acting over a section of width $\frac{a}{2}$, 10t or $\frac{b}{2}$, whichever is less

$M_z$ acting over a section of width 0.3b, 10t or $\frac{a}{2}$, whichever is less

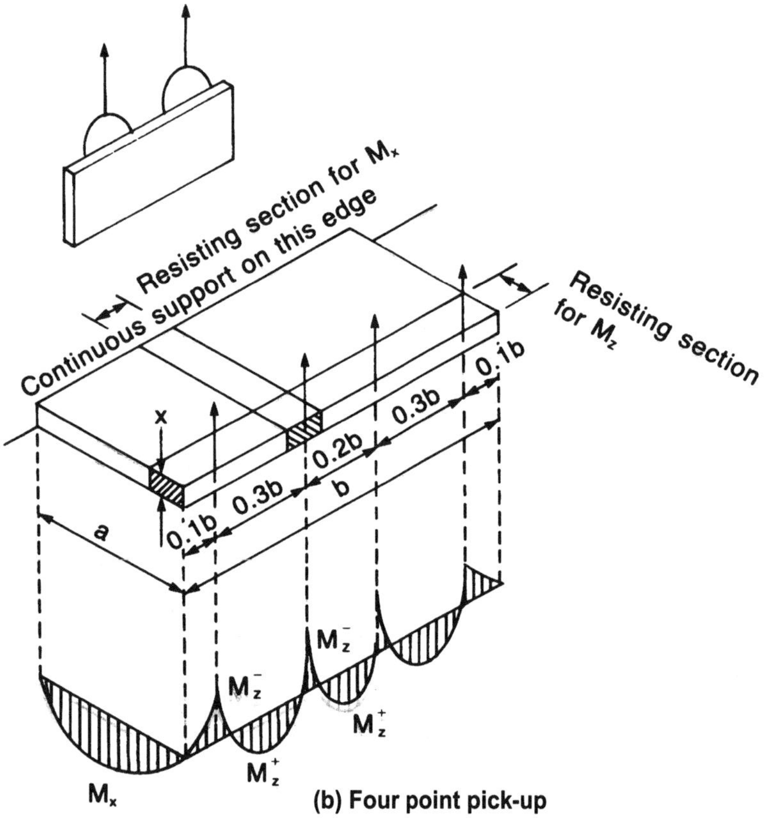


(b) Four point pick-up

Maximum moments (approximate)

$M_x = 0.0312\ wa^2b$

$M_z^- = M_z^+ = 0.0031\ wab^2$

$M_x$ acting over a section of width $\frac{a}{2}$, 10t or $\frac{b}{4}$, whichever is less

$M_z$ acting over a section of width 0.3b, 10t or $\frac{a}{2}$, whichever is less

**Fig. 3.9.3 Moments developed in panels stripped flat**

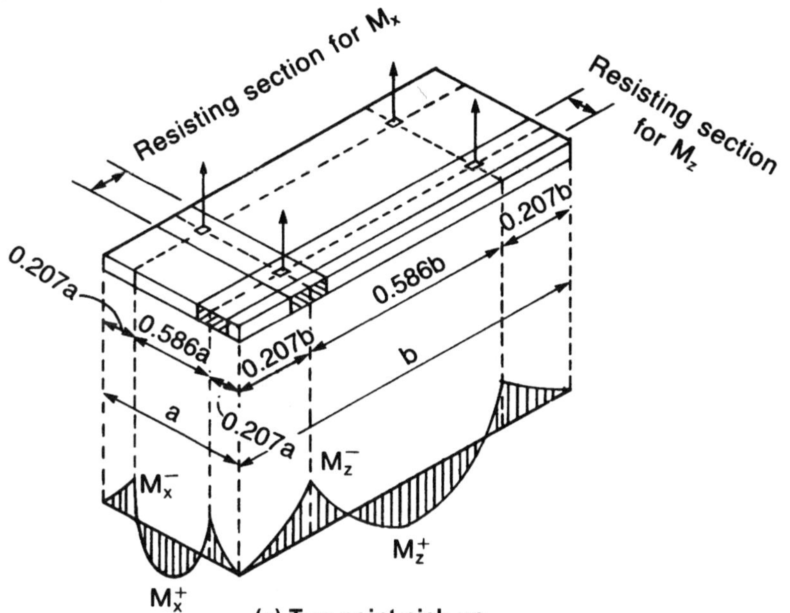


(a) Two point pick-up

Maximum moments (approximate)

$M_x^+ = M_x^- = 0.0107\ wa^2b$

$M_z^- = M_z^+ = 0.0107\ wab^2$

$M_x$ acting over a section of width 0.3a, 10t or $\frac{b}{2}$, whichever is less

$M_z$ acting over a section of width 0.3b, 10t or $\frac{a}{2}$, whichever is less

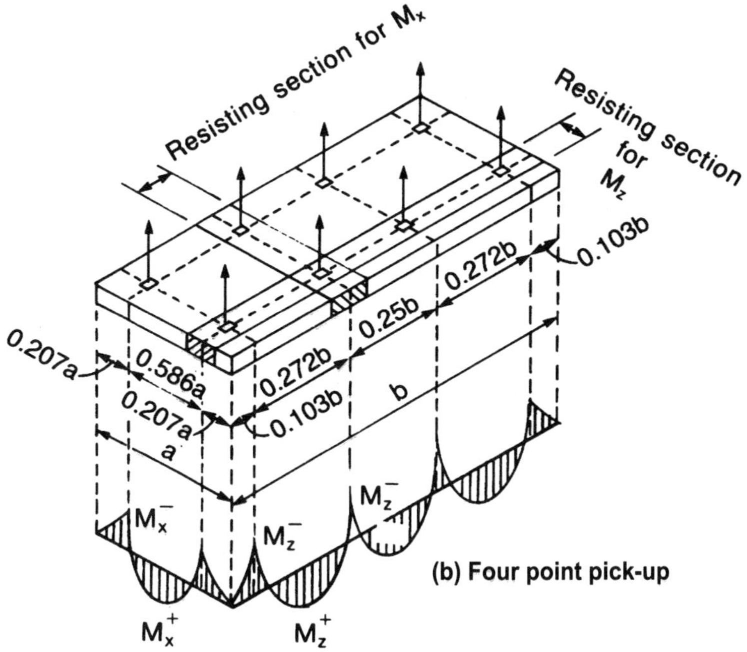


(b) Four point pick-up

Maximum moments (approximate)

$M_x^+ = M_x^- = 0.0054\ wa^2b$

$M_z^- = M_z^+ = 0.0027\ wab^2$

$M_x$ acting over a section of width 0.3a, 10t or $\frac{b}{4}$, whichever is less

$M_z$ acting over a section of width 0.15b, 10t or $\frac{a}{2}$, whichever is less

## Ribbed Panels

When a panel is ribbed or is of a configuration such that stripping by sliding or tilting is not practical, a system of vertical pick-up points on the top surface can be used. These lift points should be located so as to minimize the tension on the face of the element. However, the geometry and practical location limitations on the pick-up points may not permit a crack-free design. Provided that the in-service conditions permit, the panel in this case should be designed for the controlled cracking condition, or should be prestressed.

## Elements of Constant Cross-section

Since the section modulus of top and bottom faces will not usually be identical, the designer must determine which face is controlling when choosing position and number of lift points. He must also select the controlling design limitation:

1. Tensile stresses on both faces to be less than that which would cause cracking (Sect. 3.4.2)
2. Tensile stress on one face to be less than that which would cause cracking, with controlled cracking permitted on the other face

**Fig. 3.9.4 Stripping from a tilt table**

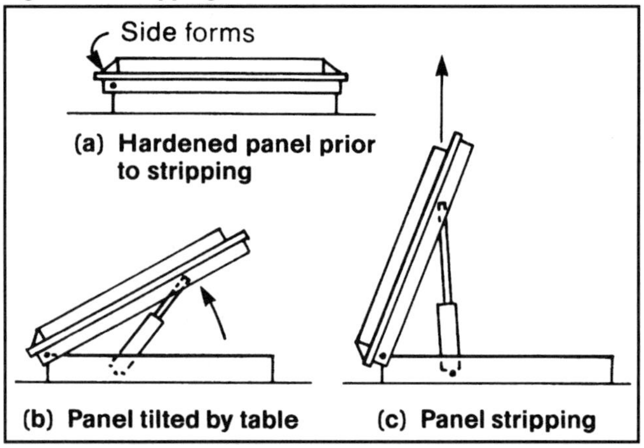


**Fig. 3.9.5 Pick-up points for equal stresses of a ribbed element**

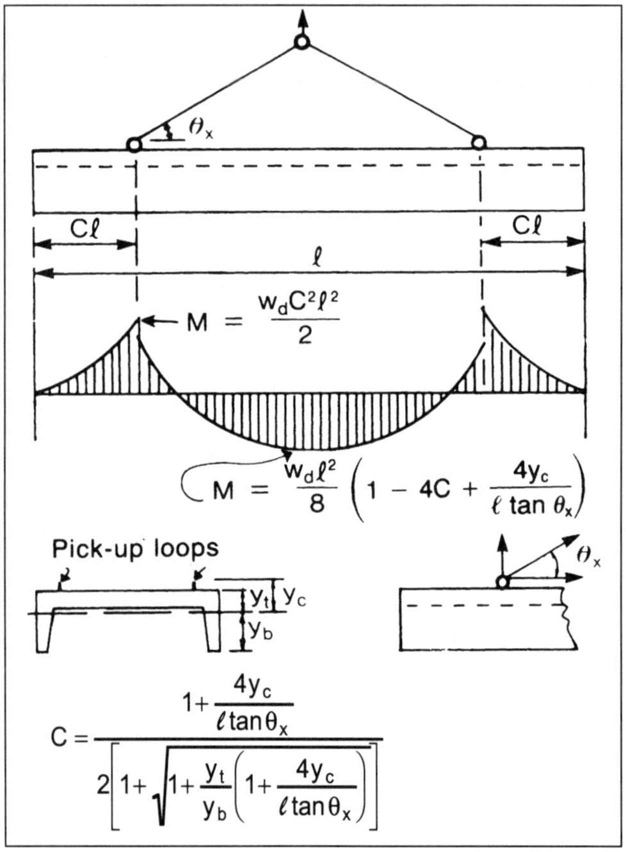


**Fig. 3.9.6 Moments caused by eccentric lifting**

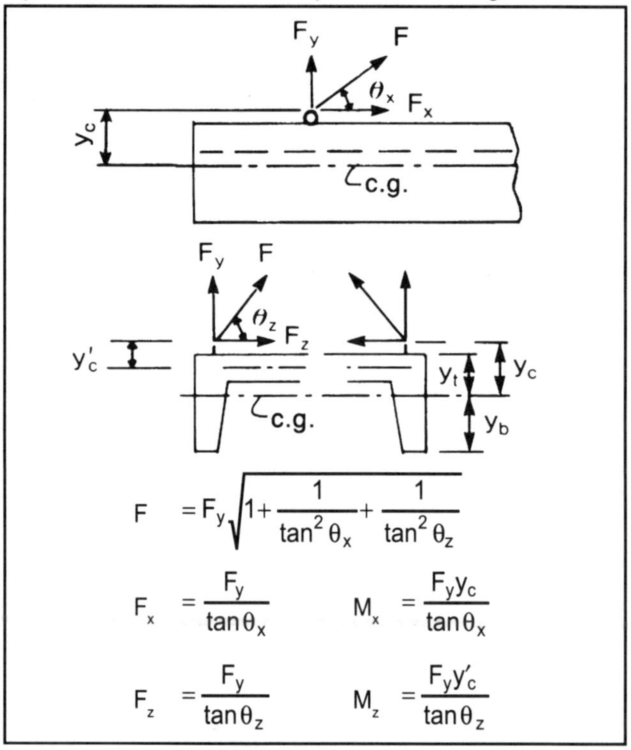


**Fig. 3.9.7 Arrangement for equalizing lifting loads**

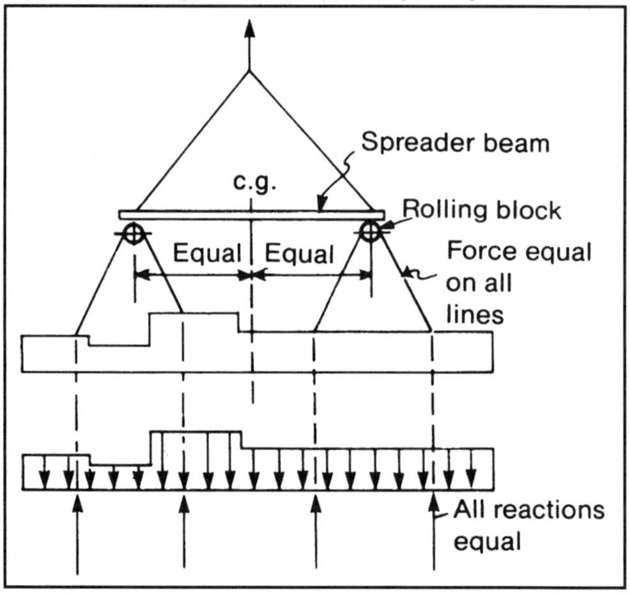


3. Controlled cracking permitted on both faces

If only one of the faces is exposed to view, then generally this is the face which will control.

The design moments for a two point lift, using inclined lines, is indicated in Fig. 3.9.5. When the angle of the lift line to the horizontal is small, the component of force parallel to the longitudinal axis may generate a significant moment. Even though the effects of non-vertical lifting lines can and should be accounted for, it is not recommended that this effect be allowed to dominate design moments. When this condition exists, consideration could be given to using spreader beams, two cranes or other mechanisms to reduce the angle of lift and thus the effect of longitudinal forces.

In addition to longitudinal bending moments, there could also be a transverse bending moment generated by the orientation of the pick-up points with respect to the transverse dimension (Fig. 3.9.6). For the particular section shown, critical conditions could result between the ribs at a relatively low magnitude of moment, due to the reduced cross-section.

**Fig. 3.9.8 Determination of force in inclined lift lines**

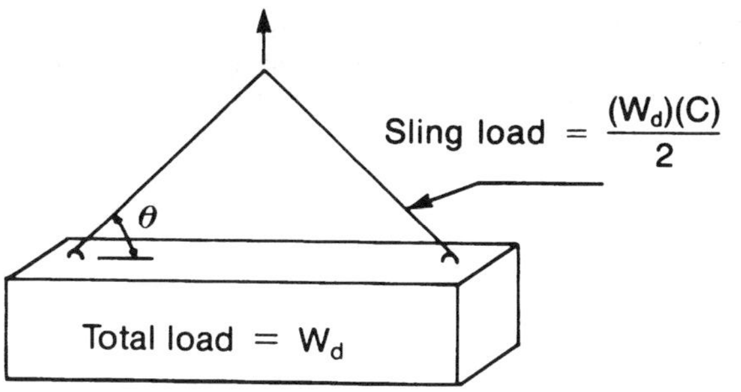


Crane line load = $W_d$

| | Correction factor "C" for the total load on the hoisting sling with an inclination = θ | | | | | |
|---|---|---|---|---|---|---|
| θ | 90° | 75° | 60° | 45° | 30° | 15°* |
| C | 1.00 | 1.04 | 1.16 | 1.42 | 2.00 | 3.86 |

*Note:* θ usually not less than 60° to 70°

*Not recommended

## Elements of Varying Cross-section

The design guidelines listed under the previous section are applicable for elements of constant cross-section. No rules can be formulated with respect to location of lift points with elements of varying cross-section; for these cases, the location of lift points requires a trial and error process to determine the position which will result in acceptable stress. In the case of long elements of varying section, rolling blocks can be employed (Fig. 3.9.7). The forces in the lifting lines will be equal, and the stress analysis can proceed by considering a beam with varying load supported by equal reactions.

The force in inclined lift lines can be determined from Fig. 3.9.8.

## Handling Devices

Lifting devices are normally strand or cable loops, projecting from the concrete, or threaded inserts.

Since lifting devices are subject to dynamic loads, ductility of the material should be part of the design requirements. Deformed reinforcing bars should not be used for lifting loops as the deformations result in stress concentrations from the shackle pin. Also, reinforcing bars are often hard-grade or rerolled rail steel, with low ductility and low impact strength at low temperature. Smooth, mild-steel bars may be used, provided adequate embedment against pullout is provided. The diameter must be such that localized failure cannot occur by bearing on the shackle pin.

Both strand and wire rope (cable) may be used for lifting loops. Ordinarily, only the straight tensile load capacity of a given product is available from the manufacturer, and a given product's capacity, when tensioned while bent over the relatively short radius of some lifting hooks, can be significantly less than its straight tensile capacity. The strand or wire rope's ability to withstand repeated bending and flexing must be considered.

Also, the bond/embedment length or embedded configuration with or without supplemental reinforcement may limit a lifting loop's capacity, and edge conditions may further limit capacity. In order to ensure that an embedded insert acts primarily in tension, a swivel plate as indicated in Fig. 3.9.9 should be used.

**Fig. 3.9.9 Swivel plate**

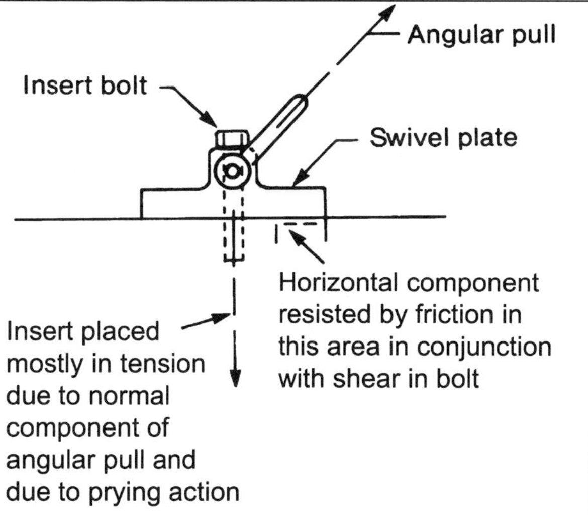


The looped (aircraft) type cable finds application in stripping small flat panels or for edge lifting of flat panels.

The following table indicates usual working loads:

| Diameter (mm) | Capacity (kN) |
|---|---|
| 5 | 5.6 |
| 6 | 8.9 |
| 8 | 13.3 |
| 9 | 22.2 |
| 13 | 44.5 |

Lifting heavy units from threaded inserts should be carefully assessed. When properly designed (for insert and concrete strength) and applied according to such criteria, threaded inserts have many advantages. Correct usage is sometimes difficult to inspect during handling operations.

Connection hardware should not be used for lifting or handling of any but the lightest units, unless carefully reviewed by the designer.

The drawings should clearly indicate which devices are intended to be used for handling.

**Example 1 - Design for handling**

*Given:*

A window unit of exposed aggregate is to be cast face down.

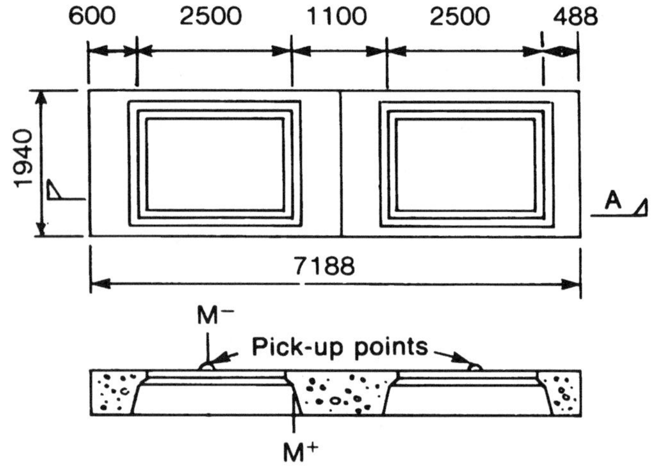


Section A

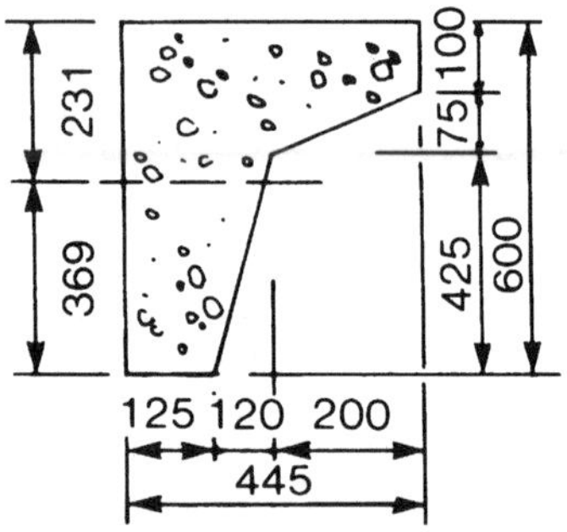


Section Properties:

$A_g = 149\ 000\,\text{mm}^2$

$I = 4291 \times 10^6\ \text{mm}^4$

$y_b = 369\ \text{mm}$

$S_t = 18\ 600 \times 10^3\ \text{mm}^3$

$S_b = 11\,600 \times 10^3\,\text{mm}^3$

*Problem:*

Locate the pick-up points to minimize tension stress in the concrete, and design handling reinforcement.

*Solution:*

Dead load of element assuming 1.6 multiplier (Fig. 3.9.1):

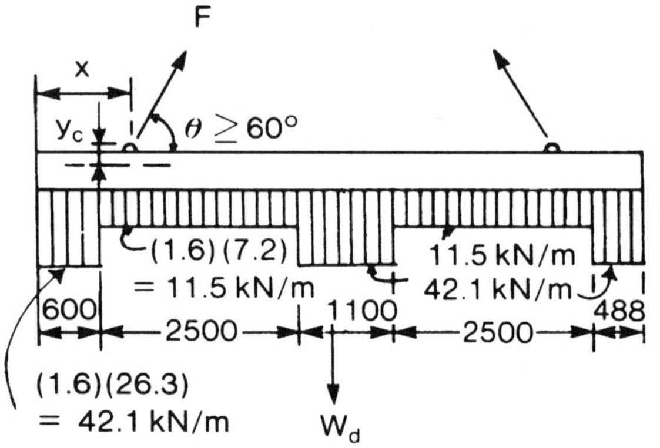


$$W_d = (5.0)(11.5) + (2.19)(42.1) = 150\ \text{kN}$$

Lifting loops should be placed symmetrically about the centre of gravity of the element. Assume critical cracking stress will occur in the narrow sections of the unit. For equal stresses on each face:

$$f_t = f_b$$

$$\frac{M^- y_t}{I} = \frac{M^+ y_b}{I}$$

$$M^- = \frac{y_b}{y_t} M^+ = \frac{369}{231} M^+ = 1.60 M^+$$

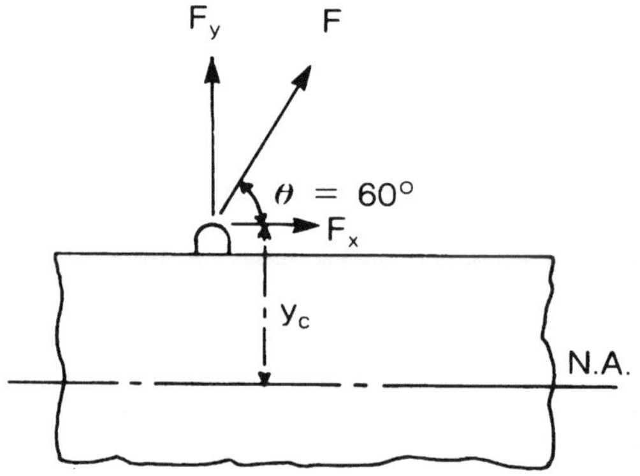


$$F_x = \frac{F_y}{\tan\theta} = \frac{150}{2\tan 60^\circ} = 43.3\ \text{kN}$$

$$y_c = y_t + 70 = 301\ \text{mm}$$

$$M = \frac{(301)}{10^3}(43.3) = 13.0\ \text{kN·m}$$

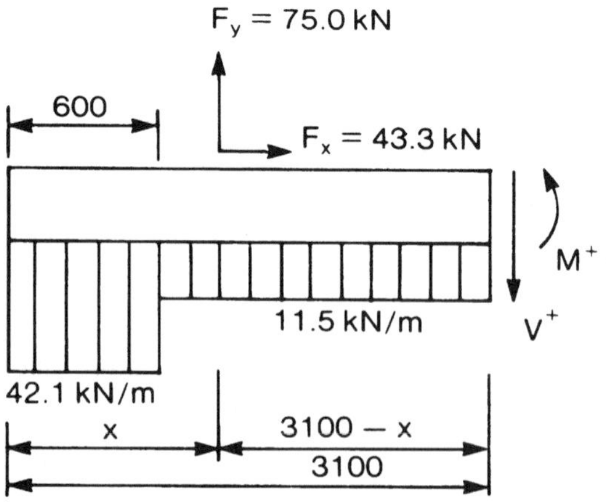


$$M^+ = (75)\frac{(3100 - x)}{10^3} - \frac{(11.5)}{2}\left(\frac{2500}{10^3}\right)^2$$

$$-(42.1)\left(\frac{600}{10^3}\right)\left(\frac{2800}{10^3}\right) + 13.0$$

$$= 139 - 75.0x$$

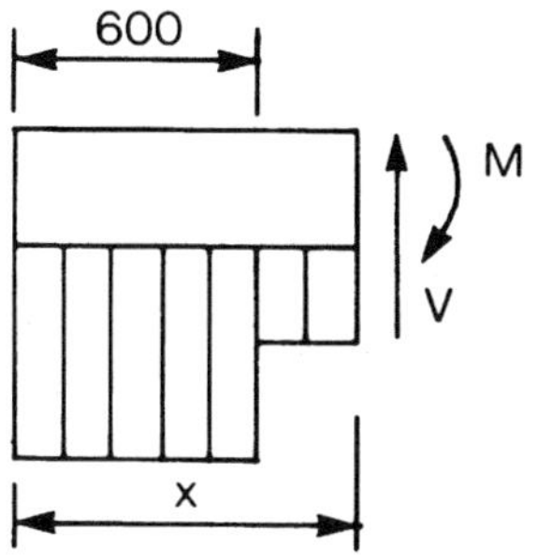


$$M^- = (42.1)\left(\frac{600}{10^3}\right)\frac{(x-300)}{10^3} + \frac{(11.5)}{2}\left(\frac{x-600}{10^3}\right)^2$$

$$= 5.8x^2 + 18.4x - 5.5$$

however:

$$M^- = 1.60M^+$$

$$\therefore\ 5.8x^2 + 18.4x - 5.5 = (1.60)(139 - 75.0x)$$

$$5.8x^2 + 138.4x - 227.9 = 0$$

$$x = 1540 \text{ mm}$$

use: $x = 1500 \text{ mm} = 1.5 \text{ m}$

$$M^+ = 139 - (75)(1.5) = 26.5 \text{ kN·m}$$

$$M^- = (5.8)(1.5)^2 + (18.4)(1.5) - 5.5 = 35.2 \text{ kN·m}$$

$$f_t = \frac{M^-}{S_t} = \frac{(35.2)(10^6)}{(2)(18\,600\times10^3)} = 0.9 \text{ MPa}$$

$$f_b = \frac{M^+}{S_b} = \frac{(26.5)(10^6)}{(2)(11600\times10^3)} = 1.1 \text{ MPa}$$

This stress would allow stripping at $f'_{ci}$ as low as 15 MPa without theoretically cracking the section. Thus, reinforcing would not be required. For illustrative purposes, determine reinforcing required for controlled cracking, assuming a permissible crack width = 0.12 mm:

$$M^- = \frac{35.2}{2} = 18.7 \text{ kN·m per mullion}$$

$$M^+ = \frac{26.5}{2} = 13.3 \text{ kN·m}$$

*Reinforcement for* $M^+$ :

$$d = 560 \text{ mm}$$

assume:

$$j_d = 0.9d = 504 \text{ mm}$$

$$k_d = 0.3d = 168 \text{ mm}$$

$$h_1 = 560 - 168 = 392 \text{ mm}$$

$$h_2 = 600 - 168 = 432 \text{ mm}$$

$$h_2/h_1 = 1.10$$

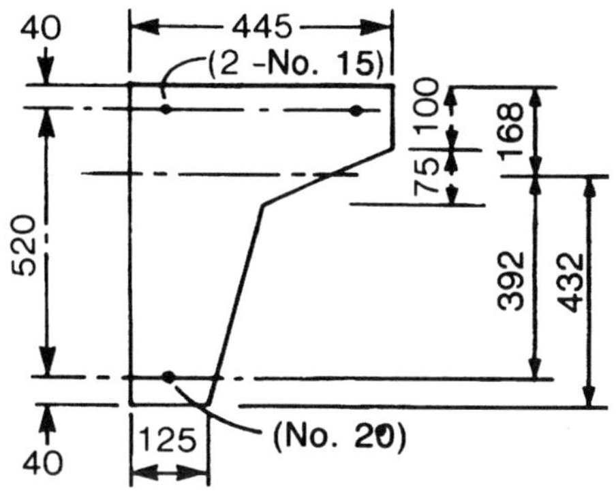


For a crack width of 0.12 mm, from Section 3.4.2:

$$w = 11\times10^{-6}\frac{h_2}{h_1}f_s\sqrt[3]{d_cA}$$

$$f_s = \frac{0.12}{(11\times10^{-6})(110)\sqrt[3]{(40)(80)(125)}} = 135 \text{ MPa}$$

$$\text{Approx. } A_s = \frac{M^+}{0.9df_s} = \frac{(13.3)(10^6)}{(0.9)(560)(135)} = 196 \text{ mm}^2$$

$$\text{Min } A_s = \frac{0.2\sqrt{f'_c}}{f_y}b_th = \frac{0.2\sqrt{35}}{400}(125)(600)$$

$$= 222 \text{ mm}^2 \text{ (controls)}$$

Use: No. 20 bar ($A_s$ = 300 mm²) at bottom of section.

*Reinforcement for* $M^-$ :

Assuming 2 top bars:

$$f_s = \frac{0.12}{(11\times10^{-6})(110)\sqrt[3]{(40)(80)(222)}} = 111 \text{ MPa}$$

$$\text{Approx. } A_s = \frac{(17.6)(10^6)}{(0.9)(560)(111)} = 315 \text{ mm}^2$$

Use: 2 – No. 15 bars ($A_s$ = 400 mm² ) at top of section.

### 3.9.3 LATERAL STABILITY

Notation Used In This Section

- $a$ = length of overhang
- $c_a$ = applied overturning moment arm
- $c_r$ = resisting moment arm
- $e_i$ = initial lateral eccentricity of the C.G. of beam (measured from supports)
- FS = factor of safety against cracking
- FS′ = factor of safety against failure or roll-over
- $h_r$ = height axis above road
- $K_O$ = sum of rotational spring constants of supports
- $\ell$ = overall length

$\ell_1$ = between supports

$M_{lat}$ = lateral bending moment at cracking

r = radius of stability = $K_o / W$

w = weight per unit length of beam

W = total weight of beam

Y = height of the C.G. of beam above the roll axis (adjusted for camber)

$Y_r$ = height of the roll axis above the C.G. of beam (adjusted for camber)

z = lateral deflection of C.G. of beam

$z_{max}$ = horizontal distance from center of vehicle to center of dual tire

$\overline{Z}_o$ = theoretical lateral deflection of the C.G. of beam with full load applied laterally

$\overline{Z}'_o$ = $\overline{Z}_o$ adjusted for cracked section at tilt angle under consideration

$\alpha$ = superelevation or tilt angle of support

$\theta_i$ = initial roll angle of rigid beam = $e_i / Y_r$

$\theta_{max}$ = tilt angle at which cracking begins (considering lateral bending)

$\theta'_{max}$ = tilt angle at the maximum factor of safety against failure

Special consideration should be given to long elements with narrow compression flanges during handling, transportation and erection. Unlike the classical buckling of steel I beams, the lateral stability of long prestressed members is dependent on the roll stiffness of the supports (i.e. lifting loops, truck, or bearing pads). The issue of lateral stability is actually an analysis of lateral bending and equilibrium, in which lateral stresses can be significant. A theory for evaluating the lateral stability of prestressed I-beams developed in Ref. R3.9.1 is summarized below.

For a hanging beam or a beam supported from below, an equilibrium diagram is used to derive a factor of safety against cracking (FS) and a factor of safety against failure or roll-over (FS′), where in general:

$$\text{factor of safety} = \frac{\text{resisting moment}}{\text{applied moment}}$$

For a hanging beam, see Fig 3.9.10,

$$FS = \frac{1}{\overline{Z}_o / Y_r + \theta_i / \theta_{max}}$$

$$FS' = \frac{Y_r \theta'_{max}}{\overline{Z}'_o \theta'_{max} + e_i}$$

For a beam supported from below, see Fig 3.9.11,

$$FS = \frac{r(\theta_{max} - \alpha)}{\overline{Z}_o \theta_{max} + e_i + Y\theta_{max}}$$

$$FS' = \frac{r(\theta'_{max} - \alpha)}{\overline{Z}'_o \theta'_{max} + e_i + Y\theta'_{max}}$$

It is recommended to use FS ≥ 1.0 and FS′ ≥ 1.5. Note that the factor of safety against failure or roll-over, FS′, is based on a cracked section and therefore need not be taken less than the calculated factor of safety against cracking, FS.

The factor of safety is a function of beam geometry and support location. For a beam with curvature due to lateral sweep and vertical camber, the centroid is moved away from the roll axis, see Fig 3.9.12.

In general, reducing the distance between supports will reduce the offset of the beam centroid from the roll axis and improve stability. The distance to the centroid can be calculated using the midspan offset for the full member length multiplied by the offset factor where,

$$\text{offset factor} = (\ell_1 / \ell)^2 - 1/3$$

The deflection of the centroid of a beam under its own weight is calculated,

$$\overline{Z}_o = \frac{w}{12E_c I_y \ell}[0.1\ell_1^5 - a^2\ell_1^3 + 3a^4\ell_1 + (6/5)a^5]$$

This deflection can be adjusted for a cracked section at the tilt angle under consideration where:

$$\overline{Z}'_o = \overline{Z}_o(1 + 2.5\theta'_{max})$$

The tilt angle at the maximum factor of safety against failure or roll-over is calculated

for a hanging beam:

$$\theta'_{max} = \sqrt{\frac{e_i}{2.5\overline{Z}_o}}$$

for a beam supported from below:

$$\theta'_{max} = \frac{z_{max} - h_r\alpha}{r} + \alpha$$

The initial stresses at the time of handling are significant to the factor of safety against cracking. Prior to applying the design loads, typically the top flange is very lightly stressed in compression or is in tension. In this case very little lateral bending can be tolerated before cracking begins. Therefore the factor of safety against cracking may need to be improved by adding compression to the top flange. Similarly, if the distance between supports is reduced to improve stability, the top flange compression due to self weight is reduced and additional compression may be required.

For beams supported from below the factor of safety can be improved by increasing the rotational stiffness of the supports. A method of determining the stiffness of vehicles, and guidelines for bearing pads are detailed in Ref. R3.9.1

For safe handling of long elements, lateral stability can be improved by several methods:

1. Move the support points inward. Decreasing the distance between supports by a small amount can significantly

**Fig. 3.9.10 Equilibrium of hanging beam**

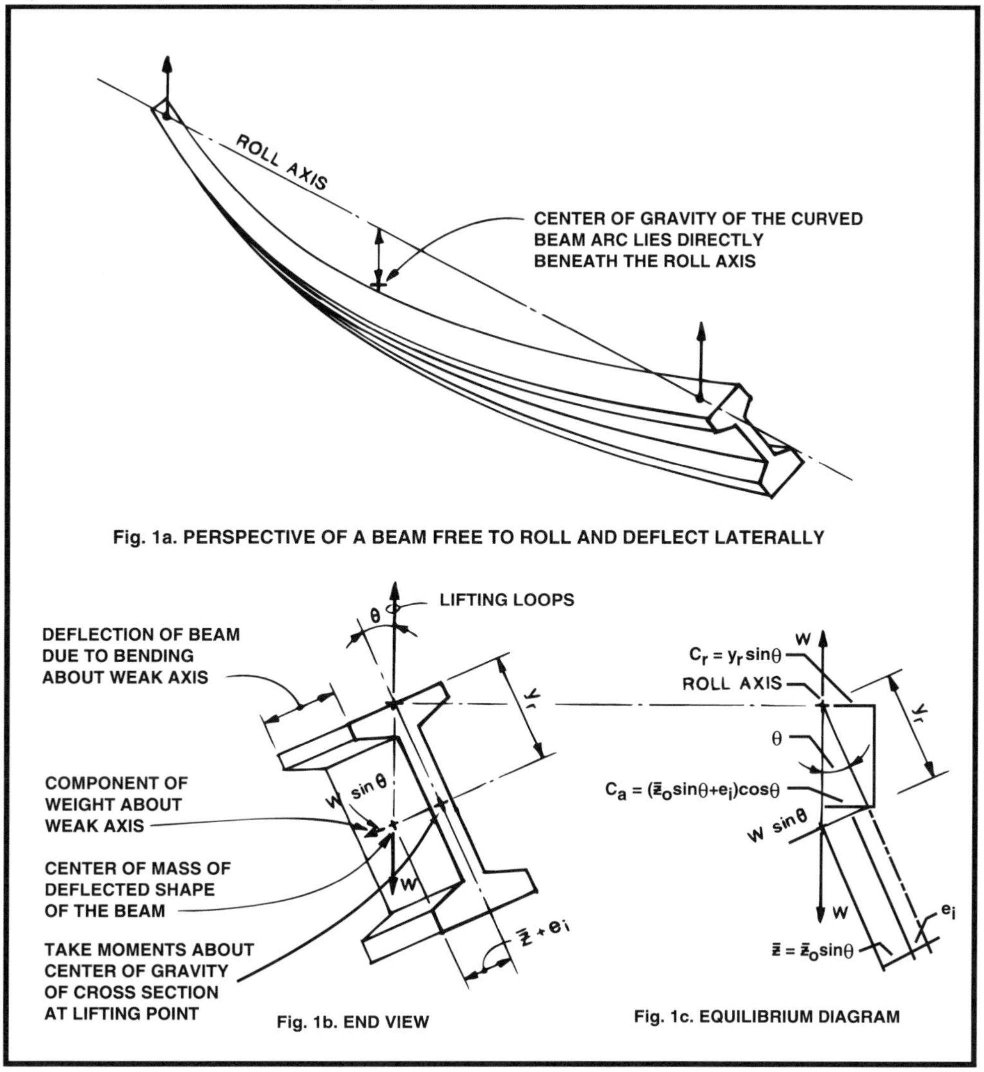


**Fig. 3.9.11 Equilibrium of beam on elastic supports**

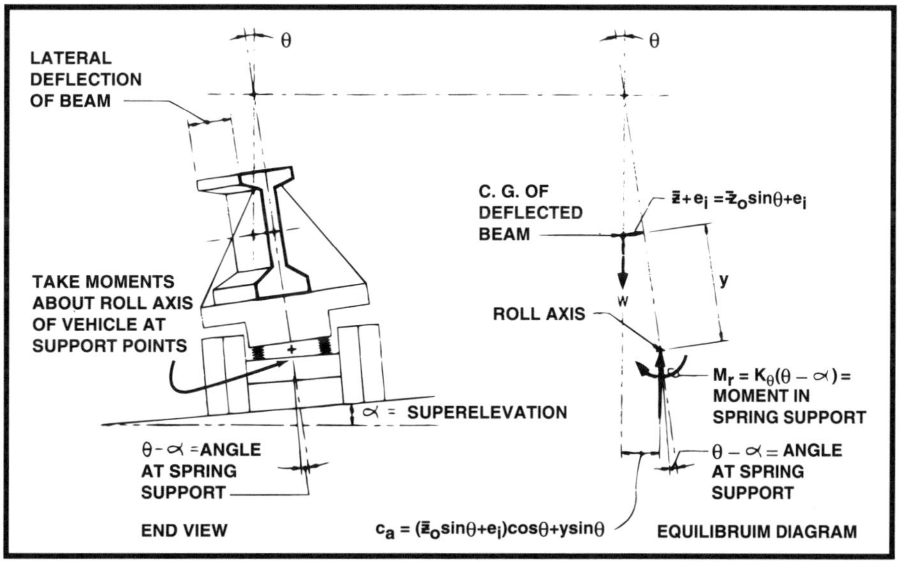

**Fig. 3.9.12 Offset of centroid at a curved arc**

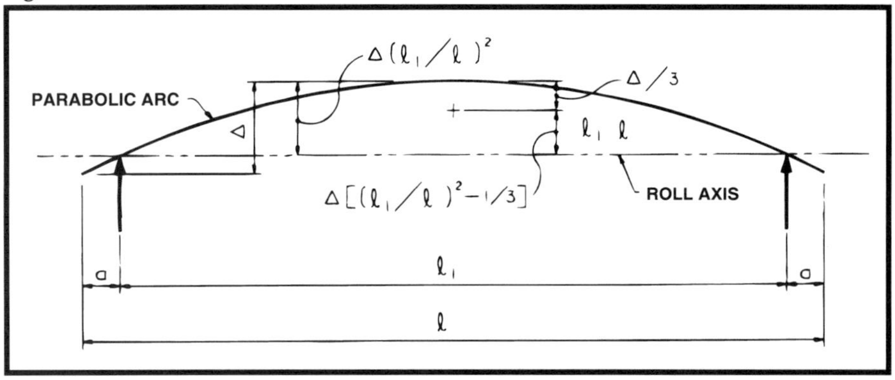


increase the factor of safety against failure or roll-over. Temporary post-tensioning can be used to improve the factor of safety against cracking.

2. Increase the distance $Y_r$ for hanging beams by using a rigid yoke at the lifting points. Decrease the distance Y for beams supported from below by using lower vehicles and bunking for transportation.
3. Increase the roll stiffness of vehicles used for transportation.
4. Attach temporary lateral stiffening in the form of strongbacks, stiffening trusses or pipe frames. Sometimes two or more units can be transported together, side by side, and tied together to provide the necessary lateral strength.
5. Increase the lateral stiffness of the members by revising the shape to increase $I_y$ or increase the concrete strength and thus $E_c$.

**Example 2 - Lateral Stability**

*Given:*

A 40 m long CPCI 1900 girder supported 2 m from each end.

$\ell$ = 40 m

a = 2 m

$\ell_1$ = 36 m

Concrete:

$f'_{ci}$ = 35 MPa

$E_{ci}$ = 28 200 MPa

$f'_c$ = 45 MPa

$E_c$ = 31 000 MPa

Loads:

w = 12.80 kN/m

W = 512 kN

Section Properties:

$y_b$ = 940 mm

$y_t$ = 960 mm

$I_y$ = 14 900 × $10^6$ $mm^4$

$b_t$ = 900 mm

Stresses at harp point (x = 16 000 mm):

$f_{tpres}$ = – 6.50 MPa (at transfer)

= – 6.00 MPa (at 40 days)

$f_{tself}$ = + 7.00 MPa for gravity moment:

$M_g$ = 1950 kN·m

*Problem:*

Evaluate the lateral stability of this beam during handling and transporting.

*Solution:*

(A) Handling (a hanging beam)

*1) Initial Eccentricity*

offset factor = $(36\,000/40\,000)^2 - 1/3$

= 0.477

lateral sweep = 20 mm

lift loop placement offset = 6 mm

$e_i$ = (20)(0.477) + 6 = 16 mm

*2) Height of Roll Axis Above C.G. of Beam*

camber = 50 mm

$Y_r$ = 960 – 50(0.477) = 936 mm

*3) Initial Roll Angle*

$$\theta_i = 16 / 936 = 0.0170 \text{ rad}$$

*4) Theoretical Lateral Deflection*

$$\overline{Z}_o = \frac{12.8 / 10^3}{(12)(28\,200)(14\,900 \times 10^6)(40\,000)}$$
$$[(0.1)(36\,000)^5 - (2000)^2(36\,000)^3 + (3)(2000)^4(36\,000) + (6/5)(2000)^5]$$
$$= 372 \text{ mm}$$

*5) Tilt Angle at Cracking*

$$f_r = 0.6\sqrt{35} = 3.55 \text{ MPa}$$

$$M_{lat} = \frac{(f_{tpres} + f_{tself} + f_r)I_y}{(b_t / 2)}$$
$$= \frac{(-6.50 + 7.00 + 3.55)(14\,900 \times 10^6) / 10^6}{(900 / 2)}$$
$$= 134 \text{ kN}\cdot\text{m}$$

$$\theta_{max} = \frac{M_{lat}}{M_g} = \frac{134}{1950} = 0.0688 \text{ rad}$$

*6) Tilt Angle at Failure*

$$\theta'_{max} = \sqrt{\frac{16}{(2.5)(372)}} = 0.1312 \text{ rad}$$

*7) Theoretical Lateral Deflection at Tilt Angle $\theta'_{max}$*

$$\overline{Z}'_o = (372)[1 + (2.5)(0.1312)] = 494 \text{ mm}$$

*8) Factor of Safety Against Cracking*

$$FS = \frac{1}{372 / 936 + 0.0170 / 0.0688} = 1.55 > 1.0 \quad \text{OK}$$

*9) Factor of Safety Against Failure*

$$FS' = \frac{(936)(0.1312)}{(494)(0.1312) + 16} = 1.52 < 1.55$$

Therefore, FS′ = FS = 1.55 > 1.5 OK

Note: If girder is supported 0.5m from each end then FS = 1.15 and FS′ = 1.15, NG.

(B) Transporting (supported from below)

*1) Radius of Stability*

rotational stiffness of vehicle = $K_o$ = 4250 kN·m

$$r = \frac{4250 \times 10^6}{512 \times 10^3} = 8300 \text{ mm}$$

*2) Initial Eccentricity*

placing offset on truck = 25 mm

$$e_i = (20)(0.477) + 25 = 35 \text{ mm}$$

*3) Height of C.G. Beam Above Roll Axis*

camber = 50 mm

height of C.G. beam above road

= 1800 + 100 + 940 = 2840 mm

height of roll axis above road

= 600 mm

$$Y = 2840 + (50)(0.477) - 600 = 2264 \text{ mm}$$

*4) Theoretical Lateral Deflection*

$$\overline{Z}_o = \frac{12.8 / 10^3}{(12)(31\,000)(14\,900 \times 10^6)(40\,000)}$$
$$[(0.1)(36\,000)^5 - (2000)^2(36\,000)^3 + (3)(2000)^4(36\,000) + (6/5)(2000)^5]$$
$$= 338 \text{ mm}$$

*5) Tilt Angle at Cracking*

$$f_r = 0.6\sqrt{45} = 4.02 \text{ MPa}$$

$$M_{lat} = \frac{(-6.00 + 7.00 + 4.02)(14\,900 \times 10^6) / 10^6}{(900 / 2)}$$
$$= 166 \text{ kN}\cdot\text{m}$$

$$\theta_{max} = \frac{166}{1950} = 0.0852 \text{ rad}$$

*6) Tilt Angle at Maximum FS′*

roadway superelevation of vehicle

= 0.0800 rad

distance center of truck to center of tire

= 900 mm

$$\theta'_{max} = \frac{900 - (600)(0.0800)}{8300} + 0.0800 = 0.1827 \text{ rad}$$

*7) Theoretical Lateral Deflection at Tilt Angle $\theta'_{max}$*

$$\bar{Z}'_o = (338)[1 + (2.5)(0.1827)] = 492 \text{ mm}$$

*8) Factor of Safety Against Cracking*

$$FS = \frac{(8300)(0.0852 - 0.0800)}{(338)(0.0852) + 35 + (2264)(0.0852)} = 0.17 < 1.0 \quad NG$$

*9) Factor of Safety Against Rollover*

$$FS' = \frac{(8300)(0.1827 - 0.0800)}{(492)(0.1827) + 35 + (2264)(0.1827)} = 1.58 > 1.5 \quad OK$$

*Note:*

If girder is supported 0.5m from each end then FS = 0.37 and FS′ = 1.44, NG.

If girder is supported 3.5m from each end then FS = 0.07 and FS′ = 1.71, NG.

The support location does not sufficiently improve the factor of safety against cracking, therefore provide temporary post-tensioning to the top flange for transporting with supports at 2m from each end. If $f_{tpost}$ = 2.70 MPa due to post-tensioning then FS = 1.13 and FS′ = 1.58, OK.

## 3.9.4 STORAGE

Wherever possible, an element should be stored on only two points of support located at or near those used for stripping and handling. Thus, the design for stripping and handling will usually control. Where points other than those used for stripping or handling are used for storage, the storage condition must be checked.

If support is provided at more than two points, and the design based on more than two supports, precautions must be taken so that the element does not bridge over one of the supports, due to differential support settlement. Particular care must be taken for prestressed elements, with consideration made for the effect of prestressing. Equalization of stresses on both faces will help to minimize deformations in storage.

The primary causes of warpage are differential temperature, differential shrinkage, creep and storage conditions. Warping in storage is minimized by resisting flexure about the strong axis of the element.

## 3.9.5 TRANSPORTATION

### Mass and Size Limitations

Governmental regulations limit the size of shipping loads. Where climatic conditions result in load restrictions during spring thaws, actual timing of the expected transportation becomes significant.

The common size of payload is 18 t with a size restriction of 2.6 m in width, 2.4 m in height, and 12.2 m in length. If a unit will fit within the confines of this size and mass restriction, the panel can be hauled on a standard trailer, without requiring permits. By using low-bed trailers, the panel height can generally be increased to about 3.0 m without requiring special permits. However, low-beds cost more to operate, and their shorter bed length may restrict the length of the panel. In some jurisdictions, total heights (roadbed to top of panel) of 4.1 m are permitted without special permit, while in others this limit is 3.7 m. On occasion, use of such heights requires special routing to avoid low overpasses and overhead wires. Restrictions generally exist for loads over 2.6 m in width, and maximum widths can vary from 3.0 m to 3.7 m to 4.3 m. Most jurisdictions allow lengths up to 17 m without special permit. Some allow loads over 21 m with only a simple permit, while others require, for any load over 17 m, special permit, escorts front and rear, and travel limited to certain times of the day. Besides lengths and widths, load restrictions vary widely. Thus, the general load limit without permit (18 t to 20 t) can in some jurisdictions be modified up to 90 t with special permit, while in others there are very severe restrictions on loads over 22 t.

One of the important factors when considering transportation, is the location of supports. Panel supports should be chosen such that the imposed tensile stresses do not exceed the permissible value of $0.4\lambda\sqrt{f'_{ci}}$ with due consideration given to the effect of dynamic loading. When an element is non-symmetrical about a bending axis of the cross-section, the location of support points to produce equal bending stresses on each face will be a function of the ratio of the distances from the bending axis to top and bottom fibre. If the tensile stresses cannot be contained within the above limiting value, auxiliary bracing, such as stiff-backs or space frames, can be attached to the element prior to loading.

Most precast firms use either flatbed or low-bed trailers, and these trailers undergo significant deformations during hauling. Thus support at more than two points on a trailer can be achieved only after considerable and expensive modification of the trailer.

When a differential strain on opposite faces can be established, it can be assumed that the panel will bend in a circular shape. From this assumption, the amount of lateral bowing can be determined, as indicated in Fig. 3.9.13.

For the type of support conditions illustrated in Fig. 3.9.14, warping can occur in both directions. By superposition, the total warpage at the maximum point can be estimated by:

$$\Delta_{max} = \frac{5 w_d \sin\theta}{384 E_c}\left(\frac{a^4}{I_x} + \frac{b^4}{I_z}\right)$$

where:

$\Delta_{max}$ = maximum deflection

$w_d$ = panel dead load per unit area

$E_c$ = modulus of elasticity of concrete

a = panel support height

b = horizontal distance between supports

$I_x, I_y$ = moment of inertia of uncracked transformed section in the respective direction, per unit width of panel

Fig. 3.9.13 Bowing due to differential strain

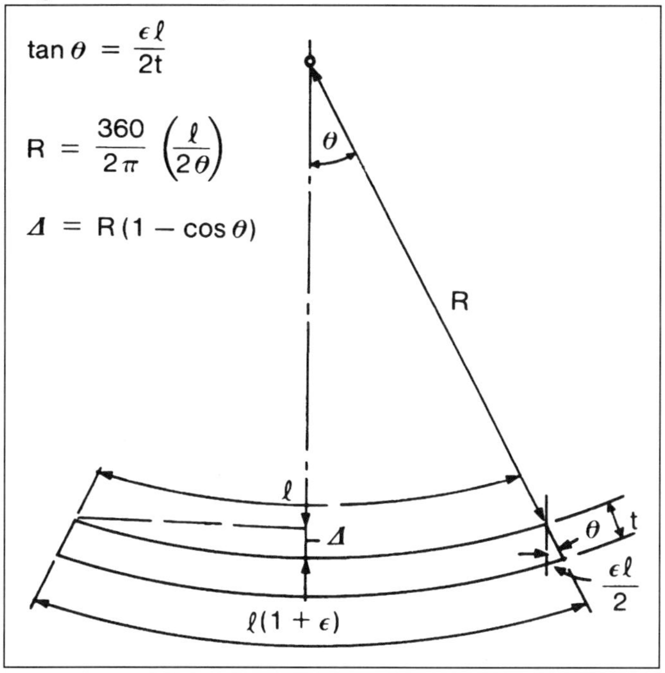


Fig. 3.9.14 Panel warpage in storage

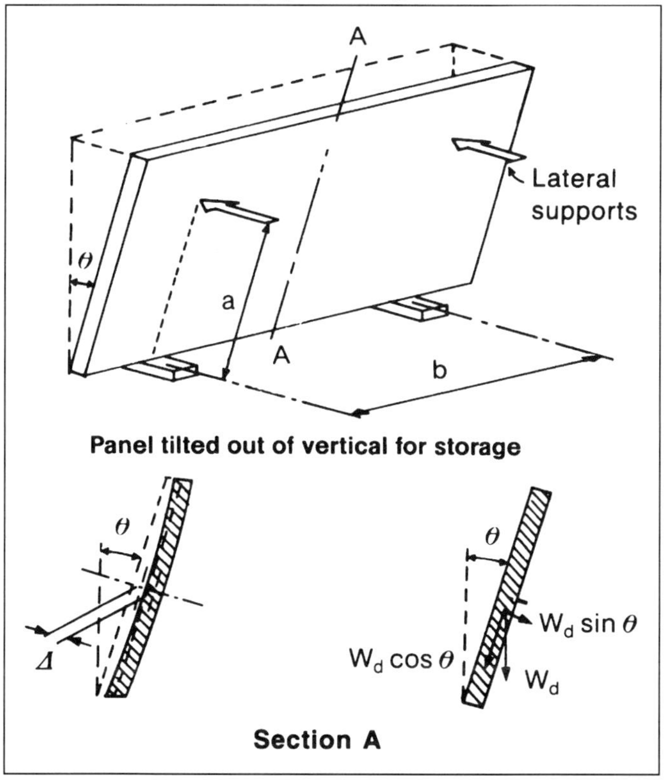


When considering warpage in storage, the time dependent creep and shrinkage effects should be considered. The total deformation will also be a function of the amount of reinforcement. The total deformation at any time can be estimated as:

$$\Delta = \Delta_i(1+\lambda)$$

where:

$\Delta$ = time dependent displacement

$\Delta_i$ = instantaneous displacement

$\lambda$ = amplification due to creep and shrinkage (Fig. 3.9.15)

$\rho'$ = $A'_s/b\,d$

Fig. 3.9.15 Effect of compression reinforcement on creep

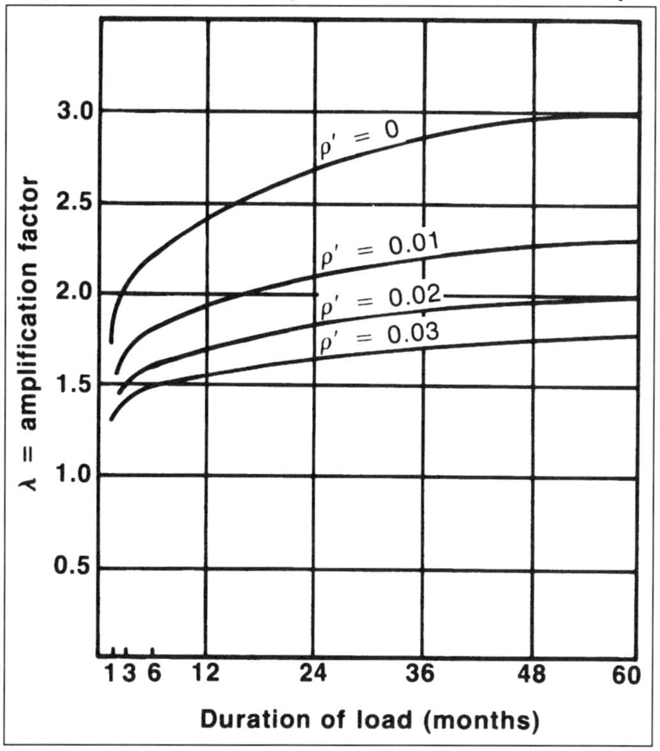


## 3.9.6 ERECTION

Generally, erection poses no particularly new problems in the design of precast elements. The stress limitations and multipliers previously discussed for stripping and handling are applicable during erection.

The centre of gravity of the element should be computed and lifting points located so as to place the centre of gravity of the loads directly below the main hook and below the lowest point of the attachment of slings, in order that the element will hang level during erection. Walls with openings should be carefully checked for erection stresses and braced for handling when necessary.

Erection is also an engineering consideration from the viewpoint of making the process fast, efficient and safe. Speed of erection is directly related to the type of connections selected (see Chap. 4) and the arrangement of the building frame. It is highly desirable that connections allow for initial setting of a panel and immediate release of crane, with final alignment completed independent of crane support.

Any specific restrictions placed on the sequence of panel erection, related to strength of the structural frame, must be indicated on the Contract Drawings. Limitations may state that loading on the structure shall be balanced, or may require that no elevation be erected more than a stated number of floors ahead of the remaining elevations, or they may involve the rigidity of the structure, requiring that walls not be erected prior to completion of floors designed to carry horizontal loads. In steel frames, it should be determined how far ahead final connections of the frame must be completed prior to panel erection. In concrete frames, it must be determined what strength of concrete is required prior to imposing loads of the precast element. The frame designer must also recognize that connections between panel and frame impose concentrated loads on the frame and that these loads may require supplementary local reinforcement. In the case of multi-storey concrete frames, consideration should be given to the effects of frame shortening due to shrinkage and creep. Delay in erecting precast elements to permit a portion of the shrinkage and creep to occur may be beneficial. For elements which are supported on a continuous bed of grout (such as in load bearing wall construction) the maximum number of floor levels which can be erected using only shims should be determined and, if critical, indicated on the Contract Drawings.

The deflection of a support is a function of the stiffness of the support. Thus, where adjacent elements are supported on different portions of the building frame with differing stiffnesses, relative deflections between adjacent elements may occur. This is often the case at building corners, where the structural arrangement may result in significantly different stiffnesses. It is also a concern where the structural frame cantilevers. If elements are attached in a manner which will tend to prevent relative displacement, element stresses must be re-evaluated.

Erection equipment will frequently influence the size of precast elements. The designer must consider access to the site to be certain that there is sufficient space in proximity to the structure to allow erection to proceed as contemplated. This requires coordination with other trades to assure that there will be no interference with crane access, such as foundation walls. It should be noted that cranes are rated by the safe capacity they will lift with the shortest boom and at the steepest boom-up angle. Maximum lifts will reduce rapidly as boom length and angle change, and the designer should, if possible, consider this during the planning stage. On very tight sites, the use of a climbing crane may be necessary. If so, this will have a significant effect on the planning of the structural frame and the sequencing of construction, which must be considered during the planning stages of the structural frame.

After erection, each element must be stable and offer resistance to wind, accidental impact, and loads which may be imposed due to other construction operations. Provision should be made both in the precast element and the support system to permit immediate bracing of the element. The arrangement of such temporary bracing should not interfere with adjacent erection and other construction processes, and the bracing must be maintained until permanent connections are accomplished. It is desirable that each precast element be braced independently of other elements, so that an element may be moved without affecting adjacent elements. Permanent connections which safely sustain loads imposed during erection may be used instead of temporary bracing.

Temporary supports or bracing should be designed with a safety factor of 2 to withstand wind loads in accordance with building code requirements, but not less than 0.7 kN/m$^2$ on projected surfaces. Experience has shown that, in bracing, tension failure of embedded inserts can be critical. Seismic forces are usually only considered when the time of unsupported erection condition will be unduly prolonged (i.e., winter shutdown), or when dislodgement could cause progressive collapse.

# 3.10 SPECIAL DESIGN SITUATIONS

## 3.10.1 GENERAL

This section outlines solutions of special situations which may arise in the design of a precast floor or roof system. Since production methods of products vary, local producers should be consulted. Also, test data may indicate that the conservative guidelines presented here may be exceeded for a specific application.

## 3.10.2 DISTRIBUTION OF CONCENTRATED LOADS

Frequently, floors and roofs are subjected to line loads, for example from walls, and concentrated loads. The ability of hollow-core systems to transfer or distribute loads laterally through grouted shear keys has been demonstrated in several published tests,[R3.10.1] and many unpublished tests. Research is continuing, and the recommendations here may be refined in the future. Based on tests, analysis and experience, the PCI Hollow-core Slab Producers Committee recommends that line and concentrated loads be resisted by an effective section as described in Fig. 3.10.1. Exception: If the total deck width, perpendicular to the span, is less than the span, modification may be required. Contact local producers for recommendations.

Load distribution of stemmed elements may not necessarily follow the same pattern, because of different torsional resistance properties.

**Example 1 – Load distribution**

*Given:*

An untopped hollow-core floor with 1220 mm wide slabs, supporting a load bearing wall and concentrated loads as shown in Fig. 3.10.2.

*Problem:*

Determine the design loads for the slab supporting the wall and concentrated loads.

**Fig. 3.10.1 Assumed load distribution**

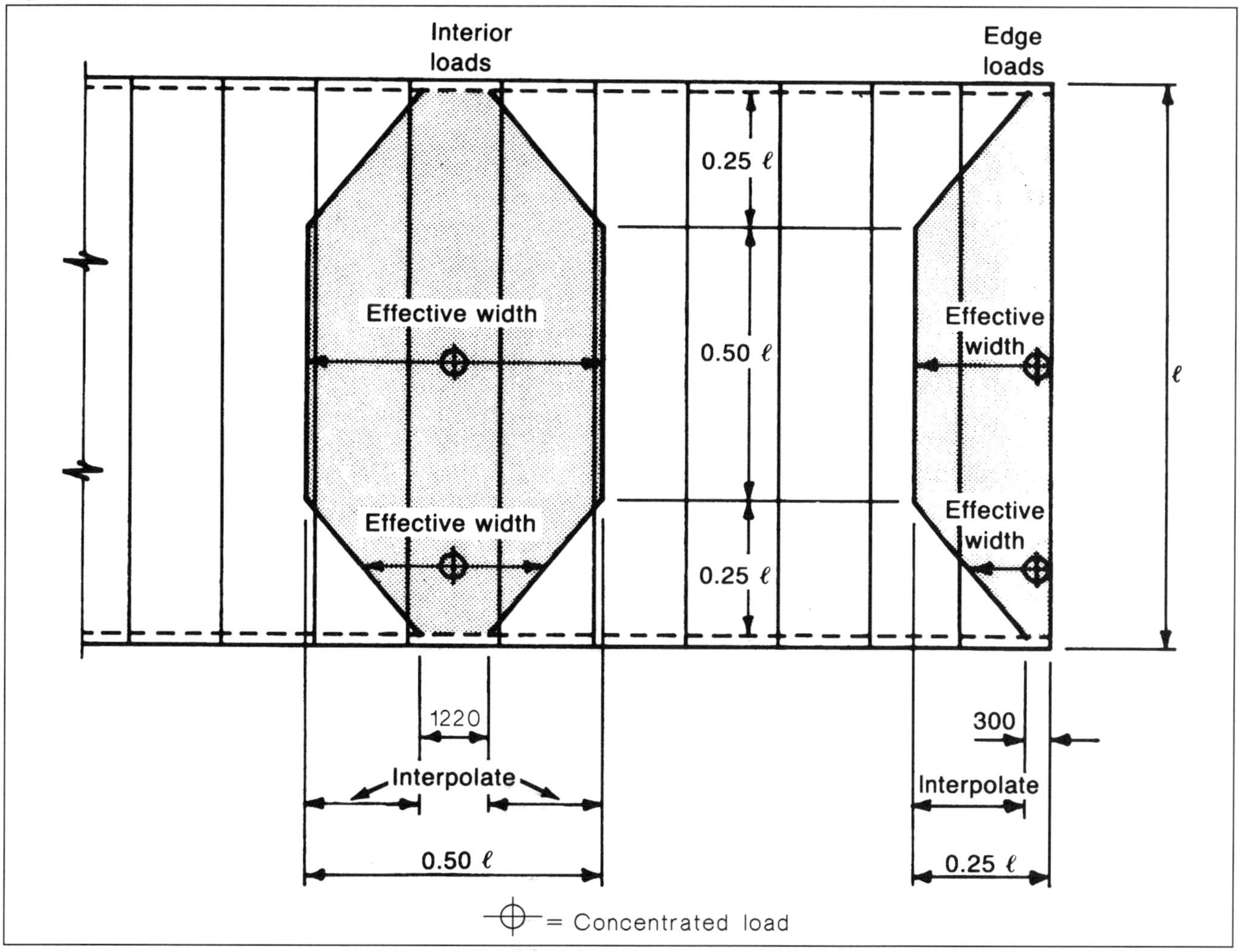


*Solution:*

(Note: Each step corresponds to a line number in table in Fig. 3.10.2):

1. Calculate the shears and moments for the non-distributable (uniform) loads:

   $w_f$ = (1.25)(2.7 + 0.5) = (1.5)(2.0) = 7.0kN/m$^2$

2. Calculate the shears and moments for the distributable (concentrated and line) loads:

   $w_f$ = (1.25)(9.5) + (1.5)(15.2) = 34.7 kN/m

   $P_{1f}$ = (1.25)(2.2) + (1.5)(4.4) = 9.4 kN

   $P_{2f}$ = (1.25)(4.4) + (1.5)(13.3) = 25.4 kN

3. Calculate effective width along the span:

   At the support: width = 1220 mm

   At 0.25ℓ (1905 mm): width = 0.5ℓ = 3810 mm

   Between x = 0 and x = 1905 mm:

   width = 1220 + (x/1905)(3810 − 1220)

   = 1220 + 1.36x

4. Divide distributable shears and moments from step 2 by the effective widths from Step 3.

5. Add the distributed shears and moments to the nondistributable shears and moments from Step 1.

Once the moments and shears are determined, the slabs are designed as described in Sect. 3.4.3.

This method is suitable for computer solution. For manual calculations, the procedure can be simplified by investigating only critical sections. For example, shear may be determined by dividing all distributable loads by 1.2 m and flexure at midspan can be checked by dividing the distributable loads by 0.5ℓ.

## 3.10.3 EFFECT OF OPENINGS

Openings may be provided in precast decks by: (1) saw cutting after the deck is installed and grouted, (2) forming (blocking out) or sawing in the plant, or (3) using short units with steel headers or other connections. In hollow-core or solid slabs, structural capacity is least affected by orienting the longest dimension of an opening parallel to the span, aligning several openings parallel to the span, or by coring small holes to cut the fewest strands.

**Fig. 3.10.2 Example 1 - Load distribution in hollow-core slab**

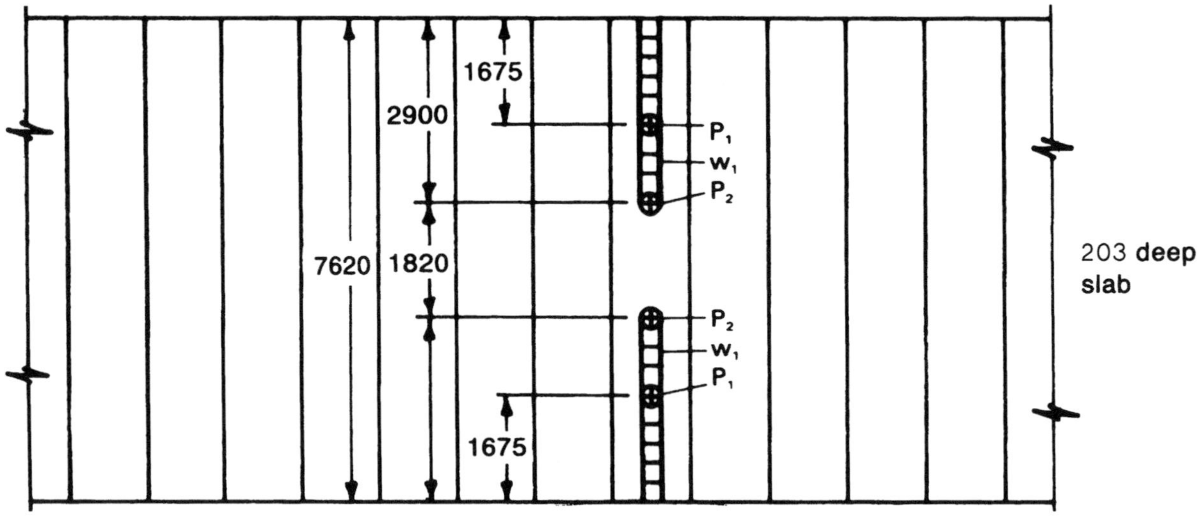


SDL = 0.5 kN/m² Slab wt. = 2.7 kN/m² $w_{1D}$ = 9.50 kN/m $P_{1D}$ = 2.20 kN $P_{2D}$ = 4.40 kN
LL = 2.0 kN/m² $w_{1L}$ = 15.20 kN/m $P_{1L}$ = 4.40 kN $P_{2L}$ = 13.30 kN

| Shears and moments | | | | | | | | | | | |
|---|---|---|---|---|---|---|---|---|---|---|---|
| Distance from support | | 0 | h/2 | 300 | 600 | 900 | 1200 | 1500 | 2250 | 3000 | 3810 |
| 1. Non-distributable loads | $V_f$ | 26.7 | 26.0 | 24.6 | 22.5 | 20.4 | 18.3 | 16.2 | 11.0 | 5.7 | 0 |
| | $M_f$ | 0 | 2.6 | 7.7 | 14.7 | 21.2 | 27.0 | 32.1 | 42.3 | 48.5 | 50.8 |
| 2. Distributable loads | $V_f$ | 135.4 | 132.0 | 125.0 | 114.6 | 104.2 | 93.8 | 83.4 | 48.0 | 0 | 0 |
| | $M_f$ | 0 | 13.4 | 39.1 | 75.0 | 107.8 | 137.5 | 164.1 | 211.4 | 235.2 | 235.2 |
| 3. Effective width (mm) | | 1220 | 1358 | 1628 | 2036 | 2444 | 2852 | 3259 | 3810 | 3810 | 3810 |
| 4. Distributed shear & moments | $V_f$ | 111.0 | 97.2 | 76.8 | 56.3 | 42.6 | 32.9 | 25.6 | 12.6 | 0 | 0 |
| | $M_f$ | 0 | 9.9 | 24.0 | 36.8 | 44.1 | 48.2 | 50.4 | 55.5 | 61.7 | 61.7 |
| 5. Design shears and moments | $V_f$ | 137.7 | 123.2 | 101.4 | 78.8 | 63.0 | 51.2 | 41.8 | 23.6 | 5.7 | 0 |
| | $M_f$ | 0 | 12.5 | 31.7 | 51.5 | 65.3 | 75.2 | 82.5 | 97.8 | 110.2 | 112.5 |

$V_f$ in kN/m; $M_f$ in kN · m/m

In some cases, it may be required to provide openings through floor and roof units. Large openings are usually made by block-outs in the forms during the manufacture; smaller ones (up to about 200 mm) are usually field drilled. Openings in flanges of stemmed elements should be limited to the "flat" portion of the flange, that is, beyond 25 mm of the edge of the stem on double tees and 75 mm of the edge of the stem on single tees. Angle headers are often used for framing large openings in hollow-core floor or roof systems.

Following are some conservative guidelines regarding design of hollow-core slabs around openings. Some producers may have data to support a different procedure:

1. An opening located near the end of the span and extending into the span less than the lesser of $0.125\ell$ or 1.2 m may be neglected when designing for flexure in the midspan region.
2. Strand development must be considered on each side of an opening which cuts strand. (see Sect. 3.4.9)
3. Slabs which are adjacent to openings which are long ($\ell/4$ or more), or occur near midspan, may be considered to have a free edge for flexural design.
4. Slabs which are adjacent to openings closer to the end than $3\ell/8$ may be considered to have a free edge for shear design.

## 3.10.4 CANTILEVERS

The most effective way to design cantilevered elements will depend on the type of product, method of production, span conditions and section properties of the element. The designer is advised to consult with local producers to determine the most effective method.

Many producers prefer to design cantilevers as reinforced concrete elements using deformed reinforcing bars or short pieces of unstressed strand to provide the negative moment resistance. When using strand which is unstressed, the stress under factored loads should be limited to 400 MPa. Pretensioned strands are sometimes used to improve the performance characteristics of cantilevers, however design, production and quality control should be rigorous to ensure proper performance.

It is suggested that concrete tensile stresses due to specified loads be limited to 0.7 MPa when prestressed strands are used and to the cracking tension $0.6\lambda\sqrt{f'_c}$ when non-prestressed reinforcement is used for negative moment resistance.

Strand development may be a problem in cantilevers, and it is suggested that when prestressed strand is used, the design stress, $f_{pr}$, be limited to one-half that indicated by Fig. 3.4.8.

### 3.10.5 COMPOSITE TOPPING WITH HOLLOW-CORE SLABS

Most precast floor and roof systems are untopped, but a composite, cast-in-place concrete topping is sometimes used in floor construction. The composite action adds stiffness and strength for gravity loads, and may be required as a diaphragm to transfer lateral loads.

Tests have shown that the normal finished surface of hollow-core and stemmed deck will develop the 0.42 MPa on the interface surface specified in Clause 17, CSA-A23.3-94, if the precast surface is thoroughly cleaned before topping is placed. Section 3.6.7 describes design procedures for horizontal shear transfer. In nearly all cases, the allowable 0.42 MPa is enough to develop the full strength of the topping on precast decks.

The strength of the topping may be determined from the design requirements for the deck. Producer's load tables usually indicate either 20 or 25 MPa. The load tables in Chapter 7 are based on a topping strength of 25 MPa.

## 3.11 SANDWICH PANEL DESIGN

### 3.11.1 FLEXURAL ANALYSIS

The flexural design of two wythe sandwich panels for stripping, handling and service loads follows the same principles as those for single wythe panels as described in Sections 3.3 and 3.4. However, the flexural resistance, flexural stiffness and axial resistance will depend upon the degree of interaction between the two wythes. For a more complete description of sandwich panel configurations and shear connectors, see Section 6.6.

If the wythes are connected so that full shear transfer can be achieved, the resistance and stiffness can be calculated using the composite section and gross moment of inertia.

In a totally non-composite panel, the resistance and stiffness are computed on the basis of the single structural wythe; in addition, the structural wythe must be designed to carry the loads from the second non-structural wythe.

Behaviour of a semi-composite panel, with partial shear transfer between wythes, falls between the above limits. For this condition, the effective moment of inertia should be determined by test or experience.

### 3.11.2 REINFORCEMENT

Non-prestressed reinforcement should be designed for handling and in-place loads. Size and spacing will be a function of strength and crack control criteria, in accordance with Sect. 3.4.2. Maximum spacing of reinforcement in any wythe exposed to the environment should be 150 mm if mesh, or 5 times wythe thickness if reinforcing bars. Except as otherwise required by analysis or experience, the minimum reinforcement in each wythe should be 0.001 of the cross-sectional area of that wythe; this amount of reinforcement should be placed in each direction, except that:

1. Wythes not exposed to the environment which are less than 1220 mm in one direction need not be reinforced in that direction;
2. Wythes exposed to the environment which are less than 610 mm in one direction need not be reinforced in that direction.

Prestressed reinforcement, if used, should provide an average prestress after all losses of not less than 1.0 MPa. Transverse reinforcement in accordance with the above should be provided. Wythes should be stressed concentrically, and consideration given to the relative stresses in the interior and exterior wythes to prevent undesirable bowing of the panel.

### 3.11.3 NON-LOAD BEARING PANEL DESIGN

Non-load bearing panels are those wall elements which are intended to resist only their own mass and those lateral forces which are directly imposed.

Unintended load transfer from or to adjacent panels should be avoided. This is accomplished by detailing joints with sufficient space so that the anticipated deformation of the supporting structure and the panel will be less than the space between elements.

The structural wythe should preferably be attached to the supporting structure such that this wythe is in compression. Hanging of panels may be acceptable, provided that the tensile loads are totally resisted by reinforcement.

For non-composite sandwich panels, the effect of the eccentricity of the non-load bearing wythe upon the forces and moments developed in the structural wythe should be considered:

1. Where the total load W is transferred through a single connector, a beam portion of the structural wythe, of width not to exceed 8t, may be assumed effective, and designed to transfer this load to the supporting structure (Fig. 3.11.1(a)).
2. Where the total load W is transferred through a series of shear connectors, a beam portion equal to the spacing between connectors, but not greater than 8t, may be assumed (Fig. 3.11.1(b)).

**Fig. 3.11.1 Effective beam widths for panel design**

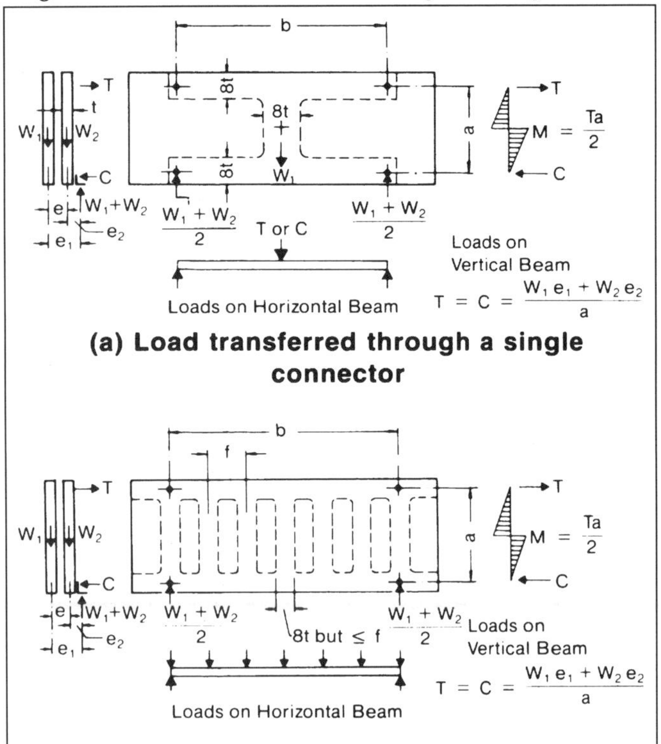


**(a) Load transferred through a single connector**

**(b) Load transferred through a series of shear connectors**

## 3.11.4 LOAD BEARING PANEL DESIGN

Load bearing panels are those elements which, in addition to their own mass, support or transfer loads originating at other locations of the structure. Generally, they will be continuously supported along their lower surface, and will uniformly transfer the superimposed loads. Shear walls, which may, but need not be, bearing walls, provide lateral load resistance for the structure.

The vertical load resistance of a load bearing panel may be determined by the design methods of Section 2.7. The design of connections for vertical and lateral load transfer is treated in Chapter 4.

**Example 1 – Non-composite, reinforced, non-load bearing panel**

*Given:*

3500 × 2500 non-composite flat panel

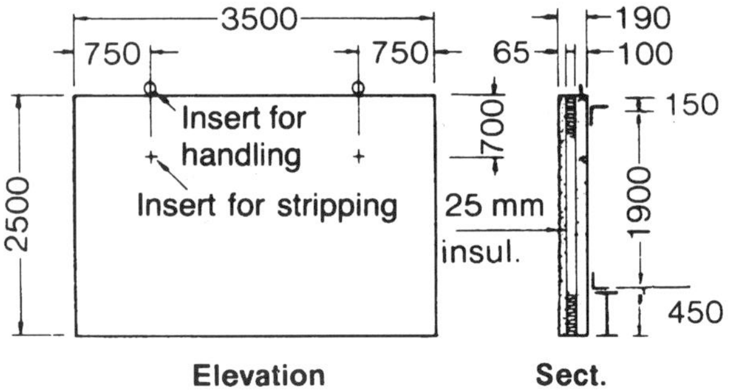


**Elevation** **Sect.**

*Concrete:*

$f'_c = 35$ MPa

$f'_{ci} = 15$ MPa

Normal Density

Section properties of structural wythe:

$A_g = 100\,000 \text{ mm}^2/\text{m}$

$I = 83 \times 10^6 \text{ mm}^4/\text{m}$

$S = 1667 \times 10^3 \text{ mm}^3/\text{m}$

Wind service load = 1.0 kN/m²

*Problem:*

Determine the flexural stresses and required reinforcement for stripping and for service conditions.

*Solution:*

1. Determine the moments during stripping:

Calculate the mass of the panel per unit area:

| | |
|---|---|
| Exterior wythe: $(0.065)(2400)(9.8)/10^3$ | = 1.5 kN/m² |
| Interior wythe: $(0.100)(2400)(9.8)/10^3$ | = 2.4 |
| Insulation: $(7)(9.8)/10^3$ | = 0.1 |
| | 4.0 kN/m² |

Assuming that the exterior non-load bearing wythe is retarded exposed aggregate cast face down, the load multiplier from Fig. 3.9.1 is 1.2.

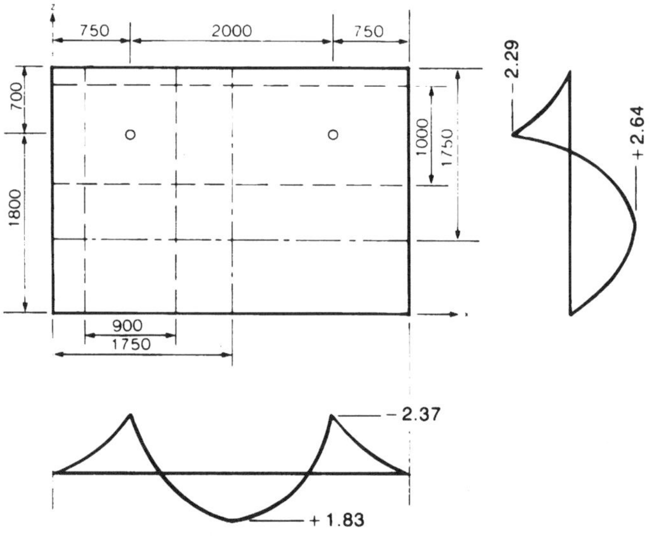


For bending about the x-x axis, the tributary width is half the panel width, namely 1750 mm, and, extrapolating from Fig. 3.9.3, the effective flexural width is the lesser of:

$(2500 - 700)/2 = 900$ mm (controls)

$10t = (10)(100) = 1000$ mm

$3500/2 = 1750$ mm

and:

$$M_x^- = -\frac{(4.0)(1.75)(1.2)(0.7)^2}{(2)(0.9)} = -2.29 \text{ kN·m/m}$$

$$M_x^+ \approx \frac{(4.0)(1.75)(1.2)(1.8)^2}{(8)(0.9)} - \frac{2.29}{2} = 2.64 \text{ kN·m/m}$$

For bending about the z-z axis, the tributary width for load on the stripping inserts is determined from statics to be 1750 mm and the effective flexural width is the lesser of:

$2000/2 = 1000$ mm (controls)

$10t = 1000$ mm

$(700)(2) = 1400$ mm

and:

$$M_z^- = -\frac{(4.0)(1.75)(1.2)(0.75)^2}{(2)(1.0)} = -2.37 \text{ kN·m/m}$$

$$M_z^+ = \frac{(4.0)(1.75)(1.2)(2.0)^2}{(8.0)(1.0)} - 2.37 = 1.83 \text{ kN·m/m}$$

Therefore, the maximum moment at stripping is $M_x^+ = 2.64$ kN·m/m.

2. Determine moments due to wind:

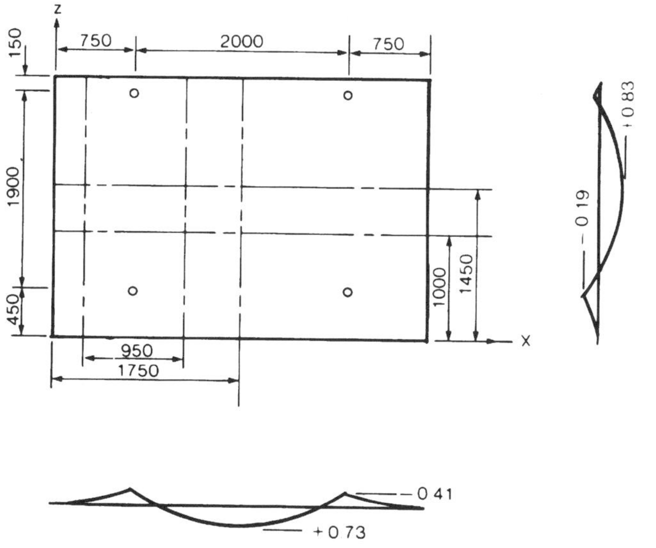


For bending about the x-x axis, the tributary width is 1750 mm and the effective width is 950 mm:

$$M_x^+ = -\frac{(1.0)(1.75)(0.45)^2}{(2)(0.95)} = -0.19 \text{ kN·m/m}$$

$$M_x^- = \frac{(1.0)(1.75)(1.9)^2}{(8)(0.95)} = 0.83 \text{ kN·m/m}$$

For bending about the z-z axis, the tributary width is 1450 mm and the effective width is 1000 mm:

$$M_z^- = -\frac{(1.0)(1.45)(0.75)^2}{(2)(1.0)} = -0.41 \text{ kN·m/m}$$

$$M_z^+ = \frac{(1.0)(1.45)(0.75)^2}{(8)(1.0)} = 0.73 \text{ kN·m/m}$$

3. Determine moments due to eccentricity of loads:

From Fig. 3.11.1, assuming that the outside wythe is supported by a single shear connector at the centre of the panel:

$W_1 = (1.5)(3.5)(2.5) = 13.2$ kN

$W_2 = (2.5)(3.5)(2.5) = 21.9$ kN

Assuming the support point for the panel connection is 50 mm from the inside face of the panel:

$e_1 = 208$ mm

$e_2 = 100$ mm

$$T = C = \frac{(13.2)(0.208) + (21.9)(0.1)}{1.9} = 2.6 \text{ kN}$$

$8t = (8)(100) = 800$ mm

$$M_x^+ = \frac{Ta}{2} = \frac{(2.6)(1.9)}{(2)(0.8)} = 3.09 \text{ kN·m/m}$$

$$M_z^+ = \frac{Tb}{4} = \frac{(2.6)(2.0)}{(4)(0.8)} = 1.63 \text{ kN·m/m}$$

4. Determine combined moments due to wind and load eccentricity:

$M_x^+ = 0.83 + 3.09 = 3.92$ kN·m/m (controls)

$M_z^+ = 0.73 + 1.63 = 2.36$ kN·m/m

5. Calculate flexural stresses:

At stripping:

$$f = \frac{(2.64)(10^6)}{(1667 \times 10^3)} = 1.6 \text{ MPa}$$

$$0.4\lambda\sqrt{f'_{ci}} = (0.4)(1.0)\sqrt{15} = 1.5 \text{ MPa} < 1.6$$

At service load:

$$f = \frac{(3.92)(10^6)}{(1667 \times 10^3)} = 2.4 \text{ MPa}$$

$$0.4\lambda\sqrt{f'_c} = (0.4)(1.0)\sqrt{35} = 2.3 \text{ MPa} < 2.4$$

6. Determine reinforcement requirements:

Assume a single layer of mesh with a wire spacing of 100 mm at mid-depth of the structural wythe:

$d = 50$ mm

$j_d = (0.9)(50) = 45$ mm

$k_d = (0.3)(50) = 15$ mm

$h_1 = 50 - 15 = 35$ mm

$h_2 = 100 - 15 = 85$ mm

$$\frac{h_2}{h_1} = \frac{85}{35} = 2.42$$

Since the interior structural wythe is not exposed to view or to the weather, use maximum w = 0.40 mm:

$$f_s = \frac{w}{(11\times10^{-6})\frac{h_2}{h_1}\sqrt[3]{dA_c}}$$

$$= \frac{0.40}{(11\times10^{-6})(2.42)\sqrt[3]{(50)(100)(70)}} = 213\,\text{MPa}$$

The controlling moment is $M_x^+$ under combined service wind load and eccentricity of the dead loads:

$$A_s = \frac{M}{0.9df_s} = \frac{(3.92)(10^6)}{(0.9)(50)(213)} = 410\ \text{mm}^2/\text{m}$$

For the non-structural wythe, use minimum reinforcement of (0.001)(65)(1000) = 65 mm²/m with maximum wire spacing of 150 mm.

7. Shear connector and tie design:

The shear connector is designed to carry the entire mass of the exterior wythe:

$V_f = (1.25)(1.5)(2.5)(3.5) = 16.4$ kN

If tension ties are provided on 600 mm centres in each direction:

$T_f = (1.25)(1.2)(1.5)(0.6)(0.6) = 0.8$ kN

See Sect. 6.6.8 for further information on wythe connector design.

**Example 2 – Composite, prestressed, non-load bearing flat panel**

*Given:*

2400 × 6000 composite prestressed flat panel

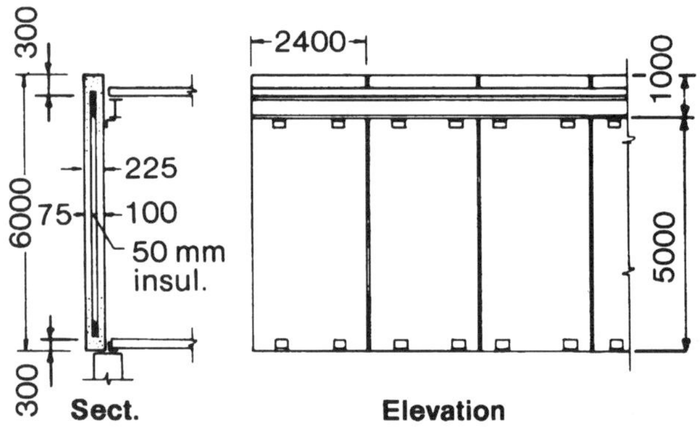


*Concrete:*

$f'_c = 35$ MPa

$f'_{ci} = 25$ MPa

Normal Density

Section properties of composite section:

$A_g = 420\,000$ mm²

$I = 2230 \times 10^6$ mm⁴

$y_b = 116$ mm

$y_t = 109$m

$S_b = 19\,200 \times 10^3$ mm³

$S_t = 20\,500 \times 10^3$ mm³

m = 1020 kg/m

w = 10.0 kN/m

Wind service load = 1.2 kN/m²

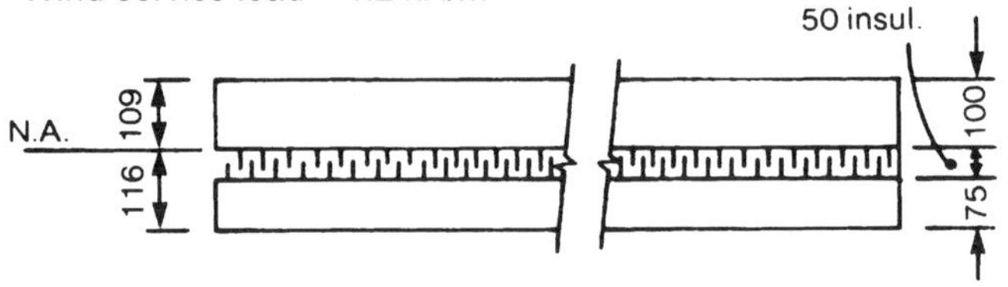


*Problem:*

Determine required prestressed reinforcement and check horizontal shear capacity between wythes.

*Solution:*

1. Stripping stresses:

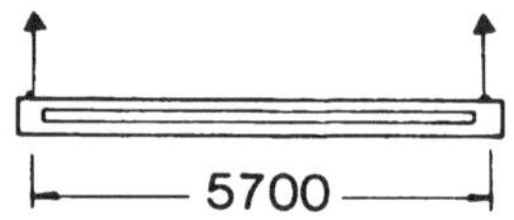


From Fig. 3.9.1, use a load multiplier of 1.2 at stripping. Stresses due to dead load of element are:

$$M = \frac{(10.0)(1.2)(5.7)^2}{8} = 48.7\ \text{kN·m}$$

$$f_b = -\frac{(48.7)(10^6)}{(19\,200\times10^3)} = -2.5\,\text{MPa}$$

$$f_t = \frac{(48.7)(10^6)}{(20\,500\times10^3)} = 2.4\,\text{MPa}$$

2. Storage stresses:

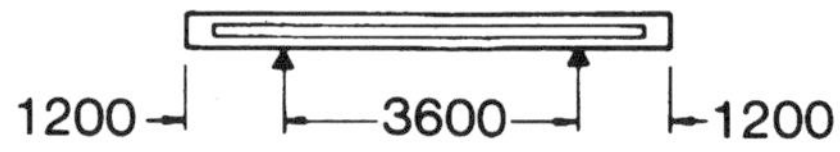


By inspection, stresses not critical.

3. Transportation stresses:

Assume that the element is supported as shown during transportation. Use a load multiplier of 1.5:

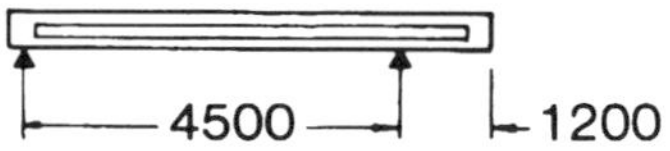


$$M^- = -\frac{(10.0)(1.5)(1.2)^2}{2} = -10.8 \text{ kN·m}$$

$$M^+ \approx \frac{(10.0)(1.5)(4.5)^2}{8} - \frac{10.8}{2} = 32.6 \text{ kN·m}$$

4. Service wind stresses:

$$M^- = -\frac{(1.2)(2.4)(1.0)^2}{2} = -1.5 \text{ kN·m}$$

$$M^+ = \frac{(1.2)(2.4)(5.0)^2}{8} = 9.0 \text{ kN·m}$$

5. Select prestressed reinforcement:

The critical loading condition from above is at stripping. By providing the same prestress force in both the interior and the thinner exterior wythe, the resulting stress will be slightly higher in the exterior wythe. This will tend to deflect the panel inward, counteracting the usual tendency to bow outward. Try providing 1.5 MPa prestress in the 100 mm wythe:

$$P = (1.5)(100)(2400)/10^3 = 360 \text{ kN}$$

Stress in 75 mm wythe:

$$f = \frac{(360)(10^3)}{(75)(2400)} = 2.0 \text{ MPa}$$

Use 5 – 11 mm 1860 MPa strands in each wythe.

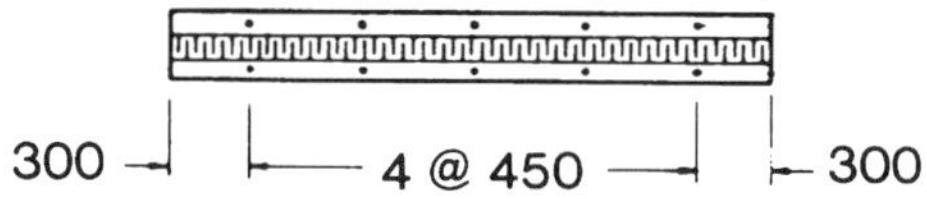


6. Check total stresses at stripping:

| Whyte | Stripping | Prestress | Total |
|---|---|---|---|
| 100 mm | + 2.4 | + 1.5 | + 3.9 OK |
| 75 mm | − 2.5 | + 2.0 | − 0.5 OK |

7. Check horizontal shear resistance:

Factored moment at stripping:

$$M_f = (1.25)(48.7) = 60.9 \text{ kN·m}$$

Conservatively, assume a moment arm of 140 mm between tension and compression forces:

$$T = C = \frac{60.9}{0.14} = 435 \text{ kN}$$

Calculate resistance of concrete rib 300 mm wide at panel ends:

$$V_r = 0.17\phi_c\sqrt{f'_{ci}}A$$
$$= (0.17)(0.65)\sqrt{25}(300)(2400)/10^3$$
$$= 398 \text{ kN} < 435$$

As the factored resistance in this case is less than the factored shear, the concrete strength must be increased

$$f'_{ci} > \left[\frac{(435)(10^3)}{(0.65)(0.17)(300)(2400)}\right]^2$$

$$f'_{ci} > 29.9 \text{ MPa}$$

Use $f'_{ci} = 30$ MPa

## 3.12 DESIGN OF POST-TENSIONED ELEMENTS

### 3.12.1 GENERAL

Post-tensioning is a method of prestressing in which tendons are stressed after the concrete has reached a pre-determined strength. It is used primarily with continuous cast-in-place construction, but also has application in precast structures for continuity connections, in segmental construction and for some special uses such as rock anchors and remedial measures.

This section primarily applies to post-tensioned flexural elements used in buildings; other applications such as bridges, tanks and nuclear containment vessels are outside the scope of this Manual.

The use of post-tensioning allows a designer to control the behaviour of a structure within a broad range of loading conditions. For example, post-tensioning forces can be applied at various stages during construction to offset permanent loads which are applied in increments, such as in transfer girders or foundation mats. Post-tensioning may be applied by either bonded or unbonded tendons.

### 3.12.2 ANALYSIS

It is usually preferable in post-tensioned design to determine the required prestressing force and, hence, the number, size, and profile of the tendons for behaviour at service loads. The flexural resistance of the element must then be checked at the critical points.

Determination of both the strength and serviceability characteristics of indeterminate post-tensioned structures depends on elastic analysis procedures. The serviceability of a structure is assessed primarily in terms of stresses and deflections, both elastic and long time.

In simple span post-tensioned beams, as in pretensioned beams, the moments induced by the post-tensioning are directly proportional to the eccentricity of the tendons with respect to the centroid of the beam cross-section. In continuous, or indeterminate post-tensioned structures, the moments due to post-tensioning are usually not directly proportional to the tendon eccentricity. The difference occurs because the deformations imposed by post-tensioning are resisted by the continuous element at the points where they are restrained by other elements in the structure. This restraint to post-tensioning deformations modifies the reactions and, hence, affects the elastic moments and shears resulting from the post-tensioning. The moments resulting from the restraints to the prestressing deformations are usually called "secondary moments." This term is used because the moments are induced by the primary post-tensioning moment, Pe, and not because the secondary moment is negligible nor necessarily smaller than the primary moment.

Primary and secondary moments, as well as the total moments due to post-tensioning, are illustrated for a two-span continuous beam in Fig. 3.12.1. This beam has a post-tensioning force, P, acting at a constant eccentricity, e. Hence the primary moment in the beam has a magnitude, Pe, as shown in Fig. 3.12.1(b). This primary moment will cause a theoretical upward deflection at the centre support of $Pe\ell^2/(2EI)$. The reactions necessary to retain the beam at its original position on the supports A, B and C are shown in Fig. 3.12.1(c). Note that the algebraic sum of the secondary reactions is zero and that, for this case, the secondary moment at the interior support is 150 percent of the primary moment and of opposite sign. It is important to note that, since the secondary moment is caused by the induced reactions, the secondary moment always varies linearly between the supports.

**Fig. 3.12.1 Primary and secondary moments**

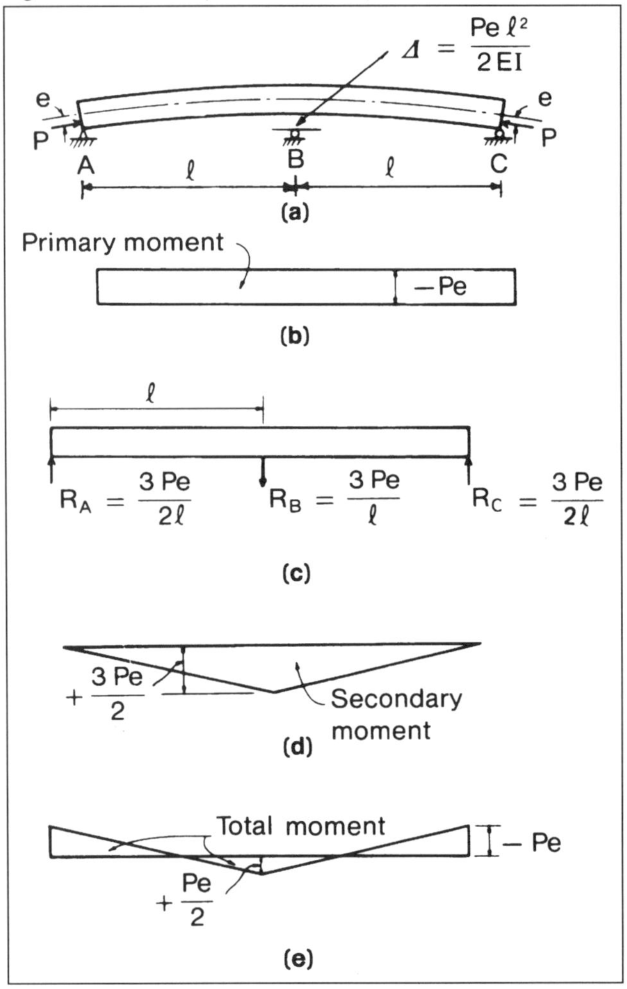


The total moment due to post-tensioning may be expressed as the superposition of the primary and secondary moments as shown in Fig. 3.12.1(e). If the profile of a tendon is such that, when stressed, it induces no secondary moments in a statically indeterminate structure, the profile is said to be concordant. A profile will be concordant if, over the entire structure, the eccentricity is proportional to the bending moment due to any applied loading on the structure. The use of a concordant tendon may reduce the amount of additional bonded non-prestressed reinforcement required in the factored load design of bending members.

### Service Load Analysis

Any method of elastic structural analysis may be applied to indeterminate post-tensioned structures. Computer programs may be required for more complex structures. Such programs are usually based on the stiffness approach and require input in the form of applied loads. The equivalent loading due to prestress can be calculated as described later. The flexibility approach is attractive for the analysis of prestressing effects in statically indeterminate structures, since input is required in the form of moments due to the loading on the corresponding statically determinate structure, namely, the primary prestressing moment. The basis of the stiffness and flexibility approaches can be found in standard texts on structural analysis. However, many simpler structures can be analyzed with the aid of a hand calculator and one of two suitable methods: namely the equivalent load method and the load balancing method.

*Equivalent load method of analysis*:

The effect of a prestressing force on an element can be determined by considering the prestressing force to be replaced by equivalent external loads. In the simple span beam shown in Fig. 3.12.2(a) containing two sets of tendons, horizontal forces $P_1$ and $P_2$ are exerted at the ends of the element along with vertical forces equal to $P_2 \tan\alpha$, which may be neglected in the design of the beam for flexure because they occur directly over a support. In addition to these loads, the curved tendon (assumed parabolic) exerts a continuous upward force on the beam along its entire length. If friction between the tendon and the concrete is neglected, the force exerted on the concrete by the tendon is normal to the plane of contact and the tension in the tendon is constant. The normal force, $w_p$, exerted by the tendon is equal to the tension in the tendon divided by its radius of curvature, as shown in Fig. 3.12.2(b).

Due to the shallow nature of most post-tensioned beams, the horizontal component of the tension in the tendon may be assumed equal to the tension. With this assumption, the horizontal component of the tendon force may be assumed constant, which allows the tendon to be held in equilibrium with the uniform vertical load as shown in Fig. 3.12.2(c).

The magnitude of the vertical load exerted by the element to hold the tendon in equilibrium may be derived in various ways. Taking one-half of the tendon as a free body, the forces are as shown in Fig. 3.12.2(d). Summing moments about support A and solving for the load, $w_p$, provides:

$$w_p = \frac{8P_2e'}{\ell^2}$$

**Fig. 3.12.2 Equivalent load analysis**

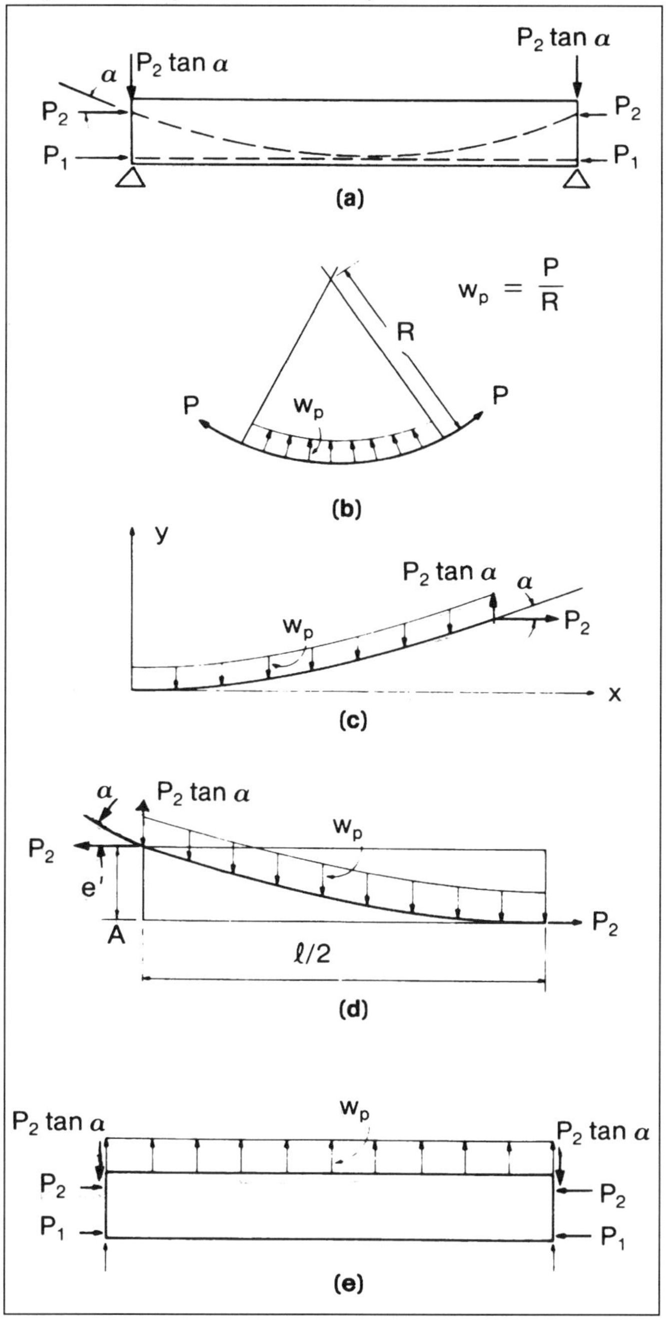


The load exerted by the tendon on the element is equal but opposite to this load, or:

$$w_p = \frac{-8P_2e'}{\ell^2}$$

Note that the sign convention considers $w_p$ to be positive when it acts downward.

Summing the loads exerted by the tendons on the simple span beam of Fig. 3.12.2(a) leads to the loads shown in Fig. 3.12.2(e). Calculating equivalent loads is probably not justified for simple span beams where the moments induced by the tendons are directly proportional to the tendon eccentricity. However, for continuous beams, the use of equivalent loads permits analysis for the total moment effects of the post-tensioning by considering a single additional loading case.

Equivalent load equations are given in Fig. 8.1.4 of this Manual. Various combinations of equivalent loads may be superimposed. For example, the moments due to end eccentricity can be added to the equivalent load moments caused by the tendon profile.

Using the equivalent load procedure, the total post-tensioning moments at supports may be obtained by calculating fixed end moments and using moment distribution or other method of indeterminate structural analysis. After the post-tensioning moments have been determined at the supports, post-tensioning moments between supports may be obtained by super-imposing the simple beam moment diagram due to the post-tensioning loads on the base line provided by connecting the total post-tensioning moments at supports.

However, it is generally simpler to obtain interior post-tensioning moments using the tendon profile. Using this approach, the interior moments may be expressed as:

$$M = Pe + M_{AB} + (M_{BA} - M_{AB})\frac{x}{\ell} - P\left[e_A + (e_B - e_A)\frac{x}{\ell}\right]$$

where $M_{AB}$ and $M_{BA}$ are the total post-tensioning moments at two adjacent supports, P is the tendon force, and $e_A$, $e_B$, and e are the eccentricities of the tendon at A, B and x, respectively. Eccentricity is taken as negative when it falls below the centroid and positive above the centroid of the cross-section. For the tendon profile in Fig. 3.12.3, the moment at x is the algebraic sum of the end moments at the point and the moment of the tendon force, P, times the distance from the tendon to the straight line connecting the ends of the tendon over the supports.

**Fig. 3.12.3 Interior post-tensioning moments**

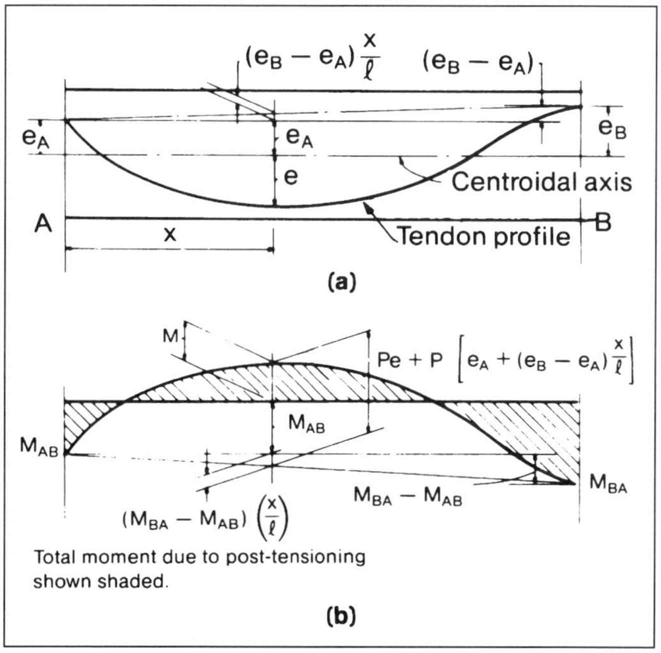

*Load balancing method of analysis:*

The basic concept of load balancing also represents the influence of the tendons by equivalent loads. The concept is illustrated in Fig. 3.12.4(a), where the tendon is selected to directly counteract the imposed loading at the indicated eccentricity, e′, that is, $w = w_p$. Since the moment induced by the tendon and the load offset each other, the net stress in the beam will be the axial compressive stress from the post-tensioning, $P/A_g$. If it is desired to design the beam for zero stress at the bottom fibre at midspan (or any other value of stress less than the modulus of rupture), it is only necessary to reduce the amount of post-tensioning provided. The net stress on the section may be calculated from $P/A_g + M_{n\ell}/s$ where $M_{n\ell}$ is the net (unbalanced) bending moment on the section. For continuous designs, the tendon geometry would be assumed as shown in Fig. 3.12.4(b).

**Fig. 3.12.4 Load-balancing**

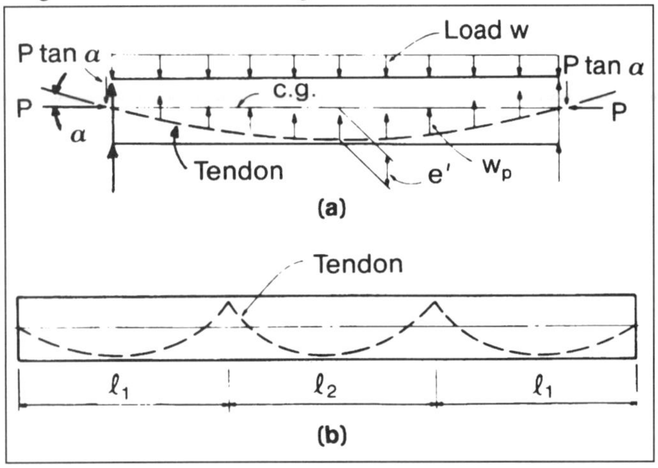


Fig. 3.12.4 serves to illustrate the salient features of load-balancing. These are that the prestressing force is selected to balance or counteract the dead load of the structure and generally some portion of the live load, and that under this loading condition, the structure will be subjected only to an axial compressive stress, $P/A_g$, from the tendon. This concept has substantial advantages in visualizing the behaviour of statically indeterminate structures. It simplifies design calculations since only the total prestressing moment induced by post-tensioning needs to be considered rather than separate consideration of primary and secondary moments.

The net moment in the structure at any point is that resulting from the load not "balanced" by the post-tensioning.

While load-balancing is an extremely powerful concept for visualizing the effect of the post-tensioning on the structure, there are some qualifications that should be kept in mind when using this technique in final detailed designs. First, it is usually not practical to install tendons with a sharp break in curvature at the supports. The stiffness of the tendons requires a reverse curvature in the tendon profile with a point of contraflexure some distance from the supports. This reverse curvature modifies the loads imposed by post-tensioning from those assumed using load-balancing. A second qualification is that the load-balancing technique should be adjusted to consider friction losses and the resulting variation of the force in the tendon along its length. Finally, load-balancing does not account directly for moments induced by tendons anchored eccentrically at ends free to rotate.

### Factored Resistance

When checking the critical sections for flexural and shear resistance, the applicable load factors of the Code should be applied to the moments and shears obtained by elastic service load analysis. A load factor of 1.0 is applied to secondary moments. Design for flexure and shear for factored loads is outlined in Sections 3.3 and 3.6 respectively, of this Manual.

## 3.12.3 DESIGN FOR SERVICE LOADS

In the design of post-tensioned elements, the response of the total structure must be considered for the instantaneous prestressing loads applied, as well as for the time-dependent volume changes and the applied service loads.

Flexural design for service loads is covered in Section 3.4 as well as in Examples 1, 2, and 3 of this section. These three examples deal with a continuous beam, a one way slab, and a flat plate respectively. Calculation of camber and deflection is given in Section 3.5.

### Preliminary Sizing of Elements

There are no set span-depth limits for post-tensioned elements, however, the values shown in Fig. 3.12.5 are provided as a guide for preliminary sizing of elements. These ratios may be increased if calculations verify that deflection, camber and vibration frequency and amplitude are not objectionable.

In the sizing of elements, concrete cover must also be considered. Fig. 1.4.1 gives details of cover for non-corrosive environments.

**Fig. 3.12.5 Approximate span-depth ratios***

| | |
|---|---|
| One way slab | 48 |
| Two way slab | 45 |
| Two way slab with drop panel (minimum drop panel L/6 each way) | 50 |
| Two way slab with two-way beams | 55 |
| Waffle (1500 × 1500 grid) | 35 |
| Beams b ≈ h/3 | 20 |
| Beams b ≈ 3h | 30 |

* These values are appropriate for elements whose live load is less than the dead load.

### Type and Placement of Tendons

Post-tensioned tendons may consist of high-strength strand, wire or bars, sizes and capacities of which are listed in Sect. 8.2. Tendon profiles are usually parabolically curved. Straight line profiles are occasionally used, particularly for strengthening existing structures.

In two way post-tensioned cast-in-place slabs, it is common practice to band post-tensioning tendons over the supports in one direction, while evenly distributing the tendons in the other direction. The advantage of this is that the expense of placing the tendons is reduced by removing the need to "weave" tendons over and under each other to achieve draped profiles in both directions. This method is especially powerful in applications with irregular column spacings. Experience has shown that a post-tensioned flat plate behaves as a flat plate regardless of where the tendons have been placed. The load balancing concept, described above, is useful in understanding the idea of the banding and distribution of tendons in this manner.

In the banded direction, the high points of the banded tendons are placed at the column locations, and the low points are placed at or near the midspan. The drape of the tendons provides an upward force along the column gridline, except for a small area on either side of the column caused by the reversed curvature of the tendon, which can be ignored. In the distributed direction, the high points of the tendons are placed at the column gridlines, while low points are placed at or near the midspan of the slab in that direction (see Fig. 3.12.4).

It is easy to see that the distributed tendons carry the loads to the column gridlines which in turn carry loads to the columns. In this way, the balanced portion of the gravity loads is carried to the columns without any flexural action of the slab.

## 3.12.4 PRESTRESS LOSSES

Post-tensioned elements are subject to loss of prestress due to elastic shortening, creep, shrinkage, and stress relaxation as described in Section 3.4.4. In addition, post-tensioned tendons are subject to losses caused by friction and anchorage set.

### Friction Losses

The stress loss due to friction is equal to ($f_{po} - f_{px}$) and may be determined using the following relationship:

$$f_{px} = f_{po}e^{-(Kx+\mu\alpha)}$$

where:

$f_{px}$ = tendon stress at distance x from jacking end

$f_{po}$ = jacking stress

$e$ = base of Naperian logarithm

$\mu$ = curvature coefficient per radian

$K$ = wobble coefficient per metre length

$\alpha$ = total angular change between tangents to the tendon at the jacking end at a distance x from the jacking end

For values of ($Kx + \mu\alpha$) less than 0.3, the above relationship can be written in the form:

$$f_{px} = f_{po} / (1 + Kx + \mu\alpha)$$

Approximate values if the coefficients, K and $\mu$, are given in Fig. 3.12.6. Actual values for a particular type of tendon and duct should be obtained from the manufacturer.

Anchorage set loss is due to the slip which occurs in anchoring the tendon at the jacking end. The effect of anchorage set loss is illustrated in Fig. 3.12.7, which shows the variation in the tendon force near the jacking end after the anchorage has been set. The value $f_{pi}$ is the stress at the jacking end of the of the tendon after the anchorage set loss. The length, y, of tendon affected by the anchorage set is given by:

$$y = \sqrt{\frac{E_p\Delta}{\beta}}$$

where:

$E_p$ = modulus of elasticity of the tendon

$\Delta$ = anchorage set

$\beta$ = slope shown in Fig. 3.12.7

**Fig. 3.12.6 Approximate friction coefficients for post-tensioned tendons**

| Type of tendon and sheath | Wobble coefficient, K per metre length | Curvature coefficient, m per radian |
|---|---|---|
| Tendons in flexible metal sheathing | | |
| Wire tendons | $(3.3 \text{ to } 5.0) \times 10^{-3}$ | 0.15 to 0.25 |
| 7 wire strand | $(1.6 \text{ to } 6.5) \times 10^{-3}$ | 0.15 to 0.25 |
| High strength bars | $(0.3 \text{ to } 2.0) \times 10^{-3}$ | 0.08 to 0.30 |
| Tendons in rigid metal sheath | | |
| 7 wire strand | $\approx 0.7 \times 10^{-3}$ | 0.15 to 0.25 |
| Pre-greased tendons in extruded plastic sheath | | |
| Wire tendons and 7 wire strand | $(1.0 \text{ to } 6.5) \times 10^{-3}$ | 0.05 to 0.15 |
| Mastic coated tendons in waxed paper | | |
| Wire tendons and 7 wire strand | $(3.3 \text{ to } 6.6) \times 10^{-3}$ | 0.05 to 0.15 |

**Fig. 3.12.7 Anchorage set prestress loss**

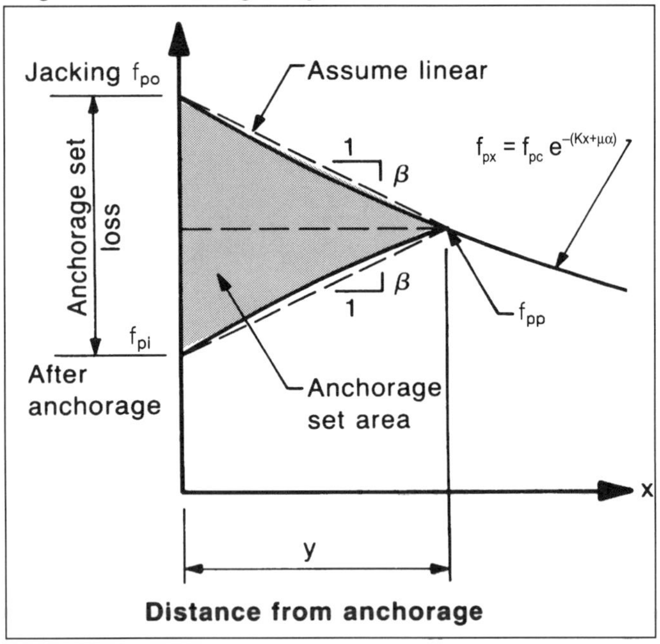


The quantity, Δ, is characteristic of the post tensioning system and has a minimum value of approximately 6 mm for prestressing strand.

If Δ is known, y and $f_{pp}$ can be computed and the level of prestress along the tendon can be determined. Alternatively, the anchorage set loss can be calculated from the shaded area, or the anchorage set area in Fig. 3.12.7, divided by $E_p$.

The values $f_{po}$, $f_{pi}$ and $f_{pp}$ (shown in figure 3.12.7) are given by CSA-A23.3, Clause 18.5. $f_{po}$ may not be greater than 0.85 $f_{pu}$, or the maximum value specified by the strand manufacturer, but not greater than $0.94f_{py}$. $f_{pp}$ may not be greater than $0.82f_{py}$, or $0.74f_{pu}$ (whichever is less). $f_{pi}$ may not be greater than 0.80 $f_{pu}$. These values apply immediately after transfer and before long term losses. Either or both of the values of jacking stress, $f_{po}$, and the anchorage slip, Δ, can be manipulated to ensure that these constraints are met.

If a value $f_{pp}$ is chosen, $f_{pi}$ is easily determined by inspection. The anchorage set area can then be determined and the value of anchorage set calculated.

Anchorage set loss can have a significant effect in a short tendon. In practice the loss is usually compensated for by temporary overtensioning, shimming of the anchor head, or control of the anchorage set.

The elongation of the tendon during jacking is given by:

$$\delta = \int_0^{\ell} \frac{f_{px}}{E_p} dx$$

where $\ell$ is the length of the tendon.

The computed elongation should be compared with the elongation measured during the actual jacking operation of the tendon.

**Long Term Losses**

The calculation of losses in post-tensioned tendons due to shrinkage, creep and relaxation is an important part of the design process. Long term losses reduce the effective stress in the tendon, affecting the service load stresses in the structural element. Refer to Section 3.4.4 for information on the calculation of these values.

## 3.12.5 DESIGN FOR FACTORED LOADS

Design for flexural resistance is covered in Section 3.3 and is illustrated in Example 1,2 and 3 in Section 3.12.6, 3.12.7 and 3.12.8 respectively.

**Fig. 3.12.8 Shear in post-tensioned element with reverse curvature in tendon**

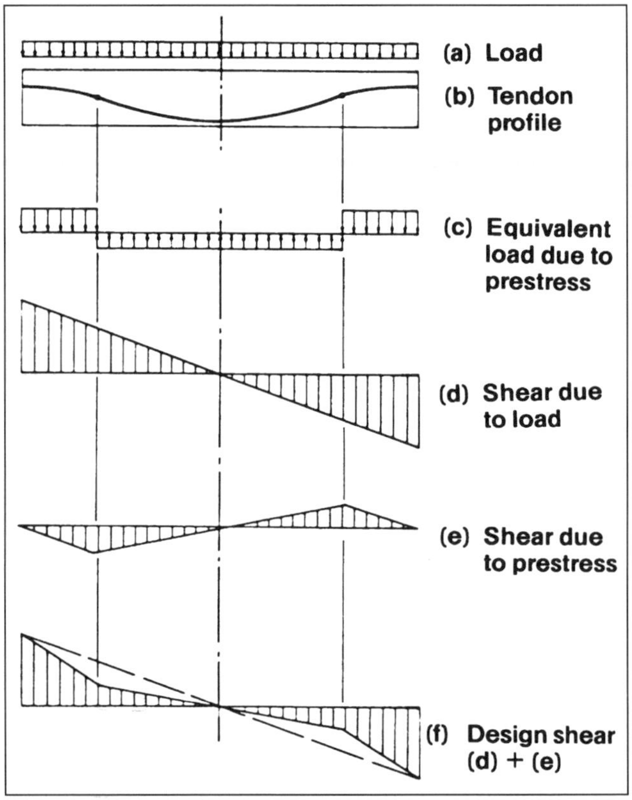


In continuous post-tensioned elements, special attention should be paid to the distribution of shear imposed by the post-tensioned tendons. This is illustrated in Fig. 3.12.8. Shear resulting from secondary reactions should be included for statically indeterminate structures.

## 3.12.6 EXAMPLE 1- TWO SPAN CONTINUOUS T-BEAM

*Given:*

Two span cast-in-place post-tensioned T-beams, spaced 8 m on centers and supported by 500 × 600 mm columns, 3000 mm long, at 17 m spacing as shown.

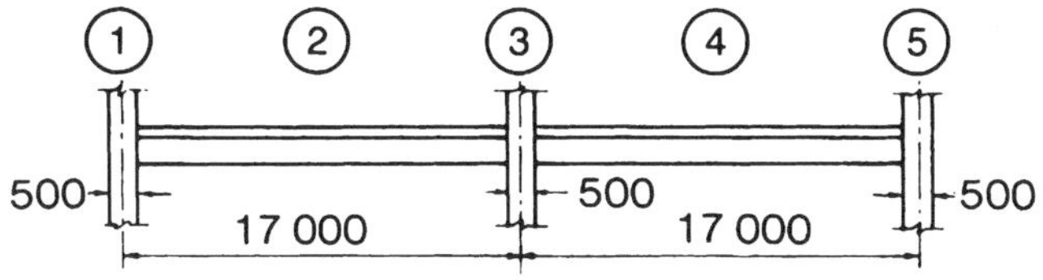


Concrete:

$f'_{ci}$ = 25 MPa

$f'_c$ = 35 MPa  $\alpha_1 = 0.798$  $\beta_1 = 0.883$

Normal Density

Prestressed reinforcement:

15 mm 1860 MPa low relaxation strand

$A_p$ = 140 mm²/strand

$f_{py}$ = 1675 MPa

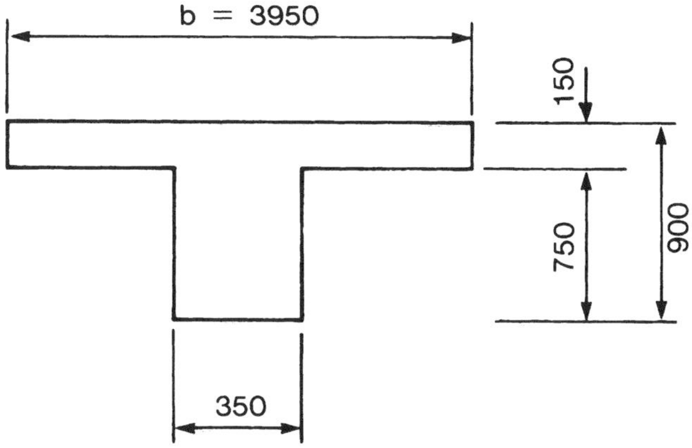


Section properties:

Although excluded by clauses 18.1.3 of CSA-A23.3 the provisions of Clause 10.3.3 are used here for a post-tensioned beam and slab. Since the exclusion of Clause 10.3.3 was meant to allow the use of many standard precast tee beams in use today, it is felt that its use is acceptable for this case. When using this method, it should be noted that the overall prestress level of the beam and the slab should be checked to ensure that it meets code minimums. This is done at the end of this example.

According to Clause 10.3.3 the overhang width on either side of the web is calculated as the lesser of:

1. 1/5 the span length for simply supported beams,
2. 1/10 the span length for continuous beams
3. 12 times the slab thickness
4. One-half the clear distance to the next web.

For this beam the effective flange width is the lesser of:

1. $1/7.5\ell_1 + b_w$ = 1/7.5 (17 000)(2) + 350 = 4883 mm
2. $(2)(12)t + b_w$ = (2)(12)(150) +350 = 3950 mm
3. $0.5\ell_2$ = (0.5)(8000) = 4000 mm

Therefore, use 3950 mm effective width.

$A_g$ = 855 000 mm²

$I$ = 50 200 × 10⁶ mm⁴

$y_b$ = 687 mm

$y_t$ = 213 mm

$S_b$ = 73 100 x 10³ mm³

$S_t$ = 235 700 × 10³ mm³

Loads:

Dead Load (including full slab) = 35.1 kN/m

Live Load = (2.5)(8.0) = 20.0 kN/m

*Problem:*

Design bonded post-tensioned flexural reinforcement.

*Service load moments*:

The dead and live load moments, with two live load patterns considered, are determined by moment distribution and are shown in the table on the following page.

*Tendon profile:*

The effects of post-tensioning are treated separately from the effects of the loads, and are dependent on the magnitude and positioning of the post-tensioning tendons, and on frame action.

Assume one end stressing and bonded strands. A parabolic tendon profile is selected as shown:

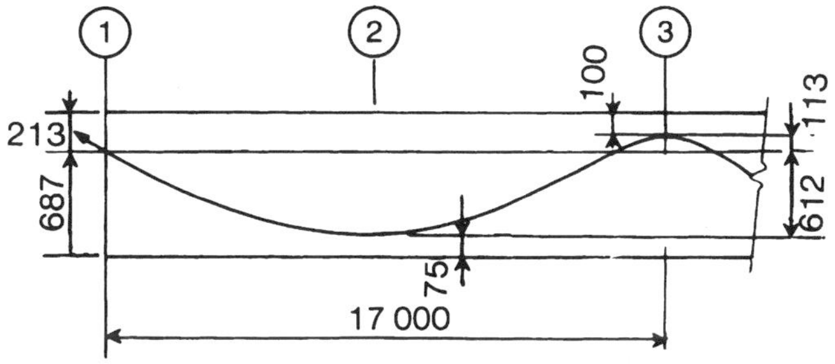


*Friction losses:*

Calculate friction losses using:

$f_{px} = f_{po}e^{-(Kx+\mu\alpha)}$

$K = 0.0023/\text{m}$

$\mu = 0.18/\text{rad}$

Assume a prestress loss of 230 MPa in addition to the loss due to an anchor set of 14 mm, with one end stressing at support ① and check later. To find point of contraflexure, equate slopes of two parabolas at distance x:

$$\frac{(2)(113)}{x} = \frac{(2)(612)}{8500 - x}$$

x = 1325 mm

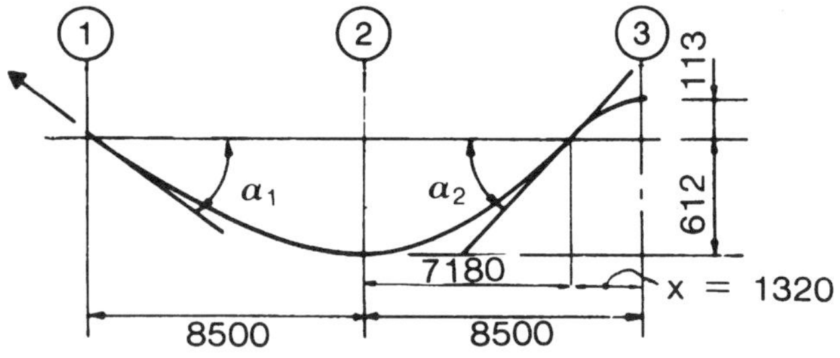


The angles $\alpha_1$ and $\alpha_2$ are determined to be:

$$\alpha_1 = \tan^{-1}\left[\frac{(2)(612)}{8500}\right] = 0.143\,\text{rad}$$

$$\alpha_2 = \tan^{-1}\left[\frac{(2)(612)}{7180}\right] = 0.169\,\text{rad}$$

Assuming a maximum temporary jacking stress of 0.80 $f_{pu}$ in the tendon at point ①, the jacking stresses, expressed as a fraction of $f_{pu}$, are determined to be:

| | | |
|---|---|---|
| ① | | = 0.800 |
| ② | $0.800\,e^{-[(0.0023)(8.5)+(0.18)(0.143)]}$ | = 0.765 |
| ③ | $0.765\,e^{-[(0.0023)(8.5)+(0.18)(2)(0.169)]}$ | = 0.705 |
| ④ | $0.705\,e^{-[(0.0023)(8.5)+(0.18)(2)(0.169)]}$ | = 0.651 |
| ⑤ | $0.651\,e^{-[(0.0023)(8.5)+(0.18)(0.143)]}$ | = 0.622 |

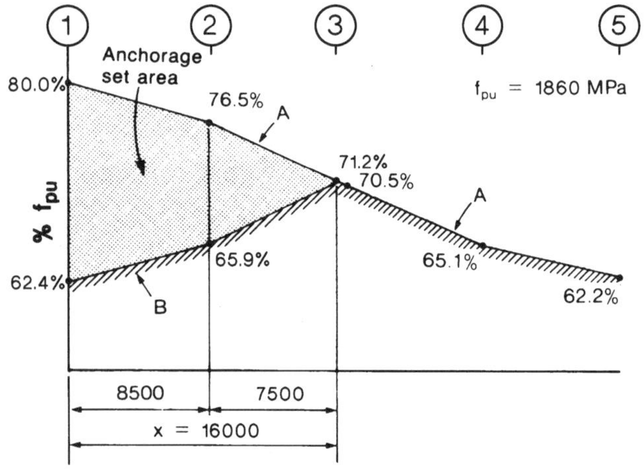


The variation of prestress along the beam at jacking is thus shown by curve A. After anchor set, the reduction in the prestress at the jacking end is shown by curve B. Note that the prestress at locations ① and ② approximately equals the prestress at locations ⑤ and ④, respectively.

The length, x, over which the anchorage set is effective is determined as follows, assuming a linear variation in prestress between supports ① and ③:

$$\beta = \frac{(1860)(0.800 - 0.705)}{(17)(1000)}$$

$$= 10.39 \times 10^{-3}\ \text{MPa/mm}$$

$$x = \sqrt{\frac{E_p \Delta}{\beta}}$$

$$= \sqrt{\frac{(190000)(14)}{10.39 \times 10^{-3}}} = 16\,000\ \text{mm}$$

and the prestress at sections ① and ② can be determined assuming equal but opposite slopes for curves A and B.

The anchorage set loss can be checked using the anchorage set area approach. The displacement of the jacking end of the tendon is given by:

$$\Delta = \frac{(f_{pu})(\text{anchorage set area})}{E_p}$$

$$= \frac{(1860)}{190\,000}\left\{\frac{(8500)}{2}[(0.8 + 0.765) - (0.624 + 0.659)] + \frac{(7500)}{2}(0.765 - 0.659)\right\}$$

$$= 15.6\ \text{mm} \approx 14\ \text{mm specified}$$

The elongation of the strand during jacking is equal to the area under the curve A divided by $E_p$, and is given by:

$$\delta = (0.5)[0.8 + (2)(0.765 + 0.705 + 0.651) + 0.622]\frac{(1860)(8500)}{190\,000} = 236\ \text{mm}$$

*Initial and final tendon stress:*

$$f_{pi} = (f_{pu})(\%\ \text{prestress})$$

$$f_{pe} = f_{pi} - 230\ \text{MPa}$$

The initial and final tendon stresses are calculated as shown in the table on Page 3-127.

*Secondary Moments:*

Secondary moments induced by post-tensioning are accounted for by considering the effects of restrained rotations.

For simplicity, a post tensioned force of 1 kN is used, giving the following primary moments:

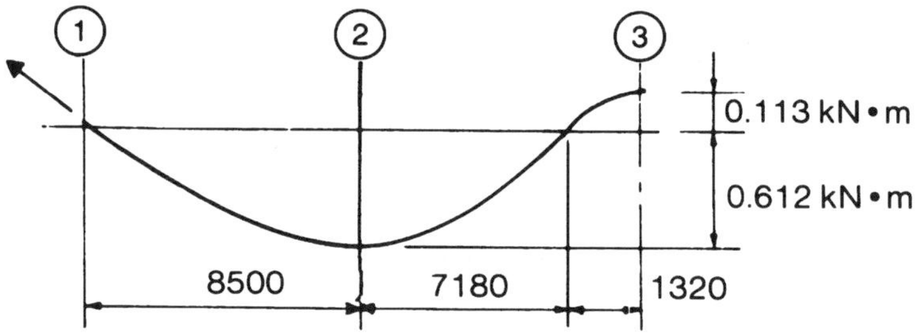

| Location | ① ⑤ | ② ④ | ③ |
|---|---|---|---|
| $M_d$ (kN·m) | – 495 | + 524 | – 1021 |
| $M_\ell$ (kN·m) | – 331 | + 349 | – 581 |
| $M_{total}$ (kN·m) | – 826 | + 873 | – 1602 |
| e (mm) | 0 | 612 | – 113 |
| Δ e (mm) | – 280 | – 320 | – 360 |
| e + Δ e (mm) | – 280 | + 292 | – 473 |
| % initial prestress | 62.2 | 65.1 | 70.5 |
| $f_{pi}$ (MPa) | 1157 | 1211 | 1311 |
| $f_{pe}$ (MPa) | 927 | 981 | 1081 |
| $P_i$ (kN) | 2106 | 2204 | 2386 |
| $P_e$ (kN) | 1687 | 1785 | 1967 |

| Prestress | Load | ① ⑤ | | ② ④ | | ③ | |
|---|---|---|---|---|---|---|---|
| | | $f_b$ | $f_t$ | $f_b$ | $f_t$ | $f_b$ | $f_t$ |
| Initial | $P / A_g$ | + 2.5 | + 2.5 | + 2.6 | + 2.6 | + 2.8 | + 2.8 |
| | $P (e + \Delta e) / S$ | – 8.1 | + 2.5 | + 8.8 | – 2.7 | – 15.4 | + 4.8 |
| | $M_d / S$ | + 6.8 | – 2.1 | – 7.2 | + 2.2 | + 14.0 | – 4.3 |
| | $M_\ell / S$ | | | | | | |
| | Stresses | + 1.2 | + 2.9 | + 4.2 | + 2.1 | + 1.4 | + 3.3 |
| | Permissible stresses | $0.6\,f'_{ci}$ | $0.6\,f'_{ci}$ | $0.6\,f'_{ci}$ | $0.6\,f'_{ci}$ | $0.6\,f'_{ci}$ | $0.6\,f'_{ci}$ |
| | | + 15.0 | + 15.0 | + 15.0 | + 15.0 | + 15.0 | + 15.0 |
| | | OK | OK | OK | OK | OK | OK |
| Final | $P_e / A_g$ | + 2.0 | + 2.0 | + 2.1 | + 2.1 | + 2.3 | + 2.3 |
| | $P_e (e + \Delta e) / S$ | – 6.5 | + 2.0 | + 7.2 | – 2.2 | – 12.9 | + 3.9 |
| | $M_d / S$ | + 6.8 | – 2.1 | – 7.2 | + 2.2 | + 14.0 | – 4.3 |
| | $M_\ell / S$ | + 4.5 | – 1.4 | – 4.8 | + 1.5 | + 8.0 | – 2.5 |
| | Stresses | + 6.8 | + 0.5 | – 2.7 | + 3.6 | + 11.7 | – 0.6 |
| | Permissible stresses | $0.6\,f'_c$ | $0.6\,f'_c$ | $0.5\sqrt{f'_c}$ | $0.6\,f'_c$ | $0.6\,f'_c$ | $0.5\sqrt{f'_c}$ |
| | | + 21.0 | + 21.0 | – 3.0 | + 21.0 | + 21.0 | – 3.0 |
| | | OK | OK | OK | OK | OK | OK |

$M_2 = (687 - 75)(1)/10^3 = 0.612$ kN·m

$M_3 = (213 - 100)(1)/10^3 = 0.113$ kN·m

The unrestrained end rotations in the span are found by the moment area method as follows:

$$\theta_1 = \frac{1}{(EI)(17.0)}\left\{+\left(\frac{2}{3}\right)(0.113)(1.32)^2\left(\frac{3}{8}\right) - \left(\frac{2}{3}\right)(0.612)(7.18)\left[8.5 - \left(\frac{3}{8}\right)(7.18)\right] - \left(\frac{2}{3}\right)(0.612)(8.5)\left[8.5 - \left(\frac{3}{8}\right)(8.5)\right]\right\}$$

$$= -\frac{3.38}{EI}$$

$$\theta_3 = \frac{1}{(EI)(17.0)}\left\{-\left(\frac{2}{3}\right)(0.612)(8.5)^2\left(\frac{5}{8}\right)\right.$$

$$-\left(\frac{2}{3}\right)(0.612)(7.18)\left[8.5+\left(\frac{3}{8}\right)(7.18)\right]$$

$$\left.+\left(\frac{2}{3}\right)(0.113)(1.32)\left[17.0-\left(\frac{3}{8}\right)(1.32)\right]\right\}$$

$$= -\frac{2.92}{EI}$$

Fixed end moments for 1 kN of post-tensioning force are:

$$\text{F.E.M. @ ①} = \frac{2EI}{\ell}(\theta_3 - 2\theta_1)$$

$$= \frac{2EI}{17.0}\left[-\frac{2.92}{EI} - (2)\left(-\frac{3.38}{EI}\right)\right] = 0.45 \text{ kN·m}$$

$$\text{F.E.M. @ ③} = \frac{2EI}{\ell}(\theta_3 - 2\theta_1)$$

$$= \frac{2EI}{17.0}\left[-\frac{3.38}{EI} - (2)\left(-\frac{2.92}{EI}\right)\right] = 0.29 \text{ kN·m}$$

Secondary moments are derived from these fixed end moments by moment distribution.

These secondary moments may be considered as effective changes in the tendon profile in accordance with the equation:

$$M' = P\Delta e$$

with the resulting values of $\Delta e$ of 280 mm at supports ① and ⑤, 360 mm at ③ and 320 mm at ② and ④, as shown in the table on Page 3-127.

*Required post-tensioning force:*

Assume an allowable tensile stress of:

$$0.5\sqrt{f'_c} = 3.0 \text{ MPa}$$

The required final prestress force, P, is determined from the equation:

$$f_{allowable} = \frac{M_d + M_\ell}{S} - \frac{P}{A_g} \pm \frac{P(e+\Delta e)}{S}$$

$$P = \frac{\dfrac{M_d + M_\ell}{S} - f_{allowable}}{\dfrac{1}{A_g} + \dfrac{(e+\Delta e)}{S}}$$

$$① \quad P = \frac{\dfrac{(826)(10^3)}{(235700\times 10^3)} - \dfrac{3.0}{10^3}}{\dfrac{1}{855000} + \dfrac{280}{(235700\times 10^3)}} = 214 \text{ kN}$$

$$② \quad P = \frac{\dfrac{(873)(10^3)}{(73100\times 10^3)} - \dfrac{3.0}{10^3}}{\dfrac{1}{855000} + \dfrac{292}{(73100\times 10^3)}} = 1729 \text{ kN}$$

$$③ \quad P = \frac{\dfrac{(1602)(10^3)}{(235700\times 10^3)} - \dfrac{3.0}{10^3}}{\dfrac{1}{855000} + \dfrac{473}{(235700\times 10^3)}} = 1195 \text{ kN}$$

Therefore, the prestress force at location ② governs.

*Determine number of strands required in tendon:*

$$n = \frac{(1729)(10^3)}{(981)(140)} = 12.6$$

∴ Use 13 strands ($A_p$ = 1820 mm²)

Determine initial and final prestress forces based on previously calculated stresses:

$$P_o = A_p f_{po}$$

$$P_e = A_p f_{pe}$$

Values for the various locations are given in the table on Page 3-125. This table shows all stresses are within permissible values.

*Factored flexural resistance:*

At exterior supports ① and ⑤.

$$M_f = \alpha_D M_d + \alpha_L M_\ell + 1.0M'$$

$$= (1.25)(-495) + (1.5)(-331) + (1.0)(1687)(0.280)$$

$$= -643 \text{ kN·m}$$

Determine value of $f_{pr}$:

$$k_p = 2(1.04 - f_{py}/f_{pu})$$

$$= (2)(1.04 - 1675/1860) = 0.28$$

$$\frac{c}{d_p} = \frac{\phi_p A_p f_{pu}}{\alpha_1 \phi_c f'_c \beta_1 b d_p + k_p \phi_p A_p f_{pu}}$$

$$= \frac{(0.9)(1820)(1860)}{(0.798)(0.60)(35)(0.883)(350)(687) + (0.28)(0.9)(1820)(1860)}$$

$$= 0.691 > 0.5 \quad \text{(Clause 18.7.2(a))}$$

Use strain compatibility to compute $f_{pr}$ = 1396 MPa and c = 408 mm.

$$a = \beta_1 c = (0.883)(408) = 360 \text{ mm}$$

Determine flexural resistance:

$$M_r = \phi_p A_p f_{pr}(d - a/2)$$

$$= (0.9)(1820)(1396)(687 - 360/2)/10^6$$

$$= 1160 \text{ kN·m} > 643 \text{ kN·m} \quad \text{OK}$$

At midspan ② and ④:

$$M_f = \alpha_D M_d + \alpha_L M_\ell + 1.0M'$$
$$= (1.25)(524) + (1.5)(349) + (1.0)(1785)(0.320)$$
$$= 1750 \text{ kN·m}$$

Check that rectangular stress block lies between within flange of T-Beam:

$$k_p = (2)(1.04 - f_{py}/f_{pu})$$
$$= (2)(1.04 - 1675/1860) = 0.28$$

$$c = \frac{\phi_p A_p f_{pu}}{\alpha_1 \beta_1 \phi_c f'_c b + k_p \phi_p A_p f_{pu}/d_p}$$

$$= \frac{(0.9)(1820)(1860)}{(0.798)(0.883)(0.60)(35)(3950) + (0.28)(0.9)(1820)(1860)/825}$$

$$= 51\text{mm} < 150\text{mm flange thickness OK}$$

Therefore, treat as a rectangular section with b = 3950 mm:

$$\frac{c}{d_p} = \frac{51.2}{825} = 0.062 < 0.5 \quad \text{OK}$$

$$f_{pr} = f_{pu}(1 - k_p c/d_p)$$
$$= (1860)\left[1 - \frac{(0.28)(51.2)}{825}\right] = 1830 \text{ MPa}$$

$$a = \beta_1 c = (0.883)(51.2) = 45 \text{ mm}$$

$$M_r = \phi_p A_p f_{pr}(d - a/2)$$
$$= (0.9)(1820)(1830)(825 - 45/2)/10^6$$
$$= 2400 \text{ kN·m} > 1750 \text{ kN·m} \quad \text{OK}$$

At interior support ③:

The code allows moments calculated at the face of supports to be used in design of beams.

Calculate shear force at centreline of column:

$$w_f = \alpha_D w_d + \alpha_L w_1$$
$$= (1.25)(35.1) + (1.5)(20.0) = 73.9 \text{ kN/m}$$

$$V_{fCL} = (73.9)(8.5) + \left[\frac{(1021 - 495)(1.25)}{17} + \frac{(581 - 331)(1.5)}{17}\right] = 689 \text{ kN}$$

Shear force at face of column:

$$V_{fCF} = 689 - (73.9)(0.25) = 670 \text{ kN}$$

Moment at column face:

$$M_f = \alpha_D M_d + \alpha_L M + 1.0M' - \left(\frac{V_{fCL} + V_{fCF}}{2}\right)(0.25)$$
$$= (1.25)(-1021) + (1.5)(-581) + (1.0)(1945)(0.36) + \left(\frac{689 + 670}{2}\right)(0.25)$$
$$= -1278 \text{ kN·m}$$

$$\frac{c}{d_p} = \frac{\phi_p A_p f_{pu}}{\alpha_1 \beta_1 \phi_c f'_c b d_p + k_p \phi_p A_p f_{pu}}$$

$$= \frac{(0.9)(1820)(1860)}{(0.798)(0.883)(0.60)(35)(350)(800) + (0.28)(0.9)(1820)(1860)}$$

$$= 0.61 > 0.5 \quad \text{(Clause 18.7.2(a))}$$

Use strain compatibility to compute $f_{pr}$ = 1582 MPa and c = 462 mm.

$$a = \beta_1 c = (0.883)(462) = 408 \text{ mm}$$

$$M_r = \phi_p A_p f_{pr}(d - a/2)$$
$$= (0.9)(1820)(1582)(800 - 408/2)$$
$$= 1546 \text{ kN·m} > 1278 \text{ kN·m} \quad \text{OK}$$

*Check stress on gross concrete area:*

Average value of $P_e$ in bay ② and ④ is 1785 kN.

Stress on gross area of concrete:

$$\frac{P_e}{A_g} = \frac{(1785)(10^3)}{855000}$$
$$= 2.09 \text{ MPa} > 0.8 \text{ MPa} \quad \text{(Clause 18.13.2.3)}$$

To complete this design, check midspan deflection, and design shear reinforcement.

### 3.12.7 EXAMPLE 2- ONE WAY SLAB

*Given:*

150 mm thick slab spanning one way over five spans between beams on 8000 mm centres.

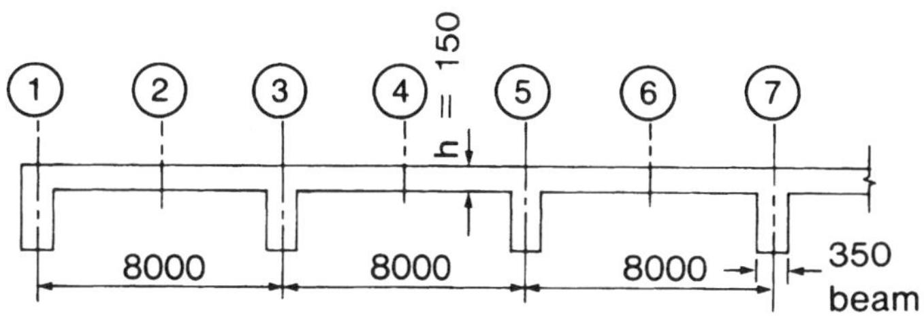

Concrete:

$f'_c$ = 35 MPa  $\alpha_1 = 0.80$  $\beta_1 = 0.88$

Normal Density

Prestressed reinforcement

15 mm - 1860 MPa low relaxation unbonded strand tendons

$A_p$ = 140 mm$^2$ / strand

$f_{py}$ = (0.9)(1860) = 1675 MPa

Section properties:

$A_g$ = 150 000 mm$^2$ / m

I = 281 x 10$^6$ mm$^4$ / m

S = 3750 x 10$^3$ mm$^3$ / m

Loads:

Dead load = 3.6 kN/m$^2$

Live load = 2.5 kN/m$^2$

*Problem:*

Design post-tensioned and non-prestressed reinforcement and check shear resistance.

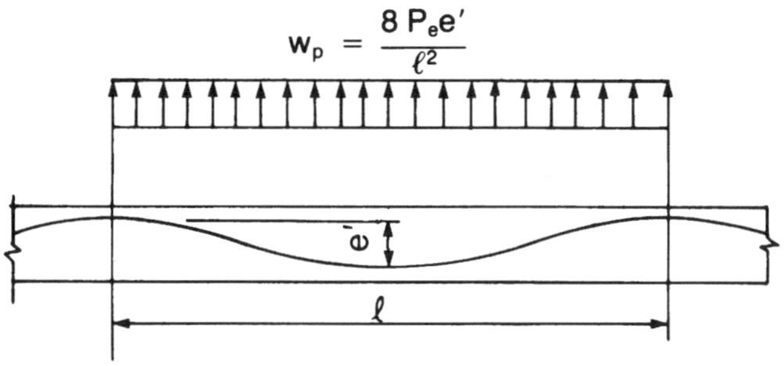


*Solution:*

Use balance load concept in design, which offers a simple approach to statically indeterminate structures:

where:

$w_p$ = uniformly distributed upward load applied to the concrete by the post-tensioning tendon

Experience shows that the required balance load by prestressing, $w_p$, is between 80 and 100% of the slab weight, when total superimposed loads are less than the slab weight.

Hence $w_p$ = 2.9 to 3.6 kN/m$^2$, use $w_p$ = 3.3 kN/m$^2$ (0.92$w_d$).

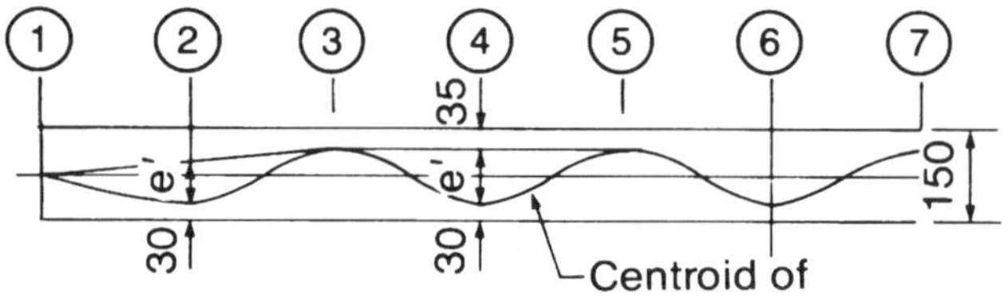


| Location | ② | ③ | ④ | ⑤ | ⑥ | ⑦ |
|---|---|---|---|---|---|---|
| $M_d$ (kN·m/m) | + 17.8 | − 24.7 | + 7.6 | − 17.8 | + 10.2 | − 19.2 |
| $M_\ell$ (kN·m/m) | + 15.8 | − 18.6 | + 12.4 | − 16.9 | + 12.4 | − 16.9 |
| $M_{bal}$ (kN·m/m) | − 16.3 | + 22.6 | − 7.0 | + 16.3 | − 9.4 | + 17.6 |
| $M_{n\ell}$ (kN·m/m) | + 17.3 | − 20.7 | + 13.0 | − 18.4 | + 13.2 | − 18.5 |
| e (mm) | + 45 | − 40 | + 45 | − 45 | + 45 | − 40 |
| $w_p$ (kN/m$^2$) | 3.3 | | 3.3 | | 3.3 | |
| e′ (mm) | 65 | | 85 | | 85 | |
| $P_e = \frac{w_p \ell^2}{8e'}$ (kN / m) | 406 | | 311 | | 311 | |
| $\frac{P_e}{A_g}$ (MPa) | 2.7 | | 2.1 | | 2.1 | |
| $M_{bal}$ (kN·m/m) | | + 22.6 | | + 16.3 | | + 17.6 |
| $P_e e$ (kN·m/m) | | + 16.3 | | + 12.2 | | + 12.2 |
| M′ (kN·m/m) | + 3.2 | + 6.3 | + 5.2 | + 4.1 | + 4.7 | + 5.4 |

Therefore, the net service load causing flexural stress is:

$$3.6 + 2.5 - 3.3 = 2.8 \text{ kN} / \text{m}^2$$

Assume a bottom cover of 20 mm and a top cover of 25 mm, and allow for a 20 mm duct diameter.

*Dead and live load moments:*

From the above information, dead and live load moments, $M_d$ and $M_\ell$, are calculated by the moment distribution method, using two live load configurations to give maximum positive and negative live load moments respectively. The dead and live load moments are summarized in the table on page 3-120.

*Balance load moments:*

The balance load moments, $M_{bal}$, due to the post-tensioning vertical component, $w_p$, are directly proportional to the dead load moments, since $w_p = -0.92\, w_d$. Values for $M_{bal}$ are shown in the table on page 3-120.

*Net load moments:*

The net load moments, $M_{n\ell}$, which cause flexure in the slab, are calculated from the equation:

$$M_{n\ell} = M_d + M_\ell + M_{bal}$$

*Service load stresses:*

Check the service load stresses on the basis of the effective prestress, $P_e$, required to produce the assumed balance load, $w_p$:

$$P_e = \frac{w_p \ell^2}{8e'}$$

The effective prestress force, $P_e$, required in the first span and in the other interior spans, is shown in the table on page 130.

The service flexural stresses can now be calculated from the equation:

$$f = \frac{P_e}{A_g} + \frac{M_{n\ell}}{S}$$

as shown in the table at the bottom of the page. Note that permissible tensile stresses are based on the provision of additional bonded non-prestressed reinforcement.

As all service load stresses are within the permissible limits, the assumed value of the effective prestress force $P_e$, is adequate.

*Required number of strands:*

Friction and anchor stress losses have to be determined as shown in Example 1.

For simplicity, an average initial prestressing level after friction losses of 0.71 $f_{pu}$ is assumed:

| | | |
|---|---|---|
| $f_{po}$ = (1860)(0.71) | = | 1321 MPa |
| Long term losses | = | – 230 MPa |
| $f_{pe}$ | = | 1091 MPa |

$P_e = (1091)(140)/10^3 = 153$ kN/strand

The required number of strands is therefore:

| | | |
|---|---|---|
| End spans: | 406/153 | = 2.66 strands/m width |
| Interior spans: | 311/153 | = 2.04 strands/m width |

The above can be achieved by spacing single strand tendons at 500 mm centres with additional tendons at 1500 mm centres in the end spans.

*Secondary moments:*

The balance moment, $M_{bal}$, is equal to the sum of primary and secondary moments. Hence, the secondary moment, which is

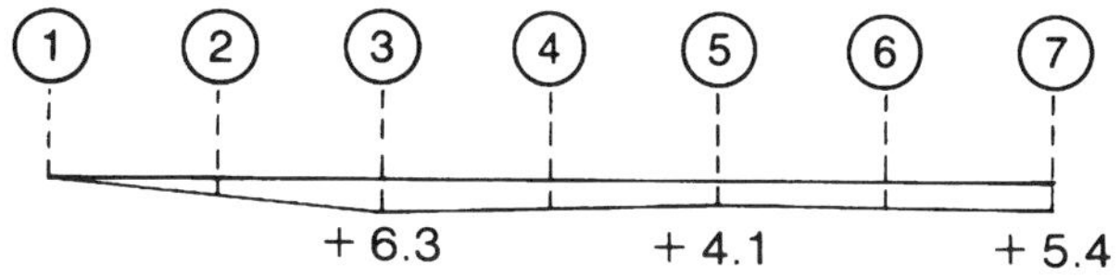


| Load | ② $P_e$ = 406 kN/m | | ③ $P_e$ = 406 kN/m | | ⑥ $P_e$ = 311 kN/m | | ⑦ $P_e$ = 311 kN/m | |
|---|---|---|---|---|---|---|---|---|
| | $f_b$ | $f_t$ | $f_b$ | $f_t$ | $f_b$ | $f_t$ | $f_b$ | $f_t$ |
| $P_e/A_g$ | + 2.7 | + 2.7 | + 2.7 | + 2.7 | + 2.1 | + 2.1 | + 2.1 | + 2.1 |
| $M_{n\ell}/S$ | – 4.6 | + 4.6 | + 5.5 | – 5.5 | – 3.5 | + 3.5 | + 4.9 | – 4.9 |
| Total | – 1.9 | + 7.3 | + 8.2 | – 2.8 | – 1.4 | + 5.6 | + 7.0 | – 2.8 |
| Stresses | $0.5\sqrt{f'_c}$ | $0.6 f'_c$ | $0.6 f'_c$ | $0.5\sqrt{f'_c}$ | $0.5\sqrt{f'_c}$ | $0.6 f'_c$ | $0.6 f'_c$ | $0.5\sqrt{f'_c}$ |
| | – 3.0 | + 21.0 | + 21.0 | – 3.0 | – 3.0 | + 21.0 | + 21.0 | – 3.0 |
| | OK | OK | OK | OK | OK | OK | OK | OK |

required for flexural resistance calculations, is the difference between balanced moment and primary moment:

$$M' = M_{bal} - P_e e$$

Values of M′ are given in the table on the previous page and are as follows:

*Factored moments:*

$$M_f = 1.25\,M_d + 1.5\,M_f + 1.0\,M'$$

② $M_f = (1.25)(17.8) + (1.5)(15.8) + (1.0)(3.2)$

$= 49.2$ kN·m/m

③ $M_f = (1.25)(-24.7) + (1.5)(-18.6) + (1.0)(6.3)$

$= -52.5$ kN·m/m

④ $M_f = (1.25)(7.6) + (1.5)(12.4) + (1.0)(5.2)$

$= 33.3$ kN·m/m

⑤ $M_f = (1.25)(-17.8) + (1.5)(-16.9) + (1.0)(4.1)$

$= -43.5$ kN·m/m

⑥ $M_f = (1.25)(10.2) + (1.5)(12.4) + (1.0)(4.7)$

$= 36.1$ kN·m/m

⑦ $M_f = (1.25)(-19.2) + (1.5)(-16.9) + (1.0)(5.4)$

$= -44.0$ kN·m/m

The minimum area of bonded reinforcement required by Clause 18.9.2.3 of CSA-A23.3 is:

$$A_s = 0.003b\frac{h}{2}$$

$$= (0.0015)(1000)(150) = 225\ \text{mm}^2/\text{m}$$

Use a minimum of No. 10 Grade 400 bars at 400 mm centres ($A_s = 250$ mm²/m).

*Factored flexural resistance:*

For location ②:

$$A_s = 250\ \text{mm}^2/\text{m}$$

$$A_p = \frac{(4)(140)}{1.5} = 373\ \text{mm}^2/\text{m}$$

Determine the value of $f_{pr}$:

$$c_y = \frac{\phi_s A_s f_y + \phi_p A_p f_{py}}{\alpha_1 \beta_1 \phi_c f'_c b}$$

$$= \frac{(0.85)(250)(400) + (0.9)(373)(1675)}{(0.798)(0.883)(0.60)(35)(1000)}$$

$$= 44\ \text{mm}$$

The variable $\ell_e$ in the equation from Clause 18.7.2(b) of CSA-A23.3 accounts for the length of the tendon over which the stress increase due to factored loading is being averaged out, and for the number of high moment regions (possible plastic hinges) in the span under consideration.

In this example, since the overall length of tendon is 40 000 mm, and the number of possible plastic hinges for the span is 2:

$$\ell_e = \frac{40\,000}{2} = 20\,000\ \text{mm}$$

$$f_{pr} = f_{pe} + \frac{8000}{\ell_e}(d_p - c_y)$$

$$= 1091 + \frac{(8000)}{20\,000}(120 - 44)$$

$$= 1121\ \text{MPa} < 1675\ \text{MPa}\ (f_{py}) \quad \text{OK}$$

$$a = \frac{\phi_s A_s f_y + \phi_p A_p f_{pr}}{\alpha_1 \phi_c f'_c b}$$

$$= \frac{(0.85)(250)(400) + (0.9)(373)(1121)}{(0.798)(0.60)(35)(1000)}$$

$$= 27\ \text{mm}$$

Determine the factored flexural resistance:

$$M_r = (\phi_s A_s f_y + \phi_p A_p f_{pr})(d - a/2)$$

$$= [(0.85)(250)(400) + (0.9)(373)(1121)]$$

$$\times (120 - 27/2)/10^6$$

$$= 49.2\ \text{kN·m} > 49.2\ \text{kN·m} \quad \text{OK}$$

For location ③, since the number of possible hinge locations is 3:

$$\ell_e = \frac{40\,000}{3} = 13\,333$$

$$f_{pr} = f_{pe} + \frac{8000}{\ell_e}(d_p - c_y)$$

$$= 1091 + \frac{(8000)}{13\,333}(115 - 42)$$

$$= 1135\ \text{MPa} < 1584\ \text{MPa}\ (f_{py}) \quad \text{OK}$$

$$a = \frac{\phi_s A_s f_y + \phi_p A_p f_{pr}}{\alpha_1 \phi_c f'_c b}$$

$$= \frac{(0.85)(250)(400) + (0.9)(373)(1135)}{(0.798)(0.60)(35)(1000)}$$

$$= 28\ \text{mm}$$

$$M_r = (\phi_s A_s f_y + \phi_p A_p f_{pr})(d - a/2)$$

$$= [(0.85)(250)(400) + (0.9)(373)(1135)]$$

$$\times (115 - 28/2)/10^6$$

$$= 47.1\ \text{kN·m} < 52.5\ \text{kN·m (req'd)}$$

Provide No. 10 bars at 200 mm centres over the first interior support, location 3:

$$A_s = 500\ \text{mm}^2/\text{m}$$

then:

$$a = \frac{\phi_s A_s f_y + \phi_p A_p f_{pr}}{\alpha_1 \phi_c f'_c b}$$
$$= \frac{(0.85)(500)(400) + (0.9)(373)(1135)}{(0.798)(0.60)(35)(1000)}$$
$$= 33\text{ mm}$$

$$M_r = (\phi_s A_s f_y + \phi_p A_p f_{pr})(d - a/2)$$
$$= [(0.85)(500)(400) + (0.9)(373)(1135)] \times (115 - 33/2)/10^6$$
$$= 54.3\text{ kN·m} > 52.5\text{ kN·m} \quad \text{OK}$$

*Check shear resistance:*

$$V_f = \frac{\ell(\alpha_D w_d + \alpha_L w_\ell)}{2}$$
$$= \frac{8.0}{2}[(1.25)(3.6) + (1.5)(2.3)] = 31.8\text{ kN/m}$$

Using the minimum value of $V_c$ specified in Clause 11.3.6.1 of CSA-A23.3:

$$V_r = 0.17\lambda\phi_c\sqrt{f'_c}\,b_w d$$
$$= (0.17)(1.0)(0.60)\sqrt{35}(1000)(0.8)(150)/10^3$$
$$= 72.4\text{ kN} > 31.8\text{ kN} \quad \text{OK}$$

To complete this design, check the flexural resistance at other locations and check the midspan deflection.

### 3.12.8 EXAMPLE 3 - TWO WAY FLAT SLAB

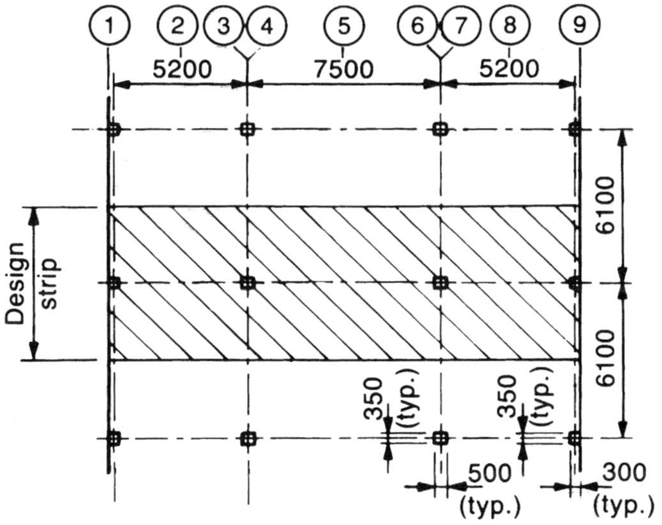


*Given:*

Flat plate floor slab supported by columns with a 2600 mm storey height.

Concrete:

$f'_{ci} = 25\text{ MPa}$

$f'_c = 35\text{ MPa}$

Normal Density

Prestressed reinforcement:

15 mm - 1860 MPa unbonded, low relaxation strands

$A_p = 140\text{ mm}^2/\text{strand}$

$f_{py} = 1675\text{ MPa}$

Non-prestressed reinforcement:

$f_y = 400\text{ MPa}$

Loads:

$w_{sd} = 1.0\text{ kN/m}^2$

$w_\ell = 1.9\text{ kN/m}^2$

*Problem:*

Design post-tensioning tendons and non-prestressed reinforcement, and check shear resistance.

*Solution:*

Determine slab thickness:

From Fig. 3.12.5: $h = \ell/45$

Longitudinal: $h = \frac{7500}{45} = 167\text{ mm}$

Transverse: $h = \frac{6100}{45} = 136\text{ mm}$

Use h = 165 mm

*Loads:*

Dead loads:

| | |
|---|---|
| 165 mm slab: (2.4)(9.81)(0.165) | = 3.9 kN/m$^2$ |
| Partitions: | = 1.0 kN/m$^2$ |
| Total dead load: | = 4.9 kN/m$^2$ |

Live loads:

| | |
|---|---|
| Occupancy: | = 1.9 kN/m$^2$ |

$$\frac{\text{Live load}}{\text{Dead load}} = \frac{1.9}{4.9} = 0.38 < 0.75$$

As per Clause 13.9.4.2 of CSA-A23.3, load all panels with full live load.

*Load balancing:*

Assume 20 mm cover from top of slab to tendon and 40 mm cover from underside of slab to tendon for fire protection. Cover to centreline of strand:

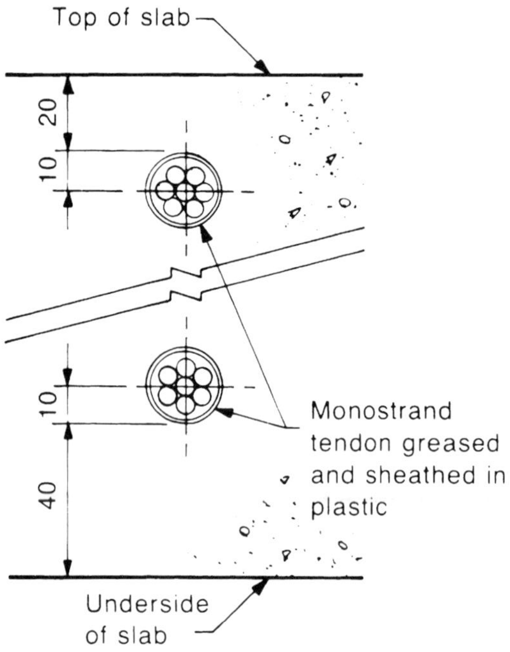


Assume a parabolic tendon profile with a force corresponding to an average compressive stress of 1.0 MPa:

$$\frac{P_e}{A_g} = 1.0\,\text{MPa}$$

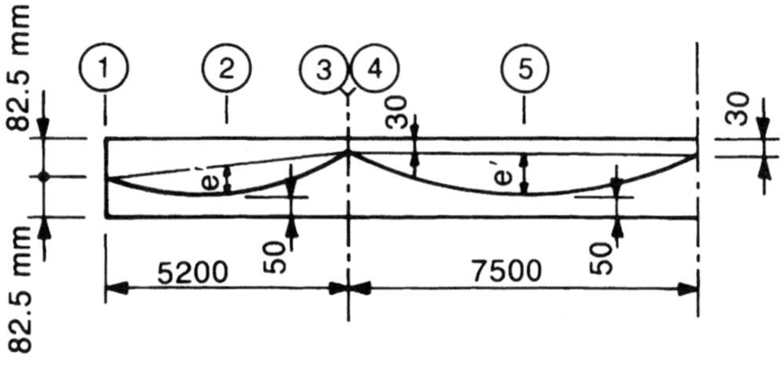


$$P_e = (10)(165)(1000)/10^3 = 165\,\text{kN/m}$$

With an average initial prestress level after friction losses of 0.7 $f_{pu}$ and an assumed long term loss of 230 MPa:

$$P_e = (0.7f_{pu} - 230)A_p$$
$$= [(0.7)(1860) - 230](140)/10^3$$
$$= 150\,\text{kN / strand}$$

For a 6.1 m bay, require:

$$\frac{(6.1)(165)}{150} = 6.7\text{, say 7 tendons}$$

thus:

$$P_e = \frac{(7)(140)}{6.1} = 161\,\text{kN/m}$$

$$\frac{P_e}{A_g} = \frac{(161)(10^3)}{(165)(10^3)} = 1.0\,\text{MPa}$$

If upward uniformly distributed load due to prestress is $w_p$ and service load (downward uniformly distributed load) is w, the net service load causing flexural stress is:

$$w_{nt} = w - w_p$$

*Equivalent frame properties:*

Moments in the slab due to net loading are determined by the equivalent frame method outlined in Clause 13.9.1 of CSA-A23.3:

At exterior column ①, 300 × 350:

$$I_c = \frac{c_2 c_1^3}{12} = \frac{(350)(300)^3}{12} = 787 \times 10^6\ \text{mm}^4$$

Height of column, floor to floor, h = 2600 mm. Centre to centre column height, $h_c$ = 2600 − 165 = 2435 mm:

$$K_c = \frac{k_c E_c I_c}{h_c} = \frac{(4)(787 \times 10^6)E_c}{2435} = 1.29 \times 10^6 E_c$$

The stiffness coefficient, $k_c$, is taken as 4.0. This is not strictly true, as the lengths of the columns within the floors are infinitely stiff. However, for two way flat plate construction, this is reasonably close to the expected value. Actual values of $k_c$ can be found in various texts on structural analysis.

As there are columns above and below:

$$\Sigma K_c = (2)(1.29 \times 10^6)E_c = 2.58 \times 10^6 E_c$$

Torsional element:

$c_1 = 300$ $h = 165$ $\ell_t = 6100$

$y = 300$ $x = 165$ $c_2 = 350$

| Location | ① | ② | ③ | ④ | ⑤ |
|---|---|---|---|---|---|
| e (mm)<br>e′ (mm) | 0 | − 32<br>58 | + 52 | + 52 | − 32<br>85 |
| w (kN/m²) | | 6.8 | | | 6.8 |
| $w_p = \dfrac{8P_e e'}{\ell^2}$ (kN/m²) | | 2.8 | | | 1.9 |
| $w_n$ (kN/m²) | | 4.0 | | | 4.9 |

then:

$$C = \left(1 - 0.63\frac{x}{y}\right)\frac{x^3y}{3}$$

$$= \left[1 - \frac{(0.63)(165)}{300}\right]\frac{(165)^3(300)}{3}$$

$$= 293 \times 10^6\ \text{mm}^4$$

$$K_t = \frac{\Sigma 9E_cC}{\ell_t\left(1 - \frac{c_2}{\ell_t}\right)^3}$$

$$= \frac{(2)(9)E_c(293 \times 10^6)}{(6100)\left(1 - \frac{350}{6100}\right)^3} = 1.03 \times 10^6 E_c$$

Equivalent column stiffness:

$$\frac{1}{K_{ec}} = \frac{1}{K_t} + \frac{1}{\Sigma K_c}$$

$$= \frac{1}{1.03 \times 10^6 E_c} + \frac{1}{2.58 \times 10^6 E_c}$$

$$K_{ec} = 0.74 \times 10^6 E_c$$

At interior columns ③ and ④, 500 × 350:

$$I_c = \frac{c_2c_1^3}{12} = \frac{(350)(500)^3}{12} = 3646 \times 10^6\ \text{mm}^4$$

$h = 2600$ mm and $h_c = 2435$ mm

$$K_c = \frac{4E_cI_c}{h_c} = \frac{(4)(3646)E_c(10^6)}{2435} = 5.99 \times 10^6 E_c$$

$$\Sigma K_c = (2)(5.99 \times 10^6)E_c = 11.97 \times 10^6 E_c$$

Torsional element: $c_1 = 500$ $\quad h = 165$ $\quad \ell_2 = 6100$

$y = 500$ $\quad x = 165$ $\quad c_2 = 300$

$$C = \left(1 - 0.63\frac{x}{y}\right)\frac{x^3y}{3}$$

$$= \left[1 - \frac{(0.63)(165)}{500}\right]\frac{(165)^3(500)}{3}$$

$$= 588 \times 10^6\ \text{mm}^4$$

$$K_t = \frac{\Sigma 9E_cC}{\ell_2\left(1 - \frac{c_2}{2}\right)^3}$$

$$= \frac{(2)(9)E_c(588 \times 10^6)}{(6100)\left(1 - \frac{300}{6100}\right)^3} = 2.02 \times 10^6 E_c$$

Equivalent column stiffness:

$$\frac{1}{K_{ec}} = \frac{1}{K_t} + \frac{1}{\Sigma K_c}$$

$$= \frac{1}{2.02 \times 10^6 E_c} + \frac{1}{11.97 \times 10^6 E_c}$$

$$K_{ec} = 1.73 \times 10^6 E_c$$

*Moment distribution:*

1. Relative stiffnesses:

At exterior column ①:

Column: $K_{ec} = 0.74 \times 10^6 E_c$

Span: $$K_s = \frac{4E_cI_s}{\ell} = \frac{(4)(6100)(165)^2E_c}{(12)(5200)}$$

$$= 1.76 \times 10^6 E_c$$

Relative stiffness, column:

$$= \frac{0.74 \times 10^6 E_c}{(0.74E_c + 1.76E_c)(10^6)} = 0.3$$

Relative stiffness, slab:

$$= \frac{1.76 \times 10^6 E_c}{(0.74E_c + 1.76E_c)(10^6)} = 0.7$$

At interior column ③:

Column: $K_{ec} = 1.74 \times 10^6 E_c$

Exterior span: $$K_{se} = \frac{4E_cI_s}{\ell} = \frac{(4)(6100)(165)^3E_c}{(12)(5200)}$$

$$= 1.76 \times 10^6 E_c$$

Interior span: $$K_{si} = \frac{4E_cI_s}{\ell} = \frac{(4)(6100)(165)^3E_c}{(12)(7500)}$$

$$= 1.22 \text{ x } 10^6 E_c$$

Relative stiffness, column:

$$= \frac{1.74 \times 10^6 E_c}{(1.74 + 1.76 + 1.22)(10^6)E_c} = 0.37$$

Relative stiffness, exterior span:

$$= \frac{1.76 \times 10^6 E_c}{(1.74 + 1.76 + 1.22)(10^6)E_c} = 0.37$$

Relative stiffness, interior span:

$$= \frac{1.22 \times 10^6 E_c}{(1.74 + 1.76 + 1.22)(10^6)E_c} = 0.26$$

2. Fixed end moments:

Exterior span:

$$\text{FEM} = \frac{w_{n\ell}\ell^2}{12} = \frac{(4.0)(5.2)^2}{12} = 9.0\ \text{kN}\cdot\text{m/m}$$

| Moment distribution of net loads | | | | | | | | | | |
|---|---|---|---|---|---|---|---|---|---|---|
| **Joint** | ① | | ③ ④ | | | ⑥ ⑦ | | | ⑨ | |
| Span | Col. | 1-3 | 3-1 | Col. | 4-6 | 6-4 | Col. | 7-9 | 9-7 | Col. |
| Rel. Stiff. | 0.3 | 0.7 | 0.37 | 0.37 | 0.26 | 0.26 | 0.37 | 0.37 | 0.7 | 0.3 |
| FEM | | 9.0 | –9.0 | | 23.0 | –23.0 | | 9.0 | –9.0 | |
| | –2.7 | –6.3 | –3.1 | | | | | | | |
| | | –2.0 | –4.0 | –4.0 | –2.8 | –1.4 | | | | |
| | 0.6 | 1.4 | 0.7 | | 2.0 | 4.0 | 5.7 | 5.7 | 2.8 | |
| | | –0.5 | –1.0 | –1.0 | –0.7 | –0.4 | –0.7 | 2.2 | 4.3 | 1.9 |
| | 0.2 | 0.3 | | | | –0.5 | | –0.7 | | |
| $M_{n\ell}$ | –1.9 | 1.9 | –16.3 | –5.0 | 21.4 | –21.4 | 5.0 | 16.3 | –1.9 | 1.9 |

Fixed end moment sense

Interior span:

$$FEM = \frac{w_{n\ell}\ell^2}{12} = \frac{(4.9)(7.5)^2}{12} = 23.0 \text{ kN·m/m}$$

3. Carry over factors: all carry over factors = –0.5

4. Moment distribution (see table above).

*Determine midspan moments:*

Exterior span:

Moment at ①: $M_1 = -1.9$ kN·m/m

Moment at ③: $M_3 = -16.3$ kN·m/m

Moment at midspan:

$$= \left(\frac{M_1 + M_3}{2}\right) + \frac{w_{n\ell}\ell^2}{8}$$

$$= \left(\frac{-1.9 - 16.3}{2}\right) + \frac{(4.0)(5.2)^2}{8} = 4.4 \text{ kN·m/m}$$

Interior span:

Moment at ④: $M_4 = 21.4$ kN·m/m

Moment at ⑥: $M_6 = 21.4$ kN·m/m

Moment at midspan:

$$= \left(\frac{M_4 + M_6}{2}\right) + \frac{w_\ell \ell^2}{8}$$

$$= \left(\frac{-21.4 - 21.4}{2}\right) + \frac{(4.9)(7.5)^2}{8} = 13.1 \text{ kN·m/m}$$

The service load moment diagram is as follows:

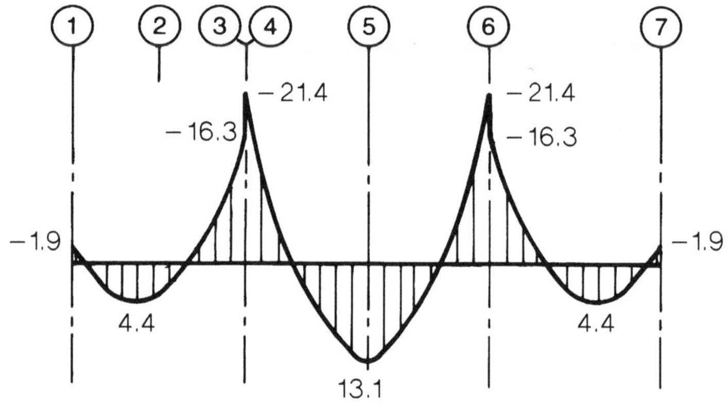


*Service load stresses:*

Flexural stresses are computed from the equation:

$$f = \frac{P_e}{A_g} \pm \frac{M}{S}$$

Note that the moment at the face of the support is used. Calculations for the interior span are shown below. Calculations are typical for all spans.

1. Determine moment and stress at ④:

$$V_{CL} = \frac{w_{n\ell}\ell_1}{2} = \frac{(4.9)(7.5)}{2} = 18.4 \text{ kN/m}$$

$$V_{CF} = V_{CL}\left(\frac{3500}{3750}\right) = (18.4)(0.933) = 17.2 \text{ kN/m}$$

Moment at column face:

$$M_{CF} = M_{CL} - \left(\frac{V_{CL} + V_{CF}}{2}\right)\left(\frac{c_1}{2}\right)$$

$$= -21.4 + \frac{(18.4 + 17.2)(0.5)}{4} = 17.0 \text{ kN·m/m}$$

Total moment in design strip:

$$M_{DS} = (-17.0)(6.1) = -104.0 \text{ kN·m/m}$$

Moment in column strip:

$$\text{Width of column strip} = (2)(0.25)\ell_1 = (2)(0.25)(6100) = 3050 \text{ mm}$$

From Clause 18.13.2.1, the column strip should be proportioned to resist 75% of the total moment:

$$M_{CS} = \frac{(-104.0)(0.75)}{3.05} = -25.6 \text{ kN·m/m}$$

Stress in column strip. Note in this calculation $P_e/A_g$ is the value for the entire section:

$$f_t = \frac{P_e}{A_g} - \frac{M}{S} = 0.98 - \frac{(25.6)(10^6)(6)}{(1000)(165)^2} = -4.7 \text{ MPa}$$

Permissible tensile stress, per Clause 18.4.2(c) of CSA-A23.3, is:

$$0.5\sqrt{f'_c} = 0.5\sqrt{35} = 3.0 \text{ MPa} < 4.7 \text{ MPa}$$

Therefore section is partially prestressed, and bonded, non-prestressed reinforcement must be provided. According to Clause 18.9.2 of CSA-A23.3:

$$A_s = (1.25)(0.0006)h\ell_n = (1.25)(0.0006)(165)(7500) = 928 \text{mm}^2$$

Therefore, provide 5 - No. 15 bars as bonded reinforcement over a width equal to the column width, $c_2$, plus 1.5 times the slab thickness beyond each side of the column (Clause 18.13.5.2):

$$c_2 + (2)(1.5)h = 350 + (3)(165) = 845 \text{ mm}$$

Thus, the spacing is:

$$= \frac{845}{5} = 169 \text{ say } 150 \text{ mm}$$

The minimum length of bonded reinforcement, according to Clause 18.10.4:

$$\ell = 2\ell_1/6 = (2)(7500)/6 = 2500 \text{ mm}$$

Provide 5 - No. 15 bars, 2500 mm long, spaced at 150 mm evenly about the column centreline.

Compressive stress:

$$f_c = \frac{P_e}{A_g} + \frac{M}{S} = 0.98 + \frac{(25.6)(10^6)(6)}{(1000)(165)^2} = 6.6 \text{ MPa}$$

Permissible compressive stress, from Clause 18.4.2(b) is:

$$0.6f'_c = (0.6)(35) = 21 \text{MPa} \quad \text{OK}$$

2. Determine stresses at ⑤:

Total moment in design strip:

$$M_{DS} = (13.1)(6.1) = 79.9 \text{ kN·m}$$

Moment in column strip:

$$M_{CS} = \frac{(79.9)(0.60)}{3.05} = 15.7 \text{ kN·m/m}$$

Tensile stress in column strip:

$$f_t = \frac{P_e}{A_g} - \frac{M}{S} = 0.98 - \frac{(15.7)(10^6)(6)}{(1000)(165)^2} = -2.5 \text{ MPa}$$

From Clause 18.9.2, permissible stress without bonded reinforcement:

$$0.5\sqrt{f'_c} = 0.5\sqrt{35} = 3.0 \text{ MPa} > 2.3 \text{ MPa}$$

use:

$$A_s = 0.004A = (0.004)(1000)(165)/2 = 330 \text{ mm}^2$$

Provide 5 - No. 15 bars in column strip of positive moment area, spaced evenly throughout:

$$\text{Spacing} = \frac{3050}{3} = 610 \text{, say } 600 \text{ mm}$$

From Clause 18.10.3, the length of bar:

$$\ell_b = \ell_1/2 = 7500/2 = 3750 \text{ mm}$$

Provide 5 - No. 15 bars, 3750 mm long, spaced at 600 mm in the column strip.

Compressive stress:

$$f_c = \frac{P_e}{A_g} + \frac{M}{S} = 0.98 + \frac{(15.7)(10^6)(6)}{(1000)(165)^2} = 4.4 \text{ MPa}$$

From Clause 18.4.2(b), permissible stress is:

$$0.6f'_c = 21.0 \text{ MPa} \quad \text{OK}$$

Results of service load analysis are in table below:

| **Location** | ① | ② | ③ | ④ | ⑤ |
|---|---|---|---|---|---|
| $M_{n\,CL}$ (kN·m/m) | −1.9 | 4.1 | −16.3 | −21.4 | 13.1 |
| $M_{n\,CF}$ (kN·m/m) | −0.4 | | −13.8 | −17.0 | |
| $M_{n\,CS}$ (kN·m/m) | −1.0 | 6.2 | −24.3 | −25.6 | 15.7 |
| $f_t$ (MPa) | 0.8 | −0.4 | −4.4 | −4.7 | −2.5 |
| $f_c$ (MPa) | 1.2 | 2.3 | 6.3 | 6.6 | 4.4 |

**Moment distribution of prestress (uplift loads)**

| Joint | ① | | ③ ④ | | | ⑥ ⑦ | | | ⑨ | |
|---|---|---|---|---|---|---|---|---|---|---|
| Span | Col. | 1-3 | 3-1 | Col. | 4-6 | 6-4 | Col. | 7-9 | 9-7 | Col. |
| Rel. Stiff. | 0.3 | 0.7 | 0.37 | 0.37 | 0.26 | 0.26 | 0.37 | 0.37 | 0.7 | 0.3 |
| FEM | | 6.3 | –6.3 | | 8.9 | –8.9 | | 6.3 | –6.3 | |
| | –1.9 | –4.4 | –2.2 | | | | | | | |
| | | | –0.1 | –0.1 | –0.1 | –0.1 | | | | |
| | | | | | 0.4 | 0.7 | 1.0 | 1.0 | 0.5 | |
| | | | | | | | | 2.0 | 4.1 | 1.7 |
| | | | | | –0.3 | –0.5 | –0.7 | –0.7 | | |
| | | | 0 | 0 | 0 | | | | | |
| $M_{bal}$ | –1.8 | 1.8 | –8.7 | –0.2 | 8.8 | –8.8 | 0.2 | 8.7 | –1.8 | 1.8 |

To complete service load analysis for this span, check stresses for middle strip, and bonded reinforcement requirements.

*Factored flexural resistance:*

Determine secondary moments, M′, by subtracting primary moments from balanced load moments since:

$$M_{bal} = Pe + M'$$

then:

$$M' = M_{bal} - Pe$$

The balanced load or uplift moments due to post-tension forces are calculated by moment distribution as were net load moments.

From the table on Page 3-134, $w_p$ for the exterior span is = 2.8 kN/m²

Fixed end moment:

$$\text{FEM} = \frac{w_p \ell^2}{12} = \frac{(2.8)(5.2)^2}{12} = 6.3 \text{ kN·m/m}$$

Also, $w_p$ for interior span is = 1.9 kN/m². Thus:

$$\text{FEM} = \frac{w_p \ell^2}{12} = \frac{(1.9)(7.5)^2}{12} = 8.9 \text{ kN·m/m}$$

Determine the factored dead and live loads:

| | | |
|---|---|---|
| Factored dead load: | (1.25)(3.9+1.0) | = 6.1kN/m |
| Factored live load: | (1.5)(1.9) | = 2.9 kN/m |
| Total factored load: | | = 9.0 kN/m |

Fixed end moments:

$$\text{Exterior span FEM}_f = \frac{w_1 \ell^2}{12} = \frac{(9)(5.2)^2}{12} = 20.3 \text{ kN·m}$$

$$\text{Interior span FEM}_f = \frac{w_1 \ell^2}{12} = \frac{(9)(7.5)^2}{12} = 42.2 \text{ kN·m}$$

**Moment distribution of net loads**

| Joint | ① | | ③ ④ | | | ⑥ ⑦ | | | ⑨ | |
|---|---|---|---|---|---|---|---|---|---|---|
| Span | Col. | 1-3 | 3-1 | Col. | 4-6 | 6-4 | Col. | 7-9 | 9-7 | Col. |
| Rel. Stiff. | 0.3 | 0.7 | 0.37 | 0.37 | 0.26 | 0.26 | 0.37 | 0.37 | 0.7 | 0.3 |
| FEM | | 20.3 | –20.3 | | 42.2 | –42.2 | | 20.3 | –20.3 | |
| | –6.1 | –14.2 | –7.1 | | | | | 7.1 | 14.2 | 6.1 |
| | | –2.8 | –5.5 | –5.5 | –3.8 | –1.9 | | | | |
| | 0.8 | 2.0 | 1.0 | | 2.2 | 4.3 | 6.2 | 6.2 | 3.1 | |
| | | –0.6 | –1.2 | –1.2 | –0.8 | –0.4 | | –1.1 | –2.2 | –0.9 |
| | 0.2 | 0.4 | 0.2 | | 0.2 | 0.4 | –0.6 | 0.6 | 0.3 | |
| | | | –0.1 | –0.1 | –0.1 | | | | –0.2 | –0.1 |
| $M_t$ | –5.1 | 5.1 | –32.9 | –6.8 | 39.8 | –39.8 | 6.8 | 33.1 | –5.1 | 5.1 |

Combine moments due to factored loads with secondary moments to determine total factored moments, $M_f$, as shown in table on this page.

1. Negative factored moment at ④:

From Clause 18.13.1, the design strip may be analyzed in its entirety for factored load conditions. As the column makes up less than 10% of the design strip, moments at the centreline of supports should be used:

Total factored moment on design strip:

$$M_f = (40.0)(6.1) = 244 \text{ kN·m}$$

Since there are 7 tendons located within the design strip:

$$A_p = (7)(140) = 980\,\text{mm}^2$$

As bonded reinforcement is provided for service load conditions, it may be utilized for factored load calculations. However, the length of the bars must be increased from the minimum allowed by Clause 18.10.4 (see Clause 18.10.1):

$$A_s = 1000\,\text{mm}^2$$

Calculate $f_{pr}$ from Clause 18.7.2(b):

$$f_{pr} = f_{pe} + \frac{8000}{\ell_e}(d_p - c_y)$$

where:

$$f_{pe} = 0.7f_{pu} - 230$$
$$= (0.7)(1860) - 230 = 1072\,\text{MPa}$$

$$\ell_e = \frac{(2)(5200) + 7500 + (2)(150)}{3} = 6067\,\text{mm}$$

$$d_p = 165 - 30 = 135\,\text{mm}$$

$$\alpha_1 = 0.85 - (0.0015)(35) = 0.80$$

$$\beta_1 = 0.97 - (0.0025)(35) = 0.88$$

$$c_y = \frac{\phi_s A_s f_y + \phi_p A_p f_{py}}{\alpha_1 \beta_1 \phi_c f'_c b}$$
$$= \frac{(0.85)(1000)(400) + (0.9)(980)(1675)}{(0.8)(0.88)(0.60)(35)(6100)}$$
$$= 20\,\text{mm}$$

then:

$$f_{pr} = 1072 + \frac{8000}{6067}(135 - 20) = 1224\,\text{MPa}$$

$$a = \frac{\phi_s A_s f_y + \phi_p A_p f_{pr}}{\alpha_1 \phi_c f'_c b}$$
$$= \frac{(0.85)(1000)(400) + (0.9)(980)(1224)}{(0.8)(0.60)(35)(6100)}$$
$$= 14\,\text{mm}$$

$$M_r = (\phi_s A_s f_y + \phi_p A_p f_{pr})(d_p - a/2)$$
$$= [(0.85)(1000)(400) + (0.9)(980)(1224)]$$
$$(135 - 14/2)/10^6$$
$$= 182\,\text{kN·m} < M_f = 244\,\text{kN·m}$$

Add additional bonded reinforcement to increase flexural resistance. Add 8 - No. 15 bars:

$$A_s = 2600\,\text{mm}^2$$

$$c_y = \frac{\phi_s A_s f_y + \phi_p A_p f_{py}}{\alpha_1 \beta_1 \phi_c f'_c b}$$
$$= \frac{(0.85)(2600)(400) + (0.9)(980)(1675)}{(0.8)(0.88)(0.60)(35)(6100)}$$
$$= 26\,\text{mm}$$

$$f_{pr} = 1072 + \frac{8000}{6067}(135 - 26) = 1216\,\text{MPa}$$

$$a = \frac{\phi_s A_s f_y + \phi_p A_p f_{pr}}{\alpha_1 \phi_c f'_c b}$$
$$= \frac{(0.85)(2600)(400) + (0.9)(980)(1216)}{(0.8)(0.60)(35)(6100)}$$
$$= 19\,\text{mm}$$

$$M_r = (\phi_s A_s f_y + \phi_p A_p f_{pr})(d_p - a/2)$$
$$= [(0.85)(2600)(400) + (0.9)(980)(1216)]$$
$$(135 - 19/2)/10^6$$
$$= 246\,\text{kN·m} > M_f = 244\,\text{kN·m} \quad \text{OK}$$

Use 13 - No. 15 reinforcing bars for factored flexural resistance.

2. Positive factored moment at ⑤:

Total factored moment on design strip:

**Factored moment calculations**

| Location | ① | ② | ③ | ④ | ⑤ |
|---|---|---|---|---|---|
| $M_{bal}$ (kN·m/m) | 1.8 | −4.2 | 8.7 | 8.8 | −4.6 |
| Pe (kN·m/m) | 0 | 5.2 | 9.0 | 9.0 | 5.2 |
| M′ (kN·m/m) | 1.8 | 1.0 | −0.3 | −0.2 | 0.6 |
| $M_f$ (kN·m/m) | −5.1 | 11.4 | −32.9 | −39.8 | 23.5 |
| $M_{fCL}$ (kN·m/m) | −3.3 | 12.4 | −33.2 | −40.0 | 24.1 |
| $M_{f\,des\,strip}$ (kN·m) | −20.1 | 75.6 | −202.5 | −244.0 | 147.0 |

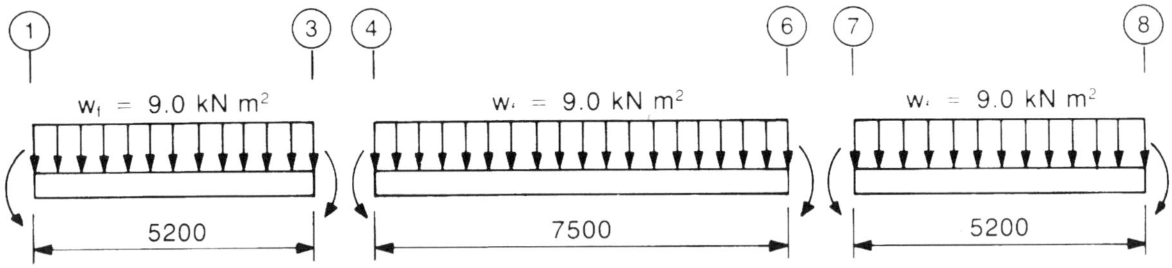


| | ① | ③ | ④ | ⑥ | ⑦ | ⑧ |
|---|---|---|---|---|---|---|
| $M_{fCL}$ (kN·m/m) | 4.0 | 32.5 | 39.2 | 39.2 | 32.5 | 4.0 |
| $R_{wf}$ (kN/m) | 23.4 | 23.4 | 33.8 | 33.8 | 23.4 | 23.4 |
| $R_{mf}$ (kN/m) | −5.5 | 5.5 | 0 | 0 | 5.5 | −5.5 |
| ΣR | 17.9 | 62.7 | | 62.7 | | 17.9 |

$M_f = 147$ kN·m

$A_p = 980\ \text{mm}^2$

$A_s = 1000\ \text{mm}^2$

$d_p = 165 - 50 = 115$ mm

$$c_y = \frac{\phi_s A_s f_y + \phi_p A_p f_{py}}{\alpha_1 \beta_1 \phi_c f_c' b}$$

$$= \frac{(0.85)(1000)(400) + (0.9)(980)(1675)}{(0.8)(0.88)(0.60)(35)(6100)}$$

$= 20$ mm

$$f_{pr} = f_{pe} + \frac{8000}{\ell_e}(d_p - c_y)$$

$$= 1072 + \frac{8000}{6067}(115 - 20) = 1197\ \text{MPa}$$

$$a = \frac{\phi_s A_s f_y + \phi_p A_p f_{pr}}{\alpha_1 \phi_c f_c' b}$$

$$= \frac{(0.85)(1000)(400) + (0.9)(980)(1197)}{(0.8)(0.60)(35)(6100)}$$

$= 14$ mm

$M_r = (\phi_s A_s f_y + \phi_p A_p f_{pr})(d_p - a/2)$

$= [(0.85)(1000)(400) + (0.9)(980)(1197)]$

$(115 - 14/2)/10^6$

$= 151$ kN·m $> M_f = 147$ kN·m OK

The post tensioning strand and bonded reinforcement is sufficient to resist the required factored moment.

*Shear considerations:*

The calculation of shear and moment transfer capacities in two way flat plates is an essential part of the design. Shear capacity is checked for factored resistance at the critical perimeters around supports.

Calculation of shear force:

$w_f = \alpha_D w_d + \alpha_L w_\ell$

$= (1.25)(4.9) + (1.5)(1.9) = 9.0\ \text{kN/m}^2$

As shown in the diagram above, reactions due to factored dead and live load and prestress are calculated, and from these reactions the factored shear forces at the centreline of supports are determined. Upward reactions are positive.

At location ③:

$$R_{wf} = \frac{w_f \ell}{2} = \frac{(9.0)(5.2)}{2} = 23.4\ \text{kN/m}$$

$$R_{mf} = \frac{M_{f3} - M_{f1}}{\ell} = \frac{32.5 - 4.0}{5.2} = 5.5\ \text{kN/m}$$

Shear at exterior column ①:

Assume a cladding load along edge of design strip of 6 kN/m acting at an eccentricity of 75 mm from the centreline of the exterior column:

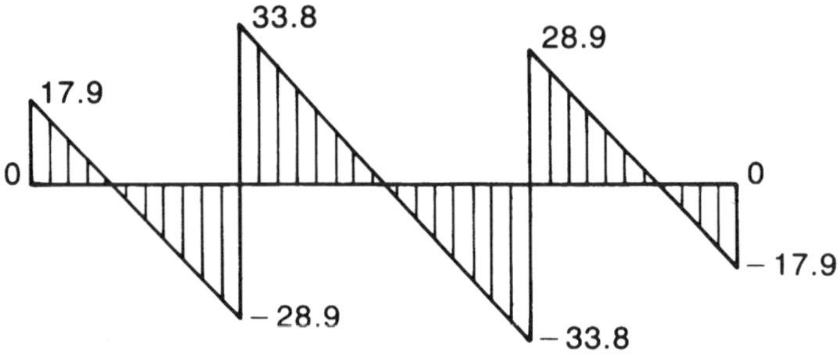


| | | |
|---|---|---|
| Shear from diagram above: | (17.9)(6.1) | = 109 kN |
| Shear due to cladding: | (1.25)(6)(6.1) | = 46 kN |
| Total Shear $V_f$: | | 155 kN |

Critical section properties:

From the diagram on page 3-141:

$c_1 = 300$ mm  $d = 132$ mm  $\frac{d}{2} = 66$ mm

$c_2 = 350\text{ mm}$

$$e_1 = \frac{(c_1 + d/2)^2}{2c_1 + c_2 + 2d} = \frac{(300+66)^2}{(2)(300)+350+(2)(132)} = 110\text{ mm}$$

$$e_2 = \frac{1}{2}(c_1 + d) = (0.5)(300+132) = 216\text{ mm}$$

$$e_3 = e_2 - e_1 = 216 - 110 = 106\text{ mm}$$

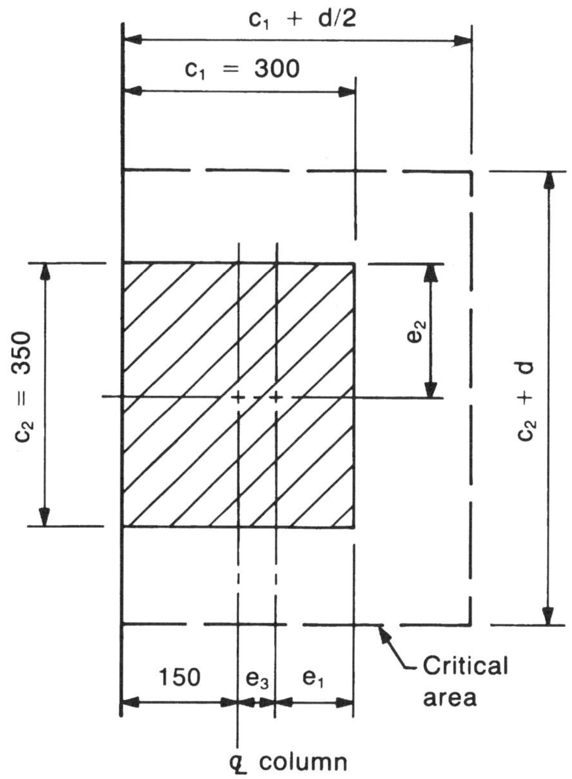


$$A_{VT} = d(2c_1 + c_2 + 2d) = (132)[(2)(300)+350+(2)(132)] = 160\,248\text{ mm}^2$$

$$J_1 = \frac{1}{6}\left[\left(c_1+\frac{d}{2}\right)d^3 + \left(c_1+\frac{d}{2}\right)^3 d\right] + (c_2+d)de_1^2 + 2\left(c_1+\frac{d}{2}\right)d\left(\frac{c_1+\frac{d}{2}}{2} - e_1\right)^2$$

$$= \frac{1}{6}[(300+66)(132)^3 + (300+66)^3(132)] + (350+132)(132)(110)^2 + (2)(132)(366)\times\left(\frac{366}{2} - 110\right)^2$$

$$= 2.5\times 10^9\text{ mm}^4$$

*Moment transferred by shear eccentricity:*

The total moment due to factored loads plus eccentric shear about centreline of critical section is found:

$$M_{vcr} = M_f + Ve_3 = (-4.0)(6.1) + (109)(106)/10^3 = 12.8\text{ kN·m}$$

From Clauses 13.4.3 and 13.4.5.3 of CSA-A23.3:

$$\gamma_f = \frac{1}{1+\frac{2}{3}\sqrt{\frac{c_1+d}{c_2+d}}} = \frac{1}{1+\frac{2}{3}\sqrt{\frac{300+132}{350+132}}} = 0.613$$

The moment transferred by eccentricity of shear:

$$M_v = \gamma_f M_{vcr} = (0.613)(12.8) = 7.8\text{ kN·m}$$

Maximum factored shear stress:

$$v_f = \frac{V_f}{A_{VT}} + \frac{M_v e_1}{J_1} = \frac{(155)(10^3)}{160\,248} + \frac{(7.8)(10^6)(110)}{2.5\times 10^9} = 1.1\text{ MPa}$$

From Eqn. 18.5 of CSA-A23.3:

$$v_c = 0.4\lambda\phi_c\sqrt{f'_c}\sqrt{1+\frac{\phi_p f_{cp}}{0.4\lambda\phi_c\sqrt{f'_c}}} + \frac{\phi_p V_p}{b_o d}$$

If the term $\phi_p V_p$ is ignored:

$$v_c = (0.4)(1.0)(0.60)\sqrt{35}\sqrt{1+\frac{(0.9)(0.98)}{(0.4)(1.0)(0.6)\sqrt{35}}}$$

$$= 1.8\text{ MPa} > 1.06\text{ MPa} = v_f \quad \text{OK}$$

To complete the shear calculation, check the shear capacity of the critical section for the interior column and provide column capitals if necessary; check shear capacity in the span. Also check that column strip section is capable of carrying factored design strip moment at column.

*Deflection:*

Calculate the live load deflection of a 1 m wide strip in the central span of the design strip.

Since the tensile stress at ④ under service load is greater than $0.5\sqrt{f'_c}$, use the effective moment of inertia according to Sect. 3.5.2, namely:

$$I_{tr} \approx I_g = \frac{(1000)(165)^3}{12} = 374\times 10^6\text{ mm}^4$$

$$d_p = 135\text{ mm}$$

$$A_p = (7)(140)(6.1) = 161\text{ mm}^2/\text{m}$$

$$n_p = \frac{(190\,000)(161)}{(28\,200)(1000)(135)} = 0.0080$$

$$C = 0.0068 \text{ from Fig. 3.5.3}$$

$$I_{cr} = Cbd_p^3 = (0.0068)(1000)(135)^3$$
$$= 16.7 \times 10^6 \text{ mm}^4$$

$$M'_{cr} = \frac{(0.6)\sqrt{35}(374)(10^6)}{(82.5)(10^6)} = 16.1 \text{ kN·m/m}$$

$$f_{ce} = \frac{(161)(10^3)}{(165)(10^3)} + \frac{(161)(10^3)(52.5)(82.5)}{(374)(10^6)}$$
$$= 2.84 \text{ MPa}$$

$$M_{dc} = \frac{(2.84)(374)(10^6)}{(82.5)(10^6)} = 12.9 \text{ kN·m/m}$$

$$M_s = \frac{(-39.8)(6.8)}{9.0} - 0.2 = -30.3 \text{ kN·m/m}$$

$$I_e = 16.7 \times 10^6 + (374 - 16.7) \times 10^6 \left(\frac{16.1}{30.3 - 12.9}\right)^3$$

$$= 300 \times 10^6 \text{ mm}^4 < I_g \quad \therefore \text{Use } I_e$$

Since tensile stress at ⑤ is less than $0.5\sqrt{f'_c}$, use gross moment of inertia throughout span.

Specified live load = 1.9 kN/m²

Live load moment at ④ $= (-39.8)\frac{(1.9)}{9.0} = 8.4$ kN·m/m

Live load moment at ⑤ $= (+24.1)\frac{(1.9)}{9.0} = 5.0$ kN·m/m

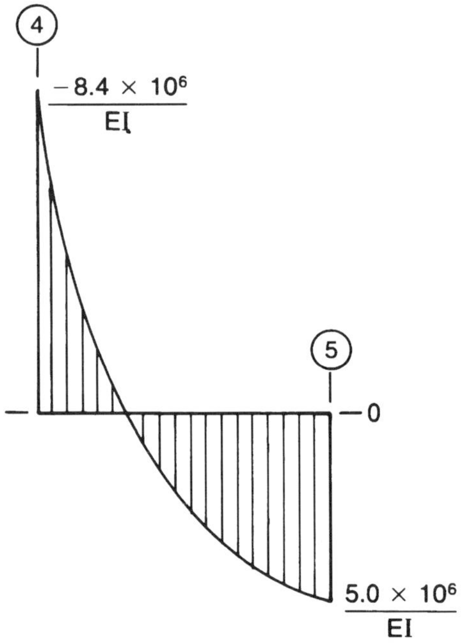


Live load bending moment over EI

*Deflection:*

Using the moment area method; deflection at midspan due to live loads:

$$\Delta_{\ell} = \frac{1}{EI}\left[\left(\frac{2}{3}\right)(5.0 + 8.4)(3750)\left(\frac{5}{8}\right)(3750) - (8.4)(3750)\left(\frac{3750}{2}\right)\right](10^6)$$

$$= \frac{19.5 \times 10^{12}}{(28\,200)(300 \times 10^6)} = 2 \text{ mm}$$

Deflection at midspan due to net dead load:

$$\Delta_{dr} = \frac{(2)(4.9)}{1.9} = 5 \text{ mm}$$

Long term deflection should also be computed according to Clause 9.8.4.4 of CSA-A23.3. This should be compared with permissible deflections specified in Table 9-1 of CSA-A23.3.

## 3.12.9 OTHER CONSIDERATIONS

### Detailing Considerations

Design and detailing of the concrete element, the post-tensioning anchorage and tendons, and the anchorage zone reinforcing require special attention.

The following items should be considered:

1. The end face of concrete elements should be large enough to comfortably contain all of the required post-tensioning anchorages.
2. The anchorage blockouts should be of sufficient size, and anchors should be adequately spaced, to allow the unimpeded operation of stressing jacks.
3. Protruding rebar dowels and any other obstruction in the stressing face of the element should be carefully detailed and placed to maintain adequate clearance for stressing operations.
4. Spacing and location of post-tensioning tendons are important considerations. At the ends of elements, care should be taken to provide adequate cover for bearing plates and other anchorage hardware. At high and low points of a tendon within the length of the element, concrete cover to the centre of gravity of the prestressing steel should take into consideration the diameter of the tendon ducts as well as the clear cover required. Individual tendons should be spaced to allow effective placing and vibration of concrete.
5. Where post-tensioning tendons conflict with the reinforcement, electrical conduits or other items included in the element, the post-tensioning tendons should take precedence.
6. Anchorage zone areas should be adequately designed and detailed for easy placement of reinforcement.
7. Anchorage blockouts may reduce the gross area of concrete where they occur within a column. Column capacities at these locations should be checked.

### Volume Change Restraints

The relative stiffness of post-tensioned beams and columns must be carefully analyzed. Restraining items such as walls which frame into columns, should be given special attention, and if necessary, properly designed control joints should be provided. When a post-tensioned beam is framed into the top of a relatively stiff column, the elastic, shrinkage, and creep shortening of the beam and slab will cause high moments and shears in the column. Furthermore, the prestress force is reduced by the amount that is diverted to column bending. In multi-storey and multi-bay frames, these effects are distributed throughout the frame.

### Anchorage Zone Reinforcement

*General:*

Post-tensioning force is applied through relatively small anchorages causing high local stresses, which decrease as the forces spread through the anchorage zone to the full cross section of the element. Usual bending theory does not apply and the works of Guyon and Leonhardt should be consulted for classical theories on anchorage zone stress.

The design of anchorage zones should consider the following steps:

1. Determine the size of the primary prism around individual anchorages.
2. Provide primary bursting reinforcement within each primary prism.
3. Provide reinforcement close to the loaded face of the anchorage zone to resist the spalling stresses.
4. Check bearing stresses behind anchorages.
5. Consider the overall equilibrium of the anchorage zone and provide secondary reinforcement as required.

Fig. 3.12.9 graphically illustrates areas to be considered. When the anchorage forces act outside the centroids of their respective $\sigma_x$ diagrams beyond the transmission length (Fig. 3.12.9(a)),

**Fig. 3.12.9 Splitting stress in anchorage zones**

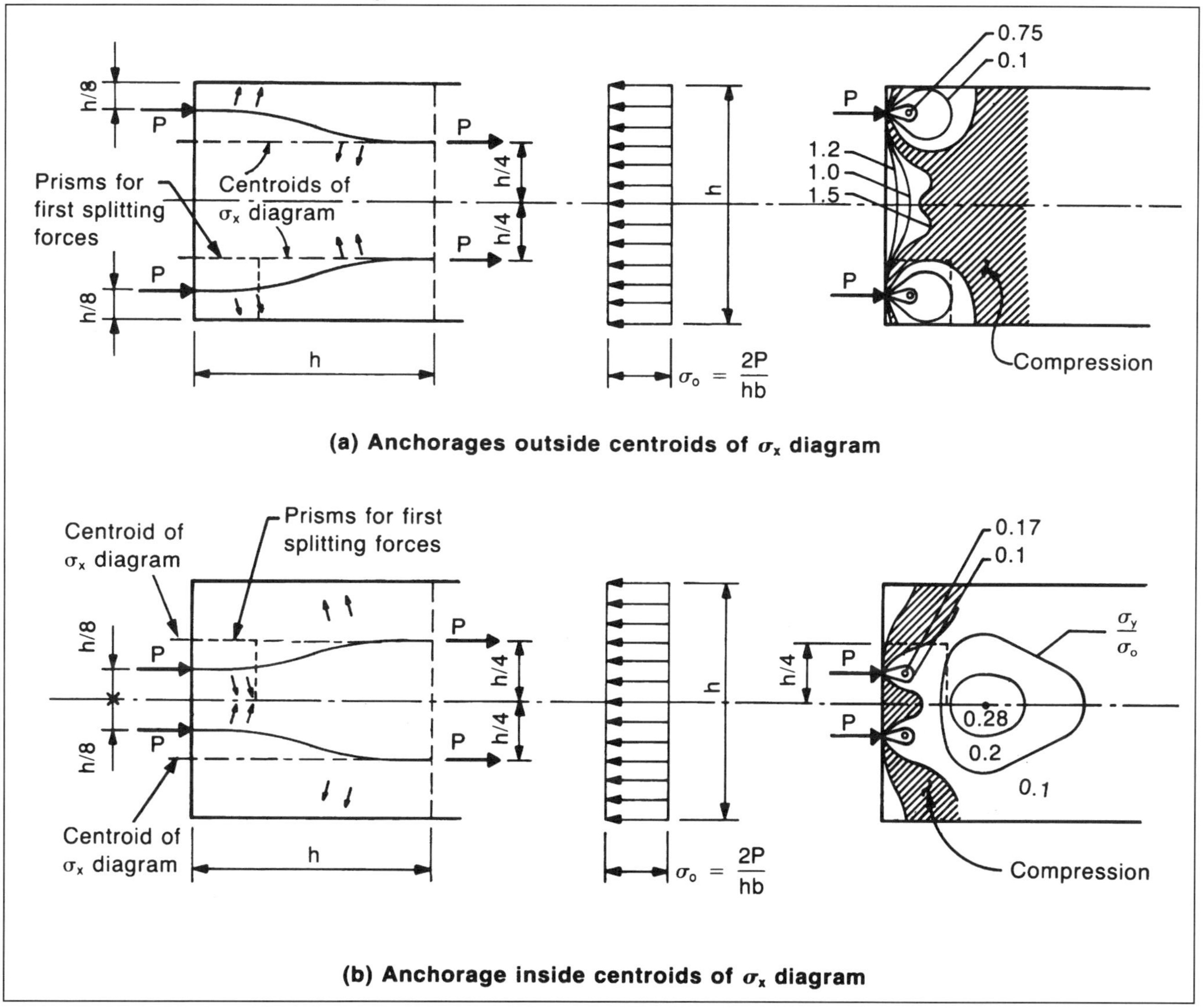

**Fig. 3.12.10 Determination of primary prisms**

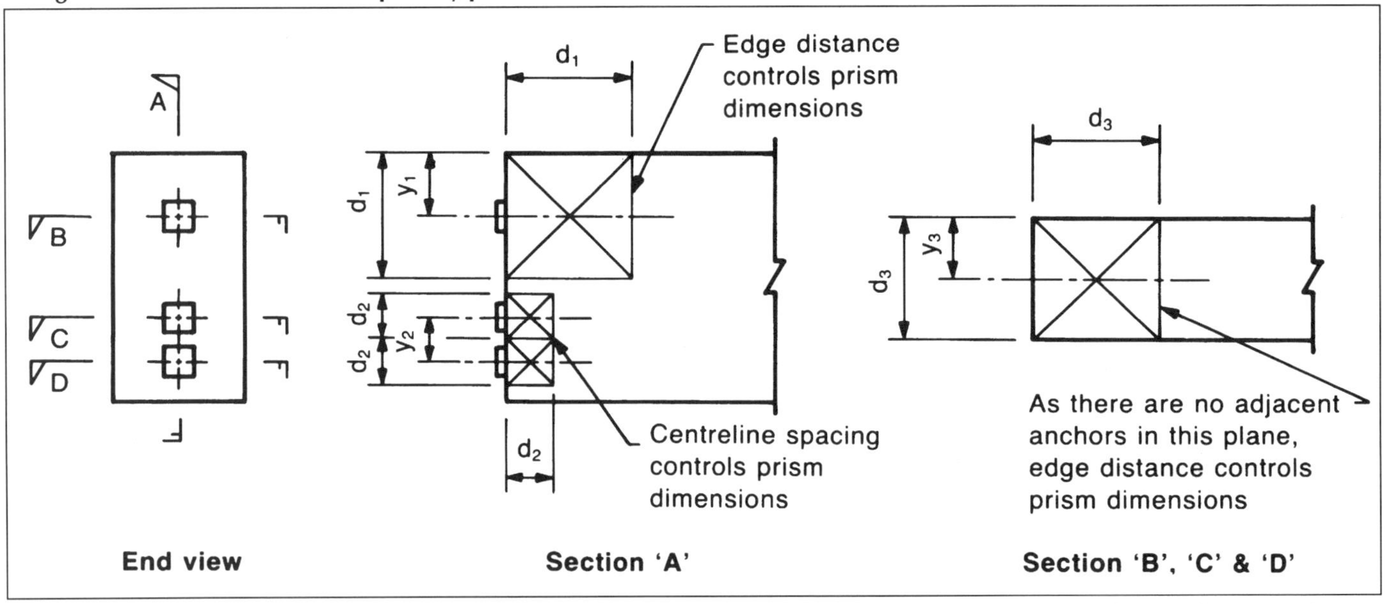


tensile bursting stresses behind the anchorages and splitting stresses between the anchorages are created. When the anchorage forces act inside the centroids (Fig. 3.12.9(b)), tensile bursting stresses behind the anchorages and further down the length of the beam, and tensile spalling stresses at the end corners of the beam are created. Typical calculations are given in Example 4.

*Determination of primary prisms:*

Before the bursting stress behind an anchorage can be determined, the dimensions of the primary prism (the area in a particular plane where bursting stresses are greatest) must be determined.

As shown in Fig. 3.12.10, for distribution in a particular plane, the depth of the primary prism, d, is taken as the lesser of:

1. The distance between centrelines of adjacent anchors.
2. Twice the distance from the centreline of an anchorage to the edge of the concrete.

When the extent of the prism is different on the two axes, or when the anchorage plate is rectangular, the distribution on each axis will also be different.

*Bursting stresses:*

Bursting stresses vary with the ratio of a/d as shown in Fig. 3.12.11.

The area under each of the curves in Fig. 3.12.11 yields the total bursting force which may be closely represented by the equation:

$$Z = 0.3P(1 - a/d)$$

where:

Z = total bursting force

P = tendon force

a = width of bearing plate

d = depth of primary prism

**Fig. 3.12.11 Transverse tensile bursting stresses in anchorage zones**

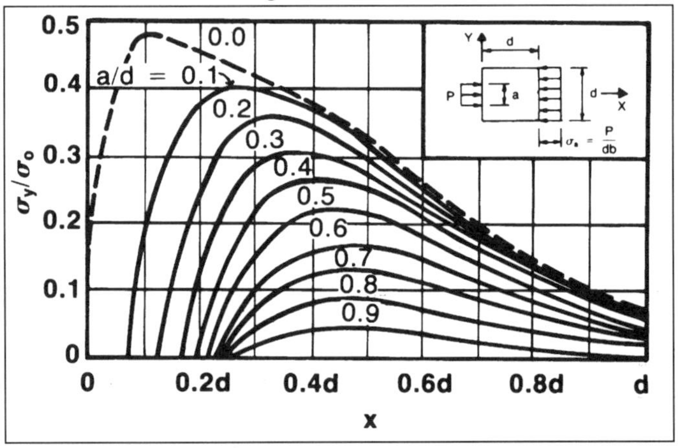


Effectively bonded reinforcement acting at a stress of 0.5 $f_y$ (but not exceeding 0.001 strain) should be distributed through the depth of the primary prisms as required. In some cases, it may be appropriate to allow some unreinforced tension in the concrete (e.g. widely spaced anchorages in large concrete sections or small low stress anchors in slabs).

Spiral reinforcement of suitable diameter is commonly used for bursting reinforcement. The reinforcement should be detailed to allow for proper placement of concrete. when the length of the primary prism is different on two axes, the longer length should be used for the length of the spiral.

*Spalling reinforcement:*

As shown in Fig. 3.12.9, zones of high tensile stress exist at the loaded face of the concrete.

Reinforcement to resist a total transverse force of 0.04P, acting at a stress of 0.5 $f_y$ (but not greater than 0.001 strain), should be placed in both directions as close to the end face of the element as possible.

Corners of concrete elements are also subject to high tensile stresses and should be adequately secured by reinforcement.

*Bearing stresses behind anchors:*

The designer of the anchorage zone is not usually concerned with the bearing stresses behind the anchorage. Proprietary anchorage designs are based on experience, tests, and usage as well as theory. Anchorages that have been successfully adopted should be considered reliable. Although this is generally true of multistrand anchorages, designers using single strand tendons may need to check anchorage stresses, as spiral reinforcement is not provided. Special projects may require consultation with post-tensioning suppliers.

*Equilibrium of the anchorage zone:*

In most cases, there is a further spread of prestress force behind the primary prisms until the stresses become fully distributed over the entire cross section of the element. It is necessary to check the overall equilibrium of the anchorage zone.

Consider an element of the end zone as shown in Fig. 3.12.12. The forces acting on opposite faces of the element produce a moment and shear force on planes parallel to the longitudinal axis of the element.

Effectively anchored reinforcement should be provided to carry the maximum moment on the assumption that the lower arm of the resisting couple is equal to half the length of the end

**Fig. 3.12.12 Freebody diagram of end zone**

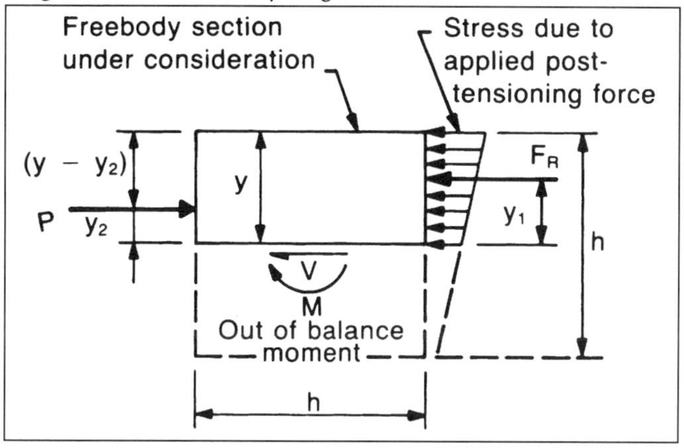


block. Depending on the direction of the out of balance moment, the reinforcement should be distributed over a distance of 0.5h from the loaded surface of the block or over a distance of 0.25h from the opposite end of the block (see Fig. 3.12.13).

The area of secondary reinforcement should not be less than 0.003 of the area of the horizontal cross-section calculated for the full length of the anchorage zone.

Since tendons are stressed sequentially, checks should be carried out to determine the worst combination of loads.

**Example 4 - Anchorage reinforcement**

*Given:*

A 400 × 1600 rectangular beam subjected to two 2500 kN post tensioning jacking forces, each on 300 × 300 anchorage plates.

**Fig. 3.12.13 Equilibrium forces in anchorage zone**

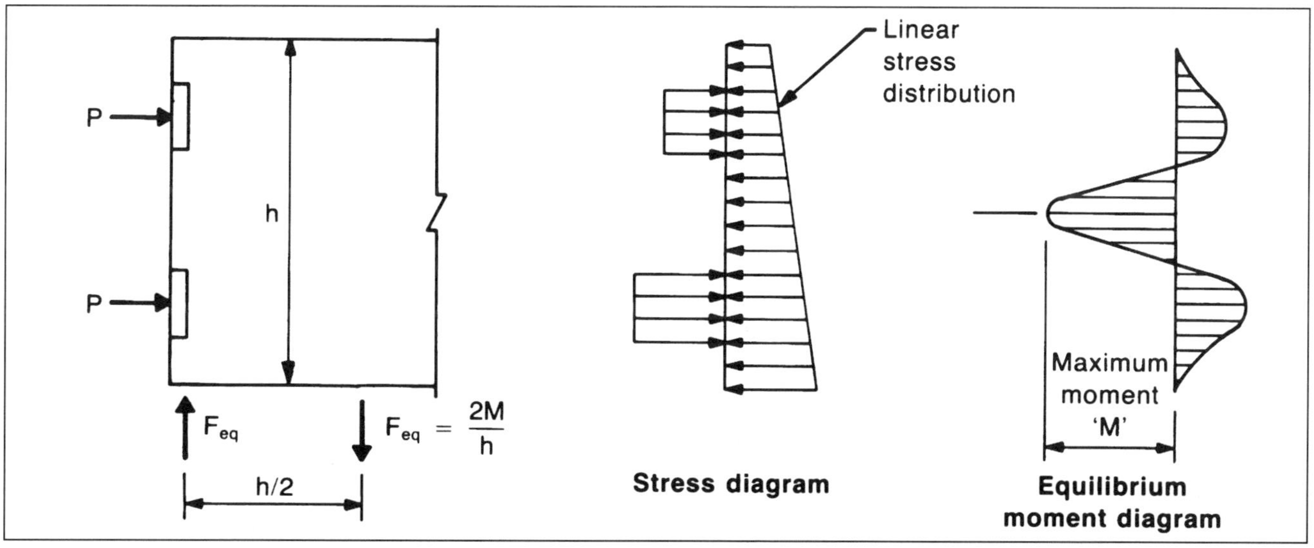

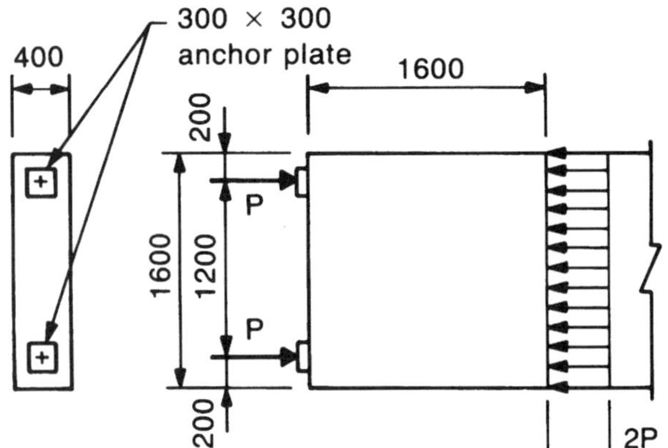


Case ①

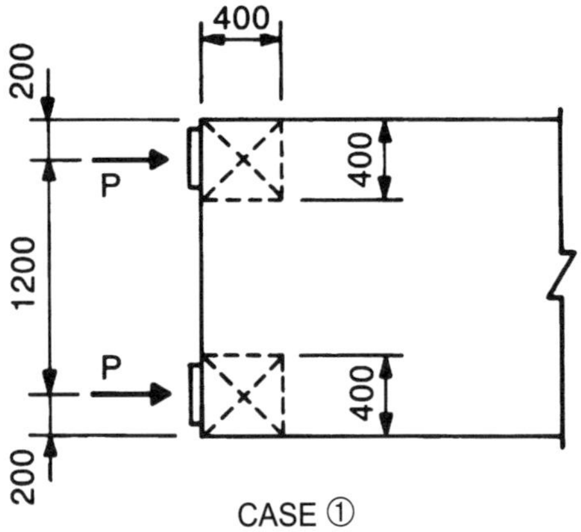


CASE ①
Vertical plane section

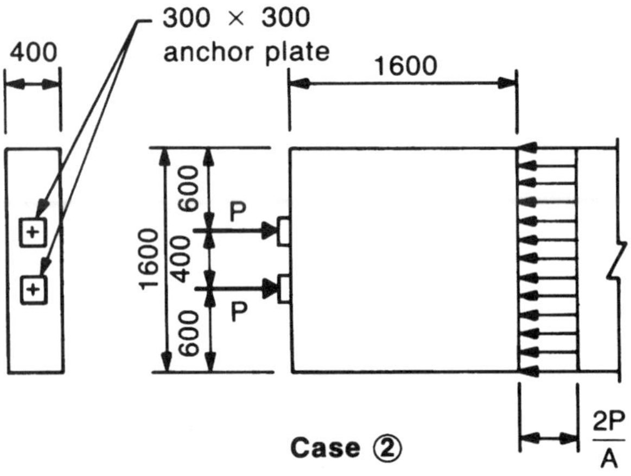


Case ②

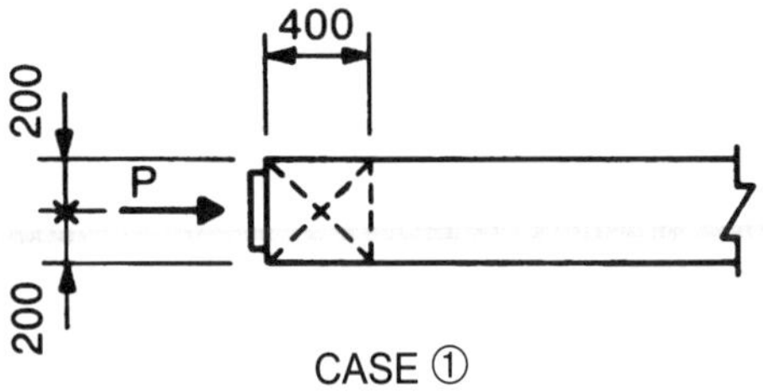


CASE ①
Horizontal plane section

*Problem:*

Determine the stresses in the end zone and design the anchorage reinforcement for the anchorages ① 1200 mm apart, and ② 400 mm apart.

*Solution:*

*Determination of primary prisms:*

The primary prisms are determined by comparing for each of the horizontal and vertical planes, the lesser of:

1. The distance between the centrelines of adjacent anchors, or,
2. Twice the distance from the centreline of an anchor to the edge of concrete.

For case ①, in the vertical plane, the depth of the primary prism may be:

1. $d_1 = 1200$ mm, or
2. $d_2 = (2)(200) = 400$ mm

Since $d_2 < d_1$, the primary prism in this plane is 400 × 400.

Similarly, in the horizontal plane, the primary prism is found to be d = 400 mm, as there are no adjacent anchors in this plane.

The same procedure for case ② gives a primary prism of 400 × 400 × 400.

*Bursting reinforcement, case ① & ②:*

$$\frac{a}{d} = \frac{300}{400} = 0.75$$

$$\sigma_o = \frac{P}{(d)(d)} = \frac{(2500)(10^3)}{(400)(400)} = 15.6 \text{ MPa}$$

From Fig. 3.12.11:

$\sigma_y = 0$ at 0.25, d = (0.25)(400) = 100 mm from bearing plate.

The maximum value of $\sigma_y$ is:

$$\sigma_{ymax} = (0.12)(15.6) = 1.9 \text{ MPa}$$

and this value occurs at a distance from the bearing plate equal to:

$$0.47\ d = (0.47)(400) = 188 \text{ mm}$$

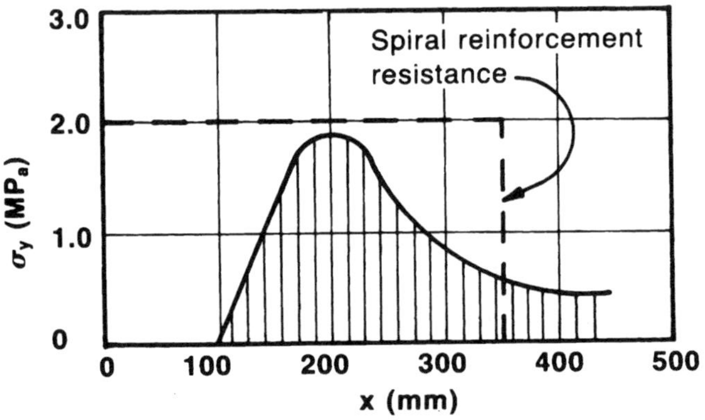

The total bursting force:

$$Z = 0.3P\left(1-\frac{a}{d}\right)$$
$$= (0.3)(2500)(1-0.75) = 188 \text{ kN}$$

$$A_s = \frac{Z}{0.5f_y} = \frac{(188)(10^3)}{(0.5)(400)} = 940 \text{ mm}^2$$

For a No. 10 spiral, 5 turns are required between x = 100 mm and x = 400 mm. Use a No. 10 spiral with 7 turns at 50 mm pitch from anchor plate to x = 400 mm:

$$\sigma_y \text{ provided} = \frac{A_s(0.5f_y)}{bd}$$
$$= \frac{(7)(2)(100)(0.5)(400)}{(350)(400)}$$
$$= 2.0 \text{ MPa} > 1.9 \text{ MPa} \quad \text{OK}$$

In this example, the lengths of the primary prisms in the vertical and horizontal planes are both equal to 400 mm. If they had been different, the longer dimension would have been used for the length of the spiral.

*Required reinforcement - spalling stresses,*

Case ① & ②:

The spalling force at the end face of the beam:

$$F_{sp} = 0.04P = (0.04)(2500) = 100 \text{ kN}$$

$$A_s = \frac{F_{sp}}{0.5f_y} = \frac{(100)(10^3)}{(0.5)(400)} = 500 \text{ mm}^2$$

Case ①:

Provide 4 - No. 15 @ 100 mm o.c. in vertical plane, and 4 - No. 15 @ 200 mm o.c. in horizontal plane:

$$A_s \text{ provided} = (4)(200) = 800 \text{ mm}^2 > 500 \quad \text{OK}$$

Case ②:

Provide 4 - No. 15 @ 100 mm o.c. vertical plane. Because spalling occurs outside anchors, 4 - No. 15 @ 100 should be placed in this location.

*Required reinforcement - equilibrium:*

Case ①:

At the horizontal plane of the top anchorage, the out of balance moment is found as follows:

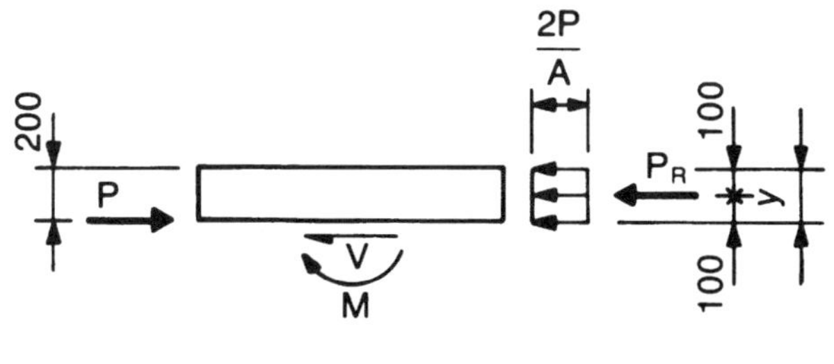


$$P_R = \frac{2Pyb}{bh} = \frac{(2)(2500)(200)(400)}{(400)(1600)} = 625 \text{ kN}$$

$$M = -P_R(0.5y) = -(625)(0.5)(200) = -63 \text{ kN·m}$$

At the horizontal plane at the mid depth of the beam:

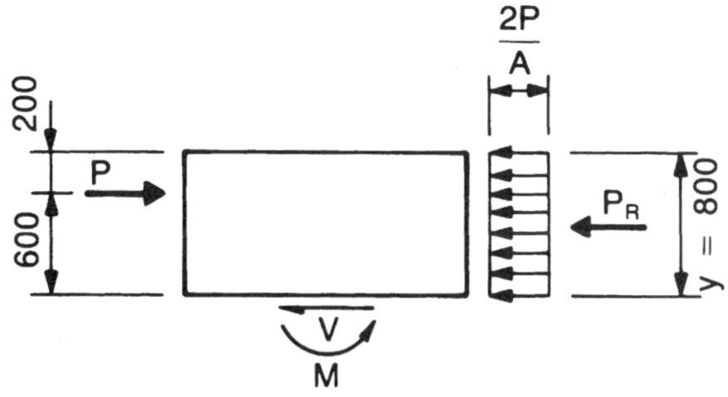


$$P_R = \frac{2Pyb}{bh} = \frac{(2)(2500)(800)(400)}{(400)(1600)} = 2500 \text{ kN}$$

$$M = -0.4P_R + 0.6P$$
$$= -(0.4)(2500) + (0.6)(2500) = 500 \text{ kN·m}$$

Since the beam is symmetric about its mid-depth, the out of balance moment for the bottom anchor is the same as for the top anchor. Thus the maximum out of balance moment is 500 kN·m.

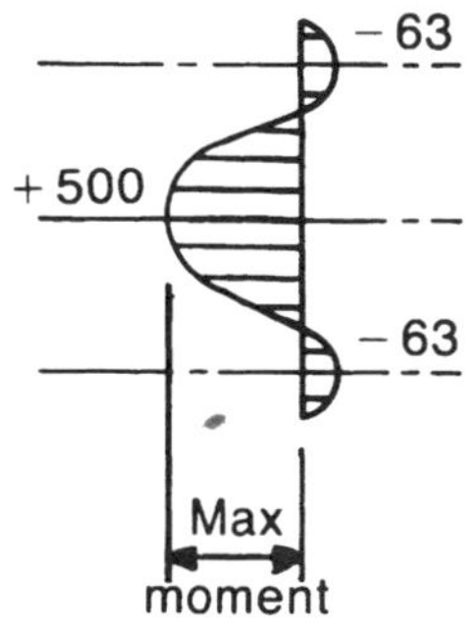


since:

$$F_{eq} = \frac{2M}{h} = \frac{(2)(500)}{1.6} = 625 \text{ kN} \quad \text{(c.f. Fig. 3.12.14)}$$

then:

$$A_s = \frac{F_{eq}}{(0.5)(f_y)} = \frac{(625)(10^3)}{(0.5)(400)} = 3125 \text{ mm}^2$$

Use 5 double No. 15 stirrups at 100 mm o.c.

Case ②:

At the horizontal plane of the top anchor:

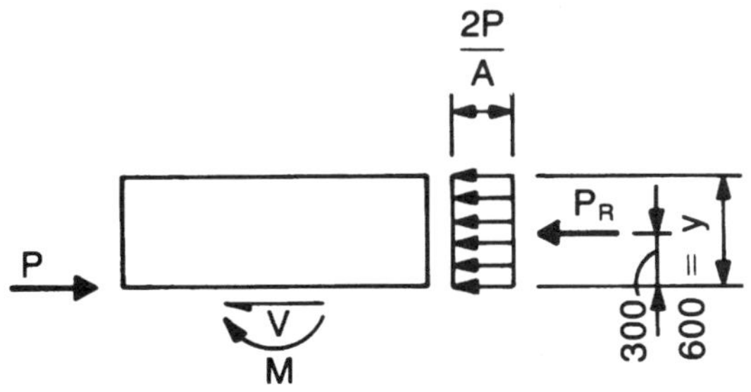

**Fig. 3.12.14 Anchorage zone reinforcement for Example**

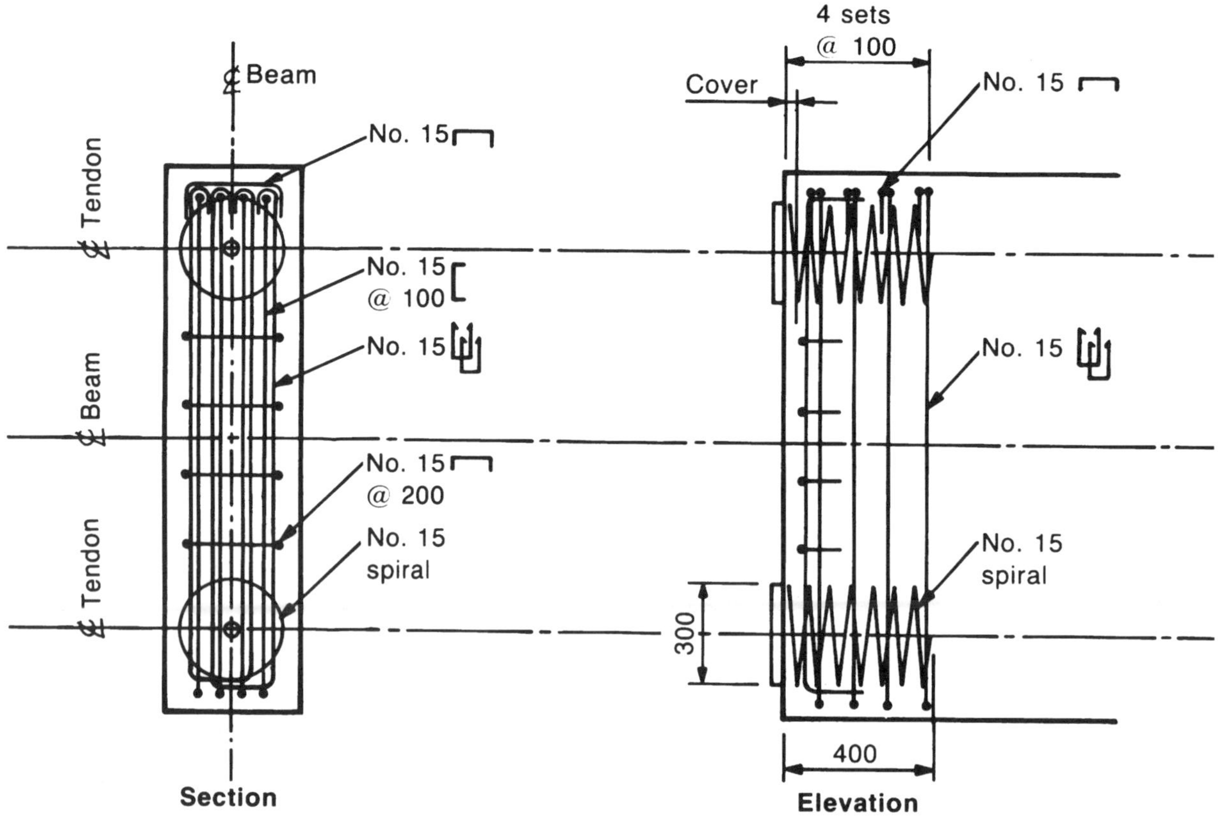


(a) Anchorage reinforcement for Case ①

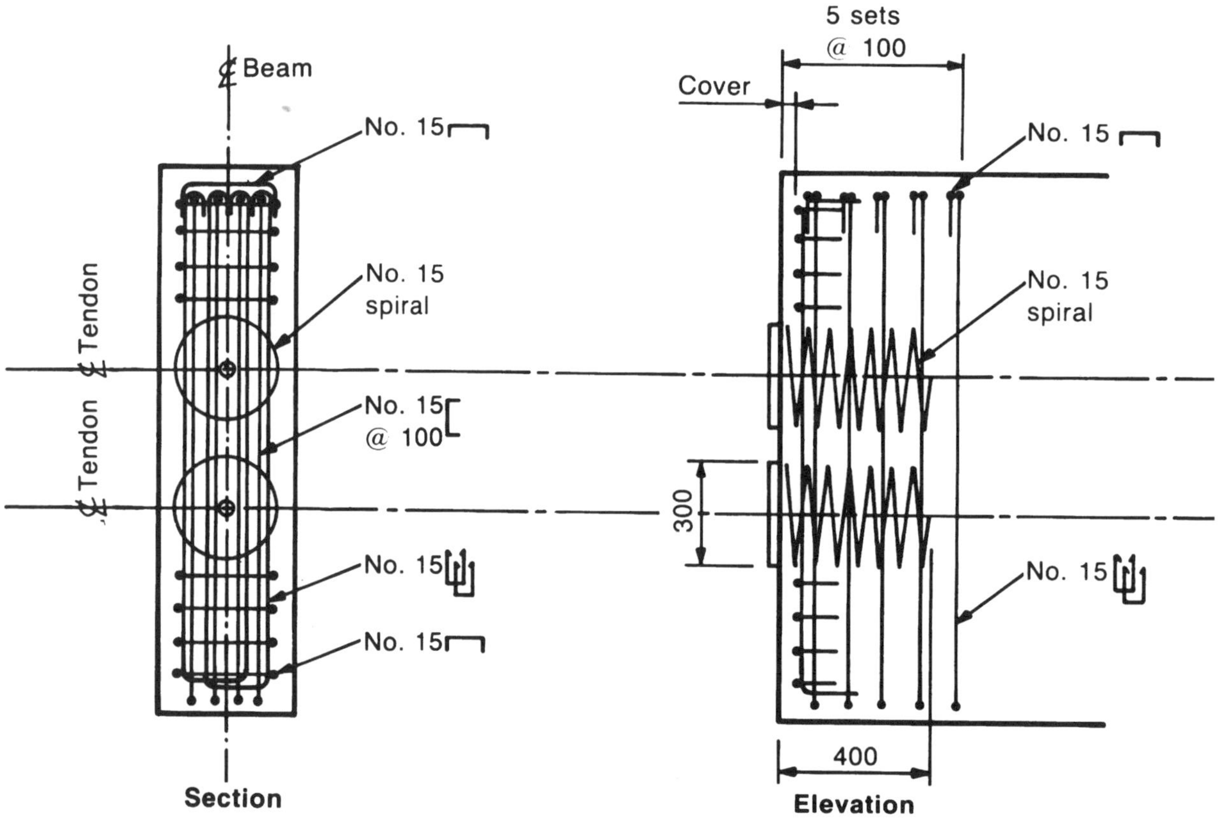


(b) Anchorage reinforcement for Case ②

$$P_R = \frac{2Pyb}{bh} = \frac{(2)(2500)(600)(400)}{(400)(1600)} = 1875 \text{ kN}$$

$$M = 0 - (1875)(0.3) = -562 \text{ kN·m}$$

At the mid-height of the beam:

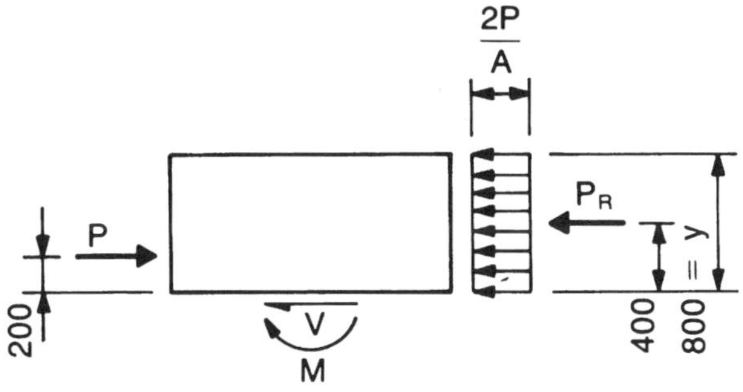


$$P_R = \frac{2Pyb}{bh} = \frac{(2)(2500)(800)(400)}{(400)(1600)} = 2500 \text{ kN}$$

$$M = 0.2P - 0.4P_R$$

$$= (0.2)(2500) - (0.4)(2500) = -500 \text{ kN·m}$$

The maximum moment is -563 kN·m. Thus:

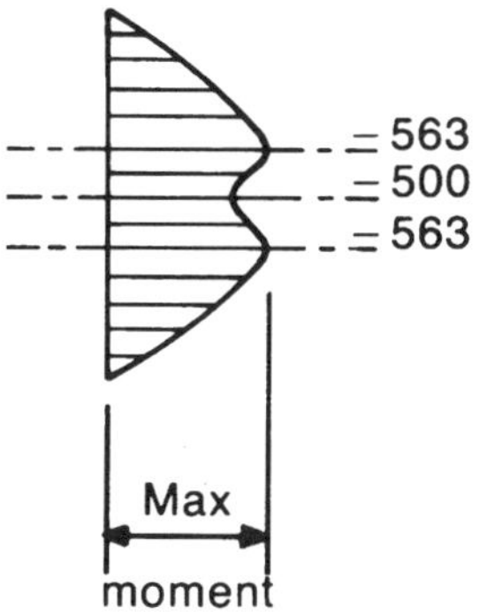


$$F_{eq} = \frac{2M}{h} = \frac{(2)(563)}{1.6} = 704 \text{ kN}$$

and:

$$A_s = \frac{F_{eq}}{0.5f_y} = \frac{704 \times 10^3}{(0.5)(400)} = 3520 \text{ mm}^2$$

Provide 5 double No. 15 stirrups @ 100 mm o.c. The required anchorage zone reinforcement for both cases ① and ② is summarized in Fig. 3.12.14.

**References**:


R3.4.1 Gergely, P. And Lutz, L.A., Maximum Crack Width in Reinforced Concrete Flexural Members, Causes, Mechanism, and Control of Cracking in Concrete, SP-20, American Concrete Institute, Detroit, 1968, pp. 1-17.

R3.4.2 Papers by Nawy, E.G. and Huang, P.T., "Crack and Deflection Control of Pretensioned Prestressed Beams," PCI Journal, May-Jun 1977, pp. 30-47, and Nawy, E.G. and Chiang, J.Y., "Serviceability Behaviour of Post-Tensioned Beam," PCI Journal, Jan-Feb 1980, pp. 74-95.

R3.4.3 Neville, A.M., Dilger, W.H., and Brooks, J.J., Creep of Plain and Structural Concrete, Construction Press, London, New York 1983.

R3.4.4 For simplicity in the calculation of the relaxation loss, it is assumed that the superimposed dead load is applied at the time of transfer. This means that the tendon stress due to $M_{sd}$ is included in $f_{pi}$ .

R3.4.5 Dilger, W.H., Creep Analysis of Prestressed Concrete Members Using Creep-Transformed Section Properties, PCI Journal, Vol. 27, No. 1, Jan-Feb 1982, pp. 98-118.

R3.4.6 Suri, V., Dilger, W.H. Crack Width in Partially Prestressed Members, Journal of the American Concrete Institute, Sep-Oct 1986. Note: As the coefficients $k_1$ have been established from tests on relatively small beams, the above value $k_1$ are somewhat higher than those listed in the paper.

R3.4.7 Hutton, S.G. , Loov, R.E., "Flexural Behaviour of Prestressed, Partially Prestressed and Reinforced Concrete Beams", ACI Journal, Dec 1966, pp. 1401-1410.

R3.5.1 Branson, D.E., "The Deformation of Noncomposite and Composite Prestressed Concrete Members," *Deflection of Concrete Structures,* SP-43, American Concrete Institute.

R3.5.2 Branson D.E. and Trost H., Application of the I-Effective Method in Calculating Deflections of Partially Prestressed Members, P.C.I. Journal, V. 27, No. 5, Sept-Oct 1982, pp 62-77

R3.5.3 It is within the precision of the calculation method and observed behaviour to use midspan d and to calculate the deflection at midspan, although the maximum tensile stress in this case is assumed at 0.4$\ell$.

R3.5.4 Martin L.D., A Rational Method for Estimating Camber and Deflection of Precast, Prestressed Concrete Members, PCI Journal, Vol. 22, No. 1, Jan-Feb 1977.

R3.5.5 Shaikh, A.F., and Branson D.E., Non-tensioned Steel in Prestressed Concrete Beams, PCI Journal, Feb. 1970.

R3.5.6 Neville, A.M. Dilger, W.H. Brooks, J.J., Creep of Plain and Structural Concrete, Longman, 1983, 392 pp.

Bazant, Z., Prediction of Concrete Creep Using Age-Adjusted Effective Modulus Method , ACI Journal 69, 1972, pp. 212-17.

R3.5.7 Tadros et al, Prestress Loss and Deflection of Precast Concrete Members, PCI Journal, Vol. 30, No. 1, Jan-Feb 1985, pp. 114-141.

R3.5.8 Suri, U.V.M., Service Load Analysis and Design of Partially Prestressed Members, Ph.D. Thesis, University of Calgary, May 1986.

R3.6.1 Collins and Mitchell, Prestressed Concrete Basics, CPCI, 1987.

R3.6.2 Loov, Robert E. and Patnaik, Anil K., Horizontal Shear Strength of Composite Concrete Beams With a Rough Interface, PCI Journal, Vol. 39. No. 1, January/February, 1994, pp. 48-69.

R3.6.3 Mattock, Alan H., Anchorage of Stirrups in a thin Cast-in-Place Topping, PCI Journal, Vol. 32, No. 6, November/December, 1987, pp. 70-85.

R3.8.1 PCI Committee on Prestressed Concrete Columns, Recommended Practice for the Design of Prestressed Concrete Columns and Walls, PCI Journal, July - August 1988.

R3.8.2 MacGregor, J.G., Oelhofen, V.H. and Hage, S., A Re-examination of the EI value for Slender Columns, Draft Paper for ACI column Symposium, Ottawa, Canada 1974.

R3.8.3 PCI Committee on Prestressed Concrete Piling, Recommended Practice for Design, Manufacture and Installation of Prestressed Concrete Piling,PCI Journal, Vol.38, No. 6, Mar – Apr 1993, pp. 14-41.

R3.8.4 Anderson, A.R., and Moustafa, S.E., Dynamic Driving Stresses in Prestressed Concrete Piles, *Civil Engineering,* ASCE, N.Y., Vol. 41, No. 8, Aug 1971, pp. 55-58.

R3.8.5 Margason, E., and Holloway, D.M., Pile Bending During Earthquakes, *Proceedings of the Sixth World Conference on Earthquake Engineering*, New Delhi, India, 1977, Vol. II, pp. 1690-1695.

R3.8.6 Sheppard, D.A., Seismic Design of Prestressed Concrete Piling, *PCI Journal, V. 28,* No. 2, Mar – Apr 1983, pp. 20-49.

R3.8.7 Park, R., and Falconer, T.J., Ductility of Prestressed Concrete Piles Subjected to Simulated Seismic Loading, *PCI Journal*, Vol. 28, No. 5, Sep – Oct 1983, pp. 112-143.

R3.8.8 Pizzano, B. A., Behaviour of Prestressed Concrete Piles Under Seismic Loading, *MSCE Thesis*, University of Washington, 1984.

R3.9.1 Mast, R.F., Lateral Stability of Long Prestressed Concrete Beams – Part 2, PCI Journal, Volume 38, No. 1, Jan – Feb 1993, pp70 – 88.

R3.10.1 LaGue, D.J., Load Distribution Tests on Precast Prestressed Hollow-Core Slab Construction, *PCI Journal*, Vol. 16, No. 6, Nov-Dec 1971.

Johnson, T. and Ghadiali, Z., Load Distribution Test on Precast Hollow-Core Slabs with Openings, *PCI Journal*, Vol. 17, No. 5, Sep-Oct 1972.

Pfeifer, D.W., and Nelson, T.A., Tests to Determine the Lateral Distribution of Vertical Loads in a Long-Span Hollow-Core Floor Assembly, *PCI Journal*, Vol. 28, No. 6, Nov-Dec 1983.

Lateral Distribution of Loads on Prestressed Concrete Decks, PCI sponsored research project at University of Washington.

# CHAPTER 4
# DESIGN OF CONNECTIONS

Page No.

## 4.0 NOTATION

$A$ = bearing area
$A_b$ = area of bar of stud
$A_{cr}$ = area of crack face
$A_{cv}$ = area of concrete section resisting shear transfer
$A_e$ = effective stress area
$A_g$ = area of gross section
$A_n$ = area of reinforcement required to resist axial tension
$A_s$ = area of reinforcement
$A'_s$ = area of vertical reinforcement near end of steel haunch and welded to the steel haunch
$A_{sh}$ = area of reinforcement for horizontal or diagonal cracks
$A_{vf}$ = area of shear friction reinforcement
$a$ = depth of equivalent concrete stress block
$B_r$ = factored bearing resistance
$b$ = effective width
$c$ = cohesion stress
C.E. = carbon equivalent
$d$ = distance from extreme compression fiber to centroid of reinforcement
$d_b$ = bar or stud diameter
$e_i$ = distance from centre of bolt to horizontal reaction
$e_v$ = eccentricity of vertical load
$f'_c$ = specified compressive strength of concrete
$F_N$ = factored frictional force
$f_{su}$ = minimum specified tensile strength of anchor
$f_y$ = minimum specified yield strength of reinforcement or structural steel
$f_{ys}$ = minimum specified yield strength of $A_{sh}$
$S_f$ = greatest sum of factored anchor bolt forces on one side of the column
$g$ = gauge of angle
$h$ = overall height of member
$j_u$ = for resisting lever arm, used in $j_u d$
$\ell_d$ = development length of reinforcement
$\ell_e$ = embedment length
$\ell_w$ = length of weld
$N$ = unfactored permenant compressive force perpendicular to shear plane
$N_{df}$ = factored dead load force normal to friction face
$N_f$ = factored horizontal or axial force
$P_f$ = factored tension force
$P_r$ = factored tension resistance
$s$ = distance between $A_s$ and $A'_s$
$t$ = thickness
$t_w$ = effective throat thickness of weld
$v_f$ = factored shear stress
$v_r$ = factored shear stress resistance of shear plane
$V_c$ = nominal shear capacity of embedded steel haunch connection without additional reinforcement
$V_f$ = factored shear force at section
$V_n$ = nominal shear capacity of embedded steel haunch connection
$V_r$ = factored resistance provided by reinforcement
$V_s$ = additional factored resistance provided by reinforcement welded to embedded steel haunch
$w$ = dimension (see specific application)
$x, y$ = surface dimensions
$x_c$ = distance from centreline of bolt to face of column
$x_o$ = base plate projection
$x_t$ = distance from centreline of bolt to centreline of reinforcement
$Z_s$ = plastic section modulus of structural steel
$\Delta$ = horizontal deformation of bearing pad
$\epsilon_1$ = principal tensile strain
$\epsilon_s$ = strain in tension reinforcement
$\theta$ = angle of assumed crack plane or angle of diagonal compression
$\theta_s$ = smallest angle between compressive strut and adjoining tension ties
$\lambda$ = coefficient for use with low density concrete
$\sigma$ = effective normal stress
$\mu$ = shear-friction coefficient
$\mu_s$ = static coefficient of friction
$P_v$ = ratio of shear friction reinforcement
$\phi_a$ = resistance factor for structural steel, 0.90
$\phi_c$ = resistance factor for concrete, 0.65 for precast concrete
$\phi_s$ = resistance factor for reinforcing bars, 0.85

## 4.1 INTRODUCTION

This part of the Manual presents concepts of analysis and equations for design of connections for precast, prestressed concrete. Design equations were developed through field experience, laboratory tests, and structural analysis. These recommendations take into consideration current design office and plant practice and are intended as reasonable guidelines for the analysis and design of connections. Other types of connections are in use and some have been extensively tested. Continuing research will lead to improved details and analyses. Therefore, engineers should not necessarily be limited to the design equations or connection details included in this part.

Practical and economical connection design must consider production of the units and construction situations pertinent to precast concrete buildings, as well as the performance of the connections for both serviceability and ultimate limit states.

## Criteria for Connections

Connections must meet a variety of design and performance criteria, and not all connections are required to meet the same criteria. Some of the items discussed in this section are self-evident. Other requirements may not be so obvious and may require special consideration or specification by the designer or owner of the structure.

### Resistance

A connection must resist the forces to which it will be subjected during its lifetime. Some of these forces are apparent, caused by dead and live gravity loads, wind, earthquake, and soil or water pressure. Others are not so obvious and are frequently overlooked. These are the forces caused by restraint of volume changes in the elements and forces required to maintain stability.

Volume changes are caused by temperature change, creep and shrinkage of the concrete. Instability can be caused by eccentric loading (intentional or unintentional), as well as lateral loads from wind and earthquake. Very often, measures taken to resist instability will aggravate the forces caused by volume changes, and vice versa.

The connection resistance can be categorized by the types of force that may be induced. These include:

1. Compression.
2. Tension.
3. Flexure.
4. Shear.
5. Torsion.

Many connections will have a high degree of resistance to one type of force, but little or no resistance to another. For example, a connection may have a high shear capacity and little or no moment capacity. It may be unnecessary, or even undesirable to provide a high capability to resist certain types of forces.

### Ductility

Ductility is defined as the ability to accommodate large deformations without failure. In structural materials, ductility is measured by the amount of deformation that occurs between first yield and ultimate failure.

Ductility in building frames is usually associated with moment resistance (rotational ductility). This is particularly true in designing for earthquake forces, where concerns over ductility are usually expressed. In concrete elements with moment-resisting connections, the flexural tension is normally resisted by steel components, either reinforcing bars or structural steel sections. First yield occurs in this steel component, and final failure may be from rupture of the steel, crushing of the concrete, or a failure of the connection of the steel to the concrete.

### Volume Change Consideration

The combined effects of creep, shrinkage and temperature differences can cause severe stresses on precast, prestressed concrete elements and their supports if the end connections restrain movement. These stresses must be considered in the design, but it is usually far better if the connection will allow some movement to take place, thus relieving the stresses.

Most of the severe problems that have been caused by restraint of volume change movements have appeared when relatively long elements, usually stemmed deck units, were welded to their supports at the bottom on both ends. When such elements are placed on bearing pads and welded only at the top, experience has shown that volume changes are adequately accommodated. Prestressed elements rarely exhibit cracking at locations further from the ends than the transfer length of the strand. Distress due to volume change constraints generally occurs at the end connections.

### Durability

Evidence of poor durability is usually exhibited by corrosion of exposed steel elements, or by cracking and spalling of concrete. Connections which will be exposed to weather should have steel elements adequately covered with concrete, or should be coated with a corrosive resistant material. All exposed connections should be periodically inspected and maintained.

Most precast, prestressed concrete is of high quality, and cracking is seldom a serious problem, provided tensile stresses in the elements and connections are kept within code limits.

### Fire Resistance

Many precast concrete connections are not vulnerable to the effects of fire and require no special treatment. For example, the bearing between slabs or stemmed units and beams do not generally require special fire protection. If the slabs or tees rest on elastomeric pads or other combustible materials, protection of the pads is not generally needed because deterioration of the pads will not cause collapse. After a fire the pads can be replaced.

Connections, in which weakening by fire would jeopardize the structure's stability, should be protected to the same degree as that required for the structural frame. For example, an exposed steel bracket supporting a beam may be weakened enough by a fire to cause the beam to collapse. Such a bracket should be protected. The amount of protection depends on the stress-strength ratio in the steel at the time of the fire and the intensity and duration of the fire.

Connections which require a fire resistance rating will usually have exposed steel elements encased in concrete. Other methods of fire protection include enclosing with gypsum wallboard, coating with intumescent mastic, or spraying with fire protection material.

There is evidence that exposed steel hardware used in connections is less susceptible to fire-related strength reduction than other exposed steel elements. This is because the concrete elements provide a "heat sink," which draws off the heat and reduces the temperature of the steel.

### Production Considerations

Maximum economy of precast, prestressed concrete construction is achieved when connection details are kept as simple as possible, consistent with adequate performance and ease of erection. Furthermore, complex connections are more difficult to control and will often result in poor fit in the field. This can contribute to slow erection and less satisfactory performance.

The following is a list of items to consider in order to improve production procedures. In many cases, some of these items must be compromised in order for the connection to serve its intended function:

1. The area of the element in which the connection is made frequently requires large amounts of additional reinforcing steel, embedded plates, inserts, blockouts, etc. It is not unusual that so many items are concentrated into one area that there is virtually no room for concrete. Precast, prestressed concrete elements are usually cast in forms with only the top accessible. If one item must be threaded in and around other items, labour costs can be significantly increased. Reinforcing bars and prestressing strands, which appear as lines on drawings, have real cross-sectional dimensions (which are larger than the nominal dimension because of the deformations). This must be considered in the design phase. Reinforcing bar bends require minimum radii, which can cause fit problems and leave some regions unreinforced. If congestion is suspected, it is helpful to draw large scale details of the area in question.

   In some cases, it may be economical to increase the element size just to avoid congestion. Also, such details as dapped or recessed ends should be avoided unless necessary. They require special reinforcement in a constricted area and are always congested.
2. Projections which require cutting through the forms are difficult and costly to place. Where possible, these projections should be limited to the top of the element as cast. Even this inhibits finishing of the top surface, especially on deck elements, double tees, hollow-core slabs as well as panels.
3. Items which are embedded in the element, such as inserts, plates, reglets, etc., require plant labour to locate precisely and attach securely. Therefore, these items should be kept to a minimum.
4. A plant casting operation is most efficient when the product can be taken directly to the storage area immediately after it is stripped from the form. Any operations which are required after stripping and before placement at the job site, such as special cleaning or finishing, or welding on projecting hardware, should be avoided whenever possible. These operations require additional handling, extra work space, and added labour, often with skilled trades.
5. Dimensional tolerances which are specified to be more rigid than industry standards, are difficult to achieve. Connections which require close-fitting parts without provision for adjustment should be avoided.
6. Wherever possible, hardware items such as inserts, studs, steel shapes, etc., should be standard items that are readily available, preferably from more than one supplier. Custom fabricated or very specialized proprietary items add cost and may cause delays.

   It also simplifies fabrication if similar items on a product or project are standardized as to size and shape. For example, if half of the inserts are required to receive 20 mm diameter bolts, and the other half 24 mm diameter, use of all 24 mm bolts will be simpler. There is also less chance of error. The same principle applies to reinforcing bars, embedded plates, studs, etc.
7. The inserts used for lifting should not be easily confused with inserts of a lesser capacity used as tiebacks or other purposes.
8. It is very desirable to repeat details as much as possible. Similar details should be identical even if it may result in a slight overdesign. Once workmen are familiar with a detail, it is easier to repeat it than to learn a new one. Also, it will require fewer form set-ups and improve scheduling.
9. Very often, a precast concrete manufacturer will prefer certain details over others. The producer should be allowed to use alternative methods or materials, provided the design requirements are met. Allowing alternate solutions will often result in the most economical and best performing connection.

### Construction Considerations

Much of the advantage of precast, prestressed concrete construction is due to the possibility of rapid erection of the structure. To fully realize this benefit, and to keep costs within reasonable limits, field connections should be kept simple.

In order to fulfill the design requirements, it is sometimes necessary to compromise fabrication and erection simplicity. The following is a list of items that should be considered during the selection, design and detailing of connections to facilitate speedy and safe erection:

1. Hoisting the precast elements is usually the most expensive and time-critical process of erection. Connections should be designed so that the element can be lifted, set, and unhooked in the shortest possible time. Before the hoist can be unhooked, the precast element must be stable and in its final position. Precast elements such as double tees and hollow-core slabs are inherently stable and require no additional connections before releasing the crane. Others, such as columns, deep beams, wall panels and single tees usually require some supplemental shoring, guying, or fastening before the hoist can be unhooked. Preplanning for the fewest and quickest possible operations that must be performed before releasing the hoist will greatly facilitate erection. Bearing pads, shims, or other devices upon which the element is to be set should be placed ahead of hoisting. Loose hardware that is required for the connection should be immediately available for quick attachment.

   In some cases, it may be necessary to provide temporary fasteners or levelling devices, with the permanent connection made after the hoist is released. These temporary devices must be given careful attention to ensure that they will hold the element in its proper position during the placement of all elements that are erected before the final connection is made.

2. A certain amount of field adjustment at the connections is always necessary. Normal fabrication tolerances preclude the possibility of a perfect fit in the field. This is true not only when the precast elements adjoin each other, but, even more so, when the precast elements must interface with other materials.

   Adjustment in the field is accomplished through the use of slotted or oversize holes for bolts and dowels, field welding, shims and grout.
3. Connections should be planned so that they are accessible to the workers from a stable deck or platform. The type of equipment necessary to perform such operations as welding, post-tensioning, or pressure grouting should be considered. Operations which require working under a deck in an overhead position should be avoided, especially for welding. Room to place wrenches on nuts and swing them in a large arc should be provided for bolts. Dry-packing column or wall panel bases in a narrow excavation is difficult.
4. All connections which serve similar functions within the building should be standardized as much as possible. As workers become familiar with the procedures required to make the connection, productivity is enhanced, and there is less chance for error.

   Some types of connections require skilled craftsmen to accomplish, for example, welding and post-tensioning. The fewer of these skilled trades required, the more economical the connection will be.
5. Whenever possible, such things as field bolts, loose angles, etc., should be of a common size and material designation for all connections. This reduces the chance for error and the time required to search for the proper item.
6. Such materials as grout, dry-pack, cast-in-place concrete, and epoxies need special provisions to be placed in cold weather. Welding is slower when the ambient temperature is low. If the connections are designed so that these processes must be completed before erection can continue, costly delays may result.
7. Reinforcing bars, steel plates, dowels, and bolts that project from the precast element will often be damaged if care is not taken in handling, requiring repair to make them fit. Anchor bolts for columns which project from cast-in-place foundations should be at least 24 mm diameter so that there is less chance of them being bent. Threads on projecting bolts should be protected from damage and rust.
8. The precast concrete manufacturer or erector may prefer certain details or procedures not anticipated by the designer. Allowing alternate solutions will often result in more economical and better performing connections.

## 4.2 LOADS, LOAD FACTORS AND RESISTANCE FACTORS

With certain exceptions, such as bearing pad design, the equations in this part are based on ultimate limit state design philosophy. Load factors and resistance factors are contained in Sect. 8.3 and 8.4 of CSA Standard A23.3-94 except that $\phi_c$ = 0.65 for concrete elements produced in a certified manufacturing plant.

Situations exist where it may be prudent for the designer to increase the load factors given in the Code. Haunches, corbels, and similar seats for flexural elements should be designed such that the flexural element's ultimate resistance would be fully developed prior to overload of the seat itself. This ensures that adequate warning of an impending flexural failure is provided as opposed to creating a situation where a sudden connection failure could occur. Increasing the factored resistance by 30% will usually achieve this desired objective.

Flexural elements seated on properly designed bearing pads should be designed for not only the vertical load at that joint but also for the horizontal load imparted to the element as a result of restraint provided by the bearing pad. This restraint force is generally small and it is usually sufficient to design for a horizontal load equal to 0.2 times the factored vertical load.

## 4.3 BEARING PADS

Bearing pads are used to distribute vertical loads over the bearing area. Some pads also reduce force build-up at the connection by permitting small displacements and rotations. Their use is encouraged wherever applicable.

There are several materials commonly available as bearing pads. In some cases, various grades are available in the same material and they exhibit widely different properties and behaviour. In case of doubt, consult the pad supplier or precast manufacturer for proper selection of the pad. Most pad manufacturers have technical brochures available to aid the designer. Bearing pads typically belong to one of the following categories:

1. Commercial grade elastomeric (Neoprene) pads are readily available and inexpensive. However, these pads exhibit wide variations in shear deformation characteristics and bearing strength. These pads are not recommended unless performance data is available.
2. Structural grade chloroprene (Neoprene) pads are those which meet the requirements of Section 11, of the Canadian Highway Bridge Design Code. They are used in places where uniform bearing is necessary or when it is desired to reduce volume change restraints. For high compression stresses and/or large horizontal displacements, laminated pads consisting of layers of elastomer bonded between steel or fiberglass plates can be used. Each layer behaves in compression like an individual pad, but the shear deformation is a function of the thickness of the total assembly.
3. Laminated fabric bearing pads composed of multiple layers of 190 g/m$^2$ cotton fabric with a high quality binder are generally used where a higher compressive strength is desired. These pads do not deform as readily as elastomeric pads, and thus provide less tolerance of horizontal movement and rotation than do chloroprene pads.

4. Pads reinforced with randomly oriented fibres have been used successfully in recent years. Vertical load capacity is higher than plain chloroprene pads, but tolerance of rotation and horizontal movement is somewhat lower than chloroprene pads. No national standard specifications are available for this material.
5. Tetrafluorethylene (TFE, trade name Teflon) bearing pads reduce horizontal stresses because of their low coefficient of friction. The TFE is normally bonded to a backing material, such as steel or chloroprene pads. These bearing pads are usually used at expansion joints. Fig. 4.3.1 shows a typical bearing detail using TFE, and Fig. 4.3.2 shows the range of friction coefficients that may be used for design.
6. A multi-polymer plastic bearing strip is manufactured expressly for bearing purposes. It is a commonly used material for the bearing support of hollow-core slabs, and is highly suitable for this application. The material has a compressive strength higher than the typical design range of concrete used in precast construction.

   Tempered hardboard strips are also used with hollow-core slabs to prevent concrete to concrete bearing. However, they should be used with caution in situations where moist conditions exist.

**Fig. 4.3.1 Typical TFE bearing pad detail**

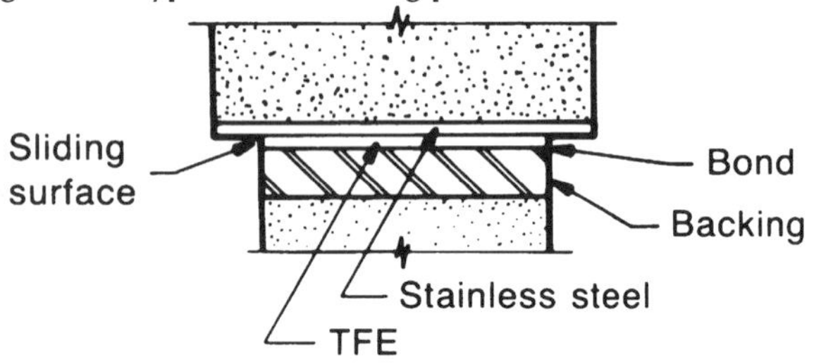


**Fig. 4.3.2 TFE friction coefficients**

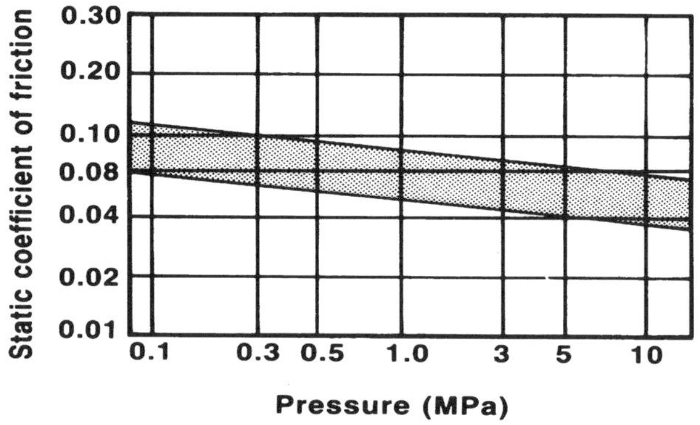


### Design Recommendations

Recent research[R4.3.1] has shown that most of the stress-relieving characteristics of elastomeric bearing pads are due to slippage instead of pad deformation. The following recommendations should be followed, along with Figs. 4.3.3 and 4.3.4, when selecting bearing pads.

1. All design loads should be service loads. Use actual contact area as bearing area for design calculations.
2. At the maximum allowable compressive stress shown in Fig. 4.3.3, 10 to 20% instantaneous strain can be expected. Non-parallel bearing surfaces could double the strain. Long-term creep of the bearing material could add another 100% if the sustained dead load is very high.
3. In order to achieve adequate stability in unreinforced pads, the length and width should be at least five times the thickness. The shape factor, S, should be greater than 2 when used under tee webs and a minimum of 3 for beam seats.

**Fig. 4.3.3 Single layer bearing pads free to slip**

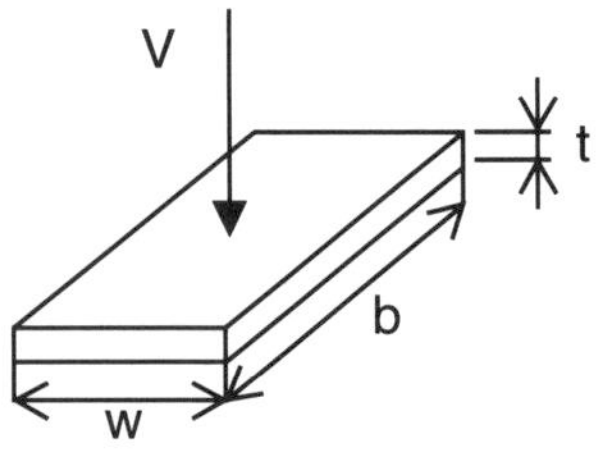


Shape Factor = $s = \dfrac{wb}{2(w+b)t}$

D = Durometer (Shore A Hardness)

Δ = Design horizontal movement at end of element

| Pad material | Allowable compressive stress (MPa) | Shore A hardness D | Recommended minimum thickness* | Recommended maximum rotation* |
|---|---|---|---|---|
| Unreinforced Chlororprene or rubber (CSA/CHBDC Sect. 11) | $\frac{DS}{36} \le 5.5$ | 50 through 70 | 1.4Δ | $\frac{0.3t}{b \text{ or } w}$ |
| Random-fiber reinforced elastomeric | 7.0+0.7S ≤ 10.0 | 80 ± 10 | 1.4Δ | $\frac{0.3t}{b \text{ or } w}$ |
| Cotton duck fabric reinforced | ≤ 14.0 | 90 ± 10 | 2.0Δ | $\frac{0.3t}{b \text{ or } w}$ |

* For movement or rotation in two directions, use the higher value. The values in the table are based on sliding criteria. If sliding is not critical or testing indicates more advantageous conditions, thinner pads may be used.

**Fig. 4.3.4 Shear resistance of bearing pads**

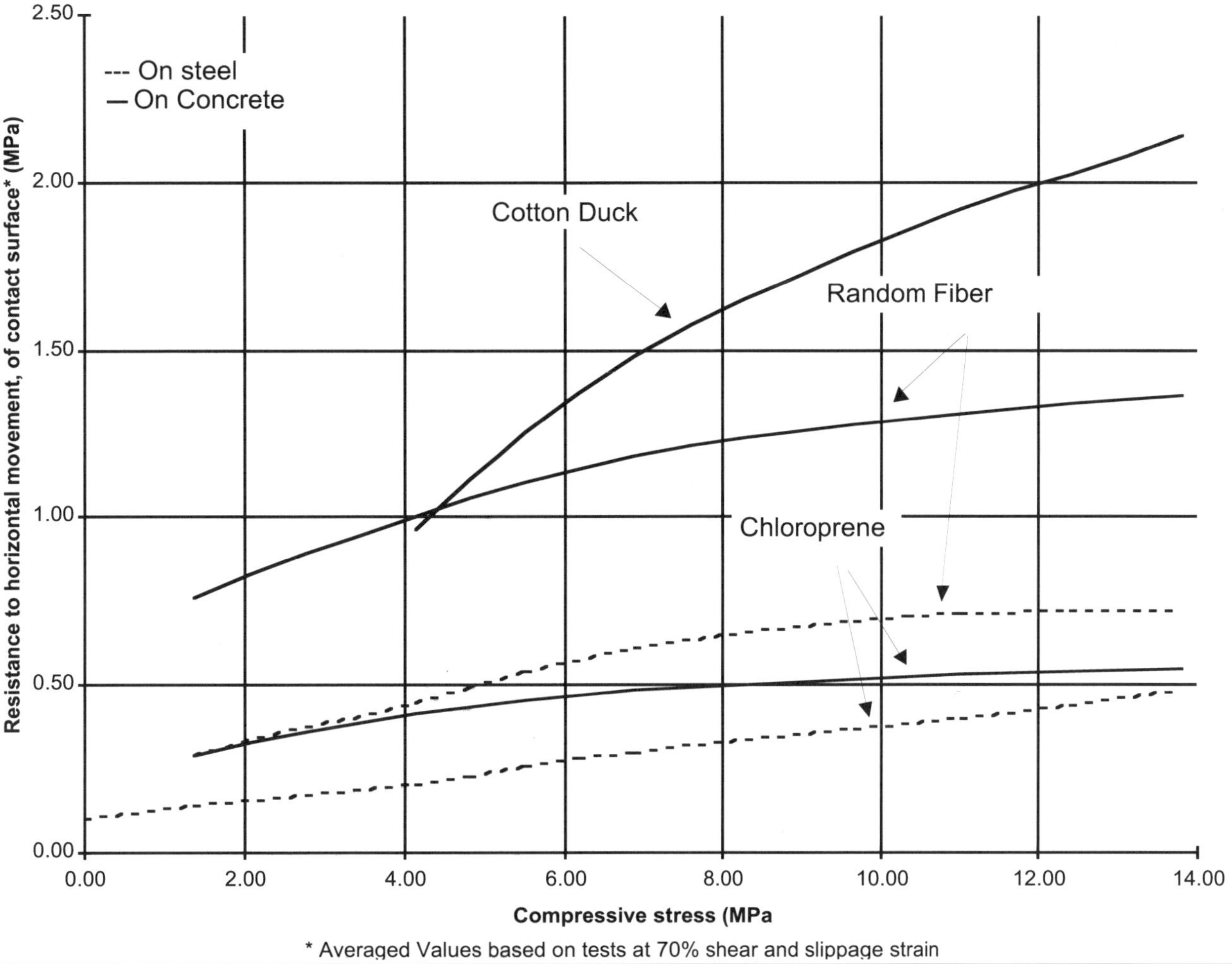


4. Bearing pads for double tee webs should be at least 6 mm thick. For beam seats, a minimum thickness of 10 mm should be used.
5. The sustained dead load on chloroprene pads should be limited to the range of 2 to 3.5 MPa.

## 4.4 FRICTION

The static coefficients of friction shown in Fig. 4.4.1 are conservative values for use in determining the upper limit of volume change forces for elements without "hard" connections. Thus, the maximum force restraint of axial movements can be determined by:

$$F_N = \mu_s N_{df}$$

where:

$F_N$ = factored friction force

$\mu_s$ = static coefficient of friction as given in Fig. 4.4.1

$N_{df}$ = factored dead load force normal to the friction face

The coefficients in Fig. 4.4.1 should be divided by 5 if friction is to be depended upon for support of temporary loads.

**Fig. 4.4.1 Static coefficients of friction of dry materials**

| Material | $\mu_s$ |
|---|---|
| Elastomeric to steel or concrete | 0.7 |
| Laminated cotton fabric to concrete | 0.6 |
| Concrete to concrete | 0.8 |
| Steel to steel (not rusted) | 0.25 |
| Concrete to steel | 0.4 |
| Hardboard to concrete | 0.5 |
| Multipolymer plastic (non-skid) to concrete | 1.2* |
| Multipolymer plastic (smooth) to concrete | 0.4* |

***Courtesy Koro Corp.**

## 4.5 SHEAR-FRICTION

Use of the shear-friction theory is recognized in Clause 11.6 of A23.3-94. This theory is applicable to situations where it is inappropriate to assume that shear stresses are uniformly distributed over the depth of an element. A crack is assumed to occur along a plane which is located in the most undesirable manner. Shear transfer across this plane is achieved by placing reinforcement across the assumed crack. If the portions of concrete on each side of the crack were to move relative to one another, a tensile strain would be created in the reinforcement which would in turn create a force normal to the shear plane. This normal force in combination with friction at the crack interface provides the shear resistance.

Where an area of shear-friction reinforcement, $A_{vf}$, is placed perpendicular to the shear plane, the factored resisting shear stress, $v_r$ is computed by:

$$v_r = \phi_c(c + \mu\sigma)$$

but shall not exceed 0.25 $\phi_c f'_c$ nor 7.0 $\phi_c$ MPa.

$c$ = cohesion stress

$\sigma$ = effective normal stress

$\mu$ = coefficient of friction

$\phi_c$ = 0.65

Values of c, $\mu$, and maximum values of $v_r$ are shown in Fig. 4.5.1.

The effective normal stress $\sigma$ is calculated by:

$$\sigma = \rho_v f_y + \frac{N}{A_g}$$

where:

$A_g$ = gross area of the section

$f_y$ = specified yield strength of reinforcement

$N$ = unfactored permanent compressive load perpendicular to the shear plane

$\rho_v$ = ratio of shear friction reinforcement perpendicular to the shear plane

The ratio of shear friction reinforcement, $\rho_v$, is:

$$\rho_v = \frac{A_{vf}}{A_{cv}}$$

where:

$A_{cv}$ = area of concrete section resisting shear transfer

$A_{vf}$ = area of shear friction reinforcement

The area of shear-friction reinforcement, $A_{vf}$, should not be less than:

$$A_{vf(min)} = \frac{0.9A_{cr}}{f_y}$$

where:

$A_{cr}$ = area of the crack interface, unless $A_{vf}$ calculated for the factored shear force is increased by one-third

Any tension acting across the shear plane should be resisted by an additional amount of reinforcement computed by:

$$A_n = \frac{N_f}{\phi_s f_y}$$

All reinforcement, on both sides of the assumed crack plane, should be properly anchored by development length or welding to angles or plates.

## 4.6 REINFORCED CONCRETE BEARING

Except for continuously grouted horizontal joints between wall elements, and for uniform bearing of hollow core and flat slabs, it is recommended that reinforcement be provided in the bearing area. This reinforcement can be designed by shear-friction as discussed in Sect. 4.5. Referring to Fig. 4.6.1, the reinforcement $A_{vf} + A_n$ across the assumed crack plane is determined to resist $N_f$ directly, and $V_f$ by shear-friction.

**Fig. 4.5.1 Values for c, $\mu$ and $v_{max}$**

| Crack Interface Condition | $\mu$ | c (MPa) | $v_{max}$ (MPa) |
|---|---|---|---|
| Concrete placed against hardened concrete with surface clean but not intentionally roughened | 0.6 | 0.25 | $0.25\phi_c f'_c$ but $< 7.0\phi_c$ |
| Concrete placed against hardened concrete with surface clean and intentionally roughened to an amplitude of 5mm | 1.0 | 0.5 | |
| Concrete placed monolithically | 1.4 | 1.0 | |
| Concrete anchored to as-rolled steel by headed studs or reinforcing bars | 0.6 | 0.0 | |

**Fig. 4.6.1 Reinforced concrete bearing**

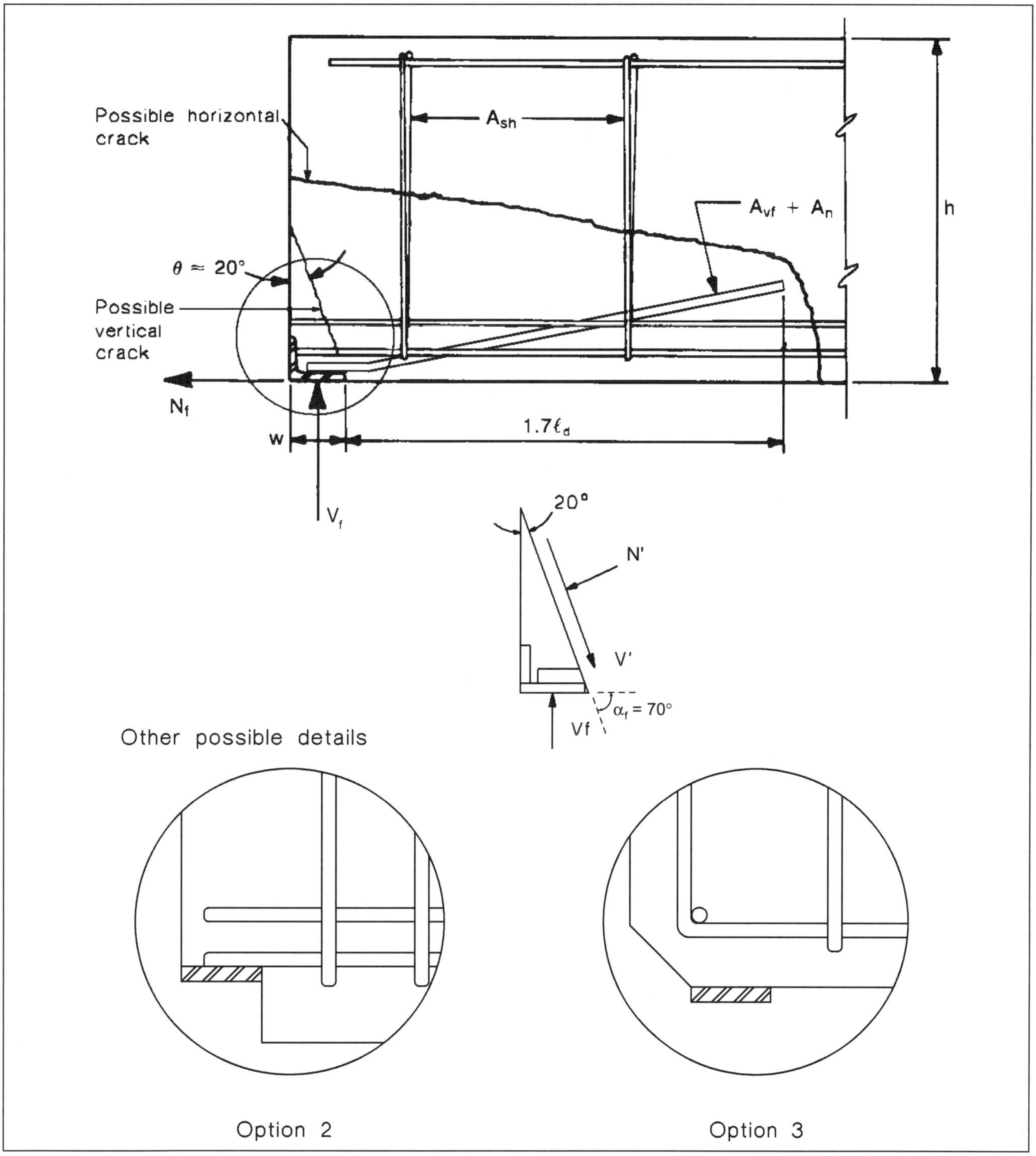


Vertical reinforcement across potential horizontal cracks can be calculated by:

$$A_{sh} = \frac{1}{f_{ys}\mu}\left(\frac{f_y(A_n + A_{vf})}{\phi_c} - cA_{cr}\right)$$

where:

$c$ = cohesion stress (Clause 11.6.2, A23.3 - 94)

$f_y$ = minimum specified yield strength of flexural reinforcement

$f_{ys}$ = minimum specified yield strength of $A_{sh}$ reinforcement

$b$ = average element width

$\ell_d$ = development length of $A_{vf}$ bars (see Clause 12.2.3, A23.3 - 94)

$A_{cr}$ = $1.7\,\ell_d\,b$

Stirrups or mesh used for diagonal tension reinforcement can be considered to act as $A_{sh}$ reinforcement.

**Example: Reinforced bearing for a rectangular beam**

*Given:*

Rectangular beam 400 × 700 × 8 m long carrying a factored UDL of 120 kN/m.

$f_y$ = 400 MPa

$f'_c$ = 35 MPa (normal density)

Bearing pad = 100 × 350

*Problem:*

Determine reinforcement requirements for the end of the element.

*Solution:*

Factored end reaction = (120)(8)/2 = 480 kN

$V_f$ = (480)(1.3) = 624 kN

$N_f$ = (0.2)(624) = 125 kN

Considering the $V_f$ only, it is equilibrated by $V'$ and $N'$ that are respectively parallel and perpendicular to the cracked surface.

$V'$ = 624 cos 20° = 586 kN

$N'$ = 624 sin 20° = 213 kN

The area of the cracked surface $A_{cr}$ is:

$A_{cr}$ = bw / sin20°
= (400)(100) / sin20°
= 117 x $10^3$ $mm^2$

Therefore the shear stress to be resisted is:

$v'_f$ = (586 × $10^3$) / (117 × $10^3$) = 5.0 MPa

Because the reinforcement is not perpendicular to the crack plane the required steel will be determined by:

$$v_r = \lambda\phi_c(c+\mu\sigma)+\phi_s\rho_v f_y \cos\alpha_f$$

where $\alpha_f$ is the angle between the shear friction reinforcement and the shear plane. Substituting for the normal stress σ gives:

$$v_r = \lambda\phi_c\left[c+\mu\left(\rho_v f_y \sin\alpha_f + \frac{N}{A_g}\right)\right]+\phi_s\rho_v f_y \cos\alpha_f$$

where:

$A_g = A_{cr}$

Setting $V_r = V'_f$ gives:

$$5.0 = (1.0)(0.65)\left[1.0+1.4\left(\rho_v(400)\sin 70^\circ+\frac{213}{117}\right)\right]$$
$$+0.85\rho_v(400)\cos 70^\circ$$

Solving for the required $\rho_v$ gives:

$$\rho_v = 0.00588$$

but $\lambda\phi_c(c+\mu\sigma)$ is limited to the lesser of $0.25\phi_c f'_c$ or $7.0\phi_c$.

(0.25)(0.65)(35) = 5.69 MPa

(7.0)(0.65) = 4.55 MPa (governs)

$$\lambda\phi_c(c+\mu\sigma)$$
$$= (1.0)(0.65)\left[1.0+(1.4)\left((0.00588)(400)\sin 70^\circ+\frac{213}{117}\right)\right]$$
$$= 4.32\text{ MPa} < 4.55\text{ MPa} \quad \text{OK}$$

therefore the steel required is:

$A_{vf}$ = (0.00588)(117 × $10^3$) = 688 $mm^2$

The minimum reinforcement required for shear friction design can be calculated by:

$$A_{vf(min)} = \frac{(0.9)(117\times10^3)}{400} = 263\text{ mm}^2 < 688\text{ mm}^2$$

(does not govern)

The additional steel required for the horizontal tension force is:

$$A_n = \frac{N_f}{\phi_s f_y} = \frac{125\times10^3}{(0.85)(400)} = 368\text{ mm}^2$$

The total amount of reinforcement required is:

$A_{st} = A_{vf} + A_n$ = 1056 $mm^2$

Use 4 No. 20 bars ($A_s$ = 1200 $mm^2$)

The reinforcement across the potential horizontal crack can be calculated by:

$$A_{sh} = \frac{1}{f_y\mu}\left[\frac{f_y(A_{vf}+A_n)}{\phi_c} - cA_{cr}\right]$$
$$= \frac{1}{(400)(1.4)}\left[\frac{(400)(688+368)}{(0.65)} - (1.0)(331\times10^3)\right]$$
$$= 569\text{ mm}^2$$

where:

$A_{cr} = 1.7\ell_d b$ = (1.7)(487)(400) = 331 × $10^3$ $mm^2$

$$A_{vf(min)} = \frac{(0.9)(331\times10^3)}{400} = 745\text{ mm}^2 \text{ (governs)}$$

Use 2 No. 15 stirrups

# 4.7 COMPRESSION STRUT AND TIE MODEL

The strut and tie model described in Clause 11.5 of A23.3-94 utilizes truss models together with concepts from the Compression Field Theory to determine the crushing resistance of the struts.

The geometry of the truss model, which consists of concrete compression struts and reinforcing tension ties, is determined by following the flow of forces from the support reaction into the body of the supported element. Once a geometry of the truss is known, the forces in the struts and ties are determined by statics.

The intersection of compressive struts with tension ties or support reactions delineates the nodal zones of multi directionally compressed concrete. Unless special confinement reinforcement is provided, the concrete compressive stresses in the nodal zones should not exceed the following limits:

1. $0.85\ \phi_c\ f'_c$ in nodal zones bounded by compression struts and bearing areas.
2. $0.75\ \phi_c\ f'_c$ in nodal zones anchoring only one tension tie.
3. $0.65\ \phi_c\ f'_c$ in nodal zones anchoring tension ties in more than one direction.

The compressive stress, $f_2$, in a compression strut must be less than the crushing limit, $f_{2max}$. The compressive stress, $f_2$, is determined by dividing the strut force by the width and the thickness of the strut. If a compressive strut is crossed by a tension tie, then $f_{2max}$ must be reduced to account for the presence of the principal tensile strain $\in_1$. This principal tensile strain may be determined from strain compatibility by conservatively assuming that the strain, $\in_s$, in the tension tie, is $f_y / E_s$, as follows:

$$\in_1 = \in_s + (\in_s + 0.002) / \tan^2 \theta_s$$

where $\theta_s$ is the angle between the tie and the strut. The crushing strength, $f_{2max}$, accounting for the reduction of strength due to the presence of the principal tensile strain, $\in_1$ is:

$$f_{2max} = \frac{\lambda \phi_c f'_c}{0.8 + 170 \in_1}$$

Reducing $\theta_s$ results in lower values of the crushing strength, $f_{2max}$. Unless the concrete is triaxially confined, $f_{2max}$ should not exceed $\lambda\ \phi_c\ f'_c$. (See Ref. 4.7.1)

The tension tie reinforcement should be chosen such that the factored resistance equals or exceeds the tie force required. The reinforcement must be effectively anchored to transfer the required tension to the nodal zone of the truss. Additionally, the tension tie reinforcement should be distributed over an effective area of concrete at least equal to the tensile tie force divided by the stress limit previously stated for compressive stresses in the nodal zone. This effective area of concrete may be assumed to be the area of concrete surrounding the tension tie reinforcement and having the same centroid as that reinforcement.

# 4.8 CORBELS

Concrete corbels must be designed in accordance with the provisions of Clause 11.7 of CSA-A23.3-94. The strut and tie models as described in Sect. 4.7 apply to the design of corbels and are illustrated by the following example.

**Example - Design of a corbel**

*Given:*
A 350 × 350 column supports a beam with a factored vertical load of 250 kN

$f'_c$ = 35 MPa (normal density)

$f_y$ = 400 MPa (weldable)

*Problem:*

Determine the corbel size and reinforcement required to support the load.

*Solution:*

Given that:

$V_f$ = 250 kN

then:

$N_f = 0.2\ V_f = (0.2)(250) = 50$ kN

*Corbel dimensions*:
Choose an overall corbel depth of 500 mm. Choose a bearing pad 300 × 110 × 10. The depth at the outside edge of the bearing should be at least 250 mm. The details of the bearing are shown in Fig. 4.8.1.

**Fig. 4.8.1 Corbel example**

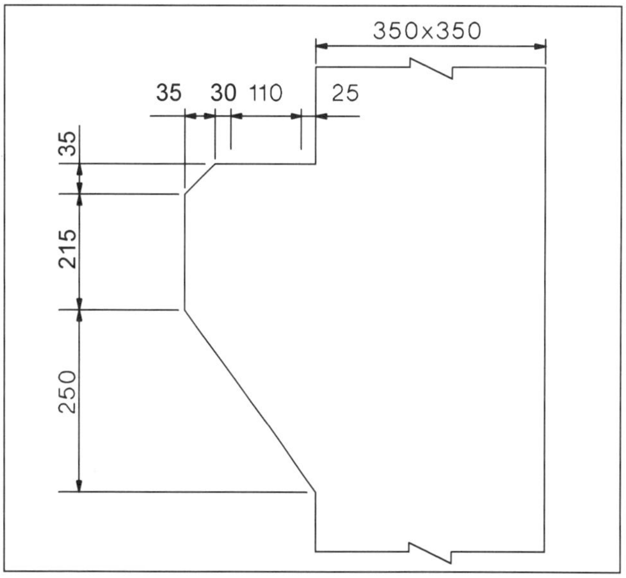


To allow for load eccentricities and erection tolerances, consider the vertical load to be placed 25 mm towards the edge of the corbel from the centre of the bearing pad.

**Fig. 4.8.2 Strut-and-tie model for corbel example**

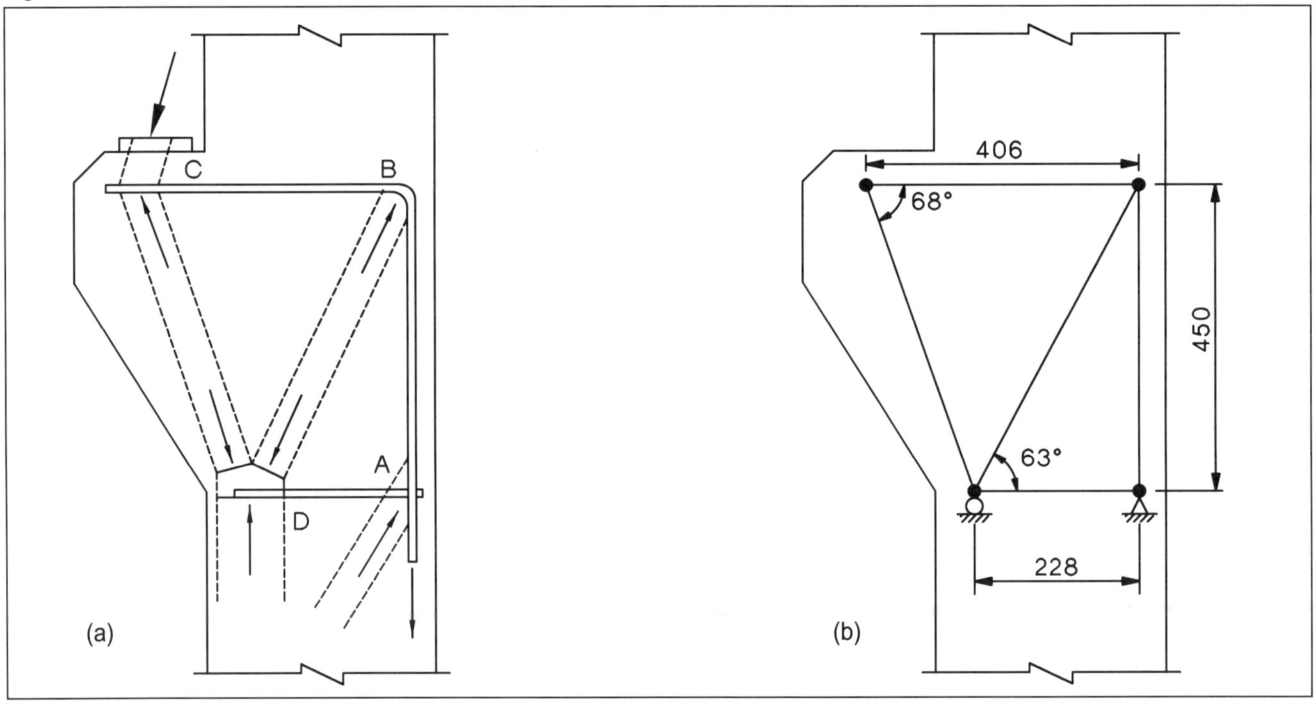


*Strut and tie model:*

The assumed compression strut, tension tie and nodal zone model for the corbel is shown in Fig. 4.8.2(a). The truss model is shown in Fig. 4.8.2(b).

Nodes A and B are located at the intersection of the center lines of the tension ties. Node C is located at the intersection of the centerline of the upper tension tie and the line of action of the applied vertical loads. Node D is located on centerline of the lower tension tie for the tension to be fully developed to equilibrate the compressive struts at the node.

The forces in the truss members and the reactions are determined from statics and are summarized below:

| Reaction | $A_x$ | $A_y$ | D |
|---|---|---|---|
| Force (kN) | 50 | 294 | 544 |

| Member | CB | BD | CD | BA | DA |
|---|---|---|---|---|---|
| Force (kN) | 149 | 329 | 269 | 294 | 50 |

Design of Tension Ties:

The reinforcement required in tie CB is:

$$A_s = \frac{(149 \times 10^3)}{(0.85)(400)} = 438 \text{ mm}^2$$

The minimum area required for the primary reinforcement is:

$$A_s = 0.04 \frac{f'_c}{f_y} bd$$

$$= 0.04 \frac{(35)}{(400)}(350)(450) = 551 \text{ mm}^2$$

therefore:

Use 3 No. 15 bars

Although tie BA has a larger tie force, the vertical reinforcement in the column would have been designed for this force. Continue the 3-No.15 bars down the column to adequately anchor them.

The reinforcement required for tension tie DA is:

$$A_s = \frac{(50 \times 10^3)}{(0.85)(400)} = 147 \text{ mm}^2$$

Use an additional No. 10 column tie at location DA.

Additional ties with an area equal to one half the primary reinforcement area must be placed within two-thirds the effective depth adjacent to $A_s$:

$$\text{Tie area} = \frac{550}{2} = 275 \text{ mm}^2$$

Use 2-No. 10 ties within $\frac{(2)(450)}{3} = 300 \text{ mm}$.

*Check compression struts:*

The stress limits for the struts are governed by the nodal limits; therefore, $f_{2max}$ is not reached.

*Design of nodal zones:*

Reinforcement at nodes A, B, C, and D is anchored around the column reinforcement. Check the anchorage conditions at the ends of the compression strut in accordance with Clause 11.5.2.2. in CSA-A23.3

The primary reinforcement can engage an area of concrete six times its diameter as long as there is room. However the maximum stress limit for the node must be within these limitations.

To satisfy the stress limits of node C the tension tie requires a depth of concrete equal to:

$$\frac{(149 \times 10^3)}{(0.75)(0.65)(35)(350)} = 25\,mm$$

This depth is available.

The final details of the corbel are shown in Fig. 4.8.3.

**Fig. 4.8.3 Reinforcement for corbel example**

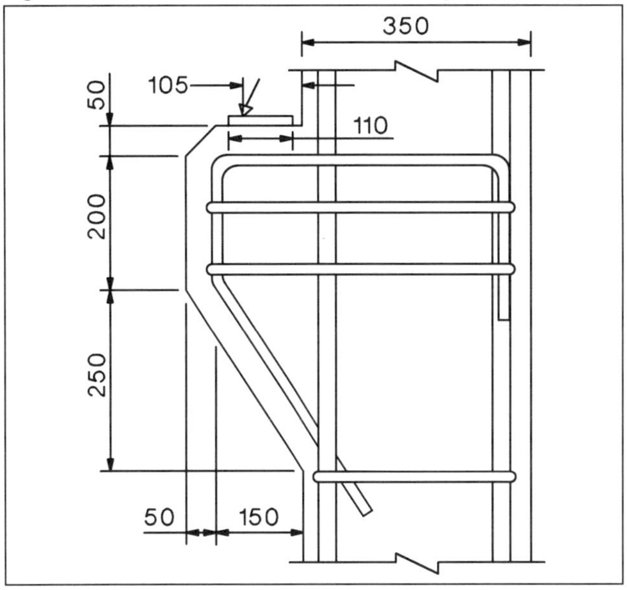


**Fig. 4.8.4 Anchorage checks example**

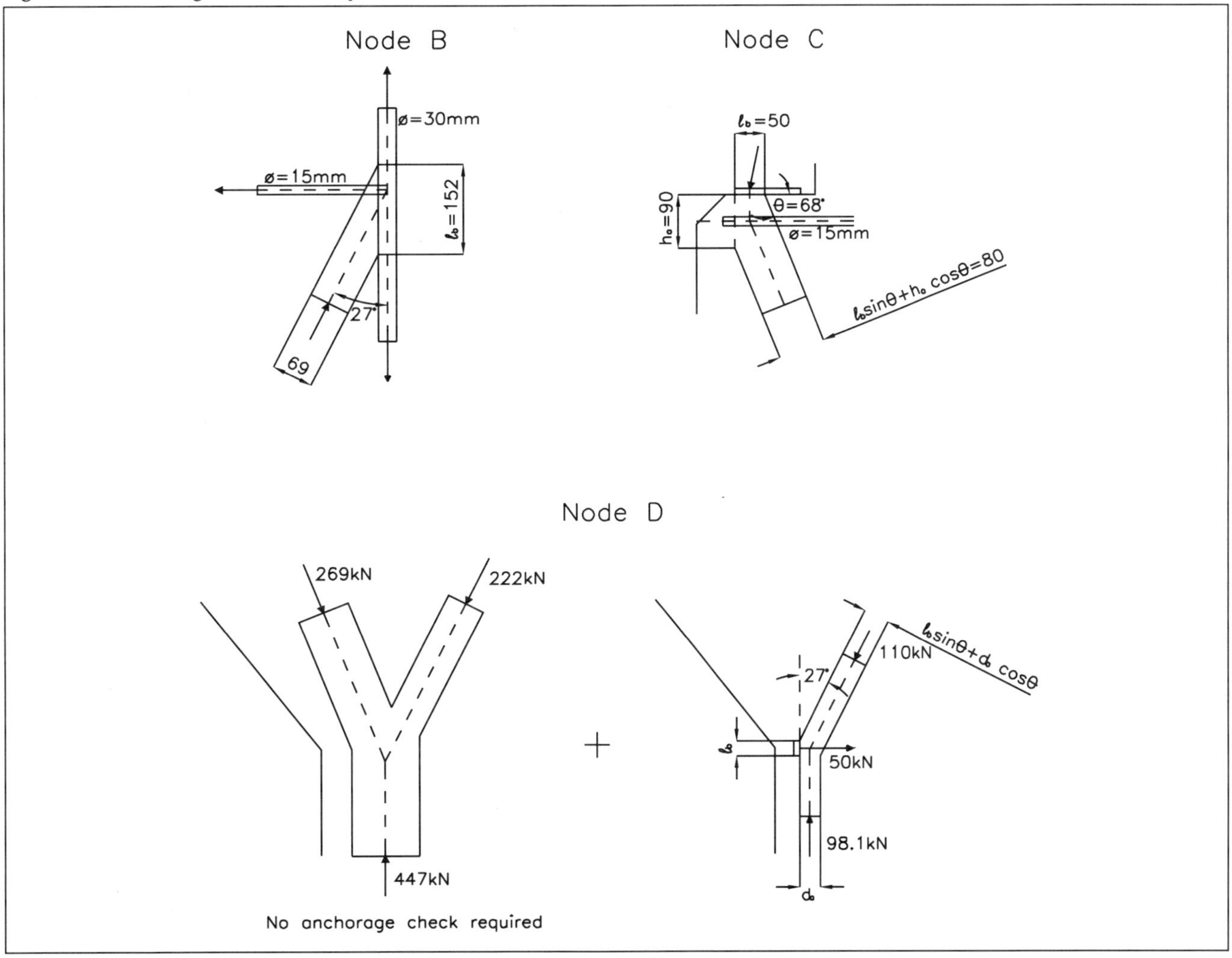

## 4.9 DAPPED END CONNECTIONS[R4.9.1]

**Fig. 4.9.1 Dapped end connection example**

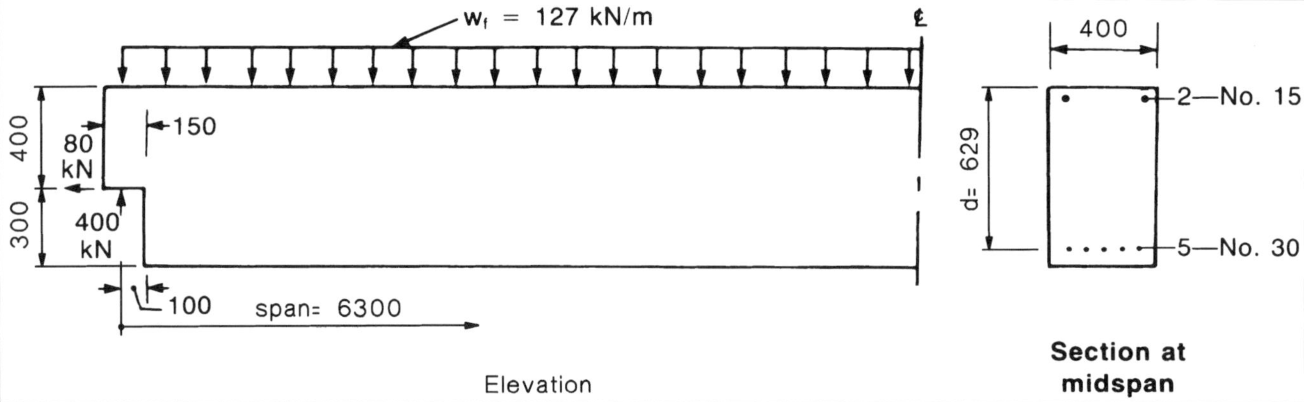


**Example - 400 mm wide beam with a dapped end as shown in Fig. 4.9.1**

*Given:*

$V_f$ = 400 kN

$N_f$ = 80 kN

$f'_c$ = 35 MPa Normal Density

$f_y$ = 400 MPa weldable

*Solution:*

Assume a 320 mm long steel angle is used at the support. The required bearing length:

$$= \frac{V_f}{0.75\phi_c f'_c b} = \frac{(400)(10^3)}{(0.75)(0.65)(35)(320)} = 73\text{ mm}$$

Use a 100 × 100 × 10 mm thick steel angle × 320 mm long.

*Strut and tie model:*

The tension tie, compression strut and nodal zone model is shown in Figs. 4.9.2 and 4.9.3.

The corresponding line drawing of the truss is shown in Fig. 4.9.4. This truss can be determined in the following manner:

1. Node A at intersection of line of action of support force and centreline of tension (No. 20 bars assumed).
2. Node B at intersection of centre line of tension hanger (assumed to be spread over 180 mm) and centre line of top longitudinal bars (No. 15 bars assumed).
3. Node C at intersection of centreline of tension hanger and centre line of bottom No. 30 longitudinal bars.
4. Node E taken to be $d_v/(2\tan\theta)$ from edge of tension hanger so that strut CE can represent the fan.

**Fig. 4.9.2 Strut- and-tie model for dapped end connection**

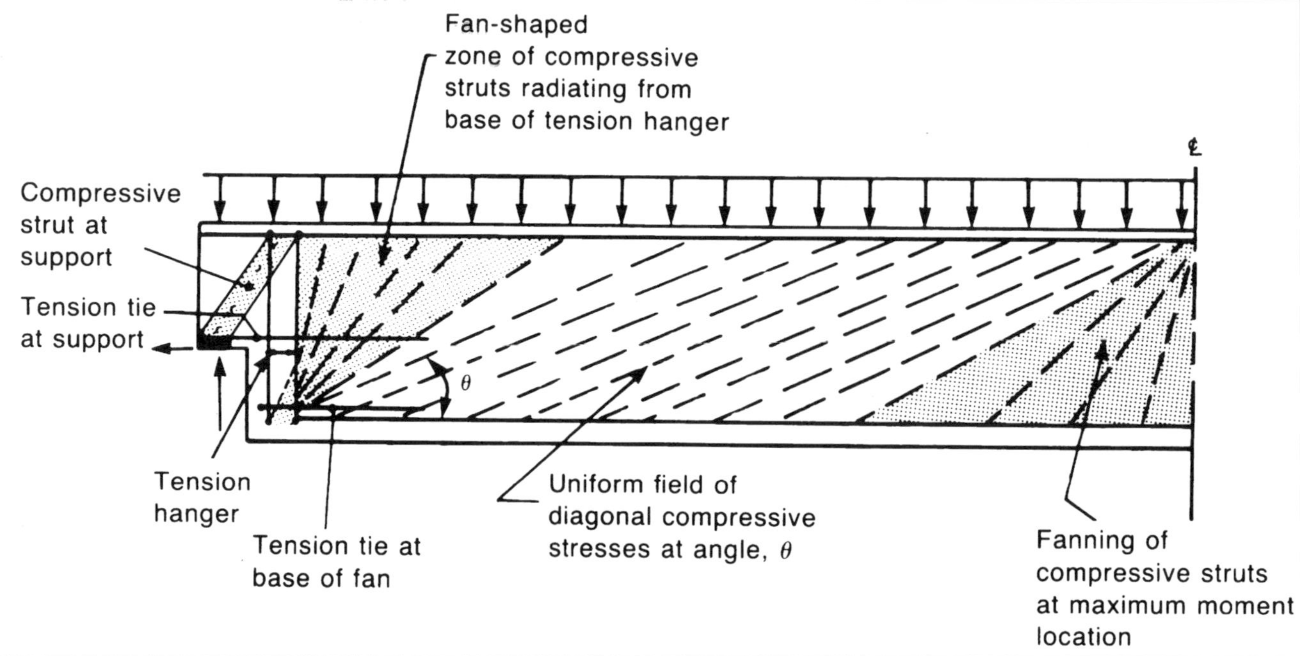

**Fig. 4.9.3 End detail of strut-and-tie model for dapped end connection example**

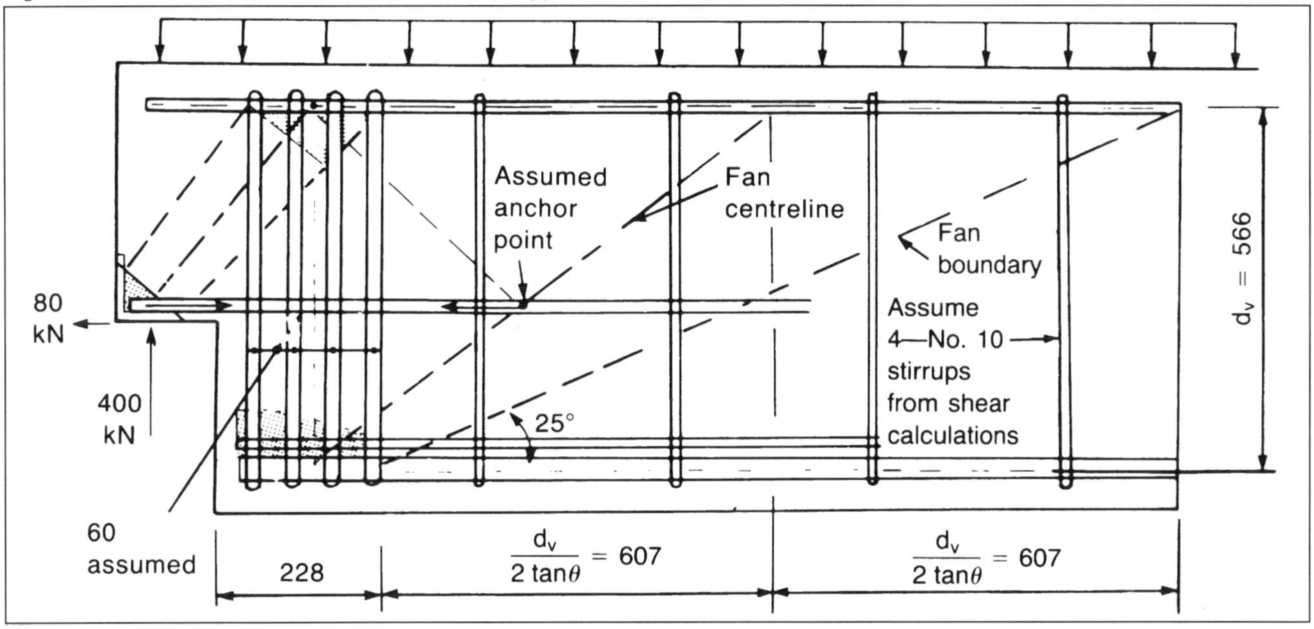


5. Node D at intersection of tension tie and strut CE to represent the anchor for this tension tie.
6. Node F directly below node E so that tie EF can represent the stirrups within the fan length $d_v/\tan\theta$.
7. The vertical load on node E represents the applied load directly carried by the fan.

The forces in the truss members can be determined by statics and are:

| Member | AB | AD | DE | DB | DC | CB | CF | BE | EF |
|---|---|---|---|---|---|---|---|---|---|
| Force | 508 | 393 | 573 | 247 | 849 | 531 | 663 | 136 | 204 |
| (kN) | C | T | C | C | C | T | T | C | T |

*Design of tension ties:*

All the tension ties will require tension reinforcement, $A_s$:

Tie AD:

$$A_s = \frac{(393)(10^3)}{(0.85)(400)} = 1156 \text{ mm}^2$$

**Fig. 4.9.4 Truss idealization for dapped end connection example**

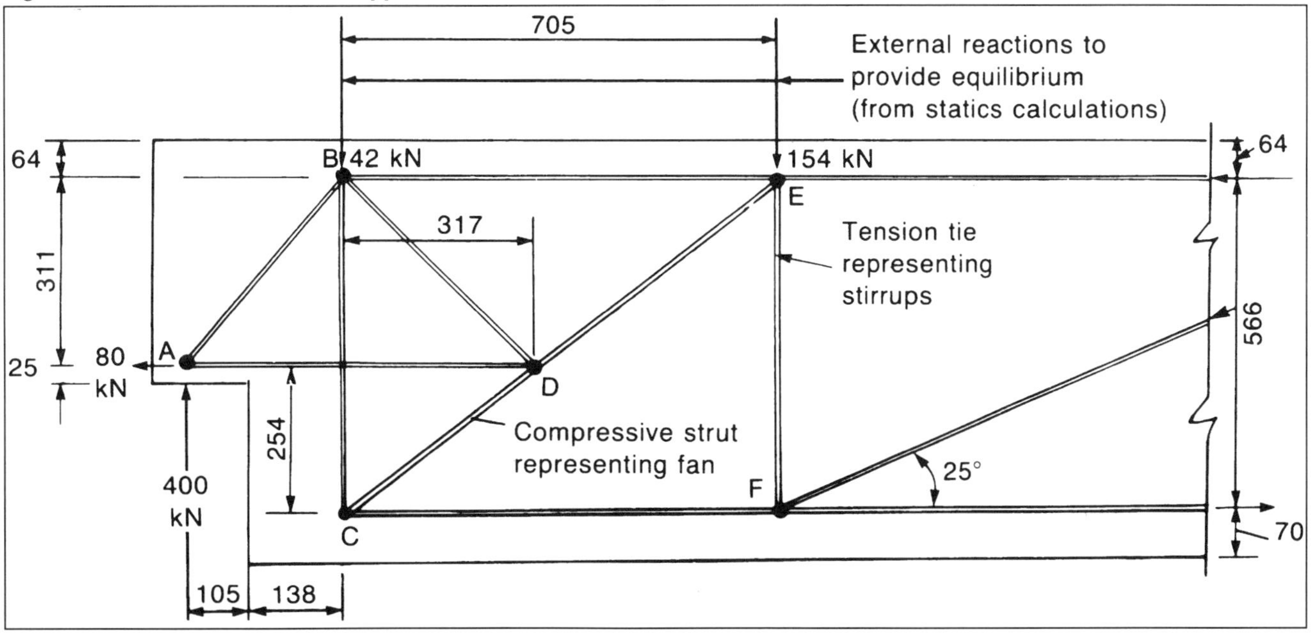

Use 4 - No. 20 bars.

Hanger CB:

$$A_s = \frac{(531)(10^3)}{(0.85)(400)} = 1562\,mm^2$$

Use 4 - No. 15 closed stirrups.

Tie CF:

$$A_s = \frac{(663)(10^3)}{(0.85)(400)} = 1950\,mm^2$$

Area of 5 - No. 30 bars is sufficient but anchorage will need to be checked.

Tie EF:

$$A_s = \frac{(204)(10^3)}{(0.85)(400)} = 600\,mm^2$$

The 4 - No. 10 U-stirrups in fan region are sufficient.

*Check compressive stresses in the nodal zones:*

At nodal zone A, the required depth of the nodal zone:

$$= \frac{(393-80)(10^3)}{(0.75)(0.65)(35)(320)} = 57\,mm$$

therefore, 100 mm provided by the steel angle is adequate.

At nodal zone B, because of a concern about spalling of the concrete cover, neglect the concrete outside the anchoring tension tie reinforcement.

The required width of the nodal zone:

$$= \frac{(531)(10^3)}{(0.75)(0.65)(35)(320)} = 97\,mm$$

provide a nodal zone width of (3)(60) + 16 = 196 mm, which is conservative.

Nodal zone C anchors two tension ties. Hence, required width:

$$= \frac{(531)(10^3)}{(0.65)(0.65)(35)(320)} = 112\,mm$$

Thus, 196 mm is still OK

Required depth of nodal zone:

$$= \frac{(663)(10^3)}{(0.65)(0.65)(35)(320)} = 140\,mm$$

To provide this nodal zone depth, provide 3 - No. 15 horizontal U-bars at 50 mm spacing above the layer of No. 30 bars.

Also check the anchorage of tension tie CF in nodal zone C.

As the No. 30 bars emerge from nodal zone C, they can resist a tensile force of approximately:

$$= \frac{196\phi_s A_s f_y}{\ell_d}$$

$$= \frac{(196)(0.85)(3500)(400)}{(726)(10^3)} = 321\,kN$$

The 3 - No. 15 U-bars will be capable of resisting a tension force of $(0.85)(3)(2)(200)(400)/10^3$ = 408 kN.

Hence, total tensile resistance at face of nodal zone = 321 + 408 = 729 kN

As 729 kN > 663 kN, anchorage OK

To obtain the proper lap length, extend the No. 15 bars at least $\ell_d$ (= 300 mm) beyond the nodal zone and far enough for the 5 - No. 30 bars to be capable of carrying the 663 kN tie force on their own (i.e., 500 mm from the end of the No. 30 bars).

Although the addition of the No. 15 U-bars will raise the location of node C somewhat, this secondary effect will be neglected.

As the compressive strut CD-DE represents a fan shaped region of radiating struts and as the nodal zone stresses at the base of the fan, which are at their highest level, have already been checked, further checks are not required.

Check compressive stress in struts which meet at node B (Fig. 4.9.5).

**Fig. 4.9.5 Nodal zone B**

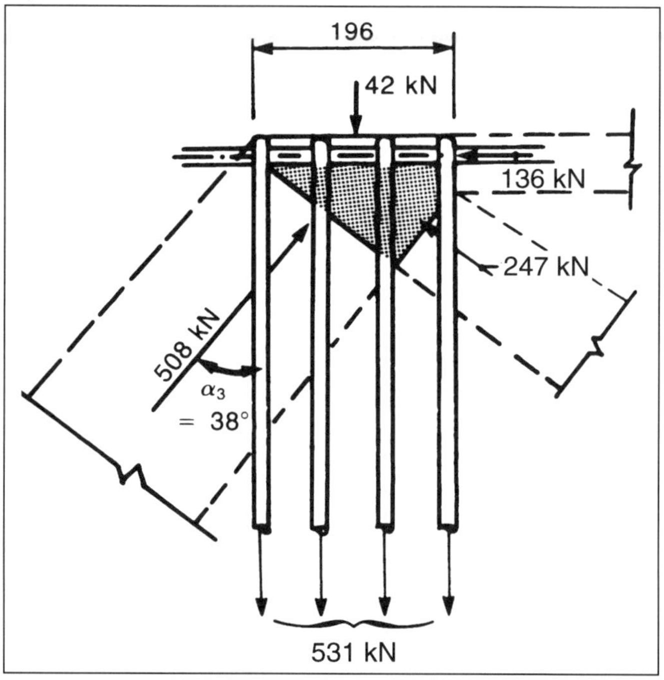


If this nodal zone is to be in equilibrium under a "hydrostatic" stress condition, the length of the faces of the nodal zone must be proportional to the loads applied to these faces and the faces must be perpendicular to the loads. Hence, width of strut AB at nodal zone B:

$$= \frac{(508)(196)}{531+42} = 174\,mm$$

Thus, stress in strut (again neglecting cover):

$$= \frac{(508)(10^3)}{(174)(320)} = 9.1\,MPa$$

As this strut is crossed by tension tie CB, its diagonal crushing strength will be reduced.

The average tensile strain in tension tie CB can be estimated as:

$$\epsilon_s = \frac{(531)(10^3)}{(8)(200)(0.85)(200)(10^3)} = 0.00195$$

The average strain will actually be somewhat smaller than this due to tension stiffening effects.

The strain $\epsilon_1$ perpendicular to the strut is:

$$\epsilon_1 = \epsilon_s + \frac{\epsilon_s + 0.002}{\tan^2\theta_s}$$

$$= 0.00195 + \frac{(0.00195 + 0.002)}{\tan^2 38^o}$$

$$= 0.0084$$

Diagonal crushing strength is:

$$f_{2max} = \frac{\lambda\phi_c f'_c}{0.8 + 170\,\epsilon_1}$$

$$= \frac{(1.0)(0.65)(35)}{0.8 + (170)(0.00842)}$$

$$= 10.2\,\text{MPa}$$

$A_s f_2 < f_{2max}$, compressive stress in strut OK

The other struts meeting at node B will have the same compressive stress but smaller values of $\epsilon_1$. Hence, they will not be critical.

To improve crack control and ductility, provide a minimum area of horizontal reinforcement parallel to the primary tensile tie reinforcement in the region above the support. If the dapped end is treated as a bracket, the required area of such additional reinforcement would be:

$$0.5\,A_s = (0.5)(4)(300) = 600\text{ mm}^2$$

Use 2 - No. 15 horizontal U-bars distributed over 2/3 of the effective depth. Extend these bars $\ell_d$ beyond the face of the dap.

To improve the support conditions for the highly stressed compressive strut AB, use two additional No. 15 top longitudinal bars in the region of node B. The final details of the dap ended beam are shown in Fig. 4.9.6.

Finally, check the anchorage conditions at the ends of the compression strut in accordance with Clause 11.5.2.2. in CSA-A23.3-94

## 4.10 BEAM LEDGES

The compression strut and tie model can be used in the design of continuous beam ledges supporting concentrated or uniformly distributed loads. The truss model is very similar to the model used in dapped end connection design.

**Example: Design of a reinforced beam ledge for a double tee leg**

*Given:*

A 125 mm wide double tee leg is resting on a L-beam as shown in Fig. 4.10.1.

$V_f$ = 100 kN
$N_f$ = 20 kN
$f'_c$ = 35 MPa
$f_y$ = 400 MPa

**Fig. 4.9.6 Reinforcement requirements for dapped end connection example**

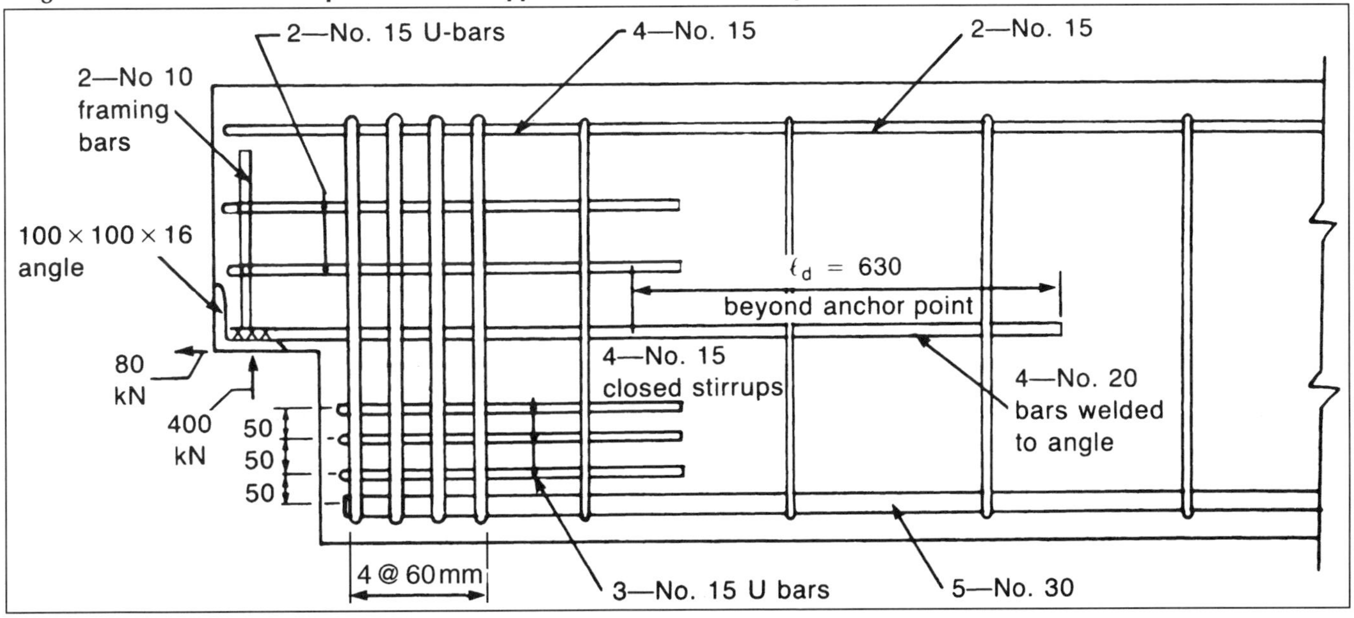

**Fig. 4.10.1 Beam ledge design example**

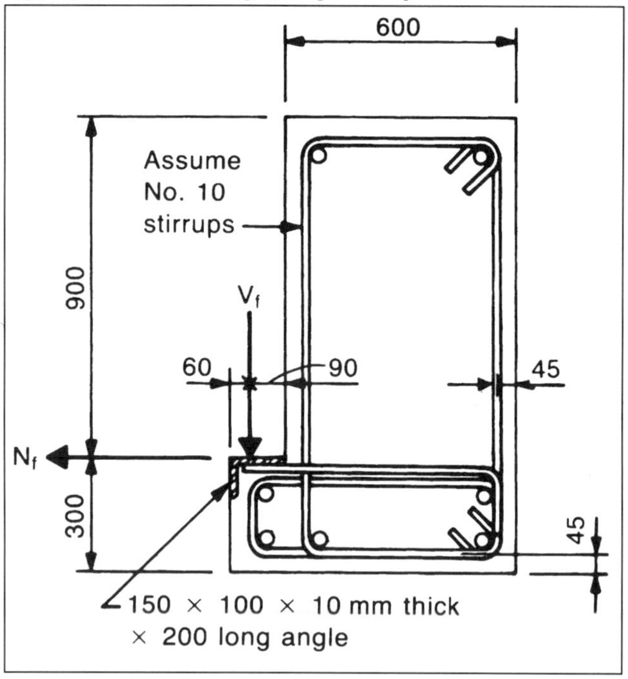


*Solution:*

Minimum bearing plate width:

$$w = \frac{V_f}{0.75\phi_c f'_c b} = \frac{(100)(10^3)}{(0.75)(0.65)(35)(150)} = 39 \text{ mm}$$

Use 150 × 100 × 10 mm thick steel plate.

*Strut and tie model:*

The flow of compressive stresses is shown in Fig. 4.10.2 and the strut and the tie model is shown in Fig. 4.10.3.

**Fig. 4.10.2 Flow of compressive stresses for beam ledge design example**

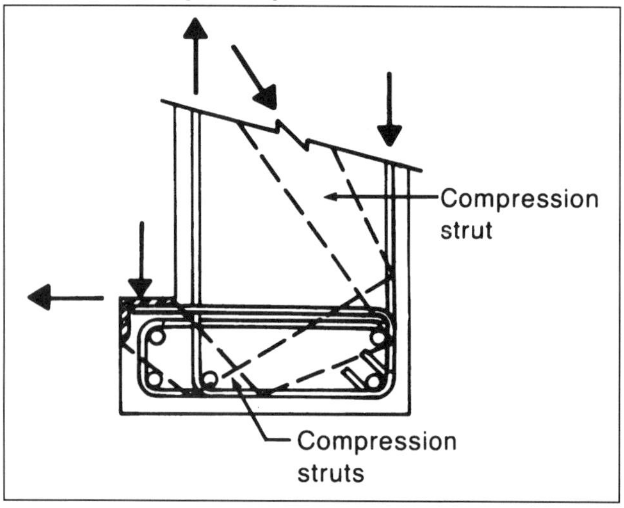


**Fig. 4.10.3 Force diagram for beam ledge example**

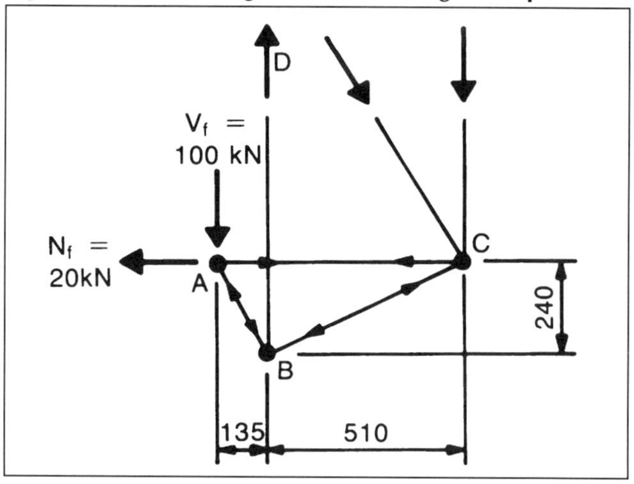


Summing moments about C:

$$DB = \frac{(100)(135+510)}{510} = 126 \text{ kN tension}$$

Summing vertical forces at A:

$AB_v$ = 100 kN

therefore:

$$AB_h = \frac{(135)(100)}{(240)} = 56 \text{ kN}$$

$$AB = \sqrt{100^2 + 56^2} = 115 \text{ kN compression}$$

Summing horizontal forces at A:

AC = 20 + 56 = 76 kN tension

*Design of tension ties:*

The reinforcement area required for tension tie AC is:

$$A_s = \frac{(76)(10^3)}{(0.85)(400)} = 224 \text{ mm}^2$$

Use 3-No. 10 bars welded to the angle.

The reinforcement area required for tension tie DB is:

$$A_s = \frac{(126)(10^3)}{(0.85)(400)} = 371 \text{mm}^2$$

Provided 4-No. 10 closed stirrups in addition to those required for shear and torsion.

*Check compressive stresses at nodal zones:*

Width of the nodal zone at A required to anchor tension tie AC is:

$$w = \frac{(76-20)(10^3)}{(0.75)(0.65)(35)(100)} = 33 \text{ mm} < 200 \text{ mm}$$

Since the compression strut AB is not crossed by a tension tie, the stresses will be OK.

Check the anchorage conditions at the ends of the compression strut in accordance with Clause 11.5.2.2. in CSA-A23.3-94. The final detail is shown in Fig. 4.10.4.

**Fig. 4.10.4 Required reinforcement for beam ledge example**

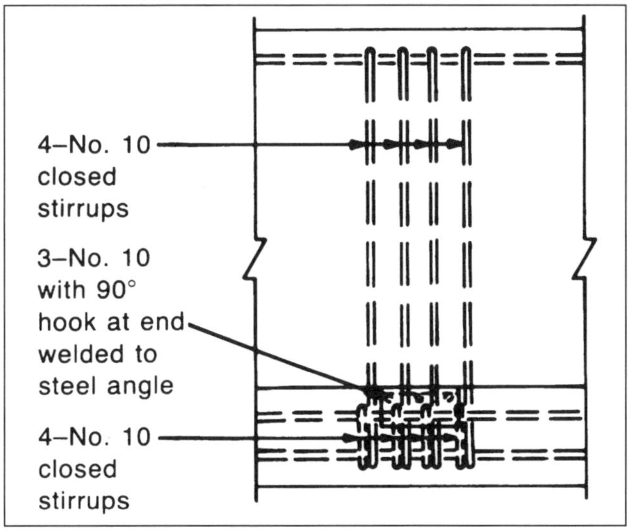


## 4.11 WELDED HEADED STUDS

Welded headed studs are designed to resist direct shear, tension or a combination of the two. Either the resistance of the concrete or of the steel may be critical, and both must be checked. The latest design philosophy is to design the embedded steel stud to ensure a tensile failure in the stud rather than a failure in the surrounding concrete. The equations are intended to apply to studs which are welded to steel plates or structural steel components, and then embedded in unconfined concrete. Confinement of the concrete, either from applied compressive loads, or from reinforcement has been shown to increase the capacity.

### 4.11.1 TENSION

The tensile resistance of the concrete surrounding a headed stud is governed by the general equation:

$$P_r = 0.30\lambda\phi_c\sqrt{f'_c}A_e$$

where:

$\phi_c$ = 0.65

$A_e$ = effective stress area

$\lambda$ = as defined in Clause 8.6.5 of A23.3

The effective stress area is the projected area of a 45° pyramid from the bearing edge of the anchor (see Fig. 4.11.1). The effective stress area should not include the bearing area of the anchor head and overlapping stress area in a stud group situation. As well, reductions must be made for the intersection of pyramids with concrete surfaces and for the overall thickness of the concrete. The above equation contains a reduction factor to account for the difference between the assumed pyramidal failure planes and the conical failure planes observed in practice.

Referring to Fig. 4.11.1, the effective stress area for a single headed stud is calculated by:

$$A = (2\ell_e + 1.58d_b)^2 - 2.5\,d_b^2$$

where:

$\ell_e$ = embedment length of stud anchor

$d_b$ = diameter of stud anchor

**Fig. 4.11.1 Effective stress area for welded headed studs**

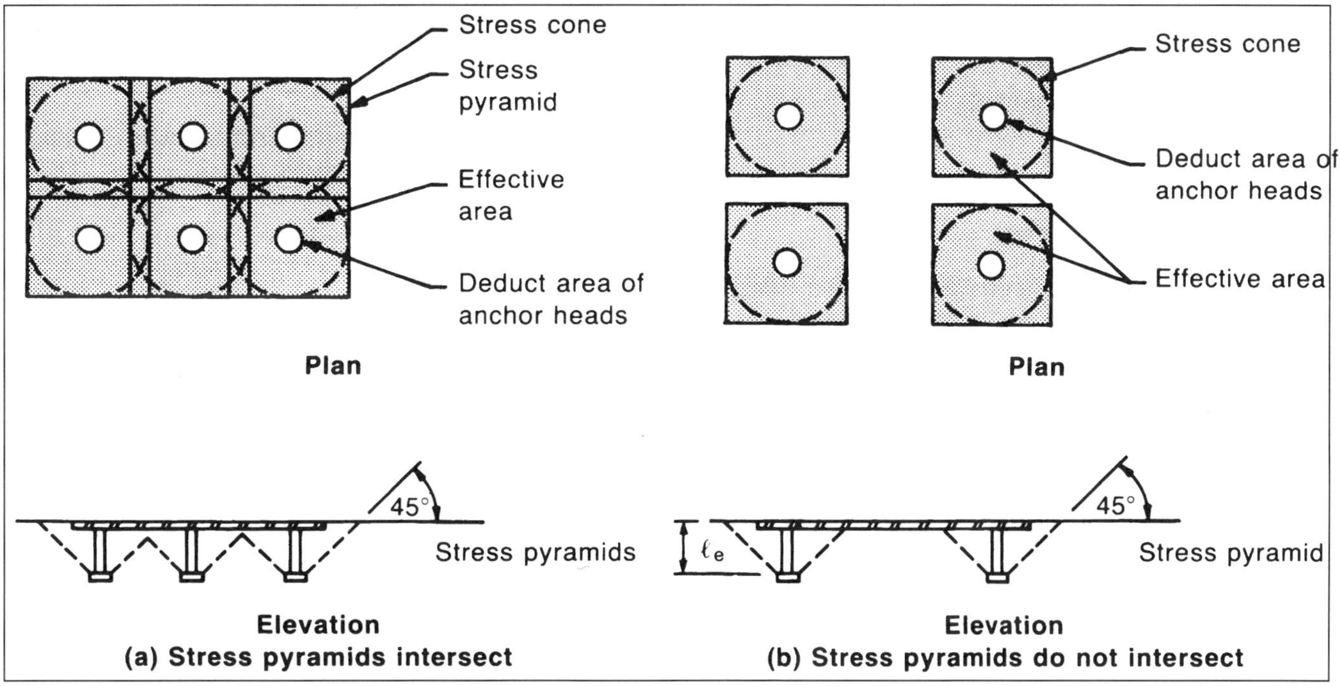

In a stud group in which the centre to centre spacing of anchors is less than two times the length of embedment plus one head diameter, the individual effective stress areas will overlap and thus reduce the factored resistance of the group (see Fig. 4.11.1).

When a stud is located near a free edge, a bursting type of failure can occur. For such cases it is recommended that the minimum cover from the anchor head to any free edge, should be:

$$m = d_b \sqrt{\frac{f_{su}}{6\sqrt{f'_c}}} \text{ but not less than 50 mm}$$

where:

$d_b$ = diameter of stud anchor

$f_{su}$ = minimum specified tensile strength of anchor

For full development of a stud group, there must be a sufficient thickness of concrete below the heads of the studs. The minimun concrete thickness (Fig. 4.11.2) should be:

$$h_{min} = \ell_e + (b / 2)$$

When the concrete thickness is less than that required for full development of the studs, an appropriate reduction must be made. Referring to Fig. 4.11.3, the effective stress area, $A_e$, is calculated as:

$$A_e = h(2a + 8\ell_e + 2b - 4h) - \text{(total area of anchor heads)}$$

**Fig. 4.11.2 Development of welded headed studs**

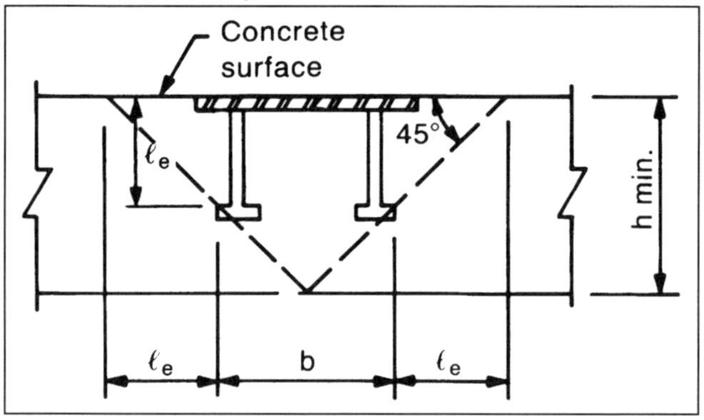


The factored resistance of a headed stud as governed by the tensile strength of the stud material is calculated by:

$$P_r = 0.8 A_b \phi_s f_{su}$$

but $0.8\,\phi_s f_{su} < 600$ MPa

where:

$A_b$ = area of stud anchor

$\phi_s$ = 0.85

$f_{su}$ = minimum specified tensile strength of anchor

The ultimate tensile strength of steel used for headed studs is typically 400 MPa. As stated previously, to ensure ductile be-

**Fig. 4.11.3 Stress area reduction for limited depth $A_r$**

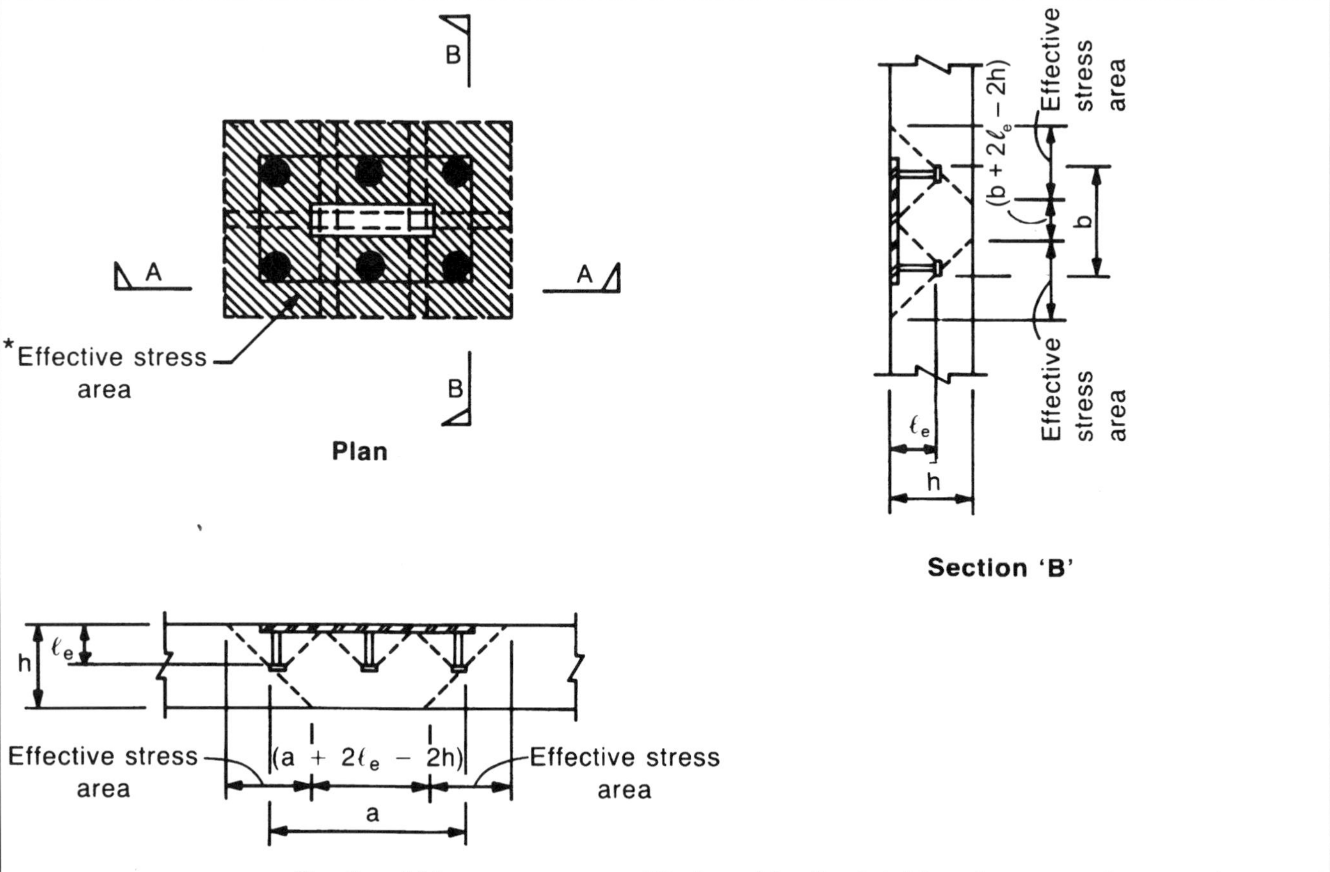


*Reduced by the total bearing area of the anchor steel

haviour, the factored resistance of the stud assembly as governed by the steel strength should exceed the factored resistance of the assembly due to concrete pullout. Factored tensile resistances of typical steel studs are shown in Fig. 4.11.4.

**Fig. 4.11.4 Factored resistance of headed studs limited by steel strength, $f_u$ = 400 MPa, $\phi_s$ = 0.85**

| Nom. dia. (mm) | Area (mm²) | $P_r$ (kN) | Min. plate thickness for full strength weld (mm) |
|---|---|---|---|
| 6 | 32 | 9 | 4 |
| 10 | 71 | 19 | 7 |
| 13 | 127 | 34 | 9 |
| 16 | 198 | 54 | 11 |
| 19 | 285 | 78 | 13 |
| 22 | 388 | 106 | 15 |

## 4.11.2 SHEAR

The concrete shear resistance of studs not located near a free edge can be analyzed by either lateral bearing of the stud or by shear-friction principles.

The factored bearing resistance of a stud is:

$$B_r = 1.4\,\phi_c A f'_c$$

where:

$\phi_c$ = 0.65

A = projected area of the stud(diameter × length)

When a stud or studs are welded to a plate, the factored shear resistance is calculated by shear-friction principles as:

$$V_r = \phi_c A_{vf} f_y \mu$$

where:

$\phi_c$ = 0.65

$A_{vf}$ = cross-sectional area of the studs

$f_y$ = yield strength of the stud, but ≤ 400 MPa

$\mu$ = 0.6λ for contact plane coincidental with the concrete surface

$\mu$ = 0.8λ for contact plane a full plate thickness below the concrete surface

The value of $V_r$ should not exceed the maximum values specified in Sect. 4.5.

When shear acts towards a free edge, the minimum distance from the anchorage to that edge should be such that the concrete does not fail in tension. The factored shear resistance is then calculated by:

$$V_r = 0.30\lambda\phi_c\sqrt{f'_c}A_e$$

where:

$A_e$ = effective stress area

For stud assemblies designed by shear-friction, the effective stress area is defined by projecting 45° planes for a line parallel to the free edge through the middle of the contact surface, and from the edges of the contact surface to the free edge (see Fig. 4.11.5).

**Fig. 4.11.5 Edge distance effect for welded headed studs**

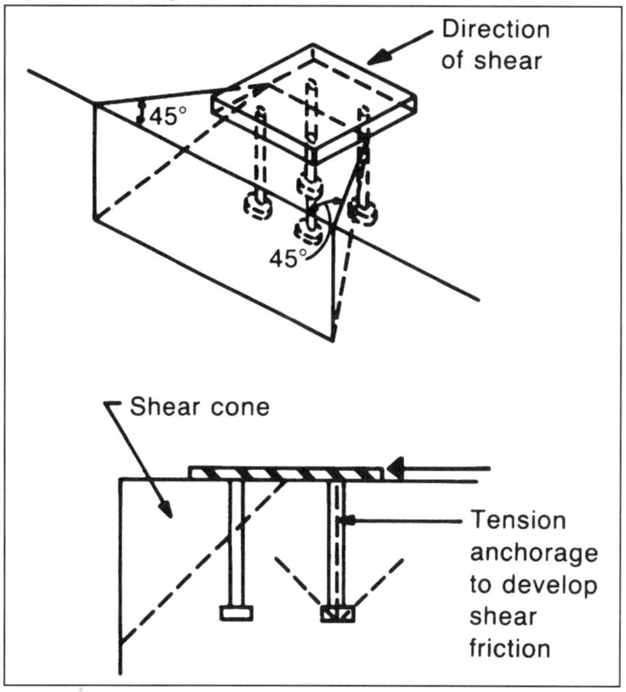


In all cases, the stud assembly should have a factored tensile resistance at least equal to the factored shear force being transferred. In an assembly designed by shear-friction, the factored tensile resistance should not be less than $\dfrac{V_r}{\mu}$.

## 4.11.3 COMBINED SHEAR AND TENSION

For stud anchor assemblies designed to resist both shear and tension, the area of steel required is additive. When the factored shear resistance is provided in accordance with shear-friction principles, the total area of steel required is:

$$A_b = A_{bt} + A_{bs}$$

where:

$$A_{bt} = \frac{P_f}{0.8\phi_s f_{su}}$$

$$A_{bs} = \frac{V_f}{\phi_c \mu f_y}$$

$P_f$ = factored tension force

$V_f$ = factored shear force

$\phi_s$ = 0.85

$f_y$ = minimum specified yeild strength of anchor

$f_{su}$ = minimum specified tensile strength of studs

$\mu$ = values in Sect. 4.11.2

Once the number of studs has been determined, select a stud spacing, embedment length, and side cover to ensure that a steel failure will occur prior to a failure of the concrete.

### Example - Capacity of welded headed studs

*Given:*

Bracket on column as shown in Fig. 4.11.6.

$f'_c$ = 35 MPa
Normal Density
Vertical specified dead load = 16 kN
Vertical specified live load = 50 kN
$f_y$ = 350 MPa
$f_{su}$ = 400 MPa

*Problem:*

Determine if the studs are adequate to resist the applied loads.

*Solution:*

$V_f = (1.25)(16) + (1.5)(50) = 95$ kN

To ensure a member failure prior to a connection failure, increase $V_f$ by 30%:

$V_f = (1.3)(95) = 123.5$ kN
$P_f = 0.2\ V_f = (0.2)(123.5) = 24.7$ kN
$C = \alpha_1\ \phi_c\ f'_c\ b\ a$
$\alpha_1 = 0.85 - 0.0015\ f'_c \geq 0.67$
$= 0.85 - 0.0015(35)$
$\alpha_1 = 0.80$
$C = (0.80)(0.65)(35)(250)\ a/10^3 = 4.55a$ kN

Summing moments about C:

$T(275 - a/2) = (123.5)(150) + (24.7)(275 - a / 2)$

$$T = \frac{(18\,525)}{(275 - a/2)} + 24.7$$

For equilibrium, C = T

$$4.55\ a = \frac{(18\,525)}{(275 - a/2)} + 24.7$$

Solving the quadratic equation gives:
a = 21 mm and, therefore, T = 94.7 kN

The top group of six studs must resist this tension and also resist the factored shear by shear-friction. For this group of studs:

$V_f$ = 123.5 kN and $P_f$ = 94.7 kN

Check combined shear and tension on the steel area:

Area of 6 - 16 mm diameter studs
$= (6)(198) = 1188\ \text{mm}^2$

$$A_{bs} = A_{bt} + A_{bs} = \frac{P_f}{0.8\phi_s f_{su}} + \frac{V_f}{\phi_c \mu f_y}$$

$$= \frac{(94.7\times10^3)}{(0.8)(0.85)(400)} + \frac{(123.5\times10^3)}{(0.65)(0.8)(350)}$$

$= 1027\ \text{mm}^2 < 1188$ OK

Check the factored tensile resistance of concrete. This must be equal to or greater than:

$$P_f + V_f / \mu = 94.7 + 123.5 / 0.8 = 249\ \text{kN}$$

$$2\ell_e + 1.58d_b = (2)(200) + (1.58)(19)$$
$$= 430\ \text{mm} > 100$$

Therefore, stress pyramids overlap:

Effective stress area, $A_e$

$= [(2)(200) + 200 + (1.58)(19)][(2)(200) + 100 + (1.58)(19)] - (6)(2.5)(19)^2$
$= 328\ 508\ \text{mm}^2$

$$P_r = 0.30\lambda\phi_c\sqrt{f'_c}A_e$$
$$= (0.30)(1.0)(0.65)\sqrt{35}(328508)/10^3$$
$$= 379\ \text{kN} > 249 \quad \text{OK}$$

Calculate the steel factored resistance. From Fig. 4.11.4:

$P_r = (6)(67) = 402$ kN

While it is desirable to have the steel factored resistance less than that of the concrete, the two values are close and a sudden connection failure is unlikely to occur since the applied loads were increased by 30%.

**Fig. 4.11.6 Welded headed stud example**

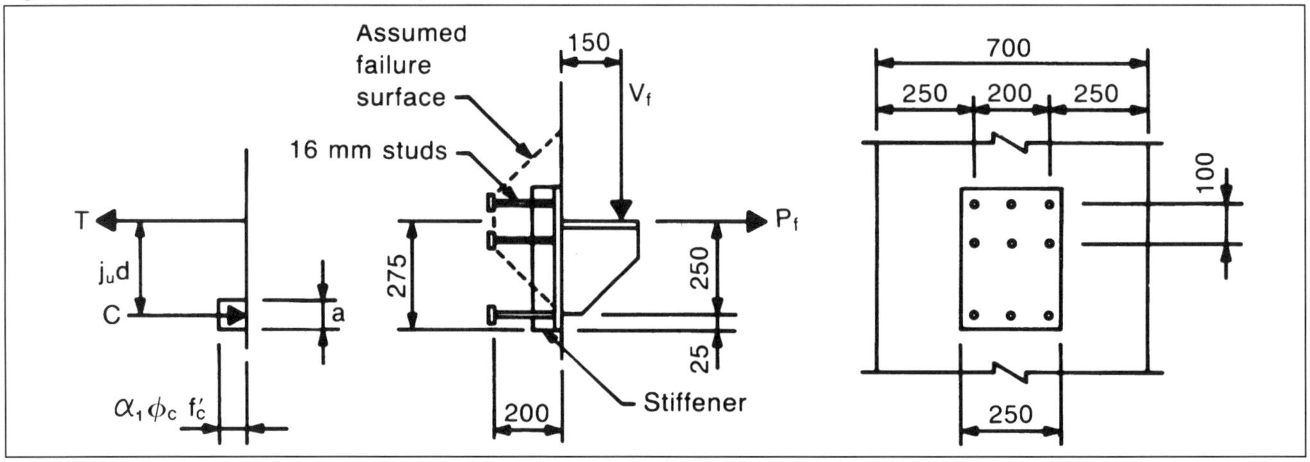

Check maximum allowable shear from Section 4.5:

Maximum $V_r = 7.0\ \phi_c\ A_{cv}$

$= (7.0)(0.65))(350)(250)/10^3$

$= 398$ kN > 123.5 OK

From Fig. 4.11.4, the minimum plate thickness is 11 mm. The plate and stiffener should also be checked to ensure that they will not yield due to the moment from the compressive force about the row of bottom studs.

## 4.12 DEFORMED BAR ANCHORS

Deformed bar anchors are available in standard diameters and lengths. They can be automatically welded to steel plates in a manner similar to headed studs. These bars are anchored in the concrete by bond and thus their required development length can be calculated in the same fashion as the development length of reinforcing bars.

## 4.13 STRUCTURAL STEEL HAUNCH

Structural steel shapes such as wide flange beams, double channels, tubes or vertical plates often serve as haunches or brackets as illustrated in Fig. 4.13.1. In lieu of a more precise analysis,[R4.13.1] when the axial load in the column does not exceed 75% of the axial resistance in the column, the capacity of these members can be calculated as follows, using the notation shown in Fig. 4.13.1:

$$V_c = \frac{\alpha_1 \phi_c f'_c b \ell_e}{1 + 3.6\, e/\ell_e}$$

where:

- $V_c$ = factored resistance of connection without additional reinforcement
- $e$ = eccentricity of resultant of vertical loads from center of embedment length
  $= a + \ell_e / 2$
- $\ell_e$ = embedment length
- $b$ = effective width of the compression block (see Fig. 4.13.2)
  $\leq b_{max}$
- $\phi_c$ = material resistance factor for concrete, 0.65

The effective width of the compression block is as shown in Fig. 4.13.2, provided that steps are taken to assure good compaction and confinement of the concrete around the section. Holes (more than 25 mm diameter) aid compaction, and additional column ties provide confinement.

It should be noted that there must be adequate concrete and/or superimposed axial dead load above and below the haunch to develop the compressive forces. If not, it may be possible to develop the force couple by using reinforcing bars in tension, in a manner similar to that shown below.

**Fig. 4.13.2 Effective width of embedded shapes**

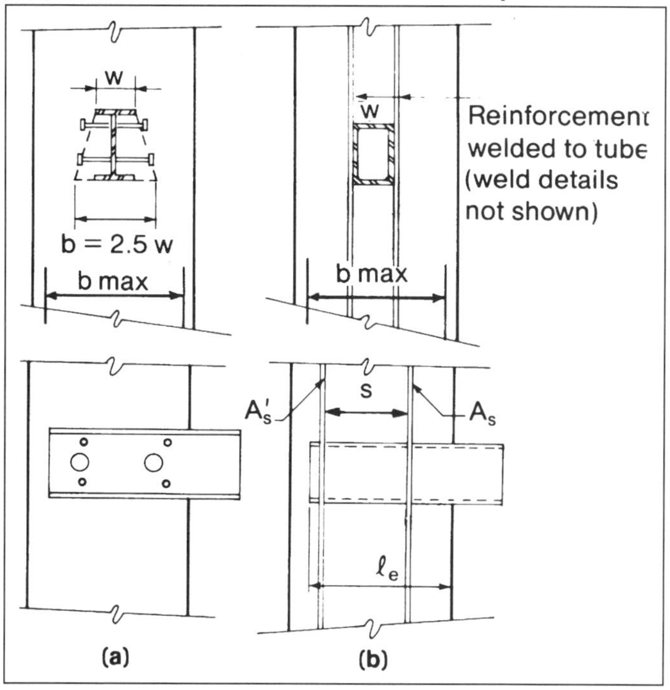


Additional capacity can be obtained by welding vertical reinforcing bars to the steel section, as shown in Fig. 4.13.2(b). If the welded reinforcement extends above and below the steel section, and if proper development lengths are provided for tension and compression, then the effective area of reinforcement can be taken as twice the area of reinforcement. In lieu of a more precise analysis, the additional capacity can be calculated as follows:

$$V_s = \frac{2 A_s \phi_s f_y}{1 + \left( \dfrac{6 e/\ell_e}{4.8 s/\ell_e - 1} \right)}$$

$$A'_s = A_s$$

where:

- $s$ = distance between $A_s$ and $A'_s$ when placed symmetrically about center of embedment length. (If reinforcement is not placed symmetrically, s = twice the distance from center of embedment to nearest reinforcement.)
- $V_s$ = additional factored resistance provided by welded reinforcement
- $A_s$ = effective area of welded reinforcement nearest to the applied load
- $A'_s$ = effective area of welded reinforcement near the end of the steel section
- $f_y$ = yield strength of the reinforcement
- $\phi_s$ = material resistance factor for reinforcing bars, 0.85

The factored resistance of the section is thus:

$$V_r = V_c + V_s$$

**Fig. 4.13.1 Embedded structural steel shape**

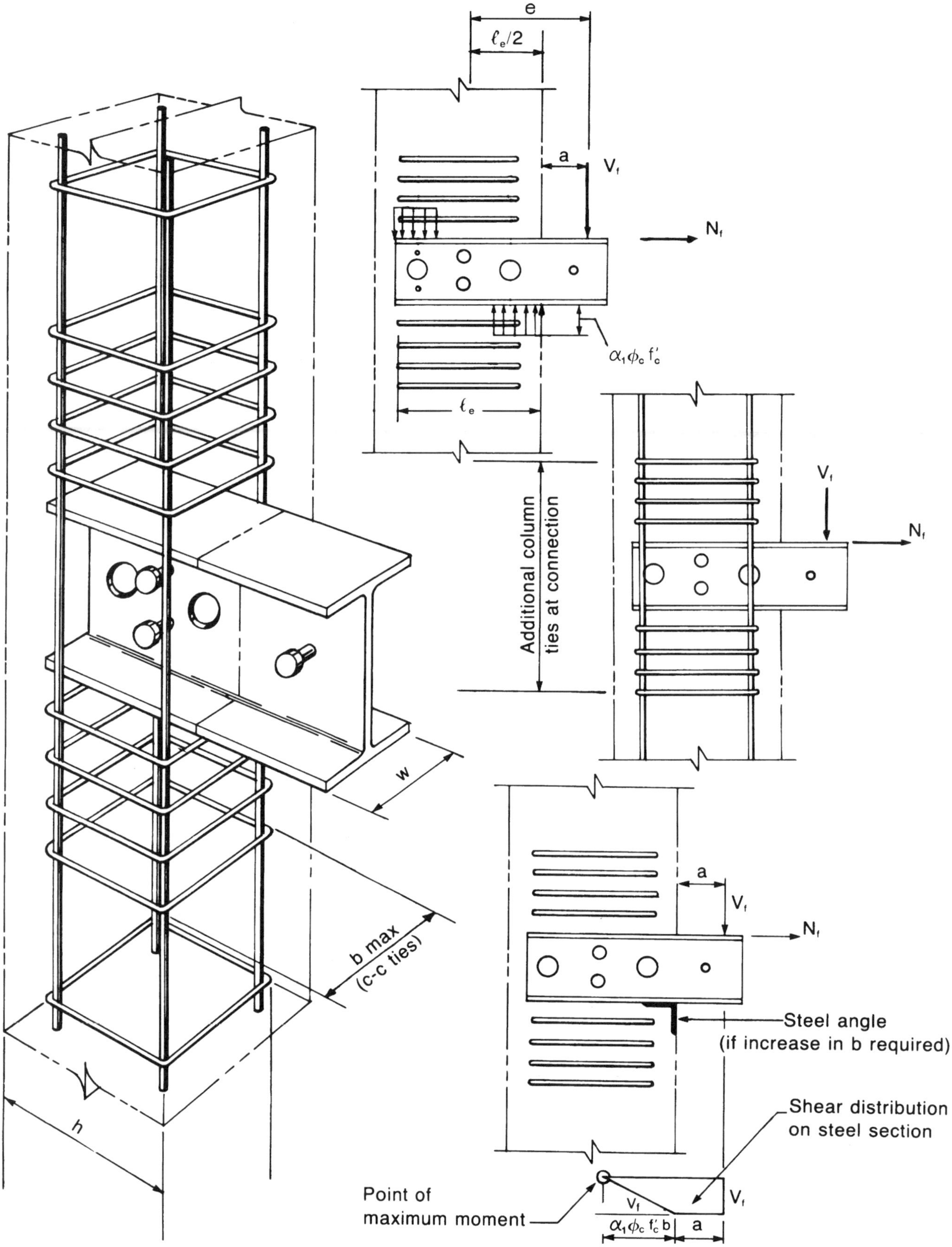

**Fig. 4.13.3 Plastic section moduli and shape factors***

| Section | Plastic section modulus, $Z_s$ | Shape factor |
|---|---|---|
| 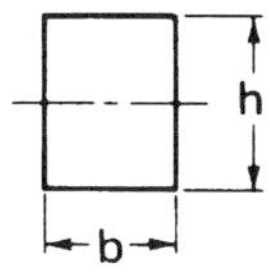  | $\frac{bh^2}{4}$ | 1.5 |
| 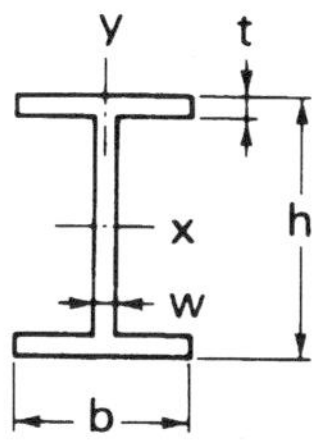  | x-x axis: $bt(h-t)+\frac{w}{4}(h-2t)^2$ | 1.12 (approx.) |
| | y-y axis: $\frac{b^2t}{2}+\frac{(h-2t)w^2}{4}$ | 1.55 (approx.) |
| 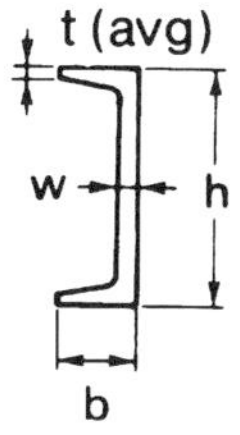  | $bt(h-t)+\frac{w}{4}(h-2t)^2$ | 1.12 (approx.) |
| 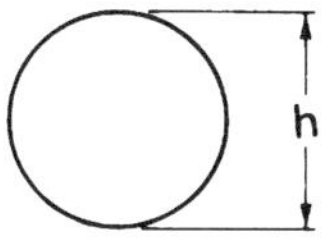  | $\frac{h^3}{6}$ | 1.70 |
| 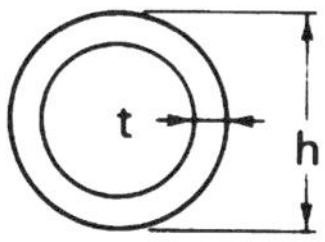  | $\frac{h^3}{6}\left[1-\left(1-\frac{2t}{h}\right)^3\right]$ <br> $t\,h^2$ for $t << h$ | $\frac{16}{3\pi}\left[\frac{1-\left(1-\frac{2t}{h}\right)^3}{1-\left(1-\frac{2t}{h}\right)^4}\right]$ <br> 1.27 for $t << h$ |
| 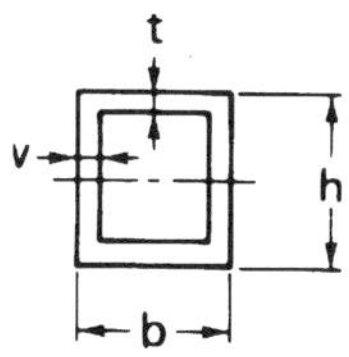  | $\frac{bh^2}{4}\left[1-\left(1-\frac{2w}{b}\right)\left(1-\frac{2t}{h}\right)^2\right]$ | 1.12 (approx.) for thin walls |
| 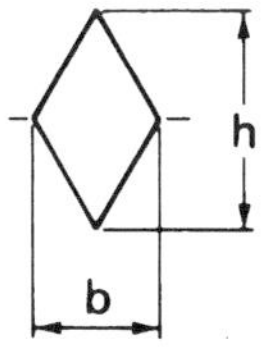  | $\frac{bh^2}{12}$ | 2 |

*For additional information on structural steel sections, see "Handbook of Steel Construction," by the Canadian Institute of Steel Construction.

The factored resistance of the steel section can be determined by:

Flexural resistance:

$$M_r = \phi_a Z_s f_y \geq M_f = V_f a + \frac{0.5V_f^2}{(\alpha_1 \phi_c f'_c\, b)}$$

Shear resistance:

$$V_r = \phi_a 0.66 f_y h t \geq V_f$$

where:

$Z_s$ = plastic section modulus of the steel section (see Fig. 4.13.3)

$f_y$ = yield strength of the steel section

h, t = depth and thickness of steel web, respectively

$\phi_a$ = 0.90

It is recommended that hollow structural steel sections be filled with concrete in order to improve the bearing condition.

For steel shapes projecting equally from each side of the element, with approximately symmetrical loading, the factored shear resistance on each side as governed by the capacity of the concrete can be calculated by:

$$V_r = \frac{\alpha_1 \phi_c f'_c b h}{2}$$

Horizontal forces, $N_f$, are resisted by bond on the perimeter of the section. If the bond stress resulting from factored loads exceeds 1.7 MPa, headed studs or reinforcing bars can be welded to the section.

**Example - Design of structural steel haunch**

*Given:*

The structural steel haunch shown.

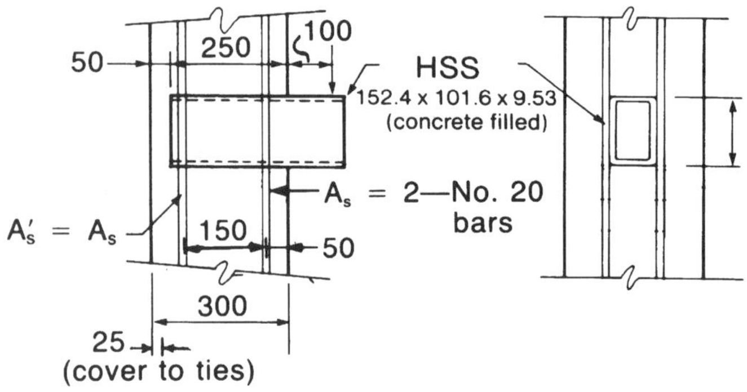


e = 100 + 250 / 2 = 225

b = 2.5 w = 254

$b_{max}$ = 300 – (2)(25) = 250

s = 150

Effective $A_s = 2\,A_s = (2)(600) = 1200$ mm$^2$

$f'_c$ = 35 MPa

$f_y$(reinforcement) = 400 MPa (weldable)

$f_y$(structural steel) = 350 MPa

*Problem:*

Find the factored resistance of the connection.

*Solution:*

$$V_c = \frac{\alpha_1 \phi_c f'_c b \ell_e}{1 + 3.6\, e/\ell_e}$$

$$= \frac{(0.80)(0.65)(35)(250)(250)/10^3}{1 + (3.6)(225/250)}$$

$$= 268 \text{ kN}$$

$$V_s = \frac{2\phi_s A_s f_y}{1 + \dfrac{6e/\ell_e}{4.8s/\ell_e - 1}}$$

$$= \frac{(2)(1200)(0.85)(400)(10^{-3})}{1 + \dfrac{(6)(225/250)}{(4.8)(150/250) - 1}}$$

$$= 211 \text{kN}$$

$$V_r = 268 + 211 = 479 \text{ kN}$$

Plastic section modulus of HSS[R4.13.2]:

$$Z_a = 205 \times 10^3 \text{ mm}^3$$

Flexural resistance of HSS (neglect concrete fill):

$$M_r = \phi_a Z_s f_y = (0.9)(205 \times 10^3)(350)/10^6$$

$$= 64.6 \text{ kN} \cdot \text{m}$$

$$M_r = V_r a + \frac{0.5V_r^2}{\alpha_1 \phi_c f'_c b}$$

$$\therefore V_r = 438 \text{ kN} < 479 \text{kN}$$

Shear resistance of HSS:

$$V_r = \phi_a 0.66 f_y h t$$

$$= (0.9)(0.66)(350)(152.4)(2)(9.53)/10^6$$

$$= 604 \text{ kN} > 438 \text{kN}$$

Therefore, flexural resistance of the HSS controls and $V_r$ = 443 kN.

## 4.14 CONNECTION ANGLES

Angles used to support light precast elements can be designed by statics as shown in Fig. 4.14.1.

In addition to the applied vertical and horizontal loads, the design should include all loads induced by restraint of relative movement between the precast element and the supporting element. The minimum thickness of non-gusseted angles loaded in shear as shown in Fig. 4.14.2 can be determined by:

**Fig. 4.14.1 Design relationships for connection angles**

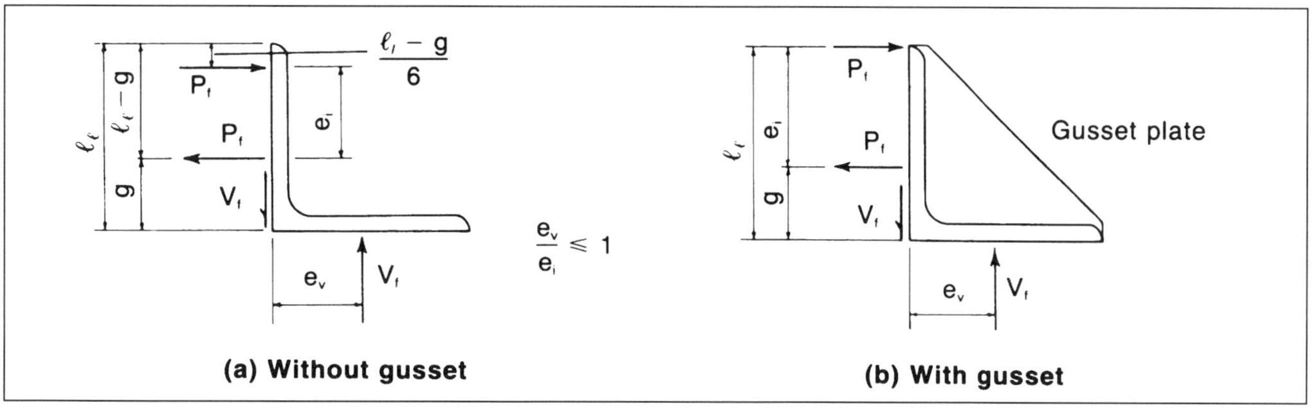


**Fig. 4.14.2 Vertical loads on connection angle**

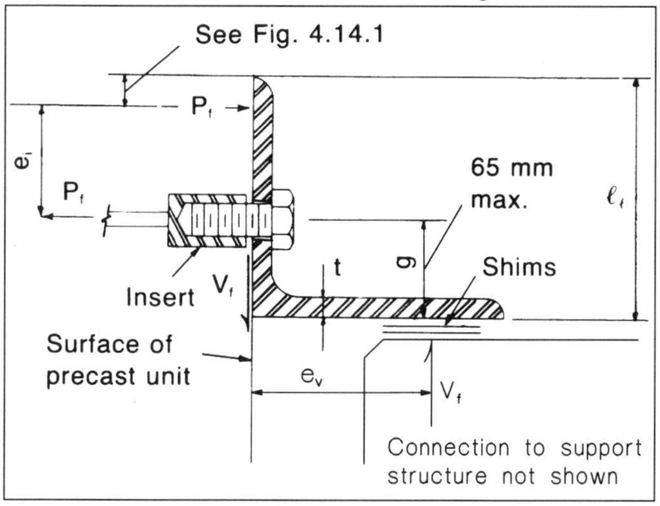


**Fig. 4.14.3 Horizontal loads on connection angle**

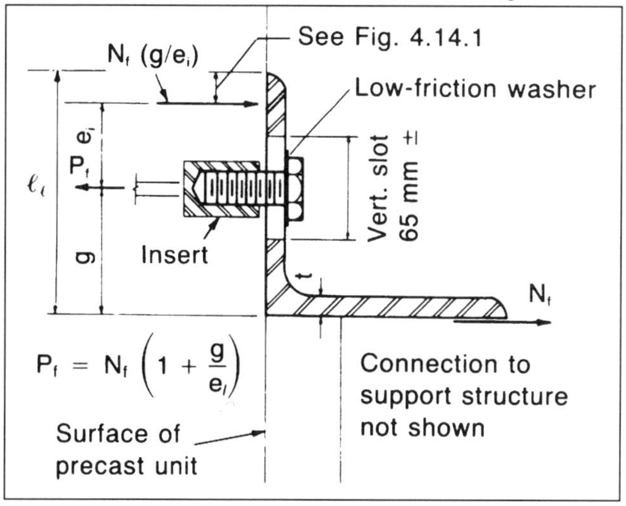


$$t = \sqrt{\frac{4V_f e_v}{\phi_a f_y b}}$$

where:

$\phi_a$ = 0.9

b = width of angle

Design $e_v$ = specified $e_v$ + 20 mm

The tension on the bolt can be calculated by:

$$P_f = V_f \frac{e_v}{e_i}$$

For angles loaded axially (Fig. 4.14.3) either in tension or compression, the minimum thickness of non-gusseted angles can be calculated by:

$$t = \sqrt{\frac{4N_f g}{\phi_a f_y b}}$$

where:

$\phi_a$ = 0.9

g = gauge of angle (see Fig. 4.14.3)

b = width of angle

## 4.15 COLUMN BASE PLATES

Column bases must be designed for both erection loads and loads which occur in service, the former often being critical. Two commonly used base plate details are shown in Fig. 4.15.1, although other details are also frequently used.

If in the analysis for erection loads or temporary construction loads before grout is placed under the plate, all anchor bolts are in compression, the base plate thickness required to satisfy bending is determined from:

$$t = \sqrt{\frac{(\Sigma F)4x_c}{\phi_a f_y b}}$$

where:

$\phi_a$ = 0.9

$x_c$, b from Fig. 4.15.1.

$f_y$ = yield strength of the base plate

$\Sigma F$ = greatest sum of anchor bolt factored forces on one side of the column

If the analysis indicates that the anchor bolts on one or both of the column faces are in tension, the base plate thickness is determined by:

**Fig. 4.15.1 Column base connections**

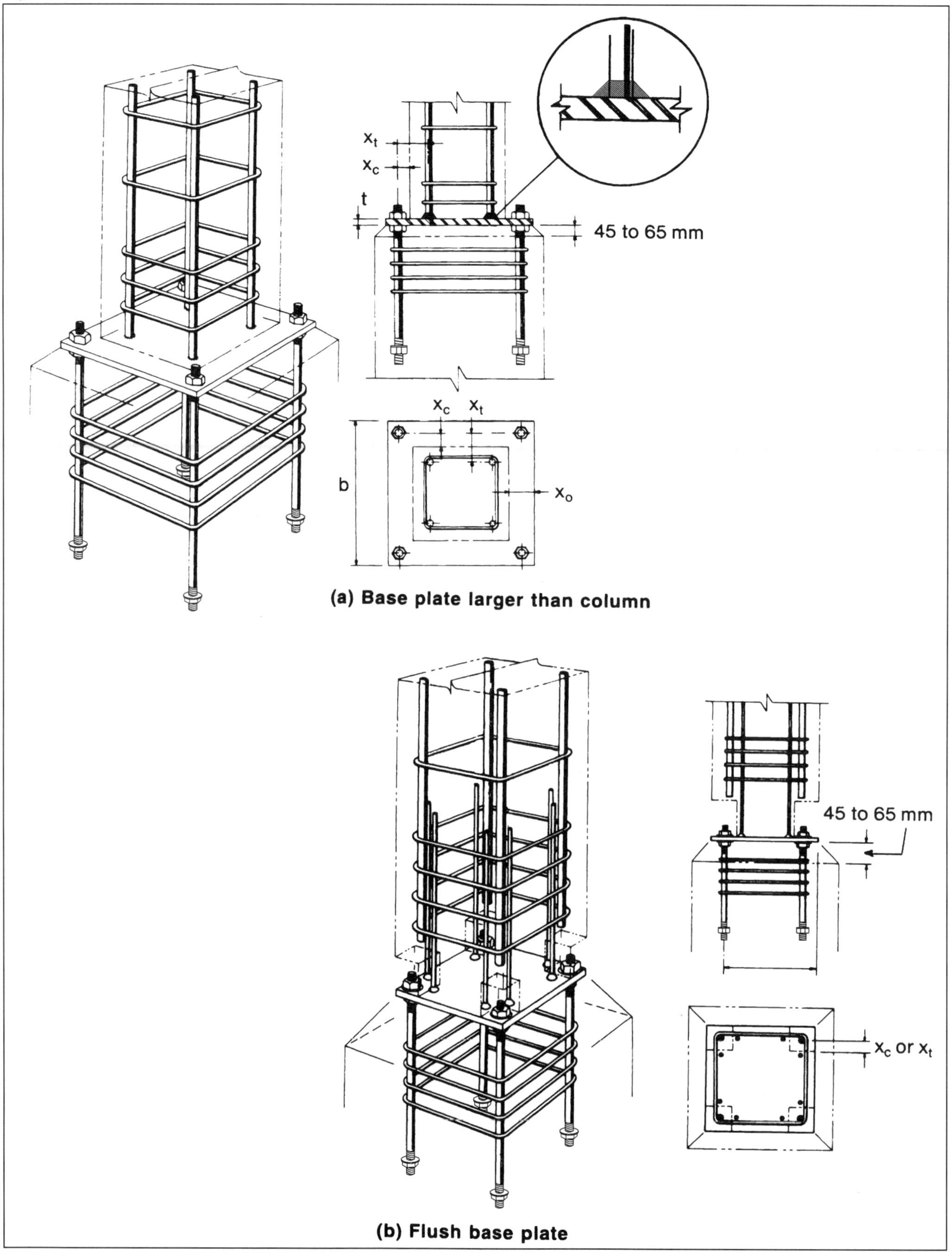

$$t = \sqrt{\frac{(\Sigma F)4x_t}{\phi_a f_y b}}$$

where:

$x_t$ from Fig. 4.15.1.

Base plate thickness may be controlled by bearing on the concrete or grout. In this case, the base plate thickness is determined by:

$$t = x_o \sqrt{\frac{2\phi_c \alpha_1 f_c'}{\phi_a f_y}}$$

in which $x_o$ is as shown in Fig. 4.15.1.

The factored shear resistance of a column base plate can be determined from:

$$V_r = \phi_a 0.66 f_y bt$$

where:

$\phi_a$ = 0.9

The anchor bolt diameter is determined as shown in Sect. 4.19. Anchor bolts may be ASTM A307 bolts or, more frequently, threaded rods of Grade 260 steel.

When the bolts are near a free edge, as in a pier or wall, the buckling of the bolts before grouting may be a consideration. Confinement reinforcement, as shown in Fig. 4.15.1 should be provided in such cases. A minimum of 4 - No. 10 ties at about 75 mm centers is recommended for confinement.

The strength of the concrete when a bolt is in tension may be critical and can be determined by assuming a shear cone pull-out failure as described for headed studs.

The length of the anchor bolts should be such that the concrete will develop the desired resistance of the bolt in bond and bearing on the hook projection or bolt head. The bearing area of bolt heads can be increased by welding a washer or steel plate to the bolt head. The bottom of the bolt should be a minimum of 100 mm above the bottom of a footing, and above the footing reinforcement.

The tensile resistance of smooth anchor bolts, limited by the bond strength to the concrete, is determined by:

$$P_r = 1.7\pi\phi_c d_b \ell_e$$

where:

$\phi_c$ = 0.65

$d_b$ = diameter of anchor bolt

$\ell_e$ = embedment length

The tensile resistance limited by the bearing strength of concrete under a bolt head or hook is:

$$P_r = \phi_c f_c' A$$

where:

$\phi_c$ = 0.65

A = net area under head or hook

Compression on anchor bolts during erection can be substantially reduced by the use of steel shims. The required area of the shims can be determined by the bearing resistance of the concrete.

## 4.16 WELDING OF REINFORCING BARS

Welding of reinforcement is a practical method of developing the force transfer in many connections, provided the weldability of the bars can be assured. Typical reinforcing bar welds are shown in Fig. 4.16.1.

CSA W186-M1990 primarily applies to the welding of reinforcing bars:

1. To each other.
2. Through splice members.
3. To structural steel members used as anchorages.

As weldability of a bar is a function of its carbon equivalent, it should be noted that unless stringent quality control procedures are present in a shop, the carbon equivalents should be limited to 0.45% for No. 25 and larger bars, and 0.55% for No. 20 and smaller bars.

Carbon equivalent (C.E.) is defined by the following formula:

$$C.E. = \%C + \frac{\%Mn}{6} + \frac{\%Ni}{20} + \frac{\%Cr}{10}$$

If mill reports are not available and if the chemical composition of the bars is also not available, then the carbon equivalent should be assumed to be above 0.75%.

Common considerations in the design and detailing of welded bar connections are that:

1. Welding should not be done on or within two bar diameters of any bent portion of a bar.
2. When welding bars to structural shapes that are embedded in concrete, allowance should be made for the thermal expansion of steel to avoid concrete spalling or cracking.

When item (2) is a concern, adequate confinement reinforcement should be provided in the immediate area on a compressible material placed around the steel plate to allow for expansion.

For further background information on the design and detailing of welded bar connections, reference should be made to CSA W186. Common welds used with reinforcing bars are shown in Fig. 4.16.1.

## 4.17 MOMENT CONNECTIONS

Moment connections are sometimes required in building frames, as discussed in Sect. 2.6. Moment resisting capacity is attained by detailing the connection so that a force couple can be developed. Tensile capacity through the connection can be attained by a variety of load transfer devices. However, particular attention should be directed to conditions which may exist that position the load transfer hardware near the "free edge" of a precast element. Discussion of this consideration can be found in Sect. 4.11.

**Fig. 4.16.1 Typical reinforcing bar welds**

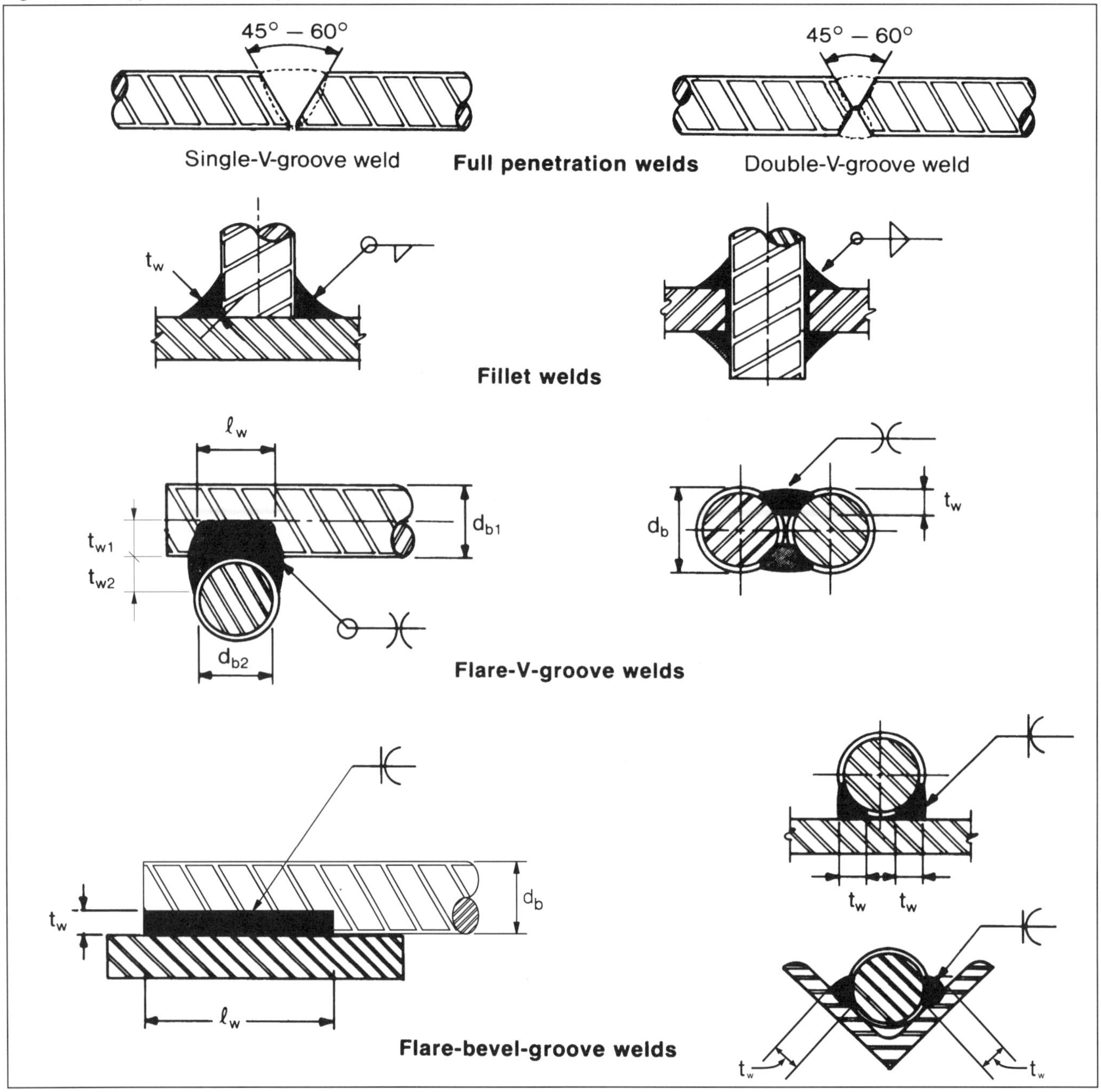


It should be noted that in designing moment resistant connections, restraint to volume changes must be considered. The designer should take particular care in specifying or understanding the erection sequence, as to whether the moment connection works only after dead loads have been applied. Shoring of slab members during pours of composite topping and joint topping reinforcement can substantially increase the moment connection capability.

## 4.18 DOWEL CONNECTIONS

Dowel connections to the support structure or to other precast elements can provide an efficient solution for certain connections. This is particularly applicable to the connecting of hollow-core wall panels to foundations of shear wall buildings. The strength of the dowels in the connection depends entirely on the embedment length and the developed bond. In the case of moment resisting frame connections, the embedment length requirements of a larger bar may prohibit such a detail from being economically achieved. It also must be realized that any adjustment after the initial set of the grout may destroy the dowel bond and reduce the connection strength accordingly.

Where continuous bearing is required, elements may be set on shims and "dry-packed" at a later date or the elements may be set on shims, dammed where appropriate and grouted with an expanding cement based material either by gravity flow or induced pressure.

### Cement Grouted Dowels

Reinforcing bars may be embedded into a dowel hole and grouted to provide the necessary connection. For compression connections, the required embedment length can be taken as the development length for the particular bar diameter (refer to Fig. 8.2.11).

However, in dowel connections subjected to tensile loads, the bond between the concrete of the element and the grout is unreliable if there is any shrinkage. Experience and tests have shown that these dowel holes should be provided using a flexible conduit within the form, as shown in Fig. 4.18.1.

**Fig. 4.18.1 Grouted tube connection**

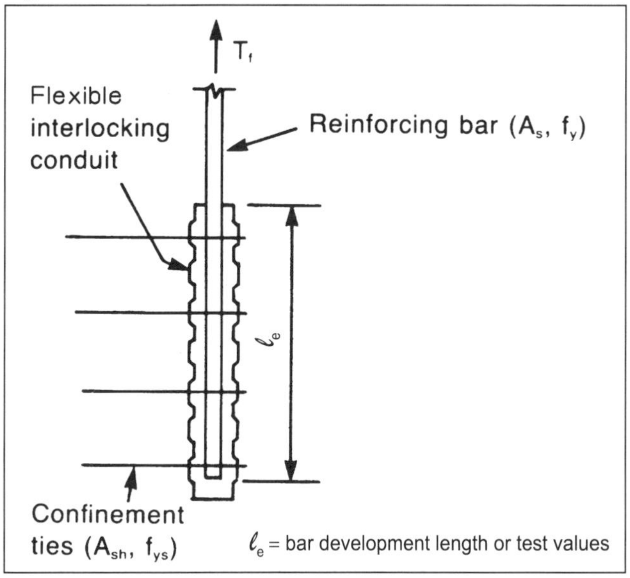


Recommended requirements to complete such a connection are as follows:

1. The minimum concrete cover to the metallic conduit should be 75 mm.
2. The conduit should have a minimum clear, internal diameter of three bar diameters. (A review of fabrication and construction tolerances may dictate an even larger opening.)
3. The grout material strength specified should parallel the choice of development length(s).
4. Confinement reinforcement consisting of a spiral or ties may be required to prevent splitting or bond failure between the conduit and the surrounding concrete, particularly at the ends of the beams in beam-column dowel connections where significant axial loads may be present.

Dowel holes, when smooth formed and approximately three times the dowel diameter, exhibit very good bond to the smooth concrete of the hole when nonshrink grout is used. Typically under failure conditions, the rebar dowel will debond from the grout before the grout plug will debond from the smooth hole. An added benefit to the grout strength is realized because the excess water of a sand cement grout is absorbed into surrounding concrete. An added benefit of a smooth hole is that the conduit thickness is eliminated, permitting a more compact connection.

The grout to concrete bond need only be 1/3 as great as the grout to rebar bond when the hole is three times as large.

### Epoxy Grouted Dowels

Epoxy adhesives can be used for anchoring dowels, bolts, bars, or similar shapes into concrete. Though the physical properties of epoxy compounds vary widely, their tensile strength is generally much larger than that of cement grout (users should refer to manufacturer's specifications for specific material properties). Designers and users should note that the thermal expansion coefficients of some epoxy compounds can be seven times that of cement grout/concrete.

Recommended requirements to complete such connections should parallel those of cement grouted dowels.

## 4.19 BOLTS AND THREADED CONNECTORS

Various types of bolts and other threaded connectors are used in connections of precast concrete elements. The advantages of these devices are rapid assembly and to avoid the attraction of volume change forces in applicable elements. When using this form of connection, attention should be directed to tolerance detailing to ensure the completion of the connection without field modification of loose hardware pieces. Whenever possible, provide over-size and/or slotted openings in hardware to be fastened to accommodate fabrication and construction tolerances.

In a large majority of connections, bolts are provided loose to the site, and installed into receptacles cast into concrete. If required, a precast element can be cast with a threaded connector projecting from the face. However, this may be undesirable as vulnerability during handling and possible form setting constraints may damage this type of connection.

The generic types of threaded connectors that are most commonly used include:

1. Standard bolts.
2. High strength bolts.
3. Threaded steel rods.
4. Coil bolts and coil rods.

Designers should attempt to standardize all loose threaded connectors of the same dimensions, to a single grade on a project-to-project basis to prevent accidental substitution during construction.

### Standard Bolts

Standard bolts as defined here are those satisfying ASTM A307.

**Fig. 4.19.1 Factored shear and tensile resistances (kN / bolt), $\phi_a = 0.90$**

| Bolt Size | | Nominal Area | Factored Shear Resistance [2] - Single Shear [3] (kN/bolt) | | | | | | Factored Tensile Resistance, $T_r$ (kN/bolt) | | |
|---|---|---|---|---|---|---|---|---|---|---|---|
| | | | Threads Excluded | | | Threads Intercepted [4] | | | | | |
| Metric [1] | Imperial | $A_b$ ($mm^2$) | A325 A325M | A490 A490M | A307 | A325 A325M | A490 A490M | A307 | A325 A325M | A490 A490M | A307 |
| | 1/2 | 127 | 42.2 | 52.8 | | 29.5 | 37.0 | | 52.7 | 66.0 | |
| | 5/8 | 198 | 65.7 | 82.4 | 32.9 | 45.9 | 57.6 | 23.2 | 82.2 | 103 | 41.2 |
| **M16** | | **201** | **67.1** | **84** | | **47.3** | **58.9** | | **83.8** | **105** | |
| | 3/4 | 285 | 94.6 | 119 | 47.3 | 66.1 | 82.9 | 33.4 | 118 | 148 | 59.3 |
| **M20** | | **314** | **105** | **131** | | **73.5** | **92.0** | | **131** | **164** | |
| **M22** | | **380** | **127** | **159** | | **88.9** | **111** | | **158** | **199** | |
| | 7/8 | 388 | 129 | 161 | 64.4 | 90.0 | 113 | 45.4 | 161 | 202 | 80.7 |
| **M24** | | **452** | **151** | **189** | | **106** | **132** | | **188** | **238** | |
| | 1 | 507 | 168 | 211 | | 118 | 148 | | 210 | 264 | |
| **M27** | | **573** | **191** | **240** | | **134** | **168** | | **239** | **300** | |
| | 1 1/8 | 641 | 187 | 267 | | 131 | 187 | | 233 | 333 | |
| **M30** | | **707** | **236** | **296** | | **165** | **207** | | **295** | **370** | |
| | 1 1/4 | 792 | 230 | 329 | | 162 | 230 | | 288 | 412 | |
| **M36** | | **1018** | **340** | **426** | | **238** | **298** | | **425** | **532** | |
| | 1 1/2 | 1140 | 332 | 474 | | 233 | 322 | | 415 | 593 | |

1 The number following the letter M is the nominal bolt diameter in millimeters
2 For joints longer than 1300 mm, use 80% of the factored shear resistance
3 For double shear, multiply tabulated values by 2.0
4 Threads are intercepted if thin material next to the nut is combined with detailing for minimum bolt stick-through (the nut)

Design of standard bolts is in accordance with the CISC Handbook of Steel Construction and applicable CSA Standards. Refer to Fig. 4.19.1 for factored resistances of bolts.

### High Strength Bolts

High strength bolts, identified as ASTM A325 and ASTM A490, were developed for friction type connections between structural steel elements. They have more than twice the tensile resistance of A307 bolts. Their application requires controlled tensioning of the fastener to develop sufficient force to prevent slipping of the connected parts. To take full advantage of high strength bolts in friction-type connections, they must be tightened in accordance with CAN/CSA-S16.1-M94.

As a result of creep and minor crushing, it is questionable whether a high-strength bolt will hold its tension when tightened against concrete. Some applications and limitations are shown in Fig. 4.19.2.

### Threaded Steel Rods

Steel rods of standard sizes and grades (both mild steel and high strength) are sometimes used in precast concrete connections. The most common applications are column base anchor bolts and beam-column connections.

### Coil Bolts and Rods

Though these threaded connectors (Fig. 4.19.3) are commonly available through various manufacturers, they are not covered by standard specifications and should not be used in primary structural connections. Their use should be restricted to temporary connections such as for precast handling or installation.

**Fig. 4.19.2 Application of high-strength bolts**

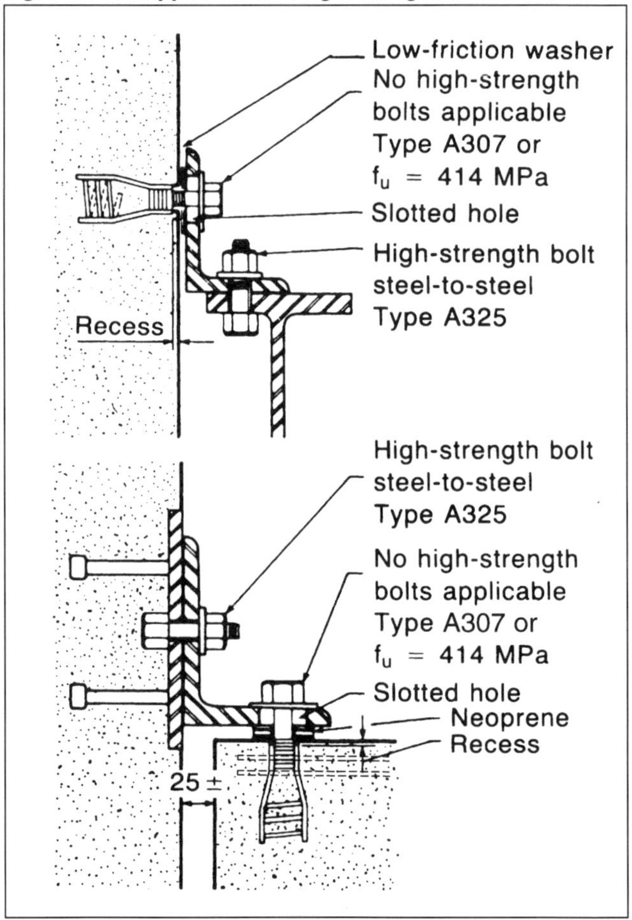

**Fig. 4.19.3 Coil bolts and coil rods**

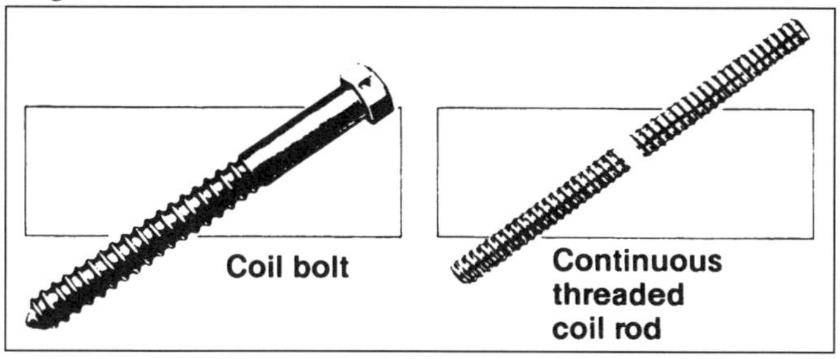


## 4.20 CONNECTIONS FOR WALL PANELS

The multitude of sizes and shapes of precast, non-load bearing panels has led to a variety of connections, nearly always custom-designed for each project, and defying any classification of standard connections.

Support systems resulting in a statically determinate force system are recommended. Indeterminate systems may cause supplementary forces of a magnitude and direction which are difficult to predict at the time of design.

Sound design procedure dictates that the mass of the entire panel be supported entirely at one level, by two connections. Lateral alignment connections, as required, should also be located near the bottom of the element. Non load bearing panels may be suspended at the top and tied with lateral connections at the bottom. Example connection details are shown in Figs. 4.20.1 to 4.20.5.

**Fig. 4.20.1 Direct seat connections**

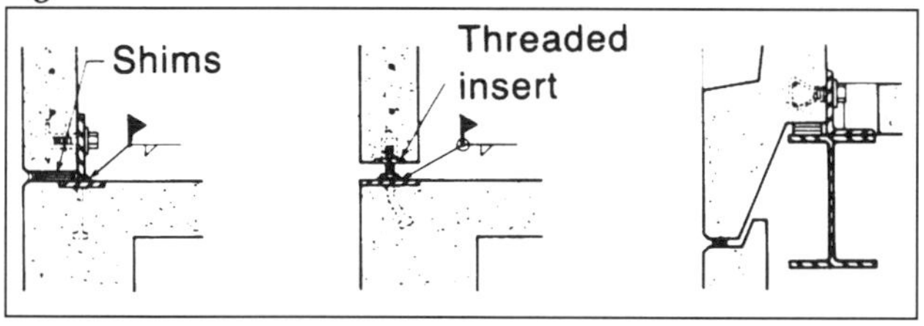


**Fig. 4.20.2 Angle seat connections**

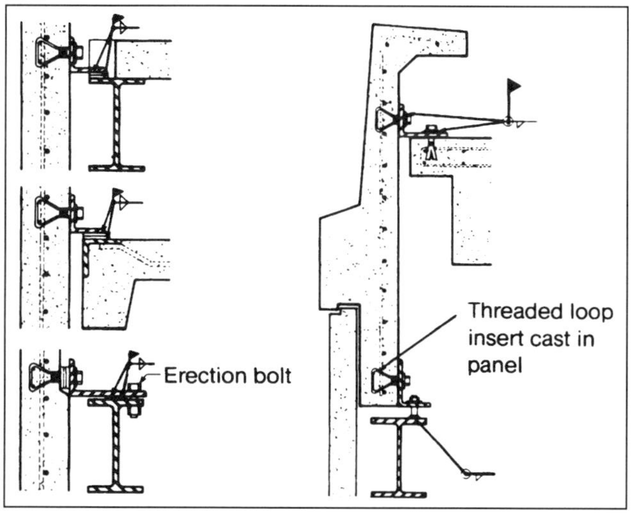


**Fig. 4.20.3 Corbel seat connections**

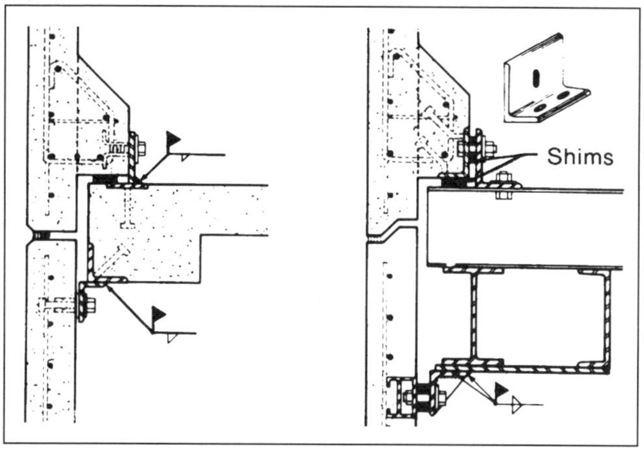


**Fig. 4.20.4 Cast-in seat connections**

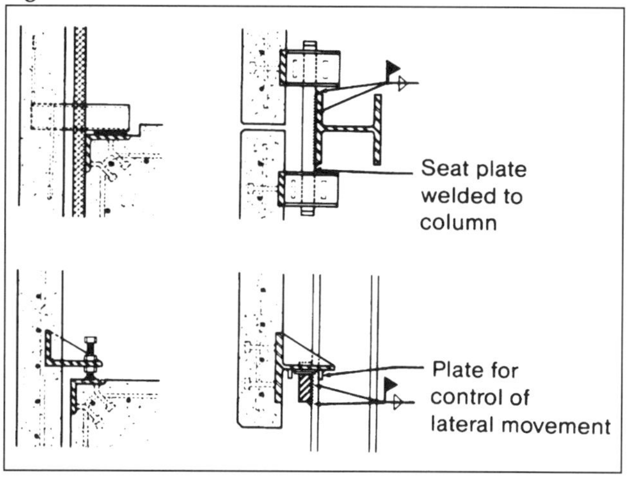


**Fig. 4.20.5 Tie-back connections**

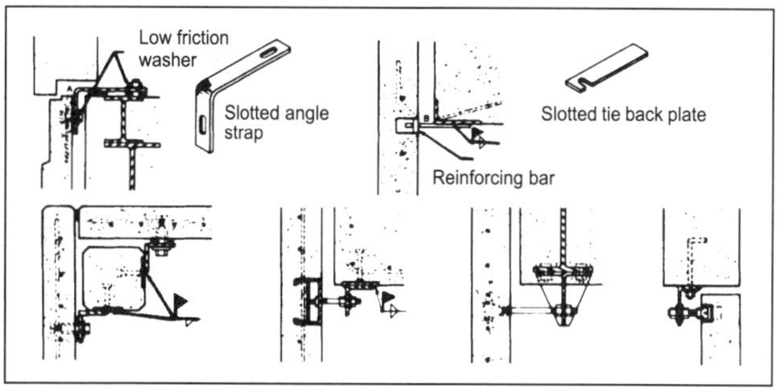


This section summarizes the connection requirements for wall panels as typically used in apartment and hotel type structures. General connection details that would be common to other precast concrete systems, such as beam and girder seats, are not covered here, and the reader is directed to other chapters for a more complete discussion.

### 4.20.1 REQUIREMENTS OF CONNECTIONS

Connections for load bearing wall panels are subject to heavier and more varied loading than for non-load bearing wall panels. These form an essential part of the structural support system, and stability of the building often depends on them.

Fig. 4.20.6 Exterior forces and joint force systems

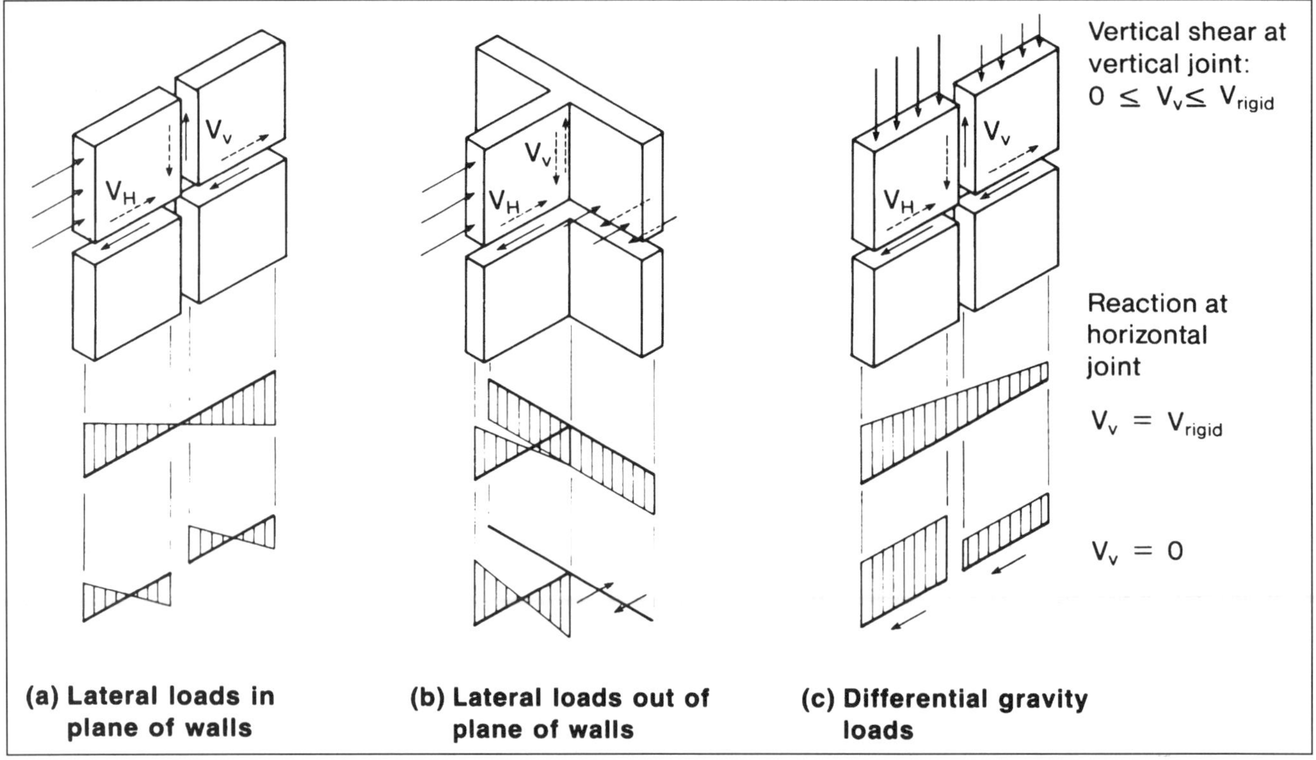


The panels are required to resist or transfer loads from one or more of the following:

1. Panel self weight.
2. Roof and floor loadings.
3. Lateral wind or seismic loads.
4. Effects of volume changes (shrinkage, creep, temperature).
5. Abnormal loads.[R4.20.1]

Erected load bearing walls may have both horizontal and vertical joints across which forces must be transferred. Fig. 4.20.6 indicates the principal exterior forces and the resulting joint force system for three separate cases. In buildings, superposition of forces and various combinations of panel and joint assemblies must be considered.

Fig. 4.20.7 Lateral force transfer in shear walls

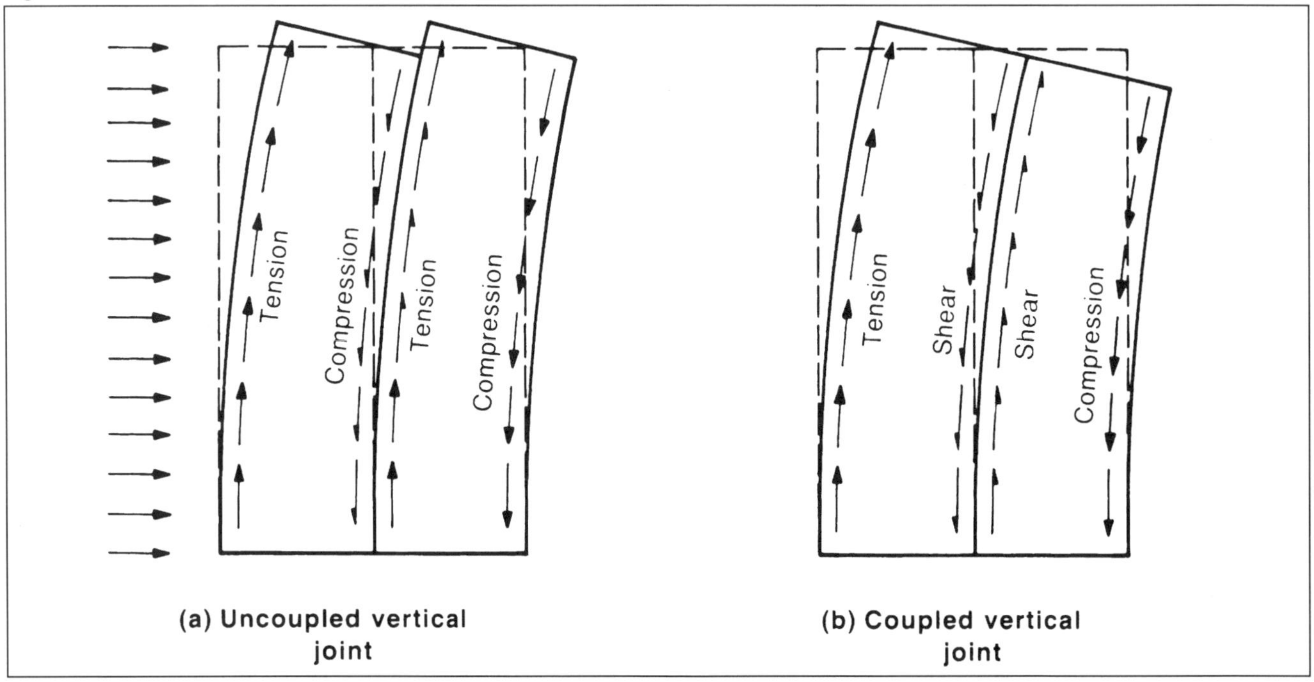

Distribution of lateral forces to shear walls depends largely on the connection of the floors to the walls. The walls carry these lateral forces by shear in combination with vertical tension and compression. Adequate connections are required in the wall panels to transfer these forces down to the foundations.

## 4.20.2 VERTICAL JOINTS

To assure lateral force transfer between adjacent wall panels, the vertical joint must be able to transfer direct tension and compression in combination with vertical shear. Fig. 4.20.7 illustrates forces at vertical joints for coupled and uncoupled wall panels. Coupled panels are preferred since they result in a more uniform distribution of forces, which is of particular importance for tall structures in high seismic areas.

There are several types of connections that may be used to connect panels at vertical joints. Mechanical fasteners are by far the most popular type of joint connection used in precast bearing wall structures and consist of cast-in steel sections or inserts at the panel joint edges that are fastened together by bolting or welding (see Fig. 4.20.8). These types of connections have low to moderate shear resistance, but are considerably less expensive to install than keyed connections. They are preferred for exposed walls where they may be hidden in grouted recesses.

**Fig. 4.20.8 Panel to panel shear connection**

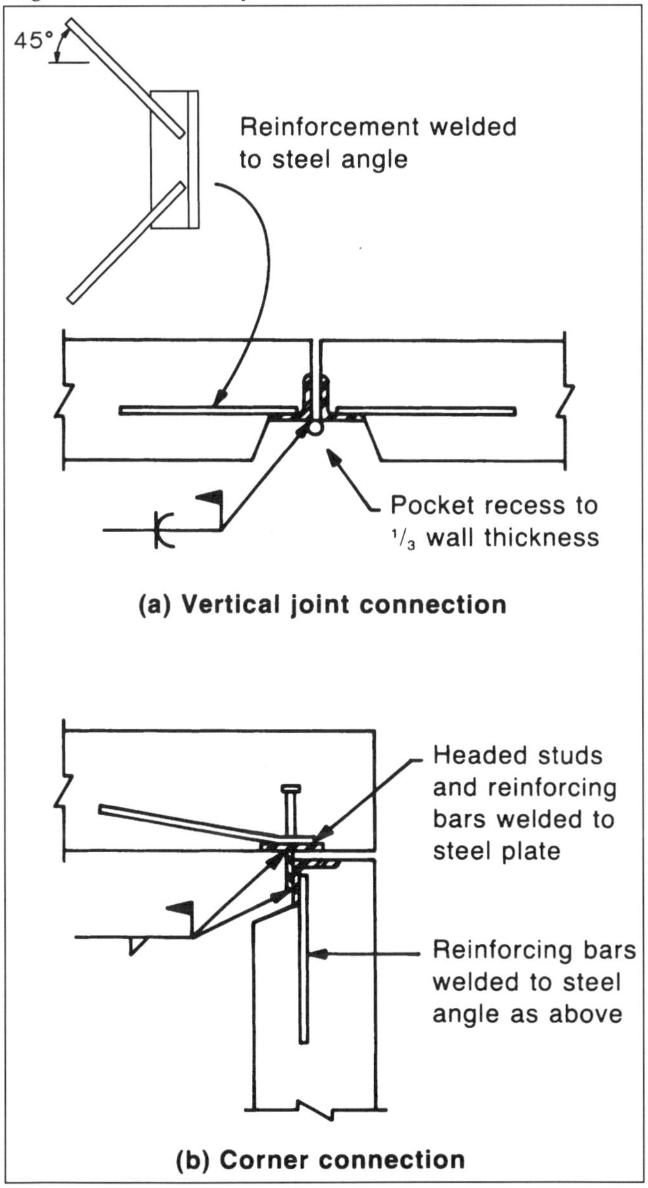


**(a) Vertical joint connection**

**(b) Corner connection**

The strength of mechanical connections is often controlled by the shear capacity of the cast-in anchor. Studded anchors exhibit reasonable static resistance but may perform poorly under cyclic loading. Deformed bar anchors (weldable steel grade) have superior strength and deformation characteristics, particularly when used in direct tension or compression with adequate embedment length.

Coil inserts and drilled-in wedge anchors are generally not suitable for vertical joint connections because of low strength and ductility properties. They should be used only to provide nominal connections for prevention of thermal warping or for other low stress effects.

The grooved connection of Fig. 4.20.9 consists of a continuous void at the vertical joint of adjoining panels with lapping horizontal steel and a vertical tie bar. A high strength, non-shrink grout is installed after erection. It is used primarily for shear transfer and can be evaluated with the shear-friction concept (Sect. 4.5) even if shrinkage, creep or temperature movements have created a crack at the wall-grout interface. The grooved connection is simple and relatively low in cost, and is often employed where mechanical connections cannot provide sufficient strength.

Keyed joints (Fig. 4.20.10) have been used where large shear forces must be transferred between wall panel units. Reinforcement increases strength, and is essential for ductility in seismic zones. The keyed joint can be expensive to install and is often avoided where other methods provide adequate resistance.

**Fig. 4.20.9 Grooved joint connections**

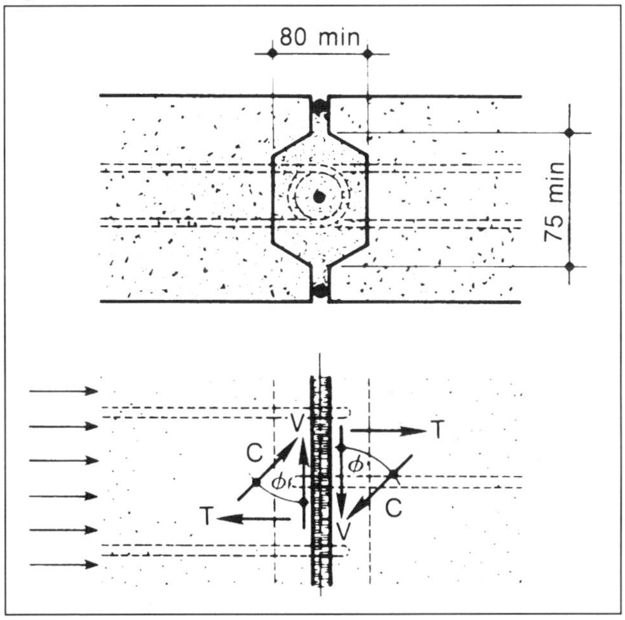

**Fig. 4.20.10 Keyed joint connection**

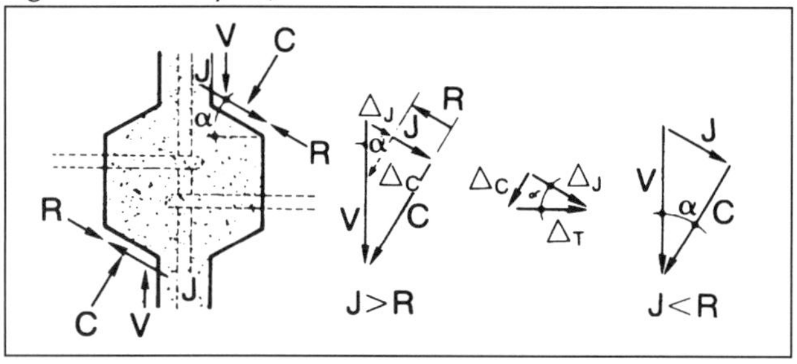


**Fig. 4.20.11 Forces in keyed joint**

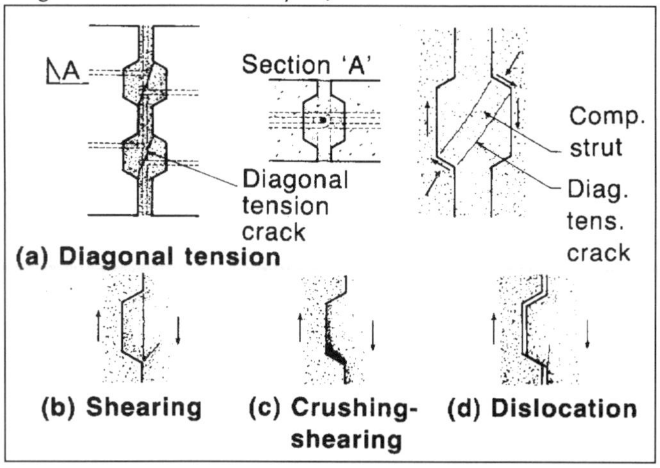


As shown in Fig. 4.20.11, the maximum shear resistance can be controlled by either:

1. Shearing of grout concrete parallel to joint.
2. Crushing of concrete or grout at key edges.
3. Slippage along contact area.

The shear friction concept would apply for reinforced keyed joints only. In this case, the key geometry should be such that reinforcement yield and joint slippage occur well before concrete crushing at the key edges. Confining ties can be effective in increasing keyed joint strength, but may be difficult to install properly. For unreinforced joints, the key shear strength often controls the design.

## 4.20.3 HORIZONTAL JOINTS

Horizontal panel joints usually occur at each floor level and at the transition to foundations or transfer beams. Connections at these joints will often perform multiple functions including support of gravity loads from floor elements, transfer of horizontal shear, and tying of tension/compression forces for wall panel overturning resistance.

The typical floor to wall panel arrangement in high-rise precast construction is illustrated in Fig. 4.20.12. Hollow-core floor slabs are supported directly on the lower wall panel unit with an end gap of 50 to 70 mm. The gap and a portion of the core is filled with a flowable grout. The upper wall unit is then placed on shim pads and supported by a continuous bed of dry pack grout.

**Fig. 4.20.12 Cross sections of interior and exterior platform type wall-to-floor connections**

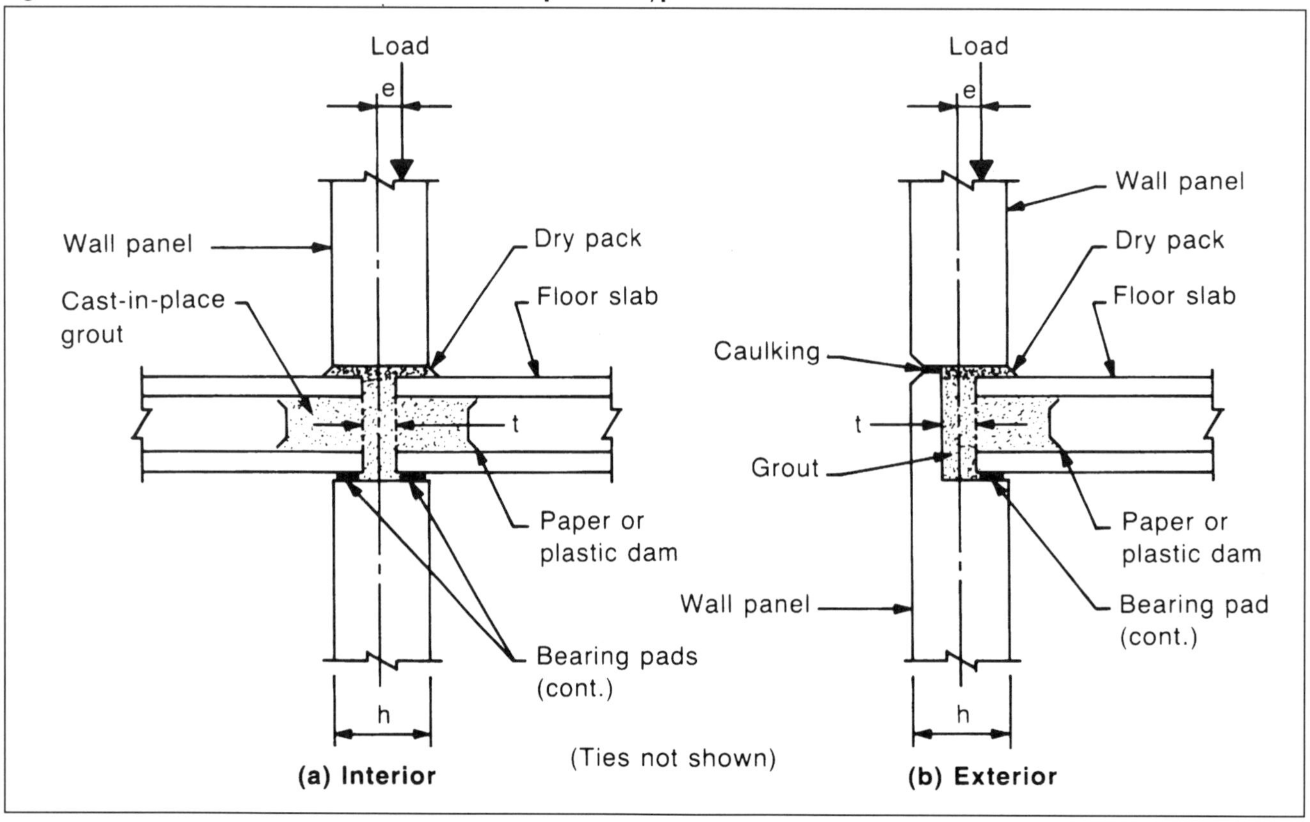

**Fig. 4.20.13 Vertical bar splice using weld angle**

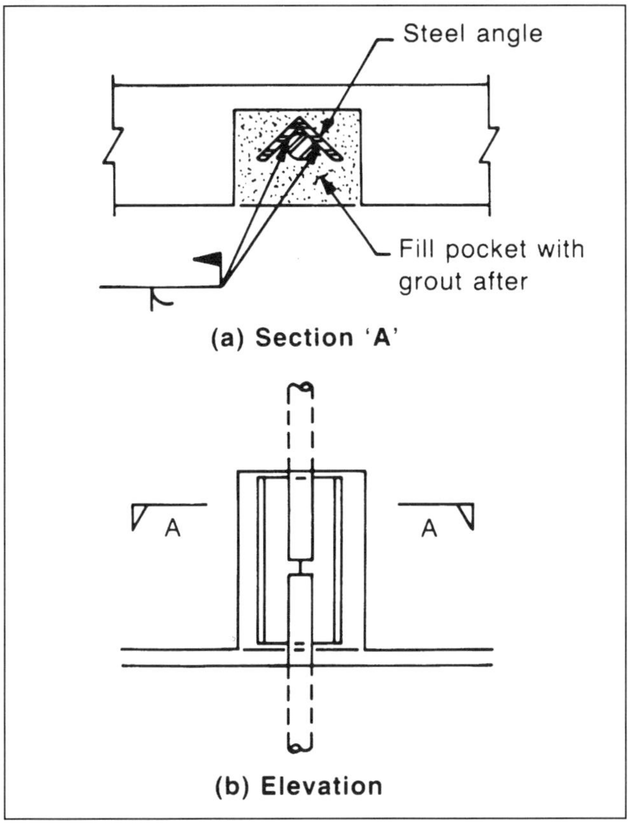


Vertical reinforcement is generally spliced at the horizontal joints as illustrated in Figs. 4.20.13 to 4.20.16. This vertical reinforcement provides overturning resistance due to in-plane shear, and also doubles as shear friction reinforcement across the joint. The detail selected will depend on physical restrictions of the building layout and on the availability of proprietary hardware.

**Fig. 4.20.15 Grouted sleeve tension tie**

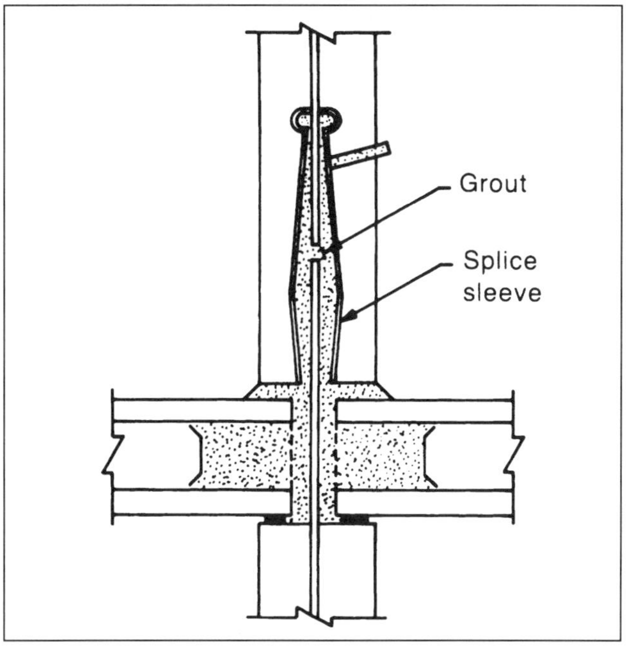


**Fig. 4.20.14 Panel vertical tie using hollow structural section**

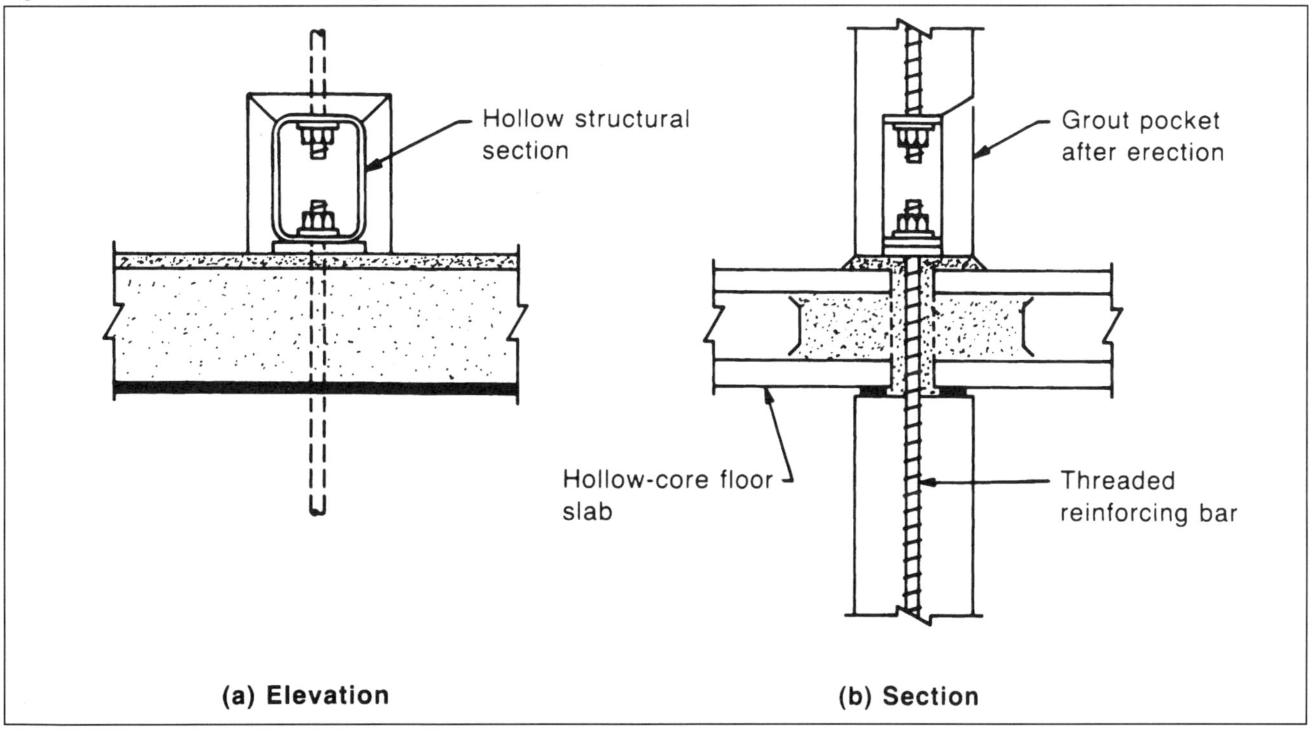

Fig. 4.20.16 Post-tensioned vertical tie

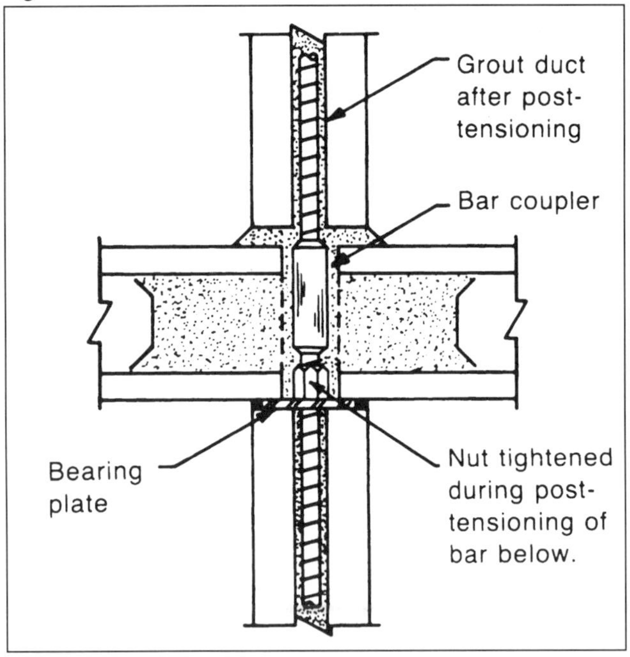


Fig. 4.20.18 Roof to bearing wall connection

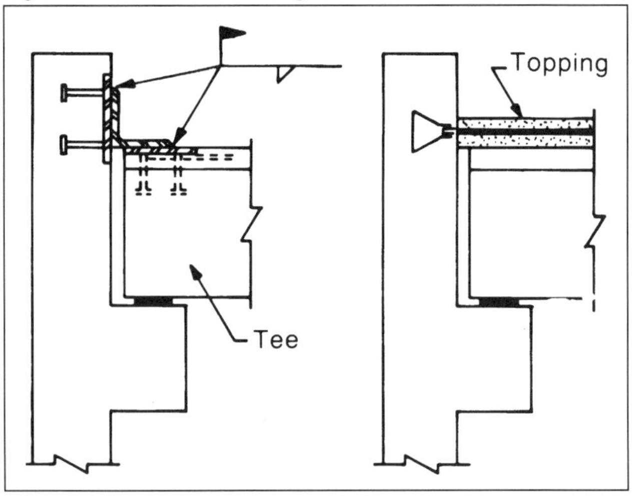


Fig. 4.20.19 Slab to shear wall connection

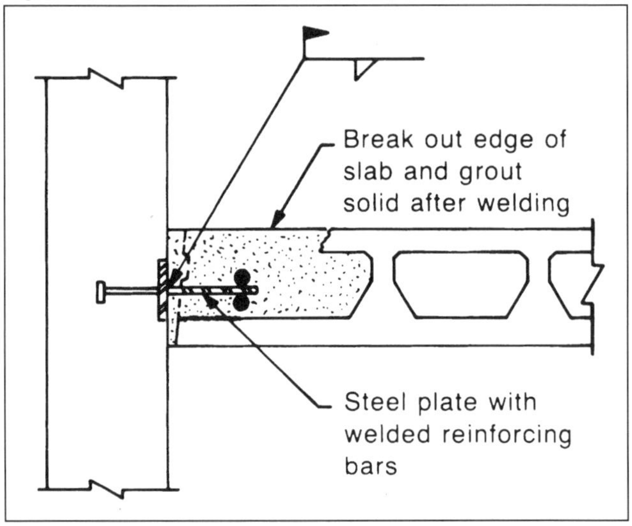


The vertical tension tie shown in Fig. 4.20.15 uses a pipe sleeve which is grouted with an expansive grout after the upper element is aligned and braced. Tests have shown that these splices develop the full strength of the bar.

Fig. 4.20.16 illustrates a typical post-tensioned bar vertical tie. Bars are tensioned at the top of each panel and threaded couplers are used to connect bars in adjacent elements.

## 4.20.4 FLOOR TO WALL SHEAR CONNECTIONS

Where floor elements are not directly incorporated into the horizontal joint of the wall panel, other methods must be used to transfer vertical load and horizontal diaphragm forces. Examples of connections used in multi storey shear wall structures are illustrated in Figs. 4.20.17 to 4.20.20.

Fig. 4.20.17 Floor to bearing wall connections

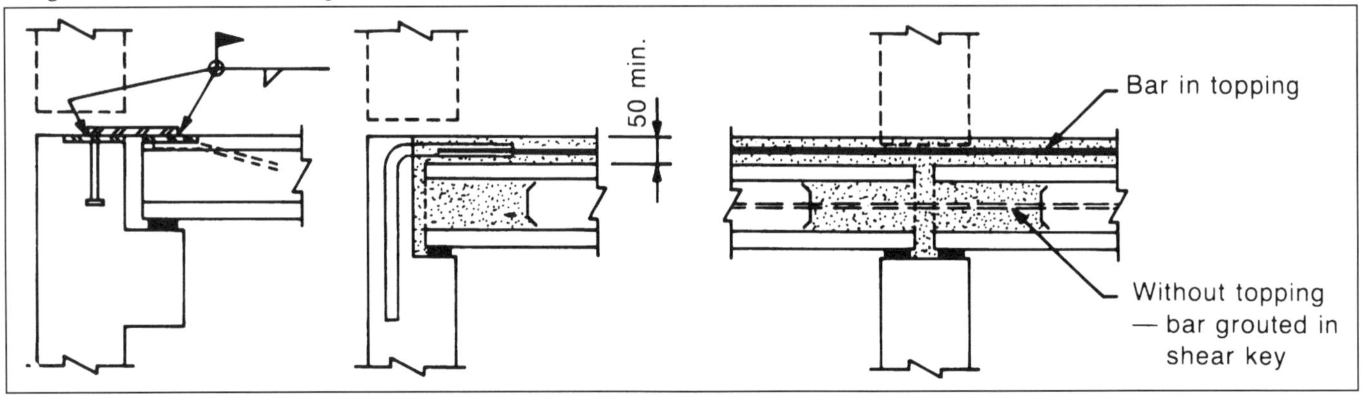

**Fig. 4.20.20 Double tee to shear wall connection**

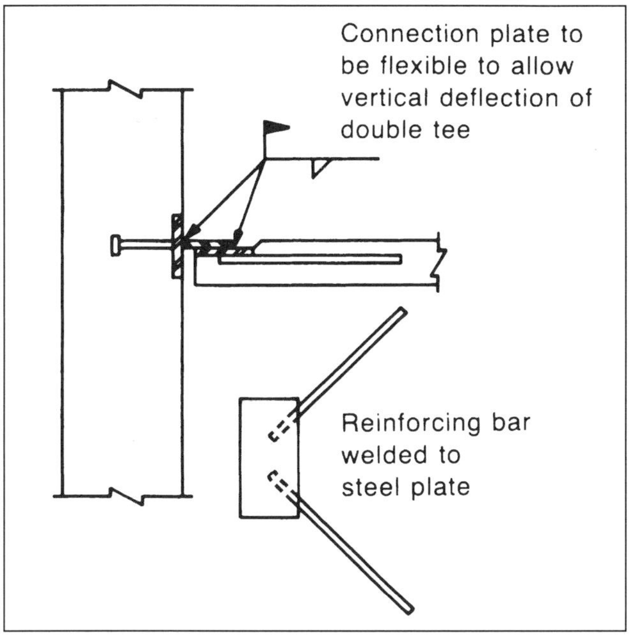


## 4.20.5 WALL TO FOUNDATION CONNECTIONS

Typical wall panel to foundation connections are often similar to the tensile connections shown in Figs. 4.20.13 to 4.20.16. Keyed joints may be necessary in earthquake zones for shear force transfer. Thickened areas or special cast-in-place infill sections may be required to provide tension and compression load resistance.

**References:**


R4.3.1 Iverson, J.K. and Pfeifer, D.W., "Criteria for Design of Bearing Pads", *Technical Report No. 4*, Prestressed Concrete Institute, 1985, pp. 111.

R4.7.1 Collins, M.P. and Mitchell, D., "A Rational Approach to Shear Design - the 1984 Canadian Code Provisions", ACI Journal, Vol.83, No.80, Nov/Dec 1986.

Collins, M.P. and Mitchell, D., "Chapter 4 - Shear and Torsion", Concrete Design Handbook, Canadian Portland Cement Association, Ottawa, 1995, pp.4-1 to 4-38.

R4.9.1 Concrete Design Handbook, Canadian Portland Cement Association, 1995.

R4.13.1 Marcakis, K. and Mitchell, D., "Precast Concrete Connections with Embeded Steel Members", *PCI Journal*, Vol. 25, No. 4, Jul-Aug 1980, pp. 88-116

R4.13.2 *Handbook of Steel Construction*, Canadian Institute of Steel Construction, Latest Edition

R4.20.1 U.S. Department of Housing and Urban Development, "Design and Construction of Large-panel Concrete Structures," Reports 1 to 6 and Supplements A, B and C, Contract H-2131R, Prepared by the Portland Cement Association, 1975-79.

# CHAPTER 5
# RELATED CONSIDERATIONS

# 5.1 THE BUILDING ENVELOPE

## 5.1.1 GLOSSARY

C = *thermal conductance*; the time rate of heat flow through the unit area of a layer of material induced by a unit temperature difference between the layer surfaces ($W / m^2 \cdot °C$).

J = *joule*; unit of energy or work, defined here as a flow of energy equivalent to the work done when the point of application of one Newton moves a distance of one metre in the direction of the force, ($N \cdot m$).

k = *thermal conductivity*; the time rate of heat flow through a homogeneous material under steady-state heat-flow conditions. A property of homogeneous materials with units of $W / (m \cdot °C)$. This is termed $\lambda$ in most documents using SI units.

Q = *total energy for heating or cooling*; expressed in joules (J) or gigajoules ($10^9$ J).

q = *rate of heat flow*; (J / s or watts).

RH = *relative humidity*; the ratio of the water vapour pressure present in air to the water vapour pressure present in saturated air at the same temperature and atmospheric pressure.

$T_s$ = *dew point or saturation temperature*; the highest temperature at which condensation of water vapour in air can occur as the air is being cooled.

U = *overall thermal transmittance* (U-value); a measure of the rate at which heat is transferred through a building assembly subject to a temperature difference - $W / (m^2 \cdot °C)$.

Where there are two or more paths of heat flow that have different thermal resistances, a sum of heat flow per unit temperature difference through each section ($U \cdot A$) is calculated and divided by the total area to yield the average rate of heat transfer per unit temperature ($U_T$).

R = *thermal resistance (RSI value);* the inverse of the overall thermal transmittance. ($m^2 \cdot °C / W$)

*Thermal resistance of an assembly* with parallel paths with different areas and thermal resistances is defined as the (areal) thermal resistance and equals:

$$R_T = \frac{A_1}{R_1 \cdot A_T} + \frac{A_2}{R_2 \cdot A_T} + \cdots + \frac{A_n}{R_n \cdot A_T}$$

where:

$R_1, R_2, \ldots, R_n$ are the thermal resistances (RSI values) of individual areas of an assembly

$A_1, A_2, \ldots, A_n$ are the areas of individual sections of an assembly

$A_T$ is the total area of the assembly

M = *water-vapour permeance*; the time rate of water vapour flow through a layer of material expressed in $ng / (Pa \cdot s \cdot m^2)$ (i.e., nanograms of water per second through unit area of the layer for each pascal of water-vapour pressure difference across the layer).

V = *water vapour resistance*; the reciprocal of M if a single layer of material is involved, or the sum of all the V values for all the layers if an assemblage of layers is involved. The unit of water vapour resistance is $(Pa \cdot s \cdot m^2) / ng$.

m = *water vapour permeability* of a material expressed in terms of unit thickness with units $ng / (Pa \cdot s \cdot m)$. The permeability of a material will vary with barometric pressure, temperature and relative humidity.

### National Energy Code Definitions

*Adjusted energy consumption* means an estimate of the annual sum of the energy consumption for lighting, service water heating and space conditioning of the proposed building design.

*Building energy target* means the annual adjusted energy consumption of a mathematical replica of the proposed building, using the same energy sources for the same functions and having the same environmental requirements, occupancy, climate data and operational schedules, but made to comply with all the applicable prescriptive requirements of the National Energy Code.

*Fenestration* means all building envelope assemblies that transfer visible light, including windows, skylights, clerestories, glass blocks, sliding or swinging glass doors and glazed inserts in doors.

*Fenestration-to-wall ratio* is calculated as the total area of fenestration divided by the sum of all exterior wall areas including the fenestration areas.

## 5.1.2 GENERAL

Some of the basic information needed to design and construct precast and prestressed concrete enclosure assemblies to meet energy-related performance criteria is presented here. This material also provides a means of assessing some of the structural implications.

The elements and assemblies that form the enclosure or envelope of a building are usually required to attain a specified thermal or structural performance.

The structural performance limits specify maximum dimensional changes, in particular, deformations from differential changes that arise from localized daily and seasonal climatic variations.

The thermal performance limits specify maximum energy flow for heating, cooling and operation of a building. The limits may or may not be prescribed by regulations or in regulatory documents. The selection of a limit for the thermal performance of a building may be affected by factors such as building occupancy, life cycle costing, etc.

Except where noted, the information and design criteria that follow are taken from or derived from:

*National Energy Code for Buildings - 1995*,
Canadian Commission on Building and Fire Codes / National Research Council of Canada.

*National Energy Code for Houses - 1995*,
Canadian Commission on Building and Fire Codes / National Research Council of Canada.

*1993 ASHRAE Handbook - Fundamentals - SI Edition*,
American Society of Heating, Refrigerating and Air-Conditioning Engineers, Inc.

It is not possible, within the limits of this section, to present a comprehensive treatment of the mechanics of heat and moisture flow, to provide all the necessary performance data, or to develop the detailed design procedures necessary to establish the so-called "optimum" level of thermal resistance.

### Moisture Control

Moisture problems in building envelopes have been attributed in large measure to moisture from outside and moisture from inside involving three transport mechanisms: rain penetration, air leakage and vapour diffusion. Often serious failures involve two or more of these mechanisms. Blown and drifted snow can enter and melt inside a building envelope.

### Rain Penetration Control

Rain penetration control strategies for walls range from face seal designs to pressure equalized rainscreen (PER) designs. The face seal approach relies heavily on the use of sealants, which need regular preventative maintenance to achieve the required service life.

The performance benefits gained by the use of a PER wall system should more than offset any extra capital cost.

PER walls consist of three elements - an exterior cladding or rainscreen, an air and vapour barrier system installed in an interior assembly and a vented air space separating the two. Vented air spaces should be designed as a drained cavity to direct any water which gets through the cladding back to the exterior. The rain penetration forces controlled by a PER wall include the kinetic energy of rain drops, surface tension and capillarity, gravity and air pressure difference. Air pressure difference has a large variable, or dynamic component because a significant amount of the air pressure difference is generated by wind. The design of a PER wall must address both static and dynamic components. It must also consider such parameters as size and location of the venting, volume and flexibility of the air space, location of compartment seals and loads on the barrier system and on the rainscreen.

A pressure equalized rainscreen wall system will prevent rain leakage without face sealing. Architectural precast rainscreen wall panels can be manufactured to provide rain penetration control using the pressure equalized rainscreen design principle. An effective air barrier system is an inherent feature of such panels. A plastic spacer material has been shown to provide a simple but effective air space. Adequate venting of panels with areas of 30 $m^2$ can be provided with a 12 mm high opening located at the bottom of the air space and extending the width of the panel. Such a design will minimize the static and dynamic pressure difference induced by the wind across the rainscreen, thus minimizing one of the rain penetration forces.

### Vapour Diffusion Control

The principal functions of a vapour barrier are to retard the passage of moisture as it diffuses through the assembly of materials in a building envelope, to control the location of the dew point in the assembly and to ensure there is a manageable flow of moisture across the assembly. The vapour barrier is usually placed toward the warm side of the insulation.

A vapour barrier is important, particularly in high humidity buildings such as computer centres, museums, hospitals and swimming pools. While a vapour barrier need not be perfectly continuous, care should be taken to minimize the occurrence of imperfections such as unsealed laps, cuts and pin holes.

### Air Leakage Control

The principal function of an air barrier is to prevent air flow from outside to inside, inside to outside, inside to inside if near the inside, or outside to outside if near the outside of a wall assembly. Moisture-laden air passing through an insulated cavity may deposit much more moisture than would diffuse through the vapour barrier at that location in the same time period.

Air leakage through openings in a building envelope is driven by one or more of the following forces:

**Stack effect** results from warmer or moister inside air having a lower density than the cooler drier outside air. This difference in density creates a slight outward positive pressure at the top of a building and a slight inward negative pressure at the base as the denser air attempts to displace the less dense air.

**Wind** can cause infiltration on the windward sides and exfiltration on the leeward side of buildings. Pressure distribution on the windward facade will vary from a maximum in the centre and diminish toward the perimeter. Pressure on walls parallel to wind flow direction is normally negative but a slight change in direction can result in a positive pressure. Flat roofs will generally experience exfiltration due to negative wind pressure above them. The difference in wind pressure across the building envelope will be greater in tall buildings since the wind velocity is dampened near the ground.

**Mechanical system pressurization** caused by the intake and exhaust of air from buildings may cause significant pressure differences across the building envelope. These will vary with the time of day according to the building ventilation requirements and with the weather.

**Air barrier system design requirements:**

1. There must be continuity throughout the building envelope.
2. The air barrier system must transfer loads to the structure and be rigid enough to avoid displacement. The insulation must be in intimate contact with the air barrier or have its own convection barrier that prevents convection.
3. The air barrier system should be virtually air-impermeable with an air flow of less than 0.1 L/s.$m^2$ at 50 Pa. Precast concrete, polyethylene, roofing membranes, gypsum board, metal and glass qualify as air-impermeable materials whereas concrete block, acoustic insulation, open cell polystyrene insulation and fibreboard do not.

4. The air barrier assembly must be durable and constructed of materials known to have excellent fatigue resistance and a long service life.

While it is preferable that an air barrier system be placed close to the warm side of an insulated assembly it is not an essential requirement (as it is with a vapour barrier). If the air barrier is positioned on the outside of the insulation, the air barrier should be 10 to 20 times more permeable to moisture than the vapour barrier material to allow any vapour to escape through the wall assembly.

A single material such as the inner wythe of a precast concrete sandwich wall panel may serve as both the air and vapour barrier in a building.

Architectural precast wall panels can provide more than an attractive, economical, long-lasting exterior cladding for buildings. With appropriate design, insulated precast panels can provide a high performance wall system.

#### Condensation Control

Condensation of water on the inside of inadequately insulated walls is another moisture-related problem. To a large extent, this problem can be eliminated by using the proper amount of insulation for energy conservation and by paying proper attention to the prevention of thermal bridging.

Reducing humidity is sometimes a procedure for reducing condensation.

#### Ventilation

The provision of adequate ventilation in buildings, particularly in those buildings designed to limit infiltration, is very important. Serious health, safety and durability problems may result if proper ventilation is not provided. Special calculation procedures may be necessary to predict ventilation under limiting conditions in buildings where natural ventilation is employed. Ventilation provided by mechanical systems can be more closely estimated with conventional calculation procedures.

#### Joint Design

**Junctions that are considered to be most critical in building construction are listed below:**

- roof / wall connection
- wall / foundation connection
- wall / window or door connection
- soffit connections
- corner-detail connections
- wall panel joint connections

One-stage joints are usually formed using field molded sealant installed close to the exterior surface of precast panels. These joints have a low initial cost but must be regularly maintained. Fatigue and exposure to weather and ultraviolet light can cause sealant failures. Even a small break in the sealant can allow water penetration.

Two-stage joints, sometimes called ventilated or pressure equalized joints, use the PER principle. A typical joint consists of a rain barrier near the exterior face and an airseal near the interior face of a panel. The rain barrier is designed to shed most of the water from the joint. The wind-barrier or airseal maintains the pressure difference across the wall assembly. In between is an equalization chamber which must be ventilated and drained to the outside. The airseal, on or near the interior of the wall, is protected from the elements (weathering and ultraviolet light).

Joints between precast wall panels and abutting materials can be caulked or sealed with a sheet membrane system. The building designer can greatly improve the ease of installation and effectiveness of the jointing system by locating the interior joints where they are easily accessible during construction.

### 5.1.3 HEATING AND COOLING ENERGY

In most buildings, the main contributors to the cost of space conditioning (heating or cooling or both) are internal heat gains, transmission heat loss, air leakage across the building enclosure and solar gains. In the design of the enclosure of any building, proper attention must be made to provide comfort, durability and the appropriate levels of thermal resistance. The relative significance of transmission and air leakage will vary from building to building, as do the cost of insulation and benefits involved.

In the sections that follow, the emphasis is on controlling transmission heat loss. However, it must be appreciated that heat loss due to air leakage is often of similar significance. Careful attention to avoiding moisture penetration from the exterior is essential. Therefore proper attention must be directed to a number of design considerations (such as the number, orientation and thermal resistance of windows) if true energy conservation is to result.

Heat loss due to transmission through precast concrete elements and assemblies comprises only a portion of the total transmission heat loss in a building envelope. Since this may constitute only a small proportion of the total space conditioning energy requirements over a year, use only the amount of insulation that can be economically justified or that is required to achieve the desired comfort and durability.

#### Economic Considerations

Energy costs can be a significant part of life-cycle costs. The owner and designer must consider the long-term impact of energy costs during the planning stage of a building. More than half of the true total costs incurred during the economic life of a building may be attributable to operating and energy costs. Life-cycle costing is a sound means of assessing the cost of all elements involved in constructing and operating a building throughout its life. It is a means of analyzing and appraising alternative solutions to design questions (such as comfort, rentability and what insulation levels to provide) so that rational economical choices can be made.

### Heating and Cooling Calculations

Heating and cooling calculations are used to provide predictions of peak loads and annual energy usage.

Peak load estimates are required to size equipment and to design distribution systems. Energy can be saved by lowering peak loads. Increasing a building's thermal mass will lower and shift peak loads which can reduce equipment size. Small equipment that runs continuously uses less energy than large equipment that is run intermittently to respond to peak loads. By lowering peak loads, energy is saved.

The energy saved as a result of reduced annual loads is also important. As design standards switch from prescriptive to performance requirements, annual load calculations will become imperative.

Calculation of both peak and annual loads in all but the simplest of buildings are relatively complex and often require special expertise and computer analysis.

### Factors to Consider

Regulatory documents may prescribe minimum requirements for thermal insulation to limit the rate of heat flow through the envelope. To comply, it is usually necessary to evaluate heat flow through each component of a building using a steady-state procedure. Thermal resistance values for floors, walls and roofs may be prescribed.

The "National Energy Code for Buildings, 1995" gives designers the option of proposing alternative measures provided they can demonstrate the building will not use more energy than if the mandatory requirements of the code had been used.

**Some factors which should be considered when calculating energy loads are:**

1. Geographic location: latitude, longitude, building exposure (landscaping).
2. Occupancy: demands, requirements, number, activities, hours of occupancy.
3. Building: site, orientation, plan, configuration, insulation, insulation location, envelope mass, mass of the construction and contents, window design, shading, surface colour and texture.
4. Climate: dry-bulb temperature, wet-bulb temperature, wind speed and direction, solar conditions, cloud cover.
5. Heat loss mechanisms: transmission, air leakage, ventilation.
6. Heat gain: transmission, solar radiation, air leakage, ventilation.
7. Internal heat gain: occupants, lights, appliances, machines, power, equipment.
8. Environmental (comfort): window area, indoor temperature, indoor relative humidity, indoor air quality.
9. Schedules: occupancy, lighting, ventilation, equipment, changes in thermostat set points, changes in humidity.

When computing energy consumption, equipment efficiency must also be considered. This is particularly important when making life-cycle cost studies involving factors such as mass, optimum insulation, glass area, environmental factors, energy type and energy costs.

### Thermal Inertia Effects

The mass of heavy materials like precast concrete wall, floor and roof construction can reduce the annual heating requirements of a building. The effect that mass and its thermal capacitance have on peak loads and annual energy requirements is primarily governed by:

1. Location and storage characteristics of the mass.
2. Effectiveness of thermal coupling between the heat source and the mass.
3. Ratio of internal heat gain to heat loss.
4. Hours when internal gains occur.
5. Solar radiation through glass.
6. Number of air changes.

### Mass and the Heating Season

For mass to contribute during the heating season, it must be inside the insulation and subjected to conditions that permit the construction materials to absorb solar heat or radiation and heat from light or equipment and occupants. Mass inside the insulation is only effective if the temperature set point of the air is allowed to vary considerably. Such a variation can have a large effect on comfort if it occurs during working hours. At night, furnaces can be turned down to allow the absorbed heat to be released to nearby cooler surfaces or space. This process of absorption and release is largely dependent upon the location of the mass in the structure relative to the space being heated, the sources of heat and the insulation. Precast sandwich wall construction is an excellent way to combine mass and insulation in walls.

### Mass and the Cooling Season

Mass on the outside of building envelopes can absorb solar gain. Heat that is absorbed by the mass is released later to the outside during cooler hours that usually occur at night. Averaging over the cooling season, the outdoor air temperature is generally below the indoor air temperature. As a result, with the exception of peak periods, the net heat flow through the walls is outward, not inward. Thermal coupling between the interior mass of the building and cool night-time outside air is also important as is forced ventilation which will flush out excess heat stored within the structure during the day. The mass of precast concrete building construction can substantially reduce the need for mechanical cooling.

### Evaluation of Mass Effects

Computer solutions are essential to fully evaluate the effect of mass on heating and cooling loads. Computerized solutions, however, are not essential to determine what reduction can be made in insulation requirements when mass is present. An explanation of the use of these methods is contained in the "Guide to Energy Efficiency in Masonry and Concrete Buildings"[R5.1.4].

### Fenestration

Fenestration is a term used for any light-transmitting opening in a building wall, floor or roof. Windows are, of course, the most important form of fenestration which designers must consider. Regulatory documents, such as the "National Energy Code for Buildings, 1995", prescribe restrictions on the amount of glass area to be provided.

Windows are, however, a dominant source of heat gain. A change in orientation can result in load reductions when thermal storage is utilized. Unshaded glazing located within 45$^0$ of South will receive maximum heat gain.

Shading devices can reduce solar radiation in the summer and allow solar radiation for winter heating. Overhangs can be designed to provide shading for windows when the summer sun is at a higher altitude and direct sun on the windows in winter when the sun altitude is much lower. Design procedures are given in Reference R5.1.4. Precast concrete is ideal for the construction of sun shades.

### Economic Considerations

Designers of building-envelope assemblies must consider the potential for providing the necessary overall energy efficiency by insulating different building assemblies in the most effective manner. The design requirements and level of insulation chosen for each assembly should be reviewed and economically evaluated.

Frequently, the cost of adding additional insulation to roofs is less than for walls. Prescriptive regulatory documents generally permit the reduction of the required thermal resistance of one assembly, providing that of another is increased so that the total calculated heat loss through the entire envelope is no greater than if the minimum prescribed values for each assembly were used.

**Fig. 5.1.1 Thermal resistance and density of various building and insulating materials**

| Material | Thermal resistance ($m^2 \cdot °C / W$) (R-value) | | Density ($kg/m^3$) | Material | Thermal resistance ($m^2 \cdot °C / W$) (R-value) | | Density ($kg/m^3$) |
|---|---|---|---|---|---|---|---|
| | Per 100 mm thickness | For thickness listed | | | Per 100 mm thickness | For thickness listed | |
| **Concrete** | | | | Concrete (semi-low density) | | | |
| Density as indicated | 0.35 | – | 1000 | 150 mm thick | – | 0.30 | 1900 |
| | 0.29 | – | 1200 | 200 mm thick | – | 0.35 | 1900 |
| | 0.22 | – | 1400 | 250 mm thick | – | 0.43 | 1900 |
| | 0.17 | – | 1600 | 300 mm thick | – | 0.46 | 1900 |
| | 0.13 | – | 1800 | | | | |
| | 0.091 | – | 2000 | **Woods** | | | |
| | 0.064 | – | 2200 | Hardwoods (maple, oak, etc.) | 0.63 | – | 720 |
| | 0.052 | – | 2320 | Softwoods (fir, pine, etc.) | 0.77 | – | 500 |
| | 0.045 | – | 2400 | | | | |
| **Insulating concretes** | | | | **Roofing** | | | |
| Density as indicated | 1.03 | – | 300 | Asbestos-cement shingles | – | 0.037 | 1900 |
| | 0.84 | – | 400 | Asphalt roll roofing | – | 0.026 | 1120 |
| | 0.68 | – | 500 | Asphalt shingles | – | 0.078 | 1120 |
| | 0.61 | – | 600 | Built-up roofing (19 mm) | – | 0.058 | 1120 |
| | 0.53 | – | 700 | Slate (12 mm) | – | 0.009 | – |
| | 0.47 | – | 800 | | | | |
| | 0.40 | – | 900 | **Siding** (flat surface) | | | |
| | | | | Asbestos-cement shingles | – | 0.037 | 1900 |
| **Hollow-core slabs**(1) | | | | Wood shingles (400 mm, 190 mm exposure) | – | 0.153 | – |
| Concrete (normal density) | | | | | | | |
| 150 mm thick | – | 0.19 | 2400 | Asbestos-cement (6 mm, | | | |
| 200 mm thick | – | 0.24 | 2400 | lapped) | – | 0.037 | – |
| 250 mm thick | – | 0.31 | 2400 | Asphalt roll siding | – | 0.026 | – |
| 300 mm thick | – | 0.34 | 2400 | Wood (11 mm, medium density) | – | 0.052 | 640 |

**Fig. 5.1.1 (Cont'd) Thermal resistance and density of various building and insulating materials**

| Material | Thermal resistance ($m^2 \cdot °C / W$) (R-value) | | Density ($kg/m^3$) |
|---|---|---|---|
| | Per 100 mm thickness | For thickness listed | |
| **Finish flooring** | | | |
| Carpet and fibrous pad | – | 0.366 | – |
| Carpet and rubber pad | – | 0.226 | – |
| Cork tile (3 mm) | – | – | – |
| Tile (asphalt, linoleum, vinyl, rubber, asbestos, ceramic) | – | 0.009 | – |
| Terrazzo (25 mm) | – | 0.014 | – |
| Hardwood (19 mm) | – | 0.12 | – |
| **Acoustical tile** | | | |
| Mineral fibreboard | | | |
| Wet-felted (13 mm) | – | 0.24 | – |
| Wet-felted (19 mm) | – | 0.36 | – |
| Wet-moulded (13 mm) | – | 0.21 | – |
| Wet-moulded (19 mm) | – | 0.31 | – |
| Wood or cane (13 mm) | – | 0.22 | – |
| Fibreboard (19 mm) | – | 0.33 | – |
| **Building board** | | | |
| Asbestos-cement board | 0.17 | – | 1900 |
| Gypsum plasterboard | 0.62 | – | 800 |
| Particleboard (mat-formed) | 0.87 | – | – |
| Plywood or wood panels | 0.87 | – | 545 |
| **Stucco** | 0.14 | – | 1860 |
| **Insulation** | | | |
| Rigid | | | |
| Glass fibre | 2.77 | – | 64-144 |
| Cane fibre | 1.73 | – | – |
| Mineral fibre | 1.82 | – | 240 |
| Cork | 2.57 | – | – |
| Perlite | 1.94 | – | 280 |
| Fibreboard | 1.94 | – | 255 |
| Phenolic foam | 3.04 | – | 60 |
| Polystyrene (expanded)(2) | | | |
| Type 1 | 2.57 | – | – |
| Type 2 | 2.77 | – | – |
| Type 3 | 2.98 | – | – |
| Type 4 | 3.47 | – | – |
| Polyurethane | 4.20 | – | 24 |
| Wood (shredded) | 1.69 | – | 350 |
| Cellular glass | 1.82 | – | 135 |
| Polyisocyanurate | 4.20 | – | – |
| Batt | | | |
| Glass fibre | 2.08 | – | – |
| Mineral fibre | 2.08 | – | – |

| Material | Thermal resistance ($m^2 \cdot °C / W$) (R-value) | | Density ($kg/m^3$) |
|---|---|---|---|
| | Per 100 mm thickness | For thickness listed | |
| **Loose fill** | | | |
| Cellulose | 4.20 | – | 37-51 |
| Mineral fibre | 2.91 | – | 130-240 |
| Perlite (expanded) | 1.94 | – | 80-130 |
| Vermiculite | 1.44 | – | 64-96 |
| Glass fibre (pouring wool) | 2.22 | – | 190 |
| Glass fibre (blowing wool) | 1.46 | – | 40 |
| **Flat glass** (winter)(3) | | | |
| Vertical panels (exterior) | | | |
| Single glass | – | 0.16 | – |
| Double glass | | | |
| 6.4 mm air space, 3.2 mm glass | – | 0.30 | – |
| 12.7 mm air space, 6.4 mm glass | – | 0.36 | – |
| Triple glass | | | |
| 6.4 mm air space, 3.2 mm glass | – | 0.45 | – |
| 12.7 mm air space, 6.4 mm glass | – | 0.57 | – |
| Single glass w/storm sash 25.4 mm to 101.6 mm air space, 3.2 mm glass | – | 0.35 | – |
| Horizontal panels (exterior) | | | |
| Single glass | – | 0.14 | – |
| Double glass | | | |
| 6.4 mm air space, 3.2 mm glass | – | 0.27 | – |
| 12.7 mm air space, 6.4 mm glass | – | 0.30 | – |
| Plastic domes | | | |
| Single wall | – | 0.15 | – |
| Double wall | – | 0.25 | – |

*(1) R-value listed is for hollow-core slabs with cores empty. Consult manufacturer for R-values of insulation filled cores and for exact mass of concrete to be used. (Core configuration may also affect the thermal resistance values listed.)*

*(2) Complying with CGSB 41-GP-14a (1972) "Thermal Insulation, Expanded, Polystyrene - June 1978."*

*(3) R-values include air films and gaps. Emissivity of uncoated glass surface = 0.84.*

## 5.1.4 CALCULATION OF RSI-VALUE

The RSI-value is a measure of the thermal resistance of a building component or assembly in a direction normal to the surface. For an assembly, this resistance is the sum of the resistances of each layer, including air gaps when they are present, and air films contiguous to each outer surface. The RSI-value is measured in units of ($m^2 \cdot °C / W$).

Thermal resistance values of selected construction materials and of air surface films and gaps are listed in Figs. 5.1.1 and 5.1.2. Examples 1, 2 and 3 demonstrate RSI-value calculations for walls, floors and roofs. These do not represent total wall systems. RSI-values for fenestration, and joint areas also have to be calculated.

**Fig. 5.1.2 Thermal resistance for air films and spaces**

| Air | Thermal resistance ($m^2 \cdot °C / W$) (R-value) |
|---|---|
| **Surface films** | |
| - Internal surface: | |
| Still air (vertical) | |
| Heat flow horizontal | 0.120 |
| Still air (horizontal) | |
| Heat flow up | 0.110 |
| Heat flow down | 0.160 |
| - External surface: | |
| Moving air (any position) | |
| - winter | 0.030 |
| - summer | 0.044 |
| **Spaces (with non-reflective surfaces)** | |
| 12 mm and greater | |
| Vertical | |
| Heat flow horizontal | 0.160 |
| Horizontal | |
| Heat flow up | 0.150 |
| Heat flow down | 0.180 |
| **Non-vented air spaces (with reflective surfaces)** | |
| 12 mm and greater | |
| Vertical (heat flow horizontal) | |
| 1 side faced | 0.330 |
| 2 sides faced | 0.450 |
| Horizontal (heat flow up) | |
| 1 side faced | 0.280 |
| 2 sides faced | 0.360 |
| Horizontal (heat flow down) | |
| 1 side faced | 0.980 |
| 2 sides faced | 1.034 |

The basic equation for calculating the rate of heat transfer through a portion of an enclosure is:

$$Q = \frac{A \cdot \Delta T}{RSI}$$

where:

- Q = heat loss rate (W)
- A = surface area of component ($m^2$)
- $\Delta T$ = difference in temperature across the assembly (°C)
- RSI = sum of the thermal resistance of the layers ($m^2 \cdot °C / W$)

### Example 1 – Rain-screen sandwich wall panel

*Given:*

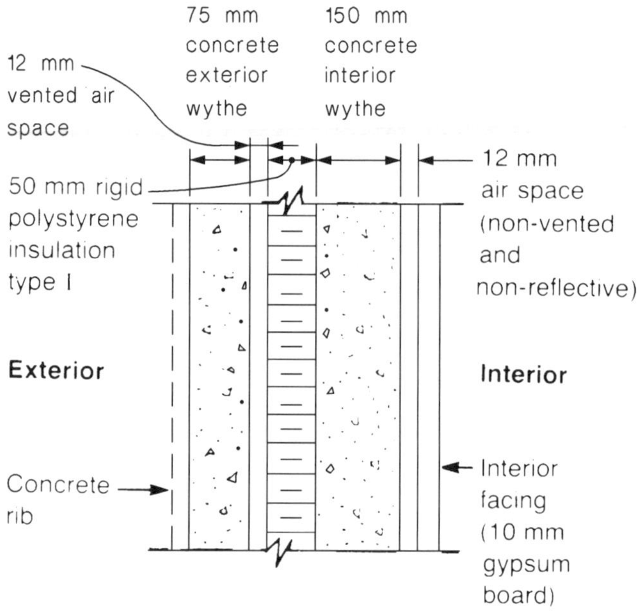


*Problem:*

Find the RSI-value of the wall assembly.

*Solution:*

| Component (layer) | R-value ($m^2 \cdot °C / W$) |
|---|---|
| Surface air film (exterior) | 0.030 |
| 75 mm normal density concrete | 0.039 |
| 12 mm air space - non-reflective (vertical) | 0.120 |
| 50 mm rigid polystyrene insulation type I | 1.285 |
| 150 mm normal density concrete | 0.078 |
| 12 mm (non-vented) air space | 0.160 |
| 10 mm gypsum board | 0.062 |
| Surface air film (interior) | 0.120 |
| Total RSI-value | 1.894 |

**Example 2 – Hollow-core slab floor system**

*Given:*

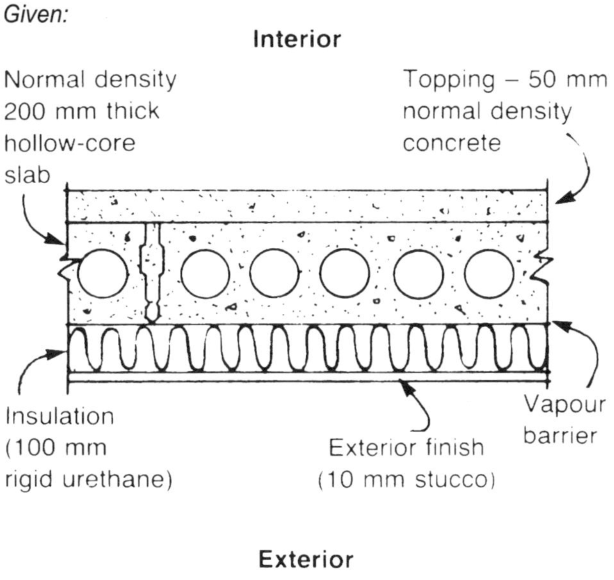


*Problem:*

Find the RSI-value of the floor assembly.

*Solution:*

| Component | R-value (m$^2$·°C /W) |
|---|---|
| Surface air film (exterior) | 0.030 |
| 10 mm stucco exterior finish | 0.014 |
| 100 mm urethane insulation | 4.200 |
| Vapour barrier | 0.000 |
| Hollow-core slab (equiv. area) | 0.240 |
| 50 mm normal density concrete topping | 0.026 |
| Surface air film (interior) | 0.160 |
| Total RSI-value | 4.670 |

**Example 3 – Double tee roof system**

*Given:*

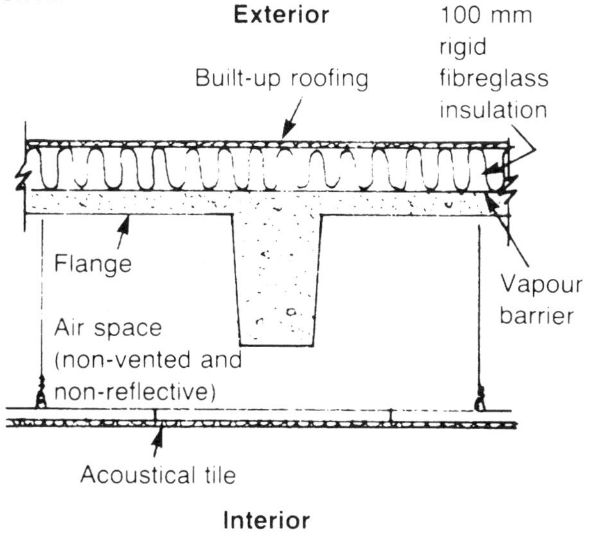


*Problem:*

Find the RSI-value of the roof assembly.

*Solution:*

| Component | R-value (m$^2$·°C /W) |
|---|---|
| Surface air film (exterior) | 0.030 |
| Built-up roofing and gravel | 0.058 |
| 100 mm rigid fibreglass insulation | 2.770 |
| Vapour barrier | 0.000 |
| 50 mm normal density tee flange | 0.026 |
| (Non-reflective) air space | 0.150 |
| Acoustical tile (13 mm wet-molded mineral fibreboard) | 0.210 |
| Surface air film (interior) | 0.110 |
| Total RSI-value | 3.354 |

## 5.1.5 CALCULATION OF TEMPERATURE GRADIENT

The calculation of the temperature gradient through a roof or wall assembly can be used to help determine whether there may be a problem with condensation or differential thermal movement. The temperature gradient alone is not sufficient to accurately locate the *dew point* (condensation point) within the assembly. However, an approximation of its location can be made which is useful in estimating where condensation can occur from exfiltrating air. For vapour diffusion the solution involves the calculation of vapour pressures within the assembly.

The solution of heat flow through walls and roofs can become very complicated. The assumed steady-state conditions are seldom reached, owing to fluctuations in the temperatures to which a building envelope is exposed and to the heat storage capacities of the building components and contents. Simplified procedures introduce inaccuracies, but the results obtained can be a valuable guide for the design of walls and roofs.

The ability to calculate the thermal gradient allows a designer to estimate the magnitude of the movements caused by external temperature changes, to estimate the location of condensation and freezing planes in the wall, and to perform an initial assessment of the suitability of wall and roof assemblies.

Fig. 5.1.3 lists dew-point temperatures for various relative humidities and several indoor temperatures. This chart can be used once the thermal gradient is determined.

**Fig. 5.1.3 Dew-point temperatures (°C)**

| Relative humidity (%) | Indoor temperature | | | | |
|---|---|---|---|---|---|
| | 15°C/59°F | 18°C/64.4°F | 21°C/69.8°F | 24°C/75.2°F | 27°C/80.6°F |
| 10 | – 14 | – 12 | – 11 | – 8 | – 7 |
| 20 | – 7 | – 4 | – 2 | 0 | + 2 |
| 30 | – 2 | 0 | + 3 | + 6 | + 8 |
| 40 | + 2 | + 4 | + 7 | + 9 | + 12 |
| 50 | + 5 | + 8 | + 11 | + 13 | + 16 |
| 60 | + 8 | + 11 | + 13 | + 16 | + 18 |
| 70 | + 10 | + 13 | + 16 | + 18 | + 21 |
| 80 | + 12 | + 14 | + 17 | + 20 | + 23 |
| 90 | + 14 | + 17 | + 19 | + 22 | + 25 |
| 100 | + 15.6 | + 18.3 | + 21.1 | + 23.9 | + 26.7 |

**Example – Thermal gradient through a wall section**

*Given:*

The wall assembly shown in Fig. 5.1.4, with an overall RSI-value of 2.33, is subjected to indoor and outdoor temperatures of 22°C and -12°C respectively (a temperature difference of 34°C).

*Problem:*

Find the thermal gradient through the wall section.

*Solution:*

By labeling each layer and each interface as shown in Fig. 5.1.4 and by tabulating the calculation procedure as shown in Fig. 5.1.5, a readily understandable record of the necessary computation can be provided. Given that the rate of heat flow (i.e., Q / A) is constant and given that Q / A = DT / RSI, then for the assembly as a whole, Q / A = 34 / 2.33. Knowing one interface temperature and the RSI value for each layer, starting from the interior, the temperature at the other interface can readily be found by determining the temperature drop across the layer, i.e. (34 / 2.33)RSI.

Assuming that the interior relative humidity is 40%, then from Fig. 5.1.3 the dew point is just above 7°C. An examination of the temperature gradient indicates that there should be no condensation on the inside wall surface since the inside surface temperature, 20.2°C, is considerably above the dew point temperature.

Without further knowledge about vapour pressures, it is safe to predict that immediately behind the wall finish, where it is 19.0°C, there should also be no condensation problem.

A dew point is likely to exist within the insulation. In this case, condensation would probably occur where there is a decrease in the rate of flow of moisture (where the resistance to moisture flow per unit distance increases). The extent of the problem will depend upon the effectiveness of the air barrier and vapour retarder.

**Fig. 5.1.4 Thermal gradient example**

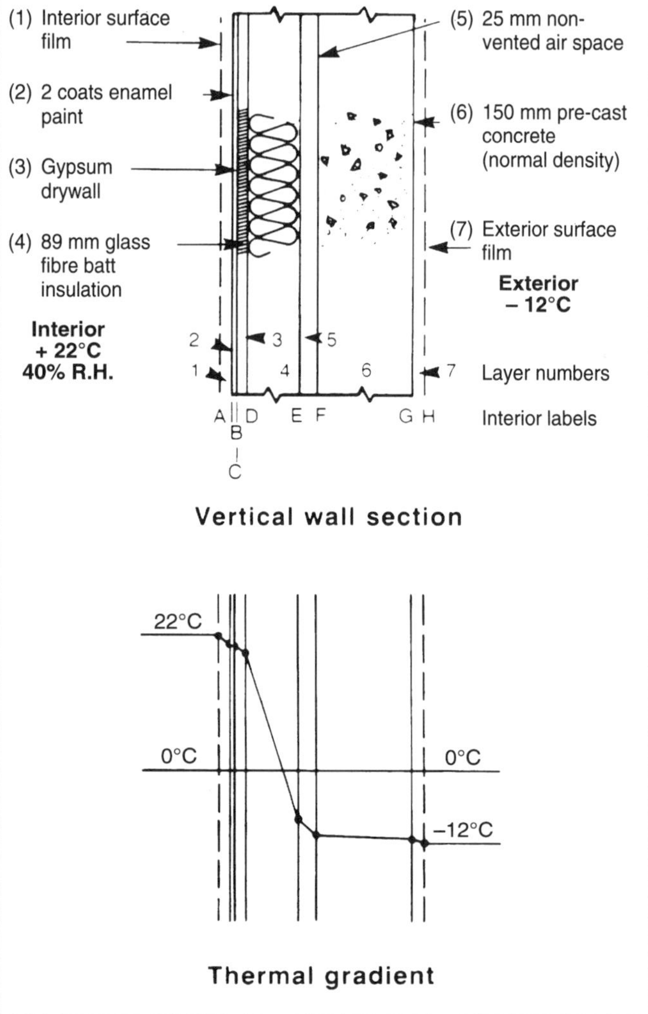

**Fig. 5.1.5 Calcualtion of thermal gradient**

| Inter-face | Layer | Nature of layer | Thick-ness (mm) | R Value ($m^2 \cdot K/W$) | Computation $(\Sigma\Delta T/\Sigma R)R = \Delta T$ (°C) | Temp. (°C) |
|---|---|---|---|---|---|---|
| A | | | | | | + 22.0 |
| | 1 | Interior surface film | – | 0.120 | (34/2.331) (0.120) = 1.8 | |
| B | | | | | | + 20.2 |
| | 2 | 2 coats enamel paint | – | 0.000 | (34/2.331) (0.000) = 0 | |
| C | | | | | | + 20.2 |
| | 3 | Gypsum drywall | 13 | 0.081 | (34/2.331) (0.081) = 1.2 | |
| D | | | | | | + 19.0 |
| | 4 | Glassfibre batt insulation | 89 | 1.851 | (34/2.331) (1.851) = 27.0 | |
| E | | | | | | – 8.0 |
| | 5 | Air space (non-vented & non-reflective) | 25 | 0.171 | (34/2.331) (0.171) = 2.5 | |
| F | | | | | | – 10.5 |
| | 6 | Precast concrete (normal density - 2320 kg/m³) | 150 | 0.078 | (34/2.331) (0.078) = 1.1 | |
| G | | | | | | – 11.6 |
| | 7 | Exterior surface film | – | 0.030 | (34/2.331) (0.030) = 0.4 | |
| H | | | | | | – 12.0 |
| Summation | | R = | 277 | 2.331 | $\Sigma\Delta T$ = 34.0 | |

## 5.1.6 TEMPERATURE INDEX

The temperature index, I, will allow the prediction of the condensation resistance of a wall assembly for other sets of operating systems (indoor relative humidity, RH, and outdoor dry-bulb temperature). The wall system should be designed to provide an I value greater than the minimum I in order to control condensation.

The temperature index, I, is defined as:

$$I = (T - T_c) / (T_h - T_c) \times 100\%$$

where:

$T$ = the temperature of the indoor surface of the vapour barrier, °C

$T_h$ = the indoor design dry-bulb temperature, °C

$T_c$ = the outside winter design temperature, °C

**Fig. 5.1.6 Relationship of temperature index and relative humidity for different winter design temperatures**

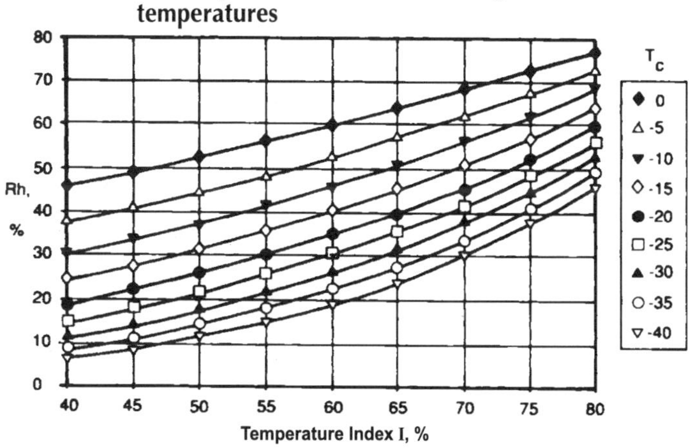


### Example

The following procedure illustrates how to select a temperature index, I, for a given assembly in a given location in Canada.

1. Select the inside relative humidity to be maintained without causing condensation.
   - Select maximum RH = 30%.
   - The indoor dry-bulb temperature is taken as 20 °C.
2. Select the outside winter design temperature, $T_c$, for the location where the assembly will be installed. This value can be found in the Supplement to the National Building Code.
   - Assume at a location $T_c$ = -15 °C.
3. Using Fig. 5.1.6:

   (a) Enter the value of relative humidity, RH = 30%, found in Step 1. on the vertical scale.

   (b) Move horizontally to the selected outside winter temperature, $T_c$ = -15 °C. Interpolate linearly for in-between values.

   (c) From the intersection of each maximum allowable RH and $T_c$, move vertically downward to find the minimum required temperature index, I, for each case.
   - Here Min. I = 48%.

Fig. 5.1.7 shows the temperature index, I, for some selected locations in Canada.

**Fig. 5.1.7 Temperature Index (I)**

| City | Inside design conditions $T_h$ (°C) | Inside design conditions RH (%) | Outside design temp. (2½%) $T_c$ (°C) | I (%) |
|---|---|---|---|---|
| Vancouver | 20 | 35 | -7 | 41 |
| Saskatoon | 20 | 35 | -35 | 71 |
| Winnipeg | 20 | 35 | -33 | 70 |
| Toronto | 20 | 35 | -18 | 58 |
| Montreal | 20 | 35 | -23 | 63 |
| Halifax | 20 | 35 | -16 | 57 |
| St. John's | 20 | 35 | -14 | 53 |

### Thermal Bridges

Paths of high conductivity called "thermal bridges" may lead to localized cold areas where surface condensation may occur. In precast concrete applications, these thermal bridges sometimes occur in sandwich panels where connections, metal shear ties or solid concrete sections penetrate the insulation layer. Because of the many variables, measured values and calculated values of heat flow through these thermal bridges differ.

Steel shear ties are often used as structural connections between the concrete wythes of precast concrete sandwich wall panels. Depending on the amount of highly conductive material used, the reduction of overall resistance values may be calculated but is often ignored if the relative area of the ties is small and the insulation is installed tight against the outline of the penetration.

# 5.2 ACOUSTICAL PROPERTIES

## 5.2.1 ARCHITECTURAL ACOUSTICS

The science of architectural acoustics is used to control sound propagation within buildings. The objective is to provide environments where occupants hear what they want to hear and are not seriously bothered by unwanted sounds.

Sound control in buildings may be divided into two categories.

1. Sound distribution within an occupancy - wanted sounds are heard properly by the recipients, without being blurred by reverberation or masked by noise. This may involve the appropriate shaping of the room boundaries and the use of sound absorbing surfaces. Typical concrete products do not provide good sound absorption, and other materials may be necessary. *This topic is rather specialized and is not dealt with here.*
2. Sound insulation between occupancies - noises originating in one occupancy do not intrude in adjacent occupancies. This mainly involves the provision of adequate barriers to sound transmission. Typical concrete elements are sufficiently massive to form effective ingredients in a sound insulation design.

This chapter deals primarily with sound insulation.

## 5.2.2 AIRBORNE SOUND

*Airborne sound* reaches a partition by propagating from the source through the air. When sound waves strike a partition, the variation in sound pressure causes the partition to vibrate. This sets the air on the other side in motion, thereby generating sound. Only a very small fraction of the vibrational energy carried by the sound waves is transferred to the partition. The amount depends on the construction of the partition.

### Sound Transmission Loss

When sound propagates between two rooms through a common partition, the principal factor controlling the sound pressure level in the receiving room (where the sound is received) is the *sound transmission loss* (TL) of the partition.

The sound transmission loss ,expressed in decibels (dB), is a measure of the ratio of the energy striking the partition relative to the energy which is transmitted through it. The greater the sound insulation provided by a partition, the higher its TL. The sound transmission loss of a partition varies with frequency, usually increasing as the frequency increases.

The sound pressure level in the receiving room also increases as the area of the common partition increases and decreases as the amount of sound absorbing material in the room increases. These two effects are secondary. Sound transmission loss is measured in specially constructed reverberation rooms in accordance with ASTM E90[R5.2.1]. Measurements can also be made in buildings by following ASTM E336[R5.2.2].

### Single-leaf Walls

The term *single-leaf partition* refers to all types of solid homogeneous panels where both faces are rigidly connected. Examples are gypsum board, plywood, concrete block, brick and concrete.

The transmission loss of a single-leaf partition depends mainly on its surface mass (mass per unit area). The heavier the partition, the less it vibrates in response to sound waves and therefore the less sound it radiates on the side opposite the sound source.

### Mass Law

The mass law is a semi-empirical expression which may be used to predict transmission loss for thin, homogeneous single-leaf panels.

The mass law is expressed as:

$$TL = 20 \log (m \cdot f) - 48 \text{ dB}$$

where:

TL = transmission loss (dB)
m = surface mass (kg / $m^2$)
f = frequency (Hz)

The mass law predicts that the transmission loss will increase by 6 dB for each doubling of the surface mass or frequency. To increase the TL by 3 dB requires an increase in the mass by a factor of 1.4. A change of less than 3 dB is considered insignificant.

Transmission loss also depends on material properties, such as stiffness. The transmission losses of two single-leaf walls are illustrated in Fig. 5.2.1. For the 16 mm gypsum board, agreement with the mass law is good below about 2 kHz . At higher frequencies, there is a dip in the TL curve, called the *coincidence dip,* where the wavelength of the flexural vibrations in the wall coincides with the wavelength of the sound in the air.

The frequency where the minimum value of TL occurs in the coincidence dip is called the critical frequency. The critical frequency depends on the material stiffness and thickness. The stiffer or thicker the layer of material, the lower the critical frequency. Mass law only applies at frequencies below about one half of the critical frequency.

For gypsum board, the critical frequency is given by $f_c = 39\,000/t$, where t is the thickness in mm. For solid, normal density concrete, the critical frequency is given by $f_c = 18\,700/t$. For example, a 150 mm thick concrete slab weighs about 345 kg/m$^2$ and has a coincidence frequency at 125 Hz. As shown in Fig. 5.2.1, the transmission loss for the 150 mm concrete slab is well below that predicted by mass law over most of the frequency range shown.

**Fig. 5.2.1 Transmission loss curves for single-leaf partitions - 15.9 mm gypsum board and 150 mm concrete.** The thick solid lines are measured data. The dotted lines are mass law predictions for each material.

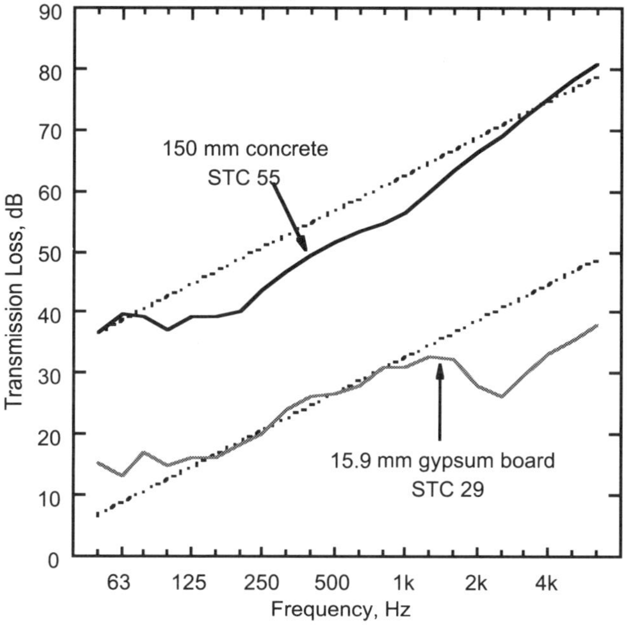


## Sound Transmission Class

To simplify acoustical design, it is convenient to replace the detailed TL data by a single-number rating known as the sound transmission class (STC)[R5.2.3].

The STC is determined by comparing the TL values in the 16 one-third octave bands from 125 to 4000 Hz with a reference contour covering the same frequency range. The reference contour is adjusted until the TL curve has an average deficiency of no more than 2 dB. (A deficiency occurs when a TL value lies below the STC contour). Additionally, no measured TL value may lie more than 8 dB below the STC contour. The sound transmission class is then given by the 500 Hz value for the fitted STC contour.

Fig. 5.2.2 shows transmission loss data and the position of the reference contour after the fitting process is complete. The higher the STC rating, the better the sound insulation provided by the partition.

Sound transmission class is intended for rating partitions for sounds such as speech, which have most of their energy in the middle and high-frequency range. Nevertheless, it is widely used for rating the airborne sound insulation of partitions, doors, and windows against other types of noise such as transportation noise, music, and machinery noise.

When the noise to be attenuated has a strong low-frequency content, the perceived noise reduction may be less than the sound transmission class indicates. In such cases, it is better to estimate the loudness of transmitted sound using the noise and transmission loss information down to the lowest frequency where data are available.

**Fig. 5.2.2 Example of fitting the STC contour to measured data for a 150 mm thick concrete slab.** Deficiencies are shown by the shaded area.

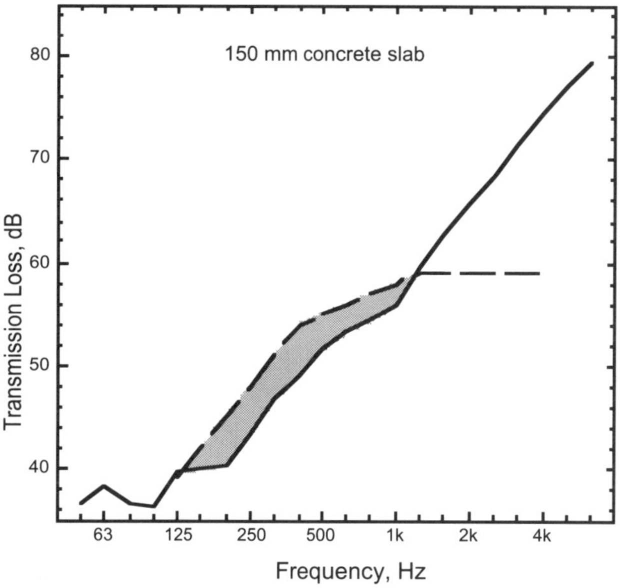


## Two Leaf Walls

A two-leaf partition contains two single-leaf layers with an airspace in between. Such partitions usually provide substantially higher sound insulation than single-leaf partitions with the same total mass. The resulting STC will not be as high as the sum of the STCs for the individual leafs.

The sound insulation provided by a two-leaf partition depends on 4 main factors:

1. *Mass per Unit Area of Component Leafs.* In general, the heavier the component leafs, the higher the sound insulation provided by a two-leaf partition.
2. *Mechanical Connection Between the Two Leafs.* Rigid mechanical connections between the two leafs permit sound to travel efficiently from one leaf to the other, reducing the sound insulation of the construction. If possible, mechanical connections should be avoided, for example, by constructing the two leafs to stand independently of each other. Where mechanical connections are required, they should be sufficiently resilient to avoid reductions in TL.
3. *Sound Absorptive Material in Airspace Between Leafs.* The addition of sound-absorptive material in the cavity of a double-leaf partition may improve the sound insulation by 10 or more points. *To achieve this improvement, the two leafs must not be rigidly connected to each other.* Where the two leafs are *rigidly* connected by structural supports, the addition of sound-absorptive material within the cavity will provide only small improvements in the sound insulation.

Other points to note are:

• Porous fibrous materials (such as glass, mineral or cellulose fibre) provide significant improvement. The type and density of the fibrous material does not influence the STC much.

• Closed-cell foam materials (such as foamed polystyrene) are poor absorbers of sound and do not improve the sound insulation.

• The airspace between leafs should be filled with about three-quarters of sound-absorptive material. Using more than this will provide little additional increase in sound insulation. The fibrous material should not be so densely compressed that it forms a solid bridge between leafs and reduces the sound insulation provided by the partition.

• The position or arrangement of the sound absorptive material inside the cavity has no significant effect provided the whole area of the partition is covered. For example, in staggered stud construction, it does not matter whether the material is against one face or zigzags between the studs.

4. *Depth of the Airspace Between the Two Leafs.* One potential disadvantage of cavity wall or floor systems is that air trapped in the cavity between the two leafs can act as a spring and transfer vibrational energy from one layer to the other. This interaction between the leafs causes a resonance called the *mass-air-mass resonance* and an associated dip in the transmission loss curve. The TL can be reduced to less than that for a single-leaf of the same total weight which in turn can lower the STC rating.

The frequency of the mass-air-mass resonance can be calculated from:

$$f_{mam} = K\sqrt{\frac{m_1 + m_2}{dm_1 m_2}}$$

where:

$f_{mam}$ = the mass-air-mass resonance frequency (Hz)
$m_1$ = the surface mass of the first layer (kg/m$^2$)
$m_2$ = the surface mass of the second layer (kg/m$^2$)
$d$ = their separation (m)
$K$ = 60 for an empty cavity
= 43 for a cavity filled with sound absorbing material

Adding sound absorbing material to the cavity will lower the resonance frequency which is why there are two values for the constant K.

The improvement in TL due to the double-layer construction begins about two-thirds of an octave above $f_{mam}$. Therefore, to maximize the improvement due to the cavity, the mass-air-mass resonance frequency should be as low as practical which implies large air spaces or heavier materials. Usually, designing for a mass-air-mass resonance of 80 Hz means that the presence of the cavity will improve the TL from 125 Hz upward, thereby ensuring an increase in the STC.

For a given total weight, the mass-air-mass resonance is lowest when $m_1 = m_2$. Having both layers with the same weight and stiffness could, however, lead to low TL values around the coincidence dip.

Many common partitions have the mass-air-mass resonance in the frequency range of common low-frequency noise sources. The deficiencies in their low-frequency transmission loss are clear when a stereophonic sound system plays in an adjacent room; the melody line is barely audible, but bass sounds are easily heard.

## 5.2.3 CONCRETE AND MASONRY FLOORS AND WALLS

Precast concrete floor and roof slabs are often supported on concrete and masonry bearing walls. Non-load bearing masonry partition walls are often installed in precast concrete structures. Information on the acoustical performance of concrete and masonry assemblies is given in this section.

**Fig. 5.2.3 STC for some common concrete floors and walls**

| | Mass, kg/m$^2$ | STC |
|---|---|---|
| **Solid slabs or walls** | | |
| 50 mm | 115 | 43 |
| 70 mm | 161 | 47 |
| 100 mm | 230 | 50 |
| 150 mm | 345 | 52 |
| 200 mm | 460 | 58 |
| **Hollow-core slabs** | | |
| 150 mm | 220 | 48 |
| 200 mm | 278 | 50 |
| 250 mm | 312 | 50 |
| **Precast double tee** | | |
| 350 mm deep with 50 mm topping | 366 | 54 |

### Single-leaf Concrete

Fig. 5.2.3 gives representative STC values for some common concrete walls or floors. Simple concrete partitions can provide STCs from about 45 to 55. Hollow-core slabs have slightly lower values than solid slabs. For STCs much greater than 55, the weight required will likely be prohibitive in modern buildings (surface mass must be doubled for an increase of 6 dB). Multi-layer partitions are often used where high STC ratings are required - a central massive wall with one or more leafs attached to each side. This is discussed below.

### Single-leaf Masonry Walls

In the thicknesses commonly used in buildings, concrete block provides approximately the same sound insulation as a solid concrete wall or brick wall of the same weight per unit area - provided all cracks, openings and voids are properly sealed.

Solid and hollow concrete blocks are manufactured in a variety of thicknesses, core sizes, aggregates, and densities. Representative values of the sound transmission class (STC) for block walls are given in Fig. 5.2.4. These values are only valid if the wall surfaces are properly sealed and the mortar joints are sound.

**Fig. 5.2.4 Sound transmission class (STC) ratings for normal and lightweight hollow block walls sealed on at least one side.** The block face dimensions are 190 × 390 mm.

| | Lightweight block | | Normal weight block | |
|---|---|---|---|---|
| Block thickness (mm) | Weight per block, (kg) | STC | Weight per block, (kg) | STC |
| 90 | 7 | 43 | 10 | 44 |
| 140 | 10 | 44 | 15 | 46 |
| 190 | 13 | 45 | 17 | 48 |
| 240 | 15 | 47 | 21 | 49 |
| 290 | 18 | 48 | 25 | 51 |

The sound insulation provided by a hollow block depends not only on the density of the block material but also on its porosity. The more porous the block, the more sound will leak through the block structure and the greater will be the improvement if the surfaces of the block are sealed, for example, by concrete paint, epoxy paint, or a skim coat of plaster. Improvements of 5 to 10 STC points are not uncommon after sealing. The porosity of acoustical materials (airflow resistivity) is a measure of the volume and velocity of air flowing through a specimen and the air pressure drop across it (ASTM C522[R5.2.4]).

A layer of gypsum board attached directly with screws or dabs of glue to the block surface is not an effective seal because it is still able to vibrate as a separate layer.

When gypsum board is used to finish a masonry wall, at least one face of the block should be sealed. In fact when gypsum board is added on resilient supports to finish the wall, this can be an advantage. Covering a flaw with gypsum board does not eliminate the detrimental effects of the flaw.

### Two-leaf Masonry Walls

Two-leaf concrete or masonry walls are capable of providing very high sound insulation because they comprise two heavy layers separated by an airspace. The maximum insulation attained may be limited by the practical difficulties of constructing two leafs that are structurally isolated.

Depending on the height, block walls may require metal ties for structural reasons or to meet applicable building codes. Ties transmit sound energy from one leaf to the other. The detrimental effects of mechanical ties can be minimized by the use of special ties with interlocking hooks or resilient couplings. During construction, the airspace between two leaf masonry walls may become filled with mortar droppings or rubble. This can bridge the gap between the walls and reduce the sound insulation. Such construction defects are usually concealed and impossible to correct after the wall is complete.

Sound energy is transmitted along floors and ceilings, along walls abutting the periphery of cavity walls, and through other parts of the structure. These flanking paths bypass the cavity wall and reduce its effective TL. This kind of flanking is illustrated in Fig. 5.2.5. Physical breaks in the floor, ceiling, and abutting walls are needed to reduce transmission along these paths.

**Fig. 5.2.5 Flanking paths under a cavity block (or concrete) wall resting on a continuous slab of concrete.** Any solid connection between the block walls will reduce the sound insulation.

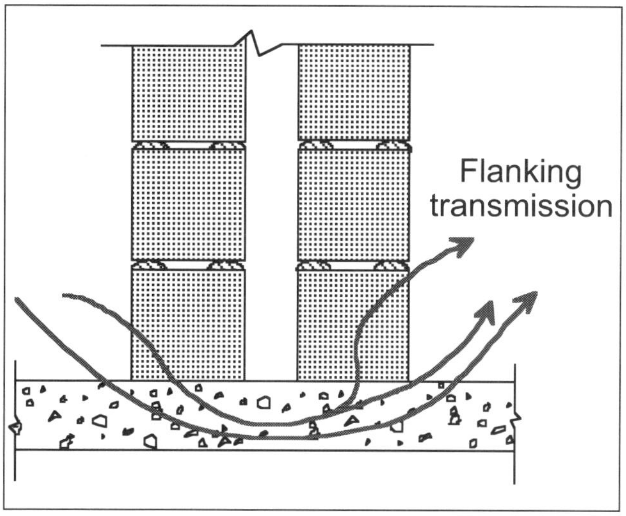


In summary, two-leaf concrete walls can provide very high sound insulation but great care and experience are needed in the overall design and construction. In most cases, it is more practical to attach gypsum board or other lightweight panels on resilient mounts as described below.

### Concrete or masonry walls with added gypsum board

A common method for improving the sound insulation of an existing concrete or masonry wall is to add a layer of gypsum board on one or both sides of the wall. It is essential to first seal masonry walls.

The gypsum board is best supported using:

- steel or wood studs not in contact with the wall
- resilient metal furring attached to the wall
- wood furring and resilient metal channels attached to the wall

Wood furring alone attached to the wall with nails, screws and other fasteners is often used to attach gypsum board to the wall. This practice is not recommended as these types of rigid connections provide a path for vibrational energy and the benefits of the added gypsum board may be lost.

**Fig. 5.2.6 Increase in STC ratings when a single layer of 16 mm gypsum board is added to one or both sides of a concrete block or poured concrete wall, with and without fiberglass batts filling the cavity between gypsum board and the concrete.** The same values may be used for 13 mm gypsum board without serious error.

| | Without fiberglass in cavity | | With fiberglass in cavity | |
|---|---|---|---|---|
| **Gypsum board attachment** | **one side** | **both sides** | **one side** | **both sides** |
| Directly on concrete block | + 0 | − 1 | | |
| On 13 mm resilient steel channels | + 2 | − 1 | + 4 | − 1 |
| On 40 mm wood furring | + 3 | + 4 | + 5 | + 9 |
| On 50 mm resilient steel furring | + 2 | + 2 | + 9 | + 14 |
| On 65 mm steel studs | + 8 | + 7 | + 10 | + 22 |
| On 75 mm resilient steel furring | + 7 | | + 11 | |

The measured improvement in sound transmission class (STC) rating which can result from the attachment of 15.9 mm gypsum board in a variety of ways to concrete and masonry walls is given in Fig. 5.2.6. The STC ratings for complete wall systems can be calculated by adding the improvements given in Fig. 5.2.6 to the STC rating of the concrete walls from Fig. 5.2.3 or the masonry walls from Fig. 5.2.4.

Note that when the cavity behind the gypsum board is too small, the STC is actually lowered relative to the bare block case. Note also the very large increases in STC when the cavity is large and filled with sound absorbing material.

### Multi-element partitions

When a partition comprises two or more different components (for example, a wall containing a door or window) the effective value of airborne sound insulation is determined by the areas and sound transmission properties of the individual components. Components having low values of sound insulation, such as windows and doors, reduce the overall sound insulation.

By selecting the areas and transmission losses of the individual components, however, the STC rating for a multi-element partition can meet the design criterion. It is usually uneconomical to require that all components have the STC rating required for the composite structure. Instead, sound transmission through the weaker component(s), is compensated for by improving the stronger components (usually the wall) beyond the overall design criterion. There are limits, however, to what can be achieved.

Calculations should be made for each of the individual TL values at each frequency band, and then the STC derived for the resultant TL curve. However, Fig. 5.2.7 can be used to calculate an approximation to the STC value of each component so the combination has the specified STC rating.

To use the table proceed as follows:

1. First, determine the fractional area of each component, i.e., $A_1/A$, $A_2/A$. where A is the total area occupied by all components and $A_1$, $A_2$, etc. are the areas of each component.
2. For each value of fractional area, find the adjustment in Fig. 5.2.7 in the column appropriate for the total number of components.
3. Determine the STC value of the component by adding the adjustment of Step 2 to the specified value of STC value for the composite partition.

**Fig. 5.2.7 Adjustments for determining component sound transmission class from fractional area and number of components in partition.**

| Fractional Area | Number of Components 2 | 3 | 4 |
|---|---|---|---|
| 0.0315 | − 12 | − 10 | − 9 |
| 0.0400 | − 11 | − 9 | − 8 |
| 0.0500 | − 10 | − 8 | − 7 |
| 0.0630 | − 9 | − 7 | − 6 |
| 0.0800 | − 8 | − 6 | − 5 |
| 0.1000 | − 7 | − 5 | − 4 |
| 0.1250 | − 6 | − 4 | − 3 |
| 0.1600 | − 5 | − 3 | − 2 |
| 0.2000 | − 4 | − 2 | − 1 |
| 0.2500 | − 3 | − 1 | 0 |
| 0.3150 | − 2 | 0 | 1 |
| 0.4000 | − 1 | 1 | 2 |
| 0.5000 | 0 | 2 | 3 |
| 0.6300 | 1 | 3 | 4 |
| 0.8000 | 2 | 4 | 5 |
| 1.0000 | 3 | 5 | 6 |

**Example**: Two components, a wall and a door, are to be combined so the STC rating of the composite partition is 50. The door occupies 0.25 of the total area and the wall occupies the remaining 0.75. From Fig. 5.2.7, the adjustment term for the door is –3. The nearest value in the table to 0.75 is 0.8 for which the adjustment term is +2. Thus the STC for the wall is 50 + 2 = 52 and the STC for the door is 50 – 3 = 47.

## 5.2.4 FLANKING TRANSMISSION

In buildings, there are many paths for the transmission of airborne, impact, and structureborne sound. Impact sound is sound that originates somewhere in the building as one or a series of impacts. The more general term structureborne sound refers to sound that originates as impacts or vibrations in the structure of the building.

Fig. 5.2.8 shows a building where all the components are rigidly connected. Sound energy can be transmitted through the ceilings, walls, and floors of the structure to reach adjacent rooms and cause annoyance. Partitions other than the common partition may be involved.

**Fig. 5.2.8 Direct (D) and flanking (F) sound paths for airborne and impact (structureborne) sound in a concrete building.**

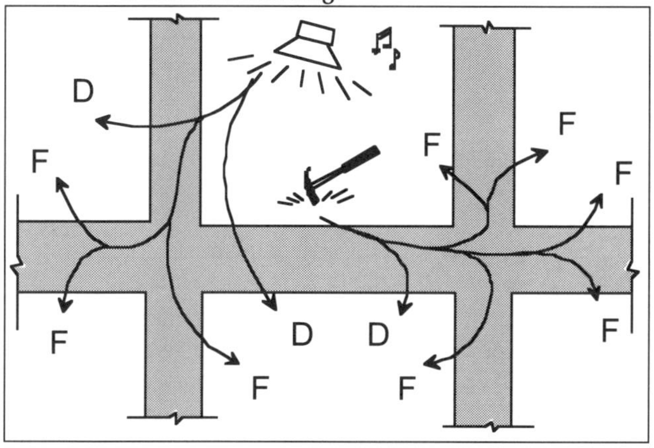


A flanking path is a path for sound transmission that allows sound to bypass the element that is intended to be the only significant sound path between two spaces. Fig. 5.2.8 indicates several flanking paths. Sound entering a structure is propagated as vibration which can travel for considerable distances. The energy lost in traveling from one point to another depends on the type of materials used and the details of construction. A vibrating wall or floor continually re-radiates energy as sound from both sides.

To provide the maximum possible sound attenuation in buildings, energy transmission along flanking paths should be reduced by introducing structural breaks (such as isolation joints) and resilient connections. Ideally in a residential building, each apartment should be independent and resiliently supported.

In practice, the ideal situation can be approached by:

1. Using independent structures wherever possible,
2. Using resilient metal channels to support gypsum board walls and ceilings, and
3. By installing floating floors.

Resiliently suspending the ceiling below the floor as in Fig. 5.2.9 reduces the sound transmission for the path directly through the floor. Flanking sound transmission to the adjacent rooms on the same level or flanking transmission down the walls are not reduced.

**Fig. 5.2.9 A resiliently suspended ceiling in a concrete building reduces direct sound but not flanking sound.**

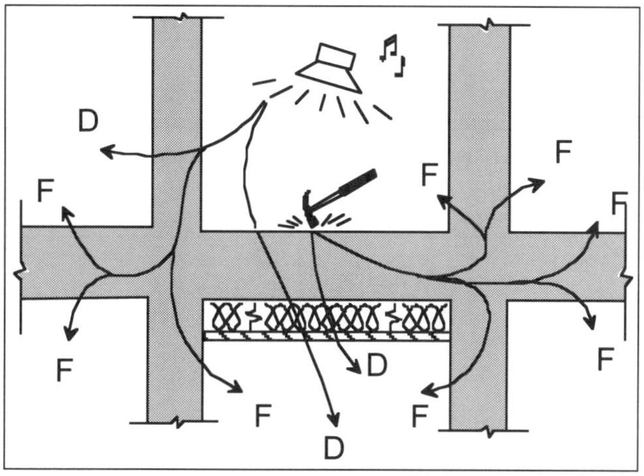


Impacts can be controlled very effectively using resilient floor coverings such as carpet. In cases where a hard floor surface is needed, the best way to reduce impact sound transmission from the floor to the building structure is to use a floating floor shown in idealized form in Fig. 5.2.10.

**Fig. 5.2.10 Use of a floating floor to reduce the transmission of impact sound to the building structure.**

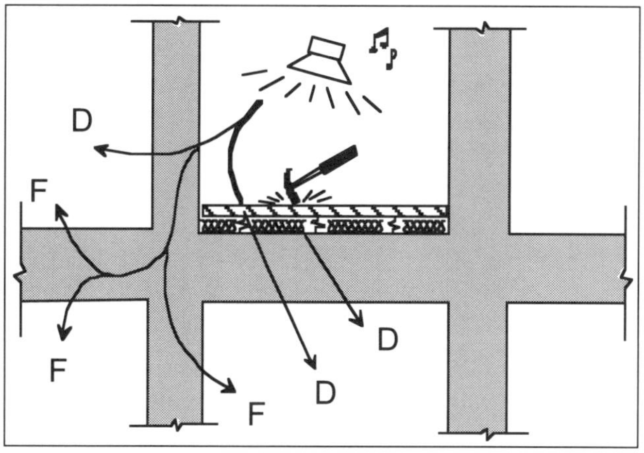


A floating floor is a slab that is supported by a structural floor but is isolated from it by a resilient support. Floating floors work well because the most effective place to control structureborne or impact noise is at its source.

Reducing the transmission of noise along the remaining direct and flanking paths from the room containing the noise sources to the adjacent rooms requires the use of resiliently or independently supported walls and a floating floor.

This kind of construction approximates the ideal condition in a building with each room not connected to the structure at all or at least only resiliently. For example, 65 mm steel studs could be used to support gypsum board on each side of a concrete or block wall.

Fig. 5.2.11 shows how this might be done in practice. Note that sound insulation measures act in both directions, i.e., the room containing the noise sources is also protected against noise coming from the adjacent rooms.

**Fig. 5.2.11 Use of floating floors and gypsum board supported independently on steel studs to attenuate direct and flanking transmission of airborne and impact sound.** Caulking is indicated by black dots.

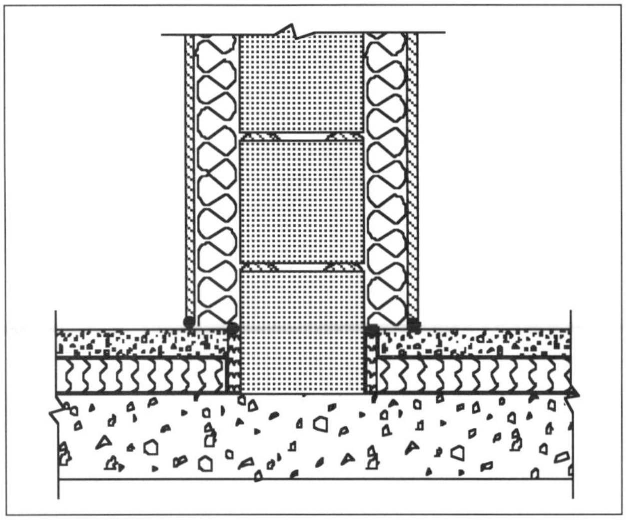

## 5.2.5 FLOATING FLOORS

Hard floor surfaces are often required (for example; kitchens, bathrooms, hospitals, ballrooms, etc.). In these cases, if excellent impact noise insulation is required, a floating floor may be used.

There are several ways of constructing a floating floor:

- The floating slab can be supported on a continuous layer of commercially available resilient material which may or may not be sound-absorptive. (see Fig. 5.2.12).

**Fig. 5.2.12 Example of a concrete slab floating on a continuous layer of resilient material.**

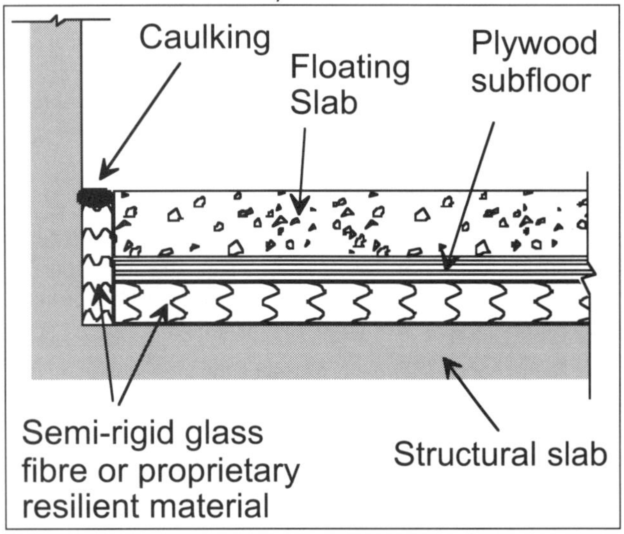


- The floating slab can rest on relatively small resilient pads or vibration isolators, spaced at intervals under the slab. The airspace between the floating slab and structural floor may or may not contain sound-absorptive material.
- In critical applications, the floating slab, usually quite thick, can be supported on commercially available "screw-jack" vibration isolators. These are placed in position on the structural floor and then used to lift the floating slab a few centimeters after it has been poured and cured. Unless special steps are taken, such floors will not have sound-absorptive material between the floating slab and the structural slab.

Floating slabs are most effective if they are relatively heavy (at least 50 mm of concrete). In large mechanical rooms, the concrete slab is often 100 mm or thicker.

Floating floors attenuate sound above a resonance frequency that is determined by the stiffness of the resilient material and the weight of the slab. The more compliant (the less stiff) the support or the heavier the slab, the lower the frequency where the floating assembly becomes effective. Adding a carpet and underpad on top of a floating slab further improves the impact sound insulation rating.

In designing a floating floor, consider the following:

- The floating slab must have adequate structural strength for the load it is to support.
- The resilient support must have load-bearing characteristics appropriate for the floating slab and other loads it is to support. At the same time, it must be compliant enough so the assembly attenuates vibration over the range of required frequencies.
- The floating slab must not contact the building structure. Fig. 5.2.12 shows an edge detail.
- Avoid penetrations of the floating slab by pipes, ducts, etc. Where a penetration is essential, ensure that it does not form a rigid connection between the floating slab and the structural slab or walls (see Fig. 5.2.13).

**Fig. 5.2.13 One method of isolating a pipe that passes through a floating floor assembly.**

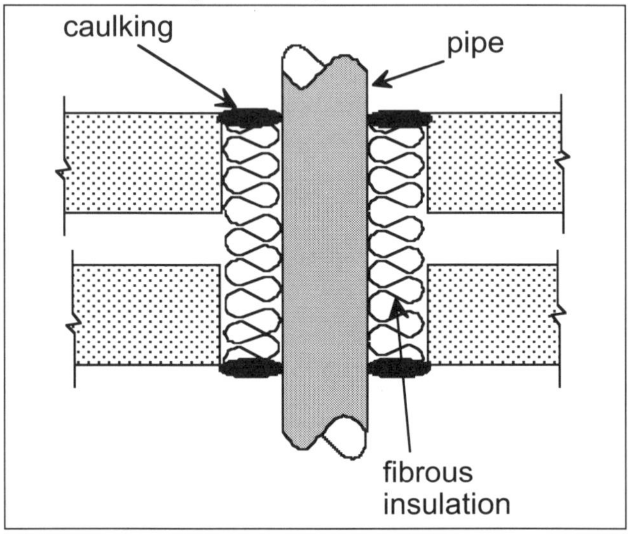

## 5.2.6 IMPACT SOUND TRANSMISSION

Noise from footsteps and other impacts is a common source of annoyance in buildings. It is more difficult to control than airborne sound because energy transfer is much more efficient when the vibrating or impacting source touches and directly excites the surfaces of the building.

Laboratory and field test methods give single number ratings for the transmission of impact sound through floors.[R5.2.5, R5.2.6] These tests use a standardized tapping machine with five steel-faced hammers that strike the floor at a total rate of 10 times per second. Sound pressure levels are measured in the room below the floor in the frequency range 100 to 3150 Hz and the resulting data are fitted to a reference contour to obtain a single number rating - impact insulation class (IIC)[R5.2.7]. Some examples of test results are shown in Fig. 5.2.14.

The higher the IIC rating, the greater the impact noise insulation provided by the construction against this kind of impact. The IIC value is determined by those values lying above the reference IIC contour once the fitting process is complete. As with the STC, the average deficiency must not exceed 2 dB and no single deficiency may exceed 8 dB.

For reasonable protection against impact noise, the IIC needs to be about 55 or more. Fig. 5.2.14 shows that a bare concrete floor, which has an IIC rating controlled by the high frequencies, may be unacceptable. Adding carpet and/or carpet with an underpad makes a very large difference to the IIC. Most of the high frequency noise is reduced and the IIC is controlled by the lower frequencies.

**Fig. 5.2.14 Tapping machine data for a 150 mm bare concrete floor tested bare (IIC 25), with a carpet (IIC 68), and with a carpet and foam underpad (IIC 86).**

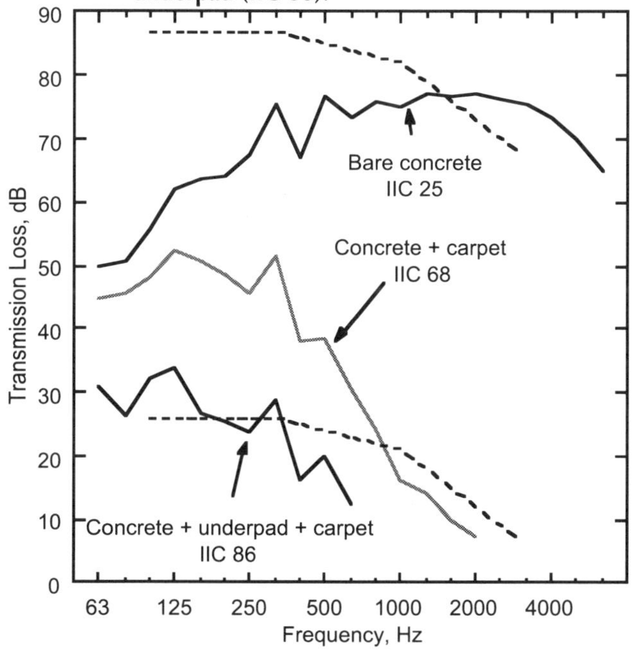


### Improving IIC Ratings for Concrete Floors

Bare concrete floors or those with hard finishes, such as tile or hardwood, provide low impact sound insulation. The thickness of the slab makes little difference. The IIC ratings are usually between 25 and 35. This is an example of the fact that a construction that provides good insulation against airborne noise reduction does not necessarily provide good insulation against impact noise. Also, good impact noise insulation does not necessarily mean good airborne sound insulation.

A soft, resilient floor covering cushions impact forces and reduces the energy transferred to the building structure. The more resilient or soft the floor covering the greater the increase in the IIC. The improvement in IIC depends on the characteristics of the floor structure, so different carpets and underpads give different results. Typical soft carpets and underpads used in the home give IIC ratings around 80. Vinyl floor coverings and similar products improve the IIC of the bare slab only about 5 to 10 points.

A ceiling supported on resilient hangers below a concrete slab will also increase the IIC rating. The improvement depends on the method of support, the cavity depth, the weight of the ceiling and the amount of sound absorbing material present in the cavity. Test data are needed to establish ratings for such floors.

Floating floors can improve the IIC ratings for a concrete slab by about 30 to 40 points. The value of IIC obtained depends on the properties of the resilient material used to support the floating layer, the depth of the cavity below the floating slab, the weight and properties of the slab and whether or not there is sound absorbing material present in the cavity. The IIC ratings are not as good as those provided by a soft carpet and underpad, but still very good for use in multifamily homes.

### Lightweight joist constructions

Wood-joist and truss floors perform quite differently than solid concrete floors. When finished with carpet and underpad, the IIC rating of lightweight wood-joist and truss floors can be quite high. However, such lightweight floors may transmit enough low-frequency noise to be judged unacceptable by many occupants.

This conflict between subjective reactions and IIC ratings arises because the impact insulation class rating system does not consider low-frequency sound below 100 Hz. The ratings for lightweight wood-joist and truss floor systems tend to be optimistic in their evaluation of impact sound insulation. Occupants living below such floors often complain of "thumps", creaking sounds, or rattling sounds when people walk on the floor above.

Similar low frequency problems can occur with relatively thin concrete floors supported on steel joists. Lightweight floors vibrate more in response to impacts and thus generate more sound.

Typically, heavy concrete floors generate about 10 dB less noise at low frequencies than do lighter joist or truss systems. For this reason, in critical applications, the more massive constructions usually are preferable.

### 5.2.7 CONTROL OF STRUCTUREBORNE NOISE IN BUILDINGS

Structureborne sound arises when a machine or vibrating device is rigidly attached to the building structure. A source that is rigidly connected to the building structure will be heard much more easily in adjacent spaces than when it is not rigidly connected. Structureborne sound can be controlled at its source, along its transmission path, and at its point of reception. Control at the point of origin is best.

To control structureborne noise effectively, the following techniques are recommended:

- Provide vibration isolation between machines and the structure by mounting the machines on soft rubber or neoprene pads, on springs or on an inertia block which rests on springs.
- Reduce the power of a source of vibration by changing its operating conditions or by selecting a different model that introduces less vibration into the structure. Well-balanced rotating machinery will generate less vibration than reciprocating equipment.
- Locate the source of vibration as far as possible from areas where low noise levels are required. For example, locate elevators and garbage chutes away from bedrooms in apartments and don't place a gymnasium above a school library.
- Use breaks (for example, expansion joints) in an otherwise solid building construction to impede the transmission of sructureborne noise.

#### Plumbing Noise

The plumbing system in a building can be considered as an extended machine and source of vibration. The same considerations given above apply. Wherever possible, the pipes and appliances should not be directly attached to the structure of the building. Resilient collars and supports should be used. Quiet faucets and toilets are available but usually cost more than the standard noisy devices. See the references for more information on this topic.

### 5.2.8 SOUND LEAKS

Common causes of sound leaks in buildings are:

- Cracked or incomplete mortar joints in a masonry wall.
- Unsealed gaps around power outlets.
- Gaps under gypsum board walls where caulking has been omitted.
- Openings around pipes where they penetrate walls.
- Cracks around the edges of doors and windows.

To eliminate sound leaks:

- Inspect and repair all mortar joints in concrete block walls and seal the surface if necessary.
- Apply non-hardening caulking to seal all holes and fissures in party walls or floors.
- Apply rubber gaskets or non-hardening caulking to seal openings around pipes and conduits. (This also prevents the pipes forming a solid link and an easy path for vibration between the two sides of a wall).
- Apply weather-stripping to windows and doors.
- Repair and seal holes made during installation of services.

Although the area of a leak can be quite small, the TL is roughly zero. Unless the area of the leak relative to that of the wall is very low, the STC is determined largely by the leak area. For example, if the leak area is 0.0001 times the wall area, the STC can never exceed 40 no matter how good the wall construction. The higher the acoustical isolation required, the more important it is to eliminate all sound leaks.

### 5.2.9 CONSTRUCTION DETAILS

The details of the construction where a wall and floor join greatly influence the sound insulation that the wall provides. Consider Fig. 5.2.15 which shows a floating thin concrete slab. The wall and the floor both provide very good sound insulation when tested individually, yet the sound insulation is poor between the rooms on each side of the wall because of the flanking transmission along the upper surface of the floor. The wall is "short-circuited" by the concrete slab.

**Fig. 5.2.15 Flanking transmission along a continuous lightweight floating slab will seriously degrade the sound insulation between the two adjacent rooms**

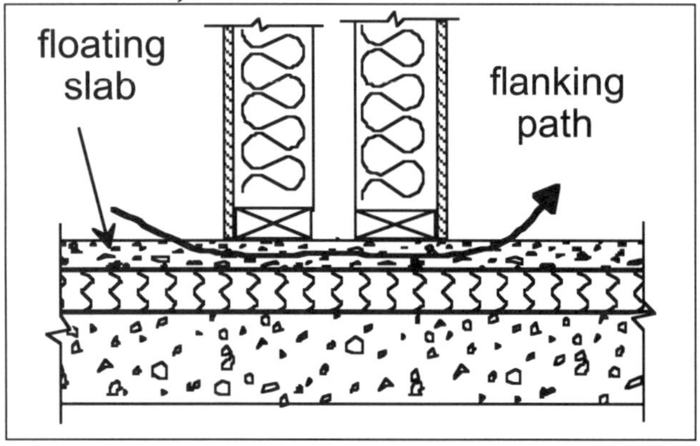


**Fig. 5.2.16 Prevention of flanking transmission using separate floating slabs in each dwelling.**

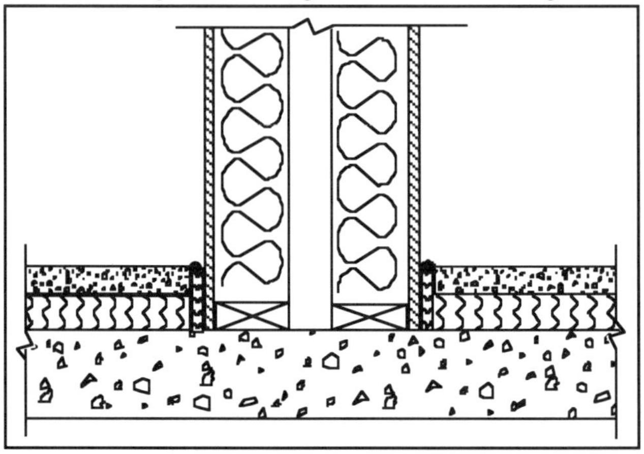

The problem may be circumvented by introducing a break in the floating slab as shown in Fig. 5.2.16. The introduction of a structural discontinuity reduces or prevents the passage of vibration in the construction. In some cases, such construction breaks cannot be employed to conform to applicable building codes or to maintain the structural integrity of the building.

## 5.2.10 SOUND ABSORPTION OF CONCRETE

Sound absorption is measured in a reverberation room following ASTM C423 [R5.2.8]. Specimen size is typically $2.4 \times 2.7$ m but can differ in certain circumstances. Measurements are made at several frequencies and an average value is calculated. For many years this average has been called the noise reduction coefficient (NRC) but this may be replaced soon by the sound absorption average (SAA).

Normal density concrete is not an efficient sound absorber. Some low density concretes made with porous aggregates (expanded shale, slag or mica) are reasonably effective absorbers, but the manufacturing process must be carefully controlled so the individual aggregate particles are not sealed by a coating of slurry. Even if this is achieved, the resulting absorption is easily destroyed if the surface is painted sufficiently to seal the porous surface.

A simpler approach is to add an absorptive surface over the concrete such as sprayed or trowelled acoustical plasters. Layers of fibrous sound absorbing materials may be attached and protected by some durable covering.

Thin layers of material attached directly to the surface of a wall are relatively inefficient at lower frequencies. To improve low-frequency absorption, thick porous layers, resonant cavity systems, porous layers behind a perforated screen, sound absorbing material mounted on furring or a suspension system that leaves space behind, all may be used. Manufacturers' trade literature usually provides sound absorption characteristics for sound absorbing materials on typical mounting systems.

## 5.2.11 SOUND INSULATION IN MULTI-UNIT DWELLINGS

Several factors determine whether noise from neighbors will bother the occupants of multi-family dwellings:

- The airborne sound transmission and impact sound transmission characteristics of party walls and floors.
- The noise level in the neighbouring space.
- The background noise level in the occupant's own dwelling.
- The sensitivity of the occupant.

The last three factors vary widely. The sound insulation value of a wall or floor partition must be selected to provide protection for most, but not all, situations. Sound insulation requirements for multi-family dwellings are usually based on experience and social surveys that have been conducted in various countries.

### Recommended Sound Insulation Criteria for Multiple Dwellings

Mandatory minimum sound insulation requirements between adjoining dwellings are generally specified in building codes and equivalent regulations. For example, the National Building Code of Canada[R5.2.9] specifies a minimum STC of 50 for party walls and floors.

A more detailed approach, which considers the sound insulation needed between various activity spaces in adjacent dwellings is embodied in Fig. 5.2.17 and Fig. 5.2.18. These tables give recommended effective sound insulation values for walls and floors between homes in terms of sound transmission class and impact insulation class ratings, respectively. The ratings may be higher than those commonly required in some national or municipal jurisdictions, but are similar to those used in many European countries and provide a quality living environment. Even higher values can be achieved with typical building ma-

**Fig. 5.2.17 Recommended minimum sound transmission class (STC) requirements for airborne sound insulation of walls and floors separating dwellings.** Values are given for separation of specific categories of rooms. If two or more of the categories are combined in one unpartitioned space the higher requirement should be used.

| | | A | B | C |
|---|---|---|---|---|
| | **Separated spaces** | Bedrooms | Living, etc. | Service spaces |
| A | Bedrooms | 50 | | |
| B | Living, dining, family rooms | 55 | 50 | |
| C | Kitchen, bathroom, hallway, foyer, storage, utility space | 55 | 50 | 45 |
| D | Service spaces common to two or more dwelling units | | | |
| | (a) Typically quiet, e.g., corridors, stairways, storage spaces | 50 | 50 | 45 |
| | (b) Typically noisy: garages, garbage disposal areas (including garbage chutes), mechanical equipment rooms, furnace rooms, laundries, squash courts, party rooms | 70 | 65 | 60 |

**Table 5.2.18 Minimum requirements for impact insulation class (IIC) floors separating dwellings.** Values are given for separation of specific categories of rooms. If two or more of the categories are combined in one unpartitioned space the higher requirement should be used.

| | | Room below | | |
|---|---|---|---|---|
| | | A | B | C |
| | Upper room | Bedrooms | Living, etc. | Kitchen, etc. |
| A | Bedrooms | 50 | 45 | 45 |
| B | Living, dining, family rooms | 55 | 50 | 45 |
| C | Kitchen, bathroom, hallway, foyer, storage, utility space | 55 | 50 | 45 |
| D | Service spaces common to two or more dwelling units | | | |
| | (a) Typically quiet, e.g. corridors, stairways, storage spaces | 55 | 50 | 45 |
| | (b) Typically noisy: garages, garbage disposal areas (including garbage chutes), mechanical equipment rooms, furnace rooms, laundries, squash courts, party rooms) | 65 | 60 | 50 |

terials without resorting to extreme designs although care is needed during the design and construction processes.

The ratings in Fig. 5.2.17 and Fig. 5.2.18 represent recommended insulation values in completed constructions. Selection of building components which have met these ratings will not guarantee that the recommended insulation will be obtained in the finished building. Flanking transmission at the junctions of the construction components can reduce noise insulation drastically. Field testing of sound insulation at an early stage while the construction is still in progress may allow changes to be made before the construction is completed.

The impact insulation class (IIC) ratings in Fig. 5.2.18 apply to floors without carpeting or other resilient coverings. Some floors that are tested covered with carpet and underpad provide quite high IIC ratings, but only because of the presence of the carpet and underpad. Owners of buildings likely to be renovated for other finishes or uses at a future time may wish to provide floors with acceptable IIC ratings not dependent on floor coverings for the required sound ratings.

### Influence of Building Layout on Sound Insulation Requirements

A simple means of controlling noise in buildings is to separate noisy areas from quiet areas as much as possible. For example, in a typical apartment the noise levels vary widely from one room to another. Kitchens are often noisy; living rooms are sometimes the source of speech or sounds from radios or TV sets. If an area where quiet is required (such as a bedroom) is located next to a noisy machine room, a high value of noise reduction will be required to achieve a quiet environment in the bedroom. A carefully chosen layout will reduce the required noise reductions between spaces and therefore reduce the sound insulation needed for the building components. Variations in activity and noise levels between rooms are taken into account in the recommendations given in Fig. 5.2.17 and Fig. 5.2.18.

## 5.2.12 INSULATION FROM OUTDOOR NOISE

Protection from outdoor noise has become a major consideration in building acoustics, particularly in the vicinity of major roads, railways and airports. The techniques already described are applicable, with modifications, to outdoor-indoor noise conditions.

Detailed procedures have been prepared by Canada Mortgage and Housing Corporation for the site planning and design of residential buildings[R5.2.10, R5.2.11]. Usually, the sound insulation of a building facade is largely determined by the performance of the windows and doors, which can be seen by calculating the effective performance of the composite wall.

## 5.2.13 HIGHWAY SOUND BARRIERS

Sound barriers are positioned along the edges of major roads to reduce the transmission of direct sound to residential areas. If the barrier is heavy enough (greater than about 20 kg/m$^2$) and does not let sound leak through it significantly, the sound that reaches the residential area will be limited mainly by diffraction over the top of the barrier.

Other factors that affect sound barrier noise reduction are reflections from the ground on either side of the barrier and the bending of sound wave paths due to wind or temperature effects. Having a sound absorptive face on the barrier which faces the traffic increases the sound attenuation. These effects and the variations in traffic density and type are sufficiently complex that computer programs[R5.2.12] are usually used to estimate the effectiveness and environmental impact of busy roads, highways and sound barriers.

Proposed barrier structures are usually tested to determine whether the sound transmission through the sound walls is low enough. Where there is a requirement to have a sound absorptive face towards the traffic, measurements of the sound absorption coefficients are made. Both of these measurements are made in reverberation rooms[R5.2.1, R5.2.8]. The criteria that have to be met may vary but typical values are STC 20 and NRC 0.7. The requirements for mounting the specimens are usually specified by the authority requiring the test results.

Sound attenuation is not the only consideration. Barriers must be structurally sound to withstand wind loads, snow plow loads, weathering, and corrosive attack. Further information about transportation noise and highway barriers can be found in journals focused on this topic[R5.2.13].

# 5.3 FIRE RESISTANCE

## 5.3.1 NOTATION

Note: Subscript $\theta$ indicates the property as affected by elevated temperatures.

| | | |
|---|---|---|
| $A_p$ | = | area of prestressed reinforcement |
| $A_s$ | = | area of non-prestressed reinforcement |
| $A_s^-$ | = | area of reinforcement in negative moment region |
| $a$ | = | depth of equivalent rectangular compression stress block |
| $b$ | = | width of element |
| $c$ | = | neutral axis depth at nominal resistance |
| $d_p$ | = | distance from centroid of prestressed reinforcement to the extreme compression fibre |
| $f'_c$ | = | specified compressive strength of concrete |
| $f_{pr}$ | = | stress in the prestressed reinforcement at nominal resistance |
| $f_{pu}$ | = | specified tensile strength of prestressed reinforcement |
| $h$ | = | total depth of an element |
| $k_p$ | = | factor for type of prestressing steel |
| $\ell$ | = | span length |
| $M$ | = | service load moment |
| $M_n$ | = | nominal flexural resistance |
| $M^+_{n\theta}, M^-_{n\theta}$ | = | positive and negative nominal flexural resistance at elevated temperatures, respectively |
| $u$ | = | distance from prestressed reinforcement to the fire exposed surface |
| $w$ | = | uniform total load |
| $w_d$ | = | uniform dead load |
| $w_\ell$ | = | uniform live load |
| $x, x_o$ | = | horizontal distances as shown in Fig. 5.3.10, |
| $x_1, x_2$ | = | 5.3.11, and 5.3.12 |
| $\alpha_1$ | = | stress block intensity factor |
| $\beta_1$ | = | stress block depth factor |
| $\theta_s$ | = | temperature of steel reinforcement |
| $\phi_c$ | = | resistance factor for concrete |
| $\phi_p$ | = | resistance factor for prestressed reinforcement |

## 5.3.2 INTRODUCTION

Precast prestressed concrete elements can be provided with any degree of fire resistance that may be required by building codes, insurance companies, and other authorities. The fire resistance of building assemblies is determined from standard fire tests defined in ULC standard CAN4-S101-M89, "Standard Methods of Fire Endurance Tests of Building Construction and Materials."

To ensure that fire resistance requirements are satisfied, the engineer can use tabulated information provided in Chap. 2 of the "Supplement to the National Building Code of Canada." That information is based on the results of standard fire tests of assemblies.

In the absence of fire test results or tabulated data, the fire resistance of precast prestressed concrete elements and assemblies can be determined in most cases by calculation. These calculations are based on engineering principles and take into account the conditions of a standard fire test. This is known as the Rational Design Method of determining fire resistance. It is based on extensive research conducted by the National Research Council, the Portland Cement Association, and many other laboratories in Europe and America.

While every attempt has been made to present information that is factual and is in usable format, the material presented herein does not have official status. Acceptance of designs based on this information rests entirely with the authority having jurisdiction.

After a discussion on fire tests and heat transmission in Sect. 5.3.3 and 5.3.4, calculations using the Rational Design Method for common situations are presented. Brief explanations of the underlying principles are also given. For additional examples, design charts, and a complete explanation of the method, refer to the CPCI manual, "Fire Resistance Ratings for Prestressed and Precast Concrete."

## 5.3.3 STANDARD FIRE TESTS

The fire resistance of building components is measured in standard fire tests defined by ULC S101. During these tests, the building assembly, such as a portion of a floor, wall, or roof or column, is subjected to increasing temperatures that vary with time as shown in Fig. 5.3.1.

In addition, ULC S101 also specifies the minimum size of assemblies, the magnitude of applied load, the region of the assembly to be exposed to the fire, and the end point criteria on which fire endurance is based.

The Standard, ULC S101, specifies the minimum sizes of specimens to be exposed in fire tests. (Much valuable data has been developed for tests on specimens smaller than the minimum sizes – see Sect. 5.3.4.). For floors and roofs, at least 16.8 $m^2$ must be exposed to fire from beneath, and neither dimension can be less than 3660 mm. For tests of walls, either

**Fig. 5.3.1 Standard time temperature curve**

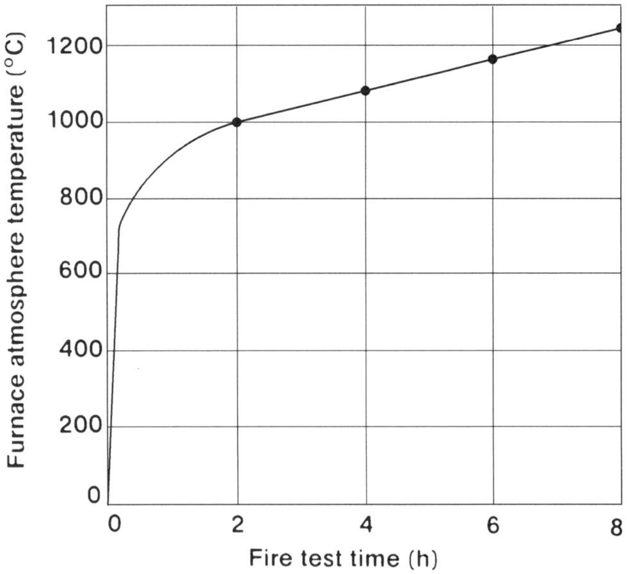


loadbearing or non-loadbearing, the minimum specified area is 9.3 $m^2$ with neither dimension less than 2750 mm. The minimum length for columns is specified to be 2750 mm, while for beams it is 3660 mm.

During fire tests of floors, roofs, beams, load-bearing walls, and columns, the maximum permissible superimposed load as required or permitted by nationally recognized standards is applied. A load other than the maximum load may be applied, but the test results then apply only to the restricted load condition.

Floor and roof specimens are exposed to fire from beneath, beams from the bottom and sides, walls from one side, and columns from all sides.

ULC S101 distinguishes between "restrained" and "unrestrained" assemblies and defines them as follows:

> "Floor and roof assemblies and individual beams in buildings shall be considered thermally restrained when (1) the surrounding or supporting structure is capable of resisting, without collapse, thermal expansion such as would be induced in an environment described by the Standard Time-Temperature Curve, or, (2) the assemblies or beams have structural continuity over supports. Constructions not complying with either of these two conditions or a combination thereof are assumed to be free to expand and rotate and should be considered as thermally unrestrained."

ULC S101 includes a guide for classifying types of construction as restrained or unrestrained. The guide indicates that cast-in-place and many precast concrete assemblies can be considered to be restrained.

### Fire Endurance, Fire-Resistance Rating, and End Point Criteria

The fire endurance of an assembly is the period of time which elapses before a prescribed condition of failure or end point is reached during a standard fire test. A fire-resistance rating is a legal term for the fire endurance required by a building code or authority having jurisdiction.

The following end point criteria are defined by ULC S101:

1. Loadbearing specimens must sustain the applied loading. Collapse is an obvious end point (structural end point).
2. Holes, cracks, or fissures through which flames or gases hot enough to ignite cotton waste must not form (flame passage end point).
3. The temperature increase of the unexposed surface of floors, roofs, or walls must not exceed an average of 140°C or a maximum of 180°C at any one point (heat transmission end point).
4. In alternate tests of large steel beams (not loaded during test), the end point occurs when the steel temperature reaches an average of 538°C or a maximum of 649°C at any one point.

Unrestrained and restrained assembly classifications can be derived from fire tests of restrained specimens. When based on results of fire tests of restrained specimens, additional end point criteria for *unrestrained* floor, roof, and beam classifications are:

1. Structural steel elements spaced more than 1220 mm on centres: temperature of the steel at any one section must not exceed an average of 593°C or a maximum of 704°C.
2. Structural steel elements spaced 1220 mm or less on centres, e.g., multiple open-web steel joists: average temperature must not exceed 593°C.
3. Concrete structural elements: average temperature of the tension reinforcement at any section must not exceed 427°C for cold-drawn prestressing steel or 593°C for reinforcing bars.

Additional end point criteria for *restrained* assembly classifications are:

4. Beams more than 1220 mm on centres: the steel temperatures in (1) and (3) above must not be exceeded for classifications of 1 h or less; for classifications longer than 1 h, these temperatures must not be exceeded for the first half of the classification period or 1 h, whichever is longer.
5. Concrete beams 1220 mm or less on centres and slabs are not subjected to steel temperature limitations.
6. Structural steel members spaced 1220 mm or less on centres: the steel temperature in (2) above must not be exceeded for classifications of 1 h or less; for classifications longer than 1 h, that temperature must not be exceeded for the first half of the classification period or 1 h, whichever is longer.

Note that there are no limiting temperatures for reinforcing steel or prestressing steel for *restrained* classifications of slabs. Also, there are no limiting steel temperatures for *unrestrained* assembly classifications derived from fire tests of *unrestrained* specimens. *Restrained* assembly classifications cannot be obtained from fire tests of *unrestrained* specimens.

Walls and partitions must meet the same structural, flame passage, and heat transmission end points described above. In addition, they must withstand a hose stream test (simulating, in a specified manner, a fire fighter's hose stream).

### Fire Tests of Prestressed Concrete Assemblies

The first fire test of a prestressed concrete assembly in North America was conducted in 1953 at the National Bureau of Standards. Since that time, more than 150 prestressed concrete assemblies have been subjected to standard fire tests in North America. Although many of the tests were conducted for the purpose of deriving specific fire ratings, most of the tests were performed in conjunction with broad research studies whose objectives have been to understand the behaviour of prestressed concrete subjected to fire. The knowledge gained from these tests has resulted in the development of (1) lists of fire resistive prestressed concrete building components, and (2) procedures for determining the fire endurance of prestressed concrete elements by calculation.

Many different types of prestressed concrete elements have been fire tested. These elements include joists, double tees, mono-wing tees, single tees, solid slabs, hollow-core slabs, rectangular beams, ledger beams, and I-shaped beams. In addition, roofs with thermal insulation and loadbearing wall panels have also been tested. Nearly all of these elements have been exposed directly to fire, but a few tests have been conducted on specimens that received additional protection from the fire by spray-applied coatings, ceilings, etc.

## 5.3.4 DESIGNING FOR HEAT TRANSMISSION

As noted above, ULC S101 imposes heat transmission criteria for floor, roof, and wall assemblies. Thus floors, roofs, or walls requiring a fire-resistance rating must satisfy the heat transmission requirements as well as the various structural criteria. The heat transmission fire endurance of a concrete assembly is essentially the same whether the assembly is tested as a floor (oriented horizontally) or as a wall (tested vertically). Because of this, and unless otherwise noted, the information which follows is applicable to floors, roofs, or walls.

### Single Course Slabs or Wall Panels

For concrete slabs and panels, the temperature rise of the unexposed surface depends mainly on the thickness, relative density, and aggregate type of the concrete. Fig. 5.3.2 shows the fire endurance (heat transmission) of concrete slabs or wall panels as influenced by aggregate type and element thickness. For a hollow-core slab, this thickness may be obtained by dividing the net cross sectional area by its width. On the graph, concretes are designated as Types L, L40S, N, and S, in accordance with the following definitions:

1. Type S concrete is that in which the coarse aggregate is granite, quartzite, siliceous gravel or other dense materials containing at least 30 per cent quartz, chert or flint.
2. Type N concrete is that type in which the coarse aggregate is cinders, broken brick, blast furnace slag, limestone, calcareous gravel, trap rock, sandstone or similar dense material containing not more than 30 per cent of quartz, chert or flint.
3. Type L concrete is that type in which all the aggregate is expanded slag, expanded clay, expanded shale or pumice.
4. Type L40S concrete is that type in which the fine portion of the aggregate is sand and low density aggregate in which the sand does not exceed 40 per cent of the total volume of all aggregates in the concrete.

**Fig. 5.3.2 Fire endurance (heat transmission) of concrete slabs or wall panels**

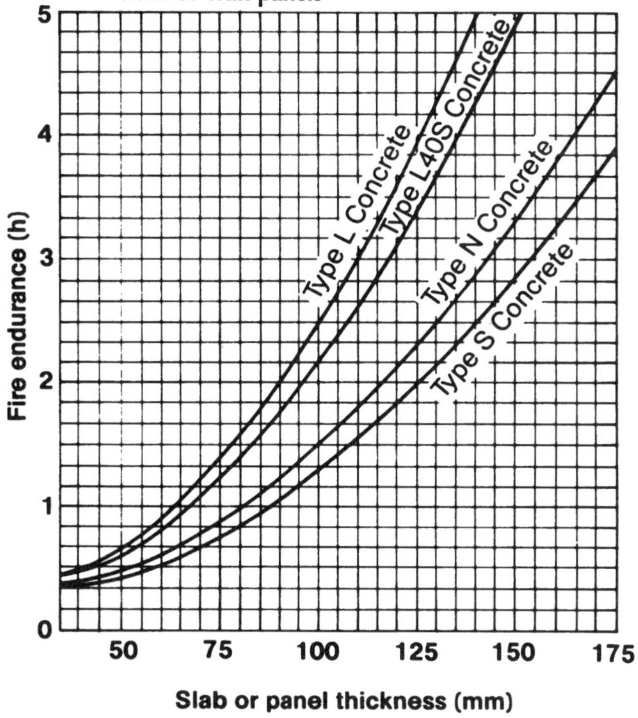


### Ribbed Panels

Heat transmission through a ribbed panel is influenced by the thinnest portion of the panel and by the panel's "equivalent thickness." Equations for calculating the equivalent thickness are given in Clause 1.6.6, Chap. 2 of the Supplement to the NBCC. An illustrated example is shown in Sect. 1.4.5 of the Canadian Portland Cement Association's Concrete Design Handbook.

### Multi-course Assemblies

Floors and roofs often consist of concrete base slabs with overlays or undercoatings of other types of concrete or insulating materials. In addition, roofs generally have built-up roofing. Walls may consist of more than one wythe, with or without an intervening air space.

The fire resistance of multi-course assemblies is outlined in Clause 2.2.3, Chap. 2 of the Supplement to the NBCC. An illustrated example is shown in Sect. 1.4.6 of the Canadian Portland Cement Association's Concrete Design Handbook.

### Sandwich Panels

Some wall panels are made by sandwiching an insulation material between two face slabs of concrete.

It should be noted that cellular plastics melt and are consumed at about 200 to 300°C. Thus, additional thickness or changes in composition probably have only a minor effect on the fire endurance of sandwich panels. The danger of toxic fumes caused by the burning cellular plastics is practically eliminated when the plastics are completely encased within concrete sandwich panels.[R5.3.1]

The fire resistance of sandwich wall panels is equivalent to the fire resistance of a solid panel with a thickness equal to the sum of the thickness of the two wythes. For further information, see Clause 2.1.2, Chap. 2 of the Supplement to the NBCC.

### Window Walls

The National Building Code of Canada limits the area of unprotected openings (windows and doors) in exterior walls which are required to be fire resistive. Percentages of unprotected opening areas are tabulated for various combinations of area of building face, height-length ratio, and spatial separation.

The percentage of openings permitted increases (1) as the spatial separation increases (2) as the area of the exposing building face decreases, and (3) as the ratio of either height-length (H/L) or length-height (L/H) increases, i.e., a greater percentage is permitted for H/L or L/H of 10:1 than for H/L or L/H of 3:1. As an example, an exposing face of an office building having an area of 350 $m^2$, an L/H = 2:1, and a limiting distance of 7 m can have a maximum of 16 percent of unprotected openings. If the ratio of L/H or H/L were 10:0 or more, the area of unprotected openings could be increased to 30 percent, or if the spatial separation were 12 m and the L/H were 10:1, the area of unprotected openings permitted is 59 percent of the exposing face.

The Code also permits a higher limit on the unexposed surface temperature if the area of unprotected openings is less than the maximum allowed, and thus somewhat thinner panels can be used.

An equivalent opening factor is applied in a formula to determine the corrected area of openings:

$$A_c = A + A_f F_{eo}$$

where:

$A_c$ = corrected area of unprotected openings including actual and equivalent openings

$A$ = actual area of unprotected openings

$A_f$ = area of exterior surface of the exposing building face exclusive of openings, on which the temperature limitation of the standard fire test is exceeded

$F_{eo}$ = equivalent opening factor

Fig. 5.3.3 shows the relation between $F_{eo}$ (as defined in the Code) and panel thickness for three types of concrete.

To illustrate the use of Fig. 5.3.3, suppose that for a particular building face, a 2 h fire-resistance rating is required and the area of unprotected openings permitted is 57 percent. Suppose also that the actual area of unprotected openings is 49 percent and that the window wall panels are made of Type N concrete. Determine the minimum thickness of the panel.

In this case, $A_c$ = 57 percent, A = 49 percent,

$A_f$ = 100 - 49 = 51 percent, hence:

$$F_{eo} = \frac{A_c - A}{A_f} = \frac{57-49}{51} = 0.16$$

From Fig. 5.3.3, for $F_{eo}$ = 0.16 at 2 h, the minimum panel thickness is 54 mm. Thus, if the panel is 54 mm thick or thicker, the code requirements will be satisfied.

### Treatment of Joints

Joints between wall panels are similar to openings, and can be treated in the manner discussed for window walls. So in many cases, joints need not be fire-protected. In other cases, openings, including joints, must be protected.

Fire tests of wall panel joints[R5.3.2] have shown that the fire endurance, as determined by a temperature rise of 180°C over

**Fig. 5.3.3 Equivalent opening factor $F_{eo}$, for concrete wall panels***

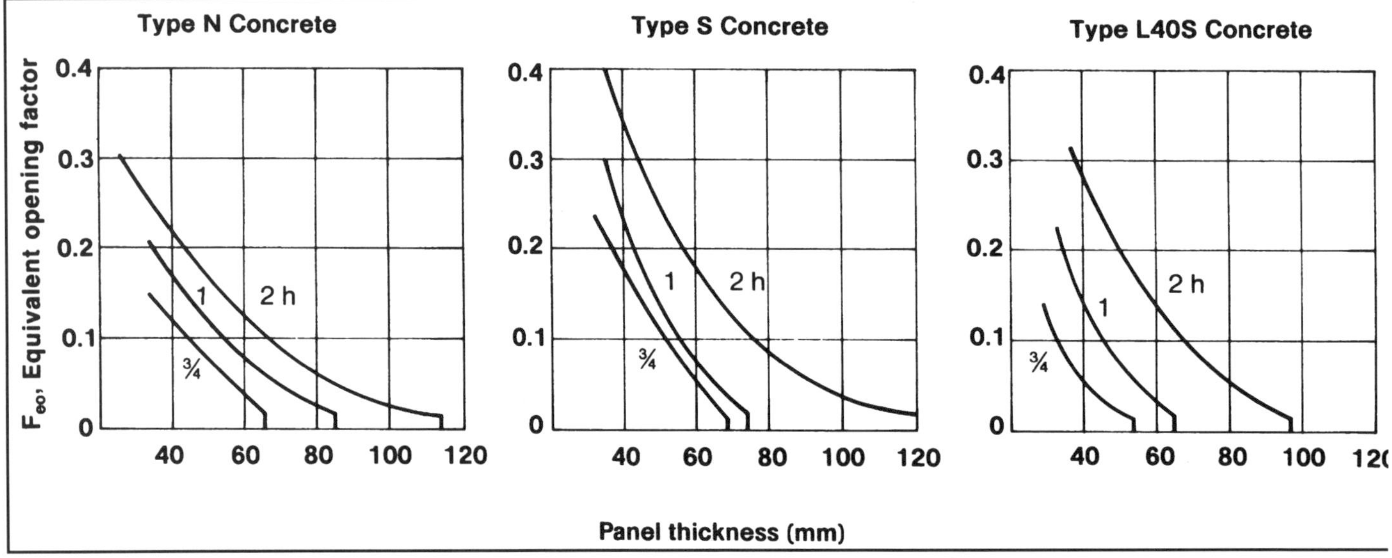


* For use with the National Building Code of Canad

the joint, is influenced by joint type, joint treatment (materials), joint width, and panel thickness. By providing the proper thickness of insulating materials within the joint, it is possible to attain fire endurances essentially as long as those of the panels.

Fig. 5.3.4 is based on results of fire tests of panels with butt joints. The tabulated values apply to one-stage butt joints and are conservative for two-stage and ship-lap joints.

**Fig. 5.3.4 Protection of joints utilizing ceramic fibre felt**

| Panel equivalent thickness* (mm) | Thickness of ceramic fibre felt (mm) required for fire resistance ratings and joints widths shown | | | | | | | |
|---|---|---|---|---|---|---|---|---|
| | Joint width = 10 mm | | | | Joint width = 25 mm | | | |
| | 1 h | 2 h | 3 h | 4 h | 1 h | 2 h | 3 h | 4 h |
| 100 | 6 | N.A. | N.A. | N.A. | 19 | N.A. | N.A. | N.A. |
| 125 | 0 | 19 | N.A. | N.A. | 13 | 58 | N.A. | N.A. |
| 150 | 0 | 0 | 32 | N.A. | 6 | 32 | 89 | N.A. |
| 175 | 0 | 0 | 0 | 25 | 6 | 19 | 51 | 96 |

N.A. = Not applicable

* Panel equivalent thicknesses are for Type N concrete. For Type S concrete change 100, 125, 150, and 175 in column 1 to "110, 135, 165, and 190," respectively. For Type L40S or L concrete change 100, 125, 150, and 175 in column 1 to "85, 105, 125, and 145."

The tabulated values apply to one-stage but joints and are conservative for two-stage and ship-lap joints as shown below.

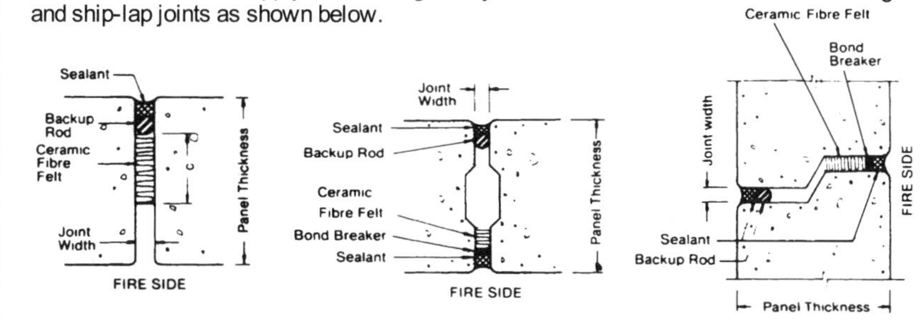


**Vertical or horizontal** **Vertical** **Horizontal**

Joints between adjacent precast floor or roof elements may be ignored in calculating the slab thickness provided that a concrete topping at least 40 mm thick is used. Where no concrete topping is used, joints should be grouted to a depth of at least one-third the slab thickness at the joint, or the joints made fire-resistive in a manner acceptable to the authority having jurisdiction.

## 5.3.5 BALANCED DESIGN

A balanced design approach to fire safety is a philosophy of fire protection based on hazard and risk assessments applied to individual buildings on an individual basis. A balanced design approach to fire safety recognizes the importance of fire resistive compartmentation, smoke detection and automatic suppression in buildings to reduce the risk of injury, loss of life and property damage from the effects of fire. The use of the components of balanced design, singularly or in combination, is dependent upon the building's occupancy, construction and type of hazards that may be present. As each of the protection components of a balanced design approach has its strengths and weakness, it is essential to understand the building's fire protection requirements to avoid redundancy in fire protection features and the appropriate model building code, to avoid compromising lives and property. This approach provides code officials and authorities having jurisdiction the latitude to determine an appropriate level of protection based on specific conditions for a particular building type and occupancy.

Multifamily residences, high-rise buildings and high hazard occupancies are examples where all the components of a balanced design approach may be necessary. Open parking structures may not require all of the balanced design approach components. The balanced design approach applies in either case as a systems concept philosophy based on hazard and risk assessments is applied on an individual basis to determine fire protection requirements.

Post earthquake fires continue to be one of the most serious threats, particularly in densely populated urban areas. Although earthquake resistant automatic sprinkler systems exist, it is probable that the public water mains supplying these systems will fail rendering the automatic suppression systems inoperable. If a trade-off between compartmentation and automatic suppression system has been made, fire can quickly spread through buildings destroying everything in its path. Fire resistant elements, designed to maintain structural integrity in the event of an earthquake, should also be included to stop or retard the spread of fire, permit safe occupant evacuation and provide an acceptable environment for fire fighting operations.

In arson related fires, automatic suppression systems can be rendered inoperable by simply closing a valve. Compartmentation and inherent fire resistant construction is very difficult if not impossible to disable.

Automatic suppression systems are not as effective as one desires for the following reasons: sprinkler valves may be shut off before or during a fire; insufficient water supply; clogged piping; system may be out of service due to freezing, earthquake, explosion, or lack of maintenance; inadequate sprinkler protection (coverage); broken piping or fittings; and sprinkler head obstructions. Therefore, permitting trade-offs between fire resistant construction or compartmentation and automatic suppression generally leads to protection imbalances as there are no established equivalency relationships between the elements of balanced design. Further, statistics show that the total dollar property loss for sprinklered buildings average about twice the dollar amount for unsprinklered buildings - probably because of trade-offs such as a reduction in fire resistance ratings, or an increase in allowable area and height of the building or permitting a larger percentage of exterior wall openings.

## 5.3.6 DESIGNING FOR STRUCTURAL INTEGRITY

It was noted above that many fire tests and related research studies have been directed toward an understanding of the structural behaviour of prestressed concrete subjected to fire. The information gained from that work has led to the development of calculation procedures which can be used in lieu of fire tests. The purpose of this section is to present an introduction to these calculation procedures. Because the method of support is the most important factor affecting structural behaviour of flexural elements during fire, the discussion that follows deals with three conditions of support: simply supported elements, continuous slabs and beams, and elements in which restraint to thermal expansion occurs.

### Simply Supported Elements

Assume that an unrestrained prestressed concrete slab is exposed to fire from below, that the ends of the slab are free to rotate, and that expansion can occur without restriction. Also assume that the reinforcement consists of straight strands located near the bottom of the slab. With the underside of the slab exposed to fire, the bottom will expand more than the top, causing the slab to deflect downward; also, the strength of the steel and concrete near the bottom will decrease as the temperature rises. When the strength of the steel diminishes to that required to support the slab, flexural collapse will occur. In essence, the applied moment remains practically constant during the fire exposure, but the resisting moment capacity is reduced as the steel weakens.

Fig. 5.3.5 illustrates the behaviour of a simply-supported slab exposed to fire from beneath, as described above. Because the strands are parallel to the axis of the slab, the nominal flexural resistance is constant throughout the span. For rectangular sections (or tee sections where the neutral axis is in the flange) containing prestressed reinforcement only:

$$M_n = A_p f_{pr} (d_p - a/2)$$

where:

$M_n$ = nominal moment resistance

$A_p$ = area of prestressed reinforcement

$f_{pr}$ = stress in the prestressed reinforcement at nominal resistance.

$d_p$ = distance from the extreme compression fibre to the centroid of the prestressed reinforcement

$a$ = depth of equivalent rectangular stress block
 = $A_p f_{pr} / (\alpha_1 f'_c b)$, where $f'_c$ is the specified compressive strength of the concrete and b is the width of the compression face.

In lieu of a more accurate determination of $f_{pr}$ based on strain compatibility, the following approximate value of $f_{pr}$ may be used for bonded tendons if $f_{pe}$ is not less than $0.6f_{py}$ and $c/d_p$ is not greater than 0.5:

$$f_{pr} = f_{pu} (1 - k_p c/d_p)$$

where:

$f_{pu}$ = tensile strength of bonded tendons

$k_p$ = factor for type of prestressing steel
 = $2 (1.04 - f_{py} / f_{pu})$

**Fig. 5.3.5 Moment diagrams for simply-supported beam or slab**

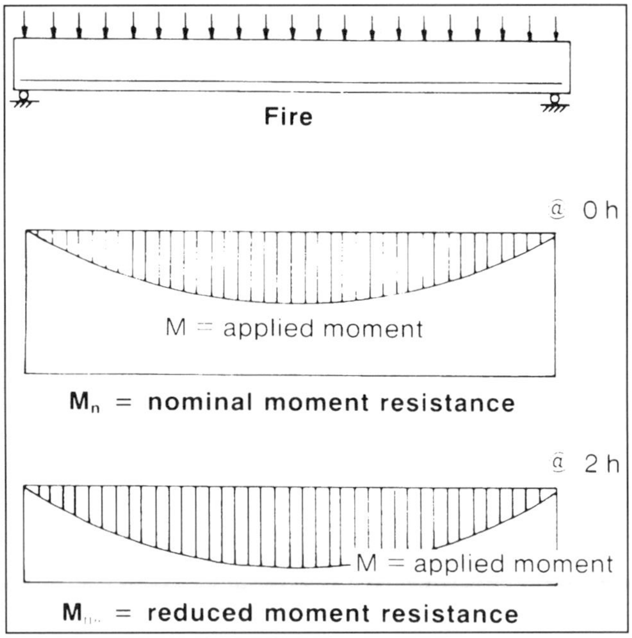


$c$ = neutral axis depth at nominal resistance

As the material strengths diminish with elevated temperatures, the retained nominal flexural resistance becomes:

$$M_{n\theta} = A_p f_{pr\theta} (d_p - a_\theta/2)$$

where:

$$f_{pr} = f_{pu\theta} (1 - k_p c_\theta/d_p)$$

$$a_\theta = A_p f_{pr\theta}/(\alpha_1 f'_c b)$$

in which θ signifies the effects of elevated temperature. Note that $A_p$ and $d_p$ are not affected, but $f_{pr}$ is reduced. Similarly $a_\theta$ is reduced, but the concrete strength at the top of the slab, $f'_c$, is generally not reduced significantly because of its lower temperature. If, however, the compressive zone of the concrete is heated above 450°C, $f'_c$ should also be reduced to calculate $a_\theta$.

Flexural failure can be assumed to occur when $M_{n\theta}$ is reduced to M. The material resistance factors, $\phi_p$ and $\phi_c$ are not applied because a factor of safety is included in the required ratings. From this expression, it can be seen that the fire endurance depends on the applied loading and on the strength-temperature characteristics of the reinforcement.

In turn, the duration of the fire before the "critical" steel temperature is reached depends on the protection afforded to the reinforcement.

To solve problems involving the above equations, it is necessary to utilize data on the strength-temperature relationships for steel and concrete, and information on temperature distributions within concrete elements during fire exposures. Fig. 5.3.6 shows strengths of certain steels at elevated temperatures, and Fig. 5.3.7 shows similar data for various types of concrete.

Fig. 5.3.6 Temperature-strength relationships for various steels

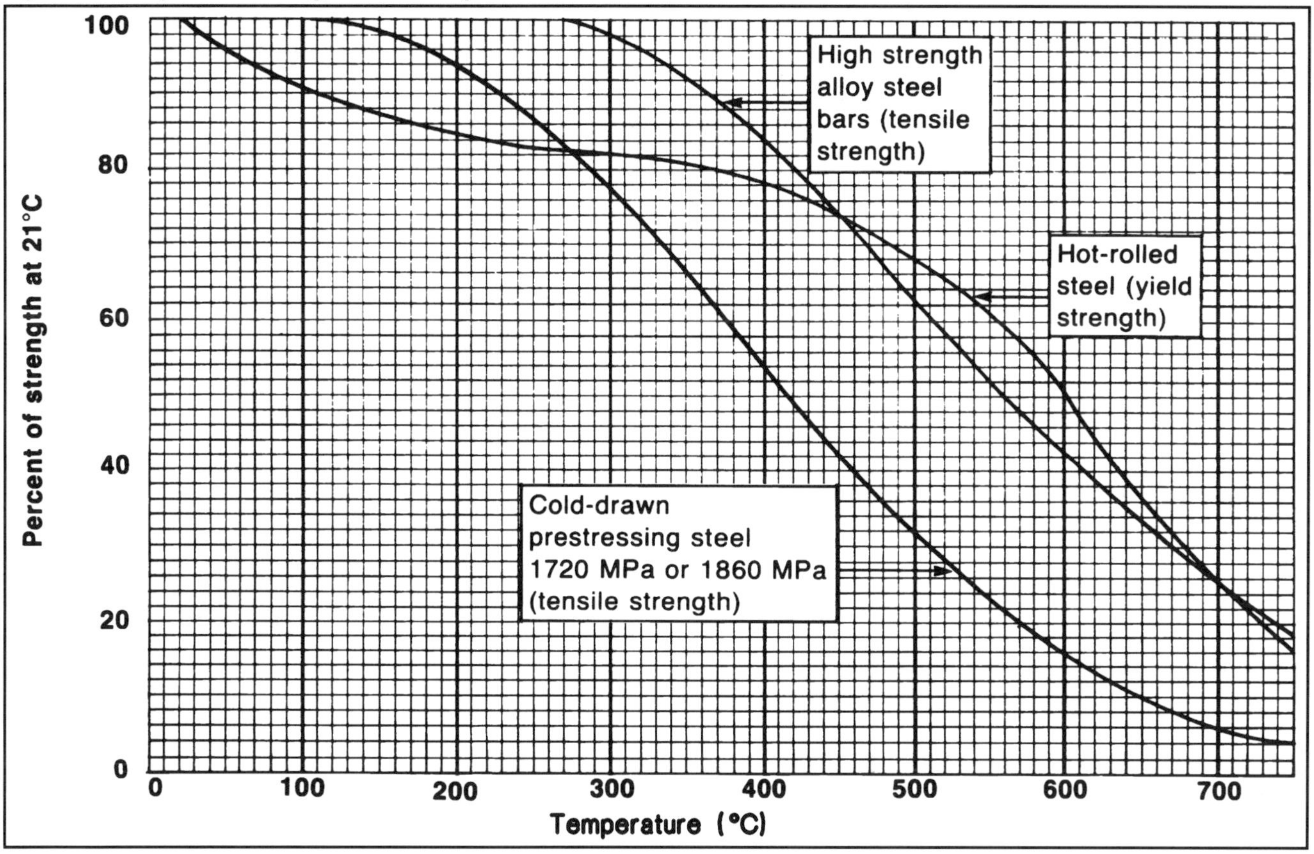


Fig. 5.3.7 Compressive strength of concrete at high temperatures

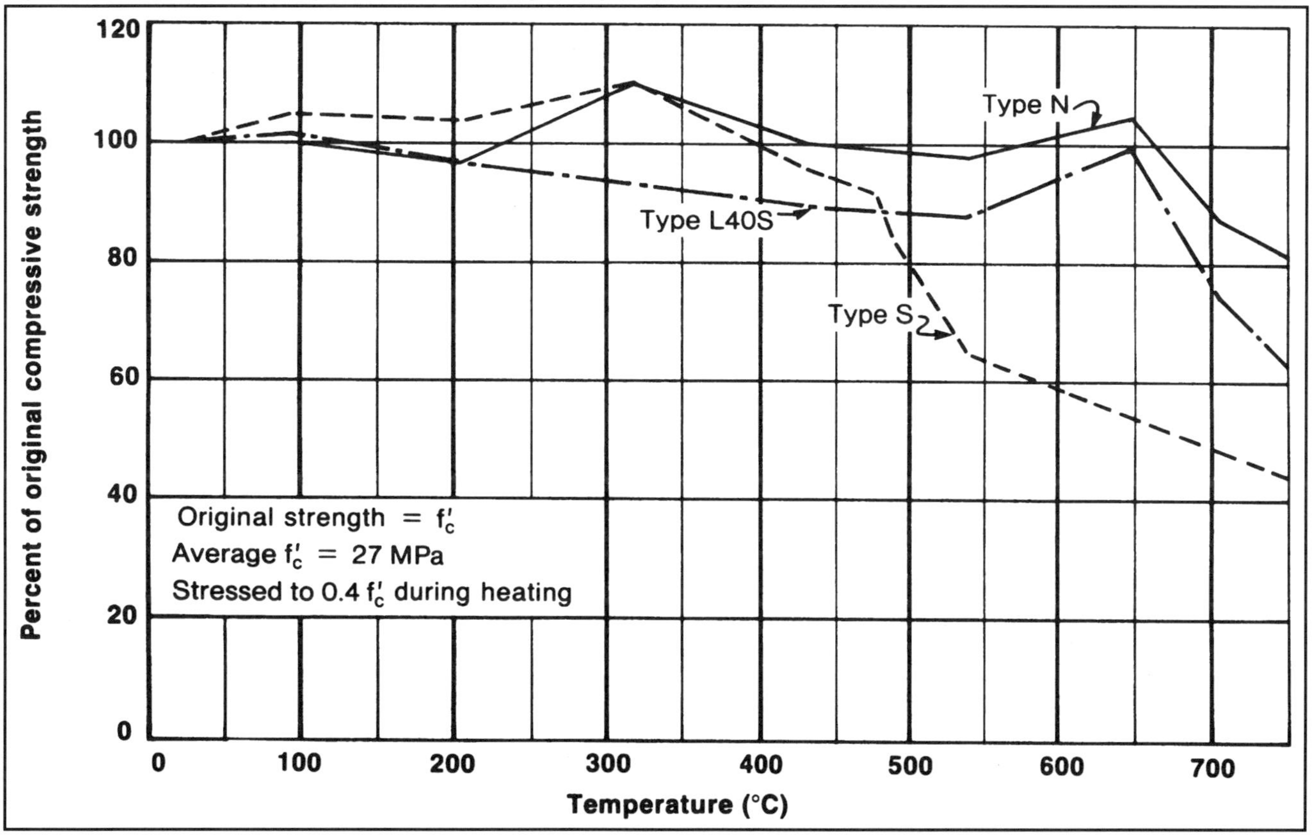

Data on temperature distribution in concrete slabs during fire tests are shown in Fig. 5.3.8. Similar data for concrete beams and joists are more complex because beams are heated from the sides as well as from beneath. Fig. 5.3.9 shows temperature data for normal density concrete joists and beams at 2 h exposure. Note that the temperatures are given for points along the vertical centreline of rectangular beams or of elements with tapered sides.

**Fig. 5.3.8 Temperatures at u mm from hot face of concrete slabs during fire tests**

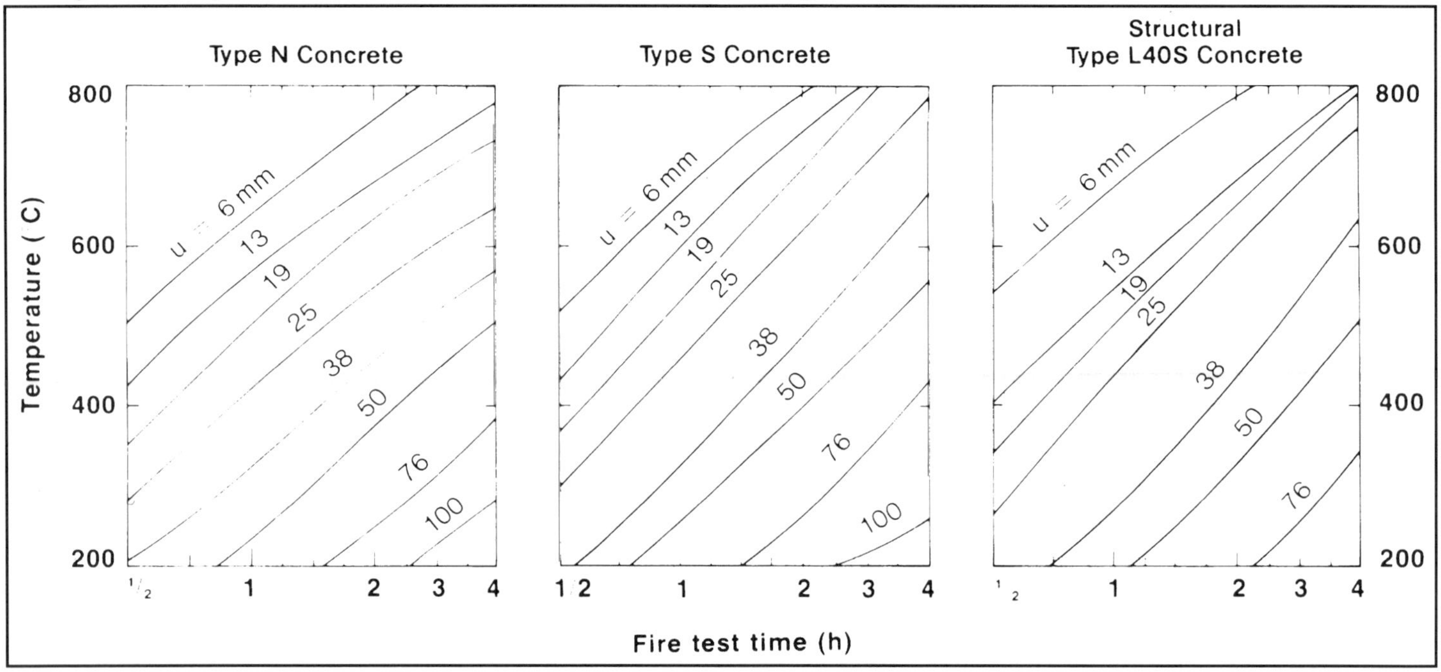


**Fig. 5.3.9 Temperature-along vertical centre lines of stemmed units at 2h of exposure***

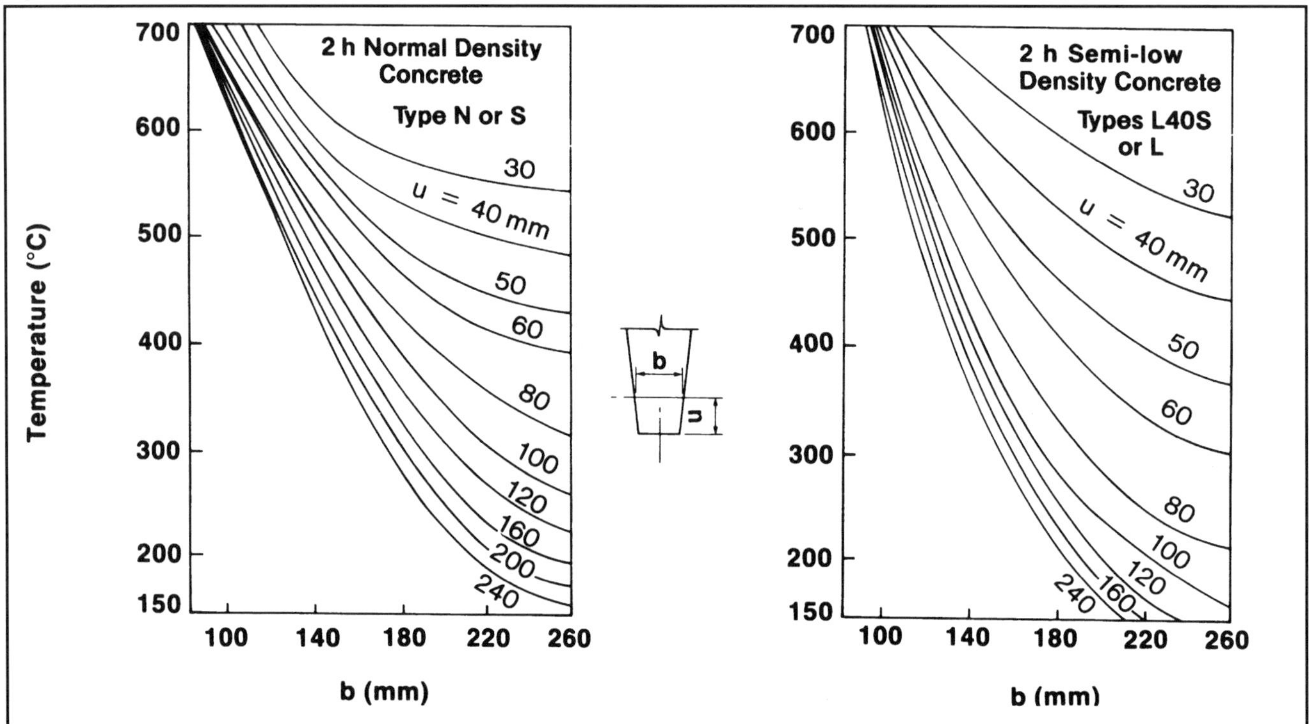


*From: FIP/CEB Report on Methods of Assessment of the Fire Resistance of Structural Members.

### Example 1 – Capacity of a hollow-core slab exposed to fire

*Given:*

A 203 mm deep hollow-core slab with a simply-supported unrestrained span of 8.5 m.

Concrete:

$f'_c$ = 35 MPa  $\alpha_1 = 0.80$  $\beta_1 = 0.88$
Normal density, Type N

Prestressed reinforcement:

7 – 13 mm 1860 MPa low-relaxation strands
$k_p$ = 0.28
$A_p$ = (7)(99) = 693 $mm^2$
h = 203 mm
u = 45 mm
$d_p$ = 158 mm
b = 1220 mm
$w_d$ = 2.6 kPa
$\ell$ = 8.5 m

*Problem:*

Determine the maximum superimposed load that can be supported after 2 h exposure to a standard fire.

*Solution:*

1. From Fig. 5.3.8, estimate strand temperature at 2 h for carbonate aggregate concrete with:
   u = 45 mm: $\theta_s$ = 400°C
2. From Fig. 5.3.6, determine:
   $f_{pu\theta} = 0.54$
   $f_{pu}$ = (0.54)(1860) = 1005 MPa
3. Determine $M_{n\theta}$ and $w_\ell$ :

$$c_\theta/d_p = \frac{A_p f_{pu\theta}}{\alpha_1 f'_c \beta_1 b d_p + k_p A_p f_{pu}}$$

$$= \frac{(693)(1005)}{(0.8)(35)(0.88)(1220)(158) + (0.28)(693)(1005)}$$

= 0.141

$f_{pr\theta}$ = (1005)[1 - (0.28)(0.141)] = 965 MPa
$a_\theta$ = (693)(965)/[(0.8)(35)(1220)] = 20 mm
$M_{n\theta}$ = (693)(965)(158 - 20/2)/$10^6$ = 99.1 kN · m
w = (8)(99.1)/[$(8.5)^2$(1.2)] = 9.1 kN/$m^2$
$w_\ell$ = w - $w_d$ = 9.1 - 2.6 = 6.5 kN/$m^2$

### Example 2 – Increasing fire endurance of a double tee by adding reinforcement

*Given:*

A 2400 × 400 double tee floor with 50 mm composite topping and a simply-supported unrestrained span of 9 m.

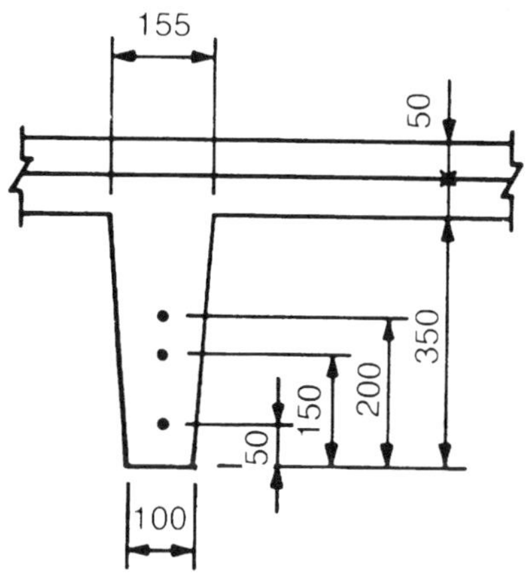


Concrete:

$f'_c$ (precast) = 35 MPa  $\alpha_1 = 0.80$  $\beta_1 = 0.88$
$f'_c$ (topping) = 20 MPa  $\alpha_1 = 0.82$  $\beta_1 = 0.92$
Normal density, Type N

Prestressed reinforcement:

6 – 13 mm 1860 MPa low-relaxation strands
$k_p$ = 0.28
$A_p$ = (6)(99) = 594 $mm^2$
$w_d$ = 3.3 kN/$m^2$
$w_\ell$ = 1.9 kN/$m^2$
u = 133 mm
$d_p$ = 317 mm

*Problem:*

Determine the amount of additional non-prestressed reinforcement required to provide a 2 h structural fire endurance.

*Solution:*

w = (2.4)(1.9 + 3.3) = 12.5 kN/m

1. Estimate strand temperature at 2 h from Fig. 5.3.9. At centroid of strand:
   $b_w$ = 100 + (133/350)(55) = 121 mm
   Avg.$\theta_s$ = 545°C
2. Estimate $f_{pu\theta}$ from Fig. 5.3.6:
   $f_{pu\theta}$ = (0.24)(1860) = 445 MPa
3. Calculate $M_{n\theta}$ and compare with M:

$$c_\theta/d_p = \frac{(594)(445)}{(0.82)(20)(0.92)(2400)(317) + (0.28)(594)(445)}$$

= 0.023

$f_{pr\theta}$ = (445) [1 - (0.28)(0.023)]
= 442 MPa
$a_\theta$ = (594)(442)/[(0.82)(20)(2400)] = 7 mm
$M_{n\theta}$ = (594)(442)(317 - 7/2)/$10^6$ = 82.4 kN · m
M = (12.5)$(9)^2$/8 = 127 kN · m > $M_{nq}$

4. Try adding one No. 25 grade 400 MPa reinforcing bar in each stem at u = 181 mm.

5. Estimate temperature and strength of the reinforcing bars:

   $b_w$ = 128 mm

   $\theta_s$ = 495°C (Fig. 5.3.9)

   $f_{y\theta}$ = (0.70)(400) = 280 MPa (Fig. 5.3.6)

   $A_s$ = (2)(500) = 1000 mm$^2$

6. Calculate $M_{n\theta}$:

$$\text{Adjusted } a_\theta = \frac{(594)(442)+(1000)(280)}{(0.82)(20)(2400)} = 14 \text{ mm}$$

$$M_{n\theta} = [(594)(442)(317 - 14/2) + (1000)(280)(269 - 14/2)]/10^6 = 155 \text{ kN}\cdot\text{m} > 127 \quad \text{OK}$$

## Continuous Elements

Continuous elements undergo changes in stresses when subjected to fire. These stresses result from temperature gradients within the structural elements, or changes in strength of the materials at high temperatures, or both.

Fig. 5.3.10 shows a two-span continuous beam whose underside is exposed to a fire test. The bottom of the beam becomes hotter than the top and tends to expand more than the top. This differential temperature effect causes the ends of the beam to tend to lift from their supports thereby increasing the reaction at the interior support. This action results in a redistribution of moments, i.e., the negative moment at the interior support increases while the positive moments decrease.

**Fig. 5.3.10 Moment diagram for two-span continuous beam**

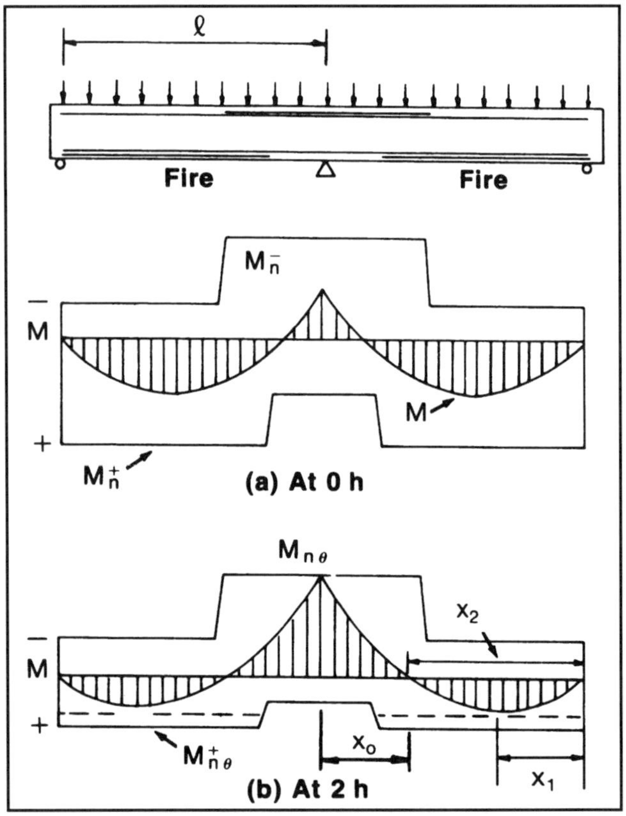


During a fire, the negative moment reinforcement (Fig. 5.3.10) remains cooler than the positive moment reinforcement because it is better protected from the fire. In addition, the redistribution that occurs is sufficient to cause yielding of the negative moment reinforcement. Thus, a relatively large increase in negative moment can be accommodated throughout the test. The resulting decrease in positive moment means that the positive moment reinforcement can be heated to a higher temperature before failure will occur. Therefore, the fire endurance of a continuous concrete beam is generally significantly longer than that of a simply supported beam having the same cover and the same applied loads.

It is possible to design the reinforcement in a continuous beam or slab for a particular fire endurance period. From Fig. 5.3.10 the beam can be expected to collapse when the positive moment capacity, $M_{n\theta}^+$, is reduced to the value of the maximum redistributed positive moment at a distance $x_1$ from the outer support.

Fig. 5.3.11 shows a uniformly loaded beam or slab continuous (or fixed) at one support and simply supported at the other. Also shown is the redistributed applied moment diagram at failure.

**Fig. 5.3.11 Uniformly loaded element continuous at one support**

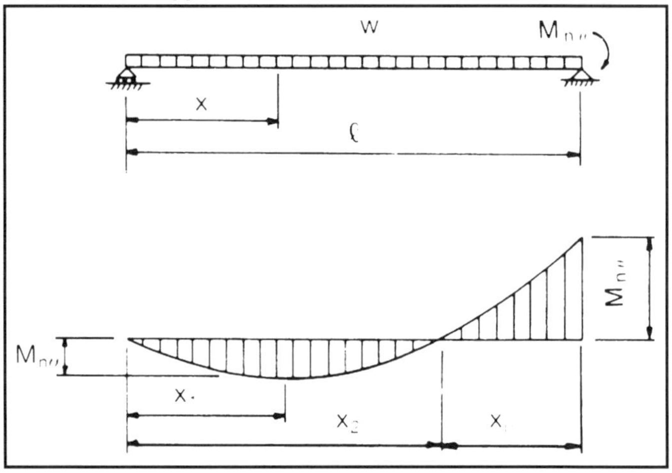


It can be shown that at the point of maximum positive moment, $x_1$:

$$x_1 = \frac{\ell}{2} - \frac{M_{n\theta}^-}{w\ell}$$

At x $= x_2$, $M_x = 0$ and $x_2 = 2\,x_1$:

$$x_o = \frac{2M_{n\theta}^-}{w\ell}$$

$$M_{n\theta}^- = \frac{w\ell^2}{2} \pm w\ell^2\sqrt{\frac{2M_{n\theta}^+}{w\ell^2}}$$

In most cases, redistribution of moments occurs early during the course of a fire and the negative moment reinforcement can be expected to yield before the negative moment capacity has been reduced by the effects of fire. In such cases, the length of $x_o$ is increased, i.e., the inflection point moves toward the simple support. If the inflection point moves beyond the cut off points of the negative moment reinforcement, sudden failure may result.

Fig. 5.3.12 shows a symmetrical beam or slab in which the end moments are equal:

$$M^-_{n\theta} = w\ell^2/8 - M^+_{n\theta}$$

$$\frac{wx_2^2}{8} = M^+_{n\theta}$$

$$x_2 = \sqrt{\frac{8M^+_{n\theta}}{w}}$$

$$x_o = \frac{1}{2}(\ell - x_2)$$

$$= \frac{\ell}{2} - \frac{1}{2}\sqrt{\frac{8M^+_{n\theta}}{w}}$$

**Fig. 5.3.12 Symmetrical uniformly loaded element continuous at both supports**

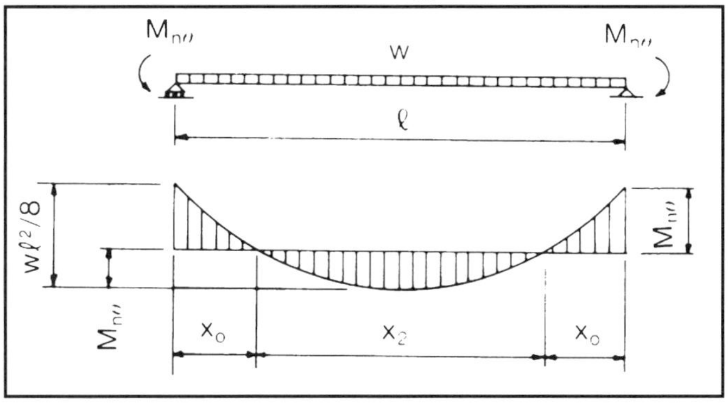


To determine the maximum value of $x_o$, the value of w should be the minimum service load anticipated, and ($w\ell^2/8 - M^-_n$) should be substituted for $M^+_{n\theta}$ in the equation:

$$x_o = \frac{\ell}{2} - \frac{1}{2}\sqrt{\frac{8M^+_{n\theta}}{w}}$$

For any given fire endurance period, the value of $M^+_{n\theta}$ can be calculated by the procedures given in the section on simply-supported elements. Then the value of $M^-_{n\theta}$ can be calculated by the use of the two preceding equations:

$$M^-_{n\theta} = \frac{w\ell^2}{2} \pm w\ell^2\sqrt{\frac{2M^+_{n\theta}}{w\ell^2}}$$

or

$$M^-_{n\theta} = \frac{w\ell^2}{8} - M^+_{n\theta}$$

and the necessary lengths of the negative moment reinforcement can be determined from:

$$x_o = \frac{2M^-_{n\theta}}{w\ell}$$

or

$$x_o = \frac{\ell}{2} - \frac{1}{2}\sqrt{\frac{8M^+_{n\theta}}{w}}$$

Use of these equations is illustrated in the following example.

It should be noted that the amount of moment redistribution that can occur is dependent on the amount of negative moment reinforcement. Tests have clearly demonstrated that, in most cases, the negative moment reinforcement will yield, so the negative moment capacity is reached early during a fire test, regardless of the applied loading. The designer must exercise care to ensure that a secondary type of failure will not occur. To avoid a compression failure in the negative moment region, the amount of negative moment reinforcement should be such that A23.3-94, Clause 18.8, Minimum Factored Flexural Resistance, is satisfied before and after reductions in $f_y$, b, d and $f'_c$ are taken into account. Furthermore, the negative moment bars or mesh must be long enough to accommodate the complete redistributed moment envelope and change in the inflection points. It should be noted that the worst condition occurs when the applied loading is smallest, such as the dead load plus partial or no live load. It is recommended that at least 20% of the maximum negative moment reinforcement extend throughout the span.

### Example 3 – Design of a continuous hollow-core floor system for a 3 h fire endurance

*Given*:

1220 × 305 hollow-core floor with 50 mm topping and a continuous span of 7.5 m.

Concrete:

$f'_c$ (precast) = 35 MPa  $\alpha_1 = 0.80$  $\beta_1 = 0.88$
$f'_c$ (topping) = 20 MPa  $\alpha_1 = 0.82$  $\beta_1 = 0.92$
Normal density, Type N

Prestressed reinforcement:

10 – 9 mm 1860 MPa stress-relieved strands
$k_p$ = 0.28
$A_p$ = (10)(55) = 550 mm²
$w_d$ = 8.4 kN/m²
$w_\ell$ = 7.2 kN/m²
u = 45 mm
$d_p$ = 310 mm

*Problem:*

Calculate the required amount of negative moment reinforcement to provide a 3 h fire endurance.

*Solution:*

From Fig. 5.3.8: $\theta_s$ = 480°C
From Fig. 5.3.6: $f_{pu\theta} = 0.36 f_{pu}$ = 670 MPa
Calculate values: $f_{pr\theta}$ = 658 MPa, $a_\theta$ = 18 mm

$M^+_{n\theta}$ = (550)(658)(310 - 18/2)/$10^6$
= 108.9 kN
M = (1.2)(8.4 + 7.2)$(7.5)^2$ / 8 = 131.6 kN · m
Req'd $M^-_{n\theta}$ = 131.6 - 108.9 = 22.7 kN · m/unit
Assume $d_\theta$ - $a_\theta$ /2 = 310 mm and $f_{y\theta}$ = 400 MPa
$A^-_s$ = (22.7)($10^6$)/[(400)(310)] = 183 $mm^2$/unit

Neglect concrete above 750°C in negative moment region, i.e. from Fig. 5.3.8, neglect bottom 13 mm. Also, concrete within compressive zone will be about 730 to 750°C, so use $f'_{c\theta}$ = 28 MPa, $\alpha_1$ = 0.81 (see Fig. 5.3.7).

Check $M^-_{n\theta}$, assuming that the temperature of the negative moment reinforcement does not rise above 90°C. If greater than 90°C, reinforcement strength should be reduced according to Fig. 5.3.6.

Use 20% $A_s$ throughout span.

Try WWF with an area of 59.8 $mm^2$/m in both directions throughout plus WWF with an area of 123 $mm^2$ /m in both directions over supports.

Over supports:
$A^-_s$ = (1.2)(59.8 + 123.0) = 220 $mm^2$ /m
$a^-_\theta$ = (220)(400)/[(0.81)(28)(1220)] = 3 mm
$M^-_{n\theta}$ = (220)(400)(330 - 13 -3/2)/$10^6$ = 28 kN · m

With dead load + 50% live load; w = 12 kN/$m^2$, M = 101 kN · m/unit and $M_r$ = 29 kN · m/unit (calculated at room temperature).
$M^+_{min}$ = 101 - 29 = 72 kN · m/unit
therefore:
max $x_o$ = (7.5/2) - (0.5)(8)(72)/[(1.2)(12)]
= 0.6 m = 600 mm

Use the lighter WWF throughout plus the heavier fabric for a distance of 900 mm in both directions from the support. The WWF must extend into walls which must be designed for the moment induced at the top.

### Elements Restrained Against Thermal Expansion

If a fire occurs beneath an interior portion of a large reinforced concrete slab, the heated portion will tend to expand and push against the surrounding part of the slab. In turn, the unheated part of the slab exerts compressive forces on the heated portion. The compressive force, or thrust, acts near the bottom of the slab when the fire first occurs but, as the fire progresses, the line of action of the thrust rises as the mechanical properties of the heated concrete change. This thrust is generally great enough to increase the fire endurance significantly.

The effects of restraint to thermal expansion can be characterized as shown in Fig. 5.3.13. The thermal thrust acts in a manner similar to an external prestressing force, which, in effect, increases the positive moment capacity.

The increase in bending moment capacity is similar to the effect of added reinforcement located along the line of action of the thrust. It can be assumed that the added reinforcement has a yield strength (force) equal to the thrust. By this approach, it is possible to determine the magnitude and location of the required thrust to provide a given fire endurance.

**Fig. 5.3.13 Axially restrained beam during fire exposure**

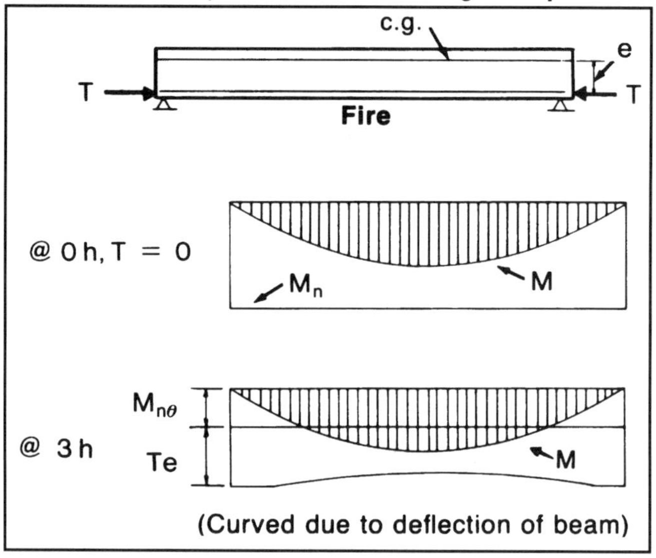


The above explanation is greatly simplified because in reality, restraint is quite complex, and can be likened to the behaviour of a flexural element subjected to an axial force. Interaction diagrams similar to those for columns can be constructed for a given cross-section at a particular stage of a fire, e.g., 2 h of a standard fire exposure.

The guidelines in ULC S101 given for determining conditions of restraint are useful for preliminary design purposes. Basically, interior bays of multi-bay floors or roofs can be considered to be restrained and the magnitude and location of the thrust are generally of academic interest only.

## 5.3.7 CPCI MANUAL, "FIRE RESISTANCE RATINGS FOR PRESTRESSED AND PRECAST CONCRETE"

The CPCI manual, "Fire Resistance Ratings for Prestressed and Precast Concrete" gives information on a number of additional topics including:

1. Cover thickness requirements for prestressed concrete slabs.
2. Anchor protection for post-tensioned tendons.
3. Cover and width requirements for prestressed concrete beams and joists.
4. Thickness requirements for precast concrete column covers.

In many instances, the CPCI manual contains more extensive data and discussion than comparable information in Chap. 2 of the Supplement to the National Building Code of Canada. The manual discusses alternative methods of achieving fire-resistance ratings such as providing added moment capacity instead of increasing the cover thickness. On the other hand, the manual gives more conservative values of cover thickness for beams and joists than Chap. 2 of the Supplement to the National Building Code.

# 5.4 VIBRATIONS IN CONCRETE BUILDINGS

## 5.4.1 INTRODUCTION

Aerobics, dancing and other rhythmic human activities have caused annoying vibrations in a number of buildings in recent years, including those of concrete and precast concrete construction.

The two main factors behind these problems are resonance (where the natural frequency of the floor structure is equal to or close to a forcing frequency of the rhythmic activity) and the presence of other occupancies in the building such as offices or restaurants where people are sensitive to the vibrations generated in the building by the rhythmic activity. Sect. 5.4.2 contains design recommendations to avoid these problems based on R5.4.1 and R5.4.2.

Floor vibrations due to walking, generally not a problem for current precast concrete construction, are discussed in Sect. 5.4.3.

Sometimes annoying floor vibrations are generated by mechanical equipment. These can usually be corrected by isolating the equipment as discussed in Sect. 5.4.4.

Design recommendations for lateral building vibrations due to wind are given in R5.4.3.

Vibration levels are generally expressed in terms of peak acceleration as a fraction of the acceleration due to gravity, a/g, where g = 9.807 m/s$^2$. In an office environment, annoyance occurs when vibration exceeds approximately 0.5%g. In an active environment such as dancing or aerobics, participants will accept much greater vibrations of approximately 10%g. Vibration limits are recommended for design in Fig. 5.4.1 for different occupancies. These limits apply for vibration frequencies between 2 and 10 Hz. Vibrations due to rhythmic human activities are within this frequency range.

**Fig. 5.4.1 Recommended Acceleration Limits for Vibrations[R5.4.1, R5.4.2]**

| Occupancies Affected by the Vibration | Acceleration Limit %g |
|---|---|
| Office and Residential | 0.4 to 0.7 |
| Dining and Weightlifting | 1.5 to 2.5 |
| Rhythmic Activity Only | 4.0 to 7.0 |

## 5.4.2 DESIGN FOR VIBRATIONS DUE TO HUMAN ACTIVITIES

To design for vibrations due to human activities it is usually necessary to avoid resonance by ensuring that the natural frequency, $f_o$, is greater than the forcing frequency, f, as given by the following formula[R5.4.1]:

$$f_o \geq f\sqrt{1+\frac{K}{a_o/g}\cdot\frac{\alpha_i w_p}{w_t}}$$

where:

$a_o/g$ = acceleration limit (see Fig. 5.4.1).

Acceleration limits for the design of floor structures for rhythmic activities can be assumed to be in the range 4 to 7 %g, and less than this if a sensitive occupancy is likely to feel vibrations which are greater than the limits in Fig. 5.4.1. The latter requires a consideration of the modal displacements in the sensitive occupancies as compared to the maximum displacement on the activity floor[R5.4.2].

$w_t$ = total dead weight supported by the floor structure expressed as a UDL.

$w_p$ = weight of participants as an equivalent UDL over the floor span. This can be estimated from the UDL of the participants over the area occupied by the participants as recommended in Fig. 5.4.2.

f = forcing frequency of the i$^{th}$ harmonic (i times the step frequency) as recommended in Fig. 5.4.2.

Only one harmonic needs to be considered for dancing whereas three harmonics need to be considered for aerobics because of its repeated impactive character. For sports events the second harmonic in Fig. 5.4.2 takes into account the impactive character of foot stamping.

**Fig. 5.4.2 Recommended Dynamic Loading for Rhytmic Events[R5.4.1, R5.4.2]**

| Activity | Forcing Frequency f (Hz) | UDL of Participants on Occupied Area $w_p$ (kPa) | Dynamic Coefficient $\alpha_i$ |
|---|---|---|---|
| Dancing | | | |
| first harmonic | 1.50 to 3.00 | 0.6 (2.5 m$^2$ per couple) | 0.50 |
| Lively Concert/Sports Event | | | |
| first harmonic | 1.50 to 3.00 | 1.5 (0.5 m$^2$ per person) | 0.25 |
| second harmonic | 3.00 to 5.00 | 1.5 (0.5 m$^2$ per person) | 0.05 |
| Aerobics | | | |
| first harmonic | 2.00 to 2.75 | 0.2 (3.5 m$^2$ per person) | 1.50 |
| second harmonic | 4.00 to 5.50 | 0.2 (3.5 m$^2$ per person) | 0.60 |
| third harmonic | 6.00 to 8.25 | 0.2 (3.5 m$^2$ per person) | 0.10 |

$\alpha_i$ = dynamic coefficient for the i'th harmonic of the step or jumping frequency, as recommended in Fig. 5.4.2

K = 1.3 for dancing, 1.7 for lively concert or sports event, 2.0 for aerobics.

To prevent resonance vibration, the above is applied to the highest harmonic.

However, if the dynamic loading for the highest harmonic, $\alpha_i w_p$, is sufficiently small compared to the mass weight, $w_t$, then resonance may be acceptable, in which case the above equation is applied to the second highest harmonic. Alternatively, a more detailed procedure is recommended in R5.4.1.

The acceleration due to harmonic resonance is obtained from[R5.4.2]:

$$a/g = \frac{1.3}{2\beta} \cdot \frac{\alpha_i w_p}{w_t}$$

where:

$\beta$ = the damping ratio

For rhythmic activities, $\beta$ can be taken as 0.04 to 0.06 for design purposes. The more people on the floor the greater the damping ratio.

The fundamental natural frequency of the floor structure, $f_o$, is the most important property of the floor structure for vibrations.

A simplified formula for estimating $f_o$ is given by:

$$f_o = \frac{18}{\sqrt{\Delta}}$$

where:

$\Delta$ = the maximum deflection (in mm) of the floor structure under the mass weight supported.

For simply-supported slabs (e.g., hollow-core or double tee slabs) on rigid supports, $\Delta$ is obtained from the simply-supported beam deflection formula. If the slabs are supported on simply-supported girders, the deflection of the girder under the weight it supports should be added to obtain the total deflection $\Delta$.

In a tall building the shortening of the supporting columns under the weight they support should also be added to obtain $\Delta$[R5.4.1]. Thus the fundamental natural frequency of the floor structure, $f_o$, is affected by the total deflection of the structure, not just the deflection of the slab itself.

This calculation procedure does not apply to two-way flat slab or two-way beam and slab construction.

## 5.4.3 WALKING VIBRATIONS

Floor vibrations due to walking are rarely a problem for concrete construction, mainly because of the large floor mass resisting the dynamic forces generated by individuals walking.

A design criterion for walking vibrations applicable to long-span floors of any material is given in R5.4.4 in terms of natural frequency, effective mass of the floor resisting the dynamic walking forces, damping ratio and sensitivity of the occupancy.

Guidelines for estimating these parameters have been developed for steel construction[R5.4.4] but not for precast concrete. Preliminary applications of this design criterion to precast concrete indicate, however, that annoying walking vibrations may occur in special situations, such as double tee slabs without topping used for office/residential occupancies. See R5.4.4 for more details.

### Example 1 Gymnasium Floor

A 2400 × 600 double tee with 50 mm topping and with a 15 m span on block walls is to be used as a gymnasium floor, aerobics being the critical rhythmic activity. There are no sensitive occupancies that might be affected by the vibrations, so the acceleration limit will be asumed as 7%g (Fig. 5.4.1).

The weight of participants is assumed to be 0.15 kPa spread out over the total span (based on 0.2 kPa over the occupied area in Fig. 5.4.2 and the size of a typical class), and the floor plus contents weigh 4.2 kPa.

From Fig. 5.4.2 the forcing frequencies for aerobics are 2.75 Hz (first harmonic), 5.5 Hz (second harmonic) and 8.25 Hz (third harmonic).

First check to see if third harmonic resonance is acceptable. Assuming a damping ratio of 0.04:

$$a/g = \frac{1.3}{(2)(0.04)} \cdot \frac{(0.1)(0.15)}{4.2} = 5.8\,\%g$$

The value 5.8 %g is acceptable, therefore the equation for $f_o$ can be applied to the second harmonic:

$$f_o \geq 5.5\sqrt{1 + \frac{2.0}{(2)(0.04)} \cdot \frac{(0.6)(0.15)}{4.2}} = 7.0 \text{ Hz}$$

The natural frequency for the double tee is determined from the simple beam deflection formula. Section properties are as given in the CPCI Design Manual, except that the modulus of elasticity of concrete is increased by 20 % to allow for expected short-term dynamic stiffness as compared to the CSA A23.3 formula. The composite section properties are used in the calculation.

$$\Delta = \frac{5}{384} \cdot \frac{w_t s L^4}{EI}$$

$$= \frac{5}{384} \cdot \frac{(4.2)(2.4)(15)^4}{(35.5)(10^6)(11.6)(10^{-3})}$$

$$= 16 \text{ mm}$$

$$f_o = \frac{18}{\sqrt{16}} = 4.5 \text{ Hz}$$

A natural frequency of 4.5 Hz is unacceptable because second harmonic resonance can occur at f = 4.5 Hz with accelerations of the order of 30%g by the application of the equation for a/g.

To obtain acceptable performance (f = 7.0 Hz), the double tee spanning 15 m must be increased to a 2400 × 1000 with 50 mm topping or the span must be shortened.

**Example 2 Stadium Seats**

The precast stadium seating shown in Figure 5.4.3 is proposed. The weight of the seating, including non-structural components, is 5 kPa on a horizontal projection. Determine the maximum acceptable span for vibration acceptability using an acceleration limit of 7%g for concerts and sports events.

**Fig. 5.4.3 Cross-Section of Proposed Stadium Seating**

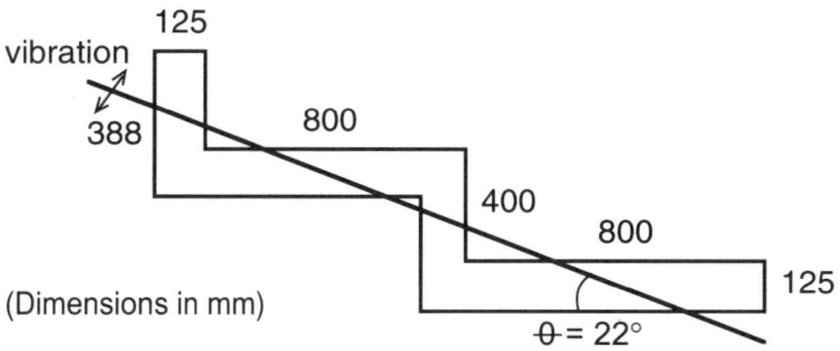


For such events, Fig. 5.4.2 recommends a UDL of 1.5 kPa for the weight of participants and two harmonic loads with forcing frequencies up to 3 and 5 Hz respectively. Because the seating vibrates in a direction normal to the plane of seating (see Figure 5.4.1) these forces are multiplied by cos 22°.

First, apply the equation for a/g to see if second harmonic resonance is acceptable:

$$a/g = \frac{1.3}{(2)(0.06)} \cdot \frac{(0.05)(1.5)\cos 22^0}{5+1.5} = 12\ \%g$$

The value of 12 %g is unacceptable. To limit accelerations to 7 %g , apply the equation for $f_o$ to the second harmonic:

$$f_o \geq 5.0\sqrt{1+\frac{1.7}{0.07} \cdot \frac{(0.05)\cos 22^0}{5+1.5}} = 5.6\text{ Hz}$$

The natural frequency for the precast seating is determined from the simple beam formula using the principle moment of inertia in the most flexible direction (see Figure 5.4.1), which is found by application of R5.4.5 to be $3.26 \times 10^{-3}$ m$^4$.

Thus:

$$\Delta = \frac{5}{384} \cdot \frac{(6.5)(1.6)L^4}{(35.5)(10^6)(3.35)(10^{-3})}$$

$$= \frac{L^4}{874} \quad (L \text{ in m}, \Delta \text{ in mm})$$

For a minimum natural frequency of 5.6 Hz (which corresponds to a Δ of 10.3 mm by application of the equation for f ), application of the above formula results in a maximum span of 9.75 m.

## 5.4.4 VIBRATION ISOLATION FOR MECHANICAL EQUIPMENT

Vibrations produced by equipment and unbalanced operating or starting forces can usually be isolated from the structure by mounting the equipment on a heavy concrete slab placed on resilient supports. This type of slab, called an inertia block, provides a low centre of gravity to compensate for dynamic forces such as those generated by large fans.

For equipment with less unbalanced weight, a "housekeeping" slab is sometimes used below the resilient mounts to provide a rigid support for the mounts and to keep them above the floor so they are easier to clean and inspect. This slab may also be mounted on pads of precompressed glass fibre or neoprene.

The natural frequency of the total mass on resilient mounts must be well below the forcing frequency generated by the equipment. The required weight of an inertia block depends on the total weight of the machine and the unbalanced force. For a long-stroke compressor, five to seven times its weight might be needed. For high pressure fans, one to five times the fan weight is usually sufficient.

A floor supporting resiliently mounted equipment must be stiffer than the isolation system. If the static deflection of the floor approaches the static deflection of the mounts, the floor becomes part of the vibrating system, and little vibration isolation is achieved. In general, the floor deflection should be limited to about 15% of the deflection of the mounts.

Simplified theory shows that for 90% vibration isolation, the equipment plus blocking mass supported on resilient mounts should have a natural frequency of about one-third the forcing frequency of the equipment. The natural frequency of this system can be calculated by[R5.4.6]:

$$f_{om} = \frac{15.8}{\sqrt{\Delta_m}}$$

where:

$f_{om}$ = natural frequency of the total mass on the mounts (Hz)

$\Delta_m$ = static deflection of the mounts due to the weight supported (mm)

From the above, the minimum required static deflection of the mounts can be determined as follows:

$$f_{om} = \frac{15.8}{\sqrt{\Delta_m}}$$

or

$$\Delta_m = \left(\frac{47.4}{f}\right)^2$$

where:

f = forcing frequency of the equipment (Hz)

$\Delta_{f\ell}$ = static deflection of the floor system at the location of the equipment caused by the weight of the equipment (including inertia block), (mm).

**Example 3 Vibration Isolation**

A piece of mechanical equipment has a forcing frquency of 15 Hz. Determine the approximate minimum deflection of the mounts due to the weight supported and the maximum deflection of the floor structure due to the eqipment.

The minimum static deflection of the mounts due to the weight supported is determined as follows:

$$\Delta_m = \left(\frac{47.4}{15}\right)^2 = 10 \text{ mm}$$

The maximum deflection of the floor structure under the weight of isolated equipment is determined as follows:

$$\Delta_{f\ell} = (0.15)(10) = 1.5 \text{ mm}$$

# 5.5 MECHANICAL, ELECTRICAL AND OTHER SUB-SYSTEMS COORDINATION

## 5.5.1 INTRODUCTION

Structural precast concrete is used in a wide variety of buildings, and its integration with lighting, mechanical, plumbing, and other services is important. Environmental demands are increasing substantially the ratio of costs for mechanical and electrical installations to total building costs.

This section is intended to show some methods which economically satisfy mechanical and electrical requirements and describe some standard methods of providing for the installation of other sub-systems.

## 5.5.2 LIGHTING AND POWER DISTRIBUTION

For many applications, designers can take advantage of the finished appearance, fire resistance and reflective qualities of precast, prestressed concrete by leaving the columns, beams and ceiling structure exposed. To achieve uniform lighting, free from distracting shadows, the lighting system should parallel the stems of tee elements.

A reflective paint and properly spaced high-output fluorescent lamps installed in a continuous strip can achieve a high level of illumination at a minimum cost. Lighting coffers can be enclosed with diffuser panels fastened to the bottom of the tee stems to provide a flush ceiling (see Fig. 5.5.1.). By applying reflective paints, these precast concrete lighting channels can be made as efficient as conventional fluorescent fixtures.

## 5.5.3 ELECTRIFIED FLOORS

The increasing use of computers, telephone and other monitoring and communication systems require adequate and flexible means of supplying electric power and cable communications. Since a cast-in-place topping is usually placed on prestressed floor elements, conduits and floor outlets can be readily buried within this topping. Shallow depth electrical systems can be provided in a reasonably thin topping (the total height of conduits can be as little as 35 mm). Most systems can easily be included in a 50 to 100 mm thick slab. Voids in hollow-core slabs can also be used as electrical raceways (see Fig. 5.5.2.).

When the system is placed in a structural composite slab, the effect of ducts and conduits must be carefully examined and their location coordinated with the reinforcing steel. Tests on slabs with buried ductwork have shown that structural strength is not normally impaired by these voids.

The high load-carrying capacity of prestressed concrete elements make it possible to locate high-voltage substations, with heavy transformers, near the areas of consumption with little or no additional expenses. For extra safety, distribution feeds can also be run between the legs of stemmed elements.

## 5.5.4 DUCTWORK

The space within stemmed elements and the voids in hollow-core slabs may be used for distribution ducts for heating, air-conditioning, or exhaust systems. In stemmed elements, three sides of the duct are provided by the bottom of the flange and the sides of the stems. The bottom of the duct is completed by attaching a metal panel to the tee stems in the same fashion as the lighting diffusers (see Fig. 5.5.1.).

Connections can be made by several means (drilled inserts, powder-activated fasteners, cast-in inserts or reglets). Field installed devices generally offer the best economy and ensure exact placement. Inserts should only be cast-in when they can be located at the design stage well in advance of casting the precast elements.

Additional duct work can be eliminated when hollow-core slabs are used. These elements have round, oval or rectangular voids of varying size which can provide ducts or raceways for various systems. Openings core-drilled in the field can provide access and distribution. The voids in the slabs can be aligned and connected to provide continuity of the system. Openings

**Fig.5.5.1 Metal panels attached to the bottoms of precast tee stems create ducts, and lighting diffuser panels provide a flush ceiling.**

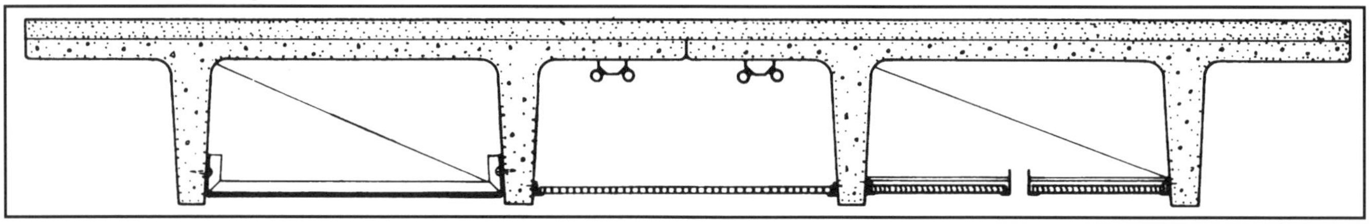

**Fig. 5.5.2 Underfloor communication and electrical ducts can be embedded within a concrete topping**

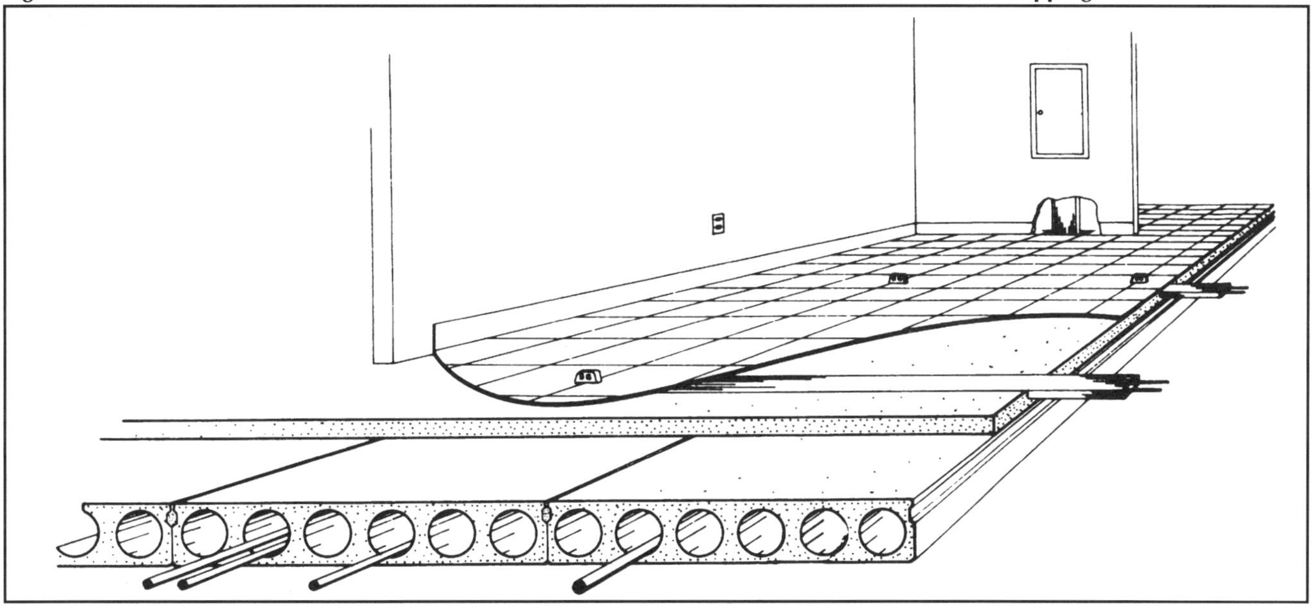

can also be provided in intermediate supporting beams, such as inverted tees, to allow duct continuity.

If high-velocity air movement is utilized, the enclosed space becomes a long plenum chamber with uniform pressure throughout its length. Diffusers are installed in ceilings to distribute the air. Branch runs, when required, can be standard ducts installed along the column lines.

When ceilings are necessary, proper selection of precast components can result in shallow ceiling spaces as shown in Fig. 5.5.3. This figure also illustrates the flexibility of space arrangements possible with long span prestressed concrete elements.

Branch ducts of moderate size can also be accommodated by providing block-outs in the stems of tees or beams. To achieve the best economy and utility in prestressed concrete elements, particularly stemmed elements, such blocks-outs should be repeated in size and location to handle all conditions demanded by mechanical, electrical, and/or plumbing runs. While this may lead to slightly larger openings in some cases, the end result will usually be more economical. Sufficient tolerance should be allowed in sizing openings to provide for field assembly.

Prestressed concrete box girders have been used to serve a triple function as air conditioning distribution ducts, conduit for

**Fig. 5.5.3 Where ceilings are required, ducts, piping and lighting fixtures can be accommodated within a shallow depth.**

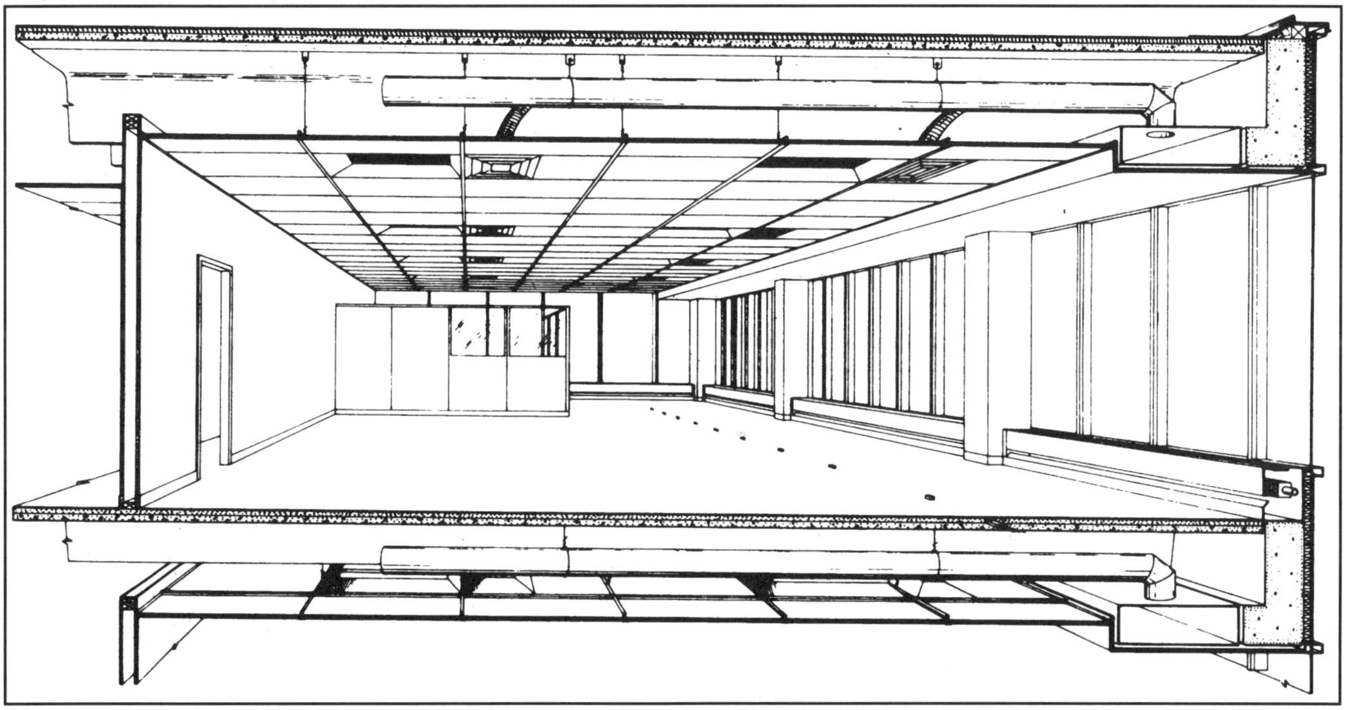

utility lines and structural supporting elements for roof deck units. Conditioned air can be distributed within the void area of the girders and introduced into the building work areas through holes cast in the sides and bottoms of the box girders. The system is balanced by plugging selected holes.

Vertical supply and return air trunks can be carried in the exterior walls, with small ducts branching out into the ceiling space. In some cases, the exterior wall cavities have been replaced with three or four sided precast boxes stacked to provide vertical runs for the mechanical and electrical systems. These stacked boxes can also be used as columns or lateral bracing elements for the structure.

These measures aid the economy of the structure by reducing the overall storey height and minimizing maintenance expenses.

## 5.5.5 OPENINGS THROUGH FLOOR AND ROOF SLABS

Large openings are usually made by block-outs in the forms during the manufacture; smaller ones (up to about 200 mm) are usually field drilled. Openings in the flanges of stemmed elements should be limited to the “flat” portion of the flange (beyond 25 mm of the edge of the stem on double tees and 75 mm of the edge of the stem on single tees). Angle headers are often used for framing large openings in hollow-core floor or roof systems (see Fig. 5.5.4.).

## 5.5.6 OTHER SUB-SYSTEMS

Suspended ceilings, crane rails, mechanical and electrical equipment and other sub-systems can be easily accommodated with standard manufactured hardware items and embedded plates as shown in Fig. 5.5.5.

Floor or window frames properly braced to prevent bowing during concrete placement can be cast into precast concrete panels. The glazing or doors can be installed prior to or after delivery to the job site. If the window or doors units are properly protected, they can often be cast into the panel at the plant.

When casting in aluminum window frames, the aluminum should be properly coated so that it will not react with the concrete. Repetition is one of the real keys to economy in precast concrete wall assemblies. Windows and doors should be located in the identical location for all panels wherever possible.

Insulated wall panels are produced by embedding an insulating materials such as expanded polyurethane or polystyrene between layers of concrete. Sandwich wall panels can provide the thermal resistance values required for modern energy-conserving buildings (see Sect. 5.1). These exterior panels are normally cast on flat beds. The inside surface of the concrete panel can be given a troweled finish at the precast plant followed by minor touch-up work at the jobsite. The interior face is completed by painting or wall papering to achieve a finished appearance. The formed surface of the panel can also be treated in a similar manner when used as the interior face.

## 5.5.7 SYSTEMS BUILDING

Many complete systems buildings have been built using precast and prestressed concrete. Interest in this method of construction will lead to more of the building sub-systems being prefabricated and precoordinated with the structure.

Those parts of the structure that require the highest labour skills can economically be prefabricated at the precast plant prior to installation in the field. The prefabricated components can be preassemblies of basic plumbing systems or electrical/mechanical systems plus lighting.

For housing systems, electrical conduits, and boxes can be cast in the precast wall panels. This process requires coordination with the electrical contractor, and savings on job site labour and time are possible. The metal or plastic conduit is usually prebent to the desired shape and delivered to the casting bed already connected to the electrical boxes. All joints and connections must be thoroughly sealed and the boxes enclosed prior to casting to prevent the system from becoming clogged. Cable TV and telephone conduits have also been cast in using the same procedure.

**Fig. 5.5.4 Large openings in floors and roof slabs made during manufacturing of the units. Small openings are field drilled. Some common types of openings are shown.**

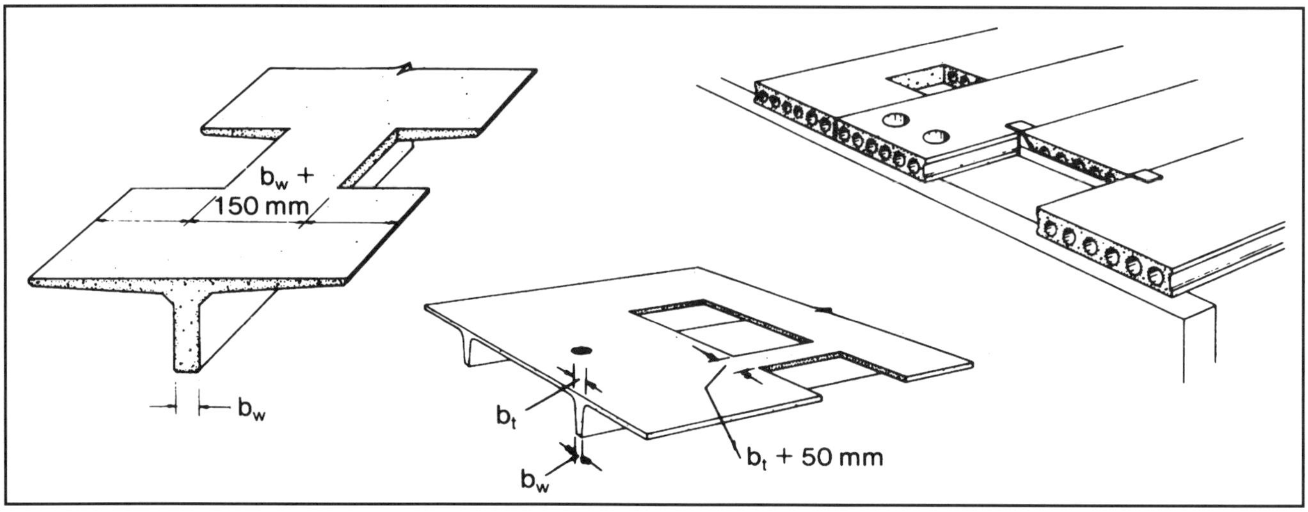

Fig. 5.5.5 Methods of attaching suspended ceilings, crane rails and other sub-systems.

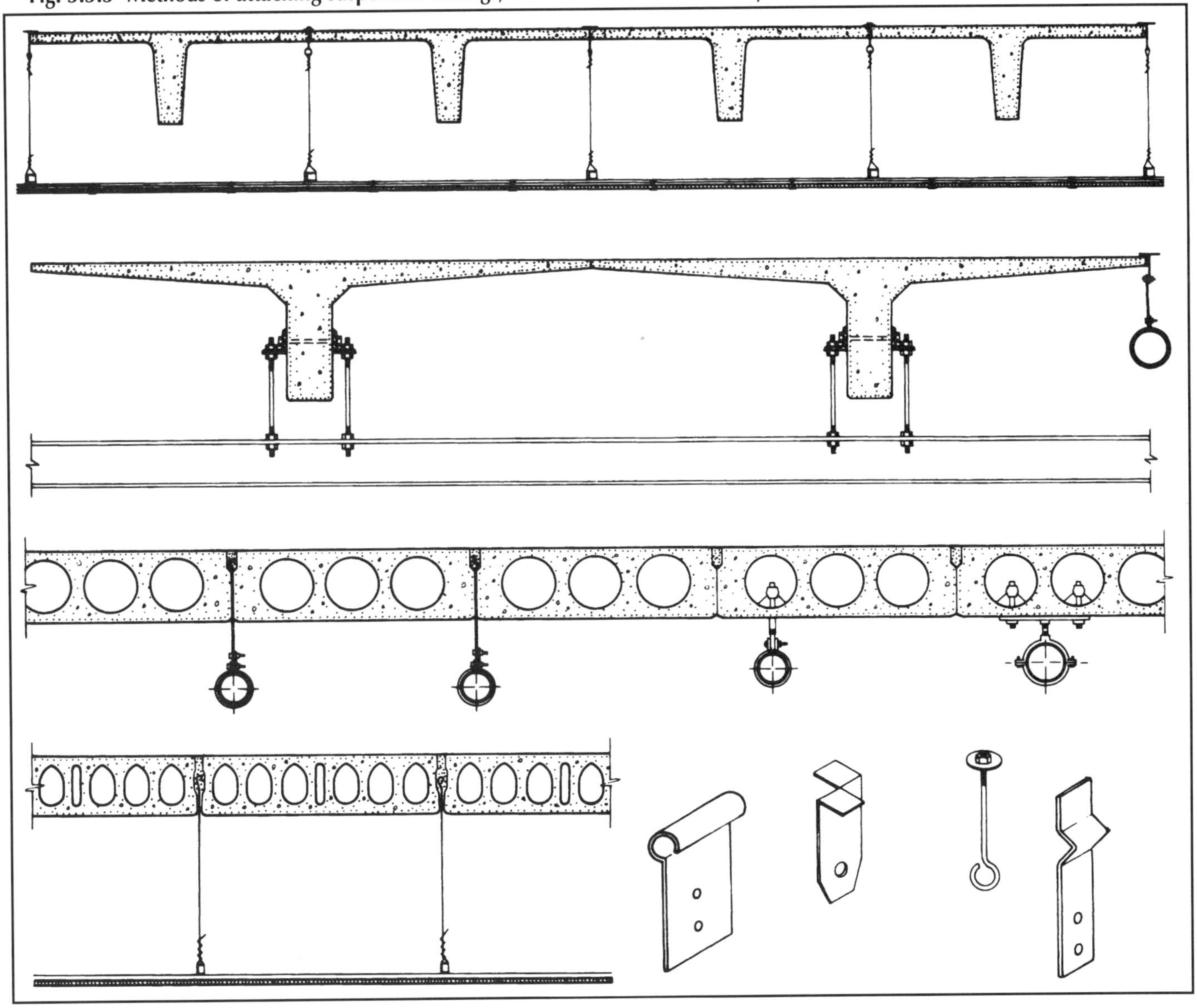

To reduce on-site labour, prefabricated bathroom units or combination bathroom/kitchen modules have been developed (see Fig. 5.5.6). Such units include bathroom fixtures, kitchen cabinets and sinks, as well as wall, ceiling, and floor surfaces.

Plumbing units are often connected and assembled prior to delivery to the job sites. These bathroom/kitchen modules can be molded plastic units or fabricated from drywall components. To eliminate a double floor, the module can be built on a structural element at the plant.

Fixtures can be wall hung if the walls of the unit can be designed strong enough. Units are placed directly on a precast floor and in multi-storey construction are stacked with one bathroom directly over the one below. A block-out for a chase is provided in the precast floor and connections are made from each unit to the next to provide a vertical plumbing stack.

Prefabricated wet-wall plumbing systems (see Fig. 5.5.7) incorporate preassembled piping system using snap-on or no-hub connections made using a variety of materials. These units only require a block-out in the prestressed flooring units and are arranged to mate with the connecting module. Best economy results when bathrooms are backed up to each other, with a common vertical run servicing two bathrooms.

Some core modules not only feature bath and kitchen components, but also HVAC components all packaged in one unit. These modules can be easily accommodated in a precast prestressed structural systems by placing them directly on the prestressed elements and shimming and grouting as required.

**Fig. 5.5.6 Kitchen/bathroom modules can be preassembled on precast prestressed slabs ready for installation into systems buildings.**

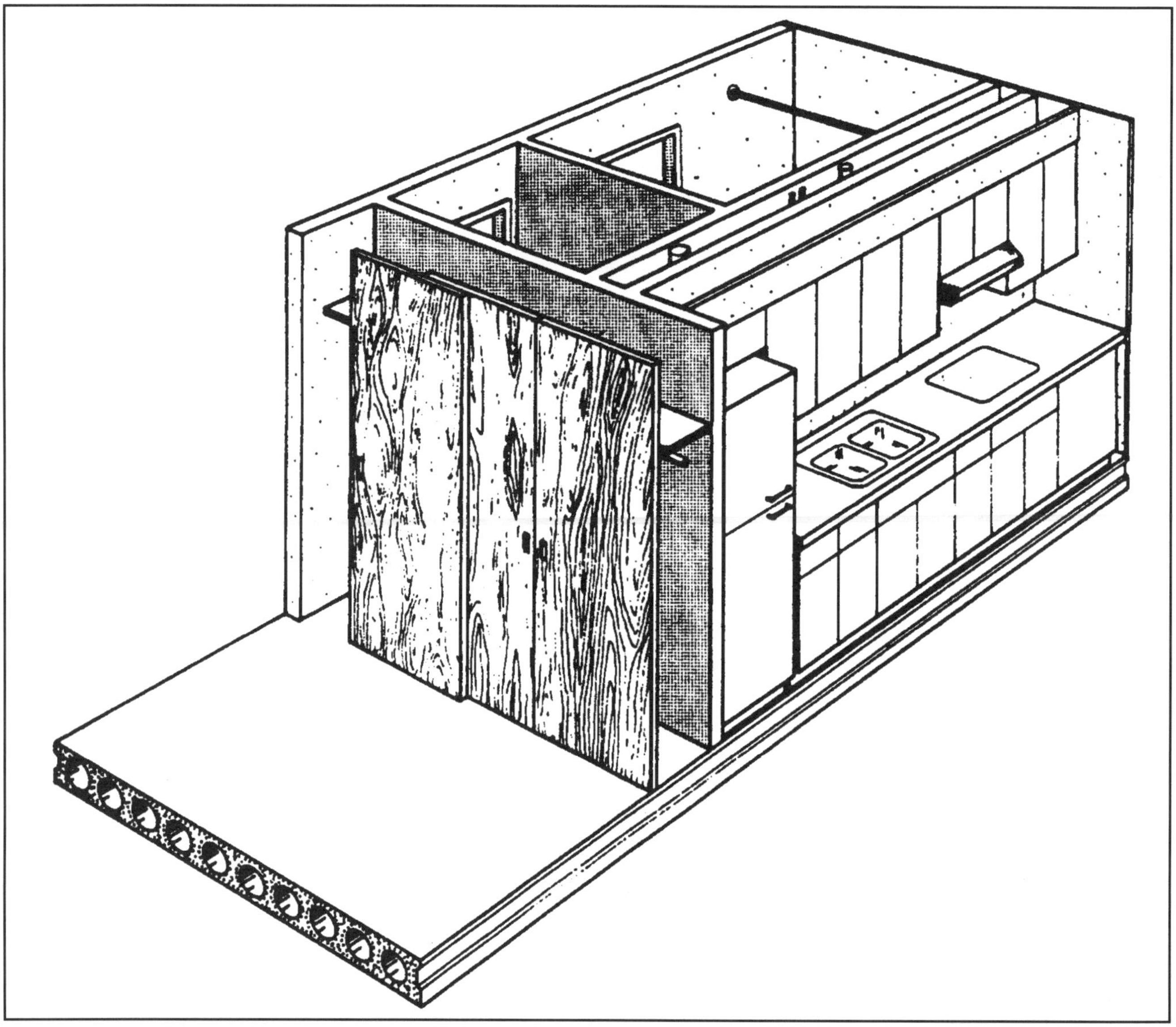


**References:**

R5.1.1 *National Energy Code for Buildings - 1995*, Canadian Commission on Building and Fire Codes / National Research Council of Canada.

R5.1.2 *National Energy Code for Houses - 1995*, Canadian Commission on Building and Fire Codes / National Research Council of Canada.

R5.1.3 *1993 ASHRAE Handbook - Fundamentals - SI Edition,* American Society of Heating, Refrigerating and Air-Conditioning Engineers, Inc.

R5.1.4 *Guide to Energy Efficiency in Masonry and Concrete Buildings*, available from the Masonry Council of Canada.

R5.1.5 Hutcheon, N.B. and Handegord, G.O., *Building Science for a Cold Climate*, Wiley, 1983.

R5.1.6 *Moisture Control in Buildings*, G. R. Treschel, Editor

R5.1.7 *Canadian Building Digests*, National Research Council of Canada, Ottawa.

R5.1.8 *Energy Conservation Design Resources Handbook,* The Royal Architectural Institute of Canada.

R5.1.9 *Architectural Details for Insulated Buildings*, Ronald Brand, Van Nostrand Reinhold (1990)

R5.2.1 ASTM E90 - Laboratory measurement of airborne sound transmission loss of building partitions, American Society for Testing and Materials, 1916 Race St., Philadelphia, PA, USA, 19103.

R5.2.2 ASTM E336 - Standard Test Method for Measurement of Airborne Sound Insulation in Buildings, American Society for Testing and Materials, 1916 Race St., Philadelphia, PA, USA, 19103.

**Fig. 5.5.7 Prefabricated wet-wall plumbing systems incorporate preassembled piping.**

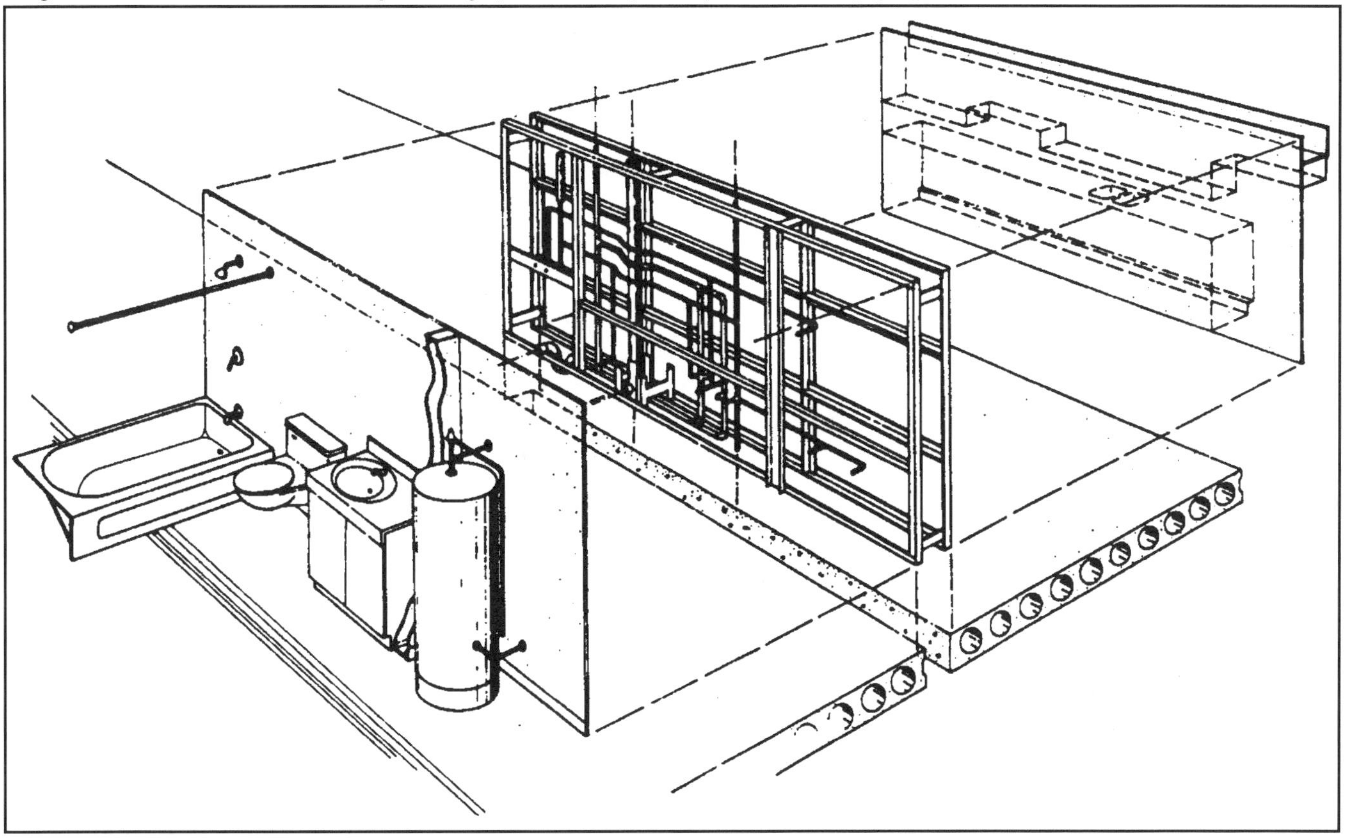


R5.2.3 ASTM E413 - Determination of sound transmission class, American Society for Testing and Materials, 1916 Race St., Philadelphia, PA, USA, 19103.

R5.2.4 ASTM C522 — Standard Test Method for Airflow Resistance of Acoustical Materials. American Society for Testing and Materials, 1916 Race St., Philadelphia, PA, USA, 19103.

R5.2.5 ASTM E492 - Laboratory measurement of impact sound transmissions through floor-ceiling assemblies using the tapping machine, American Society for Testing and Materials, 1916 Race St., Philadelphia, PA, USA, 19103.

R5.2.6 ASTM E1007, Standard Test Method for Field Measurement of Tapping Machine Impact Sound Transmission through Floor-ceiling Assemblies and Associated Support Structures, American Society for Testing and Materials, 1916 Race St., Philadelphia, PA, USA, 19103.

R5.2.7 ASTM E989, Classification for Determination of Impact Insulation Class (IIC), American Society for Testing and Materials, 1916 Race St., Philadelphia, PA, USA, 19103.

R5.2.8 ASTM C423, Standard Test Method for Sound Absorption and Sound Absorption Coefficient by the Reverberation Room Method.

R5.2.9 National Building Code of Canada, 1995, Associate Committee on the National Building Code, National Research Council of Canada. Ottawa, Canada K1A 0R6.

R5.2.10 Canada Mortgage and Housing Corporation, New Housing and Airport Noise (Metric Edition 1978).

R5.2.11 Canada Mortgage and Housing Corporation, Road and Rail Noise: Effects on Housing (1977) Metric.

R5.2.12 The US Federal Highway Administration issues a free prediction program called STAMINA.

R5.2.13 The Wall Journal, PO Box 1217, Lehigh Acres, FL 33970-1217, 813 369 0178

R5.2.14 Field Sound Transmission Loss Measurements, A.C.C. Warnock, Building Research Note 232, National Research Council Canada.

R5.2.15 Measurements of the Sound Transmission Loss of Windows, J.D. Quirt, Building Research Note 172, National Research Council Canada.

R5.2.16 Sound Transmission Loss of Masonry Walls: Tests on 90, 140, 190, 240 and 290 mm Concrete Block Walls With Various Surface Finishes, A.C.C. Warnock and D.W. Monk. Building Research Note 217, National Research Council Canada.

R5.2.17 Sound Transmission Loss of Masonry Walls: Twelve-inch Lightweight Concrete Blocks With Various Surface Finishes, T.D. Northwood and D.W. Monk, Building Research Note 90, National Research Council Canada.

R5.2.18 Sound Transmission Loss of Masonry Walls: Twelve-inch Lightweight Concrete Blocks — Comparison of Latex and Plaster Sealers, T.D. Northwood and D.W. Monk, Building Research Note 93, National Research Council Canada.

R5.2.19 Sound Transmission Loss Measurements Through 190 mm and 140 mm Blocks with Added Gypsum board and Through Cavity Block Walls, A.C.C. Warnock, NRCC Internal Report 586, 1990.

R5.2.20 Noise Control in Buildings, A Practical Guide for Architects and Engineers, Cyril M. Harris, ed., McGraw-Hill, Inc. 1994.

R5.2.21 Handbook of Acoustical Measurements and Noise Control, Third Edition, Cyril M. Harris, ed., McGraw-Hill, Inc. 1991.

R5.2.22 NIOSH Compendium of materials for noise control, R.A. Hedeen. DHEW Publication No. 80-116. US Department of Health, Education and Welfare, 4676 Columbia Parkway, Cincinnati, Ohio 45226.

R5.2.23 Catalog of STC and IIC Ratings for Wall and Floor/ceiling Assemblies, R.B. Dupree, Office of Noise control, California Department of Health Services, Berkeley, California 94704.

R5.2.24 Northwood, T.D., Dickens, H.B. and Hansen, A.T., Noise Control in Residential Buildings, NRC 9162, National Research Council of Canada 1967.

R5.3.1 Lie, T.T., Contribution of Insulation in Cavity Walls to Propagation of Fire, Fire Study No. 29, Division of Building Research, National Research Council of Canada, Ottawa, Ontario.

R5.3.2 Gustaferro, A.H. and Abrams, M.S., Fire Tests of Joints Between Precast Wall Panels: Effect of Various Joint Treatments, PCI Journal, Vol. 20, No. 5, Sept-Oct 1975, pp. 44-64.

R5.3.3 Lie, T.T., Contribution of Insulation in Cavity Walls to Propagation of Fire, Fire Study No. 29, Division of Building Research, National Research Council of Canada, Ottawa, Ontario

R5.3.4 Gustaferro, A.H. and Abrams, M.S., Fire Tests of Joints Betwen Precast Wall Panels: Effect of Various Joint Treatments, PCI Journal, Vol. 20, No. 5, Sept-Oct 1975, pp. 44-64.

R5.3.5 Fire Resistance Ratings for Prestressed and Precast Concrete, Canadian Prestressed Concrete Institute, February 1978.

R5.4.1 Commentary A on Part 4 of the National Building Code of Canada: Serviceability Criteria for Deflection and Vibration. Supplement to the National Building Code of Canada 1990. National Research Council Canada, Ottawa.

R5.4.2 Allen, D.E. Building Vibrations from Human Activities, Concrete International, 12(6), June 1990, p. 66-73.

R5.4.3 Commentary B on Part 4 of the National Building Code of Canada: Wind Load. Supplement to the National Building Code of Canada 1990. National Research Council Canada, Ottawa.

R5.4.4 Allen, D.E. and Murray, T.M. Design Criterion for Floor Vibration Due to Walking. AISC Engineering Journal, Fourth Quarter (December) 1993, p. 117-129.

R5.4.5 Timoshenko, S.P. and Gere, J.M. Mechanics of Materials (see Appendix A, p. 485 - 497). Van Nostrand Reinhold, New York, NY, 1972

R5.4.6 Harris, C.M. and Crede, C.E. Shock and Vibration Handbook, Second Edition, McGraw-Hill, New York, NY, 1976.

R5.4.7 International Standard ISO 2631-2:1989. Evaluation of Human Exposure to Whole-Body Vibration - Part 2: Continuous and Shock-Induced Vibrations in Buildings. International Organization for Standardization, Geneva, Switzerland.

R5.4.8 Bachmann, H., Ed. Vibration Problems in Structures. Birkhauser-Verlag, Basel - Boston - Berlin. 1995

R5.4.9 Vibrations of Concrete Structures. Special Publication SP-60, American Concrete Institute, Detroit, MI, 1979

R5.4.10 International Standard ISO 10137:1992. Basis for Design of Structures - Serviceability of Buildings Against Vibration. International Organization for Standardization, Geneva, Switzerland.

# CHAPTER 6
# ARCHITECTURAL PRECAST CONCRETE

Page No.

# 6.1 INTRODUCTION

Architectural precast concrete products are those precast concrete units which, through finish, shape, colour and texture, define the architectural aesthetics and function of the structure.

Architectural precast concrete is available in complex shapes which serve not only as non-load bearing (cladding) walls, but also combine their attractive appearance with the ability to serve as main structural elements.

Among the many wall applications for architectural precast concrete, a distinction could be made between load bearing and non-load bearing elements:

1. A non-load bearing (cladding) element is one which could be removed from the structure without any consequence to the integrity of the building. Although non-load bearing walls are normally provided only for the purpose of separating environments, they are subjected to externally applied loads, such as those produced by wind and earthquake.
2. A load bearing element is one which resists and transfers vertical and transverse loads applied from other elements. A load bearing element cannot be removed without affecting the structural integrity or stability of the building as a whole or any of its parts.

Non-load bearing architectural panels can also be made of glass fibre reinforced concrete. These panels consist of sprayed applications of alkali-resistant glass fibres in combination with a sand/cement slurry. They are very light and can incorporate complex shapes and face mixes with numerous variations in colour.

In composite construction, precast concrete elements may be used as forms for cast-in-place concrete. This solution is especially suitable for combining architectural and structural functions in load bearing facades.

## 6.1.1 CHECKLIST

The following is a checklist of factors the designer must consider in the design, manufacture and erection of architectural precast concrete elements.

### Architectural Requirements

1. Load bearing or non-load bearing?
2. Finishes. For full information on the many and varied types of finish available, contact CPCI architectural precast manufacturers.
3. Insulated or non-insulated panels? Refer to Section 6.6.

### Size of Panel

1. Weight limitations.
2. Production facilities.
3. Transportation weight and dimension limitations.
4. Erection feasibility and access.
5. Crack control.

### Supporting Structure

1. Points of load application
2. Overall stability.
3. Stability during erection.

### Standardization

Obtain maximum repetition of similar units to reduce overall costs.

### Design of Connections

Refer to Chapter 4 of this Manual.

### Detailed Design of Panel

Refer to Chapter 3 and Section 6.5.3 of this Manual for:

1. Concrete mix design.
2. Reinforcement:
   a) For function in final position.
   b) For stripping, storage, transportation and erection.
3. Design of connections, inserts, hardware lifting hooks, etc. for stripping, storage, transportation and erection.

### Shop Drawings

For acceptable standards refer to PCI Architectural Precast Concrete Drafting Handbook, including:

1. Erection drawings
2. Anchor layout drawings
3. Connection details.
4. Piece drawings.
5. Hardware details.
6. Storage diagrams.
7. Drawings of special handling equipment.

### Formwork

For information on the various materials used for the manufacture and design of formwork, contact CPCI architectural precast manufacturers.

### Production

For information regarding materials to be used and production methods to be followed, refer to CSA-A23.4-94.

The following items should be considered:

1. Plant capacity and cranage.
2. Placing of hardware and reinforcement.
3. Inspection of forms, hardware and reinforcement.
4. Mixing, placing and vibrating concrete.
5. Finishing.
6. Curing.
7. Stripping.
8. Storage.

### Quality Control

For full information on the requirements of quality control over the entire production sequence, refer to CSA-A23.4-94.

### Transportation

Refer to Section 3.9.5 for:

1. Types of trailers.
2. Types of frames.
3. Support material.

For further detailed information on methods, material and equipment used in handling and transporting all types of precast concrete units, contact CPCI architectural precast manufacturers.

### Erection

For the successful use of architectural precast concrete, the designer must clearly envision the erection process:

1. Unimpeded access to allow for continuous erection, working and storage space, staging area for trailers and cranes.
2. Equipment, cranes, monorail hoists, etc.
3. Lifting, turning, tilting units.
4. Survey of structural frame, location of cast-in hardware.

### Safety

Refer to local safety regulations.

### Economy

1. Design, production, transportation and erection costs.
2. Progress payments for completed units stored at plant.
3. Economy of precast concrete.

### Tolerances

1. Design.
2. In-situ concrete or steel support structure tolerances, sway, creep, differential deflection.
3. Production tolerances, dimensions, bowing, warping.
4. Erection tolerances:
    a) Precast concrete units to supporting structure.
    b) Precast concrete unit to precast concrete unit. Refer to CSA-A23.4-94.

The total effect of all tolerances will be noticed at the panel joints of the final structure.

## 6.1.2 DESIGN OBJECTIVES

The design objectives for construction, of which architectural precast concrete is a part, are similar to those objectives which apply to the design of an individual precast element.

Structural integrity of the completed structure is the primary objective. Deflections must be limited to acceptable levels, and distress that could result in instability for an individual element or for the structure as a whole, must be prevented.

Economy is an important design objective. The total cost of a completed structure is the determining factor. The designer should attempt to optimize the entire structure and consider the advantages provided by multi-functional precast panels. In this regard, the designer should be aware of the major economies offered by standardization or repetition. Consideration must also be given to the cost of large versus small panels subject to weight limitations on transportation, crane capacity and crane location.

The aim of repetition is to reduce costs by increasing productivity. Repetition means fewer moulds and a consequent reduction in mould construction cost. Production-line working can be implemented in the plant, enabling a particular casting sequence to be repeated each day, leading to improvements in efficiency through the repeated operations of familiar tasks. Handling, storage and delivery are simplified with consequent reductions in the risk of errors. Site efficiency is also improved through the repetition of familiar erection sequences. These benefits can only be achieved if there is a high degree of repetition in the design of the precast elements.

Often, in the initial design stage, a high degree of repetition appears possible. As the details are finalized, however, a very strict discipline is required on the part of the designer if the creation of a large number of non-repetitive units is to be avoided. Any budget costs given at the initial design stage should take into account the possibility that the number of different units will increase as the design progresses. If non-repetitive units are unavoidable, the increase in cost can be minimized if they can be cast from a master mould with simple modifications without the need for completely special moulds. In general, it is relatively straightforward to alter a mould if the variations can be contained within the total mould envelope, by use of bulkheads or blockouts, rather than by cutting into the mould surface.

The term "standard" is difficult to define, but panels cannot truly be described as standard unless they are identical in every respect. Even relatively minor variations, such as the positions of connections, are sufficient to make a unit non-standard (non-repetitive). However, most manufacturers expect and allow for some variations and will normally accept some minor changes to standard units without an increase in cost.

## 6.1.3 STRUCTURAL DESIGN AND ANALYSIS

Architectural precast concrete construction can be considered in three parts:

1. The precast elements individually.
2. The support system(s) for the precast element, such as the beam, slab, wall, column, foundation.
3. The connection that serves to join the precast element to its support system.

The design of the architectural precast elements and the structure of which they are a part, involves load transfer, consideration of stability and the potential for movement of the panels or the structure.

The potential movement of an element should consider shrinkage, expansion and contraction due to temperature and moisture changes, and elastic and creep deformations resulting from stresses within the element.

The potential for movement of the structure should consider the same above-mentioned causes and in addition also movement due to gravity, wind and earthquake loads, foundation effects, short term and long term deflection and differential deflection.

The potential for movement can be determined by analysis, and provisions made to accommodate this movement.

The structural design of architectural precast concrete requires the determination of all loads which will act on the precast element, and consideration of the following:

1. Stresses induced during the handling, transportation, and erection processes.
2. Acceptable crack locations and crack widths. It should be noted, however, that it is preferable to have crack-free units.
3. The strain gradients across the thickness of the panel. Stresses induced by the restraint of bowing due to thermal and moisture differentials and non-uniform curing.
4. Forces which may develop in a precast element, and its connections due to distortion of the structural frame and restrained volume change.
5. Deflection response of a precast element under the action of:
   a) Live and dead loads.
   b) Wind and earthquake loads.
   c) Thermal loads.
6. The specified tolerances for structures and those for precast installation.
7. Historical performance of various types of connections.

The designer must recognize that loads and behaviour cannot be established precisely, particularly with respect to elements continuously subjected to the environment. The imprecise nature of design loads will generally not affect the safety of the element, provided that the above factors have been considered, and provided that reasonable values have been established in accordance with the requirements of the relevant building codes.

The designer must take careful note of the allowable tolerances for the structural system. This is particularly important for isolated elements forming a long vertical line, such as column covers, where any deviation from the vertical is readily noticed.

All non-load bearing elements should be designed to accommodate movement freely and, whenever possible, without redundant supports, except where provisions are necessary to restrain bowing. It should be noted that extremely high loads can be induced if it is attempted to restrain bowing completely.

# 6.2 NON-LOAD BEARING PANELS

## 6.2.1 GENERAL

Non-load bearing panels are closure panels only, and are designed to resist wind forces, seismic forces generated from their self-weight, and forces required to transfer the weight of the panel to the support. Although service loads, wind and seismic forces do frequently produce the maximum stresses, the forces imposed during the manufacturing and erection processes will usually govern the design. The forces resulting from seismic considerations will generally govern connection design, but will usually result in panel stresses less than those imposed during manufacture.

## 6.2.2 DEFORMATION

The relationship of the deformations of the panel and the supporting structure must be evaluated, and care taken to prevent unintended restraints and imposed loads due to these deformations. Deformations of the supporting structure which may induce unintended loading will be caused by deflection due to the weight of the panel, volumetric changes of concrete frames, and torsion of spandrel beams. To prevent imposing loads on the panel, the connections must be designed and installed to permit these deformations to freely occur.

Behaviour of a series of panels supported on a flexible beam is shown in Fig. 6.2.1. The tendency for the panels with two bearing connections to deflect with the beam may cause unintended restraint forces to develop in the longitudinal direction. Alternatively, this problem could be solved by providing a single bearing connection for each panel. This illustrates the need for the panel designer to provide for vertical deformations of the supporting beam.

**Fig. 6.2.1 Deformation of panels on flexible beam**

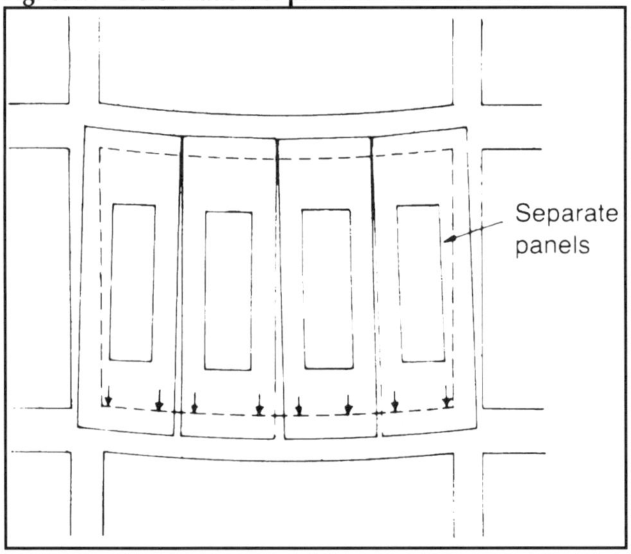


The most prevalent cause of panel deformation after placement in the structure is bowing due to thermal variations. This generally will cause panels to bow outward. The panel, if supported in a manner that will permit deformation to freely occur, and assuming a homogeneous material and linear temperature distribution, will not be subjected to stresses. However, if a panel is restrained laterally at the middle, such restraint will produce panel stresses.

Panel bowing in horizontally cast panels before erection is caused by:

1. Differential shrinkage due to evaporation from the top face of the element while the lower face is maintained at 100% humidity.
2. Deflection due to self-weight and position of support points.
3. Deflection due to difference in temperatures between faces (e.g. storage orientation).
4. Deflections due to incorrectly stacked panels, creep of the concrete, and the length of time in the storage yard.

Non-load bearing panels should be designed and installed so that they do not restrain frames from lateral translation. If such restraint is induced, significant diagonal compression may occur, as indicated in Fig. 6.2.2. To prevent this, panels which are installed between elements of the frame should be connected at the top and bottom only and left free along the sides. The space required for erection will usually be sufficient to prevent contact during lateral deformation of the frame.

**Fig. 6.2.2 Panel forces induced by frame distortion**

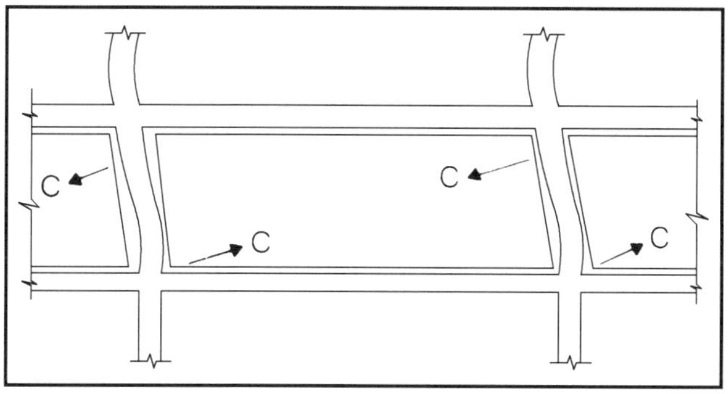


The vertical shortening of concrete columns should be considered when structures are tall. At intermediate levels, the differential shortening between two adjacent floors will be negligible, and the panel will follow the frame movement. At the lowest level, if the panel is rigidly supported at the base, the accumulated shortening of the structure above may induce unintended loading on the panel due to stacking panels on top of each other. The connections and horizontal joints widths should be designed to accommodate all column shortening.

A similar situation will result when two adjacent columns have significantly different loads.

Non-load bearing panels which contain openings may develop stress concentrations at these openings resulting from unintended loading or restrained bowing. While these stress concentrations may be partially contained by use of reinforcement, the designer should always try to eliminate the imposed restraints. Areas of abrupt change in the cross-section should be reinforced.

### 6.2.3 CRACK CONTROL

While it is most desirable to manufacture and erect panels which are crack-free, the design of non-load bearing panels will often be governed by the limiting crack widths in the panel. Sec. 3.4.2 discusses allowable tensile stresses and crack widths in concrete.

Cracking can also be controlled by introducing prestressing, using either pretensioning or post-tensioning. In either case, the prestress force should generally be concentric with the effective cross-section, to minimize out-of-plane deformations. For thin sections (75 to 100 mm) which may be subjected to undesirable bowing, experienced precasters sometimes prestress the section eccentrically to counteract the bowing. It is recommended that prestress in a panel, after all losses, be limited to the range of 1 MPa to 6 MPa. Care must be taken to avoid longitudinal splitting of the concrete collinear with the prestressing strands.

If short panels are pretensioned, care must be taken that effective bond transfer length is provided. For a discussion on transfer lengths, see Sec. 3.4.9.

## 6.3 SPANDRELS

### 6.3.1 NON-LOAD BEARING SPANDRELS

These are precast elements which are less than storey height, made up either as a series of individual units or as one unit extending between columns. Support for spandrels may be located on the floor at or near the column or on the column, and stability against eccentric loading achieved by connections to the underside of the floor (Figure 6.3.1).

**Fig. 6.3.1 Typical spandrel connections**

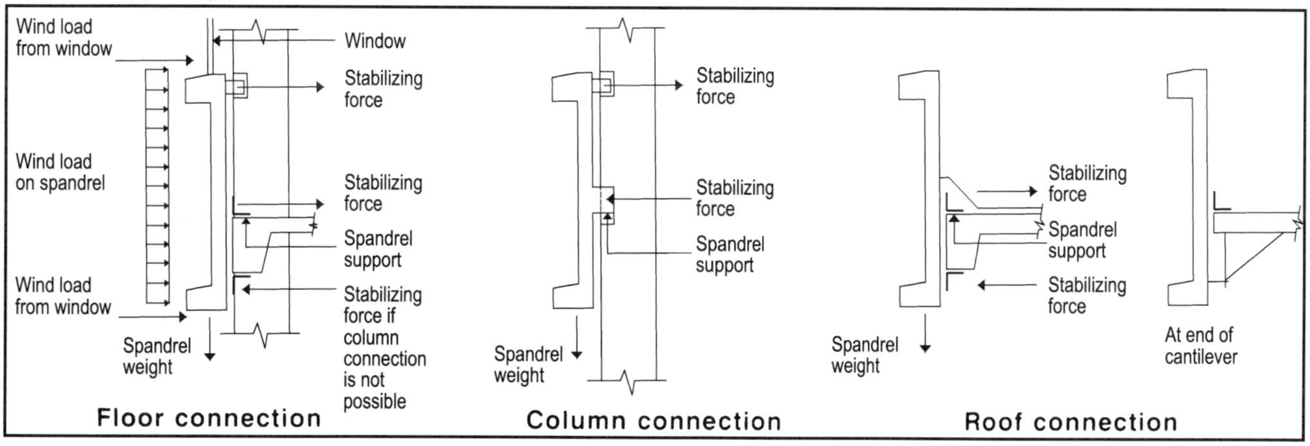

Spandrels are usually part of a window wall; thus consideration should be given to limiting the vertical deflections and rotations of the spandrel to that consistent with the requirements of the window.

Special provisions should be made for slender elements, which are defined as elements in which the spacing of lateral support of the compression flange exceeds 50 times the width of the flange. The figure 50 is based on the criteria of CSA-A23.4-94 and is conservative.

The factored resistance of the lateral support should be a minimum of 2% of the factored force in the compression flange.

### 6.3.2 EFFECT OF CANTILEVER SUPPORTS

Consideration should also be given to spandrels which are supported at the ends of long cantilevers. The designer must verify the effect of deflection and rotation of the support, including the effects of creep, and arrange the details of all attachments to accommodate this condition (Fig. 6.3.2). A particularly critical condition can occur at corners of buildings when there is a cantilever on both faces or on one face only.

**Fig. 6.3.2 Effect of cantilever supports**

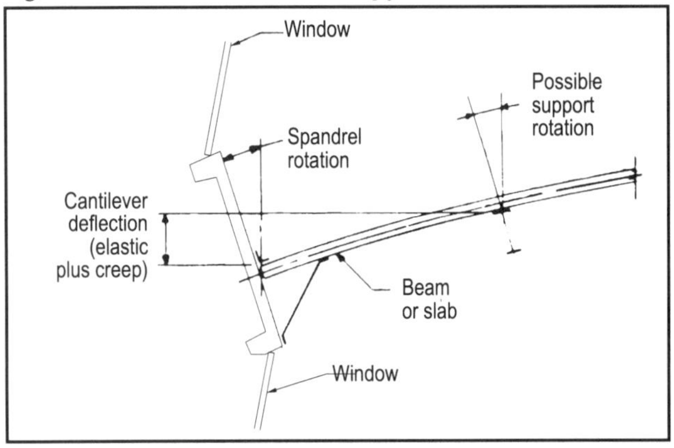


## 6.4 COLUMN COVERS

### 6.4.1 GENERAL

The use of precast concrete panels as covers over steel or cast-in-place concrete columns and beams, and as mullions, is a common method of achieving architectural expression, special shapes, and specific finishes in an economical manner. When used over steel columns and beams, covers may possibly provide the required fire-rating without resorting to further protection of the steel (see Sec. 5.3). When used over cast-in-place concrete columns and beams, it will often permit achieving a uniformity of finish in combination with a special architectural shape, all in the most economical manner. The interior surface of the cover can be readily insulated, thus minimizing the differential temperature variation through the cover thickness. Column covers can be multi-functional, providing space for services and for exterior grooves for vertical window-washing machinery. For a further discussion of beam covers (spandrels) see Sec. 6.3.

Column covers are generally supported by the structural column or the floor, and are themselves designed to transfer no vertical load other than their own weight. The vertical load of each length of column cover section is supported usually at one elevation, and tied back top and bottom to the floors for lateral load transfer and stability. Connection must allow for relative horizontal movement between floors. This may cause the column cover to rock back and forth between bearing connections. The length of a column cover will be dependent upon transportation and lifting limitations, on architectural considerations, and upon the ability of the structural column to support a specific concentrated load (panel weight) locally.

### 6.4.2 DESIGN GUIDELINES

Column covers and mullions are usually a major focal point in a structure, and aesthetic success can be achieved if careful thought is given to all facets of design and erection. Following are some items which the designer should consider:

1. The designer must take careful note of the allowable tolerances for the structural system. This is particularly important for column covers and mullions since, as isolated elements they often form a long vertical line, and any deviation from a vertical plane is readily observable. Invariably, this deviation is the result of the allowable tolerances associated with structural steel and cast-in-place concrete frames. To some degree, these deviations can be compensated for by precast connections with sufficient adjustability. This need for adjustability must be recognized during design, and clearances should be established which will reflect realistic tolerances. In any event, the designer should anticipate a planned clearance of at least 40 mm between precast elements and the structure. For steel columns, the designer should consider the clearances around splice plates and projecting bolts.
2. Support should be provided at only one level for vertical loads, and at additional levels for lateral loads and stability.
3. Column covers and mullions which project from the facade will be subjected to loads imposed from shearing winds. Connections acting to provide a moment resisting couple are required to resist these forces.
4. Effects of temperature and humidity change must be considered.
5. The width of horizontal joints between abutting precast covers should be sufficient to permit shortening of the structural frame.
6. The designer must clearly envision the erection process. Column cover connections are, by their location, often difficult to reach and, once made, difficult to adjust.
7. Insulation may be placed on the interior face of the column cover or, alternatively, it may be applied to the structural column directly. A minimum setting clearance of 20 mm is required. Connection details must be chosen to accommodate either choice.
8. Column covers or mullions should be combined with adjacent spandrels to minimize joints.

9. Where uniformity of architectural finish is required on three sides of a column cover, the designer must be aware of the manufacturing limitations of achieving this with certain types of finish, and should be guided by a certified manufacturer regarding the feasibility of his requirements. For example, to ensure uniformity of finish, it may be necessary to stage-cast U-shaped units with quirk mitre joints. Vertical legs over 300 mm high will generally require consideration for separate face down casting.

# 6.5 LOAD BEARING PANELS

## 6.5.1 GENERAL

The design and structural behaviour of exterior precast architectural concrete bearing wall panels is dependent upon the panel shape and configuration, the magnitude of applied gravity and lateral loads, the manner by which loads are transferred between panels, and the capability of individual panel components to resist shear, axial force and flexure.

The design of an architectural precast panel does not differ from the design of a two-dimensional frame, once it is isolated and taken as a free body. Accepted procedures and code requirements apply. Design examples are included in Chapter 3.

**Fig. 6.5.1 Horizontal and vertical rib panels**

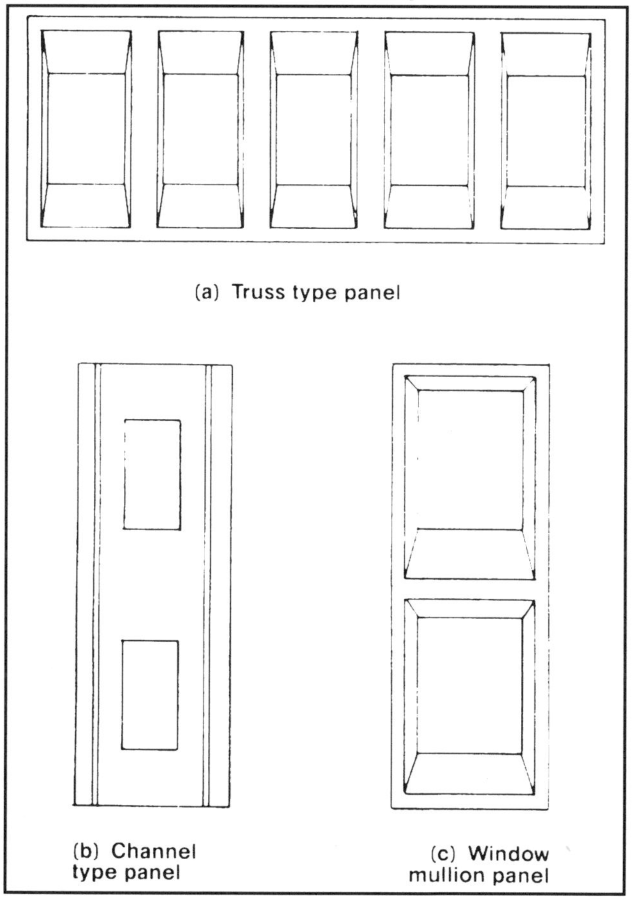


Architectural precast concrete panels used in exterior load bearing walls can be classified as either flat panels or ribbed panels. Both flat and ribbed panels may have window, door, or other openings.

Fig. 6.5.1 illustrates three types of ribbed panels. The panel of Fig. 6.5.1(a) is a horizontal Vierendeel truss window mullion panel, while the other panels are vertical window mullion panels.

Whether or not the architectural panel of the exterior wall is placed horizontally or vertically depends primarily upon handling and erection requirements and the methods or details selected for making connections. A horizontal Vierendeel truss type panel lends itself to simple handling since it is shipped in its erected position, requires vertical load transfer connections at each story level, and allows for minimum erection bracing requirements.

A two-storey vertical panel requires additional erection handling and, because of its projected height during erection, demands more sophisticated erection bracing procedures.

Shipping and erection requirements and connection design are factors which will play a part in determining the panel size and configuration. A panel can only be as large as that which can be shipped and hoisted with the available erection equipment.

## 6.5.2 DESIGN CONSIDERATIONS

Wall panels are subject to loads applied in the vertical and horizontal direction. Loads in the horizontal direction may be both parallel to and normal to the plane of the wall. Vertical loads are parallel to the plane of the wall as shown in Fig. 6.5.2, at an eccentricity influenced by the geometry of the wall, location of load, erection and manufacturing tolerances as shown in Fig. 6.5.3. Reference should be made to CSA-A23.4 for a discussion of tolerances.

Lateral loads applied normal to the wall are the result of wind and seismic activity. These loads cause lateral deflections, which may reduce the vertical load carrying capacity of the wall. Horizontal loads may be resisted by using shear walls and floor diaphragms.

Fig. 6.5.4 illustrates a special framing arrangement which provides uniform distribution of loads to panels on all sides of a

**Fig. 6.5.2 Vertical and horizontal forces on wall panels**

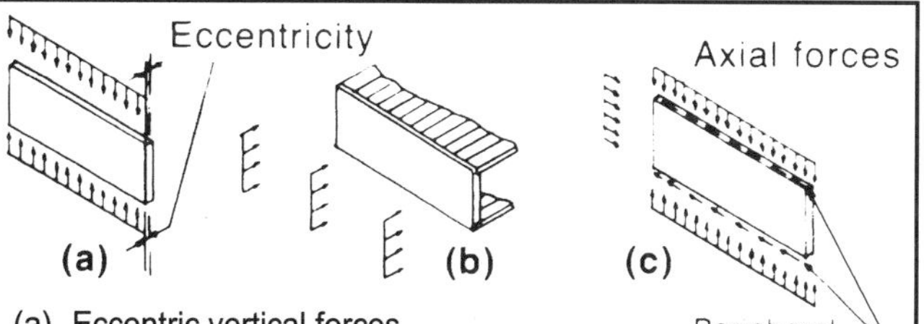


(a) Eccentric vertical forces
(b) Horizontal forces (perpendicular to the plane of the wall)
(c) Horizontal forces (parallel to the plane of the wall) along with the vertical forces

Fig. 6.5.3 Examples of accidental eccentricities

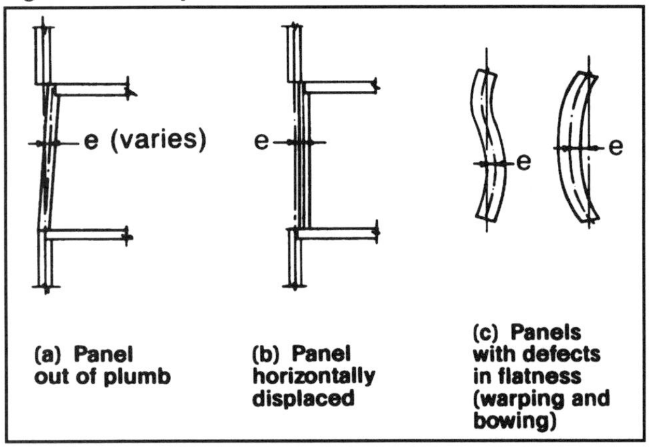


structure. The uniform loading of the perimeter by the floor and roof construction has the advantage of resisting tension forces caused by lateral loadings, and thus minimizes connection and other hold-down requirements. Further, uniform distribution of floor and roof loadings around the building's perimeter also reduces differential creep deformations of panels as discussed in Sec. 2.4.

Fig. 6.5.4 Perimeter floor loading to exterior panel wall

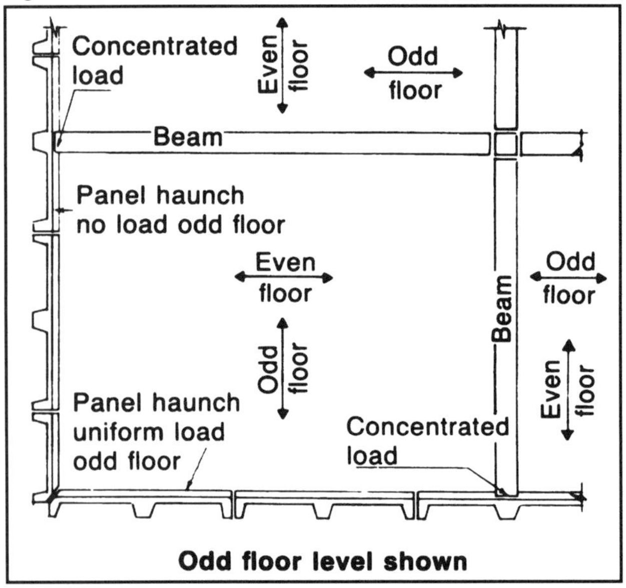


The effect of cumulative loads on connections between panels must be considered, as they become a significant factor in determining minimum panel dimensions.

The transfer of gravity loads between panels or combinations of gravity and axial loads caused by lateral loadings can become the major factor influencing the structural dimensions of a panel and the required connection details.

Often the mullion size of the panel will not be controlled by the minimum concrete structural area. Rather, minimum dimensions for grouting panels at horizontal joints and for placing reinforcement, space for locating handling devices, or space required to accommodate a variety of connection conditions can determine the mullion size.

For most precast exterior bearing wall structures, it will be found that the gravity dead and live load condition will control structural dimensions rather than load combinations which include lateral loads.

It should be noted that unless analysis or experience indicates otherwise, both load bearing and non-load bearing panels should be reinforced with an amount of reinforcing steel, as specified in the appropriate building code, and at least equal to $\rho = 0.001$.

## 6.5.3 ORDER OF SOLUTION

The order of solution can be divided into the following categories:

1. Determine the design gravity and lateral loads applied to the wall panels, and identify the panels receiving the greatest loading.
2. Select the size and shape of the wall panel. This will be affected by the details and location of the vertical and horizontal panel-to-panel connections.

   Development of connection details must consider such questions as:

   a) Do the details lend themselves to standardization?
   b) Can the details be plant-produced within required tolerances assuring proper quality and strength?
   c) Can the details be field-constructed within required tolerances assuring proper quality and strength?
   d) Once details are selected, can they be designed by an accepted method, and will volume change deformation (creep, shrinkage, and temperature) influence the detail or its design?
3. Review the preliminary size and shape of the wall panel considering production requirements for stripping, general handling, storage, shipping and erection, and architectural finish.
4. Review the creep, shrinkage and temperature behaviour of the building as a whole with respect to how the structural behaviour of the panels is influenced, and how non-structural items such as partitions, glass, and sealants are affected.
5. Check connection concepts and types of connections required to resist the various applied loads. In some cases, local practice may suggest one type of connection over another, for example, the use of bolts rather than welds. Make sure that welded connections are accessible.

## 6.5.4 LOAD BEARING SPANDRELS

### General

Load bearing spandrels are panels which support floor or roof loads. Except for the magnitude and location of these additional loads, the design consideration for load bearing spandrels is the same as for non-load bearing spandrels.

**Loads**

Load bearing spandrels support structural loads which are generally applied eccentrically to the shear centre. A typical arrangement of spandrel and supported floor is shown in Fig. 6.5.5.

**Fig. 6.5.5 Load bearing spandrel**

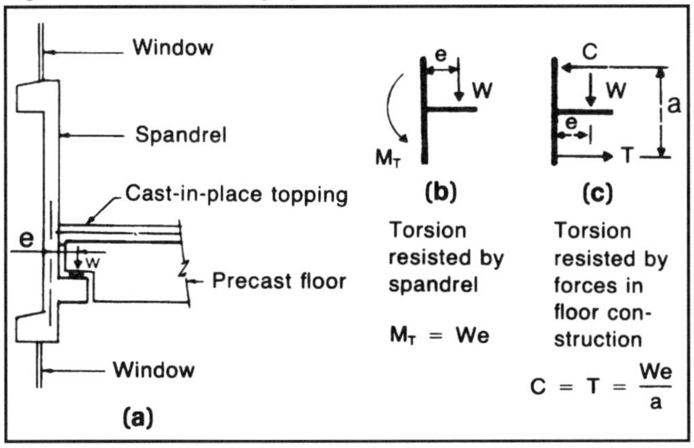


Torsion due to eccentricity must either be resisted by the spandrel, or by a horizontal couple developed in the floor construction. To take care of torsion in the floor construction, the details must provide for a compressive force transfer at the top of the floor, and a tensile force transfer at the bearing of the precast floor element. The load path of these floor forces must be followed through the structure, and considered in the design of other elements in the building. Even when torsion is resisted in this manner in the completed structure, twisting on the spandrel during erection will have to be considered.

If torsion cannot be removed by floor connections, the spandrel should be designed for the induced stresses.

# 6.6 SANDWICH PANELS

## 6.6.1 GENERAL

Sandwich panels are composed of two concrete wythes separated by a layer of insulation and, depending on the design, an air space between the exterior wythe and the insulation. If either of the concrete wythes is replaced by a material other than concrete, it may still be classified as a sandwich panel, provided that the remaining concrete layer is the major (structural) wythe and the product lends itself to production in a precast concrete plant.

In place, sandwich panels provide a means of transferring load, and insulating the structure. They may be cladding panels, in which case they support only their own weight and transfer lateral loads to the supporting structure, or they may act as beams, as bearing panels or shear walls, transferring loads imposed from remote parts of the structure.

In the past, both composite and non-composite panels have been used. Composite panels are those in which inner and outer wythes are interconnected through the insulation by means of rigid ties, or regions of solid concrete, that restrict relative movement between the wythes. The two wythes act together to resist externally imposed loads. Depending on the rigidity of the connector system (ties or ribs) wythe interaction may be total or partial. Non-composite panels are those in which one wythe is supported from the other by relatively flexible ties and/or hangers, allowing differential movement of the wythes with changing temperatures and humidity conditions. Non-composite panels with an air space allow for ventilation of the outer wythe and pressure equalization.

In general, there are three basic types of non-composite panels:

1. Panels in which both wythes participate in transferring lateral and vertical loads to the structural framework.
2. Panels in which only one wythe transfers vertical and lateral loads to the structural framework and supports the weight of the other wythe. The structural wythe is normally thicker and stiffer than the non-structural wythe, and is usually located on the interior side of the panel. Occasionally it may be the exterior wythe, particularly in the case of sculptured panels such as ribbed panels. For such cases, special attention must be paid to the details of lateral tie-back connections.
3. Air gap sandwich panels. These panels are usually a variation of those described in paragraph 2 above. The air gap removes moisture from the interior of the panel and helps to control the flow of water from the exterior to the interior of the assembly.[R6.6.1] Typical examples are panels for which the exterior wythe consists of natural stone.

Non-composite panels, and composite panels with an area less than 4 $m^2$ will generally exhibit acceptable behaviour independent of the type of reinforcement used. Composite panels larger than 4 $m^2$ should be prestressed, unless the producer is able to demonstrate satisfactory behaviour over a period of years.

In a non-composite panel (Fig.6.6.1(a)), one wythe provides the total structural function of the panel, and is analyzed assuming no interaction with the other parts of the panel, although some limited amount of shear transfer may in fact occur.

For equal overall thickness of panel, a composite element (Fig. 6.6.1(b)) will demonstrate greater lateral stiffness. However, because the deformation of the outer wythe will affect the inner wythe, experience indicates that the lateral bowing of composite panels with mild steel reinforcement is less predictable than that of non-composite panels. While the introduction of prestress in both wythes of a composite panel has no effect on thermal bowing, it can be used to induce a negative bow to counteract the tendency of the panel to bow outwards, thus improving the behaviour of the element. While this is difficult to calculate, it is a workable solution used successfully by experienced manufacturers.

Panels with continuous perimeter or interior ribs of concrete, or openings with surrounding ribs, are not recommended, because:

1. The ribs act as restraints between the two concrete layers, each of which is subjected to significantly differing deformations, thus developing forces which may lead to cracking. This is true of any composite panel. It should be noted

**Fig. 6.6.1 Non-composite and composite panels**

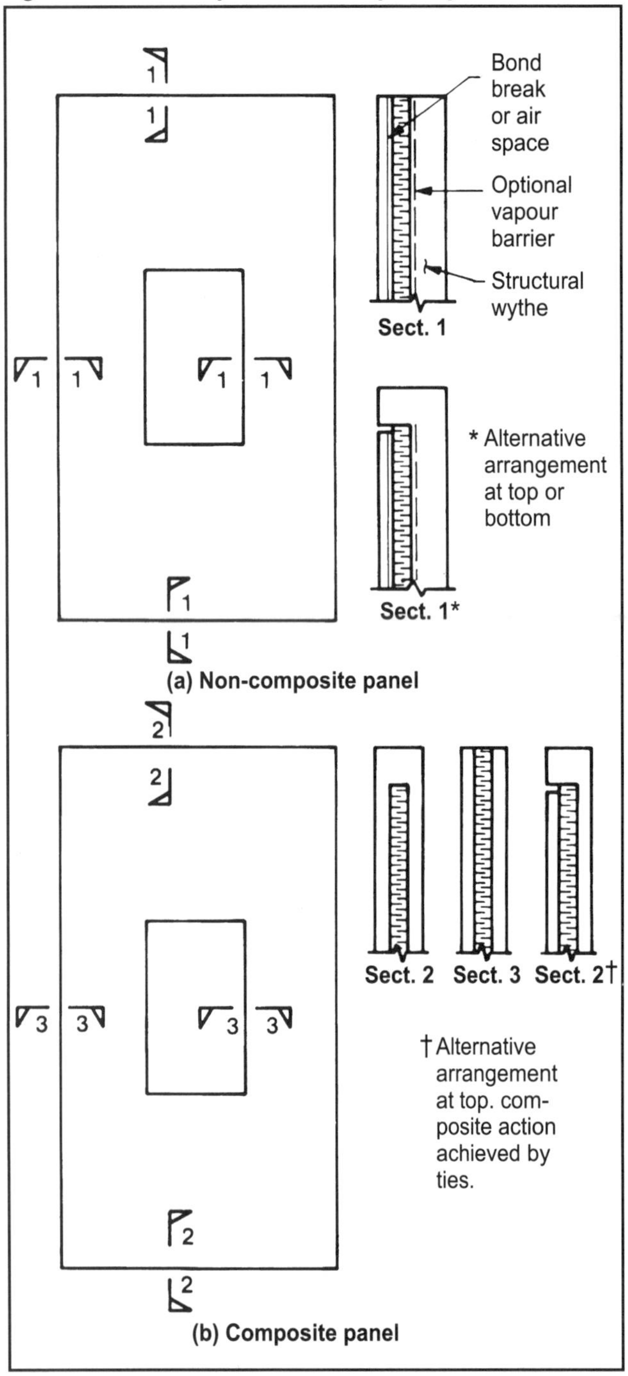


that the degree of composite action cannot be established accurately by analysis.

2. The ribs act as significant heat sinks, and will reduce the insulating effectiveness of the panel, as well as possibly causing local condensation and discolouration.

Some manufacturers have reported successful use of panels with ribs at the top and bottom only, and with no ribs on the sides. Such an arrangement provides less restraint than surrounding ribs.

A more satisfactory arrangement, where a solid concrete rib is required for bearing, is to provide freedom of movement at the other three sides.

Many manufacturers produce successful composite insulated panels. Their methods of obtaining composite action between the wythes vary widely from solid ribs to steel trusses to expanded metal, etc. The most successful composite panels are prestressed.

Manufacturers who advocate the use of non-composite sandwich panels emphasize the advantage of the structural wythe being protected by the insulation from extremes of temperature, thus minimizing the bowing of the structural wythe and eliminating thermal stresses in the structure. The exterior wythe is free to expand and contract with variations in temperature and thus will remain crack-free. For outer wythes greater than 12 $m^2$, crack control joints are recommended.

Manufacturers who advocate the use of composite panels point out that the structural wythe of a non-composite panel must carry all of the applied loads in addition to the weight of the other wythe thus increasing overall dimensions, weight and expense; whereas a composite panel can be designed to share the loads between the wythes thus reducing dimensions, weight and expense. Prestressing of composite sandwich panels can be used to lessen the effect of thermal bowing.

## 6.6.2 CURING

No special precautions are required for curing the lower cast wythe of a sandwich panel. It is protected by the insulation which retains the heat of hydration and increases the early strength of the concrete. It follows, therefore, that to ensure uniformity of the curing process, the upper wythe should be covered to minimize water loss, and insulated to produce the same curing effect for strength gain as in the lower wythe.

## 6.6.3 FLEXURAL ANALYSIS

The strength of concrete should be determined by structural and production requirements, and the need to provide a mix of high density and low shrinkage. A discussion of the principles of sandwich panel design can be found in Sec. 3.11.

## 6.6.4 INSULATION

Thermal properties of insulation and concrete are discussed in Sec. 5.1. The insulation should have low absorption or a water-repellent coating. The insulating material used may also influence the fire-rating of the wall assembly.

The thickness of the insulation will be determined by the thermal characteristics of the material and the design temperatures of the structure. A minimum thickness of 25 mm is recommended, with no limitation on maximum thickness. The deflection characteristics of the inter-wythe connectors should be considered in relation to the insulation thickness. Although one does not necessarily limit the other, the two must be designed to be compatible.

Openings in the insulation around connectors should be packed with insulation to avoid forming concrete bridges between wythes.

The maximum sizes of insulation commercially available, consistent with the shape of the panel, are recommended. This will minimize joints and the resulting cold sinks.

## 6.6.5 BONDBREAKER

The insulation itself is capable of transferring a certain amount of shear between the wythes, the value being dependent upon the thickness and properties of the insulation. It may be necessary to break the bond between the insulation and the concrete wythes of non-composite panels by physical or chemical methods to eliminate unintended restraint. This will allow relatively free movement between the wythes for the dissipation of temperature and other volume change stresses. While such bond may be destroyed in time, it is strongest at the initial stages of casting, when the concrete has its least tensile strength.

Panels may be manufactured by incorporating bond-breakers of polyethylene or reinforced paper sheets, or by applying form release oils to the insulation. If the bond-breaker has vapour retarding properties, such as polyethylene, it must be placed on the warm side of the insulation. Air gaps between the insulation and the outer wythe virtually eliminate bond between the wythes.

The use of sheeting with a single layer of insulation minimizes the danger of concrete bridges between the wythes. Holes in polyethylene sheeting at mechanical ties must be sealed to provide continuity of the vapour barrier. It should be noted that a 100 mm minimum thickness of the inner structural concrete wythes is normally regarded as a satisfactory vapour barrier, provided that it remains crack-free.

## 6.6.6 WYTHE DESIGN

The minimum thickness of the wythes is dependent upon structural requirements, finish, reinforcement protection, handling considerations and past experience.

### Non-structural Wythe (Non-composite Panels)

In order to minimize differential temperature across the thickness, the wythe should be as thin as architectural details will permit. The following limitations are applicable:

1. At the thinnest point, thickness should not be less than 50 mm, but preferably a minimum of 65 mm.
2. Thickness should be sufficient to provide proper reinforcement cover, in accordance with Code requirements.
3. Thickness should be sufficient to provide required anchorage of the wythe connector devices.
4. At no point should the thickness be less than three times the size of the maximum aggregate.

### Structural Wythe (Non-composite and Composite Panels)

The thickness of the structural wythe should be determined by structural analysis, and by the need to accommodate architectural details. The wythe should not be less than 75 mm thick. A thinner wythe may be used successfully but the amount of reinforcement and the risk of cracking and bowing must be assessed. If the wythes are prestressed, the wythe should not be less than 65 mm thick.

The other limitations listed above for the non-structural wythe also should be considered. Load bearing structural wythes are, in most cases, supported at the bottom edge. They may have a lateral tie near the top and a mid-height connection to the adjacent panels to prevent differential bowing. Non-load bearing composite or non-composite panels can be supported by hanging from suitably designed connections. It is worth noting that top hung panels eccentrically supported will bow outwards less than bottom supported units.

## 6.6.7 PANEL SIZE

The size of panel will be primarily determined by architectural considerations. Additionally, the following restrictions are applicable:

1. The maximum dimension and weight should be determined based on transportation and handling requirements.
2. a) The maximum dimension for non-composite or prestressed composite panels should be in the order of:

   $L = 48\,t$

   where

   L = maximum dimension

   t = overall thickness of panel excluding any ribs

   b) The maximum dimension for conventionally reinforced composite panels should be in the order of:

   $L = 48\,c$

   where

   L = maximum dimension

   c = overall thickness of panel minus thickness of insulation

   In addition, the area of reinforced composite panels should not exceed about 4 m$^2$. Prestressed composite panels may be much larger. The above suggested restrictions on length and area are empirical, and are based on the experience that composite reinforced panels tend to bow to a greater degree than non-composite panels. Where bowing is not a serious consideration, these figures could be exceeded by a certified manufacturer.
3. The maximum dimension from that point on a composite panel where the two wythes are rigidly connected to the furthest free edge, should not exceed approximately 4500 mm unless the panel is prestressed, or experience indicates otherwise.

Special procedures which reduce the differential shrinkage rate, or differential temperature rate will, in turn, permit larger panels. Such procedures include:

1. Use of low-shrinkage concrete.
2. Jointing of the non-structural wythe. Any joints should preferably be complete all the way to the insulation and should be provided at the corners of large openings in the panels.

### 6.6.8 WYTHE CONNECTORS

#### General Considerations

When one wythe is non-structural, its weight must be transferred to the structural wythe. This is generally accomplished by using shear connectors and tension/compression ties passing through the insulation at regularly spaced intervals. An alternative to the use of shear connectors is a solid rib at the bottom of the panel to serve as a support and to still allow the panel to move without restraint that may cause cracking. The solid rib is located at the foundation to minimize the effect on the insulation value of the panel.

#### Shear Connectors

Inclined legs of galvanized mild steel or, preferably stainless steel bars approximately 6 mm in diameter bent into configurations shown in Fig. 6.6.2 (a), (b), (c) and (f) carry the vertical shear load. These bars are usually placed along the horizontal axis of the panel. Sometimes, placement along the vertical axis is added because of panel rotation, torsional forces or special shear requirements.

Sleeve anchors, generally up to 500 mm in diameter, may be used to connect the two concrete wythes (Fig. 6.6.2 (d)). Since movements due to temperature differentials radiate away from the sleeve anchor, no restraint to the movement of the suspended wythe exists, permitting free floating of the non-load bearing wythe. The metal sleeve, in conjunction with anchorage bars, carries all applied shear loads from any direction.

**Fig. 6.6.2 Typical shear connectors**

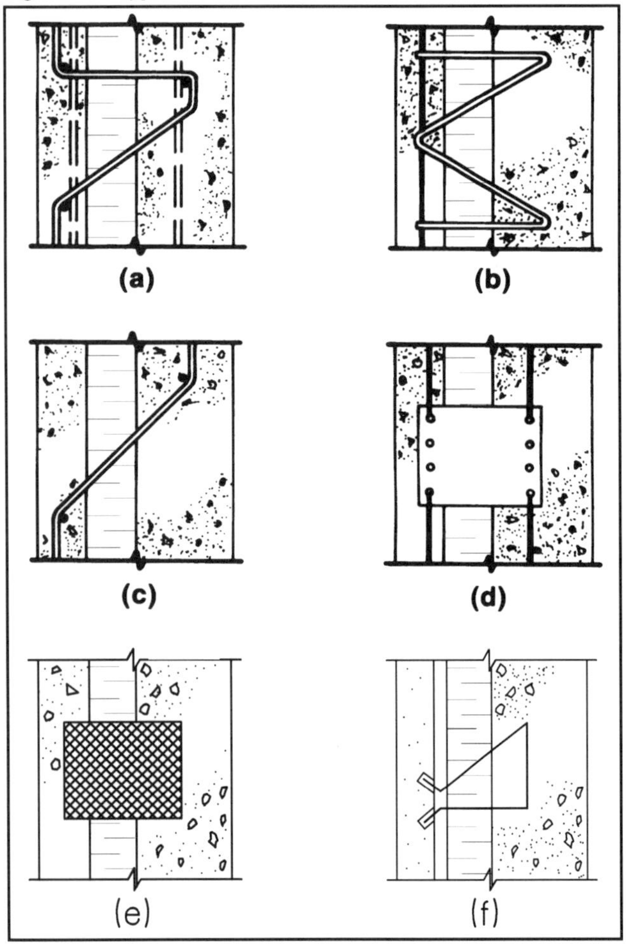


Expanded metal or a steel truss can be installed or tied to the reinforcement to connect the concrete wythes (Fig. 6.6.2 (e)). Cut sections (approximately 200 mm long) or continuous sections may be used to provide a one-directional shear capacity. Simple tie rods are also used as tension/compression ties to resist wind loads and individual layer separation.

For ribbed insulated panels, it is best to position the shear connector in the rib area of the panel (Fig. 6.6.3). By so doing, it will be easier to position and assure proper embedment depths for an effective connection between the two concrete wythes.

**Fig. 6.6.3 Shear connector for ribbed panel**

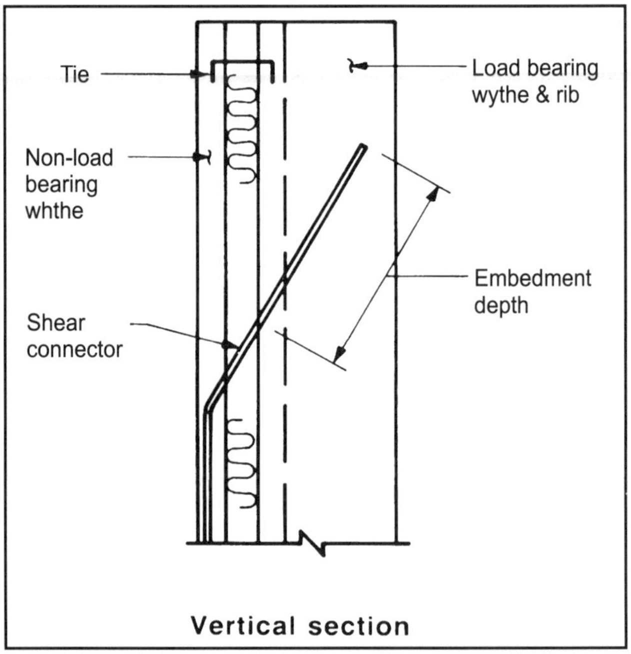


Due to expansion and contraction, it is preferable to have only one anchoring centre. In a panel with two ribs, a shear connector can be positioned in either one of the ribs and, in the other rib, a flat anchor can be positioned vertically. It should have the same vertical shear capacity as the shear connector, and should be located on the same horizontal axis. The flat anchor has little or no horizontal shear capacity and therefore restraint of the exterior concrete face is minimized (Fig. 6.6.4). In a multi-ribbed panel, the shear connector should be positioned in the rib closest to the centre, and flat anchors used in the other ribs.

#### Tension/Compression Ties

Examples of ties are shown in Fig. 6.6.5.

Wire tie connectors are generally 1.5 to 2.5 mm in diameter and are often made of stainless steel, Type 304 or 316.

**Fig. 6.6.4 Anchorage for ribbed panels**

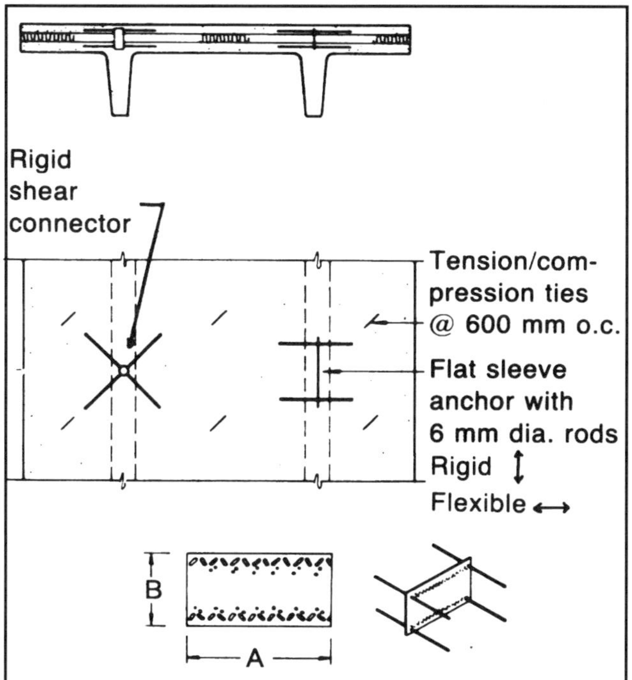


| Safe load (kN) | A (mm) | B (mm) | Note |
|---|---|---|---|
| 9 | 160 | 155 | Use a minimum of 2, |
| 18 | 320 | 155 | 6 mm diameter rods top |
| 27 | 475 | 155 | and bottom of flat sleeve |
| 36 | 640 | 155 | anchor to maintain stability. |

**Fig. 6.6.5 Tension/compression ties**

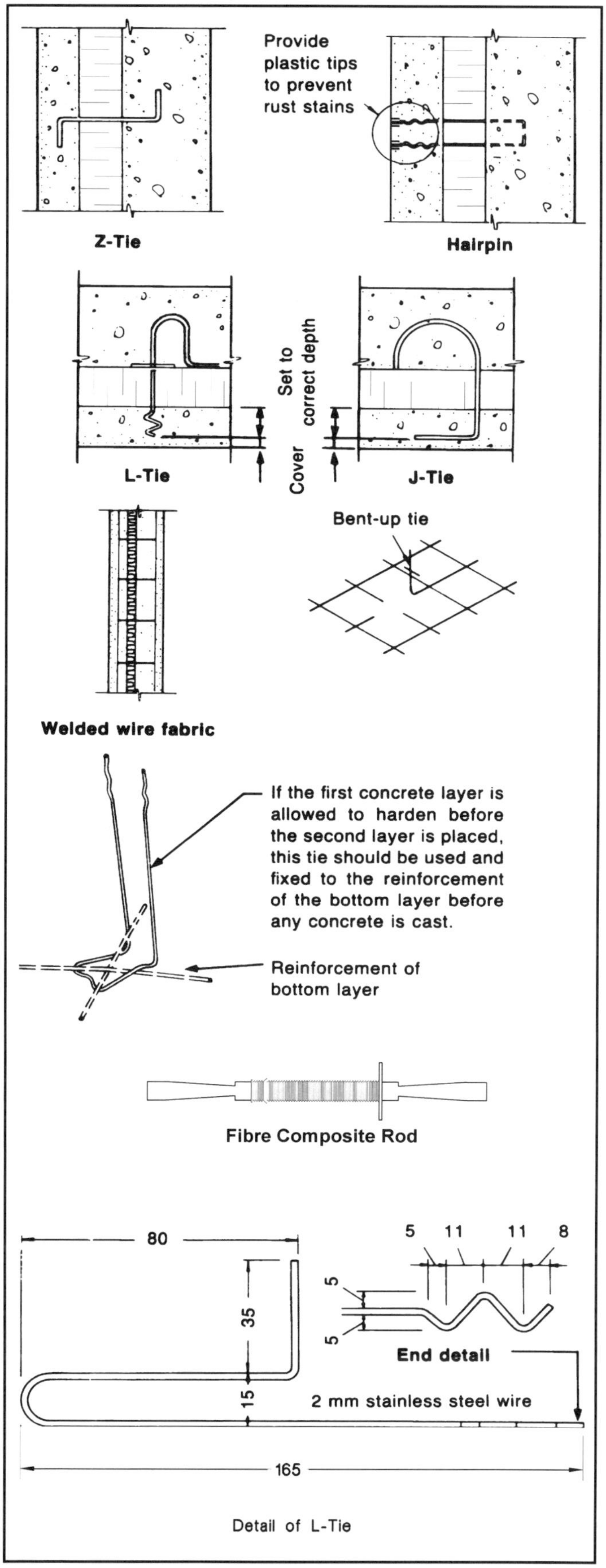


Stainless steel connectors are particularly preferred in a rainscreen panel. Galvanized metal ties are conventionally acceptable, provided that subsequent deformation does not destroy the protective coating. Reference is made to CSA A370-94 - "Connectors for Masonry" for further information. Plastic connectors are also acceptable. They minimize condensation at tie locations. Consideration may have to be given to the effect of plastic ties on the fire resistance of the wall.

Ties of welded wire fabric, conforming to CSA-G30.5, or reinforcing bars, conforming to CSA-G30.8 are acceptable. Shaped, crimped, or bent ties should be cold bent.

Connectors which are intended to accommodate differential movement between wythes should be sufficiently ductile to withstand 5000 cycles of reversal at an amplitude of 3 mm.

Tension/compression ties should be flexible enough so as not to provide significant resistance to temperature and shrinkage stresses in the direction parallel to the panel surface. They should have sufficient anchorage in each wythe to safely transfer applied loads. This can be accomplished through hooking around or tying to the reinforcement, or by bending or deforming the ends of the ties.

At time of stripping, and with the non-structural wythe cast face down, ties acting in tension will transfer the weight of the exterior wythe, the weight of the insulation, and any form suction, to the structural wythe.

During handling, erection, and in the final position, ties acting in tension and compression or in shear plus flexure, will transfer these same loads to the structural wythe. Insulation can participate in shear transfer to achieve semi-composite action.

Under wind pressure, or suction, ties will act in tension or compression, depending on the direction of force. While it is probable that the insulation will act to transfer wind compression, the possibility exists that the non-structural wythe may in time separate from the insulation. Thus it is recommended that only the ties, acting in compression, be assumed effective.

Horizontal hangers should be provided for handling and racking loads, such as inertia forces developed during seismic action. The functional behaviour of connectors is illustrated in Fig. 6.6.6.

**Fig. 6.6.6 Functional behaviour of connectors**

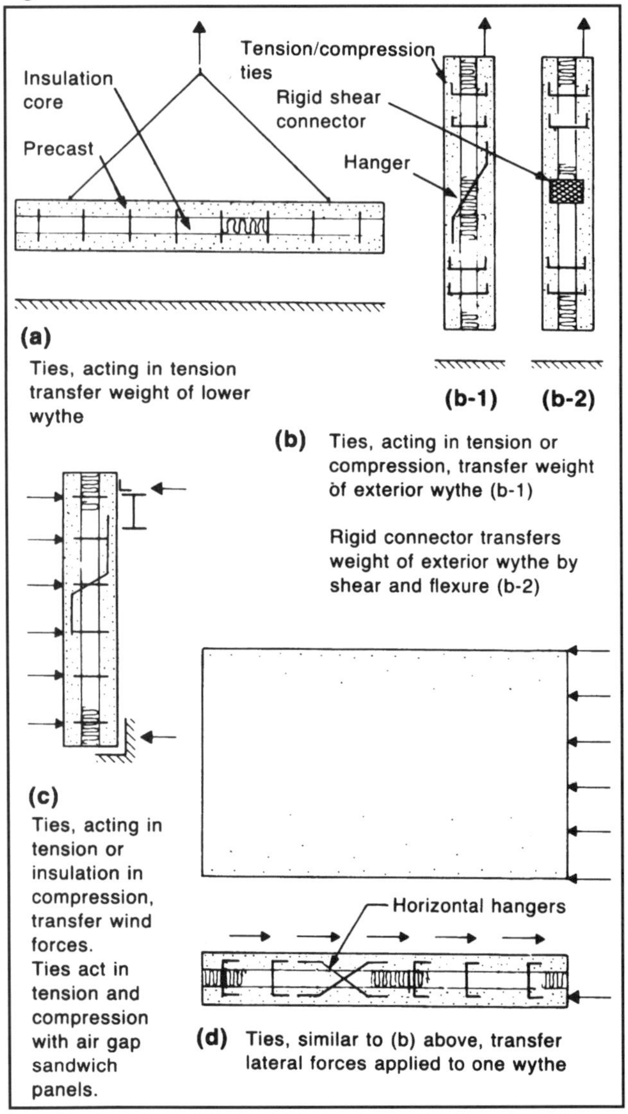


### Arrangement

Satisfactory connector design requires an assessment of panel geometry and weight, anticipated movement of the exterior wythe, and a knowledge of tie material strength and flexibility characteristics which have been established by test.

The exterior wythe will move in response to environmental changes and concrete shrinkage. To accommodate this movement without damage, the connectors should be sufficiently flexible (in the direction of movement) so that undue restraint will not develop. In the erected position, the device which transfers the weight of the non-structural wythe to the structural wythe should ideally be a single element or a closely spaced pair of elements located at the centre of the panel. This arrangement will permit the non-structural wythe to contract and expand with least total restraint (Fig. 6.6.7 (a)). The single point connector need not necessarily be at the centre of mass; however, a limitation of 3 mm maximum differential movement between wythes is suggested. The single point connector can be moved up or down the vertical centroidal axis if the panel will not be rotated during erection. As one alternative, the ties can be located along two lines at right angles, the intersection of which is coincident with the centre of mass of the panel (Fig. 6.6.7 (b)). In any event, those ties which are used solely to transfer weight during stripping (tension ties) or direct wind forces (tension or compression ties) should not unduly restrain lateral movement.

The tension and compression ties within the field of the panel should preferably be approximately 600 mm on centres, but in any event not greater than 1200 mm on centres or at least two ties per 1 $m^2$ of panel area. Around the perimeter of the panel, and the perimeter of openings larger than 600 mm, tie spacing should not exceed 600 mm (Fig. 6.6.7).

When shear connectors cannot be located so that their centre of resistance coincides with the centre of mass of the supported wythe, added forces due to torsion should be considered. Similar consideration is required when the lateral (racking) force is not coincident with the centre of resistance in that direction.

## 6.6.9 GENERAL ARCHITECTURAL DESIGN CONSIDERATIONS

The architectural design of precast concrete sandwich panels is similar to the design of single wythe precast architectural panels. However, there are some special considerations for precast concrete sandwich wall panels.

### Bowing

A recognized characteristic of composite precast concrete sandwich panels is the tendency of longer panels to bow outwards under prolonged exposure to the hot sun. In non-composite sandwich panels, there is less tendency to develop thermal bowing. The inner wythe of a temperature-controlled building is kept at a constant temperature with virtually no thermal gradient. Although the exterior wythe experiences extremes of temperature, the temperature gradient through the wythe is minimal leaving it unaffected by a steep thermal gradient through the wall system. This practically eliminates or minimizes thermal bowing, and is the main advantage of non-composite sandwich panels. Even where the bowing is recognized and considered in design, the possible failure of sealant between adjacent panels at corners or where abrupt changes in building shape occur, remains a major design and construction consideration for successful performance. The tendency of panels to bow is influenced chiefly by panel size, degree of composite action, the rigidity of connection between the wythes, and daily temperature variations on the exterior face. Differential movement between adjacent panels is not normally a problem and through good design and detailing practice, the effect of thermal bowing can be satisfactorily accommodated.

**Fig. 6.6.7 Arrangement of connectors between whythes**

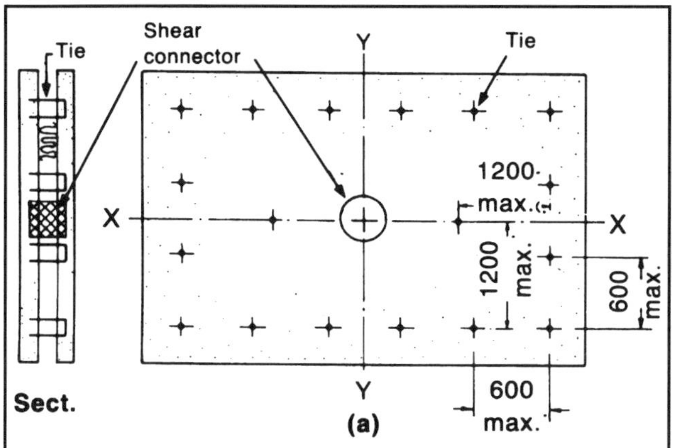


One device transfers weight of non-structural wythe to structural wythe, as well as any racking shear.

Ties spaced in the field of the panel transfer direct wind forces and stripping loads from the non-structural wythe to the structural wythe.

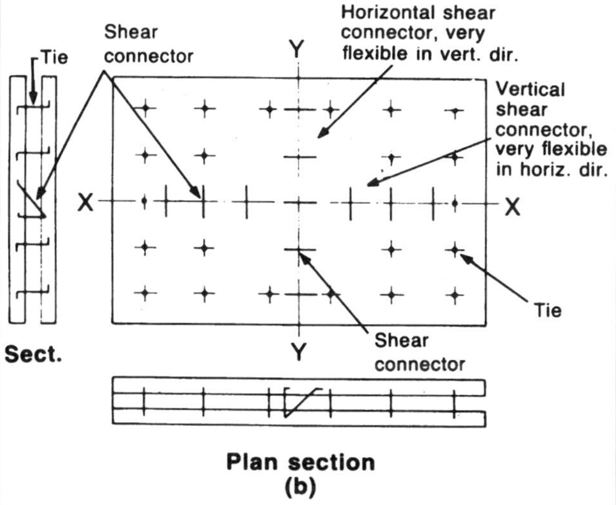


Devices spaced along the X-X axis transfer weight of non-structural wythe to the structural wythe. Similar devices along the Y-Y axis transfer racking shear from non-structural wythe to structural wythe.

Ties spaced in the field of the panel transfer direct wind loads and stripping forces from the non-structural wythe to the structural wythe.

For panels with large openings, joints in the outer wythes at the corners of such openings are desirable. These joints should preferably be completely through to the insulation layer and may subsequently be caulked or treated architecturally, in the same manner as the joints between panels.

Control joints should be provided in large panels to break the outer wythe into units which will not craze or crack due to extreme temperature changes, or shrinkage and creep of the concrete. The pattern for such control joints becomes an important architectural feature and aligning such joints with adjacent panels may present problems. These can be minimized by having the real panel joint expressed as a recess, but this may not be possible if the outer wythe is already of minimum thickness. Alternatively, the pattern may be varied and only maintained in alternate panels, so that a small misalignment will not be noticeable. The problem of crazing or cracking and the need for control joints in the outer wythe can be reduced by prestressing the panels.

### Corners

Good corner details are essential. This is where most care must be exercised with sandwich wall panels. Mitred corners, unless restrained, may not be satisfactory. Restrained corners may also cause problems unless the panels are adequately prestressed or reinforced to resist the restraint forces. Even panels with returns are not easy to weather-seal as the bowing will be in different planes. In addition, the panel with even a small return will be stiffer than its neighbour, and both joints on either side of a corner may suffer. A special corner unit, which is not necessarily flush with the adjacent panels, can be effectively used to camouflage bowing in the two different planes (Fig. 6.6.8). It should be noted that some manufacturers have successfully developed mitre corner details to suit their own product.

### Other Wall Materials

If other materials such as curtain walls are combined with precast concrete sandwich wall panels, no attempt should be made to make this material flush with the concrete surface, as it is unlikely that this material will act and bow exactly like the concrete panels. If it is essential that they are in the same theoretical plane, it is suggested that they be framed around with material which is not flush with the walls. Similar suggestions apply to corner columns. Door or window frames should be attached to the inside wythe since the movement is less than that of exterior wythe (Fig. 6.6.8).

# 6.7 VENEER FACED PANELS

## 6.7.1 GENERAL

Finishes with materials other than normal concrete aggregates, such as natural cut stone, brick, structural tile, and architectural terra cotta (ceramic veneer), provide a great variety of textures for precast units.

**Fig. 6.6.8 Architectural and structural details**

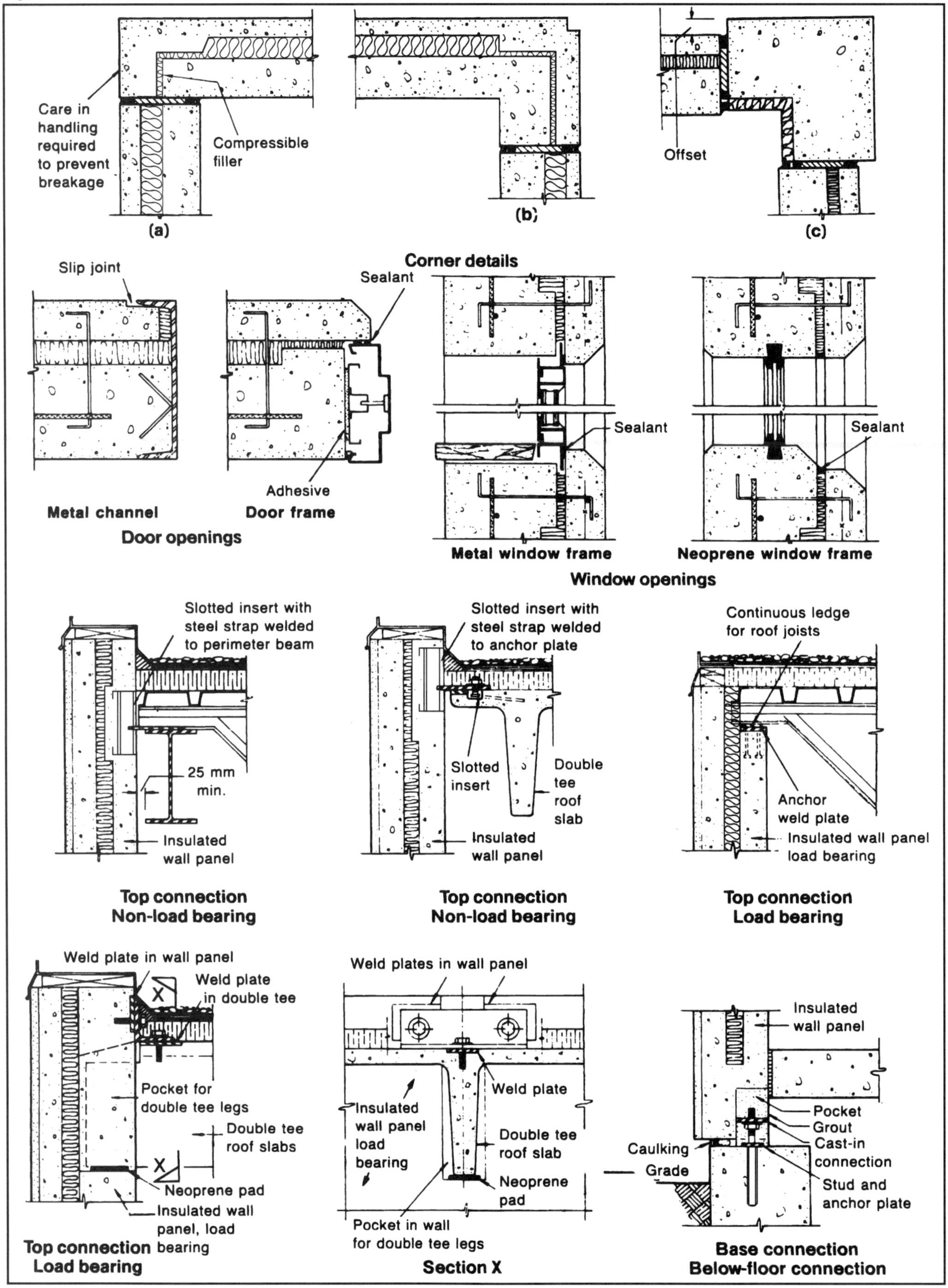

Some of the advantages found in the usage of veneer faced panels are:

1. Veneer stock can be used in thinner sections than the stock required for standard set construction.
2. Multiplane units, such as column covers, spandrels with integral soffit and sill sections, deep reveal window frames and parapet sections, which are impractical to construct conventionally, are more easily assembled with veneer units on precast elements (Fig. 6.7.1).
3. The erection of the precast units is faster and more economical than handset construction because the larger panels incorporate a number of veneer pieces and may include insulation, interior finishes and windows.

The fabrication, handling and erection of stone-faced panels follows closely that of traditional precast panels. It should be noted, however, that special handling and shipping procedures are required for units such as those in Fig. 6.7.1(a) and (b).

Interaction between the veneer and its supporting back-up present other special problems associated with their differing physical properties and structural functions. In this context, tensile, compressive and shear strength, modulus of elasticity, coefficient of thermal expansion, and volumetric change with time, are relevant. Other factors influencing interaction are concrete shrinkage, temporary and permanent applied loads, and thermal cycling. The behaviour of the assembly is also dependent on the configuration of the panel, i.e. relative thicknesses of the basic components, and the location of the insulation in the wall construction.

**Fig. 6.7.1 Applications of veneer faced precast**

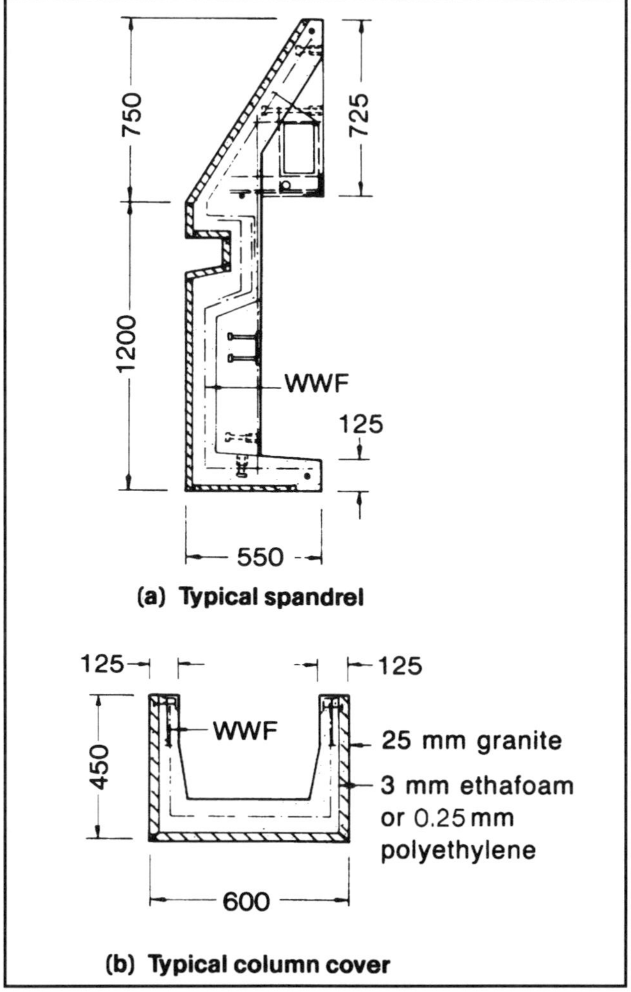


Generally, in panels of this type, the facing may be either bonded to or isolated from the concrete back-up. The use of bonding is not recommended, since it promotes excessive bending, and may lead to cracking of either wythe, and staining of the veneer. However, it may be desirable to bond heavily veined marbles and certain types of travertine and granite. Under these circumstances, stone sizes should be restricted, and mechanical anchors should be provided.

Much of the bowing arises from the drying shrinkage of the concrete. It is therefore highly desirable to minimize the water-cement ratio of the mix. The use of water-reducing agents or superplasticisers, and the sealing of the screeded surface of the concrete should be considered. Where long units are involved, the use of a cambered (concave) form might be advantageous, to compensate for the anticipated convex bowing.

There are two types of panels in which the veneer is at least partially isolated from the concrete. The first makes use of a bond breaker in the form of a polyethylene sheet, or open or closed cell pad. The second involves an essentially continuous air gap inside the veneer, and either continuous or discontinuous closed cell insulation between the veneer and the back-up. Although the bond breaker tends to render the veneer non-composite, it should be noted that the anchor arrangement should not be redundant (i.e. differential curvatures between veneer and concrete can take place). In the ideal case in which insulation is continuous behind the air gap, there is virtually no interaction. Note that there is essentially no horizontal restraint in the plane of the panel. Further, since the air gap is open to the outside atmosphere, the veneer is subjected to a significantly reduced wind load, which arises from gust effects only.

As with concrete sandwich panels, the designer should consider the use of an "air gap" between the veneer and the concrete. Connection of natural stone to the concrete should be made with mechanical anchors which can accommodate some relative movement. This is, of course, a necessity if bond-breakers (or air gap) are used. One possible exception is the limestone industry which sometimes uses rigid rather than flexible connectors.

The strength of the facing veneer material must be determined along with the anchor system, to assure adequate strength to resist flexural stresses during handling, transportation and erection.

Panel design must include the "in-service" conditions that panels will encounter in the final location on the structure. Generally, the interior surface of panels will experience a very small temperature range, while the exterior surface may be exposed to a range as great as -40°C to 70°C. Major bowing of panels can occur (magnitude of bowing could be ($\ell$ / 240) and the amount will depend on:

**Fig. 6.7.2 Typical anchor details for stone veneers for bonded and unbonded panels**

## (a) Marble

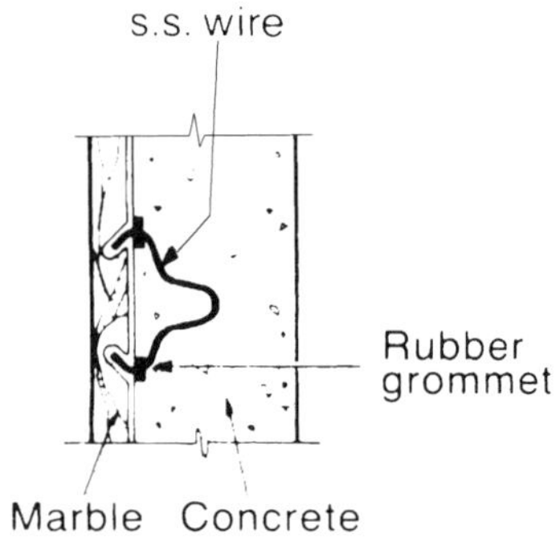


**Anchor schedule**

| Height of marble facing unit | Width of marble facing units (mm) | | | |
|---|---|---|---|---|
| | 150 to 300 | 300 to 1050 | 1050 to 1220 | over 1220 |
| Up to 1400 mm | One anchor top, one anchor bottom | Two anchors top, two anchors bottom | Two anchors top, two anchors bottom | Three anchors top, three anchors bottom |
| 1400 mm and over | Anchor details vary. Anchor placement and panel assembly should allow for differential movement between veneer and back-up concrete in plane of and perpendicular to plane of panel. | | | |

Note: *Top and bottom anchor* holes of each marble facing unit should be drilled near the top and bottom edge and should be 20 mm deep at 35 deg. angles in the back of the marble facing units and opposing each other in the vertical position to receive the performed anchors. *Anchor* holes centred in the backs of the marble facing units when necessary should be drilled in the same manner as the top and bottom anchors. All anchor holes should be fitted with epoxy or silicone to eliminate water and formation of ice. Anchors should be filled with rubber grommets at face of the concrete backup.

**Anchor test data**

| | Shear (kN)(2) | Tension (kN)(2) |
|---|---|---|
| Avg. Failure(1) | 11 | 1.9 |
| Safety Factor | 30 to 1(3) | 5 to 1(3) |

(1) Tests performed with 100% bond failure built into panel.
(2) Per anchor
(3) Based on anchor schedule.

## (b) Granite

### Anchor A

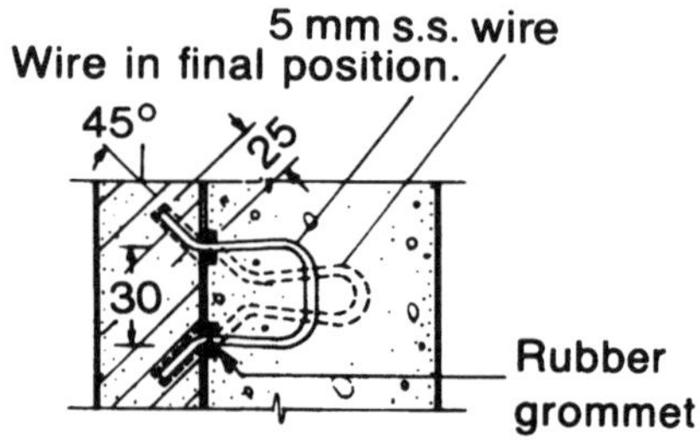


Tensile strength of one rod: 10.4 kN
Shear strength of one rod: 7.8 kN

**Test data for Anchor A**

| Type of hole | Shear parallel to anchor (kN) | Shear perpendicular to anchor (kN) | Tension (kN) |
|---|---|---|---|
| Diamond core drilled | 10.7 - 11.8 | 14.0 - 16.0 | 3.5 - 5.0 |
| Tungsten carbide drilled | 9.8 (approx.) | 13.0 (approx.) | 1.9 |

(1) Factor of safety of anchor shear and tensile capability should be a minimum of 5.
(2) Toed-out, the anchor has approximately 50% more tensile capability than a toed-in anchor.
(3) As a general rule, anchors are spaced one every 0.25 to 0.5 m$^2$.

### Anchor B

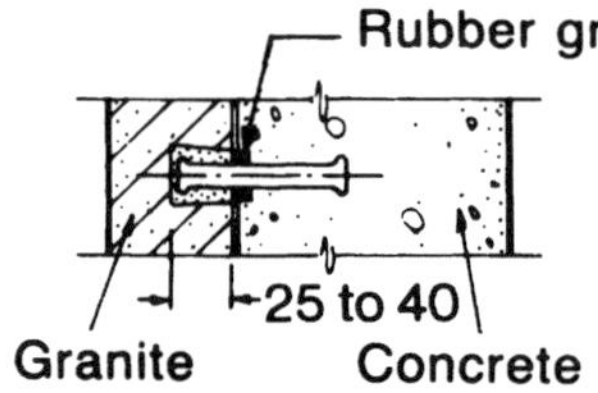


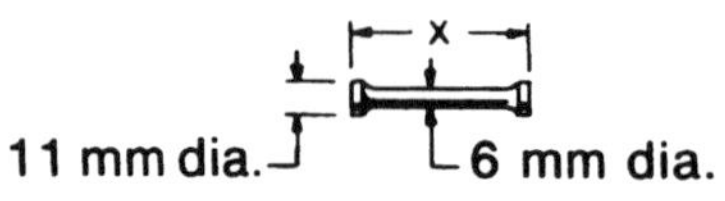


Tensile strength of anchor: 18.5 kN
Shear strength of anchor: 13.8 kN
Capacity depends on depth of holes and grout material.

**Fig. 6.7.2 (Continued)**

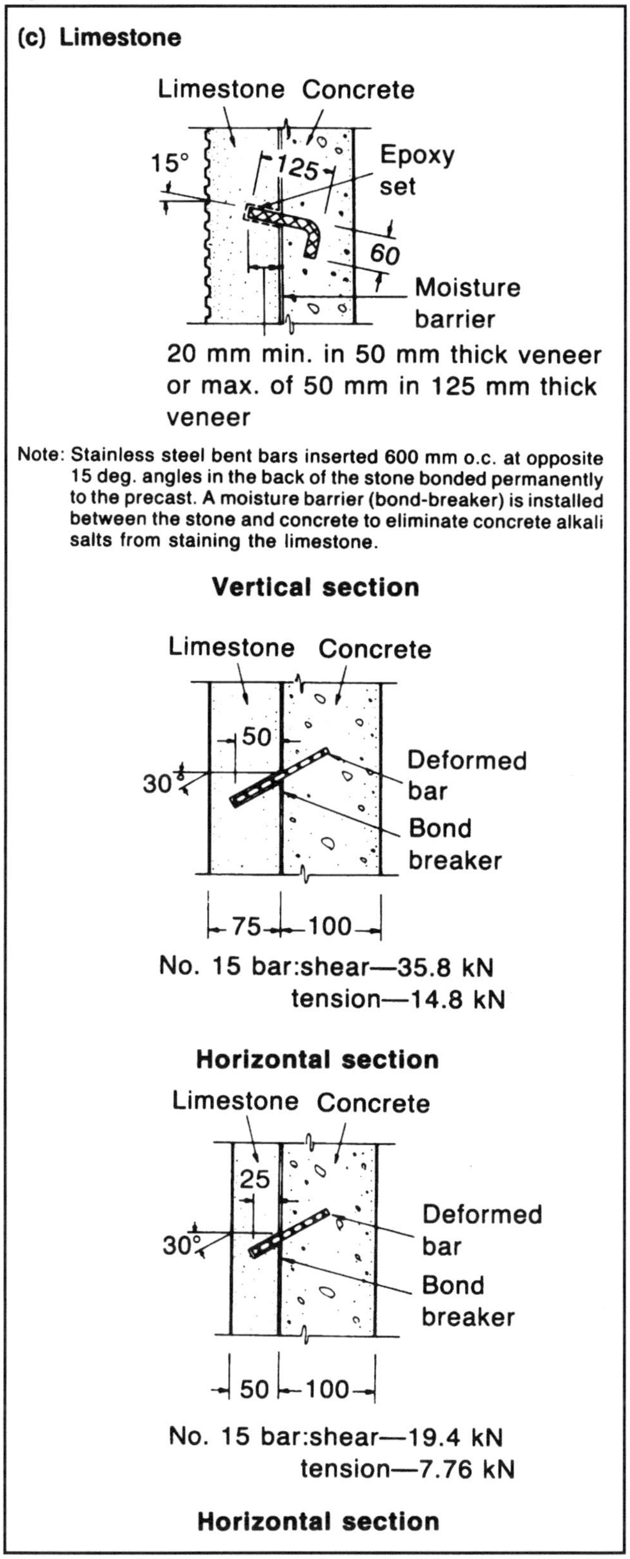


*Note: Recommended safety factor for galvanized strength deformed bar is 8.*

1. Shrinkage of the concrete.
2. The temperature differential (exterior to interior).
3. Coefficients of expansion of the materials.
4. Ratio of cross-sectional areas of the materials and their moduli of elasticity.
5. The use of prestressing or truss type reinforcement in the concrete element, to increase rigidity.
6. Type and location of connection to the structure.
7. Rigidity of connection between veneer and concrete back-up.

For cut stone finishes, emphasis on samples and mock-up units is particularly important, and supplemental tests on the behaviour of the unit for anticipated temperature changes may be required. To the degree that it is possible to test or judge from earlier performances, durability of the combined unit should be investigated.

There is now a good background of experience in the production and erection of cut stone veneer faced panels. However, it is recommended that, for new and major applications, full scale mock-up units should be manufactured to check out the feasibility of the production sequence. Tests on sample panels should be made to confirm the suitability of the cut stones and anchors, and to determine the effects of in-service loads and bowing on the unit's performance.

## 6.7.2 NATURAL STONE

Natural stone facings may be used in various sizes and regular shapes (narrow strips, small squares and rectangles, or regular sized ashlar pieces) as well as different colour combinations, to provide an infinite number of pattern possibilities.[R6.7.1]

### Sizes

Natural stone veneer stock for precast facing is usually thinner than that required for conventionally set stone. At the present time, some fabricators will supply granite with thicknesses from 22 to 50 mm. Depending on the colour range, the size of granite pieces varies from 900 mm × 900 mm to 1500 mm × 2100 mm. Reference is also made to various codes which may specify minimum thicknesses and maximum stone dimensions. Granite is an excellent material for exterior use and should be approximately 25 mm thick. Marble may be fabricated with a minimum thickness of 20 mm. Panels may be 1 to 1.5 m long and 0.5 to 1.5 m wide. Marble is softer than granite and the material should be carefully chosen for exterior use in a harsh northern climate. Limestone may have a thickness between 40 and 125 mm. Limestone at the minimum thickness will have an area of 0.4 to 1.0 $m^2$. Thicker pieces may be as large as 5500 mm high by 2400 mm wide. Minimum thickness of granite depends on sawing, handling and finishing, and type of anchors used.

Veneer materials should be sized (length and width) to a tolerance of plus zero, minus 3 mm. Flatness tolerances for finished surfaces will vary, depending on the type of finish. For example, the granite industry tolerances vary from 1.2 mm for polished surface to 4.8 mm for flame (thermal) finish when measured with a 1200 mm straightedge. While thickness variations are generally not important, since the concrete will provide a uniform back face, some consideration must be given to the possibility of individual pieces being locked in. At corner butt joints particularly, the finished edges should be gauged to within 1.6 mm of the specified thickness.

### Anchorage of Stone Facing

It is recommended that there be no concrete bonding between stone veneer and concrete back-up in order to prevent:

1. Excessive bowing of panel.
2. Cracking of panel.
3 Staining of veneer.

Mechanical anchors should be used to secure the veneer.

Where a bondbreak is desired between the veneer and the concrete to allow for independent movement, the following methods are recommended:

1. A 150 μm polyethylene sheet.
2. An air gap.
3. A 3 mm polyethylene foam pad.

Stone veneer is supplied with holes predrilled in the back surface for the attachment of mechanical anchors. Preformed anchors fabricated from stainless steel type 304 are commonly used. Generally, for sound, dense veneer materials, one anchor should be used for each approximately 0.25 m$^2$ of veneer with a maximum spacing of 450 mm between anchors and a minimum of two anchors per piece of veneer. However, four

**Fig. 6.7.3 Typical anchor details for stone veneers for insulated sandwich panels**

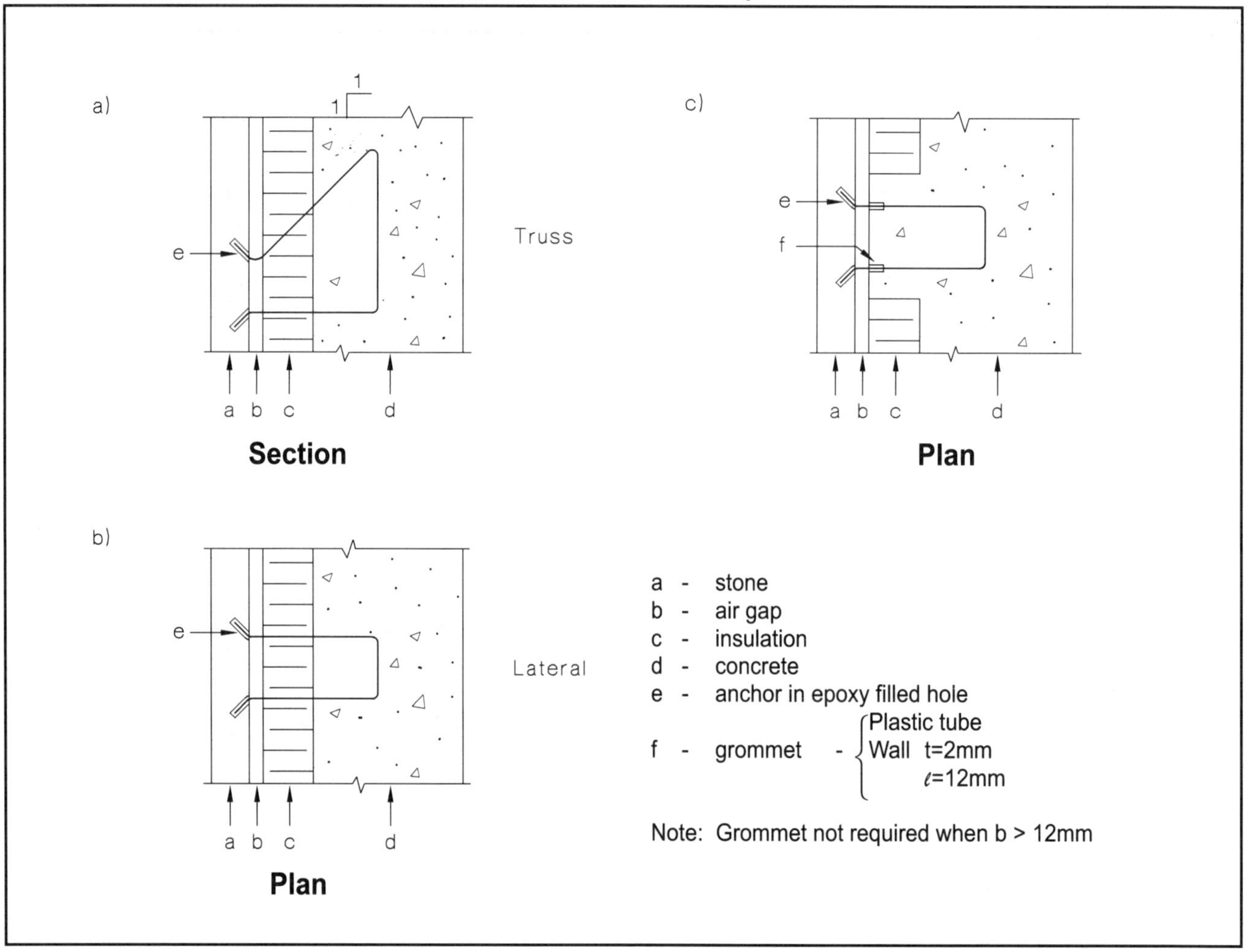

anchors should normally be used for square or substantially square veneers. Anchor size and spacing in veneers of questionable strengths or with natural planes of weakness may require special analysis. Depth of holes, drilled at an angle of 30° to 45° to the plane of the stone, should be approximately one-half the thickness of the veneer (minimum depth of 15 mm) and for granite and marble, the holes should have a diameter of 2 to 3 mm larger than the anchor to allow for stresses which could be caused by a differential in the coefficient of thermal expansion between the veneer and the concrete. A moisture resisting epoxy filler compatible with the stone is highly desirable.

Mechanical anchors are produced in a variety of sizes, shapes and materials. Typical details for bonded and unbonded conditions, test data, and schedules for marble veneers are shown in Fig. 6.7.2(a). Fig. 6.7.2(b) presents anchor test data for granite veneers and Fig. 6.7.2(c) shows typical limestone anchoring details. Note that for limestone veneers, a moisture barrier is recommended at the concrete face to eliminate a possible chemical reaction between the limestone and the concrete which could stain the exposed face.

In the case of a fully insulated panel, "Truss" anchors of the type shown in Fig. 6.7.3 (a) are used. The anchors should preferably be located in the joints in the insulation. This system has tended to replace the arrangement shown in Fig. 6.7.3(c) in which the anchor is embedded in a concrete cylinder formed by a hole drilled in the insulation. Both systems are in current use, although the thermal advantages of the truss system are clear.

The precast manufacturer should submit test reports on the stone anchoring system proposed for use. The test samples should be a typical panel section about 0.1 $m^2$ and should approximate as closely as possible actual panel anchoring conditions. A bondbreaker should be placed between stone and concrete during sample manufacture to eliminate any bond between veneer and concrete surface. Each test sample should contain at least two anchors connecting stone to concrete backup. Unless there are satisfactory historical records of similar previously successful manufacture, a minimum of ten tests should be conducted on samples to determine ultimate pullout and shear strength of each anchor.

## Veneer Jointing

Joint width between stone veneer pieces within a precast element should be a minimum of 6 mm. Joints which are to be sealed should receive a chemically neutral, resilient type, nonremovable gasket (such as a caulking back-up rod) which will not stain the veneer or adversely affect the specified type of sealant to be applied to the joints later. The gasket should be of a size and configuration that will provide a pocket to receive the sealant and also prevent any backing concrete from entering any portion of the joint between the veneer units. Shore A hardness[R6.7.2] of gasket should be less than 20. Joint width between veneer units within a precast element is a matter of the architect's design concept. It may or may not be equal to the specified joint width between precast elements.

Caulking should be non-staining to the veneer material. Caulking installation can be performed at the time the caulking between precast elements is executed, however, consideration should be given to caulking of standard units in the factory where it can be carried out efficiently and inexpensively.

## Properties

The strength of natural stone depends on several factors: the rift and cleavage of crystals, the degree of cohesion, the interlocking of crystals, and the nature of cementing materials present. The properties of a given type of stone will vary depending on the locality from which it was originally quarried; moreover, sedimentary and metamorphic rocks will exhibit differing strengths when measured parallel and perpendicular to their original bedding planes.

The range of properties for most building stones is given in Fig. 6.7.4. These may serve only as a guide for preliminary design; specific data should be obtained from suppliers and the "directionality" of the material considered in any analysis where the degree of accuracy is of consequence to a design decision. In some cases, the use of lower-bound data may prove satisfactory.

**Fig. 6.7.4 Range of natural stone properties**

| Type | Compression (MPa) | Tension (MPa) | Modulus of rupture (MPa) | Shear (MPa) | Modulus of elasticity (MPa × $10^3$) | Coef. of thermal expansion ($10^{-6}$mm/(mm·°C)) |
|---|---|---|---|---|---|---|
| Granite | 70 - 280 | 4.1 - 6.9 | 7.6 - 20.7 | 13.8 - 29.7 | 39 - 66 | 8.1 (avg.) |
| Limestone | 28 - 70 | 2.1 - 2.6 | 4.8 - 8.3 | 6.2 - 12.4 | 21 - 37 | 4.3 - 5.4 |
| Marble | 41 - 105$^{(1)}$<br>52 - 115.5$^{(2)}$ | 2.8 - 15.9 | 7.6 - 17.9<br>7.6 - 18.6 | 11.4 - 33.1<br>16.2 - 29.7 | 13 - 90<br>30 - 102 | 3.8 - 22.1 |
| Serpentine | 76 - 193 | 5.5 - 11.0 | | 17.9 - 34.5 | 33 - 66 | |
| Sandstone | 35 - 140 | 1.9 - 3.5 | 3.5 - 6.9 | 2.1 - 20.7 | 13 - 53 | 11 |
| Slate | 48 - 214 | 20.7 - 30.0 | 27.6 - 62.1 | 13.8 - 24.8 | 62 - 103 | 10 |

(1) Parallel (2) Perpendicular

## 6.7.3 REINFORCEMENT

Reinforcement of the precast concrete back-up should follow closely the recommendations for wall panels relative to design, cover and placement. Cover depth of reinforcement must be a minimum of 15 mm at the veneer surface. This cover should be maintained by non-corrosive spacers. All possible precautions should be taken to avoid any materials that could cause corrosion or rusting, which in turn could cause discolouration in the veneer.

In flat panels, it is necessary to choose the reinforcement to resist the inherent bowing that will occur. Truss systems have been very successful. Stock truss assemblies are obtainable as a catalogue item or, on large panels, they can be fabricated from reinforcing bars. Ribbing on the back of the panels is also a solution to bowing, but this requires backforming in many cases, and is more costly.

## 6.7.4 STRUCTURAL CLAY PRODUCTS

Clay products which are bonded directly to concrete include brick, structural facing tile, and architectural terra cotta (ceramic veneer). Ceramic glaze units may craze from freeze-thaw cycles or the bond may fail on exposure. Therefore, where required for exterior use, the manufacturer of the clay product should be consulted for suitable materials.

Since ceramic-glazed units have very low permeance to water vapour, it is recommended that a vapour barrier be installed on the warm side of walls enclosing areas in which the average relative humidity is expected to exceed 50%.

Clay product faced precast panels may be designed as concrete elements, neglecting, for design purposes, the structural action of the face veneer. The thickness of the precast panels is reduced by the thickness of the veneer. Design assumptions often exclude consideration of differential shrinkage or thermal expansion. However, if the panel is to be prestressed, the effect of composite behaviour and the resulting prestress eccentricity must be recognized and considered in design.

### Sizes

The selection of brick shapes and sizes depends on local availability. Brick manufacturers should be consulted early in the design stage to determine the manufacturing capability for special shapes and sizes with tightly controlled tolerances. In addition to standard shapes and sizes, brick may be available from some suppliers in veneer thickness, usually 1/3 to 1/4 standard brick thickness.

Architectural terra cotta (ceramic veneer) is a custom product and, within limitations, is produced in sizes specified for specific projects. Two thicknesses and sizes of units are usually manufactured: 30 mm thick units including dovetails which are spaced 125 mm on centres, the size may be 510 mm wide x 760 mm long; and 55 mm thick units including dovetails which are spaced 180 mm on centres, the size may be 815 mm wide x 1220 mm long.

### Bond

The nature of the surfaces of the brick are important for bond. Smooth, dense, heavily sanded or glazed surfaces are not usually satisfactory where high bond is required. Textures that may be specified to give increased bond include scored finish, in which the surface is grooved as it comes from the die; combed finish, where the surface is altered by parallel scratches; and roughened finish, produced by wire cutting or wire brushing to completely remove the smooth surface or die skin from the extrusion process.

When die skin or heavily sanded brick must be used, a mechanical bond with corrosion-resistant metal ties is required. With other brick surfaces, ties are not required. Ties may be those used in cavity wall construction, such as 5 mm diameter Z, rectangular shape, prefabricated truss, or ladder wire ties, shown in Figure 6.7.5.

**Fig. 6.7.5 Wall ties**

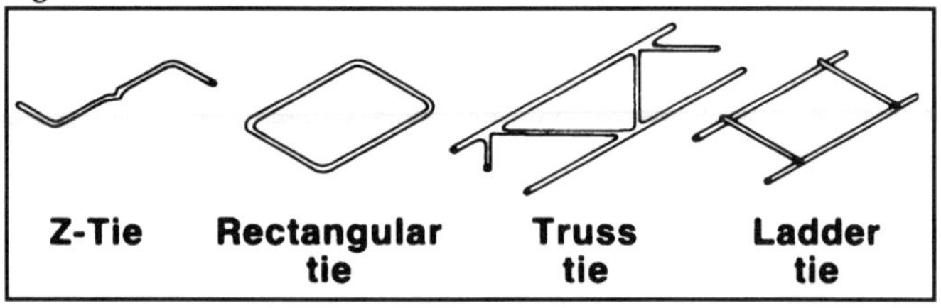


Where ties are required, there should be one for each 0.40 m$^2$ of wall area. Ties in alternate courses should be staggered. The maximum vertical distance between ties should not exceed 600 mm, and the maximum horizontal distance should not exceed 900 mm (see Fig, 6.7.6).

**Fig. 6.7.6 Spacing and staggering of metal ties (where required)**

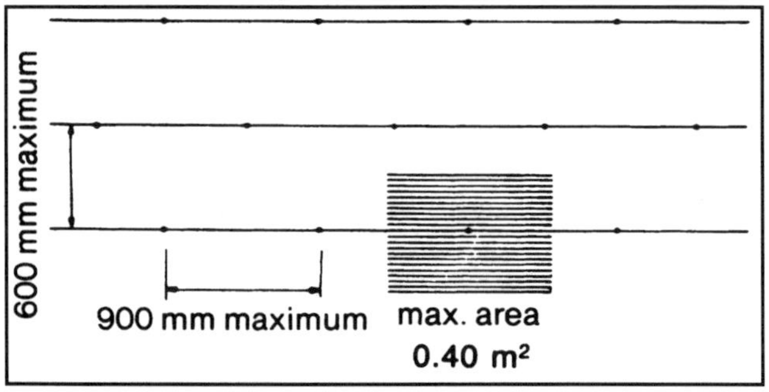


Additional bonding ties should be provided at all openings spaced not more than 900 mm apart, around the perimeter and within 300 mm of the opening. Ties should be of corrosion-resistant or non-corroding metal depending on the application. Reference is made to CSA 370-M94 "Connectors for Masonry" and CSA 371-M94 "Masonry Construction for Buildings" for guidance as to when ties should be non-corrosive or corrosion-resistant.

Corrosion resistance is usually provided by coating the metal with zinc or copper. To ensure adequate resistance to corrosion, coatings should conform to the following ASTM specifications:

1. Zinc coated ties - A 153, class B1, B2 or B3
2. Zinc coated wire - A 116, class 2 or 3
3. Copper coated wire - B 277, grade 30HS

Non-corroding ties should be made of stainless steel, ASTM type 301 or 304, or copper or bronze.

When ties are used, the brick joints are grouted and the ties placed into the horizontal joint as the wet grout is placed. The required concrete reinforcement is anchored in place after the brickwork grout has achieved its initial set. The concrete is then placed and cured.

The back side of the terra cotta units should have dovetails in order to develop adequate bond to concrete.

### Absorption

Brick with an initial rate of absorption (suction) less than 1 mg/(mm$^2$·min) net area when tested in accordance with CSA A82.2, "Methods of Sampling and Testing Brick," is not required to be wetted. However, brick with high suction or with an initial rate of absorption in excess of 1 mg/(mm$^2$·min) should be wetted to reduce the suction, and thereby improve bond, prior to placement of the concrete.

Terra cotta units must be soaked in water for at least one hour in order to reduce suction. They should be damp at the time of concrete placement.

### Properties

Engineering properties of brick vary considerably depending on the source and grade of brick[R6.7.3]. Fig. 6.7.7 shows the classification by compressive strength of brick manufactured in Canada. Fig. 6.7.8 shows the physical requirements of the brick.

As the temperature or length of burning period is increased, clays burn to darker colours, and compressive strength and modulus of elasticity are increased. The modulus of elasticity

**Fig. 6.7.7 Classification by compressive strength**

| Designation | Compressive strength (MPa) | |
|---|---|---|
| | Average of 5 brick | Individual minimum |
| Type II (not exposed) | 11 and over | 10 |
| Type I (exposed) | | |
| A | 18 to 30 | 16 |
| B | 30 to 42 | 27 |
| C | 42 to 55 | 38 |
| D | 55 to 70 | 50 |
| E | Over 70 | 63 |

Note: This classification conforms essentially to that given in the current National Building Code of Canada which relates compressive strength to the load requirements of masonry. Type I Brick, A, B, C, D, and E are all suitable for exposed masonry, and when load-bearing capacity is not critical, the selection may be made according to colour and texture.

**Fig. 6.7.8 Physical requirements**

| Designation | Minimum compressive strength (Brick flatwise) (MPa) gross area | | Maximum water absorption by 5-h boiling (per cent) | | Maximum saturation coefficient | |
|---|---|---|---|---|---|---|
| | Aver. of 5 brick | Indi-vidual | Aver. of 5 brick | Indi-vidual | Aver. of 5 brick | Indi-vidual |
| Type I Brick (Exposed) | 18 | 16 | 22.0 | 25.0 | 0.88 | 0.90 |
| Type II Brick (Not Exposed) | 11 | 10 | No Limit | No Limit | No Limit | No Limit |

* The saturation coefficient or C B Ratio is the ratio of absorption by 24-h submersion in cold water to that after 5-h submersion in boiling water.

of brick ranges from 10 000 to 35 000 MPa and Poisson's ratio from 0.04 to 0.11. In general, the modulus of elasticity of brick increases with compressive strength to a compressive value of approximately 35 MPa. After this, there is little change.

The average coefficient of thermal expansion of brick is $10 \times 10^{-6}$ mm/(mm·°C). The thermal expansion of clay units is not the same as the thermal expansion of brick-faced precast panels due to joints.

Clay bricks, when removed from the kiln after firing, will begin to permanently increase in size as a result of absorption of atmospheric moisture. The design coefficient for moisture expansion of clay bricks can very between .0002 and .0005 mm/mm and must be considered in the design of brick or tile faced concrete panels.

Relative temperature and moisture movements of concrete, brick, tile and mortar are shown in Fig. 6.7.10

Fig. 6.7.9 shows the relation of compressive strength of structural clay tile to absorption. For the same tile design, this relationship might be expected to be reasonably constant, however, the same relationship would not be expected to hold for tile of different designs.

The modulus of elasticity of structural tile varies from 12 000 to 43 000 MPa and Poisson's ratio from 0.05 to 0.10. Tile has an average coefficient of thermal expansion of $5.9 \times 10^{-6}$ mm/(mm · °C).

The compressive strength of terra cotta units usually ranges from 55 to 76 MPa, and the average coefficient of thermal expansion is $7.2 \times 10^{-6}$ mm/(mm · °C).

All clay products are subject to such local variation that the designer should obtain values from suppliers that are being considered.

Fig. 6.7.9 Compressive strength and absorption of structural clay tile

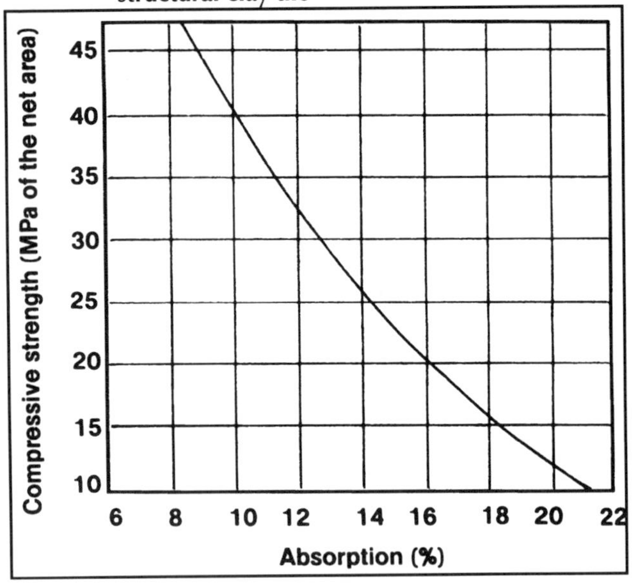


### Insulated Sandwich Panels

Clay products can be combined with insulation and back-up concrete to construct fully insulated sandwich panels similar to stone faced panels. A suggested solution for anchorage for gravity, seismic and lateral loads is shown in Fig. 6.7.11.

Fig. 6.7.10 Relative temperature and moisture movements of concrete, brick, tile and mortar

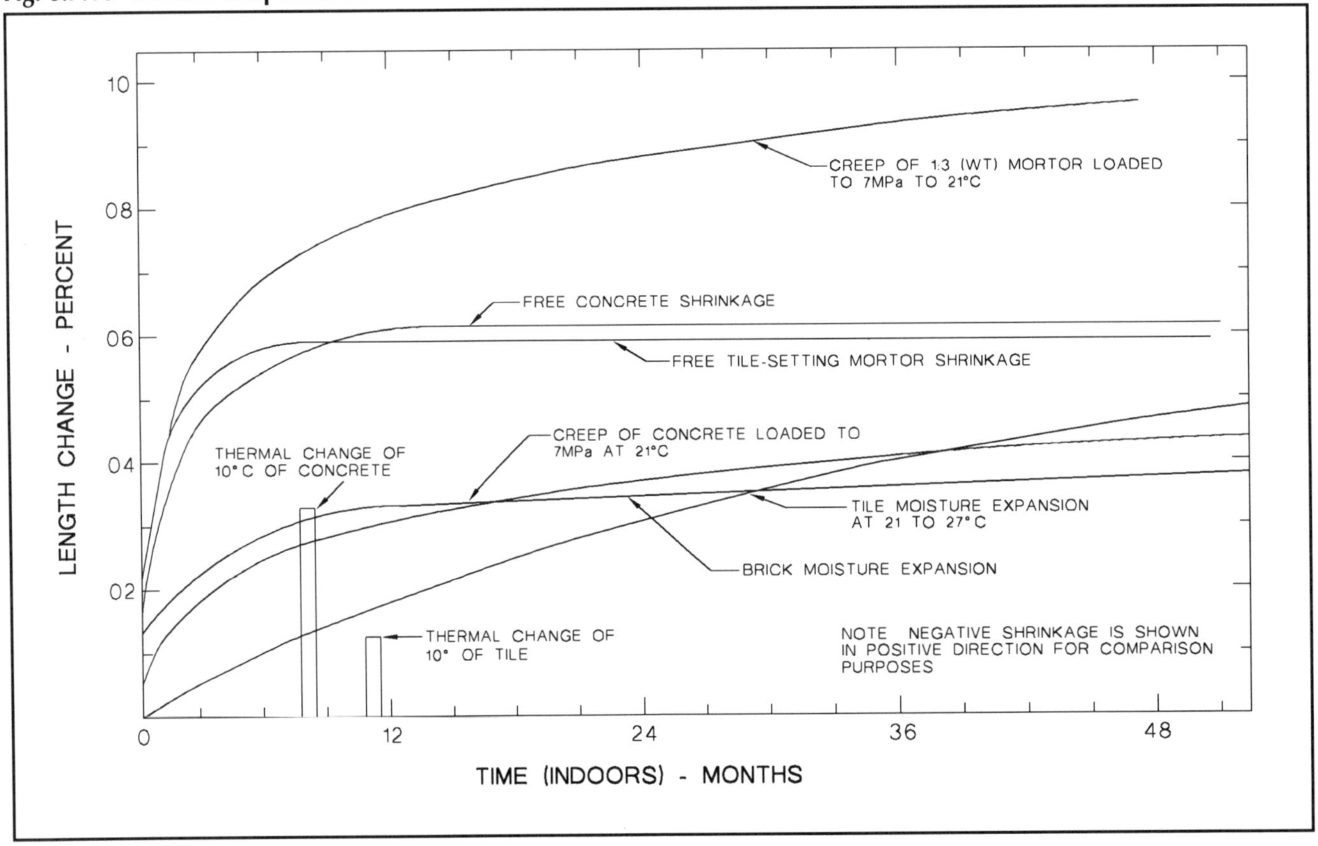

**Fig. 6.7.11 Anchor details for brick veneer for insulated sandwich panels**

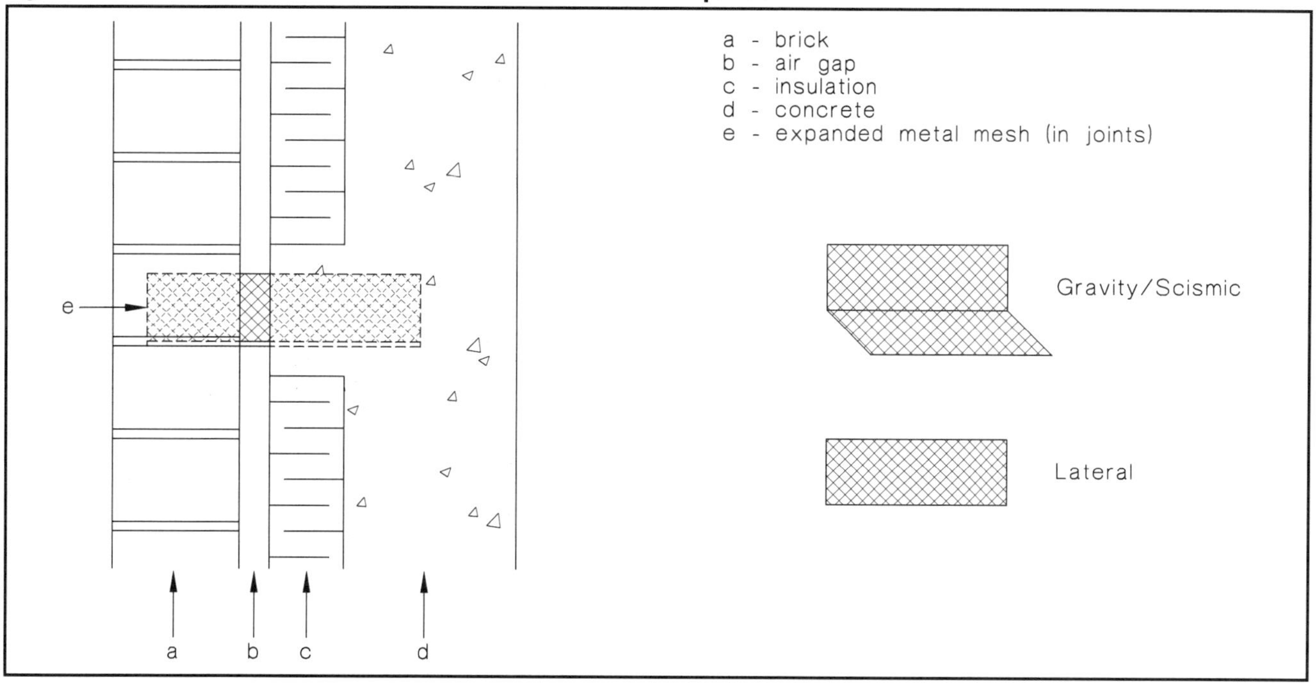


# 6.8 PRECAST CONCRETE USED AS FORMS

## 6.8.1 GENERAL

Architectural precast concrete units serving as formwork for cast-in-place concrete, provide benefits for all members of the construction team:

1. The Designers are free to shape the form, and thus the appearance of the structure, with virtually no restriction. Precast concrete, manufactured in a plant, permits rigid factory-controlled conditions, thus ensuring a uniformly high quality facade in the desired shapes, colours and textures. Precasting makes it possible to inspect the surface finish before the cast-in-place concrete is placed. Form removal, with its subsequent patching of form anchor holes and surface imperfections, and repair of other flaws, is eliminated.
2. The Designers may, by proper joint arrangement, employ the form as a contributing portion of the structural support system. They can thus readily achieve continuity and the ductility required in zones of high seismicity. Precast wall panels may be connected to form structural walls by casting in place spandrel beams and columns using the wall panels as forms.[R6.8.1]

   This will transform the usually inactive curtain walls into lateral load (seismic) resisting elements. Seismic loads will be resisted by the building's central core and by the ductile concrete exterior frame with the floor slabs acting as diaphragms.[R6.8.2]
3. Contractors will benefit from a reduction in the time required for construction. All the framework required for a structure can be manufactured in advance of concreting. This permits greater flexibility and continuity in concrete operations. Delays in concreting due to time required for preliminary curing of concrete preceding form removal and re-erection of forms can be eliminated. The precast units may be erected quickly, and the structure is complete when the concrete is placed and joints caulked.

Architectural precast concrete units used as exterior forms eliminate the need for temporary outside forms. Exterior scaffolding is not generally required, as work is carried out from inside the building, resulting in increased safety on the project. Bottom horizontal faces of spandrel beams can be formed with precast units designed as beams supporting the weight of wet concrete and construction loads above; also scaffolding or shoring normally required to support overhead removable forms, can be partly or completely eliminated, thus saving the time and cost involved in its erection.

The precast architectural concrete can be considered solely as a form, serving only decorative purposes after the cast-in-place concrete has achieved design strength. This is accomplished by providing open or compressible joints between abutting precast panels, and neglecting (or eliminating) bond at the interface of the precast and cast-in-place concrete. When the architectural precast concrete element is non-composite with the cast-in-place concrete, the reinforcing steel extending from the precast into the cast-in-place concrete should be of sufficient strength to support the formwork unit. Both quantity and anchorage should be checked. Some stiffening of the reinforcement cage may be required.

In other cases, it may be desirable to establish interaction between the precast form and the cast-in-place concrete, thus combining the strength of both. Shear transfer is accomplished by bond and/or mechanical connections. The element may then be treated as an integral unit for subsequent applied loads, designed in accordance with the composite concrete section of the Code. If the precast element is arranged vertically (such as a column or wall form) and not otherwise supported, the reinforcement that passes across the interface should be adequate to support the architectural precast unit using the shear friction concept. This reinforcement should be adequate to restrain bowing in the precast element.

## 6.8.2 DESIGN

It is recommended that, for formwork design, reference should be made to ACI Committee 347 Guide to Formwork for Concrete(ACI 347R-94) American Concrete Institute, Detroit, Michigan.

## 6.8.3 CONSTRUCTION CONSIDERATIONS

Realistic assumptions with regard to construction techniques are required for this design concept. It must be determined how and where the precast panels will be supported during concreting in order to design the proper reinforcement within the panels. A mock-up section may be necessary to test the construction procedures before the project gets underway. Such a mock-up will also assist in determining panel shape, size, finish, joint placement and connections.

**References:**

R6.3.1 Speyer, Irwin J., “Considerations for the Design of Precast Concrete Bearing Wall Buildings to Withstand Abnormal Loads,” *P.C.I. Journal,* Vol. 21, No. 2, March-April 1976, pp. 18-51.

R6.6.1 Latta, J.K., “Precast Concrete Walls - A New Basis for Design,” *Canadian Building Digest,* CBD 94, Oct. 1967.

R6.7.1 McDaniel, W. Bryant, “Marble-Faced Precast Panels,” *PCI Journal*, V, 12, No. 4, August, 1967, pp 29-37.

R6.7.2 Mantell, C.L., *Engineering Materials Handbook,* McGraw-Hill Book Co., Toronto, Ontario

R6.7.3 Plummer, Harry C., “Brick and Tile Engineering,” Brick Institute of America, McLean, Virginia, 1962.

R6.8.1 ACI Committee 347, “Precast Concrete Units Used as Forms for Cast-in-Place Concrete,” *ACI Journal,* Proceedings V.66, No. 10, October, 1969, pp 798-813.

R6.8.2 Kulka, Felix, Lin, T.Y., and Yang, Y.C. “Prestressed Concrete Building Construction Using Precast Wall Panels,” *PCI Journal*, V.20, No. 1, January-February, 1975, pp 62-72.

# CHAPTER 7
# PRODUCT INFORMATION AND CAPABILITY

# PRODUCT INFORMATION AND CAPABILITY

## 7.0 NOTATION

$A_g$ = gross area of section

$A_p$ = area of prestressed reinforcement

$A_s$ = area of non-prestressed reinforcement

b = overall width of element

$e_c$ = eccentricity of prestress force from the centroid of the section at the centre of the span

$e_e$ = eccentricity of prestress force from the centroid of the section at the end of the span

$f'_c$ = specified compressive strength of concrete

$f'_{ci}$ = compressive strength of concrete at time of initial prestress

$f_{cp}$ = average compressive stress in concrete at centroid due to effective prestress force only

$f_{pr}$ = stress in prestressed reinforcement at factored flexural resistance

$f_{pu}$ = specified tensile strength of prestressed reinforcement

$f_{pe}$ = effective stress in prestressed reinforcement after losses

$f_y$ = specified yield strength of non-prestressed reinforcement

h = overall depth

I = moment of inertia

$\ell$ = span

$M_r$ = factored flexural resistance

$M_{rb}$ = factored flexural resistance under balanced conditions

$M_{ro}$ = factored flexural resistance with zero axial load

m = mass of element

$P_r$ = factored axial load resistance

$P_{rb}$ = factored axial load resistance under balanced conditions

$P_{ro}$ = factored axial load resistance with zero eccentricity

$S_b$ = section modulus with respect to the bottom fibre of section

$S_t$ = section modulus with respect to the top fibre of section

s = strand spacing

t = thickness

$V_c$ = factored shear resistance of concrete

$V_{cw}$ = factored shear resistance provided by concrete when diagonal cracking results from principal tensile stress in the web

$V_f$ = factored shear force

$V_r$ = factored shear resistance

V/S = volume-surface ratio

w = weight expressed in force units

$y_b$ = distance from bottom fibre to centroid of section

$y_t$ = distance from top fibre to centroid of section

z = a quantity limiting distribution of flexural reinforcement

$\lambda$ = factor to account for low density concrete

$\rho$ = $A_s/bd$ = reinforcement ratio for non-prestressed reinforcement

$\phi_c$ = resistance factor for concrete

$\phi_p$ = resistance factor for prestressed reinforcement

$\phi_s$ = resistance factor for reinforcement

$\alpha_1$ = ratio of average stress in rectangular compression block to the specified concrete strength (clause 10.1.7)

# 7.1 INTRODUCTION

This part of the Manual is devoted to technical data on the shapes that are standard in the precast prestressed concrete industry today. The use of this part of the Manual as a guide for selection of a particular shape, together with information on design aids and techniques provided in other parts of this Manual, should enable the designer to quickly and expeditiously complete his design.

The load tables on the following pages show dimensions, gross section properties and engineering capabilities of the shapes most commonly used throughout the industry. These shapes include double and single tees, hollow-core slabs, solid flat slabs, beams, girders, columns, piles and wall panels. The dimensions of the shapes shown in the tables may vary among manufacturers. Adjustment for these minor variations can be made by the designer. Hollow-core slabs of different thicknesses, core sizes and shapes are available in the market under various trade names. Load tables on pages 7-18 and 7-19 are developed for non-proprietary hollow-core sections of thicknesses most commonly used in the industry.

**Designers making use of these load tables should contact the manufacturers in the geographic area of the proposed structure to determine availability and exact dimensions of products shown here. Manufacturers will usually have their own load tables for sections which are not included on the following pages.**

# 7.2 EXPLANATION OF LOAD TABLES

Load tables for stemmed deck elements, flat deck elements and beams show the safe (unfactored) superimposed load, estimated camber at the time of erection and the estimated long-time camber after the element has essentially stabilized. For the deck elements, the upper table gives the information for the element with no topping, and the lower table is for the same element with 50 mm of normal density concrete topping acting compositely with the precast section. Values in the table assume a uniform 50 mm topping the full span length, and assume the element to be unshored at the time the topping is placed. Safe loads and cambers shown in the tables are based on the dimensions and section properties shown on the page. For the calculation of concrete stresses and crack width criteria, the section properties have been transformed to account for the effects of the prestressed reinforcement. For elements with different dimensions or section properties, the safe loads and cambers will vary.

For beams, a single table is used for several sizes of elements. The values shown are based on sections containing the maximum practical number of prestressing strands, but in some cases, more strands could be used.

## 7.2.1 SAFE SUPERIMPOSED LOAD

The values for safe (unfactored) superimposed load are based on the capacity of the element as governed by the limitations tabulated below:

| Type of element | Factored flexural resistance | Factored shear resistance | Flexural stresses | Crack widths |
|---|---|---|---|---|
| Flat deck elements | • | • | • | |
| Stemmed deck elements | • | | • | • |
| Beams | • | | • | • |

A portion of the safe load shown is assumed to be dead load for the purpose of applying load factors and determining time-dependent cambers and deflections. For untopped deck elements, 0.5 kN/m$^2$ of the capacity shown is assumed as superimposed dead load, typical for roof elements. For topped deck elements, 0.7 kN/m$^2$ of the capacity shown is assumed as superimposed dead load, typical for floor elements. The capacity shown is in addition of the dead load of the topping. For beams, 50 percent of the capacity shown is assumed as dead load, normally conservative for beams which support concrete decks.

**Example:** For a 2400 × 600 double tee with an 8-D strand pattern (Page 7-9) and a 16 m span, the capacity shown is 3.8 kN/m$^2$ live.

## 7.2.2 LIMITING CRITERIA

The criteria used to determine the safe superimposed load and the strand placement are based on requirements of the Code. For design procedures, see Chap. 3 of this Manual. A summary of the code provisions used in the development of these load tables is as follows:

### Factored Flexural Resistance

Load factors: 1.25 for dead load
1.50 for live load

Resistance factors:
Precast concrete, $\phi_c = 0.65$
Prestressing tendons, $\phi_p = 0.9$
Reinforcing bars, $\phi_s = 0.85$

Calculation of moments assumes simple spans with roller supports. If the strands are fully developed (see Sect. 3.4.9), the critical moment is assumed to be at midspan in elements with straight strands; for elements with tendons depressed at midspan, the critical section is found at either 0.4 $\ell$ or 0.5 $\ell$. (Note: The actual critical point can be determined by analysis, but will seldom vary significantly from 0.4 $\ell$.) Flexural resistance is calculated using strain compatibility as discussed in Chap. 3.

#### Flexural Stresses at Transfer

The following limitations are placed on flexural stresses immediately after transfer of prestress, before long time losses; it is assumed that strands are initially tensioned to 0.75 $f_{pu}$ or less:

| | |
|---|---|
| Compression: | $0.6 f'_{ci}$ |
| End tension: | $0.5\lambda\sqrt{f'_{ci}}$ |
| Midspan tension: | $0.25\lambda\sqrt{f'_{ci}}$ |

These stresses are calculated at 50 strand diameters from the end of the element (the theoretical point of full transfer) and at midspan.

Release tension is not used as limiting criterion for beams. Supplemental top reinforcement must be provided, and designed as described in Sect. 3.4.9 of this Manual.

#### Stresses due to Specified Loads, after all Losses

| | |
|---|---|
| Compression due to sustained loads: | $0.45 f'_c$ |
| Compression due to total load: | $0.60 f'_c$ |
| Tension: | $0.5\lambda\sqrt{f'_c}$ |

The critical point for service load moment is assumed at midspan for elements with straight strands; for elements with tendons depressed at midspan, the critical section is found at either 0.4 $\ell$ or 0.5 $\ell$.

#### Crack Widths of Partially Prestressed Elements, After All Losses

Sect. 3.4.8 describes the design procedures for calculating the crack widths of partially prestressed elements. For interior exposure, the Code allows a maximum value for the crack control parameter, z, to be 20 kN/mm. This value equates to a crack width of 0.27 mm (see Fig. 3.4.1). The capacity of partially prestressed elements is limited such that this crack width is not exceeded. The critical section is assumed at midspan for elements with straight strands; for elements with tendons depressed at midspan, the critical section is found at either 0.4 $\ell$ or 0.5 $\ell$.

If it is a requirement that the allowable crack width be restricted to less than 0.27 mm, a marginal increase in the final prestress force may be required if the full superimposed load capacity in the tables is to be achieved.

#### Factored Shear Resistance

For *flat deck elements*, the capacity may be limited by the factored shear resistance. In this case, the safe superimposed load is that which will yield a factored shear force $V_f$ of no more than $V_c$ or $V_{cw}$, as permitted by the Code for slabs without shear reinforcement. See Sect. 3.6.4 for the design procedures.

For *stemmed deck elements and beams*, the factored shear resistance may exceed $V_c$ or $V_{cw}$ by providing shear reinforcement, designed as described in Sect. 3.6.4. For the majority of deck elements, however, minimum or no reinforcement may be required.

#### Span/depth Ratios

*Flat deck elements* show no value beyond a span/depth ratio of 50 for untopped elements and 40 for topped elements. These are the suggested maximums for roof and floor elements respectively, unless a detailed analysis is made.

## 7.2.3 ESTIMATED CAMBER

The estimated cambers shown are calculated to the nearest 10 mm using the multipliers shown in Sect. 3.5.4 of this Manual. *These values are estimates and should not be used as absolute values.* Non-structural components attached to elements which could be affected by camber variations, such as partitions of folding doors, should be placed with adequate allowance for error. Calculation of topping quantities should also recognize that the values can vary.

## 7.2.4 CONCRETE STRENGTH AND DENSITY

Twenty-eight day cylinder strength for concrete in the prestressed units is assumed to be 35 or 40 MPa. Tables for units with composite topping are based on the topping concrete being normal density concrete with a cylinder strength of 25 MPa. For stemmed and flat deck elements, the concrete strength at time of strand tension release is 25 MPa unless the value falls below the heavy line shown in the load table, indicating that a cylinder strength greater than 25 MPa is required. For beams, the concrete strength at transfer is 30 MPa. No values are shown when the required release strength exceeds 30 MPa. The designer should recognize that it is sometimes difficult to obtain a release strength higher than 25 MPa on a one-day casting cycle. In such cases, the cost of production will be increased and the designer should consult with prospective producers when required release strengths are above 25 MPa.

Many prestressing plants prefer to use higher strength concretes, resulting in somewhat higher allowable loads or greater spans than indicated in the load tables contained herein.

In this Chapter, all tables apply to normal density concrete only, with the density assumed as 2400 kg/m$^3$.

## 7.2.5 PRESTRESSING STRAND

Prestressing strands used in the load tables are 13 mm diameter low relaxation strands with a specified tensile strength of 1860 MPa.

Quantity, size and profile of strands are shown in the load tables under the column headed "Strand Pattern". In the double tee load tables, for example, "8-S" indicates 8-13 mm diameter 1860 MPa strands (4 per double tee stem) and the "S" indicates that the strands are straight. "8-D" indicates 8-13 mm diameter 1860 MPa strands depressed at one point in the midspan of the double tee.

For the *flat deck elements* (Pages 7-18 to 7-22) the manufacturer is allowed some flexibility in choice of strand size and tensile strength. Manufacturers should be contacted for details of strand utilization.

**Note:** For development length (see Sect. 3.4.9).

## 7.2.6 PRESTRESS LOSSES

Losses assumed in computing the required concrete strength at time of strand release are 8%. Total losses averaged 18% for normal density concrete.

For long span, heavily prestressed products, losses may be somewhat higher than these assumed values, and, for shorter spans with less prestressing, they may be lower. However, these values will usually be adequate for element selection. Additional information on losses is given in Chapter 3 of this Manual.

## 7.2.7 STRAND PLACEMENT

For *stemmed deck elements* and *beams*, the eccentricities of strands at the ends and midspan are shown in the load tables. Strands have been placed so that the stress at 50 strand diameters from the end (theoretical transfer point) will not exceed those specified above, with a concrete strength at release of 25 MPa for stemmed elements and 30 MPa for beams. For stemmed deck elements, values below the heavy line indicate a required release strength between 25 MPa and 30 MPa.

For *flat deck elements* the load table values are based on strand centred 45 mm (38 mm for solid slabs) from the bottom of the slab. Strand placement can vary from as low as 25 mm to as high as 55 mm from the bottom, which will change the capacity and camber values shown. The higher strand placements give improved fire resistance ratings (see Chapter 5 of this Manual for more information on fire resistance). The lower strand placement may require higher release strengths, or top tension reinforcement at the ends. **The designer should contact the local supplier of flat deck elements for available and recommended strand placement locations.**

## 7.2.8 COLUMNS AND LOAD BEARING WALL PANELS

Interaction curves for selected precast prestressed columns, precast reinforced columns and various types of commonly used wall panels are provided on Pages 7-32 to 7-39.

These interaction curves are for factored loads and moments and the appropriate load factors must be applied to the service loads and moments before entering the charts. Also, the curves are for *short* elements. Moment magnifiers caused by slenderness effects must be calculated and applied to the design moments before using the curves for final element selection (see Chapter 3).

The column curves are terminated at a value of 0.8 $P_{ro}$, the maximum allowable load for tied columns under the Code. Most of the wall panel curves show the lower portion of the curve only (flexure controlling). Actual design loads will rarely exceed the values shown.

The curves for double tee wall panels are shown for bending in directions causing either tension or compression in the stem.

Partial development of strands has been assumed for prestressed columns and full development for prestressed wall panels. Lateral ties have been assumed for prestressed columns and double tee wall panels and no ties assumed for hollow-core and solid wall panels. For double tee and hollow-core wall panels, the effects of strands above the neutral axis have been neglected for the calculation of the flexural resistance under zero axial load.

The curves for hollow-core wall panels are based on a generic section as shown. They can be used with small error for all sections commonly marketed for wall panel use.

## 7.2.9 PILES

The factored resistances of prestressed concrete bearing piles, based on the structural capacity of the pile alone are shown on Page 7-40. The ability of the soil to carry these loads must be evaluated by a geotechnical engineer. Values for concrete strengths up to 55 MPa are shown. Available strengths should be checked with local manufacturers. The design of prestressed concrete piles is discussed in Sect. 3.8.8 of this Manual.

Section properties and allowable service load bending moments for prestressed concrete sheet pile units are shown on Page 7-41. These units are available in some areas for use in earth retaining structures.

# DOUBLE TEE
## 2400 × 300
## Normal Density Concrete

**Strand Pattern Designation**

- No. of 13 mm strands
- S = straight; D = depressed

2-S

Superimposed loads shown include superimposed dead load of 0.5 kN/m² for untopped elements and 0.7 kN/m² for topped elements. Remainder is live load. Long-time cambers include superimposed dead load but do not include live load.

**Key**

11.1 - Superimposed (service) load, kN/m²
10 - Estimated camber at erection, mm
10 - Estimated long time camber, mm

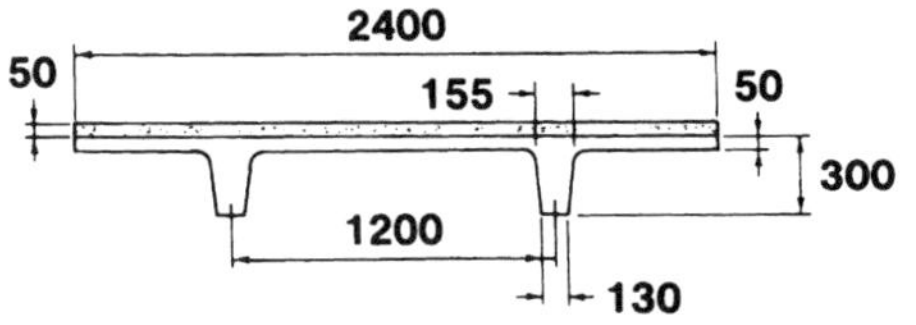


$f'_{ci}$ = 25 MPa
$f'_c$ = 35 MPa
$f_{pu}$ = 1860 MPa
$f'_{ctop}$ = 25 MPa

**Section Properties**

| | | Untopped | Topped |
|---|---|---|---|
| $A_g$ | = | 191 000 mm² | |
| I | = | 1 350 × 10⁶ mm⁴ | 2 100 × 10⁶ mm⁴ |
| $y_b$ | = | 220 mm | 257 mm |
| $y_t$ | = | 80 mm | 93 mm |
| $S_b$ | = | 6 130 × 10³ mm³ | 8 170 × 10³ mm³ |
| $S_t$ | = | 17 000 × 10³ mm³ | 22 500 × 10³ mm³ |
| m | = | 194 kg/m² | 316 kg/m² |
| w | = | 1.9 kN/m² | 3.1 kN/m² |
| V/S | = | 32 mm | |

## Table of superimposed load (kN/m²) and cambers (mm) — No Topping

| Strand Pattern | $e_e$† / $e_c$† | Span (m) 4 | 5 | 6 | 7 | 8 | 9 | 10 | 11 | 12 | 13 | 14 |
|---|---|---|---|---|---|---|---|---|---|---|---|---|
| 2-S | 170<br>170 | 9.6<br>0<br>0 | 5.6<br>10<br>10 | 3.4<br>10<br>10 | 2.0<br>10<br>0 | | | | | | | |
| 4-S | 145<br>145 | | 11.1<br>10<br>10 | 7.2<br>10<br>20 | 4.9<br>20<br>20 | 3.4<br>20<br>20 | 2.4<br>20<br>20 | 1.6<br>20<br>10 | | | | |
| 6-S | 120<br>120 | | | 7.8<br>10<br>10 | 5.4<br>10<br>20 | 3.8<br>10<br>10 | 2.7<br>10<br>10 | 1.9<br>10<br>10 | | | | |
| 6-D | 120<br>157 | | | | 8.3<br>30<br>30 | 6.0<br>30<br>30 | 4.4<br>30<br>40 | 3.3<br>40<br>30 | 2.5<br>30<br>30 | 1.8<br>30<br>10 | | |
| 8-D | 57<br>150 | | | | | | | | 3.3<br>50<br>50 | 2.5<br>50<br>40 | 1.9<br>40<br>20 | 1.5<br>30<br>– 10 |

## Table of superimposed load (kN/m²) and cambers (mm) — 50 mm Normal Density Topping

| Strand Pattern | $e_e$† / $e_c$† | Span (m) 4 | 5 | 6 | 7 | 8 | 9 | 10 | 11 | 12 | 13 |
|---|---|---|---|---|---|---|---|---|---|---|---|
| 2-S | 170<br>170 | 10.8<br>0<br>0 | 6.0<br>10<br>10 | 3.4<br>10<br>10 | 1.9<br>10<br>0 | | | | | | |
| 4-S | 145<br>145 | | | 8.1<br>10<br>20 | 5.3<br>20<br>20 | 3.5<br>20<br>10 | 2.2<br>20<br>10 | | | | |
| 6-S | 120<br>120 | | | | 6.3<br>10<br>10 | 4.2<br>10<br>10 | 2.3<br>10<br>0 | | | | |
| 6-D | 120<br>157 | | | | 9.3<br>30<br>30 | 6.6<br>30<br>30 | 4.7<br>30<br>30 | 3.1<br>40<br>20 | 2.0<br>30<br>10 | | |
| 8-D | 57<br>150 | | | | | | | | 2.8<br>50<br>30 | 1.9<br>50<br>20 | |

* Values below heavy solid line require transfer strengths higher than 25 MPa.
* Values to right of dashed line indicate that element is partially prestressed.
† $e_e$ and $e_c$ are the maximum eccentricities at end and midspan that will not violate the permissible concrete stresses at transfer.

# DOUBLE TEE
## 2400 × 400
## Normal Density Concrete

**Strand Pattern Designation**

2-S
- No. of 13 mm strands
- S = straight; D = depressed

Superimposed loads shown include superimposed dead load of 0.5 kN/m² for untopped elements and 0.7 kN/m² for topped elements. Remainder is live load. Long-time cambers include superimposed dead load but do not include live load.

**Key**
- 11.1 - Superimposed (service) load, kN/m²
- 10 - Estimated camber at erection, mm
- 10 - Estimated long time camber, mm

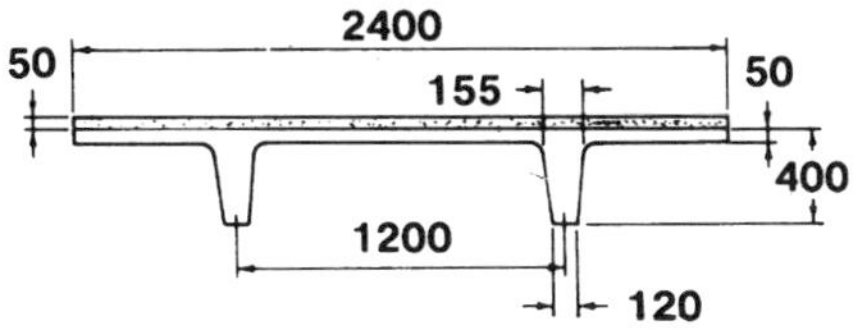


$f'_{ci}$ = 25 MPa
$f'_c$ = 35 MPa
$f_{pu}$ = 1860 MPa
$f'_{ctop}$ = 25 MPa

**Section Properties**

| | | Untopped | Topped |
|---|---|---|---|
| $A_g$ | = | 216 000 mm² | |
| I | = | 2 980 × 10⁶ mm⁴ | 4 300 × 10⁶ mm⁴ |
| $y_b$ | = | 289 mm | 333 mm |
| $y_t$ | = | 111 mm | 117 mm |
| $S_b$ | = | 10 300 × 10³ mm³ | 12 900 × 10³ mm³ |
| $S_t$ | = | 26 900 × 10³ mm³ | 36 400 × 10³ mm³ |
| m | = | 214 kg/m² | 336 kg/m² |
| w | = | 2.1 kN/m² | 3.3 kN/m² |
| V/S | = | 35 mm | |

**Table of superimposed load (kN/m²) and cambers (mm)** — **No Topping**

| Strand Pattern | $e_e$† / $e_c$† | Span (m) 5 | 6 | 7 | 8 | 9 | 10 | 11 | 12 | 13 | 14 | 15 | 16 | 17 | 18 |
|---|---|---|---|---|---|---|---|---|---|---|---|---|---|---|---|
| 2-S | 239 | 8.3 | 5.3 | 3.4 | 2.2 | | | | | | | | | | |
| | 239 | 0 | 0 | 10 | 10 | | | | | | | | | | |
| | | 0 | 10 | 10 | 0 | | | | | | | | | | |
| 4-S | 214 | | 11.1 | 7.7 | 5.5 | 4.0 | 2.9 | 2.1 | 1.5 | | | | | | |
| | 214 | | 10 | 10 | 10 | 20 | 20 | 20 | 10 | | | | | | |
| | | | 10 | 10 | 20 | 20 | 20 | 10 | 0 | | | | | | |
| 6-S | 155 | | | 9.6 | 7.0 | 5.2 | 3.9 | 2.9 | 2.2 | 1.6 | 1.5 | | | | |
| | 155 | | | 10 | 20 | 20 | 20 | 20 | 20 | 10 | 20 | | | | |
| | | | | 20 | 20 | 20 | 20 | 20 | 10 | 0 | 0 | | | | |
| 6-D | 155 | | | | 9.3 | 7.0 | 5.4 | 4.1 | 3.2 | 2.5 | 1.9 | 1.5 | | | |
| | 226 | | | | 20 | 30 | 30 | 30 | 30 | 30 | 30 | 20 | | | |
| | | | | | 30 | 30 | 30 | 30 | 30 | 20 | 10 | 0 | | | |
| 8-D | 101 | | | | | | 7.1 | 5.6 | 4.4 | 3.5 | 2.8 | 2.2 | 1.8 | | |
| | 219 | | | | | | 40 | 40 | 50 | 50 | 50 | 40 | 30 | | |
| | | | | | | | 50 | 50 | 50 | 40 | 40 | 20 | 0 | | |
| 10-D | 79 | | | | | | | | | | 3.6 | 2.9 | 2.3 | 1.9 | 1.5 |
| | 213 | | | | | | | | | | 60 | 60 | 60 | 50 | 40 |
| | | | | | | | | | | | 60 | 50 | 40 | 10 | – 20 |

**Table of superimposed load (kN/m²) and cambers (mm)** — **50 mm Normal Density Topping**

| Strand Pattern | $e_e$† / $e_c$† | Span (m) 5 | 6 | 7 | 8 | 9 | 10 | 11 | 12 | 13 | 14 | 15 | 16 |
|---|---|---|---|---|---|---|---|---|---|---|---|---|---|
| 2-S | 239 | 8.8 | 5.3 | 3.1 | 1.7 | | | | | | | | |
| | 239 | 0 | 0 | 10 | 10 | | | | | | | | |
| | | 0 | 0 | 0 | 0 | | | | | | | | |
| 4-S | 214 | | | 8.1 | 5.6 | 3.8 | 2.6 | | | | | | |
| | 214 | | | 10 | 10 | 20 | 20 | | | | | | |
| | | | | 10 | 10 | 10 | 10 | | | | | | |
| 6-S | 155 | | | | 7.5 | 5.4 | 3.9 | 2.7 | | | | | |
| | 155 | | | | 20 | 20 | 20 | 20 | | | | | |
| | | | | | 20 | 20 | 20 | 10 | | | | | |
| 6-D | 155 | | | | | 7.3 | 5.4 | 4.0 | 2.9 | 2.1 | | | |
| | 226 | | | | | 30 | 30 | 30 | 30 | 30 | | | |
| | | | | | | 30 | 30 | 30 | 20 | 10 | | | |
| 8-D | 101 | | | | | | 7.4 | 5.7 | 4.3 | 3.2 | 2.3 | | |
| | 219 | | | | | | 40 | 40 | 50 | 50 | 50 | | |
| | | | | | | | 40 | 40 | 40 | 30 | 20 | | |
| 10-D | 79 | | | | | | | | | | 3.0 | 2.2 | 1.6 |
| | 213 | | | | | | | | | | 60 | 60 | 60 |
| | | | | | | | | | | | 50 | 30 | 10 |

* Values below heavy solid line require transfer strengths higher than 25 MPa.
* Values to right of dashed line indicate that element is partially prestressed.
† $e_e$ and $e_c$ are the maximum eccentricities at end and midspan that will not violate the permissible concrete stresses at transfer.

# DOUBLE TEE
## 2400 × 500
## Normal Density Concrete

**Strand Pattern Designation**

4-S
- No. of 13 mm strands
- S = straight; D = depressed

Superimposed loads shown include superimposed dead load of 0.5 kN/m$^2$ for untopped elements and 0.7 kN/m$^2$ for topped elements. Remainder is live load. Long-time cambers include superimposed dead load but do not include live load.

**Key**
10.5 - Superimposed (service) load, kN/m$^2$
10 - Estimated camber at erection, mm
10 - Estimated long time camber, mm

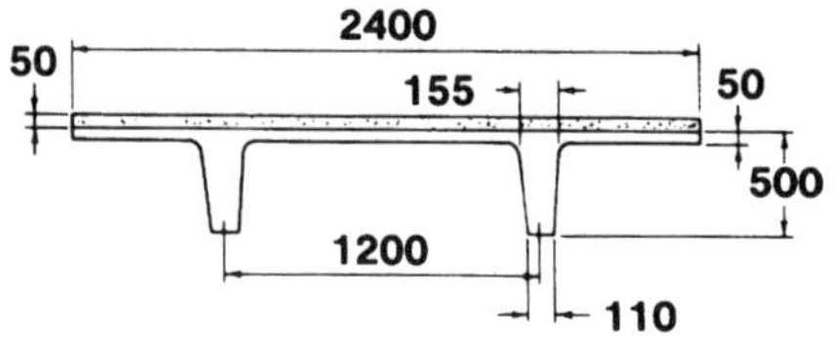


$f'_{ci}$ = 25 MPa
$f'_c$ = 35 MPa
$f_{pu}$ = 1860 MPa
$f'_{ctop}$ = 25 MPa

**Section Properties**

| | Untopped | Topped |
|---|---|---|
| $A_g$ = | 239 000 mm$^2$ | |
| I = | 5 390 × 10$^6$ mm$^4$ | 7 400 × 10$^6$ mm$^4$ |
| $y_b$ = | 357 mm | 407 mm |
| $y_t$ = | 143 mm | 143 mm |
| $S_b$ = | 15 100 × 10$^3$ mm$^3$ | 18 200 × 10$^3$ mm$^3$ |
| $S_t$ = | 37 600 × 10$^3$ mm$^3$ | 51 800 × 10$^3$ mm$^3$ |
| m = | 234 kg/m$^2$ | 357 kg/m$^2$ |
| w = | 2.3 kN/m$^2$ | 3.5 kN/m$^2$ |
| V/S = | 36 mm | |

**Table of superimposed load (kN/m$^2$) and cambers (mm)** — **No Topping**

| Strand Pattern | $e_e$† / $e_c$† | Span (m) 7 | 8 | 9 | 10 | 11 | 12 | 13 | 14 | 15 | 16 | 17 | 18 | 19 | 20 | 21 |
|---|---|---|---|---|---|---|---|---|---|---|---|---|---|---|---|---|
| 4-S | 282<br>282 | 10.5<br>10<br>10 | 7.6<br>10<br>10 | 5.6<br>10<br>10 | 4.2<br>10<br>20 | 3.1<br>10<br>10 | 2.4<br>10<br>10 | 1.7<br>10<br>10 | | | | | | | | |
| 6-S | 240<br>240 | | 10.8<br>20<br>20 | 8.1<br>20<br>20 | 6.2<br>20<br>20 | 4.8<br>20<br>20 | 3.8<br>20<br>20 | 2.9<br>20<br>20 | 2.3<br>20<br>20 | 1.7<br>20<br>10 | | | | | | |
| 6-D | 240<br>294 | | | 9.6<br>20<br>30 | 7.4<br>30<br>30 | 5.8<br>30<br>30 | 4.6<br>30<br>30 | 3.6<br>30<br>30 | 2.9<br>30<br>30 | 2.3<br>30<br>20 | 1.8<br>30<br>10 | | | | | |
| 8-D | 169<br>287 | | | | | 7.9<br>40<br>40 | 6.4<br>40<br>50 | 5.1<br>40<br>50 | 4.2<br>40<br>50 | 3.4<br>40<br>40 | 2.8<br>40<br>30 | 2.2<br>40<br>20 | 1.8<br>30<br>10 | | | |
| 10-D | 117<br>281 | | | | | | | 5.3<br>60<br>60 | 4.4<br>60<br>60 | 3.6<br>60<br>50 | 3.0<br>60<br>50 | 2.5<br>50<br>30 | 2.1<br>40<br>10 | 1.7<br>30<br>– 10 | | |
| 12-D | 90<br>275 | | | | | | | | | | | 3.0<br>70<br>60 | 2.5<br>70<br>40 | 2.1<br>60<br>20 | 1.7<br>50<br>– 10 | |

**Table of superimposed load (kN/m$^2$) and cambers (mm)** — **50 mm Normal Density Topping**

| Strand Pattern | $e_e$† / $e_c$† | Span (m) 8 | 9 | 10 | 11 | 12 | 13 | 14 | 15 | 16 | 17 | 18 | 19 |
|---|---|---|---|---|---|---|---|---|---|---|---|---|---|
| 4-S | 282<br>282 | 7.7<br>10<br>10 | 5.4<br>10<br>10 | 3.8<br>10<br>10 | 2.6<br>10<br>10 | 1.7<br>10<br>10 | | | | | | | |
| 6-S | 240<br>240 | | 8.3<br>20<br>20 | 6.2<br>20<br>20 | 4.7<br>20<br>20 | 3.5<br>20<br>20 | 2.5<br>20<br>10 | 1.8<br>20<br>10 | | | | | |
| 6-D | 240<br>294 | | | 7.4<br>30<br>30 | 5.6<br>30<br>30 | 4.3<br>30<br>30 | 3.2<br>30<br>20 | 2.4<br>30<br>20 | 1.7<br>30<br>10 | | | | |
| 8-D | 169<br>287 | | | | 8.0<br>40<br>40 | 6.3<br>40<br>40 | 4.9<br>40<br>40 | 3.9<br>40<br>40 | 3.0<br>40<br>30 | 2.3<br>40<br>20 | 1.7<br>40<br>0 | | |
| 10-D | 117<br>281 | | | | | | | 4.9<br>60<br>50 | 3.9<br>60<br>50 | 3.0<br>60<br>40 | 2.3<br>60<br>30 | 1.7<br>50<br>10 | |
| 12-D | 90<br>275 | | | | | | | | | | | 2.2<br>70<br>40 | 1.7<br>70<br>10 |

* Values below heavy solid line require transfer strengths higher than 25 MPa.
* Values to right of dashed line indicate that element is partially prestressed.
† $e_e$ and $e_c$ are the maximum eccentricities at end and midspan that will not violate the permissible concrete stresses at transfer.

# DOUBLE TEE
## 2400 × 600
## Normal Density Concrete

**Strand Pattern Designation**

4-S — No. of 13 mm strands; S = straight; D = depressed

Superimposed loads shown include superimposed dead load of 0.5 kN/m² for untopped elements and 0.7 kN/m² for topped elements. Remainder is live load. Long-time cambers include superimposed dead load but do not include live load.

**Key**

9.7 - Safe superimposed service load, kN/m²
10 - Estimated camber at erection, mm
10 - Estimated long time camber, mm

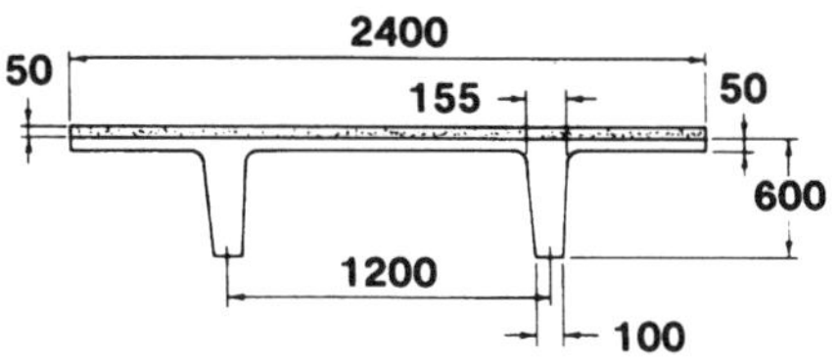


$f'_{ci}$ = 25 MPa
$f'_c$ = 35 MPa
$f_{pu}$ = 1860 MPa
$f'_{ctop}$ = 25 MPa

**Section Properties**

| | Untopped | Topped |
|---|---|---|
| $A_g$ = | 260 000 mm² | |
| I = | 8 580 × 10⁶ mm⁴ | 11 600 × 10⁶ mm⁴ |
| $y_b$ = | 424 mm | 480 mm |
| $y_t$ = | 176 mm | 170 mm |
| $S_b$ = | 20 200 × 10³ mm³ | 24 100 × 10³ mm³ |
| $S_t$ = | 48 800 × 10³ mm³ | 68 100 × 10³ mm³ |
| m = | 265 kg/m² | 387 kg/m² |
| w = | 2.5 kN/m² | 3.7 kN/m² |
| V/S = | 37 mm | |

**Table of superimposed load (kN/m²) and cambers (mm)** — **No Topping**

| Strand Pattern | $e_e$† <br> $e_c$† | 8 | 9 | 10 | 11 | 12 | 13 | 14 | 15 | 16 | 17 | 18 | 19 | 20 | 21 | 22 | 23 | 24 | 25 |
|---|---|---|---|---|---|---|---|---|---|---|---|---|---|---|---|---|---|---|---|
| 4-S | 349<br>349 | 9.7<br>10<br>10 | 7.2<br>10<br>10 | 5.5<br>10<br>10 | 4.2<br>10<br>10 | 3.2<br>10<br>10 | 2.4<br>10<br>10 | 1.8<br>10<br>0 | | | | | | | | | | | |
| 6-D | 324<br>361 | | | 9.4<br>20<br>30 | 7.4<br>20<br>30 | 5.9<br>30<br>30 | 4.7<br>30<br>30 | 3.8<br>30<br>30 | 3.1<br>30<br>30 | 2.5<br>30<br>20 | 1.9<br>20<br>10 | 1.5<br>20<br>0 | | | | | | | |
| 8-D | 224<br>354 | | | | | 8.2<br>30<br>40 | 6.7<br>40<br>40 | 5.5<br>40<br>40 | 4.5<br>40<br>40 | 3.8<br>40<br>40 | 3.1<br>40<br>30 | 2.5<br>40<br>20 | 2.1<br>30<br>10 | 1.7<br>20<br>0 | | | | | |
| 10-D | 174<br>348 | | | | | | | 7.0<br>50<br>60 | 5.9<br>50<br>60 | 4.9<br>50<br>60 | 4.1<br>50<br>60 | 3.5<br>50<br>50 | 2.9<br>50<br>40 | 2.4<br>50<br>30 | 2.0<br>40<br>10 | 1.7<br>30<br>– 20 | | | |
| 12-D | 132<br>342 | | | | | | | | | | | 4.3<br>70<br>70 | 3.7<br>70<br>60 | 3.1<br>70<br>50 | 2.6<br>60<br>40 | 2.2<br>50<br>20 | 1.9<br>40<br>– 10 | 1.6<br>30<br>– 40 | |
| 14-D | 102<br>335 | | | | | | | | | | | | | | | 2.6<br>70<br>50 | 2.2<br>70<br>20 | 1.9<br>50<br>0 | 1.6<br>40<br>– 40 |

Span (m) is given in the column headings 8–25.

**Table of superimposed load (kN/m²) and cambers (mm)** — **50 mm Normal Density Topping**

| Strand Pattern | $e_e$† <br> $e_c$† | 8 | 9 | 10 | 11 | 12 | 13 | 14 | 15 | 16 | 17 | 18 | 19 | 20 | 21 | 22 |
|---|---|---|---|---|---|---|---|---|---|---|---|---|---|---|---|---|
| 4-S | 349<br>349 | 9.7<br>10<br>10 | 7.1<br>10<br>10 | 5.1<br>10<br>10 | 3.7<br>10<br>10 | 2.6<br>10<br>10 | 1.7<br>10<br>10 | | | | | | | | | |
| 6-D | 324<br>361 | | | 9.4<br>20<br>20 | 7.2<br>20<br>30 | 5.6<br>30<br>30 | 4.3<br>30<br>30 | 3.3<br>30<br>20 | 2.5<br>30<br>20 | 1.8<br>30<br>10 | | | | | | |
| 8-D | 224<br>354 | | | | | 8.1<br>30<br>40 | 6.5<br>40<br>40 | 5.2<br>40<br>40 | 4.1<br>40<br>40 | 3.3<br>40<br>30 | 2.6<br>40<br>20 | 2.0<br>40<br>10 | | | | |
| 10-D | 174<br>348 | | | | | | | 6.8<br>50<br>50 | 5.6<br>50<br>50 | 4.5<br>50<br>50 | 3.7<br>50<br>40 | 3.0<br>50<br>30 | 2.3<br>50<br>20 | 1.7<br>50<br>10 | | |
| 12-D | 132<br>342 | | | | | | | | | | | 3.6<br>70<br>60 | 2.9<br>70<br>40 | 2.3<br>70<br>30 | 1.8<br>60<br>10 | |
| 14-D | 102<br>335 | | | | | | | | | | | | | | 2.2<br>80<br>40 | 1.7<br>70<br>10 |

Span (m) is given in the column headings 8–22.

* Values below heavy solid line require transfer strengths higher than 25 MPa.
* Values to right of dashed line indicate that element is partially prestressed.
† $e_e$ and $e_c$ are the maximum eccentricities at end and midspan that will not violate the permissible concrete stresses at transfer.

# DOUBLE TEE
## 3000 × 600
## Normal Density Concrete

**Strand Pattern Designation**

6-D — No. of 13 mm strands; S = straight; D = depressed

Superimposed loads shown include superimposed dead load of 0.5 kN/m² for untopped elements and 0.7 kN/m² for topped elements. Remainder is live load. Long-time cambers include superimposed dead load but do not include live load.

**Key**

9.4 - Superimposed (service) load, kN/m²
20 - Estimated camber at erection, mm
20 - Estimated long time camber, mm

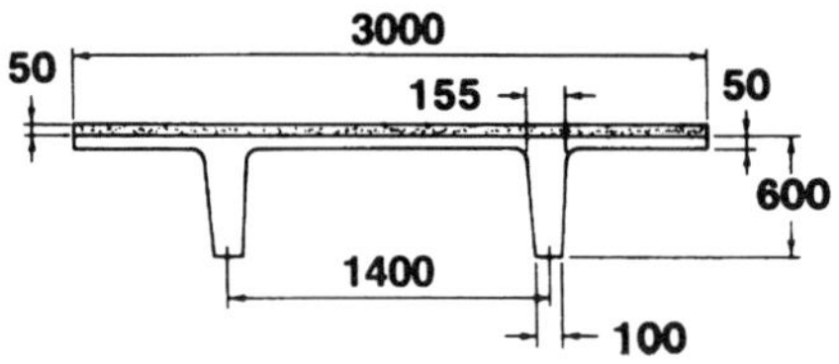


$f'_{ci}$ = 25 MPa
$f'_c$ = 35 MPa
$f_{pu}$ = 1860 MPa
$f'_{ctop}$ = 25 MPa

**Section Properties**

| | Untopped | Topped |
|---|---|---|
| $A_g$ = | 290 000 mm² | |
| I = | 9 200 × 10⁶ mm⁴ | 12 300 × 10⁶ mm⁴ |
| $y_b$ = | 440 mm | 496 mm |
| $y_t$ = | 160 mm | 154 mm |
| $S_b$ = | 20 900 × 10³ mm³ | 24 700 × 10³ mm³ |
| $S_t$ = | 57 400 × 10³ mm³ | 79 600 × 10³ mm³ |
| m = | 232 kg/m² | 352 kg/m² |
| w = | 2.3 kN/m² | 3.5 kN/m² |
| V/S = | 35 mm | |

### Table of superimposed load (kN/m²) and cambers (mm) — No Topping

| Strand Pattern | $e_e$† / $e_c$† | Span (m) 9 | 10 | 11 | 12 | 13 | 14 | 15 | 16 | 17 | 18 | 19 | 20 | 21 | 22 | 23 | 24 |
|---|---|---|---|---|---|---|---|---|---|---|---|---|---|---|---|---|---|
| 6-D | 340<br>377 | 9.4<br>20<br>20 | 7.3<br>20<br>20 | 5.7<br>20<br>30 | 4.5<br>20<br>30 | 3.8<br>30<br>30 | 2.8<br>30<br>20 | 2.2<br>20<br>20 | 1.7<br>20<br>10 | | | | | | | | |
| 8-D | 240<br>370 | | | 7.9<br>30<br>30 | 6.4<br>30<br>40 | 5.2<br>30<br>40 | 4.2<br>40<br>40 | 3.4<br>40<br>40 | 2.8<br>40<br>30 | 2.3<br>30<br>20 | 1.8<br>30<br>10 | | | | | | |
| 10-D | 180<br>364 | | | | | 6.5<br>40<br>50 | 5.4<br>50<br>50 | 4.5<br>50<br>50 | 3.7<br>50<br>50 | 3.1<br>50<br>40 | 2.5<br>50<br>30 | 2.1<br>40<br>20 | 1.7<br>30<br>0 | | | | |
| 12-D | 140<br>358 | | | | | | | | 4.6<br>60<br>70 | 3.8<br>60<br>60 | 3.2<br>60<br>60 | 2.7<br>60<br>50 | 2.3<br>50<br>30 | 1.9<br>50<br>10 | 1.6<br>40<br>− 10 | | |
| 14-D | 118<br>351 | | | | | | | | | | | | 2.8<br>70<br>60 | 2.4<br>70<br>40 | 2.0<br>60<br>20 | 1.7<br>50<br>− 10 | |
| 16-D | 96<br>345 | | | | | | | | | | | | | | | 1.9<br>70<br>30 | 1.6<br>60<br>0 |

### Table of superimposed load (kN/m²) and cambers (mm) — 50 mm Normal Density Topping

| Strand Pattern | $e_e$† / $e_c$† | Span (m) 9 | 10 | 11 | 12 | 13 | 14 | 15 | 16 | 17 | 18 | 19 | 20 |
|---|---|---|---|---|---|---|---|---|---|---|---|---|---|
| 6-D | 340<br>377 | 9.4<br>20<br>20 | 7.1<br>20<br>20 | 5.3<br>20<br>20 | 4.0<br>20<br>20 | 3.0<br>30<br>20 | 2.2<br>30<br>20 | 1.5<br>20<br>10 | | | | | |
| 8-D | 240<br>370 | | | 7.7<br>30<br>30 | 6.1<br>30<br>30 | 4.8<br>30<br>30 | 3.7<br>40<br>30 | 2.9<br>40<br>30 | 2.2<br>40<br>20 | 1.7<br>30<br>10 | | | |
| 10-D | 180<br>364 | | | | | 6.2<br>40<br>50 | 5.0<br>50<br>50 | 4.0<br>50<br>40 | 3.2<br>50<br>40 | 2.6<br>50<br>30 | 2.0<br>50<br>20 | | |
| 12-D | 140<br>358 | | | | | | | | 4.1<br>60<br>50 | 3.3<br>60<br>50 | 2.6<br>60<br>40 | 2.0<br>60<br>20 | |
| 14-D | 118<br>351 | | | | | | | | | | | | 1.9<br>70<br>30 |

* Values below heavy solid line require transfer strengths higher than 25 MPa.
* Values to right of dashed line indicate that element is partially prestressed.
† $e_e$ and $e_c$ are the maximum eccentricities at end and midspan that will not violate the permissible concrete stresses at transfer.

# DOUBLE TEE
## 2400 × 700
## Normal Density Concrete

**Strand Pattern Designation**

6-D
- No. of 13 mm strands
- S = straight; D = depressed

Superimposed loads shown include superimposed dead load of 0.5 kN/m$^2$ for untopped elements and 0.7 kN/m$^2$ for topped elements. Remainder is live load. Long-time cambers include superimposed dead load but do not include live load.

**Key**

9.0 - Superimposed (service) load, kN/m$^2$
20 - Estimated camber at erection, mm
20 - Estimated long time camber, mm

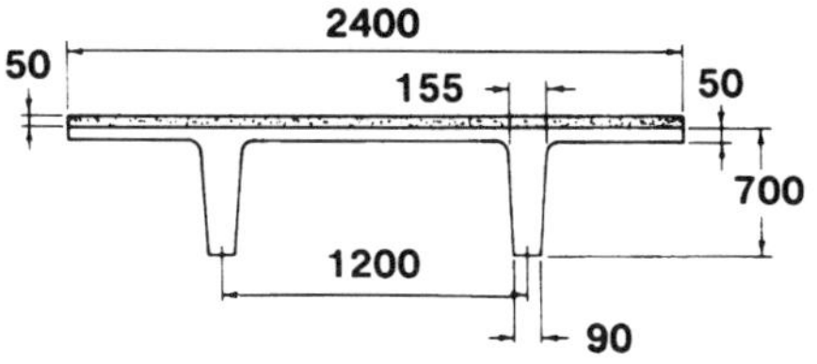


$f'_{ci}$ = 25 MPa
$f'_c$ = 35 MPa
$f_{pu}$ = 1860 MPa
$f'_{ctop}$ = 25 MPa

**Section Properties**

| | | Untopped | Topped |
|---|---|---|---|
| $A_g$ | = | 279 000 mm$^2$ | |
| I | = | 12 600 × 10$^6$ mm$^4$ | 16 600 × 10$^6$ mm$^4$ |
| $y_b$ | = | 492 mm | 554 mm |
| $y_t$ | = | 208 mm | 196 mm |
| $S_b$ | = | 25 500 × 10$^3$ mm$^3$ | 30 000 × 10$^3$ mm$^3$ |
| $S_t$ | = | 60 300 × 10$^3$ mm$^3$ | 84 800 × 10$^3$ mm$^3$ |
| m | = | 275 kg/m$^2$ | 398 kg/m$^2$ |
| w | = | 2.7 kN/m$^2$ | 3.9 kN/m$^2$ |
| V/S | = | 38 mm | |

**Table of superimposed load (kN/m$^2$) and cambers (mm)** — **No Topping**

| Strand Pattern | $e_e$† / $e_c$† | 11 | 12 | 13 | 14 | 15 | 16 | 17 | 18 | 19 | 20 | 21 | 22 | 23 | 24 | 25 | 26 | 27 | 28 |
|---|---|---|---|---|---|---|---|---|---|---|---|---|---|---|---|---|---|---|---|
| 6-D | 392<br>429 | 9.0<br>20<br>20 | 7.2<br>20<br>30 | 5.9<br>20<br>30 | 4.8<br>30<br>30 | 3.9<br>30<br>30 | 3.1<br>30<br>20 | 2.5<br>20<br>20 | 2.0<br>20<br>10 | 1.6<br>20<br>0 | | | | | | | | | |
| 8-D | 292<br>422 | | | 8.3<br>30<br>40 | 6.9<br>30<br>40 | 5.7<br>40<br>40 | 4.8<br>40<br>40 | 4.0<br>40<br>40 | 3.3<br>40<br>30 | 2.7<br>30<br>30 | 2.3<br>30<br>20 | 1.9<br>20<br>0 | 1.5<br>20<br>– 20 | | | | | | |
| 10-D | 222<br>416 | | | | | 7.3<br>50<br>50 | 6.2<br>50<br>60 | 5.2<br>50<br>50 | 4.4<br>50<br>50 | 3.7<br>50<br>50 | 3.2<br>50<br>40 | 2.7<br>40<br>30 | 2.2<br>40<br>10 | 1.9<br>30<br>– 10 | 1.5<br>20<br>– 30 | | | | |
| 12-D | 175<br>410 | | | | | | | | 5.5<br>60<br>70 | 4.7<br>70<br>70 | 4.0<br>60<br>60 | 3.4<br>60<br>50 | 2.9<br>60<br>40 | 2.5<br>50<br>30 | 2.1<br>40<br>10 | 1.8<br>30<br>– 20 | 1.5<br>20<br>– 50 | | |
| 14-D | 134<br>403 | | | | | | | | | | | | | 3.0<br>70<br>50 | 2.6<br>60<br>30 | 2.3<br>60<br>10 | 1.9<br>40<br>– 20 | 1.6<br>30<br>– 50 | |
| 16-D | 110<br>397 | | | | | | | | | | | | | | | | 2.2<br>70<br>20 | 1.9<br>60<br>– 10 | 1.6<br>40<br>– 50 |

**Table of superimposed load (kN/m$^2$) and cambers (mm)** — **50 mm Normal Density Topping**

| Strand Pattern | $e_e$† / $e_c$† | 11 | 12 | 13 | 14 | 15 | 16 | 17 | 18 | 19 | 20 | 21 | 22 | 23 | 24 | 25 |
|---|---|---|---|---|---|---|---|---|---|---|---|---|---|---|---|---|
| 6-D | 392<br>429 | 8.9<br>20<br>20 | 6.9<br>20<br>20 | 5.4<br>20<br>30 | 4.3<br>30<br>20 | 3.3<br>30<br>20 | 2.5<br>30<br>20 | 1.8<br>20<br>10 | | | | | | | | |
| 8-D | 292<br>422 | | | 8.1<br>30<br>40 | 6.6<br>30<br>40 | 5.3<br>40<br>40 | 4.3<br>40<br>30 | 3.4<br>40<br>30 | 2.7<br>40<br>20 | 2.1<br>30<br>10 | 1.6<br>30<br>0 | | | | | |
| 10-D | 222<br>416 | | | | | 7.0<br>50<br>50 | 5.8<br>50<br>50 | 4.8<br>50<br>50 | 3.9<br>50<br>40 | 3.2<br>50<br>30 | 2.6<br>50<br>20 | 2.0<br>40<br>10 | 1.6<br>40<br>– 10 | | | |
| 12-D | 175<br>410 | | | | | | | | 4.9<br>60<br>60 | 4.1<br>70<br>60 | 3.3<br>60<br>50 | 2.7<br>60<br>30 | 2.2<br>60<br>20 | 1.7<br>50<br>0 | 1.5<br>40<br>– 30 | |
| 14-D | 134<br>403 | | | | | | | | | | | | 2.6<br>80<br>40 | 2.1<br>70<br>30 | 1.7<br>60<br>0 | |
| 16-D | 110<br>397 | | | | | | | | | | | | | | | 1.7<br>80<br>10 |

* Values below heavy solid line require transfer strengths higher than 25 MPa.
* Values to right of dashed line indicate that element is partially prestressed.
† $e_e$ and $e_c$ are the maximum eccentricities at end and midspan that will not violate the permissible concrete stresses at transfer.

# DOUBLE TEE
## 3000 × 700
## Normal Density Concrete

**Strand Pattern Designation**

8-D
- No. of 13 mm strands
- S = straight; D = depressed

Superimposed loads shown include superimposed dead load of 0.5 kN/m$^2$ for untopped elements and 0.7 kN/m$^2$ for topped elements. Remainder is live load. Long-time cambers include superimposed dead load but do not include live load.

**Key**

8.0- Superimposed (service) load, kN/m$^2$
30- Estimated camber at erection, mm
30- Estimated long time camber, mm

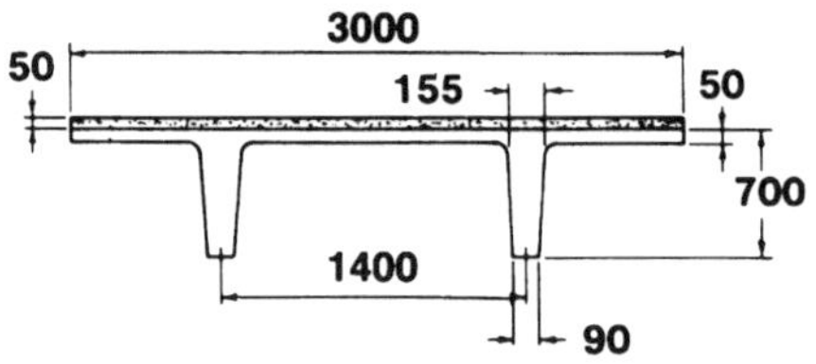


$f'_{ci}$ = 25 MPa
$f'_c$ = 35 MPa
$f_{pu}$ = 1860 MPa
$f'_{ctop}$ = 25 MPa

**Section Properties**

| | Untopped | Topped |
|---|---|---|
| $A_g$ = | 309 000 mm$^2$ | |
| I = | 13 500 × 10$^6$ mm$^4$ | 17 700 × 10$^6$ mm$^4$ |
| $y_b$ = | 510 mm | 572 mm |
| $y_t$ = | 190 mm | 178 mm |
| $S_b$ = | 26 400 × 10$^3$ mm$^3$ | 30 900 × 10$^3$ mm$^3$ |
| $S_t$ = | 70 800 × 10$^3$ mm$^3$ | 99 400 × 10$^3$ mm$^3$ |
| m = | 247 kg/m$^2$ | 367 kg/m$^2$ |
| w = | 2.4 kN/m$^2$ | 3.6 kN/m$^2$ |
| V/S = | 36 mm | |

**Table of superimposed load (kN/m$^2$) and cambers (mm)** — **No Topping**

| Strand Pattern | $e_e$ [†] / $e_c$ [†] | Span (m) 12 | 13 | 14 | 15 | 16 | 17 | 18 | 19 | 20 | 21 | 22 | 23 | 24 | 25 | 26 | 27 |
|---|---|---|---|---|---|---|---|---|---|---|---|---|---|---|---|---|---|
| 8-D | 322<br>440 | 8.0<br>30<br>30 | 6.5<br>30<br>40 | 5.4<br>30<br>40 | 4.4<br>30<br>40 | 3.7<br>40<br>40 | 3.0<br>40<br>30 | 2.5<br>30<br>30 | 2.0<br>30<br>20 | 1.7<br>30<br>10 | | | | | | | |
| 10-D | 240<br>434 | | | 6.8<br>40<br>50 | 5.7<br>40<br>50 | 4.8<br>50<br>50 | 4.0<br>50<br>50 | 3.4<br>50<br>40 | 2.8<br>50<br>40 | 2.4<br>40<br>30 | 2.0<br>40<br>20 | 1.6<br>30<br>0 | | | | | |
| 12-D | 184<br>428 | | | | | | | 4.2<br>60<br>60 | 3.6<br>60<br>60 | 3.0<br>60<br>50 | 2.6<br>60<br>40 | 2.2<br>50<br>20 | 1.8<br>40<br>10 | 1.5<br>30<br>– 20 | | | |
| 14-D | 152<br>421 | | | | | | | | | | 3.2<br>70<br>60 | 2.7<br>70<br>50 | 2.3<br>60<br>40 | 2.0<br>60<br>20 | 1.7<br>40<br>– 10 | | |
| 16-D | 122<br>415 | | | | | | | | | | | | | | 2.0<br>70<br>20 | 1.7<br>60<br>– 10 | 1.5<br>40<br>– 40 |
| 18-D | 104<br>409 | | | | | | | | | | | | | | | | 1.7<br>70<br>0 |

**Table of superimposed load (kN/m$^2$) and cambers (mm)** — **50 mm Normal Density Topping**

| Strand Pattern | $e_e$ [†] / $e_c$ [†] | Span (m) 10 | 11 | 12 | 13 | 14 | 15 | 16 | 17 | 18 | 19 | 20 | 21 | 22 | 23 |
|---|---|---|---|---|---|---|---|---|---|---|---|---|---|---|---|
| 6-D | 410<br>447 | 8.7<br>20<br>20 | 6.7<br>20<br>20 | 5.2<br>20<br>20 | 4.0<br>20<br>20 | 3.0<br>20<br>20 | 2.2<br>20<br>20 | 1.6<br>20<br>10 | | | | | | | |
| 8-D | 322<br>440 | | | 7.7<br>30<br>30 | 6.1<br>30<br>30 | 4.9<br>30<br>30 | 3.9<br>30<br>30 | 3.1<br>40<br>30 | 2.4<br>40<br>20 | 1.8<br>30<br>10 | | | | | |
| 10-D | 240<br>434 | | | | | 6.4<br>40<br>40 | 5.2<br>40<br>40 | 4.3<br>50<br>40 | 3.5<br>50<br>40 | 2.8<br>50<br>30 | 2.2<br>50<br>20 | 1.7<br>40<br>10 | | | |
| 12-D | 184<br>428 | | | | | | | | | 3.7<br>60<br>50 | 3.0<br>60<br>40 | 2.5<br>60<br>30 | 1.9<br>60<br>20 | 1.5<br>50<br>0 | |
| 14-D | 152<br>421 | | | | | | | | | | | | 2.4<br>70<br>.40 | 1.9<br>70<br>30 | |
| 16-D | 122<br>415 | | | | | | | | | | | | | | 1.9<br>80<br>30 |

* Values below heavy solid line require transfer strengths higher than 25 MPa.
* Values to right of dashed line indicate that element is partially prestressed.
[†] $e_e$ and $e_c$ are the maximum eccentricities at end and midspan that will not violate the permissible concrete stresses at transfer.

# DOUBLE TEE
## 2400 × 800
## Normal Density Concrete

**Strand Pattern Designation**

**12-D**
- **No. of 13 mm strands**
- **S = straight; D = depressed**

Superimposed loads shown include superimposed dead load of 0.5 kN/m² for untopped elements and 0.7 kN/m² for topped elements. Remainder is live load. Long-time cambers include superimposed dead load but do not include live load.

**Key**
- 10.5 - Superimposed (service) load, kN/m²
- 40 - Estimated camber at erection, mm
- 40 - Estimated long time camber, mm

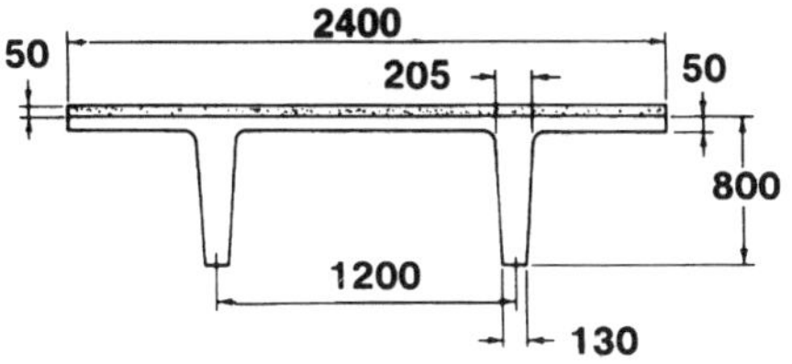


$f'_{ci}$ = 25 MPa
$f'_{c}$ = 35 MPa
$f_{pu}$ = 1860 MPa
$f'_{ctop}$ = 25 MPa

**Section Properties**

| | Untopped | Topped |
|---|---|---|
| $A_g$ = | 371 000 mm² | |
| I = | 22 800 × 10⁶ mm⁴ | 30 100 × 10⁶ mm⁴ |
| $y_b$ = | 523 mm | 588 mm |
| $y_t$ = | 277 mm | 262 mm |
| $S_b$ = | 43 700 × 10³ mm³ | 51 200 × 10³ mm³ |
| $S_t$ = | 82 500 × 10³ mm³ | 115 000 × 10³ mm³ |
| m = | 371 kg/m² | 491 kg/m² |
| w = | 3.6 kN/m² | 4.8 kN/m² |
| V/S = | 48 mm | |

### Table of superimposed load (kN/m²) and cambers (mm) — No Topping

| Strand Pattern | $e_e$† $e_c$† | 15 | 16 | 17 | 18 | 19 | 20 | 21 | 22 | 23 | 24 | 25 | 26 | 27 | 28 | 29 | 30 |
|---|---|---|---|---|---|---|---|---|---|---|---|---|---|---|---|---|---|
| 12-D | 331 | 10.5 | 8.9 | 7.5 | 6.4 | 5.4 | 4.6 | 3.9 | 3.3 | 2.8 | 2.3 | 1.9 | 1.5 | | | | |
| | 441 | 40 | 40 | 40 | 40 | 40 | 40 | 40 | 30 | 30 | 20 | 10 | 0 | | | | |
| | | 40 | 50 | 50 | 40 | 40 | 40 | 30 | 20 | 10 | – 10 | – 30 | – 50 | | | | |
| 14-D | 265 | | 10.5 | 9.0 | 7.7 | 6.6 | 5.7 | 4.9 | 4.2 | 3.6 | 3.0 | 2.6 | 2.2 | 1.8 | | | |
| | 434 | | 40 | 50 | 50 | 50 | 50 | 50 | 40 | 40 | 30 | 20 | 10 | 0 | | | |
| | | | 50 | 50 | 50 | 50 | 50 | 40 | 30 | 20 | 10 | – 10 | – 30 | – 50 | | | |
| 16-D | 216 | | | 10.1 | 8.7 | 7.5 | 6.5 | 5.6 | 4.9 | 4.2 | 3.6 | 3.1 | 2.6 | 2.2 | 1.9 | 1.6 | |
| | 428 | | | 50 | 50 | 60 | 60 | 60 | 50 | 50 | 40 | 40 | 30 | 20 | 0 | – 20 | |
| | | | | 60 | 60 | 60 | 60 | 60 | 50 | 40 | 30 | 10 | – 10 | – 30 | – 60 | – 90 | |
| 18-D | 184 | | | | | 8.2 | 7.2 | 6.2 | 5.4 | 4.7 | 4.1 | 3.5 | 3.0 | 2.6 | 2.2 | 1.9 | |
| | 422 | | | | | 60 | 60 | 60 | 60 | 60 | 60 | 50 | 40 | 30 | 20 | 0 | |
| | | | | | | 70 | 70 | 70 | 60 | 60 | 40 | 30 | 10 | – 10 | – 40 | – 70 | |
| 20-D | 153 | | | | | | | | | 5.1 | 4.4 | 3.9 | 3.3 | 2.9 | 2.5 | 2.1 | 1.8 |
| | 415 | | | | | | | | | 70 | 70 | 60 | 60 | 50 | 30 | 20 | 0 |
| | | | | | | | | | | 70 | 60 | 50 | 30 | 10 | – 20 | – 40 | – 80 |
| 22-D | 132 | | | | | | | | | | | 4.1 | 3.6 | 3.1 | 2.7 | 2.3 | 2.0 |
| | 409 | | | | | | | | | | | 80 | 70 | 60 | 50 | 40 | 20 |
| | | | | | | | | | | | | 60 | 50 | 30 | 10 | – 20 | – 50 |

Span (m) shown in column headings 15–30.

### Table of superimposed load (kN/m²) and cambers (mm) — 50 mm Normal Density Topping

| Strand Pattern | $e_e$† $e_c$† | 15 | 16 | 17 | 18 | 19 | 20 | 21 | 22 | 23 | 24 | 25 | 26 | 27 | 28 | 29 |
|---|---|---|---|---|---|---|---|---|---|---|---|---|---|---|---|---|
| 12-D | 331 | 10.3 | 8.6 | 7.1 | 5.9 | 4.9 | 4.1 | 3.3 | 2.7 | 2.1 | 1.6 | | | | | |
| | 441 | 40 | 40 | 40 | 40 | 40 | 40 | 40 | 30 | 30 | 20 | | | | | |
| | | 40 | 40 | 40 | 40 | 30 | 30 | 20 | 10 | – 10 | – 30 | | | | | |
| 14-D | 265 | | 10.3 | 8.7 | 7.3 | 6.2 | 5.2 | 4.4 | 3.6 | 3.0 | 2.4 | 1.8 | | | | |
| | 434 | | 40 | 50 | 50 | 50 | 50 | 50 | 40 | 40 | 30 | 20 | | | | |
| | | | 50 | 50 | 50 | 50 | 40 | 30 | 20 | 10 | – 10 | – 30 | | | | |
| 16-D | 216 | | | 10.1 | 8.6 | 7.3 | 6.2 | 5.2 | 4.4 | 3.6 | 3.0 | 2.4 | 1.8 | | | |
| | 428 | | | 50 | 50 | 60 | 60 | 60 | 50 | 50 | 40 | 40 | 30 | | | |
| | | | | 60 | 60 | 60 | 50 | 50 | 40 | 20 | 10 | – 10 | – 30 | | | |
| 18-D | 184 | | | | | 8.0 | 6.8 | 5.8 | 4.9 | 4.1 | 3.4 | 2.8 | 2.3 | 1.8 | | |
| | 422 | | | | | 60 | 60 | 60 | 60 | 60 | 60 | 50 | 40 | 30 | | |
| | | | | | | 70 | 60 | 50 | 50 | 40 | 30 | 10 | – 10 | – 40 | | |
| 20-D | 153 | | | | | | | | 5.3 | 4.5 | 3.8 | 3.2 | 2.7 | 2.2 | 1.7 | |
| | 415 | | | | | | | | 70 | 70 | 70 | 60 | 60 | 50 | 30 | |
| | | | | | | | | | 60 | 50 | 40 | 30 | 10 | – 20 | – 50 | |
| 22-D | 132 | | | | | | | | | | | 3.6 | 3.0 | 2.5 | 2.0 | 1.6 |
| | 409 | | | | | | | | | | | 80 | 70 | 60 | 50 | 40 |
| | | | | | | | | | | | | 40 | 20 | 0 | – 30 | – 60 |

Span (m) shown in column headings 15–29.

* Values below heavy solid line require transfer strengths higher than 25 MPa.
* Values to right of dashed line indicate that element is partially prestressed.
† $e_e$ and $e_c$ are the maximum eccentricities at end and midspan that will not violate the permissible concrete stresses at transfer.

# DOUBLE TEE
## 2400 × 900
## Normal Density Concrete

**Strand Pattern Designation**

12-D
- No. of 13 mm strands
- S = straight; D = depressed

Superimposed loads shown include superimposed dead load of 0.5 kN/m² for untopped elements and 0.7 kN/m² for topped elements. Remainder is live load. Long-time cambers include superimposed dead load but do not include live load.

**Key**
- 10.3 - Superimposed (service) load, kN/m²
- 30 - Estimated camber at erection, mm
- 40 - Estimated long time camber, mm

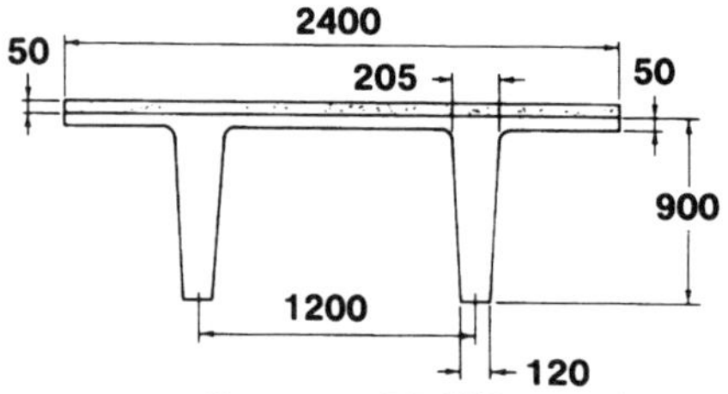


$f'_{ci}$ = 25 MPa
$f'_c$ = 35 MPa
$f_{pu}$ = 1860 MPa
$f'_{ctop}$ = 25 MPa

**Section Properties**

| | | Untopped | Topped |
|---|---|---|---|
| $A_g$ | = | 396 000 mm² | |
| I | = | 30 500 × 10⁶ mm⁴ | 39 800 × 10⁶ mm⁴ |
| $y_b$ | = | 587 mm | 656 mm |
| $y_t$ | = | 313 mm | 294 mm |
| $S_b$ | = | 52 000 × 10³ mm³ | 60 700 × 10³ mm³ |
| $S_t$ | = | 97 600 × 10³ mm³ | 135 300 × 10³ mm³ |
| m | = | 398 kg/m² | 520 kg/m² |
| w | = | 3.9 kN/m² | 5.1 kN/m² |
| V/S | = | 49 mm | |

### Table of superimposed load (kN/m²) and cambers (mm) — No Topping

| Strand Pattern | $e_e$† <br> $e_c$† | Span (m) 16 | 17 | 18 | 19 | 20 | 21 | 22 | 23 | 24 | 25 | 26 | 27 | 28 | 29 | 30 | 31 | 32 |
|---|---|---|---|---|---|---|---|---|---|---|---|---|---|---|---|---|---|---|
| 12-D | 403<br>505 | 10.3<br>30<br>40 | 8.8<br>40<br>40 | 7.5<br>40<br>40 | 6.4<br>40<br>40 | 5.4<br>40<br>40 | 4.6<br>40<br>30 | 4.0<br>30<br>30 | 3.3<br>30<br>20 | 2.8<br>20<br>10 | 2.3<br>20<br>– 10 | 1.9<br>10<br>– 20 | 1.6<br>0<br>– 40 | | | | | |
| 14-D | 322<br>498 | | 10.5<br>40<br>50 | 9.0<br>40<br>50 | 7.8<br>40<br>50 | 6.7<br>50<br>50 | 5.8<br>40<br>50 | 5.0<br>40<br>40 | 4.3<br>40<br>30 | 3.7<br>40<br>20 | 3.2<br>30<br>10 | 2.7<br>20<br>– 10 | 2.3<br>10<br>– 20 | 1.9<br>0<br>– 50 | 1.5<br>– 10<br>– 70 | | | |
| 16-D | 268<br>492 | | | 10.3<br>50<br>60 | 8.9<br>50<br>60 | 7.7<br>50<br>60 | 6.7<br>50<br>60 | 5.8<br>50<br>50 | 5.1<br>50<br>50 | 4.4<br>50<br>40 | 3.8<br>40<br>30 | 3.3<br>30<br>10 | 2.8<br>30<br>– 10 | 2.4<br>10<br>– 30 | 2.0<br>0<br>– 50 | 1.7<br>– 10<br>– 80 | | |
| 18-D | 225<br>486 | | | | 9.8<br>60<br>70 | 8.5<br>60<br>70 | 7.4<br>60<br>70 | 6.5<br>60<br>60 | 5.7<br>60<br>60 | 5.0<br>60<br>50 | 4.3<br>50<br>40 | 3.8<br>50<br>30 | 3.2<br>40<br>10 | 2.8<br>30<br>– 10 | 2.4<br>20<br>– 30 | 2.0<br>0<br>– 60 | 1.7<br>– 10<br>– 90 | |
| 20-D | 192<br>479 | | | | | | | | 6.2<br>70<br>70 | 5.4<br>70<br>70 | 4.7<br>60<br>60 | 4.1<br>60<br>50 | 3.6<br>50<br>30 | 3.1<br>40<br>10 | 2.7<br>30<br>– 10 | 2.3<br>20<br>– 40 | 2.0<br>0<br>– 70 | |
| 22-D | 164<br>473 | | | | | | | | | | 5.1<br>70<br>70 | 4.5<br>70<br>60 | 3.9<br>60<br>50 | 3.4<br>60<br>30 | 3.0<br>50<br>10 | 2.6<br>40<br>– 10 | 2.2<br>20<br>– 40 | 1.9<br>0<br>– 80 |

### Table of superimposed load (kN/m²) and cambers (mm) — 50 mm Normal Density Topping

| Strand Pattern | $e_e$† <br> $e_c$† | Span (m) 16 | 17 | 18 | 19 | 20 | 21 | 22 | 23 | 24 | 25 | 26 | 27 | 28 | 29 | 30 | 31 |
|---|---|---|---|---|---|---|---|---|---|---|---|---|---|---|---|---|---|
| 12-D | 403<br>505 | 10.0<br>30<br>40 | 8.4<br>40<br>40 | 7.0<br>40<br>40 | 5.9<br>40<br>40 | 4.9<br>40<br>30 | 4.1<br>40<br>30 | 3.3<br>30<br>20 | 2.7<br>30<br>10 | 2.1<br>20<br>– 10 | 1.7<br>20<br>– 20 | | | | | | |
| 14-D | 322<br>498 | | 10.2<br>40<br>50 | 8.7<br>40<br>50 | 7.4<br>40<br>50 | 6.2<br>40<br>40 | 5.3<br>40<br>40 | 4.4<br>40<br>30 | 3.7<br>40<br>20 | 3.1<br>30<br>10 | 2.5<br>30<br>– 10 | 2.0<br>20<br>– 30 | 1.6<br>10<br>– 50 | | | | |
| 16-D | 268<br>492 | | | | | 7.4<br>50<br>50 | 6.3<br>50<br>50 | 5.4<br>50<br>50 | 4.6<br>50<br>30 | 3.9<br>50<br>20 | 3.2<br>40<br>10 | 2.6<br>30<br>– 10 | 2.1<br>30<br>– 30 | | | | |
| 18-D | 225<br>486 | | | | | 8.5<br>60<br>60 | 7.3<br>60<br>60 | 6.2<br>60<br>60 | 5.3<br>60<br>50 | 4.5<br>60<br>40 | 3.8<br>50<br>30 | 3.1<br>50<br>10 | 2.5<br>40<br>– 10 | 2.0<br>30<br>– 30 | | | |
| 20-D | 192<br>479 | | | | | | | | 5.8<br>70<br>60 | 4.9<br>70<br>50 | 4.2<br>60<br>40 | 3.6<br>60<br>30 | 3.0<br>50<br>10 | 2.4<br>40<br>– 10 | 2.0<br>30<br>– 40 | 1.6<br>20<br>– 70 | |
| 22-D | 164<br>473 | | | | | | | | | | 4.6<br>70<br>60 | 3.9<br>70<br>40 | 3.3<br>60<br>30 | 2.8<br>60<br>10 | 2.3<br>50<br>– 20 | 1.9<br>40<br>– 50 | 1.6<br>20<br>– 80 |

* Values below heavy solid line require transfer strengths higher than 25 MPa.
* Values to right of dashed line indicate that element is partially prestressed.
† $e_e$ and $e_c$ are the maximum eccentricities at end and midspan that will not violate the permissible concrete stresses at transfer.

# DOUBLE TEE
## 2400 × 1000
### Normal Density Concrete

**Strand Pattern Designation**

12-D
- No. of 13 mm strands
- S = straight; D = depressed

Superimposed loads shown include superimposed dead load of 0.5 kN/m² for untopped elements and 0.7 kN/m² for topped elements. Remainder is live load. Long-time cambers include superimposed dead load but do not include live load.

**Key**

11.8 - Superimposed (service) load, kN/m²
30 - Estimated camber at erection, mm
40 - Estimated long time camber, mm

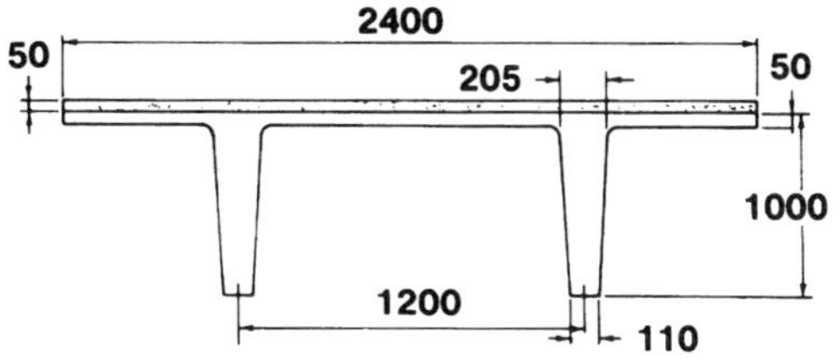


$f'_{ci}$ = 25 MPa
$f'_c$ = 35 MPa
$f_{pu}$ = 1860 MPa
$f'_{ctop}$ = 25 MPa

**Section Properties**

| | | Untopped | Topped |
|---|---|---|---|
| $A_g$ | = | 419 000 mm² | |
| I | = | 39 400 × 10⁶ mm⁴ | 50 700 × 10⁶ mm⁴ |
| $y_b$ | = | 652 mm | 725 mm |
| $y_t$ | = | 348 mm | 325 mm |
| $S_b$ | = | 60 400 × 10³ mm³ | 70 000 × 10³ mm³ |
| $S_t$ | = | 113 200 × 10³ mm³ | 156 000 × 10³ mm³ |
| m | = | 418 kg/m² | 540 kg/m² |
| w | = | 4.1 kN/m² | 5.3 kN/m² |
| V/S | = | 50 mm | |

## Table of superimposed load (kN/m²) and cambers (mm) — No Topping

| Strand Pattern | $e_e$† / $e_c$† | 16 | 17 | 18 | 19 | 20 | 21 | 22 | 23 | 24 | 25 | 26 | 27 | 28 | 29 | 30 | 31 | 32 | 33 |
|---|---|---|---|---|---|---|---|---|---|---|---|---|---|---|---|---|---|---|---|
| 12-D | 477 | 11.8 | 10.1 | 8.6 | 7.4 | 6.4 | 5.5 | 4.7 | 4.0 | 3.4 | 2.9 | 2.4 | 2.0 | 1.6 | | | | | |
| | 570 | 30 | 30 | 40 | 40 | 40 | 40 | 30 | 30 | 30 | 20 | 20 | 10 | 0 | | | | | |
| | | 40 | 40 | 40 | 40 | 40 | 40 | 30 | 30 | 20 | 10 | – 10 | – 20 | – 40 | | | | | |
| 14-D | 387 | | | 10.5 | 9.1 | 7.9 | 6.8 | 5.9 | 5.1 | 4.4 | 3.8 | 3.3 | 2.8 | 2.4 | 2.0 | 1.6 | | | |
| | 563 | | | 40 | 40 | 40 | 40 | 40 | 40 | 40 | 30 | 30 | 20 | 10 | 0 | – 10 | | | |
| | | | | 50 | 50 | 50 | 50 | 40 | 40 | 30 | 20 | 10 | 0 | – 20 | – 40 | – 60 | | | |
| 16-D | 320 | | | | 10.4 | 9.0 | 7.9 | 6.9 | 6.0 | 5.2 | 4.6 | 4.0 | 3.4 | 3.0 | 2.5 | 2.2 | 1.8 | | |
| | 557 | | | | 50 | 50 | 50 | 50 | 50 | 50 | 40 | 40 | 30 | 30 | 20 | 10 | – 10 | | |
| | | | | | 60 | 60 | 60 | 50 | 50 | 40 | 40 | 30 | 10 | 0 | – 20 | – 40 | – 70 | | |
| 18-D | 268 | | | | | 10.0 | 8.7 | 7.7 | 6.7 | 5.9 | 5.2 | 4.5 | 4.0 | 3.5 | 3.0 | 2.6 | 2.2 | 1.9 | |
| | 551 | | | | | 60 | 60 | 60 | 60 | 60 | 50 | 50 | 50 | 40 | 30 | 20 | 10 | – 10 | |
| | | | | | | 70 | 70 | 70 | 60 | 60 | 50 | 40 | 30 | 20 | 0 | – 20 | – 50 | – 70 | |
| 20-D | 232 | | | | | | | | 7.4 | 6.5 | 5.7 | 5.0 | 4.4 | 3.9 | 3.4 | 3.0 | 2.6 | 2.2 | 1.9 |
| | 544 | | | | | | | | 70 | 70 | 70 | 60 | 60 | 50 | 40 | 30 | 20 | 10 | – 10 |
| | | | | | | | | | 70 | 70 | 70 | 60 | 50 | 30 | 20 | 0 | – 20 | – 50 | – 80 |
| 22-D | 197 | | | | | | | | | | 6.1 | 5.4 | 4.8 | 4.2 | 3.7 | 3.2 | 2.8 | 2.4 | 2.1 |
| | 538 | | | | | | | | | | 70 | 70 | 70 | 60 | 60 | 50 | 40 | 30 | 10 |
| | | | | | | | | | | | 80 | 70 | 60 | 50 | 40 | 20 | 0 | – 30 | – 60 |

## Table of superimposed load (kN/m²) and cambers (mm) — 50 mm Normal Density Topping

| Strand Pattern | $e_e$† / $e_c$† | 17 | 18 | 19 | 20 | 21 | 22 | 23 | 24 | 25 | 26 | 27 | 28 | 29 | 30 | 31 | 32 |
|---|---|---|---|---|---|---|---|---|---|---|---|---|---|---|---|---|---|
| 12-D | 477 | 9.7 | 8.2 | 6.9 | 5.8 | 4.9 | 4.1 | 3.4 | 2.7 | 2.2 | 1.7 | | | | | | |
| | 570 | 30 | 40 | 40 | 40 | 40 | 30 | 30 | 30 | 20 | 20 | | | | | | |
| | | 40 | 40 | 40 | 30 | 30 | 20 | 20 | 10 | 0 | – 20 | | | | | | |
| 14-D | 387 | | | 8.6 | 7.4 | 6.3 | 5.3 | 4.5 | 3.8 | 3.2 | 2.6 | 2.1 | | | | | |
| | 563 | | | 40 | 40 | 40 | 40 | 40 | 40 | 30 | 30 | 20 | | | | | |
| | | | | 50 | 40 | 40 | 40 | 30 | 20 | 10 | 0 | – 20 | | | | | |
| 16-D | 320 | | | 10.1 | 8.7 | 7.5 | 6.5 | 5.5 | 4.7 | 4.0 | 3.4 | 2.8 | 2.3 | 1.8 | | | |
| | 557 | | | 50 | 50 | 50 | 50 | 50 | 50 | 40 | 40 | 30 | 30 | 20 | | | |
| | | | | 50 | 50 | 50 | 50 | 40 | 30 | 20 | 10 | 0 | – 20 | – 40 | | | |
| 18-D | 268 | | | | 10 | 8.7 | 7.5 | 6.5 | 5.6 | 4.8 | 4.1 | 3.4 | 2.8 | 2.3 | 1.8 | 1.6 | |
| | 551 | | | | 60 | 60 | 60 | 60 | 60 | 50 | 50 | 50 | 40 | 30 | 20 | 10 | |
| | | | | | 60 | 60 | 60 | 50 | 50 | 40 | 30 | 10 | 0 | – 20 | – 50 | – 70 | |
| 20-D | 232 | | | | | | 8.3 | 7.1 | 6.2 | 5.3 | 4.6 | 3.9 | 3.3 | 2.7 | 2.2 | 1.8 | |
| | 544 | | | | | | 70 | 70 | 70 | 70 | 60 | 60 | 50 | 40 | 30 | 20 | |
| | | | | | | | 70 | 70 | 60 | 50 | 40 | 30 | 10 | 0 | – 30 | – 50 | |
| 22-D | 197 | | | | | | | | | 5.8 | 5.0 | 4.3 | 3.7 | 3.1 | 2.6 | 2.2 | 1.8 |
| | 538 | | | | | | | | | 70 | 70 | 70 | 60 | 60 | 50 | 40 | 30 |
| | | | | | | | | | | 70 | 60 | 50 | 30 | 10 | – 10 | – 30 | – 60 |

* Values below heavy solid line require transfer strengths higher than 25 MPa.
* Values to right of dashed line indicate that element is partially prestressed.
† $e_e$ and $e_c$ are the maximum eccentricities at end and midspan that will not violate the permissible concrete stresses at transfer.

# DOUBLE TEE
## 3660 × 800
## Normal Density Concrete

**Strand Pattern Designation**

8-D
- 8 — **No. of 13 mm strands**
- D — **S = straight; D = depressed**

Superimposed loads shown include superimposed dead load of 0.5 kN/m$^2$ for untopped elements and 0.7 kN/m$^2$ for topped elements. Remainder is live load. Long-time cambers include superimposed dead load but do not include live load.

**Key**
- 10.7 - Superimposed (service) load, kN/m$^2$
- 10 - Estimated camber at erection, mm
- 20 - Estimated long time camber, mm

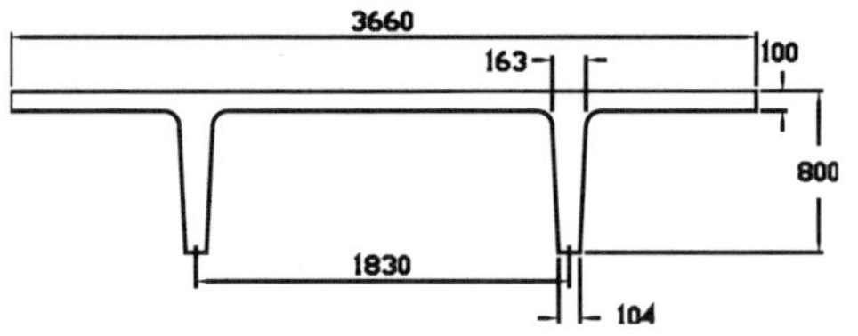


$f'_{ci}$ = **25 MPa**
$f'_c$ = **35 MPa**
$f_{pu}$ = **1860 MPa**

**Section Properties**

$A_g$ = 569 000 mm$^2$
I = 25 400 x 10$^6$ mm$^4$
$y_b$ = 623 mm
$y_t$ = 177 mm
$S_b$ = 40 800 x 10$^3$ mm$^3$
$S_t$ = 143 700 x 10$^3$ mm$^3$
m - 378 kg/m$^2$
w = 3.7 kN/M$^2$
V/S = 57 mm

### Table of superimposed load (kN/m$^2$) and cambers (mm) — No Topping

| Strand Pattern | $e_e$† / $e_c$† | 10 | 11 | 12 | 13 | 14 | 15 | 16 | 17 | 18 | 19 | 20 | 21 | 22 | 23 | 24 | 25 |
|---|---|---|---|---|---|---|---|---|---|---|---|---|---|---|---|---|---|
| | | Span (m) | | | | | | | | | | | | | | | |
| 8-D | 498 | 10.7 | 8.3 | 6.5 | 5.1 | 4.0 | 3.1 | 2.3 | 1.7 | | | | | | | | |
| | 553 | 10 | 20 | 20 | 20 | 20 | 20 | 20 | 10 | | | | | | | | |
| | | 20 | 20 | 20 | 20 | 20 | 20 | 10 | 10 | | | | | | | | |
| 10-D | 433 | | 10.9 | 8.7 | 7.0 | 5.6 | 4.5 | 3.6 | 2.8 | 2.2 | 1.7 | | | | | | |
| | 547 | | 20 | 20 | 20 | 20 | 30 | 20 | 20 | 20 | 20 | | | | | | |
| | | | 30 | 30 | 30 | 30 | 30 | 20 | 20 | 10 | 0 | | | | | | |
| 12-D | 347 | | | 10.7 | 8.6 | 7.0 | 5.7 | 4.7 | 3.8 | 3.1 | 2.4 | 1.9 | | | | | |
| | 541 | | | 30 | 30 | 30 | 30 | 30 | 30 | 30 | 30 | 20 | | | | | |
| | | | | 30 | 30 | 40 | 40 | 30 | 30 | 30 | 20 | 10 | | | | | |
| 14-D | 294 | | | | | 8.4 | 6.9 | 5.7 | 4.7 | 3.9 | 3.2 | 2.6 | 2.1 | 1.6 | | | |
| | 534 | | | | | 40 | 40 | 40 | 40 | 40 | 40 | 30 | 30 | 20 | | | |
| | | | | | | 40 | 40 | 40 | 40 | 40 | 30 | 20 | 10 | – 10 | | | |
| 16-D | 254 | | | | | | | | 5.6 | 4.7 | 3.9 | 3.2 | 2.6 | 2.1 | 1.7 | | |
| | 528 | | | | | | | | 50 | 50 | 50 | 40 | 40 | 30 | 20 | | |
| | | | | | | | | | 50 | 50 | 40 | 40 | 30 | 10 | – 10 | | |
| 18-D | 217 | | | | | | | | | | | 4.6 | 3.8 | 3.2 | 2.6 | 2.2 | 1.7 |
| | 522 | | | | | | | | | | | 60 | 50 | 50 | 40 | 40 | 30 |
| | | | | | | | | | | | | 60 | 50 | 40 | 30 | 10 | – 10 |

* Values below heavy solid line require transfer strengths higher than 25 MPa.
* Values to right of dashed line indicate that element is partially prestressed.
† $e_e$ and $e_c$ are the maximum eccentricities at end and midspan that will not violate the permissible concrete stresses at transfer.

# SINGLE TEE
## 3000 × 1200
## Normal Density Concrete

**Strand Pattern Designation**

14-D
- No. of 13 mm strands
- S = straight; D = depressed

Superimposed loads shown include superimposed dead load of 0.5 kN/m² for untopped elements and 0.7 kN/m² for topped elements. Remainder is live load. Long-time cambers include superimposed dead load but do not include live load.

**Key**

8.8 - Superimposed (service) load, kN/m²
40 - Estimated camber at erection, mm
50 - Estimated long time camber, mm

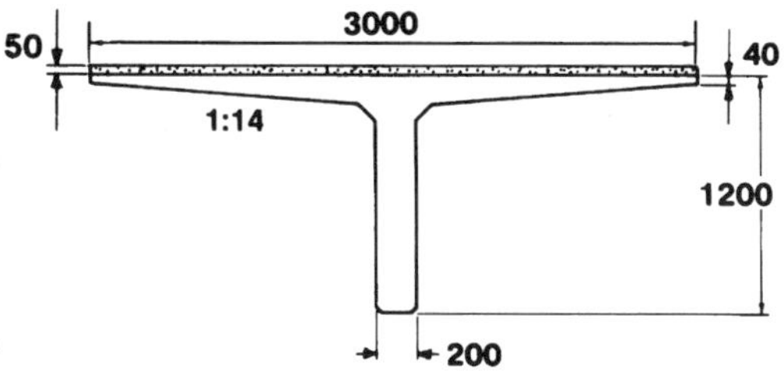


$f'_{ci}$ = 25 MPa
$f'_c$ = 35 MPa
$f_{pu}$ = 1860 Mpa
$f'_{ctop}$ = 25 MPa

**Section Properties**

| | | Untopped | Topped |
|---|---|---|---|
| $A_g$ | = | 492 000 mm² | |
| I | = | 66 300 × 10⁶ mm⁴ | 78 400 × 10⁶ mm⁴ |
| $y_b$ | = | 882 mm | 955 mm |
| $y_t$ | = | 318 mm | 295 mm |
| $S_b$ | = | 75 200 × 10³ mm³ | 82 000 × 10³ mm³ |
| $S_t$ | = | 208 000 × 10³ mm³ | 263 000 × 10³ mm³ |
| m | = | 398 kg/m² | 518 kg/m² |
| w | = | 3.9 kN/m² | 5.1 kN/m² |
| V/S | = | 60 mm | |

### Table of superimposed load (kN/m²) and cambers (mm) — No Topping

| Strand Pattern | $e_e$† <br> $e_c$† | 19 | 20 | 21 | 22 | 23 | 24 | 25 | 26 | 27 | 28 | 29 | 30 | 31 | 32 | 33 | 34 | 35 | 36 |
|---|---|---|---|---|---|---|---|---|---|---|---|---|---|---|---|---|---|---|---|
| 14-D | 510<br>793 | 8.8<br>40<br>50 | 7.7<br>40<br>50 | 6.7<br>40<br>50 | 5.8<br>40<br>50 | 5.0<br>40<br>40 | 4.4<br>40<br>40 | 3.8<br>40<br>40 | 3.2<br>40<br>30 | 2.8<br>30<br>20 | 2.4<br>30<br>10 | 2.0<br>20<br>0 | 1.6<br>10<br>– 20 | | | | | | |
| 16-D | 425<br>787 | 10.2<br>40<br>50 | 8.9<br>50<br>60 | 7.8<br>50<br>60 | 6.8<br>50<br>60 | 6.0<br>50<br>50 | 5.2<br>50<br>50 | 4.6<br>50<br>50 | 4.0<br>50<br>40 | 3.5<br>40<br>40 | 3.0<br>40<br>30 | 2.6<br>30<br>20 | 2.2<br>30<br>0 | 1.9<br>20<br>– 10 | 1.5<br>10<br>– 30 | | | | |
| 18-D | 359<br>781 | | | | 7.8<br>60<br>70 | 6.9<br>60<br>70 | 6.1<br>60<br>60 | 5.3<br>60<br>60 | 4.7<br>60<br>60 | 4.1<br>50<br>50 | 3.6<br>50<br>40 | 3.1<br>50<br>30 | 2.7<br>40<br>20 | 2.4<br>30<br>10 | 2.0<br>30<br>– 10 | 1.7<br>10<br>– 30 | | | |
| 20-D | 312<br>774 | | | | | | | | 5.4<br>70<br>70 | 4.8<br>60<br>70 | 4.2<br>60<br>60 | 3.7<br>60<br>50 | 3.3<br>50<br>40 | 2.9<br>50<br>30 | 2.5<br>40<br>10 | 2.1<br>30<br>– 10 | 1.8<br>20<br>– 30 | 1.6<br>10<br>– 60 | |
| 22-D | 272<br>768 | | | | | | | | | | 4.8<br>70<br>70 | 4.3<br>70<br>70 | 3.8<br>70<br>60 | 3.3<br>60<br>40 | 2.9<br>60<br>30 | 2.6<br>50<br>10 | 2.2<br>40<br>– 10 | 1.9<br>20<br>– 30 | 1.7<br>10<br>– 60 |
| 24-D | 240<br>762 | | | | | | | | | | | | | 3.8<br>80<br>60 | 3.4<br>70<br>50 | 3.0<br>60<br>30 | 2.6<br>50<br>20 | 2.3<br>40<br>– 10 | 2.0<br>30<br>– 30 |

Span (m): 19–36

### Table of superimposed load (kN/m²) and cambers (mm) — 50 mm Normal Density Topping

| Strand Pattern | $e_e$† <br> $e_c$† | 18 | 19 | 20 | 21 | 22 | 23 | 24 | 25 | 26 | 27 | 28 | 29 | 30 | 31 | 32 | 33 | 34 | 35 |
|---|---|---|---|---|---|---|---|---|---|---|---|---|---|---|---|---|---|---|---|
| 14-D | 510<br>793 | 9.8<br>40<br>40 | 8.4<br>40<br>40 | 7.2<br>40<br>40 | 6.1<br>40<br>40 | 5.2<br>40<br>40 | 4.4<br>40<br>40 | 3.7<br>40<br>30 | 3.1<br>40<br>30 | 2.6<br>40<br>20 | 2.1<br>30<br>10 | 1.6<br>30<br>0 | | | | | | | |
| 16-D | 425<br>787 | | 9.8<br>40<br>50 | 8.5<br>50<br>50 | 7.3<br>50<br>50 | 6.3<br>50<br>50 | 5.4<br>50<br>50 | 4.6<br>50<br>40 | 3.9<br>50<br>40 | 3.3<br>50<br>30 | 2.8<br>40<br>20 | 2.3<br>40<br>10 | 1.9<br>30<br>0 | 1.5<br>30<br>– 20 | | | | | |
| 18-D | 359<br>781 | | | | | 7.3<br>60<br>60 | 6.4<br>60<br>60 | 5.5<br>60<br>60 | 4.8<br>60<br>50 | 4.1<br>60<br>50 | 3.5<br>50<br>40 | 3.0<br>50<br>30 | 2.5<br>50<br>20 | 2.0<br>40<br>0 | 1.6<br>30<br>– 20 | | | | |
| 20-D | 312<br>774 | | | | | | | | | 4.8<br>70<br>60 | 4.2<br>60<br>50 | 3.6<br>60<br>40 | 3.1<br>60<br>30 | 2.6<br>50<br>20 | 2.2<br>50<br>0 | 1.8<br>40<br>– 10 | | | |
| 22-D | 272<br>768 | | | | | | | | | | | 4.2<br>70<br>60 | 3.7<br>70<br>50 | 3.2<br>70<br>40 | 2.7<br>60<br>20 | 2.3<br>60<br>10 | 1.9<br>50<br>– 10 | 1.5<br>40<br>– 40 | |
| 24-D | 240<br>762 | | | | | | | | | | | | | | 3.1<br>80<br>40 | 2.7<br>70<br>30 | 2.3<br>60<br>10 | 1.9<br>50<br>– 10 | 1.6<br>40<br>– 40 |

Span (m): 18–35

* Values below heavy solid line require transfer strengths higher than 25 MPa.
* Values to right of dashed line indicate that element is partially prestressed.
† $e_e$ and $e_c$ are the maximum eccentricities at end and midspan that will not violate the permissible concrete stresses at transfer.

# HOLLOW-CORE
## 1220 × 203
## Normal Density Concrete

**Strand Pattern Designation**

3-S: 3 = No. of 13 mm strands; S = straight

Superimposed loads shown include superimposed dead load of 0.5 kN/m$^2$ for untopped elements and 0.7 kN/m$^2$ for topped elements. Remainder is live load. Long-time cambers include superimposed dead load but do not include live load.

Capacity of sections of other configurations are similar. For precise values, see local hollow-core manufacturer.

**Key**

14.3 - Superimposed (service) load, kN/m$^2$
10 - Estimated camber at erection, mm
10 - Estimated long time camber, mm

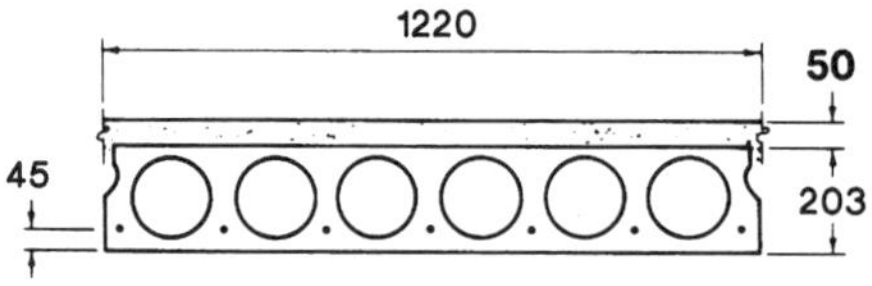


$f'_{ci}$ = 25 MPa
$f'_c$ = 40 MPa
$f_{pu}$ = 1860 MPa
$f'_{ctop}$ = 25 MPa

**Section Properties**

| | | Untopped | Topped |
|---|---|---|---|
| $A_g$ | = | 138 700 mm$^2$ | |
| I | = | 693 × 10$^6$ mm$^4$ | 1 276 × 10$^6$ mm$^4$ |
| $y_b$ | = | 101.5 mm | 134 mm |
| $y_t$ | = | 101.5 mm | 119 mm |
| $S_b$ | = | 6 800 × 10$^3$ mm$^3$ | 9 520 × 10$^3$ mm$^3$ |
| $S_t$ | = | 6 800 × 10$^3$ mm$^3$ | 10 745 × 10$^3$ mm$^3$ |
| m | = | 276 kg/m$^2$ | 393 kg/m$^2$ |
| w | = | 2.7 kN/m$^2$ | 3.9 kN/m$^2$ |
| V/S | = | 48 mm | |

### Table of superimposed load (kN/m$^2$) and cambers (mm) — No Topping

| Strand Pattern | $e_e$ / $e_c$ | Span (m) 4.5 | 5.0 | 5.5 | 6.0 | 6.5 | 7.0 | 7.5 | 8.0 | 8.5 | 9.0 | 9.5 | 10.0 |
|---|---|---|---|---|---|---|---|---|---|---|---|---|---|
| 3-S | 57 | 13.5<br>0<br>0 | 10.6<br>0<br>0 | 8.3<br>0<br>0 | 6.7<br>0<br>0 | 5.4<br>0<br>0 | 4.3<br>0<br>0 | 3.5<br>0<br>0 | 2.8<br>0<br>– 10 | 2.2<br>0<br>– 10 | 1.8<br>– 10<br>– 20 | | |
| 4-S | 57 | | 14.3<br>10<br>10 | 11.5<br>10<br>10 | 9.3<br>10<br>10 | 7.6<br>10<br>10 | 6.3<br>10<br>10 | 5.2<br>10<br>0 | 4.3<br>10<br>0 | 3.5<br>0<br>0 | 2.9<br>0<br>– 10 | 2.4<br>0<br>– 20 | 2.0<br>– 10<br>– 30 |
| 5-S | 57 | | | | 11.6<br>10<br>10 | 9.5<br>10<br>10 | 7.9<br>10<br>10 | 6.6<br>10<br>10 | 5.6<br>10<br>10 | 4.7<br>10<br>10 | 3.9<br>10<br>0 | 3.3<br>10<br>– 10 | 2.8<br>0<br>– 10 |
| 6-S | 57 | | | | | 11.2<br>20<br>20 | 9.3<br>20<br>20 | 7.9<br>20<br>20 | 6.6<br>20<br>20 | 5.6<br>20<br>10 | 4.8<br>20<br>10 | 4.1<br>10<br>10 | 3.5<br>10<br>0 |
| 7-S | 57 | | | | | | 10.5<br>20<br>30 | 8.9<br>20<br>30 | 7.5<br>20<br>30 | 6.4<br>20<br>20 | 5.5<br>20<br>20 | 4.7<br>20<br>20 | 4.0<br>20<br>20 |

### Table of superimposed load (kN/m$^2$) and cambers (mm) — 50 mm Normal Density Topping

| Strand Pattern | $e_e$ / $e_c$ | Span (m) 5.0 | 5.5 | 6.0 | 6.5 | 7.0 | 7.5 | 8.0 | 8.5 | 9.0 | 9.5 | 10.0 |
|---|---|---|---|---|---|---|---|---|---|---|---|---|
| 3-S | 57 | 13.2<br>0<br>0 | 10.4<br>0<br>0 | 8.2<br>0<br>0 | 6.5<br>0<br>0 | 5.2<br>0<br>0 | 4.1<br>0<br>– 10 | 3.2<br>0<br>– 10 | 2.5<br>0<br>– 20 | 1.9<br>– 10<br>– 30 | | |
| 4-S | 57 | | 14.2<br>10<br>10 | 11.4<br>10<br>10 | 9.3<br>10<br>10 | 7.6<br>10<br>0 | 6.2<br>10<br>0 | 5.1<br>10<br>0 | 4.1<br>0<br>– 10 | 3.3<br>0<br>– 20 | 2.7<br>0<br>– 20 | 2.1<br>– 10<br>– 40 |
| 5-S | 57 | | | 14.4<br>10<br>10 | 11.8<br>10<br>10 | 9.8<br>10<br>10 | 8.1<br>10<br>10 | 6.7<br>10<br>10 | 5.6<br>10<br>10 | 4.7<br>10<br>0 | 3.9<br>10<br>– 10 | 3.2<br>0<br>– 20 |
| 6-S | 57 | | | | | 11.8<br>20<br>20 | 9.9<br>20<br>20 | 8.3<br>20<br>10 | 7.0<br>20<br>10 | 5.9<br>20<br>10 | 5.0<br>10<br>0 | 4.2<br>10<br>– 10 |
| 7-S | 57 | | | | | | | 9.8<br>20<br>20 | 8.4<br>20<br>20 | 7.1<br>20<br>20 | 6.1<br>20<br>10 | 5.2<br>20<br>0 |

**Strand Pattern Designation**

- No. of 13 mm strands
- S = straight

5-S

Superimposed loads shown include superimposed dead load of 0.5 kN/m$^2$ for untopped elements and 0.7 kN/m$^2$ for topped elements. Remainder is live load. Long-time cambers include superimposed dead load but do not include live load.

Capacity of sections of other configurations are similar. For precise values, see local hollow-core manufacturer.

**Key**

14.8- Safe superimposed service load, kN/m$^2$
10- Estimated camber at erection, mm
10- Estimated long time camber, mm

# HOLLOW-CORE

## 1220 × 305

## Normal Density Concrete

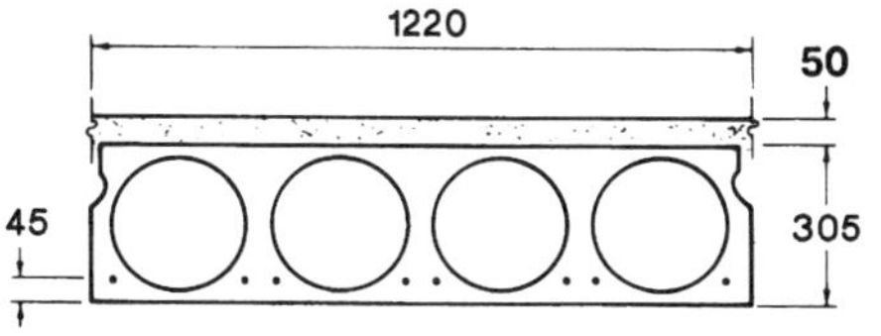


$f'_{ci}$ = 25 MPa
$f'_c$ = 40 MPa
$f_{pu}$ = 1860 MPa
$f'_{ctop}$ = 25 MPa

**Section Properties**

| | | Untopped | Topped |
|---|---|---|---|
| $A_g$ | = | 184 000 mm$^2$ | |
| I | = | 2177 × 10$^6$ mm$^4$ | 3 390 × 10$^6$ mm$^4$ |
| $y_b$ | = | 152.5 mm | 189 mm |
| $y_t$ | = | 152.5 mm | 165 mm |
| $S_b$ | = | 14 275 × 10$^3$ mm$^3$ | 17 900 × 10$^3$ mm$^3$ |
| $S_t$ | = | 14 275 × 10$^3$ mm$^3$ | 20 470 × 10$^3$ mm$^3$ |
| m | = | 363 kg/m$^2$ | 482 kg/m$^2$ |
| w | = | 3.55 kN/m$^2$ | 4.7 kN/m$^2$ |
| V/S | = | 60 mm | |

**Table of superimposed load (kN/m$^2$) and cambers (mm)** — **No Topping**

| Strand Pattern | $e_e$ / $e_c$ | 7.0 | 7.5 | 8.0 | 8.5 | 9.0 | 9.5 | 10.0 | 10.5 | 11.0 | 11.5 | 12.0 | 12.5 | 13.0 | 13.5 | 14.0 | 14.5 | 15.0 | 15.5 |
|---|---|---|---|---|---|---|---|---|---|---|---|---|---|---|---|---|---|---|---|
| | | Span (m) | | | | | | | | | | | | | | | | | |
| 5-S | 108 | 14.8 | 12.5 | 10.6 | 9.1 | 7.8 | 6.7 | 5.8 | 5.0 | 4.3 | 3.7 | 3.1 | 2.7 | 2.2 | 1.9 | | | | |
| | | 10 | 10 | 10 | 10 | 10 | 10 | 10 | 10 | 10 | 10 | 0 | 0 | 0 | – 10 | | | | |
| | | 10 | 10 | 10 | 10 | 10 | 10 | 10 | 10 | 0 | 0 | – 10 | – 10 | – 20 | – 30 | | | | |
| 6-S | 108 | | 15.2 | 13.0 | 11.2 | 9.7 | 8.4 | 7.3 | 6.3 | 5.5 | 4.8 | 4.2 | 3.6 | 3.1 | 2.7 | 2.3 | 2.0 | | |
| | | | 10 | 20 | 20 | 20 | 20 | 20 | 20 | 20 | 10 | 10 | 10 | 10 | 0 | 0 | – 10 | | |
| | | | 20 | 20 | 20 | 20 | 20 | 20 | 20 | 10 | 10 | 10 | 0 | – 10 | – 10 | – 20 | – 40 | | |
| 7-S | 108 | | | | 13.1 | 11.4 | 9.9 | 8.7 | 7.6 | 6.7 | 5.9 | 5.1 | 4.5 | 4.0 | 3.5 | 3.0 | 2.6 | 2.3 | |
| | | | | | 20 | 20 | 20 | 20 | 20 | 20 | 20 | 20 | 20 | 20 | 10 | 10 | 0 | 0 | |
| | | | | | 20 | 30 | 30 | 30 | 20 | 20 | 20 | 20 | 10 | 10 | 0 | – 10 | – 20 | – 30 | |
| 8-S | 108 | | | | | | 11.3 | 9.9 | 8.8 | 7.7 | 6.8 | 6.0 | 5.3 | 4.7 | 4.2 | 3.7 | 3.2 | 2.8 | |
| | | | | | | | 30 | 30 | 30 | 30 | 30 | 30 | 30 | 30 | 20 | 20 | 20 | 10 | |
| | | | | | | | 30 | 30 | 30 | 30 | 30 | 30 | 30 | 20 | 10 | 10 | 0 | – 10 | |
| 9-S | 108 | | | | | | | 11.1 | 9.8 | 8.7 | 7.7 | 6.8 | 6.1 | 5.4 | 4.8 | 4.3 | 3.8 | 3.3 | |
| | | | | | | | | 30 | 40 | 40 | 40 | 40 | 40 | 40 | 40 | 30 | 30 | 30 | |
| | | | | | | | | 40 | 40 | 40 | 40 | 40 | 40 | 30 | 30 | 20 | 20 | 10 | |
| 11-S | 108 | | | | | | | | 11.1 | 9.9 | 8.8 | 7.9 | 7.0 | 6.3 | 5.6 | 5.0 | 4.5 | 4.0 | 3.6 |
| | | | | | | | | | 40 | 50 | 50 | 50 | 50 | 50 | 50 | 50 | 50 | 40 | 40 |
| | | | | | | | | | 50 | 60 | 60 | 60 | 60 | 50 | 50 | 40 | 40 | 30 | 20 |

**Table of superimposed load (kN/m$^2$) and cambers (mm)** — **50 mm Normal Density Topping**

| Strand Pattern | $e_e$ / $e_c$ | 7.0 | 7.5 | 8.0 | 8.5 | 9.0 | 9.5 | 10.0 | 10.5 | 11.0 | 11.5 | 12.0 | 12.5 | 13.0 | 13.5 | 14.0 | 14.5 | 15.0 | 15.5 |
|---|---|---|---|---|---|---|---|---|---|---|---|---|---|---|---|---|---|---|---|
| | | Span (m) | | | | | | | | | | | | | | | | | |
| 5-S | 108 | 16.5 | 13.9 | 11.7 | 9.9 | 8.5 | 7.2 | 6.1 | 5.2 | 4.4 | 3.7 | 3.1 | 2.5 | 2.0 | | | | | |
| | | 10 | 10 | 10 | 10 | 10 | 10 | 10 | 10 | 10 | 10 | 0 | 0 | 0 | | | | | |
| | | 10 | 10 | 10 | 10 | 10 | 10 | 10 | 0 | 0 | – 10 | – 10 | – 20 | – 30 | | | | | |
| 6-S | 108 | | | 14.5 | 12.4 | 10.6 | 9.1 | 7.9 | 6.8 | 5.8 | 5.0 | 4.3 | 3.7 | 3.1 | 2.6 | | | | |
| | | | | 20 | 20 | 20 | 20 | 20 | 20 | 20 | 10 | 10 | 10 | 10 | 0 | | | | |
| | | | | 20 | 20 | 20 | 20 | 10 | 10 | 10 | 0 | 0 | – 10 | – 20 | – 30 | | | | |
| 7-S | 108 | | | | | | 11.0 | 9.6 | 8.3 | 7.2 | 6.3 | 5.5 | 4.7 | 4.1 | 3.5 | 3.0 | | | |
| | | | | | | | 20 | 20 | 20 | 20 | 20 | 20 | 20 | 20 | 10 | 10 | | | |
| | | | | | | | 20 | 20 | 20 | 20 | 10 | 10 | 0 | 0 | – 10 | – 20 | | | |
| 8-S | 108 | | | | | | | | 9.8 | 8.6 | 7.5 | 6.6 | 5.8 | 5.1 | 4.4 | 3.8 | | | |
| | | | | | | | | | 30 | 30 | 30 | 30 | 30 | 30 | 20 | 20 | | | |
| | | | | | | | | | 30 | 30 | 30 | 20 | 20 | 10 | 0 | 0 | | | |
| 9-S | 108 | | | | | | | | | | 8.7 | 7.7 | 6.8 | 6.0 | 5.3 | 4.6 | 4.1 | 3.5 | 3.1 |
| | | | | | | | | | | | 30 | 30 | 30 | 30 | 30 | 30 | 20 | 20 | 10 |
| | | | | | | | | | | | 30 | 30 | 20 | 20 | 10 | 0 | – 10 | – 20 | – 30 |
| 11-S | 108 | | | | | | | | | | 9.0 | 7.9 | 7.0 | 6.2 | 5.5 | 4.8 | 4.2 | 3.7 | 3.2 |
| | | | | | | | | | | | 50 | 50 | 50 | 50 | 50 | 50 | 50 | 40 | 40 |
| | | | | | | | | | | | 50 | 50 | 50 | 40 | 40 | 30 | 30 | 20 | 10 |

# SOLID FLAT SLAB

## 1220 × 100

## Normal Density Concrete

**Strand Pattern Designation**

- No. of 13 mm strands
- S = straight

3-S

Superimposed loads shown include superimposed dead load of 0.5 kN/m² for untopped elements and 0.7 kN/m² for topped elements. Remainder is live load. Long-time cambers include superimposed dead load but do not include live load.

**Key**

2.1 - Superimposed (service) load, kN/m²
−10 - Estimated camber at erection, mm
−10 - Estimated long time camber, mm

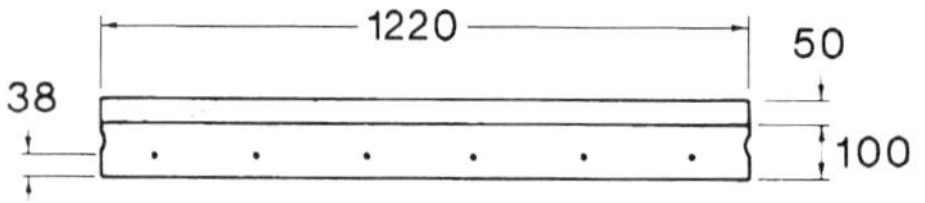


$f'_{ci}$ = 25 MPa
$f'_c$ = 35 MPa
$f_{pu}$ = 1860 MPa
$f'_{ctop}$ = 25 MPa

**Section Properties**

| | | Untopped | Topped |
|---|---|---|---|
| $A_g$ | = | 122 000 mm² | |
| I | = | 101.7 × 10⁶ mm⁴ | 316.0 × 10⁶ mm⁴ |
| $y_b$ | = | 50 mm | 72.3 mm |
| $y_t$ | = | 50 mm | 77.7 mm |
| $S_b$ | = | 2033 × 10³ mm³ | 4375 × 10³ mm³ |
| $S_t$ | = | 2033 × 10³ mm³ | 4069 × 10³ mm³ |
| m | = | 240 kg/m² | 362 kg/m² |
| w | = | 2.35 kN/m² | 3.55 kN/m² |
| V/S | = | 46 mm | |

### Table of superimposed load (kN/m²) and cambers (mm) — No Topping

| Strand Pattern | Span (m) 3 | 3.5 | 4 | 4.5 | 5 | 5.5 |
|---|---|---|---|---|---|---|
| 3-S | 8.5 | 5.8 | 4.2 | 3.0 | 2.1 | |
| | 0 | 0 | 0 | 0 | −10 | |
| | 0 | 0 | 0 | −10 | −10 | |
| 4-S | 10.4 | 7.3 | 5.5 | 3.9 | 2.8 | |
| | 0 | 0 | 0 | 0 | 0 | |
| | 0 | 0 | 0 | 0 | −10 | |
| 5-S | 12.1 | 8.6 | 6.2 | 4.5 | 3.3 | |
| | 0 | 0 | 0 | 0 | 0 | |
| | 0 | 0 | 0 | 0 | −10 | |
| 6-S | 13.3 | 9.0 | 6.6 | 4.9 | 3.6 | 2.7 |
| | 0 | 0 | 0 | 0 | 0 | 0 |
| | 0 | 0 | 0 | 0 | 0 | −10 |
| 7-S | | 9.3 | 7.0 | 5.2 | 3.9 | 2.9 |
| | | 0 | 0 | 10 | 10 | 0 |
| | | 10 | 10 | 0 | 0 | −10 |

### Table of superimposed load (kN/m²) and cambers (mm) — 50 mm Normal Density Topping

| Strand Pattern | Span (m) 3.5 | 4 | 4.5 | 5 |
|---|---|---|---|---|
| 3-S | 11.8 | 8.6 | 6.3 | |
| | 0 | 0 | −10 | |
| | 0 | −10 | −10 | |
| 4-S | | 11.4 | 8.5 | |
| | | 0 | 0 | |
| | | 0 | −10 | |
| 5-S | | | 10.1 | 7.7 |
| | | | 0 | 0 |
| | | | −10 | −10 |
| 6-S | | | 11.3 | 8.8 |
| | | | 0 | 0 |
| | | | 0 | −10 |
| 7-S | | | 12.1 | 9.8 |
| | | | 0 | 0 |
| | | | 0 | −10 |

# SOLID FLAT SLAB

## 1220 × 150
## Normal Density Concrete

**Strand Pattern Designation**

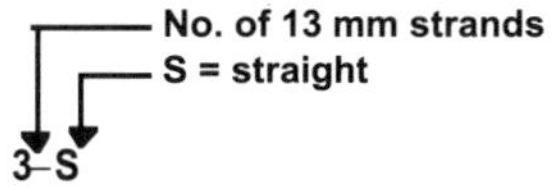


Superimposed loads shown include superimposed dead load of 0.5 kN/m$^2$ for untopped elements and 0.7 kN/m$^2$ for topped elements. Remainder is live load. Long-time cambers include superimposed dead load but do not include live load.

**Key**
- 2.0 - Superimposed (service) load, kN/m$^2$
- −10 - Estimated camber at erection, mm
- −10 - Estimated long time camber, mm

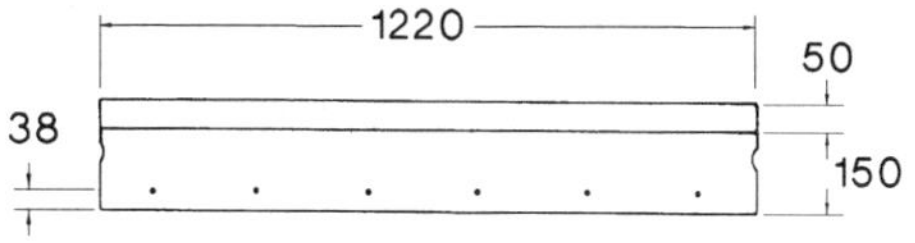


$f'_{ci}$ = 25 MPa
$f'_c$ = 35 MPa
$f_{pu}$ = 1860 MPa
$f'_{ctop}$ = 25 MPa

**Section Properties**

| | | Untopped | Topped |
|---|---|---|---|
| $A_g$ | = | 183 000 mm$^2$ | |
| I | = | 343.1 × 10$^6$ mm$^4$ | 756.0 × 10$^6$ mm$^4$ |
| $y_b$ | = | 75 mm | 97.0 mm |
| $y_t$ | = | 75 mm | 103.0 mm |
| $S_b$ | = | 4575 × 10$^3$ mm$^3$ | 7796 × 10$^3$ mm$^3$ |
| $S_t$ | = | 4575 × 10$^3$ mm$^3$ | 7339 × 10$^3$ mm$^3$ |
| m | = | 360 kg/m$^2$ | 482 kg/m$^2$ |
| w | = | 3.53 kN/m$^2$ | 4.73 kN/m$^2$ |
| V/S | = | 67 mm | |

### Table of superimposed load (kN/m$^2$) and cambers (mm) — No Topping

| Strand Pattern | Span (m) 3.5 | 4 | 4.5 | 5 | 5.5 | 6 | 6.5 | 7 | 7.5 | 8 | 8.5 |
|---|---|---|---|---|---|---|---|---|---|---|---|
| 3-S | 12.8 | 9.5 | 7.3 | 5.4 | 4.0 | 2.9 | 2.0 | | | | |
| | 0 | 0 | 0 | 0 | 0 | 0 | −10 | | | | |
| | 0 | 0 | 0 | 0 | 0 | −10 | −10 | | | | |
| 4-S | | 12.8 | 10.0 | 7.7 | 5.8 | 4.5 | 3.4 | 2.4 | | | |
| | | 0 | 0 | 0 | 0 | 0 | 0 | −10 | | | |
| | | 0 | 0 | 0 | 0 | 0 | −10 | −10 | | | |
| 5-S | | | 12.4 | 9.4 | 7.4 | 5.7 | 4.5 | 3.5 | 2.6 | | |
| | | | 10 | 10 | 10 | 10 | 0 | 0 | 0 | | |
| | | | 10 | 10 | 10 | 0 | 0 | −10 | −20 | | |
| 6-S | | | | 11.1 | 8.5 | 6.9 | 5.4 | 4.2 | 3.4 | 2.7 | |
| | | | | 10 | 10 | 10 | 10 | 10 | 0 | 0 | |
| | | | | 10 | 10 | 10 | 0 | 0 | −10 | −20 | |
| 7-S | | | | | 9.7 | 7.7 | 6.2 | 4.9 | 3.9 | 3.1 | 2.4 |
| | | | | | 10 | 10 | 10 | 10 | 10 | 0 | 0 |
| | | | | | 10 | 10 | 10 | 10 | 0 | −10 | −20 |

### Table of superimposed load (kN/m$^2$) and cambers (mm) — 50 mm Normal Density Topping

| Strand Pattern | Span (m) 4 | 4.5 | 5 | 5.5 | 6 | 6.5 | 7 | 7.5 | 8 |
|---|---|---|---|---|---|---|---|---|---|
| 3-S | 14.3 | 10.8 | 8.2 | 6.2 | 4.5 | 3.2 | | | |
| | 0 | 0 | 0 | 0 | 0 | −10 | | | |
| | 0 | 0 | 0 | −10 | −10 | −20 | | | |
| 4-S | | | 11.6 | 8.9 | 6.9 | 5.4 | 4.0 | | |
| | | | 0 | 0 | 0 | 0 | −10 | | |
| | | | 0 | 0 | −10 | −10 | −20 | | |
| 5-S | | | | 11.2 | 8.7 | 7.0 | 5.5 | | |
| | | | | 10 | 0 | 0 | 0 | | |
| | | | | 0 | 0 | −10 | −20 | | |
| 6-S | | | | 13.4 | 10.7 | 8.4 | 6.8 | 5.5 | |
| | | | | 10 | 10 | 10 | 0 | 0 | |
| | | | | 10 | 0 | 0 | −10 | −20 | |
| 7-S | | | | | 12.2 | 10.0 | 7.9 | 6.4 | 5.2 |
| | | | | | 10 | 10 | 10 | 0 | 0 |
| | | | | | 10 | 0 | 0 | −10 | −20 |

# SOLID FLAT SLAB

## 1220 × 200
## Normal Density Concrete

**Strand Pattern Designation**

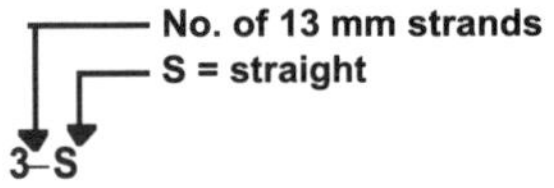


Superimposed loads shown include superimposed dead load of 0.5 kN/m² for untopped elements and 0.7 kN/m² for topped elements. Remainder is live load. Long-time cambers include superimposed dead load but do not include live load.

**Key**
2.5 - Superimposed (service) load, kN/m²
−10 - Estimated camber at erection, mm
−20 - Estimated long time camber, mm

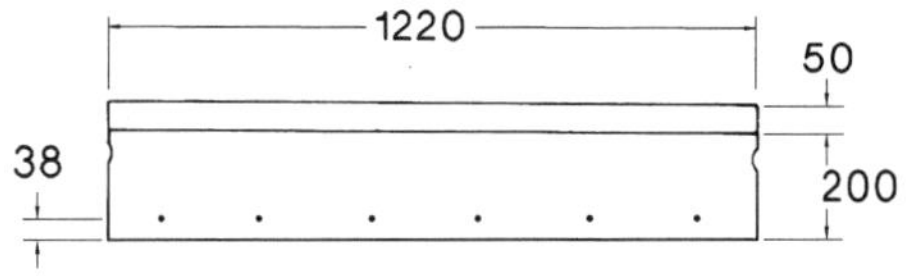


$f'_{ci}$ = **25 MPa**
$f'_c$ = **35 MPa**
$f_{pu}$ = **1860 MPa**
$f'_{ctop}$ = **25 MPa**

**Section Properties**

| | Untopped | Topped |
|---|---|---|
| $A_g$ = | 244 000 mm² | |
| I = | 813.3 × 10⁶ mm⁴ | 1489.0 × 10⁶ mm⁴ |
| $y_b$ = | 100 mm | 121.8 mm |
| $y_t$ = | 100 mm | 128.2 mm |
| $S_b$ = | 8133 × 10³ mm³ | 12 225 × 10³ mm³ |
| $S_t$ = | 8133 × 10³ mm³ | 11 615 × 10³ mm³ |
| m = | 480 kg/m² | 602 kg/m² |
| w = | 4.71 kN/m² | 5.91 kN/m² |
| V/S = | 86 mm | |

### Table of superimposed load (kN/m²) and cambers (mm) — No Topping

| Strand Pattern | Span (m) 4.5 | 5 | 5.5 | 6 | 6.5 | 7 | 7.5 | 8 | 8.5 | 9 | 9.5 | 10 |
|---|---|---|---|---|---|---|---|---|---|---|---|---|
| 3-S | 12.0 | 8.9 | 6.7 | 5.0 | 3.7 | 2.7 | | | | | | |
| | 0 | 0 | 0 | 0 | 0 | 0 | | | | | | |
| | 0 | 0 | 0 | 0 | −10 | −10 | | | | | | |
| 4-S | | 12.5 | 9.8 | 7.5 | 6.0 | 4.6 | 3.4 | 2.5 | | | | |
| | | 0 | 0 | 0 | 0 | 0 | 0 | −10 | | | | |
| | | 0 | 0 | 0 | 0 | −10 | −10 | −20 | | | | |
| 5-S | | | 12.6 | 9.9 | 7.9 | 6.2 | 4.8 | 3.9 | 3.0 | 2.2 | | |
| | | | 10 | 10 | 10 | 0 | 0 | 0 | 0 | −10 | | |
| | | | 0 | 0 | 0 | −10 | −10 | −10 | −10 | −20 | | |
| 6-S | | | | 12.1 | 9.8 | 7.9 | 6.2 | 5.0 | 4.0 | 3.3 | 2.5 | |
| | | | | 10 | 10 | 10 | 10 | 0 | 0 | 0 | −10 | |
| | | | | 10 | 10 | 10 | 0 | 0 | −10 | −10 | −20 | |
| 7-S | | | | | 11.3 | 9.4 | 7.5 | 6.2 | 5.1 | 4.2 | 3.3 | 2.6 |
| | | | | | 10 | 10 | 10 | 10 | 10 | 0 | 0 | −10 |
| | | | | | 10 | 10 | 10 | 10 | 0 | −10 | −10 | −20 |

### Table of superimposed load (kN/m²) and cambers (mm) — 50 mm Normal Density Topping

| Strand Pattern | Span (m) 5 | 5.5 | 6 | 6.5 | 7 | 7.5 | 8 | 8.5 | 9 |
|---|---|---|---|---|---|---|---|---|---|
| 3-S | 11.8 | 8.9 | 6.7 | 5.0 | 3.7 | 2.6 | | | |
| | 0 | 0 | 0 | 0 | −10 | −10 | | | |
| | 0 | 0 | 0 | −10 | −10 | −20 | | | |
| 4-S | | 12.8 | 9.9 | 7.8 | 6.2 | 4.6 | 3.6 | | |
| | | 0 | 0 | 0 | 0 | 0 | −10 | | |
| | | 0 | 0 | 0 | −10 | −10 | −20 | | |
| 5-S | | | 13.0 | 10.4 | 8.3 | 6.6 | 5.3 | 4.1 | |
| | | | 10 | 0 | 0 | 0 | 0 | −10 | |
| | | | 0 | 0 | 0 | −10 | −10 | −20 | |
| 6-S | | | | 12.7 | 10.3 | 8.5 | 6.9 | 5.6 | 4.4 |
| | | | | 10 | 10 | 0 | 0 | 0 | −10 |
| | | | | 10 | 0 | 0 | −10 | −10 | −20 |
| 7-S | | | | | 12.3 | 10.3 | 8.5 | 6.9 | 5.6 |
| | | | | | 10 | 10 | 10 | 0 | 0 |
| | | | | | 10 | 0 | 0 | −10 | −20 |

# RECTANGULAR BEAMS

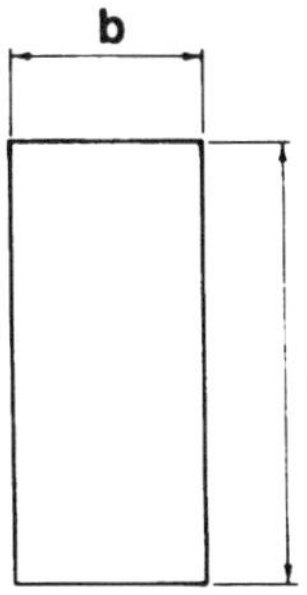


**Normal Density Concrete**

| Section Properties | | | | | | | |
|---|---|---|---|---|---|---|---|
| b (mm) | h (mm) | $A_g$ (mm$^2$) | I ($10^6$ mm$^4$) | $y_b$ (mm) | S ($10^3$ mm$^3$) | m (kg/m) | w (kN/m) |
| 300 | 400 | 120 000 | 1 600 | 200 | 8 000 | 288 | 2.8 |
| 300 | 500 | 150 000 | 3 120 | 250 | 12 500 | 360 | 3.5 |
| 300 | 600 | 180 000 | 5 400 | 300 | 18 000 | 432 | 4.2 |
| 300 | 700 | 210 000 | 8 575 | 350 | 24 500 | 504 | 4.9 |
| 300 | 800 | 240 000 | 12 800 | 400 | 32 000 | 576 | 5.6 |
| 300 | 900 | 270 000 | 18 200 | 450 | 40 500 | 648 | 6.4 |
| 400 | 600 | 240 000 | 7 200 | 300 | 24 000 | 576 | 5.6 |
| 400 | 700 | 280 000 | 11 400 | 350 | 32 700 | 672 | 6.6 |
| 400 | 800 | 320 000 | 17 100 | 400 | 42 700 | 768 | 7.5 |
| 400 | 900 | 360 000 | 24 300 | 450 | 54 000 | 864 | 8.5 |
| 400 | 1000 | 400 000 | 33 300 | 500 | 66 700 | 960 | 9.4 |

**$f'_{ci}$ = 30 MPa**
**$f'_c$ = 40 MPa**
**$f_{pu}$ = 1860 MPa**

**Key**

72.1 - Superimposed (service) load, kN/m
10 - Estimated camber at erection, mm
10 - Estimated long time camber, mm

*Superimposed loads shown include 50% dead load and 50% live load. 5.0 MPa top tension has been allowed, therefore additional top reinforcement is required.*

**Table of superimposed load (kN/m) and cambers (mm)**

| b (mm) | h (mm) | No. 13 mm strand | e (mm) | Span (m) 4 | 5 | 6 | 7 | 8 | 9 | 10 | 11 | 12 | 13 | 14 | 15 | 16 | 17 |
|---|---|---|---|---|---|---|---|---|---|---|---|---|---|---|---|---|---|
| **300** | **400** | **6** | **116** | 72.1<br>10<br>10 | 48.4<br>10<br>10 | 32.9<br>10<br>20 | 23.5<br>20<br>20 | 17.4<br>20<br>30 | | | | | | | | | |
| **300** | **500** | **7** | **157** | 119.9<br>10<br>10 | 75.6<br>10<br>10 | 51.5<br>20<br>20 | 37.0<br>20<br>20 | 27.6<br>20<br>30 | 21.1<br>20<br>30 | 16.5<br>30<br>40 | | | | | | | |
| **300** | **600** | **9** | **183** | | 114.0<br>10<br>10 | 78.0<br>10<br>10 | 56.3<br>10<br>20 | 42.2<br>20<br>20 | 32.5<br>20<br>30 | 25.6<br>20<br>30 | 20.5<br>30<br>40 | | | | | | |
| **300** | **700** | **10** | **215** | | 151.5<br>10<br>10 | 103.8<br>10<br>10 | 75.1<br>10<br>20 | 56.4<br>10<br>20 | 43.6<br>20<br>20 | 34.5<br>20<br>30 | 27.7<br>20<br>30 | 22.6<br>30<br>40 | 18.6<br>30<br>40 | | | | |
| **300** | **800** | **12** | **241** | | | 140.2<br>10<br>10 | 101.7<br>10<br>10 | 76.6<br>10<br>20 | 59.5<br>20<br>20 | 47.2<br>20<br>30 | 38.1<br>20<br>30 | 31.2<br>30<br>30 | 25.8<br>30<br>40 | 21.6<br>30<br>40 | 18.1<br>30<br>40 | | |
| **300** | **900** | **13** | **276** | | | 175.0<br>10<br>10 | 127.0<br>10<br>10 | 95.9<br>10<br>20 | 74.5<br>10<br>20 | 59.2<br>20<br>20 | 47.9<br>20<br>30 | 39.3<br>20<br>30 | 32.7<br>30<br>30 | 27.4<br>30<br>40 | 23.1<br>30<br>40 | 19.6<br>30<br>40 | |
| **400** | **600** | **12** | **183** | | 153.3<br>10<br>10 | 104.9<br>10<br>10 | 75.7<br>10<br>20 | 56.8<br>20<br>20 | 43.8<br>20<br>30 | 34.5<br>20<br>30 | 27.6<br>30<br>40 | 22.4<br>30<br>40 | 18.3<br>30<br>40 | | | | |
| **400** | **700** | **13** | **219** | | | 137.8<br>10<br>10 | 99.6<br>10<br>20 | 74.9<br>10<br>20 | 57.9<br>20<br>20 | 45.7<br>20<br>20 | 36.8<br>20<br>30 | 29.9<br>30<br>40 | 24.6<br>30<br>40 | 20.4<br>30<br>40 | 17.0<br>30<br>40 | | |
| **400** | **800** | **16** | **243** | | | | 136.8<br>10<br>10 | 103.1<br>10<br>20 | 80.0<br>20<br>20 | 63.5<br>20<br>30 | 51.3<br>20<br>30 | 42.0<br>30<br>30 | 34.7<br>30<br>40 | 29.0<br>30<br>40 | 24.4<br>30<br>40 | 20.6<br>40<br>50 | 17.4<br>40<br>50 |
| **400** | **900** | **18** | **275** | | | | 174.4<br>10<br>10 | 131.6<br>10<br>20 | 102.4<br>20<br>20 | 81.4<br>20<br>20 | 65.9<br>20<br>30 | 54.1<br>20<br>30 | 45.0<br>30<br>40 | 37.7<br>30<br>40 | 31.8<br>30<br>40 | 27.0<br>30<br>40 | 23.0<br>40<br>50 |
| **400** | **1000** | **20** | **305** | | | | | 163.4<br>10<br>20 | 127.3<br>10<br>20 | 101.4<br>20<br>20 | 82.3<br>20<br>30 | 67.8<br>20<br>30 | 56.5<br>20<br>30 | 47.5<br>30<br>40 | 40.2<br>30<br>40 | 34.3<br>30<br>40 | 29.4<br>30<br>40 |

*The above tables have been generated using an initial tension of 0.75 $f_{pu}$ resulting in fully prestressed elements with $M_r$ governing. Higher load capacities may be achieved by using more strands with a lower initial tension which may result in partially prestressed elements.*

# L-SHAPED BEAMS

**Normal Density Concrete**

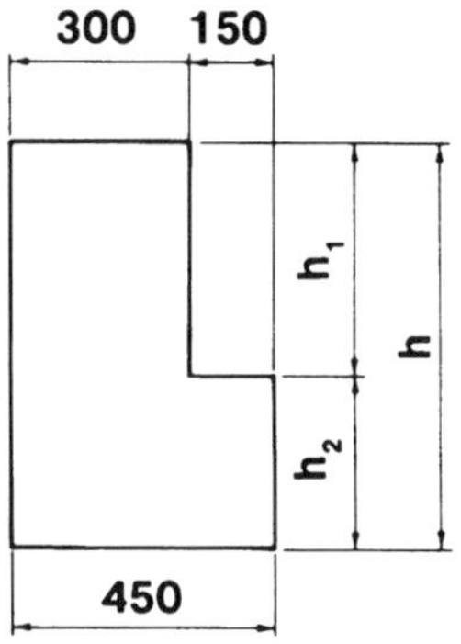


| Section Properties | | | | | | | | |
|---|---|---|---|---|---|---|---|---|
| h (mm) | $h_1/h_2$ (mm/mm) | $A_g$ ($mm^2$) | I ($10^6$ $mm^4$) | $y_b$ (mm) | $S_b$ ($10^3$ $mm^3$) | $S_t$ ($10^3$ $mm^3$) | m (kg/m) | w (kN/m) |
| 500 | 300/200 | 180 000 | 3 790 | 225 | 16 800 | 13 800 | 432 | 4.2 |
| 600 | 300/300 | 225 000 | 6 550 | 270 | 24 300 | 19 800 | 540 | 5.3 |
| 700 | 400/300 | 255 000 | 10 400 | 315 | 33 000 | 27 000 | 612 | 6.0 |
| 800 | 500/300 | 285 000 | 15 500 | 361 | 43 000 | 35 300 | 684 | 6.7 |
| 900 | 600/300 | 315 000 | 22 000 | 407 | 54 100 | 44 700 | 756 | 7.4 |
| 1 000 | 600/400 | 360 000 | 30 300 | 450 | 67 300 | 55 100 | 864 | 8.5 |
| 1 100 | 700/400 | 390 000 | 40 300 | 496 | 81 200 | 66 700 | 936 | 9.2 |
| 1 200 | 800/400 | 420 000 | 52 200 | 543 | 96 200 | 79 500 | 1 000 | 9.8 |
| 1 300 | 900/400 | 450 000 | 66 300 | 590 | 112 000 | 93 300 | 1 080 | 10.6 |
| 1 400 | 1 000/400 | 480 000 | 82 500 | 638 | 129 000 | 108 000 | 1 150 | 11.3 |
| 1 500 | 1 100/400 | 510 000 | 101 000 | 685 | 148 000 | 124 000 | 1 220 | 12.0 |

$f'_{ci}$ = **30 MPa**
$f'_c$ = **40 MPa**
$f_{pu}$ = **1860 MPa**

**Key**
122.0 - Superimposed (service) load, kN/m
10 - Estimated camber at erection, mm
10 - Estimated long time camber, mm

*Superimposed loads shown include 50% dead load and 50% live load. 5.0 MPa top tension has been allowed, therefore additional top reinforcement is required.*

**Table of superimposed load (kN/m) and cambers (mm)**

| h (mm) | No. 13 mm strand | e (mm) | Span (m) 4 | 5 | 6 | 7 | 8 | 9 | 10 | 11 | 12 | 13 | 14 | 15 | 16 | 17 |
|---|---|---|---|---|---|---|---|---|---|---|---|---|---|---|---|---|
| **500** | **8** | **143** | 137.8<br>0<br>10 | 86.8<br>10<br>10 | 59.1<br>10<br>10 | 42.4<br>10<br>20 | 31.5<br>20<br>20 | 24.1<br>20<br>30 | 18.8<br>20<br>30 | | | | | | | |
| **600** | **9** | **175** | | 122.0<br>10<br>10 | 83.2<br>10<br>10 | 59.8<br>10<br>10 | 44.7<br>10<br>20 | 34.3<br>20<br>20 | 26.8<br>20<br>20 | 21.3<br>20<br>30 | 17.1<br>20<br>30 | | | | | |
| **700** | **11** | **205** | | 172.4<br>10<br>10 | 118.0<br>10<br>10 | 85.2<br>10<br>10 | 63.9<br>10<br>20 | 49.4<br>10<br>20 | 38.9<br>20<br>20 | 31.2<br>20<br>30 | 25.3<br>20<br>30 | 20.8<br>20<br>30 | | | | |
| **800** | **12** | **244** | | | 153.5<br>10<br>10 | 111.1<br>10<br>10 | 83.6<br>10<br>10 | 64.8<br>10<br>20 | 57.3<br>20<br>20 | 41.3<br>20<br>20 | 33.7<br>20<br>30 | 27.8<br>20<br>30 | 23.1<br>20<br>30 | 19.3<br>20<br>30 | | |
| **900** | **14** | **274** | | | | 144.6<br>10<br>10 | 109.1<br>10<br>10 | 84.8<br>10<br>20 | 67.4<br>20<br>20 | 54.5<br>20<br>20 | 44.7<br>20<br>30 | 37.1<br>20<br>30 | 31.0<br>20<br>30 | 26.1<br>30<br>30 | 22.1<br>30<br>30 | 18.8<br>30<br>40 |
| **1000** | **16** | **296** | | | | | 136.1<br>10<br>10 | 105.9<br>10<br>20 | 84.3<br>10<br>20 | 68.3<br>20<br>20 | 56.1<br>20<br>20 | 46.6<br>20<br>30 | 39.1<br>20<br>30 | 33.1<br>20<br>30 | 28.1<br>30<br>30 | 24.0<br>30<br>30 |
| **1100** | **17** | **334** | | | | | 164.0<br>10<br>10 | 127.8<br>10<br>10 | 101.9<br>10<br>20 | 82.7<br>10<br>20 | 68.1<br>20<br>20 | 56.8<br>20<br>30 | 47.8<br>20<br>30 | 40.5<br>20<br>30 | 34.6<br>20<br>30 | 29.7<br>30<br>30 |
| **1200** | **19** | **361** | | | | | | 154.2<br>10<br>10 | 123.2<br>10<br>20 | 100.3<br>10<br>20 | 82.8<br>20<br>20 | 69.2<br>20<br>20 | 58.4<br>20<br>30 | 49.7<br>20<br>30 | 42.6<br>20<br>30 | 36.7<br>30<br>30 |
| **1300** | **21** | **392** | | | | | | | 147.0<br>10<br>20 | 119.8<br>10<br>20 | 99.1<br>20<br>20 | 83.0<br>20<br>20 | 70.2<br>20<br>30 | 59.9<br>20<br>30 | 51.4<br>20<br>30 | 44.5<br>30<br>30 |
| **1400** | **22** | **428** | | | | | | | 169.4<br>10<br>10 | 138.2<br>10<br>20 | 114.4<br>10<br>20 | 96.0<br>20<br>20 | 81.3<br>20<br>20 | 69.5<br>20<br>30 | 59.8<br>20<br>30 | 51.8<br>20<br>30 |
| **1500** | **24** | **457** | | | | | | | | 160.7<br>10<br>20 | 133.3<br>10<br>20 | 111.9<br>20<br>20 | 95.0<br>20<br>20 | 81.3<br>20<br>30 | 70.1<br>20<br>30 | 60.8<br>20<br>30 |

*The above tables have been generated using an initial tension of 0.75 $f_{pu}$ resulting in fully prestressed elements with $M_r$ governing. Higher load capacities may be achieved by using more strands with a lower initial tension which may result in partially prestressed elements.*

# L-SHAPED BEAMS

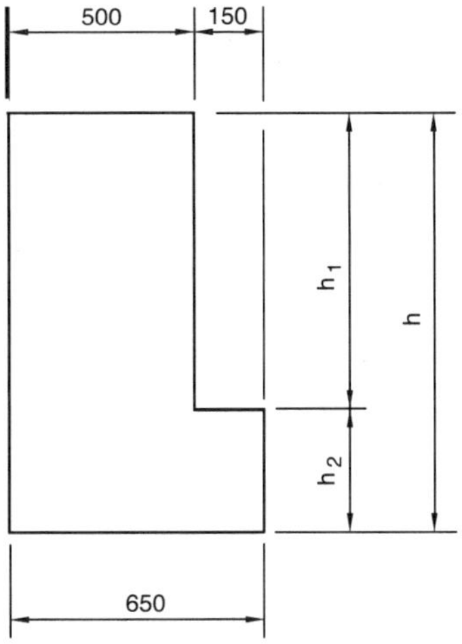


**Normal Density Concrete**

| Section Properties | | | | | | | | |
|---|---|---|---|---|---|---|---|---|
| h (mm) | $h_1/h_2$ (mm/mm) | $A_g$ ($mm^2$) | I ($10^6$ $mm^4$) | $y_b$ (mm) | $S_b$ ($10^3$ $mm^3$) | $S_t$ ($10^3$ $mm^3$) | m (kg/m) | w (kN/m) |
| 500 | 300/200 | 280 000 | 5 910 | 234 | 25 300 | 22 200 | 672 | 6.6 |
| 600 | 300/300 | 345 000 | 10 200 | 280 | 36 400 | 32 000 | 828 | 8.1 |
| 700 | 400/300 | 395 000 | 16 200 | 327 | 49 600 | 43 500 | 948 | 9.3 |
| 800 | 500/300 | 445 000 | 24 200 | 375 | 64 600 | 56 900 | 1 068 | 10.5 |
| 900 | 600/300 | 495 000 | 34 400 | 423 | 81 400 | 72 100 | 1 188 | 11.7 |
| 1 000 | 600/400 | 560 000 | 47 300 | 468 | 101 000 | 88 900 | 1 344 | 13.2 |
| 1 100 | 700/400 | 610 000 | 62 900 | 516 | 122 000 | 108 000 | 1 464 | 14.4 |
| 1 200 | 800/400 | 660 000 | 81 500 | 564 | 145 000 | 128 000 | 1 584 | 15.5 |
| 1 300 | 900/400 | 710 000 | 103 000 | 612 | 169 000 | 150 000 | 1 704 | 16.7 |
| 1 400 | 1 000/400 | 760 000 | 129 000 | 661 | 195 000 | 174 000 | 1 824 | 17.9 |
| 1 500 | 1 100/400 | 810 000 | 158 000 | 709 | 223 000 | 200 000 | 1 944 | 19.1 |

**$f'_{ci}$ = 30 MPa**
**$f'_c$ = 40 MPa**
**$f_{pu}$ = 1860 MPa**

**Key**
152.5 - Superimposed (service) load, kN/m
10 - Estimated camber at erection, mm
10 - Estimated long time camber, mm

*Superimposed loads shown include 50% dead load and 50% live load. 5.0 MPa top tension has been allowed, therefore additional top reinforcement is required.*

## Table of superimposed load (kN/m) and cambers (mm)

| h (mm) | No. 13 mm strand | e (mm) | Span (m) 4 | 5 | 6 | 7 | 8 | 9 | 10 | 11 | 12 | 13 | 14 | 15 | 16 | 17 |
|---|---|---|---|---|---|---|---|---|---|---|---|---|---|---|---|---|
| **500** | **8** | **184** | 176.1 | 110.5 | 74.9 | 53.4 | 39.4 | 29.9 | 23.0 | 18.0 | | | | | | |
| | | | 0 | 10 | 10 | 10 | 10 | 10 | 20 | 20 | | | | | | |
| | | | 10 | 10 | 10 | 10 | 20 | 20 | 20 | 20 | | | | | | |
| **600** | **9** | **230** | | 155.5 | 105.7 | 75.7 | 56.2 | 42.8 | 33.3 | 26.2 | 20.8 | 16.6 | | | | |
| | | | | 0 | 10 | 10 | 10 | 10 | 10 | 20 | 20 | 20 | | | | |
| | | | | 10 | 10 | 10 | 10 | 20 | 20 | 20 | 20 | 20 | | | | |
| **700** | **11** | **272** | | | 152.5 | 109.8 | 82.0 | 63.0 | 49.4 | 39.3 | 31.7 | 25.7 | 17.0 | | | |
| | | | | | 10 | 10 | 10 | 10 | 10 | 20 | 20 | 20 | 20 | | | |
| | | | | | 10 | 10 | 10 | 20 | 20 | 20 | 20 | 20 | 20 | | | |
| **800** | **13** | **309** | | | | 148.3 | 111.3 | 85.9 | 67.7 | 54.3 | 44.0 | 36.1 | 29.8 | 24.7 | 20.5 | 17.1 |
| | | | | | | 10 | 10 | 10 | 10 | 20 | 20 | 20 | 20 | 20 | 20 | 20 |
| | | | | | | 10 | 10 | 20 | 20 | 20 | 20 | 20 | 20 | 20 | 20 | 20 |
| **900** | **14** | **358** | | | | | 139.0 | 107.6 | 85.1 | 68.4 | 55.8 | 45.9 | 38.1 | 31.8 | 26.6 | 22.3 |
| | | | | | | | 10 | 10 | 10 | 10 | 20 | 20 | 20 | 20 | 20 | 20 |
| | | | | | | | 10 | 10 | 20 | 20 | 20 | 20 | 20 | 20 | 20 | 20 |
| **1000** | **17** | **376** | | | | | | 140.6 | 111.6 | 90.1 | 73.7 | 61.0 | 50.9 | 42.8 | 36.1 | 30.6 |
| | | | | | | | | 10 | 10 | 10 | 20 | 20 | 20 | 20 | 20 | 20 |
| | | | | | | | | 10 | 20 | 20 | 20 | 20 | 20 | 20 | 30 | 30 |
| **1100** | **18** | **427** | | | | | | 168.6 | 134.0 | 108.5 | 89.0 | 73.9 | 61.9 | 52.2 | 44.2 | 37.7 |
| | | | | | | | | 10 | 10 | 10 | 10 | 20 | 20 | 20 | 20 | 20 |
| | | | | | | | | 10 | 10 | 20 | 20 | 20 | 20 | 20 | 20 | 30 |
| **1200** | **20** | **461** | | | | | | | 162.5 | 131.8 | 108.5 | 90.3 | 75.9 | 64.3 | 54.7 | 46.8 |
| | | | | | | | | | 10 | 10 | 10 | 20 | 20 | 20 | 20 | 20 |
| | | | | | | | | | 10 | 20 | 20 | 20 | 20 | 20 | 20 | 30 |
| **1300** | **22** | **496** | | | | | | | | 157.3 | 129.7 | 108.2 | 91.2 | 77.4 | 66.2 | 56.9 |
| | | | | | | | | | | 10 | 10 | 10 | 20 | 20 | 20 | 20 |
| | | | | | | | | | | 20 | 20 | 20 | 20 | 20 | 20 | 30 |
| **1400** | **24** | **531** | | | | | | | | | 152.8 | 127.7 | 107.8 | 91.8 | 78.7 | 67.8 |
| | | | | | | | | | | | 10 | 10 | 20 | 20 | 20 | 20 |
| | | | | | | | | | | | 20 | 20 | 20 | 20 | 20 | 30 |
| **1500** | **25** | **581** | | | | | | | | | 174.2 | 145.9 | 123.3 | 105.2 | 90.3 | 77.9 |
| | | | | | | | | | | | 10 | 10 | 10 | 20 | 20 | 20 |
| | | | | | | | | | | | 20 | 20 | 20 | 20 | 20 | 20 |

*The above tables have been generated using an initial tension of 0.75 $f_{pu}$ resulting in fully prestressed elements with $M_r$ governing. Higher load capacities may be achieved by using more strands with a lower initial tension which may result in partially prestressed elements.*

# L-SHAPED BEAMS

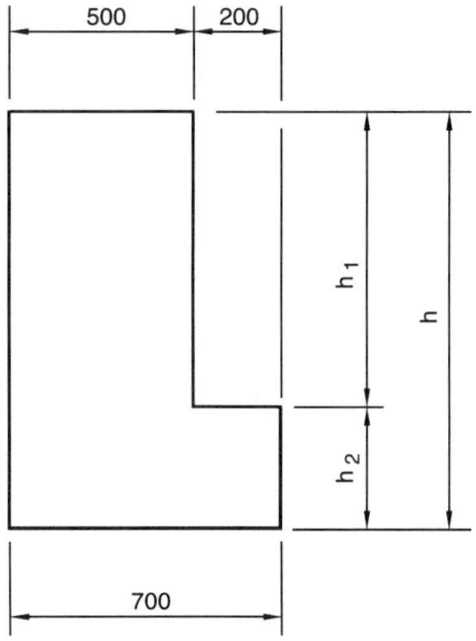


**Normal Density Concrete**

| Section Properties | | | | | | | | |
|---|---|---|---|---|---|---|---|---|
| h (mm) | $h_1/h_2$ (mm/mm) | $A_g$ (mm²) | I ($10^6$ mm⁴) | $y_b$ (mm) | $S_b$ ($10^3$ mm³) | $S_t$ ($10^3$ mm³) | m (kg/m) | w (kN/m) |
| 500 | 300/200 | 290 000 | 6 120 | 229 | 26 700 | 22 600 | 696 | 6.8 |
| 600 | 300/300 | 360 000 | 10 600 | 275 | 38 500 | 32 500 | 864 | 8.5 |
| 700 | 400/300 | 410 000 | 16 800 | 321 | 52 400 | 44 300 | 984 | 9.7 |
| 800 | 500/300 | 460 000 | 25 000 | 367 | 68 200 | 57 900 | 1 104 | 10.8 |
| 900 | 600/300 | 510 000 | 35 600 | 415 | 85 800 | 73 300 | 1 224 | 12.0 |
| 1 000 | 600/400 | 580 000 | 48 900 | 459 | 107 000 | 90 400 | 1 392 | 13.7 |
| 1 100 | 700/400 | 630 000 | 65 100 | 506 | 129 000 | 109 000 | 1 512 | 14.8 |
| 1 200 | 800/400 | 680 000 | 84 400 | 553 | 153 000 | 130 000 | 1 632 | 16.0 |
| 1 300 | 900/400 | 730 000 | 107 000 | 601 | 178 000 | 153 000 | 1 752 | 17.2 |
| 1 400 | 1 000/400 | 780 000 | 133 000 | 649 | 206 000 | 177 000 | 1 872 | 18.4 |
| 1 500 | 1 100/400 | 830 000 | 164 000 | 697 | 235 000 | 204 000 | 1 992 | 19.6 |

$f'_{ci}$ = **30 MPa**
$f'_c$ = **40 MPa**
$f_{pu}$ = **1860 MPa**

**Key**
155.1 - Superimposed (service) load, kN/m
10 - Estimated camber at erection, mm
10 - Estimated long time camber, mm

*Superimposed loads shown include 50% dead load and 50% live load. 5.0 MPa top tension has been allowed, therefore additional top reinforcement is required.*

**Table of superimposed load (kN/m) and cambers (mm)**

| h (mm) | No. 13 mm strand | e (mm) | Span (m) 4 | 5 | 6 | 7 | 8 | 9 | 10 | 11 | 12 | 13 | 14 | 15 | 16 | 17 |
|---|---|---|---|---|---|---|---|---|---|---|---|---|---|---|---|---|
| **500** | **8** | **179** | 175.9<br>0<br>0 | 110.3<br>10<br>10 | 74.7<br>10<br>10 | 53.2<br>10<br>10 | 39.3<br>10<br>20 | 29.7<br>10<br>20 | 22.9<br>10<br>20 | 17.8<br>20<br>20 | | | | | | |
| **600** | **9** | **225** | | 155.1<br>10<br>10 | 105.3<br>10<br>10 | 75.3<br>10<br>10 | 55.8<br>10<br>10 | 42.4<br>10<br>20 | 32.9<br>10<br>20 | 25.8<br>10<br>20 | 20.4<br>10<br>20 | 16.2<br>40<br>20 | | | | |
| **700** | **11** | **271** | | | 153.5<br>10<br>10 | 110.4<br>10<br>10 | 82.4<br>10<br>10 | 63.2<br>10<br>20 | 49.5<br>10<br>20 | 39.4<br>10<br>20 | 31.6<br>20<br>20 | 25.6<br>20<br>20 | 20.9<br>20<br>20 | 17.0<br>20<br>20 | | |
| **800** | **13** | **313** | | | | 151.0<br>10<br>10 | 113.3<br>10<br>10 | 87.4<br>10<br>10 | 68.9<br>10<br>20 | 55.2<br>10<br>20 | 44.8<br>20<br>20 | 36.7<br>20<br>20 | 30.3<br>20<br>20 | 25.1<br>20<br>20 | 20.8<br>20<br>20 | 17.3<br>20<br>20 |
| **900** | **14** | **361** | | | | | 141.0<br>10<br>10 | 109.1<br>10<br>10 | 86.3<br>10<br>20 | 69.4<br>10<br>20 | 56.5<br>20<br>20 | 46.5<br>20<br>20 | 38.6<br>20<br>20 | 32.2<br>20<br>20 | 26.9<br>20<br>20 | 22.6<br>20<br>20 |
| **1000** | **17** | **379** | | | | | | 142.6<br>10<br>10 | 113.1<br>10<br>10 | 91.2<br>10<br>20 | 74.6<br>10<br>20 | 61.7<br>20<br>20 | 51.5<br>20<br>20 | 43.2<br>20<br>20 | 36.4<br>20<br>20 | 30.8<br>20<br>20 |
| **1100** | **18** | **425** | | | | | | 170.0<br>10<br>10 | 135.1<br>10<br>10 | 109.3<br>10<br>20 | 89.7<br>10<br>20 | 74.4<br>10<br>20 | 62.2<br>20<br>20 | 52.4<br>20<br>20 | 44.4<br>20<br>20 | 37.8<br>20<br>20 |
| **1200** | **20** | **463** | | | | | | | 164.5<br>10<br>10 | 133.4<br>10<br>20 | 109.7<br>10<br>20 | 91.3<br>10<br>20 | 76.7<br>20<br>20 | 64.9<br>20<br>20 | 55.2<br>20<br>20 | 47.2<br>20<br>20 |
| **1300** | **22** | **498** | | | | | | | | 159.2<br>10<br>10 | 131.3<br>10<br>20 | 109.5<br>10<br>20 | 92.2<br>20<br>20 | 78.3<br>20<br>20 | 66.9<br>20<br>20 | 57.4<br>20<br>20 |
| **1400** | **24** | **536** | | | | | | | | | 155.0<br>10<br>20 | 129.6<br>10<br>20 | 109.4<br>10<br>20 | 93.1<br>20<br>20 | 79.7<br>20<br>20 | 68.7<br>20<br>20 |
| **1500** | **25** | **587** | | | | | | | | | 176.8<br>10<br>10 | 148.0<br>10<br>20 | 125.1<br>10<br>20 | 106.6<br>20<br>20 | 91.5<br>20<br>20 | 79.0<br>20<br>20 |

*The above tables have been generated using an initial tension of 0.75 $f_{pu}$ resulting in fully prestressed elements with $M_r$ governing. Higher load capacities may be achieved by using more strands with a lower initial tension which may result in partially prestressed elements.*

# INVERTED TEE BEAMS

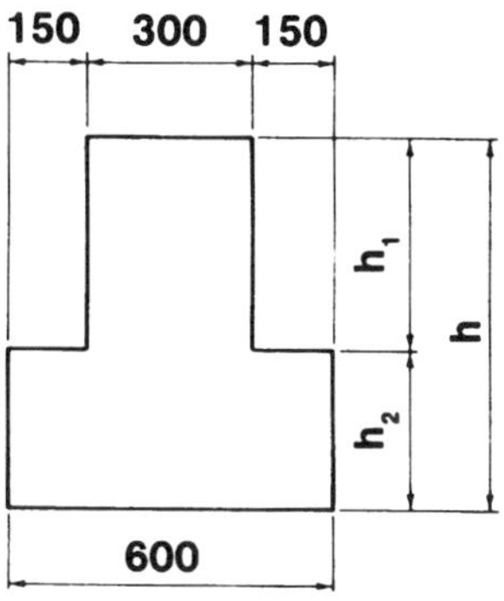


$f'_{ci}$ = 30 MPa
$f'_c$ = 40 MPa
$f_{pu}$ = 1860 MPa

**Normal Density Concrete**

| Section Properties | | | | | | | | |
|---|---|---|---|---|---|---|---|---|
| h (mm) | $h_1/h_2$ (mm/mm) | $A_g$ ($mm^2$) | I ($10^6$ $mm^4$) | $y_b$ (mm) | $S_b$ ($10^3$ $mm^3$) | $S_t$ ($10^3$ $mm^3$) | m (kg/m) | w (kN/m) |
| 500 | 300/200 | 210 000 | 4 290 | 207 | 20 700 | 14 600 | 504 | 4.9 |
| 600 | 300/300 | 270 000 | 7 430 | 250 | 29 700 | 21 200 | 648 | 6.4 |
| 700 | 400/300 | 300 000 | 11 800 | 290 | 40 600 | 28 700 | 720 | 7.1 |
| 800 | 500/300 | 330 000 | 17 600 | 332 | 52 900 | 37 500 | 792 | 7.8 |
| 900 | 600/300 | 360 000 | 25 000 | 375 | 66 600 | 47 600 | 864 | 8.5 |
| 1 000 | 600/400 | 420 000 | 34 300 | 414 | 82 800 | 58 600 | 1 000 | 9.8 |
| 1 100 | 700/400 | 450 000 | 45 700 | 457 | 100 000 | 71 000 | 1 080 | 10.6 |
| 1 200 | 800/400 | 480 000 | 59 200 | 500 | 118 000 | 84 600 | 1 150 | 11.3 |
| 1 300 | 900/400 | 510 000 | 75 100 | 544 | 138 000 | 99 400 | 1 220 | 12.0 |
| 1 400 | 1 000/400 | 540 000 | 93 500 | 589 | 159 000 | 115 000 | 1 300 | 12.8 |
| 1 500 | 1 100/400 | 570 000 | 115 000 | 634 | 181 000 | 132 000 | 1 370 | 13.4 |

**Key**
132.1 - Superimposed (service) load, kN/m
10 - Estimated camber at erection, mm
10 - Estimated long time camber, mm

*Superimposed loads shown include 50% dead load and 50% live load. 5.0 MPa top tension has been allowed, therefore additional top reinforcement is required.*

**Table of superimposed load (kN/m) and cambers (mm)**

| h (mm) | No. 13 mm strand | e (mm) | Span (m) 4 | 5 | 6 | 7 | 8 | 9 | 10 | 11 | 12 | 13 | 14 | 15 | 16 | 17 |
|---|---|---|---|---|---|---|---|---|---|---|---|---|---|---|---|---|
| 500 | 8 | 144 | 149.3 | 93.9 | 63.8 | 45.7 | 33.9 | 25.9 | 20.1 | 15.8 | | | | | | |
| | | | 0 | 10 | 10 | 10 | 10 | 20 | 20 | 20 | | | | | | |
| | | | 10 | 10 | 10 | 20 | 20 | 20 | 20 | 30 | | | | | | |
| 600 | 9 | 177 | | 132.1 | 90.0 | 64.5 | 48.0 | 36.7 | 28.6 | 22.6 | 18.0 | | | | | |
| | | | | 10 | 10 | 10 | 10 | 10 | 20 | 20 | 20 | | | | | |
| | | | | 10 | 10 | 10 | 20 | 20 | 20 | 20 | 20 | | | | | |
| 700 | 11 | 203 | | | 128.8 | 92.9 | 69.6 | 53.6 | 42.2 | 33.7 | 27.3 | 22.3 | 18.3 | | | |
| | | | | | 10 | 10 | 10 | 10 | 20 | 20 | 20 | 20 | 20 | | | |
| | | | | | 10 | 10 | 10 | 20 | 20 | 20 | 20 | 30 | 30 | | | |
| 800 | 13 | 231 | | | 171.6 | 124.1 | 93.3 | 72.2 | 57.1 | 46.0 | 37.5 | 30.9 | 25.6 | 21.4 | | |
| | | | | | 10 | 10 | 10 | 10 | 10 | 20 | 20 | 20 | 20 | 20 | | |
| | | | | | 10 | 10 | 10 | 20 | 20 | 20 | 20 | 30 | 30 | 30 | | |
| 900 | 14 | 271 | | | | 158.2 | 119.3 | 92.6 | 73.5 | 59.4 | 48.6 | 40.3 | 33.7 | 28.3 | 23.9 | 20.3 |
| | | | | | | 10 | 10 | 10 | 10 | 20 | 20 | 20 | 20 | 20 | 20 | 20 |
| | | | | | | 10 | 10 | 20 | 20 | 20 | 20 | 30 | 30 | 30 | 30 | 30 |
| 1 000 | 17 | 284 | | | | | 152.8 | 118.8 | 94.5 | 76.5 | 62.9 | 52.2 | 43.8 | 37.0 | 31.4 | 26.8 |
| | | | | | | | 10 | 10 | 10 | 20 | 20 | 20 | 20 | 20 | 20 | 20 |
| | | | | | | | 10 | 10 | 20 | 20 | 20 | 30 | 30 | 30 | 30 | 30 |
| 1 100 | 18 | 320 | | | | | | 142.8 | 113.8 | 92.3 | 76.0 | 63.3 | 53.2 | 45.1 | 38.5 | 33.0 |
| | | | | | | | | 10 | 10 | 10 | 20 | 20 | 20 | 20 | 20 | 20 |
| | | | | | | | | 10 | 20 | 20 | 20 | 20 | 30 | 30 | 30 | 30 |
| 1 200 | 20 | 352 | | | | | | 174.2 | 139.1 | 113.2 | 93.4 | 78.1 | 65.9 | 56.0 | 48.0 | 41.3 |
| | | | | | | | | 10 | 10 | 10 | 20 | 20 | 20 | 20 | 20 | 20 |
| | | | | | | | | 10 | 20 | 20 | 20 | 20 | 30 | 30 | 30 | 30 |
| 1 300 | 22 | 382 | | | | | | | 166.9 | 136.0 | 112.5 | 94.2 | 79.7 | 68.0 | 58.4 | 50.5 |
| | | | | | | | | | 10 | 10 | 20 | 20 | 20 | 20 | 20 | 20 |
| | | | | | | | | | 10 | 20 | 20 | 20 | 30 | 30 | 30 | 30 |
| 1 400 | 24 | 411 | | | | | | | | 158.4 | 131.2 | 110.0 | 93.2 | 79.7 | 68.6 | 59.4 |
| | | | | | | | | | | 10 | 10 | 20 | 20 | 20 | 20 | 20 |
| | | | | | | | | | | 20 | 20 | 20 | 20 | 30 | 30 | 30 |
| 1 500 | 25 | 450 | | | | | | | | | 149.7 | 125.7 | 106.7 | 91.3 | 78.8 | 68.4 |
| | | | | | | | | | | | 10 | 20 | 20 | 20 | 20 | 20 |
| | | | | | | | | | | | 20 | 20 | 20 | 30 | 30 | 30 |

*The above tables have been generated using an initial tension of 0.75 $f_{pu}$ resulting in fully prestressed elements with $M_r$ governing. Higher load capacities may be achieved by using more strands with a lower initial tension which may result in partially prestressed elements.*

# INVERTED TEE BEAMS

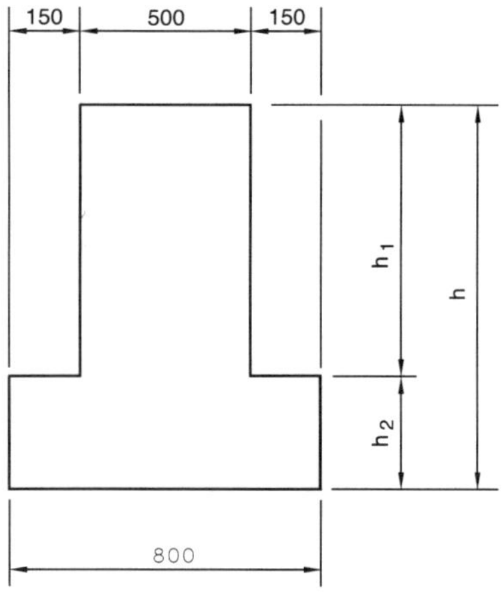


**Normal Density Concrete**

| Section Properties | | | | | | | | |
|---|---|---|---|---|---|---|---|---|
| h (mm) | $h_1/h_2$ (mm/mm) | $A_g$ ($mm^2$) | I ($10^6$ $mm^4$) | $y_b$ (mm) | $S_b$ ($10^3$ $mm^3$) | $S_t$ ($10^3$ $mm^3$) | m (kg/m) | w (kN/m) |
| 500 | 300/200 | 310 000 | 6 500 | 221 | 29 400 | 23 300 | 744 | 7.3 |
| 600 | 300/300 | 390 000 | 11 200 | 265 | 42 300 | 33 600 | 936 | 9.2 |
| 700 | 400/300 | 440 000 | 17 800 | 309 | 57 700 | 45 600 | 1 056 | 10.4 |
| 800 | 500/300 | 490 000 | 26 600 | 354 | 75 100 | 59 700 | 1 176 | 11.5 |
| 900 | 600/300 | 540 000 | 37 800 | 400 | 94 500 | 75 600 | 1 296 | 12.7 |
| 1 000 | 600/400 | 620 000 | 52 000 | 442 | 118 000 | 93 100 | 1 488 | 14.6 |
| 1 100 | 700/400 | 670 000 | 69 100 | 487 | 142 000 | 113 000 | 1 608 | 15.8 |
| 1 200 | 800/400 | 720 000 | 89 600 | 533 | 168 000 | 134 000 | 1 728 | 17.0 |
| 1 300 | 900/400 | 770 000 | 114 000 | 580 | 196 000 | 158 000 | 1 848 | 18.1 |
| 1 400 | 1 000/400 | 820 000 | 142 000 | 627 | 226 000 | 183 000 | 1 968 | 19.3 |
| 1 500 | 1 100/400 | 870 000 | 174 000 | 674 | 257 000 | 210 000 | 2 088 | 20.5 |

**$f'_{ci}$ = 30 MPa**
**$f'_c$ = 40 MPa**
**$f_{pu}$ = 1860 MPa**

**Key**

152.8 - Superimposed (service) load, kN/m
10 - Estimated camber at erection, mm
10 - Estimated long time camber, mm

*Superimposed loads shown include 50% dead load and 50% live load. 5.0 MPa top tension has been allowed, therefore additional top reinforcement is required.*

## Table of superimposed load (kN/m) and cambers (mm)

| h (mm) | No. 13 mm strand | e (mm) | Span (m) 4 | 5 | 6 | 7 | 8 | 9 | 10 | 11 | 12 | 13 | 14 | 15 | 16 | 17 |
|---|---|---|---|---|---|---|---|---|---|---|---|---|---|---|---|---|
| 500 | 8 | 171 | 175.5 | 109.9 | 74.2 | 52.7 | 38.8 | 29.2 | 22.4 | 17.3 | | | | | | |
| | | | 0 | 0 | 10 | 10 | 10 | 10 | 10 | 10 | | | | | | |
| | | | 0 | 10 | 10 | 10 | 10 | 10 | 20 | 10 | | | | | | |
| 600 | 9 | 215 | | 154.5 | 104.7 | 74.7 | 55.2 | 41.8 | 32.2 | 25.2 | 19.8 | | | | | |
| | | | | 0 | 10 | 10 | 10 | 10 | 10 | 10 | 10 | | | | | |
| | | | | 10 | 10 | 10 | 10 | 10 | 10 | 10 | 10 | | | | | |
| 700 | 11 | 259 | | | 152.8 | 109.7 | 81.8 | 62.6 | 48.9 | 38.7 | 31.0 | 25.0 | 20.2 | 16.4 | | |
| | | | | | 10 | 10 | 10 | 10 | 10 | 10 | 10 | 10 | 10 | 10 | | |
| | | | | | 10 | 10 | 10 | 10 | 20 | 20 | 20 | 20 | 20 | 10 | | |
| 800 | 13 | 304 | | | | 151.4 | 113.4 | 87.7 | 68.7 | 55.0 | 44.5 | 36.3 | 29.9 | 24.6 | 20.4 | |
| | | | | | | 10 | 10 | 10 | 10 | 10 | 10 | 20 | 20 | 20 | 10 | |
| | | | | | | 10 | 10 | 10 | 20 | 20 | 20 | 20 | 20 | 20 | 20 | |
| 900 | 14 | 350 | | | | | 141.1 | 109.0 | 86.1 | 69.1 | 56.2 | 46.1 | 38.2 | 31.7 | 26.5 | 22.1 |
| | | | | | | | 10 | 10 | 10 | 10 | 10 | 10 | 20 | 20 | 20 | 20 |
| | | | | | | | 10 | 10 | 10 | 20 | 20 | 20 | 20 | 20 | 20 | 20 |
| 1000 | 17 | 380 | | | | | | 146.5 | 116.1 | 93.6 | 76.5 | 63.2 | 52.6 | 44.1 | 37.1 | 31.3 |
| | | | | | | | | 10 | 10 | 10 | 10 | 10 | 20 | 20 | 20 | 20 |
| | | | | | | | | 10 | 10 | 20 | 20 | 20 | 20 | 20 | 20 | 20 |
| 1100 | 18 | 428 | | | | | | 174.0 | 138.1 | 111.6 | 91.4 | 75.8 | 63.3 | 53.3 | 45.0 | 38.2 |
| | | | | | | | | 10 | 10 | 10 | 10 | 10 | 10 | 20 | 20 | 20 |
| | | | | | | | | 10 | 10 | 10 | 20 | 20 | 20 | 20 | 20 | 20 |
| 1200 | 20 | 468 | | | | | | | 168.9 | 136.9 | 112.5 | 93.5 | 78.5 | 66.3 | 56.4 | 48.2 |
| | | | | | | | | | 10 | 10 | 10 | 10 | 10 | 20 | 20 | 20 |
| | | | | | | | | | 10 | 10 | 20 | 20 | 20 | 20 | 20 | 20 |
| 1300 | 22 | 504 | | | | | | | | 164.9 | 135.9 | 113.3 | 95.4 | 80.9 | 69.1 | 59.3 |
| | | | | | | | | | | 10 | 10 | 10 | 10 | 20 | 20 | 20 |
| | | | | | | | | | | 10 | 20 | 20 | 20 | 20 | 20 | 20 |
| 1400 | 24 | 539 | | | | | | | | | 161.3 | 134.8 | 113.8 | 96.8 | 82.9 | 71.4 |
| | | | | | | | | | | | 10 | 10 | 10 | 20 | 20 | 20 |
| | | | | | | | | | | | 20 | 20 | 20 | 20 | 20 | 20 |
| 1500 | 25 | 588 | | | | | | | | | | 152.9 | 129.3 | 110.2 | 94.5 | 81.6 |
| | | | | | | | | | | | | 10 | 10 | 10 | 20 | 20 |
| | | | | | | | | | | | | 20 | 20 | 20 | 20 | 20 |

*The above tables have been generated using an initial tension of 0.75 $f_{pu}$ resulting in fully prestressed elements with $M_r$ governing. Higher load capacities may be achieved by using more strands with a lower initial tension which may result in partially prestressed elements.*

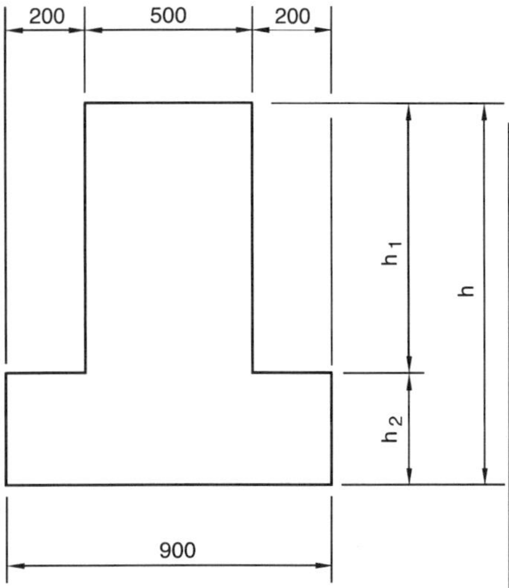


# INVERTED TEE BEAMS

**Normal Density Concrete**

| Section Properties | | | | | | | | |
|---|---|---|---|---|---|---|---|---|
| h (mm) | $h_1/h_2$ (mm/mm) | $A_g$ ($mm^2$) | I ($10^6$ $mm^4$) | $y_b$ (mm) | $S_b$ ($10^3$ $mm^3$) | $S_t$ ($10^3$ $mm^3$) | m (kg/m) | w (kN/m) |
| 500 | 300/200 | 330 000 | 6 840 | 214 | 32 000 | 23 900 | 792 | 7.8 |
| 600 | 300/300 | 420 000 | 11 800 | 257 | 46 000 | 34 500 | 1 008 | 9.9 |
| 700 | 400/300 | 470 000 | 18 800 | 299 | 62 800 | 46 800 | 1 128 | 11.1 |
| 800 | 500/300 | 520 000 | 28 000 | 342 | 81 800 | 61 200 | 1 248 | 12.3 |
| 900 | 600/300 | 570 000 | 39 800 | 387 | 103 000 | 77 600 | 1 368 | 13.4 |
| 1 000 | 600/400 | 660 000 | 54 700 | 427 | 128 000 | 95 500 | 1 584 | 15.5 |
| 1 100 | 700/400 | 710 000 | 72 800 | 471 | 154 000 | 116 000 | 1 704 | 16.7 |
| 1 200 | 800/400 | 760 000 | 94 300 | 516 | 183 000 | 138 000 | 1 824 | 17.9 |
| 1 300 | 900/400 | 810 000 | 120 000 | 561 | 213 000 | 162 000 | 1 944 | 19.1 |
| 1 400 | 1 000/400 | 860 000 | 149 000 | 607 | 246 000 | 188 000 | 2 064 | 20.3 |
| 1 500 | 1 100/400 | 910 000 | 183 000 | 653 | 280 000 | 216 000 | 2 184 | 21.4 |

**$f'_{ci}$ = 30 MPa**
**$f'_c$ = 40 MPa**
**$f_{pu}$ = 1860 MPa**

**Key**
150.6 - Superimposed (service) load, kN/m
10 - Estimated camber at erection, mm
10 - Estimated long time camber, mm

*Superimposed loads shown include 50% dead load and 50% live load. 5.0 MPa top tension has been allowed, therefore additional top reinforcement is required.*

## Table of superimposed load (kN/m) and cambers (mm)

| h (mm) | No. 13 mm strand | e (mm) | Span (m) 4 | 5 | 6 | 7 | 8 | 9 | 10 | 11 | 12 | 13 | 14 | 15 | 16 | 17 |
|---|---|---|---|---|---|---|---|---|---|---|---|---|---|---|---|---|
| 500 | 8 | 164 | 175.0<br>0<br>0 | 109.4<br>0<br>0 | 73.8<br>0<br>10 | 52.3<br>10<br>10 | 38.3<br>10<br>10 | 28.8<br>10<br>10 | 21.9<br>10<br>10 | 16.9<br>10<br>10 | | | | | | |
| 600 | 9 | 207 | | 153.8<br>0<br>0 | 104.0<br>0<br>10 | 74.0<br>10<br>10 | 54.5<br>10<br>10 | 41.1<br>10<br>10 | 31.6<br>10<br>10 | 24.5<br>10<br>10 | 19.1<br>10<br>10 | | | | | |
| 700 | 11 | 249 | | | 152.2<br>0<br>10 | 109.1<br>10<br>10 | 81.1<br>10<br>10 | 61.9<br>10<br>10 | 48.2<br>10<br>10 | 38.1<br>10<br>10 | 30.3<br>10<br>10 | 24.3<br>10<br>10 | 19.6<br>10<br>10 | 15.7<br>10<br>10 | | |
| 800 | 13 | 292 | | | | 150.6<br>10<br>10 | 112.6<br>10<br>10 | 86.6<br>10<br>10 | 68.0<br>10<br>10 | 54.2<br>10<br>10 | 43.7<br>10<br>20 | 35.6<br>10<br>20 | 29.1<br>10<br>20 | 23.9<br>10<br>10 | 19.6<br>10<br>10 | |
| 900 | 14 | 337 | | | | | 140.5<br>10<br>10 | 108.4<br>10<br>10 | 85.4<br>10<br>10 | 68.5<br>10<br>10 | 55.5<br>10<br>20 | 45.5<br>10<br>20 | 37.5<br>10<br>20 | 31.1<br>10<br>20 | 25.8<br>10<br>10 | 21.5<br>10<br>10 |
| 1000 | 17 | 377 | | | | | | 146.9<br>10<br>10 | 116.3<br>10<br>10 | 93.6<br>10<br>10 | 76.4<br>10<br>20 | 62.9<br>10<br>20 | 52.3<br>10<br>20 | 43.7<br>10<br>20 | 36.7<br>10<br>20 | 30.8<br>10<br>20 |
| 1100 | 18 | 418 | | | | | | 174.3<br>10<br>10 | 138.3<br>10<br>10 | 111.6<br>10<br>10 | 91.3<br>10<br>10 | 75.5<br>10<br>20 | 63.0<br>10<br>20 | 52.9<br>10<br>20 | 44.6<br>10<br>20 | 37.7<br>10<br>20 |
| 1200 | 20 | 458 | | | | | | | 169.1<br>20<br>10 | 136.9<br>10<br>10 | 112.4<br>10<br>10 | 93.3<br>10<br>20 | 78.2<br>10<br>20 | 65.9<br>10<br>20 | 56.0<br>10<br>20 | 47.7<br>20<br>20 |
| 1300 | 22 | 499 | | | | | | | | 164.8<br>10<br>10 | 135.6<br>10<br>10 | 112.9<br>10<br>20 | 94.9<br>20<br>20 | 80.4<br>10<br>20 | 68.5<br>20<br>20 | 58.7<br>20<br>20 |
| 1400 | 24 | 542 | | | | | | | | | 161.1<br>10<br>10 | 134.5<br>10<br>20 | 113.4<br>10<br>20 | 96.3<br>10<br>20 | 82.4<br>20<br>20 | 70.8<br>20<br>20 |
| 1500 | 25 | 587 | | | | | | | | | | 152.7<br>10<br>10 | 128.9<br>10<br>20 | 109.8<br>10<br>20 | 94.1<br>10<br>20 | 81.1<br>20<br>20 |

*The above tables have been generated using an initial tension of 0.75 $f_{pu}$ resulting in fully prestressed elements with $M_r$ governing. Higher load capacities may be achieved by using more strands with a lower initial tension which may result in partially prestressed elements.*

# I-GIRDERS

**Normal Density Concrete**

| Section Properties | | | | | | | |
|---|---|---|---|---|---|---|---|
| Designation | $A_g$ ($mm^2$) | I ($10^6$ $mm^4$) | $y_b$ (mm) | $S_b$ ($10^3$ $mm^3$) | $S_t$ ($10^3$ $mm^3$) | m (kg/m) | w (kN/m) |
| CPCI 900 | 218 000 | 19 300 | 398 | 48 600 | 38 400 | 523 | 5.13 |
| CPCI 1200 | 320 000 | 53 900 | 527 | 102 300 | 80 000 | 768 | 7.53 |
| CPCI 1400 | 413 000 | 102 600 | 635 | 161 400 | 134 200 | 991 | 9.72 |
| CPCI 1600 | 499 000 | 174 700 | 793 | 220 200 | 216 600 | 1 024 | 11.8 |

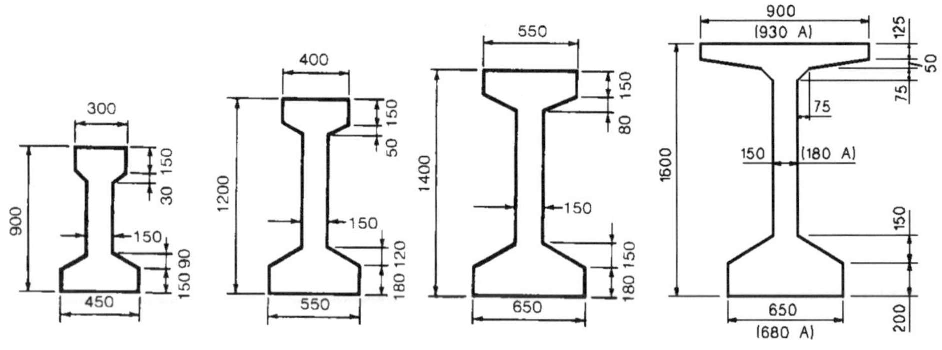


CPCI 900 CPCI 1200 CPCI 1400 CPCI 1600

$f'_{ci}$ = 30 MPa
$f'_c$ = 40 MPa
$f_{pu}$ = 1860 MPa

**Key**
102.8 - Superimposed (service) load, kN/m
10 - Estimated camber at erection, mm
20 - Estimated long time camber, mm

*Superimposed loads shown include 50% dead load and 50% live load. 5.0 MPa top tension has been allowed, therefore additional top reinforcement is required.*

**Table of superimposed load (kN/m) and cambers (mm)**

| Designation | No. 13 mm strand | $e_e$†‡<br>$e_c$† | Span (m) 8 | 9 | 10 | 11 | 12 | 13 | 14 | 15 | 16 | 17 | 18 | 19 | 20 | 21 |
|---|---|---|---|---|---|---|---|---|---|---|---|---|---|---|---|---|
| CPCI 900 | 12 | 306<br>328 | 102.8<br>10<br>20 | 80.2<br>20<br>20 | 64.1<br>20<br>20 | 52.1<br>20<br>30 | 43.0<br>20<br>30 | 36.0<br>30<br>40 | 30.4<br>30<br>40 | 25.8<br>30<br>50 | 22.1<br>40<br>50 | 19.1<br>40<br>50 | 16.5<br>40<br>60 | 14.3<br>40<br>60 | | |
| CPCI 1200 | 20 | 412<br>443 | | | 138.1<br>20<br>20 | 112.9<br>20<br>20 | 93.8<br>20<br>30 | 78.9<br>20<br>30 | 67.0<br>30<br>40 | 57.5<br>30<br>40 | 49.7<br>30<br>50 | 43.2<br>40<br>50 | 37.8<br>40<br>50 | 33.2<br>40<br>60 | 29.3<br>50<br>60 | 25.9<br>50<br>70 |
| CPCI 1400 | 26 | 496<br>554 | | | | | 132.5<br>20<br>20 | 120.3<br>20<br>30 | 110.0<br>20<br>30 | 95.8<br>30<br>40 | 83.2<br>30<br>40 | 72.6<br>30<br>50 | 63.8<br>40<br>50 | 56.4<br>40<br>50 | 50.0<br>40<br>60 | 44.5<br>50<br>60 |
| CPCI 1600 | 30 | 607<br>703 | | | | | | 136.4<br>20<br>20 | 124.5<br>20<br>30 | 114.3<br>20<br>30 | 105.6<br>30<br>40 | 98.0<br>30<br>40 | 91.4<br>30<br>40 | 84.3<br>40<br>50 | 75.0<br>40<br>50 | 67.0<br>40<br>50 |

* *Values below heavy line controlled by $V_r$.*
* *Values above heavy line controlled by $M_r$. See Limitations on Reinforcement, Sect. 3.3.1.*
† $e_e$ and $e_c$ are the maximum eccentricities at end and midspan that will not violate the permissible concrete stresses at transfer.
‡ End eccentricities may be decreased without affecting the load capacities.

# I-GIRDERS

**Normal Density Concrete**

**Section Properties**

| Designation | $A_g$ ($mm^2$) | I ($10^6\ mm^4$) | $y_b$ (mm) | $S_b$ ($10^3\ mm^3$) | $S_t$ ($10^3\ mm^3$) | m (kg/m) | w (kN/m) |
|---|---|---|---|---|---|---|---|
| **CPCI 1900** | 544 000 | 268 400 | 940 | 285 500 | 279 500 | 1 308 | 12.83 |
| **CPCI 1900A** | 601 000 | 285 600 | 940 | 303 600 | 297 600 | 1 446 | 14.18 |
| **CPCI 2300** | 604 000 | 431 800 | 1 135 | 380 300 | 370 800 | 1 453 | 14.25 |
| **CPCI 2300A** | 673 000 | 462 200 | 1 137 | 406 600 | 397 400 | 1 619 | 15.88 |

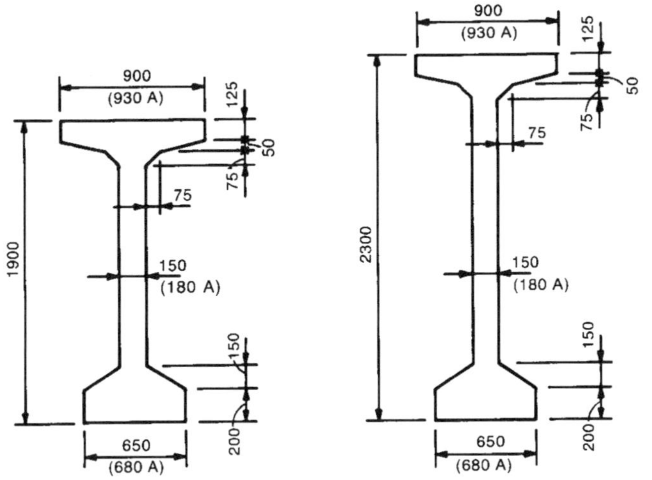


**CPCI 1900**
**CPCI 1900A**

**CPCI 2300**
**CPCI 2300A**

$f'_{ci}$ = **30 MPa**
$f'_c$ = **40 MPa**
$f_{pu}$ = **1860 MPa**

**Key**

128.1 - Superimposed (service) load, kN/m
20 - Estimated camber at erection, mm
30 - Estimated long time camber, mm

*Superimposed loads shown include 50% dead load and 50% live load. 5.0 MPa top tension has been allowed, therefore additional top reinforcement is required.*

## Table of superimposed load (kN/m) and cambers (mm)

| Designation | No. 13 mm strand | $e_e$†‡ / $e_c$† | Span (m) 17 | 18 | 19 | 20 | 21 | 22 | 23 | 24 | 25 | 26 | 27 | 28 | 29 | 30 |
|---|---|---|---|---|---|---|---|---|---|---|---|---|---|---|---|---|
| **CPCI 1900** | **30** | **838**<br>**850** | 128.1<br>20<br>30 | 119.4<br>30<br>40 | 107.4<br>30<br>40 | 95.8<br>30<br>40 | 85.8<br>30<br>50 | 77.1<br>40<br>50 | 69.6<br>40<br>50 | 62.9<br>40<br>60 | 57.1<br>40<br>60 | 51.9<br>50<br>60 | 47.2<br>50<br>60 | 43.1<br>50<br>70 | 39.4<br>50<br>70 | 36.0<br>50<br>70 |
| **CPCI 2300** | **36** | **882**<br>**1 035** | | | 132.1<br>30<br>40 | 123.9<br>30<br>40 | 116.5<br>30<br>40 | 109.9<br>30<br>50 | 100.1<br>40<br>50 | 90.8<br>40<br>50 | 82.7<br>40<br>60 | 75.4<br>40<br>60 | 69.0<br>50<br>60 | 63.2<br>50<br>60 | 58.0<br>50<br>70 | 53.4<br>50<br>70 |

* *Values below heavy line controlled by $V_r$.*
* *Values above heavy line controlled by $M_r$. See Limitations on Reinforcement, Sect. 3.3.1.*

† $e_e$ and $e_c$ are the maximum eccentricities at end and midspan that will not violate the permissible concrete stresses at transfer.

‡ End eccentricities may be decreased without affecting the load capacities.

## PRECAST PRESTRESSED COLUMNS

**Fig 7.6.1 Interaction curves for precast prestressed concrete columns**

**Criteria**

1. Minimum prestress = 1.5 MPa.
2. All strand assumed 13 mm diameter, $f_{pu}$ = 1860 MPa.
3. Curves shown for partial development of strand near end of element, where $f_{pr} \approx f_{pe}$.
4. When points of maximum stress are far from end of element and strands can fully develop, the section resistance will be greater than shown.
5. Horizontal portion of curve is the maximum for tied columns = 0.80 $P_{ro}$.

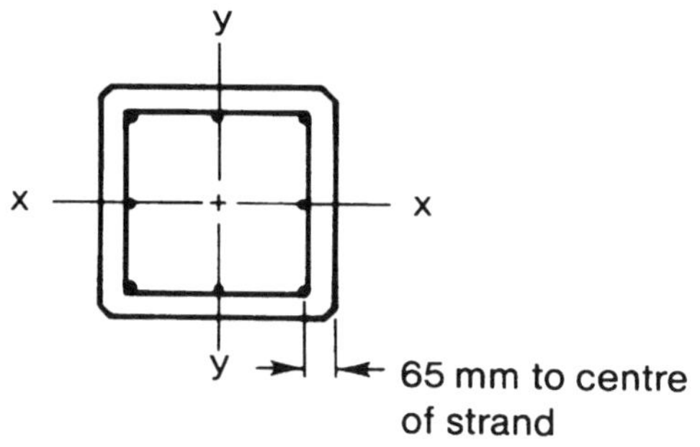


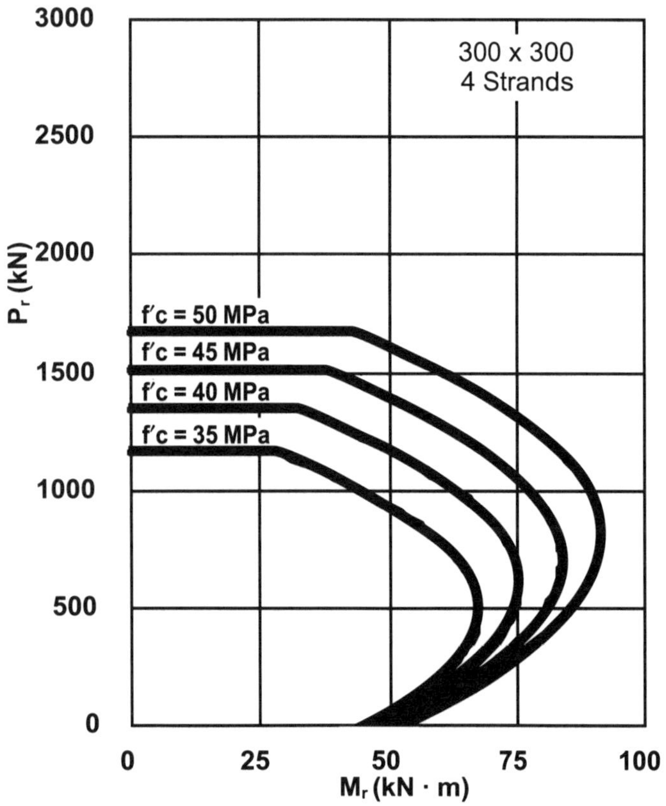


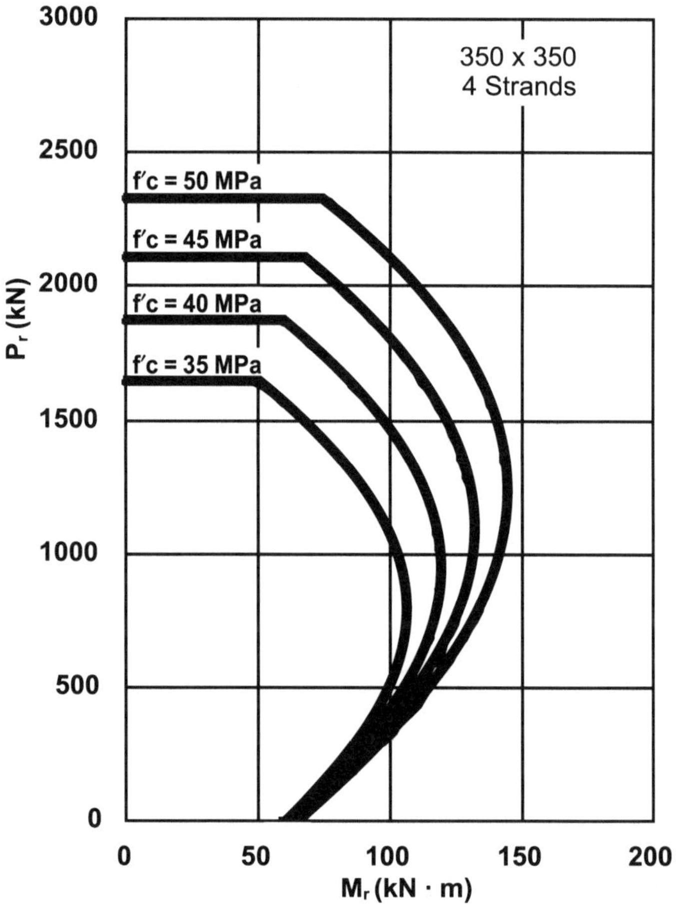

## PRECAST PRESTRESSED COLUMNS

**Fig 7.6.1 (Cont'd) Interaction curves for precast prestressed concrete columns**

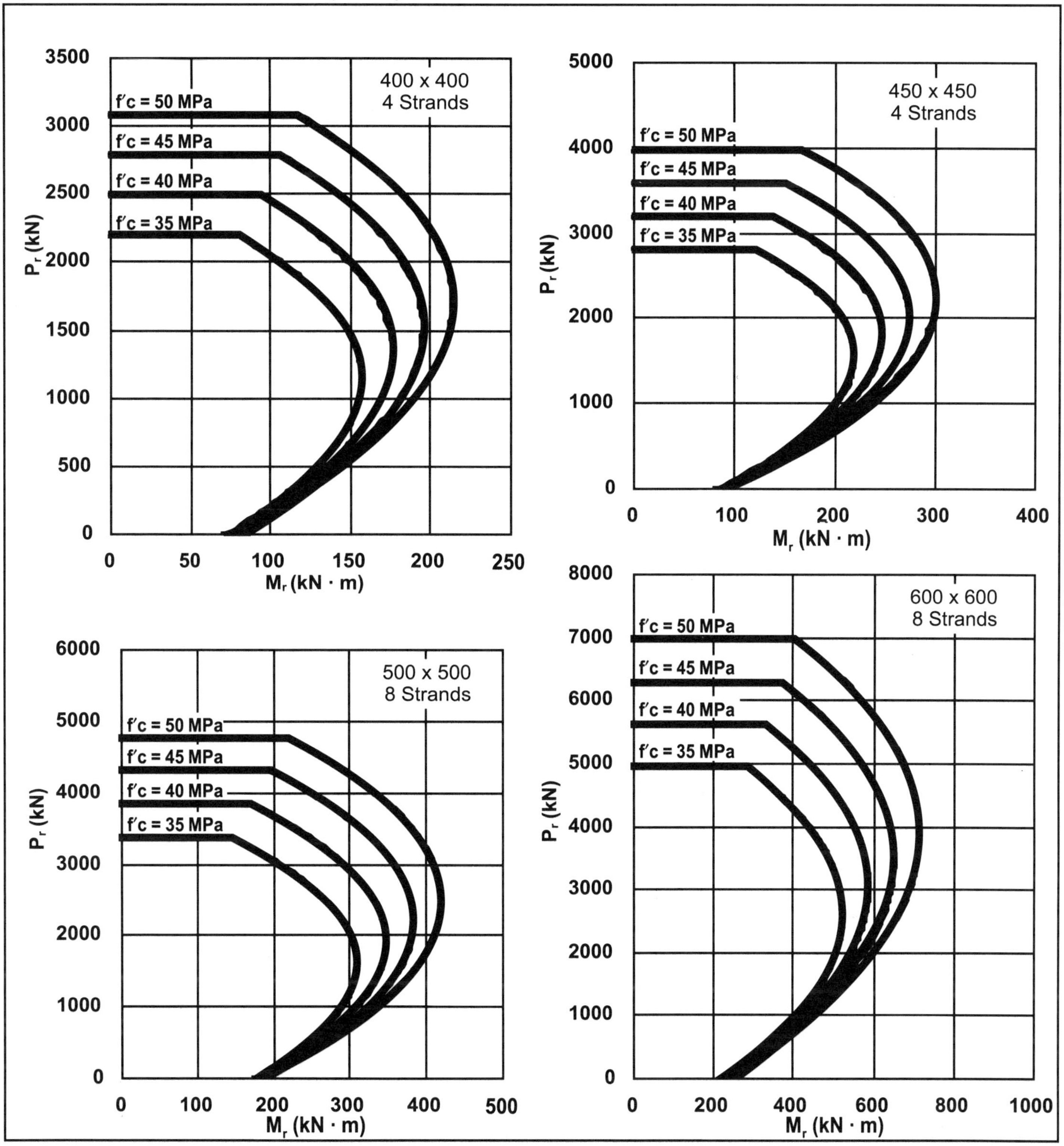

# PRECAST REINFORCED COLUMNS

**Fig 7.6.2 Interaction curves for precast reinforced concrete columns**

**Criteria**

1. Concrete $f'_c$ = 40 MPa.
2. Reinforcement $f_y$ = 400 MPa.
3. Curves shown for full development of reinforcement.
4. Horizontal portion of curve is the maximum for tied columns = 0.80 $P_{ro}$.

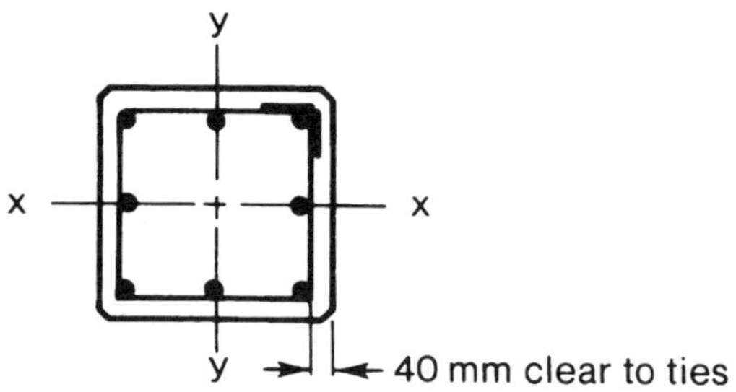


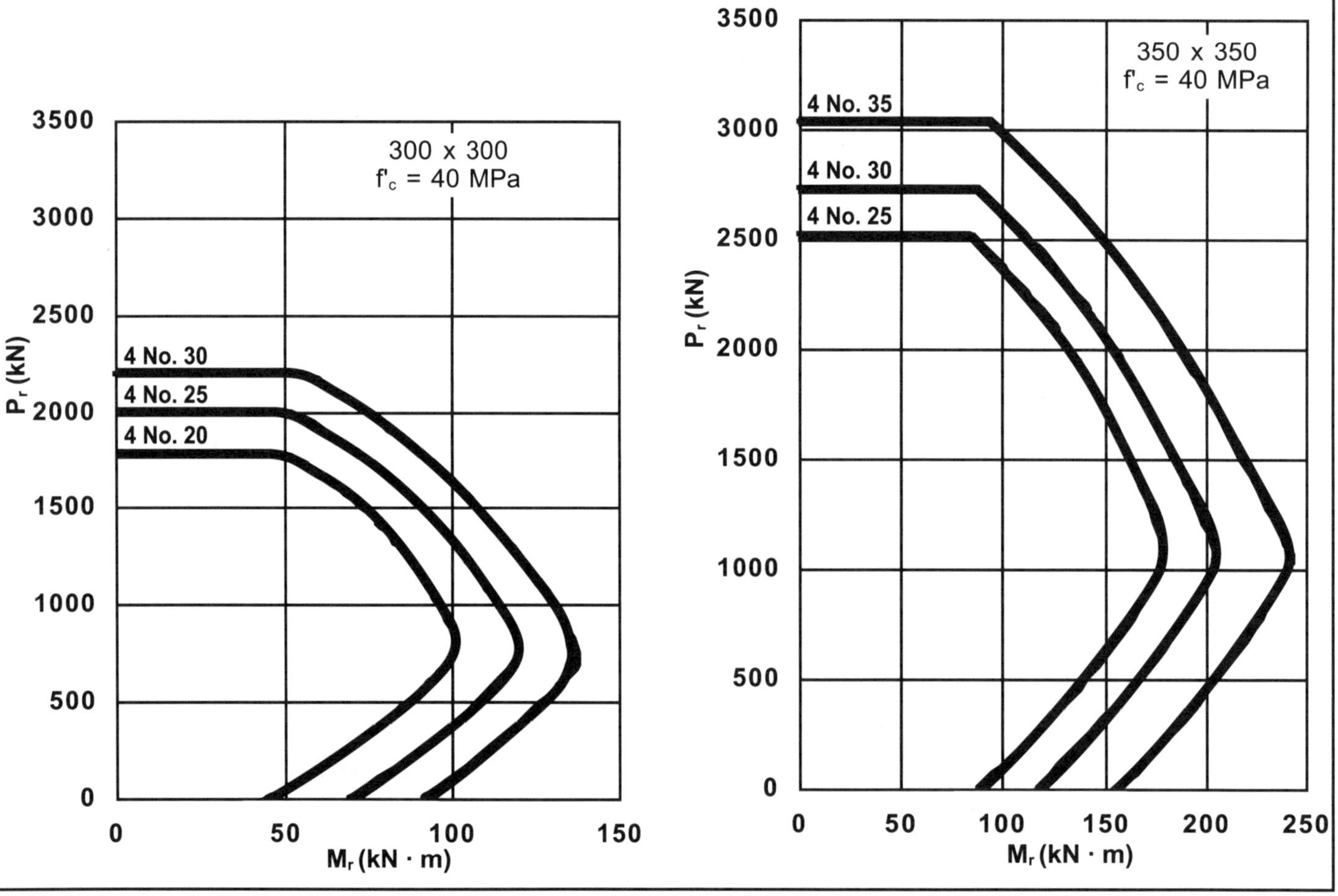

## PRECAST REINFORCED COLUMNS

Fig 7.6.2 (Cont'd) Interaction curves for precast reinforced concrete columns

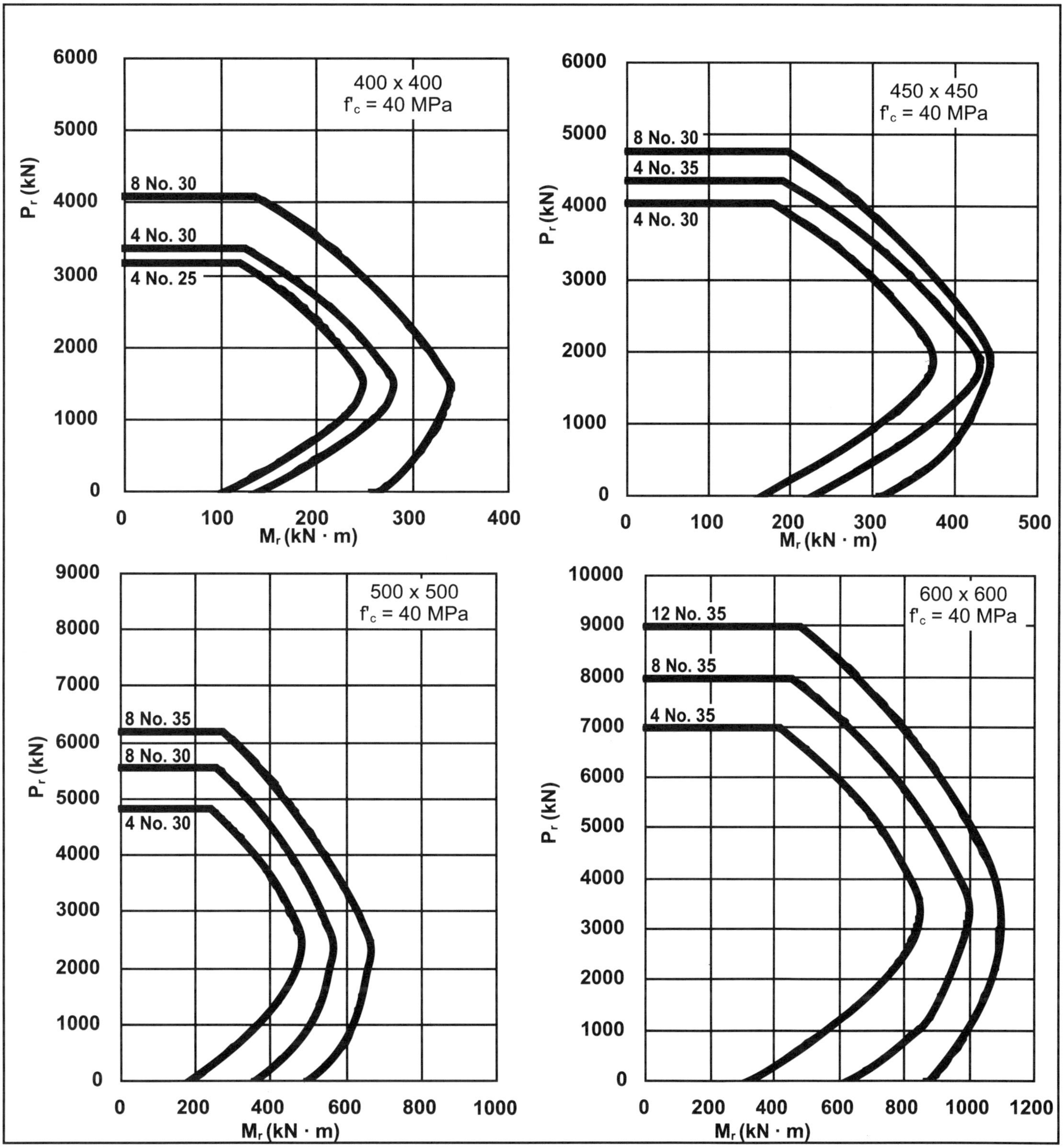

# DOUBLE TEE WALL PANELS

**Fig 7.6.3 Partial interaction curve for prestressed double tee wall panels**

| h (mm) | No. of strands | b (mm) | Interaction curve data | | | | | | |
|---|---|---|---|---|---|---|---|---|---|
| | | | | Tension in stem | | | Compression in stem | | |
| | | | $P_{ro}$ | $P_{rb}$ | $M_{rb}$ | $M_{ro}$ | $P_{rb}$ | $M_{rb}$ | $M_{ro}$ |
| 325 | 4 | 130 | 4959 | 3600 | 234 | 105 | 1120 | 154 | 63 |
| 425 | 4 | 120 | 5590 | 3830 | 362 | 135 | 1430 | 262 | 91 |
| 525 | 6 | 110 | 6020 | 3740 | 498 | 57 | 1580 | 385 | 200 |
| 625 | 6 | 100 | 6540 | 3810 | 638 | 229 | 1900 | 520 | 255 |

1. $f'_c = 40$ MPa normal density
2. Strand = 13 mm $f_{pu} = 1860$ MPa
3. Curves shown for full development of strand. Load assumed at centroid of concrete section.

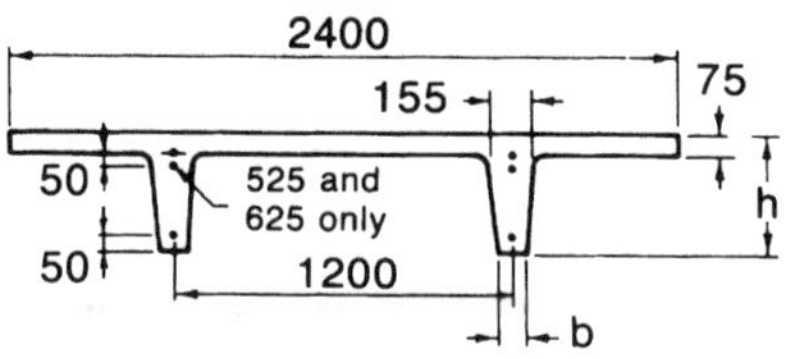


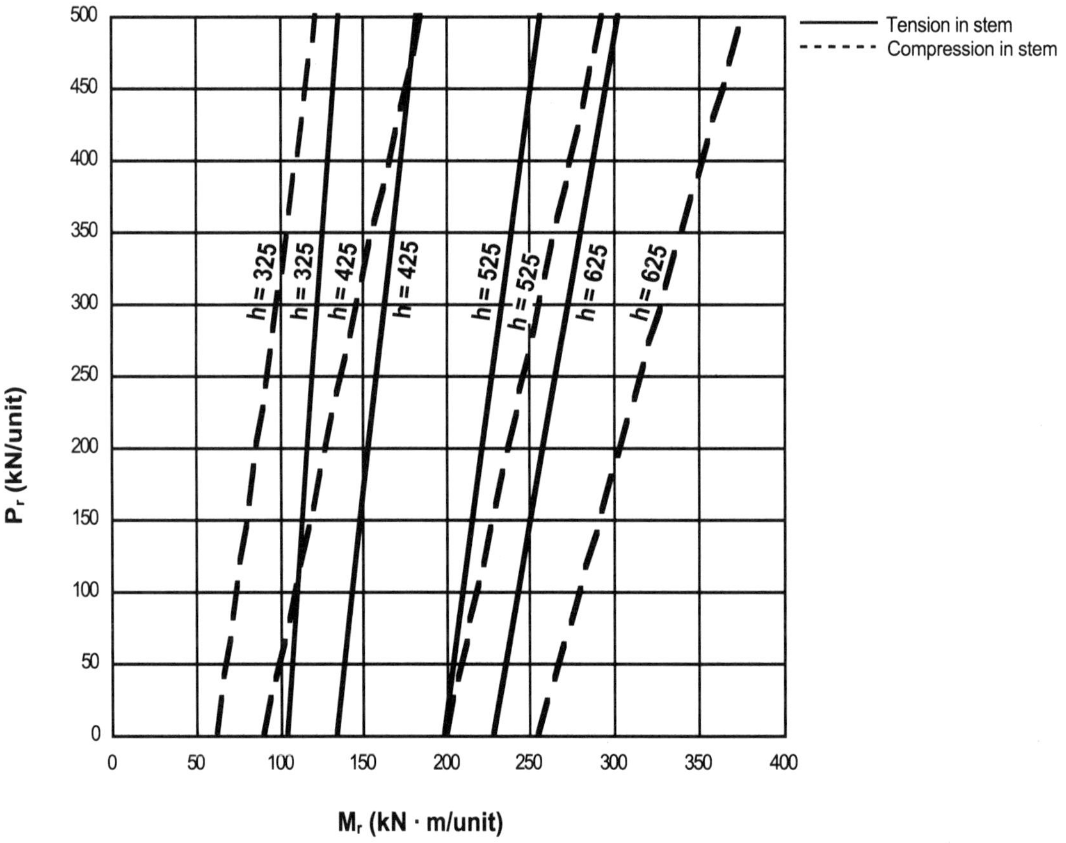

## HOLLOW-CORE WALL PANELS

**Fig 7.6.4 Partial interaction curve for prestressed hollow-core wall panels**

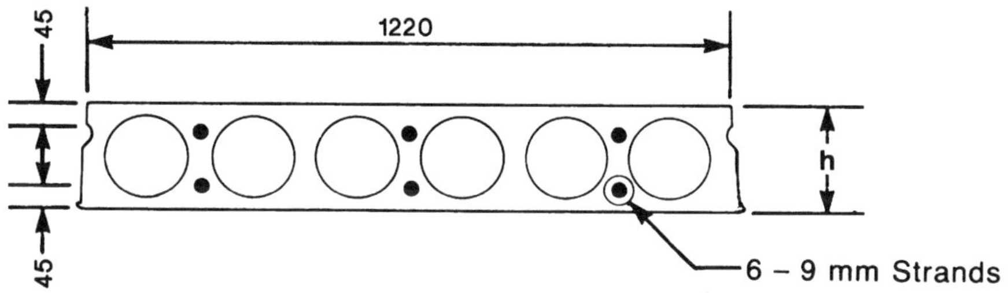


1. $f'_c$ = 40 MPa normal density
2. $f_{pu}$ = 1860 MPa
3. Curves shown for full development of strand.

| h (mm) | Interaction curve data | | | |
|---|---|---|---|---|
| | $P_{ro}$ | $P_{rb}$ | $M_{rb}$ | $M_{ro}$ |
| 203 | 2790 | 1160 | 107 | 46 |
| 305 | 3480 | 1500 | 212 | 73 |

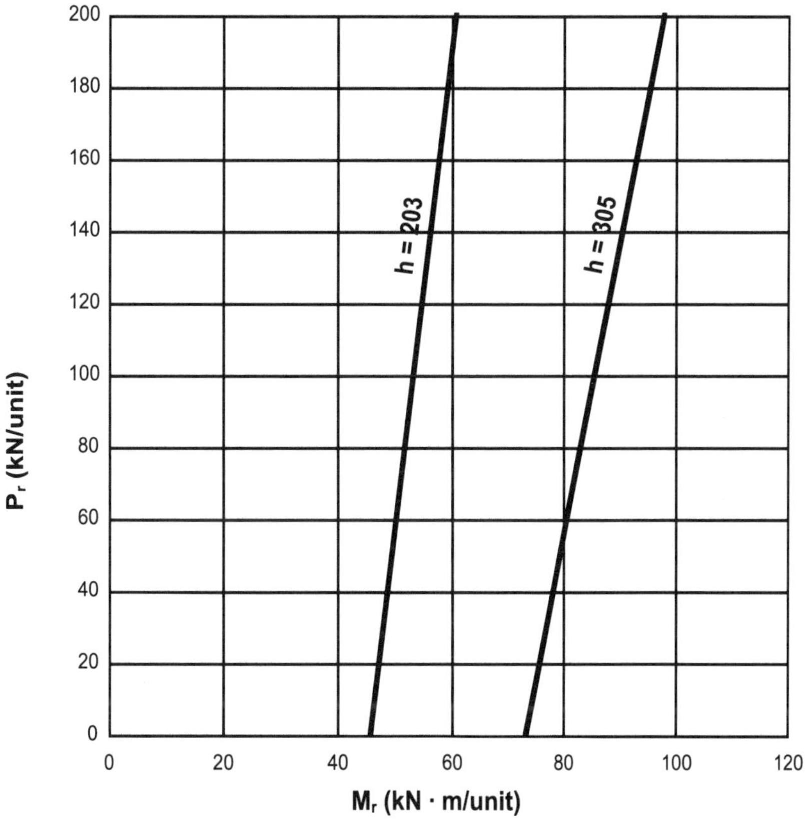

## PRECAST PRESTRESSED SOLID WALL PANELS

**Fig 7.6.5 Partial interaction curve for prestressed solid wall panels**

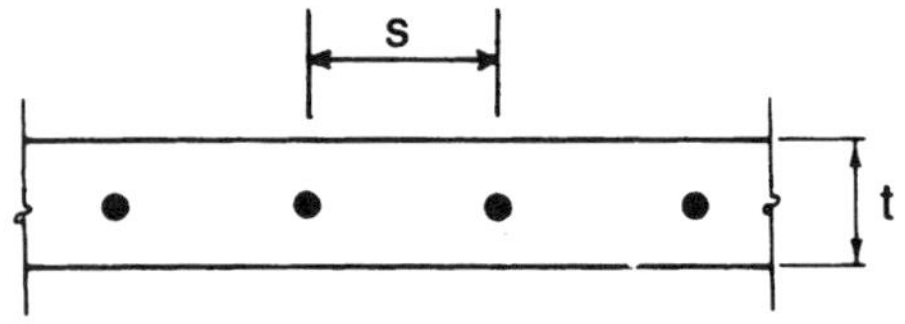


| t (mm) | s (mm) | Interaction curve data | | | |
|---|---|---|---|---|---|
| | | $P_{ro}$ | $P_{rb}$ | $M_{rb}$ | $M_{ro}$ |
| 100 | 680 | 1989 | 878 | 25.6 | 8.2 |
| 150 | 450 | 2983 | 1290 | 57.8 | 23.7 |
| 200 | 340 | 3978 | 1750 | 102.7 | 39.8 |
| 250 | 270 | 4972 | 2220 | 160.5 | 60.0 |

1. $f'_c$ = 40 MPa normal density
2. Strands = 13 mm $f_{pu}$ = 1860 MPa
3. Curves shown for full development of strand.

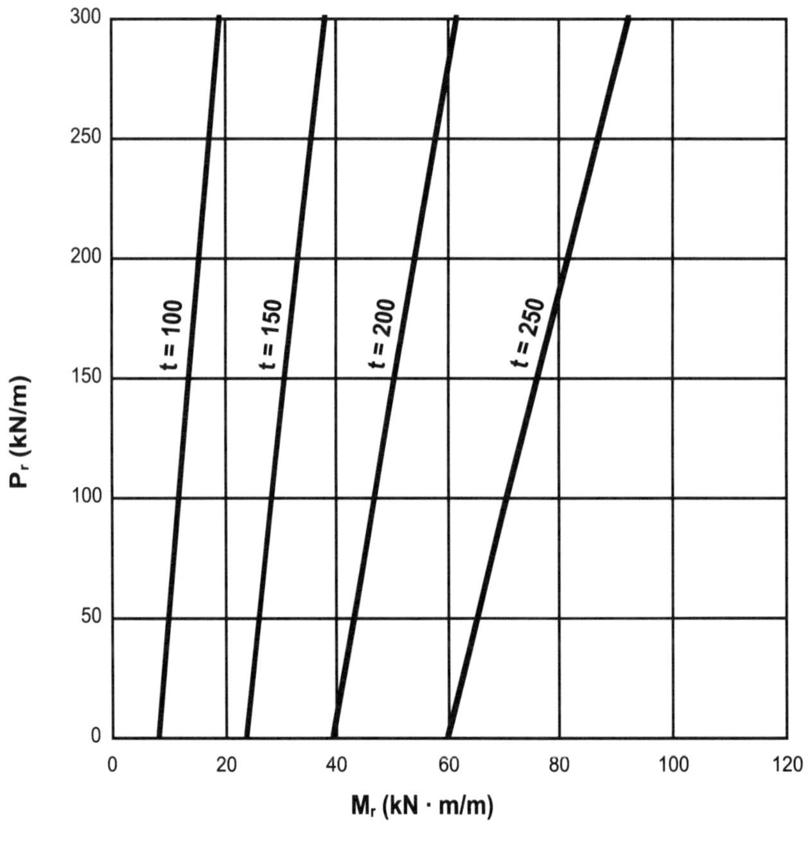

# PRECAST REINFORCED SOLID WALL PANELS

**Fig 7.6.6 Partial interaction curve for precast reinforced concrete wall panels**

| t (mm) | Interaction curve data | | | | | | | |
|---|---|---|---|---|---|---|---|---|
| | $f'_c = 35$ MPa | | | | $f'_c = 40$ MPa | | | |
| | $P_{ro}$ | $P_{rb}$ | $M_{rb}$ | $M_{ro}$ | $P_{ro}$ | $P_{rb}$ | $M_{rb}$ | $M_{ro}$ |
| 100 | 1850 | 880 | 22.6 | 2.0 | 2092 | 1035 | 25.6 | 2.3 |
| 150 | 2780 | 1370 | 51.0 | 4.5 | 3139 | 1550 | 57.8 | 5.6 |
| 200 | 3700 | 1770 | 94.9 | 7.8 | 4180 | 1970 | 106.8 | 8.7 |
| 250 | 4630 | 2160 | 149.8 | 10.6 | 5231 | 2510 | 168.2 | 11.0 |

1. $f'_c = 35$ and 40 MPa
2. $f_y = 400$ MPa
3. Curves based on minimum vertical reinforcement $\rho = 0.0012$

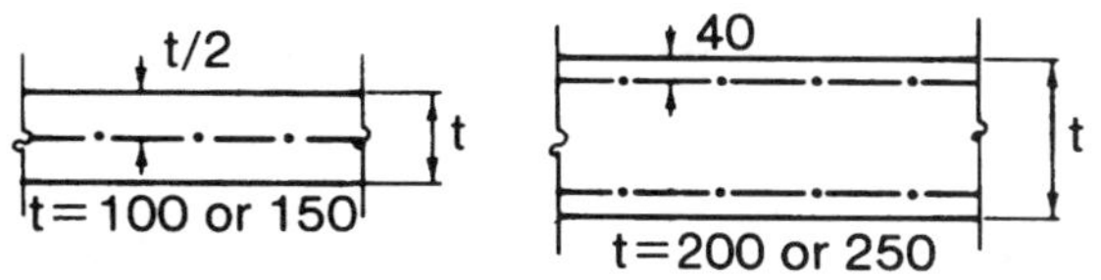


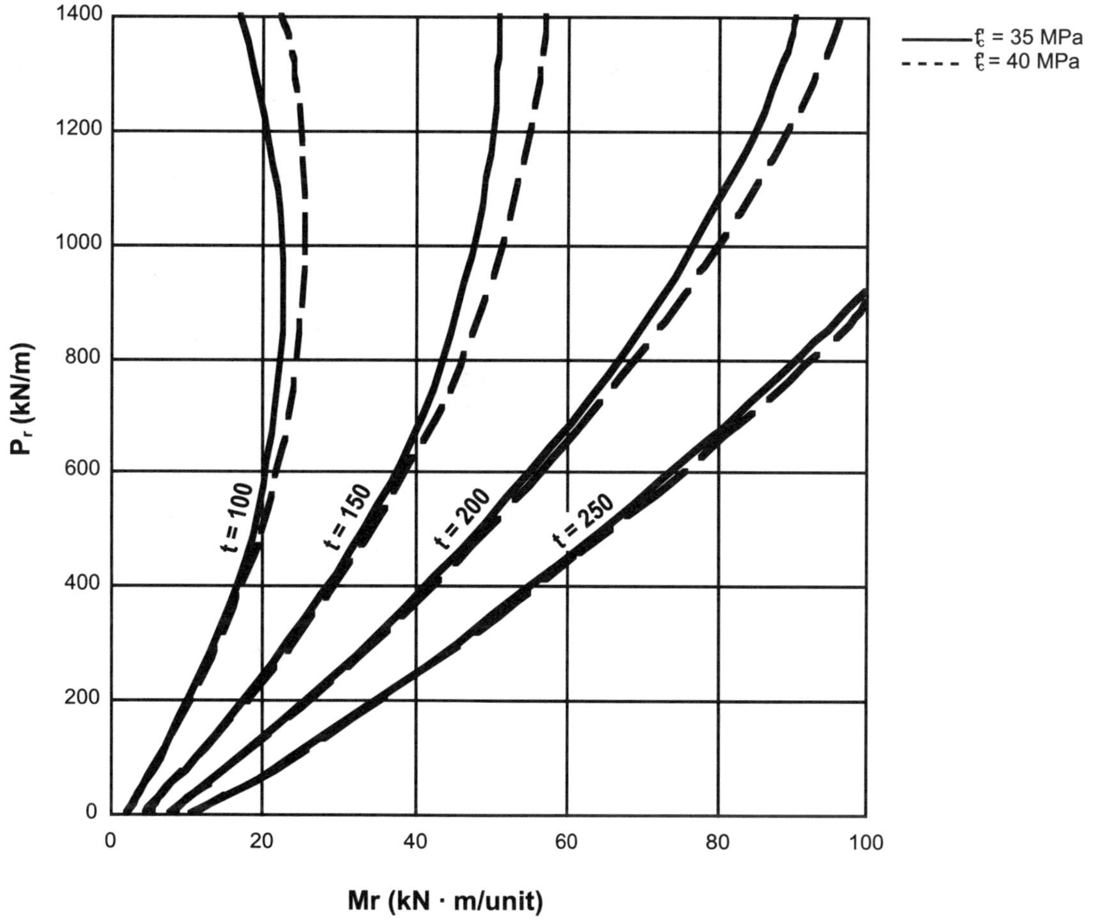

# PILES

**Fig 7.7.1 Section properties and resistance of prestressed concrete piles**

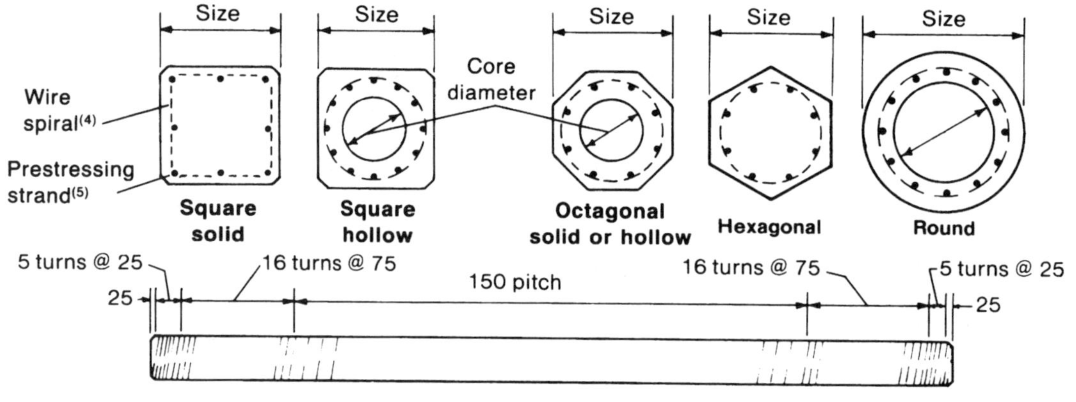


Typical elevation (4)

| Size (mm) | Core diameter (mm) | Section properties (1) Area ($mm^2$) | Mass (kg/m) | Moment of inertia ($10^6 mm^4$) | Section modulus ($10^3 mm^3$) | Radius of gyration (mm) | Perimeter (m) | Factored resistance (2)(3) (kN) $f'_c$ (MPa) 35 | 40 | 45 | 50 | 55 |
|---|---|---|---|---|---|---|---|---|---|---|---|---|
| **Square piles** | | | | | | | | | | | | |
| 250 | Solid | 63 000 | 151 | 326 | 2 610 | 72 | 1.00 | 782 | 910 | 1036 | 1159 | 1279 |
| 300 | Solid | 90 000 | 216 | 675 | 4 500 | 87 | 1.20 | 1118 | 1301 | 1480 | 1655 | 1827 |
| 350 | Solid | 123 000 | 295 | 1 250 | 7 140 | 101 | 1.40 | 1527 | 1778 | 2023 | 2262 | 2497 |
| 400 | Solid | 160 000 | 384 | 2 130 | 10 700 | 116 | 1.60 | 1987 | 2312 | 2631 | 2943 | 3248 |
| 450 | Solid | 203 000 | 487 | 3 420 | 15 200 | 130 | 1.80 | 2521 | 2934 | 3338 | 3734 | 4121 |
| 500 | Solid | 250 000 | 500 | 5 210 | 20 800 | 144 | 2.00 | 3104 | 3613 | 4111 | 4598 | 5076 |
| 500 | 275 | 191 000 | 458 | 4 930 | 19 700 | 161 | 2.00 | 2372 | 2760 | 3141 | 3513 | 3878 |
| 600 | Solid | 360 000 | 864 | 10 800 | 36 000 | 173 | 2.40 | 4470 | 5202 | 5920 | 6622 | 7309 |
| 600 | 300 | 289 000 | 694 | 10 400 | 34 700 | 190 | 2.40 | 3588 | 4176 | 4752 | 5316 | 5867 |
| 600 | 350 | 264 000 | 634 | 10 100 | 33 700 | 196 | 2.40 | 3278 | 3815 | 4341 | 4856 | 5360 |
| 600 | 375 | 250 000 | 600 | 9 830 | 32 800 | 198 | 2.40 | 3104 | 3613 | 4111 | 4598 | 5076 |
| **Octagonal piles** | | | | | | | | | | | | |
| 250 | Solid | 52 000 | 125 | 215 | 1 720 | 64 | 0.77 | 646 | 751 | 855 | 956 | 1056 |
| 300 | Solid | 75 000 | 180 | 446 | 2 970 | 77 | 0.92 | 931 | 1084 | 1233 | 1380 | 1523 |
| 350 | Solid | 101 000 | 242 | 825 | 4 710 | 90 | 1.07 | 1254 | 1460 | 1661 | 1858 | 2051 |
| 400 | Solid | 133 000 | 319 | 1 410 | 7 050 | 103 | 1.22 | 1651 | 1922 | 2187 | 2446 | 2700 |
| 450 | Solid | 168 000 | 403 | 2 260 | 10 000 | 116 | 1.38 | 2086 | 2428 | 2763 | 3090 | 3411 |
| 500 | Solid | 207 000 | 497 | 3 440 | 13 800 | 129 | 1.53 | 2570 | 2991 | 3404 | 3808 | 4203 |
| 500 | 275 | 148 000 | 355 | 3 160 | 12 600 | 146 | 1.53 | 1838 | 2139 | 2434 | 2722 | 3005 |
| 550 | Solid | 251 000 | 602 | 5 030 | 18 300 | 142 | 1.68 | 3117 | 3627 | 4127 | 4617 | 5096 |
| 550 | 325 | 168 000 | 403 | 4 480 | 16 300 | 163 | 1.68 | 2086 | 2428 | 2763 | 3090 | 3411 |
| 600 | Solid | 298 000 | 715 | 7 130 | 23 800 | 154 | 1.84 | 3700 | 4306 | 4900 | 5481 | 6050 |
| 600 | 375 | 188 000 | 451 | 6 160 | 20 500 | 181 | 1.84 | 2334 | 2717 | 3091 | 3458 | 3817 |
| **Round piles** | | | | | | | | | | | | |
| 900 | 650 | 304 000 | 730 | 23 400 | 52 000 | 277 | 2.83 | 3775 | 4393 | 4999 | 5592 | 6172 |
| 1 200 | 950 | 422 000 | 1 010 | 61 800 | 103 000 | 383 | 3.77 | 5240 | 6098 | 6939 | 7762 | 8568 |
| 1 350 | 1 100 | 481 000 | 1 150 | 91 200 | 135 000 | 435 | 4.24 | 5972 | 6951 | 7909 | 8847 | 9765 |
| **Hexagonal piles** | | | | | | | | | | | | |
| 300 | Solid | 78 000 | 187 | 486 | 3 240 | 79 | 0.90 | 969 | 1127 | 1283 | 1435 | 1584 |
| 350 | Solid | 106 000 | 254 | 900 | 5 140 | 92 | 1.05 | 1316 | 1532 | 1743 | 1950 | 2152 |
| 400 | Solid | 139 000 | 334 | 1 540 | 7 700 | 106 | 1.20 | 1726 | 2009 | 2286 | 2557 | 2822 |

(1) Form dimensions may vary with producers, with corresponding variations in section properties.
(2) Maximum factored axial load resistance, $P_{rmax} = 0.85 [\alpha_1 \phi_c f'_c (A_g - A_p) - f_{pr} A_p]$; $f_{pe}$ = 1120 MPa; $f_{cp}$ = 5.0 MPa.
(3) Resistance based on short column structural capacity only (see Sect. 7.2.9) with a maximum concrete strain of 0.002.
(4) Wire spiral varies with pile size.
(5) Strand pattern may be circular or square.

## SHEET PILES

**Fig 7.7.2 Section properties and allowable moments of prestressed sheet piles**

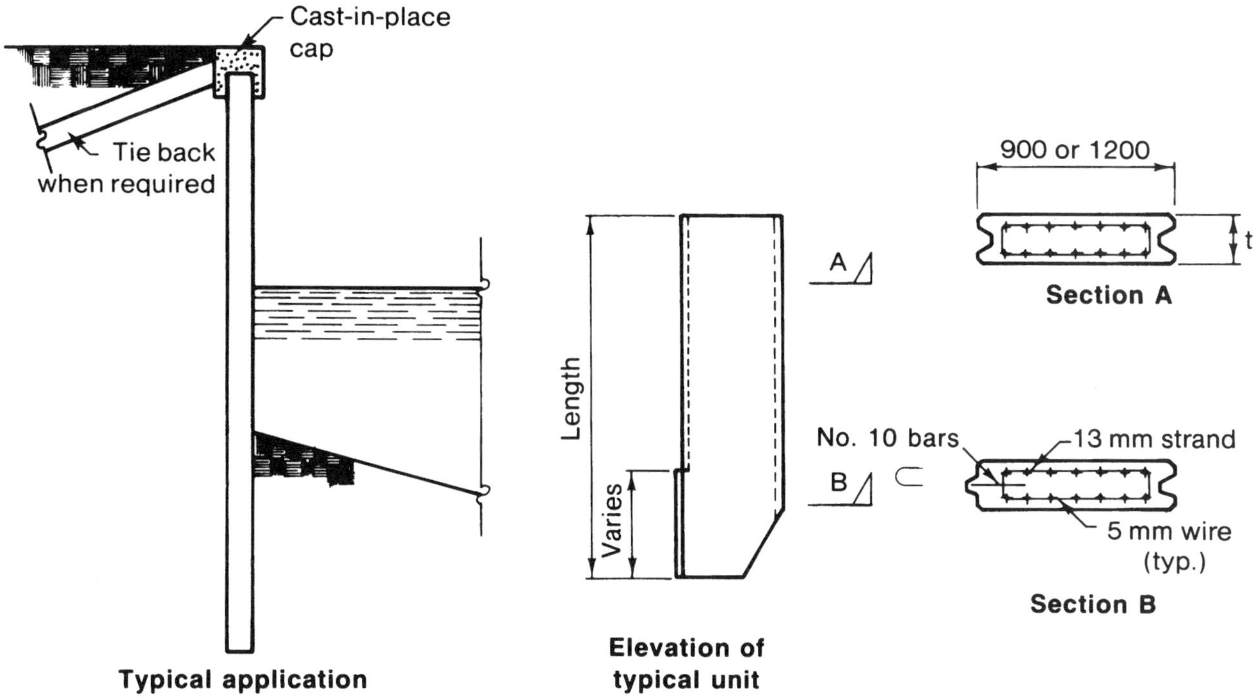


| Thickness t (mm) | Section properties per metre of width | | | | Maximum allowable service load moment (2) (kN · m/m) | |
|---|---|---|---|---|---|---|
| | Area ($mm^2$) | Mass (1) (kg/m) | Moment of inertia ($10^6 mm^4$) | Section modulus ($10^3$mm3) | $f'_c$ = 35 MPa | $f'_c$ = 40 MPa |
| 150(3) | 150 000 | 360 | 28 | 374 | 26.2 | 29.9 |
| 200(3) | 200 000 | 480 | 67 | 670 | 46.7 | 53.4 |
| 250 | 250 000 | 600 | 130 | 1 040 | 72.9 | 83.3 |
| 300 | 300 000 | 720 | 225 | 1 500 | 105.0 | 120.0 |
| 400 | 400 000 | 960 | 533 | 2 670 | 187.0 | 214.0 |
| 450 | 450 000 | 1 080 | 760 | 3 380 | 236.0 | 270.0 |
| 500 | 500 000 | 1 200 | 1 040 | 4 160 | 292.0 | 334.0 |
| 600 | 600 000 | 1 440 | 1 800 | 6 000 | 420.0 | 480.0 |

(1) Normal density concrete.
(2) Based on zero tension and maximum 0.4 $f'_c$ compression.
(3) Strand can be placed in a single layer in thin sections. Where site conditions require it, strand may be placed eccentrically.

# CHAPTER 8
# GENERAL DESIGN INFORMATION

# DESIGN INFORMATION

**Fig. 8.1.1 Dead loads of floors, ceilings, roofs, and walls**

| **Floorings** | **Force (kN/m²)** |
|---|---|
| Normal density concrete topping, per 10 mm of thickness | 0.24 |
| Semi-low density (1900 kg/m $^3$) concrete topping, per 10 mm | 0.19 |
| Low density (1900 kg/m $^3$) concrete topping, per 10 mm | 0.15 |
| 22 mm hardwood floor on sleepers clipped to concrete without fill | 0.24 |
| 40 mm terrazzo floor finish directly on slab | 0.95 |
| 40 mm terrazzo floor finish on 25 mm mortar bed | 1.49 |
| 25 mm terrazzo finish on 50 mm concrete bed | 1.79 |
| 20 mm ceramic or quarry tile on 12 mm mortar bed | 0.80 |
| 20 mm ceramic or quarry tile on 25 mm mortar bed | 1.06 |
| 8 mm linoleum or asphalt tile directly on concrete | 0.06 |
| 8 mm linoleum or asphalt tile on 25 mm mortar bed | 0.59 |
| 20 mm mastic floor | 0.45 |
| Hardwood flooring, 22 mm thick | 0.19 |
| Subflooring (soft wood), 20 mm thick | 0.13 |
| Asphaltic concrete, 40 mm thick | 0.90 |
| **Ceilings** | |
| 12.7 mm gypsum board | 0.10 |
| 15.9 mm gypsum board | 0.12 |
| 20 mm plaster directly on concrete | 0.26 |
| 20 mm plaster on metal lath furring | 0.40 |
| Suspended ceilings | 0.10 |
| Acoustical tile | 0.05 |
| Acoustical tile on wood furring strips | 0.15 |
| **Roofs** | |
| Five-ply felt and gravel (or slag) | 0.31 |
| Three-ply felt and gravel (or slag) | 0.27 |
| Five-ply felt composition roof, no gravel | 0.20 |
| Three-ply felt composition roof, no gravel | 0.15 |
| Asphalt strip shingles | 0.15 |
| Rigid insulation, per 100 mm (glass fiber) | 0.07 |
| Gypsum, per 10 mm of thickness | 0.08 |
| Insulating concrete, per 10 mm | 0.06 |

| **Walls** | **Unplastered** | **One side Plastered** | **Both sides Plastered** |
|---|---|---|---|
| 100 mm brick wall | 1.86 | 2.10 | 2.33 |
| 200 mm brick wall | 3.77 | 4.00 | 4.24 |
| 300 mm brick wall | 5.59 | 5.83 | 6.06 |
| 100 mm hollow normal density concrete block | 1.37 | 1.61 | 1.84 |
| 150 mm hollow normal density concrete block | 1.67 | 1.90 | 2.14 |
| 200 mm hollow normal density concrete block | 2.11 | 2.34 | 2.58 |
| 300 mm hollow normal density concrete block | 2.94 | 3.18 | 3.39 |
| 100 mm hollow low density block or tile | 1.08 | 1.31 | 1.55 |
| 150 mm hollow low density block or tile | 1.28 | 1.51 | 1.75 |
| 200 mm hollow low density block or tile | 1.62 | 1.85 | 2.09 |
| 300 mm hollow low density block or tile | 2.26 | 2.49 | 2.73 |
| 100 mm brick 100 mm hollow normal density block backing | 3.24 | 3.47 | 3.71 |
| 100 mm brick 200 mm hollow normal density block backing | 3.97 | 4.21 | 4.44 |
| 100 mm brick 300 mm hollow normal density block backing | 4.81 | 5.04 | 5.28 |
| 100 mm brick 100 mm hollow low density block or tile backing | 2.94 | 3.18 | 3.41 |
| 100 mm brick 200 mm hollow low density block or tile backing | 3.48 | 3.72 | 3.95 |
| 100 mm brick 300 mm hollow low density block or tile backing | 4.12 | 4.35 | 4.59 |
| Windows, glass, frame and sash | 0.38 | | |
| 100 mm stone | 2.59 | | |
| Steel or wood studs, lath, 20 mm plaster | 0.86 | | |
| Steel or wood studs, lath, 15.9 mm gypsum board on each side | 0.28 | | |
| Steel or wood studs,2 layers 12.7 mm gypsum board on each side | 0.44 | | |

# DESIGN INFORMATION

**Fig. 8.1.2 Recommended minimum floor or roof live loads**

| Uniformly Distributed Loads | |
|---|---|
| **Use of Area of Floor or Roof** | **Minimum Specified Load kN/m²** |
| Assembly Areas | |
| (a) Except for those areas listed under (b) and (c), assembly areas with or without fixed seats include | |
| Arenas | |
| Auditoria | |
| Churches | |
| Dance floors | |
| Dining Area | |
| Foyers and entrance halls | |
| Grandstands, reviewing stands and bleachers | |
| Gymnasia | |
| Museums | |
| Promenades | |
| Rinks | |
| Stadia | |
| Stages | |
| Theaters | |
| and other areas with similar uses | 4.8 |
| (b) Assembly areas with fixed seats that have backs over at least 80 per cent of the assembly area for the following uses: | |
| Churches | |
| Courtrooms | |
| Lecture halls | |
| Theaters | 2.4 |
| (c) Classrooms with or without fixed seats | 2.4 |
| Attics | |
| Accessible by stairway in residential occupancies only | 1.4 |
| Having limited accessibility so that there is no storage of equipment or material | 0.5 |
| Balconies, exterior | 4.8 |
| Balconies, interior and mezzanines that could be used for the assembly of people as a viewing area | 4.8 |

| Uniformly Distributed Loads | |
|---|---|
| **Use of Area of Floor or Roof** | **Minimum Specified Load kN/m²** |
| Corridors lobbies and aisles not more than 1200 mm in width and all upper floor corridors of residential areas only of apartments, hotels and motels (that cannot be used for the assembly of people as a viewing area) | 4.8 |
| Equipment areas and service rooms including | 3.6 |
| Generator rooms | |
| Mechanical equipment exclusive of elevators | |
| Machine rooms | |
| Pump rooms | |
| Transformer vaults | |
| Ventilating or air-conditioning equipment | |
| Exits and fire escapes | 4.8 |
| Factories | 6.0 |
| Footbridges | 4.8 |
| Garages for | |
| Passenger cars | 2.4 |
| Unloaded buses and light trucks | 6.0 |
| Loaded buses and trucks and all other trucking spaces | 12.0 |
| Kitchens | 4.8 |
| Libraries | |
| Stack rooms | 7.2 |
| Reading and study rooms | 2.9 |
| Office areas in office buildings and other buildings (not including record storage and computer rooms) located in | |
| Basement and first storey | 4.8 |
| Floors above first storey | 2.4 |
| Operating rooms and laboratories | 3.6 |
| Patients' bedrooms | 1.9 |

# DESIGN INFORMATION

**Fig. 8.1.2 (cont'd) Recommended minimum floor or roof live loads**

| Uniformly Distributed Loads | |
|---|---|
| **Use of Area of Floor or Roof** | **Minimum Specified Load kN/m²** |
| Recreation areas that cannot be used for assembly purposes including<br>Billiard rooms<br>Bowling alleys<br>Pool rooms | 3.6 |
| Residential areas<br>Sleeping and living quarters in apartments, hotels, motels, boarding schools and colleges | 1.9 |
| Residential areas | |
| Bedrooms | 1.4 |
| Other areas | 1.9 |
| Stairs within dwelling units | 1.9 |
| Retail and wholesale areas | 4.8 |
| Roofs | 1.0 |
| Sidewalks and driveways over areas and basements | 12.0 |
| Storage areas | 4.8 |
| Toilet areas | 2.4 |
| Warehouses | 4.8 |

| Specified Concentrated Live Loads on an Area of Floor or Roof | |
|---|---|
| **Area of Floor or Roof** | **Minimum Specified Concentrated Load, kN** |
| Roof Surfaces | 1.3 |
| Floors and Classrooms | 4.5 |
| Floors and offices, manufacturing buildings, hospital wards and stages | 9.0 |
| Floors and areas used by passenger cars | 11 |
| Floors and areas used by vehicles not exceeding 3600 kg gross weight | 18 |
| Floors and areas used by vehicles exceeding 3600 kg but not exceeding 9000 kg gross weight | 36 |
| Floors and areas used by vehicles exceeding 9000 kg gross weight | 54 |
| Driveways and sidewalks over area ways and basements | 54 |

Bleacher seats shall be designed for a uniformly distributed load of 1.75kN for each linear metre or for a concentrated load of 2.2 kN distributed over a length of 0.75 m, whichever produces the greatest effect on the supporting members.

Helicopter landing areas on roofs shall be constructed in conformance with the regulations for heliports established by Transport Canada.

Where the use of an area of floor is not provided for above, the specified live loads due to the use and occupancy of an area shall be determined from an analysis of the loads resulting from

(a) the weight of the probable assembly of persons,

(b) the weight of the probable accumulation of equipment and furnishings, and

(c) the weight of the probable storage of materials.

## Concentrated Loads

The specified load due to possible concentrations of load resulting from the use of an area of floor or roof shall not be less than that listed below applied over an area of 750 mm by 750 mm located so as to cause maximum effects.

# DESIGN INFORMATION

**Fig. 8.1.3 Beam design equations and diagrams**

**(1)**

**Simple Beam—uniformly distributed load**

R = V .................................. $= \frac{w\ell}{2}$

$V_x$ .................................. $= w\left(\frac{\ell}{2} - x\right)$

M max. (at centre) .................................. $= \frac{w\ell^2}{8}$

$M_x$ .................................. $= \frac{wx}{2}(\ell - x)$

$\Delta$max. (at centre) .................................. $= \frac{5w\ell^4}{384\,EI}$

$\Delta_x$ .................................. $= \frac{wx}{24\,EI}(\ell^3 - 2\ell x^2 + x^3)$

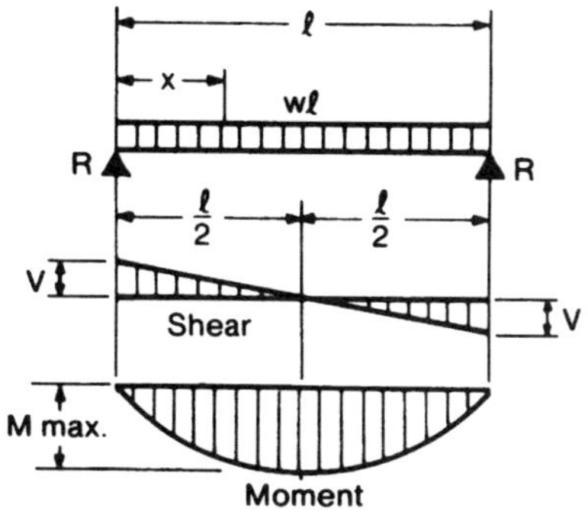


**(2)**

**Simple Beam—concentrated load at centre**

R = V .................................. $= \frac{P}{2}$

M max. (at point of load) .................................. $= \frac{P\ell}{4}$

$M_x \left(\text{when } x < \frac{\ell}{2}\right)$ .................................. $= \frac{Px}{2}$

$\Delta$max. (at point of load) .................................. $= \frac{P\ell^3}{48\,EI}$

$\Delta_x \left(\text{when } x < \frac{\ell}{2}\right)$ .................................. $= \frac{Px}{48\,EI}(3\ell^2 - 4x^2)$

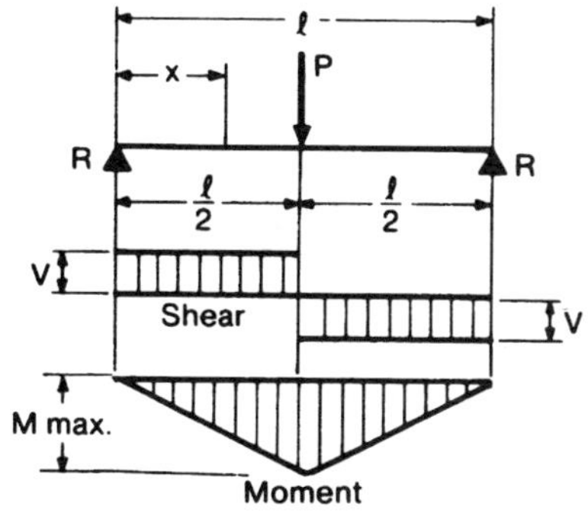


**(3)**

**Simple Beam—concentrated load at any point**

$R_1 = V_1$ (max. when a < b) .................................. $= \frac{Pb}{\ell}$

$R_2 = V_2$ (max. when a > b) .................................. $= \frac{Pa}{\ell}$

M max. (at point of load) .................................. $= \frac{Pab}{\ell}$

$M_x$ (when x < a) .................................. $= \frac{Pbx}{\ell}$

$\Delta$max. $\left(\text{at } x = \sqrt{\frac{a(a+2b)}{3}} \text{ when } a > b\right)$ .. $= \frac{Pab(a+2b)\sqrt{3a(a+2b)}}{27\,EI\ell}$

$\Delta a$ (at point of load) .................................. $= \frac{Pa^2b^2}{3\,EI\ell}$

$\Delta_x$ (when x < a) .................................. $= \frac{Pbx}{6\,EI\ell}(\ell^2 - b^2 - x^2)$

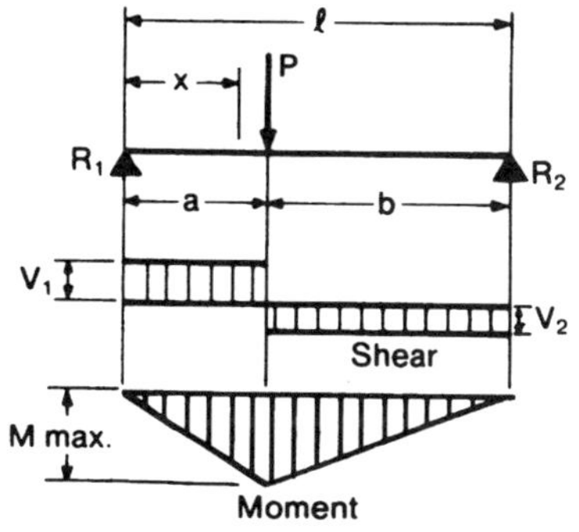


**(4)**

**Simple Beam—two equal concentrated loads symmetrically placed**

R = V .................................. = P

M max. (between loads) .................................. = Pa

$M_x$ (when x < a) .................................. = Px

$\Delta$max. (at centre) .................................. $= \frac{Pa}{24\,EI}(3\ell^2 - 4a^2)$

$\Delta_x$ (when x < a) .................................. $= \frac{Px}{6\,EI}(3\ell a - 3a^2 - x^2)$

$\Delta_x$ (when x > a and < ($\ell$ − a)) .................................. $= \frac{Pa}{6\,EI}(3\ell x - 3x^2 - a^2)$

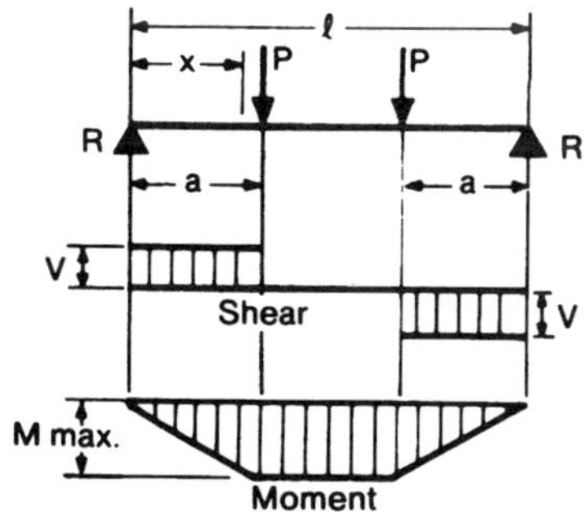

# DESIGN INFORMATION

**Fig. 8.1.3 (Cont'd) Beam design equations and diagrams**

**(5)**

**Simple Beam—two unequal concentrated loads unsymmetrically placed**

$R_1 = V_1 \ldots = \dfrac{P_1(\ell - a) + P_2 b}{\ell}$

$R_2 = V_2 \ldots = \dfrac{P_1 a + P_2(\ell - b)}{\ell}$

$V_x$ (when $x > a$ and $< (\ell - b)$) $\ldots = R_1 - P_1$

$M_1$ (max. when $R_1 < P_1$) $\ldots = R_1 a$

$M_2$ (max. when $R_2 < P_2$) $\ldots = R_2 b$

$M_x$ (when $x < a$) $\ldots = R_1 x$

$M_x$ (when $x > a$ and $< (\ell - b)$) $\ldots = R_1 x - P_1(x - a)$

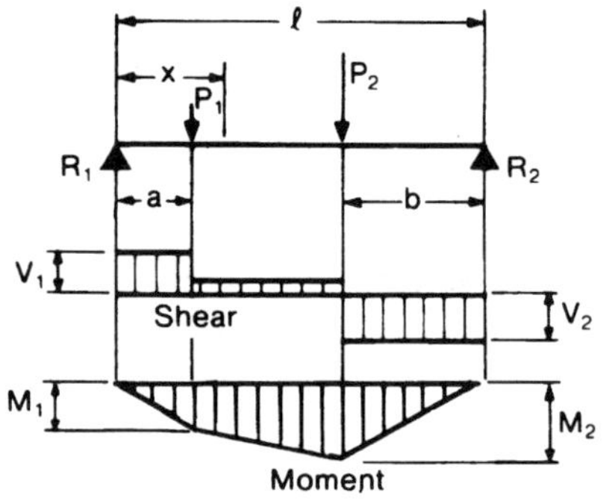


**(6)**

**Simple Beam—uniform load partially distributed**

$R_1 = V_1$ (max. when $a < c$) $\ldots = \dfrac{wb}{2\ell}(2c + b)$

$R_2 = V_2$ (max. when $a > c$) $\ldots = \dfrac{wb}{2\ell}(2a + b)$

$V_x$ (when $x > a$ and $< (a + b)$) $\ldots = R_1 - w(x - a)$

M max. $\left(\text{at } x = a + \dfrac{R_1}{w}\right) \ldots = R_1\left(a + \dfrac{R_1}{2w}\right)$

$M_x$ (when $x < a$) $\ldots = R_1 x$

$M_x$ (when $x > a$ and $< (a + b)$) $\ldots = R_1 x - \dfrac{w}{2}(x - a)^2$

$M_x$ (when $x > (a + b)$) $\ldots = R_2(\ell - x)$

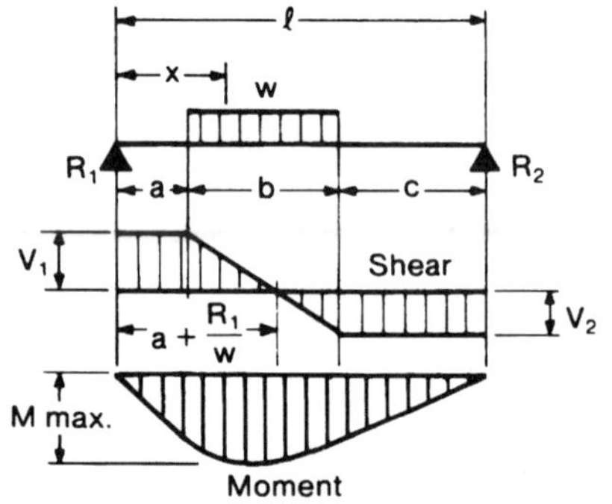


**(7)**

**Simple Beam—load increasing uniformly to one end**

$R_1 = V_1 \ldots = \dfrac{W}{3}$

$R_2 = V_2$ max. $\ldots = \dfrac{2W}{3}$

$V_x \ldots = \dfrac{W}{3} - \dfrac{Wx^2}{\ell^2}$

M max. $\left(\text{at } x = \dfrac{\ell}{\sqrt{3}} = 0.5774\,\ell\right) \ldots = \dfrac{2W\ell}{9\sqrt{3}} = 0.1283\,W\ell$

$M_x \ldots = \dfrac{Wx}{3\ell^2}(\ell^2 - x^2)$

$\Delta$max. $\left(\text{at } x = \ell\sqrt{1 - \sqrt{\dfrac{8}{15}}} = 0.5193\,\ell\right) = 0.01304\,\dfrac{W\ell^3}{EI}$

$\Delta_x \ldots = \dfrac{Wx}{180\,EI\,\ell^2}(3x^4 - 10\ell^2x^2 + 7\ell^4)$

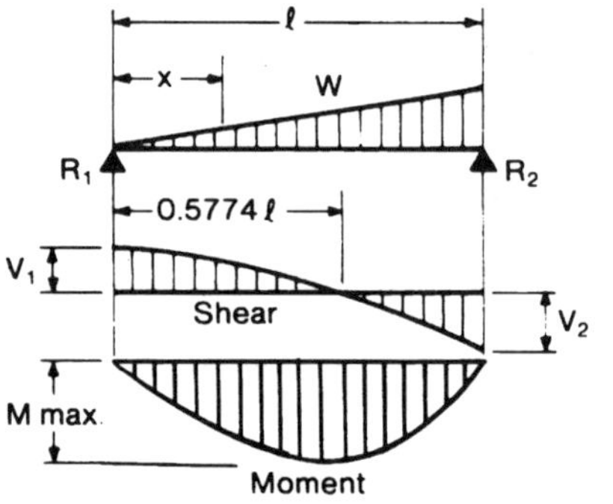


**(8)**

**Simple Beam—load increasing uniformly to centre**

$R = V \ldots = \dfrac{W}{2}$

$V_x \left(\text{when } x < \dfrac{\ell}{2}\right) \ldots = \dfrac{W}{2\ell^2}(\ell^2 - 4x^2)$

M max. (at centre) $\ldots = \dfrac{W\ell}{6}$

$M_x \left(\text{when } x < \dfrac{\ell}{2}\right) \ldots = Wx\left(\dfrac{1}{2} - \dfrac{2x^2}{3\ell^2}\right)$

$\Delta$max. (at centre) $\ldots = \dfrac{W\ell^3}{60\,EI}$

$\Delta_x \ldots = \dfrac{Wx}{480\,EI\,\ell^2}(5\ell^2 - 4x^2)^2$

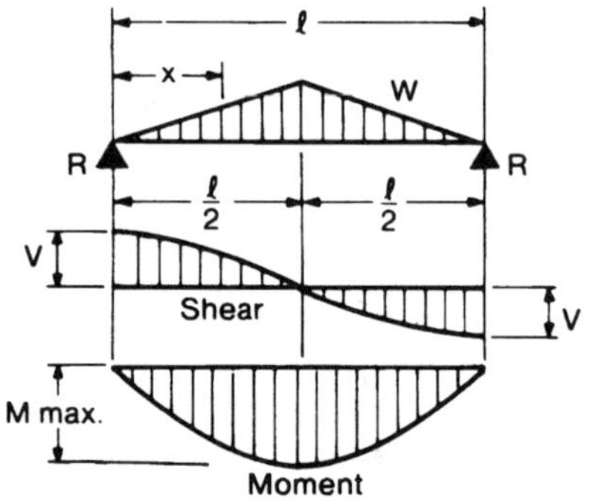

# DESIGN INFORMATION

**Fig. 8.1.3 (Cont'd) Beam design equations and diagrams**

**(9)**

**Beam overhanging one support—uniformly distributed load**

$R_1 = V_1$ . . . . . . . . . . $= \frac{w}{2\ell}(\ell^2 - a^2)$

$R_2 = V_2 + V_3$ . . . . . . . . . . $= \frac{w}{2\ell}(\ell + a)^2$

$V_2$ . . . . . . . . . . $= wa$

$V_3$ . . . . . . . . . . $= \frac{w}{2\ell}(\ell^2 + a^2)$

$V_x$ (between supports) . . . . . . . . . . $= R_1 - wx$

$V_{x1}$ (for overhang) . . . . . . . . . . $= w(a - x_1)$

$M_1 \left(\text{at } x = \frac{\ell}{2}\left[1 - \frac{a^2}{\ell^2}\right]\right)$ . . . . . . . . . . $= \frac{w}{8\ell^2}(\ell + a)^2(\ell - a)^2$

$M_2$ (at $R_2$) . . . . . . . . . . $= \frac{wa^2}{2}$

$M_x$ (between supports) . . . . . . . . . . $= \frac{wx}{2\ell}(\ell^2 - a^2 - x\ell)$

$M_{x1}$ (for overhang) . . . . . . . . . . $= \frac{w}{2}(a - x_1)^2$

$\Delta_x$ (between supports) . . . . . . . . . . $= \frac{wx}{24EI\ell}(\ell^4 - 2\ell^2x^2 + \ell x^3 - 2a^2\ell^2 + 2a^2x^2)$

$\Delta_{x1}$ (for overhang) . . . . . . . . . . $= \frac{wx_1}{24EI}(4a^2\ell - \ell^3 + 6a^2x_1 - 4ax_1^2 + x_1^3)$

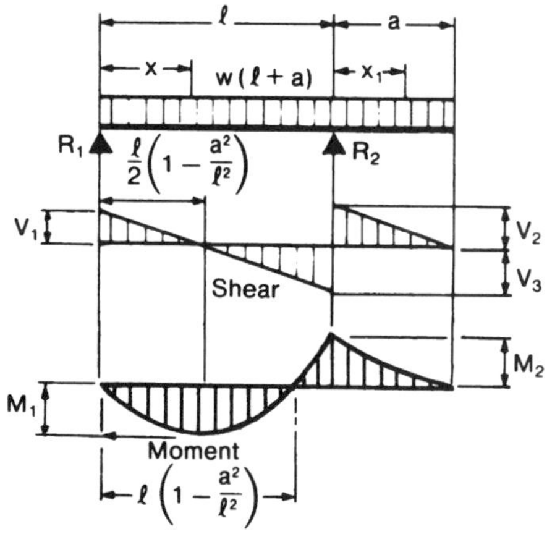


**(10)**

**Beam overhanging one support—uniformly distributed load on overhang**

$R_1 = V_1$ . . . . . . . . . . $= \frac{wa^2}{2\ell}$

$R_2 = V_1 + V_2$ . . . . . . . . . . $= \frac{wa}{2\ell}(2\ell + a)$

$V_2$ . . . . . . . . . . $= wa$

$V_{x1}$ (for overhang) . . . . . . . . . . $= w(a - x_1)$

M max. (at $R_2$) . . . . . . . . . . $= \frac{wa^2}{2}$

$M_x$ (between supports) . . . . . . . . . . $= \frac{wa^2x}{2\ell}$

$M_{x1}$ (for overhang) . . . . . . . . . . $= \frac{w}{2}(a - x_1)^2$

$\Delta$max. $\left(\text{between supports at } x = \frac{\ell}{\sqrt{3}}\right)$ . . . $= \frac{wa^2\ell^2}{18\sqrt{3}EI} = 0.03208\frac{wa^2\ell^2}{EI}$

$\Delta$max. (for overhang at $x_1 = a$) . . . . . . . . . . $= \frac{wa^3}{24EI}(4\ell + 3a)$

$\Delta_x$ (between supports) . . . . . . . . . . $= \frac{wa^2x}{12EI\ell}(\ell^2 - x^2)$

$\Delta_{x1}$ (for overhang) . . . . . . . . . . $= \frac{wx_1}{24EI}(4a^2\ell + 6a^2x_1 - 4ax_1^2 + x_1^3)$

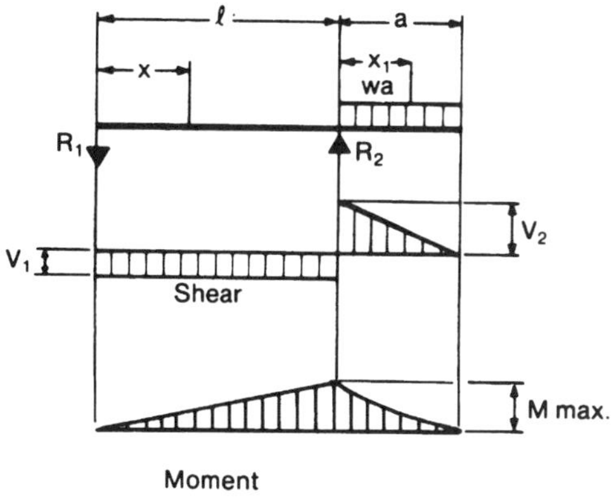


**(11)**

**Beam overhanging one support—uniformly distributed load between supports**

$R = V$ . . . . . . . . . . $= \frac{w\ell}{2}$

$V_x$ . . . . . . . . . . $= w\left(\frac{\ell}{2} - x\right)$

M max. (at centre) . . . . . . . . . . $= \frac{w\ell^2}{8}$

$M_x$ . . . . . . . . . . $= \frac{wx}{2}(\ell - x)$

$\Delta$max. (at centre) . . . . . . . . . . $= \frac{5w\ell^4}{384EI}$

$\Delta_x$ . . . . . . . . . . $= \frac{wx}{24EI}(\ell^3 - 2\ell x^2 + x^3)$

$\Delta_{x1}$ . . . . . . . . . . $= \frac{w\ell^3x_1}{24EI}$

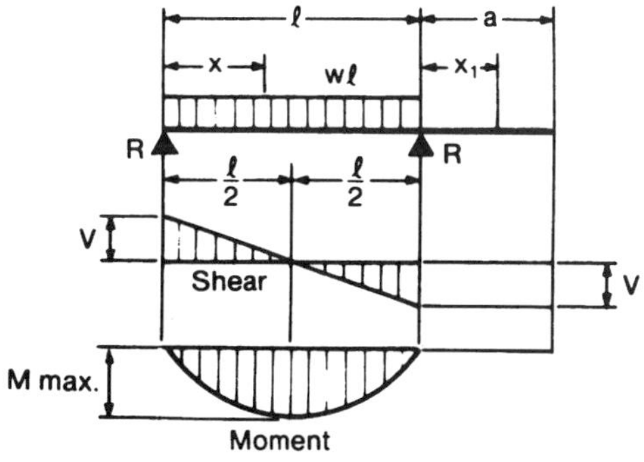

# DESIGN INFORMATION

**Fig. 8.1.3 (Cont'd) Beam design equations and diagrams**

**(12)**

**Beam overhanging one support—concentrated load at any point between supports**

$R_1 = V_1$ (max. when $a < b$) . . . . . . . $= \frac{Pb}{\ell}$

$R_2 = V_2$ (max. when $a > b$) . . . . . . . $= \frac{Pa}{\ell}$

M max. (at point of load) . . . . . . . $= \frac{Pab}{\ell}$

$M_x$ (when $x < a$) . . . . . . . $= \frac{Pbx}{\ell}$

$\Delta$max. $\left(\text{at } x = \sqrt{\frac{a(a+2b)}{3}} \text{ when } a > b\right) = \frac{Pab(a+2b)\sqrt{3a(a+2b)}}{27\,EI\ell}$

$\Delta a$ (at point of load) . . . . . . . $= \frac{Pa^2b^2}{3\,EI\ell}$

$\Delta_x$ (when $x < a$) . . . . . . . $= \frac{Pbx}{6\,EI\ell}(\ell^2 - b^2 - x^2)$

$\Delta_x$ (when $x > a$) . . . . . . . $= \frac{Pa(\ell - x)}{6\,EI\ell}(2\ell x - x^2 - a^2)$

$\Delta_{x1}$ . . . . . . . $= \frac{Pabx_1}{6\,EI\ell}(\ell + a)$

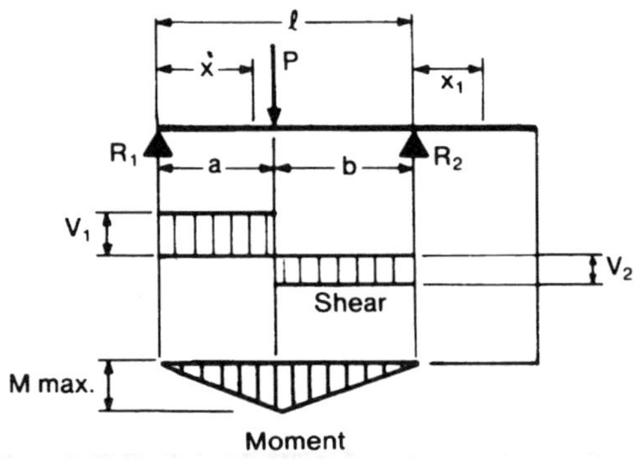


**(13)**

**Beam overhanging one support—concentrated load at end of overhang**

$R_1 = V_1$ . . . . . . . $= \frac{Pa}{\ell}$

$R_2 = V_1 + V_2$ . . . . . . . $= \frac{P}{\ell}(\ell + a)$

$V_2$ . . . . . . . $= P$

M max. (at $R_2$) . . . . . . . $= Pa$

$M_x$ (between supports) . . . . . . . $= \frac{Pax}{\ell}$

$M_{x1}$ (for overhang) . . . . . . . $= P(a - x_1)$

$\Delta$max. $\left(\text{between supports at } x = \frac{\ell}{\sqrt{3}}\right)$ . . . $= \frac{Pa\ell^2}{9\sqrt{3}\,EI} = 0.06415\,\frac{Pa\ell^2}{EI}$

$\Delta$max. (for overhang at $x_1 = a$) . . . . . . . $= \frac{Pa^2}{3\,EI}(\ell + a)$

$\Delta_x$ (between supports) . . . . . . . $= \frac{Pax}{6\,EI\ell}(\ell^2 - x^2)$

$\Delta_{x1}$ (for overhang) . . . . . . . $= \frac{Px_1}{6\,EI}(2a\ell + 3ax_1 - x_1^2)$

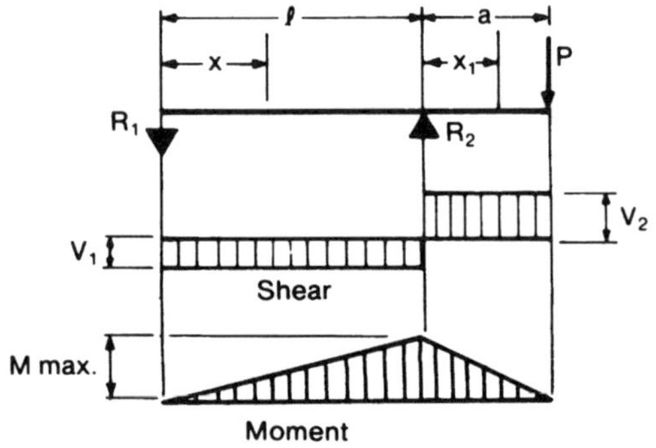

# DESIGN INFORMATION

**Fig. 8.1.3 (Cont'd) Beam design equations and diagrams**

**(14)**

**Cantilever Beam—uniformly distributed load**

R = V .................................. $= w\ell$

$V_x$ .................................. $= wx$

M max. (at fixed end) .................. $= \frac{w\ell^2}{2}$

$M_x$ .................................. $= \frac{wx^2}{2}$

$\Delta$max. (at free end) .............. $= \frac{w\ell^4}{8\,EI}$

$\Delta_x$ .............................. $= \frac{w}{24\,EI}(x^4 - 4\ell^3 x + 3\ell^4)$

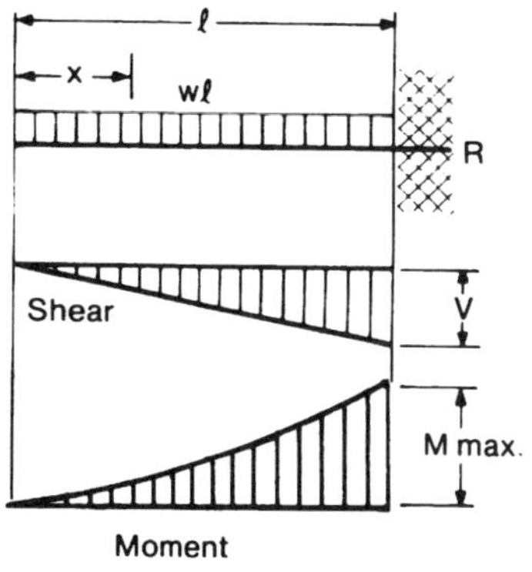


**(15)**

**Cantilever Beam—concentrated load at free end**

R = V .................................. $= P$

M max. (at fixed end) .................. $= P\ell$

$M_x$ .................................. $= Px$

$\Delta$max. (at free end) .............. $= \frac{P\ell^3}{3\,EI}$

$\Delta_x$ .............................. $= \frac{P}{6\,EI}(2\ell^3 - 3\ell^2 x + x^3)$

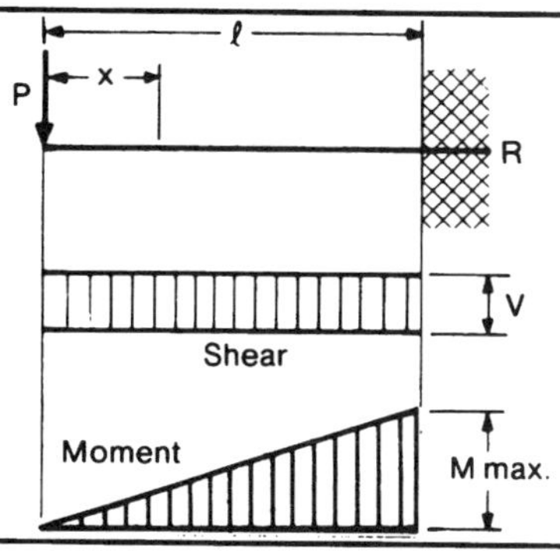


**(16)**

**Cantilever Beam—concentrated load at any point**

R = V .................................. $= P$

M max. (at fixed end) .................. $= Pb$

$M_x$ (when x > a) ..................... $= P(x - a)$

$\Delta$max. (at free end) .............. $= \frac{Pb^2}{6\,EI}(3\ell - b)$

$\Delta$a (at point of load) ............ $= \frac{Pb^3}{3\,EI}$

$\Delta_x$ (when x < a) .................. $= \frac{Pb^2}{6\,EI}(3\ell - 3x - b)$

$\Delta_x$ (when x > a) .................. $= \frac{P(\ell - x)^2}{6\,EI}(3b - \ell + x)$

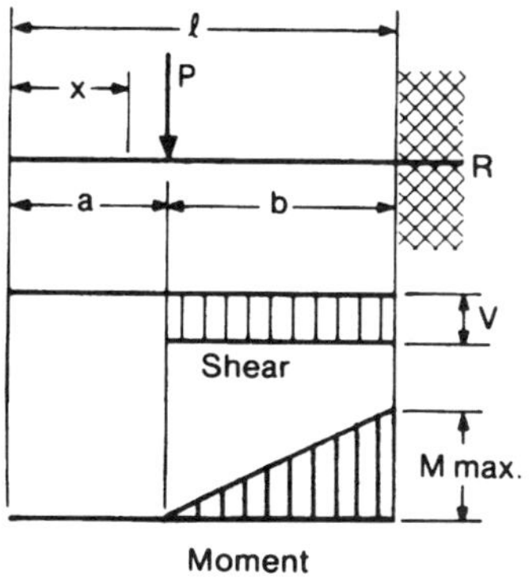


**(17)**

**Cantilever Beam—load increasing uniformly to fixed end**

R = V .................................. $= W$

$V_x$ .................................. $= W\frac{x^2}{\ell^2}$

M max. (at fixed end) .................. $= \frac{W\ell}{3}$

$M_x$ .................................. $= \frac{Wx^3}{3\ell^2}$

$\Delta$max. (at free end) .............. $= \frac{W\ell^3}{15\,EI}$

$\Delta_x$ .............................. $= \frac{W}{60\,EI\ell^2}(x^5 - 5\ell^4 x + 4\ell^5)$

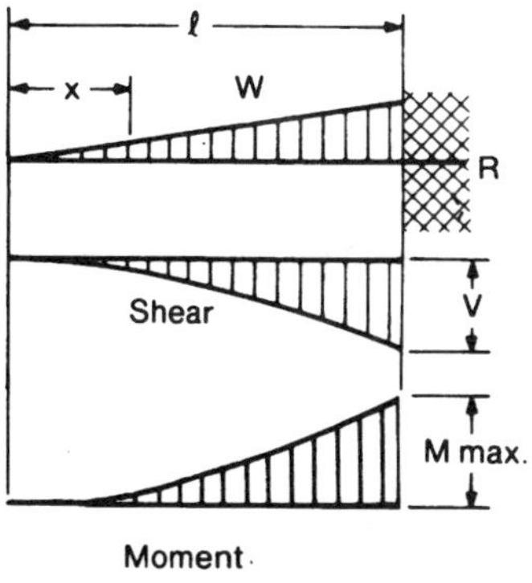

# DESIGN INFORMATION

**Fig. 8.1.3 (Cont'd) Beam design equations and diagrams**

**(18)**

**Beam fixed at one end, supported at other—uniformly distributed load**

$R_1 = V_1$ ........ $= \frac{3w\ell}{8}$

$R_2 = V_2$ max. ........ $= \frac{5w\ell}{8}$

$V_x$ ........ $= R_1 - wx$

M max. ........ $= \frac{w\ell^2}{8}$

$M_1 \left(\text{at } x = \frac{3}{8}\ell\right)$ ........ $= \frac{9}{128} w\ell^2$

$M_x$ ........ $= R_1 x - \frac{wx^2}{2}$

$\Delta$max. $\left(\text{at } x = \frac{\ell}{16}(1 + \sqrt{33}) = 0.4215\right)$ ........ $= \frac{w\ell^4}{185\,EI}$

$\Delta_x$ ........ $= \frac{wx}{48\,EI}(\ell^3 - 3\ell x^2 + 2x^3)$

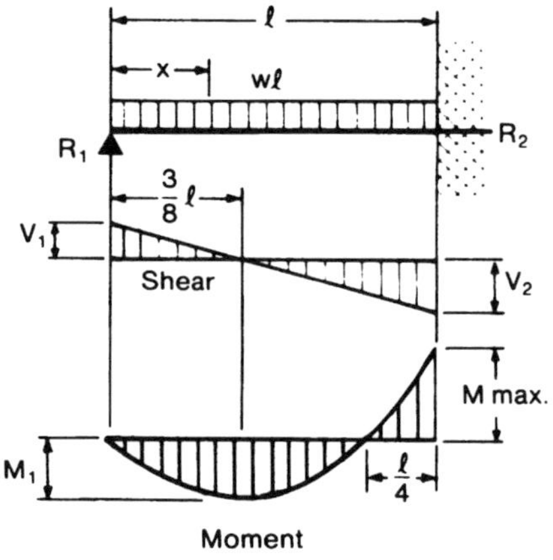


**(19)**

**Beam fixed at one end, supported at other—concentrated load at centre**

$R_1 = V_1$ ........ $= \frac{5P}{16}$

$R_2 = V_2$ max. ........ $= \frac{11P}{16}$

M max. (at fixed end) ........ $= \frac{3P\ell}{16}$

$M_1$ (at point of load) ........ $= \frac{5P\ell}{32}$

$M_x \left(\text{when } x < \frac{\ell}{2}\right)$ ........ $= \frac{5Px}{16}$

$M_x \left(\text{when } x > \frac{\ell}{2}\right)$ ........ $= P\left(\frac{\ell}{2} - \frac{11x}{16}\right)$

$\Delta$max. $\left(\text{at } x = \ell\sqrt{\frac{1}{5}} = 0.4472\ell\right)$ ........ $= \frac{P\ell^3}{48\,EI\sqrt{5}} = 0.009317\frac{P\ell^3}{EI}$

$\Delta_x$ (at point of load) ........ $= \frac{7P\ell^3}{768\,EI}$

$\Delta_x \left(\text{when } x < \frac{\ell}{2}\right)$ ........ $= \frac{Px}{96\,EI}(3\ell^2 - 5x^2)$

$\Delta_x \left(\text{when } x > \frac{\ell}{2}\right)$ ........ $= \frac{P}{96\,EI}(x - \ell)^2(11x - 2\ell)$

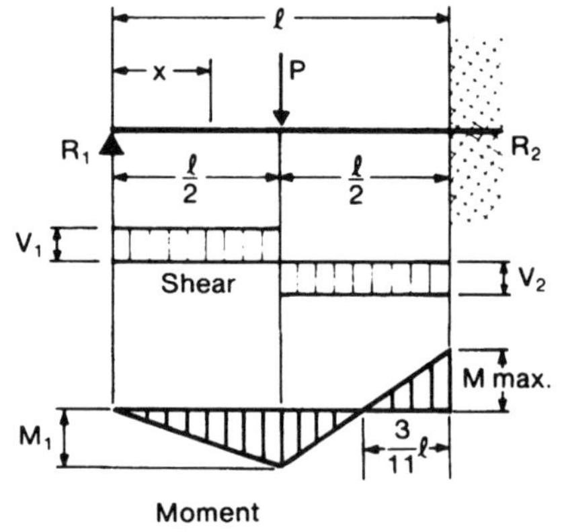


**(20)**

**Beam fixed at one end, supported at other—concentrated load at any point**

$R_1 = V_1$ ........ $= \frac{Pb^2}{2\ell^3}(a + 2\ell)$

$R_2 = V_2$ ........ $= \frac{Pa}{2\ell^3}(3\ell^2 - a^2)$

$M_1$ (at point of load) ........ $= R_1 a$

$M_2$ (at fixed end) ........ $= \frac{Pab}{2\ell^2}(a + \ell)$

$M_x$ (when $x < a$) ........ $= R_1 x$

$M_x$ (when $x > a$) ........ $= R_1 x - P(x - a)$

$\Delta$max. $\left(\text{when } a < 0.414\ell \text{ at } x = \ell\frac{\ell^2 + a^2}{3\ell^2 - a^2}\right) = \frac{Pa}{3\,EI}\frac{(\ell^2 - a^2)^3}{(3\ell^2 - a^2)^2}$

$\Delta$max. $\left(\text{when } a > 0.414\ell \text{ at } x = \ell\sqrt{\frac{a}{2\ell + a}}\right) = \frac{Pab^2}{6\,EI}\sqrt{\frac{a}{2\ell + a}}$

$\Delta a$ (at point of load) ........ $= \frac{Pa^2b^3}{12\,EI\ell^3}(3\ell + a)$

$\Delta x$ (when $x < a$) ........ $= \frac{Pb^2x}{12\,EI\ell^3}(3a\ell^2 - 2\ell x^2 - ax^2)$

$\Delta x$ (when $x > a$) ........ $= \frac{Pa}{12\,EI\ell^3}(\ell - x)^2(3\ell^2 x - a^2 x - 2a^2\ell)$

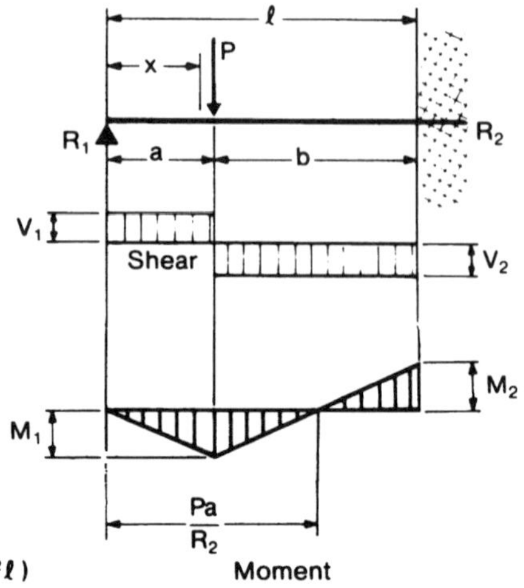

# DESIGN INFORMATION

**Fig. 8.1.3 (Cont'd) Beam design equations and diagrams**

**(21)**

**Beam fixed at both ends—uniformly distributed loads**

$R = V$ .............................. $= \dfrac{w\ell}{2}$

$V_x$ .............................. $= w\left(\dfrac{\ell}{2} - x\right)$

M max. (at ends).............................. $= \dfrac{w\ell^2}{12}$

$M_1$ (at centre).............................. $= \dfrac{w\ell^2}{24}$

$M_x$.............................. $= \dfrac{w}{12}(6\ell x - \ell^2 - 6x^2)$

$\Delta$max. (at centre).............................. $= \dfrac{w\ell^4}{384\,EI}$

$\Delta_x$.............................. $= \dfrac{wx^2}{24\,EI}(\ell - x)^2$

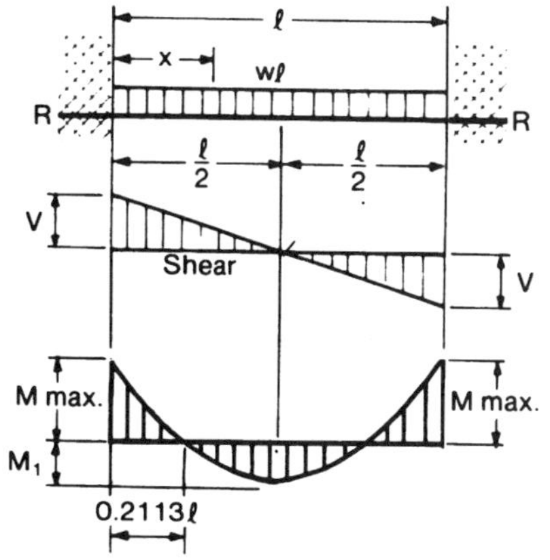


**(22)**

**Beam fixed at both ends—concentrated load at any point**

$R_1 = V_1$ (max. when $a < b$).............................. $= \dfrac{Pb^2}{\ell^3}(3a + b)$

$R_2 = V_2$ (max. when $a > b$).............................. $= \dfrac{Pa^2}{\ell^3}(a + 3b)$

$M_1$ (max. when $a < b$).............................. $= \dfrac{Pab^2}{\ell^2}$

$M_2$ (max. when $a > b$).............................. $= \dfrac{Pa^2b}{\ell^2}$

$M_a$ (at point of load).............................. $= \dfrac{2Pa^2b^2}{\ell^3}$

$M_x$ (when $x < a$).............................. $= R_1x - \dfrac{Pab^2}{\ell^2}$

$\Delta$max. $\left(\text{when } a > b \text{ at } x = \dfrac{2a\ell}{3a + b}\right)$.............................. $= \dfrac{2Pa^3b^2}{3\,EI\,(3a + b)^2}$

$\Delta_a$ (at point of load).............................. $= \dfrac{Pa^3b^3}{3\,EI\ell^3}$

$\Delta_x$ (when $x < a$).............................. $= \dfrac{Pb^2x^2}{6\,EI\ell^3}(3a\ell - 3ax - bx)$

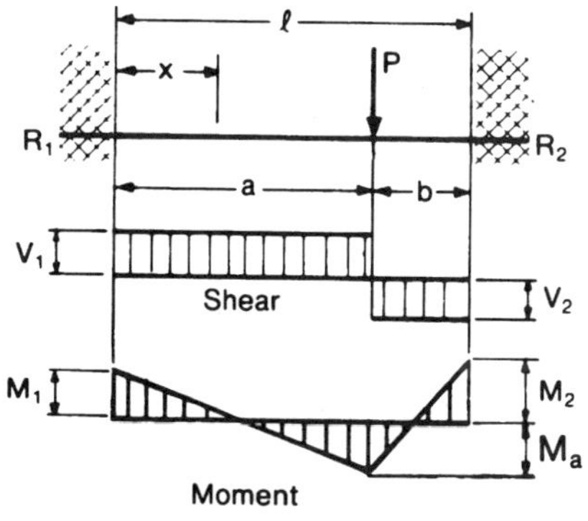

# DESIGN INFORMATION

**Fig. 8.1.3 (Cont'd) Beam design equations and diagrams**

(23) Beam-uniformly distributed loads and variable end moments.

$$R_1 = V_1 = \frac{w\ell}{2} + \frac{M_1 - M_2}{\ell}$$

$$R_2 = V_2 = \frac{w\ell}{2} - \frac{M_1 - M_2}{\ell}$$

$$V_x = w\left(\frac{\ell}{2} - x\right) + \frac{M_1 - M_2}{\ell}$$

$$M_3 \left(\text{at } x = \frac{\ell}{2} + \frac{M_1 - M_2}{w\ell}\right) = \frac{w\ell^2}{8} - \frac{M_1 + M_2}{2} + \frac{(M_1 - M_2)^2}{2w\ell^2}$$

$$M_x = \frac{wx}{8}(\ell - x) + \left(\frac{M_1 - M_2}{\ell}\right)x - M_1$$

$$b\left(\begin{matrix}\text{To locate}\\ \text{inflection points}\end{matrix}\right) = \sqrt{\frac{\ell^2}{4} - \left(\frac{M_1 + M_2}{w}\right) + \left(\frac{M_1 - M_2}{w\ell}\right)^2}$$

$$\Delta_x = \frac{wx}{24EI}\left[x^3 - \left(2\ell + \frac{4M_1}{w\ell} - \frac{4M_2}{w\ell}\right)x^2 + \frac{12M_1}{w}x + \ell^3 - \frac{8M_1\ell}{w} - \frac{4M_2\ell}{w}\right]$$

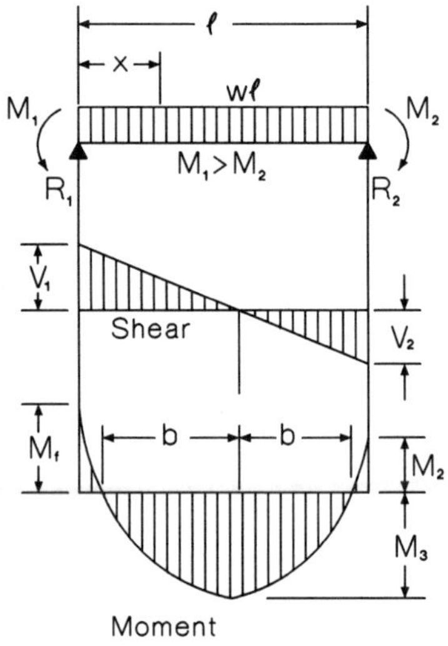


(24) Beam-concentrated load at center and variable end moments.

$$R_1 = V_1 = \frac{P}{2} + \frac{M_1 - M_2}{\ell}$$

$$R_2 = V_2 = \frac{P}{2} - \frac{M_1 - M_2}{\ell}$$

$$M_3(\text{At center}) = \frac{P\ell}{4} - \frac{M_1 + M_2}{2}$$

$$M_x\left(\text{When } x < \frac{\ell}{2}\right) = \left(\frac{P}{2} + \frac{M_1 - M_2}{\ell}\right)x - M_1$$

$$M_x\left(\text{When } x > \frac{\ell}{2}\right) = \frac{P}{2}(\ell - x) + \frac{(M_1 - M_2)x}{\ell} - M_1$$

$$\Delta_x\left(\text{When } x < \frac{\ell}{2}\right) = \frac{Px}{48EI}\left(3\ell^2 - 4x^2 - \frac{8(\ell - x)}{P\ell}[M_1(2\ell - x) + M_2(\ell + x)]\right)$$

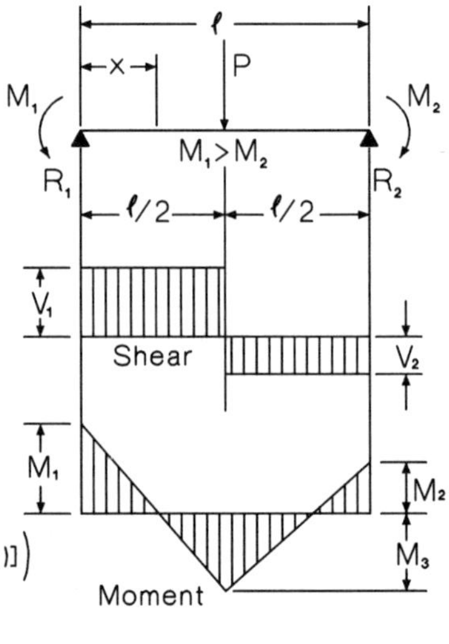

# DESIGN INFORMATION

**Fig. 8.1.3 (Cont'd) Beam design equations and diagrams**

(25) Beam fixed one end—differential settlement of supports

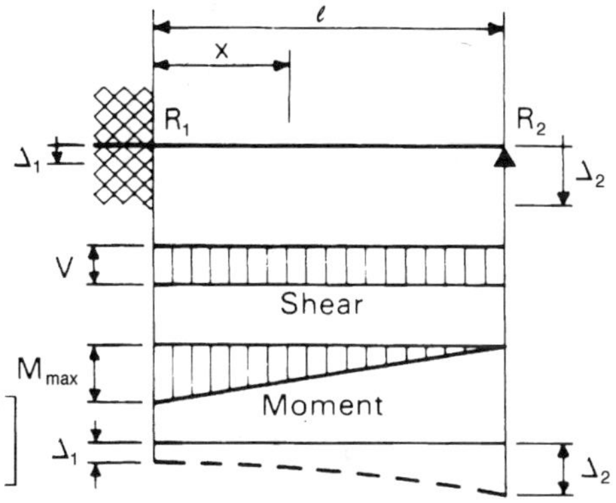


$V = R_1 = R_2 \ldots\ldots = \frac{3EI}{\ell^3}(\Delta_2 - \Delta_1)$

$M_{max} \ldots\ldots = \frac{3EI}{\ell^2}(\Delta_2 - \Delta_1)$

$M_x \ldots\ldots = M_{max}\left(1 - \frac{x}{\ell}\right)$

$\Delta_x \ldots\ldots = \Delta_1 + \frac{\Delta_2 - \Delta_1}{2}\left[3\left(\frac{x}{\ell}\right)^2 - \left(\frac{x}{\ell}\right)^3\right]$

(26) Beam fixed one end—rotation of support

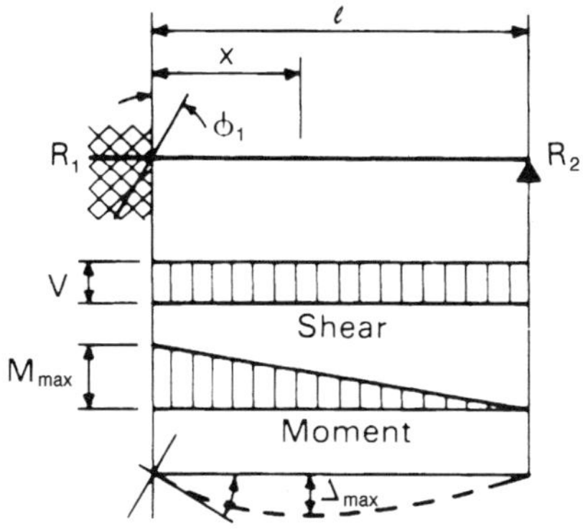


$V = R_1 = R_2 \ldots\ldots = \frac{3EI}{\ell^2}\phi_1$

$M_{max} \ldots\ldots = \frac{3EI}{\ell}\phi_1$

$M_x \ldots\ldots = M_{max}\left(1 - \frac{x}{\ell}\right)$

$\Delta_{max} \ldots\ldots = \phi_1\left[\frac{\ell}{5.196}\right]$

$\Delta_x \ldots\ldots = \phi_1\left[-x + \frac{3x^2}{2\ell} - \frac{x^2}{2\ell^2}\right]$

(27) Beam fixed both ends—differential settlement of supports

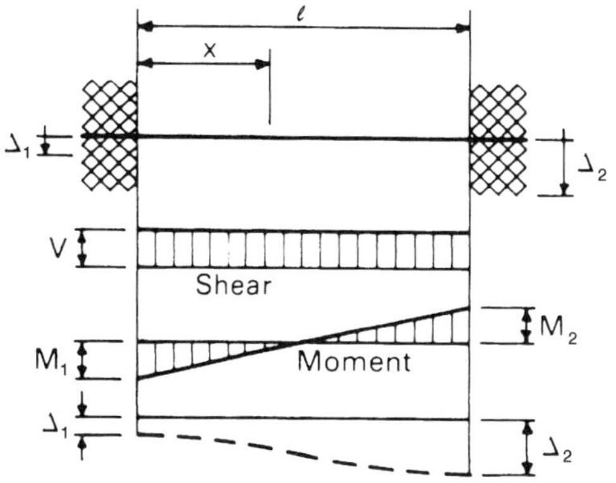


$V = R_1 = R_2 \ldots\ldots = \frac{12\,EI}{\ell^3}(\Delta_2 - \Delta_1)$

$M_1 = M_2 \ldots\ldots = \frac{6EI}{\ell^2}(\Delta_2 - \Delta_1)$

$M_x \ldots\ldots = \frac{6EI}{\ell^2}(\Delta_2 - \Delta_1)\left(1 - \frac{2x}{\ell}\right)$

$\Delta_x \ldots\ldots = \Delta_1 + (\Delta_2 - \Delta_1)\left[3\left(\frac{x}{\ell}\right)^2 - 2\left(\frac{x}{\ell}\right)^3\right]$

(28) Beam fixed both ends—rotation of support

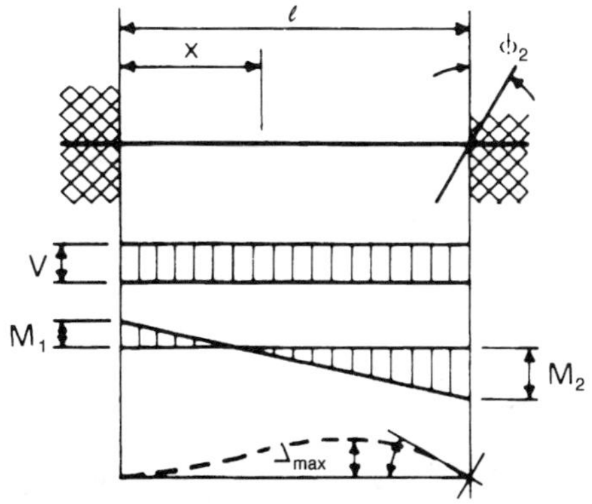


$V = R_1 = R_2 \ldots\ldots = \frac{6EI}{\ell^2}\phi_2$

$M_1 \ldots\ldots = \frac{2EI}{\ell}\phi_2$

$M_2 \ldots\ldots = \frac{4EI}{\ell}\phi_2$

$M_x \ldots\ldots = \frac{2EI}{\ell}\phi_2\left(1 - \frac{3x}{\ell}\right)$

$\Delta_{max}\ (\text{at } x = \frac{2}{3}\ell) \ldots\ldots = -\frac{4}{27}\ell\phi_2$

$\Delta_x \ldots\ldots = -\ell\phi_2\left[\left(\frac{x}{\ell}\right)^2 - \left(\frac{x}{\ell}\right)^3\right]$

# DESIGN INFORMATION

**Fig. 8.1.4 Camber (deflection) and rotation coefficients for prestress force and loads***

| Prestress Pattern | Equivalent Moment or Load | Equivalent Loading | Camber + ↑ | End Rotation (left end) + | End Rotation (right end) + |
|---|---|---|---|---|---|
| (1) 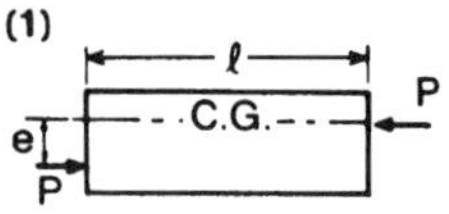 | $M = Pe$ | 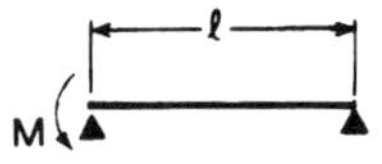 | $+\frac{M\ell^2}{16\,EI}$ | $+\frac{M\ell}{3\,EI}$ | $-\frac{M\ell}{6\,EI}$ |
| (2) 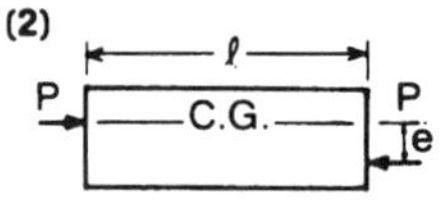 | $M = Pe$ | 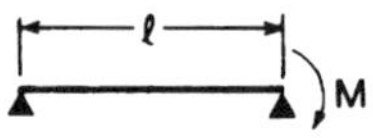 | $+\frac{M\ell^2}{16\,EI}$ | $+\frac{M\ell}{6\,EI}$ | $-\frac{M\ell}{3\,EI}$ |
| (3) 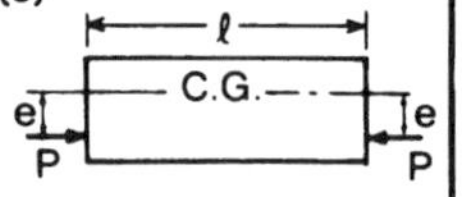 | $M = Pe$ | 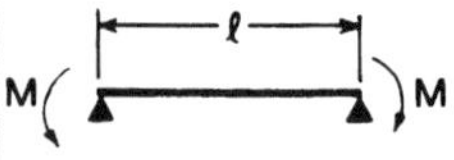 | $+\frac{M\ell^2}{8\,EI}$ | $+\frac{M\ell}{2\,EI}$ | $-\frac{M\ell}{2\,EI}$ |
| (4) 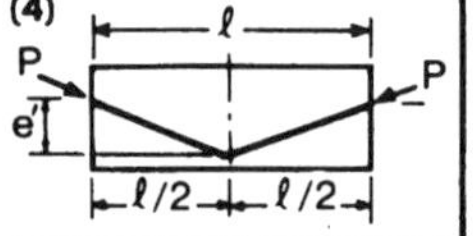 | $N = \frac{4\,Pe'}{\ell}$ | 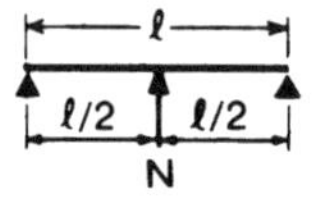 | $+\frac{N\ell^3}{48\,EI}$ | $+\frac{N\ell^2}{16\,EI}$ | $-\frac{N\ell^2}{16\,EI}$ |
| (5) 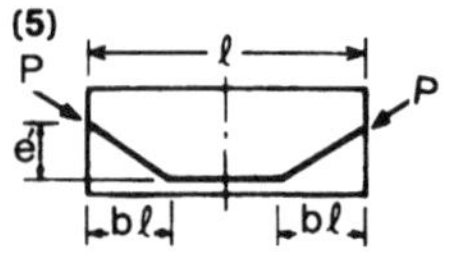 | $N = \frac{Pe'}{b\ell}$ | 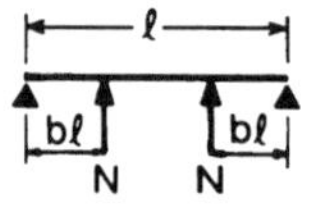 | $+\frac{b(3-4b^2)N\ell^3}{24\,EI}$ | $+\frac{b(1-b)N\ell^2}{2\,EI}$ | $-\frac{b(1-b)N\ell^2}{2\,EI}$ |
| (6) 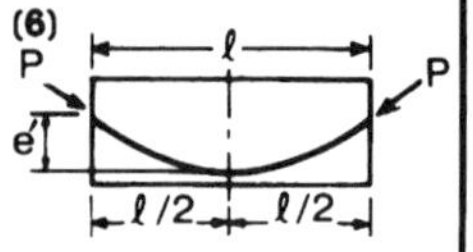 | $w = \frac{8\,Pe'}{\ell^2}$ | 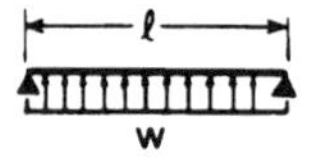 | $+\frac{5w\ell^4}{384\,EI}$ | $+\frac{w\ell^3}{24\,EI}$ | $-\frac{w\ell^3}{24\,EI}$ |
| (7) 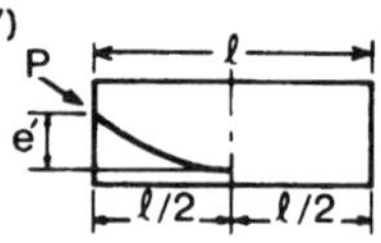 | $w = \frac{8\,Pe'}{\ell^2}$ | 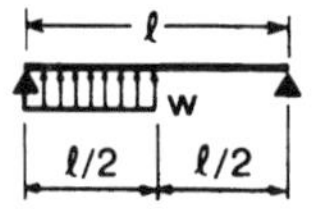 | $+\frac{5w\ell^4}{768\,EI}$ | $+\frac{9w\ell^3}{384\,EI}$ | $-\frac{7w\ell^3}{384\,EI}$ |
| (8) 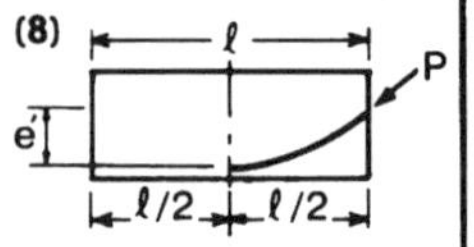 | $w = \frac{8\,Pe'}{\ell^2}$ | 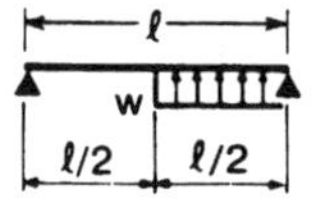 | $+\frac{5w\ell^4}{768\,EI}$ | $+\frac{7w\ell^3}{384\,EI}$ | $-\frac{9w\ell^3}{384\,EI}$ |
| (9) 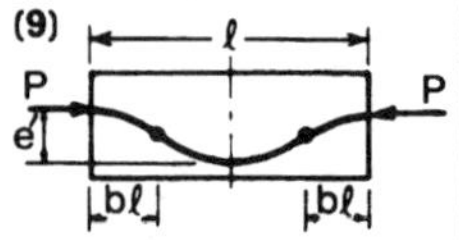 | $w = \frac{4\,Pe'}{(0.5-b)\,\ell^2}$; $w_1 = \frac{w}{b}(0.5-b)$ | 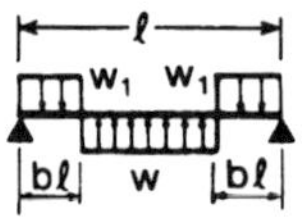 | $\left[\frac{5}{8}-\frac{b}{2}(3-2b^2)\right]\frac{w\ell^4}{48\,EI}$ | $+\frac{(1-b)(1-2b)\,w\ell^3}{24\,EI}$ | $-\frac{(1-b)(1-2b)\,w\ell^3}{24\,EI}$ |
| (10) 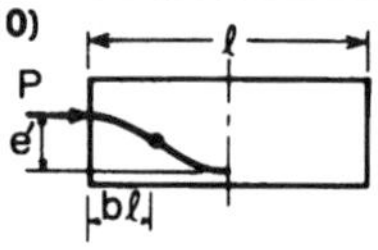 | $w = \frac{4\,Pe'}{(0.5-b)\,\ell^2}$; $w_1 = \frac{w}{b}(0.5-b)$ | 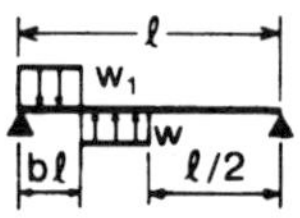 | $\left[\frac{5}{16}-\frac{b}{4}(3-2b^2)\right]\frac{w\ell^4}{48\,EI}$ | $\left[\frac{9}{8}-b(2-b)^2\right]\frac{w\ell^3}{48\,EI}$ | $\left[-\frac{7}{8}+b(2-b^2)\right]\frac{w\ell^3}{48\,EI}$ |
| (11)  | $w = \frac{4\,Pe'}{(0.5-b)\,\ell^2}$; $w_1 = \frac{w}{b}(0.5-b)$ | 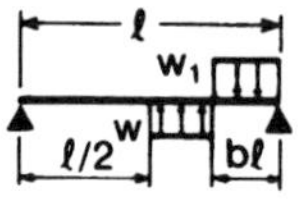 | $\left[\frac{5}{16}-\frac{b}{4}(3-2b^2)\right]\frac{w\ell^4}{48\,EI}$ | $\left[\frac{7}{8}-b(2-b^2)\right]\frac{w\ell^3}{48\,EI}$ | $\left[-\frac{9}{8}+b(2-b)^2\right]\frac{w\ell^3}{48\,EI}$ |

Add effects of end eccentricity (cases 1 to 3) as appropriate.

*The tabulated values apply to the effects of prestressing. By adjusting the directional notation, they may also be used for the effects of loads.

# DESIGN INFORMATION

**Fig. 8.1.5 Moments in beams with fixed ends**

| Loading | Moment at A | Moment at centre | Moment at B |
|---|---|---|---|
| (1) 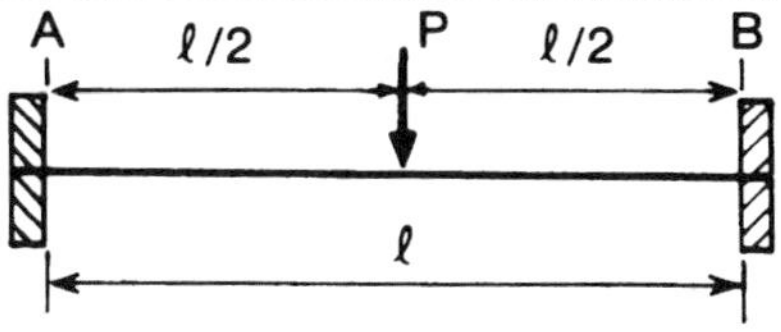  | $-\frac{P\ell}{8}$ | $+\frac{P\ell}{8}$ | $-\frac{P\ell}{8}$ |
| (2) 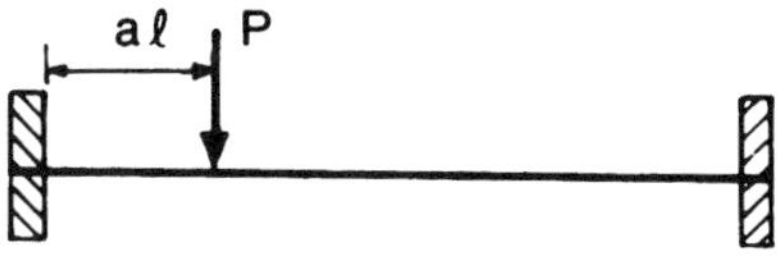  | $-P\ell a(1-a)^2$ | | $-P\ell a^2(1-a)$ |
| (3) 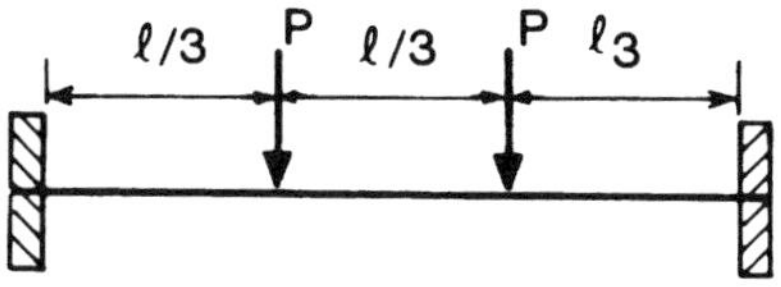  | $-\frac{2P\ell}{9}$ | $+\frac{P\ell}{9}$ | $-\frac{2P\ell}{9}$ |
| (4) 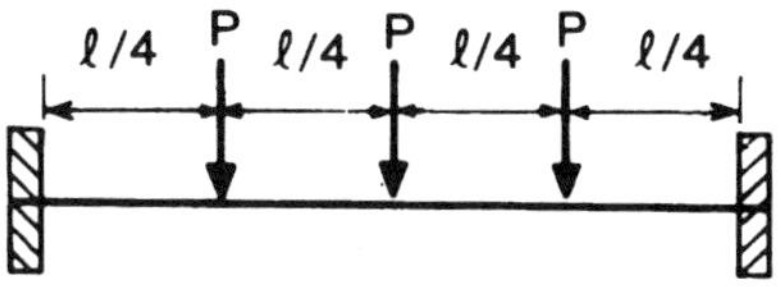  | $-\frac{5P\ell}{16}$ | $+\frac{3P\ell}{16}$ | $-\frac{5P\ell}{16}$ |
| (5) 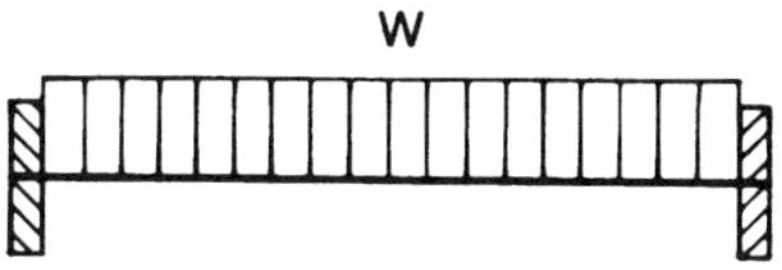  | $-\frac{W\ell}{12}$ | $+\frac{W\ell}{24}$ | $-\frac{W\ell}{12}$ |
| (6) 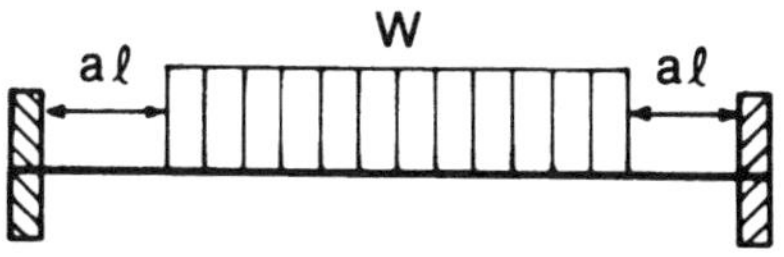  | $-\frac{W\ell(1+2a-2a^2)}{12}$ | $+\frac{W\ell(1+2a+4a^2)}{24}$ | $-\frac{W\ell(1+2a-2a^2)}{12}$ |
| (7) 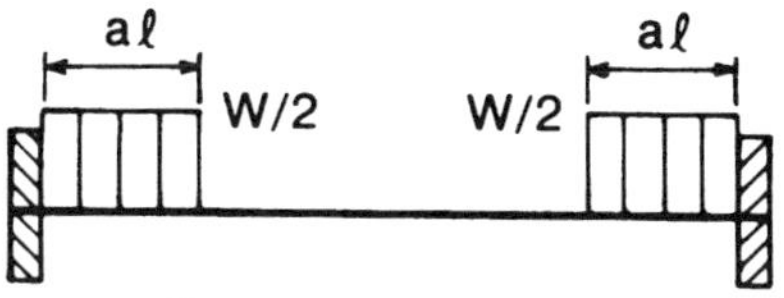  | $-\frac{W\ell(3a-2a^2)}{12}$ | $+\frac{W\ell a^2}{6}$ | $-\frac{W\ell(3a-2a^2)}{12}$ |
| (8) 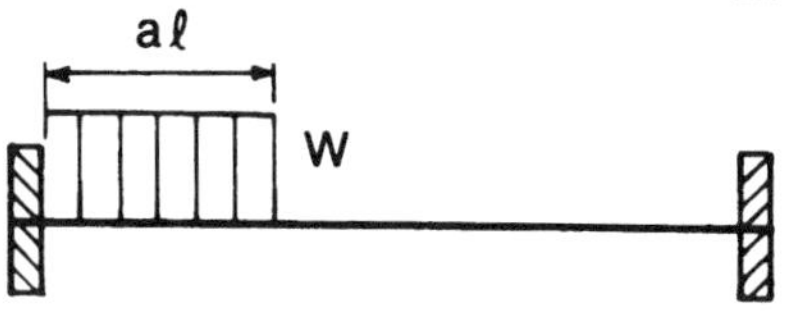  | $-\frac{W\ell a(6-8a+3a^2)}{12}$ | | $-\frac{W\ell a^2(4-3a)}{12}$ |
| (9) 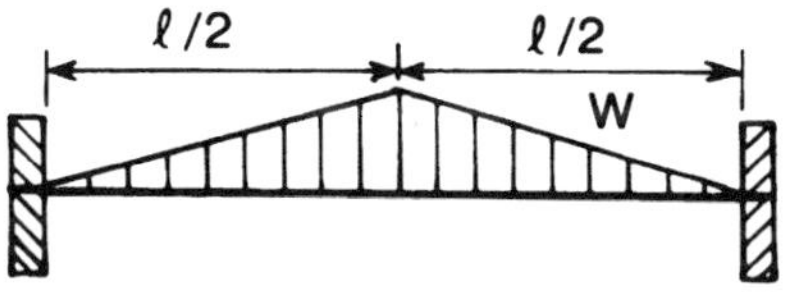  | $-\frac{5W\ell}{48}$ | $+\frac{3W\ell}{48}$ | $-\frac{5W\ell}{48}$ |
| (10) 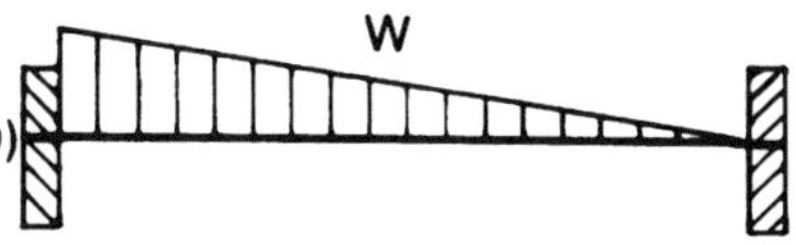  | $-\frac{W\ell}{10}$ | | $-\frac{W\ell}{15}$ |

W = Total load on beam

## DESIGN INFORMATION

**Fig. 8.1.6 Moving load placement for maximum moment and shear**

**(1) Simple Beam—one concentrated moving load**

$R_1$ max. = $V_1$ max. (at $x = 0$) ............... $= P$

M max. $\left(\text{at point of load, when } x = \dfrac{\ell}{2}\right)$ ... $= \dfrac{P\ell}{4}$

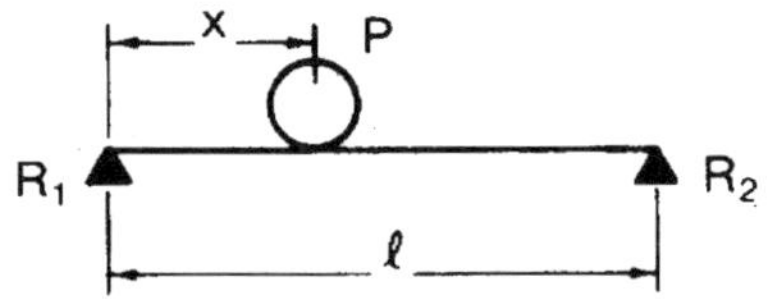


**(2) Simple Beam—two equal concentrated moving loads**

$R_1$ max. = $V_1$ max. (at $x = 0$) ............... $= P\left(2 - \dfrac{a}{\ell}\right)$

M max.:

- $\left[\begin{array}{l}\text{when } a < \left(2 - \sqrt{2}\right)\ell = 0.586\,\ell \\ \text{under load 1 at } x = \dfrac{1}{2}\left(\ell - \dfrac{a}{2}\right)\end{array}\right] = \dfrac{P}{2\ell}\left(\ell - \dfrac{a}{2}\right)^2$
- $\left[\begin{array}{l}\text{when } a > \left(2 - \sqrt{2}\right)\ell = 0.586\,\ell \\ \text{with one load at centre of span}\end{array}\right] = \dfrac{P\ell}{4}$

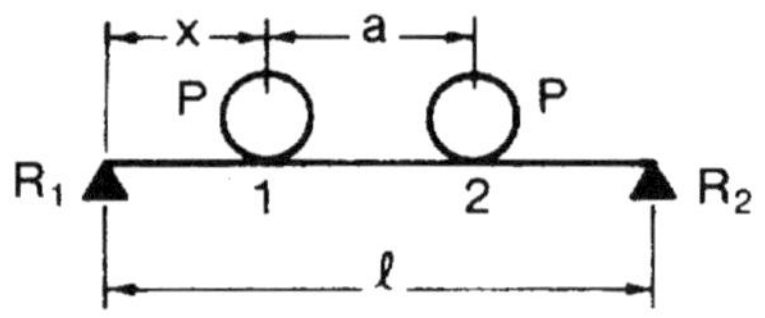


**(3) Simple Beam—two unequal concentrated moving loads**

$R_1$ max. = $V_1$ max. (at $x = 0$) ............... $= P_1 + P_2\left(\dfrac{\ell - a}{\ell}\right)$

M max.:

- $\left[\text{under } P_1\text{, at } x = \dfrac{1}{2}\left(\ell - \dfrac{P_2 a}{P_1 + P_2}\right)\right] = (P_1 + P_2)\,\dfrac{x^2}{\ell}$
- $\left[\begin{array}{l}\text{M max. may occur with larger} \\ \text{load at centre of span and other} \\ \text{load off span}\end{array}\right] = \dfrac{P_1\ell}{4}$

$P_1 > P_2$

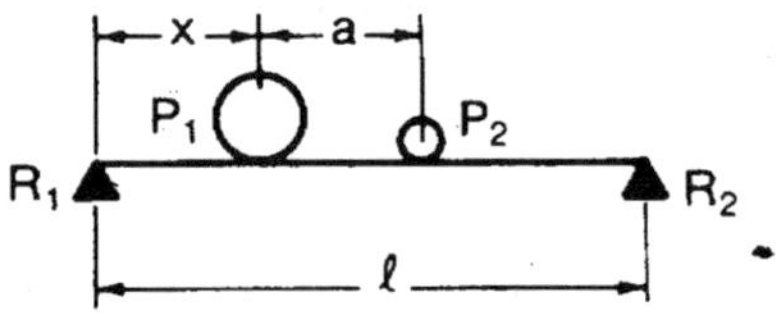

# CONCRETE MATERIAL PROPERTIES

**Fig. 8.2.1 Table of concrete stresses (MPa)**

| $f'_c$ | $0.45f'_c$ | $0.6f'_c$ | $\sqrt{f'_c}$ | $0.25\sqrt{f'_c}$ | $0.3\sqrt{f'_c}$ | $0.4\sqrt{f'_c}$ | $0.5\sqrt{f'_c}$ | $0.62\sqrt{f'_c}$ |
|---|---|---|---|---|---|---|---|---|
| 20 | 9.0 | 12.0 | 4.5 | 1.1 | 1.3 | 1.8 | 2.2 | 2.8 |
| 25 | 11.3 | 15.0 | 5.0 | 1.3 | 1.5 | 2.0 | 2.5 | 3.1 |
| 30 | 13.5 | 18.0 | 5.5 | 1.4 | 1.6 | 2.2 | 2.7 | 3.4 |
| 35 | 15.8 | 21.0 | 5.9 | 1.5 | 1.8 | 2.4 | 3.0 | 3.7 |
| 40 | 18.0 | 24.0 | 6.3 | 1.6 | 1.9 | 2.5 | 3.2 | 3.9 |
| 45 | 20.3 | 27.0 | 6.7 | 1.7 | 2.0 | 2.7 | 3.4 | 4.2 |
| 50 | 22.5 | 30.0 | 7.1 | 1.8 | 2.1 | 2.8 | 3.5 | 4.4 |
| 60 | 27.0 | 36.0 | 7.7 | 1.9 | 2.3 | 3.1 | 3.9 | 4.8 |
| 70 | 31.5 | 42.0 | 8.4 | 2.1 | 2.5 | 3.3 | 4.2 | 5.2 |
| 80 | 36.0 | 48.0 | 8.9 | 2.2 | 2.7 | 3.6 | 4.5 | 5.5 |

**Fig. 8.2.2 Concrete modulus of elasticity as affected by density and strength**

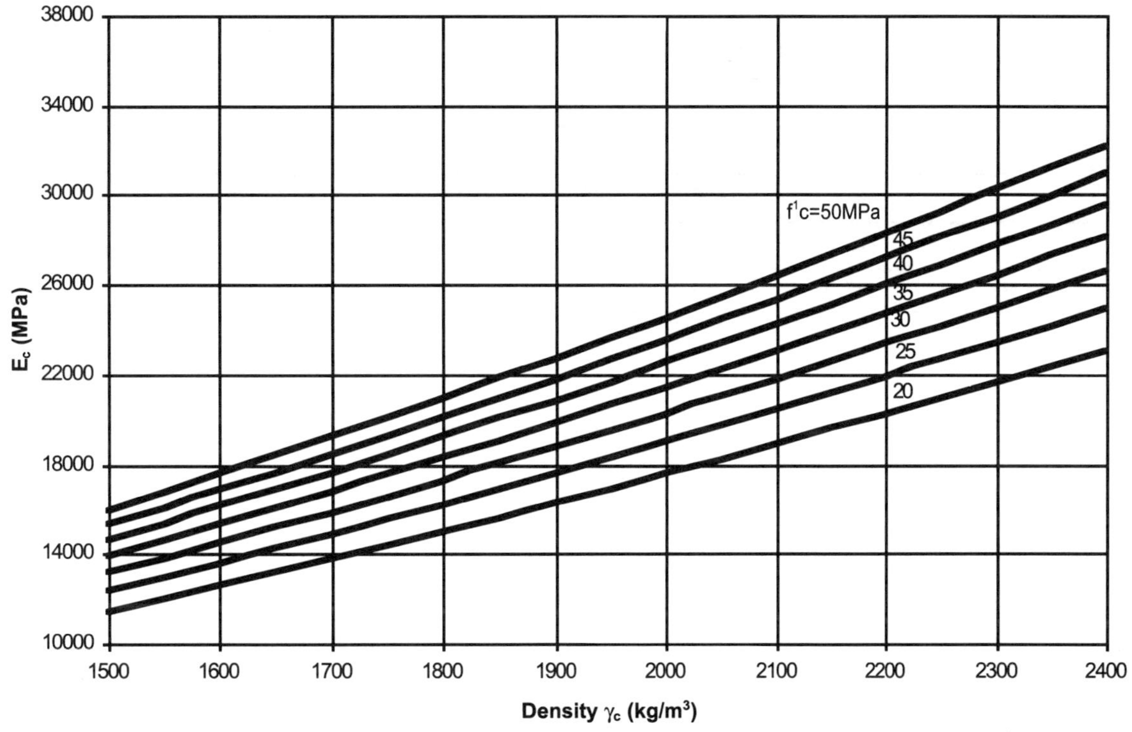

# MATERIAL PROPERTIES PRESTRESSING STEEL

**Fig. 8.2.3 Properties and resistance of prestressing strand and deformed bars**

| Seven Wire Strand, $f_{pu}$ = 1860 MPa | | | | | | |
|---|---|---|---|---|---|---|
| Size Designation | | 9 | 11 | 13 | 13 Special | 15 |
| Nominal Diameter | mm | 9.53 | 11.13 | 12.70 | 13.08 | 15.24 |
| Nominal Linear Mass | kg/m | 0.432 | 0.582 | 0.775 | 0.788 | 1.109 |
| Nominal Area | $mm^2$ | 54.8 | 74.2 | 98.7 | 107.7 | 140.0 |
| 0.7 $f_{pu}$ $A_p$ | kN | 71.3 | 96.6 | 128.5 | 140.5 | 182.3 |
| 0.8 $f_{pu}$ $A_p$ | kN | 81.5 | 110.4 | 146.9 | 160.6 | 208.3 |
| $f_{pu}$ $A_p$ | kN | 101.9 | 138.0 | 183.6 | 200.3 | 260.4 |

| Deformed Prestressing Bars | | | | | | | | |
|---|---|---|---|---|---|---|---|---|
| Grade $f_{pu}$ | MPa | 1080 | 1080 | 1030 | 1100 | 1030 | 1100 | 1030 |
| Size Designation | | 15 | 20 | 26 | 26 (1) | 32 | 32 (1) | 36 (2) |
| Nominal Diameter | mm | 15.0 | 20 | 26.5 | 26.5 | 32.0 | 32.0 | 36.0 |
| Nominal Linear Mass | kg/m | 1.44 | 2.59 | 4.48 | 4.48 | 6.53 | 6.53 | 8.27 |
| Nominal Area $A_p$ | $mm^2$ | 177 | 315 | 551 | 551 | 804 | 804 | 1018 |
| 0.7 $f_{pu}$ $A_p$ | kN | 133 | 238 | 397 | 424 | 579 | 618 | 734 |
| 0.8 $f_{pu}$ $A_p$ | kN | 153 | 272 | 454 | 484 | 662 | 707 | 839 |
| $f_{pu}$ $A_p$ | kN | 191 | 340 | 568 | 606 | 828 | 884 | 1049 |

(1) Available on special order. (2) Also available in Grade 1100.

# MATERIAL PROPERTIES PRESTRESSING STEEL

**Fig. 8.2.4 Typical stress-strain curve, 7-wire stress-relieved and low-relaxation prestressing strand**

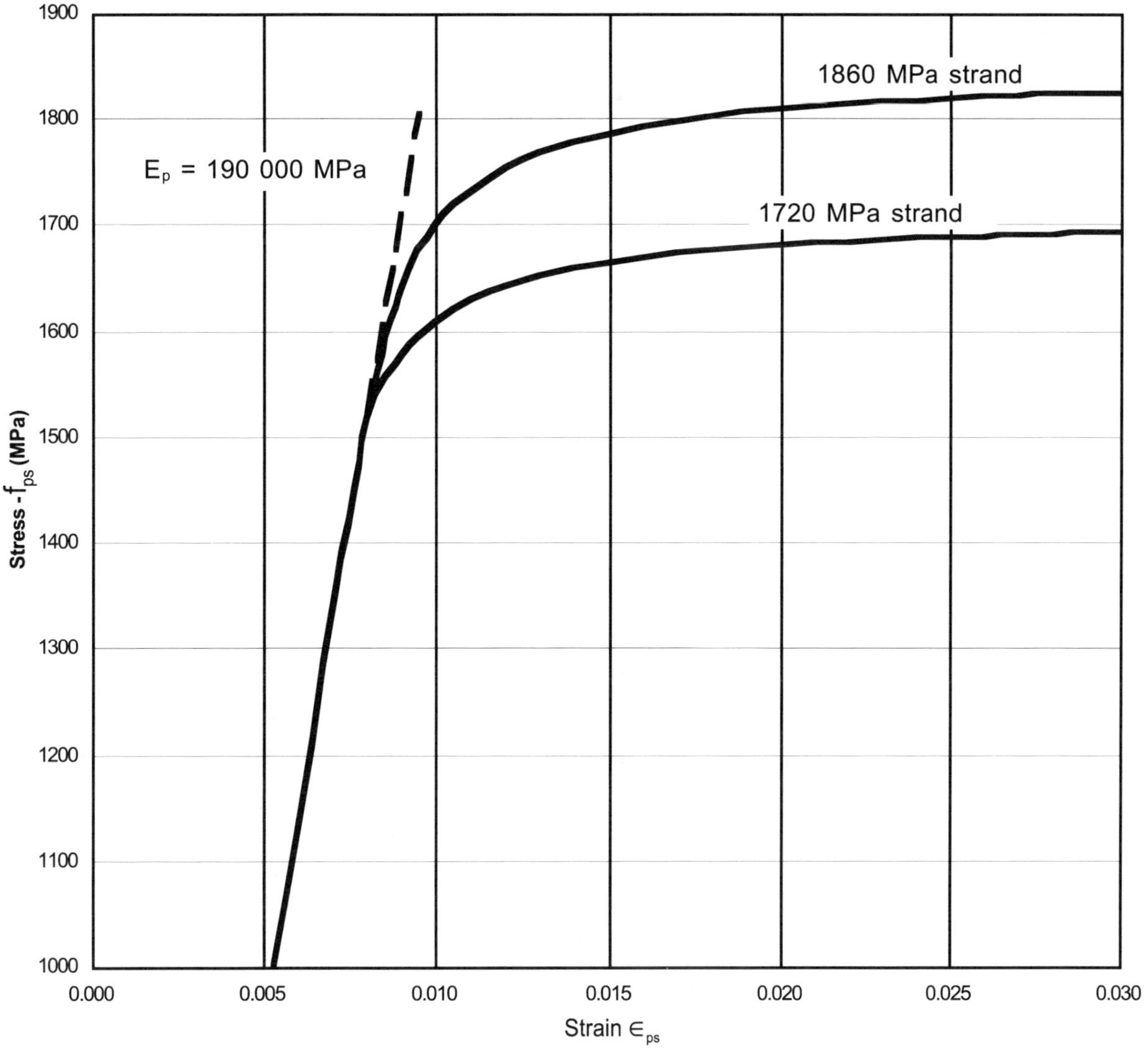


These curves are based on the following equations:

$\epsilon_{ps} \le 0.008$; $f_{ps} = 190\,000\,\epsilon_{ps}$ (MPa)

$\epsilon_{pr} > 0.008$

Grade 1720 Strand: $f_{ps} = 1710 - \dfrac{0.400}{\epsilon_{ps} - 0.006} < 0.98 f_{pu}$ (MPa)

Grade 1860 Strand: $f_{ps} = 1848 - \dfrac{0.517}{\epsilon_{ps} - 0.0065} < 0.98 f_{pu}$ (MPa)

# MATERIAL PROPERTIES POST-TENSIONING

**Fig. 8.2.5 Typical multi-strand post-tensioning anchorage details**

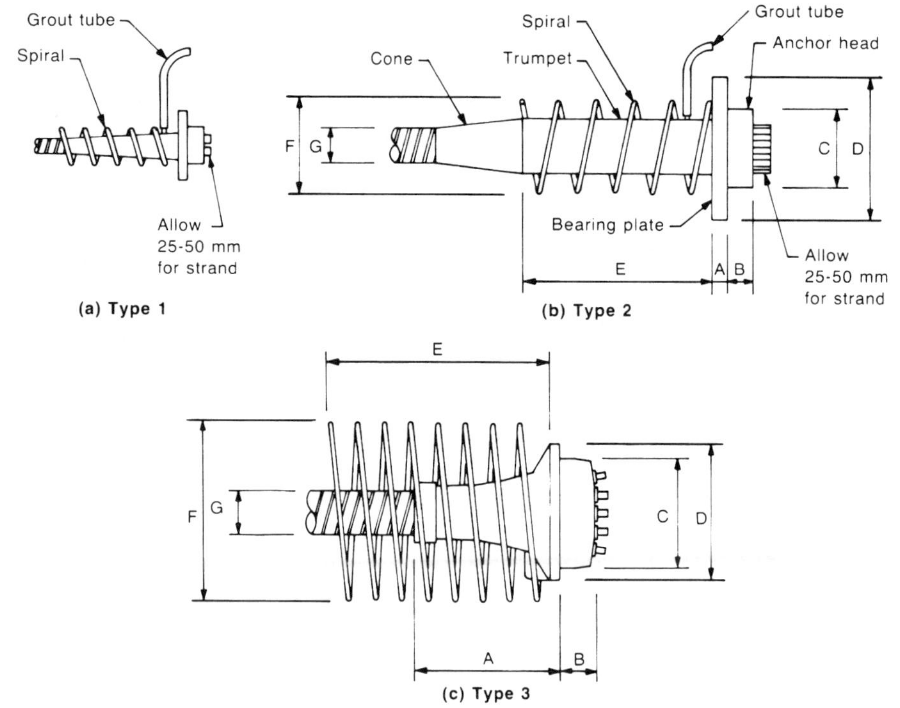


| Type | No. of strands | Strand Diameter (mm) | A (mm) | B (mm) | C (mm) | D (mm) | E (mm) | F (mm) | G I.D./O.D. (mm) | Spiral Diameter (mm) | No. of Turns |
|---|---|---|---|---|---|---|---|---|---|---|---|
| 1 | 3 | 13 | 15 | 50 | 80 | 135 | | | 40/45 | | |
| | 7 | 13 | 20 | 50 | 115 | 200 | 250 | 125 | 55/60 | 10 | 5 |
| 2 | 12 | 13 | 25 | 50 | 150 | 250 | 250 | 180 | 70/75 | 10 | 5 |
| | 19 | 13 | 30 | 63 | 185 | 330 | 250 | 215 | 85/90 | 10 | 5 |
| | 22 | 13 | 40 | 75 | 200 | 350 | 250 | 270 | 90/95 | 15 | 5 |
| | 31 | 13 | 45 | 85 | 230 | 420 | 250 | 280 | 110/115 | 15 | 5 |
| 3 | 7 | 13 | 122 | 51 | 120 | 139 | 260 | 200 | 55/60 | 10 | 7 |
| | 12 | 13 | 158 | 57 | 140 | 210 | 266 | 250 | 70/75 | 15 | 7 |
| | 19 | 13 | 190 | 60 | 185 | 245 | 266 | 320 | 85/90 | 15 | 7 |

| Type | No. of strands | Strand Diameter (mm) | A (mm) | B (mm) | C (mm) | D (mm) | E (mm) | F (mm) | G I.D./O.D. (mm) | Spiral Diameter (mm) | No. of Turns |
|---|---|---|---|---|---|---|---|---|---|---|---|
| 1 | 4 | 15 | 20 | 50 | 115 | 200 | 250 | 125 | 55/60 | 10 | 5 |
| 2 | 4 | 15 | 20 | 50 | 115 | 200 | 250 | 125 | 55/60 | 10 | 5 |
| | 7 | 15 | 25 | 50 | 150 | 250 | 250 | 180 | 65/70 | 10 | 5 |
| | 12 | 15 | 35 | 63 | 185 | 330 | 250 | 215 | 85/90 | 15 | 5 |
| | 19 | 15 | 45 | 75 | 230 | 420 | 250 | 280 | 105/110 | 15 | 5 |
| | 22 | 15 | 45 | 100 | 230 | 420 | 250 | 300 | 115/120 | 10 | 7 |
| | 31 | 15 | 64 | 100 | 292 | 500 | 600 | 500 | 130/135 | 15 | 8 |
| 3 | 4 | 15 | 122 | 51 | 120 | 139 | 266 | 200 | 55/60 | 10 | 7 |
| | 7 | 15 | 158 | 57 | 140 | 210 | 306 | 250 | 65/70 | 10 | 7 |
| | 12 | 15 | 190 | 63 | 162 | 245 | 356 | 280 | 75/80 | 15 | 7 |
| | 19 | 15 | 280 | 70 | 216 | 336 | 458 | 388 | 105/110 | 15 | 7 |

# MATERIAL PROPERTIES POST TENSIONING

**Fig. 8.2.6 Typical multi-strand fixed post-tensioning anchorage detail**

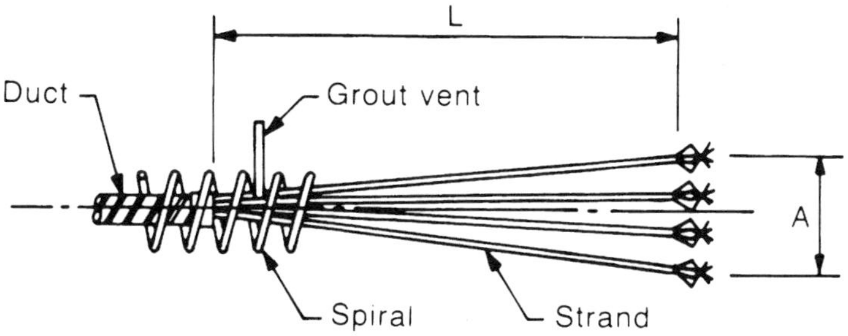


| No. of Strands | Strand Diameter (mm) | A (mm) | L (mm) |
|---|---|---|---|
| 7 | 13 | 230 | 900 |
| 12 | 13 | 300 | 1070 |
| 19 | 13 | 380 | 1200 |
| 31 | 13 | 460 | 1200 |
| 4 | 15 | 160 | 600 |
| 7 | 15 | 200 | 900 |
| 12 | 15 | 300 | 900 |
| 19 | 15 | 390 | 1200 |
| 22 | 15 | 400 | 1200 |
| 27 | 15 | 450 | 1500 |
| 31 | 15 | 450 | 1500 |

# MATERIAL PROPERTIES POST TENSIONING

**Fig. 8.2.7 Typical unbonded monostrand system details**

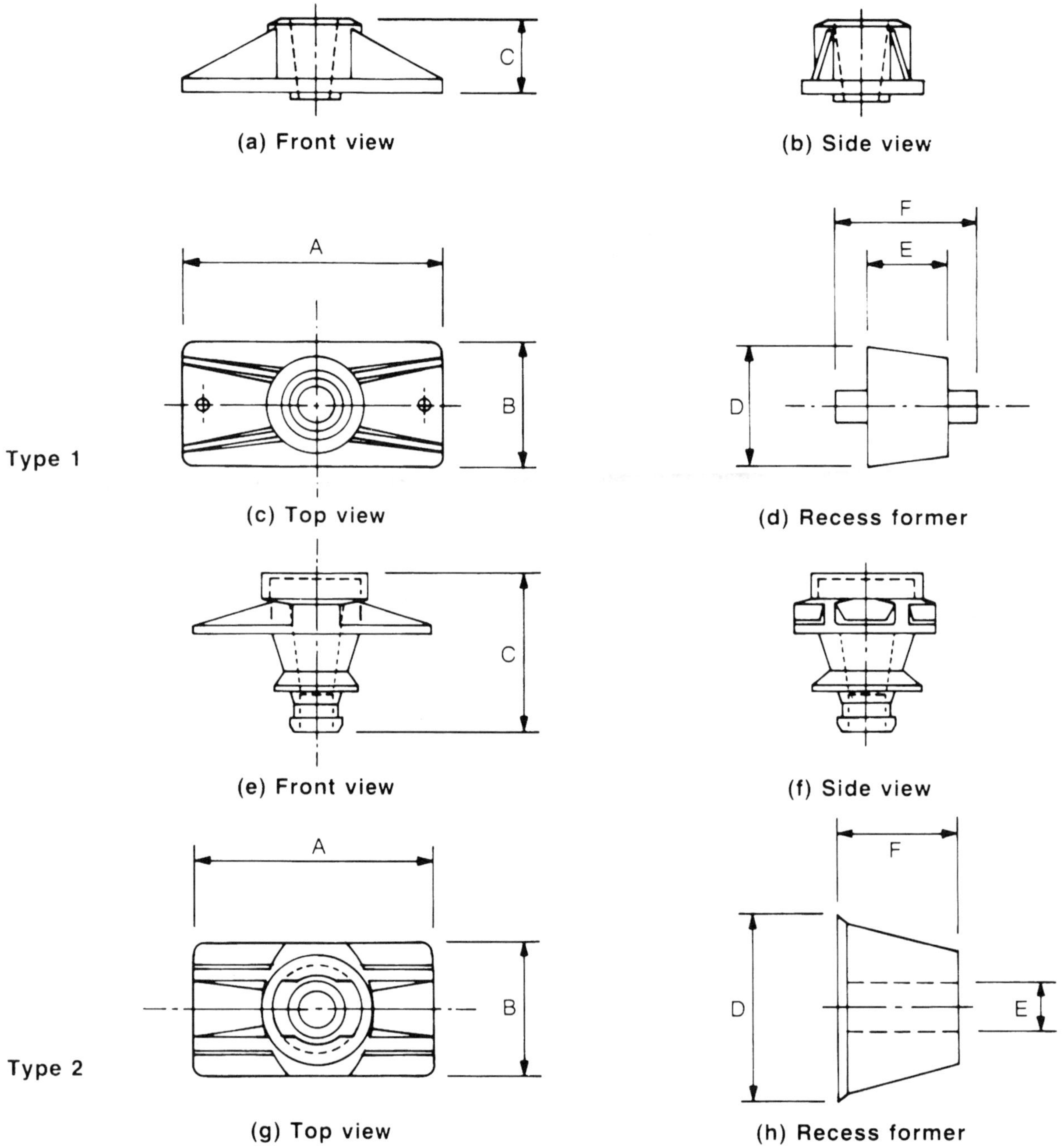


| Type | No. of Strands | Strand Dia. (mm) | A (mm) | B (mm) | C (mm) | D (mm) | E (mm) | F (mm) |
|---|---|---|---|---|---|---|---|---|
| 1 | 1 | 13 | 127 | 57 | 38 | 51 | 32-57 | 57 |
| | 1 | 13 | 133 | 73 | 38 | 51 | 32-57 | 57 |
| | 1 | 13 | 102 | 89 | 38 | 51 | 32-57 | 57 |
| | 1 | 15 | 118 | 89 | 41 | 54 | 32-57 | 64 |
| | 1 | 15 | 152 | 89 | 41 | 54 | 32-57 | 64 |
| 2 | 1 | 15 | 125 | 70 | 83 | 100 | 28 | 60 |

Note: Concrete strength at stressing to be provided by post-tensioning contractor or consulting engineer.
Generic drawing of monostrand anchor. Contact post-tensioning contractor for specific dimensions.

# MATERIAL PROPERTIES

**Fig. 8.2.8 Reinforcing bar data**

| Metric Reinforcing Bars | | | | |
|---|---|---|---|---|
| Bar Size Designation[1] | Nominal Linear Mass (kg/m) | Nominal Dimensions[2] | | |
| | | Diameter (mm) | Area ($mm^2$) | Perimeter (mm) |
| No. 10 | 0.785 | 11.3 | 100 | 35.5 |
| No. 15 | 1.570 | 16.0 | 200 | 50.1 |
| No. 20 | 2.355 | 19.5 | 300 | 61.3 |
| No. 25 | 3.925 | 25.2 | 500 | 79.2 |
| No. 30 | 5.495 | 29.9 | 700 | 93.9 |
| No. 35 | 7.850 | 35.7 | 1000 | 112.2 |
| No. 45 | 11.775 | 43.7 | 1500 | 137.3 |
| No. 55 | 19.625 | 56.4 | 2500 | 177.2 |

(1) Bar numbers are based upon the rounded of nominal diameter of the bars.

(2) The nominal dimensions of a deformed bar are equivalent to those of a plain round bar having the same linear mass as the deformed bar.

**Fig. 8.2.9 Detailing and estimating dimensions[1] (mm) for standard end hooks[2] for deformed reinforcing bars**

| Nominal Size | Steel Grade | | | | | | | | | | | |
|---|---|---|---|---|---|---|---|---|---|---|---|---|
| | 300 R | | | | 400 R or 500 R[4] | | | | 400 W or 500 W | | | |
| | 180° Hook | | | 90° Hook | 180° Hook | | | 90° Hook | 180°Hook | | | 90° Hook |
| | ID | J | G[3] | A or G | ID | J | G[3] | A or G | ID | J | G[3] | A or G |
| No. 10 | 60 | 83 | 131 | 177 | 70 | 93 | 141 | 182 | 60 | 83 | 131 | 177 |
| No. 15 | 90 | 122 | 169 | 252 | 100 | 132 | 180 | 257 | 90 | 122 | 169 | 252 |
| No. 20 | | | | | 120 | 159 | 218 | 314 | 100 | 139 | 196 | 304 |
| No. 25 | | | | | 150 | 200 | 276 | 403 | 150 | 200 | 276 | 403 |
| No. 30 | | | | | 250 | 310 | 404 | 513 | 200 | 260 | 351 | 488 |
| No. 35 | | | | | 300 | 371 | 484 | 614 | 250 | 321 | 431 | 589 |
| No. 45 | | | | | 450 | 537 | 682 | 793 | 400 | 487 | 628 | 768 |
| No. 55 | | | | | 600 | 713 | 900 | 1033 | 550 | 663 | 847 | 1008 |

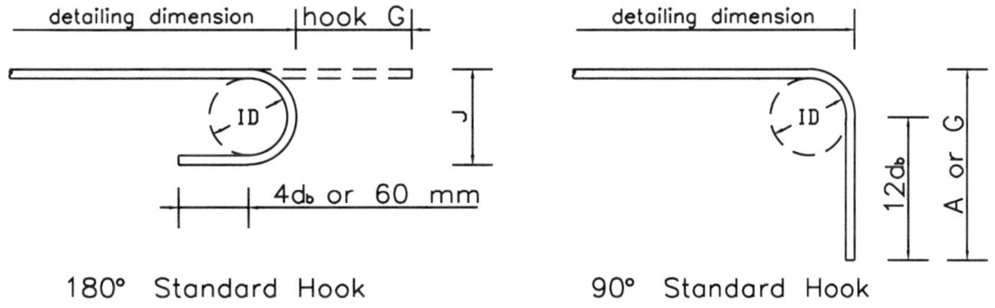


(1) The dimensions provided use the minimum bend diameters (ID) permitted in [Cl. A12.2.3 and Table A5 in A23.3 - 94]

(2) Standard hooks are defined in [Cl. A12.2.2 in A23.3 - 94]

(3) Add the additional hook dimension G to the detailing dimension to estimate the total bar length:
For 180° hooks: $G = (4d_b > 60\ mm) + \pi\,(ID + d_b) / 2 - ID / 2 - d_b$
For 90° hooks: $G \approx A = 12d_b + ID / 2 + d_b$

(4) Special fabrication is required for bends exceeding 90° for No. 45 and No. 55 bars for steel grades 400 R or 500 R

# MATERIAL PROPERTIES REINFORCING BARS

**Fig. 8.2.10 Compression lap splices**

| Bar Size | Bar Grade | Standard Lap | Within Ties[3] | Within Spirals[4] |
|---|---|---|---|---|
| 10 | 400 | 300 | 300 | 300 |
| 10 | 500 | 430 | 360 | 320 |
| 15 | 400 | 440 | 370 | 330 |
| 15 | 500 | 640 | 530 | 480 |
| 20 | 400 | 590 | 490 | 440 |
| 20 | 500 | 850 | 710 | 640 |
| 25 | 400 | 730 | 610 | 550 |
| 25 | 500 | 1070 | 890 | 800 |
| 30 | 400 | 880 | 730 | 660 |
| 30 | 500 | 1280 | 1060 | 960 |
| 35 | 400 | 1030 | 850 | 770 |
| 35 | 500 | 1490 | 1240 | 1120 |

(1) Calculated according to CSA Standard A23.3 - 94, Clause 12.14 and 12.16. The minimum length of compression lap splice shall not be less than $0.073 f_y d_b$ nor $(0.133 f_y - 24) d_b$ for $f_y$ greater than 400 MPa, nor 300 mm.

(2) When bars of different sizes are to be lap spliced in compression, the splice length shall be the larger of the compression development length of the larger bar or the splice length of the smaller bar. 45 M and 55 M bars may be lap spliced to 35 M and smaller bars.

(3) In compression members where ties have an effective area $A_v \geq 0.0015 h_s$, 0.83 times the standard lap length may be used, but not less than 300 mm.

(4) Within spirals of spiral tied columns, 0.75 times the standard lap but not less than 300 mm may be used.

# MATERIAL PROPERTIES REINFORCING BARS

**Fig. 8.2.11 Development lengths (mm) based on [Table 12-1 in A23.3 - 94][1] for heavier confined[2] deformed reinforcing bars[5] with $f_y$ = 400 MPa[3], for normal density concrete[4].**

$$\ell_d = 0.45 k_1 k_2 k_3 k_4 \frac{f_y}{\sqrt{f'_c}} d_b > 300 \text{ mm}^{(6)}$$

| Modification Factors Included | f'c (MPa) | No. 10 | No. 15 | No. 20 | No. 25 | No. 30 | No. 35 | No. 45 | No. 55 |
|---|---|---|---|---|---|---|---|---|---|
| | 20 | 322 | 483 | 644 | 1006 | 1207 | 1409 | 1811 | 2214 |
| | 25 | 288 | 432 | 576 | 900 | 1080 | 1260 | 1620 | 1980 |
| $k_1$ = 1.0 (bottom bars) | 30 | 263 | 394 | 526 | 822 | 986 | 1150 | 1479 | 1807 |
| | 35 | 243 | 365 | 487 | 761 | 913 | 1065 | 1369 | 1673 |
| $k_2$ = 1.0 (uncoated bars) | 40 | 228 | 342 | 455 | 712 | 854 | 996 | 1281 | 1565 |
| | 45 | 215 | 322 | 429 | 671 | 805 | 939 | 1207 | 1476 |
| $k_3$ = 1.0 (normal density concrete[4]) | 50 | 204 | 305 | 407 | 636 | 764 | 891 | 1146 | 1400 |
| | 55 | 194 | 291 | 388 | 607 | 728 | 849 | 1092 | 1335 |
| | 60 | 186 | 279 | 372 | 581 | 697 | 813 | 1046 | 1278 |
| | 64 | 180 | 270 | 360 | 563 | 675 | 788 | 1013 | 1238 |
| | 20 | 483 | 724 | 966 | 1509 | 1811 | 2113 | 2717 | 3321 |
| | 25 | 432 | 648 | 864 | 1350 | 1620 | 1890 | 2430 | 2970 |
| $k_1$ = 1.0 (bottom bars) | 30 | 394 | 592 | 789 | 1232 | 1479 | 1725 | 2218 | 2711 |
| | 35 | 365 | 548 | 730 | 1141 | 1369 | 1597 | 2054 | 2510 |
| $k_2$ = 1.5 (epoxy coated bars) | 40 | 342 | 512 | 683 | 1067 | 1281 | 1494 | 1921 | 2348 |
| | 45 | 322 | 483 | 644 | 1006 | 1207 | 1409 | 1811 | 2214 |
| $k_3$ = 1.0 (normal density concrete[4]) | 50 | 305 | 458 | 611 | 955 | 1146 | 1336 | 1718 | 2100 |
| | 55 | 291 | 437 | 583 | 910 | 1092 | 1274 | 1638 | 2002 |
| | 60 | 279 | 418 | 558 | 871 | 1046 | 1220 | 1569 | 1917 |
| | 64 | 270 | 405 | 540 | 844 | 1013 | 1181 | 1519 | 1856 |
| | 20 | 419 | 628 | 837 | 1308 | 1570 | 1831 | 2355 | 2878 |
| | 25 | 374 | 562 | 749 | 1170 | 1404 | 1638 | 2106 | 2574 |
| $k_1$ = 1.3 (top bars) | 30 | 342 | 513 | 684 | 1068 | 1282 | 1495 | 1923 | 2350 |
| | 35 | 316 | 475 | 633 | 989 | 1187 | 1384 | 1780 | 2175 |
| $k_2$ = 1.0 (uncoated bars) | 40 | 296 | 444 | 592 | 925 | 1110 | 1295 | 1665 | 2035 |
| | 45 | 279 | 419 | 558 | 872 | 1046 | 1221 | 1570 | 1919 |
| $k_3$ = 1.0 (normal density concrete[4]) | 50 | 265 | 397 | 529 | 827 | 993 | 1158 | 1489 | 1820 |
| | 55 | 252 | 379 | 505 | 789 | 947 | 1104 | 1420 | 1735 |
| | 60 | 242 | 363 | 483 | 755 | 906 | 1057 | 1359 | 1662 |
| | 64 | 234 | 351 | 468 | 731 | 878 | 1024 | 1316 | 1609 |
| | 20 | 547 | 821 | 1095 | 1711 | 2053 | 2395 | 3079 | 3763 |
| | 25 | 490 | 734 | 979 | 1530 | 1836 | 2185 | 2754 | 3366 |
| | 30 | 447 | 670 | 894 | 1397 | 1676 | 1994 | 2514 | 3073 |
| $k_1 \times k_2$ = 1.7 (top location and | 35 | 414 | 621 | 828 | 1293 | 1552 | 1847 | 2328 | 2845 |
| epoxy coated bars) | 40 | 387 | 581 | 774 | 1210 | 1451 | 1727 | 2177 | 2661 |
| | 45 | 365 | 547 | 730 | 1140 | 1368 | 1628 | 2053 | 2509 |
| $k_3$ = 1.0 (normal density concrete[4]) | 50 | 346 | 519 | 692 | 1082 | 1298 | 1545 | 1947 | 2380 |
| | 55 | 330 | 495 | 660 | 1032 | 1238 | 1473 | 1857 | 2269 |
| | 60 | 316 | 474 | 632 | 988 | 1185 | 1410 | 1778 | 2173 |
| | 64 | 306 | 459 | 612 | 956 | 1148 | 1366 | 1721 | 2104 |

Calculations are based on the nominal bar diameter.

(1) **Clear cover and clear spacing of bars must be at least 1.0 $d_b$ and 1.4 $d_b$, respectively.**

(2) **For lighter confinement, "(Other cases)" in [Table 12.1 in A23.3 - 94], multiply the above table values by 1.33.**

**To qualify for heavier confinement, must have either:**

**member containing minimum stirrups or ties within $\ell_d$,**

**or**

**slabs, walls, shells or folded plates having clear spacing between bars being developed not less than 2$d_b$.**

(3) **For $f_y \neq$ 400 MPa, multiply development length by $f_y$ / 400.**

(4) **Multiply table values by:**
**$k_3$ = 1.3 for structural low density concrete**
**$k_3$ = 1.2 for structural semi-low density concrete.**

(5) **The appropriate bar size factor $k_4$ has been applied.**
**For concrete strengths > 64 MPa, use development length for 64 MPa.**

(6) **After application of all modification factors, the development length must not be less than 300 mm.**

# MATERIAL PROPERTIES REINFORCING BARS

**Fig. 8.2.12 Development lengths (mm) based on [Table 12-1 in A23.3 - 94][1] for heavier confined[2] deformed wire[5] with $f_y$ = 400 MPa[3], for normal density concrete[4].**

$$\ell_d = 0.45 k_1 k_2 k_3 k_4 \frac{f_y}{\sqrt{f'_c}} d_b > 300 \text{ mm}^{(6)}$$

| Modification Factors Included | $f'_c$ (MPa) | MD9.1 | MD11.1 | MD13.3 | MD18.7 | MD25.8 | MD34.9 | MD47.6 |
|---|---|---|---|---|---|---|---|---|
| $k_1$ = 1.0 (bottom bars) | 20 | 110 | 121 | 133 | 157 | 185 | 215 | 251 |
| $k_2$ = 1.0 (uncoated bars) | 25 | 98 | 108 | 119 | 141 | 165 | 192 | 224 |
| $k_3$ = 1.0 (normal density concrete[4]) | 30 | 89 | 99 | 108 | 128 | 151 | 175 | 205 |
| | 35 | 83 | 92 | 100 | 119 | 140 | 162 | 189 |
| | 40 | 78 | 86 | 94 | 111 | 130 | 152 | 177 |
| | 45 | 73 | 81 | 88 | 105 | 123 | 143 | 167 |
| | 50 | 69 | 77 | 84 | 99 | 117 | 136 | 159 |
| | 55 | 66 | 73 | 80 | 95 | 111 | 129 | 151 |
| | 60 | 63 | 70 | 77 | 91 | 107 | 124 | 145 |
| | 64 | 61 | 68 | 74 | 88 | 103 | 120 | 140 |
| $k_1$ = 1.0 (bottom bars) | 20 | 164 | 182 | 199 | 236 | 277 | 322 | 376 |
| $k_2$ = 1.5 (epoxy coated bars) | 25 | 147 | 162 | 178 | 211 | 248 | 288 | 336 |
| $k_3$ = 1.0 (normal density concrete[4]) | 30 | 134 | 148 | 162 | 192 | 226 | 263 | 307 |
| | 35 | 124 | 137 | 150 | 178 | 209 | 243 | 284 |
| | 40 | 116 | 128 | 141 | 167 | 196 | 228 | 266 |
| | 45 | 110 | 121 | 133 | 157 | 185 | 215 | 251 |
| | 50 | 104 | 115 | 126 | 149 | 175 | 204 | 238 |
| | 55 | 99 | 109 | 120 | 142 | 167 | 194 | 227 |
| | 60 | 95 | 105 | 115 | 136 | 160 | 186 | 217 |
| | 64 | 92 | 102 | 111 | 132 | 155 | 180 | 210 |
| $k_1$ = 1.3 (top bars) | 20 | 142 | 157 | 172 | 204 | 240 | 279 | 326 |
| $k_2$ = 1.0 (uncoated bars) | 25 | 127 | 141 | 154 | 183 | 215 | 250 | 291 |
| $k_3$ = 1.0 (normal density concrete[4]) | 30 | 116 | 128 | 141 | 167 | 196 | 228 | 266 |
| | 35 | 108 | 119 | 130 | 154 | 181 | 211 | 246 |
| | 40 | 101 | 111 | 122 | 144 | 170 | 197 | 230 |
| | 45 | 95 | 105 | 115 | 136 | 160 | 186 | 217 |
| | 50 | 90 | 100 | 109 | 129 | 152 | 176 | 206 |
| | 55 | 86 | 95 | 104 | 123 | 145 | 168 | 197 |
| | 60 | 82 | 91 | 99 | 118 | 139 | 161 | 188 |
| | 64 | 80 | 88 | 96 | 114 | 134 | 156 | 182 |
| $k_1 \times k_2$ = 1.7 (top location and epoxy coated bars) | 20 | 186 | 206 | 225 | 267 | 314 | 365 | 426 |
| $k_3$ = 1.0 (normal density concrete[4]) | 25 | 167 | 184 | 201 | 239 | 281 | 326 | 381 |
| | 30 | 152 | 168 | 184 | 218 | 256 | 298 | 348 |
| | 35 | 141 | 156 | 170 | 202 | 237 | 276 | 322 |
| | 40 | 132 | 146 | 159 | 189 | 222 | 258 | 301 |
| | 45 | 124 | 137 | 150 | 178 | 209 | 243 | 284 |
| | 50 | 118 | 130 | 142 | 169 | 198 | 231 | 270 |
| | 55 | 112 | 124 | 136 | 161 | 189 | 220 | 257 |
| | 60 | 108 | 119 | 130 | 154 | 181 | 211 | 246 |
| | 64 | 104 | 115 | 126 | 149 | 175 | 204 | 238 |

(1) Clear cover and clear spacing of bars must be at least 1.0 $d_b$ and 1.4 $d_b$, respectively.

(2) For lighter confinement, "(Other cases)" in [Table 12.1 in A23.3 - 94], multiply the above table values by 1.33.

To qualify for heavier confinement, must have either:
member containing minimum stirrups or ties within $\ell_d$,
or
slabs, walls, shells or folded plates having clear spacing between bars being developed not less than 2$d_b$.

(3) For $f_y \neq$ 400 MPa, multiply development length by $f_y$ / 400.

(4) Multiply table values by:
$k_3$ = 1.3 for structural low density concrete
$k_3$ = 1.2 for structural semi-low density concrete.

(5) The appropriate bar size factor $k_4$ has been applied.

For concrete strengths > 64 MPa, use development length for 64 MPa.

(6) After application of all modification factors, the development length must not be less than 300 mm.

# MATERIAL PROPERTIES WELDED WIRE FABRIC

**Fig. 8.2.13 Standard styles of welded wire fabric**

| Metric Designation | Wire Diameter | | Wire Cross-Sectional Area | | Mass per Unit Area | Cross-Sectional Area per Metre Width | | Equivalent Imperial Styles | Equivalent Imperial Styles |
|---|---|---|---|---|---|---|---|---|---|
| | mm | in. | Long. $mm^2$ | Transv. $mm^2$ | kg / $m^2$ | Long. $mm^2$ | Transv. $mm^2$ | Gauge No. | "W" Numbers |
| 152×152 MW9.1×MW9.1 | 3.40 | 0.134 | 9.1 | 9.1 | 1.04 | 59.8 | 59.8 | 6×6 10/10 | 6×6×W1.4/W1.4 |
| 152×152 MW11.1×MW11.1 | 3.76 | 0.148 | 11.1 | 11.1 | 1.26 | 73.0 | 73.0 | 6×6 9/9 | 6×6×W1.7/W1.7 |
| 152×152 MW13.3×MW13.3 | 4.12 | 0.162 | 13.3 | 13.3 | 1.50 | 87.5 | 87.5 | 6×6 8/8 | 6×6×W2.1/W2.1 |
| 152×152 MW18.7×MW18.7 | 4.88 | 0.192 | 18.7 | 18.7 | 2.11 | 123.0 | 123.0 | 6×6 6/6 | 6×6×W2.9/W2.9 |
| 152×152 MW25.8×MW25.8 | 5.74 | 0.226 | 25.8 | 25.8 | 2.91 | 170.0 | 170.0 | 6×6 4/4 | 6×6×W4/W4 |
| 152×152 MW34.9×MW34.9 | 6.67 | 0.262 | 34.9 | 34.9 | 3.95 | 230.0 | 230.0 | 6×6 2/2 | 6×6×W5.4/W5.4 |
| 152×152 MW47.6×MW47.6 | 7.79 | 0.306 | 47.6 | 47.6 | 5.38 | 313.0 | 313.0 | 6×6 0/0 | 6×6×W7.4/W7.4 |
| 152×152 MW9.1×MW9.1 | 3.40 | 0.134 | 9.1 | 9.1 | 1.52 | 89.2 | 89.2 | 4×4 10/10 | 4×4×W1.4/W1.4 |
| 152×152 MW11.1×MW11.1 | 3.76 | 0.148 | 11.1 | 11.1 | 1.83 | 109.0 | 109.0 | 4×4 9/9 | 4×4×W1.7/W1.7 |
| 152×152 MW13.3×MW13.3 | 4.12 | 0.162 | 13.3 | 13.3 | 2.18 | 130.0 | 130.0 | 4×4 8/8 | 4×4×W2.1/W2.1 |
| 152×152 MW18.7×MW18.7 | 4.88 | 0.192 | 18.7 | 18.7 | 3.07 | 183.0 | 183.0 | 4×4 6/6 | 4×4×W2.9/W2.9 |
| 152×152 MW25.8×MW25.8 | 5.74 | 0.226 | 25.8 | 25.8 | 4.23 | 253.0 | 253.0 | 4×4 4/4 | 4×4×W4/W4 |
| 51×51 MW3.2×MW3.2 | 2.03 | 0.080 | 3.2 | 3.2 | 1.03 | 62.8 | 62.8 | 2×2 14/14 | N/A |
| 51×51 MW5.6×MW5.6 | 2.69 | 0.106 | 5.6 | 5.6 | 1.80 | 110.0 | 110.0 | 2×2 12/12 | N/A |
| 51×51 MW9.1×MW9.1 | 3.40 | 0.134 | 9.1 | 9.1 | 2.94 | 178.0 | 178.0 | 2×2 10/10 | 2×2×W1.4/W1.4 |

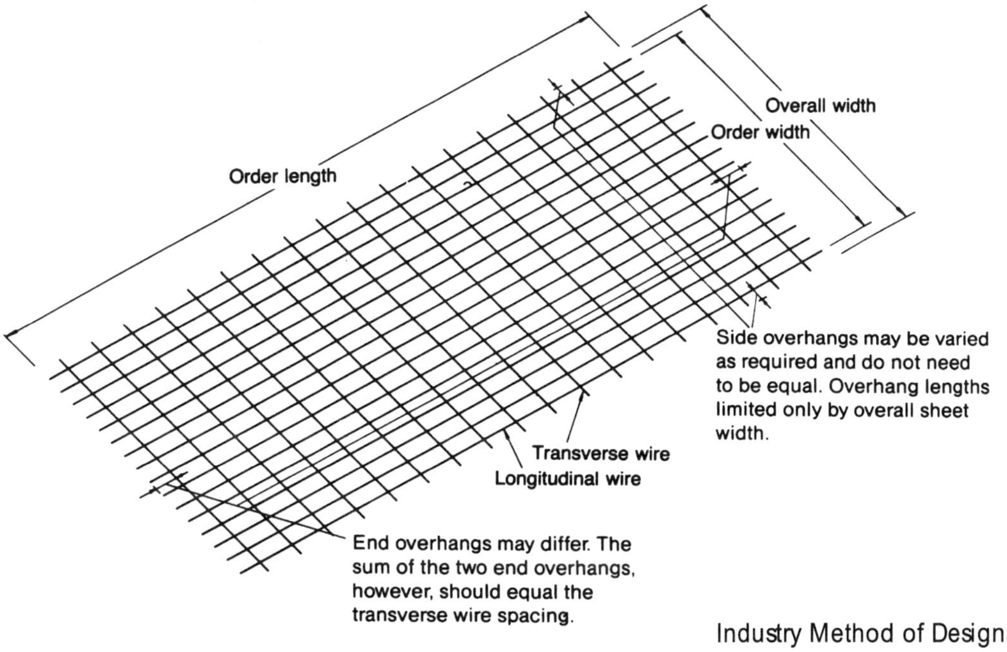


Industry Method of Designating Style:

Example: 152 × 152 MW9.1 × MW3.1

Longitudinal wire spacing — Transverse wire spacing — Longitudinal wire size — Transverse wire size

# MATERIAL PROPERTIES WELDED WIRE FABRIC

**Fig. 8.2.14 Sectional areas of welded wire fabric**

| Wire Size Smooth | Number Deformed | Nom. Diam. (mm) | Nom. Mass (kg/m) | $A_s$ (mm$^2$ / m) Centre to Centre Spacing (mm) 51 | 76 | 102 | 152 | 203 | 254 | 305 |
|---|---|---|---|---|---|---|---|---|---|---|
| MW 129 | MD 129 | 12.83 | 1.01 | 2540 | 1693 | 1270 | 847 | 635 | 508 | 423 |
| MW 116 | MD 116 | 12.17 | 0.911 | 2286 | 1524 | 1143 | 762 | 572 | 457 | 381 |
| MW 103 | MD 103 | 11.46 | 0.809 | 2032 | 1355 | 1016 | 677 | 508 | 406 | 339 |
| MW 90.3 | MD 90.3 | 10.72 | 0.708 | 1778 | 1185 | 889 | 593 | 445 | 356 | 296 |
| MW 77.4 | MD 77.4 | 9.93 | 0.607 | 1524 | 1016 | 762 | 508 | 381 | 305 | 254 |
| MW 71.0 | MD 71.0 | 9.50 | 0.556 | 1397 | 931 | 699 | 466 | 349 | 279 | 233 |
| MW 67.9 | | 9.30 | 0.531 | 1334 | 889 | 667 | 445 | 332 | 267 | 222 |
| MW 64.5 | MD 64.5 | 9.07 | 0.506 | 1270 | 847 | 635 | 423 | 318 | 254 | 212 |
| MW 61.3 | | 8.84 | 0.481 | 1207 | 804 | 603 | 402 | 301 | 241 | 201 |
| MW 58.1 | MD 58.1 | 8.59 | 0.936 | 1143 | 762 | 572 | 381 | 286 | 229 | 191 |
| MW 54.9 | | 8.36 | 0.430 | 1080 | 720 | 540 | 360 | 269 | 216 | 180 |
| MW 51.6 | MD 51.6 | 8.10 | 0.405 | 1016 | 677 | 508 | 339 | 254 | 203 | 169 |
| MW 48.4 | | 7.85 | 0.379 | 953 | 635 | 476 | 318 | 237 | 191 | 159 |
| MW 45.2 | MD 45.2 | 7.60 | 0.354 | 889 | 593 | 445 | 296 | 222 | 178 | 148 |
| MW 42.1 | | 7.32 | 0.329 | 826 | 550 | 413 | 275 | 205 | 165 | 138 |
| MW 38.7 | MD 38.7 | 7.01 | 0.304 | 762 | 508 | 381 | 254 | 191 | 152 | 127 |
| MW 35.5 | | 6.73 | 0.278 | 699 | 466 | 349 | 233 | 174 | 140 | 116 |
| MW 32.3 | MD 32.3 | 6.40 | 0.253 | 635 | 423 | 318 | 212 | 159 | 127 | 106 |
| MW 28.9 | | 6.07 | 0.228 | 572 | 381 | 286 | 191 | 142 | 114 | 95.3 |
| MW 25.8 | MD 25.8 | 5.74 | 0.202 | 508 | 339 | 254 | 169 | 127 | 102 | 84.7 |
| MW 22.6 | | 5.36 | 0.177 | 445 | 296 | 222 | 148 | 110 | 88.9 | 74.1 |
| MW 19.2 | | 4.95 | 0.152 | 381 | 254 | 191 | 127 | 95.3 | 76.2 | 63.5 |
| MW 18.7 | | 4.88 | 0.147 | 368 | 245 | 184 | 123 | 91.0 | 74.1 | 61.4 |
| MW 16.0 | | 4.52 | 0.126 | 317 | 212 | 159 | 106 | 78.3 | 63.5 | 52.9 |
| MW 13.5 | | 4.11 | 0.106 | 267 | 178 | 133 | 88.9 | 65.6 | 52.9 | 44.5 |
| MW 12.9 | | 4.06 | 0.101 | 254 | 169 | 127 | 84.7 | 63.5 | 50.8 | 42.3 |
| MW 9.7 | | 3.51 | 0.076 | 191 | 127 | 95.3 | 63.5 | 48.7 | 38.1 | 31.8 |
| MW 9.0 | | 3.40 | 0.071 | 178 | 119 | 88.9 | 59.3 | 44.5 | 36.0 | 29.6 |

Note: Wire sizes other than those listed above including larger sizes may be produced provided the quantity required is sufficient to justify manufacture.

# SECTION PROPERTIES

**Fig. 8.3 Properties of geometric sections**

### Square
Axis of moments through centre

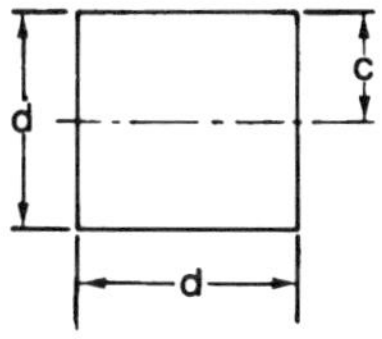


$A = d^2$

$c = \frac{d}{2}$

$I = \frac{d^4}{12}$

$S = \frac{d^3}{6}$

$r = \frac{d}{\sqrt{12}} = 0.288675\,d$

### Rectangle
Axis of moments on diagonal

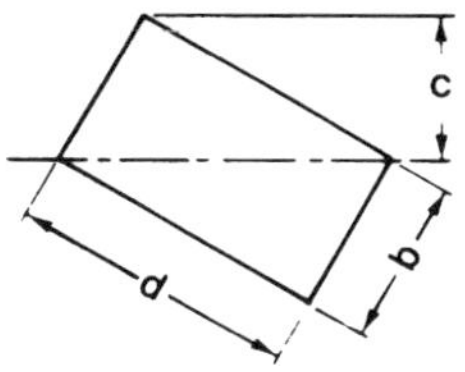


$A = bd$

$c = \frac{bd}{\sqrt{b^2 + d^2}}$

$I = \frac{b^3d^3}{6(b^2 + d^2)}$

$S = \frac{b^2d^2}{6\sqrt{b^2 + d^2}}$

$r = \frac{bd}{\sqrt{6(b^2 + d^2)}}$

### Square
Axis of moments on base

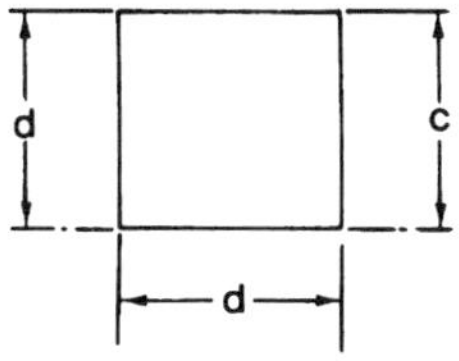


$A = d^2$

$c = d$

$I = \frac{d^4}{3}$

$S = \frac{d^3}{3}$

$r = \frac{d}{\sqrt{3}} = 0.577350\,d$

### Rectangle
Axis of moments any line through centre of gravity

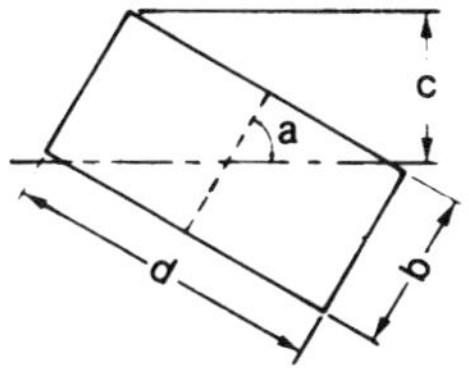


$A = bd$

$c = \frac{b \sin a + d \cos a}{2}$

$I = \frac{bd(b^2 \sin^2 a + d^2 \cos^2 a)}{12}$

$S = \frac{bd(b^2 \sin^2 a + d^2 \cos^2 a)}{6(b \sin a + d \cos a)}$

$r = \sqrt{\frac{b^2 \sin^2 a + d^2 \cos^2 a}{12}}$

### Square
Axis of moments on diagonal

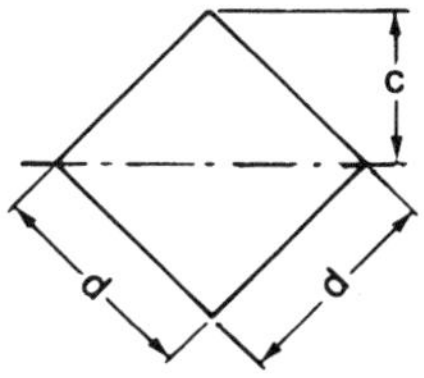


$A = d^2$

$c = \frac{d}{\sqrt{2}} = 0.707107\,d$

$I = \frac{d^4}{12}$

$S = \frac{d^3}{6\sqrt{2}} = 0.117851\,d^3$

$r = \frac{d}{\sqrt{12}} = 0.288675\,d$

### Hollow Rectangle
Axis of moments through centre

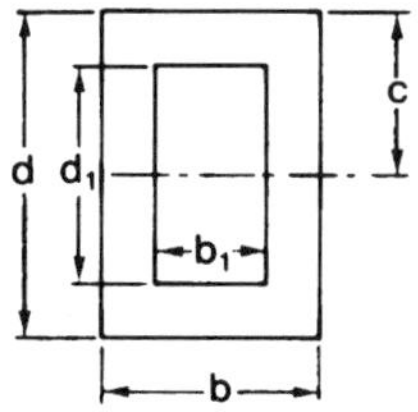


$A = bd - b_1d_1$

$c = \frac{d}{2}$

$I = \frac{bd^3 - b_1d_1^3}{12}$

$S = \frac{bd^3 - b_1d_1^3}{6d}$

$r = \sqrt{\frac{bd^3 - b_1d_1^3}{12\,A}}$

### Rectangle
Axis of moments through centre

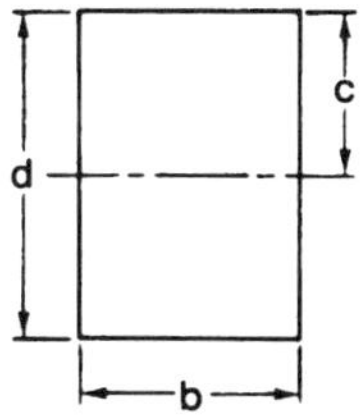


$A = bd$

$c = \frac{d}{2}$

$I = \frac{bd^3}{12}$

$S = \frac{bd^2}{6}$

$r = \frac{d}{\sqrt{12}} = 0.288675\,d$

### Equal Rectangles
Axis of moments through centre of gravity

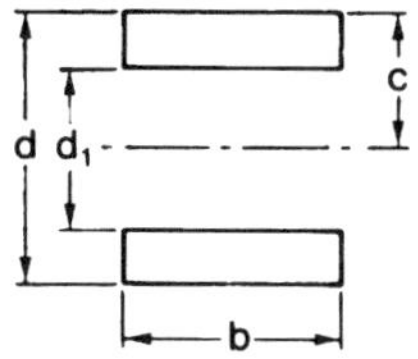


$A = b(d - d_1)$

$c = \frac{d}{2}$

$I = \frac{b(d^3 - d_1^3)}{12}$

$S = \frac{b(d^3 - d_1^3)}{6d}$

$r = \sqrt{\frac{d^3 - d_1^3}{12(d - d_1)}}$

### Rectangle
Axis of moments on base

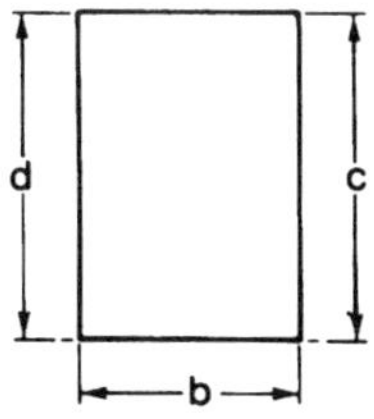


$A = bd$

$c = d$

$I = \frac{bd^3}{3}$

$S = \frac{bd^2}{3}$

$r = \frac{d}{\sqrt{3}} = 0.577350\,d$

### Unequal Rectangles
Axis of moments through centre of gravity

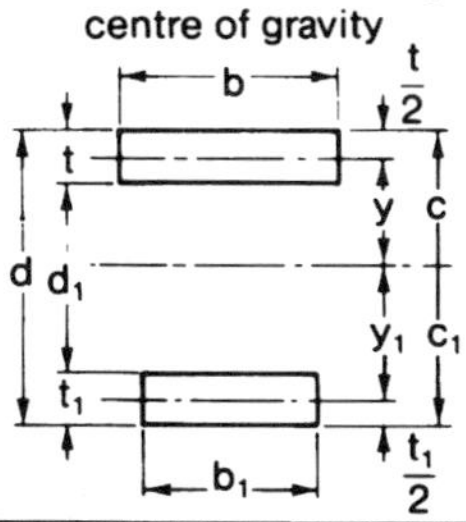


$A = bt + b_1t_1$

$c = \frac{\frac{1}{2}bt^2 + b_1t_1(d - \frac{1}{2}t_1)}{A}$

$I = \frac{bt^3}{12} + bty^2 + \frac{b_1t_1^3}{12} + b_1t_1y_1^2$

$S = \frac{I}{c} \qquad S_1 = \frac{I}{c_1}$

$r = \sqrt{\frac{I}{A}}$

# SECTION PROPERTIES

**Fig. 8.3 (cont'd) Properties of geometric sections**

### Triangle
Axis of moments through centre of gravity

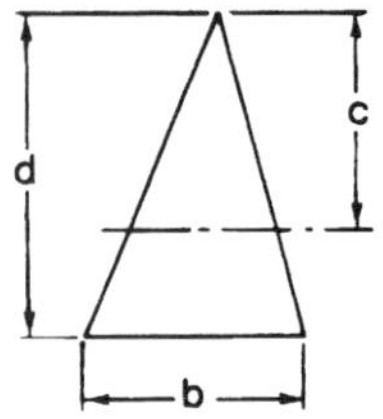


$$A = \frac{bd}{2}$$
$$c = \frac{2d}{3}$$
$$I = \frac{bd^3}{36}$$
$$S = \frac{bd^2}{24}$$
$$r = \frac{d}{\sqrt{18}}$$

### Segment of a Hollow Circle

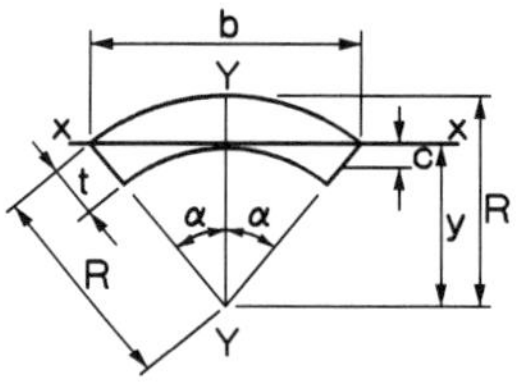


$$A = \frac{\pi\alpha t}{180}(2R-t)$$
$$c = \left(\frac{180\sin\alpha}{\pi\alpha} - \cos\alpha\right)\left(R-\frac{t}{2}\right)$$
$$n = \left(R-\frac{t}{2}\right)\sin\alpha$$
$$I_x = \left(\frac{\pi\alpha}{180} + \sin\alpha\cos\alpha - \frac{360\sin^2\alpha}{\pi\alpha}\right)\left(R-\frac{t}{2}\right)^3 t$$
$$I_y = \left(\frac{\pi\alpha}{180} - \sin\alpha\cos\alpha\right)R^3 t$$
$$v = \frac{180\sin\alpha}{\pi\alpha}\left(R-\frac{t}{2}\right)$$

### Triangle
Axis of moments on base

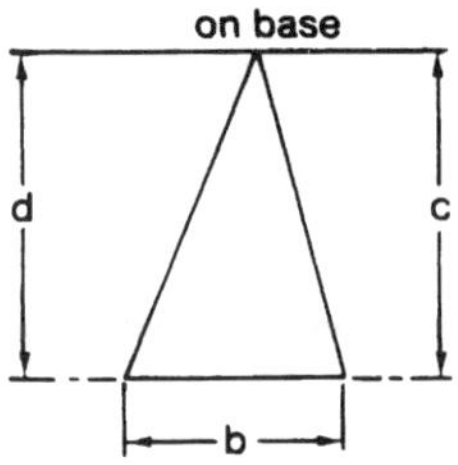


$$A = \frac{bd}{2}$$
$$c = d$$
$$I = \frac{bd^3}{12}$$
$$S = \frac{bd^2}{12}$$
$$r = \frac{d}{\sqrt{6}}$$

### Segment of a Circle
Axis of moments through circle centre

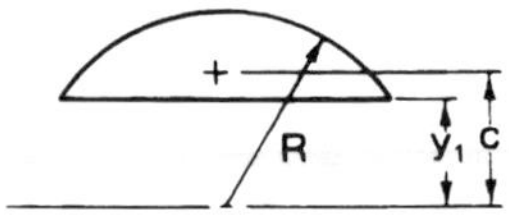


For half circle $y_1 = 0$

$$I = \frac{\pi R^4}{8} + \frac{y_1}{2}\sqrt{(R^2-y_1^2)^3} - \frac{R^2}{4}\left(y_1\sqrt{R^2-y_1^2} + R^2\sin^{-1}\frac{y_1}{R}\right)$$
$$A = \frac{\pi R^2}{2} - y_1\sqrt{R^2-y_1^2} - R^2\sin^{-1}\left(\frac{y_1}{R}\right)$$
$$c = \frac{2(R^2-y_1^2)^{3/2}}{3A}$$

### Trapezoid
Axis of moments through centre of gravity

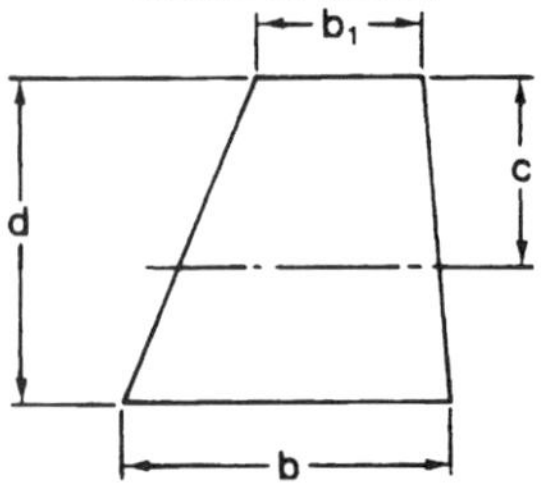


$$A = \frac{d(b+b_1)}{2}$$
$$c = \frac{d(2b+b_1)}{3(b+b_1)}$$
$$I = \frac{d^3(b^2+4bb_1+b_1^2)}{36(b+b_1)}$$
$$S = \frac{d^2(b^2+4bb_1+b_1^2)}{12(2b+b_1)}$$
$$r = \frac{d}{6(b+b_1)}\sqrt{2(b^2+4bb_1+b_1^2)}$$

### Parabola

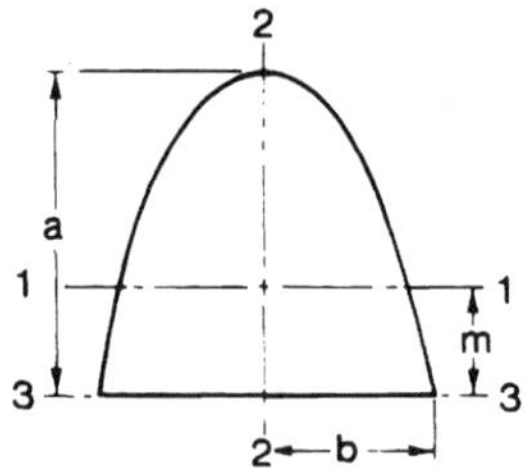


$$A = \frac{4}{3}ab$$
$$m = \frac{2}{5}a$$
$$I_1 = \frac{16}{175}a^3b$$
$$I_2 = \frac{4}{15}ab^3$$
$$I_3 = \frac{32}{105}a^3b$$

### Circle
Axis of moments through centre

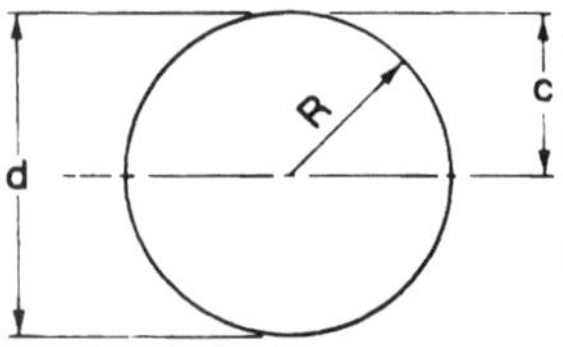


$$A = \frac{\pi d^2}{4} = \pi R^2$$
$$c = \frac{d}{2} = R$$
$$I = \frac{\pi d^4}{64} = \frac{\pi R^4}{4}$$
$$S = \frac{\pi d^3}{32} = \frac{\pi R^3}{4}$$
$$r = \frac{d}{4} = \frac{R}{2}$$

### Half Parabola

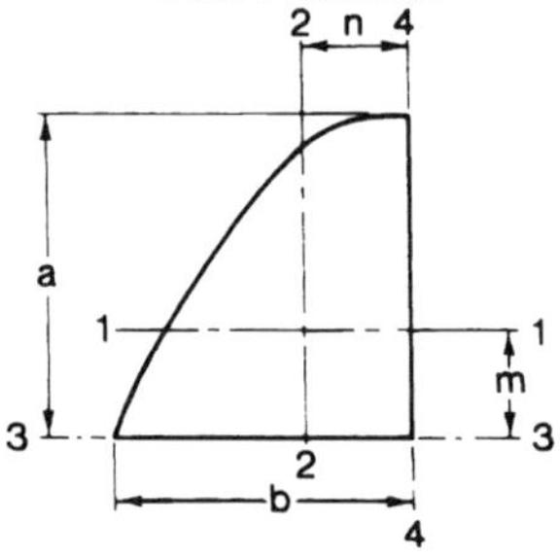


$$A = \frac{2}{3}ab \qquad I_1 = \frac{8}{175}a^3b$$
$$m = \frac{2}{5}a \qquad I_2 = \frac{19}{480}ab^3$$
$$n = \frac{3}{8}b \qquad I_3 = \frac{16}{105}a^3b$$
$$I_4 = \frac{2}{15}ab^3$$

### Hollow Circle
Axis of moments through centre

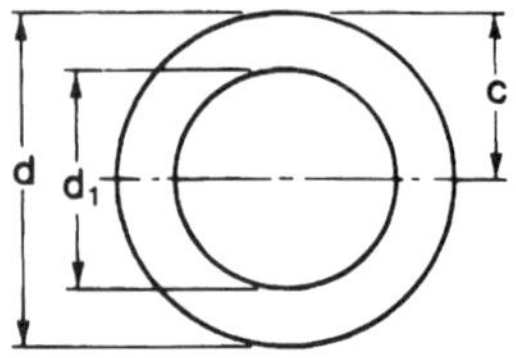


$$A = \frac{\pi(d^2-d_1^2)}{4}$$
$$c = \frac{d}{2}$$
$$I = \frac{\pi(d^4-d_1^4)}{64}$$
$$S = \frac{\pi(d^4-d_1^4)}{32d}$$
$$r = \frac{\sqrt{d^2+d_1^2}}{4}$$

### Complement of Half Parabola

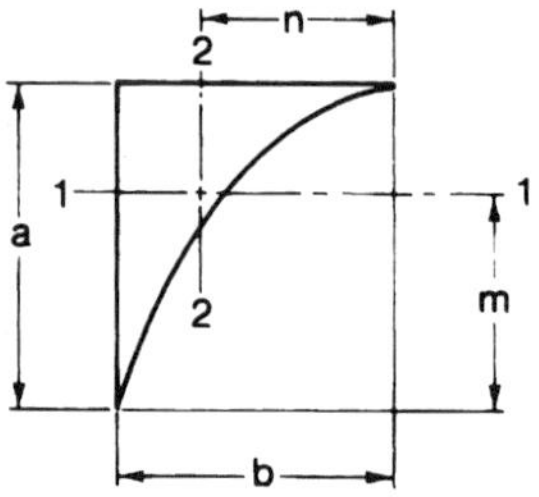


$$A = \frac{1}{3}ab$$
$$m = \frac{7}{10}a$$
$$n = \frac{3}{4}b$$
$$I_1 = \frac{37}{2100}a^3b$$
$$I_2 = \frac{1}{80}ab^3$$

# SECTION PROPERTIES

**Fig. 8.3 (cont'd) Properties of geometric sections**

### Parabolic Fillet in Right Angle

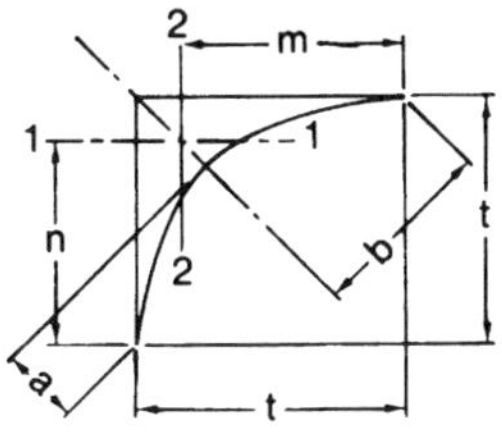


$$a = \frac{t}{2\sqrt{2}}$$

$$b = \frac{t}{\sqrt{2}}$$

$$A = \frac{1}{6}t^2$$

$$m = n = \frac{4}{5}t$$

$$I_1 = I_2 = \frac{11}{2100}t^4$$

### *Elliptic Complement

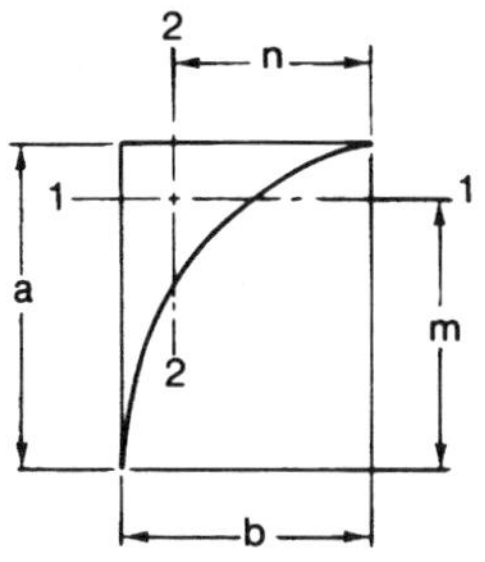


$$A = ab\left(1 - \frac{\pi}{4}\right)$$

$$m = \frac{a}{6\left(1 - \frac{\pi}{4}\right)}$$

$$n = \frac{b}{6\left(1 - \frac{\pi}{4}\right)}$$

$$I_1 = a^3b\left(\frac{1}{3} - \frac{\pi}{16} - \frac{1}{36\left(1 - \frac{\pi}{4}\right)}\right)$$

$$I_2 = ab^3\left(\frac{1}{3} - \frac{\pi}{16} - \frac{1}{36\left(1 - \frac{\pi}{4}\right)}\right)$$

### *Half Ellipse

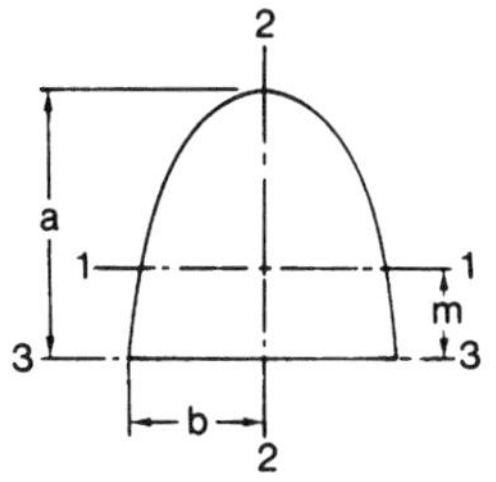


$$A = \frac{1}{2}\pi ab$$

$$m = \frac{4a}{3\pi}$$

$$I_1 = a^3b\left(\frac{\pi}{8} - \frac{8}{9\pi}\right)$$

$$I_2 = \frac{1}{8}\pi ab^3$$

$$I_3 = \frac{1}{8}\pi a^3b$$

### Regular Polygon

Axis of moments through centre

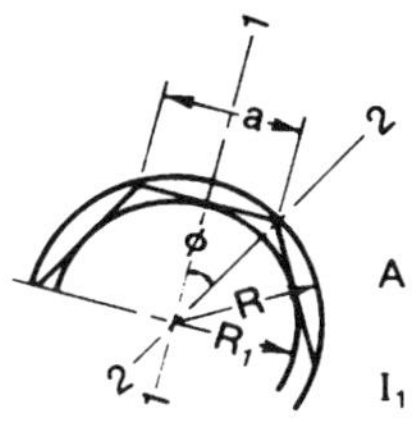


$n$ = Number of sides

$$\phi = \frac{180°}{n}$$

$$a = 2\sqrt{R^2 - R_1^2}$$

$$R = \frac{a}{2\sin\phi}$$

$$R_1 = \frac{a}{2\tan\phi}$$

$$A = \frac{1}{4}na^2\cot\phi = \frac{1}{2}nR^2\sin 2\phi = nR_1^2\tan\phi$$

$$I_1 = I_2 = \frac{A(6R^2 - a^2)}{24} = \frac{A(12R_1^2 + a^2)}{48}$$

$$r_1 = r_2 = \sqrt{\frac{6R^2 - a^2}{24}} = \sqrt{\frac{12R_1 + a^2}{48}}$$

### *Quarter Ellipse

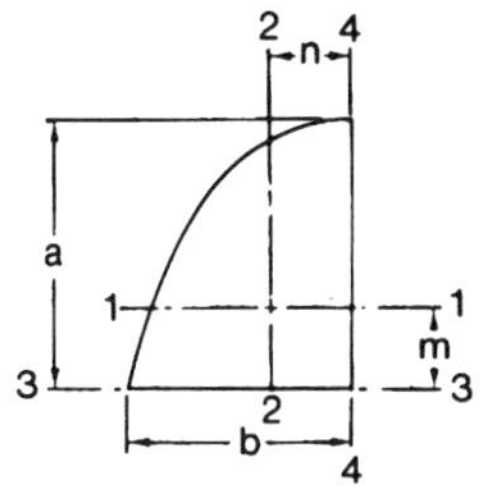


$$A = \frac{1}{4}\pi ab$$

$$m = \frac{4a}{3\pi}$$

$$n = \frac{4b}{3\pi}$$

$$I_1 = a^3b\left(\frac{\pi}{16} - \frac{4}{9\pi}\right)$$

$$I_2 = ab^3\left(\frac{\pi}{16} - \frac{4}{9\pi}\right)$$

$$I_3 = \frac{1}{16}\pi a^3b$$

$$I_4 = \frac{1}{16}\pi ab^3$$

### Beams and Channels

Transverse force oblique through centre of gravity

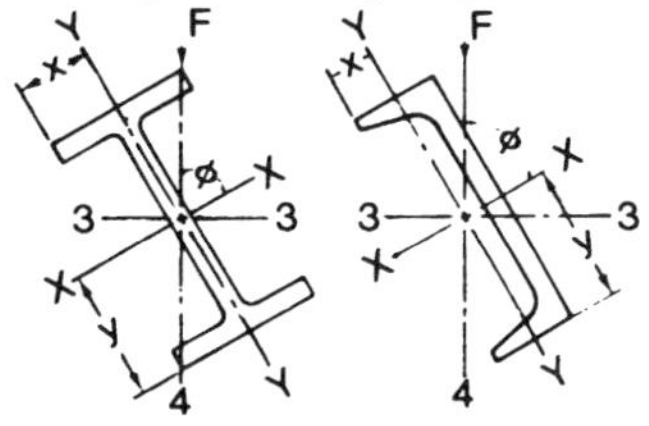


$$I_3 = I_x\sin^2\phi + I_y\cos^2\phi$$

$$I_4 = I_x\cos^2\phi + I_y\sin^2\phi$$

$$f_b = M\left(\frac{y}{I_x}\sin\phi + \frac{x}{I_y}\cos\phi\right)$$

where M is bending moment due to force F.

*To obtain properties of half circles, quarter circle and circular complement, substitute $a = b = R$.

### Angle

Axis of moments through centre of gravity

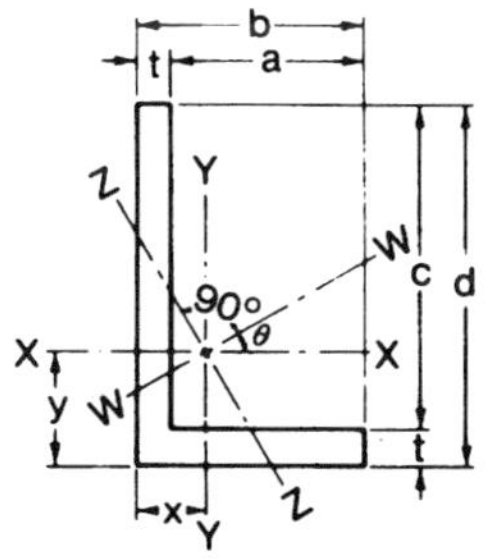


Z-Z is axis of minimum I

$$\tan 2\theta = \frac{2K}{I_y - I_x}$$

$$A = t(b + c) \quad x = \frac{b^2 + ct}{2(b + c)} \quad y = \frac{d^2 + at}{2(b + c)}$$

$K$ = Product of Inertia about X-X & Y-Y

$$= \mp\frac{abcdt}{4(b + c)}$$

$$I_x = \frac{1}{3}\left(t(d - y)^3 + by^3 - a(y - t)^3\right)$$

$$I_y = \frac{1}{3}\left(t(b - x)^3 + dx^3 - c(x - t)^3\right)$$

$$I_z = I_x\sin^2\theta + I_y\cos^2\theta + K\sin 2\theta$$

$$I_w = I_x\cos^2\theta + I_y\sin^2\theta - K\sin 2\theta$$

K is negative when heel of angle, with respect to c.g., is in 1st or 3rd quadrant, positive when in 2nd or 4th quadrant.

# METRIC CONVERSION

**Fig. 8.4.1 SI Base Units**

| Quantity | Preferred Units | Description | Base Units |
|---|---|---|---|
| Area | $mm^2$ | square millimetre | |
| Coefficient of thermal expansion | 1/Deg°C | reciprical of Degree Celcius | |
| Density, mass | $kg/m^3$ | kilogram per cubic meter | |
| Electric current | A | ampere | |
| Energy, work, quantity of heat | J | joule | 1 J = N·m |
| Force | N | newton | $1 N = kg·m/s^2$ |
| Force per unit length | N/m | newton per metre | $1 kg/m \times 9.81 m/s^2$ <br> $= (9.81 kg·m/s^2) \times 1/m$ <br> = 9.81 N/m |
| Frequency | Hz | hertz | 1 Hz = 1/s |
| Heat capacity | J/K | joule per kelvin | $m^2·kg/s^2K$ |
| Length | mm<br>m<br>km | millimetre<br>metre<br>kilometre | 1000 mm = 1 m<br>1000 m = 1 km |
| | um | micrometre | $1 mm = 1 \times 10^{-6} m$ |
| Luminus intensity | cd | candela | |
| Mass | kg<br>t | kilogram<br>tonne | <br>1000 kg = 1 tonne |
| | | | |
| Modulus of elasticity | MPa | megapascal | $1 MPa = 1 N/mm^2$ |
| Modulus, section | $mm^3$ | cubic millimetre | |
| Moment of inertia | $mm^4$ | millimetre to fourth power | |
| Moment of force | Nm | newton metre | $m^2·kg/s^2$ |
| Plane angle | deg<br>minute | °<br>' | 1 = (p/180) rad<br>1' = (p/10 800) rad |
| Power | W | watt | $1 W = 1 J/s = m^2·kg/s^3$ |
| Pressure | Pa | pascal | $1 Pa = 1 N/m^2$ |
| Specific volume | $m^3/kg$ | cubic metre per kilogram | |
| Stress | MPa | megapascal | $1 MPa = 1 N/mm^2$ |
| Temperature | °C | degree Celsius | |
| Thermal conductivity | W/m°C | watt per metre Celsius | $m^2·kg/s^3·°C$ |
| Thermodynamic temperature | K | kelvin | 0°C = 273.15 Kelvin |
| Time | s<br>min<br>h | second<br>minute<br>hour | <br>1 min = 60 s<br>1 h = 3600 s |
| Torque | kNm | kilonewton metre | |
| Volume | $mm^3$ | cubic millimetre | |

# METRIC CONVERSION

**Fig. 8.4.2 Conversion factors**

| Item | SI to Imperial | | Imperial to SI | |
|---|---|---|---|---|
| Area | 1 ha<br>1 $mm^2$<br>1 $m^2$<br>1 $m^2$ | =2.471 acre<br>=$1.55\times10^{-3}$ in.$^2$<br>=10.764 ft.$^2$<br>=1.2 yd.$^2$ | 1 acre<br>1 in.$^2$<br>1 ft.$^2$<br>1 yd.$^2$ | =0.4046856 ha<br>=645.16 $mm^2$<br>=0.09290304 $m^2$<br>=0.08361274 $m^2$ |
| Force | 1 kN | =0.225 kip | 1 kip | =4.448222 kN |
| Length | 1 m<br>1 mm<br>1 km | =3.2808 ft.<br>=0.0394 in.<br>=0.622 mile | 1 ft.<br>1 in.<br>1 mile | =0.3048 m<br>=25.4 mm<br>=1.609344 km |
| Mass | 1 kg<br>1 Mg | =2.204 lb.<br>=1.1 ton (2200 lb.) | 1 lb.<br>1 ton (2000 lb.) | =0.453592 kg<br>=0.90718474 Mg |
| Mass per unit length | 1 kg/m | =0.6719 lb./ft. | 1 lb./ft. | =1.48816 kg/m |
| Mass per Volume | 1 kg/$m^3$ | =$62.4\times10^{-3}$ lb./ft.$^3$ | 1 lb./ft.$^3$ | =16.01846 kg/$m^3$ |
| Moment of inertia | 1 $mm^4$ | =$2.4\times10^{-6}$ in.$^4$ | 1 in.$^4$ | =416231.4 $mm^4$ |
| Pressure or stress | 1 Mpa<br>1 Pa<br>1 kPa | =0.145 ksi<br>=0.02088 psf<br>=0.145 psi | 1 ksi<br>1 psf<br>1 psi | =6.894757 MPa<br>=47.8927 Pa<br>=6.894757 kPa |
| Section, modulus | 1 $mm^3$ | =$0.061\times10^{-3}$ in.$^3$ | 1 in.$^3$ | =16 387.06 $mm^3$ |
| Torque or moment of force | 1 kN·m | =0.73756 ft.kip | 1 ft.kip | =1.355818 kN·m |
| Volume | 1 L<br>1 cu m<br>1 $m^3$<br>1 $mm^3$ | =0.2199 gal (Can)<br>=219.9 gal (Can)<br>=35.314667 ft.$^3$<br>=$0.061\times10^{-3}$ in.$^3$ | 1 gal (Can)<br><br>1 ft.$^3$<br>1 in.$^3$ | =4.54609 L<br><br>=0.028317 $m^3$<br>=16 387.06 $mm^3$ |

## METRIC CONVERSION

**Fig. 8.4.3 Other conversion factors**

| To convert from | to | multiply by |
|---|---|---|
| Mass per volume (density) | | |
| pound/cubic foot (pcf) | kilogram/cubic metre ($kg/m^3$) | 16.02 |
| pound/cubic yard (pcy) | kilogram/cubic metre ($kg/m^3$) | 0.5933 |
| Bending Moment or Torque | | |
| inch-pound (in.-lb) | newton metre (N·m) | 0.1130 |
| foot-pound (ft.-lb.) | newton metre (N·m) | 1.356 |
| foot-kip (ft.-k) | newton metre (N·m) | 1356 |
| Temperature | | |
| degree Fahrenheit (° F) | degree Celsius (°C) | $t_c = (t_f - 32)/1.8$ |
| degree Fahrenheit (° F) | degree Kelvin (°K) | $t_c = (t_f + 459.7)/1.8$ |
| Energy | | |
| British thermal unit (Btu) | joule (j) | 1056 |
| kilowatt-hour (kwh) | joule (j) | 3 600 000 |
| Power | | |
| horsepower (hp) (550 ft.lb./sec) | watt (W) | 745.7 |
| Velocity | | |
| mile/hour (mph) | kilometre/hour | 1.609 |
| mile/hour (mph) | metre/second (m/s) | 0.4470 |
| Other | | |
| Section modulus ($in.^3$) | $mm^3$ | 16 387 |
| Moment of Inertia ($in.^4$) | $mm^4$ | 416 231 |
| Coefficient of heat transfer ($Btu\text{-}in./ft.^2/h/°F$) | $W/m^2/°C$ | 5.678 |
| Modulus of elasticity (psi) | MPa | 0.006895 |
| Thermal conductivity ($Btu\text{-}in./ft.^2/h/°F$) | $W·m/m^2/°C$ | 0.1442 |
| Thermal expansion in./in.°F | mm/mm/°C | 1.800 |
| Area/length ($in.^2/ft$) | $mm^2/m$ | 216.80 |

# CANADIAN STANDARDS ASSOCIATION
*A23.4-94*
# PRECAST CONCRETE – MATERIALS AND CONSTRUCTION

# CONTENTS




© Canadian Standards Association - 1994

All rights reserved.
No part of this publication may be reproduced in any form, in an electronic retrieval system or otherwise, without the prior permission of the publisher.

## TECHNICAL COMMITTEE ON PRECAST CONCRETE—MATERIALS AND CONSTRUCTION

**L. Kulcsar**
Pre-Con Company
Brampton, Ontario
*Chair*

**R.E. Loov**
University of Calgary
Calgary, Alberta
*Vice-Chair*

**M. Belanger**
Prefac Concrete Co. Ltd.t
Montréal, Québec

**P.C. Breeze**
Con-Force Structures Limited
Calgary, Alberta
*Associate*

**R. Diakiw**
Supercrete Precast Limited
Winnipeg, Manitoba

**G. Feher**
Ontario Ministry of Housing
Toronto, Ontario

**J.R. Fowler**
Canadian Prestressed Concrete Institute
Ottawa, Ontario
*Associate*

**M.J. Frye**
City of Winnipeg
Winnipeg, Manitoba

**M.V. Handa**
Canadian Standards Association
Rexdale, Ontario
*Associate*

**B. Malatesta**
Bramalea Limited
Toronto, Ontario

**M.S. Mirza**
McGill University
Montréal, Québec

**F. Nabhan**
Canadian Construction Materials Centre
Ottawa, Ontario

**T. O'Neill**
Alberta Public Works, Supply, and Services
Edmonton, Alberta

**E.C. Shelestynsky**
Procon Engineering Inspections Inc.
Inglewood, Ontario

**T. Sly**
Rocky Mountain Precast Ltd.
Calgary, Alberta
*Associate*

**R. Sneddon**
R. Sneddon & Associates Inc.
Concord, Ontario

**D. Tutty**
V.S.L. Corporation
Winterburn, Alberta

**V.N. Zubacs**
R.E. Winter and Associates
Mississauga, Ontario

**A.L. Cautillo**
Canadian Standards Association,
Rexdale, Ontario
*Administrator*

**P. McWatt**
Canadian Standards Association,
Rexdale, Ontario
*Secretary*

## PREFACE

This is the second edition of CSA Standard A23.4, *Precast Concrete—Materials and Construction*. It supersedes the first edition, published in 1978. This Standard has been updated to cover all necessary technical requirements for precast concrete including those in CSA Standard A251-M1982, *Qualification Code for Manufacturers of Architectural and Structural Precast Concrete.* In addition the Standard includes three appendices as follows:

Appendix A—Responsibilities;

Appendix B—Scope of Work; and

Appendix C—Architectural Precast Concrete—Materials and Finishes.

These three appendices are similar in content to Appendices A, B, and C in CSA Standard A251-M1982, with Appendix A and Appendix C considerably revised and expanded.

Since this Standard makes extensive reference to CSA Standard CAN/CSA-A23.1, users of CSA Standard A23.4 will need both Standards.

"Notes" to figures and tables are considered to be part of the figure or table and are therefore mandatory, but the "Notes" within the text of this Standard provide explanation and should not be considered as a mandatory part of the Standard.

This Standard was prepared by the Technical Committee on Precast Concrete—Materials and Construction under the jurisdiction of the Standards Steering Committee on Concrete and was formally approved by these Committees.

*February 1994*

Notes*:*

**(1)** *Use of the singular does not exclude the plural (and vice versa) when the sense allows.*

**(2)** *Although the intended primary application of this Standard is stated in its Scope, it is important to note that it remains the responsibility of the user of the Standard to judge its suitability for any particular purpose.*

**(3)** *CSA Standards are subject to periodic review, and suggestions for their improvement will be referred to the appropriate committee.*

**(4)** *All enquiries regarding this Standard, including requests for interpretation, should be addressed to Canadian Standards Association, Standards Development, 178 Rexdale Boulevard, Rexdale, Ontario M9W 1R3.*

*Requests for interpretation should*

*(a) define the problem, making reference to the specific clause, and, where appropriate, include an illustrative sketch;*

*(b) provide an explanation of circumstances surrounding the actual field condition; and*

*(c) be phrased where possible to permit a specific "yes" or "no" answer.*

*Interpretations are published in CSA's periodical* Info Update. *For subscription details, write to CSA Sales Promotion,* Info Update, *at the address given above.*

## GUIDE FOR THE USE OF THE STANDARD

Precast concrete construction can be referenced to this Standard as a whole, as all applicable requirements of CSA Standard CAN/CSA-A23.1 are referenced. However this Standard should not be referenced without the owner identifying, in the contract documents where applicable, the following:

(a) corrosion protection for connections (see Clause 8.2.1);

(b) insulation for sandwich wall panels (see Clause 8.3.5);

(c) concrete cover (see Clause 12.5);

(d) exposure and fire ratings (see Clause 12.5);

(e) Appendix A—Responsibilities;

(f) Appendix C—Architectural Precast Concrete—Materials and Finishes.

The following gives a fuller explanation of how the appendices should be used.

### Appendix A—Responsibilities

This Appendix provides detailed recommendations on the division of responsibilities between the owner and the precast manufacturer. It outlines three alternative options that may be adopted by the owner in preparation of contract documents (see Table A1). Consideration should be given to including Appendix A in the contract documents for a project, but if this is done, the owner must clearly specify which option has been adopted.

### Appendix B—Scope of Work Included in Precast Projects

Although this Appendix contains comprehensive advice that is intended to assist the owner in the preparation of contract documents for a precast project, it may be included in whole or in part in contract documents.

### Appendix C—Architectural Precast Concrete—Materials and Finishes

This Appendix covers all aspects of the manufacture of architectural precast concrete except for those which relate to its structural adequacy. It also covers the provision of project samples and their approval. It is therefore essential that owners include conformity with Appendix C in contract documents for any project that includes the supply of architectural precast concrete.

# PRECAST CONCRETE—MATERIALS AND CONSTRUCTION

## 1. SCOPE

### 1.1 GENERAL

This Standard covers materials to be used and methods to be followed for the manufacture, transportation, and erection of architectural, structural, and specialty precast concrete products.

*Notes:*

**(1)** *Good control over dimensions, appearance, and performance of precast concrete products can be obtained when this Standard is used.*

**(2)** *This Standard should be specified for precast concrete elements used in segmental construction.*

### 1.2 CERTIFICATION

Precast concrete elements produced and erected in accordance with this Standard shall be manufactured in plants certified in the appropriate category by a certification organization accredited by the Standards Council of Canada in the subject area of building products and structures.

*Note: Plant certification demonstrates the capability of a manufacturer to fabricate precast concrete elements to the requirements of this Standard.*

### 1.3 DESIGN BASIS

#### 1.3.1

Structural and architectural precast concrete shall be designed in accordance with CSA Standard A23.3 or S6 as specified by the regulatory authorities.

#### 1.3.2

The design of specialty precast products shall be in accordance with other applicable Standards as specified by the Owner.

*Notes:*

**(1)** *Regulatory authorities require that structural design and review of construction be carried out by a professional qualified for such activities under appropriate provincial or territorial legislation. This requirement is intended to ensure the structural adequacy of precast concrete and includes design, shop drawings, manufacture, and installation for which the manufacturer is responsible. The provision of qualified professional services is a requirement of plant certification (see Appendix A, Clause A2, Responsibility of the Manufacturer).*

**(2)** *For information on the design of precast and prestressed concrete, see the Design Manual, 3rd edition, 1996, published by CPCI.*

#### 1.3.3

Sandwich wall panels shall be designed as composite or noncomposite.

*Note: See the* Design Manual, *3rd edition, 1996, published by CPCI.*

### 1.4 SHOP DRAWINGS

This Standard outlines the requirements for precast concrete shop drawings in addition to those stated in CSA Standard A23.3.

### 1.5 ARCHITECTURAL PRECAST CONCRETE

Those aspects of architectural precast concrete falling within the scope of provincial building codes, namely those affecting its structural adequacy, are included within the body of this Standard. Architectural finishes are covered in Appendix C.

*Note: Conformity with Appendix C should be required in projects that include the supply of architectural precast concrete. (See "Guide for the Use of the Standard".)*

### 1.6 PRECEDENCE

Where this Standard states that the content of a clause shall conform to a referenced standard and additional clauses in this Standard, then, where there is a conflict the requirements of this Standard shall take precedence.

## 2. REFERENCE PUBLICATIONS, CATEGORIES, AND DEFINITIONS

### 2.1 REFERENCE PUBLICATIONS

This Standard refers to the following publications and where such reference is made it shall be to the edition listed below, including all amendments published thereto:

**CSA Standards**

CAN/CSA-A23.1-M90,
*Concrete Materials and Methods of Concrete Construction;*

CAN/CSA-A23.2-M90,
*Methods of Test for Concrete;*

CAN3-A23.3-M84,
*Design of Concrete Structures for Buildings;*

CAN/CSA-G164-M1992,
*Hot Dip Galvanizing of Irregularly Shaped Articles;*

CAN3-G279.2-M84,
*Steel Wire in Mill Coils for Prestressed Concrete Railroad Ties;*

CAN/CSA-S6-88,
*Design of Highway Bridges;*

CAN/CSA-S269.3-M92,
*Concrete Formwork;*

CAN/CSA-S413-87,
*Parking Structures;*

W47.1-92,
*Certification of Companies for Fusion Welding of Steel Structures;*

W186-M1990,
*Welding of Reinforcing Bars in Reinforced Concrete Construction.*

**ASTM* Standards**
A780-90,
*Practice for Repair of Damaged and Uncoated Areas of Hot-Dip Galvanized Coatings;*

C330-89,
*Lightweight Aggregates for Structural Concrete.*

In addition, the Appendices and "notes" of this Standard refer to the following publications and where such reference is made it shall be to the edition listed below, including all amendments published thereto:

**ACI† Publications**
214-77 (89),
*Recommended Practice for Evaluation of Strength Test Results of Concrete (MCP-2);*

309R-87,
*Guide for Consolidation of Concrete (MCP-2).*

**ASTM* Standards**
C512-87,
*Creep of Concrete in Compression;*

C641-82,
*Staining Materials in Lightweight Concrete Aggregates.*

**CPCI‡ Publications**
*Metric Design Manual, 2nd edition, 1987;*

*Fire Resistance Ratings for Prestressed and Precast Concrete.*

**PCI§ Publications**
MNL-122-89,
Architectural Precast Concrete (2nd edition);

*Report JR-271 Fabrication and Shipment Cracks in Prestressed Hollow-Core Slabs and Double Tees;*

*Report JR-316 Fabrication and Shipment Cracks in Precast or Prestressed Beams and Columns;*

*PCI Architectural Precast Concrete Color and Texture Selection Guide (1st edition);*

*PCI Special Report on Architectural Precast Concrete—Joint Details.*

**American Society for Testing and Materials.*

*†American Concrete Institute.*

*‡Canadian Prestressed Concrete Institute.*

*§Precast/Prestressed Concrete Institute.*

## 2.2 CATEGORIES

Three categories are covered in this Standard with each category further subdivided into two divisions:

(a) Precast concrete products—Architectural
  (i) Nonprestressed;
  (ii) Prestressed.

(b) Precast concrete products—Structural
  (i) Nonprestressed;
  (ii) Prestressed.

(c) Precast concrete products—Specialty
  (i) Nonprestressed;
  (ii) Prestressed.

## 2.3 DEFINITIONS

In addition to the definitions in CSA Standard A23.1, the following shall apply in this Standard:

**Bond breaker**—a substance placed on materials to prevent concrete from bonding to them.

**Bonding agent**—a substance used to bond existing concrete to a subsequent application of concrete, mortar, or grout such as a patch or composite topping.

**Camber**—

(a) The upward deflection that occurs in prestressed concrete elements due to the net bending resulting from stressing forces and self-weight. It does not include dimensional inaccuracies; and

(b) A built-in upward curvature to compensate for anticipated deflection.

**Connection**—a device for attachment of precast concrete elements to each other or to a building structure.

**Construction joint**—the surface where two sucessive placements of concrete join.

**Creep (or plastic flow)**—the time-dependent deformation under sustained stress.

**Curing**—the maintenance of humidity and temperature of freshly placed concrete during some definite period following placing, casting, or finishing to ensure satisfactory hydration of the cementitious materials. Where the curing temperature remains in the normal environmental range (generally between 10 and 30°C), this Standard uses the term normal curing. Where the curing temperature is increased to a higher range, this Standard uses the term accelerated curing.

**Debonded tendon**—a tendon that has an end zone where the tendon is permanently prevented from bonding to the concrete after stressing. This is achieved by means of a bond breaker or ducts.

**Detensioning of tendon or wire**—the transfer of tension from the bed anchorage to the concrete.

**Differential camber**—the camber difference between adjacent elements.

**Documented**—written technical substantiation that the use of a particular material, design practice, or construction method satisfies the intent of this Standard. A documented material, practice, or method shall be approved in writing prior to its use or implemention.

**Draft**—the slope of concrete surface in relation to the direction in which the precast element is withdrawn from the form. It is provided to facilitate stripping with a minimum of form breakdown.

**Draped tendon**—a tendon that is held up at specific points and held down at others to form a desired profile.

**Dunnage**—materials used for keeping concrete elements from touching each other or other materials during storage and transportation.

**Exposed aggregate concrete**—concrete manufactured so that the aggregate on the face protrudes from the paste.

**Face mix**—the concrete placed at the exposed face of architectural concrete.

**Form**—a structure or mould for the support of concrete while it is setting and gaining sufficient strength to be self-supporting. The word form is used throughout the Standard and may include mould.

**Form release agent**—a substance applied to forms for the purpose of preventing bond between the form and the concrete cast in it.

**Formed surface**—a concrete surface that has been cast against formwork.

**Grout**—a mixture of cementitious materials and water, with or without sand. It may also contain admixtures.

**Hardware**—a collective term applied to items used in connecting precast units or attaching other materials or equipment.

Note: *Hardware can be classified as follows:*

(a) *Contractor's hardware—items to be placed in the structure in order to receive the precast concrete units, eg, anchor bolts, angles, or plates with suitable anchors. These items are usually marked "C" on working or shop drawings;*

(b) *Plant hardware—items to be embedded in the concrete units themselves, either for connections and precast erector's work, or for other trades, such as the mechanical, plumbing, glazing, miscellaneous iron, masonry, or roofing trades. These items are usually marked "P" on working or shop drawings;*

(c) *Erection hardware—all loose hardware necessary for the installation of the precast concrete units. These items are usually marked "E" on working or shop drawings.*

**Holding period**—the time from the placing of concrete to the commencement of accelerated curing.

**Jig**—a device to align parts of an assembly, usually for preassembling reinforcing steel and hardware cages with a minimum of measurement and consistent accuracy from one cage to the next.

**Joint**—used in the context of erection it is the space between two adjacent precast concrete elements. Other types of joints include control joint, construction or casting joint, expansion joint, and cold joint.

**Machine cast product**—product produced by slip-form or extrusion.

**Manufacturer**—the party who produces (and usually installs) the precast concrete elements.

**Manufacturer's engineer**—the structural engineer authorized by the manufacturer to ensure the adequacy of the structural aspects of the shop drawings, manufacture, and installation for which the manufacturer is responsible.

**Mould**—a custom-made form for specific products (see definition for form).

**No-slump concrete**—concrete designed with a low water/cementing materials ratio and a low slump that is compacted by special consolidation methods such as tamping or extrusion.

**Owner**—the administrator of the requirements of this Standard or the designated representative, usually a consulting engineer or architect.

**Prestressing bed**—the platform and abutments needed to support the forms and maintain the tendons in a stressed condition during placing and curing of the concrete.

**Quirk**—a reveal or false joint between two adjoining corner surfaces to mask the mitred or cold joints.

**Sandwich wall panel**—a wall element consisting of two wythes of concrete separated by a layer of insulation.

**Sealer or protective coating**—materials used to coat precast concrete elements for the purpose of improving resistance to water penetration or for improving weathering qualities.

**Set-up**—the process of preparing forms or moulds for casting including placing of materials prior to the actual placing of concrete.

**Shop drawings**—all drawings prepared by the manufacturer for production and erection.

Note: *See Appendix A, Clause A2.3.*

**Stress-resistant form**—a form provided with suitable end bulkheads and sufficient cross-sectional strength to resist the total prestressing force.

**Stripping**—the process of removing a precast concrete element from the form in which it was cast.

**Supplier**—the party supplying materials or components to the manufacturer.

**Surface retarder**—a material used to retard or prevent the hardening of the cement paste at a concrete surface to facilitate removal of this paste after curing.

**Tendon**—

(a) a high strength steel element consisting of one or more wires, tendons or bars, or a bundle of such elements, used to impart prestress to the concrete.

(b) a group of wires laid helically over a central-core wire. A seven-wire tendon would thus consist of six outer wires laid over a single wire core.

**Tendon anchor**—sometimes called a tendon or vice. A device for holding a tendon under tension.

**Transfer strength**—sometimes called detensioning strength or release strength. The minimum concrete strength specified for the individual concrete elements before the prestressing force may be transferred to them.

**Veneered construction**—the application of two different concrete mixes such as a special architectural face mix and a standard backup mix in one element. Also the attachment of other materials, such as cut stone pieces, to a concrete panel.

# 3. CEMENTS, SUPPLEMENTARY CEMENTING MATERIALS, AND COLOURING MATERIALS

## 3.1 GENERAL

Cement and supplementary cementing materials shall conform to the requirements of Clause 3 of CSA Standard CAN/CSA-A23.1 and Clause 3 of this Standard.

*Note: Where uniformity of colour for a project or part of a project is desired, conformity with the appropriate requirements of Appendix C should be specified.*

## 3.2 COLOURING MATERIALS

Pigment used in the concrete mix or as a surface colourant shall have a proven history of colourfastness and shall not reduce the durability of the concrete.

*Notes:*

***(1)** Suitable tests or performance records are necessary to establish colour stability in the actual concrete mix and compatibility with possible retarders and finish treatments.*

***(2)** Except for exposed surfaces the use of dark pigments should be avoided.*

***(3)** The effect of colouring agents on air content, mixing water requirements, and water absorption of the face mix should be established prior to use.*

# 4. WATER

## 4.1

Water for use in precast concrete and for curing shall conform to the requirements of Clause 4 of CSA Standard CAN/CSA-A23.1.

*Note: For architectural concrete, conformity with the appropriate requirements of Appendix C should be specified.*

# 5. AGGREGATES

## 5.1 GENERAL

Aggregates for use in precast concrete shall conform to the requirements of Clause 5 of CSA Standard CAN/CSA-A23.1, and Clause 5 of this Standard.

*Note: For architectural concrete, conformity with the appropriate requirements of Appendix C should be specified.*

## 5.2 SPECIAL AGGREGATE

Special aggregate for architectural concrete such as ceramics, glass, and plastics shall not be used unless tests have proven, to the satisfaction of the owner, that they are acceptable with respect to strength, density, durability, and appearance for a minimum period of 5 years.

*Note: Investigation of special aggregates should cover the material itself and its interaction with the concrete materials and any other materials with which it may come in contact in service.*

# 6. ADMIXTURES

## 6.1 GENERAL

Admixtures for use in precast concrete shall conform to the requirements of Clause 6 of CSA Standard CAN/CSA-A23.1 and Clause 6 of this Standard.

*Note: Some admixtures may affect the colour tone of concrete, hence the same admixture should be used throughout any part of a project where colour uniformity is required.*

## 6.2 CALCIUM CHLORIDE

Calcium chloride shall not be used in precast concrete.

# 7. REINFORCEMENT

## 7.1

Reinforcement and prestressing tendons for use in precast concrete shall conform to the requirements of Clause 7 of CSA Standard CAN/CSA-A23.1 and in addition, for concrete railroad ties, CSA Standard G279.2.

# 8. HARDWARE AND MISCELLANEOUS MATERIALS

## 8.1 GENERAL

Hardware and miscellaneous materials for use in precast concrete shall conform to Clause 8 of CSA Standard CAN/CSA-A23.1 and Clause 8 of this Standard.

*Note: To prevent rust staining on finished surfaces, all exposed metal inserts, anchorages, and other hardware shall be protected with a rust preventive material.*

## 8.2 HARDWARE

### 8.2.1 Connections

Precast connection hardware shall be identified and located on the shop drawings. The owner shall specify corrosion protection adequate for the type of exposure and the design service life.

*Note: Types of exposure requiring consideration of measures such as hot dip galvanizing or the use of stainless steel include*

*(a) exposure to corrosive substances;*

*(b) exposure to the weather, condensation, or other sources of moisture;*

*(c) exposure within the air or insulation space outside the air barrier of a building.*

### 8.2.2 Tendon Deflectors

**8.2.2.1**

Tendon deflectors shall maintain the tendon location (bundled or spaced) in accordance with the design criteria and as shown on the shop drawings.

**8.2.2.2**

Tendon deflectors shall conform to the requirements of the design criteria and shall be of a type suitable for their intended use. When specified by the Owner and furnished by a supplier, test and performance data shall be documented. When manufactured by the precast concrete plant's own personnel, the design shall be approved by the Owner. Welding of hardware shall conform to the requirements of Clause 13.4 of CSA Standard CAN/CSA-A23.1.

**8.2.2.3**

Tendon reflectors used for holding the tendons in the deflected position during stressing shall have pin and roller fixtures to minimize friction at all deflection points.

### 8.2.3 Pipes, Conduits, and Other Hardware

Pipes, conduits, and other hardware shall be identified and located on the shop drawings.

### 8.2.4 Tendon Anchors

**8.2.4.1**

Tendon anchors shall be capable of anchoring the tendon positively without slippage after seating.

**8.2.4.2**

Steel casings for tendon anchors shall be proof-tested by the supplier to at least 95% of the ultimate specified strength of the tendon, without permanent deformation of the casing.

**8.2.4.3**

Proper performance and maintenance data for the anchors shall be documented by the supplier.

**8.2.4.4**

Anchors for repeated usage shall be of a type designed for reuse, while anchors used only once that remain in the product may be of the one-use only type.

**8.2.4.5**

Tendon anchors used repeatedly shall be removed from the beds and maintained regularly according to the supplier's recommendations using proper tools and materials. The frequency of maintenance will depend on the type of anchor and its exposure during use, but under no circumstance shall the same anchor be used more than 5 times without checking and maintenance.

## 8.3 MISCELLANEOUS MATERIALS

### 8.3.1 Special Facing Materials

When facing of any suitable material such as cut stone slabs, ceramic and glass products, plastics, oversized aggregates, metal sheets or sections, or wood are used as an integral part of the precast concrete element, they shall have an acceptable performance record or be tested to the satisfaction of the owner.

The design and dimensions of elements shall take into account any difference in shrinkage or thermal characteristics between the concrete and the facing material.

*Notes:*

**(1)** *Where special facing materials are used, design and testing should be based on the expected environmental conditions.*

*The test samples, preferably to full scale, should be produced, cured, and handled in a manner similar to the expected production procedures. Particular attention should be paid to the interaction of the concrete and the facing materials during humidity and temperature changes and to potentially detrimental reaction between the concrete and the facing material.*

**(2)** *Particular attention should be paid to compatibility of materials for volume changes caused by temperature or humidity changes. If materials do not have similar physical properties, the final design should include provisions for possible movement of the face material in relation to the backup concrete panel.*

**(3)** *If performance records have not already been established for these face materials for special climatic conditions including severe corrosive atmospheres, the materials should be tested for such conditions. The tests may be accelerated but should otherwise reflect the expected service condition.*

### 8.3.2 Bearing Materials

Bearing materials shall have a record of satisfactory performance, or test results that prove their adequacy shall be made available to the owner upon request. Bearing materials shall be identified on the shop drawings and shall be adequate for the loads and movements expected. The owner shall specify the bearing materials for any bearings considered to have particularly high structural significance.

#### 8.3.3 Joint Material

Materials for joints between precast concrete elements shall be as specified by the owner.

*Note: Reference should be made to the special report by the Prestressed Concrete Institute on Architectural Precast Concrete—Joint Details.*

#### 8.3.4 Weather Sealers

If weather-sealing materials are used they shall be as specified or approved by the owner.

*Note: Reference should be made to the PCI Publication,* Architectural Precast Concrete, *for a review of the merits of sealers.*

#### 8.3.5 Insulation for Sandwich Wall Panels

The owner shall specify the insulation materials for sandwich wall panels. The insulation materials shall not exhibit capillary action, otherwise, suitable vapour barriers shall be used.

*Notes:*

**(1)** *It is recommended that self-extinguishing types of insulation be used.*

**(2)** *Properties, such as insulation value, vapour transmission, and volume stability under service conditions, should be documented and this documentation should be available from the supplier.*

#### 8.3.6 Manufacturing Accessories

Form ties, inserts, bracing, spacers, and similar accessories incidental to the production of precast concrete elements shall be adequate for their intended use and shall not be detrimental to the finished product.

#### 8.3.7 Other Materials

Other materials used in the manufacture of precast concrete shall be documented. Such documentation shall be available at the plant.

# 9. STORAGE OF MATERIALS

## 9.1 GENERAL

Storage of materials for use in precast concrete shall conform to the requirements of Clause 9 of CSA Standard CAN/CSA-A23.1 and Clause 9 of this Standard.

## 9.2 HARDWARE MATERIALS

Hardware materials shall be stored according to type, grade, and size and shall be handled and stored so as to avoid distortion. The hardware materials shall be kept in a clean condition.

# 10. TOLERANCES

## 10.1 GENERAL

The tolerances in Clause 10 are for finished precast concrete work and include the installation of this work. The tolerances for precast concrete work shall conform to the requirements of Clause 10 unless otherwise specified by the owner.

#### 10.1.1 Tolerances Not Specified Elsewhere

Where tolerances are not specifically detailed in Clause 10, the allowable dimensional tolerance shall be 1:800 but not less than ±5 mm. For dimensions greater than 30 m the tolerances shall be specified by the owner.

*Notes:*

**(1)** *See Appendix C of CSA Standard CAN/CSA-A23.1 for tolerance definitions, principles, and preferred sizes.*

**(2)** *The tolerances provided in Clause 10 apply to buildings. For structures, such as bridges and wharves it is suggested that Clause 10.1.1 be used if tolerances are not already provided by the owner.*

**(3)** *Precast concrete elements not within the production tolerances provided in Clauses 10.1 to 10.4 may still be acceptable to the owner provided that they can be repaired and/or installed to the erection tolerances stated in Clauses 10.5 to 10.6 and provided that any excessive variations have been checked and approved for structural implications. The production tolerances as provided will serve to establish the responsibility of the precast manufacturer to have the elements repaired in accordance with Clause 33 and/or installed within the erection tolerances and to verify that structural performance is not impaired.*

## 10.2 TOLERANCES FOR STRUCTURAL ELEMENTS

#### 10.2.1 Floor and Roof Slabs, Prestressed or Nonprestressed

**10.2.1.1 Main Dimensions**

Allowable variations for main dimensions shall be as follows:

| | |
|---|---|
| for depths (h) | |
| ≤ 0.4 m | ±5 mm |
| > 0.4 m | ±10 mm |
| for widths (b) | |
| ≤ 2.4 m | ±5 mm |
| > 2.4 m but £ 4.8 m | ±10 mm |
| for lengths (L) | |
| < 7.2 m | ±12 mm |
| > 7.2 m but ≤ 30.0 m | ±1:600 |

For larger dimensions, the owner shall specify the allowable variations.

**10.2.1.2 Warpage**

The maximum allowable warpage of one corner out of the plane of the other three at the bottom surface shall be the greater of 1 mm per 300 mm distance from the nearest adjacent corner or 5 mm (see Figure 1). This warpage shall be measured when the slab is placed horizontally on two parallel supports or installed in its final location.

**Fig. 1 Warpage of One Corner of Slab or Panel**

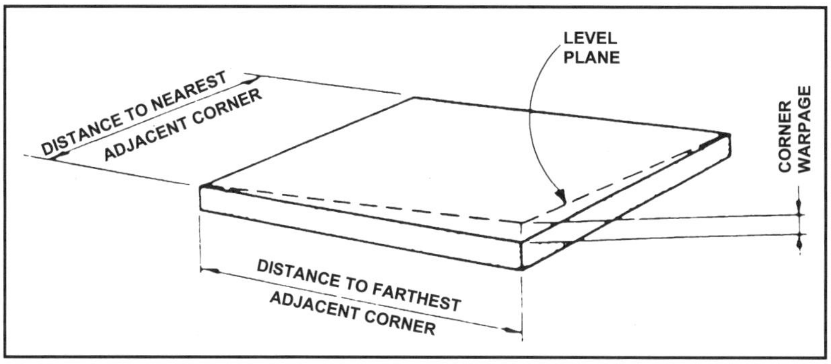


#### 10.2.1.3 Alignment of the Longitudinal Edges

Maximum allowable variation in alignment when the slab is placed horizontally on two parallel supports shall be l/600 or 5 mm, whichever is greater.

#### 10.2.1.4 Squareness and Plumbness

Variation of squareness at ends and vertical plumbness at ends with slabs supported similar to service conditions shall not exceed 1 in 100.

*Notes:*

**(1)** *The requirement in this Clause may need to be 1 in 200 where slabs are supported on narrow ledge beam seats.*

**(2)** *The ends of long and deep prestressed elements may need to be cast slightly sloped from the vertical to accommodate camber following transfer of prestress.*

#### 10.2.1.5 Camber

Maximum variation from design camber shall be l/1000, but not greater than 20 mm. Maximum differential camber variation between adjacent slabs shall be l/600 but not greater than 30 mm.

*Notes:*

**(1)** *Where roofing insulation and roofing materials are applied directly to precast slabs, feathering-out between slabs may be required when the camber differential exceeds* 10 mm. *The owner should specify the minimum differential camber above which such feathering-out by the precast erector is required.*

**(2)** *For precast slabs without subsequent floor topping being applied in the field or, where top surface conditions are otherwise important, a lower differential camber should be specified to suit the floor covering. Alternatively, remedial requirements for increased smoothness may be specified.*

#### 10.2.1.6 Flange Thickness

The allowable variation for the thickness of thin flanges up to 150 mm wide, such as single tees and double tees, shall be ±5 mm.

### 10.2.2 Girders, Beams, and Columns

#### 10.2.2.1

Allowable variations for main dimensions of prestressed or nonprestressed girders, beams, and columns shall be as follows:

for depths (h) and width (b)

| | |
|---|---|
| ≤ 0.6 m | ± 8 mm |
| > 0.6 m but ≤ 1.2 m | ±12 mm |

The tolerance for length shall be 1:1000 for columns and shall be 1:800 for beams, but not less than 12 mm in either case. Columns or beams longer than 30 m shall have the tolerances specified by the owner.

#### 10.2.2.2

Alignment of longitudinal edges shall be as specified in Clause 10.2.1.3.

#### 10.2.2.3

Squareness and plumbness at ends shall be as specified in Clause 10.2.1.4.

#### 10.2.2.4

Requirements for camber shall be similar to requirements for floor and roof slabs in Clause 10.2.1.5.

### 10.2.3 Piles

#### 10.2.3.1 General

Allowable variations for main dimensions of prestressed or nonprestressed piles shall be as follows:

Cross-section* ±12 mm

The maximum variation in pile lengths shall be 1:500 or 30 mm, whichever is greater.

**Or thickness and width in the case of sheet piling.*

#### 10.2.3.2 Alignment of Longitudinal Edges

Maximum variation in alignment when the pile is hanging vertical shall be 1:500 or 12 mm, whichever is greater, up to a maximum of 50 mm.

#### 10.2.3.3 Squareness

Variation of squareness at ends of piles shall not exceed 1 in 100.

*Note: Where squareness of the driving end of the pile is important for accurate driving the requirement of 1 in 100 should be changed to 1 in 200.*

## 10.3 WALL PANELS

### 10.3.1 Main Dimensions

The maximum variation for thickness of wall panels shall be 5 mm. The maximum variation for lengths or heights shall be 1:1000 or 3 mm, whichever is greater.

### 10.3.2 Warpage

Maximum allowable warpage of one corner out of the plane of the other three shall be 1 mm per 400 mm of distance from the nearest adjacent corner or 5 mm, whichever is greater. (See Figure 1.)

### 10.3.3 Bowing

#### 10.3.3.1

Bowing, whether concave or convex, of any part of a flat exposed surface and of the entire panel shall not exceed

$$\frac{\text{length of bow}}{400}$$

or a maximum of 20 mm for panels up to 10 m long. (See Figure 2.)

These requirements for warpage and bowing of panels may be superseded by tolerances for panels as installed with reference to joint widths, jog in alignment, and step in face as specified in Clause 10.6.

*Note: Slender panels may require larger tolerances. The major criterion for maintaining or relaxing bowing or warping tolerances will be the appearance.*

**Fig. 2 Measuring of Bowing in Panel**

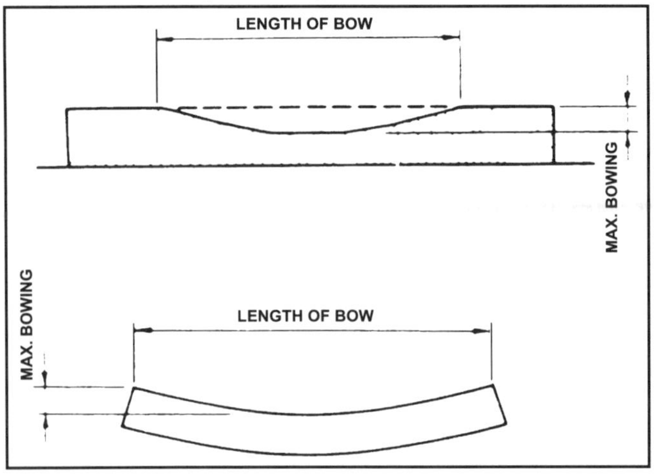


#### 10.3.3.2

Bowing of sandwich wall panels shall not exceed

$$\frac{\text{length of bow}}{400}$$

*Notes:*

**(1)** *Sandwich wall panels generally have a tendency to bow more than single wythe panels. An allowance for differential bowing between adjacent panels may be specified.*

**(2)** *Bowing allowances of this magnitude require special architectural details at corners. High panels adjacent to short panels require special attention.*

### 10.3.4 Alignment of Edges

Maximum allowable variation from a straight edge shall be 1 in 600, with a maximum of 5 mm for panels up to and including 2.4 m or 8 mm for panels above 2.4 m.

### 10.3.5 Angular Variation

The variation of the angle of abutting planes (returns) shall not exceed 1 in 100 and shall be no greater than 5 mm.

### 10.3.6 Skew

The difference between the two diagonals of a panel shall be measured. The maximum allowable difference shall be 5 mm up to and including a diagonal length of 3 m, 8 mm up to and including a diagonal length of 6 m with both parallel sides within allowable variations (see Clause 10.3.1), or 12 mm for a diagonal length greater than 6 m.

## 10.4 OTHER TOLERANCES

Tolerances for placing of reinforcement, hardware, hold-down devices, and lifting points shall be as specified in Clauses 12 and 13.

## 10.5 TOLERANCES FOR STRUCTURES ERECTED FROM PRECAST CONCRETE ELEMENTS

### 10.5.1

The actual dimensions to a vertical and horizontal reference grid system shall not vary from the dimensions on the drawings beyond the tolerances provided in Table 4 of Clause 10 of CSA Standard CAN/CSA-A23.1.

### 10.5.2

Variations in column alignment in a vertical direction shall not exceed 1 in 600 nor be greater than 40 mm, except for exterior and elevator columns where variations shall not exceed 1 in 800 nor be greater than 30 mm.

### 10.5.3

Column off-sets at each floor and beam off-sets at columns or walls shall not exceed the allowable variation for the appropriate section's dimensions as specified in Clause 10.2.2.

### 10.5.4

Tolerances for floors and roofs shall be in accordance with Clause 10.2.1 and tolerances for floor topping shall be in accordance with Clause 22 of CSA Standard CAN/CSA-A23.1.

## 10.6 TOLERANCES IN JOINTS BETWEEN ARCHITECTURAL WALL ELEMENTS

### 10.6.1 Tolerances for Joints

Precast wall elements shall be manufactured and erected so that the joints between them are within the following variations:

Joint taper — 1 in 400 with a maximum of 5 mm in either direction;

Step in face (see Figure 3) not to exceed 5 mm;

Jog in alignment of face edge (see Figure 4) not to exceed 5 mm.

### 10.6.2 Tolerances for Joint Width

Maximum allowable differential variation between joints in any one elevation shall be 5 mm.

*Notes:*

**(1)** *Maximum variation in joint width from the specified width should be either* 8 mm *or* 12 mm. *The choice depends on the specified clearance*

**Fig. 3 Section Through Face**

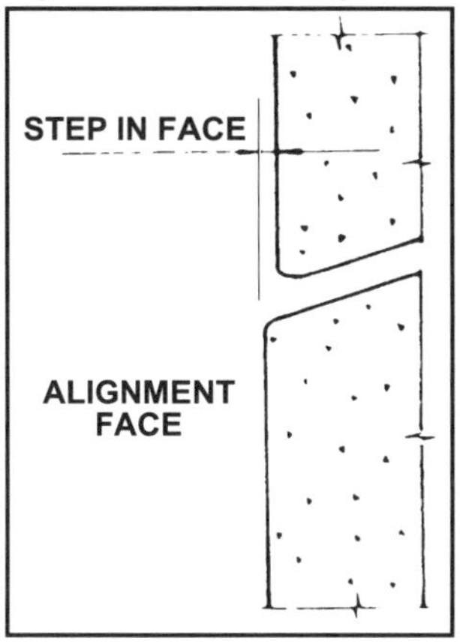


**Fig. 4 View of Elevation**

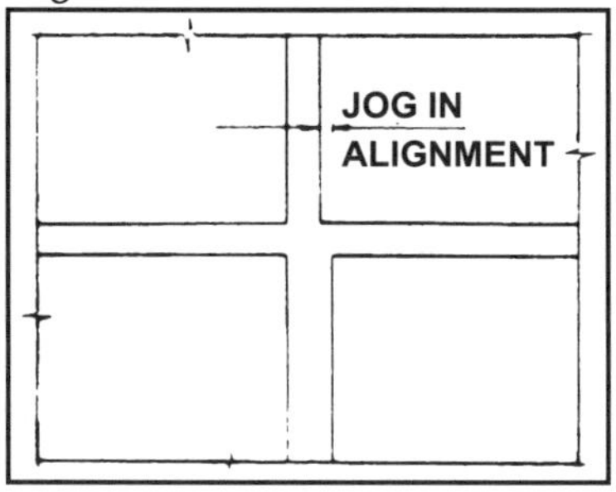


*between the structure and the panels and whether tolerances for length and height of elevations are to be accommodated solely in the joints.*

**(2)** *Small variations in width of adjacent joints may be achieved by setting out joint centrelines equally spaced along an elevation and centring panels between them.*

**(3)** *Allowable variations of joint widths from specified width may be limited to 5 mm for projects where tolerances for length and height at elevations have been designed to be accommodated as part of corner details, in expansion joints, or in overlapping details with other wall materials.*

# 11. FORMS

## 11.1 FORMS FOR PRECAST CONCRETE

### 11.1.1

Design and materials for forms shall conform to the requirements of CSA Standard CAN/CSA-S269.3 and Clause 11 of this Standard. In addition design and materials shall conform to the quality requirements for the finished product with respect to materials and dimensional controls.

### 11.1.2

Forms shall be adequate to withstand the consolidation of the concrete by external or internal vibration, or both, the hydrostatic concrete pressure, and thermal stresses during the complete cycle for either normal or accelerated curing.

*Notes:*

**(1)** *Where empty forms or forms containing concrete are to be handled during the production cycle, the forms should be designed to maintain their rigidity and the dimensional controls of the element during handling.*

**(2)** *Where external vibration is used, special design of the forms is necessary to maintain the specified tolerances. If the vibrators are fastened to the form, their position and the construction of the form should be such that uniform vibration is transmitted along any surfaces that are to be exposed.*

### 11.1.3

Parts such as bulkheads, side or top forms, or form modification pieces shall be well fitted and secured and joints shall be sealed to prevent leakage.

### 11.1.4

Forms shall be fabricated to allow stripping without damage to the precast element.

### 11.1.5

Form braces shall not touch the concrete for any finish better than smooth form finish as described in Clause 24 of CSA Standard CAN/CSA-A23.1 unless otherwise approved by the owner.

### 11.1.6

Forms shall be clean, well maintained and checked to ensure that elements of the required quality are produced.

*Note:*

**(1)** *The manufacturer should identify the specific form or forms in which an element was cast so that appearance and dimensional problems can be identified.*

**(2)** *Tolerances for forms should be one-half that given for the dimensions of the finished element in Clause 10.*

## 11.2 FORMS FOR ARCHITECTURAL PRECAST CONCRETE

Forms for architectural precast concrete shall conform to the requirements of Clause 11.1.

*Notes:*

**(1)** *For architectural concrete, conformance with the appropriate provisions of Appendix C should be specified.*

**(2)** *Because the appearance of architectural precast concrete depends on the condition of the forms, it is recommended that their design and manufacture be executed by people skilled and experienced in this work.*

## 11.3 FORMS FOR STRUCTURAL PRECAST CONCRETE

### 11.3.1

Forms for structural precast concrete shall conform to the requirements of Clause 11.1 and where prestressed reinforcement is used the following additional clauses.

### 11.3.2

Forms shall allow for shortening and movements of the precast elements during transfer of prestress and application of heat when accelerated curing is used.

### 11.3.3

Forms shall be sufficiently strong and stiff to carry concentrated loads resulting from cambering of prestressed elements following release of prestress force.

### 11.3.4

Butt joints between form units making up the length of a casting bed shall match closely to minimize form break lines on the elements.

### 11.3.5

Proper alignment of all form units shall be maintained during the entire casting operation.

### 11.3.6

Stress-resistant forms shall be so designed and constructed that they are sufficiently strong to withstand the prestressing forces and still maintain the required dimensional tolerances.

## 11.4 FORMS FOR INTERNAL VOIDS

### 11.4.1

Forms for internal voids shall have sufficient strength and stiffness to provide stability during handling and placing of concrete and to withstand hydrostatic pressures and other forces during the production cycle. However, the forms shall not be so stiff that the concrete cracks due to their restraint of shrinkage.

*Notes:*

(1) *Where new void designs are used, elements should be checked to ensure that void collapse or displacement has not taken place and that no stress raisers have been created. When voids cannot be visually checked, actual concrete quantities may be compared with theoretical quantities or the actual mass of an element may be compared with the theoretical mass.*

(2) *Storage conditions affect some void form materials and should be monitored so that a change in shape or collapse can be prevented.*

### 11.4.2

Forming material shall not be adversely affected by moisture. The adequacy of the forms shall be established prior to use.

### 11.4.3

Forms for internal voids shall be held in place during the placing and consolidation of the concrete so that their correct positions are maintained.

### 11.4.4

Anchoring devices to hold the voids in position shall be properly spaced to minimize deflection between anchor points and the devices shall not puncture or otherwise damage the forms. Where void anchors are attached to reinforcement or prestressing tendons, displacement of the reinforcement shall be prevented. Where removable templates are used they shall remain in place for a sufficient time after concrete placement so that the void is not displaced upon removal of the templates. However, the concrete shall be sufficiently plastic so that holes left by the template removal can be filled and the concrete compacted into a homogeneous mass.

### 11.4.5

The ends of void forms and the splices of void forms shall be sealed to prevent entry of cement paste or concrete.

### 11.4.6

Void forms other than the solid type shall be drained where specified by the owner. Where elements are subjected to freezing temperatures, the voids shall be provided with drains at appropriate points.

*Note: When accelerated curing is used, consideration should be given to providing vents in void forms.*

## 11.5 PREPARATION OF FORMS

### 11.5.1

Forms shall be clean and shall be assembled within the dimensional controls established for the particular set-up and element.

### 11.5.2

Form release agents shall be applied uniformly, be of a type consistent with the required finish, and be used at a rate suitable for easy release and stripping of the element. The form release agent shall be compatible with any further treatment of the concrete surface.

*Note: The choice of form release agents or surface retarders should be based on the form configuration, planned production methods, the actual curing conditions, and any further treatment of the concrete surface.*

### 11.5.3

Prestressing tendons, anchorages for miscellaneous connections, and reinforcement shall not be contaminated by form release agents whether the form release agent is applied before or after the placing of the reinforcement and hardware.

# 12. FABRICATION AND PLACEMENT OF REINFORCEMENT

## 12.1 GENERAL

### 12.1.1

Fabrication and placement of reinforcement and prestressing tendons shall conform to the requirements of Clause 12 of CSA Standard CAN/CSA-A23.1 and Clause 12 of this Standard.

### 12.1.2

The location of individual pretensioning tendons shall be shown on the shop drawings. At the ends the minimum clear spacing shall be 3 times the nominal diameter of the individual tendon but not less than 1-1/3 times the nominal maximum size of the coarse aggregate.

## 12.2 PLACING REINFORCEMENT

### 12.2.1

Reinforcement shall be placed in the forms within the tolerances specified in Clause 12.3 unless other tolerances are specified by the owner.

*Note: To maintain tolerances and concrete cover requirements in intricate shapes or small concrete sections, it is recommended that cages be produced on a jig. Tack welding will normally result in additional cage rigidity and accuracy, but the use of tack welding is restricted by CSA Standard W186. CSA Standard W186 describes the proper procedures to be followed when tack welding is done.*

## 12.3 TOLERANCES FOR PLACING

### 12.3.1

Reinforcement shall be placed within the following tolerances:

(a) for clear concrete protection or reinforcement ..... ±8 mm;

(b) for the position of primary flexural and compression reinforcement, where the depth of a flexural element, thickness of a wall or smallest dimension of a column is
  (i) 400 mm or less ........................................... ± 8 mm;
  (ii) more than 400 mm but less than 800 mm ... ±12 mm;
  (iii) 800 mm or more ......................................... ±20 mm;

(c) for the lateral position of flexural reinforcement, wall reinforcement, column ties, stirrups, temperature reinforcement, and other secondary reinforcement, where the depth of a flexural element, thickness of a wall, or the smallest dimension of a column is
  (i) 400 mm or less ........................................... ±12 mm;
  (ii) more than 400 mm but less than 800 mm ... ±20 mm;
  (iii) 800 mm or more ......................................... ±30 mm;

(d) for longitudinal locations and ends of bars, except at ends of elements ............................................. ±50 mm;

(e) for ends of bars at ends or edges of elements .. ±20 mm.

### 12.3.2

Regardless of tolerance requirements, the concrete cover shall not be reduced by more than one-fourth of the specified cover (see Clause 12.5).

### 12.3.3

Tolerances different from those specified above shall be specified by the owner in the contract documents and shall be detailed on the shop drawings.

*Notes:*

**(1)** *Smaller tolerances should only be specified after a careful evaluation for feasibility, realistic control, and need.*

**(2)** *Closer tolerances for the location of prestressing tendons of* ±5 mm *are feasible for some precast prestressed concrete elements.*

### 12.3.4

Deflection points in prestressed elements shall be located with a longitudinal tolerance of ±200 mm unless the owner specifies otherwise.

## 12.4 WELDING OF REINFORCEMENT

### 12.4.1

The welding of reinforcement for precast concrete shall conform to the requirements of CSA Standard W186.

## 12.5 CONCRETE PROTECTION OF REINFORCEMENT

### 12.5.1

Concrete cover to reinforcement shall be specified by the owner and shown on the shop drawings. Instead the owner may provide the exposure ratings and fire ratings. Where cover requirements or exposure and fire ratings are not specified by the owner then the specified cover for reinforcement shall be not less than that shown in Table 1.

*Note: For textured architectural surfaces, the cover is measured from the deepest point of the surface. The distance between the form and the reinforcement must therefore be large enough to compensate for this reduction.*

**Table 1**

| Exposure to earth or weather | Exposed | Not exposed |
|---|---|---|
| (a) For | | |
| (i) piles | | |
| principal reinforcement, No. 35 and smaller | 50 mm | |
| ties and spirals | 40 mm | |
| (ii) Beams, girders, and columns | | |
| principal reinforcement, No. 35 and smaller | 40 mm | 30 mm |
| ties, stirrups, and spirals | 30 mm | 20 mm |
| (iii) Slabs, joists, shells, and folded plates, No. 20 and smaller | 25 mm | 20 mm |
| (iv) Wall panels, No. 15 and smaller | 25 mm | 20 mm |
| (b) For bars with a diameter $d_b$ larger than listed above, the cover shall be at least, but not necessarily more than, 50 mm | 1.5 $d_b$ | 1.0 $d_b$ |
| (c) The ratio of the cover to the nominal maximum aggregate size shall be at least | 1.5 | 1.0 |
| (e) The cover for a bundle of bars shall be the same as that for a single bar with an equivalent area. | | |

### 12.5.2

Concrete in a corrosive environment shall have additional protection in accordance with CSA Standard CAN/CSA-A23.1 and the cover shall be specified by the owner. For precast concrete in parking structures, refer also to CSA Standard CAN/CSA-S413.

### 12.5.3

Where an element is required to have a specific fire rating this shall be specified by the owner.

*Note: Additional information on fire resistance can be obtained from the Canadian Prestressed Concrete Institute publication,* Fire Resistance Ratings for Prestressed and Precast Concrete.

# 13. PLACEMENT OF HARDWARE AND OTHER EMBEDDED ITEMS

## 13.1 GENERAL

### 13.1.1

Fabrication and placing of hardware and other embedded items shall conform to the requirements of Clause 13 of CSA Standard CAN/CSA-A23.1 and Clause 13 of this Standard.

### 13.1.2

Care shall be taken during handling and placing of hardware to prevent damage to corrosion protection. All damage shall be repaired. For galvanized elements this shall be done in accordance with ASTM A780.

*Notes:*

(1) *Exposed metal surfaces of hardware should be adequately protected during storage.*

(2) *Additional protection of handling hardware will depend on the exposure and treatment of such hardware following erection.*

### 13.1.3

Specified reinforcement, as shown on the shop drawings, shall not be modified, relocated, or eliminated to accommodate hardware. If hardware anchors or reinforcing steel cannot be located as shown on the shop drawings, approval for revisions shall be obtained from the manufacturer's engineer. All revisions shall be recorded on the drawings for approval by the owner.

## 13.2 TOLERANCES FOR PLACING HARDWARE

### 13.2.1

Hardware shall be placed within an allowable variation of ±12 mm.

*Notes:*

(1) *It is recommended that hardware be checked for position and realigned, if necessary, immediately after consolidation of the concrete.*

(2) *The reason for close tolerances in the case of slope variations is the difficulty in compensating for such variations when a connection is made in the field. In the case of other dimensional variations, as covered by Clause 13.3.2 of CSA Standard CAN/CSA-A23.1, it is relatively simple to compensate, if the design has allowed for sufficiently large plates, ample clearances, and proper use of shim plates.*

## 13.3 LIFTING AND HANDLING HARDWARE

Permanent connection hardware shall not be used for handling except when the approval of the owner has been obtained. It shall be designed for such additional service to prevent damage to the connection or impairment of its performance. Any resulting damage to corrosion protection shall be repaired. In the case of hot dip galvanized connections this shall be done in conformance with ASTM A780.

*Notes:*

(1) *See Clause 16.6 of CSA Standard CAN/CSA-A23.3 for the design of lifting devices.*

(2) *Lifting devices of prestressing tendon or wire cable are frequently used. If more than one tendon or wire cable is required to conform to the design requirements, special provisions should be used to ensure the combined action of all tendons or cables. Proper rigging should be used to distribute the load according to the design criteria.*

(3) *Threaded inserts and lifting devices with exposed threads should be protected.*

# 14. MIX PROPORTIONS

## 14.1 GENERAL

Mix proportioning for precast concrete shall conform to the requirements of Clause 14 of CSA Standard CAN/CSA-A23.1 and Clause 14 of this Standard.

## 14.2 ARCHITECTURAL CONCRETE

All architectural concrete exposed to the weather shall have

| | |
|---|---|
| Minimum compressive strength | 30 MPa |
| Minimum air content | 5% |
| Maximum water/cementing materials ratio | 0.45 |

## 14.3 VENEERED CONSTRUCTION

### 14.3.1

When two different concrete mixes are veneered in a precast concrete element, the two mixes shall be compatible as an integral unit.

*Note: It is important to minimize any difference in behaviour between the two veneers because dissimilar shrinkage and creep may cause unacceptable cracking or excessive bowing or warping. Accordingly it is desirable that the two mixes be reasonably similar with respect to cement content and water/cementing materials ratio.*

#### 14.3.2

Where a special face mix is used for precast concrete the thickness of the face mix following consolidation shall be at least 25 mm or a minimum of 1.5 times the maximum size of the aggregate, whichever is the larger. In the case of large aggregates hand-laid in a mould these dimensions shall govern for the concrete mix used as the matrix.

## 15. DURABILITY

### 15.1

Precast concrete shall conform to the requirements of Clause 15 of CSA Standard CAN/CSA-A23.1. Care shall be taken to protect connections and exposed hardware.

*Notes:*

**(1)** *Durable precast concrete structures are very dependent on the protection of reinforcement and hardware whether this protection is provided by concrete cover or other means for connections and exposed hardware.*

**(2)** *Where elements have exposed aggregate finishes, care should be taken in selecting the aggregate so that it can withstand the environment to which it will be exposed.*

## 16. SPECIFYING CONCRETE

### 16.1

Alternative 1 of Table 11 of Clause 16 of CSA Standard CAN/CSA-A23.1 shall be used by the owner for specifying the concrete used in precast concrete.

In addition to the items in Table 11 the Owner shall specify

(a) the allowable percentage of water absorption for architectural concrete exposed to the weather;

(b) special additives such as colour pigments for architectural concrete;

(c) the maximum allowable water/cementing materials ratio;

(d) the limits for any concrete materials passing the 315 mm sieve for special high quality architectural concrete in severe or polluted environments or when optimum weathering qualities are desirable.

## 17. CONCRETE QUALITY

### 17.1 GENERAL

Evaluation of concrete quality for precast concrete shall conform to the requirements of Clauses 17.2 to 17.6 of CSA Standard CAN/CSA-A23.1 and Clause 17 of this Standard.

### 17.2 AIR CONTENT—FREQUENCY AND NUMBER OF TESTS

#### 17.2.1

The air content of concrete shall be tested at the start of production for each mix and periodically during the daily operation if the variation of slump exceeds 30 mm or the temperature of the concrete varies by more than 5°C.

*Note: When the air content is critical it should be checked before the concrete is placed and again when strength samples are taken later during the casting.*

#### 17.2.2

Where special consolidation methods make it difficult to predict the air content in the hardened concrete, the air content shall be verified by test of the hardened concrete according to CSA Standard Test A23.2-17C, and the test shall be performed periodically and whenever the placing or consolidation methods are modified, or whenever the concrete materials are changed.

*Note: Manufacturing methods, such as machine cast extrusion techniques and some consolidation methods such as spinning machines or impact tables, cause loss of air. This may be compensated by increasing the initial air dosage unless other tests prove that the additional strength and density eliminate, to the satisfaction of the owner, the need for additional air.*

### 17.3 COMPRESSIVE STRENGTH—FREQUENCY AND NUMBER OF TESTS

#### 17.3.1

For prestressed concrete production the following minimum number of cylinders shall be taken each time a bed is used:

(a) two for determining compliance with the specified compressive strengths;

(b) two for determining compliance with the release strength at time of transfer. These cylinders shall be made from concrete placed during the latter half of bed casting; and

(c) one for each intermediate strength level, as required.

*Notes:*

**(1)** *Where several beds are cast with the same concrete mix during one working day, the testing for the specified compressive strength may be reduced to a minimum of two beds.*

**(2)** *It is advisable to increase the number of release cylinders if a high release strength is required within a short time limit.*

#### 17.3.2

Test cylinders of standard 100 x 200 mm size may be used.

#### 17.3.3

For architectural and structural precast concrete production, compliance with the specified compressive strength shall be determined by testing a minimum of two cylinders each day from each type of concrete mix, or for each 30 $m^3$ of any one mix where the daily consumption exceeds this volume.

For architectural concrete, where daily quantities involved are small, one test shall be made for every 3 $m^3$, or where daily quantities are smaller than 3 $m^3$, two tests shall be made per week.

*Notes:*

**(1)** *Project specifications may require more frequent cylinder testing, particularly where shipping or erection strengths are important.*

**(2)** *Where the stripping strength of structural or architectural precast elements is critical, one additional cylinder should be used to determine this, particularly if the mix is new or its strength gain is not well known. Otherwise, rebound hammer or other nondestructive tests may be used to indicate stripping strength.*

## 17.4 ABSORPTION—FREQUENCY AND NUMBER OF TESTS

As an indication of potential weatherstaining and durability characteristics, the absorption of all face mixes subject to weathering shall be established as the average of three tests in accordance with CSA Standard Test A23.2-11C. It shall be verified at the beginning of each production run and every 6 months thereafter, and whenever the materials or production methods are modified.

*Notes:*

**(1)** *Acceptable absorption varies according to the exposure conditions but should normally not exceed 6% (by mass).*

**(2)** *To obtain an absorption of 6% or less, thorough consolidation and curing are required as well as a low water/ cement ratio and limitations on the percentage of aggregates passing the* 315 mm *sieve.*

## 17.5 ACCELERATED CURING

### 17.5.1

During accelerated curing, temperatures within the concrete shall be monitored and recorded in accordance with Clause 21.2.2.5.

*Note:* For standard production methods air temperatures may be monitored rather than concrete temperatures.

### 17.5.2

When accelerated curing of the concrete elements takes place in accordance with Clause 21.2, all test specimens shall be cured with the concrete elements during the accelerated curing, following which the specimens shall be cured under controlled laboratory conditions in accordance with CSA Standard Test A23.2-3C.

*Notes:*

**(1)** *Test specimens should be placed so that the curing conditions simulate as closely as possible the condition of the concrete elements. Specimens should not be subjected to concentration of heat nor any heat in excess of what is likely to reach the products. During initial castings of new set-ups the ambient temperature around the test specimens should be checked in relation to the ambient temperature near the elements.*

**(2)** *Experience from precast producers indicates that the initial curing of* 28-day *cylinders with the elements during accelerated curing will normally result in* 28-day *strengths up to 8% below normally cured laboratory cylinders depending on how closely the optimum curing temperature cycle is controlled. However, this curing method is important as it is the only valid check that the accelerated curing has not been excessive or has otherwise resulted in significant strength losses of the actual concrete element below the potential* 28-day *strength.*

**(3)** *To verify the potential strength of the concrete mix it is recommended that normal laboratory cured cylinders occasionally be made and tested.*

## 17.6 INTERMEDIATE STRENGTH LEVELS

### 17.6.1

For intermediate concrete strength determination, such as strength at transfer of prestress force, test specimens made in accordance with other standards may be used provided these tests are correlated to standard cylinder testing.

*Note: Tests for intermediate strength levels may be test samples cast horizontally, if the moulds and casting procedures are in accordance with ASTM Standard C512.*

## 17.7 LOW DENSITY CONCRETE

### 17.7.1

The modulus of elasticity and the shrinkage and creep characteristics of a proposed concrete mix with low density aggregate for architectural, prestressed, or precast concrete shall be determined prior to its use and approved by the owner as meeting the requirements. This applies both to concrete where all aggregates are of low density and to concrete with only low density coarse aggregate.

*Note: Modulus of elasticity, shrinkage, and creep characteristics for low density aggregate concrete are different from those for normal density aggregate concrete. In order to assess related design and performance requirements, such as prestress losses, deflection, and camber under the proposed production methods and application, it is necessary to know the modulus of elasticity of low density aggregate concrete.*

### 17.7.2

The absorption of structural low density aggregate, as defined in ASTM Standard C330, shall not exceed 11% when such aggregate is used in architectural concrete face mixes exposed to the weather.

*Note: Low density aggregate should not be used in exposed finishes in cold or humid climates unless their performance has been verified by tests or records of satisfactory usage under similar conditions.*

## 17.8 NO-SLUMP CONCRETE

### 17.8.1

Test specimens to determine the 28 day strength shall be made in accordance with CSA Standard Test A23.2-12C.

### 17.8.2

Intermediate strength levels may be determined by other methods, provided these tests are correlated to standard cylinder testing.

*Note: Such methods may be*

*(a) testing of cores in accordance with CSA Test Method A23.2-14C;*

*(b) testing of samples sawn out of the product;*

*(c) rebound hammer or other nondestructive testing, if combined with tendon slippage recording in accordance with Clause 27.7.2.2. (Also refer to CSA Standard A23.2, Appendix A.)*

# 18. PRODUCTION OF CONCRETE

## 18.1 GENERAL

Production of concrete shall conform to the requirements of Clause 18 of CSA Standard CAN/CSA-A23.1 and Clause 18 of this Standard.

## 18.2 MIXERS

Only stationary mixers shall be used for

(a) architectural precast concrete mixes;

(b) mixes with a slump less than 50 mm.

*Notes:*

**(1)** *Force mixers are preferred for such mixes to provide the uniformity required.*

**(2)** *Mixers should be provided with timing devices to prevent under- or over-mixing of the batch.*

**(3)** *Mixers should be considered excessively worn and due for adjustment if the gap between the pan (or drum) and blades exceeds one-fourth of the maximum size of the coarse aggregate or* 6 mm, *whichever is the smaller.*

## 18.3 COLOURING MATERIALS

Colouring materials shall be carefully measured to assist in achieving colour uniformity of the finished product. Provisions shall also be made to ensure uniform dispersal of the material throughout the concrete.

*Note: When only small additions of powdered colouring materials are to be made, greater accuracy may be achieved by first dispersing the colouring material in a larger amount of an inert filler or cement.*

## 18.4 CONTROL OF SLUMP AND AIR CONTENT

The time from start of mixing to placement shall not exceed 1 h. Retempering of concrete shall not be allowed.

Where truck mixers are used, air and slump tests shall be made on every load until the truck mixer conforms to uniformity tests, at which time the frequency can be reduced to one per day for each truck.

# 19. PLACING OF CONCRETE

## 19.1 GENERAL

### 19.1.1

The placing of concrete shall conform to the requirements of Clause 19 of CSA Standard CAN/CSA-A23.1 and Clause 19 of this Standard.

### 19.1.2

The methods of mixing, conveying, spreading, consolidating, finishing, curing, and protecting the concrete shall be established by the plant in conjunction with the plant engineer and shall be documented as part of the plant procedures.

*Note: Limits in height of drop are important to prevent segregation and, in the case of mixes for architectural concrete, to maintain uniformity of appearance.*

## 19.2 DEPOSITING

### 19.2.1

The depositing of concrete shall be so scheduled that each layer is placed against concrete that has not reached its initial set.

### 19.2.2

The face mix and the backup mix shall be placed in such a manner that the backup mix bonds to but does not penetrate the face mix.

*Notes:*

**(1)** *For minimum thickness of face mix for precast concrete, see Clause 14.3.2 of this Standard.*

**(2)** *The use of separate face and backup mixes or the use of a uniform concrete mix throughout the element depends on economics, the type of finish required, the configuration of the element, and the practice of the particular precasting plant. Elements with intricate shapes and deep narrow sections generally require one uniform concrete mix throughout. See Clause 14.3.1 of this Standard.*

**(3)** *The second mix should be placed prior to the evaporation of bleedwater from the first mix, unless special bonding techniques are used.*

### 19.2.3

The backup mix shall be placed prior to the hardening of any concrete coating on the reinforcement or the coating shall be cleaned off. Alternatively the placement of the reinforcement shall follow the consolidation of the face mix.

### 19.2.4

Two-stage precasting as used for intricate elements, such as column covers or corners with large returns or elements with abrupt changes in concrete volumes, shall be made with construction joints in accordance with Clause 20.

### 19.2.5

In the construction of concrete wall panels the first wythe shall be carefully placed and consolidated to the correct thickness. The concrete shall be finished with a smooth surface to minimize air pockets between the insulation and concrete. Any pockets showing along the edges of the insulation shall be filled.

*Notes:*

**(1)** *The correct wythe thickness is normally accomplished or checked by a special template riding on the sides of the form.*

**(2)** *Some sandwich wall panels may be designed with an uninterrupted air space between the wythe exposed to the weather and the insulation. This air space should remain completely open for water to be drained from the building.*

## 19.3 CONSOLIDATION

Surface vibrators shall be of a type and shape, and shall be applied in such a manner, that suction action is prevented during use.

*Note: See ACI Publication 309R.*

## 19.4

Paste adhering to extended reinforcement shall be removed to ensure bonding of bars to concrete in a later placement.

# 20. CONSTRUCTION JOINTS

## 20.1

Construction joints shall conform with the requirements of Clause 20 of CSA Standard CAN/CSA-A23.1.

## 20.2

Where a construction joint is to be made, the surface of the set concrete shall be suitably roughened, thoroughly cleaned of foreign matter and laitance, and saturated with water and left in a damp condition with no free water on the surface immediately before new concrete is placed. Where a bonding agent is used, surface treatment shall be as recommended by the manufacturer.

*Notes:*

**(1)** *Precasting in stages is also used where small volume parts intersect larger volume parts; such practices may decrease the danger of unacceptable cracking.*

**(2)** *When selecting a bonding agent consideration should be given to the exposure conditions that will prevail during the life of the structure.*

# 21. CURING AND PROTECTION

## 21.1 GENERAL

### 21.1.1

Except as provided in Clause 21.1.2, curing and protection shall conform to the requirements of Clause 21 of CSA Standard CAN/CSA-A23.1 and to Clause 21 of this Standard.

### 21.1.2

Where the ambient temperature in the plant does not fall below 5°C and where the manufacturing facilities are protected from wind and direct rays of the sun such that the moisture evaporation is less than 0.75 kg/m$^2$h, curing and protection shall conform to the requirements of CAN/CSA-A23.1, Clauses 21.1.1 to 21.1.6, and to Clause 21 of this Standard.

*Note: CAN/CSA-A23.1 specifies additional curing and protection measures for outside casting. An estimation of the moisture evaporation can be obtained from Appendix D of that Standard.*

### 21.1.3

Immediately after placing and consolidation of concrete, provisions shall be made for moisture retention. The concrete shall not be subjected to rapid drying or excessive stresses during the curing cycle.

### 21.1.4

The curing process shall be well established, controlled, and continued until the concrete has attained its specified handling strength.

### 21.1.5

Records of concrete strength and temperature shall be maintained during the curing cycle.

## 21.2 INITIAL CURING

### 21.2.1 General

#### 21.2.1.1

The concrete strength at handling or at transfer of prestress shall be specified on the shop drawings and shall be based on design and experience.

#### 21.2.1.2

Prior to transfer of prestress or handling, the strength of the element shall be determined by test cylinders cured as closely as possible to the same conditions as the element.

*Note: For products manufactured using no-slump concrete, a rebound hammer or other nondestructive testing such as measurement of tendon slippage may be used.*

#### 21.2.1.3

Concrete strength for exposure to weather shall be shown on the shop drawings.

### 21.2.2 Accelerated Curing and Heated Concrete

#### 21.2.2.1 General

*21.2.2.1.1*

Heat treatment of concrete is used to accelerate its strength development. Two different types of heat treatment are recognized, namely

(a) **accelerated curing**, which is heating of the concrete after it has been placed in the forms; and

(b) **heated concrete,** which is heating of the concrete before it is placed in the forms.

*The provisions of Clause 21.2.2 are broadly based on the recommendations of the PCA Bulletin RD108T,* The Effect of Secondary Ettringite Formation on the Durability of Concrete: A Literature Analysis, *by R.L. Day, 1992. The potential for expansion is believed to be primarily related to cement composition and curing temperature. Air entrainment may reduce expansion.*

*21.2.2.1.2*

Where a combination of accelerated curing and heated concrete is used, the requirements set forth in Clause 21.2.2.3 shall apply.

#### 21.2.2.2 Moisture Categories

*21.2.2.2.1*

Based on the ambient conditions to which they will be exposed, concrete elements shall be classified as either

(a) moisture category dry; or

(b) moisture category damp.

*21.2.2.2.2*

Moisture category dry shall apply to concrete elements which, after curing, will not be subjected to a moist environment during service. Unless such units are air entrained, they shall not be exposed to a moist environment for a period of more than 3 months prior to service.

*21.2.2.2.3*

Moisture category damp shall apply to concrete elements that will be subjected to a moist environment during service or non-air entrained elements that will be exposed to a moist environment for a period of more than 3 months prior to service.

*Note: Examples of moisture category damp include*

***(a)*** *external elements of buildings or structures that are exposed to precipitation, surface water, or ground water, such as bridge girders, unprotected roof slabs, building facades, basement walls, railway ties, and underground vaults;*

***(b)*** *internal building elements subject to moist conditions, such as indoor swimming pool areas, laundries, and parking garages;*

***(c)*** *elements, parts of which may frequently drop below their dew point, such as chimneys, filter chambers, and internal voids in bridges.*

#### 21.2.2.3 Accelerated Curing

*21.2.2.3.1*

Accelerated curing shall be achieved by one, or a combination of the following means:

(a) Live steam applied within a suitable enclosure to retain the steam. The steam jets shall distribute the steam uniformly and shall not discharge directly onto the concrete, forms, or test cylinders.

(b) Radiant heat or electric heat within a suitable enclosure to effect an even distribution of heat. Pipes or electric heating units shall not be in direct contact with the concrete, forms, or test cylinders.

(c) Forced air within a suitable enclosure to effect an even distribution of heat. Hot air shall not discharge directly onto the concrete, forms, or test cylinders.

*21.2.2.3.2*

Throughout the entire accelerated curing cycle, exposed concrete surfaces shall be protected from moisture loss by such means as covering the surfaces with tarpaulins or plastic sheeting, using a water spray, or ensuring that 100% relative humidity is present in the curing enclosure.

*21.2.2.3.3*

Exposed concrete surfaces shall not be exposed to combustion gases during the accelerated curing cycle.

*Note: Unless exhaust flue gases from combustion heaters are vented outside the curing enclosure, concrete surfaces should be protected by formwork or tightly fitted tarpaulins or plastic sheeting.*

*21.2.2.3.4*

The accelerated curing cycle shall be as set forth in Table 2.

**Table 2. Accelerated Curing Cycle**

| | Moisture category | |
|---|---|---|
| **Heat treatment cycle** | **Dry** | **Damp** |
| Minimum holding period (delay) prior to application of heat and maximum concrete temperature during the holding period | 1 h at 30°C | 3 h at 30°C or<br>4 h at 40°C |
| Maximum heating rate* (°C/h) | 20 | 20 |
| Maximum concrete temperature (°C) | 80 | 70 |
| Maximum cooling rate† (°C/h) | 10 | 10 |

**For structural low and semilow density concrete, the heating rate shall be limited to* 10°C/h.

†*Cooling shall continue until the concrete temperature is no more than* 20°C *above the ambient temperature outside the enclosure.*

#### 21.2.2.4 Heated Concrete

*21.2.2.4.1*

Heated concrete shall be achieved by preheating the concrete materials, by heating the concrete mixture by such means as super-heated steam, or by a combination of these means.

*Note: The use of heated concrete may result in premature stiffening of the mix.*

*21.2.2.4.2*

During the heated concrete cycle, exposed concrete surfaces shall be protected from moisture loss, as specified in Clause 21.2.2.3.2.

*21.2.2.4.3*

During the heated concrete cycle, exposed concrete surfaces shall not be exposed to combustion gases, as specified in Clause 21.2.2.3.3.

*21.2.2.4.4*

The heated concrete cycle shall be as set forth in Table 3.

**Table 3. Heated Concrete Cycle**

| Heat treatment cycle | Moisture category: Dry | Moisture category: Damp |
|---|---|---|
| Maximum temperature of the compacted fresh concrete in the form (°C) | 50 | 40 |
| Maximum concrete temperature (°C) | 80 | 70 |
| Maximum cooling rate* (°C/h) | 10 | 10 |

**Cooling shall continue until the concrete temperature is no more than 20°C above the ambient temperature outside the enclosure.*

#### 21.2.2.5 Measuring and Recording the Heat Treatment Cycle

*21.2.2.5.1*

When accelerated curing or heated concrete methods are used, the governing concrete temperatures shall be those achieved within the concrete elements.

*21.2.2.5.2*

Temperatures shall be measured by one or a combination of the following means:

(a) Devices such as expendable thermometers or thermocouples cast into the concrete.

(b) Devices placed against the concrete surface under a temporary cover of heavy insulation.

(c) In the case of standard elements, temperatures may be measured within the enclosure. In such cases, the correlation between the enclosure temperature and temperatures within the concrete shall be determined and documented and the temperature within the concrete shall be used to govern the cycle.

*21.2.2.5.3*

Concrete temperatures shall preferably be recorded continuously but in no case shall the interval between temperature readings exceed 30 minutes.

*21.2.2.5.4*

The time at which concrete temperatures are measured shall be recorded.

## 21.3 ADDITIONAL CURING

### 21.3.1

Additional curing as specified in Clause 21.1.3 of CAN/CSA-A23.1 shall be used where it is important for durability and for architectural concrete. The concrete shall be kept in a surface damp condition.

*Note: For architectural precast concrete panels, additional curing should continue until the hydration of the cement on the surface is sufficient to prevent unhydrated cement from streaking the panel surfaces or being carried by rainwater to window glass areas or other surfaces that may be affected.*

### 21.3.2

Additional curing as specified in Clause 21.1.4 of CAN/CSA-A23.1 shall be used when concrete strength levels in excess of those achieved following initial curing are required for structural safety.

# 22. TREATMENT OF UNFORMED SURFACES

## 22.1 GENERAL

### 22.1.1

The finishing of plastic concrete surfaces shall conform to the requirements of Clause 22 of CSA Standard CAN/CSA-A23.1 and to Clause 22 of this Standard.

### 22.1.2

The required finish of precast concrete shall be specified on the shop drawings.

### 22.1.3

Concrete surfaces designated to have concrete placed on them shall be clean and free of laitance.

## 22.2 COMPOSITE CONSTRUCTION

Concrete surfaces designated as part of a composite design shall be finished in accordance with the requirements for scratch finish as explained in Clause 22.5.3 of CSA Standard CAN/CSA-A23.1.

# 23. TOPPINGS

## 23.1

Concrete toppings shall conform to the requirements of Clause 23 of CSA Standard CAN/CSA-A23.1.

# 24. FINISHING OF FORMED SURFACES

## 24.1 GENERAL

### 24.1.1

The required finish of formed surfaces shall be as specified by the owner and shall be detailed on the shop drawings.

*Note: For architectural finishes refer to Appendix C.*

### 24.1.2

The finish shall be established from an assessment of previous production or from samples, as described in Appendix C.

## 24.2 FINISH GRADES

### 24.2.1 Commercial Grade

For this finish concrete may be cast in forms that impart a texture to the surface but fins shall be removed and large surface voids shall be filled. Faces exposed to view shall have well defined outlines and all sharp edges visible in the finished structure shall be softened by sanding or grinding.

### 24.2.2 Standard Grade

The finish shall be as described in Clause 24.2.1 except that the forms shall not impart a pronounced texture to the finished concrete.

### 24.2.3 Finish Grade B

This finish shall be as described in Clause 24.2.2 except that all surface blemishes shall be finished to provide a smooth surface of uniform appearance if painted. Form release or other agents shall not leave a paint-rejecting coating on the concrete.

### 24.2.4 Finish Grade A

This finish shall be as described in Clause 24.2.3 except that repairs that are exposed to view shall be colour-matched within practical limits.

# 25. DAMPPROOFING AND WATERPROOFING

## 25.1 GENERAL

Preparation of concrete surfaces shall conform to the requirements of Clause 25 of CSA Standard CAN/CSA-A23.1 and Clause 25 of this Standard.

## 25.2 SURFACE SEALERS

### 25.2.1

Surface sealers and joint sealants shall be compatible with the surface on which they are to be placed.

### 25.2.2

Surface sealers specified by the owner shall be applied in accordance with the sealer manufacturer's instructions.

*Notes:*

*(1)* The PCI publication *Architectural Precast Concrete* gives some information on the merits of sealers.

*(2)* The application of some sealers should be delayed so that it follows the completion of erection and joints.

### 25.2.3

Prior to actual application on the finished product, a sealer shall be tested on reasonably sized samples of varying ages and its performance verified over a suitable period based on experience under similar exposure conditions.

# 26. POST-TENSIONING

## 26.1

Post-tensioning shall conform to the requirements of Clause 26 of CSA Standard CAN/CSA-A23.1.

# 27. PRETENSIONING

## 27.1 GENERAL

### 27.1.1

The prestress force to be applied to the tendon shall be calculated in accordance with Clause 18.6 of CSA Standard A23.3, taking into consideration the specified effective prestress force and prestress losses, and it shall be shown on the shop drawings.

## 27.2 FACILITIES

### 27.2.1

Prestressing beds shall conform to the following requirements:

(a) The design of anchorages, abutments, stress-resisting forms, and deflecting equipment shall be designed by experienced engineers and approved by the manufacturer's engineer.

(b) The design shall be based upon a stated factor of safety according to sound engineering principles taking into account the maximum forces on the bed. Such forces shall be analyzed as to magnitude, position, and frequency of occurrence.

(c) Foundations shall be sufficiently firm to prevent undesirable movements.

### 27.2.2

Information on the capacity of each bed and stress-resisting form in terms of allowable prestress force and its corresponding height of application above the bed or base shall be kept on record and made available when requested.

## 27.3 EQUIPMENT

### 27.3.1

The tensioning force shall be applied by equipment suitable for single tendon tensioning.

### 27.3.2

Hydraulic gauges, dynamometers, tension meters, load cells, or other suitable devices shall be used for controlling and measuring the prestress force and shall conform to the following requirements:

(a) they shall have an accuracy of reading within ±2%;

(b) each gauge shall be capable of indicating loads directly or be accompanied by a chart by which the reading can be converted into units of force;

(c) the indicating dials of gauges shall be at least 150 mm in diameter;

(d) gauges shall be calibrated for the jacks, hoses, and connections with which they are used;

(e) alternatively, the gauge shall be calibrated alone and friction losses in jacks, hoses, and connections established by other means shall be recorded;

(f) the calibration of each gauge shall be supervised by an experienced engineer and a certified calibration certificate for each gauge shall be signed by an engineer;

(g) the calibration shall be repeated at any time that a gauging system indicates erratic results and at intervals not greater than 6 months during regular usage or not greater than 12 months for other conditions of use.

(h) the loads to be gauged shall be not less than 1/4 nor more than 3/4 of the total graduated capacity unless calibration data clearly establish accuracy over a wider range;

(i) hydraulic gauges shall not fluctuate excessively and shall remain steady until the jacking load is released; and

(j) gauges shall be mounted near eye level and within 2 m of the operator and positioned so that readings can be obtained without parallax.

### 27.3.3

Manual hydraulic pumps shall not be allowed for stressing tendons.

### 27.3.4

Automatic hydraulic controls shall be capable of adjustment so that the jacking force will correspond to the required load. The setting accuracy of automatic cut-off valves shall be verified whenever the results are suspect.

## 27.4 STRINGING OF TENDONS

### 27.4.1

Contamination of tendons shall be prevented. After stringing and tensioning, the tendons shall be inspected and if found to be contaminated shall be cleaned.

### 27.4.2

The stressing sequence shall avoid entanglement of tendons and minimize unbalanced loads on the beds.

### 27.4.3

Binding of tendons shall not occur during the stressing operation.

### 27.4.4

Tendons cast into prestressed elements shall be free of grip marks and other damage.

*Notes:*

**(1)** *Care should be taken where lengths of tendon previously gripped by tendon vices are incorporated within the length to be stressed but outside of individual elements.*

**(2)** *Significant nicks result in local stress concentrations that may result in failure of the tendon during stressing.*

### 27.4.5

Tendon anchors shall conform to Clause 8.2.4 and shall be capable of anchoring stressing loads positively and with a minimum of slippage.

## 27.5 STRESSING

### 27.5.1 General

#### 27.5.1.1

An initial load shall be applied to the individual tendons to straighten them, eliminate slack, and provide a starting or reference point for measuring elongation. The final load shall then be applied for which elongation of tendons is computed and measured.

**27.5.1.2**

The force in the tendons shall be measured by elongation and by pressure gauge.

*Note: Stressing may be done to a calculated elongation using the force reading as a check against the calculated force or vice versa.*

**27.5.1.3**

Discrepancies between the elongation measurement and force measurement shall not exceed 5%. Neither one of these measurements shall vary by more than 5% from the required design value.

*Notes:*

**(1)** *The gauging system indicates that the proper force has been applied. A check of the elongation indicates that the correct size of tendon has been used and operational losses are within tolerance limits and provides a check on the gauging system.*

**(2)** *Inaccurate and variable stressing can result in differential camber, lateral bowing of members and reduction of the cracking load.*

**(3)** *The rotation of a tendon during stressing should be limited to one turn per* 30 m.

### 27.5.2 Initial Tensioning

Initial load shall be measured within a tolerance of ±200 N or 2%, whichever is larger.

*Notes:*

**(1)** *The same jack used for final stressing of single tendons may be used for initial tensioning, if it is provided with a proper gauging system for measuring the initial force. (See Clause 27.3.2.)*

**(2)** *In single tendon tensioning the initial and final loads may be applied in immediate succession. When this is done, the initial load is applied and held momentarily while reference marks are made for measuring elongation and slippage; the load is then increased to its full value and the tendon is seated.*

### 27.5.3 Elongation Corrections

**27.5.3.1**

Elongation calculations shall be based on the actual modulus of elasticity of the tendon determined from the stress-strain curve furnished by the tendon manufacturer.

**27.5.3.2**

Elongation calculations shall take into account the slippage of the tendon in the anchors and the seating of the chucks for both the dead and live ends. Slippage and seating adjustments shall be based on average values from tests.

**27.5.3.3**

Elongation calculations shall take into account the movement of anchorages and shortening of stress-resisting forms. Compensation for movements shall be based on tests which simulate the appropriate force and eccentricity.

**27.5.3.4**

Where rods under tension are used in multiple tendon tensioning, an allowance shall be made for the elongation of the rods.

**27.5.3.5**

Elongation calculations shall take into account the effect of temperature variations if the temperature of the steel at the time it is stressed and at the time the concrete starts to set differ by more than 15°C. The thermal coefficient of expansion of steel shall be taken as $12 \times 10^{-6}/°C$.

### 27.5.4 Measurement of Elongation

**27.5.4.1**

After completion of initial stressing, reference marks shall be established from which elongations resulting from final tensioning forces can be measured.

Reference marks shall be visible during the entire final stressing operation.

**27.5.4.2**

Elongation shall be measured and recorded to the nearest mm.

*Note: Stressing procedures and calculations should be the responsibility of the manufacturer's engineer.*

**27.5.4.3**

Where there are no intermediate bulkheads or friction points, elongation and tendon slippage shall be measured on the first and last tendon stressed and at least 10% of the remaining tendons.

### 27.5.5 Force Corrections

The following shall be taken into account in calculating the indicated hydraulic pressure.

(a) **Friction in the jacking system.**

*Note: Because of the large jacking ram and the heavy sliding or rolling anchorage to which the tendons are attached, multiple tendon tensioning methods are subject to substantial variations in friction which must be overcome by the jack. Friction in single tendon tensioning is small and can usually be ignored.*

(b) **Temperature variations.** Allowance shall be made in the stressing force for temperature variations of 15°C or more between the ambient temperature and the concrete temperature at the time of placing.

(c) **Live end seating.** Allowance shall be made for loss of force as the tendon seats into the anchor.

*Note: Refer to PCI Manual 116, Clauses 2.2.8 and C2.2.8, for assistance in stress corrections.*

### 27.5.6 Final Stressing of Straight Tendons

After application of the initial load in single tendon tensioning and the establishment of reference marks for measuring elongation and slippage, the full tendon load shall be applied and checked by at least two control measures. The frequency of elongation and tendon slippage measurement shall be as specified in Clause 27.5.4.3.

*Note: For multiple tendon tensioning, refer to Clause 2.2.9 of PCI Manual 116.*

### 27.5.7 Final Stressing of Draped Tendons

**27.5.7.1**

After application of the initial load and the establishment of the reference mark for measuring elongation and slippage, the full tendon load shall be applied and checked by at least two control measures as specified in Clause 27.5.6.

**27.5.7.2**

Final stressing shall be accomplished by one or a combination of the following two methods.

(a) **Deflection of stressed tendons**. In order to distribute friction and restraint forces at tendon deflection points, the tendon lifting and depressing shall be carried out either simultaneously or consecutively commencing at the centre of the bed and progressing alternately at points equidistant from the centre to the end.

(b) **Stressing in draped position**. Tendons shall be stressed to final value in their draped position for the full length either from one or both ends of the bed. The tendons shall pass over pin and roller fixtures that effectively minimize friction at all deflection points. Support and hold-down devices shall be of sufficient rigidity and adequate support such that the position of tendons will remain substantially unchanged under the induced loads.

*Note: If elongation is not obtained within the required 5% tolerance, the tendon may be temporarily overstressed to overcome friction. Overstress should not exceed 80% of the specified tensile strength. (Refer to PCI Manual 116, Clauses 2.2.10 and C2.2.10.)*

### 27.5.8 Recording of Stressing Operations

**27.5.8.1**

Prestress force and elongations shall be checked and recorded during the stressing operations.

**27.5.8.2**

Stressing records shall include the following items if applicable to the particular operation:

(a) date of tensioning;

(b) casting bed identification;

(c) description, identification, and number of elements;

(d) manufacturer, size, type, and ultimate strength of tendon;

(e) sequence of stressing (and detensioning if critical);

(f) straight or draped tendons (possible compensations and possible stressing from either end);

(g) identification of jacking equipment;

(h) corrections for tendon slippage (including splices), anticipated abutment movement, thermal effects, etc;

(i) required total load per tendon;

(j) initial tension;

(k) anticipated gauge pressure for each tendon or for each group of tendons stressed in one operation;

(l) anticipated elongation for each different jacking load;

(m) actual elongation or actual force measurement, depending on which operation is chosen for checking (see Clause 27.5.1.2);

(n) for stressing of draped tendons:

   (i) sequence;

   (ii) data and description; and

(o) any unanticipated problems encountered during tensioning such as wire breakage, excessive slippage, restressing, or other factors that may have an influence on the net stress.

### 27.5.9 Debonded Tendons

Bond breakers shall not alter the physical or chemical properties of the tendons or surrounding concrete. Grease or oil shall not be used.

### 27.5.10 Tendon Splices

**27.5.10.1**

Splicing of tendons shall conform to the same requirements as tendon anchors (see Clause 8.2.4).

**27.5.10.2**

A maximum of one splice per tendon shall be permitted.

**27.5.10.3**

The ends of tendons to be spliced shall be cut by shears or abrasive grinders.

### 27.5.11 Wire Failure in Strands

Failure of one wire in a 7-wire strand is acceptable provided the 5% allowable variation in prestress force for all strands in the entire element is not exceeded and provided the total area of wire failure is not more than 2.5% of the total area of tendons in any member.

## 27.6 DETENSIONING

### 27.6.1

Where concrete curing has been accelerated, the detensioning shall be done immediately following the curing period while the concrete is still warm and moist.

*Note: If the concrete is allowed to dry and cool prior to detensioning, dimensional changes take place that may cause tension stresses and cracking in the concrete.*

### 27.6.2

Forms, ties, inserts, hold-downs, or other devices that would restrict longitudinal movement of the members along the bed shall be removed or loosened, or alternatively, detensioning shall be performed in such a manner and sequence that longitudinal movement is precluded.

#### 27.6.3

The prestressing forces shall be kept nearly symmetrical about the vertical axis of the element and shall be applied in a manner that will minimize sudden loading.

#### 27.6.4

For draped tendons the sequence of release of hold-down and anchorage forces shall be established to minimize cracking of the element.

*Notes:*

**(1)** *Cracks resulting from initial release of hold-downs will generally close upon release of anchorage forces; however, cracking of concrete members is undesirable and should be avoided.*

**(2)** *Computation of the hold-down forces and comparison with the weight of the member should always be done if hold-downs are to be released prior to release of anchorage forces. Release of hold-downs without release of anchorage forces may result in concentrated vertical loads that may crack the top of the member.*

**(3)** *In general, if the sum of hold-down forces is not more than one-half the weight of the member, it is safe to release hold-downs prior to the release of anchorage forces. If the hold-down forces are less than three-fourths of the weight of the element and nominal reinforcing steel is carried in the top flange or slab, cracking is unlikely. If the hold-down forces approach the weight of the element, some cracking of upper surfaces immediately above the hold-downs is inevitable if hold-downs are released prior to the anchorage.*

**(4)** *Preloading of the element with weights or additional vertical restraint can minimize the effect of releasing hold-down forces.*

#### 27.6.5

Tendons shall be detensioned by hydraulic jack or by heat applied to the tendons at both ends simultaneously to produce a gradual release of stress prior to cutting.

## 27.7 EXTRUDED AND SLIP-FORMED PRODUCTS

### 27.7.1 General

Documented procedures for manufacturing extruded and slip-formed products shall be established by the precast concrete manufacturer.

*Note: In most cases, the supplier of such equipment has established specific operational procedures and standards that should be followed.*

### 27.7.2 Stressing and Detensioning

#### 27.7.2.1

Tendon slippage at time of detensioning in excess of 5 mm shall be investigated and evaluated. For tendon slippage in excess of this value, all slabs on the bed shall be marked with a number showing their position in the bed in order to allow an accurate evaluation of the effective prestress remaining in a particular slab following removal from bed.

*Notes:*

**(1)** *For example, a "free" (or sawn) end may show a* 100 mm *slippage of a tendon, indicating a bond loss over a length in excess of* 16 m. *If the next cut is made at* 10 m *it will show no slippage since the tendon at that point is not under tension.*

**(2)** *It is recommended that a jig be devised for measuring tendon slippage after detensioning.*

#### 27.7.2.2

Tendon slippage in excess of 2 mm following sawing and stacking of slabs not already subject to excessive "free" end tendon slippage shall be investigated and an evaluation of the capacity of such slabs made.

*Note: Refer also to PCI Manual 116, Clauses 2.3.5, 6.3.3, C2.3.5, and C6.3.3.*

# 28. STRIPPING AND HANDLING

## 28.1

Care shall be exercised in removing the elements from the forms in order to prevent damage, initial lifting shall be gradual, and equipment for initial release shall have controls allowing such operation.

*Note: Intricate shapes may require special equipment during removal from the form through the use of compressed air, vibration, hydraulic pressure, or screw jacks.*

## 28.2

The minimum concrete strength, the number and location of lifting points for handling of elements, and details of lifting devices shall be shown on the shop drawings. The number and location of the lifting points shall be compatible with the method of shipping—ie, flat or on edge.

## 28.3

Lifting equipment shall be properly secured to the element being stripped and handled, and proper rigging procedures shall be documented. All lifting equipment shall conform to established safety practices.

## 28.4

All precast elements shall be handled in positions consistent with their shape and design. Care shall be exercised to retain the required condition, quality, and appearance of all elements until finally installed.

## 28.5

Individual precast concrete elements shall be clearly marked with identification as shown on the shop drawing, including date of casting. Alternatively, precast concrete elements shall be identified with individual mark numbers and a list provided with required information correlated to mark numbers. Marking shall be done as soon as practical after stripping from forms and prior to yarding.

*Notes:*

*(1) Such markings may be made in the wet concrete, or they may be made by tagging as long as the tags are securely fastened to the elements, or they may be painted on the elements. In all cases the markings should remain legible for the longest storage period that can be anticipated.*

*(2) It is also recommended that the approximate mass of the elements be marked on them, particularly in cases where shipping or erection crews may not easily identify masses of the precast elements.*

*(3) Markings should be readable following erection for on-site quality control or the erection drawings may be marked with complete element identification including date of production.*

# 29. STORAGE

## 29.1

Products awaiting delivery shall be stored to minimize twisting, racking, bowing, or warping and to avoid damage to the product.

*Notes:*

*(1) Often the manner of storing depends on how the product is to be shipped and what limitations the element's cross-sections impose on handling.*

*(2) Wall panels should preferably be stored in vertical storage racks carefully made so as to minimize bowing or warping of the panels.*

*(3) Panels with a large vertical dimension should be designed to be turned in the air using a double crane hookup.*

## 29.2

Elements shall be stacked only when the lower elements in a stack will not be damaged and where the risk of warpage is small. Proper blocking and dunnage shall be used for separation.

## 29.3

All blocking, dunnage, and protective materials shall be of a type that will not cause damage or objectionable staining.

*Note: For exposed surfaces care should be taken that such material does not cause uneven curing that may result in permanent colour variations.*

## 29.4

Elements that are exposed in the finished structure shall be protected from soiling, staining, and contact with objectionable matter.

## 29.5

Prestressed elements shall be stored with due allowance for creep and shrinkage.

*Notes:*

*(1) The location of supports and any external loading during storage can increase or decrease camber growth. Special conditions for storage should be shown on the shop drawings.*

*(2) Preloading a pretensioned element in the plant may be considered to control excessive camber growth.*

# 30. TRANSPORTATION

## 30.1

All materials and products shall be satisfactory for delivery prior to loading.

*Note: The responsibility for this should be clearly established in the contract documents (see Appendix A).*

## 30.2

Elements shall be supported and braced during transportation to ensure they are not subjected to stresses for which they are not designed.

*Notes:*

*(1) Vibration, impact, or lateral forces occurring during transportation are significant.*

*(2) Particular care should be taken for slender elements to minimize excessive twisting or whipping.*

## 30.3

All precast elements shall be delivered to the site clearly marked in accordance with Clause 28.5. The mass of the unit shall be provided.

*Note: All orientation and identification marks should be located on the element in such a way as to be visible during loading and unloading.*

# 31. ERECTION

## 31.1

Erection procedures include

(a) erection sequence;

(b) temporary shoring, bracing, supports, guys;

(c) sequence for installation and removal of temporary shoring, etc.

## 31.2

All elements shall be erected as shown on the erection drawings to the tolerances provided in Clause 10. They shall be positioned so that cumulative dimensional errors do not exceed allowable tolerances.

## 31.3

Elements shall be erected in a sequence that takes into account the actual loading on the building, the loading on previously erected elements, attachments, weather-sealing and tolerances. Elements shall not be stacked temporarily on other elements during erection without allowance having been made for this in the design of the elements and the connections.

### 31.4

Installation shall not impose undue stresses or overloads on the structure, the elements, or the connections.

### 31.5

Connections and bearings shall be set within specified tolerances. (See Clause 10 and CSA Standard CAN/CSA-A23.1.) Elements shall be positioned to avoid any eccentric application of forces not considered in the design.

*Note: It is important that bearing surfaces make proper contact with the supports in accordance with the design criteria; otherwise the structure may be built with serious structural deficiencies.*

### 31.6

Connections shall not be modified unless approved by the engineer responsible for the connection design.

### 31.7

All bolted or welded connections shall be made in accordance with the shop drawings. Where welded connections are required, all welds shall be made in accordance with CSA Standard W47.1.

### 31.8

Temporary lifting and handling devices cast into the elements shall be removed, retained, and protected as specified on the erection drawings.

### 31.9

Where elements in partially completed structures might be damaged by the effects of weather, such as freezing water in ducts, holes, and inserts, adequate temporary protection shall be provided.

### 31.10

All discontinuities in corrosion protection on permanent connections shall be repaired. The repair method shall be specified or approved by the engineer responsible for the design of the connections.

*Note: Discontinuities in corrosion protection can result from welding or other erection procedures.*

## 32. JOINTS

### 32.1

Joints between precast elements and the jointing materials shall be as specified by the owner and shown on the shop drawings.

### 32.2

The following information shall be shown on the shop drawings:

(a) joints between precast elements;

(b) joints between precast elements and other materials;

(c) jointing materials;

(d) joint dimensions.

*Note: For further information on the design of joints refer to the PCI Manual,* Architectural Precast Concrete.

### 32.3

Joint width shall be sufficient to accommodate tolerances for precast elements, structure, and site placing of hardware. Tolerances for erected joint widths shall be as specified by the owner.

*Note: An adequate joint width is particularly important where the joints are the only means to accommodate erection tolerances and building movements. See also Clause 10.*

## 33. REPAIRS

### 33.1 GENERAL

Repairs of defects and damage to precast concrete elements shall be performed with materials and in a manner that will restore the specified quality of the product.

*Notes:*

**(1)** *A distinction is made between*

*(a) repairs of minor defects or damage normally referred to as cosmetic repairs; and*

*(b) repairs of extensive defects or damage normally referred to as structural repairs.*

**(2)** *Guidance on the prevention, significance, effect, and suggested method of repair of cracks may be found in Reports JR-271 and JR-316 published by PCI.*

### 33.2 COSMETIC REPAIRS

#### 33.2.1

Holes left by lifting or hold-down devices or other temporary inserts shall be repaired where required. Patches on the underside of elements, near the centre, or in the area of variable tensile stresses shall be bonded by an epoxy resin or other approved bonding material.

### 33.2.2

Bonding agents shall have a proven record of durability and shall be applied under conditions and by methods provided by the manufacturer.

### 33.2.3

Where it is necessary to fill air bubbles in vertical concrete surfaces or under top forms, these surfaces shall be sack-rubbed in accordance with Clause 24 of CSA Standard CAN/CSA-A23.1.

### 33.2.4

Damaged areas that will be exposed to view and other areas specified by the owner shall be repaired.

*Notes:*

**(1)** *Where repairs have to match specified architectural finishes, repair mixes should be developed early, following approval of the initial sample. A trial and error process is normally required for each newly developed face mix to effectively match colour and texture. Such trials should allow several weeks for curing and include outdoor exposure.*

**(2)** *Where repairs of grey concrete have to be colour-matched, white cement normally has to be included in the repair batch especially where curing of elements has been accelerated.*

### 33.2.5

The repair mix shall be proportioned to provide at least the strength and durability of the original mix. Where appearance is important, the proportions shall accommodate this characteristic.

*Notes:*

**(1)** *The repair mix may have to have an altered cement content (see Clause 33.2.4) or some of the larger aggregate pieces may have to be eliminated or hand-placed in the patch.*

**(2)** *Where colour-matching is desired for elements exposed to the weather, care should be taken to obtain similar absorption characteristics so that patches do not become too obvious in alternate dry or wet weather. For that reason bonding agents should normally not be included in the repair mix.*

### 33.2.6

Preparation of the repair area and the actual repair shall consist of the following steps:

(a) Remove all loose material and chip back to sound concrete.

(b) Provide cutouts at least 6 mm deep. For exposed aggregate finishes, cutouts shall be a minimum of 6 mm plus the maximum size of the exposed aggregate. Feather edges shall not be permitted.

*Note: A slight undercut is recommended.*

(c) Clean repair area by brushing or with compressed air.

(d) Saturate repair area and surrounding concrete with water to prevent absorption of water from the patching mix and to facilitate bonding.

(e) When surface water has evaporated but before complete drying, place repair mix by vibration or hand tamping for consolidation and for proper contact with the old concrete surface.

*Note: For larger repair areas it is recommended that placement of the repair batch be preceded by brushing a bond coat (cement and fine sand) onto the repair surface.*

(f) Provide proper curing, taking particular care to retain moisture and avoid excessively high or low ambient temperatures (see Clause 21).

*Note: The curing procedure should be part of the development of the trial mix particularly where colour matching is important. (See Clause 33.2.4.)*

(g) Finish as required depending on the particular finish treatment.

## 33.3 STRUCTURAL REPAIRS

### 33.3.1

Repairs of a structural nature shall not be undertaken until the manufacturer's engineer has

(a) investigated the structural implications of the defect or the damage;

(b) established the cause of the defect or the damage; and

(c) received approval of the proposed repair procedures from the owner.

### 33.3.2

A defect or damage shall be deemed to be of a potential structural nature when one or more of the following conditions occur:

(a) main reinforcement or tendons are exposed;

(b) connection hardware anchorage is exposed;

(c) cracking occurs in bearing areas;

(d) cracking extends from one face of the element through to the opposite face; and

(e) cracks in a structural element larger than 0.3 mm extend for longer than 300 mm.

### 33.3.3

The procedures for repairs shall include the requirements of Clause 33.2.6 or additional methods such as drypacking, pressure grouting, or injection of epoxy under pressure or by vacuum. The procedures shall also state whether any load test shall be required following repairs.

### 33.3.4 Drypacking

Drypacking shall be used for deep holes or any hole that has a relatively high ratio of depth to area. Drypack (approximately 1 part of cement to 2.5 parts of sand passing the 1.25 mm sieve) shall have the correct amount of water to make a sound, solid pack. It shall be placed in layers having a compacted thickness of about 10 mm. Each layer shall be compacted with a tool (hardwood stick) that will not produce too smooth a surface for proper bonding of successive layers.

#### 33.3.5 Pressure Grouting

Pressure grouting shall conform to the requirements of similar grouting of ducts for post-tensioning in accordance with Clause 26 of CSA Standard CAN/CSA-A23.1.

#### 33.3.6 Epoxy Injection

Injection of a low viscosity epoxy under pressure or by vacuum shall be done by personnel skilled in working safely with epoxies. For other than treatment of horizontal surfaces, cracks shall be sealed except for entry and vent points and care shall be taken not to trap air pockets.

*Note: It may be necessary to examine concrete cores to verify structural bond.*

#### 33.3.7

If a load test is required following repairs, such a test shall be done in accordance with Clause 20 of CSA Standard CAN/CSA-A23.3.

# APPENDIX A
# RESPONSIBILITIES

*Note: This Appendix is not a mandatory part of this Standard, but is written in mandatory language to accommodate its adoption by anyone wishing to do so.*

## A1. RESPONSIBILITY OF THE OWNER

### A1.1

The basic responsibility for the structural and aesthetic design of precast concrete, whether used architecturally or structurally, shall rest with the owner.

### A1.2

The owner shall, as applicable (see Table A1), carry out the following responsibilities:

(a) Provide clear and concise working drawings and specifications based on current Canadian standards with any exemptions clearly identified.

(b) Establish the standards of acceptability. The owner's right to approve work encompasses checking for defects and rejecting elements that would render the building or structure, or any portion of it, incapable of being used in the manner and for the purpose intended.

(c) Determine the part, if any, played by the precast concrete in the support of the structure as a whole.

(d) As part of the design, allow for the effects of material properties, stiffness, temperatures, and other characteristics or qualities that may influence the interaction of precast concrete elements with the structure.

(e) Determine all load reactions necessary for the design of both reinforcement and connections.

(f) Evaluate thermal movements as they may affect requirements for joints, connections, reinforcement, and compatibility with adjacent materials.

(g) Analyze the watertightness of exterior precast concrete wall panel systems, evaluating joint treatment, including the performance of adjacent materials for compatibility. Specify the joint treatment and the proper sealing of the precast concrete to windows and other openings.

(h) Design the supporting structure so that it will carry the mass of the precast concrete as well as any superimposed loads, including provisions for deflection and rotation of the supporting structure during and after erection of the precast concrete.

(i) If the stability of the structure is dependent on the sequence or method of erection, design the supporting structure to withstand the temporary loading conditions that might be encountered as a result of the sequence of erection and/or the sequence of loading of the structure.

(j) Make the selection of surface finishes, recognizing certain limitations in materials and production techniques in regard to uniform colour, texture, and performance, especially the limitations inherent in natural materials.

(k) Make the selection of interior finishes. Define the area of exposure and the interior appearance for occupancy requirements, and again recognize material and production limitations.

(l) Design for durable exterior walls with respect to weathering, corrosive environments, heat transmission, vapour diffusion, and moist air and rain penetration.

(m) Design sandwich wall panels for their proper structural and insulating performance.

(n) Review and approve all drawings as required in the specifications in time to meet project schedule requirements.

(o) Specify dimensional and erection tolerances for the precast concrete and tolerances for the supporting structure in accordance with A23.4.

## A2. RESPONSIBILITY OF THE MANUFACTURER

### A2.1 DESIGN CAPABILITY

#### A2.1.1

A manufacturer producing precast concrete elements in accordance with A23.4 shall have established design capability. Such capability shall be ensured by professionals qualified for such design under appropriate provincial or territorial legislation, who shall be employed or retained by the manufacturer (see Clause 1.3).

#### A2.1.2

The responsibilities of the manufacturer's design engineer in relation to the owner, shall be established in the contract documents (see Table A1).

### A2.1.3

Regardless of the option or alternative, as set forth in Table A1, that is specified in the contract documents, the manufacturer's design engineer shall be responsible for preparing and obtaining the owner's approval of shop drawings and erection drawings (see CAN3-A23.3, Clauses 12.3 and 16.4) and for translating the requirements for each project into procedures that will allow the finished job to meet the required quality.

### A2.1.4

The manufacturer shall be responsible for the following design items, if applicable:

(a) design for all loading conditions for nonstructural precast concrete elements;

(b) layout drawings, connection details, and related calculations (see Clause A2.3);

(c) design of structural precast concrete elements; and

(d) design for loads due to handling, transportation, and erection.

### A2.1.5

The manufacturer shall also be responsible for the following items during the preparation of shop drawings, the fabrication of precast concrete elements, and their installation, if applicable, in accordance with this Standard:

(a) design and protection of lifting hardware;

(b) modifications to reinforcement and hardware;

(c) conduits, pipes, wiring, and/or other appurtenances or fixtures embedded in the precast concrete;

(d) determination of the stripping strength of the concrete;

(e) curing procedures;

(f) determination of pickup devices and their locations;

(g) establishment of concrete strength for shipping; and

(h) assessment of structural damage and repairs.

### A2.1.6

The manufacturer shall clearly identify the owner's fire rating requirements and verify that instructions regarding materials and details to attain these ratings are communicated to the manufacturer's personnel (see Clause 12.5.3 of CSA Standard A23.4).

### A2.1.7

The manufacturer shall review the design of all precast concrete elements designed by others for structural soundness and feasibility with respect to finishes, connections, handling stresses, joint treatment, and tolerances both for manufacturing and installation. Any deficiencies shall be reported to the owner.

*Note: This review should not be interpreted as relieving the owner of such responsibilities as are applicable (see Table A1).*

### A2.1.8

For plants producing standard or prestressed concrete elements for which allowable load tables have been developed, the manufacturer shall be familiar with the design criteria used and the possible limitations of applications. The manufacturer's design engineer shall check the correct interpretation of such tables during normal plant operations.

### A2.1.9

The manufacturer's design engineer shall review and assess the quality control records.

## A2.2 MANUFACTURER'S ENGINEERING CAPABILITY

### A2.2.1

A manufacturer who produces prestressed concrete shall have a registered professional engineer on staff, unless special application is made to the certification agency to retain a design engineer. Written approval for such an arrangement is necessary from the certification agency. This approval can be reviewed and withdrawn if necessary.

*Note: Such applications should show how the duties of the plant engineer are to be safely carried out.*

### A2.2.2

If the manufacturer does not have a qualified professional engineer on staff, the design engineer or another retained engineer shall provide inspections of the precasting operations to ensure that the design criteria are met. Such arrangements shall be documented and the engineer's involvement in the inspection of the manufacturer's operations shall be recorded.

*Note:* The "manufacturer's engineer" as defined in A23.4 may refer to the design engineer or the plant engineer, as documented.

### A2.2.3

The manufacturer shall be responsible for the fabrication and installation of the precast concrete in accordance with A23.4, the shop drawings, and sound engineering principles, including, if applicable

(a) design of tendon deflectors;

(b) assessment of prestress losses prior to the transfer of the prestress force to the concrete;

(c) design of prestressing beds for loads applied;

(d) maintenance of records of bed capacities;

(e) maintenance and calibration of stressing equipment;

(f) establishment of clear stressing procedures which shall be recorded and available;

(g) inspection of the prestressing operations and assessment and correction of any discrepancies in measuring the prestress force;

(h) establishment of documented detensioning procedures;

(i) establishment of recorded procedures for control of tendon slippage for machine cast products;

(j) assessment of possible tendon slippage in all pretensioned products; and

(k) concrete mix designs.

## A2.3 SHOP DRAWINGS

Shop drawings and related documents prepared by the manufacturer and reviewed by the design engineer shall provide the information required by Clauses 1.2 and 16.4 of CAN3-A23.3, and include the following:

(a) Layout or general arrangement drawings showing the layout of the precast elements, their location in the building or the structure, their relation to adjacent materials, the connections, and the joint and sealing details, if within the scope of the work.

(b) Erection drawings showing any details required in addition to those in the layout drawings to describe the actual installation. These drawings shall show the markings of individual elements and special erection methods and shall indicate the sequence of erection where this is critical.

(c) Setting drawings, which shall show the location of hardware supplied by the precast manufacturer to be placed by the general contractor.

(d) Detail drawings, which shall show the dimensions and shape of individual elements, mass, reinforcement, and individual mark numbers. Such drawings shall also show special embedments such as plant and erection hardware and fastening details provided by other trades, such as reglets, cut-outs, pipe sleeves, openings, finishes, and special handling instructions.

(e) Production drawings for individual elements. These drawings shall contain enough views and sections to ensure clear interpretation by the workers involved. Information with respect to special materials, concrete mix, finish (including wet finishing), special tolerances, etc shall be shown on the production drawings. A production drawing may be a shape drawing, a set-up drawing, a reinforcing and hardware drawing, or stressing sheets.

# A3. SUMMARY OF THE RESPONSIBILITIES OF THE OWNER AND THE MANUFACTURER

Table A1 gives a summary of the responsibilities of the owner and the manufacturer.

*Note: Different options and alternatives are provided. The owner must clearly state in the contract documents which option and alternative apply to each project.*

### Table A1 Summary of Responsibilities of Owner and Manufacturer

| Contract information supplied by the owner | Responsibility of the manufacturer |
|---|---|
| **Option I** | |
| Provide complete drawings and specifications for the precast concrete detailing all of the aesthetic and functional requirements plus all of the dimensions. | The manufacturer shall make erection and production drawings (as required) with details as shown by the owner. The manufacturer may suggest modifications that would improve the economics, structural soundness, or performance of the precast installation. The manufacturer shall obtain specific approval for such modifications. Full responsibility for the precast design including such modifications shall remain with the owner. |
| **Option II** | |
| Detail all aesthetic and functional requirements but specify only the required structural performances of the precast units. Specified performance shall include all limiting combinations of loads together with their points of application. This shall be supplied in such a way that all details of a unit can be designed without reference to the behaviour of other parts of the structure. | The manufacturer has two alternatives:<br><br>(a) Submit drawings showing erection and element configurations with all necessary details and design information for the approval and ultimate responsibility of the owner.<br><br>(b) Submit drawings showing erection and element configurations for general approval and assume responsibility for the manufacturer's part of the structural design, ie, the individual units but not their effect on the building. Firms accepting this practice may either stamp (seal) drawings themselves or commission engineering firms to perform the design and stamp the drawings.<br><br>The choice between alternatives (a) and (b) shall be decided by the owner and the manufacturer prior to bidding, with either approach clearly stated in the specifications for proper allocation of the design responsibility. |
| **Option III** | |
| Cover the required structural performance of the precast units as in Option II and cover all or parts of the aesthetic and functional requirements by performance specifications. Define all limiting factors such as minimum and maximum thickness, depth, weights, and limiting dimensions. Give acceptable limits of any other requirements not detailed. | The manufacturer shall submit drawings with choices of assuming responsibilities as Option II (a) and (b). |

**Note:** *The owner shall in all cases clearly assign responsibility for the transportation and erection of the precast concrete in the contract documents. (See Appendix B, Clause B1.2.)*

# APPENDIX B

# SCOPE OF WORK

*Note: This Appendix is not a mandatory part of this Standard, but is written in mandatory language to accommodate its adoption by anyone wishing to do so.*

## B1. GENERAL

### B1.1

The Owner's specifications shall include all items that are to be included in the tender for the precast concrete work.

*Notes:*

**(1)** *Those items which may influence the quality of the overall finished precast concrete work should be included. However, specifications need not include those items which, although related, cannot practicably be carried out in sequence with the precast work or which may be affected by the actions of the general contractor or following trades, particularly when the precast manufacturer has no control over such items.*

**(2)** *The owner should stipulate the maximum time to award the precast concrete subcontract following the awarding of the general contract.*

### B1.2

Erection and transportation to job site of the precast concrete elements shall be the responsibility of the precast concrete manufacturer and executed by the manufacturer's own forces or by a qualified specialist firm acting as subcontractor to the precast concrete manufacturer.

*Note: Exception from responsibility of delivery and erection may be made for small precast elements such as window sills, lintels, stair treads, benches, covers, pavement slabs, and specialty items. Other exemptions may be made where the usage of concrete products ties in with other site operations not controllable with respect to scheduling by the precast manufacturer, such as the incorporation of precast elements into masonry work.*

### B1.3

Instructions regarding items relating to more than one trade shall be described in the precast concrete specifications and in the other related trade sections.

*Note: The general inclusion or exclusion of specific items has been attempted in this Appendix so that early budget estimates supplied to the owner may be judged on a more uniform basis. The owner should clearly outline which items are to be included for bidding purposes. These will depend on the individual job, local trade practices, and the precast manufacturer's recommendations.*

### B1.4

The cost of any specified materials testing and/or inspection, over and above the manufacturer's own quality control, as required in CSA Standard A23.4, shall be the responsibility of the owner. The manufacturer shall provide at no extra cost all materials required for testing and allow inspectors reasonable access to the manufacturer's premises.

### B1.5

The owner shall stipulate procedures for the submittal of shop drawings for approval, including the number of prints required of the erection (or layout) drawings, setting drawings for contractor's hardware, drawings of all erection details, and drawings of all typical elements.

*Note: Submittal of shop drawings for elements with modifications that are aesthetically or structurally unimportant may not be required.*

## B2. GENERAL CONTRACTOR'S WORK

### B2.1

A schedule of erection shall be mutually agreed upon between the precast manufacturer and the general contractor.

### B2.2

The general contractor shall provide and maintain access and operating space for the equipment of the precast concrete erector.

*Notes:*

**(1)** *Such access may consist of, but need not be limited to, roads, ramps, and crossings capable of supporting cranes and trucks normal to erection operations of this type, over the area required, and moving under their own power.*

**(2)** *Foundation and utility excavations should be backfilled by the general contractor where and when necessary.*

### B2.3

The general contractor shall coordinate with other trades and obtain embedded items and the locations of openings required by these trades.

### B2.4

The general contractor shall establish and permanently mark, at convenient locations, bench marks and building lines for the use of the precast manufacturer in layout and installation of the precast concrete elements. If the building frame is not precast, then the general contractor shall provide benchmarks and building lines at each floor.

### B2.5

The general contractor shall assume responsibility for the protection of the manufacturer's work after the final erection of the precast work to acceptable alignment and appearance, including all connections and joints.

*Note: Any cleaning or repairs of the precast concrete work should be carried out by the precast manufacutrer.*

# B3. HARDWARE

## B3.1 GENERAL

Connections and other embedded items required for the precast concrete work are normally called "hardware" and divided into three categories:

(a) **General contractor's hardware**, which is placed on or in the structure in order to receive the precast concrete elements. Examples are anchor bolts and angles or plates with suitable anchors;

(b) **Plant hardware**, which is placed in the concrete elements themselves, either for connections and the precast erection or for other trades, such as mechanical, plumbing, window or glazing, miscellaneous iron, masonry, or roofing trades; and

(c) **Erection hardware**, which comprises all loose hardware necessary for the installation of the precast concrete elements themselves.

## B3.2 GENERAL CONTRACTOR'S HARDWARE

### B3.2.1

Precast connection hardware to be placed by the general contractor shall be supplied to the job site by the precast manufacturer together with setting drawings for proper locating of the hardware. This hardware shall be placed in the structure to the tolerances given by CSA Standard A23.4, Clause 13.

*Notes:*

**(1)** *This practice is the most economical and practical for most projects. A different approach may be advantageous only when fastening precast concrete elements to structural steel buildings that are not fireproofed by structural concrete encasement. Brackets, angles, etc, which must be fastened to the structural steel, may be more economically supplied and shop-welded to supporting members by the structural steel supplier. In this case it becomes important to clearly indicate the supply and installation source for each item making up the connection.*

**(2)** *The contractor's hardware should be delivered to the general contractor in sufficient time to be incorporated in the structure in accordance with setting drawings supplied by the precast manufacturer and according to a predetermined and agreed upon schedule.*

**(3)** *It is recommended that the precast manufacturer check the location of the contractor's hardware prior to actual erection of the precast elements. This should in no way relieve the general contractor of the responsibility for the proper positioning of such hardware.*

### B3.2.2

The general contractor shall maintain and suitably protect the contractor's hardware (such as by capping of inserts, protection of threads clear of concrete where required) until required by the precast manufacturer.

### B3.2.3

If anchors or other structural items shown on the precast manufacturer's approved drawings cannot be accommodated owing to mislocation or unforeseen reinforcing complications or to the work of other trades, the general contractor shall obtain approval for any modifications by both the owner and the precast manufacturer to ensure the required structural adequacy of the hardware.

## B3.3 PLANT HARDWARE

### B3.3.1

All plant hardware necessary for the connection of the precast elements shall be supplied by the precast manufacturer except as noted in Clause B3.2.3.

### B3.3.2

The plant hardware necessary for the support of the work of other trades may be divided into two groups, depending on whether the hardware (or fastening) carries more or less than 500 N as follows:

(a) **Less than 500 N:** Fastenings intended to carry less than 500 N shall be supplied and installed on the site by the trade in question, unless otherwise shown on the contract drawings.

*Note: An exception to this requirement may be for delicate concrete sections where site fastening would damage the precast work. Drawings and/or specifications should then clearly specify that such hardware or inserts be placed by the precaster, but with the hardware inserts supplied to the precast manufacturer's plant free of charge by the trade in question. This procedure is common for window fastenings in precast wall panels and, if inserts are used, their locations should be given on the approved shop drawings of both trades.*

(b) **More than 500 N:** Plant hardware or inserts intended to withstand a force of more than 500 N that are shown on the contract drawings and/or described in the specifications shall be supplied by the trade in question and installed by the precast concrete manufacturer and shown on the approved shop drawings.

### B3.3.3

Other plant hardware to be placed by the precast manufacturer, such as dovetail slots for masonry anchors or inserts for flashings, shall be specified by the owner. The anchors or the flashing shall be supplied by the respective trades. Hardware of special manufacture shall be supplied by the trade requiring it.

### B3.3.4

The installation of inserts and fastenings by means of explosive-actuated or power-driven tools shall be allowed only where such installation will not damage the structural integrity of the elements by being too close to reinforcement or tendons or where they will not mar the concrete finish by spalling or chipping.

## B3.4 ERECTION HARDWARE

Erection hardware shall be supplied by the precast manufacturer when used in the connections or when forming an integral part of the overall precast installation.

## B3.5 OTHER MATERIALS

### B3.5.1

Temporary bracing stays and guy wires for the support of precast elements shall be considered erection equipment and be supplied by the precast manufacturer. Inserts or connections for such equipment may be either general contractor's hardware or plant hardware and shall be treated accordingly.

### B3.5.2

Materials for joints between precast concrete wall panels shall be supplied and installed by the precast manufacturer.

## B3.6 JOINTS

### B3.6.1

Joints between precast concrete elements and adjacent materials put in place later shall not be included in the precast concrete work.

*Note: This work is generally more efficiently carried out by the subsequent trade installing such adjacent materials.*

### B3.6.2

The contract drawings and project specifications shall clearly show the responsibility for the supply and installation of materials at the joints between precast concrete elements and other materials, erected prior to the precast installation, such as cast-in-place structures and steel members.

# B4. HOLES AND OPENINGS

### B4.1

All openings for services and equipment, with their approximate size and location shown on the contract drawings, and the same holes subsequently detailed with respect to size and location by the trades requiring the openings shall be coordinated by the general contractor and noted on the first issue of the precast manufacturer's shop drawings. The openings shall be provided by the precast manufacturer.

### B4.2

Holes of small diameter (80 mm or less) or holes for fastenings intended to carry less than 500 N shall be made at the site by the trade requiring the opening in locations approved by the precast manufacturer, except as noted in Clause B3.

*Note: Location and sizes of holes that may be made by other trades should be avoided in areas of structural importance and/or tendon locations.*

### B4.3

The procedures set forth in Clause B4.1 shall be followed whether the precast manufacturer provides openings at the plant or elects to cut and drill openings on site.

### B4.4

Openings requested after the approval of the shop drawings and production of precast elements shall be cut only if the openings can be structurally accommodated.

*Notes:*

**(1)** *These modifications should be carried out by the precast manufacturer, who has the equipment and expertise required to cut new openings.*

**(2)** *Holes in architectural wall panels for other trades should always be made at the manufacturing plant and consequently should be detailed and located on the contract drawings.*

# B5. ROOF AND FLOOR FILL

### B5.1

If roof and/or floor slab joints have grout keys that form part of the slab design, the grouting shall be executed by the precast manufacturer and the areas made ready for roofing or floor fill as required.

### B5.2

Where the joints between slabs do not have grout keys, joints may have to be taped for roofing or topping applications. This taping shall be performed by the trade (or general contractor) that subsequently carries out the roofing or floor topping operations.

### B5.3

Where a differential camber over 10 mm exists in the joint between the tops of adjacent precast concrete roof members, the joint may have to be feather-grouted to a slope of 1:12. This work shall be performed by the structural precast concrete manufacturer if requested by the general contractor. For differential camber tolerances, see CSA Standard A23.4, Clause 10.

# APPENDIX C

# ARCHITECTURAL PRECAST CONCRETE —MATERIALS AND FINISHES

*Note: This Appendix is not a mandatory part of this Standard, but is written in mandatory language to accommodate its adoption by anyone wishing to do so.*

## C1. GENERAL

Architectural precast concrete shall be manufactured, transported, and erected in accordance with the requirements of CSA Standard A23.4 and this Appendix.

## C2. MATERIALS

### C2.1 CEMENTS AND SUPPLEMENTARY CEMENTING MATERIALS

When uniformity of colour for all or part of a project is desired, each cementing material used shall be of one type and from the same mill. This applies whether a single cement or a mixture of cementing materials are used.

*Notes:*

**(1)** *The characteristics of special cements should be investigated prior to use.*

**(2)** *Some suppliers will provide cement that falls within guaranteed percentage colour variations.*

**(3)** *Where a mixture of white and grey cements is used, uniformity normally improves by increasing the percentage of white, but the grey colour tends to dominate.*

### C2.2 WATER

Water for architectural concrete mixtures and curing shall be free of impurities or colour that may affect the concrete finishes.

### C2.3 FINE AGGREGATE

#### C2.3.1 General

Fine aggregate for architectural concrete shall conform to the following requirements in addition to those in CSA Standard A23.4, Clause 5.1.

*Note:* Where fine aggregate is likely to be a major colouring agent in architectural concrete, this material should be controlled even more closely by using manufactured one-size fine aggregate.

#### C2.3.2 Grading

To control the uniformity in appearance of architectural concrete, the amount of material passing 160 μm sieve shall either be limited to a maximum of 2% or be maintained within ±1 percentage point of the amount established in the mix design.

*Note: To obtain a low absorption, the percentage of aggregate passing the 315 μm sieve, when tested in accordance with CSA Standard Test A23.2-11C, may have to be lower than the limits specified in CSA Standard CAN/CSA-A23.1.*

#### C2.3.3 Limits for Deleterious Substances

Fine aggregate for architectural concrete with exposed aggregate finishes exposed to the weather shall conform to the limits prescribed in Table C1 when tested in accordance with CSA Standard CAN/CSA-A23.2-3A, 4A, and 5A unless at least 5 years' exposure under similar environmental conditions indicates acceptable durability and appearance.

**Table C1. Limits for Deleterious Substances in Fine Aggregates for Architectural Concrete with Exposed Aggregate Finishes Exposed to the Weather**

| Substances | Maximum % by mass of total sample* |
|---|---|
| Clay lumps | 1.0 |
| Coal or lignite | 0.5 |
| Material finer than 80 μm sieve | 2.0 |

**Refer also to Clause 5.6 of CSA Standard CAN/CSA-A23.1.*

#### C2.3.4 Organic Impurities

Fine aggregate tested in accordance with CSA Test Method A23.2-7A, 9A, and 16A shall not produce a darker colour than the standard colour.

### C2.4 COARSE AGGREGATES—ARCHITECTURAL CONCRETE

#### C2.4.1 Requirements

Coarse aggregate for architectural concrete shall conform to the following requirements in addition to those in CSA Standard A23.4, Clause 5.1.

#### C2.4.2 Grading

Coarse aggregate for architectural concrete shall conform to the grading requirements of CSA Standard CAN/CSA-A23.1 except when a special architectural appearance is required and when at least 5 years' exposure under similar environmental conditions indicates the satisfactory performance of the concrete mix.

#### C2.4.3 Deleterious Substances and Physical Properties

Coarse aggregates for architectural concrete with exposed aggregate finishes exposed to the weather shall conform to the limits prescribed in Table 2 when tested in accordance with CSA Standard CAN/CSA-A23.2, unless at least 5 years' exposure under similar environmental conditions indicates acceptable durability and appearance.

#### C2.4.4 Staining Particles

Aggregates with iron compounds or other potentially staining materials shall not be used in architectural concrete that will

**Table C2. Limits for Deleterious Substances and Physical Property Requirements of Coarse Aggregates for Architectural Concrete with Exposed Aggregate Finishes Exposed to the Weather**

| Maximum % by mass of total sample | | | | |
|---|---|---|---|---|
| Clay lumps | Coal particles | Material finer than 80 μm | Magnesium sulphate soundness loss | Abrasion loss* |
| 0.25 | 0.5 | 1.0 | 12 | 35 |

*Refer also to Clause 5.6 of CSA Standard CAN/CSA-A23.1.

be exposed to the weather. When tested in accordance with ASTM C641, such aggregates shall show a stain index of less than 20.

*Note: An additional quick test for the staining potential of gravels or crushed carbonate rock is to immerse newly fractured or sandblasted aggregate particles in a lime slurry. If staining particles are present, a blue-green gelatinous precipitate will form within 5 to 10 min. This will rapidly change colour to brown on exposure to light and air. The reaction should be complete within 30 min. If no brown gelatinous precipitate is formed when the suspect aggregate is placed in the lime slurry, there is little likelihood of staining taking place on the surface of the exposed concrete.*

### C2.4.5 Structural Low Density Aggregate

The absorption of structural low density aggregate, as defined in ASTM Standard C330, shall not exceed 11% when such aggregate is used in architectural face mixes exposed to the weather.

*Note: Low density aggregate should not be used in exposed finishes in cold or humid climates unless their performance has been verified by tests or they have demonstrated satisfactory performance under similar climatic conditions.*

## C3. FORMS

### C3.1

Forms shall be made so that water and cement paste are not lost from the concrete.

*Note: Where the form is assembled from several components, compressible seals may be required between components, or the joints should be sealed prior to placing the concrete.*

### C3.2

Joints in the form material shall be made so that they will not be reflected in the concrete surface and mar the appearance of the product.

### C3.3

Repeated casting, consolidation of concrete, and stripping of elements shall not affect the dimensions or the planes of the elements beyond the allowable tolerances.

*Notes:*

**(1)** *It is important that window openings be made accurately. For glazing directly to the concrete, the plane of the opening should be held to close tolerances during the entire casting and curing cycle.*

**(2)** *Corners of precast concrete elements that are exposed to view should normally be rounded. Lines and shapes should be true and uniform.*

**(3)** *Whenever possible, forms should have sufficient draft to facilitate stripping with a minimum of form breakdown.*

**(4)** *It is often advantageous to precast architectural elements with large returns in stages in order to cast exposed surfaces horizontally and thus obtain a more uniform texture.*

### C3.4

Forms shall permit fixed positioning or jigging of hardware and allow for accurate positioning of the reinforcing cage while maintaining the specified cover to the reinforcement. Spacers shall not be used between the reinforcing and those surfaces which will be exposed to view.

*Notes:*

**(1)** *When spacers are necessary for unusual products, their use and type should be approved by the owner.*

**(2)** *Forms may be of wood, steel, concrete, plastic, rubber, or a combination of these materials. The choice of material should be the manufacturer's, providing this does not compromise the quality of the finished product specified by the owner.*

### C3.5

When preparing forms for concrete, form release agents and surface retarders shall be applied uniformly.

## C4. ARCHITECTURAL FINISHES

*Note: Many architectural finishes are now available and new ones are constantly being developed. The more common finishes and the means of achieving them are described in detail in references such as the PCI's* Architectural Precast Concrete *and* Architectural Precast Concrete Color and Texture Selection Guide.

### C4.1

Finishes that include materials not normally associated with concrete mixes, such as cut stone, ceramics, brick, tile, ferrous or nonferrous metal products, plastics, or other materials, shall be tested, as required by CSA Standard A23.4, Clause 5.2, to establish their performance under the service condition expected for the product.

*Notes:*

**(1)** *Particular attention should be paid to the compatibility of materials for volume changes caused by temperature and humidity changes. If materials do not have similar physical properties, the final design should include* *provisions for possible differential movement of the face material and the backing concrete.*

**(2)** *If performance records have not already been established for these face materials for special climatic conditions, including severe corrosive atmospheres, the materials should be tested for such conditions. The tests may be accelerated but should otherwise reflect the expected service conditions.*

### C4.2

Elements having more than one finish shall have features incorporated into the casting to ensure proper demarcation between the finishes involved.

*Note: When a precast element is exposed to view from all sides as in a column, it is impossible to match completely the formed surfaces with the unformed surfaces with respect to colour and texture. This difference may be made more acceptable by placing such elements in the finished structure in a way that hides the unformed surfaces or by orienting these surfaces so that they are viewed from one direction only. For the best possible match, the highest degree of smoothness of formed and unformed surfaces should be specified. This will require a high degree of hand-finishing. (See also CSA Standard A23.4, Clause 24.1.2.)*

## C5. PROJECT SAMPLES AND APPROVALS

### C5.1

The appearance of architectural precast concrete shall be determined by sample approval prior to regular production. The finishes shall be defined by such samples and by specifications for performance requirements such as durability, strength, and water absorption.

*Note: Samples for architectural precast concrete should be regarded only as a standard for performance within the variation of materials and workmanship to be expected. If the colour or appearance of the cement or aggregates is likely to vary significantly, samples showing the expected range of variations should be supplied for approval by the owner.*

### C5.2

Project samples shall be submitted with sufficient information for the owner to assess the quality and expected performance of the concrete in the sample.

### C5.3

All finishes shall be carefully assessed and detailed with regard to controlling the effects of weathering.

*Note: For special weathering considerations, refer to the PCI publication* Architectural Precast Concrete.

### C5.4

Samples shall bear a shape that resembles the actual casting. Flat samples shall only be used for flat castings. The size of the sample shall relate to the maximum size of the aggregate to be used and allow for realistic placing of the concrete and the accurate expression of detail.

*Note: Unavoidable segregation in finishes that use mixes having a large colour difference between the coarse aggregate and the sand and cement matrix will produce noticeable colour and texture variations.*

### C5.5

The concrete placement and consolidation method used to make samples shall be representative of the intended production method.

### C5.6

Samples shall be supplied for each of the different finishes for a project and all samples shall be clearly identified.

### C5.7

The first production units shall be approved by the owner before regular production continues. Such elements shall be properly marked and serve as acceptance criteria for the balance of the project. They shall remain identifiable, even on the building, until final acceptance of the project.

## C6. JOINTS

Joints between architectural precast concrete elements shall be constructed as shown on the contract drawings, and joint materials shall be specified or approved by the owner.

*Notes:*

**(1)** *It is recommended that samples of the joint materials for architectural precast concrete be approved at the same time as the concrete samples.*

**(2)** *It is often advantageous to precast architectural elements with large returns in stages in order to cast exposed surfaces horizontally and thus obtain a more uniform texture. The joints between exposed aggregate match-cast surfaces may be hidden by locating them in a groove or a quirk return (for mitred joints).*

## C7. STORAGE

Any soiling or weathering of architectural concrete panels that would result in a pattern different from normal weathering shall be avoided during storage.

# INDEX